Lecture Notes on Data Engineering and Communications Technologies

Volume 73

Series Editor

Fatos Xhafa, Technical University of Catalonia, Barcelona, Spain

The aim of the book series is to present cutting edge engineering approaches to data technologies and communications. It will publish latest advances on the engineering task of building and deploying distributed, scalable and reliable data infrastructures and communication systems.

The series will have a prominent applied focus on data technologies and communications with aim to promote the bridging from fundamental research on data science and networking to data engineering and communications that lead to industry products, business knowledge and standardisation.

Indexed by SCOPUS, INSPEC, EI Compendex.

All books published in the series are submitted for consideration in Web of Science.

More information about this series at http://www.springer.com/series/15362

Kavita Khanna · Vania Vieira Estrela ·
Joel José Puga Coelho Rodrigues
Editors

Cyber Security and Digital Forensics

Proceedings of ICCSDF 2021

Springer

Editors
Kavita Khanna
The NorthCap University
Gurugram, India

Joel José Puga Coelho Rodrigues
Federal University of Piauí (UFPI)
Teresina-PI, Brazil

Instituto de Telecomunicações
Aveiro, Portugal

Vania Vieira Estrela
Telecommunications Department
Federal Fluminense University (UFF)
Duque de Caxias, Rio de Janeiro, Brazil

ISSN 2367-4512 ISSN 2367-4520 (electronic)
Lecture Notes on Data Engineering and Communications Technologies
ISBN 978-981-16-3960-9 ISBN 978-981-16-3961-6 (eBook)
https://doi.org/10.1007/978-981-16-3961-6

This Springer imprint is published by the registered company Springer Nature Singapore Pte Ltd.
The registered company address is: 152 Beach Road, #21-01/04 Gateway East, Singapore 189721, Singapore

Preface

International Conference on Cyber Security and Digital Forensics (ICCSDF 2021) was held on April 3–4, 2021, at the NorthCap University, Gurugram, India. The conference was organized with the aim to bring together the leading academic scientists, researchers, industrialists to exchange and share their experiences and research results in all aspects of cybersecurity and digital forensics. This international conference was conceived after analyzing the increase in the prevalence of cybercrime attacks on business organizations, government infrastructures, and individuals. With this conference, we encouraged the community of multinational researchers to showcase the research work done in their field of cybersecurity. The conference provided an opportunity to researchers on an international forum to learn about the latest developments through scientific information interchange in the field of cybersecurity and digital forensics. The topics were categorized into four tracks, namely cryptology and its applications, cyber and digital forensics, network and mobile security, blockchain, and software technologies. We received a total of 205 submissions; each submission was anonymously reviewed by three reviewers. After extensive reviews and shepherding, 50 papers were accepted, and this proceeding contains revised versions of all accepted papers.

In addition to the presentations of the accepted papers, the occasion was graced by the presence of Honorable Chief Guest **Dr. Gulshan Rai**, *Former National Cyber Security Coordinator, at the Prime Minister Office, Government of India*, and four keynote speakers, namely **Prof. (Dr.) Gregory Conti**—*Co-founder and Principal at Kopidion, New York, USA*, **Ms. Vandana Verma**—*Security Solutions Architect at IBM India Software Labs and Member of the OWASP Global Board of Directors*, **Mr. Shree Parthasarathy**—*Leader South Asia Cyber and Strategy Services, APAC Cyber-Innovation, Deloitte*, and **Mr. Santosh Khadsare**—*Cyber Forensics Expert, Government of India*. The conference featured respective keynote talks, namely "The current and future state of Cybersecurity: Deflecting the Trajectory," "Careers in AppSec: Lead with Grit," "Digital Forensics," and "Digital Era and its impacts".

A number of people contributed to the success of ICCSDF 2021. The organizers are truly grateful to all the committee members for smoothly completing their assigned responsibilities and also thankful to the external reviewers for assisting in the reviewing process and their in-depth discussions. We must mention that the selection of the papers was an extremely challenging task. We would like to thank everyone who contributed directly or indirectly in making this conference a success and ensured its smooth running. The Conference was organized and facilitated by **Dr. Mehak Khurana** as Convener and **Dr. Pooja Sapra** as Co-Convener.

The support of technical partners, Open Web Application Software Project (OWASP), Spoken Tutorial, Infosecgirls, Women In Application (WIA), Computer Society of India (CSI), The Institution of Engineers (India) IEI is also appreciatively acknowledged.As per standard, editor's names and their locations (city and country) have been inserted in Preface. Please check and amend if necessary.

Gurugram, India — Kavita Khanna
Duque de Caxias, Brazil — Vania Vieira Estrela
Teresina, Brazil — Joel José Puga Coelho Rodrigues

Contents

Section–C

About the Editors

Prof. Kavita Khanna has done M.Tech. Computer Engineering, MDU, Rohtak, and Ph.D. in Computer Graphics and Soft Computing Techniques, GGSIPU, Delhi. With 21 years of teaching, administration, and research experience, currently she is associated with The NorthCap University, Gurugram, as Professor and Head of the Department (CSE Department). Her interests include artificial neural networks, computer graphics, cryptographic techniques, artificial intelligence: metaheuristic techniques, neural networks, machine learning, and analysis and design of algorithms. She has more than 60 publications in reputed journals, conferences, and chapters. She has guided many M.Tech. and Ph.D. students. Prof. Khanna is receiving a research grant from the Department of Science and Technology (DST), India, and NTRO for research projects. She has also organized various conferences, workshops, FDPs and has been Reviewer, Session Chair, and Committee Member of conferences and journals.

Vania Vieira Estrela B.S. degree from Federal University of Rio de Janeiro (UFRJ) in Electrical and Computer Engineering (ECE); M.Sc. from the Technological Institute of Aeronautics (ITA), Brazil; M.Sc. degree in ECE at Northwestern University, USA; and Ph.D. in ECE from the Illinois Institute of Technology (IIT), Chicago, IL, USA. Taught at DePaul University, USA, and Universidade Estadual do Norte Fluminense (UENF), Brazil. She was visiting professor at the Polytechnic Institute of Rio de Janeiro (IPRJ)/State University of Rio de Janeiro (UERJ) in Brazil and currently working at UFF's Department of Telecommunications. Research interests include signal/image/video processing, inverse problems, computational & mathematical modeling, stochastic models, multimedia, communications, motion estimation, machine learning, and geoprocessing. She reviews several journal/magazine articles, and she is deputy editor of the Neuroscience Informatics journal (Elsevier). She is engaged in technology transfer, STEAM education, environmental issues, and digital inclusion. Editor of several books and special issues. Member of IEEE and ACM.

Joel José Puga Coelho Rodrigues [Fellow, AAIA and IEEE] is a professor at the Federal University of Piauí, Brazil; and senior researcher at the *Instituto de Telecomunicações*, Portugal. Prof. Rodrigues is the leader of the Next Generation Networks and Applications (NetGNA) research group (CNPq), an IEEE Distinguished Lecturer, Member Representative of the IEEE Communications Society on the IEEE Biometrics Council, and the President of the scientific council at ParkUrbis—Covilhã Science and Technology Park. He was Director for Conference Development—IEEE ComSoc Board of Governors, Technical Activities Committee Chair of the IEEE ComSoc Latin America Region Board, a Past-Chair of the IEEE ComSoc Technical Committee (TC) on eHealth and the TC on Communications Software, a Steering Committee member of the IEEE Life Sciences Technical Community and Publications co-Chair. He is the editor-in-chief of the International Journal of E-Health and Medical Communications and editorial board member of several high-reputed journals (mainly, from IEEE). He has been general chair and TPC Chair of many international conferences, including IEEE ICC, IEEE GLOBECOM, IEEE HEALTHCOM, and IEEE LatinCom. He has authored or coauthored about 1000 papers in refereed international journals and conferences, 3 books, 2 patents, and 1 ITU-T Recommendation. He had been awarded several Outstanding Leadership and Outstanding Service Awards by IEEE Communications Society and several best papers awards. Prof. Rodrigues is a member of the Internet Society, a senior member ACM, and Fellow of AAIA and IEEE.

Section–A

A Systematic Approach for Analyzing Log Files Based on String Matching Regular Expressions

Keshav Kaushik, Gargeya Sharma, Gaurav Goyal, Asmit Kumar Sharma, and Ashish Chaubey

Abstract In the past few years, there has been a tremendous increase in cyberattacks and cybercrimes. Technology is changing at a very fast pace, thus inviting more advanced cyberattacks. Any event that is triggered on the system is recorded in the form of logs in log files. It may be any warning, any alert, and information, and all the things get stored in the logs. Therefore, from the security point of view, analyzing the logs plays a crucial role in the forensic investigation or for analytics purposes also. This paper highlights a systematic approach for analyzing the log files using a string-matching algorithm and regular expressions. Thus, it helps in log analysis, management, and analytics for future reference. Analyzing logs in a systematic way is always crucial in digital forensics, and it will help in the smooth conduction of forensic investigations.

Keywords Log analysis · String matching · Regular expressions · Cyberattacks

1 Introduction

We live in a world where data surrounds us in every direction. Where there is technology, some amount of data is always involved. One of the many data sources that accumulate around us is in the form of log files. These log files are automatically produced files that contain the record of events from various software and operating systems. According to statistics, an average enterprise accumulates up to 4 GB of log data a day from all their servers and other network devices. Over 95% of that data within those log files are the entries that record every successful/unsuccessful event or transaction taking place in those systems. For example, it can be a server crash, user logins, logouts, applications start-up and shutdown, file access, etc. These log files are unnoticed by many people working under that enterprise because their production is a regular process and relatively holds less priority and attention than other tasks. However, it is the active and complete

K. Kaushik (✉) · G. Sharma · G. Goyal · A. K. Sharma · A. Chaubey
University of Petroleum and Energy Studies, Dehradun, India

K. Khanna et al. (eds.), *Cyber Security and Digital Forensics*,
Lecture Notes on Data Engineering and Communications Technologies 73,
https://doi.org/10.1007/978-981-16-3961-6_1

analysis of these log data that creates the differences among organizations, how they handle failures, changes to adapt by looking at the behavior of daily analysis periodically, etc. So, underestimating these data files is not advised for the smarter functioning of any organization and its client interactions.

This paper focuses on the same fundamental of not ignoring these log files and analyzing them in a much faster and easier way by the owners of these log files. Data analysis is getting more popular in the current few years than it has ever been. Since, we are achieving some invaluable information using analysis on the unstructured data present around us with the help of big data analytics, machine learning, and other scientific techniques. Analyzing logs is also a part of them. They contain some very intuitive information about the users, services, machines, their patterns, and various relationships between them, which can help us formulate techniques to take advantage of such information and provide more personal and better services. In addition to better services, this also mitigates a large amount of risk involved in the process. With predictive and descriptive analysis, organizations can build a more secure and effective network, which in turn prevents losses and increases their overall profits. Due to its increased popularity and attention, log analysis has spread its role in various directions. For example, correlational analysis: This is where you find the correlation between data that is not visible in a single log, especially since multiple logs are relevant over a single incident. For instance, if an attack is experienced, then you can perform correlation analysis using logs generated by servers, firewall, network devices, and other possible sources involved and find those logs that are relevant to that particular attack. During such conditions, log files play an important role as a piece of evidence as their analysis helps to find the cause and suspect of the attack.

There are other useful use cases as well with log analysis, one of which is anomaly detection. Through this, we can overlook the usual less important logs and only focus on unusual and unique log entries for further investigation. This paper focused on achieving anomaly detection on IPs through their number of occurrences either overall or in a particular period. Other than just detection of any existing anomalies, the feature to block such IP addresses if the user wishes to be also included, while adding all the blocked IPs in a separate file to maintain for reporting or future references. All the functionalities are further discussed in detail in the working methodology section. This paper fulfills the intention of providing enterprises and organizations with a tool that will help reduce the extensive searching and crawling through the entire log files related to the required situation and makes the process easier for further investigation and analysis. The implementation for this paper worked on a command–line interface and built using C language on top of the Linux operating system. With this, many users can be covered in the industry, as most of the servers out there are Linux kernel-based, contributing toward speeding the process of analyzing logs for various use cases.

2 Related Work

The extensive quantity of research proposing different technologies that can be used for analysis is increasing every year. Among them, many are focusing on the automated systems for the analysis of log files. Pinjia He in [1] reveals his dataset "log hub" that contains 17 datasets including stand-alone software, web server, mobile, and operating systems logs. These can be used for AI-powered log analytics which focuses on anomaly detection. It is an approach for log analysis in combination with machine learning. In this, regularly generated usual logs are ignored, and the unique or different ones in the log files are recognized as anomalies and treated as potential threats, respective to how the model was trained earlier. Zengchao NI and Honqui Liu in [2] discussed the idea of dealing with massive web system logs by extracting log templates based on the label recognition method effectively solving the problem of insufficient log format. Effective log monitoring allows us to do behavioral analysis. Xiaojian Liu and Yi Zhu in [3] talk about generating valued genealogy information that helps in generating attributes for specialized profiles. Specialized profiles lead to personalized recommendations, and thus biased monitoring and blocking of users are attained. Analyzing the data stored in logs of web search services provides a great view of the information searching tactics of the user. This can help in developing information systems, design, interface development, and information architecture construction for content collection. Robust learning of web search logs is quite important. Laura M. Koesten in [4] describes the drawback of web search engines in finding the most relevant datasets that meet the user's query. They suggested the first query analysis for dataset search containing logs of four open data portals. Logs from that reveal that queries stated on the data portal are quite different from those issued to web search engines in their length and structure, which finally draws the result that portals are used explorative rather than answer focused questions. Nevertheless, the exploration does not stop here as researchers are using log datasets to find anomalies in the system. Traditionally, anomalies are detected by using keyword search and regular expression match that is the root idea of our project. In modern times, many machine learning-based detection mechanisms have been proposed as discussed in [1]. Qimin Cao and Yinrong Qiao in [5] carry the above-stated idea by implementing a two-level machine learning algorithm and using a decision tree model to classify between normal and anomalous datasets, thus adding future inspiration for our work.

3 Working Methodology

The log file of any server consists of multiple records, and each record takes up one row in the log file. These log files could be in terabytes or more, thus it becomes very difficult to filter out required data. Each row of a log file could contain multiple columns, for example, the log file of a web server contains, source IP address, date, time, time zone, status code, response code, etc. These columns depend on a server that generates them. Thus, we often require a tool that can filter out required data whenever needed. Our project aims to create a tool that can filter the log files based on the requirement. As soon as the tool is executed, it displays a menu to the user that lists all the features provided by the tool. While writing the code, we kept the basic requirements in mind.

This tool provides five features to deal with logs as of now. These features are as follows:

I Convert the log file into CSV format.
II Search for a particular keyword in the log file.
III Block and unblock any IP address from the server.
IV Visualize the log files.
V Filter the logs in a particular time frame (Fig. 1).

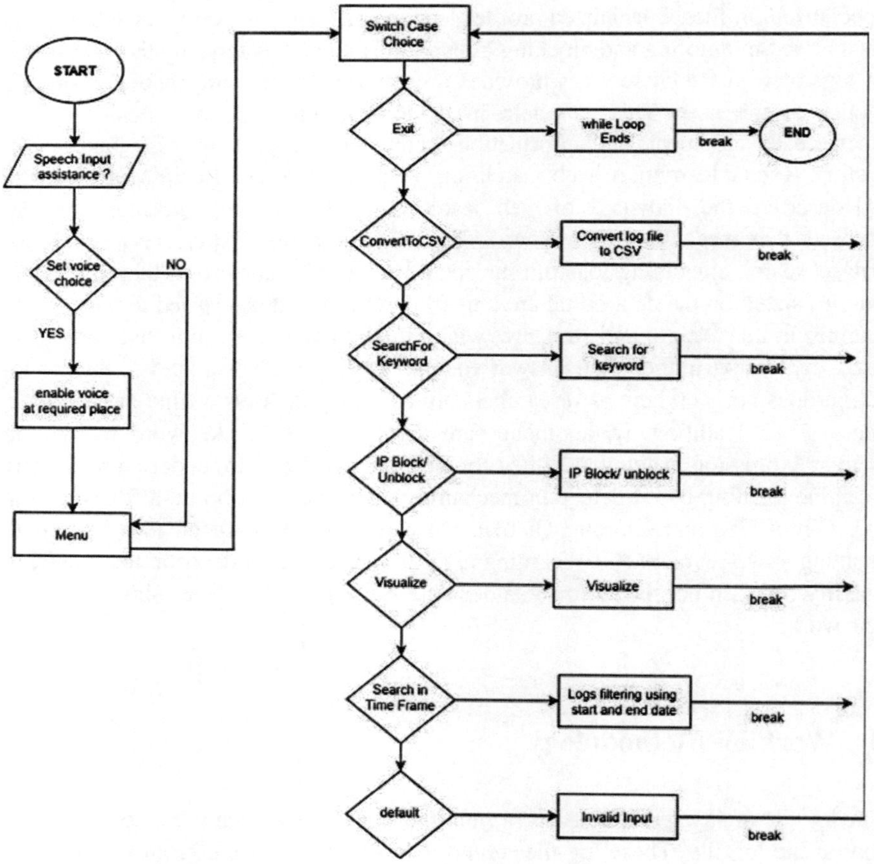

Fig. 1 Flow diagram explaining the working methodology

3.1 Converting the Log File into CSV Format

Log files could be very difficult to read most of the time, and this is why we need to convert them into such a format so that further processing can be done with ease. We wrote the code to convert the log file into comma-separated values (CSV) format. Nevertheless, why did we choose the CSV format specifically? Performing data cleaning, data processing, and data analysis can be done easily on a CSV file. We used Linux utilities like "sed", "awk", and regular expressions to convert the log [6] file into CSV format. We used "awk" to separate fields of a log file and write the output into a separate file. Then we used "sed" to remove unwanted special characters and separate different columns with a comma (",").

3.2 Search for a Particular Keyword in the Log File

Sometimes or the other, we might run into a situation where we want to search a pattern or keyword in a log file [7]. We implemented this feature using a basic pattern-matching algorithm. We used the "strstr()" method from the "string.h" header file of the C programming language. This algorithm has the quadratic worst case time complexity with respect to the searched time. Through the median of this tool, we are providing five options to search for a keyword. The user can search for either an IP address, date, HTTP method, status code, response code, or other. These options are specifically provided to deal with the web server logs, though this can be tweaked accordingly.

3.3 Block and Unblock an IP Address

This is a vital feature of any log analysis tool. While dealing with logs, we might need to blacklist certain IP addresses belonging to various users trying to access the server. To block and unblock the IP address, we use the "firewall-cmd" command–line tool to do the same. The program initially creates a new file "block.txt" which will store all the blocked user's IP addresses. We take the IP addresses to be blocked or unblocked from the user.

- **Command to block the IP**
  "firewall-cmd –permanent –add-rich-rule=rule family='ipv4' source address={input IP here} reject"
- **Command to unblock the IP**

  "firewall-cmd –permanent –remove-rich-rule=rule family='ipv4' source address={input IP here} reject"

The input IP address is concatenated within the command. The command is then executed to block/unblock the file. However, this is not it; we will be maintaining the list of blocked IP addresses (block.txt) so that our tool is unambiguous. Before blocking the IP address, the “block.txt” is traversed and if the IP address, is already blocked, then it will prompt an alert to the user. Similarly, while unblocking the IP address, it should be present in the “block.txt”, only then it will be unblocked.

3.4 *Visualizing Log Files*

The detailed working of this feature will be explained in the next section. In short, we will take a list of IP addresses from the user as input and then plot a graph that will represent the number of occurrences of an IP address in the log file. This representation is not limited to occurrences of IP addresses only; it can also be used to visualize the number of requests in a day that were made to the server.

3.5 *Search for Logs in a Time Frame*

This feature provides an option to search for logs [8] in a particular period (e.g., we can filter out logs that were recorded between January 1, 2021, to January 5, 2021). Both beginning and end dates were taken from the user. To print the output, we traverse the log file line by line. If the beginning date is found in any line, then the “flag” variable is set to “true”. Then if the “flag” variable is set to “false”, the end date is found in any line. Within each iteration, we print out the corresponding line if the “flag” is set to “true”. This will print out logs from the beginning date to the end date (exclusive). Now we have to print only those records of the end date. This can be easily done by traversing the log file and print only those records that contain the end date. Pattern matching is again the core algorithm for this feature.

4 Results and Discussion

After the experimentation for further convenience and better understanding, we visualized the final output in the form of a bar graph. As mentioned above, we provided functionality in our tool for visualizing IP addresses (refer to Sect. 3.4). We know that visualizing statistics [9] in a proper way provides additional intuition for a better understanding of the data, which often leads to a better examination of that same data. The below-represented bar graph depicts the visualization of the number of occurrences of different IP addresses required by the user in the required time frame in order to examine if any one of them shows abnormal behavior or in simple words shows a huge increase in its occurrences which could be directing

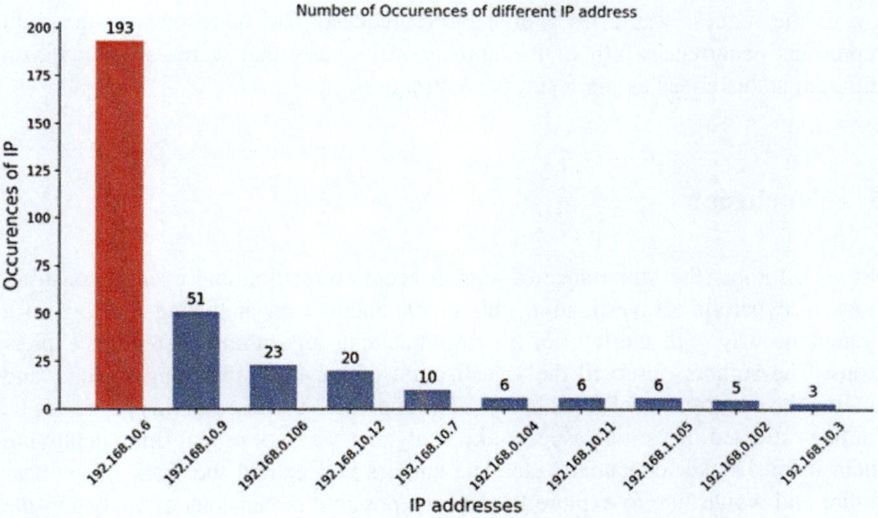

Fig. 2 Occurrences of IP addresses in the log file

toward some kind of malicious behavior or signs of malicious activity taking place. In Fig. 2, we can see that the first IP address (192.168.10.6) that is selected by the user within the selected time frame occurred the most (193 times to be precise) which is undoubtedly a high number for any IP address. This is also the case, due to the current threshold being set at a limit of 100 occurrences within a period, which implies that there is some issue or unethical activity is being performed. (Note: The threshold can be changed/set according to the requirements of the user). In Fig. 3, two pie plots are presented, one of the left represents the visualization over the comparison of occurrences of the different types of requests that were in the dataset,

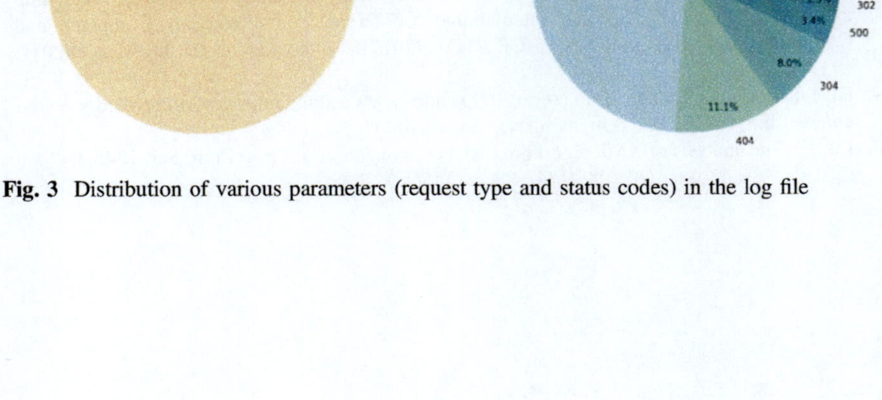

Fig. 3 Distribution of various parameters (request type and status codes) in the log file

i.e., if the request was a GET or a POST request. The other plot on the right represents occurrences [10] of the number of requests that were associated with different status codes as the result of that request.

5 Conclusion

As we all know the importance of logs in event correlation and event reconstructions in cybercrime investigation, this article focuses on analyzing the logs in a systematic way with the help of a string-matching algorithm and regular expressions. The authors observed the visualization of log files, searching of logs, and finding the occurrences of IP addresses. The approach discussed in this paper can be further extended in various aspects like analyzing the logs in real time, deploying them using Docker containers, etc. The authors will extend the work in the near future and would like to explore the various possible dimensions of analyzing the logs.

References

1. He, S., Zhu, J., He, P., Lyu, M.R.: Experience report: system log analysis for anomaly detection. In: 2016 IEEE 27th International Symposium on Software Reliability Engineering (ISSRE), Ottawa, ON, pp. 207–218 (2016). https://doi.org/10.1109/ISSRE.2016.21
2. Ni, Z., Liu, H., Chen, Y., Wu, D.: Research and implementation of a method for web log analysis template extraction. Procedia Comput. Sci. **162**, 673–681 (2019). ISSN: 1877-0509
3. Liu, X., Zhu, Y., Ji, S.: Web log analysis in genealogy system. In: 2020 IEEE International Conference on Knowledge Graph (ICKG), Nanjing, China, pp. 536–543 (2020). https://doi.org/10.1109/ICBK50248.2020.00081
4. Kacprzak, E., Koesten, L., Ibáñez, L., Simperl, E., Tennison, J.: A query log analysis of dataset search. In: Lecture Notes in Computer Science, pp. 429–436 (2017). https://doi.org/10.1007/978-3-319-60131-1_29
5. Cao, Q., Qiao, Y., Lyu, Z.: Machine learning to detect anomalies in web log analysis. In: 2017 3rd IEEE International Conference on Computer and Communications (ICCC), Chengdu, pp. 519–523 (2017). https://doi.org/10.1109/CompComm.2017.8322600
6. Best Practices: Event Log Management for Security and Compliance (2021). https://www.whatsupgold.com/resources/best-practices/event-log-management
7. Jansen, B.: The methodology of search log analysis. In: Handbook of Research on Web Log Analysis, pp. 100–123 (2009). https://doi.org/10.4018/978-1-59904-974-8.ch006
8. Sultana, N., Paira, S., Chandra, S., Alam, S.: A brief study and analysis of different searching algorithms. In: 2017 Second International Conference on Electrical, Computer and Communication Technologies (ICECCT) (2017). https://doi.org/10.1109/icecct.2017.8117821
9. IBM Knowledge Center.: Ibm.com (2021). https://www.ibm.com/support/knowledgecenter/en/ssw_ibm_i_72/rzasp/rzasp_regularexpression.html
10. Lu, X.: The analysis of KMP algorithm and its optimization. J. Phys. Conf. Ser. **1345**, 042005 (2019). https://doi.org/10.1088/1742-6596/1345/4/042005

An Efficient Detection and Prevention Approach of Unknown Malicious Attack: A Novel Honeypot Approach

Aatif Sarfaraz, Atul Jha, Avijit Mondal, and Radha Tamal Goswami

Abstract In this modern era, security has gotten to be the foremost broadly concerned in each domain as recently approaching malware postures a danger to the systems. So our fundamental concern is to identify and anticipate a malware assault on the system. As the polymorphic worm postures, an enormous challenge to identify as they have more than one occurrence and exceptionally expansive endeavors is required to detain every occurrence and to generate signatures. This work proposes malicious attack detection and prevention using honeypot technology. We have proposed a double-honeynet framework, which can distinguish and avoid modern worms. We apply system call analysis to detect malware which mostly focuses on the polymorphic and metamorphic virus rather than utilizing a signature-based approach.

Keywords Polymorphic worm · Honeypot · Double honeypot · Sticky honeypot · System call analysis

1 Introduction

Nowadays, malicious attacks became the biggest threat to network security. Malicious attacks include viruses, worms and trojans. In each of these attacks, some codes are written called malicious code or malware. Malware could be a sort of computer program outlined to require over or harm a computer's working framework without the client's information or endorsement. Worms are being the major

A. Sarfaraz (✉) · A. Jha
Department of Computer Science, Techno International Batanagar, Kolkata, India

A. Mondal
Makaut, Kolkata, India
e-mail: avijit.mondal@tib.edu.in

R. T. Goswami
Techno International New Town, Kolkata, West Bengal, India
e-mail: rtgoswami@tict.edu.in

K. Khanna et al. (eds.), *Cyber Security and Digital Forensics*,
Lecture Notes on Data Engineering and Communications Technologies 73,
https://doi.org/10.1007/978-981-16-3961-6_2

threat among all malicious attacks, because of their self-replicating nature [1]. It is capable of copying itself and sending it over the network to the computers on the network without any human interaction. Unlike viruses, worms did not need to attach themselves to any program in order to propagate. As there is a particular behavior, blocking software program is there which coordinates with the working framework and monitors the program behavior in genuine time for malicious activities. With the evolution of network security tools and techniques, malware is getting to be more intelligent each day, and polymorphic malware is the most recent participants in this calamitous amusement of overcoming the adversary. Polymorphic worm being one of the latest and effective among all [2]. A polymorphic worm changes its appearance with every occurrence [3] such that each occurrence of which receives a totally distinctive personality from its parent, which makes it difficult to gets detected.

Network intrusion detection system (NIDS) must be established in order to protect the network from malware, and it acts as an alarm that notifies the network admin every time malware is detected. IDSs basically are of two types: host-based and network-based. Host-based IDSs inspect data held on discrete computers in the network which serves as a host, whereas network-based IDSs inspect data interchanged between computers.

Security specialists physically create the IDS signature by studying the network traces after the modern worm has been discharged. Tragically, this work takes a parcel of time. Analysts have, as of late, given consideration to computerizing the era of marks for IDS to coordinate worm activity.

Our research is based on the honeypot technique which provides a solution for the above-mentioned problem. A honeypot may be a trap that mimics an authentic network resources, a self-contained secure and monitored area [4, 5]. Its essential objective is to bait and detect malicious attacks and intrusions. It can be used for surveillance and early warning; it can also benefit security researchers to understand emerging threats. Honeypots can be categorized as high and low interaction honeypot [6]. A high interaction honeypot like honeynet works as a real operating system, whereas a low interaction honeypot like honeyd imitates one or more than one real system.

Our research proposes a double-honeynet architecture which gathers all possible instances of the polymorphic worm and sends them for signature generation. In our research, we have used system call analysis approach instead of signature-based approach for the detection of the polymorphic worm inside the signature generator. Our system makes it possible to gather all the instances of the polymorphic worm and then forward those instances to the signature generator which then generates the signature.

This paper is divided into the following segments. Segments 1.1 and 1.2 give a brief idea about worms and polymorphic worms. Segment 2 reviews the associated work for the automatic signature generation framework. Segment 3 presents our proposed framework to describe the problem encountered by the current automatic signature generator. The signature generator algorithm for polymorphic and metamorphic virus using system call analysis will be discussed in Segment 4. The algorithm for our proposed architecture is discussed in Segment 5. Segment 6 concludes the paper, and in Segment 7 we will discuss some future work that we intend to perform in the future.

1.1 Worms

"A worm is a self-reproducing program that can be created with the capabilities to perform any kind of task," for example, deletion of files or sends documents via e-mail. There can be an adverse effect on network traffic due to the worm self-replicating property:

A worm:

(a) may install a pathway also called a backdoor for unprotected access by the attacker in the infected system.
(b) vulnerability introduces the worm into the system.
(c) may infect one system and by the property of self-replication, it can spread all over the system network.

1.2 Polymorphic Viruses

Polymorphic infections assume numerous forms by scrambling code in an unexpected way with every infection. This term caught in within the mid-1990s with devices that risen to produce thousands of modern slight variations of code based on transformed routines to subvert-signature innovation utilized by an anti-virus computer program.

A polymorphic infection incorporates an encrypted infection body and a decryption routine, to begin with controlling the computer and later decodes the infection body [7].

However, a polymorphic infection moreover includes a mutation engine that produces randomized decryption routines that alter each time an infection infects an unused program. The mutation engine and virus body were both encrypted in a polymorphic infection.

As soon as the users run an infected program, following activities will take place:

- The framework is beneath the control of the decryption routine that can decode both the infection body and the mutation engine [8].
- Next, the control has been shifted from the system to the virus through the decryption routine which finds a new program to infect.
- Next the infection replicates itself along with the mutation engine in RAM.

- At this point, mutation engine gets invoked by the virus which then arbitrarily produces a new decryption routine that can decrypt the virus. However, it marks a very little or no likeness to any earlier decryption routine
- At this point, the current copy of the virus body and the mutation engine gets encrypted by the virus [9, 10].
- Ultimately, the virus adds the current encryption routine, together with the recently encrypted virus body and mutation engine, onto the latest program.

As a consequence of the above-mentioned process, the virus body is encrypted as well as virus decryption routine differs from malware to malware. With no predetermined signature to scan for and no predetermined decryption routine to malware, it cannot look the same.

2 Literature Review

Levin et al. [9] in his paper outlined the technique of how the honeypot pulls out the characteristics of worm exploits, which can be examined for signature generation. The drawback of his paper is that the signature for the worm exploit is detected manually, and this process of manual signature generation for the worm takes a lot of time which gives worms more time to propagate and infect other hosts in the network. Our paper uses double-honeynet architecture which overcomes the above-mentioned problem by consequently producing signatures for worms. Our framework reduces the time and effort required for signature generation.

Honeycomb was one of the first proposed frameworks for signature generation designed by Kreibich and Crowcroft [11]. Honeycomb can be implemented as a honeyd plugin, and it produces signatures from the detected activities at honeypot. The honeycomb system is based on the LCS algorithm, mainly focuses on the "longest-shared-byte" sequence between different sets of packets. To match almost all kinds of worm instances, a honeycomb generates a signature that comprises a single or contiguous substring of worm payload. In this paper, the major drawback is the signature which is generated by the honeycomb fails to capture all the instances of the polymorphic worm with low false negative and low false positive. Our paper uses double-honeynet architecture which is based on system call analysis that overcomes the above-mentioned drawback by capturing almost all instances of polymorphic worms. The system call analysis is used in our framework over the signature-based approach as it focuses mostly on polymorphic and metamorphic worms which help in capturing almost all kinds of polymorphic worm instances.

Goswami et al. [12] in their paper described how to automatically generate the signature for a polymorphic worm using double-honeynet framework which is built on the PCA, which uses the most noteworthy information shared between each and every polymorphic worm instances as the signature. The drawback of this framework is that it may not distinguish all the occasions of the polymorphic worm.

Singh et al. [13] in their paper outlined a system called the Earlybird system for the signature generation to identify worm. This architecture calculates packet content prevalence at a single observation point like a network demilitarized zone. The Earlybird system differentiates benign repetitions from outbound content, by counting the number of definite starting and endpoints related to the string that repeats frequently in the payload. Similar to honeycomb, Earlybird also generates a signature which comprises a single-contiguous substring of a payload to coordinate most of the worm occasions. In this paper, the major drawback is the signature which is generated by the Earlybird system fails to catch all the occurrences of the worm with low false negative and low false positive. This paper uses a system call analysis-based double honeynet framework which focuses more on polymorphic and metamorphic worms that overcomes the above-mentioned paper's drawbacks.

3 System Architecture

A polymorphic worm could be a sort of worm which can alter its appearance each time it propagates, so a polymorphic worm has numerous instances. After infecting every host, a new occurrence of polymorphic worm is produced for propagation. A single honeynet is capable of detecting only one instance of it. Therefore, we need a double-honeynet framework which will catch all the conceivable polymorphic worm instances. The proposed double-honeynet model establishes a loop that accumulates all polymorphic worm occurrences as shown in Fig. 1. For all of the above-mentioned reasons, we propose a double honeynet architecture for the detection of zero-day polymorphic worm automatically.

The process begins when the incoming traffic approaches the gate translator and passes through it. The gate translator then separates the suspicious incoming traffic and swerve them to honeynet 1. "The gate translator consists of publically attainable addresses, which represent wanted amenities." Traffics linked to any additional addresses, other than the publicly attainable addresses, were considered suspicious and are swerve to the honeynet 1 by gate translator.

In the next step, the unwanted suspicious traffic which arrived at honeynet 1 will strive to make an outbound connection. An internal translator that is implemented in the router and associated with each honeypot separates the honeypot from the entire network. All the outbound connections that are made from honeynet 1 will be obstructed by the internal translator and are swerve to honeynet 2. Honeynet 2 does the same forming a loop.

When adequate occurrences of worm payloads are accumulated by honeynet 1 and honeynet 2, then they are passed through a sticky honeypot which moderates down the worm proliferation rate so that it will be easy for the signature generator to generate signatures. Afterward, the signature generator generates signatures automatically by system call static analysis that will be discussed in the next section. All the malicious traffic will go through the system call analysis, and a corresponding signature will be generated. The generated signatures are forwarded

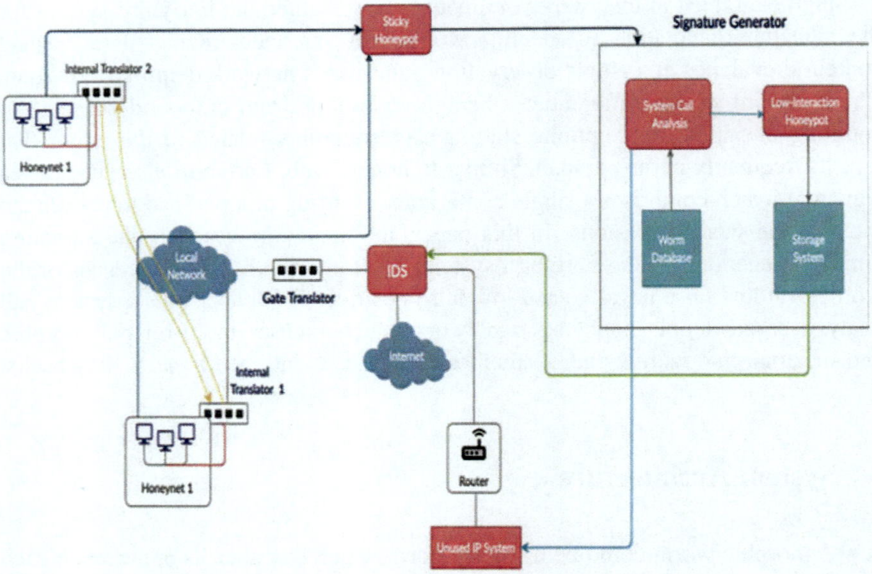

Fig. 1 System architecture

to the centralized storage system via a low interaction honeypot. Then the IDS can get all the information so that by analyzing it can protect against any attack. If the system call analysis is incapable of detecting any signature, then the payload will be automatically transferred to the unused IP system.

4 System Call Analysis

It is utilized to distinguish malware, generally focuses on the polymorphic and metamorphic infections. This strategy works based on the presumption that all malware variations share a common core signature—a combination of many highlights of the programming code [14]. In this technique, two fundamental steps were included: To start with, the portable executable (PE) decompressed and passed through a parser, this parser makes a list of Windows APIs calling grouping. Minute, this API grouping will be compared against the worm database, and a similarity measure was utilized to conclude the analyzed file [15]. In the event, if the similarity is more noteworthy than a certain limit, then at that point detection is triggered.

5 Proposed Algorithm

Step 1: The gate translator swerves the traffic coming from the Internet toward honeynet 1.
Step 2: Internal translator that is implemented in the router disconnects honeynet 1 from the entire network.
Step 3: All the outbound connection from honeynet 1 was obstructed by the internal translator and swerves them to honeynet 2. Honeynet 2 does the same forming a loop.
Step 4: When an adequate amount of worm instances are acquired by looping through honeynet 1 and honeynet 2, then they are swerved to the sticky honeypot to slow down the worm propagation rate.
Step 5: Sticky honeypot transfers the worm instances to the signature generator to generate their signatures.
Step 6: Signature generator consists of system call analysis, worm database, low interaction honeypot, and centralized storage system.
Step 7: In system call analysis, portable executable file is decompressed and passed through a parser which creates a list of windows API calling sequences.
Step 8: These API sequences will be compared against the worm database, if the similarity is greater than a certain threshold, then detection is triggered.
Step 9: All these worm signatures which are detected by system call analysis will be transferred to a centralized storage system through low interaction honeypot by which IDS can get all the payload information.
Step 10: In case, the system call analysis is incapable of detecting any signature, then that payload will be automatically redirected to the Internet through an unused IP system.

6 Conclusion

We have proposed an algorithm to generate the signature of newly emerged polymorphic worms automatically. In this paper, we have proposed a new detection technique "double honeynet" for the detection of newly emerged polymorphic or metamorphic worms. The Framework also includes a sticky honeypot which slows down the worm propagation rate. The system is based on the system call analysis that compares windows API sequences with the worm database; if the similarities are greater than a certain threshold, the detection is triggered.

7 Future Work

As honeypots are moderately a modern innovation and have a great scope for future applications, we plan to execute this proposed framework on an experimental test bed.

In the future, we intend to propose a new architecture which will be based on sandboxing. Using a sandboxing for modern malware discovery gives another layer of assurance against modern security threats—zero-day polymorphic worm. And whatever happens within the sandbox remains within the sandbox—avoiding system failures and keeping computer program vulnerabilities from spreading.

"A sandbox in a cybersecurity domain is a secluded environment on a network that imitates an end user working environment. Sandboxes are used to execute malicious code to analyze the results without having any adverse effect on the rest of the network. Sandbox environment gives a proactive layer for network security defense against threats."

The key objective of our research is to lower the false alarm rates and to produce high-quality signatures for the polymorphic worm.

References

1. Spitzner, L.: Honeypots: Tracking Hackers. Addison Wesley Pearson Education, Boston (2002)
2. Tang, Y., Chen, S.: An automated signature-based approach against polymorphic ınternet worms. IEEE Trans Parallel Distrib Syst **18**(7), 879–892 (2007)
3. Cavallaro, L., Lanzi, A., Mayer, L., Monga, M.: LISABETH: automated content-based signature generator for zero-day polymorphic worms. In: Proceedings of the Fourth İnternational Workshop on Software Engineering for Secure Systems, Leipzig, Germany, May 2008, Lissa (2008)
4. Mohhammed, M.M.Z.E., Anthony Chan, H., Ventura, N.: Honeycyber: automated signature generation for zero-day polymorphic worms. In: Proceedings of the IEEE Military Communications Conference, MILCOM (2008)
5. Fogla, P., Sharif, M,. Perdisci, R., Kolesnikov, O., Lee, W.: Pol polymorphic blending attacks. In: Proceedings of the 15th Conference on USENIX Security Symposium, Vancouver, B.C., Canada (2006)
6. Li, Z., Sanghi, M., Chen, Y., Kao, M.Y., Chavez, B.: Hamsa: fast signature generation for zero-day polymorphic worms with provable attack resilience. In: Proceedings of the IEEE Symposium on Security and Privacy, Oakland, CA, May (2006)
7. Kim, H.-A., Karp, B.: Autograph: toward automated, distributed worm signature detection. In: Proceedings of 13 USENIX Security Symposium, San Diego, CA, Aug. (2004)
8. Kreibich, C., Crowcroft, J.: Honeycomb–creating intrusion detection signatures using honeypots. In: Workshop on Hot Topics in Networks (Hotnets-II), Cambridge, Massachusetts, Nov. 2003 (2003)
9. Levine, J., La Bella, R., Owen, H., Contis, D., Culver,B.: The use of honeynets to detect exploited systems across large enterprise networks. In: Proceedings of IEEE Workshops on Information Assurance, New York, June 2003, pp. 92–99 (2003)
10. Gusfield, D.: Algorithms on Strings, Trees and Sequences. Cambridge University Press, Cambridge (1997)

11. Krebitch, C., Crowcroft, J.: Honeycomb: creating intrusion detection signatures using Honeypots. ACM SIGCOMM Comput Commun Rev **34**(1), 51–56 (2004)
12. Goswami, R. T., Mondal, A., Mishra, B. K., Mahanti, N. C.: Handbook of Information Security. John Wiley & Sons, Inc., Hoboken, New Jersey
13. Singh, S., Estan, C., Varghese, G., Savage, S.: https://www.lastwatchdog.com/internets-40th-anniversary-timeline-milestones/
14. https://epdf.pub/official-isc2-guide-to-the-sscp-cbk-second-edition.html
15. Salah, E.D., Aslan, H.K., El-Hadidi, M.T.: A detection scheme for the SK Virus. In: Proceedings International Federation for Information Processing (2002)

Analysis of Risk and Security Within Fog Computing-Enabled e-Healthcare System in Uttarakhand

Naveen Tewari and Sandeep Kumar Budhani

Abstract With the advent of fog computing, various e-governance services get influenced as fog provides new service delivery models and new ways to interact with the citizens. Uttarakhand, as a state of 86% hilly region and 65% of forest area, got geographical conditions that are not so favored for cloud-enabled technologies, because cloud needs regular and high bandwidth Internet connectivity. Fog computing can be a key player, in terms of providing e-Governance service. E-Health is the major service provided under the e-governance platform. With the application of Internet-enabled and IoT-based e-Healthcare systems in the state like Uttarakhand, there can be a drastic improvement in health services. E-Health services require real-time processing, low latency, high consistency, and high data rate, and these all parameters are fulfilled by fog computing. There are many kinds of research that describe how fog can be used in e-Healthcare systems. In this research paper, we discuss fog computing in the context of Uttarakhand. Our main concerns are security issues and challenges faced by fog computing while using the E-Healthcare system in hilly areas of Uttarakhand. In studying those challenges and security issues, the technologies related to fog computing are also discussed.

Keywords Fog computing · Internet of Things · Cloud computing · e-Health care · e-Governance

1 Introduction

Uttarakhand is a state of the Himalayan region that has very different geographical conditions than other states in India. Many services and applications use Internet of Things (IoT) today, due to a variety of sensor availability, low-cost Internet, and cloud services [1]. Technology like IoT and fog can be installed within the people's

N. Tewari (✉)
School of Computing, Graphic Era Hill University, Bhimtal, India

S. K. Budhani
Computer Sciences and Engineering, Graphic Era Hill University, Bhimtal, India

K. Khanna et al. (eds.), *Cyber Security and Digital Forensics*,
Lecture Notes on Data Engineering and Communications Technologies 73,
https://doi.org/10.1007/978-981-16-3961-6_3

range and serve them with different offerings. One such important e-Governance service provided by fog computing is in e-Health care. It also raises the patient's life quality by providing real-time health data analysis [2].

IoT extends the services drastically that are traditionally given by Internet-based technology. With the advancement in sensors used in IoT, the data collection techniques raised their standards significantly. CISCO forecasts that the size of data accumulated by IoT devices may reach 600 zeta bytes per year by the year 2020 [3]. When this large amount of health data gets transferred between devices and the cloud, it creates the problem of low latency and system idleness. As we are concern about e-Health implementation, the above-stated issue represents some difficulties in how to deal with a large number of accumulated health datasets to keep low latency in real-time applications [2].

Fog computing solves some of the difficulties raised by the cloud of things (CoT). Fog computing has the potential to give reasonable answers to all issues related to the use of IoT and cloud in e-Healthcare applications. In the E-Health solution, all the data gathered from IoT sensors must not be transferred to the cloud, some can be interpreted at the data generation point, and the processed information is then transferred to cloud or any other device for further usage. This can only be done with the help of fog computing [1].

E-Healthcare platform consists of applications and services used to gather and provide clinical data [4]. Our main focus in this research is to figure out the risk factors involved in fog-enabled e-Health system in Uttarakhand. The cost will also cut down because the patient does not have to come to the hospital for monitoring purposes, they can be monitored remotely with the help of fog architecture [5].

The present research gives the details of fog-enabled e-Healthcare systems. This research also gives the main characteristics of an e-Health system combined with fog computing. Some of the main objectives of this paper include

i. An overview of e-Healthcare systems in Uttarakhand.
ii. An investigation of how e-Health applications profit by the fog computing.
iii. Present the various terms in a presentable way, so that it can be understood why these technologies are collaborating.
iv. Finding the risk of involving fog in the e-Healthcare system in a state like Uttarakhand and analyzing them.

2 Methodology

This is a descriptive study in which several published reports, research studies, articles, e-books, vulnerabilities notes, and government notifications are used as the secondary data source to assess and analyze the security vulnerabilities in popular web browsers. To gather the secondary data, Google Scholar (https://scholar.google.com/), Directory of Open Access Journals (https://doaj.org/) and Connecting Repositories (https://core.ac.uk/) are used. Appropriate search keywords are used to find the requisite data on the above-used databases, such as e-Health, fog computing, cloud-based e-Health system, fog-based e-Health system, e-Health in Uttarakhand, etc. Further, the data is reviewed and analyzed accordingly to the objectives of the study.

3 Overview of Key Technologies

E-Health care is the utilization of data and correspondence innovative technology in health care and is viewed as basic for an advanced, financially savvy health administration system that is prepared to address difficulties and challenges. Various technologies are combined to provide the foundation for an e-Healthcare system. In this section, these technologies are discussed as follows:

Internet of Things—In recent years, devices are developed with the inbuilt feature of Internet communication. These smart devices or things can communicate with each other via the wireless network. Devices can be hardware, software, or even physical objects that can talk to each other [6]. With its rapid advancement and low cost, IoT will be very beneficial if included in the healthcare environment. Using data gathered with IoT sensors, one can advance his decision-making abilities, and when used in health care, doctors or health managers can take quick and decisive measurements for the betterment of patients. The patients also take part in the system actively and contribute by checking and maintaining their health records and IoT devices [5].

Cloud Computing—Cloud computing as defined by NIST is a pool of shared computing resources accessed on-demand on a network [7]. Computing resources can be shared with the help of three service models (SaaS, PaaS, and IaaS) and four deployment models (public, private, hybrid, and community) [7]. Scalability, rapid deployment, and low cost are the key characteristics of cloud computing. Consumers can use, switch, or end any computing service automatically according to its need [2]. One of its main advantages is the availability of enormous data storage capacity.

Fog Computing—Fog computing was characterized by Cisco as an amplification of cloud computing where data is stored at the edge to raise the standard of services with low latency and high data analysis [1, 8]. It has the same features as implemented by the cloud, i.e., computation, storage, and services, etc. Fog processes data and can segregate and transmit it after analyzing, resulting in time and resource saving [9].

The Network Technology Used in IoT, Cloud, and Fog Computing—The network is also a vital player in using IoT, cloud, or fog. Some of the important network technologies used are discussed here:

(i) *RFID*—The short-range communication can be done by using radio frequency identification (RFID) for each device used in the network [10]. RFID can be very well suited for healthcare fog devices, as these are cheap, reliable, and trackable.

(ii) *Bluetooth*—Short-wavelength UHF radio waves are used in this standard to exchange data between mobile and fixed devices from 2.402 to 2.480 GHz [11]. It can provide speed up to 3 Mbps over a range of 100 m.

(iii) *WiFi*—WiFi is one of the main network technology used in IoT. This is the method through which millions of devices are connected to offer Internet-based services [12]. Some WiFi exceeds the speed of 1 Gbit/s over a range of 100–150 m. Most of the healthcare environments in hospitals have a WiFi network available. WiFi can be accessed with mobile devices as well as wearable sensor devices, which makes it a highly commendable option for fog computing.

(iv) *WiMax*—WiMax provides a range of up to 50 km. It is best suited for communication between fog nodes that are in different parts of a city. Different standards of WiMax starting from IEEE 802.16a to IEEE 802.16c provide flexibility in choosing bandwidth range from 2 to 66 GHz [13].

(v) *Mobile Network* (*2G, 3G, 4G, LTE, 5G,* etc.)—The use of mobile in IoT and fog is a necessity without which the concept does not work. 2G, 3G, 4G, LTE, and 5G are the cellular technology known to use for long-range connectivity wirelessly [14]. E-Healthcare facility uses a mobile device and network to authenticate and share information and services. Long-range connectivity is also needed for healthcare facilities in Uttarakhand state. Cost-effective development in smart devices in recent years allows common people to use its services and connect to the world with the latest technology [14].

(vi) *WSN*—Wireless sensor network or WSN is a network of geographically detached sensors that are used to take care of physical environments [15]. Information assortment and transmission are the essential activities of WSNs [16]. And, like IoT, fog is dependent on their sensors for data collection; therefore, WSN has the advantage to be used in these technologies. The advancement of detecting IoT devices makes WSNs continuously develop into a classic detecting stage, which can give information mindful administrations to a variety of uses [17].

4 Background Study

This section is to study the work done by other researchers to get an understanding of implementing fog computing-based e-Healthcare systems in rural areas of Uttarakhand. Fog computing is becoming an important delivery model of Internet-based services. Implementing fog gives benefits as well as some disadvantages to be discussed. This section will be focused on getting and reviewing the work in the last few years implemented in this area. The aim of getting the gray area (security challenges) will also be done through this literature review.

Kuo [18] described and focused on the use of cloud computing in the e-Healthcare environment. In the research paper, Kuo elaborated on the challenges of cloud computing some are lack of trust by the user, cultural resistance, not defining proper service level agreement by the service provider, data lock-in, low latency, and virus in distributed cloud systems, etc. Many other security issues are discussed in the paper with a detailed explanation.

Alharbi et al. [19] study the role of cloud base e-Healthcare applications in the context of Saudi Arabia. The study provides a business perspective on the healthcare model and shows its cost effectiveness. Habiba et al. [20] give the security issues involved in the cloud computing identity management paradigm. They analyze various cloud identity management systems and discussed their security issues and also provide ways to solve these vulnerabilities.

Ouedraogo et al. [21] address the lack of transparency in security models of cloud, whereas Hashizume et al. [22] discussed the differences between vulnerabilities and threats and created a relationship between them. Then they also proposed new security techniques to mitigate these threats. Wang et al. [23] dissect the productivity, avg. frame rate, delivering execution bottleneck of the cloud delivering framework, and set forward an extraordinary boundary change technique to improve framework execution, by advancing related server and delivering machine setup. They conclude that an increase in the number of users directly increases the average response time, i.e., low latency.

Mitton et al. [24] include sensors in cloud computing in their research. They coined the name sensing and actuation as a service (SAaaS) for this architecture. This architecture is used to provide sensors and actuators as a service to the user. And the cloud is involved to give the computation and data storage benefits. We can also say this as CoT, i.e., cloud of things.

Ibrahim et al. [25] defined the term fog health as the inclusion of fog computing into the healthcare paradigm. They also recapitulate the issues in the domain of fog computing and health care. From their literature review, they found that low latency and slow data analysis are the main issues in distant healthcare monitoring systems. Silva and de Aquino Junior [26] also consider low latency in cloud computing as one of the main issues that can cause a failure of the healthcare system. Fog-enabled healthcare systems are liable to solve the problem of low latency. Therefore, the authors suggested the use of fog computing in this area for solving much of its issues. They also figure the lack of well-demonstrated architecture of fog-enabled e-Healthcare systems. Vilela et al. [2] present a review of the application and challenges of the fog-enabled e-Healthcare system. They summarize the already available solution to the problems raised within the system like latency, security, privacy, etc. They also study communication protocols used between edge and cloud computing.

By introducing fog computing for various e-Health services that are previously delivered by cloud computing, most of the vulnerability (issues) can be solved. But whether these services are beneficial and practical if implemented in rural and hilly regions of Uttarakhand, we will study the challenges that have to come in those distant areas.

5 Healthcare Scenario in the Himalayan Region (Uttarakhand)

Admittance to medical services in the rural parts of Uttarakhand keeps on being poor. Given the limitations of territory and geology and the small and dissipated nature of the provincial settlements, expanding access represents a significant challenge. As most of the people living in hilly areas are poor, they cannot bear the cost of the significant expense of private clinical consideration. There is an intense lack of different healthcare human resources in the state-run Primary Health Centers (PHC). Almost 90–94% of physician's posts are vacant in the state. In hilly areas, there are only two PHC which are operational per 100,000 peoples [27]. This shows that the ratio between health facilities and patients is very low. This scarcity leads to the unsuccessful implementation of health services provided by the government [27, 28]. For raising the standard of healthcare facilities and improving the service availability to common and poor people, there should be a mechanism that will address the problem in hilly and rural areas of Uttarakhand.

In the Himalayan state, Uttarakhand, initiatives are been taken by the state government to fulfill the need of people for e-Healthcare services. Many projects are going on successfully like [29].

The e-Health services (Table 1) that are started by the Uttarakhand Government are unproductive due to lack of Internet infrastructure and awareness. Expanding and reinforcing the quantity of health facilities focus that could work nonstop, particularly in rural zones, would go far in diminishing the health problems in the state. This can be done if health facilities are incorporated with fog computing. Using fog computing for providing e-Health services can help in the timely recognition and avoidance of diseases.

Several healthcare services can be provided by incorporating fog computing in e-Health systems like smart health monitoring systems/smartphone-based e-Health solutions [30], home healthcare systems [5], collecting health data in real time [9], automatics dialing of emergency numbers, hospital e-Management [5], etc.

Table 1 Various e-Healthcare services running in Uttarakhand

e-Healthcare service	Benefit/Use of service
e-Parchi	Outpatient department (OPD) registration system
e-Attendance	Attendance tracking of medical staff
e-Aushadhi	Record of medication data from providers to the stockroom to healthcare offices to patients
e-RaktKosh	Online facility to collect blood, plasma, and platelets
108 Services	24 × 7, toll-free emergency telephone number for calling Healthcare ambulance from any location within the state

6 Security and Challenges

This section gives a clear idea of security and challenges that must be covered to build a reliable, user-friendly, manageable, and secure fog-implemented e-Healthcare system. Fog provides many advantages to authorities and patients, also enhances the quality and effectiveness of the various services provided by healthcare systems. In a region where demographic conditions are not-so-good, fog-based e-Healthcare systems can provide an expert solution to the health problems of the citizen. As there are several merits, fog also has some issues that must be addressed before applying it to healthcare systems. Fog-based healthcare system manages very sensitive health datasets, so security is the main concern while using it.

Important security concerns/challenges in fog-based e-Healthcare system are as follows:

Safety of Healthcare Data—Safety of data is the main concern in the healthcare system. Patient-related data that is generated through IoT sensors is very much sensitive and should be handled carefully [9]. Data security concern contains issues like illegitimate data access, data changing and removing issues, ownership changing issues, and data sharing with other users. With the said issues, there is a very high risk of illegal file/database access in the fog system. Data should be encrypted with secured keys and transferred via secured socket layer (SSL) for maintaining data security. Ownership access should be managed securely so that the user cannot view the health dataset of other people.

Wireless Security Issues—Generally, a wireless network is used between fog devices for data communication [31]. This network is also vulnerable to various attacks like data infringe, losing data in-between communication, IDimitation, communication response attacks, etc. This can result in low privacy, inadequate accuracy, and also lower the trustworthiness of the system. Encrypted communication is a way to overcome this type of attack. Proper authentication and secure data routing will enhance wireless security features.

Computer Virus—These issues are related to computer viruses like Trojans, malware, worms, spyware, etc., which can degrade the performance of the system and can also spoil the health data permanently. Various antivirus programs are available that can be included in the system, for getting rid of these computer viruses. Data backups should be taken periodically to overcome any such vulnerability.

Although monitoring the health of old age people become very easy if we include fog computing into the system, this system is tracked with the help of a mobile phone. There is also a problem with aged people as they are almost unaware of modern gadgets like smartphones, etc., and they are not able to use them efficiently and accurately. This is a major challenge in applying smartphone-based technology in health care.

7 Conclusion

Despite various improvements in the health infrastructure in Uttarakhand, there is a shortage of health personals and primary health care centers. The geographical condition is also not very favorable for infrastructural development. An insufficient number of trained health personals give rise to the system that enables to monitor the health of people without much human intervention. Fog computing-based e-Healthcare system provides a way to enhance the capability of health monitorial without including trained health personals, as almost all the monitoring is done by the IoT and fog-based sensors. So this technology enhances the option of giving better e-Health solutions to the distant people of Uttarakhand. There are a good number of benefits seeing we can suggest fog computing as a technology that can change the scenario of health facilities in Himalayan regions. But there are some security issues and challenges also that must be entertained before proceeding further. For future work, we are working on an architectural model of a fog computing-based e-Health system that includes RFID, NFC, LWPAN, and other WiFi technologies so that the healthcare department gets benefitted from these advancements.

References

1. Caiza, G., Saeteros, M., Oñate, W., Garcia, M.V.: Fog computing at industrial level, architecture, latency, energy, and security: a review. Heliyon **6**(4) (2020). https://doi.org/10.1016/j.heliyon.2020.e03706
2. Vilela, P.H., Rodrigues, J.J., Righi, R.D., Kozlov, S., Rodrigues, V.F.: Looking at fog computing for e-Health through the lens of deployment challenges and applications. Sensors **20**(9), 2553 (2020). https://doi.org/10.3390/s20092553
3. Networking, C.V.: Cisco Global Cloud Index: Forecast and Methodology, 2015–2020. White Paper; Cisco Public, San Jose, CA, USA (2016)
4. Bisio, I., Estatico, C., Fedeli, A., Lavagetto, F., Pastorino, M., Randazzo, A., Sciarrone, A.: Brain stroke microwave imaging by means of a newton-conjugate-gradient method in Lp Banach spaces. IEEE Trans. Microw. Theory Tech. **66**, 3668–3682 (2018). https://doi.org/10.1109/TMTT.2018.2849060
5. Ahmadi, H., Arji, G., Shahmoradi, L., Safdari, R., Nilashi, M., Alizadeh, M.: The application of Internet of things in healthcare: a systematic literature review and classification. Univ. Access Inf. Soc. **18**(4), 837–869 (2018). https://doi.org/10.1007/s10209-018-0618-4
6. Jung, S., Myllyla, R., Chung, W.: Wireless machine-to-machine healthcare solution using Android mobile devices in global networks. IEEE Sens. J. **13**(5), 1419–1424 (2013). https://doi.org/10.1109/jsen.2012
7. Mell, P., Grance, T.: The NIST Definition of Cloud Computing; Computer Security Division. Information Technology Laboratory, National Institute of Standards and Technology Gaithersburg, Gaithersburg, MD, USA (2011)
8. Tewari, N., Datt, G.: Towards FoT (fog-of-Things) enabled architecture in governance: transforming E-Governance to smart governance. In: 2020 International Conference on Intelligent Engineering and Management (ICIEM) (2020). https://doi.org/10.1109/iciem48762.2020.9160037

9. Khan, S., Parkinson, S., Qin, Y.: Fog computing security: A review of current applications and security solutions. J. Cloud Comput. **6**(1) (2017). https://doi.org/10.1186/s13677-017-0090-3
10. Yue, Z., Sun, W., Li, P., Rehman, M.U., Yang, X. (eds.): Internet of things: architecture, technology and key problems in implementation. In: 2015 IEEE 8th International Congress on Image and Signal Processing (CISP) (2015)
11. Gentili, M., Sannino, R., Petracca, M.: BlueVoice: voice communications over Bluetooth low energy in the Internet of things scenario. Comput. Commun. **89–90**, 51–59 (2016). https://doi.org/10.1016/j.comcom.2016.03.004
12. Lee, I., Kim, M.: Interference-aware self-optimizing Wi-Fi for high efficiency Internet of things in dense networks. Comput. Commun. **89–90**, 60–74 (2016). https://doi.org/10.1016/j.comcom.2016.03.008
13. Zemrane, H., Abbou, A.N., Baddi, Y., Hasbi, A.: Wireless sensor networks as part of IOT: performance study of WiMax—Mobil protocol. In: 2018 4th International Conference on Cloud Computing Technologies and Applications (Cloudtech) (2018). https://doi.org/10.1109/cloudtech.2018.8713351
14. Basir, R., Qaisar, S., Ali, M., Aldwairi, M., Ashraf, M.I., Mahmood, A., Gidlund, M.: Fog computing enabling industrial Internet of things: state-of-the-art and research challenges. Sensors **19**(21), 4807 (2019). https://doi.org/10.3390/s19214807
15. Alhalafi, A., Sboui, L., Naous, R., Shihada, B.: GTBS: A green task-based sensing for energy efficient wireless sensor networks. In: 2016 IEEE Conference on Computer Communications Workshops (INFOCOM WKSHPS) (2016). https://doi.org/10.1109/infcomw.2016.7562060
16. Akber, S.M.A., Khan, I.A., Muhammad, S.S., Mohsin, S.M., Khan, I.A., Shamshirband, S., Chronopoulos, A.T.: Data volume based data gathering in WSNs using mobile data collector. In: Proceedings of the 22nd International Database Engineering and Applications Symposium on—IDEAS 2018, pp 199–207. ACM Press, New York (2018). https://doi.org/10.1145/3216122.3216166
17. Chen, Y., Liu, W., Wang, T., Deng, Q., Liu, A., Song, H.: An adaptive retransmit mechanism for delay differentiated services in industrial WSNs. EURASIP J Wireless Commun. Netw. **2019**(1) (2019). https://doi.org/10.1186/s13638-019-1566-2
18. Kuo, A.M.: Opportunities and challenges of cloud computing to improve health care services. J. Med. Internet Res. **13**(3), e67 (2011). https://doi.org/10.2196/jmir.1867
19. Alharbi, F., Atkins, A., Stanier, C.: Understanding the determinants of cloud computing adoption in Saudi healthcare organizations. Complex Intell. Syst. **2**(3), 155–171 (2016). https://doi.org/10.1007/s40747-016-0021-9
20. Habiba, U., Masood, R., Shibli, M.A., Niazi, M.A.: Cloud identity management security issues and solutions: a taxonomy. Complex Adapt. Syst. Model. **2**(1) (2014). https://doi.org/10.1186/s40294-014-0005-9
21. Ouedraogo, M., Mignon, S., Cholez, H., Furnell, S., Dubois, E.: Security transparency: The next frontier for security research in the cloud. J. Cloud Comput. **4**(1) (2015). https://doi.org/10.1186/s13677-015-0037-5
22. Hashizume, K., Rosado, D.G., Fernández-Medina, E., et al.: An analysis of security issues for cloud computing. J. Internet Serv. Appl. **4**, 5 (2013). https://doi.org/10.1186/1869-0238-4-5
23. Wang, R., Zhang, B., Wu, M., Zhang, J., Guo, X., Zhang, X., Li, H., Jiao, D., Ma, S.: Performance bottleneck analysis and resource optimized distribution method for IoT cloud rendering computing system in cyber-enabled applications. EURASIP J Wireless Commun Netw **2019**(1) (2019). https://doi.org/10.1186/s13638-019-1401-9
24. Mitton, N., Papavassiliou, S., Puliafito, A., Trivedi, K.S.: Combining cloud and sensors in a smart city environment. EURASIP J Wireless Commun Netw **2012**(1) (2012). https://doi.org/10.1186/1687-1499-2012-247
25. Ibrahim, W.N.H., Selamat, A., Krejcar, O., Chaudhry, J.A.: Recent advances on fog health—a systematic literature review. In: Fujita, H., Herrera-Viedma, E. (eds.) New Trends in Intelligent Software Methodologies, Tools and Techniques—Proceedings of the 17th International Conference SoMeT_18, Granada, Spain, 26–28 Sept 2018; Frontiers in

Artificial Intelligence and Applications, vol. 303, pp 157–170. IOS Press, Amsterdam, The Netherlands (2018). https://doi.org/10.3233/978-1-61499-900-3-157
26. Silva, C.A., de Aquino Junior, G.S.: Fog computing in healthcare: a review. In: 2018 IEEE Symposium on Computers and Communications (ISCC) (2018). https://doi.org/10.1109/iscc.2018.8538671
27. Directorate of Economics and Statistics, Planning Department, Government of Uttarakhand (2018). https://des.uk.gov.in/files/uttarakhand_human_development_report_.pdf
28. Joshi, M., Tewari, N., Budhani, S.K.: Security challenges in implementing a secured hybrid cloud model for e-health services. In: 2020 9th International Conference System Modeling and Advancement in Research Trends (SMART), Moradabad, India, pp. 3–7 (2020). https://doi.org/10.1109/SMART50582.2020.9337096
29. Uttarakhand Health Services Dashboard. Retrieved 27 Oct. 2020, from https://healthdashboard.uk.gov.in/
30. Kharel, J., Reda, H. T., Shin, S.Y.: An architecture for smart health monitoring system based on fog computing. J. Commun. (2017). https://doi.org/10.12720/jcm.12.4.228-233
31. Sen, S. (ed.): Invited-context-aware energy-efficient communication for IoT sensor nodes. In: Proceedings of the 53rd Annual Design Automation Conference. ACM (2016)

Android Malware Detection Using Extreme Learning Machine Optimized with Swarm Intelligence

Rahul Gupta, Aviral Agarwal, Devansh Dua, and Ankit Yadav

Abstract Android devices remain vulnerable to an increasing number of unidentified Android malware that has greatly compromised the efficacy of traditional security measures. Most classifiers are programmed to use a training process to learn from the data itself, since full expert insight to evaluate classifier parameters is difficult or impossible. This paper proposes a methodology which is a hybrid of machine learning and swarm intelligence. This methodology combines the successful exploration algorithm called the particle swarm optimization (PSO) with the extreme learning machine (ELM) classifier. ELM is a single-hidden layer feed-forward neural network (FFNN) consisting of large number of hidden layer neurons, which has proved to be an excellent classifier. In this research, the optimum values of input weights and biases for the ELM classifier are determined using PSO, and it further improves the classifier's accuracy. The dataset consists of 1700 benign and 418 malicious Android applications from which over 15,000 features of different types were extracted, and feature selection techniques were applied. The aforementioned dataset was used to experiment our proposed model, and significant results were achieved.

Keywords Android security · Malware analysis · Machine learning · Swarm intelligence · Static analysis

1 Introduction

Smartphones are ubiquitous today, serving as a portable way to enter personal data, accounts, contacts, and communication services. The Android platform has become the most dominant mobile OS since 2012. Android privacy is seriously jeopardized because of Android's open nature due to which innumerous adware, ransomware, etc., are disguised in a large number of non-malicious apps. Given this, smartphones have been tempting victims of cyber-attacks over the years. During

R. Gupta · A. Agarwal (✉) · D. Dua · A. Yadav
Delhi Technological University, New Delhi, India

K. Khanna et al. (eds.), *Cyber Security and Digital Forensics*,
Lecture Notes on Data Engineering and Communications Technologies 73,
https://doi.org/10.1007/978-981-16-3961-6_4

installation, Android device users who do not regularly check or understand app permissions allow malwares to be installed and gain access to sensitive resources, even without understanding the risk. Users' private data, such as IMEI, SMS data, call logs, contact list, and application usage statistics, is the main target for intruders, posing a significant threat to mobile users' confidentiality and security. Therefore, immediate measures are needed to identify and deal with malware for Android devices.

In this research article, we present a machine learning (ML)-based solution for classifying applications into benign and malware classes using an extreme learning machine (ELM) classifier whose parameters are optimized by particle swarm optimization (PSO) algorithm. We had a collection of 2118 Android applications (1700 benign + 418 malicious) as APK files [1]. We generated a dataset by unpacking these APK files into source code and extracted static information including permissions, actions, intents, services, and receivers from AndroidManifest.xml files. Using this, we created a feature set consisting of over 15,000 features. Then, we preprocessed our dataset using feature selection techniques before feeding it into the aforementioned model for classification of Android applications into benign and malicious. The explained architecture is depicted in Fig. 1.

Fig. 1 Architecture of the proposed model

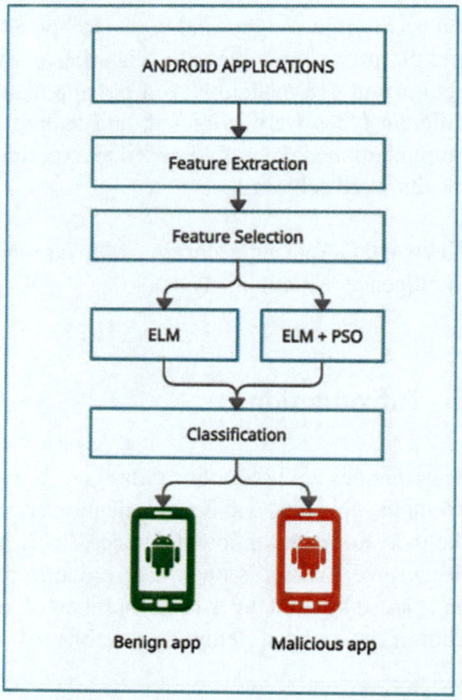

2 Related Work

Substantial research has been devoted in applying different machine learning and deep learning algorithms to classify malicious Android apps based on features which are extracted from static as well as dynamic analysis. In [2], authors have implemented a bio-inspired hybrid intelligent method for detecting Android malware (HIMDAM). This research classifies Android apps by applying extreme learning machine (ELM) as a classifier. Further, the model's accuracy was increased by using evolving spiking neural networks (eSNNs). In this paper [3], an engine for Android malware detection (DroidDetector) based on deep learning is proposed which uses deep belief network (DBN) architecture giving 96.76% testing accuracy. A feature set consisting of 192 features was extracted using static and dynamic analysis of Android applications. This work [4] proposes DroidDeep which merges the concept of static analysis and deep learning. Firstly, more than 30,000 static features were extracted and then fed into a deep learning model based on DBN to reduce the number of features. Finally, the new feature set was put into a support vector machine (SVM) model to detect Android malware. In this paper [5], detection of probable anomalies in two malware datasets was done using a high-performance extreme learning machine (HP-ELM). The results show that this approach achieved a maximum accuracy of 95.92% for top three features. In this study [6], an efficient model is introduced which is based on feature extraction and ELM classifier to detect Android malware. This model utilizes three different feature extraction methods including independent component analysis (ICA), Karhunen–Loeve transform (KLT), and principal component analysis (PCA) and then finally compiles the results using various stacking methods. In [7], authors have proposed a fully connected deep neural network approach for malware detection resulting with maximum accuracy of 94.65%. The dataset used here consists of 331 features including information about permissions in API. This paper [8] presents an effective machine learning approach based on Bayesian classification model. The features were extracted through static analysis, and total of 58 features were selected. DroidMat [9] is a model that collects static information including API calls, required permissions, etc., from AndroidManifest.xml file and then enhances the malware detection efficiency using K-means algorithm. Singular value decomposition (SVD) was used to determine the cluster's number, and finally, KNN algorithm was used to classify apps as benign or malicious.

3 Methodology

Static analysis for the identification of mobile malware is a fast and inexpensive approach. This method tests an application without any code execution and detects malware before the inspected application is executed. Dynamic analysis detects malware after the program under review is run or during it. Hybrid analysis consists of static as well as dynamic analysis. We are concentrating on the static analysis of mobile applications in this research.

The proposed system consists of three phases—feature extraction, feature selection and detection using machine learning models. All the phases are discussed in detail in this section.

3.1 Creating the Dataset (Feature Extraction)

We received a total of 2118 Android applications consisting of 1700 benign and 418 malicious applications [1]. The corresponding APK files were unpacked to their source code using APK tool. For static analysis, AndroidManifest.xml files were fetched. For each application, this XML file was parsed using Python 3 modules, and information related to permissions, actions, features, and receivers was gathered. These parameters would become the features of the dataset which would be used to train and test the classifier to detect malware in Android applications. Figure 2 diagrammatically represents the feature extraction process.

In this way, after parsing all the applications, we obtained a total of 15,939 features consisting of the aforementioned parameters. Note that any feature is Boolean or binary, meaning that its feature value is 1 whenever a feature appears in an application; otherwise, it is 0.

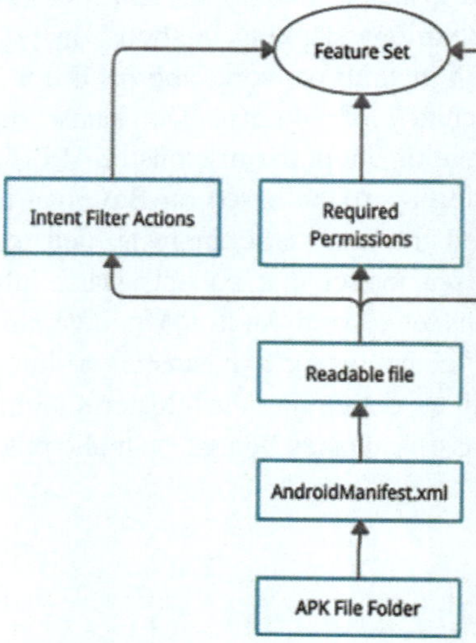

Fig. 2 Feature extraction process for an Android application

3.2 Feature Selection

One possible solution to reduce the amount of features is to use small datasets, resulting in a smaller feature set. A second approach for reducing the number of the features is to consider a limited set of possible features, e.g. permissions for Android only. A third approach is to use methods of feature ranking and feature selection, e.g. univariate statistical tests (chi-square), evaluation of correlation-based features, information gain, gain ratio, and community-based identification and elimination of recursive features. The strength of the relationship between an individual feature and the response variable is determined by these methods. Thus, a single score indicating this relationship is obtained by each feature. Then, according to these scores, all the features in the feature vector are ranked, and the top k features are considered.

For this research, we have applied the chi-square (χ^2) [10] test to find out the correlation among the features and to select the necessary features from these 15,939 features.

For a relationship with a given number of variables and sample size, the chi-square method compares the differences between the actual and the predicted results. Degrees of freedom (DOF) is considered for these experiments to determine whether, based on the cumulative number of variables and samples, a specific null hypothesis can be rejected. χ^2 provides a way to assess how well a data sample fits the characteristics of the wider population the sample is supposed to represent. If the sample data does not match the population's predicted properties that we are interested in, then we may not want to use this sample to draw conclusions about the larger population.

We apply the sklearn's select KBest function for feature selection and use the chi-square test to evaluate the best features. We receive the scores and alpha values (p_values) for each feature. The rejection region lies when the value to alpha drops below 0.05. This indicates that the features with p_values < 0.05 are not useful for the classification of our results, and we can discard them.

After selecting the features using chi-square, we finally get a dataset with 2654 features which will be fed to the machine learning classifiers in the next phase.

3.3 Model Design

Computational models are built which are capable of generalizing a concept when a machine learning algorithm is trained with a sample of input data. In our situation, trained models generalize whether or not an application is malicious or not. Therefore, when a new unidentified dataset is used by this model, it should be able to synthesize it, understand it, and provide us a reliable result.

Firstly, we are going to split the dataset into two groups: 80% for training and 20% for testing. Next, to understand its efficiency, we will run the primitive ELM.

To compare the proposed ELM with particle swarm optimization outcomes, the outcomes of the native ELM will be regarded as the gold standard [11–13].

3.4 Extreme Leaning Machine (ELM)

Machine learning methods have gained interest in artificial intelligence in recent years, primarily because of their contributions to various disciplines. A specific case of machine learning techniques is the extreme learning machine. ELM can be labelled as supervised learning algorithm competent of figuring out linear and nonlinear problems of classification. ELM is described by conventional artificial neural network architectures as the single layer feedforward neural network (SLFN), where it is not necessary to compute input weights and hidden layer biases iteratively [11]. An example of an ELM model is shown in Fig. 3.

For each hidden neuron, let IW be $(L \times n)$ input weights and B be $(L \times 1)$ bias values and O be $(C \times L)$ output weights. In a problem with N observations for a multi-category classification (K-distinct classes) $\{X_i,\ T_i,\ i = 1,\ 2,\ \ldots,\ N-2,\ N-1, N\}$, the ELM network outputs with L hidden layer neurons are identified as:-

$$y_k = \sum_{j=1}^{L} O_{kj} F_j(\mathrm{IW},\ B,\ X_i), \quad k = 1, 2, \ldots, K-1, K \tag{1}$$

For the jth hidden neuron, the output is $F_j(.)$, and the corresponding activation function is $F(.)$.

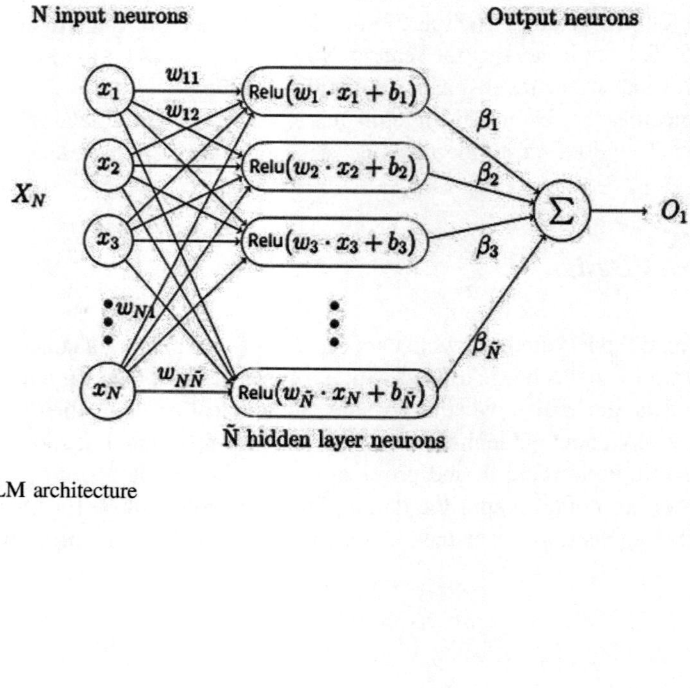

Fig. 3 ELM architecture

The target t is defined as

$$t_i^p = \begin{cases} 1 & \text{if} \quad k_i = p \\ 0 & \text{Otherwise} \quad k = 1,\ 2,\ \ldots,\ K-1, K \end{cases} \tag{2}$$

In the matrix representation, it can be written as—$Y = OY_l$.
where Y_l is an $L \times n$ matrix, which is represented as

$$Y_l = \begin{bmatrix} F_1(\mathrm{IW},\ B,\ X_1) & \cdots & F_1(\mathrm{IW},\ B,\ X_N) \\ \vdots & \ddots & \vdots \\ F_L(\mathrm{IW},\ B,\ X_1) & \cdots & F_L(\mathrm{IW},\ B,\ X_N) \end{bmatrix} \tag{3}$$

In our case, $K = 2$ for binary classification. For a given number of hidden layer neurons in the ELM algorithm, the input weights (IW) and bias (B) are initialized with random values. The output weights (O) are determined analytically by considering the classifier output (Y) equal to the Boolean class label (t):

$$O = YY_l^+ \tag{4}$$

where Y_l^+ is defined as the Moore–Penrose pseudo-inverse of matrix Y_l.

3.5 *Optimization Technique—Particle Swarm Optimization (PSO)*

PSO belongs to the group of techniques of swarm intelligence, influenced by the behavioural interactions of animals that huddle together such as ants, fireflies, and bees. It is an algorithm based on the interaction of individuals which consist of a population, in a search space to evaluate for promising regions. The action of an individual is influenced by either the best personal value achieved or the best global value achieved. Using fitness parameters just like that of evolutionary algorithms, the performance of each individual is determined. The population is called as a swarm, and particles are referred as individuals. The particles remember their best position in the past in the PSO and the optimal global position the particles have ever reached. This property lets particles search more rapidly in multidimensional space [12].

Let us consider a swarm with S number of individuals. Every P_i ($i = 1, 2, \ldots, S-2, S-1, S$) particle in the population is defined by

(i) Its current $p_i(k)$ position, referring to a probable solution to the optimization problem at iteration k;
(ii) Its $v_i(k)$ velocity;
(iii) The optimal $\mathrm{Pbest}_i(k)$ position attained in its previous path.

Let Gbest(k) be the optimal global position that the particles of the swarm have found across all trajectories. Optimality of position is determined by using one or more fitness parameters specified in accordance with the optimization problem considered. The particles travel according to the following equations during the search process:

$$v_i(k+1) = Wv_i + c_1r_1(\text{Pbest}_i(k) - p_i(k)) + c_2r_2(\text{Gbest}_i(k) - p_i(k)) \quad (5)$$

$$p_i(k+1) = p_i(k) + v_i(k+1) \quad (6)$$

These steps are shown as a flow chart in Fig. 4. The values of constants in the above equations such as W, c_1 and c_2 should be calculated before the experiment for the velocity updating process. The parameters c_1 and c_2 denote the weighting of the

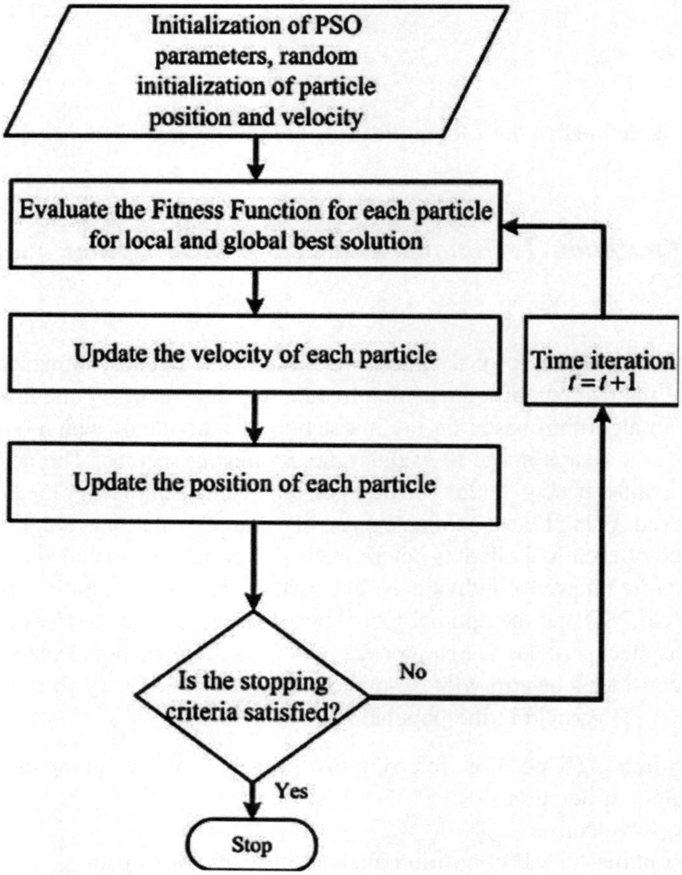

Fig. 4 Native PSO flow chart

terms of stochastic acceleration pulling each particle towards the positions of Pbest_i and Gbest. The parameters r_1 and r_2 are random variables in the range of [0, 1].

3.6 PSO Algorithm for ELM Optimization

The following major steps need to be taken to create an ELM classifier which uses PSO to optimize its parameters. These steps are as follows:

1. Randomly generate a swarm with population of S particles.
2. Initialize the vectors of velocity parameter v_i (i = 1, 2, 3, …, S) for each particle with random values.
3. Train an ELM classifier for each particle position of the particle p_i (i = 1, 2, 3, …, S), and calculate the respective accuracy/fitness function.
4. Initialize the Pbest_i of the particle with its starting position configuration.
5. Detect Gbest representing the maximum value of the fitness function observed over all paths pursued.
6. Update the velocities and the positions of the particles using the equations mentioned in the previous section.
7. For every candidate particle p_i, create an ELM model, train the classifier, and calculate the accuracy.
8. Update the particle's best position (Pbest_i) if the current accuracy is greater than the previous attained maximum value for Pbest_i.
9. If maximum number of iterations have been completed, then proceed to step 10, else go to step 5.
10. Select the best global value from the swarm of these particles, and use its values for input weights and biases to train the final ELM classifier which will have the best accuracy from all the models that PSO will have trained.
11. Use this ELM classifier to classify applications into benign and malware classes (Fig. 5).

4 Results

After finalizing our dataset by reducing the feature set to 2654 features using chi-square feature selection method, the proposed experimental model was divided into two major experiments.

The first experiment consisted of training a native ELM classifier using the dataset for different values of various parameters. The second experiment was the main focus of this research work. The previously developed ELM classifier was optimized using particle swarm optimization (PSO) and was trained using the same dataset for same parameter settings as the previous experiment and the results were compared.

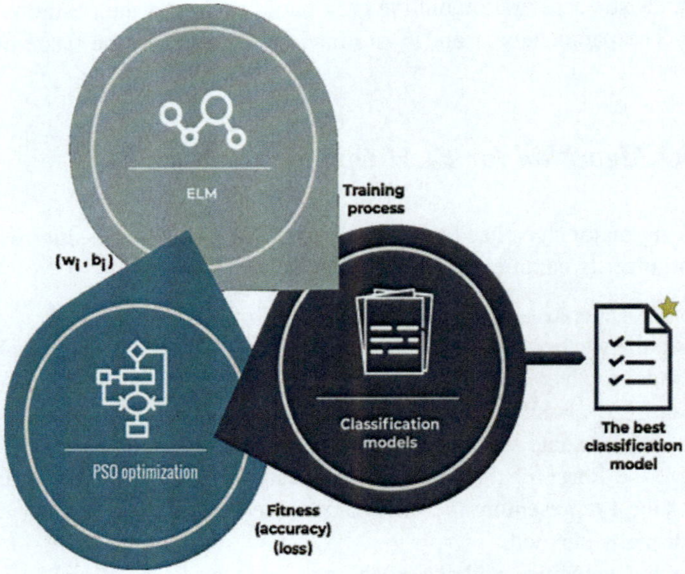

Fig. 5 Method representation used in the proposed PSO-ELM. ELM is trained and produces a learning model. This model offers accuracy and loss, all of which are used as the fitness by the PSO algorithm to be evaluated. Finally, for the ELM, the PSO algorithm measures new weights and biases. This approach works when a stop condition is met.

These experiments were performed using the Python environment on a system with the specifications—Core i5-9300H @2.40 GHz with 8 GB memory.

4.1 Experiment 1

An extreme learning machine (ELM) classifier was implemented in Python using various modules like NumPy, SciPy, etc. The whole experiment was conducted twice using two activation functions—Relu and Sigmoid [13]. In each experiment, we trained a number of ELM classifiers with different number of hidden neurons starting from 100 till 1000 neurons with a step increase of 100 neurons. For a specific number of hidden neurons, ELM classifier was tested 31 times, and the mean of these accuracies was considered.

The maximum average accuracy achieved using Relu as an activation function was **93.28**% at 600 neurons in hidden layer, whereas Sigmoid achieved a maximum average accuracy of **94.12**% at 500 neurons in hidden layer.

Table 1 Values for PSO constants

Constant Parameters	Values
Swarm population	100
No. of iterations	100
C_1—Self confidence	2
C_2—Swarm confidence	2
W—Velocity scaling factor	0.5

Table 2 Comparison of accuracies

No. of hidden neurons	ELM—Accuracy (%)		ELM + PSO—Accuracy (%)	
	Relu	Sigmoid	Relu	Sigmoid
100	90.07	90.93	96.93	97.17
200	91.96	92.86	97.88	98.35
300	92.85	93.53	97.88	98.35
400	92.94	93.93	97.88	**98.82**
500	93.27	**94.12**	98.35	98.11
600	**93.28**	93.71	**98.58**	98.35
700	92.66	93.21	97.40	98.35
800	92.09	92.86	97.64	97.87
900	91.39	91.49	96.46	97.65
1000	90.12	89.59	95.51	96.69

4.2 Experiment 2

Particle swarm optimization (PSO) was used to improve the performance by optimizing the values of input weights and biases of ELM classifiers. This experiment was also conducted twice with the same activation functions and same iterations for the number of hidden neurons. But at each step, ELM classifier was optimized using PSO algorithm explained in Sects. 3.5 and 3.6. The parameter settings used in the PSO algorithm for this experiment are mentioned in Table 1.

The maximum accuracy achieved using Relu as an activation function was **98.58%** at 600 neurons in hidden layer, whereas Sigmoid achieved a maximum accuracy of **98.82%** at 400 neurons in hidden layer.

The complete results of both the experiments are described in Table 2, and their comparison is visualized in Fig. 6.

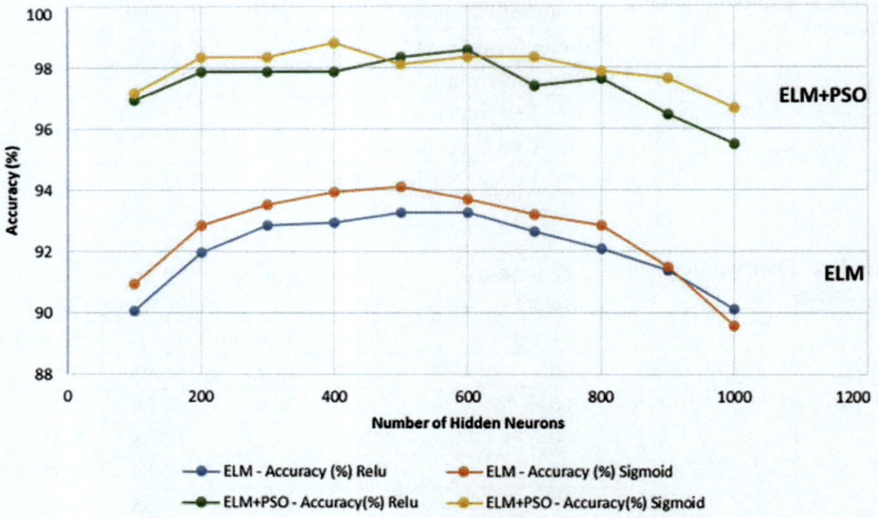

Fig. 6 Comparison of accuracies for various classifiers with different parameters

5 Conclusion and Future Work

Malware poses a serious threat to Android applications in today's world because data security and data privacy are of important concern to users. To tackle this problem, in this research, we came up with a modified approach to classify malicious applications from benign ones using machine learning techniques and optimized it with swarm intelligence. We extracted a total of 15,949 features which were used to create the dataset. After applying feature selection techniques, we reduced the dataset to achieve better results using our proposed model. Extreme learning machine (ELM) was used to classify the applications, which is a native method that provided us average results. Particle swarm optimization (PSO) was used to improve the performance of the classifier which gave us a maximum accuracy of **98.82%** which is far superior than the traditional algorithms. The proposed idea gives us a fine-tuned model which can tackle the problem efficiently and can be of great help to expert users.

For further enhancements, we plan to upgrade this model using other machine learning and deep learning techniques and aim to optimize it further using different swarm intelligence algorithms. Work done in this research is of static analysis, and we plan to update our dataset by adding new features extracted through dynamic analysis.

References

1. Lashkari, A.H., Kadir, A.F.A., Taheri, L., Ghorbani, A.A.: Toward developing a systematic approach to generate benchmark android malware datasets and classification. In: Proceedings —International Carnahan Conference on Security Technology 2018-Oct 50 (2018). https://doi.org/10.1109/CCST.2018.8585560
2. Demertzis, K., Iliadis, L.: Bio-inspired hybrid intelligent method for detecting android malware. Adv. Intell. Syst. Comput. **416**, 289–304 (2016). https://doi.org/10.1007/978-3-319-27478-2_20
3. Yuan, Z., Lu, Y., Xue, Y.: Droiddetector: Android malware characterization and detection using deep learning. Tsinghua Sci. Technol. **21**, 114–123 (2016). https://doi.org/10.1109/TST.2016.7399288
4. Su, X., Zhang, D., Li, W., Zhao, K.: A deep learning approach to android malware feature learning and detection. In: Proceedings—15th IEEE International Conference on Trust, Security and Privacy in Computing and Communications, 10th IEEE International Conference on Big Data Science and Engineering and 14th IEEE International Symposium on Parallel and Distributed Processing, pp. 244–251 (2016). https://doi.org/10.1109/TrustCom.2016.0070
5. Shamshirband, S., Chronopoulos, A.T.: A new malware detection system using a high performance-ELM method. arXiv (2019)
6. Hutchison, D., Mitchell, J.C.: Advances in neural networks—ISNN 2015. In: Theoretical Computer Science, pp. 166–173 (2015). https://doi.org/10.1007/978-3-319-25393-0
7. Sandeep, H.R.: Static analysis of android malware detection using deep learning. In: 2019 International Conference on Intelligent Computing and Control Systems, ICCS, pp. 841–845 (2019). https://doi.org/10.1109/ICCS45141.2019.9065765
8. Suleiman, Y., Sezer, S., McWilliams, G., Muttik, I.: New android malware detection approach using Bayesian classification. In: Proceedings—International Conference on Advanced Information Networking and Applications, AINA, pp. 121–128. https://doi.org/10.1109/AINA.2013.88
9. Wu, D.J., Mao, C.H., Wei, T.E., et al.: DroidMat: android malware detection through manifest and API calls tracing. In: Proceedings of the 2012 7th Asia Joint Conference on Information Security, Asia JCIS, pp. 62–69 (2012)
10. Tallarida, R.J., Murray, R.B.: Chi-square test. In: Manual of Pharmacologic Calculations, pp. 140–142. Springer, New York, New York, NY (1987)
11. Huang, G., Zhu. Q.Y., Siew, C.K.: Extreme learning machine: theory and applications. Neurocomputing **70**, 489–501 (2006). https://doi.org/10.1016/j.neucom.2005.12.126
12. Marini, F., Walczak, B.: Particle swarm optimization (PSO): a tutorial. Chemom. Intell. Lab. Syst. **149**, 153–165 (2015). https://doi.org/10.1016/j.chemolab.2015.08.020
13. Ratnawati, D.E., Marjono, Widodo, Anam, S.: Comparison of activation function on extreme learning machine (ELM) performance for classifying the active compound. In: AIP Conference Proceedings, p 140001. American Institute of Physics Inc. (2020)

Asymmetric Image Cryptosystem Based on Chaotic Zone Plate Phase Mask and Arnold Transform

Mehak Khurana and Hukum Singh

Abstract An optical asymmetric cryptosystem built on chaotic zone plate phase mask (CZPPM) has been proposed. Here the pixels of an image are shuffled by employing Arnold transform (AT^{ω}) and is then modulated with the CZPPM featuring in the gyrator Transform domain (GT). This increases randomness and adds chaotic parameters that make the system highly secure. The proposed system strengthens the security of the cryptosystem and does not permit the attacker to retrieve the initial image without the expertise of keys. The robustness of the projected cryptosystem has been investigated and validated based on an extra degree of freedom by simulating on MATLAB 9.9.0 (R2020b), and investigational outcomes have been shown to emphasize the efficacy of the algorithm.

Keywords Fresnel zone plate · Chaotic phase mask · Arnold transform · Gyrator transform

1 Introduction

With substantial innovations in transmission and information processing technologies, it has become a challenge to convey the information with security. An optical 4*f* system of DRPE in Fourier domain proposed by Refregier and Javidi was a great accomplishment in this arena in 1995 [1]. This set off the path for many researchers for constructing other cryptosystem centered on DRPE with superior security and elevated noise diminution. These systems were still exposed to many attacks known as known plaintext attack (KPA) [2, 3], chosen plaintext or ciphertext attack (CPA, CCA), etc. [4] An idea has been taken in the direction to

M. Khurana (✉)
Department of Computer Science, The NorthCap University, Gurugram, India
e-mail: mehakkhurana@ncuindia.edu

H. Singh
Department of Applied Sciences, The NorthCap University, Gurugram, India
e-mail: hukumsingh@ncuindia.edu

K. Khanna et al. (eds.), *Cyber Security and Digital Forensics*,
Lecture Notes on Data Engineering and Communications Technologies 73,
https://doi.org/10.1007/978-981-16-3961-6_5

model an asymmetric system [5] that deliver solution to insecure transmission concern and make it resilient from numerous attacks by boosting the randomness and improving chaotic mask key parameters.

2 Key Generation: Chaotic Zone Plate Phase Mask

Chaotic random phase masks (CRPM) [6] are used to increase randomness and to add chaotic parameters which makes the system highly secure in terms of private keys. Logistic map is one dimension (1D) non-linear chaotic maps and is used to generate the randomness in the CRPM. It can be expressed [7, 8].

$$x_{n+1} = px_n(1 - x_n) \tag{1}$$

It is iterated n times. Where p is bifurcation parameter and lies between $0<p<4$, x_0 is an initial value and $x_n \in [0, 1]$ is an iterative value where n varies from 0 to $M \times N$. M and N are size of a chaotic random mask in pixels. The 1D sequence $X = x_1, x_2,\ x_3, x_4, \ldots,\ x_{M \times N}$ that is produced by Eq. (1) is rearranged into 2D matrix as $Y = y_{ij}$, where i varies from $\{1, 2, \ldots, M\}$ and j varies from $\{1, 2, \ldots, N\}$ and $y_{ij} \in (0, 1)$. CRPM a 2D matrix is expressed as

$$\mathrm{CRPM}(x, y) = \exp\left(i2\pi y_{ij}(x, y)\right) \tag{2}$$

The Fresnel lens is built on quadratic phase change and the efficacy of zone plates. It is given [3] by

$$L_{\lambda,f}(r) = \exp(-i\pi r^2/\lambda f) \tag{3}$$

where r is the lens radius, f is the focal length, and λ is the wavelength of incident light. Now, CZPPM(x, y) is obtained by multiplying above two functions $L_{\lambda f}(r)$ and CRPM(x, y), i.e., Equation (1) and (2) and can be stated as

$$C(x, y) = \exp\left\{ i\pi \left[2y_{ij}(x, y) + \frac{r^2}{\lambda f} \right] \right\} \tag{4}$$

3 Proposed Cryptosystem

Asymmetric encryption technique built on CZPPM and AT$^{\omega}$ [9, 10] is proposed in this paper. The AT$^{\omega}$ scrambles the image $I(x, y)$ of size $M \times M$ by shuffling the pixels arbitrarily as in Eq. (5). Then the obtained result is transformed in gyrator domain [10], and phase is truncated to produce $G(u, v)$ as in Eq. (6). The encrypted image is generated by multiplying $G(u, v)$ with arbitrary phase mask in gyrator domain and further truncating its phase as in Eq. (7). The CZPPM and AT^{ω} applied through encryption operates as supplementary key because it comprises number of parameters as well as expands the randomness which further enhances the security.

$$I(x', y') = \mathrm{AT}^{\omega}(I(x, y)) \times C_1(x, y) \tag{5}$$

$$G(u, v) = \mathrm{PT}[\mathrm{GT}^{\alpha}(I(x', y'))] \tag{6}$$

$$E(x, y) = \mathrm{PT}\left[\mathrm{GT}^{-\beta}[G(u, v) \times C_2(x, y)]\right] \tag{7}$$

Two decryption keys DK_1 and DK_2 were generated in the process where amplitude truncation is performed as shown in Eqs. (8) and (9)

$$\mathrm{DK}_1 = \mathrm{AT}[\mathrm{GT}^{\alpha}(I(x', y'))] \tag{8}$$

$$\mathrm{DK}_2 = \mathrm{AT}\left[\mathrm{GT}^{-\beta}[G(u, v) \times C_2(x, y)]\right] \tag{9}$$

The current graph of the planned encryption stated is displayed in Fig. 1.

The decipherment procedure is the reversal process of the encryption which retrieves the original image $I(x, y)$. Following are the equations to decipher the encoded image $E(x, y)$.

$$G(u, v) = \left[\mathrm{IGT}^{-\beta}[E(x, y) \times \mathrm{DK}_2\ ] \times C_2^*(x, y)\right] \tag{10}$$

$$I(x', y') = \mathrm{IGT}^{\alpha}[G(u, v) \times \mathrm{DK}_1] \times C_1^*(x, y) \tag{11}$$

$$I(x.y) = \ \mathrm{AT}^{\omega}(\ I(x', y')) \tag{12}$$

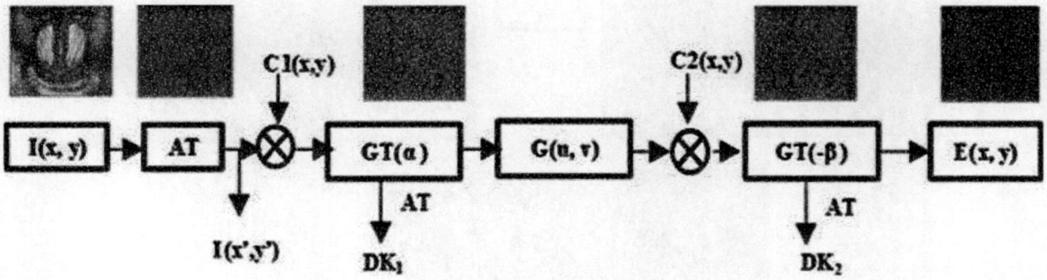

Fig. 1 Flow graph for proposed encryption scheme

4 Simulations Results

This paper demonstrates the statistical analysis and simulation results verifies the sensitivity of the indicated procedure and reveals that it accomplishes superior performance of recuperating a good superiority image. Findings also confirms the protection assessment based on correlation coefficient, noise attack, and key sensitivity. Contemplate initial image of baboon with size 256 × 256.

For simulation values use are $p = 3.96$, $x_0 = 0.35$, $\lambda = 6328$ Å, $f = 200$ mm, $\omega = 5$ and orders of GT are $\alpha = 0.4\pi$ and $\beta = 0.7\pi$. Figure 2 shows the CZPPM key generated from CRPM and Fresnel zone plate (FZP).

4.1 Statistical Analysis

For statistical analysis, entropy [11] of the image has been analyzed, and the obtained value for baboon image is 6.319 which is close to standard value that guarantees failure of information is nearby zero. The information is homogenously scattered, and it does not deliver any valuable information to the attacker. It can be confirmed from the value obtained above that the projected system is extremely efficient.

Peak signal to noise ratio (PSNR) [12] achieved 23 dB after first level of phase retrieval and 28 dB attained after the second level of phase retrieval.

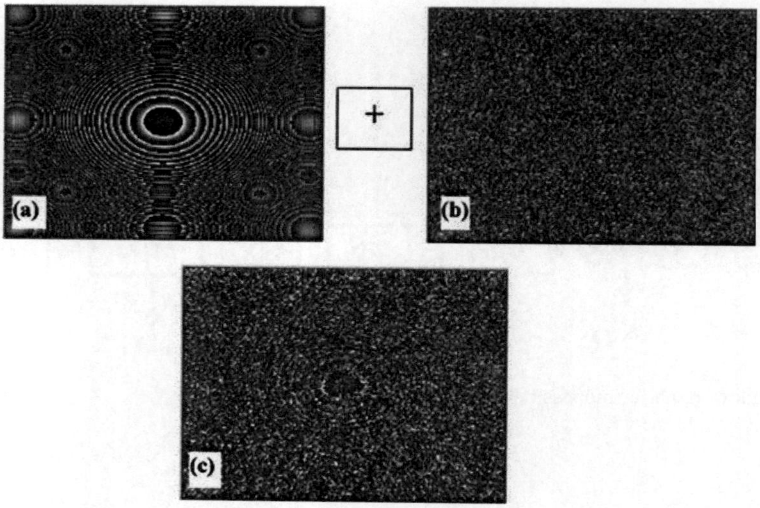

Fig. 2 **a** Fresnel zone plate, **b** chaotic random phase masks, **c** chaotic zone plate phase mask

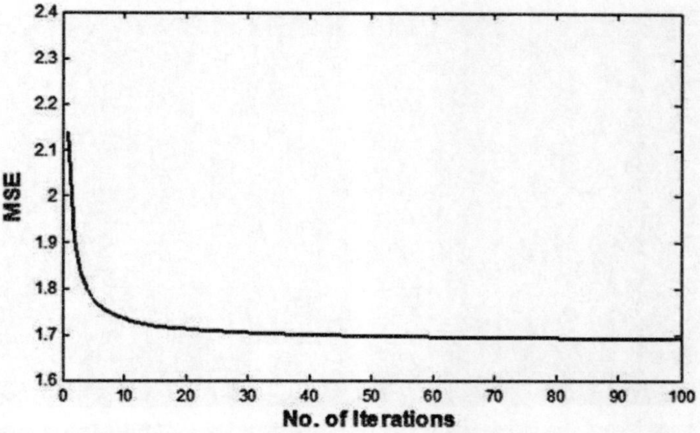

Fig. 3 MSE versus number of iteration graph

4.2 Performance Analysis

The mean square error (MSE) [13] determined for baboon image is 5.6346×10^{-22}. Figure 3 presents the number of iterations versus MSE. Here, MSE drops with the increase in number of iterations and after 50 iterations, MSE becomes constant therefore image randomization stabilizes.

4.3 Robustness Analysis

The proposed algorithm is verified against occlusion attack [14, 15] where the encrypted image has been occluded for different sizes, but the images are still visible from human eye till 75% which proves the system is robust. 45% and 70% occluded encrypted images and its retrieved images are presented in Fig. 4.

4.4 Robustness Analysis

This system has been investigated against KPA, CPA, and special attack to verify its strength. Sets of initial and cipher images have been picked to uncover the authentic key, the erroneous value of parameter headed to incorrect generation of key. To verify the conjunction of the iteration procedure, MSE versus iterations are employed to recover the decryption keys. It can be viewed from MSE plot that decipherment key cannot be recovered even with large iteration number as shown in Fig. 5.

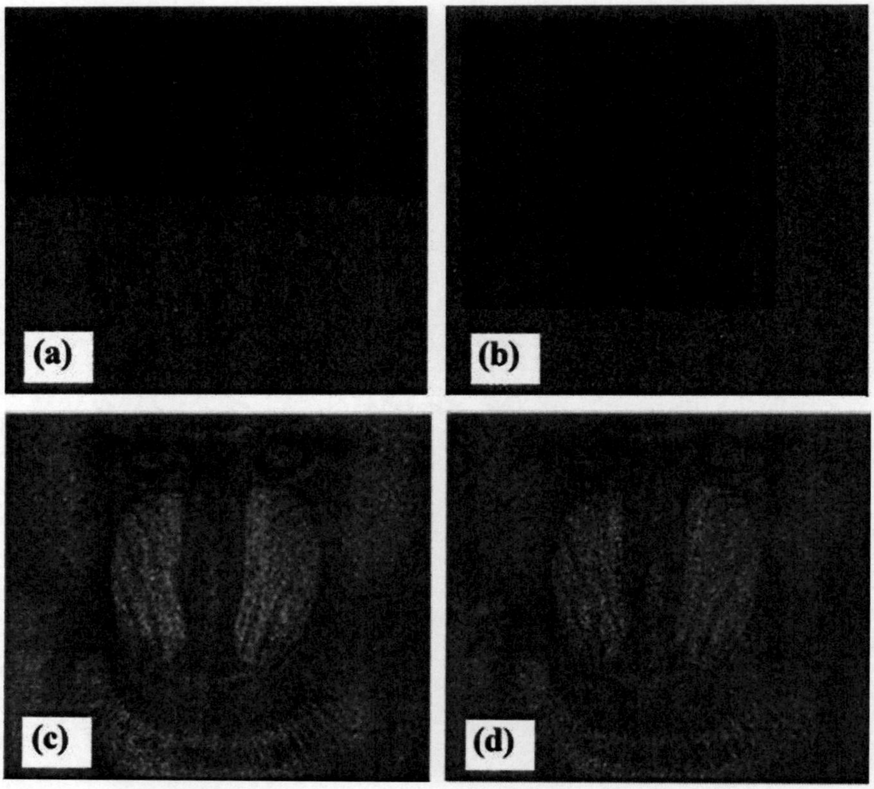

Fig. 4 **a** 45% occluded encoded image, **b** 70% occluded encoded image, **c** recovered image after 45% of occlusion, **d** recovered image after 70% of occlusion

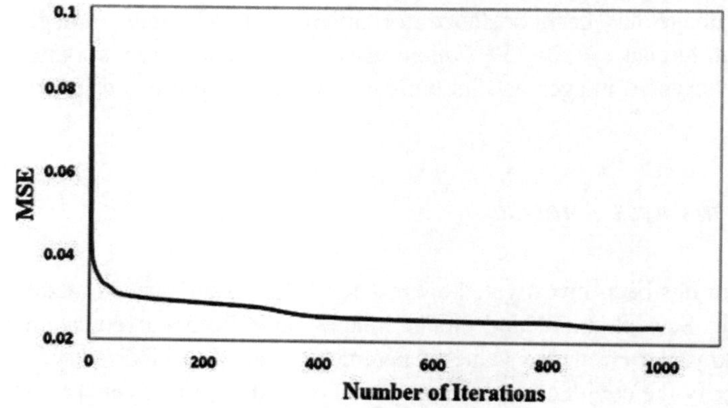

Fig. 5 MSE plot versus number of iterations for decipherment key generation with proposed scheme for 1000 iterations

5 Conclusion

The proposed asymmetric cryptosystem improves the security of the scheme as chaotic mask shifts the phase of the retrieved output and scrambling creates dispersion in the system. The introduced masking accomplishes superior execution of PSNR, MSE, occlusion, and noise as equated to existing systems. The CZPPM is further protected and cannot recapture initial image without the knowledge of all the parameters utilized for producing key. Experimental result reveals the viability and robustness of asymmetric cryptosystem.

References

1. Refregier, P., Javidi, B.: Optical image encryption based on input plane and Fourier plane random encoding. Opt. Lett. **20**(7), 767–769 (1995)
2. Peng, X., Zhang, P., Wei, H., Yu, B.: Known-plaintext attack on optical encryption based on double random phase keys. Opt. Lett. **31**(8), 1044–1046 (2006)
3. Gopinathan, U., Monaghan, D.S., Naughton, T.J., Sheridan, J.T.: A known-plaintext heuristic attack on the Fourier plane encryption algorithm. Opt. Express **14**(8), 3181–3186 (2006)
4. Zhang, C., Liao, M., He, W., Peng, X.: Ciphertext-only attack on a joint transform correlator encryption system. Opt. Express **21**(23), 28523–28530 (2013)
5. Qin, W., Peng, X.: Asymmetric cryptosystem based on phase-truncated Fourier transforms. Opt. Lett. **35**(2), 118–120 (2010)
6. Abuturab, M.R.: Asymmetric multiple information cryptosystem based on chaotic spiral phase mask and random spectrum decomposition. Opt. Laser Tech. **98**, 298–308 (2018)
7. Barrera, J.F., Henao, R., Torroba, R.: Optics encryption method using toroidal zone plates. Opt. Commun. **248**, 35–40 (2005)
8. Liansheng, S., Bei, Z., Xiaojuan, N., Ailing, T.: Optical multiple-image encryption based on the chaotic structured phase masks under the illumination of a vortex beam in the gyrator domain. Opt. Express **24**(1), 499–515 (2016)
9. Chen, L., Zhao, D., Ge, F.: Image encryption based on singular value decomposition and Arnold transform in fractional domain. Opt. Commun. **291**, 98–103 (2013)
10. Liu, Z., Chen, H., Liu, T., Li, P., Xu, L., Dai, J., Liu, S.: Image encryption by using gyrator transform and Arnold transform. J. Electron. Imaging. Proc. SPIE. **20**(1), 013020 (2011)
11. Khurana, M., Singh, H.: A spiral-phase rear mounted triple masking for securing optical image encryption based on gyrator transform. Recent Patents Comput. Sci. **12**(2), 80–94 (2019)
12. Khurana, M., Singh, H.: Two-level phase retrieval in fractional Hartley domain for secure image encryption and authentication using digital signatures. Multimedia Tools Appl. **79**, 13967–13986 (2020)
13. Wang, S., Meng, X., Yin, Y., Wang, Y., Yang, X., Zhang, X., Peng, X., He, W., Dong, G., Chen, H.: Optical image watermarking based on singular value decomposition ghost imaging and lifting wavelet transform. Opt. Lasers Eng. **114**, 76–82 (2019)
14. Liansheng, S., Xiao, Z., Chongtian, H., Ailing, T., Krishna Asundi, A.: Silhouette-free interference-based multiple-image encryption using cascaded fractional Fourier transforms. Opt. Lasers Eng. **113**, 29–37 (2019)
15. Singh, H., Khurana, M.: An asymmetric optical cryptosystem of double image encryption based on optical vortex phase mask using Gyrator transform domain. Recent Patents Comput. Sci. **13**(3), 672–685 (2020)

Authentication of Digital Media Using Reversible Watermarking

Geeta Sharma, Vinay Kumar, and Kavita Chaudhary

Abstract Digital watermarking is a technique to hide and transmit the data in such a manner that attackers cannot perceive it. The watermarks can be used for authentication, and the creator/owner of the digital data can claim the rights, in case of any dispute. The reversible digital watermarking ensures the reusability of the cover media. In this research paper, we are suggesting a more robust method of watermarking using the combination of LWT-SVD. The digest of the message is generated using MD5 algorithm, and a quantum representation is used as the trap door. The results are measured for the existing quantum technique, LSB quantum watermarking, and the proposed algorithm. We have compared the results for MSE, SSIM, and PSNR in the analysis section. The improvements can be seen in the results persistently.

Keywords Digital watermarking · DWT · Singular value decomposition · Quantum representation

1 Introduction

Multimedia, Internet of things and the Web of digital devices around us are rapidly generating digital data. This data need to be protected in such a manner that no unintentional receivers can even perceive the very presence of the data. Digital watermarking provides a strong method for information hiding as data is stored in the cover in non-perceivable manner and the process can be reversed by the intended users with security information provided by the sender. A digital water-

G. Sharma (✉)
IP University, Delhi, India
e-mail: geeta.sharma@jimsindia.org

V. Kumar
NIC, Delhi, India

K. Chaudhary
Jagannath University, Jaipur, India

K. Khanna et al. (eds.), *Cyber Security and Digital Forensics*,
Lecture Notes on Data Engineering and Communications Technologies 73,
https://doi.org/10.1007/978-981-16-3961-6_6

Fig. 1 Three pillars

marking technique has three major parameters that are robustness, capacity, and security as in Fig. 1. Robustness indicates that any kind of attacks will not affect the watermark much. Capacity is the maximum information that the cover can hold. Security, as name suggests, refrains the unauthorized access. Another important parameter is reversibility of the watermarking process that allows the use of cover image at the receiver's end.

The lifting system is an efficient way of implementing wavelet filtering, which also increases the wavelet transformation process speed. The prediction stage of the lifting scheme for the Haar transform predicts that the each odd element will be equal to its even element. The dissimilarity between the even value (the predicted element) and the original value of the odd value restores an odd element. The basics of the work is already discussed in [1].

2 Literature Review

Digital watermarking technique/algorithms can be characterized on the basis of many factors. The cover that is used for the information hiding can be text, image, sound, video, etc. A watermark can be visible or invisible, according to capacity of human eye to perceive. An invisible watermark can be fragile and robust both. Since fragile means imperceptible, it can be further divided into fragile and semi-fragile categories. The process of data hiding can be reversible and irreversible. In reversible data hiding, the cover can be used by the receiver after extracting the watermark. Watermarking is divided into two main groups: spatial domain and transform domain. Now, we will discuss few papers that helped in understanding and formed a base of the research.

For image steganography, the suggested technique is a mixed method that combines discrete wavelet transform (DWT) along with singular value decomposition (SVD) and lifting wavelet transform (LWT) techniques.

2.1 Discrete Wavelet Transform

In digital images, the discrete wavelet transform is used. There are several DWTs accessible. The most acceptable one should be used, depending on the application. Integer wavelet transformation can be used to cover text details. When DWT is applied to an image, it divides it into four sub-bands: LL, HL, LH, and HH. The first section (LL) consists of the most critical features. So if the data is hidden in the LL section, the compression or other changes do not affect the stego image to great extent. In the stego picture, distortion can sometimes be created and now other sub-bands can be used. The expanded function is discrete (i.e., a number sequence), and the resultant coefficients are named discrete wavelet transformation (DWT).

"Thakkar and Srivastava" [2]: "A blind image watermarking technique based on DWT and SVD was developed in this paper. The DWT technique was used in the medical image's Region of Interest. On the low-frequency sub-band LL-ROI, the program Block-SVD was used to obtain separate singular matrices. A pair of elements with identical values was discovered using either the left singular value matrix of all of the selected blocks. Using some criterion to embed a bit with watermark information, their properties of these pairs are changed. In order to achieve that imperceptibility of the medical image with watermark material, acceptable thresholds were selected."

"Zear et al." [3]: "Throughout this study, an algorithm with multiple watermarking based on discrete wavelet transformations (DWT), discrete cosine transformation (DCT), and singular value decomposition (SVD) was proposed for healthcare applications. The suggested application uses a total of three watermarks: a medical 'Lump picture watermark,' a code for the doctor's signature, and the patient's diagnosis details as text watermarks for identity authentication. Back Propagation Neural Network (BPNN) was used on both the derived watermark i.e. image format and the image watermark to reduce the potential impact on the watermarked image and to increase the image watermark's robustness."

2.2 Quantization-Based Watermarking

In 1984, Bennett and Brassard presented the first quantum key distribution protocol. Since then it was used in many theories-based and/or practical research. In the traditional method, a computer stores the pixel representation of the image with the necessary color information and the corresponding coordinates of each point. The same method was used by Shahrokh and Mosayeb [4], and the picture information is converted into the quantum state. The basic unit of quantum information is known as quantum bits (qubits).

Nezhadarya et al. [5]: "This paper proposed a robust picture watermarking framework based on quantization, referred to as quantization-based authentication path watermarking (GDWM), with a focus on uniform gradient vector direction quantization. The watermark bits become embedded in GDWM through quantifying vector angles and critical gradient vectors on many wavelet scales. The current technique had the following key features: (1) increased invisibility of both embedded watermarks, owing to the fact that watermarks were embedded in major gradient vectors, (2) strength against amplitude scaling attacks, owing to the fact that watermarks were embedded across the entire gradient vector angles, and (3) increased watermarking capability, owing to the method's ability to allow multiple-scale embedding. Throughout recognition of both the discrete wavelet transform (DWT) coefficients, that gradient vector at one pixel was expressed. The DWT coefficients were updated to quantize that gradient direction mostly on resultant relationship here amongst changes during the coefficients or the modification in the gradient direction."

Sachdeva and Kumar [6]: "Digital images are most prevalent cover files used only for steganography would those be. In this article, a current format of steganography called JMQT were introduced based on an updated table of quantization. It method of steganography has been contrasted with the JPEG-JSteg method of steganography. Two output parameters are being contrasted, namely capability and stego scale. Existing customers and stego scale represents the amount. Therefore, JMQT offers better power and JPEG-JSteg offers better stego-size."

2.3 Singular Value Decomposition

Through the discernment of image processing, a picture can be interpreted as a matrix of non-negative scalar entries. The SVD method for decomposing a rectangular matrix "A" into an orthogonal matrix U, a diagonal matrix S, and transposing an orthogonal matrix V is an efficient linear algebra numerical analysis method. Through image processing, SVD decomposes a given image A of size MN as decomposition (SVD) of A, which is interpreted as a matrix of non-negative scalar entries. The SVD method for decomposing a rectangular matrix "A" into an orthogonal matrix U, a diagonal matrix S, and transposing an orthogonal matrix V is an efficient linear algebra numerical analysis method. A given image A of size MN is decomposed as decomposition (SVD) of A using SVD.

"Ali et al." [7]: "By analyzing multiple scaling variables in image watermarking, a Differential Evolution (DE) algorithm has been used to balance the trade-off between robustness and imperceptibility in this article. To begin, the original image was divided into blocks, which were then transformed into discrete cosine transformation domains (DCT). The DC coefficients from each block were used to create a low-resolution approximation image, which was then subjected to the Singular Value Decomposition (SVD) algorithm. After that, the watermark was merged by replacing the singular values of both watermarks with the singular values of both watermarks."

"Du et al." [8]: "In order to ensure the copyright of a color image, a robust image watermarking approach based on 'Tensor-Singular Value Decomposition (T-SVD)' was proposed. The color image was created by transforming the third-order tensor with T-SVD to produce three orthogonal tensors and a diagonal tensor in order to establish clear associations between the three RGB color image channels. Because the main energies of that color picture are contained in the diagonal tensor, robustness can be used and the watermark can be embedded. Second, three color image channels were decomposed into four sub bands using discrete wavelet transformation (DWT), but only the LL sub band of each channel was used to create an approximate image. In addition, the approximation picture was divided into non-blocks of 4 × 4 dimensions, and the first, second, and third diagonal matrices were calculated using TSVD from each block. Finally, singular value decomposition was used to decompose the grayscale watermark (SVD)."

3 Method Proposed

Based on the standard existing algorithm, this proposed algorithm is similar to performing. The embedding is performed on the images, generally a loss less file format is used. The DWT method is combined with the benefits of SVD for embedding the image. The quantum representation is used as the trap door to fetch the watermark without changing the picture information of the cover. So the embedding ability and image quality are enhanced effectively by this algorithm. The ability for embedding is doubled over and the graphical output is greatly enhanced. The recommended reversible image watermarking algorithm largely helps increase the embedding rate of the watermark while retaining better visual quality than other algorithms.

3.1 Embedding Algorithm

This function extracted the watermark without damaging the cover image, using the cover image, watermark/hidden message, key, signature as an input. The flow of the function is explained in Fig. 2.

Step 1: Import cover image and watermark (image).
Step 2: For each channel apply following to the cover image:

(a) Resize image to 512 × 512 using bilinear algorithm.
(b) Assign a random key (secret key).
(c) Get the non-negative integer seed key.
(d) Assign 10 to QR block size.

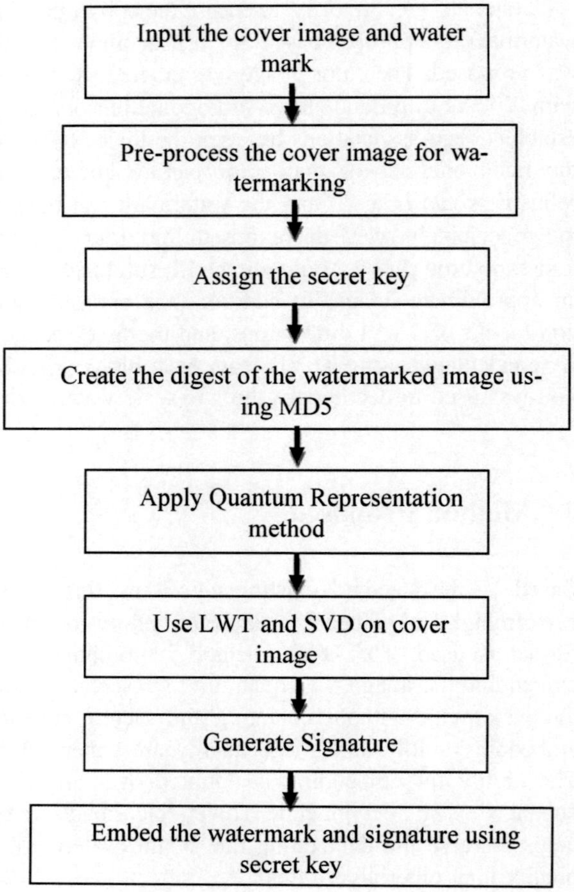

Fig. 2 Flowchart of the embedding algorithm

Step 3: Apply quantum representation on secret image using MD5.
Step 4: Convert watermark to B&W and apply following operations on watermark:

(a) Decompose the watermarked image or signal into four sub-bands, using Haar wavelet.
(b) Apply DWT on LL band up to fourth level.
(c) Also perform single value decomposition (SVD) to the high-frequency band (HH) of the watermark.

Step 5: Also decompose the watermarked image using SVD and replace values of HH band with SV of the watermark.
Step 6: Generate signature using values from watermark and key, and then embed the signature in LL band of the cover.
Step 7: After getting the updated HH band of cover image, finally acquire the watermarked image by applying inverse LWT.

3.2 Extraction Algorithm

Step 1: Import watermarked image (cover image + watermark embedded as single unit).
Step 2: Apply following operations on the watermarked image:

(a) Convert the watermarked image into four sub-bands, using Haar wavelet.
(b) Apply LWT on LL band up to fourth level.
(c) Also perform single value decomposition (SVD) to the high-frequency (HH) band.

Step 3: Generate signature using matrices of the watermarked image.
Step 4: Reconstruct the signature from the fourth level LL an HH band.
Step 5: Compare these signatures and follow these steps if the user is authenticated, else the extraction process should be stopped:

(a) Apply SVD to the HH band.
(b) Get the singular values from the HH band.
(c) Now, get the watermark image using orthogonal matrices and singular values.

Step 6: The watermark and the cover image both are in useable forms.

4 Analysis

Authentication and copyright management are two of the most common uses for reversible digital watermarking. In the field of image steganography, many studies have been conducted using the discrete wavelet, lifting scheme, and singular value decomposition. During our research, we examined the outcomes of various variations before settling on the most promising technique. We have taken into account the results of quantum steganography, LSB quantum watermarking, and LWT-QR-MD5 for this analysis.

The image Fig. 4 is clearly showing the high correlation between the input image and the resultant image. This high correlation indicates that embedding of the watermark to the cover image is not making much difference in the actual image that further ensures the better imperceptibility. As we discussed earlier, if a watermark is not able to perceive easily than it automatically reduce the number of attacks. It is also visible in the histogram of the image in Fig. 3. Now, we will use the algorithm on the basic three images Lena, Barbara, and Peppers and see the results.

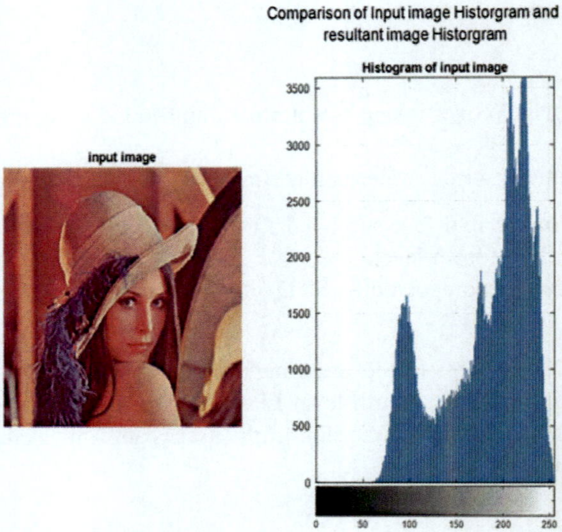

Fig. 3 Comparison of original and resultant image "Lena"

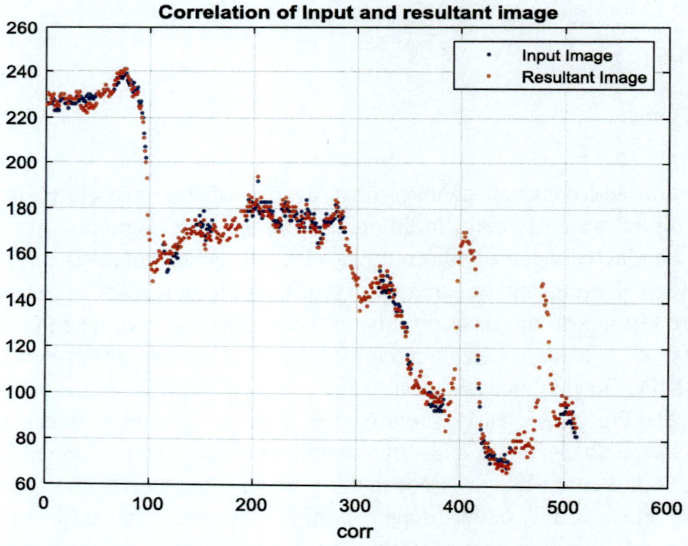

Fig. 4 Depiction of correlation

4.1 PSNR

It is used to check and compute the strength of the signal [9]. Figure 5 is showing the PSNR results after improvement in absolute term and percentage have been

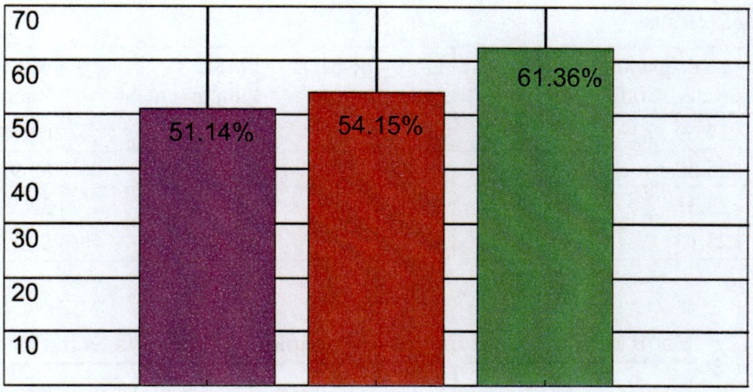

Fig. 5 Mean value of PSNR

Table 1 PSNR results

Image	Quantum steganography	LSB quantum watermarking	LWT-QR-MD5	PSNR improvement	% Improvement
Lena	51.1789	54.24	62.69	8.44	15.6
Barbara	51.0889	54.10	59.94	5.85	10.8
Peppers	51.1549	54.13	61.44	7.31	13.5
Average	51.1409	54.156	61.356	7.2	13.3

found that mean value of the PSNR applied on the various images like Lena, Barbara, and Peppers. The comparison was made for the three algorithms, i.e., quantum steganography, LSB watermarking, and the proposed algorithm. If we look at the average improvement of the resultant images, there is a clear increase of more than 13% that is a significant improvement (Table 1).

4.2 *MSE*

It is used to see how closely the original image and the watermarked image resemble each other [10]. This discrepancy is then squared, according to Wikipedia. Yi denotes the pixel position in the original image and Yi denotes the pixel value in the watermarked image at the same location. The MSE value is then computed by multiplying the difference by two [11]. Table 2 presented the mean square error results after applied proposed algorithm with LSB quantum watermarking, LWT-QR-MD5, MSE improvement, % improvement with three images. It was found that image Lena was showed that LSB quantum watermarking was 24.69, LWT-QR-MD5 8.14, MSE improvement 16.55, and it increased up to 67.02%, respectively (Fig. 6).

Table 2 MSE results

Image	LSB quantum watermarking	LWT-QR-MD5	MSE improvement	% Improvement
Lena	24.69	8.14	16.55	67.02
Barbara	25.50	11.75	13.74	53.91
Peppers	25.32	17.09	8.24	32.52
Average	25.17	12.32	12.84	51.15

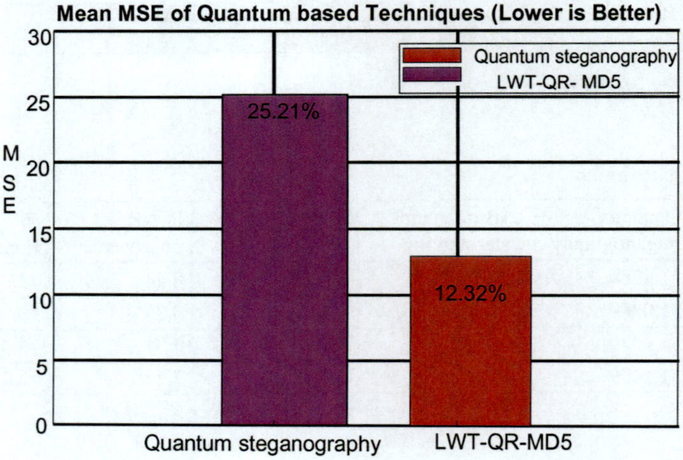

Fig. 6 Mean of MSE

It was found that image Barbara was showed that LSB quantum watermarking was 25.50, LWT-QR-MD5 11.75, MSE improvement 13.74, and it increased up to 53.91%, respectively. It was found that image Peppers was showed that LSB quantum watermarking was 25.32, LWT-QR-MD5 17.09, MSE improvement 8.24, and it increased up to 32.52%.

4.3 SSIM

Structural similarity (SSIM) was used to calculate the similarity between the filtered residual images and original images [12]. From Table 3 presented, we can observe the improvements in SSIM improvements. The three images were observed under LSB quantum watermarking, LWT-QR-MD5, SSIM improvement, and % improvement. It was found that Lena was showed that LSB quantum watermarking was 81.53, LWT-QR-MD5 93.71, SSIM improvement 12.18, and it increased up to 14.94%, respectively. It was found that Barbara was showed that LSB quantum watermarking was 84.33, LWT-QR-MD5 95.83, SSIM improvement 11.50, and it

Table 3 SSIM results

Image	LSB quantum watermarking	LWT-QR-MD5	SSIM improvement	% Improvement
Lena	81.53	93.71	12.18	14.94
Barbara	84.33	95.83	11.50	13.64
Peppers	82.69	89.88	7.19	8.70
Average	82.85	93.14	10.29	12.29

increased up to 13.64%, respectively. It was found that Peppers was showed that LSB quantum watermarking was 82.69, LWT-QR-MD5 89.88, SSIM improvement 7.19, and it increased up to 8.70%, respectively.

5 Conclusion and Future Scope

Watermarking is typically used to include evidence of digital data control. Generally, that is done by integrating certain copyright knowledge into digital details. Since it can be found on the World Wide Web for automated tracking of copywrite products. The algorithm suggested is capable of retrieving the secret message without making significant changes to the cover image. For authentication and copyright management, reversible automated watermarking is largely utilized. Through the usage of quantum watermarking, LWT-QR-MD5 techniques, this loss less approach ensures the improvement in the existing algorithms.

LWT promises faster computation while consuming less memory. Furthermore, the MD5 algorithm ensures better security and improves the algorithm's robustness against various attacks. In the steg analysis, this reversible watermarking technique produces better results. In future, we are planning to apply the same algorithm on more images and will also execute some attacks on the watermarked image.

References

1. Sharma, G., Kumar, V.: Review of different parameters for digital reversible watermarking. Int. J. Res. Anal. Rev. (2018). E-ISSN 2348–1269, P- ISSN 2349–5138
2. Thakkar, F.N., Srivastava, V.K.: A blind medical image watermarking: DWT-SVD based robust and secure approach for telemedicine applications. Multimedia Tools Appl. **76**(3), 3669–3697 (2017)
3. Zear, A., Singh, A.K., Kumar, P.: A proposed secure multiple watermarking technique based on DWT, DCT and SVD for application in medicine. Multimedia Tools Appl. **77**(4), 4863–4882 (2018)
4. Shahrokh, H., Mosayeb, N.: A Novel LSB based quantum watermarking. Int. J. Theor. Phys. Springer Science + Business Media New York (2016)

5. Nezhadarya, E., Wang, Z.J., Ward, R.K.: Robust image watermarking based on multiscale gradient direction quantization. IEEE Trans. Inf. Forensics Secur. **6**(4), 1200–1213 (2011)
6. Sachdeva, S., Kumar, A.: Colour image steganography based on modified quantization table. In: 2012 Second International Conference on Advanced Computing and Communication Technologies, pp. 309–313. IEEE (2012)
7. Ali, M., Ahn, C.W., Pant, M.: A robust image watermarking technique using SVD and differential evolution in DCT domain. Optik **125**(1), 428–434 (2014)
8. Du, M., Luo, T., Li, L., Xu, H., Song, Y.: T-SVD-based robust color image watermarking. IEEE Access **7**, 168655–168668 (2019)
9. Jadhav, A., Kolhekar, M.: Digital watermarking in video for copyright protection. In: 2014 International Conference on Electronic Systems, Signal Processing and Computing Technologies (ICESC), pp. 140–144. IEEE (2014)
10. Lou, D.C., Chou, C.L., Tso, H.K., Chiu, C.C.: Active steg analysis for histogram-shifting based reversible data hiding. Optics Commun. **285**(10–11), 2510–2518 (2012)
11. Botta, M., Cavagnino, D., Pomponiu, V.: A modular framework for color image watermarking. Sig. Process. **119**, 102–114 (2016)
12. Song, C., Sudirman, S., Merabti, M., Llewellyn-Jones, D.: Analysis of digital image watermark attacks. In: IEEE CCNC 2010 Proceedings (2010)
13. Le, P.Q., Iliyasu, A.M., Dong, F., Hirota, K.: A flexible representation and invertible transformations for images on quantum computers. In: Ruano, A.E., Várkonyi-Kóczy, A.R. (eds) New advances in intelligent signal processing. Studies in Computational Intelligence, vol. 372. Springer, Berlin, Heidelberg (2011). https://doi.org/10.1007/978-3-642-11739-8_9

Automatic Test Case Generation and Fault-Tolerant Framework Based on N-version and Recovery Block Mechanism

Seema Rani and Amandeep Kaur

Abstract A fault is any type of bug or error or failure that occurs in the system when any hardware is failed and requires replacement or strained to reboot or software fails to give the result. System failure occurs when the fault in system is not revealed and corrected in time. Software fault tolerance is the major research area in software development industry. Software fault tolerance enables a system to operate efficiently if any fault exists in the system and makes 011.0210 s a system capable to protect against any type of accidental or malicious destruction of the information in system. Many techniques of fault tolerance and recovery have been proposed in work of literatures in order to ensure efficient working of software. The main objective of this paper is to propose fault-tolerant framework based on N-version and recovery block mechanism for automatic test case generation. Simulation results show that proposed framework improves the reliability in terms of elapsed time, classification accuracy, and mean square error.

Keywords Software testing · Software fault prediction · Software fault tolerance · Software fault tolerance techniques · Real-time systems

1 Introduction

In the specific area of testing software, the main goal is to find the maximum number of faults or failures with the different test cases. The systematic and step-wise recognition of faults or errors with less time and less effort is the main aim of software testing [1]. With the growth and popularity of computer technology, the recent development is inclined to be more intelligent, complex, and automated. Software fault prediction is the process to recognize software modules containing the faults prior to the actual software testing process starts. It is essential to detect early fault prediction. One-third of the faults in software development is to be

S. Rani (✉) · A. Kaur
Computer Science and Engineering Department, Maharishi Markandeshwar University, Ambala, Haryana, India

K. Khanna et al. (eds.), *Cyber Security and Digital Forensics*,
Lecture Notes on Data Engineering and Communications Technologies 73,
https://doi.org/10.1007/978-981-16-3961-6_7

identified early to minimize the rework. A fault prediction is helpful in developing high-quality software with optimized efforts, time, and cost. It also helps in the maintainability of the software [2]. So there is a requirement of fault tolerance techniques that can handle different types of faults. Fault tolerance is a process where the system tries to work even if numbers of faults exist in the system. To handle these kinds of problems, our research focuses on software fault tolerance. In modern industrial practice, real-time data-based fault detection methods like a prediction of faults have been used [3]. For these types of problems, software fault tolerance is a proficient and competent method. Due to incapability to produce fault-free complex systems, software fault tolerance is the major issue. Software fault tolerance is mainly required in designing of a highly reliable system that can work under undesirable and difficult situations.

The main objective of software fault prediction is to predict where the fault might exist in the future and to quantify the importance of various factors [4]. Software fault tolerance makes a system proficient to work properly under any faulty conditions. It protects the system from any malevolent demolition of information. Software fault techniques can be categorized in two ways: single version and multiversion. Single-version techniques deal with single software, whereas multiversion deals with multiple numbers of versions of the same software to make sure that faults in one version do not cause a system failure. In this research, we compared the N-version and recovery block techniques from multiversion fault-tolerant techniques here. In earlier research, an efficient framework is designed for the automatic generation of test cases and prioritization. The research paper is well-thought-out as Sect. 1 is related to the introduction of the research work. Section 2 deals with the literature review. Section 3 describes software fault tolerance and deals with software fault tolerance techniques. Section 4 describes the proposed software fault-tolerant framework for automatic test case generation. Section 5 shows the simulation results. Section 6 gives the conclusion of the research.

2 Literature Review

D'Ambros et al. [5] compared bug prediction approaches. They compared the explanative and extrapolative power of bug prediction approaches in the form of a public dataset. Chinnaiah and Niranjan [6] projected an approach with an algorithm that provides the best candidate (fault-tolerant) for software having a critical configuration and developed techniques for the frequency of configuration interactions (IFrFT), and characteristics and frequency of interactions (ChIFrFT) to achieve fault tolerance and reliability with minimum cost. Choudhary and Khan [7], with industry data proposed a structure for fault tolerance and validation for the system.

Gao et al. [8] proposed a component analysis method based on enhanced kernel principal by using the theory of indiscernibility and compared conventional techniques for fault prediction. Sumra et al. [9] describe a comparative study of techniques of software fault-tolerant and surveyed to find out the issues of implementing SFT techniques in a project are: quality items missing, quality of code, project duration, team problem, lack of SFT techniques, skills and training, high cost, duplication, more space, code complexity, critical application, selection algorithm, difficult maintenance, need of SFT, work load, and ambiguous specification of requirements. A solution of some of the issues has also been suggested in this research work, i.e., quality items missing, quality of code, project duration, team problem, lack of SFT techniques, skills and training, high cost, selection algorithm, difficult maintenance, work load, and ambiguous specification of requirements.

Perälä [10] analyzed the probable drawbacks of automatic test implementation and provided an approach to handle and recover bugs that occurred in the system under test (SUT). Rathore and Kumar [11] described the review for software fault tolerance into three classes: fault prediction techniques, software metrics, and data quality which helps to deal with various elements and issues related to the fault prediction process. Alam [12] discussed the development of software fault tolerance techniques and their characteristics. Sharma and Yadav [13] projected a structure for fault prediction in a real-time system which is reliable and validated by using the simulation process.

3 Software Fault Tolerance

Software fault tolerance is a rising and emergent concern for long-run applications. Software fault tolerance makes a system able to function perfectly in any flawed situation and protects the system from any erroneous outputs and the destruction of information [14]. Software fault tolerance is very important in real-time and critical applications like flight control systems, medical systems, etc. So in software testing, researchers try to develop the software fault tolerance system continuously. Despite its popularity, it is extremely difficult to develop fault-free software. Varieties of approaches are proposed for software fault prediction and tolerance [15]. High software reliability and high quality are the main target in the software testing. Software reliability and software quality are the possibility of fault-free software for a specific period and environment. The reliability of any element can be calculated as Eq. (1):

$$\text{Reliability} = \frac{n(t)}{N} = \frac{\text{Failure free elements}}{\text{Numbers of elements at time} = 0} \tag{1}$$

The main purpose of software fault tolerance is to provide an alternate solution to fault problems where fault can be detected and the recovery to normal operation is carried out as internal functions of the system itself. A system cannot be truly fault-tolerant until software fault-tolerant techniques are applied to the system. Almost 60–90% of system failures are recognized as software failures. These failures can be controlled by applying software fault tolerance techniques that resolve the problems related to cost and time [16]. Software fault tolerance techniques are of two types, i.e., single-version and multiversion techniques. In this work, multiversion techniques, i.e., N-version programming (NVP) and recovery block (RcB) are used. The basic difference between NVP and RcB is the type of decision algorithm used. In NVP decision algorithm, delivering agreement/disagreement decision is implemented whereas in RcB acceptance test specific to the application program is implemented. From the same initial requirements, NVP has defined as an independent generation of $N \geq 2$ software modules, known as versions.

4 Proposed Software Fault-Tolerant Framework for Automatic Test Case Generation

The software fault tolerance concept is very popular these days. The significant growth and application of fault tolerance is an essential and unique architectural attribute of software testing. Nowadays, automated software fault-tolerant techniques are used [17]. The primary function of automation is to reduce the necessity of human intervention.

Automated software fault-tolerant techniques detect the faults in the system without any interaction of the human users. It also provides solutions to these faults and repairs them without human interaction. In the proposed framework, N-version and recovery block software fault-tolerant techniques applied to automated test case generation and prioritization framework using IGA proposed by Rani et al. [18]. With the addition of this software fault tolerance, the proposed framework minimizes the expenditure, to lessen time and minimize the effort for the development of better-quality software.

The existing software fault tolerance technique recovers the software or system from an error and prevents system failure. The key idea is to replace the erroneous state with a stable state. For this N-version and recovery block fault-tolerant techniques are used in the proposed framework. To enhance the reliability, the proposed framework declares an error condition and provides correct output with a very high probability. The proposed framework is compared based on elapsed time, classification accuracy, and mean square error (MSE) here. For all these parameters, the genetic algorithm evaluates the test cases by the program execution with the inputs.

The detailed process of fault tolerance and comparison of both techniques for the proposed framework of automated test case generation and test case prioritization is described in two phases, i.e., training phase, recovery, and classification phases below (Figs. 1 and 2).

```
/*Training Phase- Algorithm*/
Step A: Generating Training Models
1. For each Fl in Uploaded File: (i=1:1000 test cases)
a. Measure
i. Cosine Similarity
ii. Soft Cosine Similarity
iii. Cosine + Soft Cosine Similarity (Hybrid)
b. Vary the Similarity Index to create N-versions (N=200)
c. Find the average of N-version's
i. Cosine Similarity
ii. Soft Cosine Similarity
iv. Cosine + Soft Cosine Similarity (Hybrid)
Step B: Training
1. Train The system Model
2. Classify each sample in its accurate class
Step C: Measuring Statistics in Training
1. Calculate elapsed time for the completion of each case
2. Calculate Classification Accuracy
----------------* End of Training Phase*------------------
```

Fig. 1 Algorithm for training phase

```
/* Recovery and Classification Phase */
Step A: Getting Similarity Metrics
1. For each Fl in Uploaded File:
Measure:
i. Cosine Similarity
ii. Soft Cosine Similarity
iii. Cosine + Soft Cosine Similarity (Hybrid)
2. Getting the BASIS bias for applying Clustering and classification:
 - The SIMILARITY INDEXES and the SIMILARITY MATRIX be evaluated
Step B: Classification
1. Apply k-means clustering to divide the data in two segments.
2. The basis or the measure in k-means clustering is SIMILARITY INDEX.
3. Calculate MSE(Mean Squared Error) for each group in k-means clustering.
4. MSE = 1/n ((Sum(Ele)* Group1 - Sum(Ele)*Group2)^2)
 Where n: is the total number of calculated similarity
5. if(MSE.group1>MSE.group2)
{Classify it as Faulty Class- Group 1}
Else{Classify it as Faulty Class- Group 2}
Step C: Training and Calculating Metrics
1. For each Group in each cluster
    i. Train
    ii Classify and Evaluate Class Accuracy
    iii Calculate Elapsed Time
```

Fig. 2 Algorithm for recovery and classification phase

5 Simulation Results

To appraise the performance, the proposed framework is implemented on MATLAB-R2015 software with Java code-based files. The main focus of this work is to analyze the efficiency of the proposed framework using a genetic algorithm in terms of fault prediction and fault tolerance. Parameters used to compare the performance of both fault-tolerant techniques, N-version (NVP) and recovery blocks (RcB), are elapsed time, classification accuracy, and mean square error. For analysis, 1000 test cases are generated by a designed framework with the use of the genetic algorithm. The test units and the tables are listed with a difference of 10 cases. So for 1000 cases, 100 sample values have been evaluated for software fault tolerance purpose on the proposed framework. Simulation results of the proposed framework are described below.

Elapsed Time (EL): Elapsed time in simple terms is the amount of time that passes from the start of an event to its end. In the proposed work, elapsed time is done for the fault tolerance. Evaluation results of the proposed implementation are as given in Table 1 and Fig. 3.

In proposed framework, elapsed time for recovery block is better than the N-version. The proposed framework has achieved 78% improvement as compared to existed framework.

Classification Accuracy (CA): Classification accuracy can be defined as the number of total correct classifications or predictions divided by the total number of classified elements or predictions made and to find the percentage multiplied by 100. So it is the rate of correct classifications for an independent test dataset. We had calculated the classification accuracy for both N-version (NVP) and recovery blocks (RcB). Evaluation results of the proposed implementation are as given below in Table 2 and Fig. 4. The formula used for calculating classification accuracy is:

$$\text{Classification Accuracy(CA)} = \frac{\text{Total Correct Classified}}{\text{Total Classified Elements}} * 100 \tag{2}$$

Table 1 Calculated results of NVP and RCB with parameter elapsed time

S. no.	N-version	Recovery block
1	0.32684	0.440169
2	0.471102	0.201985
3	0.179231	0.484462
4	0.407456	0.422244
5	0.615325	0.188306
:	:	:
99	0.744868	0.446134
100	0.75522	0.490227

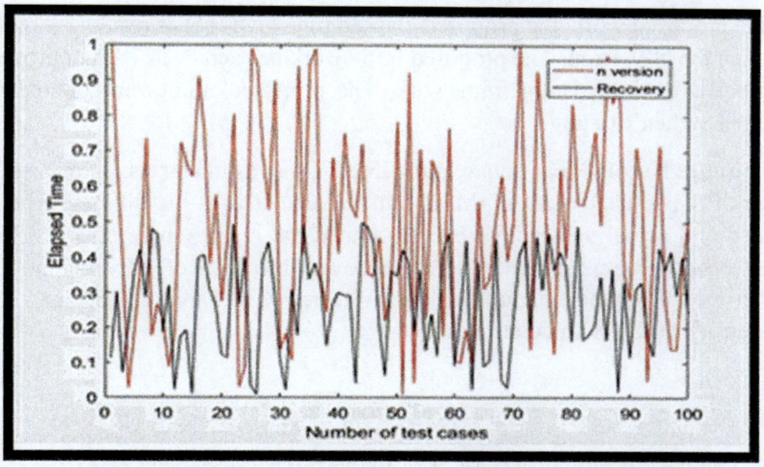

Fig. 3 Comparison result of NVP and RCB with parameter elapsed time

Table 2 Calculated results of NVP and RCB with parameter classification accuracy

S. no.	N-version	Recovery block
1	76.01701	82.59435
2	78.80281	81.04229
3	77.82223	83.65898
4	75.68417	82.32762
5	78.95729	81.88432
:	:	:
:	:	:
99	80.57971	81.84588
100	76.01328	83.2355

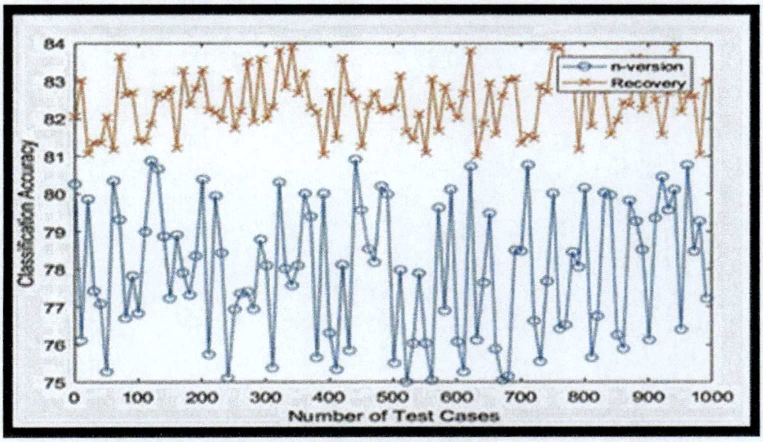

Fig. 4 Comparison result of of NVP and RCB with parameter classification accuracy

In the proposed work, the proposed elapsed time gives better results for recovery block than the N-version. The proposed framework has achieved 78% improvement as compared to the existing framework. The proposed framework offered better performance than existing.

Mean Square Error: Mean square error (MSE) is an estimator which measures the average of the square of errors. MSE can be calculated as the difference between estimated and actual values. We had calculated the mean square error (MSE) for both N-version programming (NVP) and recovery blocks (RcB). Evaluation results of the proposed implementations are as given below in Table 3 and Fig. 5. Mean square error (MSE) is calculated as:

Table 3 Calculated results of NVP and RCB with parameter classification accuracy

S. no.	MSE non-faulty N-version	MSE faulty N-version	MSE non-faulty recovery	MSE faulty recovery
1	2.075012	3.248718	1.538479	1.806141
2	2.007644	3.279142	1.536538	1.85851
3	2.015366	3.282688	1.530096	1.838388
4	2.076627	3.237418	1.518986	1.86465
5	2.06386	3.259207	1.532529	1.898895
:	:	:	:	:
:	:	:	:	:
99	2.019203	3.277696	1.58645	1.833358
100	2.031861	3.225238	1.520007	1.806901

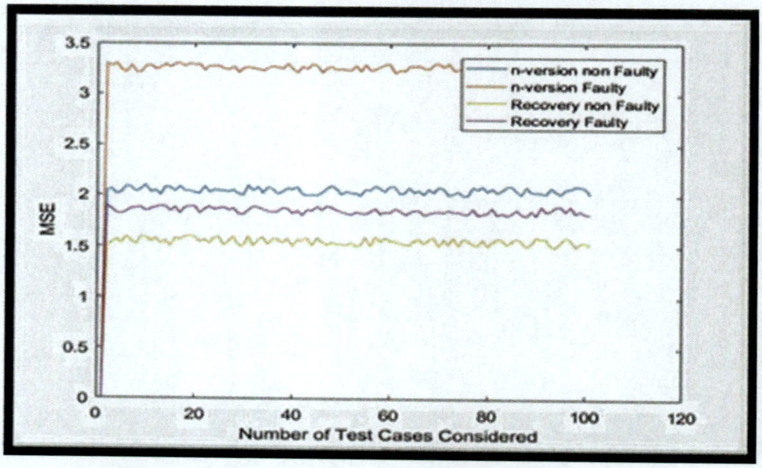

Fig. 5 Result of NVP and RCB with parameter mean square error

$$\text{MSE} = \frac{1}{n}\sum_{i=1}^{n}\left(Yi - \widetilde{Yi}\right)x^2 \tag{3}$$

In proposed work, the proposed elapsed time gives better results for recovery block than the N-version. The proposed framework has achieved 78% improvement as compared to existed framework. The proposed framework offered better performance than existed.

6 Conclusion

The purpose of software testing is the systematic detection of different types of faults with minimum effort and time. In this work, software fault tolerance techniques, N-version (NVP), and recovery blocks (RcB) are implemented for the proposed framework. To analyze the performance of the proposed framework, implementation is done in the. Net platform and various parameters are calculated with different experimentation. To analyze the performance of the proposed framework, in terms of fault prediction and fault tolerance, a set of different numbers of files is uploaded and results are calculated with three parameters elapsed time, classification accuracy, and mean square error (MSE). It concludes that the proposed efficient automated test case generation and test case prioritization framework is a good fault-tolerant framework. From the results, it is tested and proved that the proposed framework is an effective and reliable fault-tolerant framework within the budget of software. It also proved that from both used fault tolerance techniques the recovery block (RcB) gives better result as compared to N-version (NVP). Recovery blocks (RcB) on the proposed framework show better performance as compared to N-version programming (NVP). The experimental result shows the performance with parameters elapsed time, classification accuracy, and mean square error and concludes the effectiveness and reliability of the proposed framework for fault tolerance.

References

1. Rani, S., Gupta, D.: A comparative study of different software testing techniques: a review. J. Adv. Shell Program. **5**(1), 1–8 (2018). ISSN: 2395-6690 (Online)
2. Rani, S., Kaur, A.: Software fault prediction using boundary value analysis testing and loop testing techniques. In: 8th International Conference on Advancements in Engineering and Technology (ICAET-2020), Bhai Gurdas Institute of Engineering and Technology, Sangrur (Punjab), India, pp.192–196 (2020). ISBN: 978-81-924893-5-3
3. Baraldi, P., Cadini, F., Mangili, F.: Model-based and data-driven prognostics under different available information. J. Prob. Eng. Mech. **32**(4), 66–79 (2013)

4. Kamei, Y., Shihab, E.: Defect prediction: accomplishments and future challenges. In: 23rd International Conference on Software Analysis, Evolution, and Reengineering (SANER), IEEE, Publons, United States (2016). https://doi.org/10.1109/SANER.2016.56
5. D'Ambros, M., Lanza, M., Robbes, R.: An extensive comparison of bug prediction approaches. In: 7th IEEE Working Conference on Mining Software Repositories (MSR 2010): Co-located with ICSE, Cape Town, South Africa, pp. 31–41 (2010). https://doi.org/10.1109/MSR.2010.5463279. ISBN: 978-1-4244-6803
6. Chinnaiah, M.R., Niranjan, N.: Fault tolerant software systems using software configurations for cloud computing. Springer J. Cloud Comput. Adv. Syst. Appl, 1–17 (2018). https://doi.org/10.1186/s13677-018-0104-9
7. Choudhary, R.K., Khan, R.A.: Testing software fault tolerance techniques—future direction. ACM SIGSOFT Softw. Eng. Notes **36**(3), 1–5 (2011). https://doi.org/10.1145/1968587.1968604
8. Gao, Q., Liu, W., Zhao, X., Li, J., Yu, X.: Research and application of the distillation column process fault prediction based on the improved KPCA. In: International Conference on Mechatronics and Automation (IEEE ICMA 2017), Takamatsu, Japan (2017)
9. Sumra, M., Qadri, S., Fahad, M.: Issues and challenges of automated software fault tolerance techniques. Int. J. Nat. Eng. Sci. **9**(3), 39–44 (2015). ISSN: 1307-1149. E-ISSN: 2146-0086
10. Perälä, J.: Improving TTCN-3 test system robustness using software fault tolerance. In: First International Conference on Advances in System Testing and Validation Lifecycle, Porto, Portugal. IEEE (2009). ISBN: 978-1-4244-4862-3
11. Rathore, S.S., Kumar, S.: A study on software fault prediction techniques. Artif. Intell. Rev. **51**(2), 255–327 (2019). https://doi.org/10.1007/s10462-017-9563-5
12. Alam, M.J.: Analysis of different software fault tolerance techniques. J. Eng. Sci. (JES). **1**(1) (2009). ISSN: 2078-6174
13. Sharma, G., Yadav, A: Fault tolerance in real time distributed system. Rev. Comput. Eng. Res. **5**(2), 20–24 (2018). https://doi.org/10.18488/journal.76.2018.52.20.24
14. Bhat, A., Samii, S., Rajkumar, R.: Practical task allocation for software fault-tolerance and its implementation in embedded automotive systems. In: IEEE Real-Time and Embedded Technology and Applications Symposium (RTAS), Pittsburgh. PA, pp. 87–98 (2017). https://doi.org/10.1109/RTAS.2017.33
15. Sheng, J., Dong, S., Liu, Z.: Fault feature extraction method based on local mean decomposition Shannon entropy and improved kernel principal component analysis model. J. Clim. Change **8**(8), 625–635 (2016)
16. Egwutuoha, I.P., Levy, D., Selic, B., Chen, S.: A survey of fault tolerance mechanisms and checkpoint/restart implementations for high performance computing systems. Springer J. Soft Comput. **35**, 1302–1326 (2013)
17. Rani, S., Kaur, A.: A literature survey on automatic generation of test cases using genetic algorithm. Wesleyan J. Res. (UGC Care Listed) **13**(2), 65–76 (2020). ISSN: 0975-1386
18. Rani, S., Kaur, A.: Efficient framework for fully automatic test case generation and prioritization using genetic algorithm in software testing. J. Comput. Theor. Nanosci. **17**(11), 5198–5204 (2020)

Chatbot to Map Medical Prognosis and Symptoms Using Machine Learning

Himani Aggarwal, Saniya Kapur, Varun Bahuguna, Preeti Nagrath, and Rachna Jain

Abstract Computer-aided system is a subject of great importance and extensive requirement. Nowadays, deep learning and machine learning have gained quite a knack in people's eye and widely used among them. Gone are the occasions when the products were utilized for complex count issues or graphical portrayal alone. And Chatbots are proven revolutionary in our day-to-day lives where they are present in health, career, insurance and customer care support. In this paper, we have built up a Health-Bot using RNN network and Keras classifier. During such a pandemic period when there is an enormous crowd present in hospitals, people can get themselves checked at their homes with this interactive language system. Neural network adds more exactness to our work and reactions. And we further implemented our model on StreamLit which is an open-source framework for machine learning and deep learning.

Keywords Health-Bot · Machine learning · Deep learning · Keras model · Recurrent neural networks · Sequential model

1 Introduction

There are poverty-stricken regions in our country where the necessities are neglected or are not provided. These issues cannot be resolved by a bot. There comes a time back in 1966 when the first-ever Chatbot was created 'ELIZA' [1].

H. Aggarwal · S. Kapur · V. Bahuguna (✉)
Electronics and Communication Engineering, Bharati Vidyapeeth's College of Engineering, New Delhi, India

P. Nagrath · R. Jain
Computer Science and Engineering, Bharati Vidyapeeth's College of Engineering, New Delhi, India
e-mail: preeti.nagrath@bharatividyapeeth.edu

R. Jain
e-mail: rachna.jain@bharatividyapeeth.edu

K. Khanna et al. (eds.), *Cyber Security and Digital Forensics*,
Lecture Notes on Data Engineering and Communications Technologies 73,
https://doi.org/10.1007/978-981-16-3961-6_8

It was capable of mimicking human conversation using pattern matching. Be that as it may, a bot can assist by improving a circumstance. The genuine intensity of a Chatbot is to give data just on your order, which could help one lead a better life, a healthier life and install people with basic knowledge of proper health care. A sufficient number of populations need information about safe sex and have no mindfulness about an infection that is explicitly transmitted, because it is as yet considered as a No–No in the family to discuss sex. According to a WHO report, very nearly 11 million individuals infuse drugs, out of which 1.3 million people are surviving with HIV. It is additionally realized that a great region of the total populace does not have a clue about the right use of essential medications and anti-infection agents, also according to WHO [2] approximately 31 million people across the globe each year suffer from drug use disorder, which at later stage prompts clinical maltreatment and by implication cause to be in certain conditions the infused therapy more or less ineffective. In a recent report, it is proved that artificial intelligence is playing a crucial role in battling against these chronic diseases which start at a small level and grow up to be life threatening [3]. Internet work and the entrance to enormous chunks of medical resource are proved revolutionary to medical science and artificial intelligence helping to curb and cure people to live a substantial and better life. Bots are intelligent agent formed by deep learning and machine learning as a parent domain by this train and test our dataset to derive our accurate results and predict the accuracy of different models imposed on it. In this study, we have built a Health-Bot using Keras classifier and RN network to help others in the health sector [4]. The primary domain of our bot is in healthcare domain to provide people with their necessities.

Organization of the paper is as follows: Sect. 1 includes previous work in this field aka related work; in Sect. 2, we discuss various methodologies, designs, activation functions and neural networks applied; Sect. 3 walks us through features offered by our model; in Sect. 4, we talk about result generated methodology. Next, this work is concluded and future prospects of the presented project.

2 Related Work

Chatbots, or Health-Bots, interact with humans in a humane way. The presence of Chatbot has been increased significantly and has a great potential to be used in education [5], entertainment [6] and the public sector like this one. The development of Chatbot has increased by many folds. Chatbot respond to the doubt or query related to health which a person is afraid to discuss; recently due to the progress in development, there are many methods to implement Health-Bot visually by using web frameworks such as Django. Bots designed using deep learning [7] have a vast amount of data to train. In this, we have used recurrent neural networks (RNN) for the understanding of both encoding and [8] decoding. Eliza was the first Chatbot created by the Massachusetts Institute of Technology. On the off chance that a patient said ‘my head hurts,’ Eliza would react with, ‘Why do you say your head hurts?' Eliza, with just 200 lines of code, works like a [1] therapist. It is a typical inclination to have a human association in practically the entirety of our day-by-day exercises; be that as it may, innovation enhances our capacities. Today, on account of AI and NLP, we can utilize Chatbot innovation to give practically human-like discussions. The discussions are fueled by AI (man-made reasoning). A decent human services foundation is the basic for any country's metro life, and the must wipe out be conceded an indigenous arrangement to get to better social insurance, without the need to hang tight for quite a long time or months only for a little while. These calculated issues can easily be solved with the help of the Internet work and access to large chunks of medical resources—which are primarily free. These intelligent personal assistants (IPA) on our phones suddenly become definite responses for certain needs which are supported by machine learning and neural networks. It is close to unfeasible to determine these troubles with a bot, be that as it may, a bot can assist by improving the circumstance. The genuine favorable position of these Chatbots is the capacity to give appropriate direction and data to have a sound existence, as there are numerous individuals who despite everything do not have the essential information on legitimate social insurance [9]. AI is the parent space from which profound learning is inferred, which when joined with calculations of structure and working of the human mind clears approach to fake neural systems. Profound learning design comprises neural systems comprised of neurons, initiation capacities and loads that learn on their utilizing learning calculations [10]. Bots are only savvy operators dwelling on a worker to speak with people or different bots to make the human assignment a lot simpler, without the need of a particular convention or APIs nor with any 'ace bots,' for example, Google Assistant [11]. They convey in plain English, and profound learning makes them more precise in tossing the proper reaction to the given inquiry. In this examination, we have manufactured a relevant Chatbot utilizing TensorFlow [10] and Python to contribute to the well-being area. Our bot is equipped for diagnosing the medical problem, recommending a suitable doctor, giving updates about medicine and making an online meeting with the doctor. The most significant and essential advantage of Chatbots in the medicinal services area is the preeminent capacity to give exhortation and data to a solid life to help those individuals who need fundamental information on social insurance (Fig. 1).

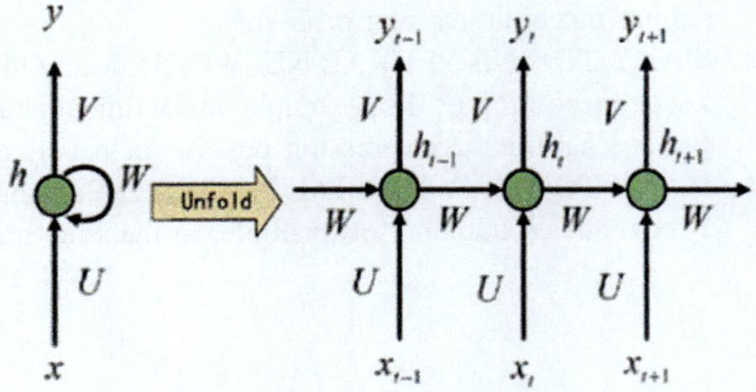

Fig. 1 Recurrent Neural Networks

3 Methodology and Design

3.1 Neural Networks

Neural networks have a main working concept which is to allow distribution of activity throughout the link with the help of a learning algorithm that is similar to the working of the human brain. Neural networks are trained and tested by a simulator [1]. It works along with the definition of neural network topologies. Neural units are of three types: input, hidden or output units. The relation that is connecting them is unidirectional, even though it supports recurrent links. There are many algorithms used for training a dataset that are available. Standard back-propagation and momentum back-propagation are only used. A pattern file directly receives the values of the input. Result file gets passed the output unit values, or values from pattern files can be received by it. Many input values and linked output values are contained in a pattern file. The simulator can also support a dynamic mode [13]. N-gram model can be implemented using these features. Values from the previous units can be received by units and hidden units. Further changes to that value are specified by the output function.

3.2 Recurrent Neural Network

A sequence or a collection of most frequently occurring consecutive words are fed into an RNN. It analyzes the data with the technique of finding the words occurring more frequently and creates a model that predicts the next or upcoming word in the sentence, i.e., it auto-fills the most probable data [14].

Recurrent neural networks are preferred because in feed forward neural network, it only considers the current input and cannot memorize previous outputs. So, in RNN [15] it memorizes what is going on the hidden layers and that produces a data to feed into the next one. Therefore, it allows to handle sequential data.

Types of RNN

1. ONE TO ONE NEURAL NETWORK: one to one type of RNN is also known as the most basic (Vanilla) form of artificial neural network. It is required for regular machine learning problems.
2. MANY TO ONE NEURAL NETWORK: Many to one neural network takes in a sequence of inputs. For example, in sentiment analysis where a given sentence can be classified as expressing positive or negative sentiments.
3. MANY TO MANY NEURAL NETWORKS: Many to many networks takes in a sequence of outputs, for example, in machine translation.

3.3 Label Encoding

Sometimes, our label is not a number but a string. We want to convert these strings to numbers that start from zero and one. If the classification is three classes, then the label is zero, one and two. Label encoder can help encode labels with a value between zero and n_classes − 1.

3.4 ReLU Activation Function

The ReLU function also can be expanded as rectified linear unit function. It is signified as follows:

$$R(x) = \max(0, x) \tag{1}$$

ReLU function basically avoids and solves vanishing gradient and removes the negative part [12].

$$\Rightarrow y = \max(0, o_i) \tag{2}$$

$$\Rightarrow \arg\max f(x) \tag{3}$$

$$\Rightarrow i \in 1, 2, \ldots, N \tag{4}$$

3.5 Softmax Activation Function

Softmax activation function is similar to the sigma function. A sigmoid clamps the value in between zero and one, but it does not represent the probability of something happening. In this function, we take the sum of the wakes times the previous output and plug that into the sigmoid function. This sum that we just quoted is like before we put it in an activation function. So, there is a need to find an activation function that can deal with real probability. That is where Softmax activation function comes handy. It does not really have a proper graph [16].

Softmax work is equivalent to the exponential of the component at position 'k' partitioned by the whole of the exponentials of all components of the vector. It is unique in relation to other actuation capacities which is on the grounds that while the other enactment capacities get an info esteem and change it paying little mind to different components, and the Softmax considers the data about the entire set of numbers one has. In this sense, Softmax is special because the output depends on the entire set of elements of input. A key part of Softmax transformation is that the estimations of yield are in the range from 0 to 1 and their aggregate is one [17]. The point of the Softmax transformation is to transform a lot of discretionarily big or small values that come out of previous layers and fit them into a valid probability distribution. This makes everything so intuitive and useful that the Softmax activation function is often used as the activation of the final output layer in classification problems [12].

$$p_k = \frac{\exp(o_k)}{\sum_{k=0}^{n-1} \exp(o_k)} \tag{5}$$

3.6 Keras

Keras is a Python-based deep learning framework which is widely used. It basically runs on top of TensorFlow and is very simple to work in as building models in Keras are as simple as stacking layers and later connecting these graphs. It is an open source project which is actively developed by developers and contributors across globe, also the documentation offered is vast and new features are added almost daily. It reduces the cognitive load which ensures that that the APIs are simple and consistent [12]. Keras provides clear feedback upon occurrence of any error, and this minimizes the number of user actions required for the majority of the common use cases. Keras also provides high flexibility to all of its developers by integrating with the lower-level deep learning framework languages like TensorFlow or Theano. This ensures that you can implement anything that you actually built in your base language. Keras supports multi-platform and lets its users work with multiple back ends. It feels like a tailor-made API for framework. The code can be run on the CPU or the GPU as well. Producing models on Keras is very effective and beneficial as it has total support to run with TensorFlow serving, GPU acceleration, example CUDA, native support to develop Android and iOS apps using TensorFlow and core ML and a full-blown support to work with Raspberry Pi as well.

Working Principle

Features: Computational graphs are used for expressing complex expression as a combination of simple operations for Keras to work with. It is mainly useful for calculating the derivatives during the phase of back-propagation and hence it makes it easier to implement distributed computing on a whole [18]. So, all it takes is to specify the inputs and outputs and to make sure that the graph is connected throughout (Fig. 2).

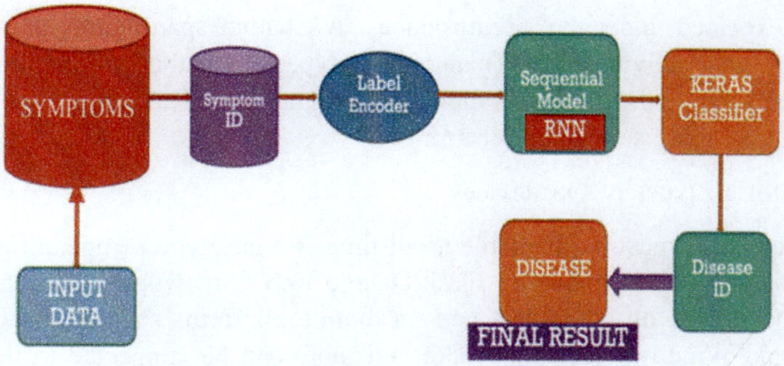

Fig. 2 Data Pipelining Model

3.7 Sequential Model

The working of the sequential model is like a linear stack of layers. This model is majorly useful for building simple classification network and encoder/decoder model. So, here we treat every layer as an object that feeds into the next layer and so on [19]. Features: model.fit() is used to train network. Bunch size is the quantity of training models in one forward and in reverse pass, so higher the batch size, the more memory you require.

Functional Model: It is a widely used model and holds good for about 95% of use cases. This model supports multi-input, multi-output and arbitrary graph topology. It has branches so wherever there is a complex model, it is folded onto two or more branches based on the requirement.

4 Features

1. Build an interactive real-time chat system [20]

Chatbots give customers a more amicable encounter. It has been very much wanted that individuals lean toward the feeling of discussion and association in which Chatbot furnishes overlooking with a mouse and snap. Regardless of whether it is something profound established in our mankind that we anthropomorphize, the truth of the matter is that individuals feel more joyful with a correspondence experience by means of a bot than something else. Clients have to search a ton and bother around which is disappointing and tedious; however with Chatbots, they get a human language-type communication. This is significantly more agreeable for patients as the cooperation is refined, giving them a customized proactive encounter. Through Chatbots, answers are gotten rapidly and productively. Who has the opportunity to be required to be postponed on the telephone or hustled starting with one office then onto the next? None of us. We need the data as fast as could be expected under the circumstances. A Chatbot spares time, letting loose patients for other activities [20]. There are two types of Chatbot's unintelligent ones that act using predefined conversation flows written by people and intelligent AI Chatbot's [21] that use machine learning.

2. Made for all types of OS devices

The services [22] must be available at all times on any type of operating system. Hence, this app will be made on Flask/Django web framework which will intend that it can be used on computers and on mobile platforms such as macOS, iOS, Android and Windows such that more and more can be connected to this application and can be benefitted from it.

3. Effective symptom-based disease prediction

As we all know that each disease at some later point shows its unique kind of problem which might convert into a life-threatening disease if not treated or judged early, the most common diseases can be easily identified by analyzing the symptoms. The symptoms can be anything like headache, itching, etc. So, by reading the symptoms and analyzing them, any possible health problem can be predicted [23], if any. If a person's body is analyzed periodically, it is possible to predict [24] any possible problem even before they start to cause any damage to the body.

4. Easily integrable and updatable

The system should be integrated [24], which signifies it has many individual modules since we use different types of modules which is used in performing a task, so they can be upgradeable individually. This will be helpful in increasing the efficiency of the system which will help to predict the data better.

5 Result-Generated Methodology

Initially, we created a custom dataset by consulting various healthcare practitioners where we asked them about commonly occurring illnesses and diseases and created a dataset out of it. We feature engineered our data as in machine learning one of the challenges is selecting the best features which are most appropriate and suitable for.

Figure 3 gives us the graphical representation of the dataset how it is mapped to identify the symptom with the help of its ID. This shows the plots of various symptoms ID (i.e., 1, 2 and 3) on the *Y*-axis and Disease ID on *X*-axis. This gives us the graphical representation of the dataset how it is mapped to identify the symptom with the help of its ID.

The type of algorithm you are trying to model, also many times the existing features, is not enough which results in re-engineering new features to be used to train the machine learning model which can be called feature engineering (FE). Then we label encoded our mapping of the symptoms to diseases [Sect. 3.3]. Once we got our data label encoded, then we implemented the recurrent neural network

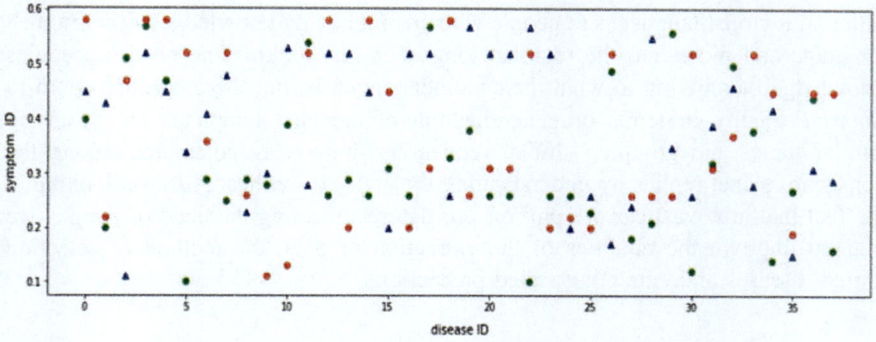

Fig. 3 Scatter plot representation of symptom ID w.r.t Disease ID

model. Recurrent neural network model [Sect. 3.2] is a type of neural network, and they require data to learn. The more the data is provided, more accuracy will be given back. On RNN, we used Keras framework [Sect. 3.6] which runs on top of TensorFlow. In this RNN model, we used activation functions like ReLu and Softmax. This function is attached to each neuron in the network and determines whether it should be activated ('fired') or not, based on whether each neuron's input is relevant for the model's prediction. Also in recurrent neural network (RNN), we used sequential model as it is used in problems related to classification. Here we treat every layer as an object that feeds into the next layer and so on. Once the symptoms were mapped with the disease, corresponding precautions and descriptions were displayed. Then using JSON file, we implemented a chat function to engage with the user. By doing this, the Chatbot will calculate the similarity between the trained text sequence and the user's input. We integrated our chat application with StreamLit which is an open-source framework for machine learning and deep learning models to make it effective for real-world users.

6 Conclusion

In this paper, we have effectively actualized a powerful Health-Bot which is where we also were successfully able to implement a train test model from sklearn model selection, where 20% of the model was given for testing and the rest 80% for training our model. Then we were able to implement cross-validation where we resample the procedure to assess the AI model on a constrained information test. Hence, we also used the K-fold method where k represents the number of folds. Further, we were successfully able to implement the Keras framework on the sequential model. In the coming years, prospects of Chatbot or Health-Bot specifically are very high by looking at the present needs of the people. The way this industry is prospering it is more prominent to be seeing it in other people's life. In addition, there are further prospects that Chatbots should be offering setup with different regional languages to people who are not having knowledge of the English language and more into the regional tongue. As on looking the need of peoples knowledge of replying to what their mother tongue is, but there is much room to improve quality in terms of generalization of the chat templates by clustering similar topics and grouping similar replies and improving coherence among the consecutive chat replies by understanding the styles of replies. Also, on looking at the fact that how vast can we pull off our datasets meeting the need of people, we plan to improve the qualities of the extraction of data, the method of selecting varied diseases and extracting suited precautions.

References

1. Szymczak, A.: Introduction to chatbots in healthcare. https://blog.infermedica.com/introduction-to-chatbots-in-healthcare/. Accessed July 2017
2. World Health Organization: Management of substance abuse advice for the public. Available https://www.who.int/substance_abuse/facts/en/. Accessed 23 June 2020
3. Alqudah, A.M., Qazan, S., Alqudah, A.: Automated systems for detection of COVID-19 using chest X-ray images and lightweight convolutional neural networks (2020). https://doi.org/10.21203/rs.3.rs-24305/v1
4. Kumar, A., Shanmugvadivu, P.: Space of RGB-H-CMYK. **1**(Feb). Spring Singapore (2019). https://doi.org/10.1007/978-981-13-1708-8
5. Mckie, I.A.S., Narayan, B.: Enhancing the academic library experience with Chatbots: an exploration of research and implications for practice. J. Aust. Lib. Inf. Asso. (2019). https://doi.org/10.1080/24750158.2019.1611694
6. Melián-González, S., Gutiérrez-Taño, D., Bulchand-Gidumal, J.: Predicting the intentions to use chatbots for travel and tourism. Curr. Issue Tour. (2019). https://doi.org/10.1080/13683500.2019.1706457
7. Emima, Y., Rajesh, M., Rao, K.S.: Experimental investigation on performance and exhaust emission characteristics of diesel engine using eesame blends with diesel and additive. Int. J, Rec. Technol. Eng. **8**(1), 6–11 (2019)
8. Pati, B., et al. (eds.): Progress in advanced computing and intelligent engineering. In: Advances in Intelligent Systems and Computing, vol. 713. https://doi.org/10.1007/978-981-13-1708-8_10
9. Kucherbaev, P., Bozzon, A., Houben, G.: Human-aided bots. IEEE Internet Comput. **22**(6), 36–43. https://doi.org/10.1109/MIC.2018.252095348
10. Singh, R., Paste, M., Shinde, N., Patel, H., Mishra, N.: Chatbot using TensorFlow for small Businesses. In: 2018 Second International Conference on Inventive Communication and Computational Technologies (ICICCT), Coimbatore, India, pp. 1614–1619. https://doi.org/10.1109/ICICCT.2018.8472998
11. López, G., Quesada, L., Guerrero. L.A.: Alexa vs. Siri vs. Cortana vs. Google assistant: a comparison of speech-based natural user interfaces. In: Nunes, I. (eds.) Advances in Human Factors and Systems Interaction. AHFE 2017. Advances in Intelligent Systems and Computing, vol. 592. Springer, Cham (2018). https://doi.org/10.1007/978-3-319-60366-7_23
12. Agarap, A.F.: Deep learning using rectified linear units (relu). arXiv preprint arXiv:1803.08375 (2018)
13. Marques, N.C., Lopes, G.P.: A neural network approach to part-of-speech tagging. In: Proceedings of the 2nd Meeting for Computational Processing of Spoken and Written Portuguese, pp. 21–22 (1996)

14. Yousif, J.: Neural computing based part of speech tagger for Arabic language: a review study. Int. J. Comput. Appl. Sci. IJOCAAS **5**(1) (2018)
15. Ramadevi, R., Sheela Rani, B., Prakash, V.: Role of hidden neurons in an Elman recurrent neural network in classification of cavitation signals. Int. J. Comput. Appl. **37**(7), 9–13 (2012)
16. Chen, J., Jing, H., Chang, Y., Liu, Q.: Gated recurrent unit based recurrent neural network for remaining useful life prediction of nonlinear deterioration process. Reliab. Eng. Syst. Saf. **185**, 372–382 (2019)
17. Shim, K., Lee, M., Choi, I., Boo, Y., Sung, W.: Svd-Softmax: Fast Softmax approximation on large vocabulary neural networks. In: Proceedings of the 31st International Conference on Neural Information Processing Systems, pp. 5469–5479 (2017)
18. Abdul-Kader, S.a., Woods, J.: Survey on Chatbot design techniques in speech conversation systems. School of Computer Science and Electronic Engineering/University of Essex Colchester/UK
19. Vidnerova, P., Neruda, R.: Evolving Keras architectures for sensor data analysis. In: 2017 Federated Conference on Computer Science and Information Systems (FedCSIS), pp. 109–112. IEEE (2017)
20. Choi, K., Joo, D., Kim, J.: Kapre: On-gpu audio preprocessing layers for a quick implementation of deep neural network models with Keras (2017). arXiv preprint arXiv:1706.05781
21. Atanasov, A.: Dynamic Working Memory in Recurrent Neural Networks
22. Awwalu, J., Garba, A., Ghazvini, A., Atuah, R.: Artificial intelligence in personalized medicine application of AI algorithms in solving personalized medicine problems
23. Chantarotwong, B.: The learning Chatbot. Fall (2006)
24. Kelly, J.E.: Computing, cognition and the future of knowing how humans and machines are forging a new age of understanding. IBM Research and Solutions Portfolio

Cloud Security: The Future of Data Storage

Parv Bajaj, Ritika Arora, Mehak Khurana, and Shilpa Mahajan

Abstract In this modern era, all the organizations seem to be shifting their data and services to cloud because of the increase in information leaks and data thefts. It has become easier for intruders to break into the organization's data stored locally. As cloud provides a significant hike in security levels, it is slowly turning into the future of data storage. It has become crucial to acknowledge the issues regarding cloud security. Cloud security is the security of all the services provided by the cloud—storage, servers, networking, and databases. In this paper, the issues related to the same are covered along with the mitigation proposed to deal with those problems. The key to the overall security of the cloud is identity access management (IAM). IAM is inherently more secure than a simple username and password combinations because of the profile of information IAM collects. It can make access to data and networks a much more convenient process.

Keywords Identity and access management · Cloud service models · Authorization · Cloud security

1 Introduction

Cloud security refers to the security dedicated towards protecting cloud computing systems from intruders. This involves keeping data safe and private across online applications, and platforms. Ample amount of security is provided by the cloud providers as end-users trust them with their personal data. Client's trust is a key component in their business. Cloud security methods are used to keep client's data safe and private. It will be wrong to say cloud security entirely depends on the cloud provider because some part of it is certainly in the end user's hand.

P. Bajaj (✉) · R. Arora · M. Khurana · S. Mahajan
The NorthCap University, Gurugram, India
e-mail: mehakkhurana@ncuindia.edu

S. Mahajan
e-mail: shilpa@ncuindia.edu

K. Khanna et al. (eds.), *Cyber Security and Digital Forensics*,
Lecture Notes on Data Engineering and Communications Technologies 73,
https://doi.org/10.1007/978-981-16-3961-6_9

Cloud security consists of:

- IAM
- Data security
- Policies
- Data retention
- Legal compliance.

The security guidelines focus on these three phases:

Phase 1: Understanding cloud usage and risk assessment.
Phase 2: Protecting the cloud.
Phase 3: Responding to cloud security issues.

These guidelines are related to protecting the hosts running the computer instances and the network these instances are connected to. This is where IAM plays the most important role because in order to secure the network, managing users and their access privilege are necessary. The protection of cloud data assets is done through encryption of all the data present in cloud storage.

The currently used cloud environments are:

- Public cloud services are offered by third-party datacenter provider to end-user. Public cloud offers resource pooling, self-service, service accounting, multi-tenancy to manage the solutions, deployment, and securing the resources and applications.
- Private clouds are deployments set up within the organization's firewall (on-premises datacenters) and traditionally run by on-site servers. Some of the benefits of a public cloud computing environment, such as elastic on-demand capacity, service-based access, and self-service provisioning are offered [1]. Private cloud is suitable when the traditional requirements, such as control, security, and resiliency, are more emphasized by an organization with the restricted and designated user access and authorization.
- A hybrid cloud is a combination of an interoperating public and private cloud. This is the model where consumer takes the noncritical application or information and compute requirements to the public cloud while keeping all the critical information and application data in control. It is an intermediate step in the evolution process, providing businesses an onstage from their present IT conditions into the cloud environment. A hybrid cloud provides you with resources of public cloud for small projects at a cheaper rate than if you use data center's IT infrastructure. With this, you do not invest much in the resources you need on a temporary basis.
- A community cloud is the cloud managed by groups of people, communities, and agencies especially government to have common interests—such as maintaining the compliance, regulation, and security parameters—working on the same mission.
- Shared private cloud is a shared compute capacity with variable usage-based pricing to business units that are based on service offerings, accounts

datacenters. It requires an internal profit center to buy infrastructure made available through account consolidations.

2 Cloud Service Models

Before we dive into cloud security, it is necessary to study about cloud service models. Clouds will transform the IT industry and profoundly affect how we live and how businesses operate. Cloud computing:

1. Offers the scalable compute model to be accessed from anywhere.
2. Offers you with simplified service delivery.
3. Disaster recovery.
4. Provides dynamic infrastructure for upcoming technologies data center.

Some say it is grid or utility computing or software-as-a-service, but it is all three of those combined—software as a service (SaaS), platform as a service (PaaS), and infrastructure as a service (IaaS).

2.1 SAAS

In software as a service model, it is not required to install an application instead you can reliably access it via the Internet. It saves you from complex software and hardware management. A third-party provider hosts it on the internet, and it can be used by various customers. These applications are not actually managed by your company but by the software provider itself. It is more convenient as it relieves you from a pestering of software maintenance, network security, data availability, infrastructure management, and all other operational issues involved with keeping applications up and running. SAAS model onboard has the largest market share in cloud security.

2.2 PAAS

Platform as a service (PaaS) is more of a complete development and deployment environment in cloud security. In this platform-as-a-service (PaaS) model, the developers hire almost everything they need to design an application, depending on a cloud provider for development tools, infrastructure, and operating systems. The PaaS provider makes available everything like servers, networks, storage, operating system software, databases at their centralized data center. The user can purchase the resources as per need from a cloud service provider on a pay-as-you-go basis and access them over a secure Internet connection.

2.3 IAAS

Infrastructure as a service model hosts applications on the public cloud and private cloud rather than in a traditional on-premises data center. The application is made available to users on-demand while it is being fully controlled by the cloud service provider (CSP) itself. The enterprises hire servers for computation and storage in the cloud environment. Infrastructure as a service is highly flexible as you need to purchase only the components you need according to your needs and demand and scale them up or down based on your business needs. Infrastructure as a service model is identical to utility computing, the user gets the resources on the rental basis and pays only for the resources he uses like power, data storage space, etc., on your business needs (Fig. 1).

3 Threats in Cloud Security

3.1 Data Breaches

The risk of a data breach is not new to cloud security, but it consistently ranks as a top threat to cloud customers. Common data breach exposures incorporate personal information, such as social security numbers, credit card numbers, and healthcare histories of customers, as well as corporate information, such as customer lists and software's source code [2].

Mitigation:

You can encrypt the data using data-at-rest and data-in-transit security. Traffic monitoring technique (log management to identify unreliable IP addresses) is also

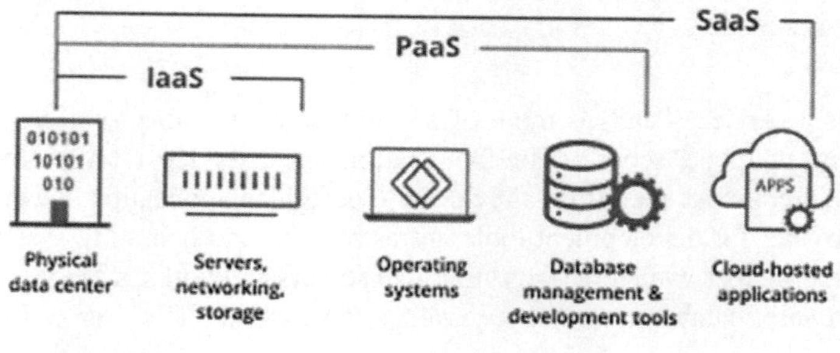

Fig.1 Cloud services models

advisable to prevent data breaches. Implementation of data access governance and establishing data remediation workflows to secure data.

3.2 *Insecure APIs*

The APIs are delivered by the cloud service providers to software developers to build the interfaces and with these interfaces; they can interconnect to the cloud services. Unique characteristics of API leave the door wide open for threats. The difficulties like anonymous access, reusing passwords, or tokenization could hamper the cloud services and could restrict logging, monitoring capabilities, and API dependencies which could result in DDoS attack or repudiation attack.

Mitigation:

Testing techniques that mimic an outsider attack which targets specified API endpoints and aim to break the security and get access to company's sensitive data. API keys must be protected and not reused.

3.3 *Denial-Of-Service Attacks (DoS/DDoS)*

A denial of service (DOS/DDoS) attack is an attempt to tie up a website's resources so that users who need to access the site cannot do so. It generally makes a website service unavailable by overwhelming it with traffic from multiple sources. These are usually launched by bots or malware from hundreds of infected hosts. Application-level DoS/DDOS attacks can easily manifest themselves as high-volume Web page reloads, XML* web services requests [3]. This attack consumes all available resources in the Web server such as memory, CPU, space in disk and indirectly abuses the functioning of the website.

Mitigation:

Implement firewalls and allowing only legitimate IP addresses or blocking ones from known attacker is one of best ways to mitigate DDoS attacks. These attacks can be nullified by rate-limiting the amount of traffic available to a specific network interface controller. The impact of DDoS attack can be reduced by filtering requests upstream, long before it reaches the target network.

3.4 *Misconfigured Cloud Storage*

Misconfigured cloud storage is when there are critical gaps in your cloud security that leave your organizational data at risk. It happens when you make errors while configuring the security controls, or you forget to implement them at all. One of the repeated misconfigurations grants public access to storage buckets. These buckets are often vulnerable because of authentication methods such as passwords, making them available to everyone and prone to attacks. Although, access to these storage buckets is one of the examples of the type of misconfiguration. Organizations face innumerable types of misconfigurations as they drift to public IaaS cloud environments [4].

Mitigation:

Cloud security configurations must be double-checked upon setting up a particular cloud server. This should be understood that configurations are part of security. A third-party security tool should be used that can look at configurations constantly. It provides a constant check and alerts you when things are misconfigured. Outside security testers should be hired to ensure that everything is configured correctly. Sometimes audits can find things that a client may overlook. The stuff being uploaded on the storage should be checked for hidden malwares.

4 Identity and Access Management

Identity and access management refers to the ability to manage user identities and their privileges to access the resources. IAM is the first step to make an organization effective in communication, reliability and securing sensitive data. Identity is the necessity for the foundation of identity and access management. It lubricates the path of providing user's privacy and secure guarding their sensitive information from data theft using identity. Identity plays a crucial role in this heterogeneous cloud environment. IAM strategies have been an important facet of IT platform as it helps to make work possible. IAM as a framework consists of policies which authenticate and authorize user to access a specific resource.

IAM ensures that you as an identity are accessing the appropriate resource at the right point of time securely and you are right person to access it with the help of these two components:

- Authentication.
- Authorization.

Within **authentication**, applications recognize you as a person by looking at your identity cards, digital certificates, etc.

Within **authorization**, applications check what permissions you have to perform actions in that specific environment considering roles, attributes of your identity, or other affiliations to decide if you are authorized to have access to a particular resource. Both these calls help to reduce the load of help desks and maintain the user's privacy. Main functionalities required from an IAM system are:

- An individual may have multiple accounts requiring PINs or passwords.
- Authentication of the user is done by a unique ID provided by IAM system.
- For authentication purpose, complexity of passwords are determined; SSO techniques and OTPs are used to prevent stealing of passwords.
- IAM manages log activity so no suspicious activity passes unnoticed resulting in preventing identity thefts.
- IAM lets individuals manage their personal and confidential data and lets organizations modify data of their employees.

IAM stops threatening activities with the help of techniques like log management, machine learning, and other algorithms. It maintains policies to make users comply with security agreement of the companies.

5 Authentication

The methods to determine that someone is who they claim to be. The most commonly used authentication method is the log-on credentials. But using passwords alone for authentication is increasingly regarded as insecure since passwords are easy for hackers to crack using various attacks (Fig. 2).

5.1 Single Sign-On (SSO)

SSO is an Internet access management tool that allows a user to log in to one of an organization's portal and automatically be logged in to a designated set of other properties. For example, when you log in to Google, you are automatically logged in to your Gmail and YouTube accounts. For users, since they do not have to keep track of different credentials for every application, SSO reduces friction. For organizations, SSO helps in collecting valuable insights about client's behavior and

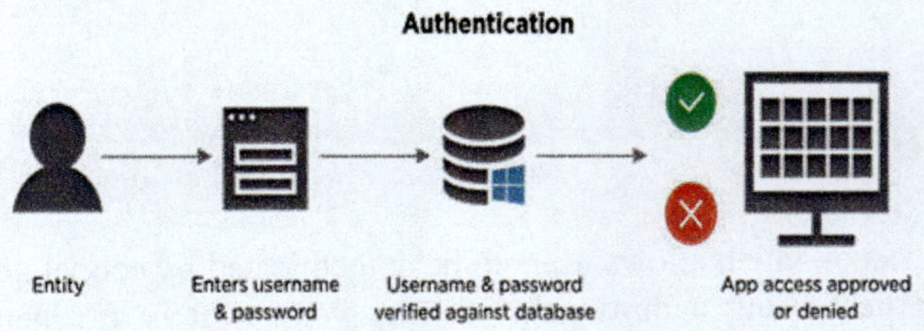

Fig. 2 Authentication basic concept

preferences since it tracks them as they move from one application to another, connected by a single login.

5.2 *Multi-Factor Authentication (MFA)*

Implementing multi-factor authentication is crucial to protect your organization's data and crucial information from malicious intrusions. Every IAM platform offers some form of MFA. However, it is equally crucial to customize MFA with the appropriate level of security. IN B2C contexts, you need to consider UX (user experience) and try not to create unnecessary friction for users who do not want to be subjected to heightened scrutiny every time they log in. For IAM in workplace, you may want more stringent multi-factor authentication, since the consequences of an unauthorized party gaining access to your private network can be so devastating. A modern IAM solution will allow you to implement MFA only when it is needed. This can be accomplished through step-up authentication or adaptive authentication in which users only trigger MFA if they are trying to access sensitive data or their behavior is flagged as risky.

5.3 *Anomaly Detection*

In the past few years, identity has become a preference for hackers to break into systems. Credential stuffing attacks, brute-force attacks and even highly targeted phishing campaigns are all attempts by hackers to break in through an organization's front door: the login box. These attacks can be devastating, leading to legal fallout, huge spikes in traffic that crash applications and most of all, stolen data. IAM solutions are designed to get in front of those issues through anomaly detection. There are multiple ways identity access management systems can help to detect and mitigate malicious attacks. For example, by detecting attacks by monitoring signals such as the velocity of traffic, detection of login patterns that differ from a user's routine (such as location and browser), use of devices and IP addresses with a poor reputation, or use of a breached password.

5.4 *OpenID*

OpenID is a protocol which allows users to be authenticated by cooperating sites—RP (relying parties) using a third-party identity vendor. It is an open standard authentication protocol. A relying party (RP) is a service that depends on a third-party identity provider to identify and authenticate a user who is requesting access to a digital resource [4]. OpenID supports single sign-on services by allowing users to sign in to multiple websites and web services using just one identity.

5.5 *Federated Identity*

While in single sign-on, the tool lets users log in to different properties or brands owned by a single organization; federated identity does the same thing across multiple organizations. The example of federated identity we see in real life is through social login, in which you can use your Google, Facebook, or Apple ID to log in to a wide range of apps. Federation is built on trust, so when you order from a food delivery app with your Samsung pay ID, you are not ordering from Samsung but indicating that the app trusts Samsung enough to take their word that you are who you claim to be.

5.6 *SAML*

SAML works by transferring the client's identity from the identity provider to the service provider. This is done via an exchange of digitally signed XML documents [5].

5.7 *OAuth*

OAuth offers one-way authentication. It is an open-standard authorization framework that allows a user to grant a third-party website access to the user's protected resources, without necessarily revealing their long-term credentials or even their identity [6]. Commonly used by consumer applications and services, so users do not have to sign up for a new username and password. Some examples of OAuth in real world are when you go to log on to a website and it offers one or more ways to log on using another website's logon—"Sign in with Google" or "Log in with Facebook".

6 Authorization

Authorization is the module that is used to determine users' privileges for a particular resource which usually includes system data, files, network services, etc. To explain it better, we can look at some examples: privilege to access any service at all, to access any service with a well-stated write access, to access only a part of the network service or permission to get into whole cloud console. Authorization in

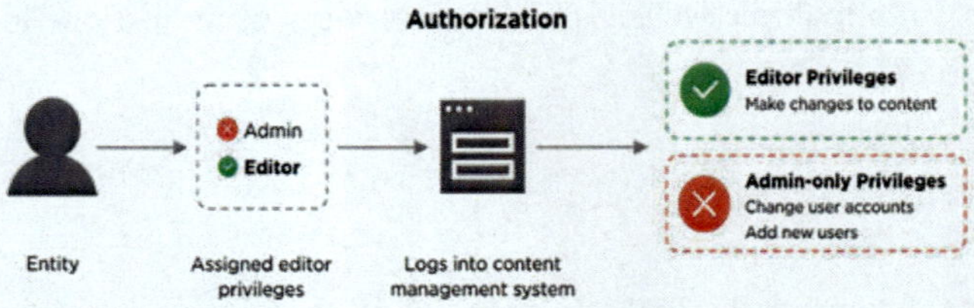

Fig. 3 Authorization basic concept

organizations is usually achieved by access control policies or access right delegations. Access control policies can be defined as—only the authorized clients will be able to access both the resources and services. Authorization is based on complex access control mechanisms which keep record of user attributes like session id, personal information, etc. It helps in safeguarding sensitive information and reducing security risks (Fig. 3).

6.1 *Access Control Mechanisms*

6.1.1 MAC (Mandatory Access Control)

Mandatory access control sets up access policies and states security attributes for the clients where the system provides users with access rights. The access policies are handled by the director, and the users are not given the rights to set, alter, or repeal permissions. When a user tries to access any program, the operating system examines the user's security attributes and decides whether the access can be granted or not. The system verifies the credentials of every individual when any resource is accessed by them. All the security measures are forced by the kernel itself. The system provides access to users based on their clearance level. This type of access control is implemented on systems with high security such as army systems or bank systems. In mandatory access control model, security labels are assigned to every file system object (determines whether it is confidential, secret, etc.). The classification label is maintained by the operating system. If a client wants to access a confidential or a secret file, he must have a secret clearance or high clearance level [5].

6.1.2 DAC (Discretionary Access Control)

In the discretionary access control model, power lies in the hands of owner of the resource; he decides who can have access to a specific object. In DAC, controls are discretionary as the users can access a file on identity basis. The file owner has the authority to change the permissions of the file. The owner can grant access to other clients as per the requirements. Discretionary access control offers a high level of security to data networks of company [7]. DAC minimizes threats and attacks by setting up a firewall with a well-organized IP-tables rules. Unauthorized users are blind to resource characteristics, such as file size, file name, and directory path. All operating systems like Windows, UNIX, and Linux use discretionary access control. DAC provides more flexibility as compared to MAC and is labor-intensive.

6.1.3 RBAC (Role-Based Access Control)

Role-based access control is a way for organizations to manage and assign access privileges across the network in a structured way. RBAC grants permissions based on the employee groups and their subsequent roles. The duties for those groups of employees can be segregated and only the amount of access the groups needs to perform their jobs can be granted. Users may be assigned multiple roles as required. In RBAC, an employee's position determines the permissions they are granted and ensures that lower-level employees are not able to access sensitive information or perform high-level tasks. Organizations that utilize RBAC are better able to secure their sensitive data and critical application [8]. RBAC provide the organizations with some benefits like:

- The process of adding or changing roles of the employees is automated resulting in less paperwork, therefore, increasing in efficiency.
- In RBAC, the data access is appropriately managed resulting in overall better control of compliance efforts.
- RBAC reduces the possibility of data theft, breaches, and information leaks by ensuring that only authorized users have permission to access certain areas of the system.

6.1.4 ABAC (Attribute-Based Access Control)

Attribute-based access control provides access rights based on any type of attributes —environmental attributes, resource attributes, and user attributes.

- Environmental attributes are those that relate to environmental conditions. It includes the location of the data, time of access, and current organizational threat levels [9].
- Resource attributes can be used to enable access control that relates to a particular resource, such as an operating system or application. It includes things like resource owner, creation date, file name, and data sensitivity [10, 11].
- User attributes include things like the user's role, name, ID, and security clearance [12].

7 Conclusion

We can conclude that for digital success in the future, IAM with strong privilege access management is necessary. This paper summarized every security issue that might occur in cloud environment, proposed mitigation for those issues, and potential threats are discussed with an emphasis on IAM. If all these threats are acknowledged and provided mitigation methods are performed, the transfer from local storage to cloud storage can be carried out very smoothly. As with proper implementation of IAM, cloud is going to become the most secure place to keep confidential data, and all the organizations will be shifting to cloud for storage resulting in cloud becoming the future of storage.

References

1. Singh, A., Chatterjee, K.: Identity management in cloud computing through claim-based solution. In: 2015 Fifth International Conference on Advanced Computing and Communication Technologies (2015). https://doi.org/10.1109/acct.2015.89
2. Rowe, N.S.: The future of identity management (2018). Retrieved from https://techvisionresearch.com/
3. Mogull, R., Arlen, J., Gilbert, F.: Cloud Security Alliance. Security Guidance for Critical Areas of Focus in Cloud Computing, V2.1 (2009). http://www.cloudsecurityalliance.org/
4. Sharmaa, H.D., Dr. Dhoteb, C.A., Poteyc, M.: Identity and access management as security-as-a-service from clouds. In: 7th International Conference on Communication, Computing and Virtualization (2016)
5. Bresz, F., Renshaw, T., Jeffrey R., Torpey, W.: Identity and access management (2007). Retrieved from https://chapters.theiia.org/montreal/ChapterDocuments/GTAG%209%20%20Identity%20and%20Access%20Management.pdf
6. Dragoş, M.M.: Cloud identity and access management—a model proposal. Account. Manag. Inf. Syst. **11**(3), 484–500 (2012)
7. Indu, I., Anand, P.M.R.: Identity and access management in cloud environment: mechanisms and challenges (2015). Retrieved from https://www.researchgate.net/publication/325336543
8. Zhu, Y., Huang, D., Hu, C.-J., Wang, X.: From RBAC to ABAC: constructing flexible data access control for cloud storage services. IEEE Trans. Serv. Comput. (2015). https://doi.org/10.1109/tsc.2014.2363474
9. Mohammed, H.K., Hassan, A., Yusuf, D.M.: Identity and access management system: a web-based approach for an enterprise. Path Sci. **4**(11), 1–11 (2018)
10. Khurana, M., Singh, H.: Two level phase retrieval in fractional Hartleydomain for secure image encryption and authentication using digitalsignatures. Multimed. Tools. Appl. **79**(19), 13967–13986 (2020)
11. Khurana, M., Singh, H.: An asymmetric image encryption based onphase truncated hybrid transform. 3D Res. **8**, 1–17 (2017)
12. Rohilla, A., Khurana, M., & Singh, L.: Location privacy usinghomomorphic encryption over cloud. Proquest. Int. J.Comput. Netw. Inf. Secur. **09**(08), 32–40 (2017)

Curbing Criminal Acts on Mobile Phone Network

Olasina Jamiu Rotimi, Sanjay Misra, Akshat Agrawal, Ezenwoke Azubuike, Rytis Maskeliunas, and Robertas Damasevicius

Abstract It is no longer a story that the criminal act is now the order of the day in Nigeria with the way mobile phones and network applications are being used. A lot of Nigerian citizens have in one way or the other fallen prey to this crime through online fraud, hacking into the person's account, retrieving vital information, and the likes. It is, therefore, necessary to detect some of these crimes, reveal them, and proffer possible solutions on how to avoid them, and possibly, how to solve such problems. To do this, the most recent papers were reviewed to create awareness of such criminal act detection and how to curb such activities to create a friendly environment for the usage of mobile phones and network applications without any panic. It is also evident that with the stated methods in this paper, such criminal acts would have been greatly mitigated.

Keywords Mobile phones · Criminal act · Crime · And network

O. J. Rotimi · S. Misra (✉) · E. Azubuike
Covenant University, Ota, Nigeria
e-mail: sanjay.misra@covenantuniversity.edu.ng

E. Azubuike
e-mail: azu.ezenwoke@covenantuniversity.edu.ng

A. Agrawal
Amity University, Gurgaon, Hairiyan, India

R. Maskeliunas · R. Damasevicius
Vytautas Magnus University, Kaunas, Lithuania
e-mail: rytis.maskeliunas@vdu.lt

R. Damasevicius
e-mail: robertas.damasevicius@vdu.lt

K. Khanna et al. (eds.), *Cyber Security and Digital Forensics*,
Lecture Notes on Data Engineering and Communications Technologies 73,
https://doi.org/10.1007/978-981-16-3961-6_10

1 Introduction

Criminal offenses on mobile phones vary from one level of occurrence to another. This act is ranging from virus attacks, malware activities, hacking, and the likes to mobile phones. It is a criminal offense to illicitly have access to data such as contacts, files, private information (such as Bank Verification Number (BVN), passwords, account details, e-mails, and others) and also to induce destructive mobile software such as malware in the system. It is evident that mobile phone is a necessity, and as such, it makes life easier for people in the way we do businesses, carries out researches, fast-tracking of online processes and activities and for these; it has therefore become what we cannot do without. A paper iterated the outrageous development the Internet use and widespread acceptance [1]. The paper proved the increase in security threats and Nigeria of today, cybercrimes are committed in many ways, and these include frauds on e-mails, wrong identity, hacking, criminal harassment, spamming, spoofing on the use of ATM, piracy acts, phishing, and many more. Cybercrime is no doubt a threat to our immediate environment, and people who used their mobile phones to connect to the Internet are the major points of reference [2]. Information and communication (ICT) trends aided these activities by its enormous tools to perpetrate cybercrimes [3].

Viruses are illicitly introduced software into the system to disrupt the system activities, but malware includes SMS fraud—where users unknowingly subscribed to the service, and at the later end, a cruel way was used to charge users without their knowledge, Spyware—it steals data from users without prior permission of the owner, and so on. [4]. Also, the echo reflection of mobile phones has made them very important means of exchanging information, in the field of communication. Furthermore, there also exists a limitation to how to convey criminal activities to law agencies for quick and needed actions [5, 5]. According to [7], the growth of the telephone network and its availability to the Voice over Internet Protocol (VoIP) has made it a major contribution to its flexibility and also the very easy-to-use artifact for end users. This also contributed significantly to the increase in cyber-criminal activities, and these criminals use emergent technologies to conduct illegal and suspicious activities. Crime analysis is a law enforcement function, and it involves systematic analysis for patterns and trends identification and analysis in crime and disorder. Information acquired from this pattern can then be used in aiding the activities of detectives in identifying and apprehending criminals [8].

The likelihood that a crime is detected and its offender is appropriately charged is a central component of the standard economic model of crime. It is also critical to the incapacitation channel, by which societies can prevent hardened criminals from reoffending. Yet, the economics literature has barely devoted any attention to studying the determinants of crime detection in detail. Typical approaches include examining whether police numbers, police composition, or high visibility patrolling are associated with lower crime rates. The implicit assumption is that a change in these variables can lead to higher chances of catching offenders, which has an immediate deterrence effect as well as an incapacitation effect over longer horizons

[9]. As part of the study, it is necessary to incorporate the action of the police force and other law enforcers to the needed technologies to apprehend the offenders who are cybercriminals. At times, the police may visit the perpetrators at home for questioning as the data of the ownership of the phone in question would be automatically captured for necessary actions. Due to the rapid need for telephone by the service it renders, criminals are now abusing its use to perpetrate various cybercrime attacks, and it is therefore necessary to logically detect and study some of the already existing solutions to these problems and also propose a new solution as there have never been any of the existing solutions without its setback.

In recent years, the number of online crimes, such as the various emerging scams and criminal schemes, has tremendously increased, and also, the online crime suspects utilize the anonymous nature of the Web to disguise their identity through various methods to evade detection and surveillance from law enforcement agencies. The prime suspect's communication methods and identifying numbers have become so hard to be verified under the electronic surveillance warrant for the Law Enforcement Monitoring Facilities (LEMF). When this is compared with traditional telephone and mobile phone communication records, online communication is more uncertain to be tracked, and its techniques require a sophisticated system [10]. Also, nowadays, providing information and security for mobile phones and also processing it in mobile network data is a great issue of importance and interest. As mobile phones and computers are provided with broader functions, the geometrical growth of vulnerabilities is also increasing in that regard. Today, an unauthorized person who is a criminal can make a call and eavesdrop on them, text message, drop malicious programs like malware, spyware, also steal money, and put the system out of order to his advantage without prior knowledge of the owner [11].

Mobile devices have become an integral part of our daily life, and there is no doubt about that. These have therefore proven to be an advantageous and almost the most successful scientific resolution of our time that fills personal and business needs in a very efficient manner [12]. In this era, the availability of mobile services has significantly increased because of the rich variety of mobile devices and essential applications provided by mobile device manufacturers as has been rightly emphasized in the previous chapter. At the same time, numerous mobile security issues and data privacy threats are challenging the use of this device in various aspects of human endeavor. Therefore, mobile devices are an ideal target for various security issues and data privacy threats in a mobile ecosystem [13]. Also, the impact of cybercrime poses a negative signal over the state of any system, and this can be felt on every individual, economy of a nation, international prospect, and integrity of a nation [1]. This paper, therefore, looks into threat detection and prevention, and also, necessary steps and algorithm to alert the security agencies are put up as a proposed framework and architectural modeling for the future most prominent solution to the problem at hand.

2 Background Information and Literature Review

Researchers have gone far as to carry out a lot of studies and investigations on the issues of threats posed by the use of mobile phones. Particularly, vulnerabilities of this amazing device to illegal act by cybercriminals have necessitated the prompt need of all researchers in the chosen area of the framework to be on their toes as the terror of criminal act has become the order of the day. Any attack on the use of the mobile phone that makes life uneasy for the user is indeed an action that has put such a user in a state of great concern. Therefore, any such challenges should be dealt with. Cybercrime issues encompass stakeholders such as the offender; its target/victim; technological; society and the law, and the architectural model was proposed. This does for the future reservoir of resources for mitigation of cybercrime [14]. The threat on the mobile phone includes (1) viruses, (2) malware, (3) spyware, and (4) limited way to convey information to the needed security agency. Authors [13] fully explained that the threat to the mobile device is multiple and it requires multiple means of protection and restrictions. These researchers proposed a defensive mechanism architecture that can simultaneously handle various threats on the use of mobile phones, but the security measure to alert the police was not put in place. Authors [15] explained that advanced persistent threat (APT) attack is a carefully planned attack that involved both social engineering and malware, where spear phishing is the most popular method that has been used by the attacker. The attacker will send an email to the targeted victim by including link (s) to the targeted Web or malicious attachment to carry out their illegal activities, but this was taken care of by the adoption of smartphones and implementation of security on bringing your device (BYOD). In [16, 17], authors launched preventive measures such as passwords, firewalls, encryption, and detection mechanisms such as tripwires, configuration-checking tools, and anomaly detection systems [18]. In their paper established a helping mechanism in investigating agencies in the detection and identification of criminal acts by the use of methods that involved six approaches—data extraction (DE), data preprocessing (DP), GSM technology, Google map representation, advanced embedded system (ES), and preprocessing of the dataset to safeguard its originality before its cluster analysis. Authors in [19] their paper posed that mostly threat is always from the Internet and therefore tackled by setting up Apache Hadoop cluster with Apache Ambari. Where master and slave nodes represent a Mesintempur as a cluster. Apache Hive was used to aggregate data from raw data for the attack type incorporated with related mechanism, deeper analysis was made, and the resultant attack data was converted to the dataset. The mining was done by the use of the SPMF tool. Besides, authors in [20] studied the Web site, using six different metrics, such as speed, SEO, security, broken links, updates, and availability to come to 54% of good Web sites of Web clustering.

Authors [21] in their paperwork validated the factual establishment of the generated results of their analytically solved problem that was based on leveraging phone numbers analysis in improving the world understanding of the underground markets, illegal activities with computers and software, and cybercrime in general and by these emphatically stated that scam activities with phone numbers were often and more stable over the period than email addresses. This was backed up by the combination of graph analysis and geographical home location register (HLR) lookup. Authors in [22] went further in their work to gather dataset related to the criminal activities from the cloud, by making the criminal-related information available to the law enforcer to speed up their investigation of criminal identification by developing the police Android application and the general user mobile application.

In [23], authors stated algorithms and mechanisms such as statistics-based algorithm, decision tree-based algorithm, rule-based algorithm, Bayesian classification model to detect fraud in automobile insurance, Naïve Bayesian visualization for data analysis and interpretation, the classifier predictions, and the use of ROC curves data assessment. All these are used for data mining. As a proper check and unbiased technique, authors in [24] proposed discrimination prevention degree (DPD), discrimination protection preservation (DPP), misses cost (MC), and ghost cost (GC). The DPD and DPP measured the success of the proposed method in discrimination prevention which ideally should be 100%. The MC and GC, therefore, measured the degree of information loss that forms the impact on data quality, and it should ideally be 0%. Authors in [25] studied that the best-recommended cryptography application that can be used on IM is encrypt, followed by AES-Crypto, EnDe-Crypto, and Kryptokaz as it is a good way for securing data.

Nevertheless, authors [11] in his work studied three asymmetric encryption techniques: RSA, ElGamal, and elliptic curves to select the most efficient and appropriate algorithm for data transmission. Elliptic curves became the best, and RSA and ElGamal algorithms are not good for the text message at all. These comparisons were done on a carefully designed platform Java 2 Micro Edition (J2ME) using wireless messaging PI (API). The application runs on a mobile phone with an ARM9 processor with a frequency of 219 MHz and 10 Mb of internal memory and the Java virtual machine (JVM) with enabled just-in-time (JIT).

The key parameter of interest is to establish a framework to address spyware, malware, viruses, unauthorized access, and bad Web sites, as there is no single system or model which can take care of all the already existing cybercrimes at the present. Besides, there is no mobile-based app for this function, and in this work, HTML5, CSS, PHP, API-VIRUS TOTAL, and MYSQL would be greatly employed for the system design. One can find other security issues in mobile in various researches [26–29].

Motivation: The existing techniques are not on a mobile app, and besides, no existing solution used multiple solutions to solve the criminal acts. This system was proposed to use multiple techniques of different algorithms to mine data, detect unwanted data, and curb Internet crimes, and reporting to the necessary law enforcement agent using mobile phones at the discretion of the user is the required task. Also, the driven force centered on the gap created by the past researchers. This then enables us to come up with this tremendous model as a mobile app.

Aim and Objectives: The aim is to establish and build a mobile app to compliment the already existing algorithms, using different techniques of the framework to prefer solutions to the cybercrimes. The objective is for the proposed framework to deliver real data to the receiving-end phone; send details of the sending-end phone handler to the law enforcement agent; and block any unwanted data from disrupting or intercepting the system.

3 Proposed Model

In this study, we propose an upgrade of existing technologies to curb cybercrimes on the use of mobile phones. In the upgrade, a Web-based application that will interrupt the activities going-on on a mobile phone will automatically be linked to the phone number of the police or law enforcer agencies for proper action. Besides, any bad Web would be kicked out of action, and only the good one will go through. We, therefore, propose the architectural framework that will detect, identify, and solve multiple problems of cybercrimes using a mobile app. The proposed model is shown in Fig. 1 below.

The model is divided into four categories, and these categories include the sending-end section, intruder section, mobile app section, and receiving-end

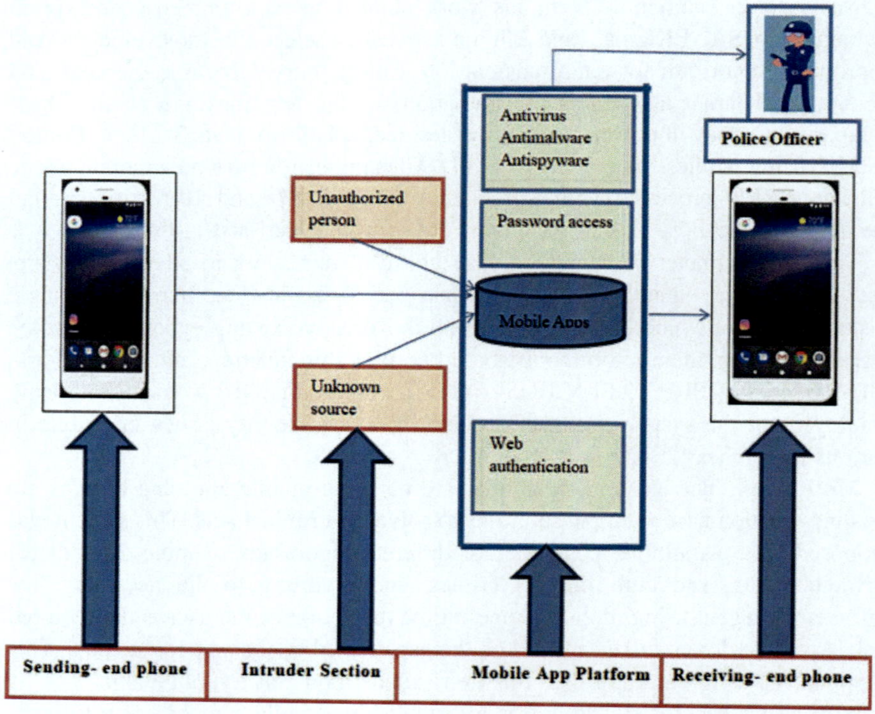

Fig. 1 Proposed architectural framework for the model

section. At the sending-end section, data which would either be a text message or multimedia message and so on is to be sent to the receiving-end phone section. But before then, such data would have to go through the mobile app section for necessary actions before it is finally received at the receiving-end phone. For the scope of this work, data from the sending-end phone is called real data, while data from unauthorized access and unknown sources, virus, malware, spyware, and the like is called unwanted data. The main function of the mobile app section is to detect unwanted data and block it, and a platform is provided to dial security code for necessary action or sanction of the culprit. The app will send personal information of the culprit to the law enforcer agent that is with such a phone identity number. The algorithm that explains the operational flow of the proposed framework and flowchart is shown in Fig. 2.

0 Start.
1 Log in your detail to open the APP (APP Launching).
2 If no details, create one (Create One).
3 Check if there is any data from an unauthorized person (Duk).
4 Check whether a good Web site (GWS) or bad Web site.
5 Check for any software threat (THREAT).
6 Allow the real data on the device (REAL DATA).
7 Block any unwanted data (UD) from having access to the recipient phone.
8 Forward the address of the sending-end phone to the closest law enforcer.
9 Stop.

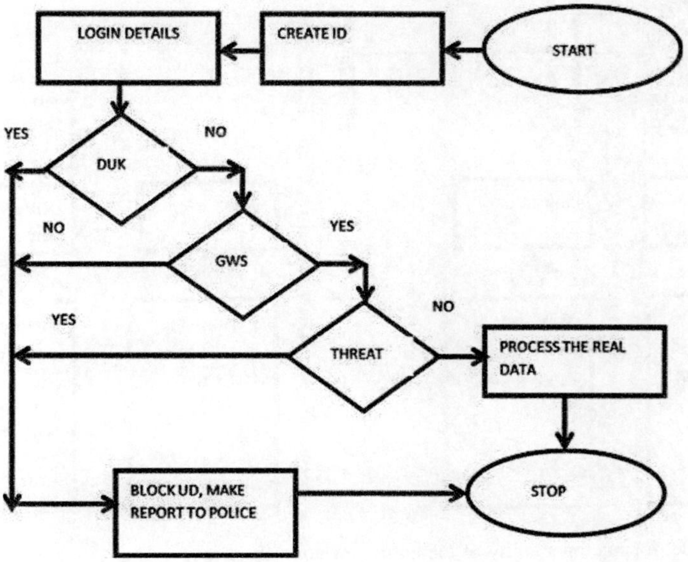

Fig. 2 Flowchart

3.1 *Implementation of the Proposed Model*

The process is divided based on the aim of the proposed work, and this goes down the line from specification to the evaluation across tasks to be handled by the mobile app section. The app should synchronize between the incoming data and the receiving-end phone. See Fig. 3, being a model for interoperability of the entire development process. The process is discussed below:

Specifications: The app will remove viruses, malware, spyware, and block bad Web sites and unwanted data. It will relay the good data to the intended receiver and channel details of any detected criminal act to the police agency if necessary by the user.

Design: The design was done by the use of HTML5, CSS, JQUERY, and PHP. These software tools were used to design both the front end and back end. The database that was used is MYSQL, and the Android-based application that was used for virus detection function is API-VIRUS TOTAL.

Implementation: The design as it has been listed is implemented following all algorithms incorporated in the mobile app model, and the necessary results were generated.

Evaluation: The mobile app was evaluated through deployment to check if any aspect will require visitation for the betterment of the system.

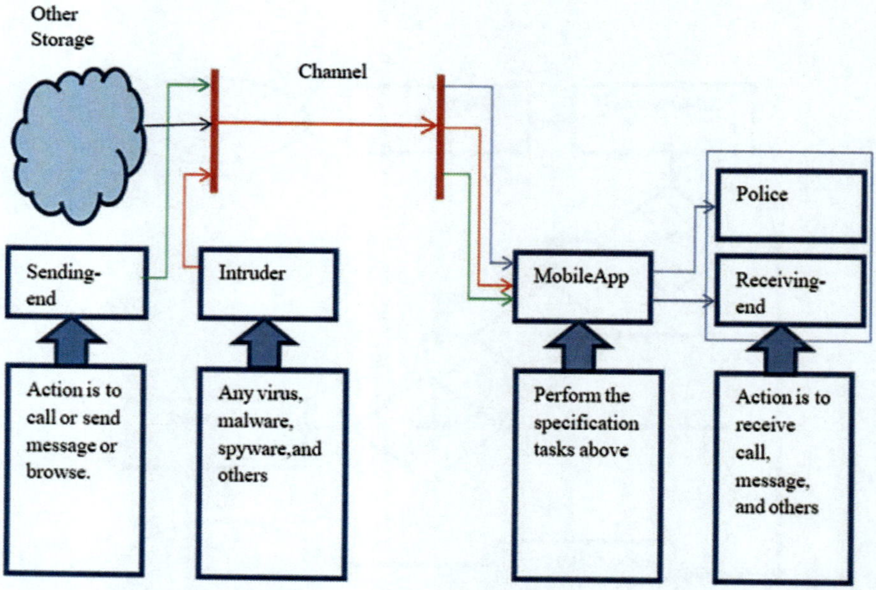

Fig. 3 Model for interoperability of the entire development process

4 Experimental Results

The results obtained are demonstrated in cropped screenshot pictures. The interface for each stage is shown under appropriate sections. See Figs. 6, 7, and 8, for the front-end view of options for test link, test files, file report, and help. The help option will guide any new user on how to use the app, for there are links that such users will follow to help with how it works.

When the mobile app is launched, it will display the front end as shown in Fig. 4 below. As soon as the requirement for logins is satisfied, it will launch the front end in Fig. 5, and from there, the option will be chosen for the link test, file test, file report, and help. The test link option will help to check if the link is a good one or not. The file option will check if the file is free from viruses and the data does not contain any malicious program. The file report option helps to show the environment where the user may decide to let law enforcer such as the police be aware.

Figure 6 is the front end that provides a space for testing any link whether such link is free of any software problem.

Figures 7 and 8 show the interfaces for the help menu and test file, respectively. At the help menu interface, the new user or any first-timer will be able to learn about the app and get familiar with how it works. On the other end, the test file platform provides a space to test a particular file and establish its state whether it is a problem-free file or not.

Fig. 4 Link for login

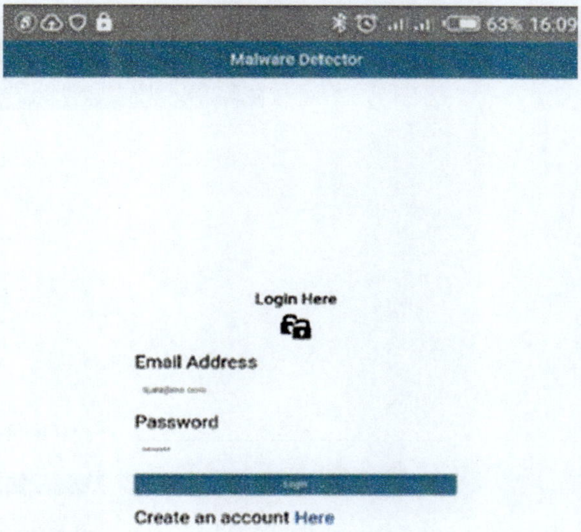

Fig. 5 Links for necessary actions

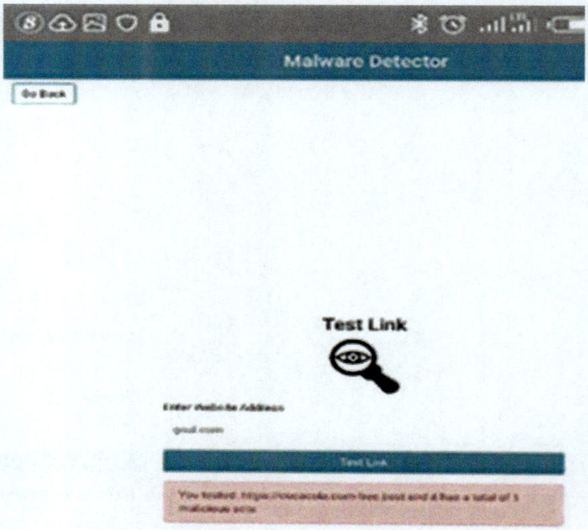

Fig. 6 Test link

Fig. 7 Help

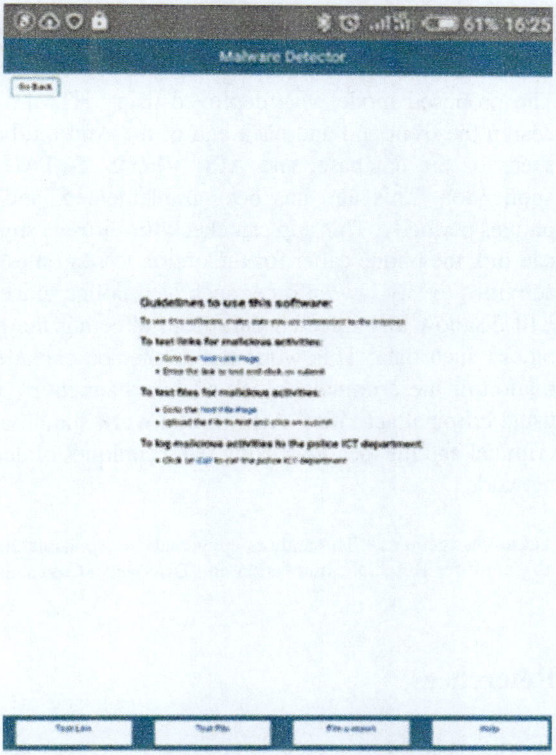

Fig. 8 Test file

5 Conclusion and Future Work

The proposed model was deployed using HTML5, CSS, JQUERY, and PHP to design the front end and back end of the Android-based mobile app, MYSQL was used for the database, and API- VIRUS TOTAL was used for virus detection application. This app has been implemented, and it works on Android mobile phones perfectly. This app can check for viruses, spyware, and malware, and also, it can link the phone caller for the option to send messages or details of any criminal activities to any law enforcer such as a police officer for necessary action. Also, it will disallow any infected data from affecting the receiver system, and it equally blocks such data. This work can further be extended by capturing detailed information of the criminals, block all his channels by which he was carrying out his usual criminal activities. Also, future work should address the identification of the criminal and his locations using the techniques of deep learning and artificial neural network.

Acknowledgements The authors appreciate the sponsorship from Covenant University through its Centre for Research, Innovation and Discovery, Covenant University, Ota Nigeria.

References

1. Omodunbi, B.A., Odiase, P.O., Olaniyan, O.M., Esan, A.O.: Cybercrimes in Nigeria: analysis, detection and prevention. FUOYE J. Eng. Technol. **1**(1), 37–42 (2016)
2. Ishwarya, T.A.S.K.: Cyber crime : prevention and detection **4**(3), 45–48 (2015) https://doi.org/10.17148/IJARCCE.2015.4311
3. Tanui, D.K.: Use of ICT in the detection and prevention of crime in Kenya **6**(9), 62–71 (2016)
4. Rashid, A.M., Al-oqaily, A.T.: Detect and prevent the mobile malware. Int. J. Sci. Res. Publ. IJSRP **5**(5), 9–11 (2015)
5. Agangiba, W.A., Agangiba, M.A.: Journal of Computing: mobile solution for metropolitan crime detection and reporting. J. Emerg. Trends Comput. Inf. Syst. Univ. Cape Town, South Africa **4**(12), 916–921 (2013)
6. Agangiba, W.A., Agangiba, M.A.: Journal of computing: mobile solution for metropolitan crime detection and reporting. J. Emerg. Trends Comput. Inf. Sci. (2019)
7. Bordjiba, H.E., Karbab, E.B., Debbabi, M.: Data-driven approach for automatic telephony threat analysis and campaign detection. Digit. Investig. **24**, S131–S141 (2018). https://doi.org/10.1016/j.diin.2018.01.016
8. Umamaheswari, B., Nithya, P., Chandran, N.S.: Survey on web crime detection using data mining technique **5**(1), 177–184 (2016)
9. Blanes, J., Kirchmaier, T.: The effect of police response time on crime detection ∗ 1–47 (2015)
10. Chen, C., Chen, W., Wang, Y., Lo, C.: Procedia engineering a real-time crime detection system based on lawful interception—a case study of msn messenger (2011). https://doi.org/10.1016/j.proeng.2011.08.304.
11. Starikovskiy, A.: Text messages protection system text messages protection system text messages protection system text messages protection system. Procedia Comput. Sci. **123**, 457–466 (2018). https://doi.org/10.1016/j.procs.2018.01.070

12. Dong, Y., Pinelli, F., Gkoufas, Y., Nabi, Z., Calabrese, F., Chawla, N.V.: Inferring unusual crowd events from mobile phone call detail records **2**, 474–492 (2015) https://doi.org/10.1007/978-3-319-23525-7
13. Khan, J., Abbas, H., Al-Muhtadi, J.: Survey on mobile user's data privacy threats and defense mechanisms. Procedia Comput. Sci. **56**(1), 376–383 (2015). https://doi.org/10.1016/j.procs.2015.07.223
14. Singh, M.M., Bakar, A.A.: A systemic cybercrime stakeholders architectural model a systemic cybercrime stakeholders architectural model. Procedia Comput. Sci. **161**, 1147–1155 (2019). https://doi.org/10.1016/j.procs.2019.11.227
15. Zulkefli, Z., Singh, M.M., Mohd Shariff, A.R., Samsudin, A.: Typosquat cyber crime attack detection via smartphone. Procedia Comput. Sci. **124**, 664–671 (2017). https://doi.org/10.1016/j.procs.2017.12.203
16. Smith, J.L., Smith, M., Smith, J.L.: The perpetration and prevention of cybercrimes. Available at SSRN 1123743
17. Prasanthi, M.M.L.: Cyber crime: prevention and detection. Ijarcce **4**(3), 45–48 (2015). https://doi.org/10.17148/ijarcce.2015.4311
18. Sathish, A., Prathyusha, M., Priyanka, M.: Crime detection and criminbal identification using IOT **4**(1), 2–4 (2018)
19. Hidayanto, B.C., Muhammad, R.F., Kusumawardani, R.P., Syafaat, A.: Network intrusion detection systems analysis using frequent item set mining algorithm FP-max and Apriori. Procedia Comput. Sci. **124**, 751–758 (2017). https://doi.org/10.1016/j.procs.2017.12.214
20. Rakhmawati, N.A., Ferlyando, V., Samopa, F., Astuti, H.M.: A performance evaluation for assessing registered websites. Procedia Comput. Sci. **124**, 714–720 (2017). https://doi.org/10.1016/j.procs.2017.12.209
21. Costin, A., Isacenkova, J., Balduzzi, M., Antipolis, S.: The role of phone numbers in understanding cyber-crime schemes
22. Dabhere, A., Kulkarni, A., Kumbharkar, K., Chhajed, V., Tirth, S.: Crime area detection and criminal data record **6**(1), 510–513 (2015)
23. Bhowmik, R.: Journal of digital forensics, security and law data mining techniques in fraud detection data mining techniques in fraud detection **3**(2) (2008)
24. Hajian, S., Domingo-Ferrer, J., Martinez-Balleste, A.: Discrimination prevention in data mining for intrusion and crime detection. IEEE SSCI 2011 Symp. Ser. Comput. Intell. CICS 2011 IEEE Symp. Comput. Intell. Cyber Secur. 47–54 (2011). https://doi.org/10.1109/CICYBS.2011.5949405
25. Liwandouw, V.B., Wowor, A.D.: The existence of cryptography: a study on instant messaging. Procedia Comput. Sci. **124**, 721–727 (2018). https://doi.org/10.1016/j.procs.2017.12.210
26. Jambhekar, N.D., Misra, S., Dhawale, C.A.: Mobile computing security threats and solution. Int. J. Pharm. Technol. **8**(4), 23075–23086 (2016)
27. Osho, O., Mohammed, U.L., Nimzing, N.N., Uduimoh, A.A., Misra, S.: Forensic analysis of mobile banking apps. Lecture Notes Comput. Sci. **11623**, 613–626 (2019
28. Alhassan, J.K., Oguntoye, R.T., Misra, S., Adewumi, A., Maskeliūnas, R., Damaševičius, R.: Comparative evaluation of mobile forensic tools. Adv. Intell. Syst. Comput. **721**, 105–114 (2018)
29. Osho, O., Musa, F.A., Misra, S., Uduimoh, A.A., Adewunmi, A., Ahuja, R.: AbsoluteSecure: a tri-layered data security system. Commun. Comput. Inf. Sci. **1078**, 243–255 (2019

Designing of Fuzzy Logic-Based Intrusion Detection System (FIDS) for Detection of Blackhole Attack in AODV for MANETs

Ruchi Makani and B. V. R. Reddy

Abstract Mobile Ad hoc networks (MANETs) are wireless/infrastructure-less and resource-constraint, having collection of nodes with high mobility feature (Ramanathan and Redi in IEEE Commun Magaz 40(5) 2002). It is a challenge to have efficient intrusion detection system (IDS) for such wireless and mobile architecture of systems. Researchers have presented in their research that the fuzzy logic-based intrusion detection systems are more adoptable to MANET's application because behavior of any mobile node may be visualized in fuzziness characteristics. It is required to design robust IDS system which can sustain and can work efficiently in MANET environments. The work presents the selection of suitable protocol features and fuzzy rules generation which exhibits substantial role for precision of the fuzzy logic-based intrusion detection system (FIDS). Here, set of fuzzy rules have been proposed to protect network against blackhole attack. These set of rules are created using three AODV critical attribute which are rate of *RREQ*, *RREP* and *Sequence number* value. The proposed FIDS, thereafter, evaluated using ns2 simulator and are found efficient to detect and isolate the attacker node from the network. The deployment of FIDS has resulted in increase of throughput of the network.

Keywords AODV · MANET · Fuzzy-logic · Intrusion detection · Blackhole

1 Introduction

Aim of an intrusion detection system (IDS) is to ascertain attack or malicious activities in a standalone system or in networked systems, by continuously monitoring the audit trail of traffic. IDS largely use soft-computing techniques to monitor the 'arriving at' and 'passing by' of traffic packets to detect intrusion. Fuzzy

R. Makani (✉) · B. V. R. Reddy
University School of Information, Communication & Technology,
Guru Gobind Singh Indraprastha University, Delhi, India
e-mail: ruchichaudhary@nic.in

K. Khanna et al. (eds.), *Cyber Security and Digital Forensics*,
Lecture Notes on Data Engineering and Communications Technologies 73,
https://doi.org/10.1007/978-981-16-3961-6_11

logic-based intrusion detection model uses fuzzy rules or fuzzy classifiers to detect various intrusive behaviors by creating more abstract and flexible patterns for intrusion detection and provides significant advantages over other techniques [2, 3]. The use of fuzziness helps to understand the abrupt separation in normal and abnormal behavior of the node distinctly and also provides a measure of the degree of normality or abnormality of an event. Section 2 of this paper has presented general overview of fuzzy logic-based intrusion detection systems. Section 3 covers the critical filed of AdHoc on-demand distance vector (AODV) protocol used in MANETs. Section 4 is presenting proposed fuzzy rules for IDS designing in detail. Sect. 5 covering the simulation of FIDS for MANETs, and the performance of AODV has been recorded with and without blackhole attack. Section 6 concludes the finding of the work.

2 Overview of Fuzzy Logic-Based Intrusion Detection

Only fuzzy logic is a computational paradigm that builds a set of user-defined rules which are converted into mathematical equivalents [4–6]. Moreover, it has the advantage to offer valuable flexibility for reasoning that considers inaccuracies and uncertainties [7–10]. It can also manage approximate reasoning instead of fixed reasoning and able to handle imprecise and incomplete data. In standard set theory, each element is either completely a member of a category or not a member at all. In contrast, fuzzy set theory allows partial membership in sets or categories. In fuzzy logic, the truth value of any event can be termed between '0' and '1'. Fuzzy logic-based intrusion detection (FLID) classifies the nodes into 'trusted nodes' and 'malicious nodes' after evaluating the nodes' behavior (i.e., degree of reliability) by utilizing user-defined fuzzy inference rules [11–13]. Primarily, an FLID consists of three processing steps: fuzzification, inference system and defuzzification.

Fuzzification: In fuzzification, input data values are obtained and converted into fuzzy set by using fuzzy linguistic variables and membership functions. In inference processing system, a set of fuzzy rules are created (*in form of 'IF–THEN' decision statements to encode an expert's knowledge of known patterns of attack and system vulnerabilities*) by the user, depending on the application environment/requirement.

Inferences System: Inferences are established after the processing of inputs based on fuzzy rules. The inferences are then combined to compute the fuzzy output distribution.

Defuzzification: In defuzzification, the resulting fuzzy output distribution is mapped back to a crisp output value using the membership functions. The detailed processing involved in each step of fuzzy-based intrusion system is available in [14, 15]. However, a broad flow chart of the FLID is shown in Fig. 1.

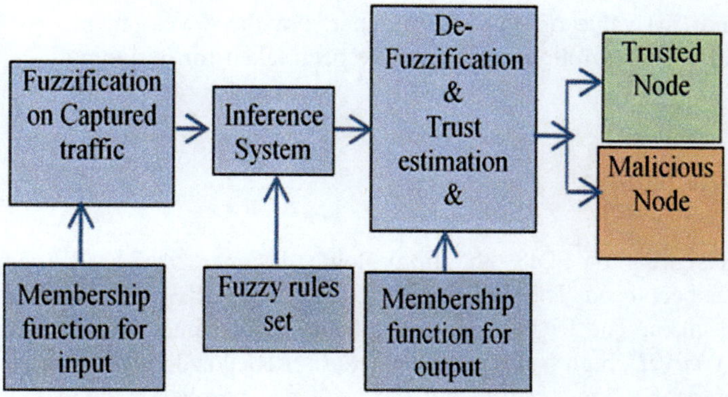

Fig. 1 Flow chart of fuzzy-logic-based IDS

3 Critical Fields of AODV Routing Protocol

AODV [16] is one of the most popular and widely used reactive routing protocols in MANETs. AODV ensures loop-free, single path and hop-by-hop distance vector routing [17]. AODV operates in two sub phases: *Route discovery* and *Route maintenance*. It uses four types of message packets to communicate among each other—route request (RREQ) and route reply (RREP) messages are needed in route discovery subphase, and route error (RERR) messages and HELLO messages are needed in route maintenance subphase. In each AODV routing packet, some critical fields such as control packet ID, type, hop count, sequence numbers of source and destination, IP addresses of source and destination, flags, lifetime are essential for correct protocol execution and routing [18, 19]. However, attackers may launch attack by advertising altered routing information to mislead correct routes (*known as route logic compromising attacks*) or by intently dropping the packets (*known as packet distortion attacks*). Any misuse or alteration of these critical fields by the attacker can cause AODV routing protocol to malfunction resulting in network performance degradation. Various publications relating to attacks of AODV and its effect on the network are available in [18, 19].

4 Proposed Fuzzy Rules for IDS Designing

In the above Sect. 3, the critical field of AODV protocols have been discussed which may be modified by attackers to deteriorate network performance. Here, efforts have been made to design a robust fuzzy logic-based intrusion detection (FIDS) system which is capable of detecting a blackhole attacker node in a network. For this, three attributes have been used to build up the fuzzy rules for fuzzification: (i) rate of forwarding RREQ control packet, (ii) rate of forwarding the RREP packets and (iii) value of sequence number. For the development of robust and effective FIDS, three following steps have been taken for designing.

4.1 Fuzzification

For the designing of FDIS, the fuzzy logic designer MATLAB (version 9.7) toolbox has been used. The three inputs, i.e., RREQ, RREP and sequence number, have been taken. The RREQ input have been divided into three categories, viz., very high (*reqVH*), high (*reqH*) and low (*reqL*). RREP is divided into two datasets, viz., high (*repH*), low (*repL*), and the sequence number is divided into three datasets, viz., high (*nauH*), medium (*nauM*), low (*nauL*). The designed FIDS's block diagram is shown in Fig. 2. The output value (*named as 'trust'*) has been estimated from the FIDS.

The membership function for output 'trust' has been divided in five sets ranging from 0 to 1 as very high (*VH*), high (*H*), medium (*M*), low (*L*) and very low (*VL*), as depicted in Fig. 3. Data is converted into fuzzy set by using fuzzy linguistic

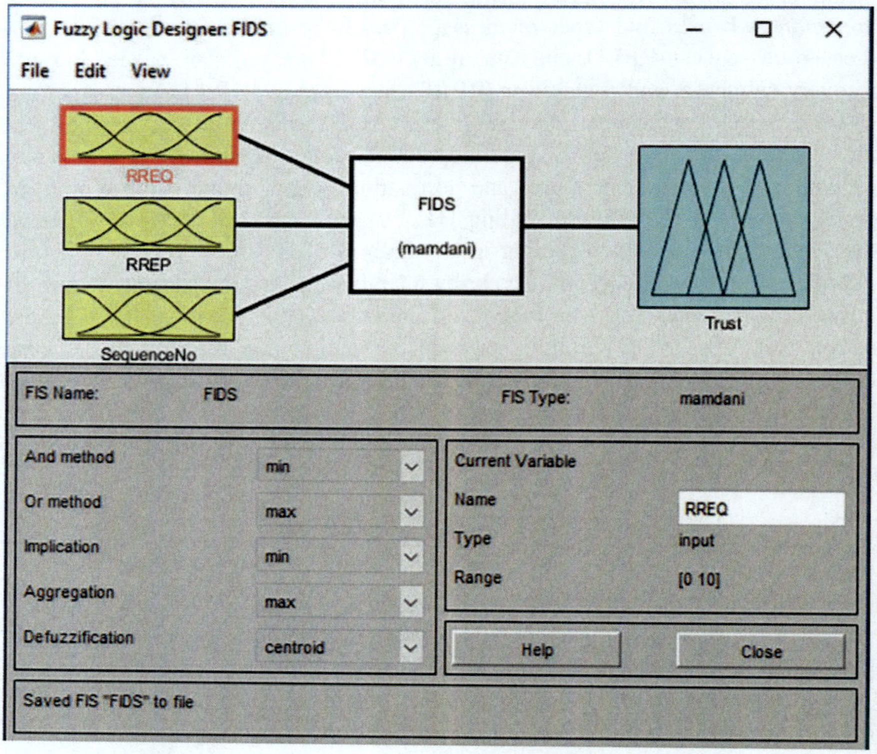

Fig. 2 FIDS designer

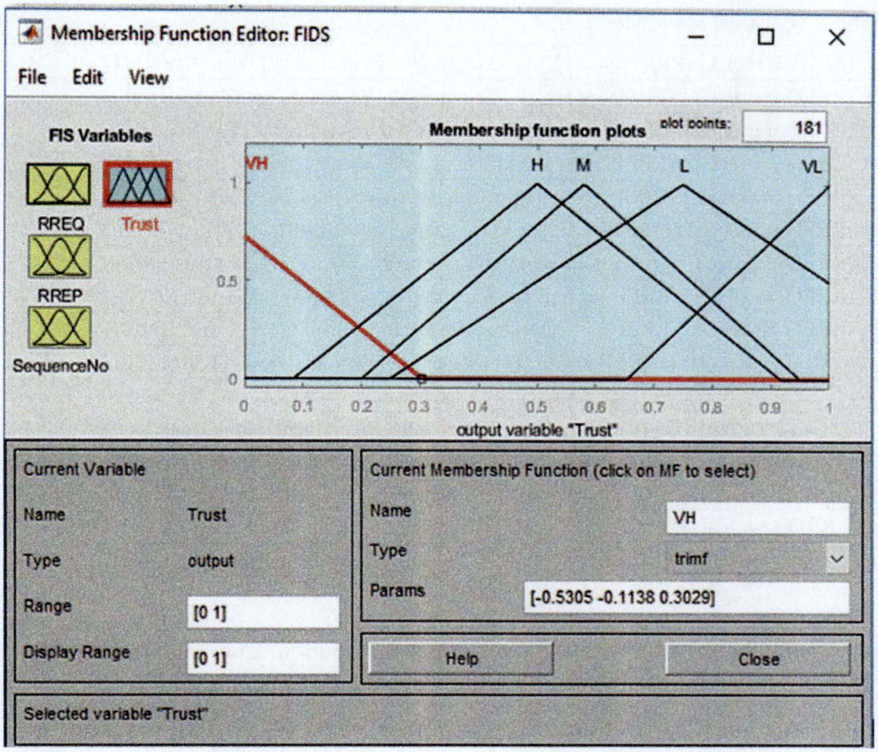

Fig. 3 Output fuzzy membership function plot

variables and membership functions. Here, triangular function (*trimf*) have been used for fuzzification. For the evaluation of the proposed FIDS, an input data value is obtained using simulations of MANET network in network simulator software (version ns 2.35) and has been discussed in the next section.

4.2 Inferences System

A set of 16 fuzzy rules have been created (in the form of 'IF–THEN' decision statements) to discern the behavior of node for the detection of a blackhole attacker (malicious node) in a given network. The blackhole attacker node broadcasts the route reply packet against the response of route request packet send by node with high sequence number to present the freshness of the loop. Based on this theory, following 16 rules have been included in FIDS. The details of the rules are listed in Table 1, and their inclusion in the fuzzy toolbox is shown in the Fig. 4. Inferences are derived after the processing of inputs based on fuzzy rules. The inferences are

Table 1 Proposed fuzzy rules

1. (REQ == reqL) | (RREP == repL) | (SequenceNo == nauL) => (Trust∼=VH) (1)
2. (RREQ == reqH) | (RREP == repL) | (SequenceNo == nauM) => (Trust=H) (0.1)
3. (RREQ == reqL) | (RREP == repH) | (SequenceNo == nauL) => (Trust=M) (0.1)
4. (RREQ == reqH) | (RREP == repL) | (SequenceNo == nauL) => (Trust=M) (0.1)
5. (RREQ == reqH) | (RREP == repH) | (SequenceNo == nauL) => (Trust=M) (0.1)
6. (RREQ == reqVH) | (RREP == repL) | (SequenceNo == nauL) => (Trust=M) (0.1)
7. (RREQ == reqVH) | (RREP == repH) | (SequenceNo == nauM) => (Trust=M) (0.1)
8. (RREQ == reqL) | (RREP == repL) | (SequenceNo == nauM) => (Trust=M) (0.1)
9. (RREQ == reqL) | (RREP == repH) | (SequenceNo == nauM) => (Trust=L) (0.1)
10. (RREQ == reqH) | (RREP == repH) | (SequenceNo == nauM) => (Trust=L) (0.1)
11. (RREQ == reqVH) | (RREP == repL) | (SequenceNo == nauM) => (Trust=L) (0.1)
12. (RREQ == reqVH) | (RREP == repL) | (SequenceNo == nauH) => (Trust=L) (0.1)
13. (RREQ == reqL) | (RREP == repL) | (SequenceNo == nauH) => (Trust∼=VL) (0.1)
14. (RREQ == reqL) | (RREP == repH) | (SequenceNo == nauH) => (Trust∼=VL) (0.5)
15. (RREQ == reqH) | (RREP == repH) | (SequenceNo == nauH) => (Trust∼=VL) (0.5)
16. (RREQ == reqVH) | (RREP == repH) | (SequenceNo == nauH) => (Trust∼=VL) (0.5)

then combined to compute the fuzzy output distribution. The graphical representation of the inference obtained after the computation is depicted in Fig. 5.

4.3 *Defuzzification*

The resulting fuzzy output distribution is mapped back to output value using the same membership functions which was used during fuzzification. Here, triangular function (*trimf*) has been used. In the following Table 2 are a few test data which were used in simulation to compute the defuzzification.

5 Simulations and Result Discussion

To establish the importance of intrusion detection systems in a mobile network in Sect. 5.1, performance of AODV has been recorded with and without blackhole attack. In Sect. 5.2, performance of FIDS has been evaluated in presence of blackhole attack.

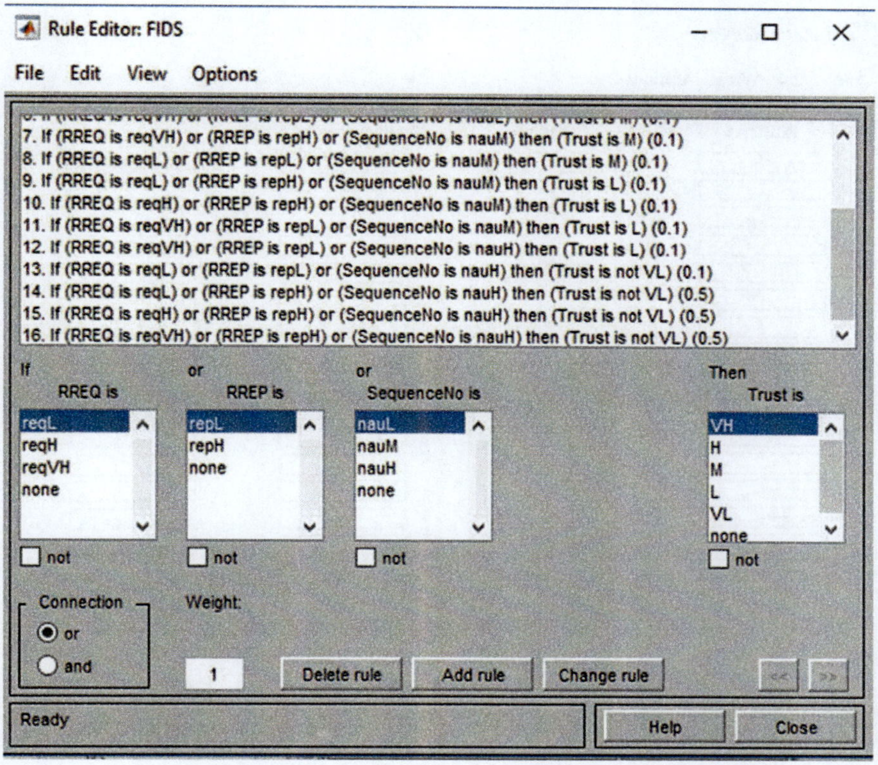

Fig. 4 Fuzzy rules (if-then-else form)

5.1 *Performance Evaluation of AODV Protocol Under Attack*

Firstly, within AODV protocol source codes, blackhole attacks have been simulated using network simulator software (ns2.35). The general simulation parameters are given in Table 3. The AODV performance is recorded in terms of packet delivery ratio, throughput, energy consumption and routing overhead in the presence of blackhole attack and without attack.

It is observed that packet delivery ratio decreases when there is a one blackhole node attack in a network. Further, this reduces by 60% when there is two blackhole node present in a network, and simulation results are shown in Fig. 6. It is observed that throughput is reduced in the presence of one blackhole node attack in a network. This reduces more significantly approximately 60% in the presence of two blackhole nodes in a network, and simulation results are shown in Fig. 7. Further, it is recorded that network energy consumption increases by 50% in the presence of two blackhole nodes in a network, and simulation results are shown in Fig. 8. Similarly, it is found that routing overhead in a given network also increases in the presence of two blackhole nodes in a network, and simulation results are shown in Fig. 9.

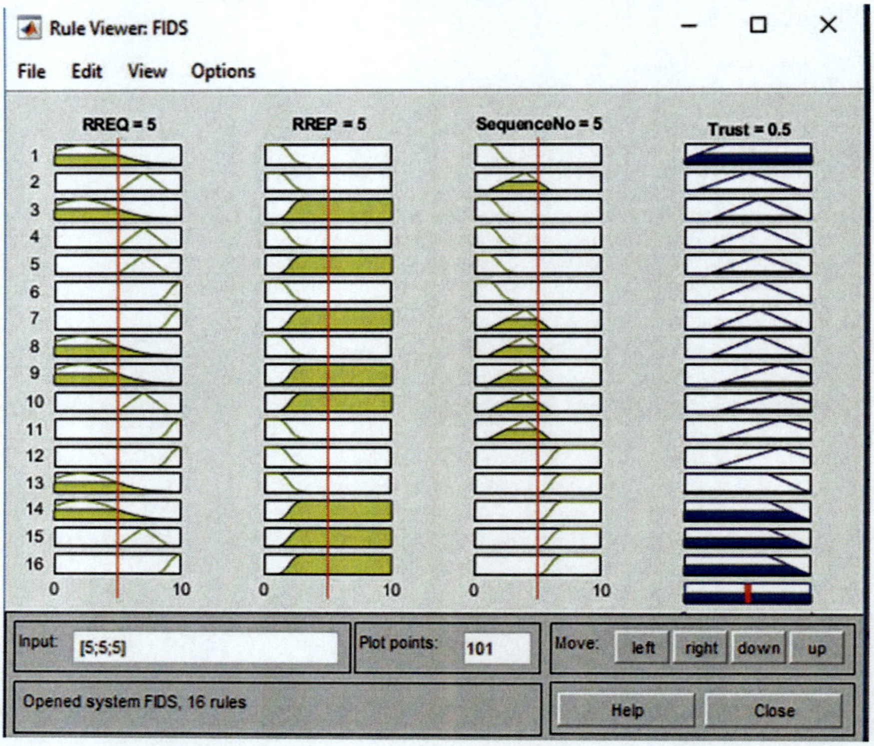

Fig. 5 Graphical representation of fuzzy rules

Table 2 Inference computation

Rate of RREQ	Rate of RREP	Sequence No	Computed trust	Inferences
5	5	5	0.5	Normal
0.301	5	5	0.544	Normal
0.301	9.46	5	0.544	Normal
0.958	9.46	9.22	0.458	Malicious
0.958	0.542	9.22	0.544	Normal
0.958	0.542	0.663	0.546	Normal
7.41	0.904	0.663	0.549	Normal
9.1	0.904	0.663	0.551	Normal
9.34	0.542	9.22	0.544	Normal
9.34	4.4	9.22	0.458	Malicious

Table 3 Simulation parameters for MANETS

Parameters	Value
Simulation tool	NS-2.35
Network nodes	20
Grid area	500 × 500
Routing protocol	AODV
Antenna	Omni directional
MAC type	802.11
Traffic type	CBR
Number of blackholes	2 nodes
Simulation time	500 s

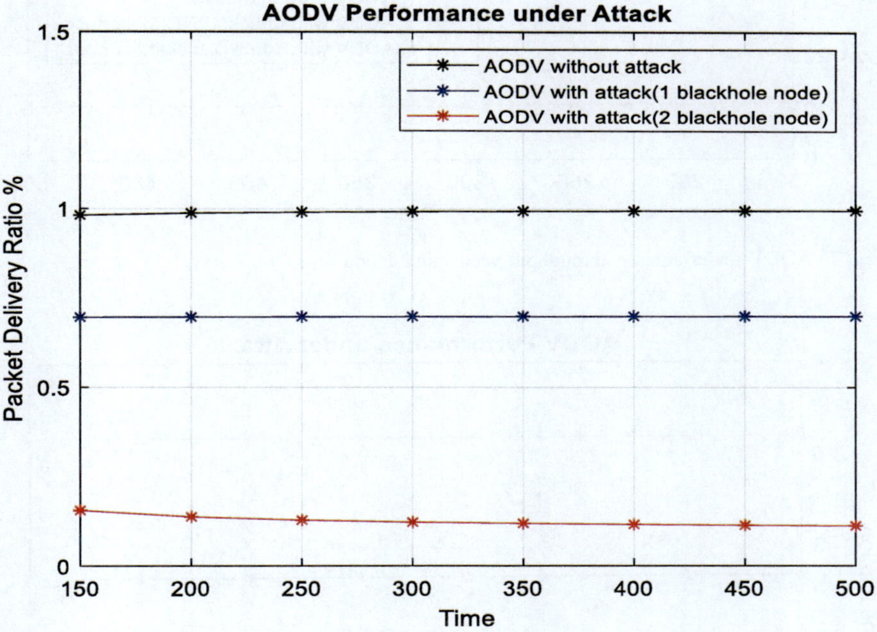

Fig. 6 AODV under attack—PDR versus simulation time

5.2 Performance Evaluation of Proposed FIDS in AODV Protocol

For the evaluation of the proposed FIDS, AODV source codes in ns 2.35 software have been modified and fuzzy-based intrusion detection module has been added. Further, the multiple blackhole attacks have been simulated in a given network.

It is observed that the performance of the network increases on the incorporation of proposed FIDS, as shown in the Figs. 10 and 11. Both PDR and throughput of the network are found to increase when FIDS has been included in AODV protocol.

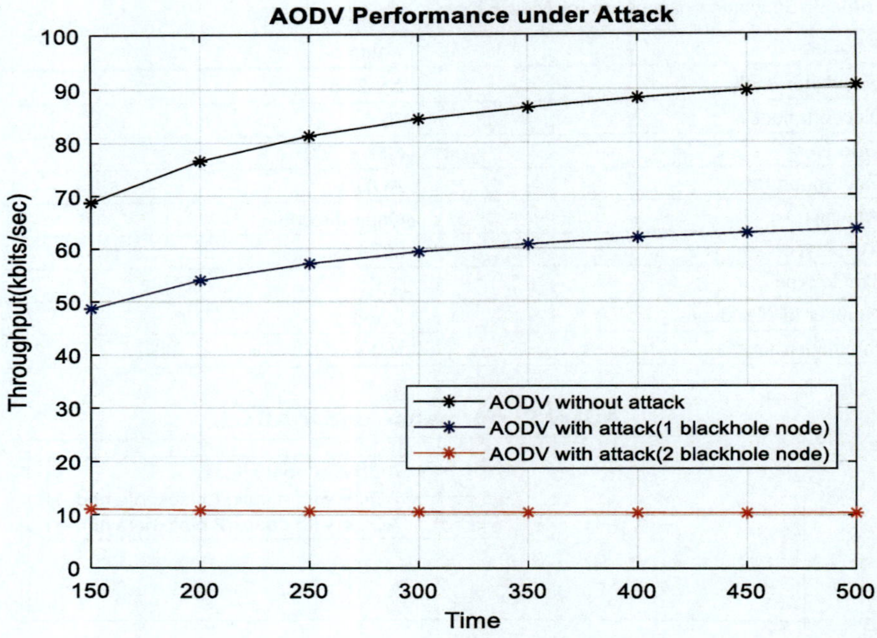

Fig. 7 AODV under attack—throughput versus simulation time

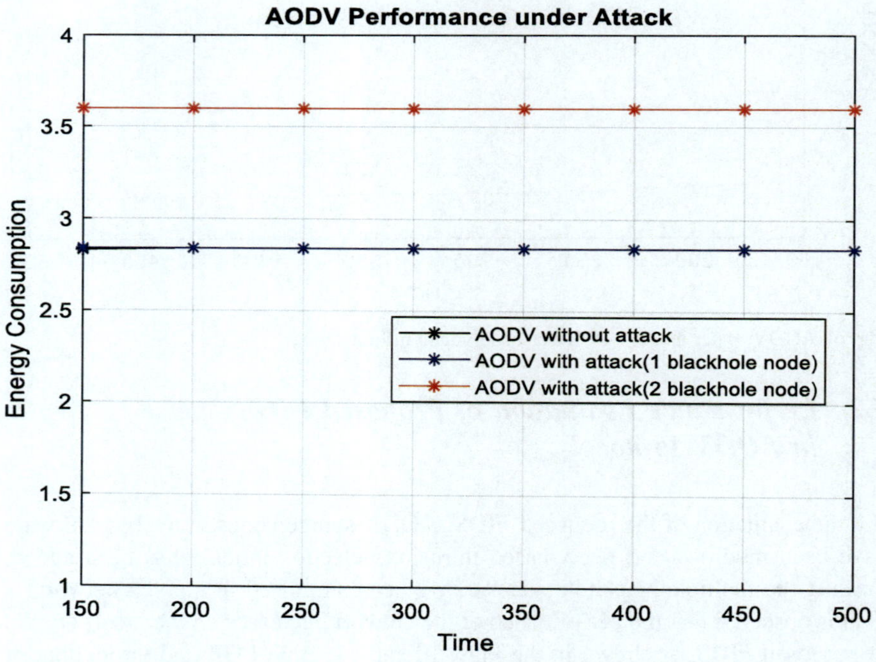

Fig. 8 AODV under attack—energy consumption versus simulation time

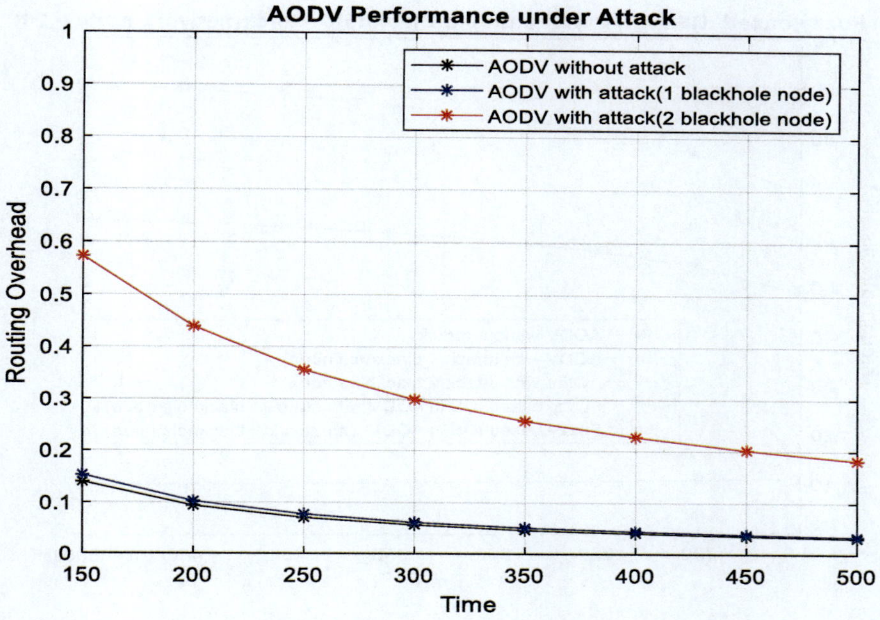

Fig. 9 AODV under attack—routing overhead versus simulation time

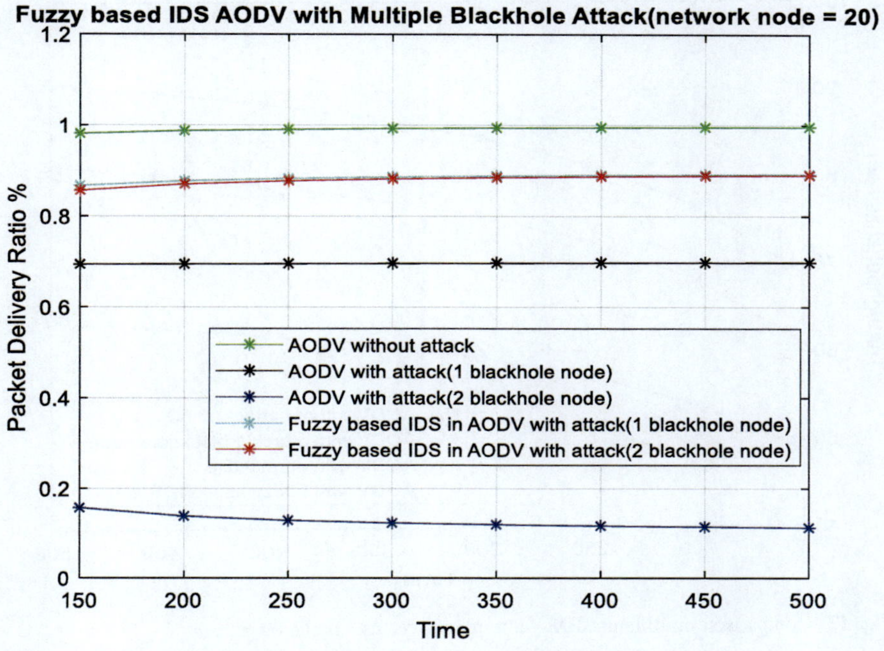

Fig. 10 FIDS performance under attack—PDR versus simulation time

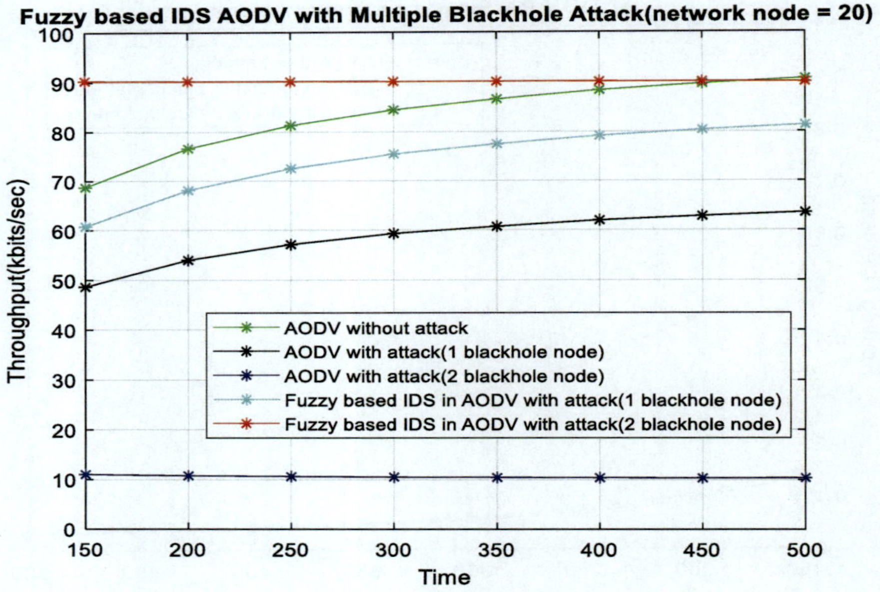

Fig. 11 FIDS performance under attack—throughput versus simulation time

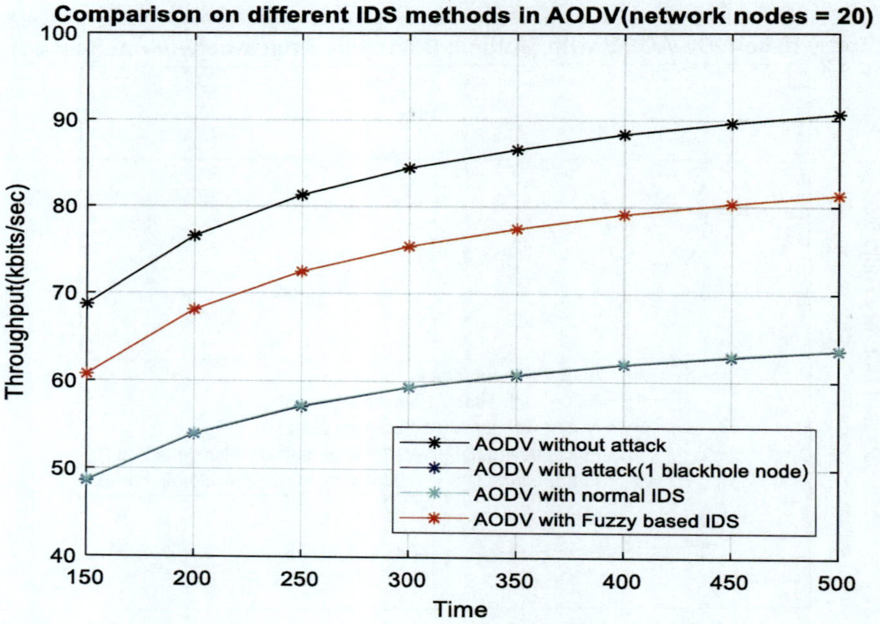

Fig. 12 Comparison on different IDS—throughput versus simulation time

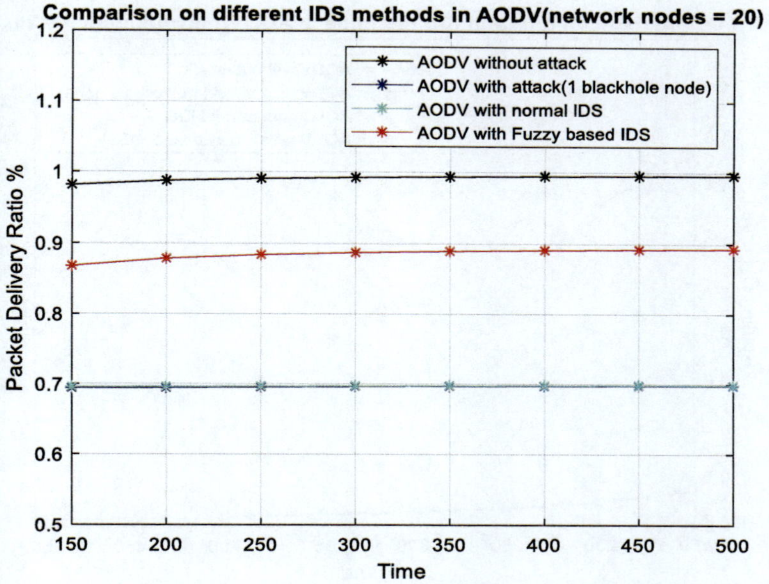

Fig. 13 Comparison on different IDS—PDR versus simulation time

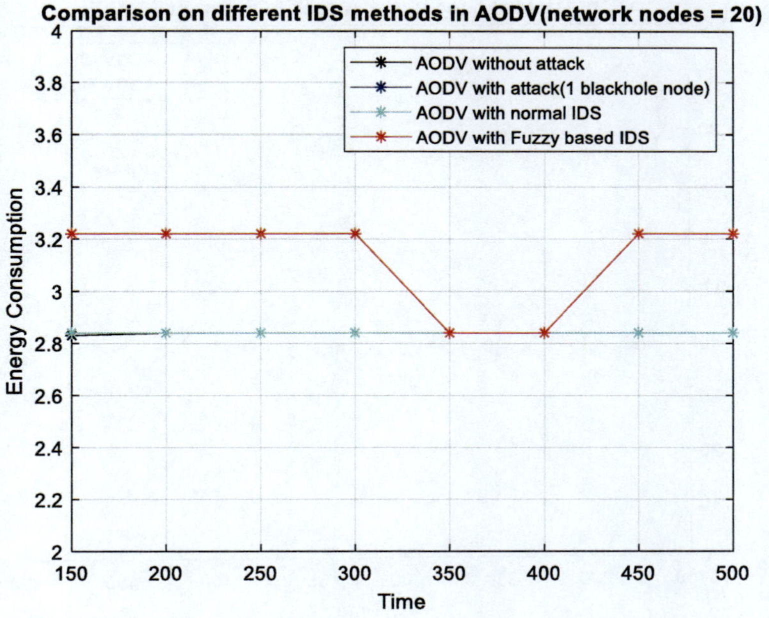

Fig. 14 Comparison on different IDS—energy consumption versus simulation time

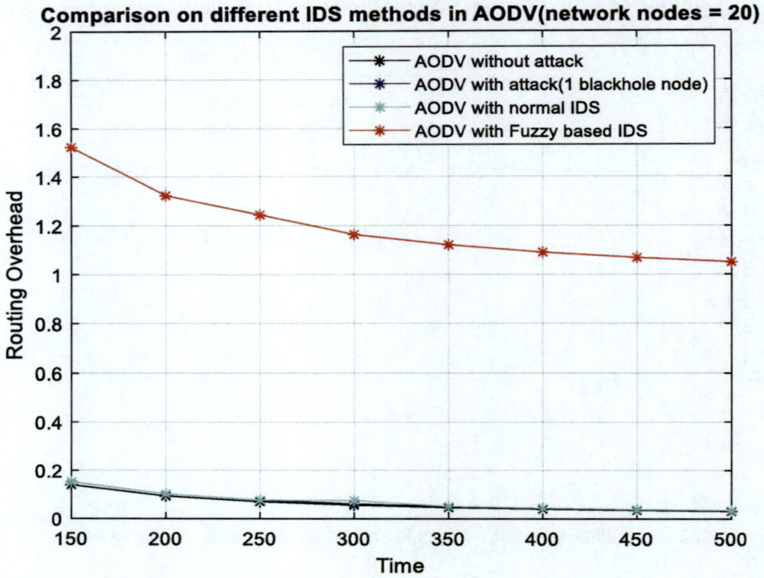

Fig. 15 Comparison on different IDS—routing overhead versus simulation time

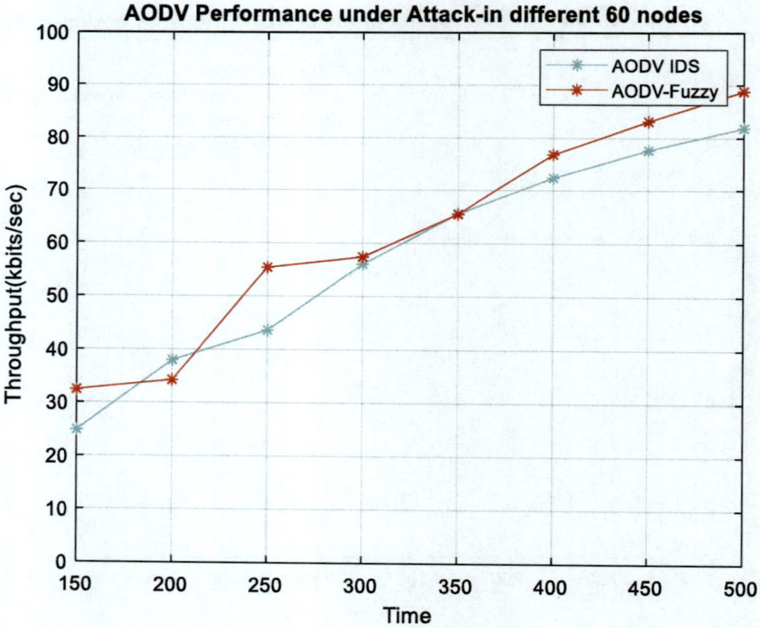

Fig. 16 Comparison on IDS with different node numbers—throughput versus simulation time

Moreover, in the presence of multiple blackhole attacks, similar results have been observed. In addition to this, the performance of FIDS is compared with the another IDS which was proposed by author Dokurer [20]; here, it is named as normal IDS which works on the principle of ignoring the first established route to reduce the effects of the blackhole attack because blackhole node always responds with a fake reply without making delay.

The evaluation of the proposed FIDS has been carried out by calculating the performance metrics, i.e., throughput, packet delivery ratio, energy consumption and routing overhead in the Figs. 12, 13, 14 and 15, respectively. At each time, it is being compared with the normal IDS. It is observed that in the proposed FIDS has increased throughput and PDR which results in better network performance. However, slight increase in energy consumption and routing overload is found because computation processing is involved in FIDS working. Further, results are also validated with different number of nodes present in a network, i.e., 60 nodes. Similar results like better throughput as compared to normal IDS are recorded and presented in the following Fig. 16.

6 Conclusion

Although MANETs are more vulnerable to inside and outside attacks than conventional wired networks, they are increasingly being used in many applications because they provide low-cost mobile connectivity solutions. Researchers are constantly focusing on developing or evolving methods for preventing, detecting and response mechanism for MANETs. Fuzzy-based intrusion detection has been identified as a suitable technique for dynamic environment (i.e., MANETs) by various researchers. It emerges that, to secure a network against the unknown attacks, the fuzzy-based intrusion detection may be the appropriate technique to tackle attacks in MANETs. Here, in this work, its applicability in MANETs has been presented. An effective set of fuzzy rules for inferences is necessary to be identified by making use of the fuzzy rule learning strategies, which would contribute more effectively for detecting intrusion in MANETs. Here, under the scope of this paper, the fuzzy rules (*set of 16 rules*) have been proposed for the detection of blackhole attack in a network. These fuzzy rules are developed on three critical attributes of AODV which are rate of RREQ, RREP and sequence number. Subsequently, the proposed FIDS has been evaluated for the detection of blackhole attack using different simulation parameters of MANETs. And it is observed that fuzzy logic-based intrusion detection systems performs better than the other method-based intrusion detection methods. However, here in the proposed solution, only limited features have been taken into account for the detection of blackhole attack ONLY, to use this model for different attacks, more features are required to be added for accuracy in detection of attack that may results in more computation processing on node which limits the operational philosophy of MANETs. As a result, IDS in MANETs remain a demanding, complex and challenging subject for researchers.

References

1. Ramanathan, R., Redi, J.: A brief overview of ad hoc networks: challenges and directions. IEEE Commun. Magaz. **40**(5) (2002)
2. Vydeki, D., Bhuvaneswaran, R.S.: Effect of clustering in designing a fuzzy based hybrid intrusion detection system for mobile ad hoc networks. J. Comput. Sci. **9**(4), 521–525, ISSN: 1549-3636 (2013)
3. Poongothai, T., Duraiswamy, K.: Cross layer intrusion detection system of mobile ad hoc networks using feature selection approach. Wseas Trans. Commun. **13** (2014)
4. Introduction to fuzzy logic. http://www.francky.me/doc/course/fuzzy-logic.pdf
5. Lectures on Fuzzy. http://ce.sharif.edu/courses/92-93/1/ce957l/resources/root/Lectures/Lecture6&7.pdf
6. Zadeh, L.A.: Fuzzy logic—computing with words. IEEE Trans. Fuzzy Syst. **4**, 103–111 (1996)
7. Ruchi, M., Reddy, B.V.R.: Taxonomy of machine leaning based anomaly detection and its suitability. In: International Conference on Computation Intelligence and Data Science (ICCIDS 2018), Procedia Computer Science, vol. 132, pp. 1842–1849, Elsevier (2018)
8. Garcia Teodora, P., Diaz Verdejo, J., MaciaFarnandez, G., Vazquez, E.: Anomaly based network intrusion detection: techniques, systems and challenges. J. Comput. Secur. **28**(1), 18–28 (2009)
9. Shelly, X.W., Wolfgang, B.: The use of computational intelligence in intrusion detection systems: a review. Appl. Soft Comput. Appl. Soft Comput. **10**, 1–35 (2010)
10. Izakian, H., Pedrycz, W.: Agreement-based fuzzy c-means for clustering data with blocks of features. Neurocomputing **127**, 266–280 (2014)
11. Animato, M.E., Kim, H., Kim, K.: Another fuzzy anomaly detection system based on ant clustering algorithm. Kumamoto, Japan (2016)
12. Mkuzangwe, N.N.P., Nelwamondo, F.V.: A fuzzy logic based network intrusion detection system for predicting the TCP SYN flooding attack. Springer International Publishing, Part II, LNAI 10192, pp. 14–22 (2017)
13. Kulbhushan, Singh, J.: Fuzzy-logic-based intrusion detection system against blackhole attack AODV in Manet. IJCA Special issue on "Network Security and Cryptography", vol. NSC, no. 2, pp. 28–35 (2011)
14. Mandal, S.N., Pal Choudhury, J., Bhadra Chaudhuri, S.R.: In search of suitable fuzzy membership function in prediction of time series data. Int. J. Comput. Sci. Issues **9**(3), 3 (2012)
15. Chaudhary, A., Kumar, A., Tiwari, V.N.: A reliable solution against Packet dropping attack due to malicious nodes using fuzzy Logic in MANETs. IEEE Int. Conf. Optimiz. Reliab. Inf. Technol. 178–181 (2014)
16. Perkins, C.E., Royer, E.M.: Ad-hoc on-demand distance vector routing. In: Proceedings 2nd IEEE Workshop Mobile Computer System and Applications, pp. 90100 (1999)
17. Ning, P., Sun, K.: How to misuse AODV: a case study of inside attacks against mobile ad-hoc routing protocols. In: Proceedings of the 2003 IEEE Workshop on Information Assurance, United States Military Academy, West Point, NY (2003)
18. Rajya Lakshmi, G.V., Anusha, K.: Detection of anomaly network traffic for mobile ad-hoc network using fuzzy logic. Int. J. Emerg. Res. Manag. Technol. (2013)
19. Chaudhary, A., Tiwari, V.N., Kumar, A.: Analysis of fuzzy logic based intrusion detection systems in mobile ad hoc networks. BIJIT—BVICAM s Int. J. Inf. Technol. **6**(1) (2014)
20. Dokurer, S., Ert, Y.M., Acar, C.E.: Performance analysis of adhoc networks under blackhole attacks. In: Southeast Con, 2007, Proceedings IEEE, pp. 148–153 (2007)

Detection of Phishing Websites Using Classification Algorithms

Sumathi Ganesan

Abstract Phishing is a form of online fraudsters try to mimic genuine websites and get the user's sensitive and personal information. Such malicious users post phishing links and advertisements which may transmit malwares and viruses. Not only individual users are affected by such fraudsters but also several organizations and corporate those depend on internet for services and sales. Such malicious users adopt social engineering skills such as sending emails or by online pop-up advertisements. A number of solutions for detecting phishing websites are mentioned such as non-technical methods, method of black listing and white listing and machine learning techniques. In this paper, data mining techniques in machine learning have been applied to detect phishing websites. One of the data mining techniques is a classification which seems to have a high potential in detecting phishing websites. Here, bagging, C4.5 (J48) and random forest classifiers are tested on the phishing dataset. The dataset is taken from UCI repository which has 1353 instances. The C4.5 classification model has the highest accuracy rate of 90.8%.

Keywords Phishing websites · Data mining · Classification algorithms · Hackers

1 Introduction

Today's world has become impossible without Internet; right from booking a ticket to bank transactions and scientific research the information gained from the internet has played a very crucial role in human life. It is important for the users to know about certain illegal websites which were built with a motive to steal user's sensitive information for their benefit. Website phishing [1, 2] can be defined as imitating a trusted website to obtain sensitive information from the user. The fraudsters aim in gaining information from the user or sometimes try to destroy certain information from the users account. There are types of phishing based on the

S. Ganesan (✉)
Department of Mathematics, CEG Campus, Anna University, Chennai-25, TamilNadu, India
e-mail: sumisundhar@auist.net

K. Khanna et al. (eds.), *Cyber Security and Digital Forensics*,
Lecture Notes on Data Engineering and Communications Technologies 73,
https://doi.org/10.1007/978-981-16-3961-6_12

types of users they aim at. Deceptive phishing websites are online threats which appear to be a copy of an honest organization's but aims at acquiring personal and sensitive details of the users such as username, password, transaction details, images, etc. These websites are lookalikes of the original legitimate websites and are hard to be detected by common man. Spear phishing aims at a specific user by customizing their attacks through mails making the receiver believe that they received it from a genuine sender. The mail may contain URL which redirects them into malicious sites which will obtain all their personals. CEO Phishing is similar to spear phishing but with a difference that these phishers send mails to targeted users. Pharma phishing tries to hijack a website's domain name and use to redirect it to a malicious website. Using DNS cache poisoning, the fraudster targets a DNS server and changes the IP address associated with the alphabetical website name. In drop box phishing, the fraudster sends emails that claim to come from drop box and asks the user to click to secure their account. In Google Docs phishing, the attackers mimic the Google account login screen to obtain user's information.

Website phishing [3, 4] has become a very serious problem because the number of such websites is increasing enormously, and hence, difficult for even experts to detect them. One of the most common methods used by the fraudsters is hiding the phishing URL. While visiting some websites, the Web browsers may assure us that the website is secure, but the user may end up landing into a malicious website. Such URLs may contain Unicode standard characters which resemble normal alphabets. So when a request to such a website is given, it directs us to a different domain and the user end up believing it is a genuine website. Such phishing URLs or links are sent through email. The hackers try to counterfeit the legitimate website, and the users are misled.

The phishing dataset forms the training dataset which has the class defined with it. Mining algorithms are applied to form association rules from the training data set. The features are selected in such a way that the learning process is done in a faster way. The efficiency of the rules formed depends upon the algorithm used and the input dataset. When a URL is requested within a Web browser, certain features are extracted from the URL and are applied to rules discovered by association classification to find out whether the requested website is genuine or not.

In this paper, the WEKA tool is used to conduct the experiments. The dataset was taken from UCI repository, and several classification algorithms are applied to the training dataset to form rules that could be used for a test data and the most efficient algorithm is determined to detect the phishing websites.

2 Related Works

Priya and Meenakshi [5] made a detailed study on phishing websites using C4.5 algorithm. They used a training dataset that contains 750 instances and a tree was generated. The trained model was used in the testing process and an accuracy was found to be 82.6%.

Babagoli et al. [6] proposed the technique for the detection of phishing website that uses a meta-heuristic-based non-linear regression algorithm. The dataset contained 30 attributes which was reduced using feature selection methods. Comparative study between two meta-heuristic algorithms was performed. The non-linear regression-based on harmonic search showed an accuracy of 94.13% for training data and 92.8% for test data.

Şentürk et al. [7] used J48 algorithm in WEKA tool to detect phishing in emails. They have used decision tree to build the classification model which has produced a success rate of 89%.

Ibrahim and Hadi [8] used several classification techniques in WEKA tool to detect phishing websites. A number of mining algorithms such as Prism, artificial neural networks, naïve Bayes and K star have been applied to compare their effectiveness in terms of accuracy. They concluded that artificial neural networks showed the highest accuracy of 94.8%.

Zaman and Sharmin [9] did an analysis on spam detection in social media. They used a number of machine learning algorithms such as naïve Bayes, K nearest neighbour, bagging and support vector machine. The performance measures have been analysed as well as performance of ensemble classifier over single classifier algorithm has been analysed. The analysis was done using the WEKA tool. In the study, it was observed that ensemble classifier and naïve Bayes gave better result in most of the cases.

Ramanathan and Wechsler [10] employed latent dirichlet allocation for semantic analysis and AdaBoost for classification. The dataset used for the analysis consisted of 47,500 phishing websites and 52,500 legitimate websites. The phishing website classifier is built using distribution probabilities for attributes latent dirichlet allocation and AdaBoost voting. The true positive rate and F-measure obtained was 99%.

Hu et al. [11] experimented on the legitimate website server of a financial company. Phishing websites may ask for resources from the user which is used to find whether the website is phishing or not from the logs. It was efficient and had high accuracy.

Abdelhamid et al. [12] proposed a method to detect phishing websites using multilayer classifier-based associative classification (MCAC). In associative classification, rules are formed in the training phase, and later, a classification model is constructed after removing useless and redundant rules. They have compared the performance MCAC with CBA, PART, C4.5 and JRip. They found that the MCAC algorithm had the highest accuracy of more than 90%.

3 Dataset Description

The dataset is taken from UCI repository. It contains 1353 instances, ten attributes and contains no missing values. There are three classes namely legitimate, phishing and suspicious. Out of the 1353 instances 702 are phishing, 103 are suspicious and

548 are legitimate. The ten attributes of phishing dataset are server form handler (SFH), popUpWindow, SSL Final_State, Request_URL, URL_of_anchor, web_-traffic, URL_Length, age_of_domain, having_IP_Address and Result [13].

4 Methodologies

The dataset must be pre-processed to remove noisy and inconsistent data before applying classification algorithm. The knowledge discovery process will become difficult if the dataset contains redundant values. A pre-processed dataset will give more accurate results than a data which is not pre-processed. Redundant attributes are found using correlation and covariance analysis, and such attributes are removed. Correlation and covariance are two similar measures used to identify how to attributes are related. Once all outliers are rectified, the data could be reduced in such a way that no information is lost. Finally, the dataset could be normalized or discretized if it contains numerical values.

Classification is a step-by-step procedure, used to classify the instances in a data set into their respective classes depending on the attribute values they take. In data mining, classification can be done either by building a model or by generation classification rules since the number of instances are too large to be classified individually. In both the methods, a training set, i.e. instances for which the class label is known, is taken and algorithms are applied to build the model/rules. These model/rules are used to classify the rest of the unclassified instances.

There are four types of test options, namely using training set, supplied test set, cross-validation and percentage split. Here, cross-validation and supplied test set are used for experiment.

4.1 Classification Algorithms

Input training data is utilized by the classification algorithms in machine learning to predict the likelihood that ensuing data can be classified as one of the predetermined categories. Perhaps the most well-known uses of classification is filtering emails into spam or non-spam. The following classification algorithms are applied to the pre-processed dataset.

C4.5 Algorithm Decision tree produce human interpretable models. J48 algorithm is used to generate decision tree to make predictions. It is the extended version of ID3 algorithm. It aims at generalizing until it is flexible and accurate.

This algorithm involves the following steps. The potential info or entropy of each attribute is found and info gain for each attribute is calculated. The attribute with the highest gain is selected and chosen to be the node and leaf node returns the class. C4.5 can deal with both discrete and continuous values. Missing values are also handled. Pruning which generalizes the tree is also done. It constructs the tree based on the attribute value [14, 15].

$$\text{Entropy}(\vec{y}) = -\sum_{j=1}^{n} \frac{|y_i|}{|\vec{y}|} \log\left(\frac{|y_i|}{|\vec{y}|}\right) \tag{1}$$

$$\text{Entropy}(j|\vec{y}) = -\frac{|y_i|}{|\vec{y}|} \log\left(\frac{|y_i|}{|\vec{y}|}\right) \tag{2}$$

$$\text{Gain}(\vec{y}, j) = |\text{Entropy}(\vec{y}) - \text{Entropy}(j|\vec{y})| \tag{3}$$

Bagging Algorithm Bagging or bootstrap aggregation is an ensemble algorithm used in classification and to reduce variance. An ensemble method is a technique that combines the predictions from multiple machine learning algorithms together to make more accurate predictions than that of any individual model. Bootstrapping is a method in which a sample is taken from a large dataset and determines various model and statistic accuracy. The model is made efficient after the evaluation of each sample. In bagging, each sample model is overviewed, and the final model is the aggregated version of the individual sample models. By doing this, the efficiency is highly increased and reduces the variance that is over fitting the data [14, 15].

Random Forest Algorithm Random forest is a tree classifier which uses ensemble method to classify. Hence, it has a better predictive accuracy. It is a combination of many tree predictors, and each tree relies upon the value of a random vector tested independently. These random vectors govern the growth of each tree in the ensemble. The advantages of random forest are that it is non-parametric and has a high classification accuracy rate. It is also used to calculate missing values and can be used for several data analysis like classification, regression, survival analysis and unsupervised analysis. A random tree constructs several decision trees and applies it to the training data and output the class which is the mode or mean of each tree [14, 15].

5 Experimental Results

WEKA provides a number of classification techniques that includes fuzzy logic, Bayesian, decision tree and support vector machine. This paper analyses chosen three algorithms and infers the best-suited algorithm for the phishing dataset. WEKA provides different test options to provide classification results.

The experiment is conducted in WEKA tool. Here, three algorithms namely C4.5, random forest and bagging are used. The efficiency of these algorithms is calculated in terms of accuracy. The experiment was conducted using cross-validation with ten-folds and using supplied test set. Both the methods are compared and evaluated.

5.1 Pre-processing

The dataset has numerical attributes like request URL, URL anchor, web traffic, URL length and age of domain. These values were classified into different range groups; for example, the attribute URL length is divided into three groups: less than 54 characters, between 54 and 75 characters and greater than 75 characters. The dataset took for the study consists of ten attributes and 1353 instances. Figure 1 shows the sample of pre-processed results.

5.2 Evaluation Metrics

The results of various classification algorithms were evaluated by using the following metrics [11].

Accuracy is the percentage of the test tuples that are correctly classified by the classifier.

$$\text{Accuracy} = \frac{(TP + TN)}{(P + N)} \tag{4}$$

Precision is defined as the percentage of positive tuples that are correctly classified.

$$\text{Precision} = \frac{TP}{(TP + FP)} \tag{5}$$

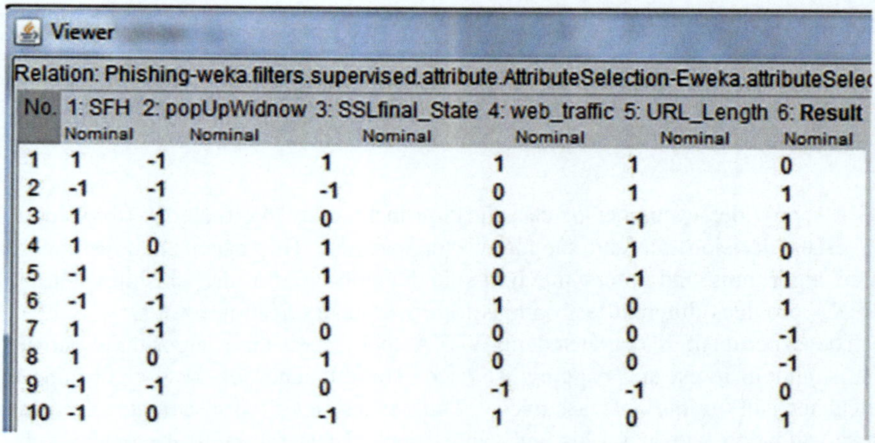
Viewer

Relation: Phishing-weka.filters.supervised.attribute.AttributeSelection-Eweka.attributeSelec

No.	1: SFH	2: popUpWidnow	3: SSLfinal_State	4: web_traffic	5: URL_Length	6: Result
	Nominal	Nominal	Nominal	Nominal	Nominal	Nominal
1	1	-1	1	1	1	0
2	-1	-1	-1	0	1	1
3	1	-1	0	0	-1	1
4	1	0	1	0	1	0
5	-1	-1	1	0	-1	1
6	-1	-1	1	1	0	1
7	1	-1	0	0	0	-1
8	1	0	1	0	0	-1
9	-1	-1	0	-1	-1	0
10	-1	0	-1	1	0	1

Fig. 1 Sample of pre-processed result

Recall is nothing but the sensitivity of the classifier or TP rate.

$$\text{Recall} = \frac{TP}{(TP + FN)} \tag{6}$$

Specificity or FP rate is the percentage of negative tuples that are correctly classified.

$$\text{Specificity} = \frac{TN}{(TN + FP)} \tag{7}$$

F-measure is the method of combining both precision and recall. It is found with the harmonic mean of the values of both precision and recall.

$$F\text{-Measure} = 2 * \frac{(\text{Recall} * \text{Precision})}{(\text{Recall} + \text{Precision})} \tag{8}$$

The entire dataset containing 1353 instances is used to build a model using cross validation with ten-folds. Classification is done with C4.5, bagging and random forest. The results of the various classification algorithms are shown in the Table 1.

The confusion matrix of C4.5 algorithm is shown in the Table 2. The diagonal elements of the above confusion matrix are the correctly classified instances.

Hence, from Table 2, it is visible that the instances of 96, 489 and 643 are correctly classified into the classes suspicious, legitimate and phishing, respectively. Therefore, it has correctly classified 1228 instances.

Similarly, Tables 3 and 4 show the correctly classified instances of bagging and random forest, respectively. It shows that the bagging classifier has correctly classified 1215 instances and random forest has correctly classified 1217 instances.

Now, the experiment was done using supplied test set. For this, dataset is split into two: one with 1253 instances and one with 100 instances. The test dataset which contains 100 instances has 55 phishing, 36 legitimate and 9 suspicious instances. Table 5 summarizes the results of the supplied test set test.

From Tables 6, 7 and 8, the inferred instances of 92, 91 and 89 are correctly classified by C4.5, bagging and random forest, respectively, with the supplied test set containing 100 instances. The C4.5 algorithm which has the highest number of correctly classified instances has correctly classified all the suspicious websites. Among the 36 legitimate websites, 33 were correctly classified, and 55 phishing websites out of which 50 were predicted correctly by the model.

Table 1 Result with cross validation test

	Weighted average of C4.5	Weighted average of bagging	Weighted average of random forest
TPR	0.908	0.898	0.899
FPR	0.070	0.077	0.078
PRECISION	0.908	0.898	0.900

Table 2 Confusion matrix of C4.5

Class	0	1	−1
0	96	2	5
1	10	489	49
−1	9	50	643

Table 3 Confusion matrix of bagging

Class	0	1	−1
0	85	9	9
1	9	491	48
−1	12	51	639

Table 4 Confusion matrix of random forest

Class	0	1	−1
0	88	7	8
1	9	489	50
−1	9	53	640

Table 5 Result with supplied test set

	Weighted average of C4.5	Weighted average of bagging	Weighted average of random forest
TPR	0.920	0.910	0.890
FPR	0.060	0.048	0.084
PRECISION	0.921	0.918	0.892
RECALL	0.920	0.910	0.890
F-MEASURE	0.920	0.911	0.890

Table 6 Confusion matrix of C4.5

Class	0	1	−1
0	9	0	0
1	0	33	3
−1	1	4	50

5.3 *Performance Comparison*

The main objective of this study is to categorize websites into three classes namely suspicious, legitimate and phishing. This is a supervised classification problem in which the classifier C4.5 and two ensemble methods of bagging and random forest are used. Tables 9 and 10 show the accuracy with cross validation and supplied test set.

From Fig. 2, it is concluded that C4.5 algorithm has the highest accuracy and is quiet efficient in both the cases: cross validation and supplied test set.

Table 7 Confusion matrix of bagging

Class	0	1	−1
0	8	1	0
1	0	35	1
−1	2	5	48

Table 8 Confusion matrix of random forest

Class	0	1	−1
0	7	1	1
1	0	33	3
−1	1	5	49

The other three metrics of precision, recall and F-measure are also used to evaluate the results of C4.5, bagging and random forest classification algorithms. Tables 11 and 12 shows the precision, recall and F-measure values with cross validation test and supplied test set.

Figures 3 and 4 show the performance comparison of C4.5, bagging and random forest classification algorithms with cross validation test and supplied test set. This comparison is analysed with the evaluation metrics of precision, recall and F-measure.

From the above figures, it is concluded that C4.5 algorithm has the highest values for precision, recall and F-measure with the cross validation test and supplied test set. It has been found that C4.5 model has the highest accuracy. And in most of the cases, the accuracy rate is above 80%. The C4.5 model correctly classified 93.20% while using the full training set, 90.76% while using cross validation with ten-folds, 92% while using supplied test set of 100 instances. Since the attribute values are discrete which are 1, 0 and 1, decision tree model has proven to be efficient.

Figure 5 shows the pruned tree of J48 algorithm. The size of the tree is 84 and total number of leaves is 55.

Table 9 Classifier accuracy results of cross validation

Testing mode	Ten-fold cross validation	
Classifier	Correctly classified	Incorrectly classified
C4.5	90.76	9.24
Bagging	89.8	10.20
Random forest	89.95	10.05

Table 10 Classifier accuracy results of supplied test set

Classifier	Correctly classified	Incorrectly classified
C4.5	92	8
Bagging	91	9
Random forest	89	11

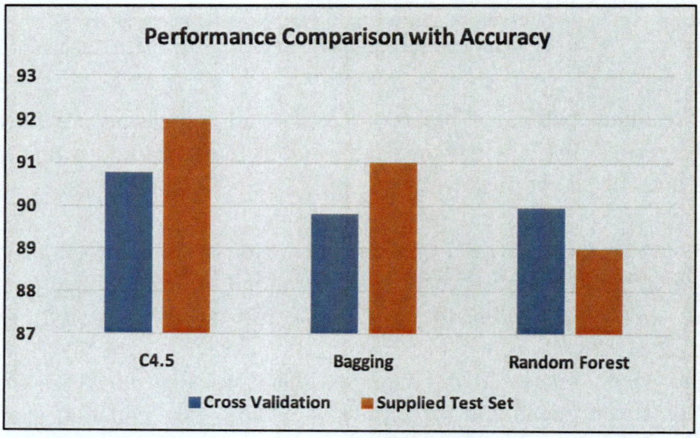

Fig. 2 Performance comparison with accuracy

Table 11 Result with cross validation test

Classifier	PRECISION	RECALL	F-MEASURE
C4.5	0.908	0.908	0.908
Bagging	0.898	0.898	0.898
Random forest	0.900	0.899	0.900

Table 12 Result with supplied test set

Classifier	PRECISION	RECALL	F-MEASURE
C4.5	0.921	0.920	0.920
Bagging	0.918	0.910	0.911
Random forest	0.892	0.890	0.890

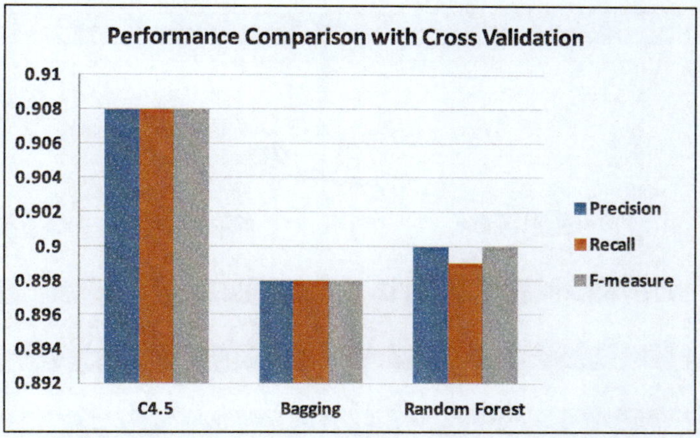

Fig. 3 Performance comparison with cross validation test

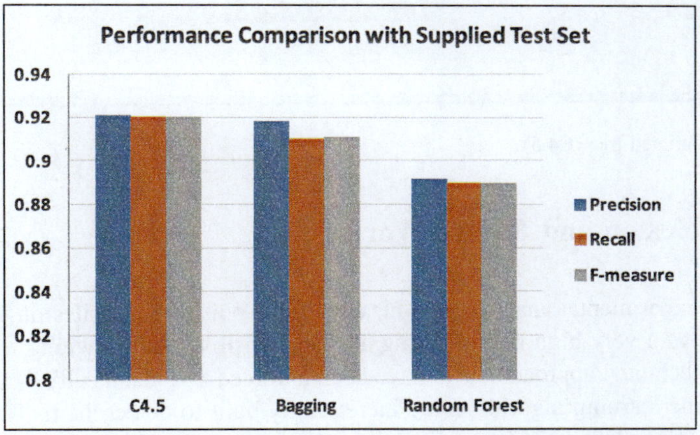

Fig. 4 Performance comparison with supplied test set

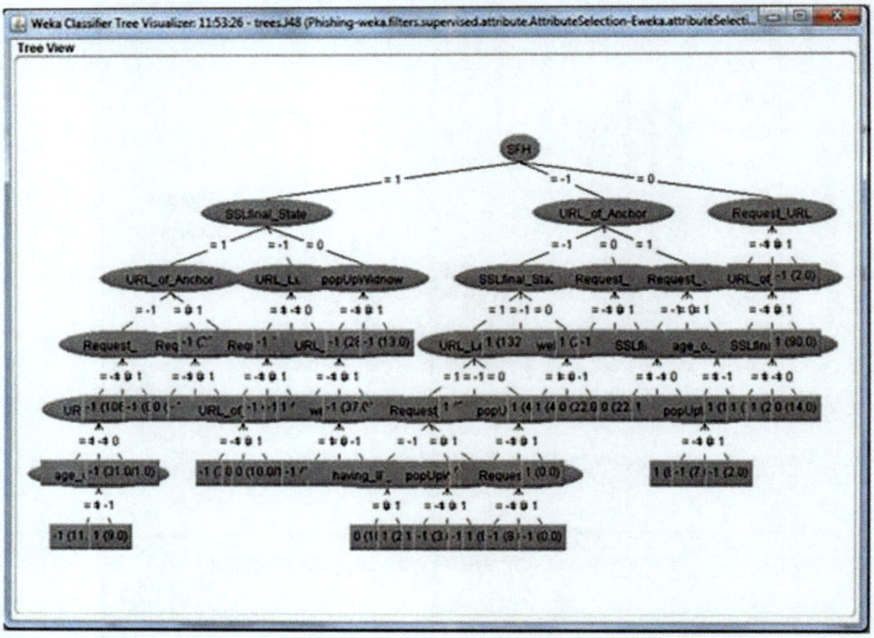

Fig. 5 J48 pruned tree (C4.5)

6 Conclusion and Future Work

From the experimental analysis, it is clear with the phishing websites on Web that people are at a very high risk of losing all their sensitive data to wrong hands. As from non-technical approach, laws have been enforced and technically data mining and machine learning algorithms are increasingly built to detect them. The experiments concluded that C4.5 algorithm has the highest accuracy of 90.76%. This paper has analysed and found the most accurate of three different algorithms which can be used by researchers to find the best classifying approach.

References

1. Mohammad, R.M., Thabtah, F., McCluskey, L.: An assessment of features related to phishing websites using an automated technique. In: 7th International Conference for Internet Technology and Secured Transactions Proceedings, pp. 492–497. IEEE (2012)
2. Sumathi, R., Prakash, M.R.V.: Prediction of phishing websites using optimization techniques. Int. J. Modern Eng. Res. (IJMER) 341–348 (2012)
3. Pan, Y., Ding, X.: Anomaly based web phishing page detection. In: 22nd Annual Computer Security Applications Conference Proceedings, pp. 381–392. IEEE (2006)

4. Zhang, Y., Hong, J.I., Cranor, L.F.: Cantina: a content-based approach to detecting phishing web sites. In: 16th International Conference on World Wide Web Proceedings, pp. 639–648 (2007)
5. Priya, A., Meenakshi, E.: Detection of phishing websites using C4. 5 data mining algorithm. In: 2nd IEEE International Conference on Recent Trends in Electronics, Information and Communication Technology Proceedings, pp. 1468–1472 (2017)
6. Babagoli, M., Aghababa, M.P., Solouk, V.: Heuristic nonlinear regression strategy for detecting phishing websites. Soft. Comput. **23**(12), 4315–4327 (2019)
7. Şentürk, Ş., Yerli, E., Soğukpınar, İ.: Email phishing detection and prevention by using data mining techniques. In: International Conference on Computer Science and Engineering Proceedings, pp. 707–712 (2017)
8. Ibrahim, D.R., Hadi, A.H.: Phishing websites prediction using classification techniques. In: International Conference on New Trends in Computing Sciences Proceedings, pp.133–137 (2017)
9. Sharmin, S., Zaman, Z.: Spam detection in social media employing machine learning tool for text mining. In: 13th International Conference on Signal-Image Technology and Internet-Based Systems Proceedings, pp. 137–142 (2017)
10. Ramanathan, V., Wechsler, H.: Phishing website detection using latent Dirichlet allocation and Adaboost. In: IEEE International Conference on Intelligence and Security Informatics Proceedings, pp. 102–107 (2012)
11. Hu, J., Zhang, X., Ji, Y., Yan, H., Ding, L., Li, J., Meng, H.: Detecting phishing websites based on the study of the financial industry webserver logs. In: 3rd International Conference on Information Science and Control Engineering Proceedings, pp. 325–328 (2016)
12. Abdelhamid, N., Ayesh, A., Thabtah, F.: Phishing detection based associative classification data mining. Expert Syst. Appl. **41**(13), 5948–5959 (2014)
13. Website Phishing Dataset. https://archive.ics.uci.edu/ml/datasets/Website+Phishing. Accessed 23 Dec 2021
14. Abernethy, M.: Data mining with WEKA, Part 2: classification and clustering. IBM Developer (2010)
15. Han, J., Jian, P., Micheline, K.: Data mining: concepts and techniques. Elsevier, Third Edition (2011)

IoT-Based Smart Security System for Agriculture Fields

Sukhwinder Sharma, Puneet Mittal, and Anuradha

Abstract Farmers face well-known challenges of crop protection from insect pests, diseases, and weeds along with protection from contrary weather conditions like hail or frost. However, they face another important challenge of protecting crops from wild animals that may cause severe damage to their grown crops by feeding on plant parts or trampling over the crops. As most of farmers stay away from their fields, continuous monitoring of fields is not possible due to distance as well as costs involved to appoint manpower for this purpose. Present technologies have made it possible to provide low cost, easy to install, and user friendly solution to such problems. This paper aims to develop and install IoT-based security system for agriculture fields capable to detect and communicate presence of wild animals using PIR sensor and GSM module. It generates SMS alerts on the farmer's mobile phone whenever some animal crosses specific area. It helps farmers to take timely action for crop protection. The security system is deployed in real environment to validate its applicability and possible future extensions.

Keywords Agriculture · GSM · IoT · PIR · Security system · Sensor

1 Introduction

Farmers plan their potential yields before the beginning of farm seasons. They prefer crops which grow well in their present environment and weather conditions as well as give financial benefits at the end. Even with their best possible efforts and planning, farmers face various challenges and obstacles that hamper their crop productivity and resulting financial outcomes. Protecting crops from insect pests,

S. Sharma (✉) · P. Mittal
Mangalore Institute of Technology and Engineering, Mangalore, India

Anuradha
The NorthCap University, Gurugram, India

K. Khanna et al. (eds.), *Cyber Security and Digital Forensics*,
Lecture Notes on Data Engineering and Communications Technologies 73,
https://doi.org/10.1007/978-981-16-3961-6_13

diseases, and weeds are usually considered most important challenges, and a prior planning and timely usage of pesticides help to avoid any damages to crops [1]. Also, protection from contrary weather events such as hail or frost is done through immediate possible solutions. However, they face another important challenge of protecting crops from wild animals that may cause severe damage to their well grown crops by feeding on plant parts or trampling over the crops. As most of farmers stay away from their fields, continuous monitoring of fields is not possible due to distance as well as costs involved to appoint permanent manpower for this purpose. Therefore, the wild animals may cause serious yield losses and induce additional financial problems. This challenge necessitates deployment of some security system to protect crops from wild animals. Present technologies like Internet of Things (IoT) [2–6] have made it possible to provide low cost, easy to install, and user friendly solutions to such problems. Hence, this paper aims to develop and install IoT-based security system for agriculture fields capable to detect and communicate presence of wild animals using passive infrared (PIR) sensor and global system for mobile communications (GSM) module. It generates short message service (SMS) alerts on the farmer's mobile phone whenever some animal crosses specific area. It helps farmers to take timely action for crop protection. The security system is deployed in real environment to validate its applicability and possible future extensions.

The rest of paper is organized as follows. The existing security systems are discussed in Sect. 2. Section 3 presents proposed IoT-based smart security systems including its hardware and software requirements. Results are discussed in Sect. 4, while Sect. 5 gives conclusions and future research directions.

2 Existing Methods of Security for Agriculture Fields

Traditionally, farmers make frequent visits to their agriculture fields to check intrusions of wild animals and rush them out of fields if present. It fails to ensure complete protection as the animals may come back in absence of farmer. Some farmers even install wire fences, plastic fences, or electric fences to keep wild animals away from their fields. This method needs to fulfill guidelines set by local bodies, governments and regulatory bodies, and quite costly for larger field sizes. Another method is to use burglar alarm (see Fig. 1) having PIR sensor, to detect the presence of wild animals while they cross a particular area, and a buzzer to alert when an intrusion takes place. These kinds of burglar alarms are available in market and are useful only when the farmer is around the system within few meters to hear the alarm and respond. Such security systems are widely used for intrusion detection in smart homes and offices to protect against theft or damages [7–9]. Also, some solutions are proposed having Wi-Fi module to communicate within limited ranges [10].

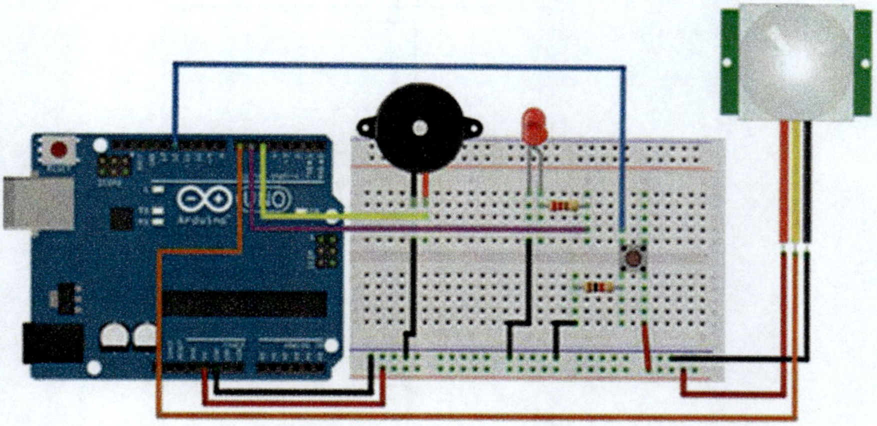

Fig. 1 Burglar alarm with buzzer [11]

These existing systems to protect crops in agriculture fields are not fruitful due to their cost and limited range resulting need of a better protection system having real-time communication and response to intrusion of wild animals. The proposed system is a low cost, portable, and easy to use solution to existing systems.

3 Proposed System and Its Implementation

The proposed system consists of device(s) having Arduino Uno microcontroller [12, 13], GSM module, and PIR sensor to detect and communicate presence of wild animals in agriculture fields. The microcontroller provides required control mechanism to set parameters for sensing, processing, and communication of information from this device to the farmer's mobile phone. The PIR sensor detects the presence of an object possibly an animal crossing the sensing area covered by the sensor. The microcontroller processes this object detection and turns on the GSM module. The GSM module makes the device capable to send an announcement in the form of preset short message service (SMS) message to the farmer's mobile phone. A buzzer is also provided to alert the farmer within audible range as well as rush the animals out of field through human like voices. Such devices can be easily installed on poles or trees depending upon the initial budget and can be further extended as per requirement. The block diagram of proposed smart security system for agriculture fields is shown in Fig. 2.

The proposed system development requires following hardware components:

(1) Arduino Uno
(2) Bread Board

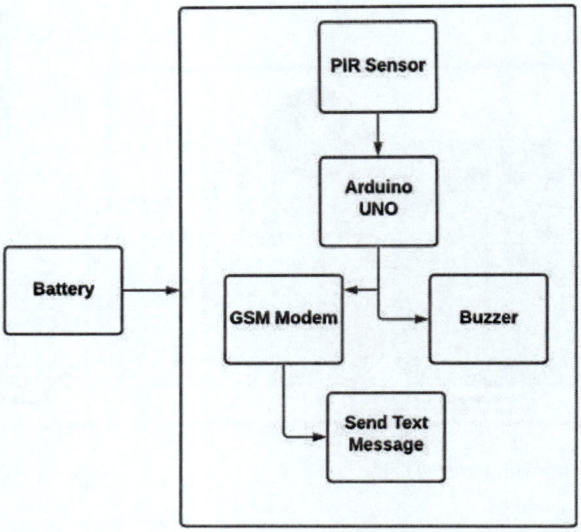

Fig. 2 Block diagram of security system

(3) PIR Sensor
(4) GSM module
(5) Piezo Buzzer
(6) Jumper Wires (as required)
(7) Batteries

The device, having all components connected together, is shown in Fig. 3.

Fig. 3 Hardware implementation of security system

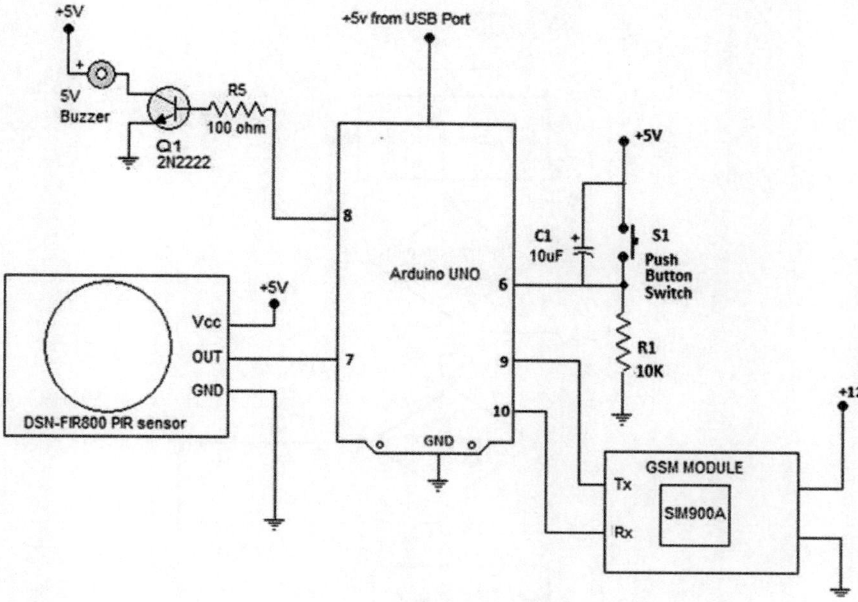

Fig. 4 Circuit diagram

The most important phase in development of proposed system is to place components at right position and connect their pins to microcontroller pins correctly in accordance with the circuit diagram. The circuit diagram for the proposed system is shown in Fig. 4.

After connecting all the components, next important thing is to program the microcontroller to get the desired functions. In general, Arduino Uno microcontroller uses Arduino IDE to program it. It can also be programmed using C programming language. Being friendly to use, C programming language is used in the development of proposed system.

4 Working of Proposed System

The above implemented devices may be installed on trees and poles in the agriculture fields at the points of possible intrusions and may be powered by some power source or replaceable batteries. The device will detect the intrusions of wild animals crossing the PIR sensor sensing area, and process and communicate it in form of predefined SMS to the farmer through GSM module. It also activates the buzzer for programmed duration to alert farmer as well as scares the wild animals to run out of agriculture fields. The working of proposed security system is shown in Fig. 5.

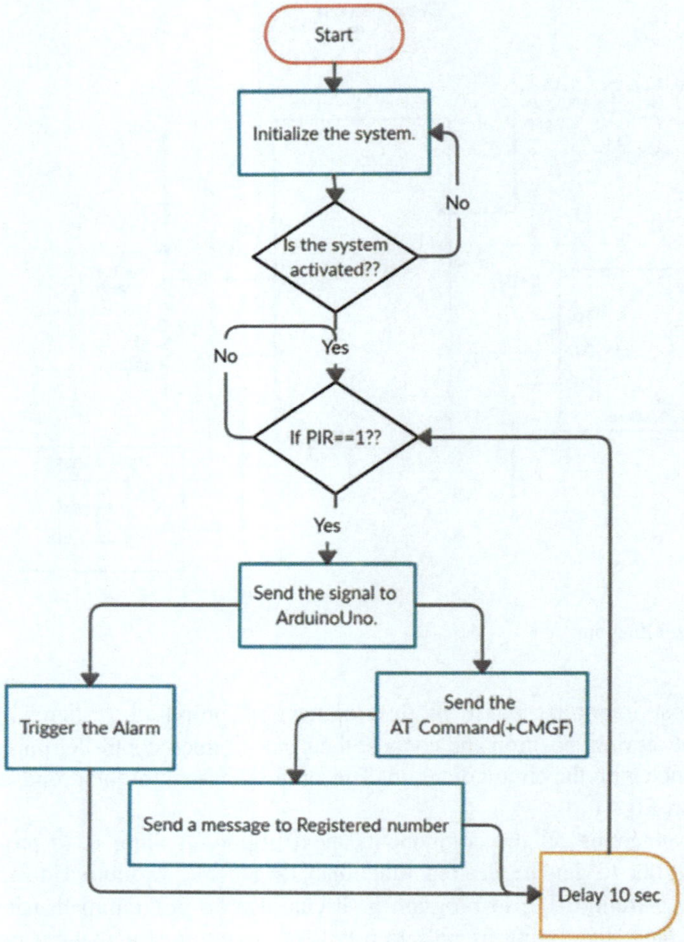

Fig. 5 Flowchart of security system

A PIR sensor is the most essential component of the proposed device to detect presence of an object and inform the microcontroller to perform the desired security system functions. The range and detection accuracy of the device depends upon the quality of PIR sensor used as well as programming and processing capabilities of microcontroller in use. PIR sensor uses inferred radiation to detect the motion produced by humans, animals, and other moving objects. After detecting an object, it will send the signals to the microcontroller. The microcontroller turns on the GSM module and communicates the predefined message to the farmer's mobile number written into the program. It can also be programmed to make a phone call to the farmer's mobile to overcome chances of delayed checking of message by the farmer. A typical PIR sensor can detect an object within a range of around 30 feet or 6 m. A warmup time of duration 20–60 s is required for proper functioning of

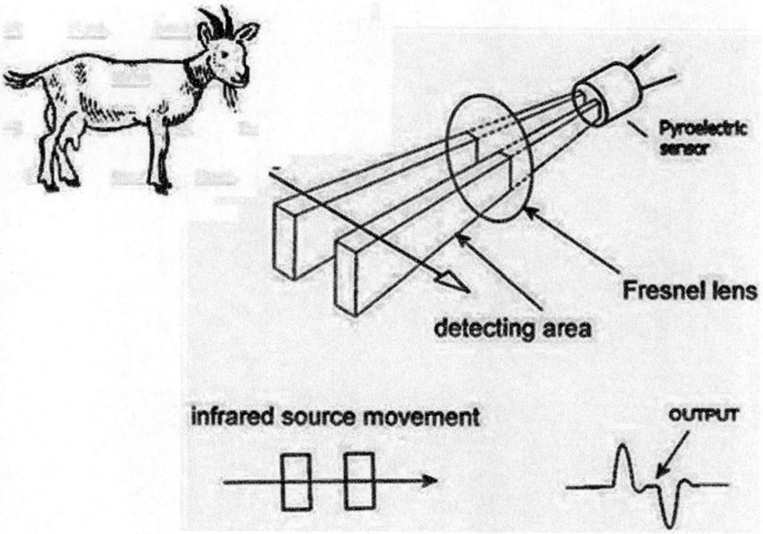

Fig. 6 Working of PIR sensor [14]

the PIR sensor. Actually, this is the settling time for consistent sensor calibration in its environment and to stabilize the IR detector. There are usually 3-pin connectors in PIR modules modules: +Vcc, Ground, and Output. There may be different pinouts, and it is generally powered through power supply of 5–12 V having own on-board transformer. When it senses some motion, the output becomes high. There is a three pin jumper selection for single/continuous trigger output modes. There are two labels—L and H for these two positions. The sensor is repeatedly re-triggered if the jumper is on H position, while output remains in high state. If the position is L, the output will be high and low whenever the sensor is in triggered state. So, during this mode, there will be endless motion giving repeated high low pulses. There is a Fresnel lens, for focusing the infrared onto sensor element, in front part of PIR sensor module. The working of PIR sensor to detect presence of an object is shown in Fig. 6.

The microcontroller turns on the GSM module and sends the predefined message to the farmer's mobile phone. The intrusion detection alert SMS received on farmer's registered mobile number is shown in Fig. 7. After receiving the message, the farmer may immediately reach the agriculture fields to rush away the wild animals. After sending the message, the microcontroller turns on the buzzer for predefined duration considering power requirement and usability.

Fig. 7 SMS received on mobile phone

5 Conclusions and Future Scope

A smart security system for crop protection in agriculture fields is proposed, developed, and implemented to provide user friendly, low cost, and portable solution to the farmers for protection of their crops from wild animals. It helps in reducing implementation cost through decreased deployment of manpower as well as self-inspections, while the cost of device is limited to few hundred Indian Rupees. It is easy to use due to availability of simple mobile phone interface to access notifications like SMS. It is portable in the sense that the device can be relocated depending upon the particular requirements like field area where the crops are present at a particular time, while the farmer can access notifications on the move. The installation of the device in real field demonstrated working capabilities as well as limitations for possible implementation as well as improvements. As the device detects any moving object like wild animals as well as humans and vehicles, it results in false notifications many a times. A camera can be installed on the device to take photographs of object being detected and can be communicated to farmer as a picture or object name through image processing at device level. The device may be powered through some power source or batteries which require availability of power source or timely replacement of batteries. Uninterrupted power supply or effective utilization of available batteries is another bottleneck which may be considered for future work. Overall, the proposed system gives an initial prototype having wide possibilities and future scope.

References

1. Rao, K.S., Maikhuri, R.K., Nautiyal, S., Saxena, K.G.: Crop damage and livestock depredation by wildlife: a case study from Nanda Devi Biosphere Reserve, India. J. Environ. Manag. **66**(3), 317–327 (2002)
2. Gubbi, J., Buyya, R., Marusic, S., Palaniswami, M.: Internet of things: a vision, architectural element, and future directions. Future Gener. Comput. Syst. **29**(7), 1645–1660 (2013)
3. Vermesan, O., Friess, P.: Internet of Things: Converging Technologies for Smart Environments and Integrated Ecosystem. River Publishers, Denmark (2013)
4. Kodali, R.K., Jain, V., Bose S., Boppana, L.: IoT based smart security and home automation system. In: 2016 International Conference on Computing, Communication and Automation (ICCCA), pp. 1286–1289. Noida (2016)
5. Salman, L., Salman, S., Jahangirian, S., Abraham, M., Germah, F., Blair C., Krenz, P.: Energy efficient IoT-based smart home. In: 2016 IEEE 3rd World Forum on Internet of Things (WF-IoT), pp. 526–529. Reston, USA (2016)
6. Chhabra, J., Gupta, P.: IoT based smart home design using power and security management. In: 2016 International Conference on Innovation and Challenges in Cyber Security (ICICCS-INBUSH), pp. 3–5. Noida, India (2016)
7. Indumathi, J., Asha, N., Gitanjali, J.: Smart security system using IoT and mobile assistance. In: Krishna, P.V., Obaidat, M. (eds.) Emerging Research in Data Engineering Systems and Computer Communications, Advances in Intelligent Systems and Computing, vol. 1054, pp. 441–453. Springer, Singapore (2020)
8. Prasetyo, T.F., Zaliluddin, D., Iqbal, M.: Prototype of smart office system using based security system. In: 2017 4th International Seminar of Mathematics, Science and Computer Science Education, pp. 1–8. Bandung, Indonesia
9. Furdik, K., Lukac, G., Sabol, T., Kostelnik, P.: The network architecture designed for an adaptable IoT-based smart office solution. Int. J. Comput. Netw. Commun. Secur. **1**(6), 216–224 (2013)
10. Jha, J., Dubey, P.R., Pradhan, P., Pai, S.N.: IoT-Based home security system with wireless communication. In: Hassanien, A., Bhatnagar, R., Darwish, A. (eds) Advanced Machine Learning Technologies and Applications, Advances in Intelligent Systems and Computing, vol. 1141, pp. 525–533. Springer, Singapore (2020)
11. Arduino Project Hub. https://create.arduino.cc/projecthub. Accessed on 25 Dec 2021
12. Andriansyah, M., Subali, M., Purwanto, I., Irianto S.A., Pramono, R.A.: e-KTP as the Basis of home security system using Arduino Uno. In: 2017 4th International Conference on Computer Applications and Information Processing Technology (CAIPT), pp. 1–5. Kuta Bali, Indonesia (2017)
13. Monk, S.: Programming Arduino, 2nd edn. The McGraw Hill, USA (2012)
14. How Infrared motion detector components work. http://www.glolab.com/pirparts/infrared.html. Accessed on 25 Dec 2021

Efficient Clustering of Transactional Data for Privacy-Preserving Data Publishing

Vartika Puri, Parmeet Kaur, and Shelly Sachdeva

Abstract Transactional data is set-valued data which is generated from retail store, healthcare, etc. The data needs to be published to extract useful information from the data. The data contain some sensitive information about the individual, and if leaked, then it will cause serious implications to the privacy of an individual. Therefore, it is required to protect the user's privacy on the published data while ensuring the data should remain useful for analysis purpose. The paper proposes efficient clustering method using ant-colony-based clustering algorithm to bring similar transactions in same equivalence class/cluster. Finally, we can achieve privacy with minimal information loss. The approach has been tested on INFORMS dataset and compared with the Disassociation. The result shows that the more information is preserved as compared to Disassociation approach.

Keywords Ant colony clustering · Clustering · Privacy-preserving data publishing · Relative error

1 Introduction

Privacy-preserving data publishing (PPDP) is the emerging area where data is published to the third party while preserving privacy of individuals whose data is contained in the published data. Transactional data is the real-world dataset generated by the widely used applications such as retail store and healthcare.

V. Puri (✉) · P. Kaur
Department of Computer Science & Engineering and Information Technology, Jaypee Institute of Information Technology, Noida, India
e-mail: vartika.puri@jiit.ac.in

P. Kaur
e-mail: parmeet.kaur@jiit.ac.in

S. Sachdeva
Department of Computer Science, National Institute of Technology, Delhi, India
e-mail: shellysachdeva@nitdelhi.ac.in

K. Khanna et al. (eds.), *Cyber Security and Digital Forensics*,
Lecture Notes on Data Engineering and Communications Technologies 73,
https://doi.org/10.1007/978-981-16-3961-6_14

The dataset is of the form {id: a_1, a_2...a_n} where id denotes the identity of user and a_1, a_2...a_n denote the set of attributes belongs to the user.

The primary requirement in PPDP is the protection of identity disclosure [1]. k^{m}-anonymity [1] is the model which ensures the protection of identity disclosure in transactional data with minimal information loss. K^{m}-anonymity model ensures every m no. of items should occur in k transactions. There are numerous methods are available to achieve k^{m}-anonymity. Disassociation [2] is the method which is based on bucketization to achieve k^{m}-anonymity while incurring less information loss. There are three phases in Disassociation—(i) Horizontal partitioning, (ii) Vertical partitioning, and (iii) Refining.

In first phase, the similar transactions are put into one cluster. In the second phase, the cluster is converted into k^{m}-anonymous record chunks by placing infrequent item combinations in different record chunks. In any of the privacy-preserving data publishing methods, there are two steps, first, make equivalence classes of similar records, and second, apply anonymization method to anonymize the records in the equivalence class. Therefore, if there are more similar records in a cluster, then there will be less modifications to achieve the desired privacy level and data utility will be maintained. Thus, creation of good equivalence classes/clusters is the main step of PPDP.

Ant colony optimization (ACO) is the technique used by insects existing in adjacent colonies in the search for food. If a source of food is found by any ant team/colony, then some teams of ants follow diverse paths searching this food, leaving behind pheromone trail, a chemical usually excreted by animals, and is of great importance for insects. The pheromone trail directs the other ants, and with its help, other ants follow the way laid down by the ants moving in front of them. Few ant teams will reach the food source prior to the other teams due to the fact that they would have traversed the shortest path, and then they will follow the same path to go back to their colony before the other ant teams. Now this shortest path will have the pheromone trails as the team have traversed this path and came back before other team following another path; therefore, probability of other teams taking the same path over other paths is much higher, lest some other paths (better) are discovered by another teams. Pheromone trail of the shortest path is expected to be more concentrated than the other paths.

Inspiration of the ant-based clustering algorithm comes from the clustering of corpses and larval sorting events found in real ant colonies. Deneubourg et al. [3] have first started the study in this field. He has proposed a basic model in which objects in clusters are randomly moved, picked up, and dropped as per the similarity found in surrounding objects. LF algorithm proposed by Lumer and Faieta [4] which is an extension of basic model, which is applicable for numerical datasets. In this algorithm, ants are considered as agents who travel in a four-sided grid in a random fashion. These agents pick up, transport, and drop the data items scattered within this environment. Operations (picking and dropping) are executed as per the similarity and density of the data items found in the ants' neighborhood: either isolated or data items surrounded by dissimilar ones are likely to be picked up by ants, and ants have a tendency to drop them near the comparable ones. This is how elements are clustered and sorted in the grid. The ant colony clustering algorithm are more flexible, robust, and decentralized [5–7] than traditional methods.

The paper proposed the use of ant colony clustering algorithm on transactional dataset for making optimized clusters of similar transactions.

The rest of the paper is structured as follows: Sect. 2 presents the related work in this domain. Section 3 proposes application of ant colony clustering algorithm for efficient clustering of transactional data. The results of the implementation of the proposed algorithm and its comparison with related approach are discussed in Sect. 4. Lastly, we summarize and conclude the work followed by future work.

2 Related Work

The application of ACO to solve the clustering problem was introduced by Shelokar et al. [8]. Firstly, the sample data is represented by each string element, and its content signify the cluster number which the sample data allotted to. Each ant in the ACO at that time builds a solution on the basis of string representation. As per [9], the ant algorithm can be segregated into two sets to achieve clustering, ant-based sorting, and ACO based clustering. Ant-based sorting algorithm uses 2D grid. As per that algorithm, foremost the objects are scattered randomly. Afterward, objects dissimilar to its neighborhood are picked up by artificial ants and transfer it to the cluster containing similar objects. The proposed solution was also used in the studies [10–13]. Though a defined cluster number is not required in the beginning by ant-based sorting, the processing time will be high as it requires post-processing to recognize the generated clusters [9]. This was proven in few prior studies where the analysis of the cluster number should be done visually once the clustering is completed [12]. ACO based clustering is another ant algorithm for clustering which uses the same idea of solution string to denote the clustering solution. The solution string is built on each iteration and assessed by the objective function to discover the most optimal one. Although a defined cluster number is a prerequisite, ACO based clustering is more efficient in computation than ant-based sorting. Also, once the clustering is done, it does not require post-processing [9]. Apart from ACO, some of the proposed clustering algorithms also practice the same concept of solution string as ACO based clustering [14–16]. ACOC [17] is the first implementation of ACO based clustering. After that, ACOC has been enhanced in some studies such as [18] which revised the original ACOC by keeping the identified best solution as the initial solution for the next iteration and adding the ability to determine the optimal cluster number automatically using Jaccard index. The study has demonstrated that the algorithm takes more time to run. The research [17] have adopted another methodology by combining the ACOC with k-means algorithm. In this, the ACO explores the initial solution generated by k-means. However, the algorithm was only tested on financial services data processing. Besides, in research [19], ACO based clustering concept was used; however, it builds the classification model according to the training dataset which is clustered using ACO. The fast ant colony optimization for clustering (FACOC) improves the efficacy of computation in ACOC [20]. In FACOC, the threshold value is used to define whether a cluster number turn out to be common for an object once it is being selected for multiple times. If a cluster number for an object turns out to be common, then that cluster number will be selected without computing the probability in the next iteration for that particular object. With this, the redundant computations can be reduced, enhancing the execution time. In addition, local search will not affect the object with common cluster number. However, the result indicates that FACOC outputs have inferior clustering quality than ACOC.

3 Proposed Approach: Application of Ant Colony Clustering Algorithm for Efficient Clustering of Transactional Data

We propose an algorithm in which efficient partitions are created using ant colony clustering algorithm and then utilizes VERPART algorithm [21] to finally achieve k^{m}-anonymity. The clusters are initialized using HORPART algorithm [21]. HORPART algorithm selects the most frequent item "a" in the dataset and splits the dataset into two partitions, the records which contain "a" come in one partition and the rest of the records come in another partition. The process of splitting like this will continue till we get the partitions of predefined size, say P_1, $P_2 \ldots P_n$.

At the beginning, an ant m is associated to a partition P_I, and during the iterations, the ant will select the most dissimilar transaction d_i of partition P_I, and another partition P_J is selected at random using a roulette-wheel with probability p_{IJ}, where p_{IJ} depends on the pheromone trail and a local heuristic. Ant will assign d_i to the partition P_J. The value of the pheromone trail is modified according to the rule.

$$\tau_{xy} = (1 - \rho)\tau_{xy} + \Delta\tau_{xy}^{k} \tag{1}$$

where τ_{xy} is the amount of pheromone deposited for a state transition from partition x to partition y, ρ is the pheromone evaporation coefficient where $\rho \in [0, 1]$ and $\Delta\tau_{xy}^{k}$ is the amount of pheromone deposited by k^{th} ant.

$$\Delta\tau_{xy}^{k} = \begin{cases} 1 & \text{if ant } k \text{ transfer a transaction from } x \text{ to } y \\ 0 & \text{otherwise} \end{cases} \tag{2}$$

The local heuristic or short-term visibility is defined as.

$$\eta_{IJ} = \frac{A \cap B}{A \cup B} \tag{3}$$

where A and B are the transactions. If the transaction A is more similar to the transaction B of particular partition, then it gives a big value in order to influence in the probability of assigning it to the partition. If ant m is at partition I, partition J is chosen with probability

$$p_{XY}^{k} = \frac{\left(\tau_{XY}^{\alpha}\right)\left(\eta_{XY}^{\beta}\right)}{\sum_{z \in \mathrm{allowed}_X}\left(\tau_{XZ}^{\alpha}\right)\left(\eta_{XZ}^{\beta}\right)} \tag{4}$$

To find whether the obtained solution is better than previous or not, the Jaccard similarity is calculated.

$$B(P) = \sum_{I=1}^{n} JS(P_I) \tag{5}$$

Then, dissimilar transaction d of partition I is assigned to the partition J.

```
Algorithm: Proposed algorithm

Input:  number of iterations tmax, the number of ants M; the initial
value of pheromone τ0; α; β; ρ; Original dataset D.
Output: Anonymized dataset D'.
Initialize τIJ= τ0;
Calculate η according to (3)
Initialize the probabilities p
Initialize partitions P1,…,Pm using HORPART
For n=1 to tmax:
 For t=1 to M do:
   For each p in P:
     Find the most dissimilar transaction di in the partition p
     Select a partition PJ according to (4)
     Assign di to the partition PJ
   End For
   Calculate the cost B(P)according to (5),
   if B(P) > B(PI-1) then
      keep B(P)
   else
       keep B(PI-1)
   Update τ according to (1) and (2)
 End For
End For
For each partition p∈P do:
Apply VERPART[21] on p.
```

Fig. 1 Relative error

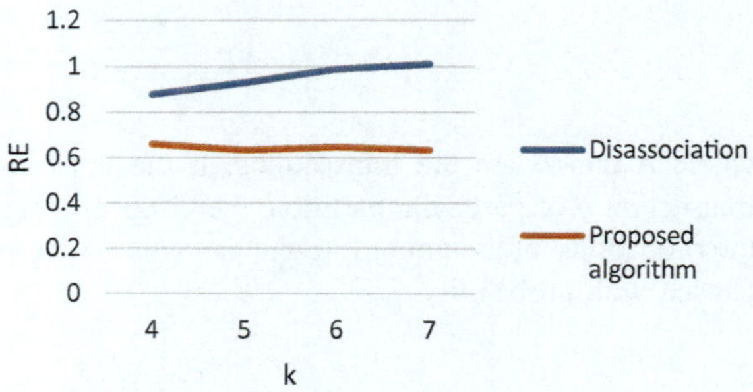

4 Implementation and Results

The proposed approach is implemented in Python and tested on INFORMS dataset.[1] In ACO, there are five parameters that needs to be fixed. In the literature [22], following conditions have been specified on selecting these values:

- β has to be larger than α; so that destination cluster should be chosen based on local heuristic, i.e., similarity with the destination cluster instead of deposited pheromone.
- α, $\beta <= 1$ is better than α, $\beta > 1$;
- $\rho = 0.8$ is better than $\rho = 0.7$ and $\rho = 0.9$ decided by set of experiments.

Considering the above conditions, the following values are considered:

- M = number of clusters
- $t_{\max} = 100$
- $\alpha = 0.8$
- $\beta = 1$
- $\rho = 0.8$
- $\tau_0 = 0.001$

The results are analyzed in terms of information loss while achieving privacy-preserving data publishing. To evaluate information loss, we have used relative error measure [21] which measure the loss in the association of items occurred while anonymization shown in Fig. 1. The relative error is calculated for different values of k (= 4, 5, 6, 7) and for the anonymized data using proposed algorithm and Disassociation algorithm, respectively. The result shows that if data is anonymized using the proposed algorithm, it gives lower values of relative error for each k than Disassociation algorithm. The lower value of relative error shows that more items are still associated in the anonymized data and, thus, preserves data utility.

[1] https://sites.google.com/site/informsdataminingcontest/data.

5 Conclusion

The paper has clearly shown that if we can create the equivalence classes/clusters which have similar records result in less information loss due to anonymization process. Thus, data utility maintained. The proposed algorithm uses ant colony optimization to further refine the equivalence classes/clusters; it shows the significant improvements in equivalence class and, thus, reduces the information loss cause by anonymization. The proposed approach has been tested on INFORMS dataset, and it gives lower relative error than Disassociation algorithm. In future, the applicability of other nature-inspired algorithm can be tested and compared.

References

1. Terrovitis, M., Mamoulis, N., Kalnis, P.: Privacy-preserving anonymization of set-valued data. VLDB Endowment, 115–125 (2008)
2. Loukides, G., Liagouris, J., Gkoulalas-Divanis, A., Terrovitis, M.: Disassociation for electronic health record privacy. J. Biomed. Inform. **50**, 46–61 (2014)
3. Deneubourg, J.L., Goss S., Franks, N., Sendova-Franks A., Detrain, C., Chretien, L.: The dynamics of collective sorting: robot-like ants and ant-like robots. In: Meyer, J.A., Wilson, S. (eds) Proceedings of the 1st International Conference on Simulation of Adaptive Behaviour: From Animals to Animals, pp. 356–365, MIT Press, Cambridge, Mass, USA (1991)
4. Lumer, E., Faieta, B.: Diversity and adaptation in populations of clustering ants. In: Proceedings of the Third International Conference on Simulation of Adaptive Behaviour: From Animals to Animals, pp. 501–508, MIT Press, Cambridge, Mass, USA (1994)
5. Bonabeau, E., Dorigo, M., Theraulaz, G.: Swarm intelligence: from natural to artificial system. Oxford University Press, New York, NY, USA (1999)
6. Ghosh, A., Halder, A., Kothari, M., Ghosh, S.: Aggregation pheromone density based data clustering. Inf. Sci. **178**(13), 2816–2831 (2008)
7. Gao, W., Yin, Z.X.: Modern intelligent bionics algorithm and its applications. Science Press, Beijing, China (2011)
8. Shelokar, P., Jayaraman, V.K., Kulkarni, B.D.: An ant colony approach for clustering. Anal. Chim. Acta **509**(2), 187–195 (2004)
9. Jabbar, A.M., Ku-Mahamud, K.R., Sagban, R.: Ant-based sorting and ACO-based clustering approaches: a review. In: 2018 IEEE Symposium on Computer Applications and Industrial Electronics (ISCAIE) (2018)
10. Gao, W.: Improved Ant Colony Clustering Algorithm and Its Performance Study. Computational Intelligence and Neuroscience **2016**(19) (2016)
11. Yang, Y., Kamel, M.S.: An aggregated clustering approach using multi-ant colonies algorithms. Pattern Recogn. **39**(7), 1278–1289 (2006)
12. Kuo, R.J., Wang, H.S., Hu, T.L., Chou, S.H.: Application of ant K-means on clustering analysis. Comput. Math. Appl. **50**(10–12), 1709–1724 (2005)
13. Korurek, M., Nizam, A.: A new arrhythmia clustering: technique based on ant colony optimization. J. Biomed. Inform. **41**(6), 874–881 (2008)
14. Tao, W.A., Ma, Y., Tian, J.H., Li, M.Y., Duan, W.S., Liang, Y.Y.: An improved ant colony clustering algorithm. In: Zhu, R., Ma, Y. (eds) Information Engineering and Applications, vol. 154. Lecture Notes in Electrical Engineering, pp. 1515–1521, Springer, London, UK (2012)
15. Inkaya, T., Kayaligil, S., Ozdemirel, N.E.: Ant colony optimization based clustering methodology. Appl. Soft Comput. J. **28**, 301–311 (2015)

16. Chaturvedi, A., Green, P.E., Carroll, J.D.: k-Modes clustering. J. Classif. **18**(1), 35–55 (2001)
17. Handl, J., Knowles, J., Dorigo, M.: Ant-based clustering and topographic mapping. Artif. Life **12**(1), 35–62 (2006)
18. Wu, B., Zheng, Y., Liu, S., Shi, Z.Z.: CSIM: a document clustering algorithm based on swarm intelligence. In: Proceedings of the Congress on Evolutionary Computation (CEC '02), pp. 477–482, Honolulu, Hawaii, USA (2002)
19. Yang, Y., Kamel, M.S., Jin, F.: Topic discovery from document using ant-based clustering combination. In: Web Technologies Research and Development—APWeb of Lecture Notes in Computer Science, pp. 100–108, Springer, Berlin, Germany (2005)
20. Ramos, G.N., Hatakeyama, Y., Dong, F., Hirota, K.: Hyperbox clustering with ant colony optimization (HACO) method and its application to medical risk profile recognition. Appl. Soft Comput. J. **9**(2), 632–640 (2009)
21. Terrovitis, M., Liagouris, J., Mamoulis, N., Skiadopoulos, S.: Privacy preservation by disassociation. VLDB Endowment **5**, 944–955 (2012)
22. Trejos, J., Murillo, A., Piza, E.: Clustering by Ant Colony Optimization. Classification, Clustering, and Data Mining Applications (2004)

Passive Video Forgery Detection Techniques to Detect Copy Move Tampering Through Feature Comparison and RANSAC

Jatin Patel and Dr. Ravi Sheth

Abstract With the advancement of innovation, these days respectability of digitized information has been addressed from various perspectives. With the assistance of innovation, the integrity level of advanced video can be disturbed from various perspectives. Video tempering performs with two different ways: one is frame level and other is changing the sequence of frames to hide or highlight the specific aim of the original video. In some cases, the original information of video frames is altered and afterward pasted at some other area of the same video. The proposed framework completed two stages of calculation to distinguish the doctored frames. Feature extraction is executed as the initial steps of the proposed framework. Correlation of separated features will assist with distinguishing the altered frames. In the subsequent advance, each frame will be compared with one before and after frames with the assistance of extricated highlights to check the measure of changes in the tempered frames all through the video stream. To check the uprightness of a video, include correlation and RANSAC strategies are utilized and it shows huge outcomes distinctive when checked with the video which altered and not altered. With the utilization of this strategy, we can ensure the originality of the advanced video when it assumes a significant part as proof or verification in specific conditions.

Keywords Video forgery detection · Copy-move forgery · Spatiotemporal tampering · Feature comparison · RANSAC

J. Patel (✉) · Dr. R. Sheth
Rashtriya Raksha University, Gandhinagar, Gujarat, India
e-mail: jatin.patel@rru.ac.in

Dr. R. Sheth
e-mail: ravi.sheth@rru.ac.in

K. Khanna et al. (eds.), *Cyber Security and Digital Forensics*,
Lecture Notes on Data Engineering and Communications Technologies 73,
https://doi.org/10.1007/978-981-16-3961-6_15

1 Introduction

Because of wide progression in advanced correspondence, now daily's video information has gotten more mainstream. Henceforth, numerous frameworks depend on the integrity of such video correspondence. In the present advanced time accessibility of practical equipment and straightforward video preparing, programming instruments have made a video recording and altering easy. Ordinarily content accessible is not accurate and more often than not, a few recordings are altered deliberately to pass on wrong data. Checking the integrity level of any video is a challenge for an examiner or any layman.

Altered recordings can make sway on political perspectives, lawfulness, and contention among nations just as it may create social strains in a tranquil society. In such cases, it is hard to distinguish an altered recording. Subsequently, proving the authenticity of the video dataset is a challenging task in today's era [1].

Video forgery is classified into two categories: one is an interframe forgery and another one is intraframe forgery. In the case of interframe forgery, the frame, in general, goes through an altering cycle, while in intraframe, only a specific portion of the video frames is altered. The first category of video tempering is further characterized by deletion, insertion, duplication, rearranging, and recompression of video frames. Frame deletion manages the occasion's evacuation from the recording by eliminating the edges concerned. While in frame insertion, the specific portion of the video replicated from the video and pasted into another location or maybe in a different recording. Duplication of the frames includes replicating a specific portion of the video and embedding at other fleeting areas of a similar file. Frames rearranging is a subcategory of edge replica where the replicated video frames are rearranged transiently before addition. Insertion, duplication, and rearranging of the video frames can be utilized to fill the hole of erased content of the original video. Altering any unique video can be performed at two distinct stages which are tampering during recording or tampering in the wake of recording.

Video altering detection is a child category of forensics that looks at the content adjustments of digital data and may find the spatial or temporal areas of video tampering [2]. These forensics methodologies are subdivided into active and passive dependent on the accessibility of earlier data of the video viable. Passive or blind video altering identification techniques do not need earlier data for ordering a video as altered or not. Anyway on account of active methodology, authenticator takes the help of advanced watermark or digital signature to identify tempering in advanced recordings. Here, advanced watermarking or digital signature or hash esteem is an equipment-based element and accordingly it is restricted to less number of users.

The rest of the paper is organized as follows: Sect. 2 quickly clarifies the different kinds of assaults in the copy-move type of forgery. In Sect. 3, survey of existing techniques was carried out, Sect. 4 characterizes the recent concerns in the current framework, and Sect. 5 features the proposed strategy, at long last in Sect. 6 outcome examination and area Sect. 7 conclusion.

2 Types of Video Tampering

Altering of any video file should be possible differently. Copy-move forgery is one of the common techniques of video tempering, where the specific subpart of the video frames is adjusted or the changes are performed in the succession of video frames. First category of forgery is called spatial-temporal while the second one is known as temporal altering [3, 4].

2.1 Spatial Tampering

An attacker can alter source recordings spatially by changing the pixel esteems in a specific frame only or in the next available frames [5, 6]. In such type of imitation, following prospects are considered:

- A specific portion of the frame is copied and is glued in different areas of a similar casing to shroud objects.
- A specific portion of the one frame might be pasted in a different frame of similar group of picture (GOP) or other GOP.

2.2 Temporal Tampering

Temporal-temporal altering kinds of fabrication fundamentally influence the arrangement of a specific scene caught by the camera [5, 6]. Such kind of forgery has the following prospects:

- Video frames are reordered in a similar GOP.
- Video frames are reordered in other GOP.
- It is additionally conceivable that the sequence of frames is changed and afterward pasted at some other area.

2.3 Spatiotemporal Tampering

Such type of altering is considered as mixing of spatial and temporal altering. The sequence of the frames is adjusted just as the visual substance of the edges is altered in a similar video [5, 6]. Figure 1 shows the pictorial representation of the above-mentioned video forgery attacks.

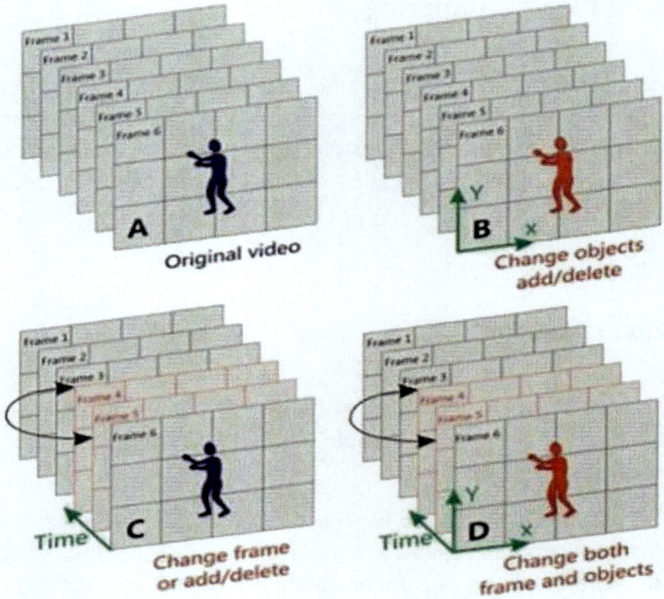

Fig. 1 Graphical representation of video forgery types [7]

3 Literature Review

To identify the research gap, here we have analyzed various standard research papers from well-known journals. The following papers define their capability and limitations.

C. Long, A. Basharat, and A. Hoogs have proposed a profound learning approach to define the duplication of video frame location at both edge level and video level, just as for the video transient limitation. An I3D network is utilized to deliver a sequence-to-grouping framework and decide the competitor frame arrangements at the coarse-search stage. In the next step, a Siamese network is used to define the fine-level search to define whether frame duplication exists or not. In the last phase, an irregularity finder is applied to additionally recognize copied outlines from chosen outlines. Proposed procedure cannot work on multistream 3D neural organizations for both edge drop, outline duplication [8].

In another methodology, M. Raveendra and K. Nagireddy have proposed a DNN-based moth search optimization strategy in which at first the input video is decompressed to identify the double compression. Double compression measurement based on the Markov insights is applied to the video. The frames which have a double compression are then given as a contribution to the SLIC superpixel. This approach based on SLIC is utilized to play out the division cycle in double compressed frames. Here, these extracted frames are portioned into various locales by SLIC superpixel division measure. Now the extracted frames are given as an input to the Gabor scale channel, which uses the scale data for highlight extraction. In the last phase, highlighted frames are passed into DNN for falsification detection. In light of these extricated highlights, the fashioned casings are distinguished by this DNN classifier. The exactness of this proposed forgery discovery measure is discovered to be 94.37% [9].

In the proposed strategy [10], the input video is divided into multiple frames in the pre-handling stage. Frames are classified into specific groups as per the extracted features. The probabilities are arrived at the midpoint of the post-preparing stage to get the end product. The excess parts are built a similar route as to when the information is a picture. In the pre-preparing stage, faces are identified and scaled to 128×128. Creator has utilized a piece of the VGG19 network to separate the dormant highlights, which are the contributions to the case network. Now they have taken the yield of the third max pooling layer rather than three yields before the ReLU layers. In the proposed network, there are three primary capsules and two output capsules, one for original data and another for counterfeit pictures. The inactive highlights separated by part of the VGG-19 network are the information sources, which are conveyed to the three primary capsules. The yields of the three capsules ($uj|i$) are powerfully directed to the output (vj) for reemphases utilizing the algorithm. The proposed technique is not compatible to oppose antagonistic machine assaults, and it cannot forestall against blended assaults.

An interframe falsification location conspires dependent on 2D stage congruency, and *k*-implies bunching was proposed for reconnaissance video. Creator has determined 2DPC for each edge initially. At that point, the connection coefficients of adjoining outlines and the variety of continuous relationship coefficients are obtained. At long last, the discontinuous focus brought about by altering is distinguished by utilizing *k*-implies grouping calculation. Here, deleting outlines show up toward the start or the finish of the video, and the location technique is inconceivable [11].

S. Jia, Z. Xu, H. Wang, C. Feng, and T. Wang determined optical flows (OF), and their high and stable relationship in duplicate move altered recordings offers a reason for effective recognition. For computation cost reasons, the consistency of OFs is dissected first to find speculated altered positions. This cycle will assist with lessening numerous counts and examinations of relationship lattices, however, may prompt all the more bogus locations. Fine discovery dependent on OF relationship is then proposed to define the set of copied frames, and a decrease of false-positive rate dependent on approval checks will be directed further for exactness. In the event that vitality, the copied succession can be separated and erased to acknowledge video recuperation. The proposed strategy may not distinguish some copy-move forgery that does not affect optical flow integrity consistency, for example, altered recordings with a generally static scene and different kinds of care-completely arranged control [12].

S. Kingra, N. Aggarwal, and R. D. Singh depend on the way that expectation leftover and optical progression of back-to-back P frames fluctuate extraordinarily at the area where outline altering has been performed. Wanted highlights are carried out after the extraction of frames (*I* and *P*) from the video. These registered highlights are then contrasted with limits which create spikes for bigger extents of PRG and OFG highlights. Now in the next step, performance on the total number of count and progression in these spikes to unmistakable unique and manufactured recordings. The strategy for figuring chosen highlights and technique of identifying falsification by contrasting these chosen highlights and edges utilizing two unique algorithms. On the off chance that progression check gives negative outcomes in Algorithm 1, at that point concerning video succession is retested utilizing Algorithm 2. On the off chance that the video creates broken spikes, at that point that shows that the video is authentic-spasm. Non-stop spikes, nonetheless, show the location of imitation. Interframe falsification with normal precision is of 83%. Substitute piece rates for recording and reproducing a video do not influence the identification consequences of this strategy except if the objective piece rate is incredibly high or very low [13].

L. Su, C. Li, Y. Lai, and J. Yang proposed algorithm which principally comprises the accompanying four phases: (1) highlight extraction, (2) block coordinating, (3) finding the tempered location in the current frame, and (4) following the produced zones in the ensuing edges. The principal phase of the calculation removes EFMs from each square utilizing an IM-demonstrated component extraction strategy. The subsequent stage applies another square coordinating strategy to look for likely coordinating squares, which altogether lessens the computational cost. In the third stage, the PVS technique is intended to take out falsely coordinated combines and find the altered zones in the current edge. In the last stage, AFCT calculation is adjusted to gain proficiency with the altered zones recognized in the current casing and track the produced districts in ensuing edges. Proposed technique is incapable to identify more perplexing kinds of video forgery or combine video falsification [14].

I. Bozkurt, M. H. Bozkurt, and G. Ulutaş proposed a strategy for forgery discovery dependent on the forgery line. The technique separates binarized DCT highlights from the casings and speaks to the closeness between them to be a connection picture. Components of the connection picture give a thought regarding the similitude between comparing outlines. The technique at that point researches a line called a falsification line on the connection picture, to decide the area of the fabrication line in a coarser way. The Hough change is applied to the picture to decide the area of the line. Besides, the proposed technique recommends two new methods, contracting and extending, to find the imitation line in a better way. The number esteemed focuses on the line show the comparing record estimations of manufactured edges proposed strategy cannot work adequately against outline reflecting assault [15].

4 Current Issues

The quick innovative progression makes the advanced information assume an essential part in moving and tending to thoughts all the more agreeably. This has driven toward finding a fitting component for looking at and recognizing picture and video fraud in this computerized period.

Numerous researchers chipping away at recognition methods accept that extra noise, surface and shading changes would have been made in the additional locale and these regions can be recognized by contrasting the areas and the whole picture. Still epic falsification location strategies cannot adapt up to the most developed sort of malevolent exercises that are done by counterfeiters by changing the calculation of the picture highlights. Subsequently, there is a requirement for the most reformist discovery method to recognize the despiteful activities on advanced proof [9].

Recordings are in compacted design. Subsequently, when a unique video goes through certain sorts of object-based fraud, the first step is to decompress it into an arrangement of individual frames and each frame can be viewed as a still picture. At that point, the frames in the selected sections of the succession are altered while the rest frames stay immaculate. After all the modifications are finished, the subsequent framing arrangement is repacked to produce a manufactured form. For this situation, the specialist needs to recognize the frames which went through some noxious movement and separate the frame [16].

- **High calculation multifaceted nature**. Straightforwardly or in a roundabout way connection or dependent on pixel approaches required high computational recognition component to discover forgery. Now and again it is tedious to look at countless frames in recordings with an immense measure of information far more prominent than that of still pictures.
- **Unstable discovery result**. Various strategies are accessible which are dealt with picture highlights, including surface, shading modes, commotion, and pixel dark qualities. A significant number of them are helpless against normal assaults or post-handling on recordings, as an illustration auxiliary pressure and added substance clamor. Here not many existing discovery approaches consider the location power as a record, and for the most part set fixed static boundaries for identification of forgery.
- **Limited pertinence**. Hardly any techniques work on a particular sort of falsification as it were. They are not sufficiently able to distinguish the complex type of forgery. A strategy that is competent to recognize entomb frame fraud may not distinguish intra frame forgery from object-based forgery. Indeed, even the technique which can recognize the frame removing cannot distinguish the framing insertion [17].

As mentioned above, there are various techniques available to check the integrity of the video but they may have some limitations to define the complex or dynamic forgery of digital video. The proposed method is capable enough to define the exact tampering location with the types of video forgery, i.e., temporal tampering or spatiotemporal tampering. In the proposed method, we have defined the copy-move forgery at frame level and the entire flow of the video.

5 Proposed Method of Forgery Detection

The first step of the proposed system is to extract frames from the input video and then features are extracted from the given video. To extract the features from the input video, here we have implemented a new transformation function which is applied on the partitioned block of the given video.

Discrete transformation assists with changing over spatial domain into the frequency of the picture. This operation is performed while utilizing the orthogonal polynomial [18, 19]. Here, the survey of video signals is performed in specific domains and provides a huge move to its high computing capacity to investigate the pieces of shifted signals [18, 20]. The discrete Tchebichef change (DTT) and discrete Krawtchouk change (DKT) zone unit made upheld the OP portions of the KP and TP, severally. As of late, DTT and DKT polynomials are encouraged by analysts, explicitly Krawtchouk-Tchebichef polynomials (KTP) [21] and Tcheb-ichef-Krawtchouk polynomials (TKP). KTP and TKP portions region unit acclimated change the sign from the special area to the second space, explicitly discrete Krawtchouk-Tchebichef change (DKTT) and Tchebichef-Krawtchouk change (DTKT), severally [22].

The steps of the proposed method are shown in Fig. 2.

The output of the KTP and TKP sets indicated momentous attributes as far as to highlight extraction and localization properties contrasted and different existing genuine transformations [23]. DTKT and DKTT can separate highlights from a

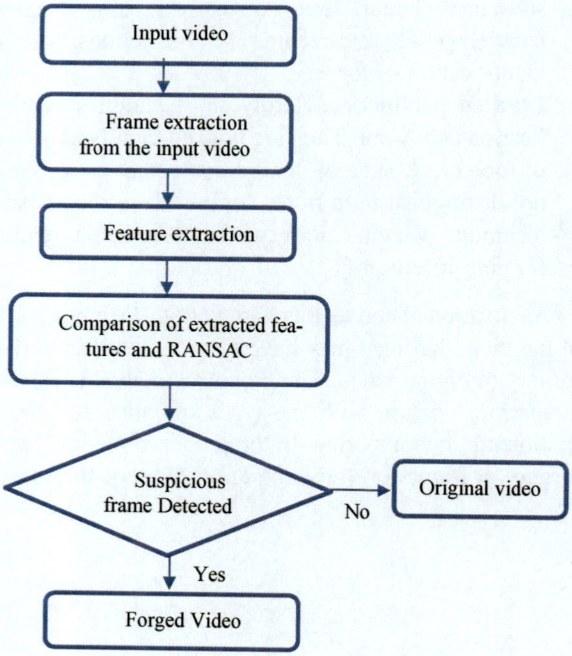

Fig. 2 Flowchart of the proposed method

particular segment of the picture by the benefits of the confinement property in existence spaces.

The overall expenses of computation for separating the predominant features from the input image in the DKTT is not as much as compared to DTKT. Consequently, in this paper, we have used DKTT to extricate highlights from the operated video frame.

As mentioned in the equations, we have performed the mathematical calculation of KTP. Here, Rn(x) is the nth order of KTP and it is given by [21]:

$$\mathrm{Rn}(x) = \sum_{i=0}^{N-1} Ki(n;p,N-1)\cdot ti(x) \tag{1}$$

$$x,n = 0,1,\ldots,N-1,$$
$$N > 0;\ \ p \in (0;\ 1)$$

where $ti(x)$ is the TP of the ith order [24].

In conventional strategy, the frame I of size $N1 * N2$ pixels is separated into number square size $B1 * B2$. All out number of squares is $v1 * v2$, where $v1 = N1/B1$ and $v2 = N2/B2$. Here, $BL_{i,j}$ is considered as a picture block in the I and j course. Presently, the change work is applied on each square with the accompanying equation:

$$\Psi\ BL_{i,j} = R_{B1}BL_{i,j}R_{B2}^{T} \tag{2}$$

$$i = 1,2,3\ldots v1,$$
$$j = 1,2,3\ldots v2$$

where $RB1 \in ROrd * B1$ and $RB2 \in ROrd * B2$ are the change grids that define the transformation function. ()T characterizes the render of the lattice, and Ord is the sequence for the transformation. Here, $\Psi\ BL_{i,j}$ is determined for each edge of the video. To separate the features of the whole video document, the previously mentioned computation is rehashed for $v1 * v2$ times which is not successful from the speed and cost viewpoint.

To beat the previously mentioned issue, another technique for feature extraction from the video record is performed while utilizing one bunch of matrix multiplication ($P1\ I\ P2T$). The change highlights of the whole frame are joined in one matrix.

$$\Psi = \begin{bmatrix} R_{B1}BL_{1,\ 1}R_{B2}^{T} & R_{B1}BL_{1,\ 2}R_{B2}^{T} & \cdots & R_{B1}BL_{1,\ v2}R_{B2}^{T} \\ R_{B1}BL_{2,\ 1}R_{B2}^{T} & R_{B1}BL_{2,\ 2}R_{B2}^{T} & \cdots & R_{B1}BL_{2,\ v2}R_{B2}^{T} \\ & & \vdots & \\ R_{B1}BL_{v1,\ 1}R_{B2}^{T} & R_{B1}BL_{v1,\ 2}R_{B2}^{T} & \cdots & R_{B1}BL_{v1,\ v2}R_{B2}^{T} \end{bmatrix} \tag{3}$$

Here, the output worth $\Psi \in R\mathrm{Ord}\cdot v1 * \mathrm{Ord}\cdot v2$ is considered as the separated element for the whole casing. As demonstrated in Eq. (3), $RB1$ and $RB2$ are two orthogonal polynomial and they are multiplied with each square of image ($BL_{i,\ j}$). Here, while utilizing Kronecker item Eq. (3) can be revamped as:

$$\Psi = \left(\underset{[v1 * v1]}{\begin{matrix} 1 & 0 \\ 0 & 1 \end{matrix}} \otimes \underset{[\mathrm{Ord} * B1]}{R_{B1}} \right) * \left(\underset{[N1 * N2]}{I} \right) * \left(\underset{[v2 * v2]}{\begin{matrix} 1 & 0 \\ 0 & 1 \end{matrix}} \otimes \underset{[\mathrm{Ord} * B1]}{R_{B2}^{T}} \right) \tag{4}$$

where $\begin{matrix} 1 & 0 \\ 0 & 1 \end{matrix}$ is the identity matrix, and $\otimes$ is the Kronecker product. Equation (4) defines the transposition property of Kronecker [25], and it can be written as follows:

$$\underset{[Ord \cdot v1 * \mathrm{Ord} \cdot\ v2]}{\Psi} = \left(\underset{[\mathrm{Ord} \cdot\ v1 * v1 \cdot B1]}{\begin{matrix} 1 & 0 \\ 0 & 1 \end{matrix} \otimes R_{B1}} \right) * \left(\underset{[N1 * N2]}{I} \right) * \left(\underset{[\mathrm{Ord} \cdot\ v2 * v2 \cdot B1]}{\begin{matrix} 1 & 0 \\ 0 & 1 \end{matrix} \otimes R_{B2}} \right)^{T} \tag{5}$$

$$= \underset{\mathrm{Ord} \cdot\ v1 * N1}{P_{B1}} * \underset{[N1 * N2]}{I} * \underset{\mathrm{Ord} \cdot\ v2 * N2}{P_{B2}^{T}} \tag{6}$$

In the above equation, P_{B1} and P_{B2} are two matrices that are constructed from R_{B1} and R_{B2}, respectively. Matrix P_B is an orthogonal matrix that satisfies the orthogonality condition [19]:

$$\sum_{x \in X} P_B(n, x) P_B(m,\ x) = \delta_{nm} \tag{7}$$

$$n, m = 1, 2, \mathrm{Ord}$$

Here, δnm is defined as the Kronecker delta and it is defined as a symbol of the orthonormal representation of orthogonal polynomials and is defined by:

$$\delta_{nm} = \left\{ \begin{matrix} 1 & n = m \\ 0 & n \neq m \end{matrix} \right\} \tag{8}$$

Image reconstruct is performed using two orthogonal matrices P_{B1} and P_{B2}, and it is represented as follows:

$$I = P_{B1}^{T} \Psi P_{B2} \tag{9}$$

From Eq. (6), it is extremely evident that the future extraction from the given info video is performed for each square of the picture without isolating the picture into various squares. Here, we have utilized just one bunch of matrix multiplication.

Presently, in the following period of usage extracted highlights were put away and determined the invariant highlights of the image. Highlights are localized and separated, keeping exclusively the individuals who are without a doubt to remain stable over relative changes, having satisfactory differentiation, and do not appear to be on edges. At long last, the central issue descriptor is made by inspecting the sizes and directions of the picture angle in a fix around the distinguished element. Here, each info picture is changed over into a grayscale picture. Extracted highlights arranged as dubious highlights and trustworthiness highlights of the video frames.

5.1 Comparing of Features Extraction with Suspicious Frame

Features of the suspicious frames are given as input training data. Extracted features of each picture contrasted and the test input information. So here RANSAC homography comparison method is utilized for ascertaining the coordinating distance between the test input and the current edge of the video individually. Here, score has been created dependent on the coordinating substance of the video with the dubious features. Edge which contains most elevated coordinating with the dubious features is to be highlighted as produced outline in a succession.

5.2 Comparison of Extracted Features Using RANSAC

For *i*th ($I = 1{:}\ N$) assessment arbitrarily pick four correspondences. Here we have checked if these focuses are collinear. If it is there, retry the above advance. Presently, the holography is computed. Now distance di was calculated and the Hcurr by standardized DLT generated from the four-point sets for each putative correspondence. At that point, the standard deviation (*Sd*) of the inlier distance curr, std is calculated. Here the total number of inliers m is counted which has the overall distance di less than the threshold value.

5.3 Decision Making Based on the Feature Matching

The image which has the highest matching with the extracted feature of the forged image is removed from the sequence. Such an image is defined as a most probable forge or tempered image into the video file.

While the proposed strategy does not require picture dividing, nor amassing the outcome, here there is a requirement of one set of matrix convolution. Subsequently, any kind of transmission, for example, orthogonal polynomials, can be actualized utilizing the proposed method for the feature extraction, and RANSAC utilized as an ultimate conclusion to characterize the tempering dependent on the matching of the extracted features.

5.4 Details of Forgery

The original frame sequence number 109–123 is shown in Fig. 3. Figure 4 shows a similar edge grouping with the spatiotemporal forgery. Here as appeared in Fig. 4, a picture of someone else is added to characterize the item-based fraud just as edge grouping is likewise changed which characterizes the fleeting altering of the source video.

Edge number 23–37 of a similar video was duplicated and it was stuck at frame number 109–123. Figure 5 characterizes the recurrence dispersion of unique videos. Figure 6 characterizes the beginning and finishing point of the tempered area.

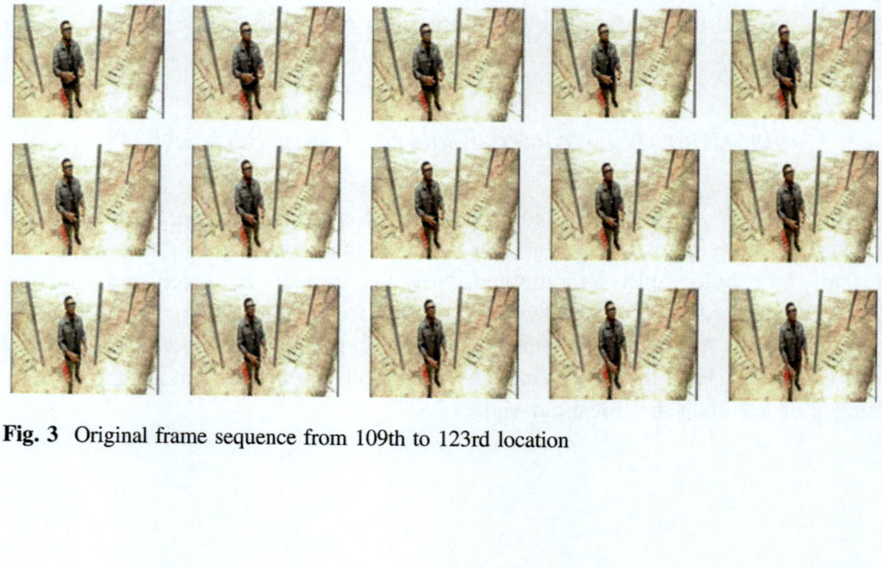

Fig. 3 Original frame sequence from 109th to 123rd location

Fig. 4 Spatiotemporal tempering frame sequence from 109th to 123rd location

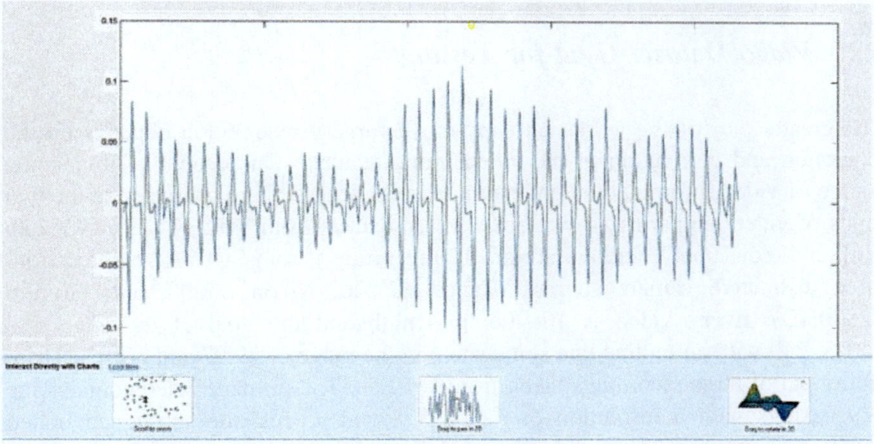

Fig. 5 Frequency distribution of original video frames

6 Experimental Result

As appeared in Fig. 6, our proposed strategy characterizes unexpected pick an incentive in the diagram which characterizes the beginning and finishing area of the manufacture casing of the video. Here, we can characterize the specific area of the manufacturing district of the video. After concluding the produce district of the picture, separated highlights contrasted and the current edges of a similar video. Highlight correlation of dubious framing assists us with distinguishing the kinds of video forgery.

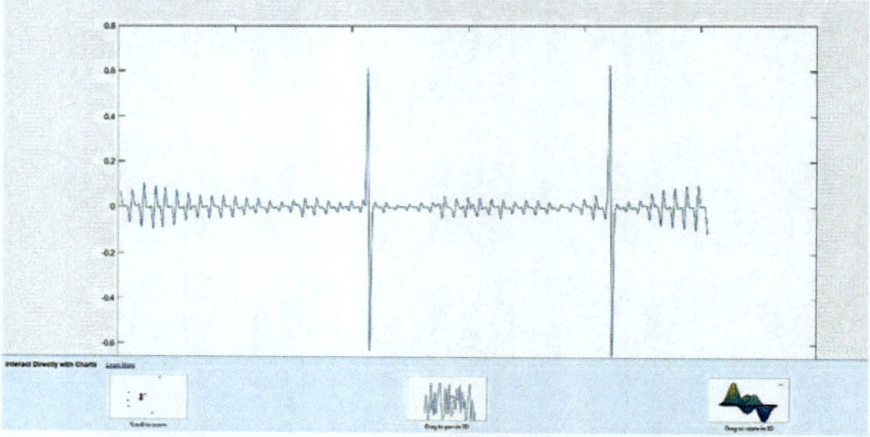

Fig. 6 Frequency distribution of forge video frames

6.1 Video Dataset Used for Testing

The greater part of the works on copy-move forgery recognition depends on the extraction and correlation of the video frame features. On account of this single source of video informational collection was not accessible to characterize the two kinds of video tempering so we took a video dataset from SULFA [25], IVY Lab [26], and YouTube. There are around 120 recordings from SULFA gathered from three distinctive camera sources (Canon SX220, Nikon S3000, and Fujifilm S2800HD). Every video is 10–20 s in length and the goal of the edge was 320 × 240 with an outline rate somewhere in the range of 29.97 and 30 fps. There are a total of three recordings taken from Ivy Lab. The primary video contains the 729 casings with a resolution of 722 × 578 and 25 fps; the subsequent video contains a total of 662 frames with a resolution of 767 × 578 and 30 fps; the third video has 708 edges with a resolution of 704 × 576 and 30 fps.

6.2 Robustness of the Proposed Method

To legitimize the precision level of the proposed technique, the true positive (TP) rates (e.g., forged is identified as forged), false positives (FP) (e.g., forged is distinguished as credible), true negatives (TN) (e.g., true is recognized as false), and false negatives (FN) (for example genuine is distinguished as valid) are determined. Our strategy is precise under double compressor with false alert 2.8%, fixed GOP with false caution 2.7%, and differing GOP with false caution 6.3%.

Table 1 shows the details of forgery detected on the tested database.

Table 1 Details of video forgery

Video	Types of forgery	Total frames	Forgery location and details
1	Temporal	729	Frame number 129th–164th copied and pasted at 225th–260th location
2	Temporal	663	Frame number 223th–254th was copied and pasted at 495th–526th location
3	Temporal	256	Frame number 229–248 copied and pasted at 125th–144th location
4	Spatiotemporal	255	The object of the 3rd frame is copied and pasted at 16th–19th location
5	Spatiotemporal	219	The object of the 239th frame is copied and pasted at 349th. Another object from 236th was copied and pasted at the 389th location
6	Spatiotemporal	432	An object from 165th frame copied and pasted at 234th–256th location

7 Conclusion and Future Work

In this paper, we have proposed feature extraction and examination-based tempering identification method for copy-move transient and spatiotemporal altering location. The proposed system utilizes feature extraction while utilizing transform matrices (*RB*1 and *RB*2) created by the latest transform which is DKTT and RANSAC to discover the copy-move frame tempering in test video succession. We have tried different sports videos with the most extreme movement action. For every video, at least one edge is reordered at different areas in the succession, and objects are reordered at different areas in the video. Other than that, few frames are altered and pasted at other locations in a similar grouping. The results of the proposed simulation demonstrate that the proposed method is very proficient; it can identify all tempered locations and frames sequence with 100% precision. The proposed strategy is additionally equipped for locating the tempered frames in a given video grouping. In the future, we will test our strategy with various sorts of recordings with countless altered frames to check the robustness of the technique. Additionally, execution boundaries have not been investigated in this examination as the dataset is restricted. The work will be stretched out to address different sorts of unpredictable and dynamic assaults. We will try to develop a robust method to detect all the above-mentioned video tempering in a single digital video.

References

1. Wahid, A., et al.: Passive video forgery detection techniques : a survey. Statistical correlation of video feature international conference on information assurance and security (lAS). In: 2014 10th International Conference on Information Assurance and Security, pp. 29–34 (2014)
2. Yao, Y., Cheng, Y., Li, X.: Video objects removal forgery detection and localization. In: Proceedings of NICOGRAPH International 2016, NicoInt 2016, p. 137 (2016)
3. Gavade, J.D., Chougule, S.R.: Passive blind forensic scheme for copy-move temporal tampering detection. In: 2018 International Conference on Advanced in Communication and Computer Technology ICACCT 2018, pp. 155–160 (2018)
4. Patel, J.J., Bhatt, N.: Review of digital image forgery detection. Int. J. Recent Innov. Trends Comput. Commun. **5**(7), 152–155 (2017)
5. Upadhyay, S., Singh, S.K.: Video authentication: issues and challenges. Int. J. Comput. Sci. Issues **9**(1), 409–418 (2012)
6. Upadhyay, S., Singh, S.K.: Video authentication—an overview **2**(4), 75–96 (2011)
7. Al-Sanjary, O.I., Sulong, G.: Detection of video forgery: a review of literature. J. Theor. Appl. Inf. Technol. **74**(2), 207–220 (2015)
8. Long, C., Basharat, A., Hoogs, A.: A Coarse-to-fine Deep Convolutional Neural Network Framework for Frame Duplication Detection and Localization in Forged Videos (2018)
9. Raveendra, M., Nagireddy, K.: Dnn based moth search optimization for video forgery detection. Int. J. Eng. Adv. Technol. **9**(1), 1190–1199 (2019)
10. Huy, I.E., Nguyen, H., Yamagishi, J.: Capsule-Forensics: Using Capsule Networks to Detect Forged Images and Videos, and Isao Echizen SOKENDAI (The Graduate University for Advanced Studies), Kanagawa , Japan National Institute of Informatics, Tokyo, Japan The University of Edinburgh, ed. IEEE International Conference Acoustics Speech Signal Process, pp. 2307–2311 (2019)
11. Li, Q., Wang, R., Xu, D.: An inter-frame forgery detection algorithm for surveillance video. Information **9**(12) (2018)
12. Jia, S., Xu, Z., Wang, H., Feng, C., Wang, T.: Coarse-to-fine copy-move forgery detection for video forensics. IEEE Access **6**, 25323–25335 (2018)
13. Kingra, S., Aggarwal, N., Singh, R.D.: Video inter-frame forgery detection approach for surveillance and mobile recorded videos. Int. J. Electr. Comput. Eng. **7**(2), 831–841 (2017)
14. Su, L., Li, C., Lai, Y., Yang, J.: A fast forgery detection algorithm based on exponential-fourier moments for video region duplication. IEEE Trans. Multimed. **20**(4), 825–840 (2018)
15. Bozkurt, I., Bozkurt, M.H., Ulutaş, G.: A new video forgery detection approach based on forgery line. Turkish J. Electr. Eng. Comput. Sci. **25**(6), 4558–4574 (2017)
16. Chen, S., Tan, S., Li, B., Huang, J.: Automatic detection of object-based forgery in advanced video. IEEE Trans. Circuits Syst. Video Technol. **26**(11), 2138–2151 (2016)
17. Deshpande, R.G., Ragha, L.L.: Performance analysis of various video compression techniques. Int. J. Sci. Res. **2**(8), 335–338 (2013)
18. Mahmmod, B.M., Ramli, A.R., Abdulhussian, S.H.: Low-Distortion MMSE Speech Enhancement Estimator Based on Laplacian Prior, pp. 1–15 (2017)
19. Abdulhussain, S.H., Ramli, A.R., Mahmmod, B.M., Jassim, W.A.: Image edge detection operators based on orthogonal polynomials. Int. J. Image Data Fusion **00**(00), 1–16 (2017)
20. Abdulhussain, S.H., Ramli, A.R., Mahmmod, B.M., Jassim, W.A.: Methods and challenges in shot boundary detection: a review, pp. 1–42 (2018)
21. Mahmmod, B.M., Abdulhussain, S.H., Al, S.A.R., Jassim, W.A.: Signal compression and enhancement using a new orthogonal-polynomial-based discrete transform, pp. 129–142 (2017)
22. Abdulhussain, S.H.: A fast feature extraction algorithm for image and video processing. In: 2019 International Joint Conference on Neural Networks, pp. 1–8 (2019)

23. Saripan, M.I., Al-Haddad, S.A.R., Jassim, W.A.: Shot boundary detection based on orthogonal polynomial (2019)
24. Abdulhussain, S.H., et al.: A fast feature extraction algorithm for image and video processing. In: Proceedings of International Joint Conference on Neural Networks, pp. 1–8 (2019)
25. Zhang, H., Ding, F.: On the Kronecker Products and Their Applications (2013)
26. Sohn, H., Neve, W.D., Ro, Y.M.: Privacy protection in video surveillance systems: analysis of subband-adaptive scrambling in JPEG XR. IEEE Trans. Circuits Syst. Video Technol. **21** (2), 170–177 (2011). Available at http://ivylab.kaist.ac.kr/demo/vs/dataset.htm

Evaluation and Performance Analysis of Apache Pulsar and NATS

Vanita Jain, Aarush Ahuja, and Dharmender Saini

Abstract Messaging systems are a crucial part of many distributed systems. Various implementations of open-source message brokers, message queues, and messaging protocols are in wide use to facilitate highly available and reliable messaging. We evaluated the architectures of two popular open-source message brokers used in the cloud: Apache Pulsar and NATS, studying qualitative differences like broker distribution, client architecture, messaging features, etc., and benchmark the two message brokers in realistic deployments.

Keywords Message-oriented middleware · Performance evaluation · Benchmarking

1 Introduction

Message brokers allows independent applications on a distributed system to communicate with each other using messages. They are useful for processing streaming data like analytics, IoT sensors, microservice communication, remote procedure calls, cloud messaging, command and control, telemetry, etc. The software architecture of message brokers consists of named message queues and protocols to transfer data efficiently. Message brokers offer various messaging transmission semantics like at-least-once, at-most-once, etc., making them reliable for critical systems. They are often deployed as a distributed system, a cluster of highly available, reliable, and fault tolerant systems. We aim to compare two message brokers: Apache Pulsar [1] and NATS [2], with NATS Streaming. A technical com-parison of Apache Pulsar and NATS is pursued on the basis of design and performance, while both are message brokers and essentially promise similar functionality, i.e., a publish/subscribe messaging model. The underlying design in both is wildly different, Pulsar is a heavyweight messaging system with a variety of

V. Jain (✉) · A. Ahuja · D. Saini
Bharati Vidyapeeth's College of Engineering, New Delhi, India
e-mail: vanita.jain@bharatividyapeeth.edu

K. Khanna et al. (eds.), *Cyber Security and Digital Forensics*,
Lecture Notes on Data Engineering and Communications Technologies 73,
https://doi.org/10.1007/978-981-16-3961-6_16

features, and NATS is a lightweight system focusing only on providing the messaging system. We compare the architecture of Apache Pulsar and NATS on the basis of messaging protocol, partitioning, persistence, clustering, clients, messaging semantics, etc. Both of them are also open source and heavily documented, we aim to compare the design and implementation of both the message brokers objectively. Using the OpenMessaging standard, an industry standard for messaging systems, we aim to identify the key performance indicators (KPIs) for message brokers and benchmark the performance of Apache Pulsar and NATS for a quantitative comparison in similar configurations.

2 Related Work

Work on evaluation of message-oriented middleware since the 1990s has produced research focusing on qualitatively and quantitatively comparing legacy messaging systems. Eugster et al. [3] sets up a conceptual base for publish-subscribe message patterns [4] presenting its various forms such as topic-based pub/sub, content-based pub/sub, type-based pub/sub implemented via RPC, message passing, notifications, etc. It presented a qualitative comparison of two early message-oriented middleware TIBCO's TIB/RV and IBM's MQ series on factors like software architecture, message persistence, reliability, etc. Maheshwari et al. [5] evaluate TIBCO's TIB/RV and progress SonicMQ, high-lighting the major difference in TIB/RV's P2P-based architecture and SonicMQ's broker-based architecture support java messaging system (JMS). It studies publish-subscribe and point-to-point messaging on both middleware benchmarking them on message throughput, message latency, and resource utilization. Sachs et al. [6–11] present SPECjms2007 and SPECjms2009-PS standardized benchmarks for JMS-based message-oriented middleware. It simulates a real-world application employing the middleware throughout a supply chain modeled after a supermarket's supply chain where orders, shipments, and sales constantly go through the system. The SPECjms2007 benchmarks effectively tested point-to-point messaging capabilities in JMS-based middleware but lacked in testing publish-subscribe messaging. SPECjms2009-PS was developed with focus on testing publish-subscribe messaging capabilities in JMS-based middleware. Folkerts et al. [12] lay out the requirements for benchmarking and evaluation of cloud infrastructure posing many essential question toward building a benchmark for the cloud. It identifies the requirements of a benchmark to be portable, fair, highly configurable, scalable, representing the real-world and applies the requirements to the layers of a cloud considering example platforms like an online gaming platform deployed in the cloud and analytical platforms deployed in the cloud. It suggested metrics important for the cloud like 99%ile wait time and 99%ile dropped requests. Importantly, the research considered factoring in pricing, elasticity, SLAs, and scalability into development of benchmarks for cloud infrastructure. John et al. [13] consider the case of two distributed message-oriented middleware RabbitMQ [14] and Apache Kafka. It presents a study of AMQP and Kafka comparing their software design, messaging capabilities and benchmarking the two based on latency and message throughput. The paper concluded Apache Pulsar to be a more effective system compared to RabbitMQ in terms of both latency and throughput.

3 Architectural Overview of Apache Pulsar and NATS

Apache Pulsar and NATS are two popular open-source distributed message-oriented middleware developed to fulfill large-scale computing requirements in the cloud where scalability, performance, and security are most important. The software also highlights two different open-source software engineering philosophies as Apache Pulsar is under the Apache Software Foundation, while NATS is hosted under the Cloud Native Computing Foundation [15]. Even though both software provide similar features like Publish/Subscribe, wide client compatibility and multi-tenancy across clusters, their underlying design brings out differences in terms of clustering, topic storage, security, etc.

In the following sections, we qualitatively describe and compare the design of Apache Pulsar and NATS on the basis of messaging patterns, distribution capabilities, software architecture, and areas of applications. In the case of NATS, we will also include NATS Streaming to be part of the system as the features provided by NATS Streaming enables a fairer comparison with Apache Pulsar.

3.1 Apache Pulsar

Apache Pulsar is a multi-tenant messaging solution for server-to-server messaging hosted under the Apache Software Foundation. It is developed focusing on large-scale clustering capabilities with geo-replication across clusters, introducing very low overhead and latency for messaging publishing, scaling to millions of topics streaming data, and providing wide client compatibility with clients in many languages like Java, Go, Python, and C++. Apart from typical messaging features like topic-based messaging, Pulsar also offers a special Pulsar Functions framework to build a serverless computing system for streaming data in and out of Pulsar [1]. Apart from typical messaging features like topic-based messaging, Pulsar also offers a special Pulsar Functions framework to build a serverless computing system for streaming data in and out of Pulsar [1].

3.2 System Architecture

The Pulsar system is a software package utilizing many open-source software underneath to implement the messaging functionalities. An instance of Pulsar consists of one or more Pulsar clusters, each Pulsar cluster consists of broker instances, Apache ZooKeeper [16] instances for configuration, and coordination and Apache BookKeeper [17] instances for storage. The Pulsar broker is implemented in Java. The Zookeeper cluster is instrumental in the system as it handles configuration and coordination, with tasks like geo-replication dependent on its availability. A Pulsar Broker is a stateless application running a HTTP server for administration and a dispatcher (a TCP server) talking Pulsar's binary protocol. Zookeeper is used for metadata storage, cluster configuration, and coordination, enabling Pulsar to be globally consistent among tenants and instances. The architecture uses BookKeeper as the storage backend for the messages, and this allows Pulsar to provide the guarantee of a message to be always delivered as it is persisted in long-term storage. Using this architecture also enables Pulsar to have geo-replication of messages. Due to the high horizontal scaling capacity of Zookeeper, Pulsar Brokers and BookKeeper, Pulsar can provide massive multi-tenancy among clusters and namespaces efficiently. Storage in Pulsar supports tiered storage, where older messages can be moved to a cheaper storage system like S3 off from the file system. Pulsar has community and official client implementations in many languages including Java, Go, Python, C++, etc. (Figs. 1 and 2).

The Client APIs wrap complex functionality and provide useful interfaces to the user. Clients provide features like transparent reconnection, connection failover handling, and internal message queuing.

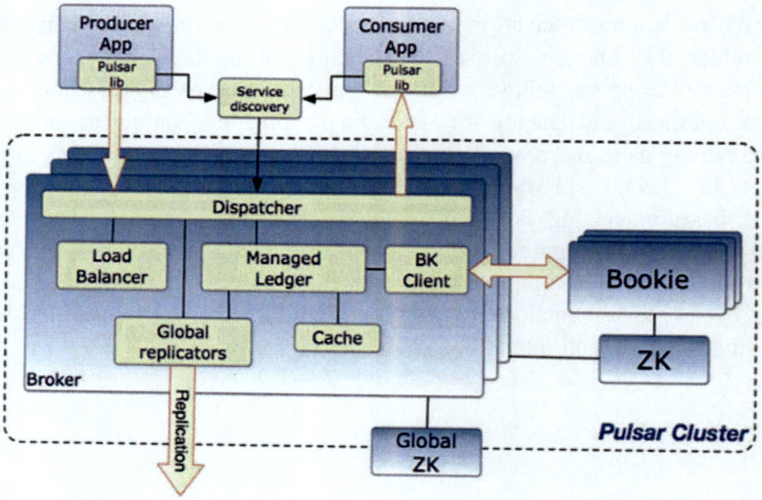

Fig. 1 System architecture of a Pulsar cluster

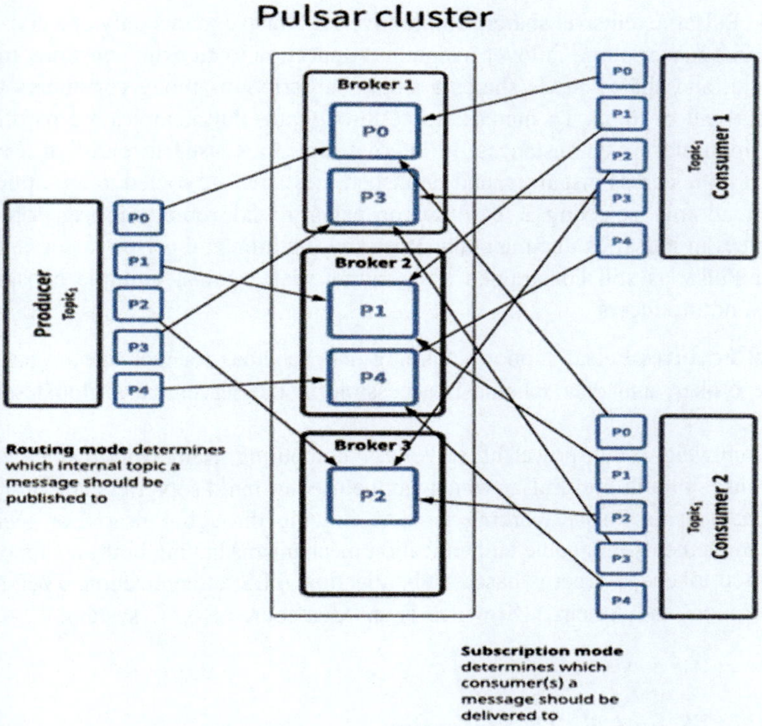

Fig. 2 Topic partitioning in an Apache Pulsar cluster

Messaging Pulsar is built to fulfill publish-subscribe requirements, using the familiar terminologies of producers which publish messages to topics and consumers which can subscribe to those topics. Messages can be processed by implementing clients for a target programming language and subscribing to a topic as a consumer. Messages in Pulsar are always retained in storage unless acknowledged by a consumer; if a consumer fails, then Pulsar will stream messages to them on reconnection. Messages are the basic level of abstraction in Pulsar, messages are published in topics, and consumers receive messages on subscription. Pulsar messages are made to transport raw bytes along with support for typed data schema. A producer can publish messages with inbuilt compression options to save bandwidth. Pulsar Consumers receive messages from the Pulsar Broker and send acknowledgments back to the broker, and acknowledgments can be either for a single message or cumulative for many messages. Messages can be negatively acknowledged to make the broker redeliver the message. Topics in Pulsar are named channels for producers to publish message to and for consumers to subscribe to. Pulsar provides both persistent and non-persistent topics which can be divided into namespaces distributed across clusters.

Namespacing topics are a handy system for defining topics across different tenants and clusters. Three major message subscription modes are provided in

Apache Pulsar: exclusive, shared, and failover. Exclusive grants only one consumer access to a subscription, failover maintains connection to an extra consumer in case of failure, and shared mode shares a single subscription among consumers transmitting to all of them. To increase topic throughput, Pulsar topics are partitioned across multiple broker instances, N internal topics are used to partition a single external topic across instances, and published messages are routed to a partition by either randomly selecting a partition or using round robin. Message ordering guarantees in Pulsar can be maintained per-key-partition and per-producer, and this part of Pulsar is still constrained as it cannot scale against multiple consumers, topics, and producers.

Special Features Pulsar supports defining data schemas for messages transmitted via the broker, and data schema is accessible to the server via a global schema registry.

A lightweight but powerful serverless computing framework called Pulsar Functions is inbuilt into Pulsar which can be used to build serverless stream-native workloads inside Pulsar working on data flowing through Pulsar. For security, Pulsar implements pluggable authentication mechanisms having built-in support for role-based tokens, Kerberos-based authentication, TLS authentication, JWT-based authorization, and Athenz [18] which is an open-source RBAC system.

3.3 NATS and NATS Streaming

NATS is a fire-and-forget messaging system built to fulfill the requirements of large-scale distributed computing. NATS Core is a simple and lightweight messaging solution hosted under the Cloud Native Computing Foundation. It is aimed to be applied in the areas of telemetry collection, building service meshes and microservices, deployment on the edge, etc. Similar to Apache Pulsar, NATS offers high performance, availability, wide client compatibility, etc. NATS itself is very limited in terms of what it offers, a sibling system NATS Streaming extends the core NATS API and enhances its capabilities bringing message delivery guarantees, message persistence, etc., bringing it closer to be compared to a typical messaging system like Apache Pulsar. Comparable to Apache Pulsar, NATS also has client implementations for many programming languages including Python, JavaScript, Golang, Rust, etc.

NATS, also known as NATS Core, is the essential component of the middleware responsible for listening to and publishing messages and provides core capabilities like topics, publish-subscribe messaging, message acknowledgments, load balancing, etc. NATS abstracts messaging topics as subjects, these are strings which represent names a publisher, and subscriber can use to identify themselves. In NATS, subscribers connect to subjects to listen for messages and publishers send messages to a subject which is received by any active subscriber. NATS also supports the request-reply messaging pattern where the server uses the publish-subscribe system to create unique "inboxes" for subscribers to publish replies to. NATS provides a built-in client-side load balancing technique called "Queue Groups" which is an implementation of a distributed queue. Multiple subscribers can connect to a single "Queue" forming a queue subscription, and NATS will balance messages between the subscribers of this queue, scaling messaging for the client becomes as simple as adding more subscribers (i.e., client processes) to the system. NATS Core supports clustering and forms a mesh overlay between all the nodes by using a gossip protocol, and this clustering support can be used to create a HA configuration of the NATS server. NATS messaging follows at-most-once semantics and to ensure message delivery in unreliable networks, and NATS provides built-in request timeouts and message acknowledgements. These simple features bring reliability in NATS messaging letting the sender implement various messaging patterns like request-reply, scatter–gather, point-to-point, etc. Compared to Apache Pulsar, which is a heavyweight compared to NATS with various dependencies and extraordinary features, NATS is a minimal open-source message broker [19] implemented in Golang only focusing on high performance topic-based messaging rather than providing a plethora of features. Implementation in Golang also lets NATS be released as a single compiled binary.

NATS Streaming extends NATS Core and brings advanced messaging capabilities to NATS. NATS Streaming is open source under the Apache-2.0 license and adds fault-tolerance, high availability, data replication, message and subscription persistence, topic partitioning, etc. The underlying messaging protocol is based upon protocol buffers [20] transmitted as binary payloads over the NATS Core messaging protocol. NATS Streaming is essentially a client to a NATS Sever, and a NATS Streaming binary embeds the NATS Server through which all messaging happens with NATS Streaming clients. NATS Streaming implements topics-based messaging via channels extending subjects in NATS Core. Each channel is a FIFO queue, thus creating a message log for all messages received to the subject. A special type of durable subscriptions is possible in NATS Streaming which enables resuming message consumption by a subscriber in the case of server or subscriber failure. Channels support at-least-once message delivery semantics using NATS message acknowledgements with the ability to perform redelivery of messages as required (Fig. 3).

NATS Streaming nodes are clustered using the RAFT consensus algorithm [21] to manage message and channel replication among nodes, and clusters can be pre-configured or an initial "seed" node can start the service discovery process to let any new nodes join the cluster. Similar to topic partitioning in Apache Pulsar, NATS Streaming supports channel partitioning which is incompatible running simultaneously with the clustering mode as clustering requires replication of channels across all nodes in the cluster, channel partitioning can share the load of channels among various NATS Servers, every server is assigned a cluster ID in a NATS Network, and a group of NATS Servers with the same cluster ID can share the network handling their own partition of channels. NATS Streaming also has

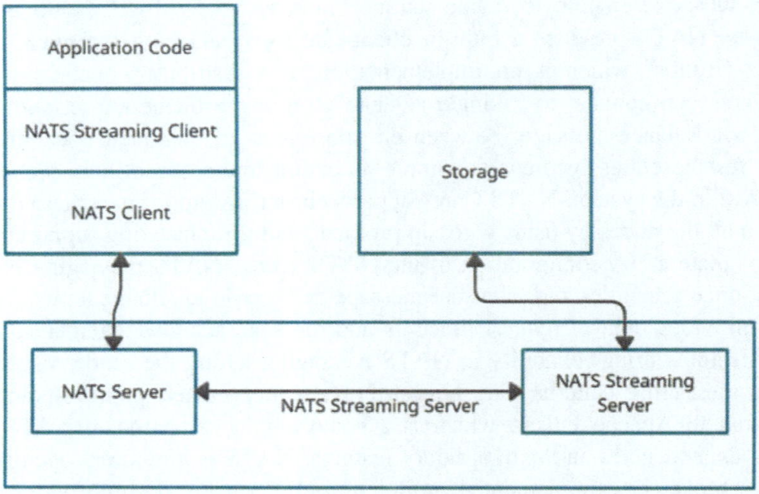

Fig. 3 NATS streaming is an advanced messaging system built on NATS core

inbuilt REST API endpoints for monitoring the system thus allowing ingestion of metrics into any monitoring framework with appropriate data collection tools (Fig. 4).

4 Benchmarking and Results

We designed the benchmark to objectively measure the performance of Apache Pulsar and NATS in realistic cloud deployments. For the server, we deployed NATS, NATS Streaming, and Apache Pulsar on a DigitalOcean Kubernetes cluster with 12 virtual CPUs and 24 GB of RAM running Kubernetes 1.16. For the client, A DigitalOcean server with 2 virtual CPUs and 4 GB RAM running Ubuntu 18.04 was used. The benchmark is based on an open-source project bench [22] which was modified to include NATS [19] v2.14, NATS Streaming [23] v0.17 and Apache Pulsar [24] v2.5.0 clients. The benchmark was run with Go clients of NATS [25] v1.9.1, NATS Streaming [26] v0.6.0, and Apache Pulsar (which are bindings for the C++ client) [24]. We found the Go clients for NATS and NATS Streaming easier to use compared to Apache Pulsar's, especially due to its non-portable nature being bindings for the C++ Apache Pulsar client. Also, TLS was enabled for communication on both NATS and Apache Pulsar. The benchmark involves publishing and subscribing messages from the client server to message brokers deployed in the Kubernetes cluster over the Internet simulating a client of the system. We benchmarked the message brokers considering three major factors: message payload size, message throughput rate, and round-trip latency. For latency benchmarks, we varied the payload size from 256 B up to 1 MB fixing the

Features	Apache Pulsar	NATS Streaming
System Architecture	Mutiple Moving components(Broker, BK, ZK)	Single executable server
Replication and Clustering	Using BookKeeper and ZooKeeper	Using Raft
Messaging Format	Custom Binary Protocol	Google Protobuf based protocols based on top of NATS Protocol
Security	TLS Authentication, JWT and Athenz, Kerberos	TLS Authentication, PKI
Multi Tenancy	Individual Authentication and Authorized streams via namespaces	Using Decentralized JWT authentication and authorization
Message Ordering Guarantee	Ordering guarantee per key partition and per producer	No guarantee of order in case of multiple subscribers
Client Libraries	C++ library and bindings for Python, Golang, etc.	Native Clients for Golang, JS, Rust etc.
Partitioning	Partitioning and replication can Work together	Either only Partitioning or Replication can work

Fig. 4 Feature comparison between subject frameworks

maximum message rate at 1500 messages/s. For throughput benchmarks, we varied the payload size from 256 B up to 50 KB and measured the publish and subscribe throughput of 50,000 messages.

The NATS Cluster was deployed with three nodes, each node being a single Kubernetes pod of NATS Server [19] containers. A Kubernetes service for the NATS cluster was exposed to the Internet by a Digitalocean Kubernetes load balancer. Similarly, a NATS Streaming cluster of three nodes was deployed with channel persistence and data replication. Both NATS and NATS Streaming clients are connected to the cluster via the load balancer.

Apache Pulsar was deployed in a minimal configuration due to the restrictions of the cluster with 1 ZooKeeper instance, 2 BookKeeper instance, and 2 Pulsar Broker instances. Similar to the NATS approach, the broker was exposed to the Internet using a load balancer and a Kubernetes service. Pulsar is meant to be a server-to-server first messaging solution and is thus heavyweight in the number and resource requirements of components, individual ZooKeeper, BookKeeper, and broker instances require a lot of CPU time, and the JVM applications consume a lot of memory. This especially, when compared to NATS which can work directly as a compiled native binary, puts Pulsar at a disadvantage for developer environments with limited resources. To deploy Apache Pulsar in Kubernetes, nodes with large amounts of RAM, powerful CPUs, and expansive disk are recommended.

We take a look at publish-subscribe round-trip latency between NATS, NATS Streaming, and Apache Pulsar varying the payload sizes between 256 bytes upto 1 megabyte apt for representing workloads like transmitting minimal RPC messages and large data files across the brokers. We should notice that both Pulsar and NATS Streaming brings both topic persistence and replication, whereas NATS is a fire-and-forget system. NATS for smaller payload sizes: 512 B and 1 KB had 99%ile latencies of 2 ms and 7 ms, respectively, with tail latencies worsening up to 15 ms and 25 ms, respectively. For larger payload sizes upto 1 MB, the latencies increased to 21 ms at 99%ile and further tail latencies upto 50 ms. NATS Streaming, for small payload sizes: 512 B and 1 KB, had 99%ile latencies of 245 ms and 170 ms, respectively, with further tail latencies up to 270 ms, respectively, for both, we note the increased latencies compared to NATS due to disk persistence and replication. For larger payload sizes up to 1 MB, the latencies increased to 351 ms at 99%ile and further tail latencies up to 820 ms. Apache Pulsar, for small payload sizes: 512 B and 1 KB, had 99%ile latencies of 360 ms and 260 ms, respectively, with further tail latencies up to 650 ms and 350 ms, respectively. For larger payload sizes up to 1 MB, the latencies increased to 740 ms at 99%ile and further tail latencies up to 980 ms. In conclusion, NATS is especially performant in terms of latencies at both smaller and larger payloads.

Apache Pulsar and NATS Streaming both suffer in latency due to presence of disk-based persistence and data replication, while NATS only uses in-memory persistence of messages (Figs. 5 and 6).

We compare message throughput in NATS and NATS Streaming which highlights the overhead disk-based persistence incurs compared to in-memory persistence. We vary the payload sizes from 256 bytes up to 50 KB and transmit 50,000 messages. For small payload sizes, 512 bytes and 1 KB, NATS turned out far superior having a throughput of 78,000 messages/s and 45,000 messages/s, respectively, compared to NATS Streaming having 13,000 messages/s and 7000 messages/s. Though, on a larger 50 KB payload size, the overhead of disk-based persistence becomes less visible with NATS having a throughput of 1000 messages/s and NATS Streaming with a throughput of 700 messages/s (Table 1).

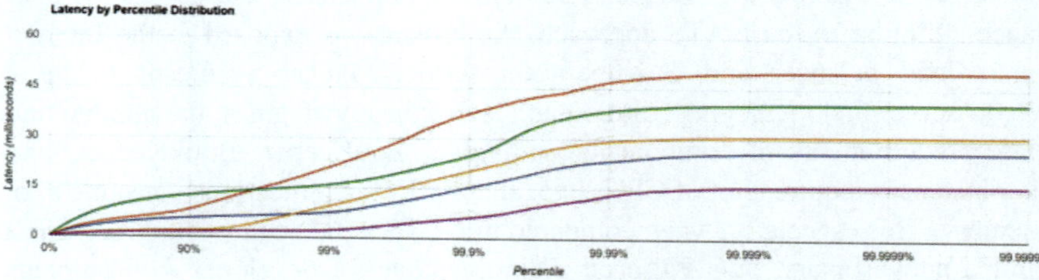

Fig. 5 NATS latency

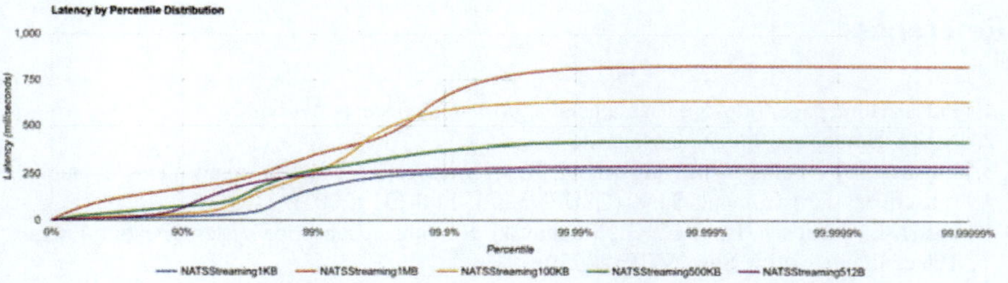

Fig. 6 NATS Streaming latency

Table 1 Throughput with different message sizes

Size	0.5 KB	1 KB	100 KiB	500 KiB	1 MiB
NATS	564.21	560.72	441.15	61.17	
NATS streaming	74.57	69.44	47.73	22.39	9.45
Apache Pulsar	102.85	103.1	63.35	21.14	13.23

5 Conclusion

The benchmarks show clear performance differences in the three systems, namely NATS (Core), NATS Streaming, and Apache Pulsar. Using file-based persistence to store messages and replication capabilities in the messaging system introduces a trade-off for both latency and throughput as writing to files is an expensive operation, which can be possibly improved by using faster disks in the form of storage-optimized instances in deployments. In our benchmarks, while NATS, obviously, was superior to NATS Streaming and Apache Pulsar in terms of throughput and latency. In a fairer comparison, NATS Streaming and Apache Pulsar can be seen to perform similarly in terms of the presented latency benchmark. NATS, extendable by NATS Streaming, is a very lightweight messaging option giving flexibility for both development and production environments, giving the choice between both NATS and NATS Streaming as requirements change, high throughput, low latency, wide client support (especially, first-class Go client support, in our case). Apache Pulsar is a heavyweight system designed for high performance server-to-server messaging systems with large amounts of resources, and this high resource requirement also provides expansive capabilities compared to NATS and NATS Streaming in the form of its serverless function framework Pulsar Functions, extensive topic partitioning and namespacing options, large-scale geo-replication support, and a scalable configuration due to its modular battle-tested components like ZooKeeper individually scalable at will.

References

1. Pulsar Homepage. https://pulsar.apache.org/docs/en/concepts-overview/
2. NATS Homepage. https://docs.nats.io
3. Eugster, P.T., Felber, P.A., Guerraoui, R., Kermarrec, A.-M.: The many faces of publish/subscribe. ACM Comput. Surv. (CSUR) **35**(2), 114–131 (2003)
4. Fidler, E., Jacobsen, H.A., Li, G., Mankovski, S.: Publish/subscribe system. Feature Interact. Telecommun. Softw. Syst. **VIII** (2005)
5. Maheshwari, P., Pang, M.: Benchmarking message-oriented middleware: TIB/RV versus SonicMQ. Concurrency Comput. Pract. Exp. **17**(12), 1507–1526 (2005)
6. Appel, S., Sachs, K., Buchmann, A.: Towards benchmarking ofAMQP. In: Proceedings of the Fourth ACM International Conference on Distributed Event-Based Systems, pp. 99–100 (2010)
7. Sachs, K., Appel, S., Kounev, S., Buchmann, A.: Benchmarking publish/subscribe-based messaging systems. In: International Conference on Database Systems for Advanced Applications, pp. 203–214. Springer, Berlin, Heidelberg (2010)
8. Sachs, K., Kounev, S., Appel, S., Buchmann, A.: Benchmarking of message-oriented middleware. In: Proceedings of the Third ACM International Conference on Distributed Event-Based Systems, pp. 1–2 (2009)
9. Sachs, K., Kounev, S., Bacon, J., Buchmann, A.: Performanceevaluation of message-oriented middleware using the SPECjms2007 benchmark. Perform. Eval. **66**(8), 410–434 (2009)
10. Sachs, K., Kounev, S., Appel, S., Buchmann, A.: A performance test harness for publish/subscribe middleware. In: Sigmetrics/Performance (2009)
11. Sachs, K., Kounev, S., Bacon, J., Buchmann, A.: Workload characterization of the SPECjms2007 benchmark. In: European Performance Engineering Workshop, pp. 228–244. Springer, Berlin, Heidelberg (2007)
12. Folkerts, E., Alexandrov, A., Sachs, K., Iosup, A., Markl, V., Tosun, C.: Benchmarking in the cloud: what it should, can, and cannotbe. In: Technology Conference on Performance Evaluation and Benchmarking, pp. 173–188. Springer, Berlin, Heidelberg (2012)
13. John, V., Liu, X.: A Survey of Distributed Message Broker Queues. arXivpreprint arXiv:1704.00411 (2017)
14. RabbitMQ Homepage. rabbitmq.com
15. CNCF Homepage. https://www.cncf.io/
16. Zookeeper Homepage. https://zookeeper.apache.org/
17. Bookkeeper Homepage. https://bookkeeper.apache.org/
18. Athenz IO. https://www.athenz.io/
19. Nats-Io Server Github. http://github.com/nats-io/nats-server
20. Protocol Buffers. https://developers.google.com/protocol-buffers
21. RAFT Consensus Algorithm Paper. https://raft.github.io/raft.pdf
22. Tylertreat GitHub. https://github.com/tylertreat/bench
23. Nats Streaming Server GitHub. https://github.com/nats-io/nats-streaming-server
24. Apache Pulsar GitHub. https://github.com/apache/pulsar
25. Nats-Io Github. https://github.com/nats-io/nats.go
26. Stan-Go GitHub. https://github.com/nats-io/stan.go

Problems of Providing Access to a Geographic Information System Processing Data of Different Degrees of Secrecy

Vitaly Gryzunov and Darina Gryzunova

Abstract Geographic information systems contain data of varying degrees of secrecy. Access to data is regulated by different documents. The data is loaded and processed in real time and booked in several hundred layers. On the other hand, geographic information systems are integrated into almost all existing information systems and tend to evolve into spatial data infrastructure. This circumstance imposes strict requirements on ensuring the integrity, confidentiality, and availability of data, makes it extremely difficult to make a decision on granting access of subjects to data, on the allocation of appropriate access rights. The article analyzes the problems of providing access to geographic information systems and suggests approaches to their solution.

Keywords Secure geographic information systems · Security of spatial data · Safe processing of spatial data · Geographic information system · Information security

1 Introduction

Geographic information systems penetrate all spheres of human activity: investigation of crimes and creation of safe conditions for the existence of the population [1], automation of delivery vehicles to consumers and the transition to Industry 4.0 [2, 3], the use of weapons, and other closed technologies.

Some authors believe that in the near future, all information systems will contain geographic information systems (GIS) or be based on them [4].

GIS is an information system operating with spatial data [5]. A high degree of GIS integration with other information systems, a variety of tasks to be solved, and

V. Gryzunov (✉)
Department of Information Technology and Security Systems, Russian State Hydrometeorological University, Saint-Petersburg, Russian Federation

D. Gryzunova
Legal Regulation of Economic Activity, Financial University Under the Government of the Russian Federation, Moscow, Russian Federation

K. Khanna et al. (eds.), *Cyber Security and Digital Forensics*,
Lecture Notes on Data Engineering and Communications Technologies 73,
https://doi.org/10.1007/978-981-16-3961-6_17

a large number of heterogeneous users suggests the following features of the GIS application:

- work in real time or close to real time;
- a persistence increase in the number of layers provided by technical systems of different ministries, departments, commercial companies, and personal devices of individual: today GIS has several hundred layers;
- layers available in GIS contain information with varying degrees of secrecy;
- GIS users have different rights to access GIS layers;
- use of cloud computing and application of fog computing [6];
- GIS elements are both stationary and mobile means;
- usually, GIS itself is not used and does not perform any actions, but provides data for analysis, allocation of money, geomonitoring, investigation of computer crimes, etc. [7, 8], so in fact, it is an element of the support and decision-making system [9, 10] or "smart card" [11].

These features allow us to say that GIS is being transformed into a flexible infrastructure of spatial data [12], continuously changing its structure and functions, which, firstly, opens up GIS for a wide range of users and dramatically increases the amount of processed data, and secondly, makes GIS an object of malicious actions on the part of computer hooligans, criminal gangs, special services, etc. Both of these destabilize the work of GIS, so reduces the likelihood of achieving the set goals and leads to financial, environmental, social, and moral damage [13]. The foregoing determines the relevance of issues of providing access to information and GIS services.

The purpose of this study is to analyze the existing problems of providing access to information in GIS and propose ways to solve them. The analysis is carried out on the basis of the current legislation of the Russian Federation and experience in the design, implementation, and operation of complex information systems.

Let's consider who exactly is the consumer of GIS services and what information the GIS works with.

2 Consumers of GIS Services and the Information that GIS Works With

The following information is processed in GIS with different requirements in access control: cartographic data; coordinates of objects and routes of movement; coordinate of users, search queries, preferences, etc. Then properties of objects: financing, cost, size, number, etc.; personal data of users of all categories; medical data on foci of diseases and rates of their spread; meteorological situation; data protected by copyright; other data.

Classification of GIS information regarding the access mode according to the legislation of the Russian Federation is shown in Fig. 1.

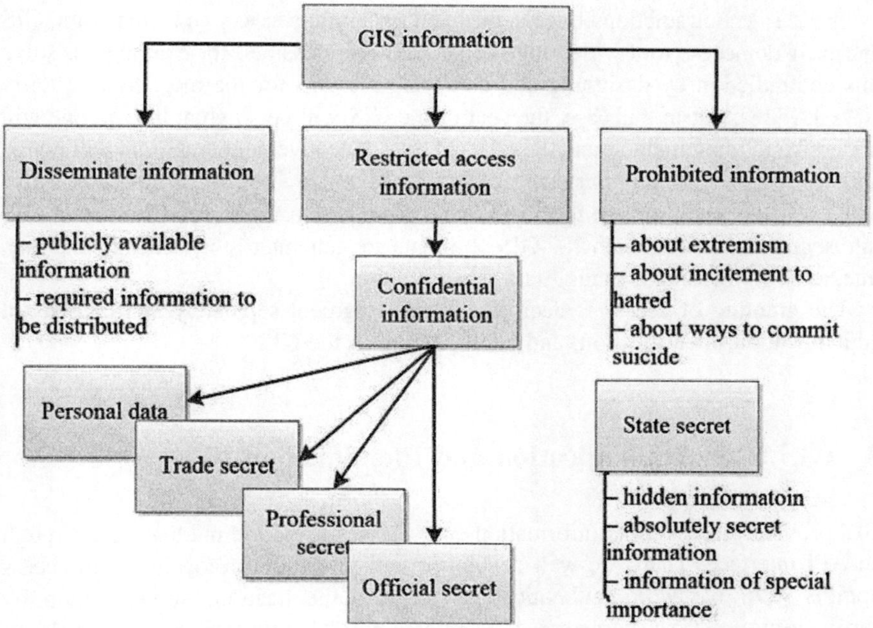

Fig. 1 Classification of GIS information by access mode

Each type of information has its own requirements for access control, which determines the specific requirements for GIS.

From the point of view of organizing information security and providing access to information, the following types of GIS are interesting: state and non-state organizations; public and private; commercial and free.

Information security in state GIS and public GIS is strictly regulated by regulatory documents.

A public information system is an information system in which participants in electronic interaction make up an indefinite circle of persons and the use of which cannot be denied to these persons [14].

Commercial GIS organize information security at their own discretion, if the processed information does not belong to government agencies.

Individuals only have access to the level of information security provided by the GIS owners.

Everyone can attack GIS to use it illegally [15].

In GIS as such, all information is processed, which means that for each specific GIS implementation, there will be its own set of information security threats, its own set of protective measures, its own implementation of the mechanism of the access control. For example, for GIS that fall under the definition of public information systems, the most important is accessibility [16]; for GIS containing information constituting a state secret, confidentiality is primarily required. It is important for all GIS to ensure the authenticity of the incoming data [17].

There is a contradiction between the need to delimit access to information in GIS and provide access to a wide range of people. It is possible, for example, to solve this contradiction by designing and certifying the GIS for the most rigid security class 1A [18]; but in this case, the cost of the GIS will be so great that no one will undertake to implement it, and the GIS will be so inconvenient that users will not be able to operate normally Services.

Classically, the problem is solved either from above, by forced division of GIS into segments, or from below,—GIS segments are generated stochastically and then integrated into a single spatial data infrastructure.

The granting of access is decided for each segment separately and consists of authentication, identification, and authorization in the GIS.

3 GIS User Authentication and Identification

GIS provides access to its information and services using a thin client and through an API interface. There are well-developed authentication protocols for all access options [19]: password authentication; forms authentication; authentication by certificates; one-time password authentication; authentication by access keys; authentication by tokens; using mandates; many others.

The main problem with all the described protocols is that they are suitable for a centralized system. This means that there is an element determining who can use the system and with what authority and who cannot.

If the GIS is segmented and there are several such centers, then relations of trust and delegation of authority are set up between them. One option is to create a GIS public key infrastructure and use certification authorities. In this case, the following methods of spreading trust between segments are available:

(1) determination of the main element in all GIS and the subordination of all others to this element;
(2) cross-certification. GIS segment owners decide with whom and what trust to establish. Moreover, if A trusts B, and B trusts C, then this does not mean that A trusts C. Or that C trusts B, or B trusts A. Each trust is established separately;
(3) bridge scheme. The central element is formally assigned. Technically, trust relationships are established by option 2 and legally by option 1;
(4) distribution of trust through CTL (certificate trusted list). Each segment has a master electronic key that determines who and what with authority can work in the system. Keys of all segments are collected in a single trusted list that is distributed across all GIS segments.

Schemes 1 and 3 have been tested and proved by authors [20] their worth when creating a network of trusted certification centers of the Federal Tax Service of Russia, the Pension Fund of the Russian Federation, electronic trading platforms, and customs. The FTS network included more than 400 certification centers and several million users. The main element of the network is the Ministry of Communications and Mass Media. It seems reasonable to apply the already worked out scheme to create the state infrastructure of spatial data.

It is worth noting that with the expansion of GIS to the international level, problems arise associated with the recognition of the countries participating in the GIS of the applied cryptographic information protection tools, as well as with the existing restrictions on the export of such tools in some countries.

Identification issues while using this scheme are also well-developed for a large number of users with different types of identifiers and do not entail any particular difficulties.

The problem is the identification of GIS services and information. The existing solutions for mapping resources or using attribute certificates and Privilege Management Infrastructure [21] seem to be rather cumbersome for practical implementation and require further study.

4 GIS Users Authorization

User authorization is to allow information access the user with requested permission or reject the access request, if the requested permission exceeds the allowed.

The access mode and, therefore, the degree of secrecy of information are determined by the user-owner of the information, with the exception of the following cases: required information to be distributed; prohibited information, here, often, we are not talking about protecting information from the user, but about protecting the user from information; personal data; professional secret.

For these cases, there is a specific list of what is a subject to protection in accordance with the requirements of current legislation.

This means that the GIS must have mechanisms for differentiating access for all types of information, connected: (1) as directed by the user; (2) automatically, upon detection of compulsory protected information.

In the first case, the responsibility for the differentiation of access lies with the user in terms of deciding on the access mode and on the GIS in terms of implementing the decision made by the user.

In the second case, a settling module is needed, where all the information is initially received, and then, a decision is made what type it belongs to, and its further processing is prohibited or allowed. And the information itself must have clear signs by which it can be attributed to some of the classes without human participation.

Access to information—the ability to obtain information and use it [22]. Classically, GIS provides two types of access: read and write.

The main models for controlling access and the possibilities of their application are analyzed in the works [23, 24]: (1) discretionary model; (2) mandate model; (3) role model; (4) attributed model.

Even with the ideal implementation of the access control model, situations of security breaches of GIS information are possible.

For example, the wrong choice of security policy in a GIS can contribute to a breach of confidentiality. The cultural heritage protection authorities need to preserve the confidentiality of the boundaries of territories in relation to which there are grounds to assume the presence of archeological heritage objects or objects with signs of an archeological heritage object in these territories [25]. And the GIS security policy is to issue open areas of the map without objects of archeological heritage and closed ones containing objects—not. An attacker can query the GIS for all open areas. And those map elements that the GIS does not provide contain some kind of closed objects.

The recommendations set out in [26] will help to avoid a large number of typical mistakes when creating GIS. This is some quintessence of the experience of information security specialists. A similar approach to creating secure GIS is used in other countries [27].

Some problems found during the operation of complex information systems (IS):

- tracking uncontrolled transfer of rights during the operation of IS;
- adding new users and resources;
- removal of unnecessary users and resources and the appointment of new owners;
- giving legal significance to the processed information;
- granting access to special services;
- appointment of administrators of the unified system;
- technical and organizational features of GIS safety control;
- GIS interoperability when integrating with other information systems [28];
- a change in the worth of information, entailing both an increase and a decrease in the secrecy level of the processed information.

5 Conclusion

GIS processes a wide variety of different information types. For some information types, the degrees and access mode is determined by the owner of the information, for some it should be done automatically. GIS should provide both options. Authentication and identification for a decentralized GIS are done using a public key infrastructure. The ideal implementation of access control models does not always guarantee the protection of information, so the development of new models is necessary. Many access control problems are solved by GIS segmentation.

Acknowledgements The reported study was funded by Russian Ministry of Science (information security), project № 08/2020.

References

1. Kudelkin, V.A., Denisov, V.F.: Experience of integration of distributed information systems. IT Standard **1**, 24–30 (2017)
2. Stoletov, O.V., Chikharev, I.A., Moskalenko, O.A., Makovskaya, D.V.: Geoinformation support of the mediterranean branch of the silk road. In: Geoinformatical and Cartographical Security of Ecological, Economic and Social Aspects of Sustainable Development, Part 25, vol. 1, pp. 102–113 (2019)
3. Schwab, K.: Globalization 4.0. new architecture for the fourth industrial revolution. Eurasian Integr. Econ. Law Polit **1**(27), 79–84 (2019)
4. Kolbanev, M.O., Palkin, I.I., Tatarnikova, T.M.: The challenges of the digital economy. Hydrometeorol. Ecol. Sci. Notes RSHMU **58**, 156–167 (2020)
5. GOST R 52438-2005 Geographical Information Systems Terms and Definitions
6. Gryzunov, V.V.: Dynamic aggregation of pools in military computing systems. Inf. Control Syst. **1**, 13–20 (2015)
7. Ananiev, Y.S.: Geoinformation Systems: Tutorial. TPU, Tomsk, Russia (2003)
8. Boltachev, E.F.: Visualization of data describing information technologies crimes. In: Crime in the Sphere of Information and Telecommunication Technologies: Problems of Prevention, Disclosure and Investigation of Crimes, vol. 1, no. 4, pp. 14–19 (2018)
9. Voronin, A.V., Zatsarinny, A.A.: Geoinformation system as the most important component of management decision making system. Highly Available Syst. Sistemy vysokoy dostupnosti **15**(3), 27–33 (2019)
10. Lisitsky, D.V., Katsko, Y.S.: User segment of a single territorial geoinformation space. Bull. SGUGiT **4**(36), 89–100 (2016)
11. Yankelevich, S.S., Radchenko, L.K., Antonov, Y.S.: From multi-purpose cartographic resource to "smart map". Bull. SGUGiT **23**(1), 142–155 (2018)
12. Koshkarev, A.V.: Laws and standards for the development of geomatics in the digital economy era. Vestnik of North-Eastern Federal University. Earth Sci. **3**(15), 46–54 (2019)
13. Nesterowskij, I.P., Yazov, Y.K.: Possible approach to assessment of damage suffered from a threat to security of information being processed in federal information systems. Cybersecur. Issues **2**(10), 20–25 (2021)
14. Federal Law of 06.04.2011 N 63-FZ (Revised 10.07.2012) "About the Electronic Signature"
15. Gryzunov, V.V.: Model of purpose aggressive actions on the information-computing system. In: Third International Conference on Human Factors in Complex Technical Systems and Environments (ERGO)s and Environments (ERGO), pp. 300–305 (2018) https://ieeexplore.ieee.org/document/8443814
16. Order of FSTEC of Russia of August 31, 2010 N 489 "About Protection of Information Contained in Public Information Systems"
17. Gryzunov, V.V., Ukraintseva, D.A.: Appearance of the data authenticity protection system of the geoinformation system. In: Information Security of Russian Regions. Materials of the XI St. Petersburg Interregional Conference, pp. 172–173 (2019). http://spoisu.ru/files/ibrr/ibrr2019/ibrr2019_materials.pdf
18. GOST R 50922-2006. Information Security Basic Terms and Definitions
19. Overview of Authentication Methods and Protocols in Web Applications. https://habr.com/ru/company/dataart/blog/262817/
20. Certificate of Registration of a Computer Program of the Russian Federation. No. 20077613135. Software Complex Information Resource of the Certification Center ("IRUC")/V.V. Gryzunov (RF), G.M. Dranishnikov (RF), A.V. Kirpichnikov (RF), A.S. Sotenko (RF). No. 2007612892; Declared 07/12/2007; Registered July 25, 2007
21. Sazonov, A.V.: Subject and technology rights management infrastructure in transboundary space. Inf. Secur. Issues **3**(98), 83–87 (2012)

22. Guidance Document. Automated Systems. Security Against Unauthorized Access to Information. Classification of Automated Systems and Requirements for Information Protection. Approved by the Decision of the Chairman of the State Technical Commission Under the President of the Russian Federation of March 30, 1992
23. Overview and Comparison of Existing Access Control Methods. Institute of Computing Technologies SB RAS [Electronic Resource]. http://www.ict.nsc.ru/ws/YM2003/6312/
24. Kalimoldayev, M.N., Biyashev, R.G., Rog, O.A.: Analysis of the Methods for Attribute-Based Access Control. https://doi.org/10.17223/20710410/44/4
25. Asset 9. Federal Law No. 73-FZ of 25.06.2002 (Revised 24.04.2020) "About Objects of Cultural Heritage (Historical and Cultural Monuments) of the Peoples of the Russian Federation"
26. FSTEC. Methodology for Determining Threats to Information Security in Information Systems
27. Ageev, V.O., Shilov, A.K.: Protection sis in the foreign and domestic system. In: VI All-Russian School-Seminar 'Perspective-2015', pp. 313–315, Russia (2015)
28. GOST R 55062-2012. Information Technologies (it). Industrial Automation Systems and Their Integration. Interoperability. Fundamentals

Security Augmented Symmetric Optical Image Cryptosystem Based on Hybrid Transform Employing Rear Mounted Technique Using Three Different Complex Masks

Priyanka Maan, Hukum Singh, and A. Charan Kumari

Abstract This paper presents the utilization of the novel rear mounted triple phase masking procedure to increase the security of traditional double random phase encoding (DRPE) scheme with a variation of hybrid transform. Two chaotic random phase masks and a deterministic phase mask are used in the scheme to enhance its security as opposed to traditional DRPE which uses random phase masks. From the previous studies on cryptanalysis of DRPE, it is clear that the second lens used in it proves to be irrational which leads to cryptanalytic attacks on DRPE. The enhanced framework discards the invalidation of the second lens and therefore reinforces DRPE security. The hybrid transform is a combination of fractional Hartley transform and Gyrator transform that improves the quality of the cryptosystem. The carried out simulations exhibit the efficacy of the designed symmetric cryptosystem.

Keywords Chaotic random phase mask · Deterministic phase mask · Fractional Hartley transform · Gyrator transform · Rear mounted triple phase masking

P. Maan (✉)
Department of Computer Science Engineering, The NorthCap University, Sector 23 A, Gurugram 122017, India

H. Singh
Department of Applied Sciences, The NorthCap University, Gurugram, India
e-mail: hukumsingh@ncuindia.edu

A. Charan Kumari
Department of Electrical Engineering, Faculty of Engineering, Dayalbagh Educational Institute, Agra 282005, India

K. Khanna et al. (eds.), *Cyber Security and Digital Forensics*,
Lecture Notes on Data Engineering and Communications Technologies 73,
https://doi.org/10.1007/978-981-16-3961-6_18

1 Introduction

In data security, cryptography, which scrambles the private and secret data into an encrypted form before transmission to keep away from data revelation, plays a critical role. Due to inherent advantages like multi-dimension encryption qualities, high parallel processing and speed optical cryptosystems are preferred over the digital ones [1–3]. DRPE is a traditional run of the optical image encryption strategy that uses two random phase masks (RPM) for the encoding of an image into a stationary noise [4]. From there on, various variations of the DRPE scheme came in picture based on different transforms and different masks. Every scheme has its own highlights and challenges [5, 6]. The cryptanalysis study of DRPE discovered that it is vulnerable to several attacks like known plaintext, chosen plaintext, chosen cipher text, etc. So to strengthen the security perspective of the DRPE technique, various advancements come in picture like fully phase encoding [7, 8], chaos-based mask generation [9, 10], pixel randomization, extending DRPE in different transforms domain like fractional Fourier transform (FrFT) [11, 12], Fresnel transform (FrT) [13, 14], Gyrator transform (GT) [15, 16], etc., and various asymmetric schemes [17, 18]. Some schemes also make use of combination of different transforms and some new additional techniques like singular value decomposition (SVD), equal modulus decomposition (EMD), etc., for performing image encryption. Later, the encryption is performed on phase images [19, 20] while some use the technique of performing encryption in spectrum domain on spectral images [21] and other uses sandwiched phase code to diffuse random phase masks to perform encryption and decryption [22]. Various researchers have also presented the use of logistic and chaotic map in the field of image encryption [23]. But dismally, these proposed schemes have their drawbacks too and can be cryptanalyzed like DRPE is recently cryptanalyzed using deep learning technique [24]. As known, DRPE is the elementary and the simplest technique, enhancing its security with lesser complexity is yet an appealing issue.

In this paper, the security improvement of the DRPE is strengthened by employing rear mounted triple phase masking technique in the fractional Hartley transform (FHT) and Gyrator transform domain as the encryption strategy and also to overcome the invalidation of second lens in the traditional DRPE. As identified [25, 26], the invalidation of second lens used in Fourier domain in DRPE leads to the cryptanalysis of the encrypted image. So to overcome the effect of this concept, the introduction of one more phase mask called rear mounted phase mask at the output plane is proposed in the discipline of hybrid transform-based DRPE. The three masks leading to triple phase masking scheme are two chaotic random phase masks and one deterministic phase mask which leads to strengthening the security of the proposed method. The proposed algorithm extends the security by exploiting the potential of second lens by adding an extra layer of rear mounted mask. The three complex masks used in the scheme are much more secure and prove to be a great advantage in enhancing its security potential as they are difficult to regenerate without proper information than the random phase mask used in the traditional

DRPE scheme. The use of fractional Hartley transform and Gyrator transform has a reason of providing computational ease and convenient implementation optically. Simulation results provide the validation and efficacy of the proposed cryptosystem.

2 Mathematical Background

The proposed system uses chaotic random phase mask, deterministic phase mask in the FHT, and GT domain employing rear mounted scheme.

2.1 Chaotic Random Phase Mask

Chaotic random phase mask can be formed with the combination of random phase mask and chaotic function. RPMs are used as a secretive key in DRPE-based image encryption. Chaos theory indicates that when a small error in the input calculations can generate a largest possible error in the output, then it can be classified as a chaotic function. Chaotic function portrays nonlinear framework from organized to confused state. A chaotic map is a map that displays some kind of chaotic behavior. Chaotic map can generate tremendous number of random iterative values exhibiting ergodicity, non-correlation, and pseudo randomness kind of nature. RPM can also be likewise built from chaotic function.

A logistic map can be defined as a one-dimensional nonlinear chaos function using the equation below:

$$f(x) = bx(1 - x) \tag{1}$$

where b represents the bifurcation parameter having value in the range $0 < p < 4$. The iterative version of Eq. (1) can be asserted as:

$$x_{n+1} = bx_n(1 - x_n) \tag{2}$$

where x_0 represents the initial value and x_n represents the iterative value with the range $x_n \in [0,1]$.

For generating chaotic random sequence to be use in the encryption process, we first need one-dimensional random value sequence. Using Eq. (2), the one-dimensional random value sequence can be depicted as:

$$X = \left\{x_1, x_2, x_3 \ldots x_{(M \times N)+k}\right\} \tag{3}$$

where k represents any chosen value of integers and $x_i \in (0,1)$. As images are a two-dimensional function, so the two-dimensional sequence can be generated by rearranging the X as Y which can be depicted as:

$$Y = \{y_{i,j}, | i = 1, 2, \ldots, M + k, j = 1, 2, \ldots, N + k\} \tag{4}$$

where $y_{i,j} \in (0,1)$. Now, the CRPM can be finally generated using the equation below:

$$\text{CRPM}(x, y) = \exp[i2\pi y_{i,j}(x, y)] \tag{5}$$

where (x, y) represents the coordinated of CRPM. CRPM comprises of three keys, namely $x_{0,}$ b and n. In the proposed system, two CRPMs are generated CRPM1 and CRPM2 which are depicted in Fig. 3a, b.

2.2 Deterministic Phase Mask

In the proposed system, we have used a deterministic phase mask as the rear mounted key. To generate a DPM [27, 28], firstly we need to set the value of order of encryption parameter 'm' whose value can be m = 2, 3, 4. Now, we decide the number of subkeys (NS) using the parameter m with the help of equation below.

$$\text{NS} = 2^m \times 2^m \tag{6}$$

Then, we split the input image in NS equal sub-blocks. The size of each sub-block can be represented using parameter 'd' and can be calculated using the equation below.

$$d = \frac{\text{dim}}{2^m} \tag{7}$$

where dim represents input image size.

DPM can now be generated using the continuous combination of sub-masks ($M_{i,j}$) using the following equation:

$$\text{DPM} = \sum_{i=1}^{2^m} \sum_{j=1}^{2^m} M_{i,j}(d \times d) \tag{8}$$

where $M_{i,j}$ represents the number of sub-blocks required for the generation of DPM. $M_{i,j}$ can be defined as:

$$M_{i,j}(p, q) = \exp[i2\pi(u_k.p + v_k.q)] \tag{9}$$

where u_k and v_k range lies between $[1, d]$ in which they are randomly generated. The value of k lies in the given interval [1, 2^m]. In our proposed scheme, we have used $m = 2$ for deterministic phase mask generation and for simplicity of the scheme. Figure 3c depicts the DPM used.

3 Methodology

Based on the cryptanalytic attacks possible on the elementary DRPE scheme, an upgraded hybrid transform-based DRPE cryptosystem in the light of rear mounted triple phase masking has been proposed. A supplementary phase mask called rear mounted phase mask has been used which is characterized as DPM, i.e., deterministic phase mask and the other two random phase masks in traditional DRPE has been replaced with chaotic random phase masks, namely CRPM1 and CRPM2. DPM fills in as extra secret key and is used to phase modulate the cipher data again that is generated after the application of hybrid transform-based DRPE. Equation (10) below shows the output of the encryption scheme. The decryption is the reverse of the encryption procedure, and Eq. (11) given below depicts the calculation of the decryption output.

$$C(x, y) = \mathrm{DPM.GT}^{-\alpha}(\mathrm{CRPM2.FHT}(p, q)(I(x, y).\mathrm{CRPM1})) \tag{10}$$

$$I(x, y) = \mathrm{CRPM1}^{*}.\mathrm{FHT}(-p, -q)(\mathrm{CRPM2}^{*}.\mathrm{GT}^{\alpha}(C(x, y).\mathrm{DPM}^{*})) \tag{11}$$

where $I(x, y)$ and $C(x, y)$ are the original image and cipher image, respectively. * denotes the conjugate operation. DPM denotes the deterministic phase mask, and CRPM1 and CRPM2 denote the two chaotic random phase masks. GT^{α} represents Gyrator transform with α as rotation angle [14]. FHT(p, q) denotes the fractional Hartley transform with two order parameters p and q.

The encryption and decryption mechanism has been depicted in Fig. 1a, b, respectively. During encryption, the input image $I(x, y)$ is introduced to the encryption algorithm, and the final output we get is represented by $C(x, y)$. Decryption process is the reverse of the encrypted one, where we apply the decryption algorithm on the encrypted image $C(x, y)$ which has been sent to the receiver after encryption. Firstly, we multiple $C(x, y)$ to the conjugate of rear mounted phase mask DPM*, and then the product is subjected to Gyrator transform with rotation angle α. Then we multiply the resultant with the conjugate of second chaotic random phase mask CRPM2*. On the output of this step, we apply fractional Hartley transform with parameters ($-p$, $-$ q). Finally to get the original image $I(x, y)$ back, we subject the resultant of previous step to the conjugate of first chaotic random phase mask CRPM1*. In the presented paper, the security potential of the second lens used in the DRPE has been enhanced by the introduction of one more mask at the output plane for the prevention of the realization of the cipher.

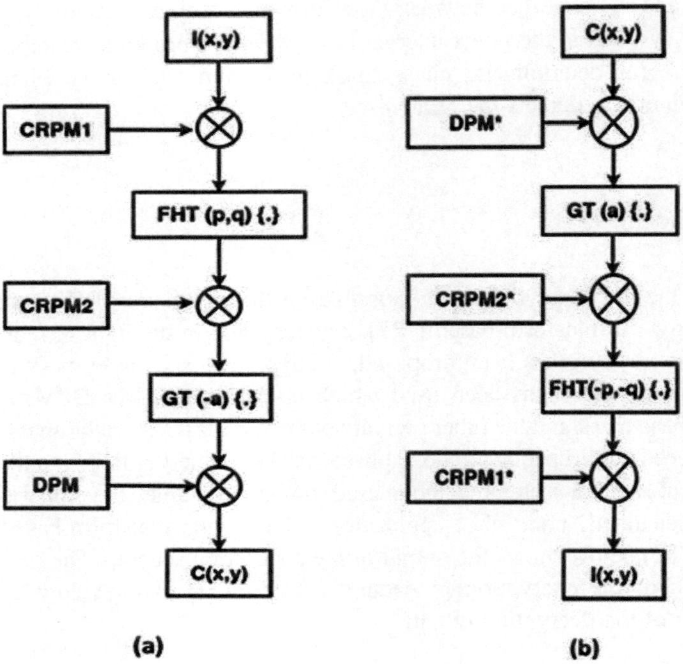

Fig. 1 **a** Encryption flowchart; **b** decryption flowchart

4 Simulation Results

MATLAB (R 2019a) 5(9.6.0.117 49 12), L.4066 4799 has been used to carry out the simulation, and the results have been interpreted. Figure 2a depicts the original input image 'Peppers.png,' a gray scale image with size 256 × 256 which is to be encoded. Figure 2b, c shows the encrypted and decrypted images with respect to the original image by employing the presented scheme. After applying encryption algorithm, we get the encrypted image which is a random image from which it is difficult to identify the original image corresponding to it. After applying decryption algorithm with correct parameters, we get the decrypted image. Figure 3a–c depicts the masks that act as the keys during the encryption and decryption process. The presented scheme includes three keys as phase masks, i.e., two chaotic random phase masks (CRPM1, CRPM2) and one deterministic phase mask (DPM) which acts as rear mounted phase mask.

Figure 4a–c depicts the 3D views of original, encrypted, and decrypted images.

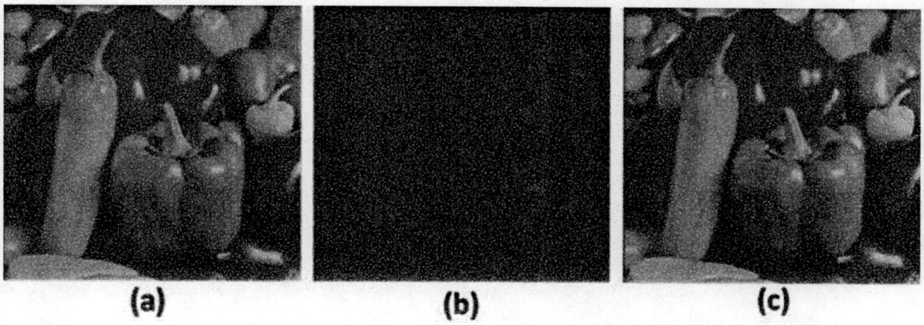

Fig. 2 **a** Original image; **b** encrypted image; **c** decrypted image

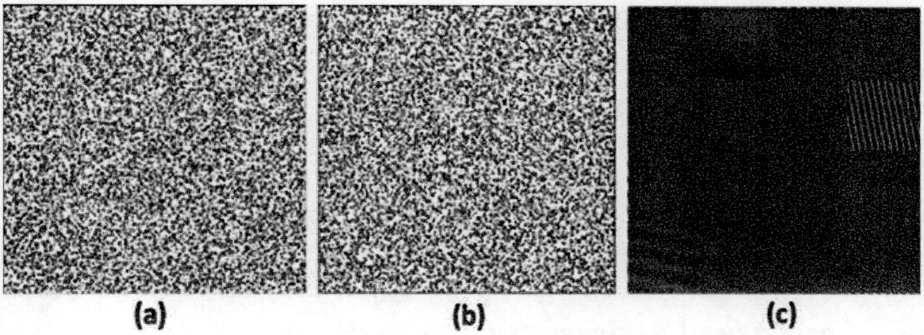

Fig. 3 **a** CRPM1; **b** CRPM2; **c** rear mounted phase mask (DPM)

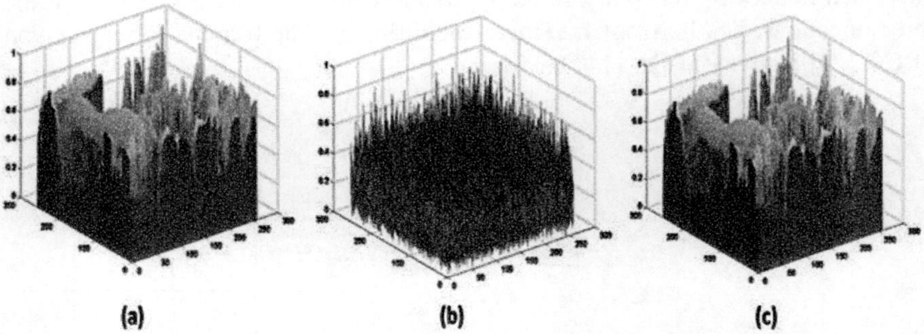

Fig. 4 3D views of **a** peppers image; **b** encrypted image; **c** decrypted image

5 Performance and Security Analysis

5.1 Error Analysis

To check the competency and strength of the proposed algorithm, we calculate the mean square error (MSE) and peak signal to noise ratio (PSNR) [18] values using the formula are given below:

$$\text{MSE} = \frac{1}{M \times N} \sum_{x=1}^{N} \sum_{y=1}^{N} |O(x, y) - D(x, y)|^2 \tag{12}$$

$$\text{PSNR} = 10 \times \log_{10}\left(\frac{(N-1)^2}{\text{MSE}}\right) \tag{13}$$

where M and N are the length and width of the images. $O(x, y)$ is the original input image, and $D(x, y)$ is the decrypted image. MSE and PSNR values calculated using the given scheme are 1.50×10^{-24} and 286.34, respectively. Figure 5a, b depicts the plot of MSE and PSNR curves versus fractional order.

5.2 Correlation Coefficient (CC) Analysis

The correlation coefficient analysis has been done for both the original and encrypted images by randomly selecting 10,000 pairs of adjacent pixels in all the three directions, i.e., horizontal, vertical and diagonal. The formula for calculation of CC [28] is given in Eq. (14) below:

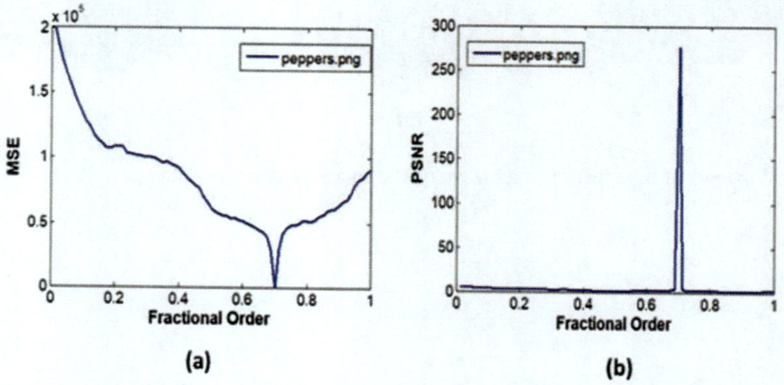

Fig. 5 **a** MSE versus fractional order plot; **b** PSNR versus fractional order plot

$$\mathrm{CC} = \frac{\sum_{i=1}^{N}(x_i - \bar{x})(y_i - \bar{y})}{\sqrt{\left(\sum_{i=1}^{N}(x_i - \bar{x})^2\right)\left(\sum_{i=1}^{N}(y_i - \bar{y})^2\right)}} \tag{14}$$

$$\bar{x} = \frac{1}{N}\sum_{i=1}^{N} x_i \tag{15}$$

$$\bar{y} = \frac{1}{N}\sum_{i=1}^{N} y_i \tag{16}$$

where $\bar{x}$ and $\bar{y}$ represent the mean values of x_i and y_i, respectively. Table 1 and Fig. 6 represent the CC analysis done for the original and encrypted images.

The CC distribution values of encrypted image in all the three directions as given in the table below show that they are very less correlated as compared to the values of the original image. The scatter plot of encrypted image also depicts that it has random distribution, and after encryption, it is difficult for intruder to identify the original image.

5.3 Key Sensitivity Analysis

To check the strength and robustness of the proposed algorithm, sensitivity analysis has been done for different parameters. Figure 7a displays the output when we use wrong second mask, i.e., CRPM2 for decryption. If we exchange the first and second chaotic random phase masks during decryption, then we get the output as displayed in Fig. 7b. When we use wrong rear mounted phase mask, i.e., DPM for decryption, then also we are not able to get the original image while performing decryption as shown in Fig. 7c. If the rotation angle of Gyrator transform used is wrong, i.e., $\alpha = 0.05$ * pi instead of the original one, i.e., $\alpha = 0.05$ * pi, then we get the decrypted output as shown in Fig. 7d. When we take the wrong values for order parameters (p, q) of fractional Hartley transform, then we get the result as shown in Fig. 7e, f. The original values used in the presented scheme are $p = 0.7$ and $q = 0.4$.

From the analysis, it is positively implicit that the use of wrong parameters produces wrong results. The key space of the presented scheme constitutes rotation

Table 1 Horizontal, vertical, and diagonal directions values of correlation coefficient of original and encrypted image

Image	Original image			Encrypted image		
	Horizontal	Vertical	Diagonal	Horizontal	Vertical	Diagonal
Peppers	0.9398	0.8869	0.8274	0.0247	0.0825	0.0575

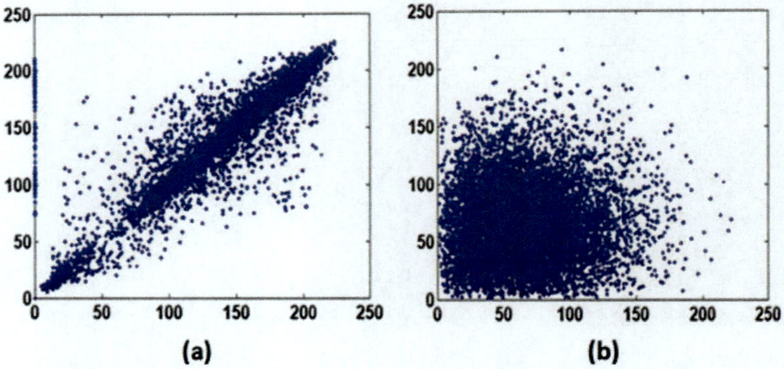

Fig. 6 **a** Correlation distribution of original image; **b** correlation distribution of encrypted image

angle (a) of Gyrator transform, order parameters (p, q) of fractional Hartley transform, and 3 phase masks used. It is evident from the above discussion that even the slight variation in these key parameters produces scrambled result from which we even cannot identify a part of the original image. Thus, it proves the robustness of proposed framework.

6 Conclusion

A security enriched image encryption method in light of fractional Hartley and Gyrator-based hybrid transform utilizing rear mounted triple phase masking has been proposed to overcome the invalidation of second lens in DRPE. The key space constitutes: rotation angle of Gyrator transform (a), 2 order parameters (p, q) of fractional Hartley transform, 2 logistic map parameters, and 3 masks, i.e., CRPM1, CRPM2, and DPM. The enlarged key space resists the brute-force attack on the proposed algorithm. The utilization of DPM as a mask in the third layer of cryptosystem enhances the security potential of the presented scheme. The usage of complex masks and the enhancement in the number of security parameters of the stated algorithm makes it immune against chosen plaintext attack, known plaintext attack, etc. The scheme has been verified on different factors like correlation coefficient analysis, MSE, and PSNR to check its adequacy and strength. The key sensitivity analysis and numerical simulation results mark the security enhancement of the scheme in comparison to the beforehand existing schemes. The developed cryptosystem has an advantage of being a symmetric one which makes it less complex and fast in comparison to the asymmetric ones. This technique highlights the use of rear mounted triple phase masking to build the key space, enhancing the security potential of second lens and makes the scheme troublesome for an eavesdropper to recuperate the original image which affirms the security improvement.

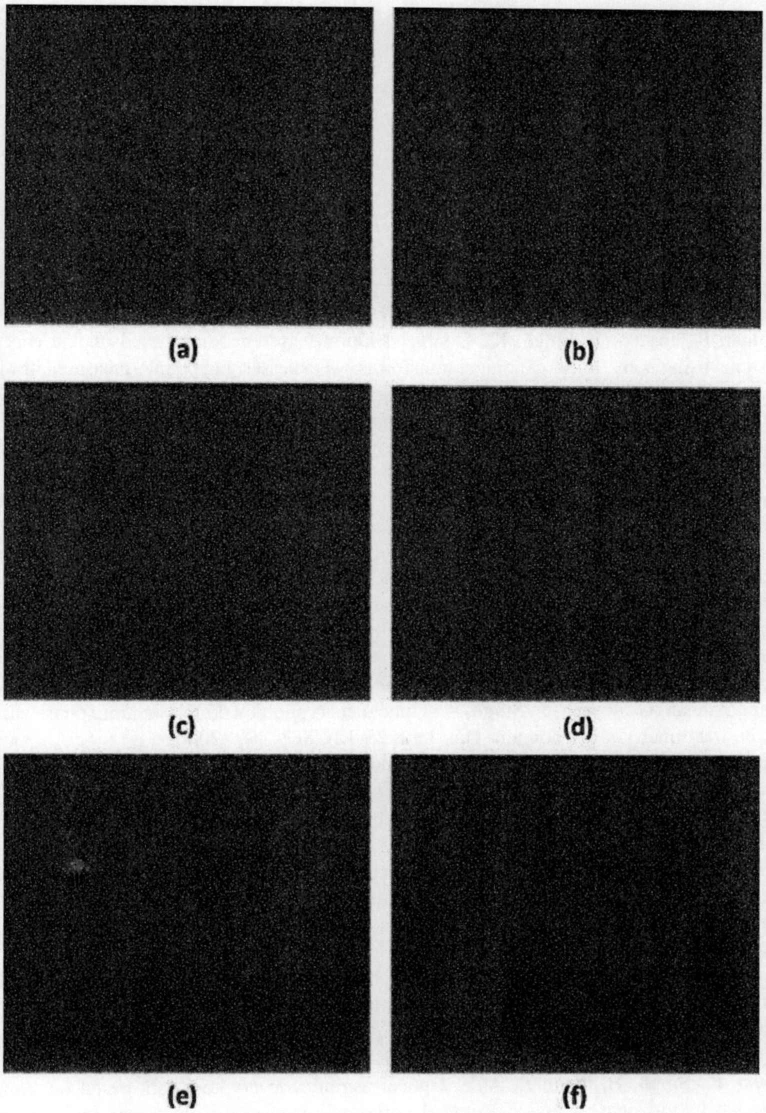

Fig. 7 **a** When second mask (CRPM2) is wrong; **b** when both CRPM are exchanged; **c** when DPM is wrong; **d** when Gyrator angle is wrong ($\alpha = 0.05 * \text{pi}$); **e** when $p = 0.2$ first fractional order is wrong; **f** when $q = 0.2$ s fractional order is wrong

References

1. Matoba, O., Nomura, T., Perez-Cabre, E., Millan, M.S., Javidi, B.: Optical techniques for information security. Proc. IEEE **97**(6), 1128–1148 (2009)
2. Chen, W., Javidi, B., Chen, X.: Advances in optical security systems. Adv. Opt. Photonics **6** (2), 120–155 (2014)
3. Javidi, B., Carnicer, A., Yamaguchi, M., Nomura, T., Pérez-Cabré, E., Millán, M.S., Nishchal, N.K., Torroba, R., Barrera, J.F., He, W., Peng, X.: Roadmap on optical security. J. Opt. **18**(8), 083001 (2016)
4. Refregier, P., Javidi, B.: Optical image encryption based on input plane and Fourier plane random encoding. Opt. Lett. **20**(7), 767–769 (1995)
5. Kumar, P., Joseph, J., Singh, K.: Double random phase encoding based optical encryption systems using some linear canonical transforms: weaknesses and countermeasures. In: Linear Canonical Transforms, pp. 367–396. Springer (2016)
6. Maan, P., Singh, H.: A survey on the applicability of Fourier transform and fractional Fourier transform on various problems of image encryption. In: Communication and Computing Systems, pp. 475–480 (2017)
7. Singh, H., Yadav, A.K., Vashisth, S., Singh, K.: Fully phase image encryption using double random-structured phase masks in gyrator domain. Appl. Opt. **53**(28), 6472–6481 (2014)
8. Nishchal, N.K., Joseph, J., Singh, K.: Fully phase encryption using fractional Fourier transform. Opt. Eng. **42**(6), 1583–1589 (2003)
9. Abuturab, M.R.: Securing multiple information using chaotic spiral phase encoding with simultaneous interference and superposition methods. Opt. Lasers Eng. **98**, 1–16 (2017)
10. Enayatifar, R., Abdullah, A.H., Isnin, I.F.: Chaos-based image encryption using a hybrid genetic algorithm and a DNA sequence. Opt. Lasers Eng. **56**, 83–93 (2014)
11. Unnikrishnan, G., Joseph, J., Singh, K.: Optical encryption by double-random phase encoding in the fractional Fourier domain. Opt. Lett. **25**(12), 887–889 (2000)
12. Tao, R., Xin, Y., Wang, Y.: Double image encryption based on random phase encoding in the fractional Fourier domain. Opt. Express **15**(24), 16067–16079 (2007)
13. Situ, G., Zhang, J.: Double random-phase encoding in the Fresnel domain. Opt. Lett. **29**(14), 1584–1586 (2004)
14. Khurana, M., Singh, H.: Asymmetric optical image triple masking encryption based on Gyrator and Fresnel transforms to remove silhouette problem. 3D Res. **9**(3), 1–17 (2018)
15. Rodrigo, J.A., Alieva, T., Calva, N.L.: Gyrator transform: properties and applications. Opt. Express **15**, 279–284 (2007)
16. Singh, H.: Hybrid structured phase mask in frequency plane for optical double image encryption in gyrator transform domain. J. Modern Opt. **65**(18), 2065–2078 (2018)
17. Maan, P., Singh, H.: Non-linear cryptosystem for image encryption using radial Hilbert mask in fractional Fourier transform domain. 3D Res. **9**(4), 1–12 (2018)
18. Maan, P., Singh, H., Kumari, A.C.: Optical asymmetric cryptosystem based on kronecker product, hybrid phase mask and optical vortex phase masks in the phase truncated hybrid transform domain. 3D Res. **10**(1), 1–18 (2019)
19. Singh, P., Yadav, A.K., Singh, K.: Phase image encryption in the fractional Hartley domain using Arnold transform and singular value decomposition. Opt. Lasers Eng. **91**, 187–195 (2017)
20. Chen, H., Du, X., Liu, Z.: Optical hyperspectral data encryption in spectrum domain by using 3D Arnold and gyrator transforms. Spectroscopy Lett. **49**(2), 103–107 (2016)
21. Han, C.: An image encryption algorithm based on modified logistic chaotic map. Optik **181**, 779–785 (2019)
22. Hasheminejad, A., Rostami, M.J.: A novel bit level multiphase algorithm for image encryption based on PWLCM chaotic map. Optik **184**, 205–213 (2019)

23. Rakheja, P., Vig, R., Singh, P.: An asymmetric hybrid cryptosystem using hyperchaotic system and random decomposition in hybrid multi resolution wavelet domain. Multimedia Tools Appl. 1–26 (2019)
24. Hai, H., Pan, S., Liao, M., Lu, D., He, W., Peng, X.: Cryptanalysis of random-phase-encoding-based optical cryptosystem via deep learning. Opt. Express **27**(15), 21204–21213 (2019)
25. Chen, J., Zhang, Y., Li, J., Zhang, L.B.: Security enhancement of double random phase encoding using rear-mounted phase masking. Opt. Lasers Eng. **101**, 51–59 (2018)
26. Khurana, M., Singh, H.: A spiral-phase rear mounted triple masking for secure optical image encryption based on gyrator transform. Recent Pat. Comput. Sci. **12**(2), 80–94 (2019)
27. Zamrani, W., Ahouzi, E., Lizana, A., Campos, J., Yzuel, M.J.: Optical image encryption technique based on deterministic phase masks. Opt. Eng. **55**(10), 1–9 (2016)
28. Girija, R., Singh, H.: A cryptosystem based on deterministic phase masks and fractional Fourier transform deploying singular value decomposition. Opt. Quantum Electron. **50**(5), 1–24 (2018)

Security Considerations in the Design of IEEE 802.15.4 Transceiver: A Review

K. Vivek Raj, P. Dinesha, and S. I. Arpitha Shankar

Abstract As internet of things (IoT) is extending internet connectivity beyond standard devices; the secure data transmission between IoT devices becomes more challenging. However, most of the traditional upper layer security schemes are computationally complex and increase the latency. Moreover, the security provided at upper layer is software implemented, and its strength depends on complexity of the encryption algorithm. This is a bottleneck situation for low-power IoT applications. As IoT uses many remote sensors which operate on battery power; IEEE 802.15.4 standard is gaining attention because of low-power consumption. An 802.15.4 protocol defines medium access control (MAC) and physical layer (PHY) specifications, and is designed to allow low-power, low-cost short range communication. Basic encryption and authentication in 802.15.4 are provided by link layer. Hence, considerable attention is required to study 802.15.4 specifications that can be used to provide alternative methods of security. So this paper is motivated to study the importance of 802.15.4 PHY. Firstly, we review the different physical layer security (PLS) schemes. Secondly, we present the concept of physical layer encryption (PLE) and further we analyze and compare the implementation of different PLE schemes in wireless standards. Later, we will try to give insights of 802.15.4 security standards and bring out the drawbacks of AES based link-layer security. Finally, we present the design and implementation aspects of 802.15.4 transceiver hardware architecture by considering performance and security.

K. Vivek Raj (✉)
Department of Electronics and Telecommunication Engineering, Dayananda Sagar College of Engineering, Bengaluru, Karnataka 560078, India

P. Dinesha
Department of Electronics and Communication Engineering, Dayananda Sagar College of Engineering, Bengaluru, Karnataka 560078, India

S. I. Arpitha Shankar
Department of Electronics and Telecommunication, GSSS Institute of Engineering and Technology for Women, Mysore, Karnataka 570016, India

K. Khanna et al. (eds.), *Cyber Security and Digital Forensics*,
Lecture Notes on Data Engineering and Communications Technologies 73,
https://doi.org/10.1007/978-981-16-3961-6_19

Keywords IEEE 802.15.4 · Computational security · Physical layer security · Internet of things · Physical layer encryption · Transceiver · Medium access control · Authentication

1 Introduction

In recent years, wireless communication is advancing with respect to its infrastructure and services with an aim of meeting rapidly growing demands. Wireless networks are being used for many applications. The International telecommunications Union [ITU] facts and figures in 2020 indicates that in 104 countries, half of the population is now using internet of which 70% are the youth [1]. Extending internet connectivity beyond conventional mobile phones and computers leads to evolution of new technology called internet of things. Currently, IoT is gaining more importance and it is playing important role in most of the fields such as health monitoring, automation, e-cities and so forth. However, secure data communication on IoT technology is more challenging scenario due to many constraints like power consumption, less storage memory and availability of processing power etc. [2]. Due to this, existing cryptographic techniques that are provided at the upper layer are not suitable because of their power consumption and the complexity in implementation. The IEEE 802.15.4 is designed to provide very low-cost communication between nearby devices in wireless personal area network (WPAN). Low-power wireless connectivity of 802.15.4 attracts wide variety of applications especially which operates on battery power. Since IoT employs different sensors which rely on battery, the use of 802.15.4 becomes most common in IoT applications. The structure of 802.15.4 protocol stack is shown in Fig. 1. It defines PHY and MAC layer specifications. PHY sits at the bottom of the protocol stack and transmit data packets. Direct sequence spread spectrum (DSSS) allows two physical layer configurations one works at 868/915 MHz and provides data rate of 20–40 kbps, second operates at 2.4 GHz with 250 kbps.

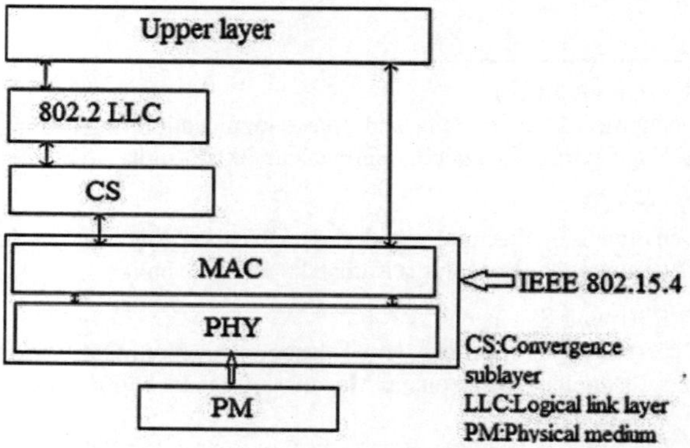

Fig. 1 IEEE 802.15.4 protocol stack

However, further revision in the standard allows the use of different modulation schemes. MAC layer performs different functions including channel access, frame delivery, synchronization, flow control, error control and it also provides interface between PHY and application layer. 802.15.4 is gaining importance as demand for IoT is increasing and security becomes a major concern. The PHY security technology combines the communication PHY and the security layer; hence, it is important to study the fundamental ability of the 802.15.4 PHY to offer secure wireless communication. In recent times, PLS and PLE are in much attention. Both of these approaches exploit the properties and characteristics of the PHY to provide the security. PLS provide information-theoretic security that is impossible to break by computing power. On the other hand, PLE is based on computational complexity and uses distribution of secret key. Unlike conventional cryptography, encryption in PLE is performed during the modulation process. That is in PLE, the security is provided at signal level. Further, PLE takes advantage of effect of the channel and noise. Hence, it increases the strength of underlying algorithm as it makes difficult in receiving the encrypted text itself. However, there is a clear difference between PLS and PLE.

Some of the characteristics which differentiates PLE from PLS are listed below.

1. Unlike PLS, PLE can guarantee secrecy, even if the channel capacity of eavesdropper is better than legitimate channel.
2. PLE has low-computational complexity, provides low latency with minimum power consumption and provides long lifetime hence it is best suited for IoT applications.
3. PLE provides security at signal level. It uses Boolean algebra therefore it gives much more functions to design and construct secure encryption algorithm.
4. In contrast to PLS which only focuses on eavesdropping attack, PLE uses randomness function which can be used to prevent linear and plaintext attacks.

Since our focus is on 802.15.4 protocol, further we concentrate more on the security aspects in the design of 802.15.4 transceiver. Table 1 provides the comparison between PLS, traditional cryptography and PLE and shows that PLE is different than PLS and conventional cryptography.

Table 1 Comparison of PLS, cryptography and PLE

Type	Distribution of key	Type of channel	Type of security	Operate domain
PLS	NO	Noise channel	Prefect secrecy, depends on channel	Complex vector
Traditional cryptography	YES	Perfect channel	Computationally secure	Bit level
PLE	YES	Noise channel	Computationally secure, uses characteristics of channel	Complex vector and signal level

Most of the available upper layer encryption schemes are computationally complex and increases the latency and its strength completely depends on complexity of the encryption algorithm. However, PLS schemes can provide security at the physical layer, but its implementation faces serious problems. PLS cannot guarantee secrecy if the legitimate node has weak channel capacity than eavesdropper or if no channel information of eavesdropper is available. Power consumption is large in PLS as it uses more number of relays or antenna systems. From Table 1, we have seen that PLE schemes have many advantages over other two. It is more practical to use PLE in the design and construction of secure 802.15.4 transceiver.

Rest of the paper is organized as follows. Section 2 presents overview of PLS. In Sect. 3, PLE and its implementation in different wireless systems are discussed. In Sect. 4, different security suits of 802.15.4 protocol is compared. Section 5 gives hardware implementation of 802.15.4 transceiver, and finally Sect. 6 provides conclusion.

2 Physical Layer Security

PLS provides unbreakable, quantifiable and provable secrecy from the information-theoretical point of view. PLS takes an advantage of wireless channel characteristics for improving the reliability and security. In this subsection, we discuss about different PLS schemes such as information-theoretic security, artificial noise generated secrecy, secure beam forming and diversity-assisted security.

2.1 Information-Theoretic Security

It is a cryptosystem whose secrecy is derived mainly from an information theory. Basic principle of security schemes was developed by focusing on the properties of mathematical structures. Claude Shannon proposed a security principle as a mathematical transformation of valid plaintext into another set of reliable cryptograms. Here, each transformation is done by encrypting the message, using secret keys. This cryptosystem is unbreakable even though attacker has unlimited computing power. One-time pad is an example of such a cryptosystem. This secrecy system was designed for protection against eavesdropping attacks. Since Shannon model uses distribution of key, key management becomes difficult for wireless networks with no fixed infrastructure [3]. Wyner improved Shannon's model without secret keys and analyzed the performance of a discrete memory less wiretap channel and achieved perfect secrecy, provided if the capacity of a channel of the link spanning from an sender to receiver must be greater than the wiretap link between the sender and the adversary [4] and he proved that secure communications without keys can be realized with channel quality between intended nodes which is better than the adversary link [5]. However, the use of information theoretic security is not practical because of difficulty in key generation and distribution. Further, secrecy capacity of wireless networks is extremely attenuated because of time varying fading effect. Usually, fading degrades the signal received at the legitimate receiver, which decreases the capacity of the valid channel, and reduces the secrecy capacity.

2.2 Artificial Noise Aided Security

Artificial noise (AN) has been exploited to enhance the security performance by degrading the capacity of the adversary, termed as artificial noise aided security. Basic principle of the noise injection method is to simultaneously transmit the message and the generated noise in order to reduce the performance of the adversary [5]. In AN pre-coding, the source node splits the transmission power between data transmission to legitimate receiver and the noise transmission to the eavesdropper. The transmitter is designed in such a way that only adversary channel is selectively degraded. With this model, certain minimum rate of security can be achieved. However, secrecy cannot be guaranteed, if the eavesdropper channel is better than the recipient [6, 7]. AN pre-coding gives provable security at the PHY, but it comes with an extra cost of additional energy requirements. In this approach, a fraction of transmission power of message signal is taken to produce the AN. It depletes transmission power that can be utilized to improve the capacity of a channel and receiver signal-to-noise ratio (SNR). Hence, it is a compromise between the transmission power and secrecy rate [8].

2.3 Security Oriented Beam Forming

This technique is extensively used in the relay systems; it helps in improving the SNR at the receiver end. It has an ability to control the direction of transmission, it only focuses energy in specified direction and suppresses energy in other direction. This in turn increases the energy efficiency of a system. Consider a model which consists of A, B and C, where A is the transmitter, B is the receiver, and C is the eavesdropper. When A sends information to B in the presence of C, and if C is within the coverage area then C can easily intercept the message sent by A to B. Thus, this can be avoided by using beam forming. Beam forming creates a beam only in the direction of B to maximize the SNR ratio and suppresses the transmission or reception in the direction of the C. One of the main features of beam forming techniques is spatial filtering. The spatial filtering helps in distinguishing the secure and insecure locations for the transmission of information. Beam forming helps in utilizing the wireless medium in order to provide better service with respect to error performance and bit rate at the PHY [7].

However, beam forming faces issues such as fraction of the power being spread through minor side lobes, even though signal is directed toward legitimate node and the transmitted power is concentrated in the beam of main lobe. It creates a loophole for an eavesdropper who is in the coverage area to decode the transmitted information. It is also observed that beam forming only concentrates on improving the quality of the main channel and it neglects the possibility of having favorable channel by eavesdropper. Further, beam forming is computationally expensive and difficult to implement [8].

2.4 Security Diversity Techniques

The diversity technique is usually employed to enhance the transmission reliability which in turn improves the wireless security and also to reduce the duration of fading, experienced at the receiver. The PLS can be improved by various diversity methods which includes MIMO, multiuser and cooperative diversity [9].

MIMO Diversity: In this technique, multiple antennas are used for transmitting and receiving the signal. When the information bearing signal is sent through the channel, it will be transmitted through more than one antenna and, while receiving, it will be received through multiple antennas. Basically, MIMO is considered as an effective method of overcoming wireless fading and thus increases the channel capacity. However, there is possibility that eavesdropper could also utilize the structure of MIMO to improve the capacity of wiretap channel.

Multiuser Diversity: This is obtained by the user scheduling it either at the transmitter or at the receiver. Basically, it uses OFDMA and TDMA. So at any point of time, transmitter will select the best user among different receivers based on the quality of their channel and throughput. Disadvantage of this method is that if the user is far away from the base station and experiences deep fading and worst propagation loss, then the user will not get a chance to access channel. Hence, user fairness needs to be maintained and should provide guarantee that each user will get opportunity to use the channel [10].

Co-operative Diversity: Cooperative diversity is one of the important beam forming techniques against eavesdropping. It is a multiple antenna technique which is employed to improve the legitimate channel capacities for a given set of bandwidth [9]. Cooperative network includes source (s), X relays (r), destination (D) and an eavesdropper (E). X relays (r) is used to help the message transmission between source and destination. The source first transmits the message to X relays that then relay (r) sends the received signal to the destination.

2.5 Physical Layer Secret Key Generation

Secret key is extracted by exploring the physical layer characteristics such as channel randomness, independent channel variation over space, channel reciprocity. The randomness is extracted either from amplitude or phase of wireless fading. Key is generated by alternatively sending probe signal and estimating the channel state information (CSIs). Intended nodes can convert their CSI into the same bit strings. The bit discrepancies are corrected using privacy amplification techniques and key reconciliation [11]. Basically, two methods are used to generate key streams, one way is by using received signal strength (RSS) where power of received signal is used and in the other scheme phase of the received signal is used to extract the common randomness. Even though this can provide an alternative approach to the conventional key generation algorithms there are certain limitations which requires an attention. RSS provides a low-key bit generation rate and it faces scalability issues however signal phase based scheme gives good performance the implementation is difficult since it requires analog-to-digital converter which increases hardware complexity [11].

Even after researchers have proposed significant number of mathematical models, algorithms, and solutions, PLS faces challenges in its implementation. Most of the problems which are faced by PLS completely depend on the channel. PLS, cannot guarantee security if the channel capacity of eavesdropper is better than legitimate channel or unavailable eavesdropper's CSI. Increased power requirement in PLS due to the use of MIMO and relay systems [12] is another issue. In addition to this, all the existing work related to PLS is only concentrated on improving the security against eavesdropping attack by completely neglecting the various wireless PHY attacks. Hence, it is important to search alternative techniques to improve security at the PHY.

3 Physical Layer Encryption

PLE is yet another method of providing security at the PHY. Unlike PLS, PLE rely on computational complexity and uses distribution of secret key. PLE has no strict requirement on the channel conditions and the number of antennas. Compared to upper layer security schemes PLE resists the influence of noise and the effect of the channel in order to give reliability along with security. PLE provides security at the signal level and makes use of channel error to enhance the secrecy level. PLE schemes are more modulation intended; uses joint design of encryption and modulation, and it varies along with wireless technology. Based on processing of plain text PLE can be categorized into two types: Stream and Block PLE.

Stream PLE: It uses an encryption unit which encrypts binary message sequence using pseudo-random cipher key streams. Key streams are generated by using pseudo-random complex sequence generation function. The encrypted symbol is a function of message binary sequence and complex sequence. And the complex sequence is calculated based on the initial key. In stream PLE, each plain text symbol is encrypted with the corresponding key symbol. Receiver jointly performs decryption and demodulation to get plaint text back. Security in stream PLE depends on encryption function and the complex sequence. This scheme provides low-propagation error and latency. Disadvantage is low diffusion and lacks symbol overlap. Hence, stream PLE is mostly used in simple and high-speed applications.

Block PLE: This scheme encrypts a fixed size plaintext symbols as one block. Block PLE uses mapping functions which maps fixed block of an input sequence to complex vector based on corresponding key. Different types of mapping functions can used to design PLE and they can be random. In block, PLE encrypted output depends on block of binary sequence, mapping function and a key. Designing a suitable mapping function plays key role. Advantages of block PLE are; it provides high diffusion and is immune to tampering, it suffers with error propagation and slow encryption process.

In recent years, many researchers had proposed PLE implementation techniques in various communication systems like OFDM, MIMO and 802.15.4 so on. In this subsection, we try to present PLE schemes used in different wireless standards. As 5G is in focus, MIMO technology is evolving significantly and there is a need for security. New PLE method is proposed to improve the security in MIMO with spatial modulation [13]. Adding spatial modulation to MIMO provides high energy and spectral efficiency. This paper presents a chaotic-antenna-index-3D modulation and rotated constellation points PLE which effectively makes use of spatial modulation as well as chaotic theory. Key generation algorithm is designed based on chaotic theory. Security can be achieved by protecting the spatial constellation diagram, for this the chaotic sequence generated antenna index is used. The simulation results clearly show that even with infinite MIMO system, eavesdropper cannot recover plaintext. In [14], PLE implemented at the PHY using OFDM is discussed. The encryption method reserves a part of OFDM subcarriers which transmits dummy data used to hide information at subcarrier level and provides randomness. This makes information about subcarriers unclear. Obfuscation of subcarrier makes the transmission secure. Along with obfuscation it also uses re-sequencing of training symbols. The reserved subcarriers used for re-sequencing provides protection to entire packet in physical layer without affecting synchronization and channel estimation between intended user meanwhile it prevents the eavesdropper from doing all these operations. This scheme is implemented on 802.11 OFDM and compared the results with key rate, complexity and search space. Calculated results show that entire data search space is (48!)38 and a search space for entire packet is 2.47×10^{173} with this eavesdropper will take 3.74×10^{121} years to break. Key streams are generated using stream ciphers. Key rate can be adjusted by varying 's' OFDM symbol and 'k' reserved subcarrier. With increase in s and k value increases the search space which in turn improves security.

Further Li et al. [15] presented both stream and block PLE for 802.11 OFDM in their work. Two PLE framework designs are developed which can provide security against known plaintext and chosen-plaintext attacks (CPA). In design framework, reliability and security are considered together. Stream PLE unit have of two parts; pseudo-random complex number generator (PRCNG) which is used to generate pseudo-random complex binary sequences (PRCBS). Both the legitimate user generates same PRCBS using a key. Next part is the design of encryption function, which performs mapping of plaintext symbols into constellation points (complex signal). The mapping is done according to PRCBS. Constellation distance and confusion are taken into account, while designing the mapping function. 3D constellation scheme maps 2 bit message to 3D constellation point and these points are distributed over spherical surface. After mapping, 3D rotation is used to disturb the constellation. This disturbance creates confusion which helps in improving the security. In block, PLE key generation algorithm produces three sub keys K1, K2, K3 which are used at three different stages. Three stages of PLE are bit change stage, modulation stage and block change stage. Bit change disturbs and creates confusion among the binary sequence using first sub key K1. Modulation uses K2 to map confused sequence to complex vector. Block change stage is a function used to make symbols confuse and interlace so that it will be difficult for eavesdropper to get plaintext. At this stage, K3 maps 2 complex vector spaces. Table 2 shows the comparison between different PLE methods used for 802.11 OFDM. Search and key space, CPA security, throughput, BER performance and design complexity are considered for performance analysis. All three PLE schemes give similar results when it comes to bit penalty, key and search space. The subcarriers obfuscate scheme uses a part of transmission power to send dummy bits so its throughput decreases. And since it uses two stream ciphers CPA security relies on stream cipher which it uses. Block PLE and stream PLE provide better throughput performance however CPA security in stream PLE depends on PRCNG which is used to produce PRCBS using keys. All three are linearly complex and these PLE can be software or hardware implemented.

Huo et al. [16] presented a new generalized phase encryption scheme which can be applied to any of the wireless communication standard independent of modulation scheme. XOR encryption uses bitwise XOR between data bits and the corresponding key bit to give encrypted output. Unlike XOR encryption, phase encryption is performed on modulated sequences. Basically, phase encryption

Table 2 Comparison between different PLE methods

Type	Bit penalty	Throughput decrease	Search space	Key space	CPA security	Complexity
Subcarrier obfuscate [15]	NO	Depends on reserved subcarrier	High	High	Relies on stream cipher	Linear
Block PLE [16]	NO	NO	High	High	Good	Linear
Block PLE [16]	NO	NO	High	High	Relies on PRCNG	Linear

neither depends on system or on specific modulation scheme. Security functions provided at upper layers makes added data and the headers of succeeding layers unprotected and it is vulnerable to traffic analysis attacks. Authors proposed PHY structure using phase encryption. Here, two bits are used to encrypt one modulated symbol. First bit represents real part and another bit is used for imaginary part. Modulated symbol consists of $\log_2 M$ bit information where M is the size of constellation. It shows phase encryption is not one to one mapping of plaintext and secret key. N bit symbols are encrypted by a key stream of 2-bits. Further, 1 bit key stream is enough to encrypt modulated symbol in case if it consists of only real component. Once phase encryption is done the modulated discrete symbols and encrypted sequence are given to digital to analog converter. Obtained analog part is up converted into carrier frequency later it is transmitted over a channel. In phase encryption, the cipher text is a complex number and relation between plaintext and cipher text depends on channel coding, source coding and type of modulation is used. Since the encryption is done at physical layer just before the transmission can prevent the traffic analysis attack. Comparison results of both XOR and phase encryption are given in Table 3. From the result, it is shown that phase encryption gives higher encryption efficiency and can be used as a substitution for XOR encryption.

Generalized phase encryption is extended to design the PHY of 802.15.4. [17]. Since PLE is modulation specific considerable attention is required for each wireless standard. Even though different PLE methods are already available, most of them are for OFDM systems [14, 15]. And some schemes are implemented by rotating the constellation points [13]. Paper [16] presented the phase encryption to mitigate traffic analysis. All the above-mentioned methods are concentrated in providing security to 802.11 OFDM systems. But all these schemes are not suitable when it comes to the security of IEEE 802.15.4 because of different operating conditions of its devices. In general, security services like confidentiality, integrity of 802.15.4 is provided via MAC. These security primitives increase the computational energy which cannot be neglected [18]. Hence in [17] authors proposed an efficient phase encryption scheme for PHY of 802.15.4 and analyzed it against energy depletion and traffic analysis attack. During an encryption, phase of the modulated symbol changes with respect to key. The size of key is depends on the underlying modulation method. Each one of the modulated symbols consists of 2

Table 3 Comparison between XOR and phase encryption

Type	Encryption	Encryption efficiency	Cipher text	Mapping
XOR encrypted	Before modulation	Low	Bit level	Bit to bit
Phase encrypted	After modulation	High	Complex number	2 bits used to encrypt 1 modulated symbol

message bits and it is the form (I, Q). Values of I and Q are from the set of $\{1, -1\}$. Hence I and Q of key stream also take its values from the set of $\{1, -1\}$ .cipher text is produced by multiplying the corresponding components of the modulated symbol and key streams.

4 IEEE 802.15.4 Security Suits

In recent years, 802.15.4 standard is becomes popular and an association with an IoT makes its applications even broader.802.15.4 protocol defines low-power, low-complexity and low-data rate communication in WPAN. An 802.15.4 application includes smart cities, home and industrial automation, health monitoring and military surveillance so on. All these applications need secure transmission of information. Particularly, when it comes to a health and military applications, security is the utmost concern. Hence, it is very important to study the security aspects of IEEE 802.15.4. This section presents the IEEE 802.15.4 security specification. Security requirements of 802.15.4 are frame integrity, confidentiality and access control. Integrity resists the modification of frames; Confidentiality guarantees that only intended nodes can transmit the secret message. Access control protects the frames against unauthorized users. Sastry et al. [19] extensively discussed about security provisions. In 802.15.4 link-layer security provides confidentially, integrity and access control. The 802.15.4 uses two types of packets; data packet which uses a flag to indicate type of packet, security enable and addressing modes. At last 2 bytes of CRC are used for error correction. Acknowledgment packet is used to send acknowledgment by the recipients. In 802.15.4, MAC layer controls the security. In case of security requirement, the application has to explicitly specify that using different control parameters. Application can choose different security schemes which control the security for transmission of message. Security suits of 802.15.4. Specification can be broadly categorized into two types: secure and unsecure mode. Table 4 gives information of security suits supported by 802.15.4. Each scheme offers different aspects of security. Null suit is unsecure where it defines no security. AES-CTR only performs encryption and gives confidentiality. AES-CBC-MAC provides data integrity which comes with 3 variations

Table 4 Comparisons between various IEEE 802.15.4 security suits

Security suit	Description
Null	Unsecure
AES-CTR	Only encryption
AES-CBC-MAC-128,64,32	Authentication only. Flexible with different MAC sizes: 128,64, 32 bits
AES-CCM-128,64,32	Authentication and encryption flexible with different MAC sizes: 128, 64, 32 bits

based on size of the MAC bit. Size can be 32, 64 or 128 bits. Each one is considered as separate security suit. Higher the size of MAC bits lower the risk of adversary, but it increases the size of packets. Whereas AES–CCM first provides integrity using CBC-MAC later it encrypts the data by using AES-CTR.

However, the above-mentioned security suits faces serious problems. Implementation of AES-CTR is unsafe because, encryption without authentication introduces significant risk of vulnerability at protocol level. Also it is shown that AES-CTR is more prone to denial of service attacks. Further, none of these security suits gives data integrity to acknowledgment packets. With jamming, this loophole can be used to stop the delivery of packets. This makes acknowledgments untrustworthy. A new method of security enhancement in 802.15.4 Zigbee is proposed [20]. This is done by altering the MAC. This paper also presents security enhancement in application and network layer, and it is implemented by using RFID detector and application gateway. In Zigbee, security is maintained by external service provider interface, which work on every layer. In order to enhance the security, MAC addressing, authentication, security unit is added at MAC layer. RFID detector is integrated at network layer, and APL security is included at application layer. Riverbed modeler is used to implement the zigbee protocol and simulation results shows with the proposed work MAC can block unauthorized devices and performs authentication, network layer blocks illegal packet and application layer block adversary data. Since performance and security compete for same CPU, memory, energy and bandwidth, it is very much necessary to analyze the resource consumption, while giving security. Hence, further we study the impact of security on memory, energy consumption and network performance [21]. To test security, authors used 2Tmotesky motes with 48 kbyte ROM and MSP430 microcontroller operates at 8 MHz with RAM of 10 kbytes. In order to analyze the network performance, full function device (FFD) and reduced function device (RFD) are consider. FFD acts as PAN coordinator which manages network and security. RFD acts as a sender and continuously transmits protected data to the PAN coordinator using a common secret key shared between RFD and FFD. After receiving data packets, PAN coordinator verify data frame and sends back ACK. Memory consumption is compared with and without security sub layer. Obtained results shows that without security FFD function takes 33.2 Kbytes that is (69% of total memory) and the same FFD with security consumes 39.31 Kbytes (81.9%) overall there is around 13% increase.

Similarly for RFD takes 34.43 Kbytes (71.7%) without and 39.85 Kbytes (81.1%) with security, 9.4% increase. By adding security layer the size and complexity of PAN coordinator increases. Table 5 shows how different security suits affect the transmission of frames. When compared to null security suit, 26.2% reduction in frames when only encryption (AES-CTR) is provided. With only authentication (CBC-MAC) 28 to 33.5% of frames are reduced (varies with size of MAC bits) and adding both encryption and authentication (AES-CCM) reduces 28.2 to 33.9% of frames. Size of each frame increase with different security suits as it adds overhead. Cost of energy consumption when implemented security suits is shown in Table 6. Overall energy consumption is split into energy for secure one

Table 5 Security impact on frame transmission

Security suit	Frames	Decrease in frame (%)	Frame size (bytes)
Null	7685.5	–	27
AES-CTR	5671.6	26.2	37
AES-CBC-MAC-4,8,16	5534.8, 5371.6, 5110.3	28, 30.1, 33.5	41, 45, 53
AES-CCM-4,8,16	5514.6, 5374.4, 5082.1	28.2, 30.1, 33.9	41, 45, 53

Table 6 Energy consumption

E (μJ)	V (v)	I	t (ms)	Device
$E1 = 240.54$	3.6	17.4 mA	3.84	CC2420
$E2 = 150.34$	3.6	17.4 mA	2.40	CC2420
$E3 = 2.66$	3	600 μA	1.48	MSP430

frame transmission (E1), energy for encryption and authentication (E2) and energy for security management (E3). As 802.15.4 protocol gaining importance for its low-power, low-complexity nature, and all these parameters discussed above plays an importance role hence one has to look in this direction in the design of efficient and secure 802.15.4 protocol.

After analyzing related work it is clear that there is less research work done toward the implementation of efficient and secure 802.15.4 protocol. As for as IoT applications are concerned the existing security suits using AES is unsuitable. Hence, it is important to search for alternative security methods which meet the low power and low-complexity requirements.

5 Hardware Implementation of IEEE 802.15.4

IEEE 802.15.4 defines PHY and MAC layer specifications. PHY is designed to transmit data packets. Physical layer operates at 868/915 MHz and 2.4 GHz. Further revisions in 802.15.4 allow the use different modulation schemes including BPSK, QPSK, O-QPSK, GFSK and UWB. In 802.15.4 data link consists of two parts; link control and MAC. Link control is standard used in all 802 protocols logic. MAC layer is designed to perform various operations like channel access, frame delivery, synchronization; flow control and error control etc. In this sub-section, various hardware implementation techniques of IEEE 802.15.4 are discussed. A small subset of 802.15.4 MAC protocol is designed to provide a point to point communication [22]. Author implemented 2.4 GHz protocol design. Mapping of bit to symbol is done by mapping 4 bit of LSB to first symbol and 4 bit of MSB to next symbol. Later each and every symbol are mapped to 32-chip sequence.

Mapped chip sequences now modulated by using O-QPSK. Modulated signals are then added with 16-bit cosine and sine values. Summed signal is finally transmitted. At the receiver sine and cosine are subtracted. Demodulation is later performed to get original signal. Verilog code is used to design the proposed protocol. Virtex-4 pro FPGA is used for hardware implementation which has a clock speed of 100 MHz. Since the design has to work only at 250 kHz; this hardware is not suitable for optimal zigbee transceiver design. However, clock divider circuit can be used, but it will add additional hardware. 2.4 GHz zigbee receiver is implemented using sparton-3E XC3S500E FPGA kit, which supports clock frequency of 250 kHz and 2000 kHz [23]. The receiver consists of O-QPSK demodulator, chip-synchronization and de-spreading block which are designed using Verilog. Further to optimize, the design all the blocks are integrated into single Verilog module. The proposed system gives data rate up to 250 kbps. In [24], MQAM modulation scheme is used to design 802.15.4 transceiver which operates at 2400 MHz. Proposed system consists of FIR filter, serial in parallel out and parallel in serial out shift registers, up and down samplers and a chip generator block. Verilog is used as a designing language and it is implemented over sparton-3 XC3S200E FPGA. Model–sim is used to simulate the waveforms of transceiver. The clock speed used is 1000 kHz and 8 MHz, even though 802.15.4 standard is designed for very low-power applications, designing energy aware transceiver is very much needed. Elmiligi et al. [25] implemented energy scalable 802.15.4 transceiver using a BPSK modulation which uses 868 MHz frequency band. VHDL is used to design zigbee transceiver and AMIRIX AP1000 for implementation. Recently, a design of fully integrated 802.15.4 zigbee transceiver is implemented on ARTIX-7 using Verilog [26]. Table 7 gives the comparison results of the implementation techniques discussed above.

All the research work discussed above only concentrated on the efficient hardware implementation of 802.15.4 transceiver design and they failed to implement security services to 802.15.4 transceiver along with performance. Nain et al. [17]

Table 7 Comparison between different implementation techniques

References	[23]	[24]	[25]	[26]
Design approach	Verilog	Verilog	VHDL	Verilog
FPGA family	Sparton-3E XC3S500E	Sparton-3 XC3S200E	AMIRIX AP1000	Atrix-7
Modulation scheme	O-QPSK	MQAM	BPSK	O-QPSK
Operating frequency	2.4 GHz	2.4 GHz	868 MHz	2.4 GHz
Clock frequency	250 kHz and 2 MHz	1 MHz and 8 MHz	105.502 MHz	270.1 MHz
LUTs	3,228	2526	284	428
FFs	2993	60	227	179
Slice registers	–	320	229	224

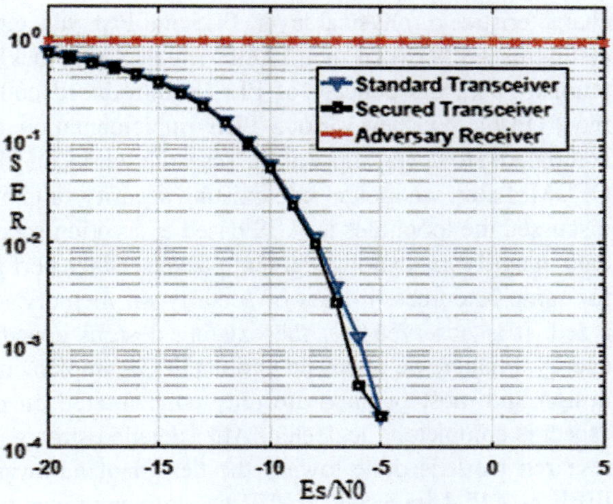

Fig. 2 Comparison of SER

designed and implemented a secure IEEE 802.15.4 transceiver which provides a protection against multiple attacks. Along with the confidentiality and message integrity, proposed scheme provide security to brute-force, cryptanalysis and traffic analysis attacks. Proposed secure transceiver uses a stream PLE scheme using phase encryption which provides a security at physical layer during the modulation. The PHY preamble here uses 8 symbols/256 chips which will be converted to 128 complex samples when it is modulated. Since QPSK is used, one of the 4 phases is used to rotate each sample according to key. This gives 4^{128} possible set of key for 1 preamble. This proposed system resists brute-force attack. Secure 802.15.4 transceiver is designed in Xilinx environment using Verilog. After simulation, results are analyzed in terms of security. Figure 2 gives the comparison of symbol error rate (SER) between standard and proposed 802.15.4 transceiver in noisy environment. The result shows that there is no degradation in SER performance at legitimate node and a very high SER degradation at adversary which increases the difficulty for an attacker to get the plain text.

Proposed system is implemented on Kintex-7 FPGA and ASIC UMC. Synthesis results show that proposed secure PLE based 802.15.4 transceiver uses 132,046 gates compared to the gate count of 104,477 in standard design. 26% increase in the gate count is observed and the implementation on FPGA uses 6507 slices and 15,954 LUTs. However, this resource overhead is because of RC4 cipher.

6 Conclusion

With the growing demand for the IoT applications, IEEE 802.15.4 protocol is gaining much attention in recent times. In this paper, we presented the importance of 802.15.4 PHY and discussed how security can be achieved by exploiting the characteristics of physical layer. PLS and PLE schemes are studied extensively and distinguished between them. Different PLS schemes are reviewed with their limitations. Next, the concept of PLE is introduced and discussed about stream and block PLEs. Further various PLE implementation techniques are analyzed and compared. However, most of the research works on PLE have been done on 802.11 OFDM. Later, we looked into existing security suits of IEEE 802.15.4 standard and discussed the loopholes in AES based encryption schemes as they failed to provide data integrity to acknowledgment packets. Since performance and security compete for same resources, impact of security on memory, energy and network are analyzed. It is observed that the existing designs consume more memory space and energy. Finally, we reviewed various FPGA implementation of 802.15.4 transceiver design and most of them are only concentrated on performance and the security aspect is completely neglected. After detailed survey, it is observed that a very less research work is done toward the design of hardware based secure and efficient IEEE 802.15.4 transceiver. PLE has low-computational complexity, provides low latency with small power consumption and provides longer lifetime. Future scope of this paper proposes the integration of PLE scheme along with the efficient lightweight key generation algorithm for the design of 802.15.4 transceiver system which can greatly enhance the security and efficiency thereby providing the security to IoT applications.

References

1. ITU, ICT Facts and Figures, 2020. Available on-line at https://www.itu.int/en/ITU-D/Statistics/Pages/facts/default.aspx
2. Stankovic, J.A.: Research directions for the internet of thing. IEEE Internet Things J. **1**, 3–9 (2014)
3. Zou, Y., Zhu, J., Wang, X., Hanzo, L.: A survey on wireless security: technical challenges, recent advances, and future trends. Proc. IEEE **104**, 1727–1765 (2016)
4. Wyner, A.D.: The wire-tap channel. Bell Syst. Tech. J. **54**, 1355–1387 (1975)
5. Sun, L., Du, Q.: A review of physical layer security techniques for internet of things: challenges and solutions. In: Entropy (2018)
6. Hyadi, A., Rezki, Z., Alouini, M.: An overview of physical layer security in wireless communication systems with CSIT uncertainty. IEEE Access **4**, 6121–6132 (2016)
7. Goel, S., Negi, R.: Guaranteeing secrecy using artificial noise. IEEE Trans. Wirel. Commun. **7**, 2180–2189 (2008)
8. Sanenga, A., Mapunda, G.A., Jacob, T.M.L.: An overview of key technologies in physical layer security. In: Entropy, Multidisciplinary Digital Publishing Institute (2020)
9. Johansson, M.: Benefits of multiuser diversity with limited feedback. In: 4th IEEE Workshop on Signal Processing Advances in Wireless Communications—SPAWC, Rome, Italy, pp. 155–159 (2003)
10. Zou, Y., Zhu, J., Wang, X., Leung, V.C.M.: Improving physical-layer security in wireless communications using diversity techniques. IEEE Netw. **29**, 42–48 (2015)
11. Ren, K., Su, H., Wang, Q.: Secret key generation exploiting channel characteristics in wireless communications. IEEE Wirel. Commun. **18**, 6–12 (2011)
12. Trappe, W.: The challenges facing physical layer security. IEEE Commun. Mag. **53**, 16–20 (2015)

13. Wang, S., Li, W., Lei, J.: Physical-layer encryption in massive MIMO systems with spatial modulation. China Commun. 159–171 (2018)
14. Zhang, J., Marshall, A., Woods, R., Duong, T.Q.: Design of an OFDM physical layer encryption scheme. IEEE Trans. Veh. Technol. **66**, 2114–2127 (2017)
15. Li, W., Mclernon, D., Lei, J., Ghogho, M., Zaidi, S.A.R., Hui, H.: Cryptographic primitives and design frameworks of physical layer encryption for wireless communications. IEEE Access **7**, 63660–63673 (2019)
16. Huo, F., Gong, G.: Physical layer phase encryption for combating the traffic analysis attack. In: IEEE International Symposium on Electromagnetic Compatibility (EMC), Raleigh, NC (2014)
17. Nain, A.K., Bandaru, J., Zubair, M.A., Pachamuthu, R.: A Secure phase-encrypted IEEE 802.15.4 transceiver design. IEEE Trans. Comput. **66**, 1421–1427 (2017)
18. Razvi Doomun, M., Sunjiv Soyjaudah, K.M., Bundhoo, D.: Energy consumption and computational analysis of rijndael-ES. In: 3rd IEEE/IFIP International Conference in Central Asia on Internet, Tashkent, pp. 1–6 (2007)
19. Sastry, N., Wagner, D.: Security considerations for IEEE 802.15.4 networks. In: Proceedings of the 3rd ACM Workshop on Wireless Security, Association for Computing Machinery, New York (2004)
20. Biddut, M.J.H., Islam, N., Sultana, R.S., Sarker, A., Rahman, M.M.: A new approach of ZigBee MAC layer design based on security enhancement. In: IEEE International Conference on Telecommunications and Photonics (ICTP), Dhaka, pp. 1–5 (2015)
21. Daidone, R., Dini, G., Tiloca, M.: On experimentally evaluating the impact of security on IEEE 802.15.4 networks. In: International Conference on Distributed Computing in Sensor Systems and Workshops (DCOSS), Barcelona, pp. 1–6 (2011)
22. Bhat, N.S.: Design and implementation of IEEE 802.15.4 mac protocol on FPGA. In: Innovative Conference on Embedded Systems, Mobile Communication and Computing (ICEMC2) (2011)
23. Ahmad, R., Sidek, O., Shukri, M.: Implementation of a Verilog-based digital receiver for 2.4 GHZ Zigbee applications on FPGA. J. Eng. Sci. Technol. **9**, 135–152 (2014)
24. Supare, V.P., Sayankar, B.B., Agrawal, P.: Design & implementation of MQAM based IEEE 802.15.4/ZigBee tranceiver using HDL. In: International Conference on Smart Technologies and Management for Computing, Energy and Materials (ICSTM), Chennai, pp. 455–458 (2015)
25. Elmiligi, H., El-Kharashi, M.W., Gebal, F.: Design and implementation of BPSK MODEM for IEEE 802.15.4/ZigBee devices. In: IEEE Canadian Conference on Electrical and Computer Engineering (CCECE), Vancouver, BC, pp. 1–5 (2016)
26. Guruprasad, S.P., Chandrasekar, B.: Design and implementation of 802.15.4 transceiver for wireless personal area networks (WPANs) on FPGA. Int. J. Innovative Technol. Exploring Eng. (2020)

An Enhanced Security Framework for Robotic Process Automation

K. Murugappan and T. Sree Kala

Abstract Robotic process automation (RPA) is an emerging field in any industry that takes care of automation of the monotonous jobs. This can help to optimize the resources utilization for the organization, save the cost, time and improve the accuracy and quality of the jobs performed. However, lack of security in the implementation and the management of these RPAs shall affect the business in an adverse way. In the current scenario, privacy has also taken important role, and allowing bots (RPA) to have an access to these privacy applications can lead to regulatory compliance issues and invite heavy penalty to the company, and at certain times, it may result in business shutdown. Hence, in this paper, we explored various security risks associated with the bots automation and provided a proposal to build a holistic security framework for the RPA environment.

Keywords Robotic process automation · Bots · Noncompliance · Regulatory · Policy compliance

1 Introduction

Robotic process automation (RPA) is a software program that imitates human actions when interacting with a computer systems application and accomplishing automation of repetitive, conditional-based processes. This shall be called as a software robot or a bot. These bots also can leverage artificial intelligence and machine learning technologies to improve the experience of organization workforce and customers. While doing so, the bots may have an access to the organization's critical applications and that can be misused by the unauthorized users. Also, mis-configuration and lack of management controls can result in security breaches, data leakage and financial impacts. To counter these potential risks, there is no single end-to-end comprehensive framework available in the IT industry. Since this RPA solution is being embraced by organizations around the world to carry out

K. Murugappan (✉) · T. Sree Kala
VISTAS, Chennai, India

K. Khanna et al. (eds.), *Cyber Security and Digital Forensics*,
Lecture Notes on Data Engineering and Communications Technologies 73,
https://doi.org/10.1007/978-981-16-3961-6_20

crucial processes across multiple business functions, this paper would help in providing an insight to those security risks and provide a framework to safeguard the organization and its customer assets.

2 Problem Identification

When RPA is interacting with multiple applications (in-house or off-the-shelf), it increases the attack surface. Other problems are manual override of the bot set-up, unauthorized changes, software licence misuses where generic ID is used, weak credentials storage, security incident response is not aligned or designed to cope up with the bot's speed and volume of transaction, no accountability established for the bots used in the network, regulatory noncompliance, a corrupt bot can access sensitive data and move laterally in the network or destroy high value information, security breach can result to disclosure of sensitive information to external parties, a rogue robot can create security vulnerabilities for data at rest or in motion, and Service denials.

3 Proposed Solution

As per a 2018 report by Ernst and Young for RPA implementations, organizations should consider the technical, process and people elements of the entire robotics ecosystem. A secure implementation should be in accordance with the entire product lifecycle starting from requirements, architecture to the ongoing operations. This is well captured in the below sections.

3.1 Security Framework—Key Pillars

In Fig. 1, key pillars for building a security framework are provided. This shall be tailored as per the organization's requirement.

Governance: The organization should ensure a governance framework which will build strategy and security requirements from an RPA perspective. This shall explain the management support and its commitment to ensure the data security [1].

Risk Management: Any risks that shall obstruct the RPA objective of the business have to be mitigated based on the priority.

Product and Software Security: Organizations should perform a product architecture risk assessment both internally and externally.

Access Management: Role-based access control is one of the most crucial features to keep in mind while opting for an RPA solution [2]. A credential

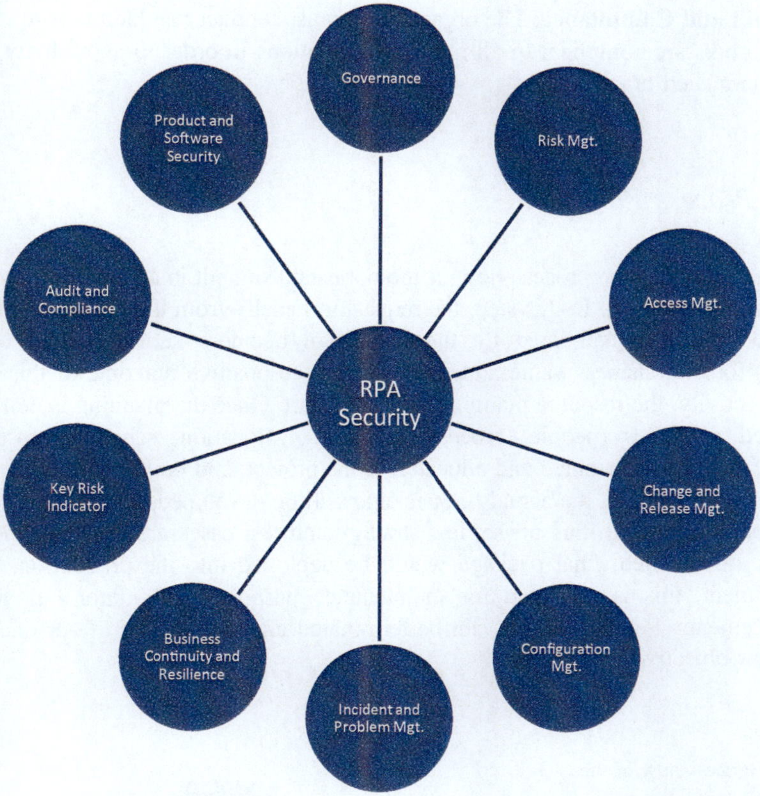

Fig. 1 Security framework

management process must be put in place for the bots to store credentials in a vault and access it as and when needed as per assigned privilege.

Change and Release Management: Any changes to the RPA systems and applications are authorized and implemented accordingly to avoid unauthorized or unnecessary downtime or security breaches.

Configuration Management: All deployed configurations are backed up and version controlled to ensure the timely recovery in case of any operational failure or application or hardware issues [3].

Incident and Problem Management: Focus on any operational and security incident to be addressed and avoid any repeated issues.

Business Continuity Planning (BCP) and Resilience: This will support the business resumption in case of a disaster.

Key Risk Indicators (KRI): This will support the RPA business to take a quick proactive action and avoid any significant impact to the business.

Audit and Compliance: The organization must conduct regular audits to ensure that the bots are compliant to all industry regulations in order to avoid hefty fines and a tarnished brand image.

3.2 *Process Flow*

Figure 2 provides the process flow of the bot establishment in a network that starts from *initiate* step [4]. In this step, the requestor, usually from the business, raises a request, and it is validated by the respective business subject matter expert (SME) for the business value added. Based on the positive outcome of this qualifying activity, the resource planning will kick start. Once the planning is done, the required resources (people, process, technology, location, schedule, cost and quality) will be determined and allocated to the project, and build phase will begin.

During this phase, the actual bot package will be developed, and then, it moves into testing phase. In this phase, test strategy and test cases are tested. If the test results are satisfied, final package would be deployed into the production. After deployment, this is managed for maintenance purposes and minor bug fix or enhancements. Finally, the bot can be terminated after the end of life or once the business objective is achieved.

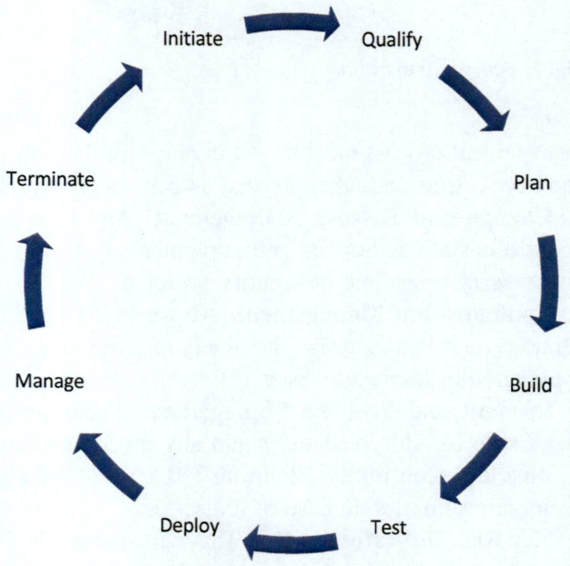

Fig. 2 Process flow of the proposed work

4 Control Requirements in Each Phase

The control requirements in each phase are listed below. The audit and compliance are applicable for every phase.

Phase	Includes	Control requirement
Initiate	Request creation is restricted to authorized users/limited users only	• Limit the request creation based on job role
Qualify	Requirement is captured or understood clearly from the business	• Risk management • RPA vendor or platform selection • Business and security requirement sign-off
Plan	Proper planning for development estimate, detailed process breakdown, value stream map and estimated savings	• Policy adherence • Return on investment is calculated and accounted for every resource • Project/change request is created
Build	• Code/script development • Bot ID creation request • FTE redeployment • Communication and escalation • Package creation • Version control • Bot reuse • Best practices establishment	• Secure coding or vendor recommended security practices are followed
Test	• Unit test • System integration test (SIT) • Quality assessment test (QAT) • User acceptance test (UAT) • Business sign-off	• Every test case is tested • All significant and high rated errors are fixed • Positive and negative tests conducted • Functional and non-functional tests are performed
Deploy	• Infra requirements (virtual machine/ physical system) • Bot installation and configuration • Code deployment to production • License request and management • ID creation in authentication servers, applications and e-mail systems • Bot console management • Scheduling • SMOKE test • Saving confirmation • Handover to operation • Project closure	• Dedicated deployment team • Deployment team not to have an access to development and testing environment • Release ticket is created • Version control/configuration management
Manage	• User access management • Communication plan • Upgrades (operating system, applications and databases) • Change request (functional /technical)	• Unique account for every bot (attended and unattended) • Log monitoring • Bot task monitoring and traceability • Segregation of duties (SOD)

(continued)

(continued)

Phase	Includes	Control requirement
	• Hot fix • Patch management (OS, applications) • Issue management • Monitoring • Logs management • Backup and archival • Bot inventory and its mapped applications	• Privileged account management • No direct write access to the database • Strong passwords • Least privilege and need to know basis access management • Periodic vulnerability assessment • Penetration test performed at least once in a year or whenever significant change in the environment • Software and bot license management • Regulatory compliance • KRI is established and monitored • Periodic backup • Periodic restoration test • BCP/DR • Security incident and problem management
Terminate	Bot decommissioning	• Periodic bot reconciliation and review process established • Unnecessary or unwanted bots de-commissioned

5 Technical Security Requirements

Data Flow: End-to-end data flow is identified for the bot, and blueprint is created with upstream and downstream applications or interfaces [5].

Data Flow Security: In each stage of the bot, data flow security is ensured. This shall include and not be limited to access control and encryption for achieving confidentiality, hashing for achieving data integrity and backup and recovery control for achieving availability [6].

Encryption: Minimum encryption shall be AES 256 and can be used in data at rest, data in transit and processing stages. The hashing algorithm shall be MD5 or SSH latest version.

Audit Trail: Every activity of the bot is logged and tracked with timestamps [7].

Passwords: Strong passwords for the bots and secured with password/credential vault.

Single Sign-On: Centralized authentication is enabled, and no local authentication is recommended.

Alert Management: In case of security breach, an alert is trigged to the concerned stakeholders/security administrator.

Vulnerability Management: Periodic vulnerability scanning and Threat modelling exercises are established to identify technical vulnerabilities and process gaps.

Security Testing: Identify bot security vulnerabilities by performing static and dynamic testing.

Penetration Testing: If any of the bots is exposed to Internet, then penetration testing is mandatory.

Remote Control: All remote console accesses are to be safeguarded with two-factor authentication.

Security by Design: It is better to ensure the security in each stage of the bot development instead of trying to fix gaps at the later stages.

Accountability: Every bot needs to have unique ID and mapped to a business owner so that accountability is established.

Sensitive Data: Sensitive data are to be wiped out before reassigning the bot to any other business.

Regulatory and Policy Compliance: Every Bot needs to be in compliance with the organization policy, standards and regulatory requirements. For example, Sarbanes–Oxley Act (SOX) and Health Insurance Portability and Accountability Act (HIPAA).

PKI Solution: Public key infrastructure shall be used to secure the bot communication process.

6 Comparative Analysis

In the current industry, there are many commercial RPA platforms available, every player has their own way of developing, implementing and managing the RPA, and this might lead to a platform dependent security and inconsistency in following the security measures when the organization wants to implement the RPA with multiple vendors. This paper provides the comprehensive and holistic approach in implementing the RPA security framework, and consistency shall be maintained across the organization even if multivendor environment exists.

7 Conclusion

The need for RPA is increasing and proliferating in every industry; however, implementing the RPA should not impact the existing systems and services in the adverse way. To strikeout the balance, we need to really consider security versus stability. In this paper, the key pillars, process flow and control requirements are discussed for each phase of the bot lifecycle, and these can help to establish the strong security framework for the organization when implementing the RPA solution.

References

1. Deborah, G.: Robotic Process Automation (RPA) Within Federal Identity Management (2019). https://www2.deloitte.com/content/dam/Deloitte/us/Documents/public-sector/us-gps-robotic-process-automation.pdf. Accessed, 1 March 2021
2. Gautam, R.: 4 Security Must-Haves for a Safe RPA Solution (2020). https://www.automationanywhere.com/company/blog/product-insights/four-security-must-haves-for-a-safe-rpa-solution. Accessed, 1 March 2021
3. CemDilmegani.: Technical Buyer's 11 Point RPA Checklist: In-Depth Guide (2021). https://research.aimultiple.com/rpa-technology. Accessed, 1 March 2021
4. UIPATH Best Practices for IT Compliant RPA Implementation (2021). https://www.uipath.com/resources/automation-whitepapers/information-technology-compliant-rpa-implementation. Accessed, 1 March 2021
5. Enríquez, J.G., Jiménez-Ramírez, A., Domínguez-Mayo, F.J., García-García, J.A.: Robotic process automation: a scientific and industrial systematic mapping study. IEEE Access **8**, 39113–39129 (2020). https://doi.org/10.1109/ACCESS.2020.2974934
6. Lewin, A.R.W., Edwards, P.P.: Open-Source Robotics and Process Control Cookbook, USA, pp. 222–225 (2005)
7. Sumit, S.: RPA Implementation: Key Considerations (2018). https://www.pwc.in/assets/pdfs/publications/2018/rpa-implementation-key-considerations.pdf. Accessed, 1 March 2021

Analysis of the Trust and Resilience of Consumer and Industrial Internet of Things (IoT) Systems in the Indian Context

Akaash R. Parthasarathy

Abstract The Internet of Things (IoT) connects every device possessing some element of computer technology or a digital interface. These devices constitute a global interconnected network that bridges the gap between the physical and virtual worlds. Today, there are two major applications for IoT—Consumer Internet of Things (CIoT), concerned with interactions between consumers and IoT devices, and industrial Internet of Things (IIoT), focussed on the utilisation of IoT for designing industrial systems. With the proliferation of IoT devices for myriad applications, it is becoming increasingly important to investigate and understand the factors essential to securing them against external threats. These factors directly influence the design, functionality and the standards and regulations for IoT devices. This paper defines the trust and resilience of IoT systems and provides unambiguous definitions for key factors (security, privacy, safety, recoverability, reliability and scalability) that directly influence the trust and resilience of IoT systems. Based on the results of a survey conducted amongst IoT consumers and experts, this paper ranks each of these factors in the order of their importance in determining the trust and resilience of CIoT and IIoT systems. These rankings are generated using the analytic hierarchy process (AHP) and a pairwise analysis of the collected data.

Keywords Internet of things · Trust and resilience · Security

1 Introduction

The International Telecommunication Union Telecommunication Standardisation Sector (ITU-T) defines the Internet of Things (IoT) "as a global infrastructure for the information society, enabling advanced services by interconnecting (physical and virtual) things based on existing and evolving interoperable information and communication technologies" [1]. IoT broadly refers to the worldwide system of

A. R. Parthasarathy (✉)
The Shri Ram School—Aravali, Gurgaon, Haryana 122002, India

K. Khanna et al. (eds.), *Cyber Security and Digital Forensics*,
Lecture Notes on Data Engineering and Communications Technologies 73,
https://doi.org/10.1007/978-981-16-3961-6_21

devices that are connected to the Internet and can communicate and exchange data with each other. Today, IoT connects every device that possesses some element of computer technology or a digital interface. These include devices such as Amazon Echoes, which function as smart speakers and assistants, and fitness bands, which track your activity and monitor your health. IoT is responsible for bridging the gap between the physical and virtual world and has been extensively employed in myriad applications ranging from home automation to facility management.

These applications depend on the collection of data through sensors, which can measure physical quantities in the surrounding environment. IoT devices, composed of a multitude of these sensors, are integrated with powerful IoT platforms capable of organizing and manipulating the collected data to perform specific tasks [2]. For example, using IoT technology, smart lights containing proximity sensors are able to turn on or off when they detect the presence or absence of people nearby. More often than not, IoT devices and platforms are enabled with artificial intelligence (AI) to effectively handle and detect patterns in the enormous amounts of incoming data, allowing for enhanced convenience [3]. Using AI, the same smart lights can learn your sleep and work patterns and automatically adjust the lighting to suit your needs.

Historically, consumer devices, home control systems and industrial machines have been offline and not connected to any network. These entities were inherently secure since they could only be compromised through physical access. At the turn of the century, however, there was a significant explosion in computing power and techniques for data analysis, enabling the shift of security and safety systems to virtual platforms [4]. This shift came with its associated risks, both in terms of data leakages and reductions in the integrity of these systems. The past decade saw the emergence of the concept of IoT, associated with an increase in the connectivity of physical devices with each other and with virtual systems. In fact, Gartner forecasts that over 25 billion connected devices will be in use around the world by 2021 [5]. This ever-rising number of IoT devices, which is expected to soon surpass the number of people on the planet, has brought into question their trust and resilience. The immense scale of implementation and the intricacy of the computer systems involved in developing IoT devices leave them vulnerable to malicious hackers and cyber attacks [6]. Each unsecured endpoint serves as a potential location for attacks that can cripple entire IoT systems.

In 2017, Ronen et al. [7] discovered a vulnerability in the Zigbee protocol for IoT devices, allowing them to develop a self-replicating Zigbee worm to exploit Philips Hue smart lamps. This worm could spread to other lamps based on their wireless connectivity and physical proximity. Their research had large implications since similar attacks could be carried out across entire cities, leading to city-wide blackouts. More recently, in February 2020, it was demonstrated that several Philips Hue smart lamps could be hacked with the assistance of drones.

It is evident that device interoperability, trusted communication and the secure sharing and management of data are key aspects that need to be addressed when designing and analysing IoT systems. Thus, research into developing standards for and enhancing the resilience of IoT systems has gained considerable traction in recent years.

2 Consumer and Industrial Internet of Things

IoT can be categorically divided into consumer IoT (CIoT) and industrial IoT (IIoT). CIoT is the more widely known variant of IoT and broadly encompasses IoT devices used to meet consumer needs and increase consumer convenience. CIoT is mainly focussed on residential and consumer interactions with IoT devices such as smart home appliances and wearable technology. IIoT, on the other hand, is concerned with using IoT devices for industrial applications such as synchronisation of manufacturing equipment and operation of integrated supply chains. IIoT makes use of a combination of sensor-driven computing, data analytics and intelligent machines to promote the efficiency of industrial processes and increase enterprise productivity [8].

Since IIoT systems involve the transmission of vast amounts of confidential data, unauthorized access to this data has far-reaching impacts. Despite being less popular and prevalent, IIoT has made significant progress towards standardisation with the help of industrial consortiums dedicated to the advancement of machine-to-machine communication and the promotion of open standards for security and interoperability [4]. These standards have continuously been revisited and updated for several years.

CIoT devices are independently developed by smart device and application providers through the use of traditional interfaces, which emphasise usability and functionality over trust and resilience [9]. Mechanisms to ensure security, privacy and safety are often incorporated solely on the basis of present consumer needs without appropriate planning for subsequent integrations. These concerns arise due to the absence of well-documented standards for CIoT systems, and as a result of the fact that CIoT devices are marketed directly to consumers with limited knowledge of security protocols. For example, in September 2019, a couple's smart home was compromised by a hacker, who took control of their cameras, played disturbing music and manipulated the heat levels in their house by accessing the Google Nest thermostat. This is just one of the many attacks that have been carried out on CIoT systems.

With the widespread adoption of both CIoT and IIoT devices, it is becoming increasingly important to investigate the issues and challenges related to their trust and resilience. The impact of the various characteristics of trust and resilience on the buying and adoption decisions of consumers and industries is of great significance.

3 Literature Survey

Recently, much research has been conducted in order to identify the security and privacy risks associated with IoT devices. Atlam and Wills [10] analysed the security, privacy and safety requirements for IoT systems. They reviewed the challenges faced in IoT security and privacy and presented a case study revolving around the security threats affecting smart cities. Additionally, they provided details regarding the implementation of security and privacy by design and listed the types of cyber and physical attacks that can affect IoT systems. Papp et al. [11] wrote a primer on hacking the hardware and software of IoT systems, where they analysed the most commonly employed methods and scientific research on IoT hacking.

Prior research has focussed on the development of malware to test the resilience of IoT systems. A few of these attempts have been successful in depicting threats to IoT protocols on a global scale [7, 12]. In an attempt to enhance the recoverability of IoT systems, researchers have also proposed self-recoverable IoT architectures, which employ time synchronisation combined with a novel algorithm to achieve formerly unobtainable results [13].

Within the field of IoT, there has also been research into the specific security threats and concerns related to CIoT and IIoT, with individual analyses having been performed for both CIoT [14, 15] and IIoT [16] systems. In order to achieve significantly higher levels of security and privacy, researchers have explored the viability of applying blockchain to CIoT and IIoT security [17]. Additionally, Wurm et al. [9] analysed and contrasted the security features and concerns in CIoT and IIoT devices. Techniques for enhancing the reliability and scalability of CIoT and IIoT systems have also similarly been investigated.

Research has also be conducted into the standardisation of IoT systems. Reference [4] is a comprehensive document analysing and outlining regulations and standards related to the security framework of IIoT systems from both business and implementational viewpoints. Additionally, [4] defines crucial terms related to the trustworthiness of IIoT systems. Taking a major step in the right direction, the European Telecommunication Standards Institute released the first global standard for CIoT devices, which outlines baseline requirements for internet-connected consumer products, in early 2019. The most recent iteration of this standard was released in June 2020 [18].

3.1 Research Gap and Contributions

It is clear that substantial research has been conducted in the field of IoT. Whilst a majority of this research has focussed on individual factors such as security, privacy and safety that impact the functioning of IoT devices, these variables or characteristics have not been comprehensively analysed in terms of their cooperative functioning. Additionally, these factors have usually been examined for IoT frameworks in general and have not been compared and contrasted across the CIoT and IIoT spaces.

This paper investigates the factors affecting the trust and resilience (see Sect. 3.2) of IoT systems. This paper initially identifies and provides definitions for key factors that influence the trust and resilience of IoT systems according to the current research in the field. Next, it assesses the importance of each of these factors in determining the trust and resilience of CIoT and IIoT systems (both separately and combined) based on the results of a survey conducted amongst IoT consumers and experts. The paper ranks the factors against each other based on a pairwise analysis of the collected data. Using the survey results, this paper makes recommendations to enhance the trust and resilience in both the CIoT and IIoT spaces, thereby providing logical steps to follow as the IoT market enlarges.

The remainder of this paper is organised as follows: Sect. 3.2 introduces the concept of trust and resilience and presents key definitions of certain factors that describe the trust and resilience of IoT systems. Section 4 describes the methodology of the study utilised to analyse the impact of each of these factors on the trust and resilience of both CIoT and IIoT systems. The results of the study are presented and analysed in Sect. 5. Conclusions are drawn in Sect. 6, and the future scope of the work is discussed in Sect. 7.

3.2 *Trust and Resilience of IoT Systems*

IoT systems, like all other information systems, have key features and elements that define their trust and resilience (T&R). T&R of IoT systems is particularly important due to our continuously increasing dependence on them. A system is said to be resilient to a fault if its core capabilities are unhindered in the presence of that fault. T&R of an IoT system can be defined by its capacity to "withstand a major disruption within acceptable degradation parameters and to recover within an acceptable time" [19]. This encompasses resistance to both external attacks and internal failures, and adaptation to continual change in global policies and standards.

Whilst no system can be fully trustworthy and resilient, laws and standards assist in maintaining their T&R to a certain degree by creating a balance between functionality and compliance. Although creativity and innovation can lead to the development of novel products and services, improper maintenance and failure to adhere to standards can result in issues such as data and identity theft. It is imperative to establish thorough guidelines and standardisation techniques in order to streamline processes related to the collection and transmission of data, and enhance the interoperability of IoT devices.

However, developing effective regulations necessitates an understanding of the different facets of the subject you are dealing with. In this subsection, six key characteristics that comprise T&R of IoT systems—security, privacy, safety, recoverability, reliability and scalability—have been defined based on the current research in IoT [4]. All further mentions of these characteristics will adhere to the definitions in this subsection.

Security. Collins Dictionary defines security as “all the measures that are taken to protect a place, or to ensure that only people with permission enter it or leave it” [20]. A similar definition can be extended to IoT systems: Security in an IoT system refers to the countermeasures that can be put in place to prevent any individual or group from exploiting the system. Systems have vulnerabilities or security risks that may be exploited and thus, need to be kept secure and protected from unauthorised access. The security of a system is defined by the CIA triad: confidentiality, which deals with unauthorised disclosure of information, integrity, which deals with unauthorised modification or deletion of data, and availability, which deals with reliable access to systems and data by authorised personnel [4].

Privacy. For the purposes of this paper, Adam Moore’s [21] definition of privacy, wherein privacy is defined as an “access control right over oneself and to information about oneself”, has been adopted. Privacy is the right of an individual or a group to decide how information concerning them should be utilised. This includes control over who has access to this personal information and the methods involved in collecting, processing and storing the information [4]. In terms of IoT, privacy refers to preventing the unauthorised access of personal information of an individual or a group in an IoT system. Privacy is ensured if this data is handled by an entity and in a manner that said individual or group has lawfully agreed to. With the growing popularity of IoT and the spike in the amount of personal information being handled and analysed by IoT systems and devices, manufacturers and service providers are required to become increasingly sensitive to consumer privacy and data protection. As such, major steps are being taken towards redefining the current techniques for ensuring privacy and aligning them with global standards [22].

Safety. An IoT system is said to be safe when it can operate without putting people at risk beyond specified acceptable limits. An IoT system must operate without endangering the lives of or causing physical harm to its users [4]. Device malfunctions and transmission of incorrect data may affect health, cause bodily harm or even be life threatening. Safety of IoT systems encompasses the measures taken to prevent such occurrences.

Recoverability. The recoverability of an IoT system is its ability to be restored to a stable state once the failures acting on it cease [23]. Recoverability is closely related to fault tolerance which aims to ensure the attainment of system goals even in the presence of unfavourable conditions and errors [24]. They are focussed on providing certain service level guarantees despite the occurrence of faults. A system is said to be recoverable to a fault if “there exists a control law such that the post-fault system satisfies the design specifications” [24].

Reliability. Reliability is the ability of an IoT system to consistently perform as it is expected to. Reliability determines whether an IoT system is capable of performing its assigned tasks for an extended period of time [4]. Reliability is applicable not only to IoT devices and their data collection techniques but also to the utilised communication frameworks. It is an essential factor in building trust in both commercial and industrial applications.

Scalability. As defined by Gupta et al. [25], scalability is "the ability of a device [or system] to adapt to changes in the environment and meet changing needs in the future". In terms of IoT, scalability refers to the capability of IoT systems to "support an increasing number of connected devices, users, application features and analytics capabilities, without any degradation in the quality of service" [26]. In this increasingly virtual and hyper-connected world, ecosystems must possess the ability to scale and plan for unusual spikes in requests. It is important for IoT systems to adapt to changing volumes of work due to factors such as seasonal demands.

4 Methodology

In order to analyse the influence of each of the T&R factors on the T&R of CIoT and IIoT systems, an online survey was conducted amongst more than 90 consumers and industry and corporate experts from across the world (however, a majority of the survey respondents were from India) using the SurveySparrow platform. The survey explicitly defined each of the T&R factors in order to reduce any ambiguity resulting from the wording of the questions.

The survey was divided into two sections—one for CIoT systems (see Fig. 1) and one for IIoT systems. Both of the sections contained the same questions, but were specific to the respective system. For each pair of factors, survey respondents were asked which factor they believed is more important in determining the T&R of the relevant type of IoT system. These choices were to be made under the assumption that the remaining 4 factors were stable or perfectly implemented. Respondents were additionally asked to take into account the impact of the remaining 4 factors on the factors under consideration, whilst making their choices. Figure 2 details the instructions for answering the questions in the CIoT section of the survey (Fig. 3).

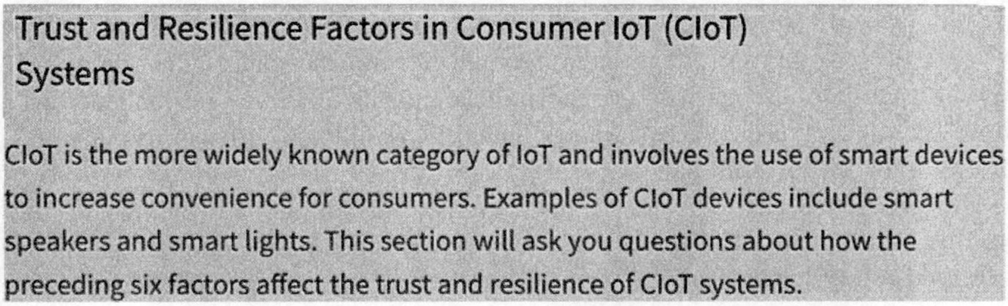
Trust and Resilience Factors in Consumer IoT (CIoT) Systems

CIoT is the more widely known category of IoT and involves the use of smart devices to increase convenience for consumers. Examples of CIoT devices include smart speakers and smart lights. This section will ask you questions about how the preceding six factors affect the trust and resilience of CIoT systems.

Fig. 1 Survey section on T&R factors in CIoT systems. *Source* SurveySparrow

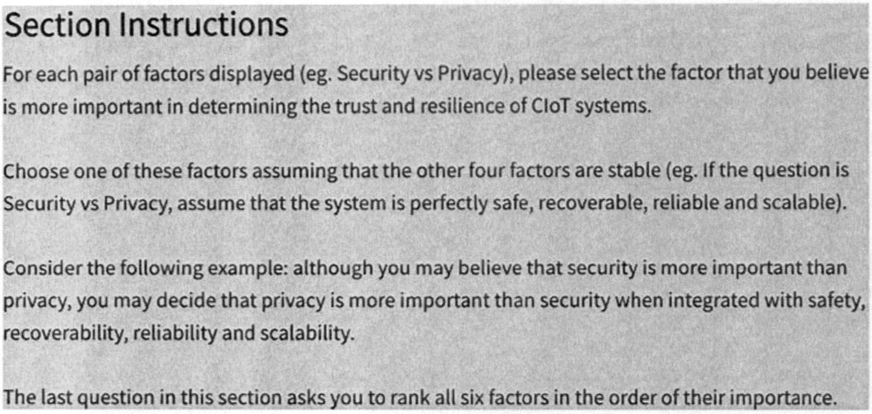

Fig. 2 Section instructions for survey section on T&R factors in CIoT systems. *Source* SurveySparrow

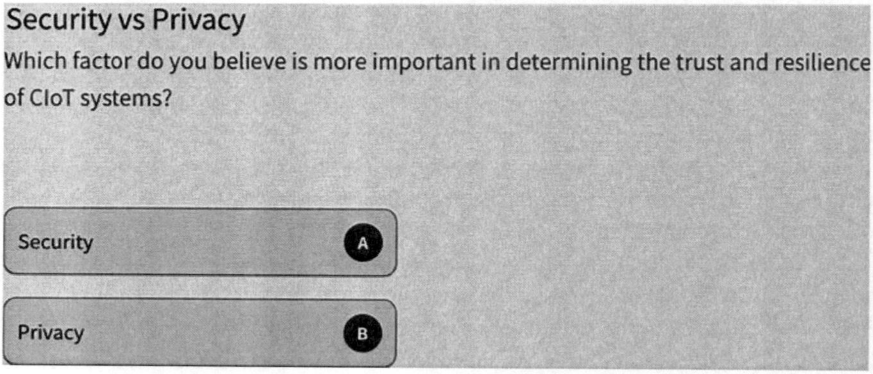

Fig. 3 A sample question from the CIoT section of the survey. *Source* SurveySparrow

The collected data was examined using the SpiceLogic AHP Software v2.2 [27]. The technique used to analyse the data was the analytic hierarchy process (AHP) for multi-criteria decision making (MCDM). Initially, as shown in Fig. 4, the objectives for the AHP were added, with the aim to maximise each T&R factor in the IoT system. Then, a pairwise analysis was conducted on the data, taking into each account each possible pair of factors. The percentages of survey respondents who chose each factor in the survey were inputted as weights in order to calculate the priority trade-off for each pair.

This procedure was followed for the responses for both CIoT systems and IIoT systems. Additionally, a pairwise analysis was performed on the combined data for both types of systems.

Objective	Attribute Type	Range	Relative Priority
1. Maximize Security	Subjective	0 to 100	24.77%
2. Maximize Privacy	Subjective	0 to 100	19.51%
3. Maximize Safety	Subjective	0 to 100	24.46%
4. Maximize Recoverability	Subjective	0 to 100	9.26%
5. Maximize Reliability	Subjective	0 to 100	17.84%
6. Maximize Scalability	Subjective	0 to 100	4.15%

Fig. 4 Maximisation objectives and their relative priorities. *Source* SpiceLogic Analytic Hierarchy Process Software v2.2

5 Results

The pairwise analysis conducted on the survey responses yielded a chart of the relative priority numbers of percentages of each of the T&R factors. These percentages indicate the importance of each of the T&R factors to the respective IoT system when stacked against each other. Based on the results from the AHP software, bar charts were created using Excel for the relative priorities of the T&R factors for CIoT systems, IIoT systems and the two when considered as a whole (Fig. 5).

As depicted in Fig. 6, security assumed the highest priority for CIoT systems with a relative priority number of nearly 25%, followed closely by safety. The subsequent levels of importance were occupied by privacy and reliability, respectively, whilst recoverability and scalability were considered to be the least important T&R factors by a margin of almost 9%.

As shown in Fig. 7, security was similarly assigned the highest relative priority number for IIoT systems. The generated relative priority number for safety was

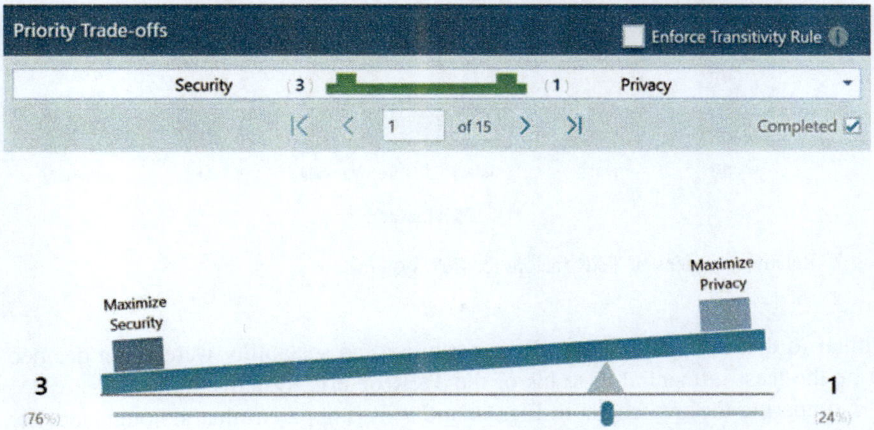

Fig. 5 Priority trade-off for maximisation of security and maximisation of privacy. *Source* SpiceLogic Analytic Hierarchy Process Software v2.2

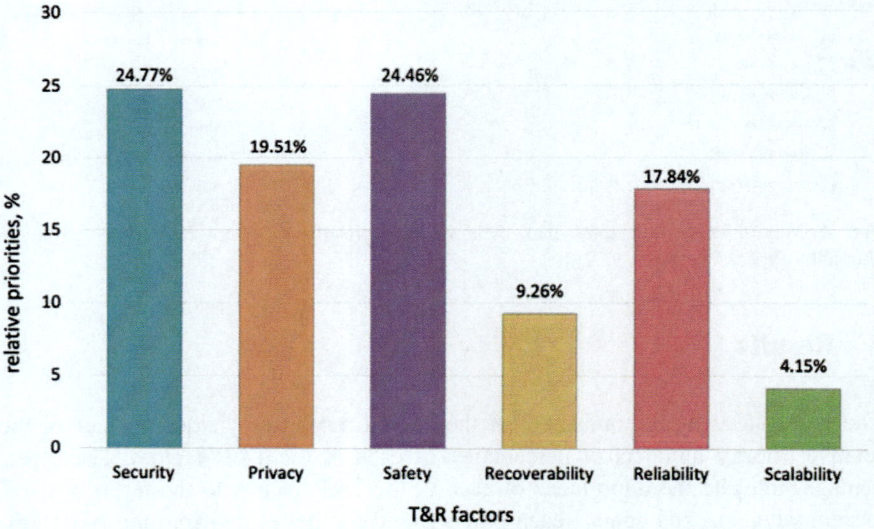

Fig. 6 Relative priorities of T&R factors for CIoT systems

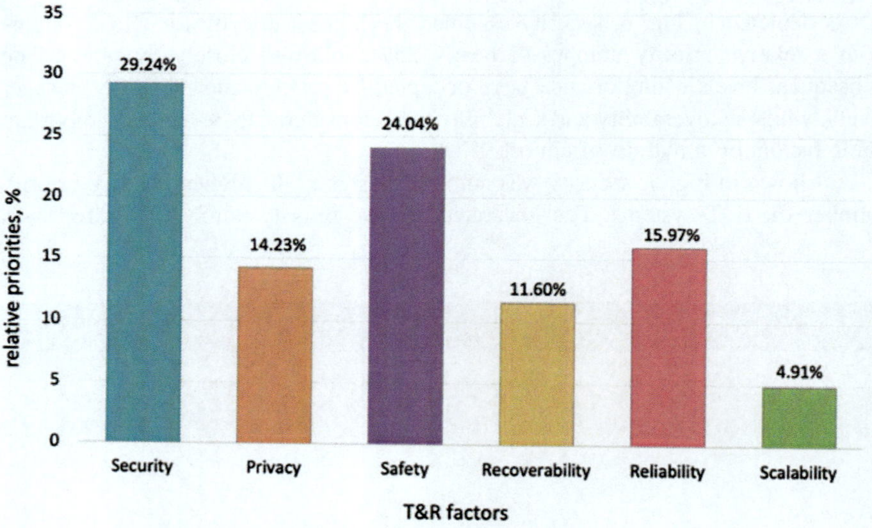

Fig. 7 Relative priorities of T&R factors for IIoT systems

similar to that in CIoT systems. Recoverability and scalability were again deemed to be the least influential in terms of the T&R of IIoT systems.

Comparing the two charts in Figs. 6 and 7, it is apparent that although security and safety occupied the highest two ranks for both types of systems, survey respondents considered security more important in determining the T&R of IIoT systems as opposed to CIoT systems. This is reasonable given the fact that IIoT involves access to greater amounts of confidential data and the coordination of many more physically adjacent systems. These systems must additionally be accessible to authorised staff at all times, failing which severe consequences such as the shutting down of entire factories could be observed.

Privacy and reliability were found to be less important to IIoT systems than to CIoT systems. One reason for this could be concern surrounding personal data in view of relatively recent data leaks such as the Facebook-Cambridge Analytica data scandal. Further, although privacy was considered more important than reliability for CIoT systems, reliability was considered more important in relation to IIoT systems.

Recoverability was more important to IIoT systems, indicating the need for the development of more fault-tolerant IIoT systems, whilst scalability remained the least important T&R factor with no major changes in its relative priority number across IoT systems.

On performing a combined pairwise analysis (see Fig. 8) of T&R factors for both CIoT systems and IIoT systems, a similar trend was observed. An important detail to note is that security was indisputably assessed to be the most important T&R factor across IoT systems. Further, despite the aforementioned reflection regarding reliability and privacy, overall, reliability fared marginally better than privacy.

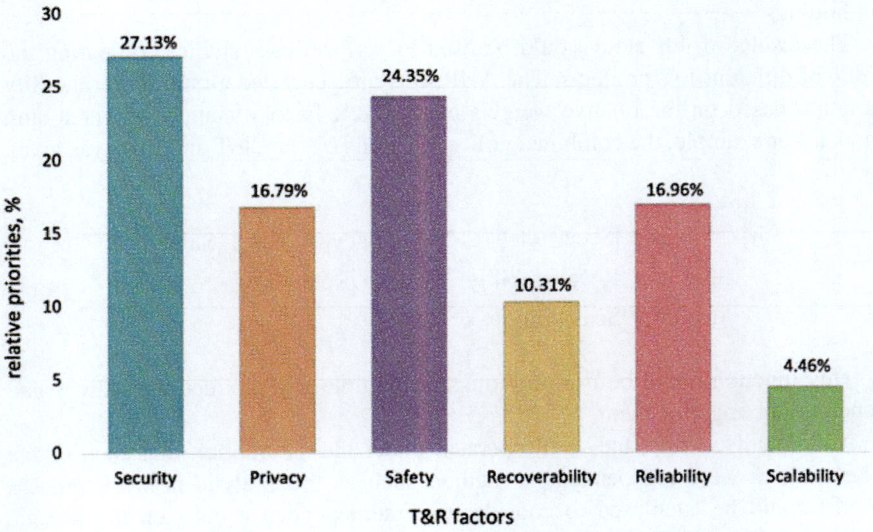

Fig. 8 Relative priorities of T&R factors for combined analysis (both CIoT and IIoT systems)

6 Conclusion

With the recent surge in IoT, many concerns have been raised as to its long-term feasibility. It is becoming increasingly necessary to address the issues surrounding the security and privacy of IoT, amongst other factors. This paper collectively laid down concise definitions for six key terms (security, privacy, safety, recoverability, reliability and scalability) influencing the T&R of IoT systems with the aim of analysing these factors with respect to the CIoT and IIoT spaces. Specifically, it utilised pairwise comparisons as part of the Analytic Hierarchy process to rank the factors according to their importance in determining the T&R of IoT devices. It compared and analysed results, obtained from a survey conducted amongst IoT experts, between CIoT and IIoT systems. It established that security is the most important factor influencing the T&R of Io systems and, in the future, special emphasis must be placed on enhancing the security features of IoT systems in order to defend against the onset of ever-increasing cyber attacks.

7 Future Scope

Although this study was limited to a small sample and the results may not be representative of the general consensus regarding the T&R of CIoT and IIoT systems, going forward the study could be extended to a larger sample to validate its findings.

The results of this study could be used to perform case studies comparing the T&R of different IoT products. The AHP software generated a multi-criteria utility function based on the relative weights of the T&R factors using a weighted sum model. For example, the combined utility function (U) for CIoT and IIoT was given by:

$$\begin{aligned} U = {} & 0.27 * [\text{Security}] + 0.17 * [\text{Privacy}] + 0.24 * [\text{Safety}] \\ & + 0.10 * [\text{Recoverability}] + 0.17 * [\text{Reliability}] \\ & + 0.04 * [\text{Scalability}] \end{aligned} \tag{1}$$

This function could be used to compare multiple IoT product alternatives and generate rankings for them.

Additionally, whilst the AHP operated under the assumption that each of the T&R factors were independent, a method such as the analytic network process (ANP) could be later used to consider the interdependence between the factors, whilst developing IoT systems.

Acknowledgements I would like to thank Dr. Dinesh Likhi, adjunct professor at the Indian Institute of Technology, Roorkee, for guiding me throughout this research, Dr. Akanksha Upadhyaya, Associate Professor, Rukmini Devi Institute of Advanced Studies, New Delhi for her review and comments on the paper, and Dr. Kavita Khanna, Associate Professor and the Head of the Department of Computer Science and Engineering, The NorthCap University, Gurgaon for guiding me and giving me the opportunity to present my paper at the ICCSDF 2021 conference at The NorthCap University. I would also like to thank my parents for their constant guidance and support.

References

1. International Telecommunication Union.: ITU-T Y.4000/Y.2060 "ITU-T recommendations". http://handle.itu.int/11.1002/1000/11559. Last accessed 20 Jan 2021
2. IBM.: What is the Internet of Things (IoT)? https://www.ibm.com/blogs/internet-of-things/what-is-the-iot/. Last accessed 26 Jan 2021
3. Norton.: What is the Internet of Things? How the IoT Works, and More. https://us.norton.com/internetsecurity-iot-what-is-the-internet-of-things.html. Last accessed 26 Jan 2021
4. Industrial Internet Consortium.: IIC:PUB:G4:V1.0:PB:20160919 "Industrial Internet of Things Volume G4: Security Framework". https://www.iiconsortium.org/pdf/IIC_PUB_G4_V1.00_PB.pdf. Last accessed 24 Jan 2021
5. Gartner.: Gartner Identifies Top 10 Strategic IoT Technologies and Trends. https://www.gartner.com/en/newsroom/press-releases/2018-11-07-gartner-identifies-top-10-strategic-iot-technologies-and-trends. Last accessed 20 Jan 2021
6. Ngo, Q.-D., Nguyen, H.-T., Le, V.-H., Nguyen, D.-H.: A survey of IoT malware and detection methods based on static features. ICT Express (2020). https://doi.org/10.1016/j.icte.2020.04.005
7. Ronen, E., Shamir, A., Weingarten, A.-O., OFlynn, C.: IoT goes nuclear: creating a ZigBee chain reaction. In: IEEE Symposium on Security and Privacy (SP), pp. 195–212. IEEE (2017). https://doi.org/10.1109/sp.2017.14
8. Panchal, A.C., Khadse, V.M., Mahalle, P.N.: Security issues in IIoT: a comprehensive survey of attacks on IIoT and its countermeasures. In: IEEE Global Conference on Wireless Computing and Networking (GCWCN), pp. 124–130. IEEE (2018). https://doi.org/10.1109/gcwcn.2018.8668630
9. Wurm, J., Hoang, K., Arias, O., Sadeghi, A.-R., Jin, Y.: Security analysis on consumer and industrial IoT devices. In: 21st Asia and South Pacific Design Automation Conference (ASP-DAC), pp. 519–524. IEEE (2016). https://doi.org/10.1109/aspdac.2016.7428064
10. Atlam, H.F., Wills, G.B.: IoT security, privacy, safety and ethics. In: Digital Twin Technologies and Smart Cities, pp. 123–149. Springer, Cham (2020). https://doi.org/10.1007/978-3-030-18732-3_8
11. Papp, D., Tamás, K., Buttyán, L.: IoT hacking—a primer. Infocommun. J. **11**(2), 2–13 (2019). https://doi.org/10.36244/ICJ.2019.2.1
12. Vaccari, I., Cambiaso, E., Aiello, M.: Remotely exploiting AT command attacks on ZigBee networks. Secur. Commun. Netw. 1–9 (2017). https://doi.org/10.1155/2017/1723658
13. Qiu, T., Liu, X., Han, M., Li, M., Zhang, Y.: SRTS: a self-recoverable time synchronization for sensor networks of healthcare IoT. Comput. Netw. **129**, 481–492 (2017). https://doi.org/10.1016/j.comnet.2017.05.011
14. Alladi, T., Chamola, V., Sikdar, B., Choo, K.-K.R.: Consumer IoT: security vulnerability case studies and solutions. IEEE Consum. Electron. Mag. **9**(2), 17–25 (2020). https://doi.org/10.1109/mce.2019.2953740

15. Loi, F., Sivanathan, A., Gharakheili, H.H., Radford, A., Sivaraman, V.: Systematically evaluating security and privacy for consumer IoT devices. In: Proceedings of the 2017 Workshop on Internet of Things Security and Privacy, pp. 1–6 (2017)
16. Bakhshi, Z., Balador, A., Mustafa, J.: Industrial IoT security threats and concerns by considering Cisco and Microsoft IoT reference models. In: 2018 IEEE Wireless Communications and Networking Conference Workshops (WCNCW), pp. 173–178. IEEE (2018). https://doi.org/10.1109/wcncw.2018.8368997
17. Wang, Q., Zhu, X., Ni, Y., Gu, L., Zhu, H.: Blockchain for the IoT and industrial IoT: a review. Internet Things **10**, 100081 (2020). https://doi.org/10.1016/j.iot.2019.100081
18. European Telecommunication Standards Institute: ETSI EN 303 645 "ETSI Releases World-Leading Consumer IoT Security Standard". https://www.etsi.org/newsroom/press-releases/1789-2020-06-etsi-releases-world-leading-consumer-iot-security-standard. Last accessed 27 Jan 2021
19. IGI Global: "What is System Resilience?" https://www.igi-global.com/dictionary/cyber-threats-to-critical-infrastructure-protection/51260. Last accessed 26 Jan 2021
20. Collins Dictionary.: Definition of 'Security'. https://www.collinsdictionary.com/dictionary/english/security. Last accessed 26 Jan 2021
21. Moore, A.D.: Defining privacy. J. Soc. Philos. **39**(3), 411–428 (2008)
22. Tawalbeh, L., Muheidat, F., Tawalbeh, M., Quwaider, M.: IoT privacy and security: challenges and solutions. Appl. Sci. **10**(12), 4102 (2020). https://doi.org/10.3390/app10124102
23. Nykyri, M., Kuisma, M., Karkkainen, T.J., Hallikas, J., Jappinen, J., Korpinen, K., Silventoinen, P.: IoT demonstration platform for education and research. In: IEEE 17th International Conference on Industrial Informatics (INDIN), vol. 1, pp. 1155–1162. IEEE (2019). https://doi.org/10.1109/indin41052.2019.8972280
24. Yang, H., Jiang, B., Staroswiecki, M., Zhang, Y.: Fault recoverability and fault tolerant control for a class of interconnected nonlinear systems. Automatica **54**, 49–55 (2015). https://doi.org/10.1016/j.automatica.2015.01.037
25. Gupta, A., Christie, R., Manjula, R.: Scalability in Internet of Things: features, techniques and research challenges. Int. J. Comput. Intell. Res. **13**(7), 1617–1627 (2017)
26. Tata Consultancy Services: "Build a Scalable Platform for High-Performance IoT Applications". https://www.tcs.com/content/dam/tcs/pdf/discover-tcs/Research-and-Innovation/Build_a_Scalable_Platform_pdf.pdf. Last accessed 26 Jan 2021
27. SpiceLogic Inc.: Analytic Hierarchy Process Software v2.2. https://www.spicelogic.com/Products/ahp-software-30/

A Comprehensive Study on Vulnerabilities and Attacks in Multicast Routing Over Mobile Ad hoc Network

Bhawna Sharma and Rohit Vaid

Abstract MANET is an autonomous collection of mobile devices. They need the features of infrastructure-less network, flexibility, random mobility, and they do not require any base station or centralized device for the communication process. Rather than this, each device in MANET acts as a client and server. So it becomes a hot research topic among researchers. Communication between nodes is completed by intermediate nodes. Sometimes the intermediate nodes act as malicious nodes by implementing any abnormal function. So we would like to guard the traditional nodes. Therefore, we examine some routing attacks, and how they drastically affect the MANET communication process.

Keywords Network security · Ad hoc network · Denial of service · Route request · Route reply · Attacks · Secure routing protocols

1 Introduction

Mobile Ad-hoc networks (MANET) are the organizations of portable processing gadgets joined remotely with no help of fixed cooperation. There are a few attributes of MANET, which are as per the following:

- No requirement of fixed street and rail organization.
- Network of the organization is dynamic.
- Two nodes be in contact straightforwardly on the off chance that they are inside radio reach.
- Less secure than wired organization.
- MANET is an independent arrangement of portable nodes. It can work in disengagement or may have doors to and interfaces with a fixed organization.
- There are bandwidth constraints and energy constraints.
- Distributed nature of action for security, controlling, and have arrangements.

B. Sharma (✉) · R. Vaid
Department of Computer Science and Engineering,
MMEC, MM (Deemed To Be University), Mullana, Ambala, India

K. Khanna et al. (eds.), *Cyber Security and Digital Forensics*,
Lecture Notes on Data Engineering and Communications Technologies 73,
https://doi.org/10.1007/978-981-16-3961-6_22

- More adaptable than fixed network.
- High client thickness and enormous degree of client portability.
- Nodal network is irregular.

In Fig. 1, design of MANET has been appeared in which a bunch of cell phones is associated together to shape a portable impromptu organization. The gadget with high calculation capacity and more battery force can be chosen as the gathering chief, who is dependable, the general administration of gathering correspondence inside the organization.

In MANET, there are different types of routing—unicast routing and multicast routing. The unicast routing is used for one-to-one communication, whereas multicast routing is used for one-to-many communications [2]. Broadcast conveys a message to all or any hub inside the organization. Multicast conveys a message to a bunch of hubs that demonstrate revenue in accepting the message. Anycast conveys a message to anybody out of a bunch of hubs, as a rule of the one nearest to the source. Geocast conveys a message to a geological area [3] (Fig. 2).

2 Multicasting

Multicasting correspondence fills in as one basic activity to help numerous uses of mobile Ad hoc networks (MANETs) that accomplishes bunch correspondence as opposed to sets of people. Multicast steering conventions turns out to be progressively significant in MANETs since they adequately arrange a lot of nodes [4]. Moreover, it gives viable coordinating to blended media applications, for instance, video social occasions, military, and rescue errands (Fig. 3).

2.1 Routing Protocols

There are many routing protocols in MANET. At whatever point a hub needs to talk with target hub, it broadcasts its current status to neighbors. Guiding shows can be arranged into proactive, reactive, and hybrid directing show.

Fig. 1 Structure of mobile Ad hoc network [1]

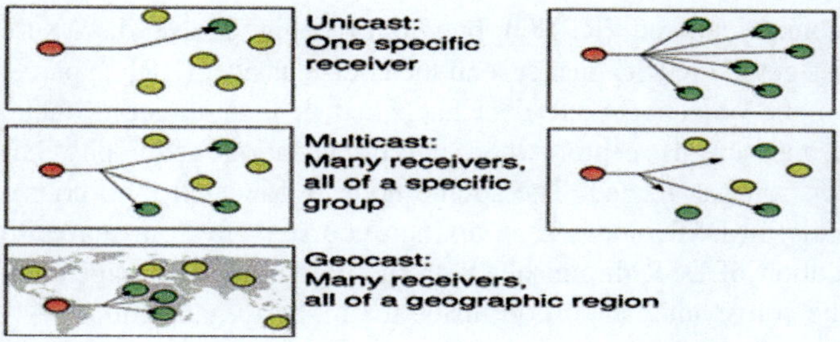

Fig. 2 Different types of routing

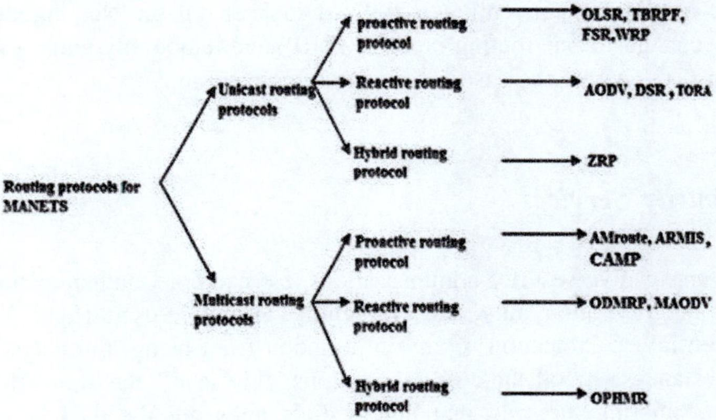

Fig. 3 Classification of routing protocols in MANET [5]

Proactive Routing Protocol:
This is a table-driven coordinating show. Each hub keeps a coordinating table which not only contains record of bordering hubs and reachable hubs, but also the amount of hops. If the size of association extends, the overhead furthermore increases which achieves decline in execution. Target sequenced distance vector (DSDV) and optimized interface state coordinating (OLSR) are proactive shows.

Reactive Routing Protocol:
This convention is likewise approached as request directing convention. At the point when a node needs to send information bundle, the responsive convention began. The preferred position of this convention is that squandered data transmission incited from consistently broadcast gets decreased. The primary weakness of this convention is that it prompts bundle misfortune. Ad hoc on-request distance vector (AODV) and dynamic source routing (DSR) are the cases of responsive directing convention. In AODV, every node records the data of next bounce in its steering table. The course revelation measure is executed at the point when the objective node cannot be reached from source node. The source node

communicates the course demand (RREQ) bundle to begin course disclosure measure. All the nodes get the RREQ packet send the course answer (RREP) parcel to the source node if the objective node data happened in their directing table. Course maintenance measure is begun when the organization geography has changed or the association has fizzled. The source node is educated by a course mistake (RRER) bundle. In DSR, nodes keep up their course store from source to objective node. Execution of DSR diminishes with the portability of organization builds, a lower bundle conveyance apportion inside the higher organization.

Hybrid Routing Protocol:
This convention contains the upsides of proactive, what is more, responsive convention. Proactive convention is utilized to accumulate the new steering data. At that point, responsive convention is utilized to keep up the steering data when geography changes. Zone routing protocol (ZRP) and temporally requested routing calculation (TORA) are the cases of crossover convention.

2.2 Security Services

MANETs are to give security administrations, for example, authentication, confidentiality, integrity, anonymity, and availability, to mobile users [5].

Confidentiality: Protection of any information from being introduced to unintended substances. In off-the-cuff associations, this is all the more difficult to achieve, considering the way that intermediate hubs get the packs for various recipients, so they can without a doubt tune in the information being coordinated.

Availability: Services should be available at whatever point required. There should be an affirmation of survivability, paying little heed to a denial of service (DOS) attack. On physical and media access control layer, the assailant can use adhering techniques to intrude with correspondence on real channel. On association layer, the attacker can upset the coordinating show. On higher layers, the attacker could chop down raised level organizations.

Authentication: Assurance that an element of concern or the cause of a correspondence is the thing that it professes to be or from. Without which, an aggressor would mimic a node in this manner, picking up unapproved admittance to asset and touchy data, and meddling with activity of different nodes.

Integrity: Message being sent is rarely adjusted.

Non-disavowal: Ensures that sending and getting gatherings can never deny truly sending or getting the message.

3 Literature Review

Jhaveri [6] proposed an MR-AODV convention which is an adjustment of R-AODV. MR-AODV not just distinguishes the dark opening and dim opening hubs, but additionally builds up free from any danger course for information transmission during the course disclosure measure.

Dhurandher et al. [7] proposes GAODV convention which is an altered AODV convention. Here, the presence of dark opening can be identified by utilizing critical control parcels CONFIRM, REPLYCONFIRM, and CHCKCNFRM. The source hub communicates RREQ message, and the middle hubs send RREP message to source, and afterwards, they unicast CONFIRM bundle to objective hub.

Karthikkannan et al. [8] proposed the grouping number distinguishing proof technique to keep away from the dark opening assaults in MANET. Here, an extraordinary grouping number will be given to every data parcel and the new bundle should have an arrangement number more noteworthy than that of pervious parcel.

In MANET, major focus was on increasing performance parameter values by developing new and updated mechanisms, and for this, several methodologies were offered. But, along with performance, security is also an important concern that must be taken care of [9]. In MANET, several attacks were found out due to which security of information can be compromised. Unauthenticated or malicious nodes are performing their attempts to be successful so that vulnerabilities can be found out in system, and accordingly, attack can also be imposed on network [10]. Each layer faces distinctive sort of assaults. Table 1 shows the normal assaults on different layers of MANETs [11].

4 Classification of Security Attacks on MANET

Making sure about MANETs is an exceptionally testing issue inferable from its existing engineering weaknesses. Assaults can be focused at steering conventions or even at security instruments conveyed in networks. Traded-off nodes can be

Table 1 Type of attacks on layers [13]

Layer	Attacks
Physical layer	Jamming, interceptions, eavesdropping
Data link layer	Traffic analysis, monitoring
Network layer	Wormhole, black hole, gray hole, message tempering, Byzantine, flooding, resource consumption, location disclosure attacks
Transport layer	Session hijacking, SYN flooding
Multiple layer	Denial of service (DoS), man-in-the-middle attack

available outside also as within the organization. Assailants can disturb typical organization steering, confine node(s), may burn through imperative assets.

4.1 Internal Attacks

This sort of assaults are started by approved (real) nodes inside an organization. An inside node may get undermined by an outer aggressor, or it might carry on egotistically to spare its assets. Inward assaults are extremely difficult to recognize.

Ex: Byzantine attacks.

4.2 External Attacks

This sort of assault is started by non-approved (non-legitimate) nodes which are not a piece of the organization. Outside bargained nodes can seriously upset organization's directing and can cause blockage in different pieces of the organization (Fig. 4).

Ex: eavesdropping.

4.3 Passive Attacks

In this assault, an aggressor just tunes in or monitors information of data that is being moved between two parties. No change and manufacture is finished. Instances of latent assaults are snooping and traffic analysis. Assailants can undoubtedly get all the data about the organization that is helpful in

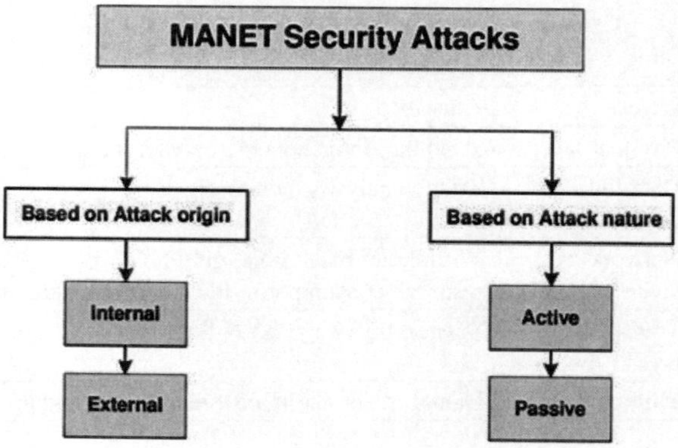

Fig. 4 Classification of security attacks in MANET

commandeering or infusing an assault in the network. It is very difficult to identify inactive assaults when contrasted with dynamic assaults [12].

Ex: eavesdropping, traffic monitoring and analysis.

4.4 Active Attacks

In this assault, an aggressor endeavors to adjust or modify the information being traded in the organization. It might disturb the ordinary working of the organizations. In dynamic assault, the interlopers can change the bundles, infuse the parcels, drop the parcels, or it can utilize the different component of the organization to dispatch the assault.

Ex: spoofing, denial of services, wormhole, black hole, sinkhole, Sybil, etc.

Wormhole Attack: In this assault, an assailant records parcels at one area in the organization and passages them to another area. This passage between two plotting assailants is alluded as wormhole. Directing can be disturbed when steering control message are burrowed [14]. Wormhole assault is utilized against on-demand routing protocol the assault could forestall the disclosure of any courses other than through the wormhole. Tunneling is used by the attacker [15].

Black-hole Attack: In this assault, a black opening is a vindictive node that erroneously answers for course demands without having a functioning course to the objective and endeavors the directing convention to promote itself as having a most brief course to objective. By promoting the most limited course, source station begins sending information through the black opening node, and it become the dynamic component in the course (Fig. 5).

Byzantine Attack: In this attack, a sabotaged temporary hub works alone, or a lot of haggled center hubs works in plan and complete attacks. These assailant hubs make controlling circles, sending groups through non-ideal ways, or explicitly dropping packs, which achieves interference or debasement of the guiding organizations.

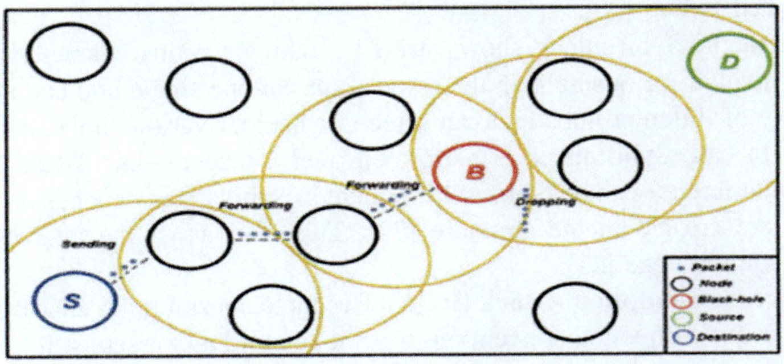

Fig. 5 Illustration of black-hole attack in MANET [16]

Traffic Monitoring and Analysis: In MANET composing, it is moreover named as location disclosure attack. In this kind of attack, the noxious hub screens, the conveyed groups, and examinations in this traffic which may reveal information, for instance, zone of sender–gatherer, sender collector pair, network topography, network coordinating structure, traffic rate, presence, zone of other genuine hubs, etc. A couple of association gadgets exist in the Web which can be used, thus, for instance, NetStumbler. Using this divulged information, other malicious hubs may similarly configure further attack circumstances in coordination. The attacker can even record, change, and retransmit changed packages to other veritable hubs remaining absolutely vague. Spillage of such information can be wrecking in security fragile conditions.

Eavesdropping: In this type of assault, the malevolent node captures the bundles sent or got, and it may uncover some classified data, for example, area of sender/beneficiary, mystery keys, passwords, and so on which might be generally left well enough alone during the correspondence between approved clients [17]. This is an aloof type of assault which owes itself because of simple tapping of remote nature of correspondence medium in MANETs.

Gray Hole Attack: In this sort of assault, a scornful node does not take an interest in course revelation instrument that is started by different nodes and is consequently not a piece of dynamic course. Such contemptuous nodes would build the course revelation disappointment and damage the general organization execution [18]. Another goal of such assailants is to moderate their energy by deciphering the message planned for them just and else they do not help out different nodes, which at last debase the presentation of the organization.

Jellyfish Attack: In this assault, the vindictive node first turns into a piece of the organization, and afterward, it might reorder the arrangement of got bundles, create undesirable postponements in bundle sending, or drop parcels [19]. This assault is like black-hole assault in any case; here, recognition is more troublesome in view of inclination of assailant to act as per convention rules. This makes the making trouble node yield very good quality to-end delay, high jitter and fundamentally influences the throughput of the organization.

Impersonation Attack: In impersonation attack, attacker node impersonates itself as authentic hub and sends bogus directing data and veils itself as sending from confided in hub [20].

Sybil Attack: Sybil attack shows itself by faking various characters by professing to involve various hubs in the association. So one single hub can anticipate the capacity of different hubs and can screen or hamper various hubs at the same time [21]. In case Sybil attack is performed over a blackmailing attack, by then degree of interference can be high. Achievement in Sybil attack depends on how the characters are created in the structure [22]. This may assist the aggressor with breaking required edge [23].

Resource Consumption Attack (RCA): Resource consumption attack (RCA) is against on-request directing convention. It is the one of DOS assaults, in which the aggressor abuses the course revelation process. During the course disclosure measure when the source node sends the RREQ parcel, at that point assailant node kept this bundle with an alternate ID, to adjust the cycling ID of every node ceaselessly and devour its restricted energy of asset, memory, and bandwidth is appeared. The primary reason for RCA is to burn through the energy of genuine hubs and to locate the accessible connection all through [24].

Flooding Attack: Flooding assault is dispatched by flooding the organization with counterfeit RREQ's or information bundles prompting the blockage of the organization and decreases the likelihood of information transmission of the approved hubs [25]. The identification of assault is exceptionally hard, and it debilitates the organization assets (Table 2).

Table 2 Summary table

S. No.	Name of attack	Attack effect
1	Wormhole attack	• Packet drain/rope methods • MAD convention and OLSR convention • Directional reception apparatuses • Multi-dimensional scaling calculation (versatility) • Using nearby neighborhood data • DAWWSEN convention • Designing appropriate steering conventions (grouping-based and topographical steering conventions) • Leveraging worldwide information
2	Black-hole attack	• Approval and monitoring • Redundancy • Using another course • Multipath steering
3	Byzantine attack	• Prevent the route establishment • Create loops, forwards packets through non optimal paths [26]
4	Traffic monitoring and analysis	• Access control • Reduction in detected information subtleties • Distributed handling • Strong encryption methods • Sending faker bundles persistently and normal checking
5	Eavesdropping	• Access control • Reduction in detected information subtleties • Distributed preparing • Access limitation • Strong encryption procedures
6	Gray hole attack	• Cautious instruments of black-hole assault, aside from excess also, utilizing worldwide information
7	Jellyfish attack	• Compliance with all data and control protocols • Affects mainly closed-loop flows [27]
8	Impersonation attack	• Strong and legitimate verification methods • Using solid information encryption

(continued)

Table 2 (continued)

S. No.	Name of attack	Attack effect
9	Sybil attack	• Certificate authority (CA) and using personality endorsements • Limiting the quantity of hub's neighbors • Physical insurance of gadgets • Changing key consistently • Resetting gadgets and changing meeting keys (network layer) • Authentication, interface layer encryption, and worldwide shared key procedures [28]
10	Resource consumption attack	• Consumes the energy of legitimate nodes and to find the available link throughout [24]
11	Flooding attack	• Customer puzzles • AODV (Ad hoc on-request distance vector) convention • Limiting the quantity of hub's associations • Routing access restriction • Key the board

5 Conclusion

Security is the standard concern in MANETs. Because of their basic properties, for instance, dynamic topography, nonattendance of central position, confined resources and open access medium Remote exceptionally named associations are introduced to being attacked or harmed. These basic credits familiarize new troubles with interference disclosure advancement, so it is difficult to achieve security in Ad hoc network when stood out from wired organizations. In this paper, we first briefly summed up the MANET and mainstream steering conventions in it. At that point, kinds of assaults alongside a most recent review of existing arrangements are examined. Various creators have given different expert throbs for discovery and counteraction of vindictive assault in MANET, yet every methodology has its own restriction. The malignant assault is as yet a functioning research zone in MANET. In the future, assessment fuses intend to develop such a security computation, which will be presented in header of each center point that helps in acknowledgment and expectation of malicious attacks.

References

1. Chander, D., Kumar, R.: Analysis of scalable and energy aware multicast routing protocols for MANETs. Ind. J. Comput. Sci. Eng. (IJCSE) **8**(3) (2017)
2. Jain, S., Agrawal, K.: Prevention against rushing attack in mobile Ad hoc networks. Int. J. Comput. Sci. Technol. (IJCST) (2014)
3. Gridher, V., Jain, S.: Review paper on an optimized approach for attack detection and prevention in wireless sensor networks. Int. J. Comput. Sci. Technol. (IJCST) (2017)

4. Qabajeh, M.M., Abdalla, A.H., Khalifa, O., Qabajeh, L.K.: A tree-based QoS multicast routing protocol for MANETs. In: 4th International Conference on Mechatronics (ICOM) (2011)
5. Sliman, K.A., Yaklaf, A., Abdurrezagh, S.E., Ekreem, N.B., Abosdel, A.A.M.: Security routing protocols in Ad hoc networks: challenges and solutions. In: Proceedings of the International Conference on Recent Advances in Electrical Systems, Tunisia (2016)
6. Jhaveri, R.H.: MR-AODV: a solution to mitigate blackhole and grayhole attacks in AODV based MANETs. In: 2012 Third International Conference on Advanced Computing & Communication Technologies, pp. 254–260. IEEE (2012) (978–0–7695–4941)
7. Dhurandher, S.K., Woungang, I., Mathur, R., Khurana, P.: GAODV: a modified AODV against single and collaborative black hole attacks in MANETs. In: 2013 27th International Conference on Advanced Information Networking and Applications Workshops, pp. 357–362. IEEE (2013) (978–0–7695–4952)
8. Karthikkannan, P., Lavanya Priya, K.P.: Reduction of Delays in Reactive Routing Protocol for Unobservable Mobile Ad-Hoc Networks. IEEE (2013)
9. Kumar, A.: Security attacks in MANET—a review. In: IJCA Proceedings on National Workshop-Cum-Conference on Recent Trends in Mathematics and Computing (2011)
10. Kumar, A., Singh, J.: Security attacks in mobile Ad hoc networks (MANET): a literature survey. Int. J. Comput. Appl. **122**(20), 31–35 (2015)
11. Mohammad, S.N.: Security attacks in MANETS (survey prospective). Int. J. Eng. Adv. Technol. (IJEAT) **6**(3) (2017). ISSN: 2249–8958
12. Dobhal, N., Pundir, D.: An investigative survey of different security attacks in MANETs. Int. J. Comput. Appl. (0975–8887) **126**(1) (2015)
13. Goyal, M., Poonia, S.K., Goyal, D.: Attacks finding and prevention techniques in MANET: a survey. Adv. Wirel. Mobile Commun. **10**(5), 1185–1195 (2017). ISSN 0973–6972
14. Rajkumar, K., Prasanna, S.: Complete analysis of various attacks in MANET. Int. J. Pure Appl. Math. **119**(15), 1721–1727 (2018)
15. Majumder, S., Bhattacharyya, D.: Mitigating wormhole attack in MANET using absolute deviation statistical approach. In: 2018 IEEE 8th Annual Computing and Communication Workshop and Conference (CCWC), pp. 317–320. IEEE (2018, January)
16. Yasin, A., Abu Zant, M.: Detecting and isolating black-hole attacks in MANET using timer based baited technique. In: Hindawi Wireless Communications and Mobile Computing Volume 2018, Article ID 9812135
17. Goyal, U., Gupta, M., Kaur, K.: Meliorated detection mechanism for the detection of physical jamming attacks under AODV and DSR protocols in MANETs. IJAIEM **3**(10) (2014)
18. Alkatheiri, M.S., Liu, J., Sangi, A.R.: AODV routing protocol under several routing attacks in MANETs. IEEE (2011) 978–1–61284–307–0/11
19. Sachdeva, S., Parneet Kaur, M.: Routing attacks and their countermeasures in MANETs: a review. Int. J. Adv. Res. Comput. Sci. **7**(4) (2016)
20. Latha, R., Sasikala, S.: A survey of routing attacks in Manet. In: Proceedings of the UGC Sponsored National Conference on Advanced Networking and Applications, 27th March 2015
21. Singh, R., Singh, J.: A novel Sybil attack detection technique for wireless sensor networks. IEEE J. Sel. Areas Commun. **10**, 185–202 (2017)
22. Saha, H.N., Bhattacharjee, D.: Different Types of Attacks in Mobile Ad hoc Network: Prevention and Mitigation Techniques. https://arxiv.org/ftp/arxiv/papers/1111/1111.4090.pdf
23. Rajakumar, P., Prasanna, V.T., Pitchaikkannu, A.: Security attacks and detection schemes in MANET. In: 2014 International Conference on Electronics and Communication Systems (ICECS), pp. 1–6. IEEE (2014, February)
24. Jain, S., Agrawal, K.: The impact of resource consumption attack on signal-stability based adapting routing protocol in MANET. In: International Conference on Recent Developments in Science, Engineering and Technology (IJST) (2016)
25. Nithya, S., Prema, S., Sindhu, G.: Security issues & challenging attributes in mobile ad-hoc networks. Int. Res. J. Eng. Technol. (IRJET) **03**(01), 1083–1087 (2016)

26. Manohar, B., Kumar, M.: Review on Byzantine attack in MANET and solution to avoid. Int. Res. J. Eng. Technol. (IRJET) **6**(1) (2019). e-ISSN: 2395–0056
27. Kaur, M., Rani, M., Nayyar, A.: A comprehensive study of jelly fish attack in mobile Ad hoc networks. Int. J. Comput. Sci. Mob. Comput. **3**(4), 199–203 (2014). ISSN 2320–088X
28. Newsome, J., Shi, E., Song, D., Perrig, A.: The Sybil attack in sensor networks: analysis & defenses. Cent. Comput. Commun. Secur. (2004)

An Offensive Approach for Hiding Malicious Payloads in an Image

Keshav Kaushik and Sneha Surana

Abstract Steganography is the oldest technique that is been used from century, steganography purpose has not changed, i.e., all these techniques aim at hiding data or protecting data. With the help of steganalysis, the media can be analyzed to check for the presence of any secret information. Nowadays, attackers are making the use of advanced steganography approaches to conceal the secret information and communicate in a stealth manner. In this paper, the authors have discussed about the novel approach to hide malicious payload into image metadata. Therefore, metadata is a data that describes about the image rights and its administration. Hacker generally uses this metadata to perform various malicious attacks such embedding malicious script inside the image metadata and many more.

Keywords Steganography · Steganalysis · EXIF · Payload · Metadata · Image steganography · Stager · Cybersecurity · Digital forensics · Cyber forensics

1 Introduction

With the increase of Internet usage, people uploading and downloading pictures activity have increased. Now hackers are taking advantage of these activities and performing attacks. Once such type of attack is hiding payload in the image, it is actually steganography. Steganography is the art of concealing or hiding data within another object. The message can be hidden in physical objects or in reference of electronic, it can be an image, video, audio. Now the message is payload for hackers, they generally try to embed the malicious script when the image is downloaded on the victim system, it will execute on the system and run the process or delete the files or spread of malware, viruses, etc., ultimately leading to damage

K. Kaushik (✉)
Department of Systemics, School of Computer Science, University of Petroleum and Energy Studies, Dehradun, India

S. Surana
University of Petroleum and Energy Studies, Dehradun, India

K. Khanna et al. (eds.), *Cyber Security and Digital Forensics*,
Lecture Notes on Data Engineering and Communications Technologies 73,
https://doi.org/10.1007/978-981-16-3961-6_23

or loss of the victim. Image steganography is majorly classified based on two main methods that are spatial domain and transform domain. Others types of steganography includes masking, filtering, and distortion techniques.

2 Related Work

In [1], research paper describes an approach a technique named hue–saturation–intensity (HSI) color space, which depend upon the least significant bit (LSB). This technique is assessed both on subjective and objective part. In the technique proposed, using LSB method, the message that is to be hidden is embedded in I-Plane of HIS color space model. The outcome image is retransformed to RGB color model to develop the stego-image. This technique adds the extra level security to image making difficult for an attacker to detect the whether the data is hidden in the image or not.

In [2], research paper discusses about the technique for hiding an image in audio file using least significant bit (LSB) substitution techniques for implementing image and audio steganography. This technique proposed converts the normal colorful image into grayscale image and use a 16-bit AES algorithm to encrypt a grayscale image. The secret audio is embedded in the grayscale image and 16 bits key in any frame in the video file in which data is to be hidden with the help of LSB substitution technique. The audio and video on which steganography is performed can easily transmit to receiver over the network.

The Android malware clustering is identified through mining of malicious payload in [3]. The research paper discusses keeping the malicious code separate from legitimate library code from malware in Android. It implements robust technique to determine if malware samples having the same version as malicious payload. The method uses tradition hierarchical clustering and fuzzy hashing fingerprint. The techniques improve the performance and scalability on examining of large malware families, and accuracy is also maintained. It can determine that if library is having old version and whether it shares the name with a bogus library.

The STEGANOGAN [4], end-to-end model that is based on deep learning, mitigates vanishing gradient problem for image steganography. Using this model, a 4.4 bits per pixel relative payload is achieved and evades the detection of steganography in the image. This technique is efficacious on images using multiple datasets. STEGANOGAN is an open-source library to enable fair comparisons. This technique [5] uses generative adversarial networks to hide binary data in images to enhance the quality of images produced by this model.

3 Working Methodology

The authors will add a malicious script to the image EXIF [6] metadata. The malicious script (Payload) will be directly embedded into the EXIF metadata using the tool. The image will be available online, and it will be easily accessible. Generally, when the image is open and the payload will execute itself, but, in this case, the stager will be required to execute the payload in the image. So firstly, the stager will download the image from the Web site, and after that, it will extract the payload from image EXIF metadata. Once the payload is extracted, the stager will execute it. Therefore, a stager can be written in short length let us say 100 characters and it will not take much space on disk.

Following tools will be used like curl, Exiftool, bash, and grip.

First step, the image is downloaded, as it is the medium for attack. The image will not be saved on the system or target computer. For downloading the image, the tool will use named Wget [7] that is non-integrated tool network downloader that even download the file from the server whether user is logged on to the system or not. Wget tool can work without hampering the process that is currently working in the background. Wget has various flags like -h flag that displays the information about other various flags, and here, the authors will use -O flag to write documents to file and will save the file to particular directory say in /tmp directory.

Here, the image is being downloaded and saved that file with name image.jpeg in tmp folder (Fig. 1).

Now the **second step** is generating payload, in reference to a cyberattack [8], it is that component of the attack that harms the target system. It can be any virus, worms, malware, or any malicious payloads that can drastically affect the target computer. In this step, the utility named Mate calculator in Kali Linux will be trying to open. The steps are performed, so that it can be run through the terminal. Therefore, whenever payload is executed, Mate calculator will be opened through stager. Here, any type of bash script can be used to perform any operations. Here, the combination printf, base64, and tr commands will be used to encode the payload. The printf command displays the passed formatted text on a terminal in Linux. The base64 is the command in Linux that encodes the text the base64 and decodes base64 encoded text to normal text. The tr is a utility in Linux that translates and

Fig. 1 Downloading image for steganography

deletes characters and supports transformations like uppercase to lowercase, deleting characters. The command be written as this (Fig. 2).

The **third step** will payload embedding into the image, and here, the authors will be modifying the EXIF metadata [9] of the image to embed the above results in it using a tool named Exiftool. Exiftool is an open-source tool for manipulating and writing audio, image, video, and PDF metadata. Now, install that tool in Linux [10] using this command (Fig. 3).

Now, after the installation, it clears all the EXIF metadata that can be in an image using this command (Fig. 4).

Now, the payload is added to the metadata of the image by using Exiftool tag—and certificate tag is only for demonstration purpose (Fig. 5).

Now to verify whether certificate tag is added to the image or not, the authors will use below displayed command (Fig. 6).

See certificate tag in the image in the above image. Now **fourth step** is uploading image on the Web site. It might look easy, but one should upload the image by taking the following factors in mind:

Web Traffic Irregularities—The image should be uploaded on the Web site [11] that is regularly visited by the target. For example, if the target visited a particular news Web site daily, it would not appear suspicious to sysadmin, which

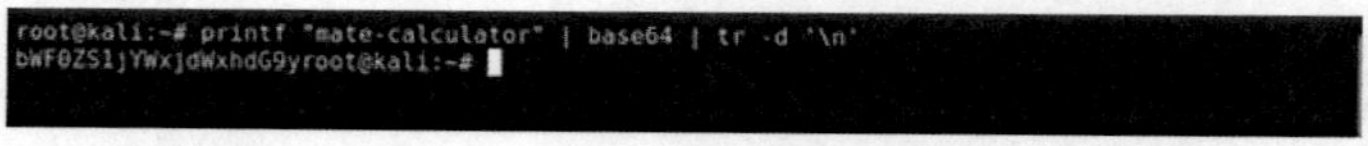

Fig. 2 Preparing payload and encoding it

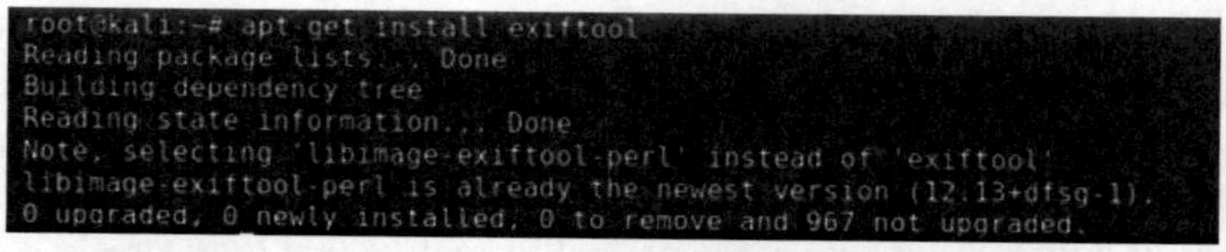

Fig. 3 Updating the packages and installation of Exiftool

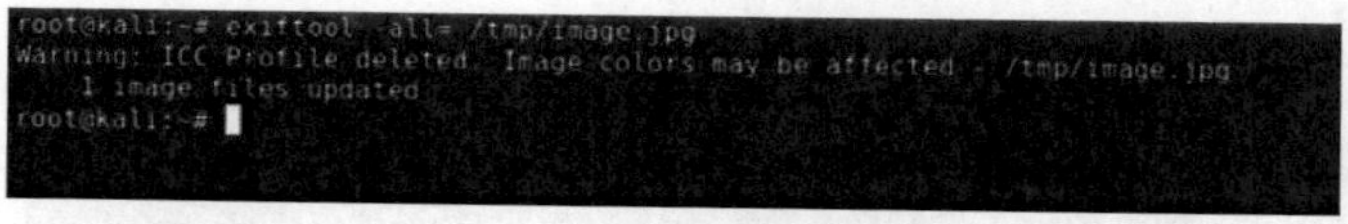

Fig. 4 Cleaning metadata of the image using Exiftool

Fig. 5 Embedding of payload to the image using Exiftool

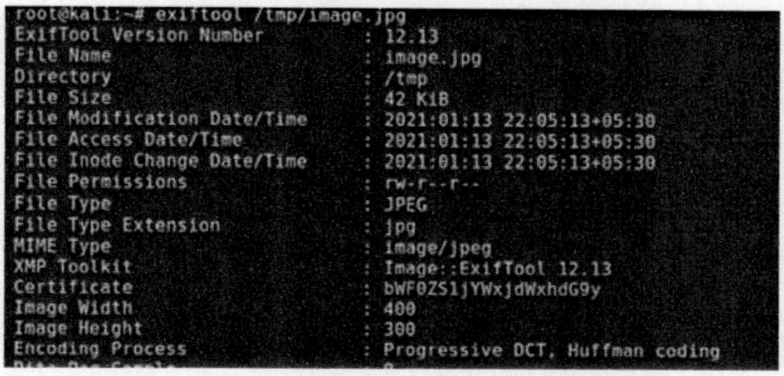

Fig. 6 Payload is embedded using Exiftool

monitor network traffic. The main idea is to make traffic look normal or ordinary to the victim Web behavior.

Avoid EXIF Data Sanitization [12]—Generally, when images are uploaded on popular Web site like Twitter, Instagram, etc., image metadata is deleted to protect the user GPS coordinates that are coincidentally uploaded with photographs, and these GPS coordinate can be source for a cyberstalker to perform attacks or harass people. Therefore, uploading the image on a mainstream Web site [13] that contain payload be a waste, as EXIF metadata will be deleted.

Website Encryption—The Web site on which image will be hosted should be HTTPS, as this will prevent sysadmin network monitor to analyze the GET request with surgical precision. Therefore, transport layer security is a necessary part in making this attack obscure.

Now, the last step is generating the stager; in this step, stager will download the image from the Web site, and then, it will extract the payload from the image and execute it (Fig. 7).

Therefore, this last step is further broken into further steps to understand the role of stager easily:-

- **p = $(….)**: This variable that is used to store the results of commands [14] that are used are executed and it is known as "p," p refers to payload. This is done, so that the image is not directly saved to Linux hard drive. Since the payload is stored in the variable after the commands like to curl, grep, sed, and base64 are being executed, so that it can be payload that contain certain scripts that executed further.
- **curl -s** https://127.0.0.1/image.jpg**-**: Curl is the utility in Linux that uses several protocols, so that data can be transferred from the server. Here, the curl will be used with flag "-s," so that the image is downloaded silently, so victim do not know that script is executed. After the flags, the URL have added for image downloading purpose. For demonstration purpose the image is been hosted on

Fig. 7 Generation of stager

localhost server. The output of curling image from the server will be in the binary form. Now output of this command is piped to grep command.

- **grep certificate -a**: grep command is used to search a mentioned word in input files or in the standard input. As default, grep prints the lines that are matched. The output of the curling image from the server will be in the form binary form. If grep is used with –a, it used to process or search the given input in the binary files like these binary files are containing text. Therefore, in the above command mentioned, grep will search the input keyword certificate in the binary output that has been obtained from binary files. Now output of this command is piped to sed command.
- **sed 's/<[^>]*>//g'**: sed is a basic text editor that is used on input stream that be a file or pipeline input. So as text editor, it performs operations like finding, searching, and replacing. The output of grep command is surrounded by xml data type [15], so to extract payload from this output, it is necessary to use sed command. Here, the "s" in the single quotes in this command stands for substitution operation. The "/" refers to delimiters, and "/g" refers to substitute flag that replaces all occurrences of a string in the line. So here, the command used is replacing all xml data types with leaving empty, and only payload data are extracted by using sed command. Now output of this command is piped to base64 command.
- **base64 -d -i**: base64 is utility in the command line that is used for encoding a text into base64 encoding and decode the base64 encoding to normal text. Here, base64 command is being used to decode the payload that extracted from the image. Here "-d" flag stands for decode data, and "-I" flag stands for ignoring garbage like non-alphabet characters while decoding. The final output from this command is stored in variable "p."
- **eval $p:** eval is utility in Linux that concatenates the arguments into a single string and take it as input and execute the commands. This command is being used to execute the payload which is extracted from the image on the terminal.

The results can be verified that Mate calculator utility is open on the screen while executing the payload. For the demonstration purpose, an open Mate calculator is opened from terminal by executing the payload. The Mate calculator is utility, i.e., present in Kali Linux (Fig. 8).

4 Result and Discussion

The authors are successful in implementing execute the payload which was embedded in the image by using stager. The payload was first encoded with base64 encoding, and then, using Exiftool, all metadata of the image was cleared. The encoded payload was added to the image using Exiftool to the metadata of an image by assigning tag to payload in this tag was named as "certificate." After that stager was written, curl the steganography [16] image from the server and extracted

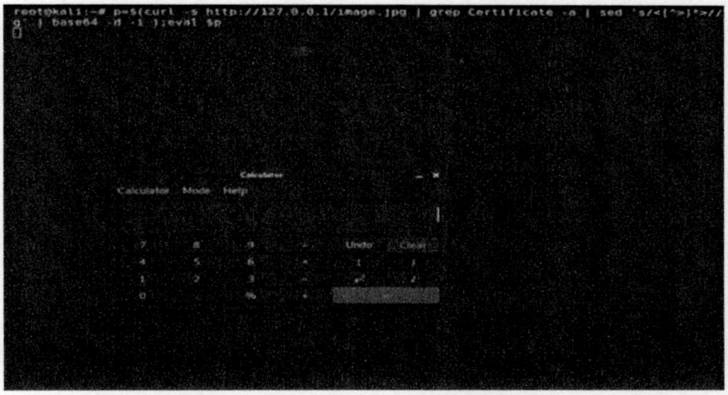

Fig. 8 Triggering Mate calculator after execution of payload using stager

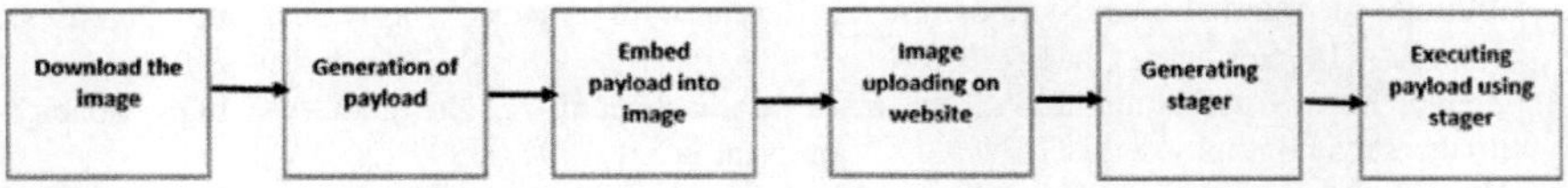

Fig. 9 Procedure of embedding the payload and its execution

encoded payload from binary image data by using grep and sed utility in Linux. The encoded payload was decoded and executed it by assigning payload to a variable and executing variable using eval command in Linux and further executed that variable. For demonstration purpose, the Mate calculator (utility) was opened. We can use various types of payload such as it consists of malware or script that count keystroke or to open software. Once these payloads are executed, they are successful in implementing the things that were assigned to them. It tells the importance of image sanitization and network sanitization to prevent these kinds of attacks in future (Fig. 9).

5 Conclusion

This paper has highlighted the use of steganography in cyberattacks and cybercrimes. Although, steganography is an ancient technique, it is predominant in many cyberattacks with some advanced mechanism and approaches. This paper also focused on the hiding of some malicious payload in the metadata of an image. Likewise, any image can be used for this purpose and later on can be used for performing specialized type of cyberattack. The main motive of this paper was to make the readers and cybersecurity enthusiasts aware about such type of approaches that are in practice in the current scenario. Furthermore, the payload may contain any type of costumed malicious scripts that may be capable of damaging the entire organization with the help of an image only. In future aspect, the metadata of the image can also be used for hiding some payment related QR codes that may trigger some financial transaction from some mobile app. The authors will also look forward to extend this work in near future.

References

1. Muhammad, K., Ahmad, J., Farman, H., Zubair, M.: A novel image steganographic approach for hiding text in color images using HSI color model. arXiv preprint arXiv:1503.00388 (2015)
2. Rawde, M., Kumbhare, M., Chaudhari, S., Bhongade, S., Bhagat, V.: Novel approach towards higher security using crypto-stego technology. Int. J. Emerg. Trends Technol. Comput. Sci. (IJETTCS) **4**(1) (2015)
3. Li, Y., Jang, J., Hu, X., Ou, X.: Android malware clustering through malicious payload mining. In: International Symposium on Research in Attacks, Intrusions, and Defenses, pp. 192–214. Springer, Cham (2017)
4. Zhang, K.A., Cuesta-Infante, A., Xu, L., Veeramachaneni, K.: SteganoGAN: high capacity image steganography with GANs. arXiv preprint arXiv:1901.03892 (2019)
5. Hosam, O., Ben Halima, N.: Adaptive block-based pixel value differencing steganography. Secur. Commun. Netw. **9**(18), 5036–5050 (2016)
6. Alwan, I.M., Mohammed, F.J.: Image hiding using discrete cosine transform. J. College Edu. Women **27**(1), 393–399 (2016)
7. Hamid, I.: Image steganography based on discrete wavelet transform and chaotic map. Int. J. Sci. Res. (IJSR) **7** (2018)
8. Deshpande, S., Mallayyanavarmath, S.: Complete study on steganography [online]. Ijraset.com. Available at: https://www.ijraset.com/fileserve.php?FID=8366 (2017). Accessed 11 Jan 2021
9. Hacking Macos: How to hide payloads inside photo metadata [online]. Available at: https://null-byte.wonderhowto.com/how-to/hacking-macos-hide-payloads-inside-photo-metadata-0196815/ (2019). Accessed 11 Jan 2021
10. Kernel.org.: The linux man-pages project [online]. Available at: https://www.kernel.org/doc/man-pages/ (n.d.). Accessed 11 Jan 2021
11. En.wikipedia.org.: Steganography [online]. Available at: https://en.wikipedia.org/wiki/Steganography (n.d.). Accessed 11 Jan 2021
12. Data Sanitization Terminology And Definitions—International Data Sanitization Consortium [online]. Available at: https://www.datasanitization.org/data-sanitization-terminology/ (n.d.). Accessed 11 Jan 2021
13. LSB Based Image Steganography Using MATLAB—Geeksforgeeks [online]. Available at: https://www.geeksforgeeks.org/lsb-based-image-steganography-using-matlab/ (2020). Accessed 11 Jan 2021
14. The Types and Techniques of Steganography Computer Science Essay [online]. Available at: https://www.ukessays.com/essays/computer-science/the-types-and-techniques-of-steganography-computer-science-essay.php (n.d.). Accessed 11 Jan 2021
15. Kaur, S., Bansal, S., Bansal, R.K.: Image steganography for securing secret data using hybrid hiding model. Multimedia Tools Appl. 1–21 (2020)
16. Zhou, Z., Mu, Y., Wu, Q.J.: Coverless image steganography using partial-duplicate image retrieval. Soft. Comput. **23**(13), 4927–4938 (2019)

A Review of Anti-phishing Techniques and its Shortcomings

Bhawna Sharma and Parvinder Singh

Abstract Phishing has become one of the most common activities observed over the Internet quite often. To investigate the methods through which phishing can not only be detected but can also be controlled, a lot of researchers have contributed and have opened gates for the industry. This paper illustrates the types of phishing attacks and ways to optimize the anti-phishing architecture. The highlights of this paper are listing down the ways to detect phishing activities over web services. The analyzed techniques are compared on the basis of suitable comparative parameters listed in reputed articles.

Keywords Information security · Phishing · Phishing countermeasures · Phishing identification · Assessment measurements

1 Introduction

The revolution of the Internet can be analyzed from our daily life. Due to the wide use of digitized systems, data security is of main concern. Security means that users should take necessary precautions while using technologies. Phishing is a social engineering attack. Phishers maliciously gather user's confidential data like credit card numbers, bank account details, and passwords [1]. The phisher introduced several attacks every day to attack the authentic site and misled the user by sending malicious code. Phishing can be done on a single user or a number of users. If phishing is done on specific users, then it is termed spear phishing. Today, phishing is a critical challenge. However, several methods have been introduced by the

B. Sharma (✉)
Deenbandhu Chhotu Ram University of Science and Technology,
Murthal, Sonepat, Haryana 131039, India

P. Singh
Department of Computer Science and Engineering, Deenbandhu Chhotu Ram University of Science and Technology, Murthal, Sonepat 131039, India
e-mail: parvindersingh.cse@dcrustm.org

K. Khanna et al. (eds.), *Cyber Security and Digital Forensics*,
Lecture Notes on Data Engineering and Communications Technologies 73,
https://doi.org/10.1007/978-981-16-3961-6_24

researchers to prevent phishing attacks or anti-phishing techniques. Phishing may prompt money-related misfortune, information misrepresentation, sensitive data loss, spoliation of reputation, and reduction of trust in the web [2]. Phishing smartly removes the security measures taken by any organization to prevent phishing activity from the hacker. Phishing and legitimate sites seem to be visually alike. Phishing attacks mainly target financial organizations. Phishing is carried out via emails, websites, and instant messengers.

This paper starts by showing the phishing issue in Sect. 2 and portraying distinctive anti-phishing techniques in Sect. 3. Section 4 focuses on evaluation metrics. The conclusion is drawn in Sect. 5 which concludes this paper.

2 The Phishing Problem

A. **History**
The term phishing came into account in 1996 in America. Fishers (e.g., attackers) use a trap (e.g., fake email tempting users for entering sensitive information) to catch a fish (i.e., to befool a user). **Ph**one **ph**reaking was the earliest form of hacking. So the character "f" of fishing was replaced by "ph," and the term "phishing" came into the picture [3]. "Phishing" takes after "angling" in a way that fishers (e.g., attackers) utilize a snare (e.g., socially designed messages) to angle (i.e., to take client's delicate data). The oldest kind of hacking was known as phone phreaking. It was utilized for phone systems.

B. **Definition**
Phishing is an assault in which the aggressor misuses social designing methods to perform data fraud. It is the illegal activity that hackers take to steal crucial data of the user such as details of bank, passwords, unique identification, and other sensitive data.
It occurs when the attacker concealments his identity and acts as an authentic person and fools the users by fake message or text, including malicious connection. This attack consequence in respect of ingathering funds and steal personal details.

Figure 1 displays the mechanism of the phishing attack to steal crucial data of the authentic user using phishing website when users log in to the original website. It mainly happens when the user clicks on the malicious link that makes by the phisher, and the user redirects to the phishing website, so the phisher takes control over it. Phishing customarily works by sending an email, mirroring an online bank, closeout, or installment destinations, controlling clients to a fake website page, which is cautiously intended to resemble the login to the certified webpage. Phishing means gathering touchy and personal information by imitating an authentic entity [4]. Portray a phishing assault in three different ways: (1) a simple site must be picked; (2) a phishing site must be made so that it has a comparative look-and-feel as that of the real site, and (3) delicate data of the client must be

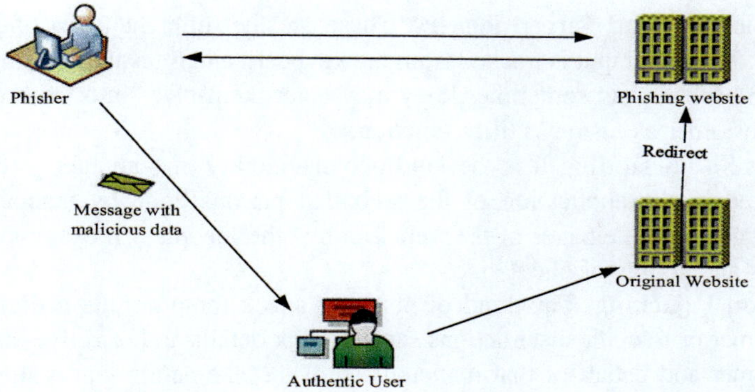

Fig. 1 Mechanism of phishing attack

gathered. Phishing assaults could have genuine ramifications for their exploited people, for example, the loss of protected innovation and sensitive data, monetary misfortune, and the trade-off of national security, just as general debilitating trust.

Different types of phishing attacks as discussed by authors are as follows (Fig. 2):

1. **Deceptive Phishing**: It is a kind of phishing. It happens by sending emails and links to the malicious website that makes to tackle the user and theft the sensitive data of the user. Mainly, these phishing attacker occurs by requesting to users to verify the details of the account, re-enter data like password or logins details and steal the sensitive data of the authentic user.
2. **Malware-Based Phishing**: This phishing attack happens through malicious software on the system of the user. The attacker sends the attachment or as a downloadable document manipulating vulnerabilities of security. It is more problematic for medium and small businesses who unsuccessful in updating their software of the system.
3. **SMiShing**: It utilizes text messages or SMS on cellular mobile or smartphone to tackle the user for the purposes of stealing data of the user. In this process, the attacker tries to trick the customer into providing them crucial data of the customer through the text message.
4. **Vishing**: It is the form of the attack that is derived from phishing and voice. It is the experience of leveraging technologies of IP-based voice messaging to engineer the intended user into giving financial, personal, socially, or other crucial data for financial reward. This attack utilizes the phone calls to mislead the user or trick to user into providing details of the user.
5. **Web Trojan**: It is the kinds of phishing attack that involves malicious code or software that seems to be authentic to steal the sensitive data of the customer. It pops up when the consumer wants to log in to a significant website. But this is not visible to the customer and store the identity details of the user and send to the hacker.

6. **Key loggers and Screen loggers**: These are the different types of malware attack in which input is tracked from the keyboard and transmit this crucial data to the hacker. And sometimes, they may enter themselves into the browser of the customer as a small utility function.
7. **Cross-Site Scripting**: It is the kind of vulnerability of computer privacy discovered in the application of the web that permits code by malicious web customers into the pages of the web seen by other people, for example, HTML codes and scripts of client side.
8. **Session Hijacking**: This kind of phishing attack monitors the activity of the customer or track the user actions such as bank details and sensitive data of the customer and behalf of that malicious software take action that is illegal, like transferring funds without customer knowledge.
9. **Denial of service (DoS) Attack**: In this, the hacker floods the system with so much traffic to steal the crucial data of the customer by misdirecting them. The main purpose of this attack is to restrict the services of the authentic user by so much traffic.
10. **Clone Phishing**: Phisher makes a cloned mail via getting data or addresses of the recipient from the authentic mail that was transferred earlier, then the attacker transfers the similar email with the malicious connection. Phisher utilized a previously sent legitimate email to tackle the customers to misdirect into giving them confidential data.
11. **Content-Injection Phishing**: In this, then, the hacker inserts the malicious code into the authentic website, and this code redirects the customer to another site or installs malware on the system of the user. This phishing attack replaces the content of the authentic site with false data to misdirect the customer into providing their sensitive data or account details.
12. **Man-in-The-Middle Phishing**: This is a kind of phishing attack that is introduced by phishers by inserting themselves between the website and users to record the data of the customer without knowing the customer. It is complicated to identify compared to other forms of attack. In this, the phisher intercepts the connection of the user or website, sits in the middle, and stores data that is entered by the user but does not show the effect on the transaction of the customer.

C. **The Phishing Existence Cycle**

The different periods of existence of phishing site are depicted in Fig. 3 [1] and are explained below:

Period 1: A fake webpage is created by a phisher that is visually similar to a genuine webpage.

Period 2: The fake webpage is sent to the user via email.

Period 3: The fake webpage is opened by the client, and personal information of the client is collected.

Period 4: The client's personal information is gathered by the phisher and is misused by him.

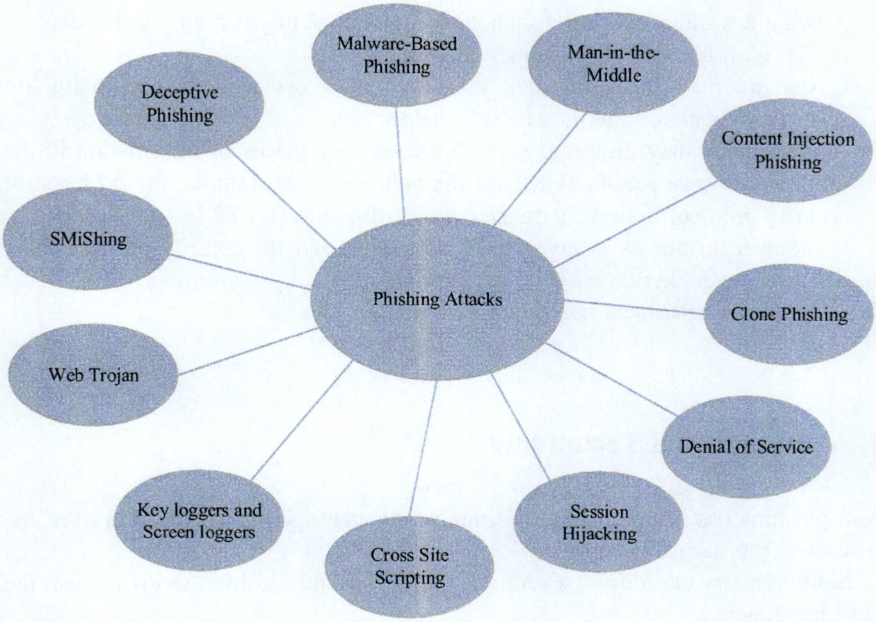

Fig. 2 Phishing attack

Fig. 3 Phishing existence cycle (*1* Phisher creates a fake webpage. *2* Phisher sends the fake webpage to the user. *3* User clicks the link to the phishing webpage and enters sensitive information into it. *4* Phisher gets the personal information of the user and uses it to perform fraud)

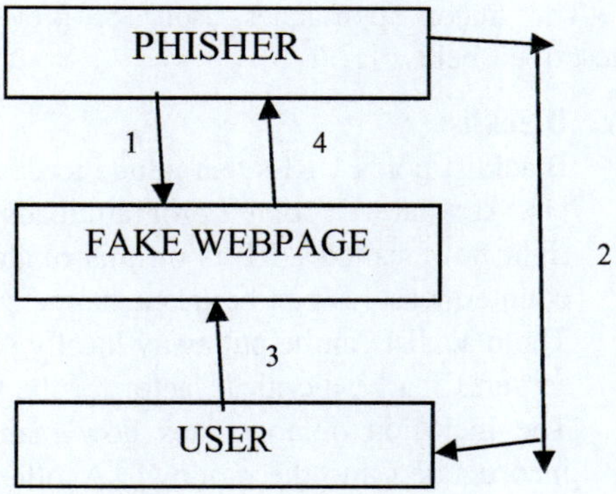

The clients are defenseless to phishing assault because of the accompanying reasons:

1. A few people may need basic information existing on the web dangers.
2. Only a few clients may have a decent comprehension of what does PC infections, programmers, and extortion mean and how to shield themselves from these dangers; they may not be well-known of what does phishing implies. Subsequently, they cannot sum up what they knew to new dangers.

3. Albeit a few clients are careful about being victims of phishing attacks; they do not have an idea about phishing countermeasures.
4. Customers may focus on their essential endeavors while concentrating on security snippets of data is seen as a helper task.
5. A few clients may disregard some basic security pieces of information in the uniform resource locator (URL) of the webpage, for example, the presence of security protocol **h**yper**t**ext **t**ransfer **p**rotocol **s**ecure (HTTPS).
6. Customers do not have any idea of protocols used for security purpose (e.g., SSL: secure socket layer).
7. A couple of customers may ignore warnings.

3 Anti-phishing Techniques

Anti-phishing has been utilized to recognize the attack of the hacker. It protects the customer from illegal access.

Many authors introduced several techniques of anti-phishing to distinguish the phishing attack.

The anti-phishing method may also have an inbuilt feature to identified illegal access from the unauthorized party, such as one-time password, captcha, hypertext transfer protocol secure.

The major approaches used for preventing users from phishing scam are described below [5, 6, 3]:

A. **Blacklist**

Blacklist holds URLs that allude to destinations that are viewed as malevolent [1]. At whatever point a program loads page, it questions blacklist to decide if right now visited URL is on this rundown. Assuming this is the case, proper countermeasures can be taken. Something else, the page is viewed as real.

The blacklist can be put away locally at the customer or facilitated at the focal server. Clearly, a critical factor for the viability of the blacklist is its inclusion. The inclusion demonstrates how many phishing pages on the Internet are incorporated into the rundown. Another factor is the nature of the rundown.

The quality shows how many non-phishing locales are inaccurately included in the rundown. For each off-base section, the client encounters a bogus cautioning when she visits a genuine site, undermining her trust in the convenience and accuracy of the arrangement.

At long last, the last factor that decides the adequacy of blacklist-based arrangement is the time which it takes until a phishing site is incorporated. This is on the grounds that numerous phishing pages are brief, and a large portion of the harm is done in the time length between going on the web and evaporating. Notwithstanding when a blacklist contains numerous passages, it is not viable when it takes excessively long until new data is incorporated or achieves the customers.

B. **Whitelist**

A white list is a list containing all genuine URLs. White list is used to classify URLs as legitimate or phishing.

If a given URL is available in the white list, then declare the URL as legitimate, else the URL is declared phishing.

The drawback of using the whitelist approach is that the updation process of the white list is time-consuming. One solution is automated individual-white list [7].

C. **Heuristics**

Ordinarily, phishing countermeasures depend on URL blacklists or on methods for separating the basic highlights (attributes) from a site [8, 9]. As opposed to blacklist approaches, which principally rely upon the client's encounters, the highlights-based methodologies are moderately progressively advanced and requires profound examinations. Highlight-based systems right off the bat gather a lot of discriminative highlights that can isolate phishing sites from real ones, at that point, train an AI model to anticipate phishing endeavors dependent on these highlights, lastly utilize the model to perceive phishing sites in reality [10]. Selecting an appropriate set of attributes or features (feature set) is of primary concern (Fig. 4).

D. **Visual Similarity**

This approach is freethinker to the essential development that provides the visualization to the customers' eyes [11]. Phishing sites give a similar look-and-feel as given by legitimate sites. For instance, phishing sites may use similar images as used in legitimate sites [12]. GooglePay, Amazon, Flipkart, banks, and other financial institutions are the main target websites that are attacked by phishers. Users should visit these sites only after assuring security.

E. **Machine Learning**

Like heuristic tests, machine learning-based strategies can direct phishing strikes, which makes them beneficial when they appeared differently in relation to blacklists. Inquisitively, machine learning frameworks are in like manner prepared for structure their very own gathering models by examining far-reaching courses of action of information [1, 3]. ML methods have the accompanying preferences when compared with the heuristic approach:

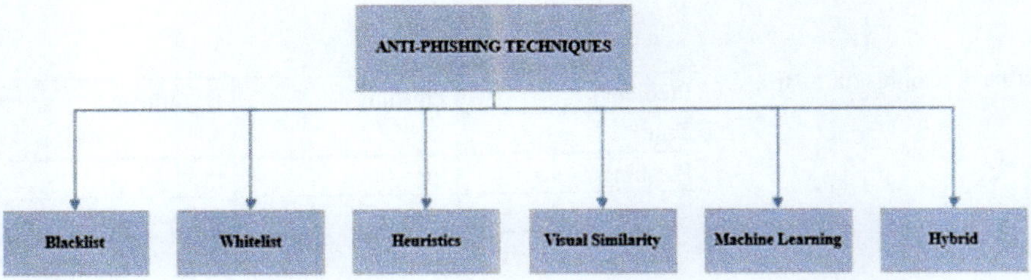

Fig. 4 Anti-phishing techniques

- It is conceivable to build viable characterization models when vast informational collection tests are accessible, without the requirement of physically dissecting information for finding complicated connections.
- As phishing endeavors create, machine learning classifiers can thus progress through learning. On the other hand, it is likewise conceivable to intermittently develop more up to date arrangement models by basically retraining the student with refreshed example informational collections. The examination of various phishing discovery strategies is presented in Table II.

F. **Hybrid**
The hybrid approach utilizes a blend of previously described anti-phishing approaches (list-based approaches, heuristics, visual similarity, and machine learning).

4 Evaluation Metrics

A. **Stability**
The steadiness estimates how stable every calculation performs as far as recognition precision. Standard deviation (SD) is used to evaluate security. A vast SD esteem infers the vacillation in discovery exactness and precariousness of the execution.

B. **Speedup**
The speedup is defined as the ratio of the runtime of a sequential algorithm for solving a problem to the time taken by the parallel algorithm to solve the same problem on multiple processors.

C. **Performance Measures**
Here, PP is a number of correctly identified phishing occurrences. LP is a number of incorrectly identified legitimate occurrences. PL is a number of incorrectly identified phishing occurrences. LL is a number of correctly identified legitimate occurrences. Table 1 shows the confusion matrix. Table 2 shows different performance measures [3]. The anti-phishing algorithms are compared in Tables 3, 4, 5, and 6.

Table 1 Confusion matrix

Predicted / True	Phishing	Legitimate
Phishing	PP	PL
Legitimate	LP	LL

Table 2 Performance measures used in anti-phishing techniques

S. no.	Measure	Definition	Formula
1	False-positive rate (*FPR*)	Quantifies the incorrectly identified legitimate instances relative to the total number of legitimate instances	FPR = LP/(LL + LP)
2	True-positive rate (*TPR*)	Quantifies the correctly identified phishing instances relative to the total number of phishing instances	TPR = PP/(PP + PL)
3	False-negative rate (*FNR*)	Quantifies the incorrectly identified phishing instances relative to the total number of phishing instances	FNR = PL/(PP + PL)
4	True-negative rate (*TNR*)	Quantifies the correctly identified legitimate instances relative to the total number of legitimate instances	TNR = LL/(LL + LP)
5	Accuracy (*ACC*)	Quantifies the correctly classified instances relative to the total number of instances	ACC = (LL + PP)/(LL + LP + PL + PP)
6	Precision (*P*)	Quantifies the correctly identified phishing instances relative to the number of instances that are identified as phishing	P = PP/(LP + PP)
7	Recall (*R*)	It is same as true positive rate	R = TP
8	*f1* score	It is equal to the *H.M. (harmonic mean)* between *precision* and *recall*	F1 = (2 * P * R)/(P + R)

Table 3 Comparison of different anti-phishing algorithms

Detection technique	Advantages	Disadvantages	Remarks
Auto-updated white list [7]	It can detect a zero-hour phishing attack	Only hyperlink features are considered for classification	More highlights can be included alongside the hyperlinks, yet it will build the run-time multifaceted nature of the framework
Collaborative approach [13]	Classification accuracy is high	Complex and time-consuming	Not efficient
Security protocols [14]	Prevents users from phishing	Many users are unaware of security protocols	Effective from the users who have an idea of security protocols. User education can improve the efficiency of this method
Routine activity theory [15]	It helps to increase the chances of cyber-crime detection	Attackers cannot be caught easily. They are smart	It focuses on the methods to prevent cyber-crime

(continued)

Table 3 (continued)

Detection technique	Advantages	Disadvantages	Remarks
User education [16]	Simple technique	Cannot prevent all phishing attacks	Simple phishing attacks can be easily
URL-based phishing detection [2]	Simple and easy to implement	Classification accuracy is low, a large number of rules are needed to achieve high accuracy, but it will increase the run-time complexity	URL features are explored to prevent phishing
LinkGuard algorithm [4]	A robust algorithm, fewer false positives	Time-consuming	In this algorithm, the actual and visual links are compared. If the DNS names of both the links are the same, then the links are declared as legitimate
Features-based Approach [17]	Efficient	The feature set can be improved	Twenty-eight features are used as critical indicators of phishing. More features can be added to improve efficiency
Machine learning (SVM learning algorithm)-based phishing detection using features extractable from URLs [18]	Efficient (high detection rate, fewer false positives)	The feature set can be improved	Eighteen features are used as critical indicators of phishing. More features can be added to improve efficiency
Visual similarity assessment [11]	High accuracy	Time consuming	It highlights the main features of the webpages and calculates the similarity of the current website with the genuine webpage on the basis of design and style

Table 4 Comparison of different anti-phishing algorithms

Detection technique	Advantages	Disadvantages	Remarks
CANTINA [8]	Simple and easy to implement	High false positives	It is a content-based way that helps distinguishing phishing sites from genuine ones by using TF-IDF (term frequency-inverse document frequency) algorithm. It is a tedious methodology
CANTINA+ [9]	Low FP rate, good run-time speed	Cannot deal with image data	Feature set can be improved for better results
Intelligent phishing URL detection using association rule mining [10]	High detection rate	The feature set can be improved for better results	Spotlights on perceiving the huge highlights that separate among real and fake URLs. The main highlights are extracted. The standards acquired are deciphered to underscore the highlights that are increasingly predominant in phishing URLs
A secure and practical authentication scheme using personal devices [19]	High detection rate	High run-time complexity	Enables clients to dispose of numerous issues, for example, retaining usernames and passwords for various sites and frameworks. Lightweight cryptographic procedures or visual unscrambling and mark check can improve this plan
Effective defense schemes for phishing attacks on mobile computing platforms [20]	It does not depend upon HTML code	Less run-time speed	Utilizes OCR, which can precisely remove content from the screen capture of the login interface with the goal that the guaranteed personality can be confirmed
New rule-based phishing detection method [21]	High accuracy	Feature sets are totally relying upon the website page content	Cannot deal with phishing that is done through images

(continued)

Table 4 (continued)

Detection technique	Advantages	Disadvantages	Remarks
A new fast associative classification algorithm for detecting phishing websites [22]	High classification accuracy	Classification accuracy depends upon the feature selection method	The associative classification was used as a powerful administered learning approach to anticipate inconspicuous occurrences
Phishing threat avoidance behavior: An empirical investigation [23]	Designed a game for URL classification	More areas, identifying fake content, should be included to teach users	The main motive was to train users how to identify phishing URLs, which is one of many ways of identifying a phishing attack

Table 5 Comparison of different anti-phishing algorithms

Detection technique	Advantages	Disadvantages	Remarks
Statistical Twitter spam detection demystified: performance, stability, and scalability [24]	>90% of accuracy was achieved using C5.0 and random forest algorithms	Training dataset can be updated by adding more tweets	Random forest performed was found to be the most stable among all algorithms
A survey of phishing email filtering techniques [25]	This overview improves the comprehension of phishing messages issue, current arrangement space, and future phishing email recognition procedures	Discusses the detection techniques proposed till 2013	Discusses advantages and limitations of phishing email detection techniques proposed till 2013
Email classification research trends: review and open issues [26]	Analyzed email classification area, data sets, feature space, techniques used for classifying emails, and the significance of performance measures	Feature selection and email classification tools were not examined	Reviewed articles on the classification of e mails
Phishing susceptibility: An investigation into the processing of a targeted spear phishing email [27]	Provides a research model for email processing	Sample data is that of university students. The sample should be more representative	This is the first study that presents an empirical model for phishing and concludes the importance of two factors: visceral triggers and deception indicators

(continued)

Table 5 (continued)

Detection technique	Advantages	Disadvantages	Remarks
You are probably not the weakest link [28]	An automated model to predict user susceptibility	"Selection of features" can have some limitations. For example, security training is a high-level feature, the content of training should also be considered	Automates the prediction of user susceptibility to phishing attacks. Uses machine learning algorithms (random forest and logistic regression) for implementation
PhishStorm: detecting phishing with streaming analytics [29]	Classifies URL as fake or genuine	This technique cannot be applied to all types of URLs	Uses an approach based on intra-URL relatedness to classify URLs as phishing or legitimate. This technique is implemented using machine learning algorithms. The performance of the random forest algorithm comes out to be the best

Table 6 Comparison of different anti-phishing algorithms

Detection technique	Advantages	Disadvantages	Remarks
Mobile device affordance: Explicating how smartphones influence the outcome of phishing attacks [30]	Concludes that device affordances (uses of mobiles) enhance habituation	The number of design elements can be increased	Describes that device affordances stimulate heuristic processing and that uses of mobiles enhance habituation
PhiDMA: A phishing detection model with multi-filter approach [31]	Visually challenged people can utilize this model	This paper focuses only on phishing detection capabilities. Research should be done on phishing detection usabilities as well	Uses a hybrid approach to solve the phishing problem. It makes use of whitelist approach; heuristics approach, content-based approach, and string matching algorithm

(continued)

Table 6 (continued)

Detection technique	Advantages	Disadvantages	Remarks
Phishing websites features classification based on extreme learning machine [32]	Extreme learning machine came out to be the most accurate method when compared with traditional machine learning methods	Feature set should be improved	This study used 30 features and an extreme machine learning approach for classifying websites
Inside a phisher's Mind: Understanding the anti-phishing ecosystem through phishing kit analysis [33]	Analyzes main components of the anti-phishing ecosystem	The database used by APWG is partial for its partners	Pointed an image of an anti-phishing ecosystem from the viewpoint of culprits
Performance comparison of classifiers on reduced phishing website dataset [34]	Feature selection is used so as to reduce the dataset	Feature set should be improved to get better results	This study compared different anti-phishing algorithms on the basis of their performance on a reduced data set
Using URL shorteners to compare phishing and malware attacks [35]	Analyzed that phishing attacks have a shorter timespan than malware attacks	Flagged URLs no longer record new clicks	URL shortener click analytics is used to compare the lifecycle of phishing and malware attacks

5 Conclusion

This study explored various countermeasures of phishing programming procedures. A portion of the essential angles in estimating phishing countermeasures is as follows: The most recent phishing attack, which is termed as "zero-hour phishing attack," is not caught by all countermeasures; a framework with a high false-positive rate may cause more damage than anything else. There are a number of approaches used as anti-phishing methods such as blacklist, whitelist, heuristics, content-based approach, visual similarity method, and machine learning algorithms. The machine learning method shows the best performance among all methods.

References

1. Mohammad, R.M., Thabtah, F., McCluskey, L.: Tutorial and Critical Analysis of Phishing Websites Methods, pp. 1–24. Elsevier (2015)
2. Waziri Jr., I.: Website forgery: understanding phishing attacks & nontechnical countermeasures. In: IEEE 2nd International Conference on Cyber Security and Cloud Computing, 2015
3. Khonji, M., Iraqi, Y., Jones, A.: Phishing detection: a literature survey. IEEE Commun. Surv. Tutorials **15**(4) (2013)
4. Shekokar, N.M., Shah, C., Mahajan, M., Rachh, S.: An ideal approach for detection and prevention of phishing attacks. Proc. Comput. Sci. 82–91 (2015) (Elsevier)
5. Nazreen Banu, M., Munawara Banu, S.: A comprehensive study of phishing attacks. Int. J. Comput. Sci. Inf. Technol. **4**(6), 783–786 (2013)
6. Gupta, S., Singhal, A., Kapoor, A.: A literature survey on social engineering attacks: phishing attack. In: International Conference on Computing, Communication and Automation, 2016
7. Jain, A.K., Gupta, B.B.: A novel approach to protect against phishing attacks at client side using auto-updated white-list. EURASIP J. Inf. Secur. 1–11 (2016)
8. Zhang, Y., Hong, J., Cranor, L.: CANTINA: a content-based approach to detecting phishing web sites. In: ACM Proceedings of 16th International Conference on World Wide Web, pp. 639–648, May 2007
9. Xiang, G., Hong, J., Rose, C.P., Cranor, L.: CANTINA+: a feature-rich machine learning framework for detecting phishing web sites. ACM Trans. Inf. Syst. Secur. (TISSEC) **14**(2), 1–32 (2011)
10. Carolin Jeeva, S., Rajsingh, E.B.: Intelligent phishing url detection using association rule mining. SpringerOpen Hum. Centric Comput. Inf. Sci. 1–19 (2016)
11. Liu, W., Deng, X., Huang, G., Fu, A.Y.: An antiphishing strategy based on visual similarity assessment. IEEE Comput. Soc. 58–65 (2006)
12. Chen, K.-T., Huang, C.-R., Chen, C.-S.: Fighting phishing with discriminative keypoint features. IEEE Comput. Soc. 56–63 (2009)
13. Nirmal, K., Janet, B., Kumar, R.: Phishing—the threat that still exists. In: IEEE International Conference on Computing and Communications Technologies, 2015
14. Iuga, C., Nurse, J.R.C., Erola, A.: Baiting the hook: factors impacting susceptibility to phishing attacks. Hum. Cent. Comput. Inf. Sci., SpringerOpen (2016)
15. Choo, K.K.R.: The cyber threat landscape: challenges and future research directions. Comput. Secur. **30**, 719–731 (2011) (Elsevier)
16. Parsons, K., McCormac, A., Pattinson, M., Butavicius, M., Jerram, C.: The design of phishing studies: challenges for researchers. Comput. Secur. **52**, 194–206 (2015) (Elsevier)
17. Montazer, G.A., Yarmohammadi, S.A.: Identifying the critical indicators for phishing detection in iranian e-banking system. In: IEEE 5th Conference on Information and Knowledge Technology, 2013
18. Chu, W., Zhu, B.B., Xue, F., Guan, X., Cai, Z.: Protect sensitive sites from phishing attacks using features extractable from inaccessible phishing URLs. In: IEEE Communication and Information Systems Security Symposium, 2013
19. Alhothaily, A., Chunquiang, Hu., Alwaris, A., Song, T., Cheng, X., Chen, D.: A secure and practical authentication scheme using personal devices. IEEE Access **5**, 11677–11687 (2017)
20. Wu, L., Du, X., Wu, J.: Effective defense schemes for phishing attacks on mobile computing platforms. IEEE Trans. Veh. Technol. (2015) (Accepted for publication)
21. Moghimi, M., Varjan, A.Y.: New rule-based phishing detection method. Expert Syst. Appl. (2016) (Accepted for publication)
22. Hadi, W., Aburub, F., Alhawari, S.: A new fast associative classification algorithm for detecting phishing websites. Appl. Soft Comput. **48**, 729–734 (2016) (Elsevier)
23. Arachchilage, N.A.G., Love, S., Beznosov, K.: Phishing threat avoidance behaviour: an empirical investigation. Comput. Hum. Behav. **60**, 185–197 (2016)

24. Lin, G., Sun, N., Nepal, S., Zhang, J., Xiang, Y., Hassan, H.: Statistical twitter spam detection demystified: performance, stability and scalability. IEEE Access, Special Section on Big Data Analytics in Internet of Things and Cyber-physical Systems **5**, 11142–11154 (2017)
25. Almomani, A., Gupta, B.B., Atawneh, S., Meulenberg, A., Almomani, E.: A survey of phishing email filtering techniques. IEEE Commun. Surv. Tutorials **15**(4) (2013)
26. Mujtaba, G., Shuib, L., Raj, R.G., Majeed, N., Al-Garadi, M.A.: Email classification research trends: review and open issues. IEEE Access **5**, 9044–9064 (2017)
27. Wang, J., Herath, T., Chen, R., Vishwanath, A., Raghav Rao H.: Phishing susceptibility: an investigation into the processing of a targeted spear phishing email. IEEE Trans. Prof. Commun. **55**(4), 345–362 (2012)
28. Heartfield, R., Loukas, G., Gan, D.: You are probably not the weakest link: towards practical prediction of susceptibility to semantic social engineering attacks. IEEE Access **4**, 6910–6928 (2016)
29. Marchal, S., François, J., State, R., Engel, T.: PhishStorm: detecting phishing with streaming analytics. IEEE Trans. Netw. Serv. Manage. **11**(4) (2014)
30. Vishwanath, A.: Mobile device affordance: explicating how smartphones influence the outcome of phishing attacks. Comput. Hum. Behav. **63**, 198–207 (2016)
31. Sonowal, G., Kuppusamy, K.S.: PhiDMA—a phishing detection model with multi-filter approach. J. King Saud Univ. Comput. Inf. Sci. 1–14 (2017)
32. Sönmez, Y., Tuncer, T., Gökal, H., Avcr, E.: Phishing web sites features classification based on extreme learning machine. In: 6th International Symposium on Digital Forensic and Security, pp. 1–5, 2018
33. Oest, A., Safei, Y., Doupé, A., Ahn, G.-J., Wardman, B., Warner, G.: Inside a phisher's mind: understanding the anti-phishing ecosystem through phishing kit analysis. In: APWG Symposium on Electronic Crime Research, pp. 1–12, 2018
34. Karabatak, M., Mustafa, T.: Performance comparison of classifiers on reduced phishing website dataset. In: IEEE, pp. 1–5, 2018
35. Le Page, S., Jourdan, G.-V., Bochmann, G.v., Flood, J., Onut, I.-V.: Using URL shorteners to compare phishing and malware attacks. In: IEEE, pp. 1–13, 2018

Assessment of Open Source Tools and Techniques for Network Security

U. Guru Prasad, R. Girija, R. Vedhapriyavadhana, and S. L. Jayalakshmi

Abstract Providing network security with an open source has been a huge concern today. Data transmitted over the network shall not be assumed to be stable. There are numerous attacks such as phishing, spoofing, and sniffing. This paper provides a summary of the tools available to fight against these network attacks and threats. There are several resources in the open source that are built to use to handle these threats. Resources including OpenSSH, OpenSSL, NMap, digital certificate, and IP tables were discussed in this article.

Keywords OpenSSH · OpenSSL · NMap · Digital certificate · IP tables

1 Introduction

Networks are easily vulnerable to multiple threats due to the exponential growth of the Internet and e-commerce [1]. Undoubtedly, the Internet has created a big infrastructure in the world that makes and supports both personal and technical statements around the world. Wide use of the Internet work by consumers has opened the door to a number of threats. The effects of the attacks are rising rapidly. There are a lot of data breaches going on in the Internet world. Network security consists of measures implemented to avoid and track the denial of entry to the computer network, permitted entry, manipulation, and misuse of the computer. It is the method of utilizing applications and physical prevention. Measures are to

U. Guru Prasad (✉) · R. Girija · R. Vedhapriyavadhana · S. L. Jayalakshmi
Vellore Institute of Technology, Chennai Campus, Chennai, India
e-mail: guru.prasad2018@vitstudent.ac.in

R. Girija
e-mail: girija.r@vit.ac.in

R. Vedhapriyavadhana
e-mail: vedhapriyavadhana.r@vit.ac.in

S. L. Jayalakshmi
e-mail: jayalakshmi.sl@vit.ac.in

K. Khanna et al. (eds.), *Cyber Security and Digital Forensics*,
Lecture Notes on Data Engineering and Communications Technologies 73,
https://doi.org/10.1007/978-981-16-3961-6_25

safeguard the underlying networking system against unwanted entry, misuse, degradation, change, loss, or inappropriate disclosure, providing a safe computing interface.

The purpose of network protection is to have permission to access data on a network. Network protection begins with verification and is supported with a username and password. Another means of delivering protection uses a firewall. Firewall enforces access rules such as which resources should be reached by the network administrator. Thus, security of the computer is vulnerable (weakness in the network) and leads to misuse of the network by the hackers [2]. To prevent the network from various attacks such as phishing, spoofing, and sniffing, various security tools are used. There are several security tools in the open source that are built to use to handle these threats and attacks. This review gives an overview of a few security tools which are highly used by network professionals. Some tools used include OpenSSH, OpenSSL, NMap, digital certificate, and IP tables [3].

2 Network Security Tools: Overview

2.1 OpenSSH

OpenSSH (popularly known as OpenBSD Secure Shell) is a package of secure networking utilities based on the secure shell protocol [4]. It is commonly used to provide a secure channel, for multiple parties, over an unsecured network using client–server architecture. Unlike most GUI based tools, OpenSSH is not a single computer program, but a bundle of programs that serve as a better alternative to the various unencrypted connectivity protocols like Telnet and FTP. OpenSSH is built-in into several operating systems, mostly macOS and almost all Linux operating systems by default. As it is a terminal-based tool, this is expected. OpenSSH or SSH, as it is generally referred to, has a variety of operations available. We can execute shell commands in the same way as we would if we were physically operating the remote computer.

SSH utilizes various encryption algorithms like—Symmetric Encryption, Asymmetric Encryption, and Hashing. It can easily transfer files using the associated SSH protocols such as SSH file transfer protocol (SFTP) or secure copy protocol (SCP). SSH works by trying a client–server model to allow for authentication of two remote systems and thus encryption of the data that will pass between them. SSH operates on TCP port 22 by default (however, this can be changed if needed, as demonstrated in the presentation). The host (server) listens on port 22 for any inbound connections. It starts the secure connection by first authenticating the client and only then opening the correct shell environment if the verification is successful. Now, there are two major stages in establishing a connection: First, both the systems must agree upon an encryption standard, so that the unknown client can be added to the known clients' list for future reference. Second, the user must authenticate themselves [5].

2.1.1 Output

Become root user in the client system: Command: sudo -i, (Option: -i login).

By becoming a root user, the user can accomplish any task on the remote system as the root user group has all permissions and access rights enabled. The root user can also add other users to the group. It can also allow for other client systems to connect to the server system.

Debug Mode ON: SSH-ing to the server system through client system to get a detailed explanation of how the systems are reacting to form the SSH connection, Command: ssh -v fv@127.0.0.1 -p 2222

Options: -v—Verbose-list out all background details, fv—Server System Name, 127.0.0.1—Server IP address, -p—To set custom port number , 2222—Custom Post Number (Fig. 1).

```
debug1: identity file /home/sagarika/.ssh/id_ed25519 type -1
debug1: key_load_public: No such file or directory
debug1: identity file /home/sagarika/.ssh/id_ed25519-cert type -1
debug1: Local version string SSH-2.0-OpenSSH_7.6p1 Ubuntu-4ubuntu0.3
debug1: Remote protocol version 2.0, remote software version OpenSSH_8.0
debug1: match: OpenSSH_8.0p1 Ubuntu-6ubuntu0.1 pat OpenSSH* compat 0x040
debug1: Authenticating to 127.0.0.1:2222 as 'fv'
debug1: SSH2_MSG_KEXINIT sent
debug1: SSH2_MSG_KEXINIT received
debug1: kex: algorithm: curve25519-sha256
debug1: kex: host key algorithm: ecdsa-sha2-nistp256
debug1: kex: server->client cipher: chacha20-poly1305@openssh.com MAC: <
debug1: kex: client->server cipher: chacha20-poly1305@openssh.com MAC: <
debug1: expecting SSH2_MSG_KEX_ECDH_REPLY
debug1: Server host key: ecdsa-sha2-nistp256 SHA256:+ZFYPOk+sH1pDVF+jrnR
debug1: Host '[127.0.0.1]:2222' is known and matches the ECDSA host key.
debug1: Found key in /home/sagarika/.ssh/known_hosts:4
debug1: rekey after 134217728 blocks
debug1: SSH2_MSG_NEWKEYS sent
debug1: expecting SSH2_MSG_NEWKEYS
debug1: SSH2_MSG_NEWKEYS received
debug1: rekey after 134217728 blocks
debug1: SSH2_MSG_EXT_INFO received
debug1: kex_input_ext_info: server-sig-algs=<ssh-ed25519,ssh-rsa,rsa-sha
debug1: SSH2_MSG_SERVICE_ACCEPT received
debug1: Authentications that can continue: publickey,password
debug1: Next authentication method: publickey
debug1: Trying private key: /home/sagarika/.ssh/id_rsa
debug1: Trying private key: /home/sagarika/.ssh/id_dsa
debug1: Trying private key: /home/sagarika/.ssh/id_ecdsa
debug1: Trying private key: /home/sagarika/.ssh/id_ed25519
debug1: Next authentication method: password
fv@127.0.0.1's password:
debug1: Authentication succeeded (password).
Authenticated to 127.0.0.1 ([127.0.0.1]:2222).
debug1: channel 0: new [client-session]
```

Fig. 1. Displays the output of SSH

SSH Custom Key Generation: Command: ssh-keygen

Option: -keygen—To generate custom key, Passphrase is optional.

Web sites such as GitHub and Heroku often ask for an SSH public key, so that we can push/deploy code without entering a password. This can be generated using the ssh-keygen command. The RSA key pair is stored in the home directory under . ssh directory (Fig. 2).

Service Command to check the status: Command: service ssh status.

The service command can be used to check the status of the service of your choice. Apart from status, it can also be used to start, stop and restart the said service through the terminal with the following syntax.

service ssh stop, service ssh restart, service ssh stop.

The status screen also shows the last interactions with the particular service that took place in the current login session. It gives a variety of information such as Process Name, Main Process ID, Tasks performed, Memory consumed, Doc files or associated Man pages to name a few.

SSH Secure Copy: Command: scp filename username@hostname: destination_path.

Alternatively, scp filename from_destinaton to_destination, can also be used.

Fig. 2 Service command

2.2 *Open SSL*

OpenSSL might be termed as a strong, business-grade, and feature-rich tool-set used for the transport layer security (TLS) and secure sockets layer (SSL) protocols [6]. It is also a general-purpose cryptography library. OpenSSL may be a cryptography software library or tool set that creates connections between computer networks safely. This is a command-line based tool for using various cryptographic functions of OpenSSL's cryptolibrary from the shell. OpenSSL is a software library for applications that protect connections over computer networks against intrusion or need to recognize the party at the other end. It is extensively employed by Internet servers, including the bulk of HTTPS Web sites [7]. The core library, written within the C programming language, executes basic cryptographic functions and provides various utility functions.

2.2.1 Output

Public key cryptography: In this type of cryptography, two keys are used—one key for encryption and the other key for decryption. It is used for authentication and confidentiality [8]. The first key is a public key that is available to everyone, and the second key is the private key which is kept as secret and known only to the owner who generated the key pair (Fig. 3).

Command for Creating Private Keys: openssl genrsa –out keypairA.pem 2048.

Command for Creating Public keys: openssl rsa –in keypairA.pem –pubout –out publicA.pem.

Command for Creating Public keys: openssl rsa –in keypairA.pem –pubout –out publicA.pem, openssl rsautl –decrypt –in received –out msg –inkey keypairB. pem (Figs. 4 and 5).

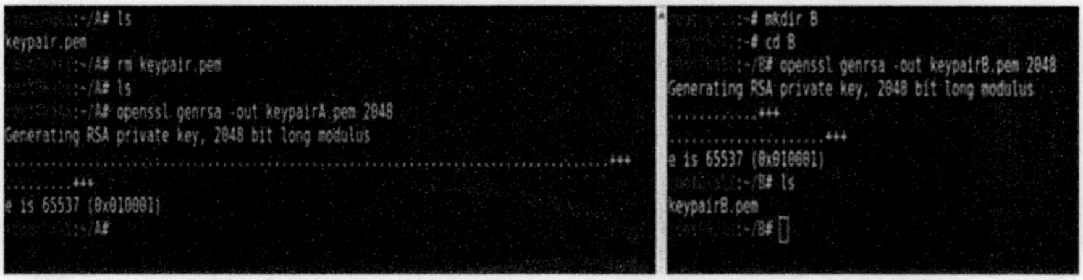

Fig. 3 OpenSSL—key pair generation

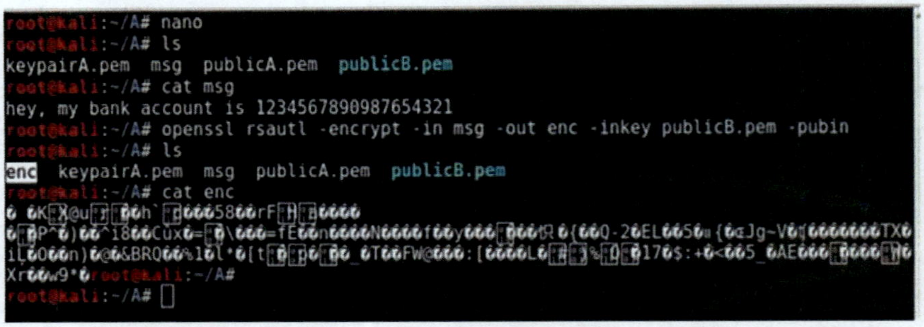

Fig. 4 Encryption in OpenSSL

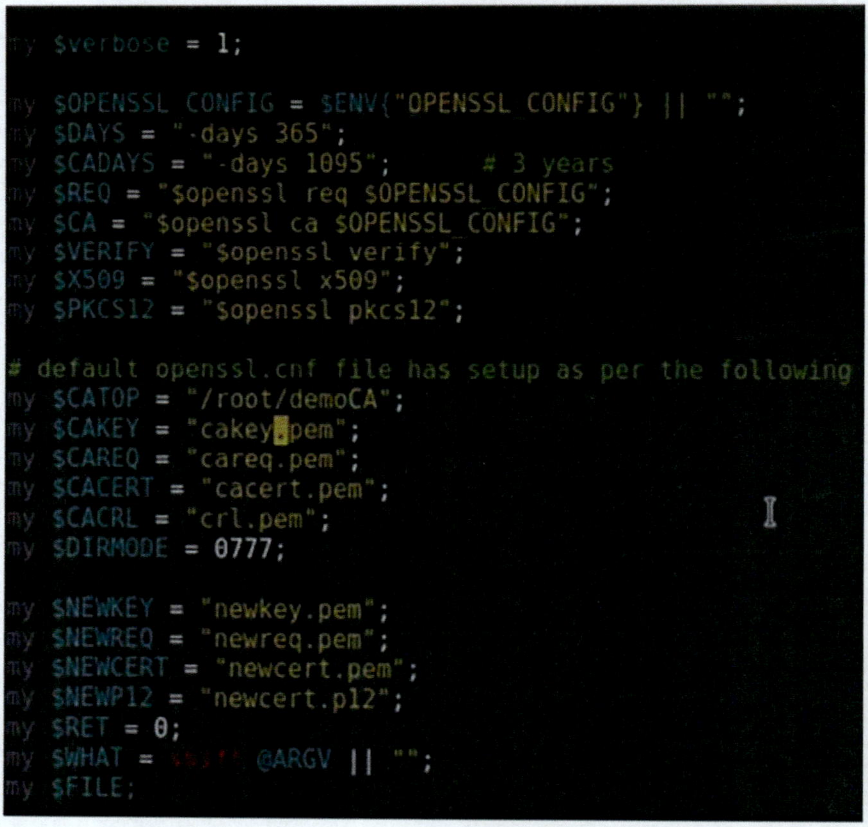

Fig. 5 Setup—OpenSSL

2.3 N-Map

It is a network scanner used to find hosts and services on a network by sending packets and analyzing the response [9]. Network administrators use this to discover the devices running on their systems, identify hosts and the services, discovering open ports, and identifying security risks. It is used to look after single hosts and vast networks which in turn look after hundreds of thousands of devices and subnets. It is free and open source and extremely flexible, but deep down it is a port scanning tool. It used to gather information on whether the ports are opened or closed by sending raw packets to system ports as well as listens for responses. Likewise, it can also be called port discovery or enumeration [10].

2.3.1 Output

There are four types of output formats in N-Map in which the interactive output is saved to a file. N-Map output may be manipulated by text processing software, enabling the user to form customized reports [11].

Interactive: When a user runs N-Map from the terminal, it is updated in real-time. Various options are entered during the scan to facilitate monitoring. XML: A format that will be further processed by XML tools. It will be converted into an HTML report using XSLT.

Grepable: Output that is tailored to line-oriented processing tools like grep, sed, or awk. Normal: The output as seen while running N-Map from the terminal, but saved to a file.

N-Map [9] present Scans: Intense scan - Command: Nmap -T4 -A -v, this scan is reasonably quick and is used to scan the most used TCP ports, which will help in determining the OS type and what services and their versions are running. Intense scan plus UDP - Command: Nmap -sS -sU -T4 -A -v, this command is similar to the regular intense scan, the difference is that it will also scan the UDP ports (-sU). Intense scan, all TCP ports—Command: Nmap -p 1–65,535 -T4 -A -v, this command checks all the TCP ports without leaving anything unchecked. Ping scan Command: Nmap -sn, This command is used to ping on the target network and does not do a port scan. Quick scan—Command: Nmap -T4 -F, this command is used to scan the TCP ports faster than the intense scan by constraining the number of TCP ports to be scanned to only the top 100 most used TCP ports. Quick traceroute—Command: Nmap -sn –traceroute, This option is used when you need to list hosts and routers in a network scan. It will traceroute and ping all hosts defined within the target. Regular scan—Command: Nmap, this command is used for TCP SYN scan for the most used 1000 TCP ports by using an ICMP echo request (ping) to detect the host. Slow comprehensive scan—This command is used to detect the host with an immense amount of effort without giving up even if the first ping request fails. TCP, UDP, and SCTP are the protocols used to detect the host. Quick scan plus—Command: Nmap -sV -T4 -O -F.

Fig. 6 N-Map

–Version-light+, this command is used to detect the operating system and the version (Fig. 6).

Port Scan—To scan a single port—use -p: Nmap –p 53 192.168.1.8,

To scan a single port when the port number is specified: Nmap -p T:7777, 973 192.164.0.1,

To scan a range of ports—separate it by a hyphen: Nmap -p 76–973 192.164.0.1

To scan the top port of a scan—use –top-ports flag: Nmap –top-ports 10 scanme.nmap.org (Fig. 7).

2.4 IPTables

It is a utility program that is used by the system administrator to set the IP packet filtering rules in the Linux kernel firewall, and it is also implemented as various Netfilter modules [12]. All these filters are arranged in different tables, which has the rules on how to filter network traffic packets. Various kernel modules and programs are used now for different protocols like IPv4 uses iptables, IPv6 uses ip6tables, ARP uses arp tables, and Ethernet frames uses the ebtables. They need high privileges to operate and should be executed by the root user, otherwise, it might fail to function. On most Linux kernel, iptables are installed in /usr/sbin/iptables and documented in man pages. These pages can be opened using iptables after installing. It can also be found in /sbin/iptables, because iptables is like a service rather than an "essential binary," the preferred location remains /usr/sbin. iptables that allow the system administrator to define the tables which have the rules on how to treat the packets. Each table is linked with a different way of treating the packet which is processed by sequentially following the rules. A rule in a chain can cause a jump to another chain, and this can be repeated to any level depending on the desired nesting. (A jump is like a "call," i.e., the point that was jumped from is remembered.) Every packet that is entering or leaving the device will traverse at least one chain. The source of the packet shows which chain it traverses in the

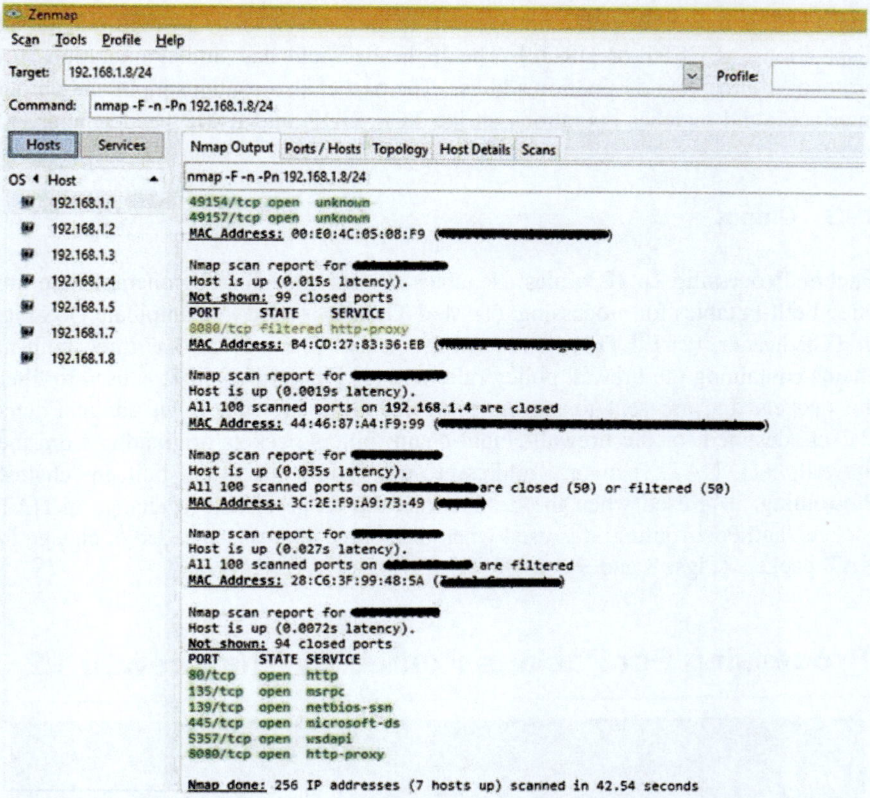

Fig. 7 Port scan—network map

beginning. There are five types of pre-defined chains which has a policy, like DROP, which is added to the packet when it reaches the end of the rule chain. The system administrator has the privilege to create as many chains as they want. These chains do not have a policy; if a packet reaches the end of the chain and then returns to the initial chain, this type of chain might be empty.

2.4.1 Features

It uses the concept of having a separate rule table for various types of packet processing functionality. The three types of primary modules are the rule filter table, the NAT table, and the packet-handling mangle table. Module extensions are dynamically loaded when it will be first referenced unless you have built them directly into the Linux kernel for each of these table modules. The filter table is the default table, and other tables are specified by a terminal option. The features in the filter table are chain-related operations on the built-in chains that are INPUT,

OUTPUT, and FORWARD and on user-defined chains that are help and target disposition (ACCEPT or DROP). The IP header field has similar operations for protocol, source, and destination address. The packet also contains input and output interfaces, and matches operations on the TCP, UDP, and ICMP header fields.

2.4.2 Output

Packet Processing In IP tables: IP tables are complex for beginners. There are three built-in tables for processing: (1) MANGLE: it is used to manipulate QoS bits in TCP header, (2) FILTER: it is used for packet filtering and has three built-in chains containing the firewall policy rules and (3) Forward chain: it is used to filter the packets that are sent to servers protected by a firewall, Input chain: Filters packets destined for the firewall- Input chain: Filters packets originating from the firewall, (4) NAT: network addresses translation has two built-in chains. Pre-routing: it is used when the destination address is needed to change in NAT packets, and Post-routing: it is used when the source address is needed to change in NAT packets (Figs. 8 and 9).

Processing For Packets Routed By The Firewall 1/2

Queue Type	Queue Function	Packet transformation chain in Queue	Chain Function
Filter	Packet filtering	FORWARD	Filters packets to servers accessible by another NIC on the firewall.
		INPUT	Filters packets destined to the firewall.
		OUTPUT	Filters packets originating from the firewall
Nat	Network Address Translation	PREROUTING	Address translation occurs before routing. Facilitates the transformation of the destination IP address to be compatible with the firewall's routing table. Used with NAT of the destination IP address, also known as **destination NAT** or **DNAT**.

Fig. 8 IP config table

Processing For Packets Routed By The Firewall 2/2

		POSTROUTING	Address translation occurs after routing. This implies that there was no need to modify the destination IP address of the packet as in pre-routing. Used with NAT of the source IP address using either one-to-one or many-to-one NAT. This is known as **source NAT**, or **SNAT**.
		OUTPUT	Network address translation for packets generated by the firewall. (Rarely used in SOHO environments)
Mangle	TCP header modification	PREROUTING POSTROUTING OUTPUT INPUT FORWARD	Modification of the TCP packet quality of service bits before routing occurs (Rarely used in SOHO environments)

Fig. 9 IP config table (continued)

3 Conclusion

The reliability of the network is the biggest concern these days. While the open source is deemed stable, the data exchanged over the network is still not safe. This paper reviews some network security tools for handling various types of attacks and threats on the Internet. This paper discusses various tools with their features and their structure with respect to different types of network attacks.

References

1. Kumar, S.N.: Review on network security and cryptography. Int. Trans. Electr. Comput. Eng. Syst. **3**(1), 1–11 (2015)
2. Dowd, P.W., McHenry, J.T.: Network security: it's time to take it seriously. Computer **31**(9), 24–28 (1998)
3. Kartalopoulos, S.V.: Differentiating data security and network. In: Conference on Differentiating Data Security and Network, pp.1469–1473, 19, 23 May 2008
4. Hofstede, R., Hendriks, L., Sperotto, A., Pras, A.: SSH compromise detection using NetFlow/IPFIX. ACM SIGCOMM Comput. Commun. Rev. **44**(5), 20–26 (2014)
5. Alsaadi, H.H., Aldwairi, M., Taei, M.A., Albuainain, M.: Penetration and security of OpenSSH remote secure shell service on raspberry pi. In: International Conference on New Technologies, Mobility & Security (NTMS'18), February 2018
6. Viega, J., Messier, M., Chandra, P.: Network Security with OpenSSL: Cryptography for Secure Communications. O'Reilly Media, Inc. (2002)

7. Chordiya, A.R., Majumder, S., Javaid, A.Y.: Man-in-the-middle (MITM) attack based hijacking of HTTP traffic using open source tools. In: IEEE International Conference on Electro/Information Technology (EIT), 2018
8. Großschädl, J., Kizhvatov, I.: Performance and security aspects of client-side SSL/TLS processing on mobile devices. In: International Conference on Cryptology and Network Security, pp. 44–61, 2010
9. Calderon, P.: Nmap: Network Exploration and Security Auditing Cookbook. Packt Publishing Ltd. (2017)
10. Sarmiento, O.P., Guerrero, F.G., Argote, D.R.: Basic security measures for IEEE 802.11 wireless networks. Revista Ingeniería E Investigación **28**(2), 89–96 (2008)
11. Bagyalakshmi, G., Rajkumar, G., Arunkumar, N., Easwaran, M., Narasimhan, K., Elamaran, V., Ramirez-Gonzalez, G.: Network vulnerability analysis on brain signal/image databases using Nmap and Wireshark tools. IEEE Access **6**, 57144–57151 (2018)
12. AL-Musawi, B.Q.M.: Mitigating DoS/DDoS attacks using iptables. Int. J. Eng. Technol **12**(3), 101-111 (2012)

A Detailed Comparative Study and Performance Analysis of Standard Cryptographic Algorithms

Chetan Rathod and Atul Gonsai

Abstract Nowadays the most important issue we face during data transmission and exchanging of data is security of data. The cryptographic algorithms play very significant role to secure data. By using different algorithms, it improves security of data by making data unreadable, only the authenticate users can read data after decryption using keys. All algorithms perform same work, but consume different volume of computing properties such as time of CPU, memory utilization, throughput time, encryption and decryption time, simulation etc. Also different algorithms use various size of keys, size of block and cipher type. So we need to compare the cryptographic algorithms to choose the best one. In this paper, we have done the evaluation of both block cipher (AES, DES, 3DES, Blowfish) as well as stream cipher (SALSA20) cryptographic algorithms is shown by taking different size of audio files. This comparison of different algorithms has been conducted to evaluate parameters such as encryption/decryption time, throughput time, memory utilization, scalability and ratio. Simulation results are given to demonstrate the efficiency of each. This research contributes to identify state of the art cryptographic technique.

Keywords Encryption · Decryption · Cryptography · Cipher · AES · 3DES · Blowfish · SALSA20

1 Introduction

Encryption means coding and decryption means decoding, by this technique, we can protect our information through transmission so that only authorized user can read the data and data can be prevented from being misused. If data goes to the unauthorized user, there can be a chance to read and misuse it [1]. That is why we translate our data into unreadable form by encryption and in decryption it will be convert into readable form. In other word, we can say that by encryption, plain text

C. Rathod (✉) · A. Gonsai
Department of Computer Science, Saurashtra University, Rajkot, India

K. Khanna et al. (eds.), *Cyber Security and Digital Forensics*,
Lecture Notes on Data Engineering and Communications Technologies 73,
https://doi.org/10.1007/978-981-16-3961-6_26

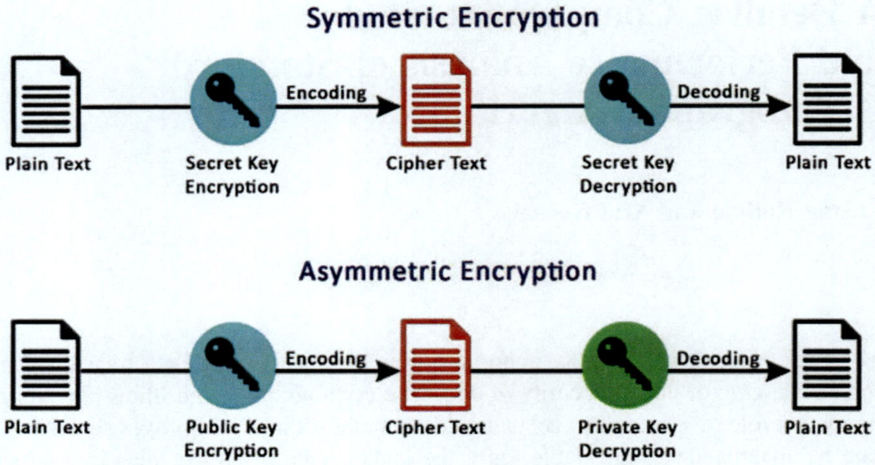

Fig. 1 Symmetric versus asymmetric encryption [2]

will be converted into cipher text and in decryption, cipher text will converted into plain text. Only authorized user who have the key, they can read the message, e.g., original message is "ABCD" after encryption it will be "1234," if someone got the message through transmission, they will got "1234," not "ABCD" but if someone has key then they can convert your encrypted message ("1234") in original form ("ABCD"). This method is known as encryption/decryption or cryptography [2, 3].

Figure 1 shows that there are no. of algorithms available for encryption/decryption [4]. Two different techniques available for cryptography: Symmetric and Asymmetric. In symmetric, only one key (private) is used and in asymmetric two keys (private and public) are used [5]. There are two types of cipher text: block cipher and stream cipher [6]. Apart from that different parameters like methods (structures), no. of rounds, different kind of P-Arrays and S-Boxes will be considered to select the best algorithm for encryption/decryption. Each have its own unique features and specialty like speed, memory utilization, security level, throughput time, encryption ratio etc. these parameters are also considered for selecting algorithm.

2 Literature Review of Standard Algorithms

2.1 Aes: [3, 6–8]

AES means advanced encryption standard is a block cipher algorithm in symmetric category for encryption as well as decryption. In 1998, it was established by Daemon and Vincent Rijmen. It is a commonly used encryption algorithm. AES perform encryption on plain text and decryption on ciphertext of 128 bits with same key at both side. It is 128 bits block size algorithm which has variable length key size (128, 192, 256 bits). Key size decides the rounds for encryption and decryption.

- 128 bits = 10 cycles
- 192 bits = 12 cycles
- 256 bits = 14 cycles

2.2 *Blowfish: [3, 6–8]*

In 1993, Bruce Scheier designed Blowfish algorithm for encryption. It is a symmetric block cipher technique. It has key size 32–448 bits with same key at both side and 64 bits block size. Blowfish has fiestel network of 16 rounds with XOR operation in each round. It has key expansion technique. This algorithm contains P-Array and S-Boxes. P-Array consists 18 sub keys each of 32 bits. S-Boxes contain 256 entries each of 32 bits. Blowfish is faster and secure algorithm.

2.3 *Des: [3, 6, 7]*

In March 1975, data encryption standard developed by IBM later adopted in 1977 by NIST. For encryption, DES is mostly used algorithm. DES is block cipher algorithm with 64 bits key size. It is symmetric algorithm for input it takes 64 bits plain text and gives 64 bits ciphertext as an output. DES contains 16 Round. Each block is divided into each of two 32 bits block L and R (Left and Right). This algorithm is not robust enough. Some attacks noted against DES algorithm.

2.4 *Triple DES: [7, 9]*

It is an advanced form of DES created by IBM in 1978. With long key size, it is created by using same rules of DES. Performance of triple DES is trice it means it has 48 rounds that is why it is slower than DES, but it is more secure than DES. 3DES is block cipher algorithm including key size of 192 bits and block size of 64 bits. In terms of power consumption, its performance is low compared to DES. It is symmetric block cipher algorithm. Several attacks also recorded against it.

Table 1 Comparison of algorithms with various parameters

Algorithms	Key size	Block size	Cipher type	Rounds	Structure	No of S-Boxes
AES	128–192–256 bits	128 bits	Block	10, 12, 14	Substitution-permutation network	1
DES	64 bits	64 bits	Block	16	Feistel	8
3DES	192 bits	64 bits	Block	48	Feistel	8
Blowfish	32–448 bits	64 bits	Block	16	Feistel	4
SALSA20	256 bits	N/A	Stream	20	ARX	0

2.5 Salsa20: [10, 11]

Salsa20 is come in stream cipher category's algorithm developed by Daniel J. Bernstein for encryption submitted to eSTREAM. Salsa20 used the key of 256 bits, 64 bits nonce and a 64 bits position of stream to a 512 bits block of the keystream. It used pseudorandom function based on add rotate xor(ARX) operations 32 bits addition, bitwise addition (XOR) and rotation operations. The user can seek to any position in the keystream constant time. Some attacks noticed.

In this paper, we have compared various parameters like key size, block size, cipher type, no. of rounds, no. of S-boxes and structure of these algorithms. Following Table 1 shows details of this.

Nowadays in the era of multimedia security of data transmission is very annoying task in any format of data required more secure algorithm they may be in a form of text, image, audio or video. In this paper, we used different cryptographic algorithms for audio files.

3 Research Methodology

Here, we used five most common security algorithms in this program for encoding and decoding (Blowfish, AES, DES, Triple DES and SALSA20). A computer having Intel® Core™ i5-4210U CPU @1.70 GHz 2.40 GHz with 4 GB RAM and 64-bit operating system (Windows 10 Home). Python 3.7.2 (64 Bit) used to developed and run programs for the algorithms. Blowfish, AES, DES, Triple DES and SALSA20 cryptographic algorithms were implemented using Python programming language on the same programming environment. Program consists of three phases: (1). Key generation, (2). Encryption and (3). Decryption. It also checks encryption/decryption time, memory utilization during encryption/decryption, scalability of algorithms along with throughput and encryption ratio of all algorithms. The Python program took as individual inputs for five different size audio file, the sizes (463.77, 777.52 KB, 2.08, 11.2 and 24.03 MB) of mp3 format for each algorithm. The time of encryption and decryption, memory utilization, throughput scalability and ratio for each algorithm and for each size of audio files were recorded. We also compared various parameters (key size, block size, cipher type, no. of rounds, no. of S-boxes and structure of these algorithms.

3.1 Outcome of Research: A Comparative Study

Tables 2, 3, 4, 5 and 6 show the detailed comparative study of the audio encryption using different standard algorithms. As the tables showcases that there are mainly seven parameters and five standard algorithms using which we have carried out comparison. The results are as follows.

Table 2 Algorithms with various parameters for audio file of 463.77 KB

Algorithms	Enc. Time (s)	Dec time (s)	Memory utilization (Enc.)	Memory utilization (Dec.)	Loss	Throughput (MB/s)	Enc. ratio
AES	0.007943392	0.196688175	1,544,192	385,024	No	58.38400166	1
DES	0.036031008	0.039975405	1,011,712	221,184	No	12.87133024	1
3DES	0.040014982	0.079990149	466,944	294,912	No	11.58983396	1
Blowfish	0.0119977	0.020001411	274,432	135,168	No	38.65465966	1
SALSA20	0.004013062	0.053800344	626,688	487,424	No	115.5643883	1

Table 3 Algorithms with various parameters for audio file of 777.52 KB

Algorithms	Enc. time (s)	Dec time (s)	Memory utilization (Enc.)	Memory utilization (Dec)	Loss	Throughput (MB/s)	Enc. ratio
AES	0.011958599	0.087993145	487,424	294,912	No	65.01806726	1
DES	0.039977074	0.040005445	1,482,752	270,336	No	19.44927251	1
3DES	0.095997572	0.152122021	991,232	573,440	No	8.099423603	1
Blowfish	0.01997304	0.028001785	282,624	544,768	No	38.92872665	1
SALSA20	0.011981726	0.020003796	532,480	294,912	No	64.89257223	1

Table 4 Algorithms with various parameters for audio file of 2.08 MB

Algorithms	Enc. time (s)	Dec time (s)	Memory utilization (Enc.)	Memory utilization (Dec)	Loss	Throughput (MB/s)	Enc. ratio
AES	0.028020382	0.092093229	3,837,952	5,763,072	No	74.28521158	1
DES	0.083986282	0.088006973	3,551,232	2,228,224	No	24.78380923	1
3DES	0.192012787	0.196011305	2,424,832	2,269,184	No	10.84042388	1
Blowfish	0.060003757	0.132009983	1,019,904	1,990,656	No	34.68949425	1
SALSA20	0.040008545	0.043998003	3,416,064	2,375,680	No	52.02638596	1

Table 5 Algorithms with various parameters for audio file of 11.2 MB

Algorithms	Enc. time (s)	Dec time (s)	Memory utilization (Enc.)	Memory utilization (Dec)	Loss	Throughput (MB/s)	Enc. ratio
AES	0.144010544	0.196668625	16,207,872	11,038,720	No	77.78214499	1
DES	0.392023563	0.384027004	11,993,088	4,034,560	No	28.57340743	1
3DES	0.969902039	0.961265564	11,988,992	847,872	No	11.54905192	1
Blowfish	0.272023439	0.272011042	12,107,776	5,242,880	No	41.17824929	1
SALSA20	0.152007341	0.159998655	8,339,456	11,202,560	No	73.69018429	1

Table 6 Algorithms with various parameters for audio file of 24.03 MB

Algorithms	Enc. time (s)	Dec time (s)	Memory utilization (Enc.)	Memory utilization (Dec)	Loss	Throughput (MB/s)	Enc. ratio
AES	0.461263657	0.428479433	17,313,792	17,940,480	No	52.10560307	1
DES	1.001013756	0.976339817	4,653,056	8,269,824	No	24.01008064	1
3DES	2.445178032	2.23591423	2,805,760	5,480,448	No	9.82931332	1
Blowfish	0.768339396	0.742202282	5,570,560	18,096,128	No	31.28099527	1
SALSA20	0.464538336	0.504232883	7,483,392	17,801,216	No	51.73829402	1

4 Conclusion

According to the result of this program (Tables 2, 3, 4, 5 and 6), in encryption AES takes least time and then the less time consuming algorithm is Salsa20, but Salsa20 is stream cipher so according to cipher text Blowfish is second less time taking algorithm in block cipher. In decryption, Salsa20 takes least time, but according to cipher text Blowfish is the least time consuming algorithm and AES stands after blowfish in less time consuming algorithm in block cipher. In memory utilization, the best algorithm is Blowfish in both encryption and decryption and then after 3DES comes. According to throughput of above result the fastest algorithm is AES and then, followed by Salsa20, but when we consider according to block cipher, Blowfish will stand after AES in less time consuming algorithm.

References

1. Mota, A.V., Azam, S., Shanmugam, B., Yeo, C., Kannoorpatti, K.: Comparative analysis of different techniques of encryption for secured data transmission. In: IEEE International Conference on Power, Control Signals and Instrumentation Engineering, pp. 231–237, 2017
2. Dhanalaxmi, B.: Multimedia cryptography—a review. In: 2017 IEEE International Conference on Power, Control Signals and Instrumentation Engineering, pp. 764–766, 2017
3. Survey, R.C.A., Princy, P.: A comparison of symmetric key algorithms DES, AES, Blowfish. Int. J. Comput. Sci. Eng. Technol. **6**(5), 328–331 (2015)

4. Access, O.: International journal of computer sciences—comparative analysis on different parameters of encryption algorithms for information security. JCSE. Eng. **4**, 76–82 (2014)
5. Jeeva, A.L.: Comparative analysis of performance efficiency and security measures of some encryption algorithms. IJERA **2**(3), 3033–3037 (2012)
6. Thakur, J., Kumar, N.: DES, AES and Blowfish : symmetric key cryptography algorithms simulation based performance analysis. IJETAE **1**(2), 6–12 (2011)
7. Nadeem, A., Javed, M.Y.: Performance Comparison of Data. IEEE, Sept 2005
8. Mahamat, Y., Othman, S.H., Siraj, M., Nkiama, H.: Comparative study Of AES, Blowfish, CAST-128 and DES encryption algorithm international organization of scientific research international organization of scientific research. IOSR J. Eng. **6**(6), 1–7 (2016)
9. Jun, Y., Na, L., Jun, D.: A design and implementation of high-speed 3DES algorithm system. IEEE, pp. 175–178 (2009). https://doi.org/10.1109/FITME.2009.49
10. Afdhila, M.D., Nasution, S.M., Azmi, F.: Implementation of stream cipher Salsa20 algorithm to secure voice on push to talk application. In: IEEE Asia Pacific Conference on Wireless and Mobile Implement, pp. 137–141, 2016
11. Panda, M., Nag, A.: Plain text encryption using AES, DES and SALSA20 by java based bouncy castle API on windows and linux. In: Proceedings of 2015 2nd IEEE International Conference on Advance Computing and Communication Engineering, ICACCE 2015, pp. 541–548, 2015. https://doi.org/10.1109/ICACCE.2015.130

Secured Communication Using Virtual Private Network (VPN)

Paul Joan Ezra, Sanjay Misra, Akshat Agrawal, Jonathan Oluranti, Rytis Maskeliunas, and Robertas Damasevicius

Abstract The evolution and era of the latest programs and services, collectively with the enlargement of encrypted communications, make it difficult for site visitors within a safety enterprise. Virtual private networks (VPNs) are an instance of encrypted communique provider that is becoming famous, as a way for bypassing censorship in addition to gaining access to offerings which are geographically locked. This paper reviews the layout of an IP security, VPN. The Cisco Packet lines platform is used for the simulation, evaluation and verification. It uses a virtual connection to carry the records packets from a non-public network to remote places.

Keywords Virtual private network · Authentication · Security · Confidentiality

1 Introduction

Individuals who use the Internet are highly exposed to social media exploitation where they are victims of attacks. Due to the various attack and vulnerability that data are exposed to when been transmitted from a sender to a receiver, a protection mechanism ought to be provided to address several safety assaults on statistics transmission through the Internet. There are different attacks over the Internet, such

P. J. Ezra · S. Misra (✉) · J. Oluranti
Center of ICT/ICE Research, Covenant University, Ota, Nigeria
e-mail: Sanjay.misra@covenantuniversity.edu.ng

J. Oluranti
e-mail: jonathan.oluranti@covenantuniversity.edu.ng

A. Agrawal
Amity University, Gurgaon, Haryana, India

R. Maskeliunas · R. Damasevicius
Silesian University of Technology, Gliwice, Poland
e-mail: rytis.maskeliunas@polsl.pl

R. Damasevicius
e-mail: robertas.damasevicius@polsl.pl

K. Khanna et al. (eds.), *Cyber Security and Digital Forensics*,
Lecture Notes on Data Engineering and Communications Technologies 73,
https://doi.org/10.1007/978-981-16-3961-6_27

as the denial-of-service attack which makes the network service unavailable by flooding network traffic to the target which exhausts the processing power of the target [1, 2]; information has been changed either accidentally or by malicious attack affects the integrity of the data or creates false information. Eavesdropping on data containing confidential information, such as the location, keys and even passwords of the node, can be redirected to another location. Many security mechanisms have been reviewed to protect data integrity, confidentiality, availability, authenticity and non-repudiation. Cell users want to get entry to assets from their company or domestic network in an efficient but relaxed manner which is done with the help virtual private network (VPN) connections. VPN is a virtual connection routed through the Internet on a public network, from the sender's private network to the receiver. VPN aims to initiate a secure communication path among different networks. It is usually created across the public network [3]. VPN tunnels are used to maintain the privacy of statistics shared over the physical network connection protecting packet-level encryption, consequently making it very hard to become aware of the programs strolling through these VPN services [4].

Authors in [5] showed that a current survey indicated that almost 50% of agencies would adopt the preceding idea by 2025. VPN provides privacy which prevents intermediated users from eavesdropping, altering or deleting the data, authentication which validates that the packet sent by the authorized sender, checks that the data is not altered and prevent intermediate users from copying and resending the information. A VPN tunnel is created for the information to be secured over the physical community connection, maintaining packet-stage encryption, making it very hard to discover the software passing through the VPN offerings. This paper focuses on secure communication using a VPN.

VPNs continue to develop with an increasing number of options that is frequently used in both big and small organization. They also have an advantage of flexibility, connectivity and security at cheap cost. Organizational gains from VPN are reduction in cost and increases in scalability and productivity without compromising the security [6]. This study covers the simulation, evaluation and verification with the help of a packet tracer simulator.

The main aim of the present work is to design a simple system that uses a VPN to secure wireless communication. The following are the main objectives of the presented work.

1. To show how to protect data from being attack over the Internet.
2. To enable communication to be kept private between only the receiver and sender.
3. To show how VPN is over other security mechanisms such as firewall defense.

A brief knowledge of the work is given in this section. Section 2 presented related works in the field of secured communication. Section 3 outlines the method that is used for design and implementation and results. Section 4 describes the conclusion and future work.

2 Related Works

We have reviewed the related works in several databases. The summary of all those important and selected studies is given in Table 1.

There are several other related works [8, 23, 24] available in the literature but due to limitation of work, we are not providing details.

3 Methodology and Results

CISCO packet tracer is used for the design and the simulation of the proposed network using VPN. Only the authorized user will be able to communicate with the other network. The routers will be configured with advance encryption standard to protect data and privacy, Hash-sha tool for IP security authentication, ISAKMP protocol to ensure that two hosts agree on how to build a security association.

3.1 Design and Implementation

Any device connected to the Internet has an IP address which is a sequence of number; a VPN will mask the IP address. An IP address identifies address and location, and a VPN erases IP address from been detected, encrypts your data and keeps your activities private but they do reduce the speed due to the extra security.

For a system to have a working VPN, the following must be configured.

1. Access-list to permit corresponding traffic that will go over the tunnel.
2. ISAKMP policy and ISAKMP key. It is used to set up key authentication and tunnel.
3. IP sec transform-set. It provides authentication and integrity.
4. Crypto map. The crypto map should be applied to the interface.

VPN tunnel must have a security license on the router. The encryption algorithm that was used is the advanced encryption standard (AES) with a key of 256, to protect data and ensure privacy. The IPsec message integrity used is the HMAC-SHA which defines the key size to support different encryption key size. The pre-shared authentication key was used to require VPN devices on each end to configure with the identical mystery key.

Figure 1 shows a conceptual diagram of the VPN network within an organization with all configured interfaces. If the interfaces are not connected to an IP address, there cannot be any form of communication, secured or not secured. This IP address is a unique identifier that indicates the location of a device and governs the way data is sent over the Internet. In the fig above, router 3 interface is

Table 1 Summary of the literature review

Summary of the literature review		
Authors	Work done	Result
Liyanage and Gurtov [7]	VPN architecture for LTE backhaul was addressed. The Internet key exchange model 2(IKEv2) and host identity protocol (HIP) were used as the safety key	Provided secured backhaul traffic during DoS, DDoS and TCP reset attacks
Deshmukh and Iyer [5]	A secure VPN for remote access was addressed. The advanced encryption standard (AES) algorithm was used for data security. The work was carried out on a packet tracer	The smart devices can securely get linked with users on the Internet as if they had been part of the equal non-public network
Busschbach [4]	Enhancing the QoS of a VPN by using an asynchronous transfer mode (ATM) and multiprotocol label switching is studied	An MPLS-based VPN using next-technology IP switches as the best method to enhance the QoS for VPNs
Jaha et al. [8]	A VPN formula that relies on remote access connections requirements and a site-to-site VPN formula that relies on site-to-site connections requirements	Provide a basis while creating business enterprise WAN which connects websites and customers using VPN technology
Azhar et al. [9]	A social media detection on SMS and camera by using application programming interface (API) and permissions was addressed	Identify measures for API and permission for SMS and camera and possible methods for API and permission SMS and API exploitation
Chze and Leong [10]	Proposed a secure multi-hop (SMRP) to merge the routing and authentication processes	Result suggests that the SMRP produces a secured multi-hop IoT communication network without performance degradation in comparison with the well-known optimized link nation routing protocol (OLSR)
Das and Islam [11]	A remarks-based dynamic computation version that could correctly come across unexpected strategic alteration in malicious behavior with the additional feature of balancing workload among service provider was addressed	Strategic behavior of a malicious agent was detected. They provided a mathematical definition of secured trust
Sarika et al. [12] Wu et al. [13] Dinesh et al. [14] Zhou and Hass [15]	Studied the vulnerabilities, attacks and security mechanisms for mobile ad hoc networks (MANETs)	Gives more understand on MANETs, their traits, uses, the criterion for the network security

(continued)

Table 1 (continued)

Summary of the literature review		
Authors	Work done	Result
Manvi and Tangade [16]	The authentication schemes in vehicular ad hoc networks (VANETs) were studied	Security was based on cryptography systems and digital signature. VANET became famous in transport because of the broadcast of safety messages between vehicles
Assadhan et al. [17]	Monitoring and analyzing of a botnet by the command and control (C2) communication traffic	The result achieved was by evaluating a periodogram of the packet and address count sequences
Lan et al. [18]	The hardware implementation of a lightweight Chaskey algorithm using different implementation scheme was studied	A hardware implementation of 3334.33 gate equivalent is achieved with an operating clock frequency of 1 MHz
Liu [19]	Hash-based message authentication codes (HMAC) was studied to achieve authenticity and integrity without the support of a digital signal	Prove that the HMAC is cheap and easy to implement, and it can be used on websites
Draper-gil et al. [3]	Two standard machine learning algorithms, C4.5 and KNN, were used as classification techniques for time-related features	The results prove that the proposed set of time-related features are good classifiers, reaching accuracy levels above 80%. C4.5 and KNN had a similar performance in all experiments; C4.5 has achieved better results
Padmavathi [20]	Zone routing protocol (ZRP) with wireless transport layer security (WTLS) models was addressed to provide security	The proposed work provided security in both routing and transport layer of MANET
Liang et al. [21] Kobayashi and Shitz [22] Nawej and Technologiae [6]	Secured communication over fading channel was addressed They evaluated the impact of VPN on network performance. The network performance covers the hypertext transfer protocol (HTTP), file transfer protocol (FTP) and constant bit rate (CBR)	They establish a capacity state of the parallel broadcast channel, parallel Gaussian broadcast channel and an optimal power location A simulation was done on NS2 for network on VPN and no VPN. network with VPN always delivers better results

having an IP address of 209.165.100.2 and 209.165.200.2, router 4 209.168.100.1 and 192.168.1.1 and router 5 209.165.100.2 and 192.168.3.1

Figure 2 shows that the router does not have a security license. Without this security license VPN encryption, secure collaborative encryption, dynamic multipoint VPN is impossible. The securityk9 can be checked by using the "show version" command in the privilege mode. The security license has been configured and shown in Figs. 3 and 4

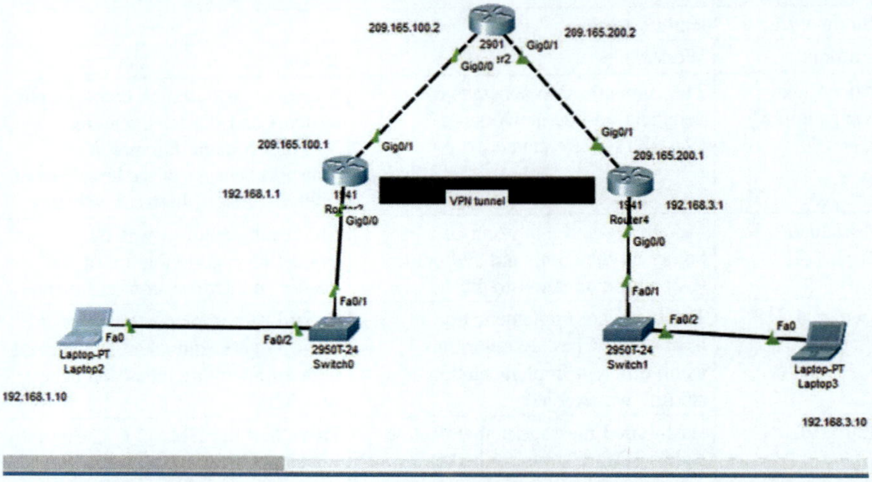

Fig. 1 Theoretical diagram of a VPN network

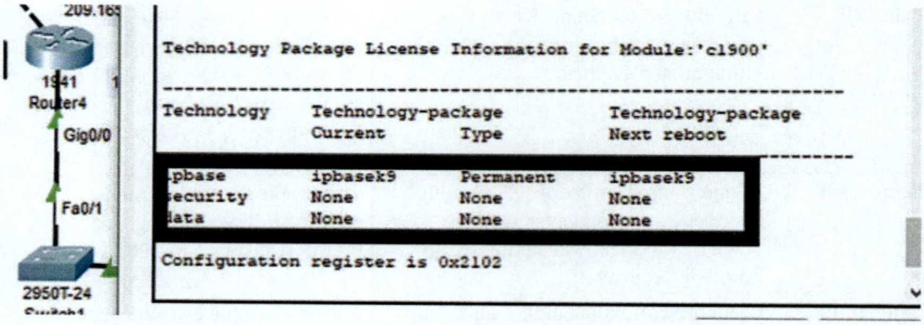

Fig. 2 Router without security license

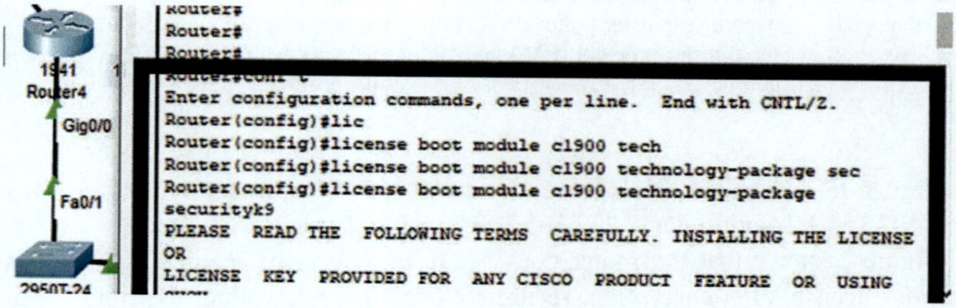

Fig. 3 Configuration of the security license

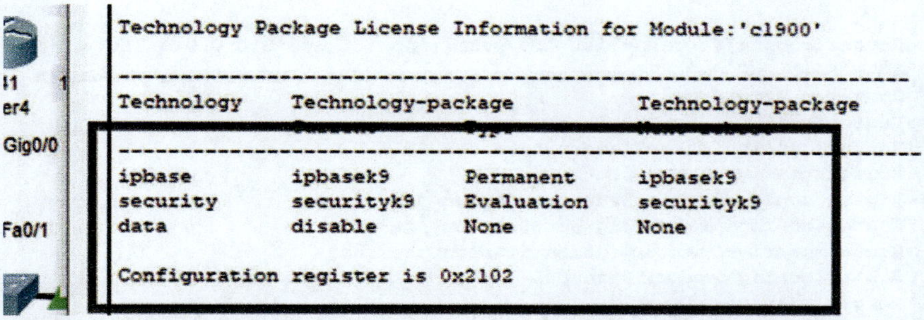

Fig. 4 Router with security license

Figure 5 shows the access-list configuration. The access-list grants permission to allow traffic from one network to the other through the tunnel. The access-list only allow listed IP addresses to communicate across the tunnel.

The policy and key enable the router to utilize IP security as showed in Fig. 6. Every ISAKMP coverage is assigned a unique precedence number among 1 and 10,000. The coverage with precedence number 1 is considered the highest priority policy.

Figure 7 shows the IP sec transform-set configuration, which verifies authentication and integrity. A transform set is a merger of an IP sec transforms designed to enact a particular protection coverage for data traffic

Figure 8, shows the crypto mapping configuration. A crypto map is a configuration entity that select data flow that needs security processing. A crypto map must be named. In the configuration above, the crypto map name is "IPSEC-MAP". Figure 9 shows interface of the crypto map.

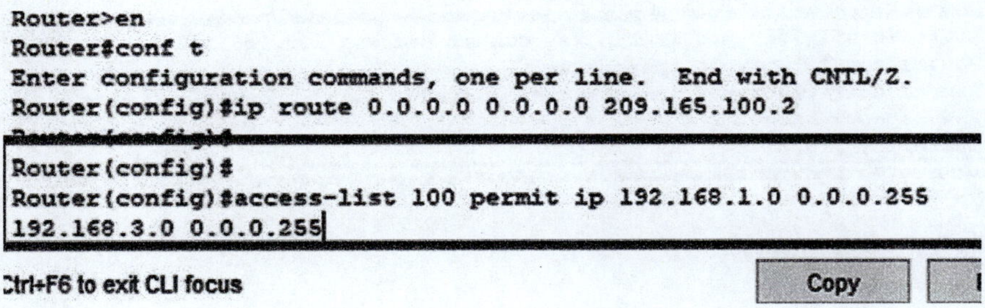

Fig. 5 Access-list

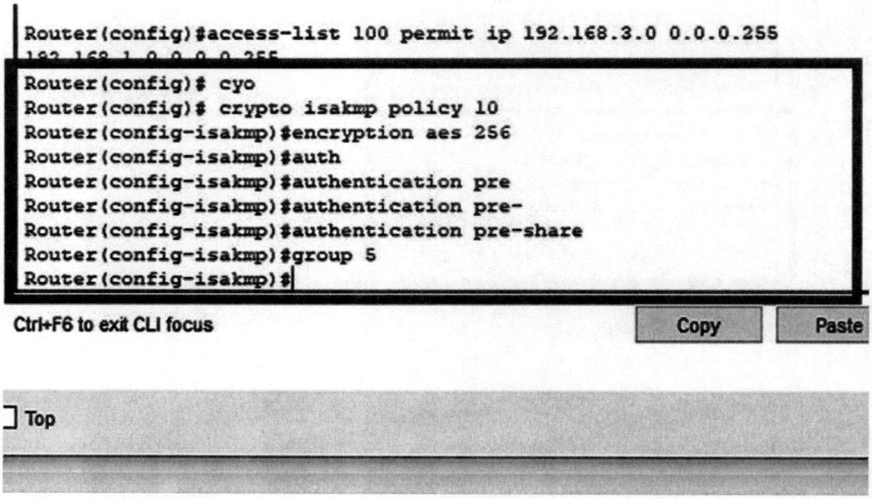

Fig. 6 ISAKM policy and ISAKM key

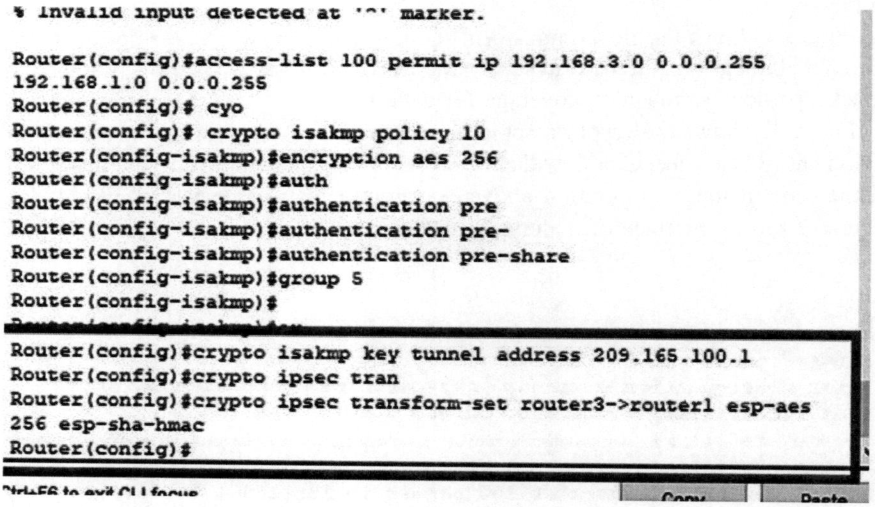

Fig. 7 IPsec transform-set

3.2 *Result and Discussion*

When using the real-time mode to check for the communication process, it is observed that laptop 2 could communicate with laptop 3 without router three been aware of the network; this process is seen using the simulation mode as shown in Fig. 10. Information about the VPN is checked from the inbound PDU details; it is

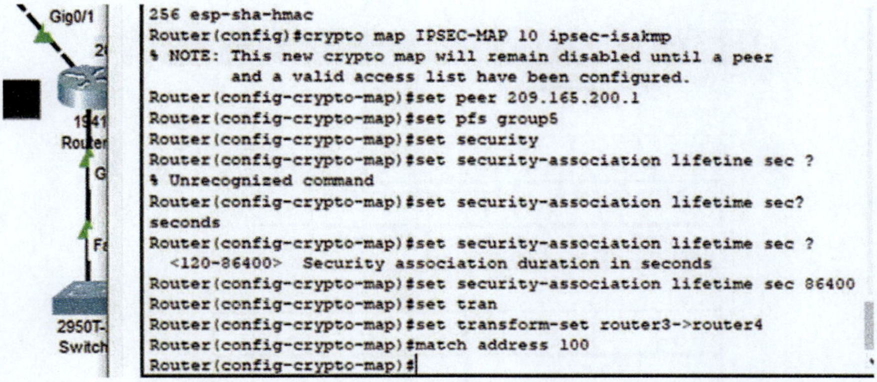

Fig. 8 Crypto map

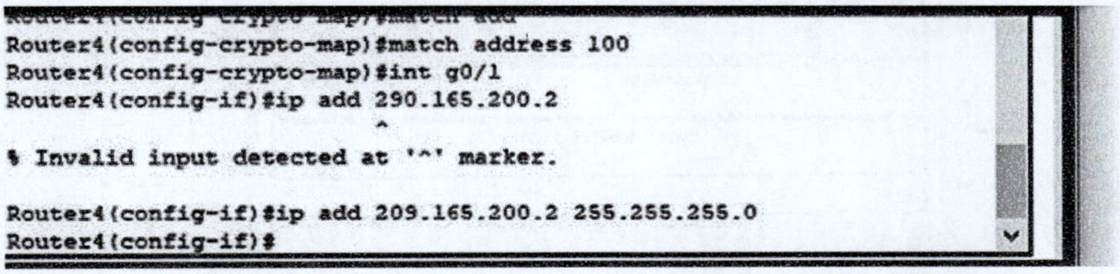

Fig. 9 The interfaces applied to the crypto map

noticed that router 3 had no idea about router 4 and router 5 but they are pinging across router 3 because of the VPN. From the simulation result below, only the source IP address 192.168.1.10 and the destination IP addresses 192.168.3.10 are seen, but the path through which the packet goes through is not recognized.

4 Conclusion and Future Work

This paper presented a VPN architecture within an organization that proposed solution to secure traffic through authentication, authorization, payload encryption and privacy protection. Simulation result on cisco packet tracer verifies that they provide secured traffic communication. This paper proposed a simple VPN solution that can be used in an organization. They also have the advantage of flexibility, connectivity and security at cheap cost. Organizational gains from VPN are increased in the scalability and productivity. Future work can be carried out by

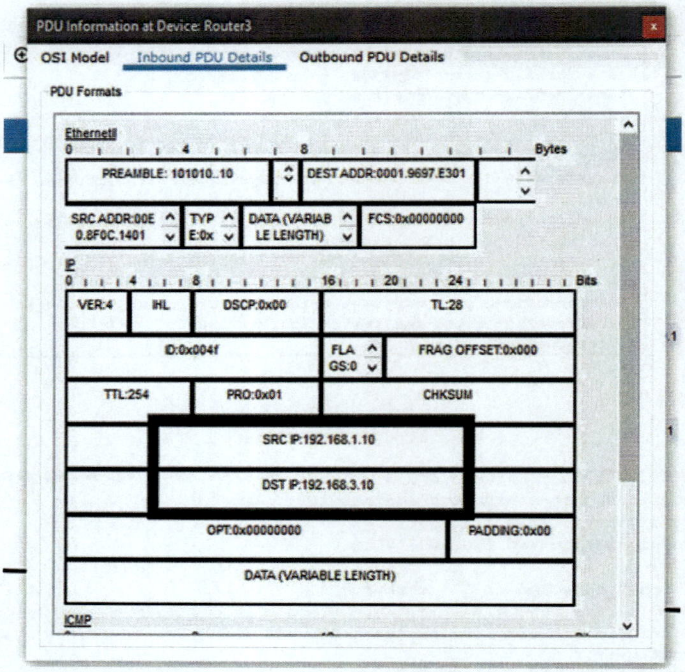

Fig. 10 Simulation result

using other simulation packages order than cisco packet tracer for a simple VPN connection within an organization, and also, the model can also be expanded by using VPN connections across multiple countries.

Acknowledgements The authors appreciate the sponsorship from Covenant University through its Center for Research, Innovation and Discovery, Covenant University, Ota Nigeria.

References

1. Odusami, M., Misra, S., Adetiba, E., Abayomi-Alli, O., Damasevicius, R., Ahuja, R.: An improved model for alleviating layer seven distributed denial of service intrusion on webserver. J. Phys.: Conf. Ser. **1235**(1), 012020 (2019)
2. Odusami, M., Misra, S., Abayomi-Alli, O., Abayomi-Alli, A., Fernandez-Sanz, L.:. A survey and meta-analysis of application-layer distributed denial-of-service attack. Int. J. Commun. Syst. **33**(18), e4603 (2020)
3. Draper-gil, G., Lashkari, A.H., Saiful, M., Mamun, I., Ghorbani, A.A.: Characterization of encrypted and VPN traffic using time-related features. In: Proceedings of the 2nd International Conference on Information Systems Security And Privacy (ICISSP), pp. 407–414, 2016
4. Busschbach, P.B.: ♦ Toward QoS-capable virtual private networks. Bell Labs Tech. J. **3**(4), 161–175 (1998)

5. Deshmukh, D., Iyer, B.: Design of IPSec virtual private network for remote access. In: 2017 International Conference on Computing, Communication and Automation (ICCCA), pp. 716–719. IEEE, 2017
6. Nawej, M.C., Technologiae, M.: Evaluation of virtual private network impact on network performance (2016)
7. Liyanage, M., Gurtov, A.: Secured VPN models for LTE backhaul networks. In: 2012 IEEE Vehicular Technology Conference (VTC Fall), Sept 2015, pp. 1–5. IEEE
8. Jaha, A.A., Ben Shatwan, F., Ashibani, M.: Proper virtual private network (VPN) solution. In: Proceedings of 2nd International Conference on Next Generation Mobile Applications, Services, and Technologies, NGMAST 2008, pp. 309–314, 2008
9. Azhar, M.A., Saudi, M.M., Ahmad, A., Bakar, A.A.: Detection of social media exploitation via SMS and Camera. IJIM **13**(4), 61–78 (2019). Last accessed 01 Mar 21. https://www.learntechlib.org/p/208525/paper_208525.pdf
10. Chze, P.L.R., Leong, K.S.: A secure multi-hop routing for IoT communication. In: 2014 IEEE World Forum on Internet of Things, WF-IoT 2014
11. Das, A., Islam, M.M.: SecuredTrust: a dynamic trust computation model for secured communication in multiagent systems. **9**(2), (2012)
12. Sarika, S., Pravin, A., Vijayakumar, A., Selvamani, K.: Security issues in mobile ad hoc networks. Proc. Comput. Sci. **3**(5), 1022–1024 (2014)
13. Wu, B., Chen, J., Wu, J., Cardei, M.: COUNTERMEASURES IN
14. Dinesh, D., Kumar, A., Singh, J.: Security attacks in mobile adhoc networks (MANET): a literature survey. Int. J. Comput. Appl. **122**(20), 31–35 (2015)
15. Zhou, L., Haas, Z.J.: Securing ad hoc networks. IEEE Netw. **13**(6), 24–30 (1999)
16. Manvi, S.S., Tangade, S.: A survey on authentication schemes in VANETs for secured communication. Veh. Commun. (2017)
17. Assadhan, B., Moura, J.M.F., Lapsley, D., Jones, C., Strayer, W.T.: Detecting botnets using command and control traffic, 4, 156–162 (2009)
18. Lan, J., Zhou, J., Liu, X.: An area-efficient implementation of a message authentication code (MAC) algorithm for cryptographic systems. In: IEEE Reg. 10 Annual International Conference Proceedings/TENCON, pp. 1977–1979, 2017
19. Liu, Z., Lallie, H.S., Liu, L., Zhan, Y., Wu, K.: A hash-based secure interface on plain connection, 1236–1239 (2011)
20. Padmavathi, G., Subashini, P., Aruna, M.D.D.: ZRP with WTLS key management technique to secure transport and network layers in mobile adhoc networks. Int. J. Wirel. Mob. Netw. **4**(1), 129–138 (2012)
21. Liang, Y., Poor, H.V., Shamai, S.: Secure communication over fading channels. IEEE Trans. Inf. Theory **54**(6), 2470–2492 (2008)
22. Kobayashi, M., Shitz, S.S.: Secured communication over frequency-selective fading channels : a practical vandermonde precoding, 2009 (2009)
23. Azeez, N.A., Salaudeen, B.B., Misra, S., Damaševičius, R., Maskeliūnas, R.: Identifying phishing attacks in communication networks using URL consistency features. Int. J. Electron. Secur. Digit. Forensics **12**(2), 200–213 (2020)
24. Osho, O., Musa, F.A., Misra, S., Uduimoh, A.A., Adewunmi, A., Ahuja, R.: AbsoluteSecure: a tri-layered data security system. Commun. Comput. Inf. Sci. **1078**, 243–255

Survey for Detection and Analysis of Android Malware(s) Through Artificial Intelligence Techniques

Sandeep Sharma, Kavita Khanna, and Prachi Ahlawat

Abstract Artificial intelligence techniques have been intensively used for android malware detection and analysis in the last past few years. The proposed methodologies do not suffice the requirement while characteristics of malwares are changing so rapidly and evolving new complex malwares. Therefore, it is a very complex task to classify and identify these malwares. This paper presents an organized and comprehensive survey for the detection techniques of android malware(s) in chronological order. These detection and analysis techniques are elaborated in two core categories: statics and dynamic analysis and hybrid analysis with machine learning or artificial intelligence. The core contributions of this paper are: (1) explaining a methodical, chronicle and organized summary of the existing techniques of android malware detection, (2) exploring the major elements and challenges in the detection methods and (3) explaining the importance of artificial intelligence for android malware detection. The detection approaches are explained in a manner that new approaches are compared with the old ones based on their features. The advantages and disadvantages of each approach are discussed. This study facilitates researchers and academics to have a wide-ranging conception of the field of android malware detection and provides a platform to enhance the fundamental knowledge to implement the new idea and subsequent improvement further in existing techniques.

Keywords Android · Malware detection · Static · Dynamic · Hybrid · Artificial intelligence

S. Sharma (✉) · K. Khanna · P. Ahlawat
The Northcap University, Gurugram, Haryana, India
e-mail: kavitakhanna@ncuindia.edu

P. Ahlawat
e-mail: prachi@ncuindia.edu

K. Khanna et al. (eds.), *Cyber Security and Digital Forensics*,
Lecture Notes on Data Engineering and Communications Technologies 73,
https://doi.org/10.1007/978-981-16-3961-6_28

1 Introduction

The smart mobile phones are attracting all due to their tremendous features and have provided online banking, online marketing, online study, etc., in addition to fundamental telephony services. These mobile terminals are outfitted with huge processing power and ample storage that holds a variety of sensitive and critical data like contacts, pictures, passwords, cookies, credit card details, location information, etc. As per statista's report, the global figure of subscribers of smart mobile phone has reached 4.01 billion in 2020 and is expected to reach 4.7 billion in 2022 [1]. However, this popularity of smartphones has attracted the attention of hackers who may steal the data stored in smartphones and further may deduce the code to unlock the mobile to compromise the device or manipulate popular services. The various Android Operating Systems (OS) is the most accepted operating systems for the existing smartphones (82.8%) [2]. On one hand, the Android OS is an open-source software system to provide a wide range of accessibility to users and on the other hand, Android OS is more susceptible to cyber-attacks. The malicious apps or malwares are developing so rapidly worldwide that was never observed in the past. Normally, android apps and games are downloaded and installed from the Play store verified by play protect developed by Google. However, android apps and games can also be installed from other non-authenticated and not verified stores where apps and games are integrated with malicious codes that contaminate the android phones and exfiltrate the critical data to hackers. The protected Play store is also not secured from the hackers even after their continuous effort to monitor and remove infected applications. Moreover, several variants of Android Operating System are available in multiple manufacturer's devices and all variants are needed patches and updates regularly from the respective developer immediately. There are some major aspects like diverse ARM processors, limited RAM, and power battery restriction, etc. that create the complexity in detection of malicious codes. TrendMicro declared in the first quarter of 2020 that, mobile cyber espionage operations have increased by 1400%, distributed among multiple firmware(s) and operating systems from 2015 to 2019 [3].

Several anti-malware techniques, frameworks and solutions have been developed to protect android phones. These analysis frameworks are divided into static and dynamic features analysis. The static features analysis is based on the reverse engineering of.apk file of apps. Applications are scanned for malicious code instead of executing them. This is helpful for detecting the malicious codes that only run in explicit circumstances like rebooting; however, the same techniques are not able to identify the encrypted and dynamically loaded malicious codes of android. In the dynamic feature analysis, runtime activities of apps are analyzed and encrypted and dynamically loaded malicious codes of android are identified. Since, all execution paths cannot be examined, which limits overall and complete analysis of the code/apk. The Hybrid techniques combine the advantage of static and dynamic features analysis; however, these techniques have several limitations that force the researchers to develop new hybrid analysis technique based on artificial or machine learning to achieve high detection rates. All analysis techniques have some advantages and weaknesses for the detection of android malwares. It is also observed that there are several deficiencies in the surveyed presented in the research paper. A quantity of papers has not considered the latest articles in the comparison and analysis. On the other hand, some surveys have not classified the detection techniques systematically for their research work. To overcome these flaws, this paper presents an organized literature survey on the latest android malware detection and analysis frameworks. The main contributions of this paper and study are:

(a) resenting a systematic, chronically and categorized study of the current approaches to android malware detection.
(b) Exploring the architecture and elements of noteworthy android malware detection methods and challenges in these methods.
(c) Demonstrating the importance of deep learning and artificial intelligence for android malware detection and analysis techniques.

The paper is organized as follows. "Review of Android Malware Analysis Techniques" presents a systematic assessment of the existing techniques for detection and analysis of android malwares. In "Discussion," a brief tabulated comparison of malware detection techniques is presented for ready reference. A comparison chart for the efficiency of major android malware detection methods is presented which can work as a benchmark level for new research works. At last, "Conclusion" exhibits the crux of the same study.

2 Review of Android Malware Analysis Techniques

The existing android malware detection and analysis framework are elaborated based on their basic architecture, feature extraction module, test data set, efficiency and advantages and disadvantages. The technique is divided into two parts. The first basic techniques including static and dynamic analysis that are base for any artificial intelligence technique are discussed in chronological order. Then, the latest artificial intelligent techniques for the analysis of android malware are analyzed in detailed.

2.1 *Integrated Static and Dynamic Malware Analysis Techniques*

There are several android malware analysis techniques that implement the combination of static and dynamic analysis to improve the accuracy of detection of malware. The general architecture of an integrated static and dynamic analysis framework is shown in Fig. 1. Blaising [4] demonstrated an android application (AA) sandbox in 2010 that detects the malicious activities in android apps through

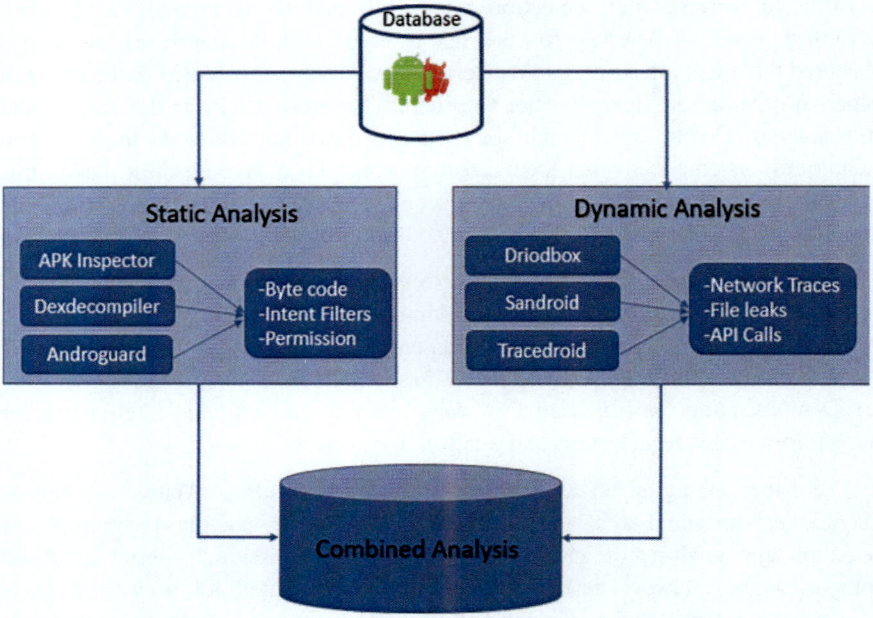

Fig. 1 Statics and dynamic combined malware analysis

analysis both statically and dynamically. In the section of static analysis, android api/code is decompiled by the Baksmali module. Then, the decompiled smali files are scanned and static features are extracted. In the section of dynamic analysis, the apk/code is run through the emulator devices. The Monkey module is also integrated to activate the apps with testing parameters like gestures and clicks. This framework is loaded in the kernel under a fully controlled environment for execution of the apk/code that generates the logs files by capturing the system calls. Vectors are defined based on the log of static and dynamic behavior analysis and these same vectors are inserted into a framework for monitoring of malicious activities of apps and further detection. The main constraint of AASandbox was that it could not be implemented in a large volume of android devices as the root privilege is required for capturing system calls. Authors did not calculate the accuracy of the proposed framework.

Zhou et al. [5] developed a new framework DroidRanger in 2012 for detection of unknown and known malwares. In first section, DroidRanger examines and detects the essential permissions which are used by the malwares to execute the malicious actions that segregate the malicious apps for further analysis in second stage. This data is then collaborated with the defined behavioral rules to distinguish, filter and analysis of identified malwares. However, detection of unknown malwares is achieved in two segments: In the first segment, the filter is defined based on heuristic algorithms. Researcher has implemented two heuristics algorithms in this technique. In the first heuristic technique, the code from the remote server is loaded dynamically. In the other heuristic technique, the native code is loaded dynamically and the behavior of app is monitored. If app is not stored in default directory, then this behavior is noticeable. Heuristic technique facilitates the framework to determine the malwares which exploit the OS kernel and get root access and provides a capability to detect the DroidDreamLight [6]. Dynamic executions are performed to monitor the runtime behavior of applications. A data set of 204040 android applications was collected in which 75% and 25% were taken from the official and other alternative android markets, respectively. The limitation of this framework is that it developed for only official android markets (Google Playstore) and alternative android markets (eoe market, gfan, alcatelclubm mmoovv, etc.), not for android devices.

Talha et al. [7] developed a new framework for android malware analysis, i.e., APK Auditor in 2015, that uses permissions features for detection of malware behavior. This framework comprises of following segments: (1) database for signatures: store signatures of all the apps; (2) android client: offers the facility of malware analysis to clients; (3) server for processing, which offers a connection between android client and database and processing. The server executes the analysis without installing the apps on the device and optimizes the available assets. APK Auditor monitors all the permissions as called by the apps and determines the value of permission malware score (PMS). Thereafter, this tool classifies the apps as malware if the value of PMS apps scores higher than the threshold value. The outcome demonstrated that APK Auditor accomplished 92.5% accuracy but cannot identify the malicious payloads which installed dynamically.

Abraham et al. [8] developed a 2-Hybrid malware detection framework in 2016 which executes the detection and analysis of android malwares on the server installed at a distant location. The extractor module of framework extracts and accumulates the parameters of the apps and classifies the apps as malware or benign. This framework does not conclude the categorization of apps only based on host analysis but also sent to the server installed at a distant location for exhaustive analysis. If malicious activities are observed in app then server forwards the same data to local device for future detection. As a test set, 39 malicious samples were collected and 69 permissions were monitored from these samples. The results were satisfactory; however, authors did not compare the efficiency of this framework with other frameworks.

Sun et al. [9], developed a new framework MONET in 2017, that detects the malicious codes of an already acknowledged malware family. This framework extracts the static parameters, disassemble the codes, and monitors the dynamic activities of apps for uncovering of malwares. MONET uses client-end applications that run on the user device for analysis of application and design its specific signature. The sets of signatures are stored in the server and a signature matching algorithm is applied. A test set of 3723 and 500 malwares and legitimate apps, respectively, were used for experiments. MONET demonstrated 99% accuracy of detection of malware but shown lacks of resource efficient.

2.2 *Hybrid Malware Analysis with Artificial Intelligence*

The general architecture of the hybrid malware analysis technique with artificial intelligence is shown in Fig. 2. Wang et al. [10] developed a hybrid malicious code analysis framework in 2015 which performs both static and dynamic analysis and implement machine learning for training the framework. It implements signature-based detection to detect the malwares. In first stage, the manifest file is used for static features extraction and android packaging tools are used for disassembling the dex files [11]. In second stage, Cuckoo Droid [12] and Robotium [13] are used for dynamic features extraction. These static and dynamic features are converted into vectors and mapped into vector space. Different feature extraction and selection methods are implemented in this method to increase the accuracy and susceptibility for misuse and anomaly detection. In misuse detection, SelectKBest method is applied where k highest scoring features are selected and Chi2 is used as a scoring function. After feature selection, SVC [14] and SVM classifier [15] are applied for the classification of known and unknown malware, respectively. Based on the probability of signature matching, apps are categorized as malwares and training database is updated accordingly. In case, if there is any uncertainty, then only anomaly detection is applied and if any abnormal behavior is detected, then app is classified as unknown malware and training database is updated accordingly. A test set of 12000 and 5560 for benign and malware apps are collected,

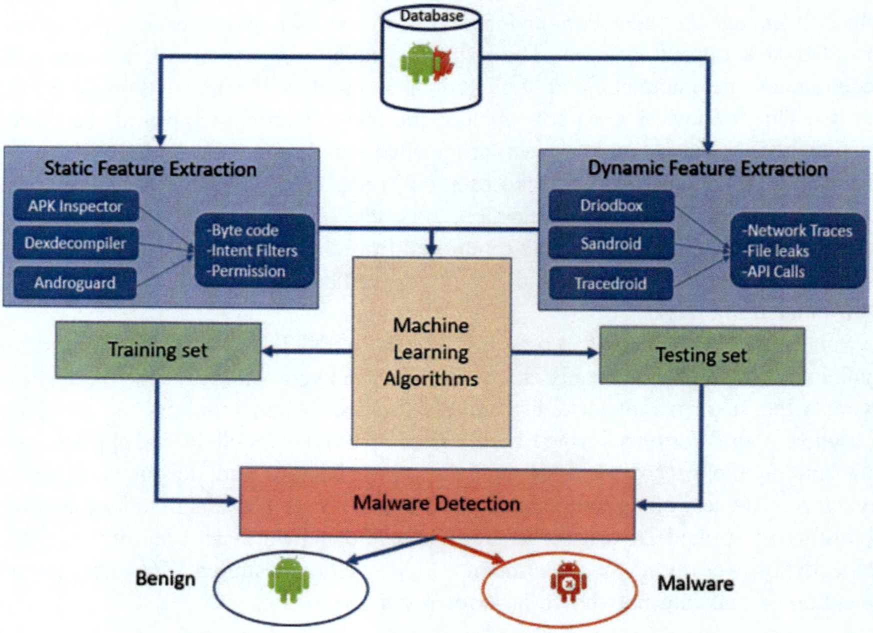

Fig. 2 Hybrid malware analysis using artificial intelligence

respectively, from various App stores. The Outcome demonstrated that the framework can detect the malwares with an accuracy of 98.79% from the used test set.

Sahijo et al. [16], developed an integrated framework for malware detection and analysis based on machine learning in 2015. This framework assimilates the static analysis and dynamic analysis for detection and analysis with training of modules of machine learning. Framework extracts printable strings information as static feature from app and then sorted as per the rate of incidence in each file. After extraction, feature selection is performed such that PSI features with the rate of incidence above benchmark are selected. Selected features are then enlisted into global list named feature list. This feature list is used for creating static feature vector. In next stage, Cuckoo platform is implemented and API call logs are extracted for dynamic feature extraction of apps. API call grams are generated for all codes and sorted as per the rate of incidents. The vectors for static and dynamic features are designed and inserted into two separate machine learning modules SVM [17, 18] and random forest [19] for classification. This framework demonstrated 98.7% accuracy on a test set of 997 malicious and 490 benign apps; however, huge storage and high processing power are required for the same framework.

Yuan et al. [20] developed an online framework DroidDetector in 2016 to detect malwares online. In this framework, both the static and dynamic features extraction tools are implemented in the servers hosted remotely and then machine learning modules are integrated for discriminating between malwares and benign apps. Major permissions and sensitive API are monitored to determine the static features. DroidBox is applied for a specific period of time to monitor the dynamic activities of apps. The static and dynamic features are mapped into vectors and these features are inserted into the machine learning module for learning of detection of malicious apps. A test set of 20,000 apps of a combination of both legitimate and malware apps is used to evaluate the framework of DroidDetector. Experimental results demonstrated the achievement of 96.7% accuracy for the detection of malwares. The key flaw of this framework is that the various malicious apps can be escaped from the detection of framework if malicious behavior of apps is not observed during the observation time-period.

Yu et al. [21] developed a hybrid automatic static-dynamic switch framework in 2016 to counter the transformation technique adopted by the malwares. This framework can distinguish the malicious codes of apps by either static or dynamic detection and analysis technique. In first section, apps are decompiled through Apktool [22] and accordingly manifest and Smali files are created and static malware analysis is applied. The extracted static features are mapped into the vectors and these vectors are inserted into machine learning modules for detection of the apps as malware. However, if apps are not properly decompiled due to transformation techniques, then automatic dynamic malware analysis is executed on the apps. In the same way, dynamic malware analysis is performed on apps and extracted dynamic features are mapped with vectors. These vectors are also inserted into machine learning modules like kNN and Naive Bayes, etc., for training of framework for detection. Framework demonstrated the 99% accuracy for static detection and 90% accuracy for dynamic detection. However, the main flaw of this framework is that only static malware analysis process is executed if the app is decompiled properly. Framework does not have capability of detection of dynamically loaded codes in the app. Therefore, app that is not able to execute accurately cannot be analyzed properly and accordingly cannot be classified as malware or benign app.

Saracino et al. [23], developed a host-based novel framework MADAM in 2016 which suspicious the apps based on their misbehavior and malicious activities. The framework segregates the malware into a number of behavioral modules and each module execute specific misbehavior that is tagged with the same malware. Framework extracted and correlated features at four stages: package, application, user, and kernel. MADAM structural design consists of four main pillars that are App Risk Assessment, Global Monitor, App Monitor, and User Interface and Prevention. MADAM executes App Risk Assessment process and generates a list of suspicious apps. Framework applies the static and dynamic analysis to determine the risk of app. The Global monitor process monitors the system call, user activity, message and activity logger and generates feature vectors. These vectors are inserted in a machine learning module like kNN and are used to train the framework for legitimate and malicious behaviors. App monitor process decompiled the malicious codes by analyzing the behavior of malwares at API and kernel level. App Monitor continuously monitors the (a) Background apps with administrator privileges, (b) Automatic SMS send-receive apps, and (c) Foreground apps. Lastly, the user interface warns about the malicious apps and assistance for blocking and removing these apps. The proposed framework is tested on a test set collected from virus share and demonstrated 96.9% detection rate. However, MADAM framework has one flaw that a root privilege is required on the device to execute the detection and this cannot be executed in the mass market.

Hou et al. [24] developed a heterogeneous information network, i.e., Hindroid for android malware analysis in 2017. This framework created higher-level semantics in place of application programming interface (API) calls. HinDroid framework created a heterogeneous information network (HIN) of android applications and a meta-path to provide connection to these applications. All meta-path is required to calculate a similarity measure over android applications and automatically weighted by the learning algorithm to make predictions. The authors claimed that HinDroid shows better performance than other android malware detection systems; however, accuracy and efficiency were not provided.

Karbab et al. [25] developed an automatic framework MalDozer using deep learning in 2018 for android malware analysis. MalDozer automatically extracts the features and learns the new patterns from the actual samples to detect the malwares. MalDozer can be deployed not only on remote servers but also on android mobile agents. MalDozer offered many advance features like automatic features extraction in the training phase and minimal preprocessing power. A test sets of 33,000 malwares and 38,000 benign apps are used to evaluate the performance of framework. MalDozer demonstrated that malware can be correctly detected with an F1-Score of 96–99%.

Li et al. [26] introduced a new framework in 2018 for malware analysis system based on Significant Permission IDentification (SigPID). This framework is designed in three systematic stages of pruning approach with machine learning techniques to identify the most significant permissions so that framework can be utilized effectively. SigPID demonstrated that only 22 permissions out of 132 permissions are significant and improve the runtime performance by 85.6%. This framework also demonstrated that 90% of precision can be achieved by the support vector machine (SVM). A test set of 1,000 malwares was applied for testing and SigPID technique demonstrated 93.62% and 91.4% efficiency in detection of known malware and unknown/new malware samples, respectively.

Feng et al. [27] proposed an innovative ensemble learning-based EnDroid in 2019 for android malware detection. This framework integrated multiple types of dynamic features and analysis techniques and achieved very precise malware detection accuracy. These analysis techniques include the monitoring of malicious behavior of system and application level like stealing of critical data, uploading on a remote server, and update of firmware with malicious codes. Two datasets were taken for experiments and result demonstrated that stacking accomplished the best detection performance and exhibits 96.56% efficiency in android malware detection.

Zhou et al. [28] presented a new framework incorporating the control flow graph with machine learning algorithms in 2019 for android malware analysis. In this framework, the applications are decompiled and a control flow graph is constructed to obtain the API information. Three types of system API uses data sets, (1) API calls, (2) API frequency, and (3) API sequence based on control flow graphs are constructed to develop three detection models. The accuracy of all three models is compared using Precision, Recall and F-score metrics. The Framework demonstrated 98.98% detection precision on a test set of 10,683 malicious and 10,010 benign applications.

Mehtab et al. [29] proposed an innovative approach, AdDroid in 2019 to detect android malwares based on a variety of permutations of artifacts. The rules/artifacts designate the activities of codes of android device like establishing a connection to the ISP through the internet, secretly uploading personal data to the pre-defined server, updating malicious package/patches of firmware, etc. AdDroid uses ensemble-based machine learning algorithms, i.e. Adaboost to train the model for static analysis of android apps. AdDroid is able to extraction and selection of statics feature and then able to recognizing malicious applications based on the most unique rules. This machine learning framework is trained and developed by applying a test set comprising 1,420 apps including 910 malicious and 510 benign android apps. The framework demonstrated an accuracy of 99.11% on a similar test set. The Authors did not include dynamic features for machine learning.

Ma et al. [30] proposed the deep learning-based framework Droidetec in 2020, for android malware detection and exact localization of malicious codes in the apps. This framework implemented an innovative feature extraction method to monitor behavior sequences from malicious apps. Extracted behavior sequences are represented as a vector that automatically scrutinizes the semantics of sequence of fragment and determines the malicious code. A test set of 9,616 malicious and 11,982 benign programs is used to evaluate the capability of Droidetec framework and the result demonstrated a precision of 97.22% for detection.

Mohammed et al. [31] proposed a deep learning framework, i.e., DL-Droid in 2020 to detect the malicious apps through dynamic features and generation of stateful input. DL-Droid framework is implemented with 30,000 benign and malware apps on real android terminals. Result demonstrates that the performance of detection of this framework with deep learning module is better than existing traditional methods. DL-Droid achieved accuracy of 97.8% for detection with dynamic features only and accuracy of 99.6% for detection with dynamic and static features. However, self-adaptation for intrusion detection systems is not available to improve the performance of model for malware detection.

Su et al. [32] proposed DroidPortrait in 2020, an approach of construction of multi-dimensional and vertical behavioral representation for detection of android malwares. In the analysis, the behavior of android malware is monitored and static and dynamic behavior as dataset are extracted. In the next phase of analysis, different kinds of behaviors are segregated based on android malware and a specific and unique behavioral tag is attached with its signature. Machine learning (ML) algorithms are implemented to correlate these behavioral tags with specific malwares automatically. A high performance machine learning algorithm, i.e., random forest algorithm is very suitable and easily integrates with the basic framework to detect the android malware. The result demonstrated that DroidPortrait framework can illustrate the behavior appearances of android malware with high accuracy. The efficiency level was 90% which is lower and can be increased further.

Mahindru et al. [33] proposed MLDroid in 2020, the web-based model to analyze the android apps as malware and benign. Authors implemented feature selection techniques and trained the framework by these techniques. These selected features developed an innovative model by implementing different machine learning algorithms. A test set of 500,000 plus android apps are used for the experiment and the four distinct machine learning algorithms (1) deep learning, (2) first and farthest clustering, (3) YMLP, and (4) nonlinear ensemble decision tree forest is applied in parallel. Experiment results exhibit that framework developed by considering all the algorithms can achieve an accuracy rate of upto 98.8% for detection of malware from android apps.

Zhang et al. [34] proposed an automatic framework TC-Droid in 2020, for android applications analysis based on text classification technique. In this framework, the text sequences of APPs are analyzed by AndroPyTool and generated analysis reports are feed in deep learning algorithms. A machine learning algorithm, i.e., convolutional neural network (CNN) is used to extract considerable information instead of manual feature engineering from the same analysis report. A variety of well-known samples were collected for evaluation, and it is demonstrated that the performance of TC-Droid is better than other classic algorithms, i.e., NB, LR, KNN, RF, etc. However, actual data for accuracy and efficiency were not provided for comparison.

Thongsuwan et al. [35] demonstrated a hierarchical approach in 2021 using extraction of authorization-sensitive feature and implementing deep learning algorithms to design an android malware detection framework. This framework extracts four sensitive features: basic blocks, permissions, api, and key functions used for authorization. An innovative machine learning model, i.e., convolution neural network and eXtreme Gradient Boosting (CNNXGB) are implemented for training, learning and detection of malware. This framework sequences the key functions as per the timing of API calls and collects a similar section that confines the global semantics of malware family. Permissions and API calls were extracted from 1,330 android test samples to drill the model by XGBoost. The result demonstrated that efficiency increased upto 98% by the CNNXGB model. However, as the input data for analysis is increased, the number of the convolution layers and complexity increases.

McDonald et al. [36] developed a framework based on manifest permission and machine learning for android malware detection in 2021. In this framework, four different machine learning algorithms (1) random forest, (2) support vector machine, (3) Gaussian Naïve Bayes and (4) K-Means are applied in conjunction with features selected from android manifest file permissions to distinguish the apps as malicious or benign. A test set of 5,243 samples are used to test the framework and it was demonstrated that random forest ML algorithm performed the best with 82.5% precision and 81.5% accuracy.

Above mentioned techniques are briefly explained and tabulated in Table 1 for ready reference for researchers.

Table 1 Overview of the existing framework and research gaps

S.no	Reference	Year	Methodology/ contribution	Research gap
Integrated static and dynamic malware analysis techniques				
	Android application (AA) sandbox [4]	2010	Uses static and dynamics analysis	Root privilege is needed to capture the system calls
	DroidRanger [5]	2012	Uses static and dynamic analysis	Developed for official android and alternative android markets only
	APK Auditor [7]	2015	Uses static analysis with permission-based malware detection	Framework is installed at central server, and therefore, internet connection is required for android terminal for the malware analysis at server

(continued)

Table 1 (continued)

S.no	Reference	Year	Methodology/ contribution	Research gap
	Novel hybrid android malware detection [8]	2016	Static and dynamic features of the applications, are extracted at the server configured at distant. If app is distinguished as risk, database is updated for this risk signature	Efficiency and accuracy of framework were not determined and compared with another existing framework
	MONET [9]	2017	The specific signatures for respective malware are generated and forwarded to server for matching and detection	This framework works only with android DVM; however, latest android supports only ART
Hybrid malware analysis with artificial intelligence				
	A novel anomaly and misuse hybrid mobile malware detection system [10]	2015	Static features from manifest file and dynamic features through CuckooDroid are used with SVC classifier in misuse detection	Comparison table for detection with other frameworks are not computed
	Detection and mitigation of android malware through deep learning [16]	2015	Static features from printable strings information (PSI) and dynamic features from Cuckoo are extracted. These features are inserted in machine learning modules for classification	Power and storage consumption are intense
	DroidDetector [20]	2016	The extracted static and dynamic features through different tools are applied in deep learning algorithm for classification as malware	Malwares are escaped from the detection system during the dynamic monitoring time-interval
	A hybrid automatic static-dynamic switch framework [21]	2016	Analysis is conditional-based. If properly decompiled the app, extracted static vectors are inserted into deep learning algorithms like SVM, kNN and Naïve Bayes. However, in case, app is not decompiled properly then only dynamic are inserted into machine learning for classification	Only performs one analysis, i.e., static or dynamic. Framework will not be able to detect the malicious codes if app is decompiled properly

(continued)

Table 1 (continued)

S.no	Reference	Year	Methodology/ contribution	Research gap
	MADAM [23]	2016	Features are extracted at four stages: (1) package, (2) application, (3) user and (4) kernel. These features are mapped with vectors and inserted as input in machine learning kNN classifier for training	It runs only on rooted device, therefore, not supported in the large users. Moreover, pre-loaded apps/games cannot be analyzed by this framework
	Hindroid [24]	2017	Created heterogeneous information network (HIN) of android applications is used with machine learning algorithms	Accuracy and efficiency were not provided
	MalDozer [25]	2018	Automatically extracts the features and learns the new patterns from the actual samples through machine learning algorithms	Less effective in malware family attribution
	Significant permission IDentification (SigPID) [26]	2018	Instead of all android permissions, only most significant permissions SigPID are extracted and utilized for machine learning models	The analysis can be performed only on the rooted devices
	EnDroid [27]	2019	Dynamic features with system-based behavior trace and common apps-based malicious behaviors are used for deep learning	Power consumption is very high
	A control flow Graph-based android malware detection including machine learning [28]	2019	An ensemble of three detection models is created based on control flow graph of three data sets for android malware detection and analysis	Capability to establish the malware family is not presented
	AdDroid [29]	2019	Only static features are extracted and inserted into the ensemble-based machine learning model that is trained by the Adaboost	Artifacts called rules were defined for only static analysis. Dynamic analysis is not integrated with the model

(continued)

Table 1 (continued)

S.no	Reference	Year	Methodology/ contribution	Research gap
	DroidDetec [30]	2020	An innovative behavior sequences feature is extracted for analysis of syntax of linked segments and finding of malicious code	Not accurate framework for further classification into various malware families
	DL-Droid [31]	2020	Dynamic feature extraction with stateless approach is used with deep learning algorithms for analysis of malicious android apps	Self-adaptation for Intrusion detection systems is not available to improve the performance of model
	DroidPortrait [32]	2020	Extracted static and dynamic behavior and create an informative behavior dataset that includes specific behavior tag for specific android malware	The efficiency level was 90% which can be increased further
	MLDroid [33]	2020	Static and dynamic features are extracted and inserted into 04 different machine learning algorithms in parallel	This framework has capability for only detection the app as malware or benign. Detection rate is lower
	TC-Droid [34]	2020	Convolution neural network (CNN) algorithm is used for extraction of significant tags instead of manual features	Data for accuracy and efficiency were not provided
	ConvXGB [35]	2021	Authorization-sensitive features are extracted for machine learning algorithm of convolution neural network and eXtreme gradient boosting (CNNXGB)	Number of the convolution layers is increased, depending on the input data for analysis
	Manifest permission (MP) and machine learning (ML)-based framework [36]	2021	Static features are selected and inserted into four different machine learning algorithms for grading the apps as malicious or benign	Other remaining static features can also be explored further to form a greater feature set

3 Discussion

Based on the review of all frameworks, an overall comparative analysis among all malware detection frameworks based on static and dynamic features and hybrid features using artificial intelligence are analyzed. We discover that the hybrid method using artificial intelligence has the best accuracy in comparison with statics and dynamic malware analysis detection approaches. The accuracy efficiency of each framework is analyzed and it is observed that all frameworks have an accuracy efficiency higher than 80%. It is also examined that the DL-Droid malware analysis framework has maximum accuracy efficiency that is 99.6% [31], and machine learning-based android malware detection framework using manifest permissions has minimum accuracy efficiency that is 81% [36].

4 Conclusion

A methodical and chronically literature investigation of the detection and analysis frameworks and techniques for android malware are explained. The work done by researches were reviewed and investigated and existing android malware analysis frameworks were categorized into two categories: (1) static and dynamic malware analysis and (2) hybrid malware analysis using artificial intelligence. These malware analysis frameworks were compared and analyzed according to their specific features and technique. The advantage and disadvantage of each analysis framework were deliberated. It is essential to develop the frameworks which automatically learn without the intervention of human and detect the zero-day malwares. Therefore, artificial intelligence techniques surfaced as a potential solution to handle the different types of malwares. This state-of-the-art research survey is a fundamental instrument for further any research which can make a tremendous change in the development of a new framework.

References

1. Gartner says worldwide sales of smartphones grew in the first quarter of 2020 [Online]. Available: http://www.gartner.com/newsroom/id/3609817 (2020)
2. Number of smartphone users in India in 2015 to 2020 with a forecast until 2025 [Online]. Available: https://www.statista.com/statistics/467163/forecast-of-smartphone-users-in-india/ (2020)
3. Trend micro threat report | review, refocus and recalibrate, the 2019 mobile threat landscape [Online]. Available: https://www.trendmicro.com/vinfo/hk-en/security/research-and-analysis/threat-reports/roundup/review-refocus-and-recalibrate-the-2019-mobile-threat-landscape (2019). Accessed 25 Mar 2020

4. Bläsing, T., Batyuk, L., Schmidt, A.D., Camtepe, S.A., Albayrak, S.: An android application sandbox system for suspicious software detection. In: 2010 5th International Conference on Malicious and Unwanted Software, October 2020, pp. 55–62. IEEE
5. Zhou, Y., Wang, Z., Zhou, W., Jiang, X.: Hey, you, get off of my market: detecting malicious apps in official and alternative android markets. In: NDSS, February 2012, vol. 25, no. 4, pp. 50–52
6. Balanza, M., Abendan, O., Alintanahin, K., Dizon, J., Caraig, B.: Droiddreamlight lurks behind legitimate android apps. In: 2011 6th International Conference on Malicious and Unwanted Software, October 2011, pp. 73–78. IEEE (2011)
7. Talha, K.A., Alper, D.I., Aydin, C.: APK auditor: permission-based android malware detection system. Digit. Investig. **13**, 1–14 (2015)
8. Rodriguez-Mota, A., Escamilla-Ambrosio, P.J., Morales-Ortega, S., Salinas-Rosales, M., Aguirre-Anaya, E.: Towards a 2-hybrid Android malware detection test framework. In: 2016 International Conference on Electronics, Communications and Computers (CONIELECOMP), February 2016, pp. 54–61. IEEE (2016)
9. Sun, M., Li, X., Lui, J.C., Ma, R.T., Liang, Z.: Monet: a user-oriented behavior-based malware variants detection system for android. IEEE Trans. Inf. Forensics Secur. **12**(5), 1103–1112 (2016)
10. Wang, X., Yang, Y., Zeng, Y., Tang, C., Shi, J., Xu, K.: A novel hybrid mobile malware detection system integrating anomaly detection with misuse detection. In: Proceedings of the 6th International Workshop on Mobile Cloud Computing and Services, September 2015, pp. 15–22
11. Android aapt—eLinux.org [Online]. Available: http://elinux.org/Android_aapt (2016). Accessed 13 Aug 2016
12. CuckooDROiD: Dalvik monitoring framework for CuckooDroid (2020)
13. Ghahrai, A.: 10+ open source mobile test automation tools. Retrieved September, 16, 2016 (2014)
14. Gunn, S.R.: Support vector machines for classification and regression. ISIS Tech Rep **14**(1), 5–16 (1998)
15. Li, K.L., Huang, H.K., Tian, S.F., Xu, W.: Improving one-class SVM for anomaly detection. In: Proceedings of the 2003 International Conference on Machine Learning and Cybernetics (IEEE Cat. No. 03EX693), November 2003, vol. 5, pp. 3077–3081. IEEE (2003)
16. Shijo, P.V., Salim, A.J.P.C.S.: Integrated static and dynamic analysis for malware detection. Proc. Comput. Sci. **46**, 804–811 (2015)
17. Andrew, A.M.: An introduction to support vector machines and other kernel-based learning methods. Nello Christianini and John Shawe-Taylor, Cambridge University Press, Cambridge, 2000, xiii+ 189 pp., ISBN 0–521–78019–5 (Hbk,£ 27.50). Robotica **18**(6), 687–689 (2000)
18. Andrew, A.M.: An introduction to support vector machines and other kernel-based learning methods. Kybernetes (2001)
19. Dittman, D., Khoshgoftaar, T.M., Wald, R., Napolitano, A.: Random forest: a reliable tool for patient response prediction. In: 2011 IEEE International Conference on Bioinformatics and Biomedicine Workshops (BIBMW), November 2011, pp. 289–296. IEEE (2011)
20. Yuan, Z., Lu, Y., Xue, Y.: Droiddetector: android malware characterization and detection using deep learning. Tsinghua Sci Technol **21**(1), 114–123 (2016)
21. Liu, Y., Zhang, Y., Li, H., Chen, X.: A hybrid malware detecting scheme for mobile android applications. In: 2016 IEEE International Conference on Consumer Electronics (ICCE), January 2016, pp. 155–156. IEEE (2016)
22. Winsniewski, R.: Apktool: a tool for reverse engineering android apk files. https://ibotpeaches.github.io/Apktool/ (2012). Visited on 27 July 2016
23. Saracino, A., Sgandurra, D., Dini, G., Martinelli, F.: Madam: effective and efficient behavior-based android malware detection and prevention. IEEE Trans. Dependable Secure Comput. **15**(1), 83–97 (2016)

24. Hou, S., Ye, Y., Song, Y., Abdulhayoglu, M.: Hindroid: an intelligent android malware detection system based on structured heterogeneous information network. In: Proceedings of the 23rd ACM SIGKDD International Conference on Knowledge Discovery and Data Mining, pp. 1507–1515. ACM (2017)
25. Billah, K.E.M., Mourad, D., Abdelouahid, D., Djedjiga, M.: MalDozer: automatic framework for android malware detection using deep learning. Digit Investig **24**, S48–S59 (2018)
26. Li, J., Sun, L., Yan, Q., Li, Z., Srisa-An, W., Ye, H.: Significant permission identification for machine-learning-based android malware detection. IEEE Trans. Industr. Inf. **14**(7), 3216–3225 (2018)
27. Feng, P., Ma, J., Sun, C., Xu, X., Ma, Y.: A novel dynamic android malware detection system with ensemble learning. IEEE Access **6**, 30996–31011 (2018)
28. Ma, Z., Ge, H., Liu, Y., Zhao, M., Ma, J.: A combination method for android malware detection based on control flow graphs and machine learning algorithms. IEEE Access **7**, 21235–21245 (2019)
29. Mehtab, A., Shahid, W.B., Yaqoob, T., Amjad, M.F., Abbas, H., Afzal, H., Saqib, M.N.: AdDroid: rule-based machine learning framework for android malware analysis. Mobile Netw. Appl. **25**(1), 180–192 (2020)
30. Ma, Z., Ge, H., Wang, Z., Liu, Y., Liu, X.: Droidetec: Android malware detection and malicious code localization through deep learning. arXiv preprint arXiv:2002.03594 (2020)
31. Alzaylaee, M.K., Yerima, S.Y., Sezer, S.: DL-Droid: deep learning based android malware detection using real devices. Comput. Secur. **89**, 101663 (2020)
32. Su, X., Xiao, L., Li, W., Liu, X., Li, K.C., Liang, W.: DroidPortrait: android malware portrait construction based on multidimensional behavior analysis. Appl. Sci. **10**(11), 3978 (2020)
33. Mahindru, A., Sangal, A.L.: MLDroid—framework for android malware detection using machine learning techniques. Neural Comput. Appl. 1–58 (2020)
34. Zhang, N., Tan, Y.A., Yang, C., Li, Y.: Deep learning feature exploration for android malware detection. Appl. Soft Comput. **102**, 107069 (2021)
35. Thongsuwan, S., Jaiyen, S., Padcharoen, A., Agarwal, P.: ConvXGB: a new deep learning model for classification problems based on CNN and XGBoost. Nucl. Eng. Technol. **53**(2), 522–531 (2021)
36. Mcdonald, J., Herron, N., Glisson, W., Benton, R.: Machine learning-based android malware detection using manifest permissions. In: Proceedings of the 54th Hawaii International Conference on System Sciences, January 2021, p. 6976

Section–B

A Blockchain-Based Secure Car Hiring System

Sonakshi and Seema Verma

Abstract Currently, the blockchain technology is a boom in creating a secure and distributed environment for many real-life applications. Here, the focus is to introduce a car hiring system so that the car owners will get the fare without any significant loss and consumers can hire the car with variety of preferences. The proposed system is designed with blockchain technology; anyone can join the system by entering user's information and is added to the specified block, computing hash for that required block by applying proof of work and storing hash of previous block in it. Thus, the system works on the complexity of blockchain with proof of work. The proposed model is implemented in Python and aims at removal of organizational role from car hiring system using blockchain technology.

Keywords Blockchain · Hash · Nonce · Timestamp · Proof of work

1 Introduction

All the car hiring system these days run under an organization where all a user can do is hire a car and use it for some period of time. As an example, in zoom cars, there is an organizational hold on all the cars and profit from the car, you just have to register yourself on their portal, hire a car, pay and go on. As a conclusion, all the profit goes into the pocket of the organization. After going through the existing systems like decentralized e-Voting portal using blockchain [1], A blockchain-based shopping cart [2], blockchain technology in farmer's portal [3] and securing e-FIR data through blockchain [4], our proposed system provides a distributed system to manage the renting of cars. To conquer this, we are planning to design something which will remove the role of organization between the car and customer. The user for this application can either be a car owner or the renter. For the sake of making this system more secure, we will be using blockchain technology, in which details of each car (owner details and list of customers who have hired the

Sonakshi (✉) · S. Verma
Manav Rachna International Institute of Research and Studies, Faridabad, Haryana, India

K. Khanna et al. (eds.), *Cyber Security and Digital Forensics*,
Lecture Notes on Data Engineering and Communications Technologies 73,
https://doi.org/10.1007/978-981-16-3961-6_29

car) would be assigned a unique hash and stored in a block. Each block would be identified on the basis of that hash value. We are not using any pre-deployed blockchain like Ethereum or hyperledger but developing our own blockchain in Python. This gives us advantage of adding features to the block as per our own requirements [5]. The architecture, working, framework and implementation of basic blockchain using Python are discussed in this paper.

The structure of the paper follows as—blockchain technology is described with its architecture, applications and challenges in next section. The proposed system of car hiring is presented in Sect. 3 followed by the implementation results and analysis.

2 Blockchain

Basically, blockchain [6] provides resistance to change of data stored in it. As the name suggests, it is a chain of blocks. It is also known as distributed ledger technology (DLT), and it is a time-stamped record of unchangeable data. It is an infrastructure behind cryptocurrency. The most common example of blockchain is Bitcoin [5]. Dejan Vujičić, Dijana Jagodić, Siniša Ranđić in their work "blockchain technology, Bitcoin and Ethereum: A Brief Overview" [7] helped in increasing understanding about blockchain and basic terms related to it. Blockchain consists of different kind of data like in Bitcoin the data field which contains information regarding the receiver, sender and number of Bitcoin to be transferred. Nowadays, blockchain is highly used in IOT field to make smart decisions [8, 9].

2.1 Blockchain Architecture and Working

Blockchain architecture consists of nodes, i.e., a computer or user that has a complete copy of block chain ledger, a block, i.e., the data structure that will be storing the data [10]. Block is the smallest building block of blockchain system. A block mainly consists of index, data, block's hash, previous block's hash, nonce and timestamp.

Hash: Hash is encrypted string of data. If the data changes, hash also changes. Each block of a blockchain is assigned a unique hash. If someone tries to change the data of the block, its hash changes and the blockchain comes to know that something is corrupt as somewhere in between the "block's hash" will no more match "previous block's hash" of the next block; hence, any malicious activity gets highlighted.

Previous Hash: Previous hash of a block contains hash of block preceding the current block. It allows us to traverse through the blockchain. And also any malicious activity or attempt to corrupt the blockchain can be encountered with the use of previous hash (Fig. 1).

```
class block:
    def __init__(self, index, timestamp, data, previous_hash, nonce):
        self.index=index
        self.timestamp=timestamp
        self.data=data
        self.previous_hash=previous_hash
        self.nonce=0
        self.hash=self.proof of work()
```

Fig. 1 Data of a single block

```
def proof_of_work(self):
    found=False
    self.nonce=0
    while found==False:
        block_hash=str(self.hash_block())
        if block_hash.startswith('0000'):
            found=True
        self.nonce+=1
        block_hash=str(self.hash_block())
    return block_hash
```

Fig. 2 Procedure and working of proof of work

Data: Data is different for every blockchain. It may be text, audio, video, image or even smaller blockchain. Hash value of a blockchain depends upon the value of data (Fig. 2).

Proof of Work: In this general architecture of blockchain, anyone can corrupt the data of block and calculate the hash for preceding blocks within a few seconds as calculating hash is just a matter of 1–2 s, solution to this is something called proof of work [11, 12]. Working of proof of work can be explained as finding a hash which starts from a specified number of 0's, finding the hash which matches our requirement and takes computational time as the hash is computed again and again until a hash which starts with required number of zeros are found. So with proof of work, if someone tries to corrupt a block, it will take much more time as compared to a general blockchain. A Bitcoin blockchain takes about 8–10 min to add a new block [5].

3 Proposed System (Car Hiring System Using Blockchain)

When we think of hiring a car, all the organizations providing car rental facility are run by organizations that own all the cars. We are providing a connected network of car owners and renters where any car owner can lend their car to someone who is in need and earn profit from their idle car. In this paper, we are focusing on implementation of such a network of cars. The role of blockchain in this network comes into play when we talk about non changeable or corruptible form of data, transaction or user history. Whenever the user will get involved in the transaction his caroins (used as cryptocurrency in this work later), all other necessary information would be encrypted in the form of hash using proof of work and will be added to the blockchain.

3.1 Framework

The framework of the system which would be working on blockchain follows the algorithm of a simple blockchain. The only difference is that the data field consists of car owner and renter's name and car number. Other than this, everything is very much same such as the nonce, timestamp, hash and previous block hash. Whenever a new block is created, it goes through the assignment of hash to the block. The block gets its data through the user input which is depicted in the flowchart below. Each block is stored in a secure way using the proof of work mechanism on the localhost server.

In Fig. 3, initially, nonce = 0. As explained in the earlier part of paper. Proof of work increases the computing time by calculating hash again and again until a requirement is met. Hence, it makes the task of interfering with the blocks difficult and time consuming for any fraudulent activity. As depicted by the flowchart as per

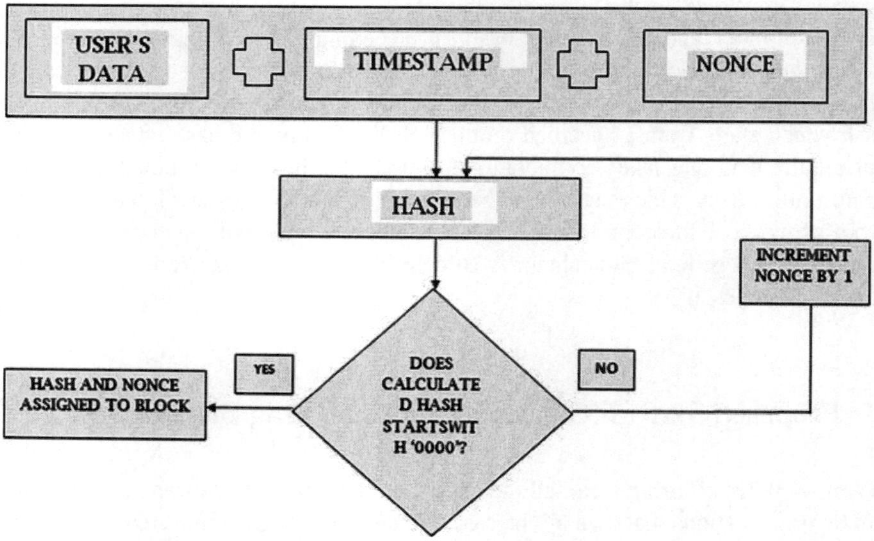

Fig. 3 Framework of hash calculation using proof of work

the difficulty of the developer, the time increases as finding a hash value which starts with some specified number of zeros and is not an easy task. In our paper, we are using a constant difficulty of 4, i.e., only those hashes would be taken which starts with 4 numbers of zeros. From the above flowchart, hash is calculated and assigned to the block. The user is asked to give his/his information and that is stored in the block.

Figure 4 depicts all the relevant information stored in the blockchain for the proposed system. Based on the type of user, the data to be stored in blockchain varies which is shown in the flowchart below. If the user is a car owner, car's detail and user's name are stored in the block. Likewise, if the user is a customer, details of car the customer are renting, customer's name, amount based on user's decision for hiring with or without fuel is stored in the blockchain.

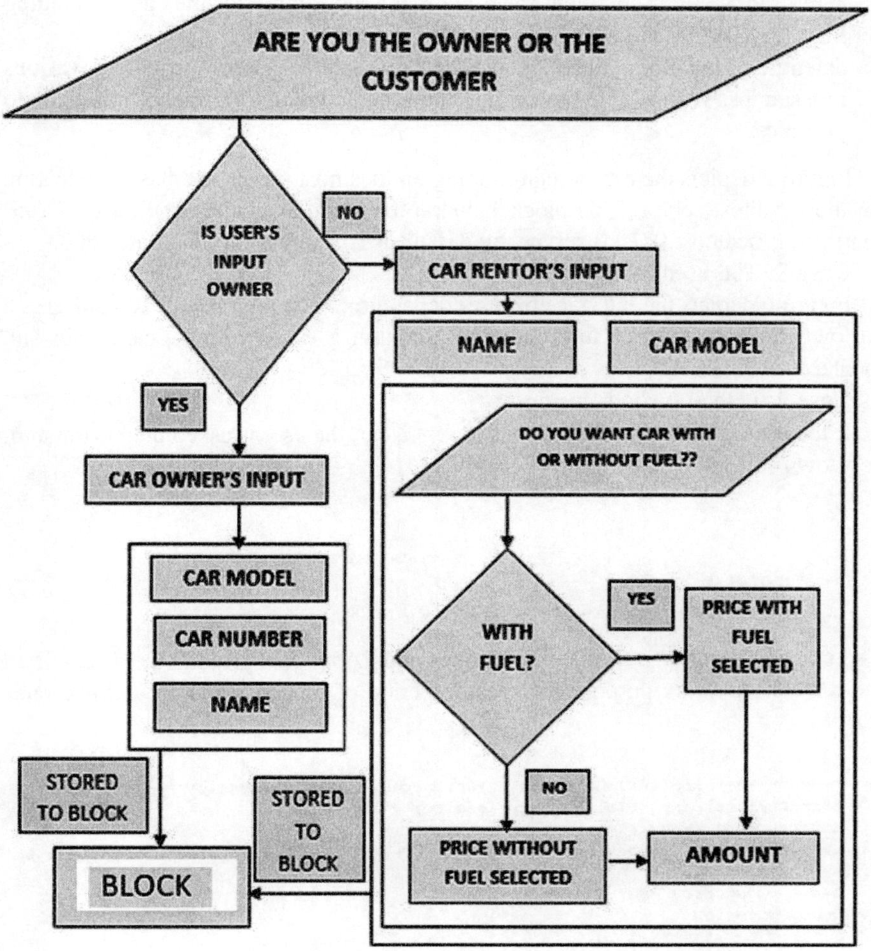

Fig. 4 Structure of block in our framework

3.2 Implementation and Result

Proposed car hiring system is implemented using Python 3.x. Firstly, the user will be identified as renter or owner based on the identification they would be asked to give his/her information which would be used in calculation of hash, and the user will get added in the blockchain running on the localhost server.

Python libraries used in its implementation are as follows:

1. Flask: This library is used for running web applications using Python. It is easy to use and have many applications.
2. sqlite3: It is used to run the database for keeping the records of various cars with its prices and availability.
3. hashlib: It is used for generating hash values for the blocks of blockchain. We here used SHA256 although there are many algorithms for the same including SHA1, SHA224 and SHA384, etc.
4. datetime: This library can be used in multiple ways for providing various timestamps. We used this to capture the time and date of a transaction made to hire a car.

Figure 5 depicts the blockchain running on localhost server as a flask application on the localhost server. This blockchain can be made available externally by running it on localhost:0.0.0.0 server, by default it is running on localhost:5000.

Case 1: The user is an owner.

Figure 6 depicts the interface for user depending upon their needs. If the user is a car owner, all the relevant information of that user is asked with the car details and availability.

Case 2: If the user is a customer.

If the user is a customer and wants to rent a car, the user is asked the information as shown in Fig. 7.

3.3 Analysis

The existing rental system like Yolo bikes and Zoom cars provide their vehicle to the renters for riding purpose for specific period of time at some pre-defined rates

```
=============== RESTART: C:\Users\aa\Desktop\blockchain\block.py ===============
 * Serving Flask app "block" (lazy loading)
 * Environment: production
   WARNING: This is a development server. Do not use it in a production deployme
nt.
   Use a production WSGI server instead.
 * Debug mode: on
 * Running on http://127.0.0.1:5000/ (Press CTRL+C to quit)
```

Fig. 5 Starting the blockchain server

```
==================== RESTART: C:\sqlite3\car_interface.py ====================
Do You Want To Earn From Your Idle Car???(y/n)y
Enter your namesonakshi
(car_ID, car_name)
(1, 'Hyundai Creta AT')
(2, 'TATA Tiago')
(3, 'Maruti Swift')
(4, 'Mahindra Scorpio')
(5, 'Hyundai Creta')
(6, 'Mahindra Marazzo')
(7, 'nano')
Choose Car ID corresponding to your Car Model if not available enter N5
('Hyundai Creta',) registered
Enter Your Car NumberDLDHE325
Enter Your Citydelhi
Enter Your Mobile Number1234567890
Enter Your Email Idsona@gmail.com
Enter The Week's day when your car is availablewednesday
Enter the duration (Eg:2:00PM-7:00PM)1:00-4:00PM
```

Fig. 6 User interface for case 1

```
==================== RESTART: C:\sqlite3\car_interface.py ====================
Do You Want To Earn From Your Idle Car???(y/n)n
enter your namehimani
If you want to rent a car these are the available cars with their price
('car_id','car_name','price without fuel','price with fuel')
(1, 'Hyundai Creta AT', 5163, 8580)
(2, 'TATA Tiago', 4623, 2782)
(3, 'Maruti Swift', 5700, 3400)
(4, 'Mahindra Scorpio', 6533, 3900)
(5, 'Hyundai Creta', 6500, 3940)
(6, 'Mahindra Marazzo', 7400, 4500)
(7, 'nano', 4500, 4500)
(8, ' WagonR', 5500, 5500)
Enter the car's ID in which you are interested
5
(5, 'Hyundai Creta', 6500, 3940)
Hyundai Creta selected
enter 1 for car with fuel and 2 for car without fuell
you have to pay 6500
enter your license numberSJDFIJE3432
enter your adhaar numberA45WEJ3
enter your phone number2545965
enter email IDhimani@gmail.com
```

Fig. 7 User interface for case 2

and fine (in case of any damage or exceeding the deadline for returning of their vehicle). In such system, all the vehicles are under one organization that has a centralized hold on the vehicles and profit made out of them. Taking an instance of OLA or Uber Cabs, if a person needs to go to airport, suppose he books a cab from his home to the airport, but after that, he has to pay for another cab from airport to hotel on the other side. Additionally, if someone's car returns with a scratch or dent, or if the owner finds that the car has violated any traffic rule and "challan" has been filed in his name the customer shall be liable to pay it and should be blocked to hire

Table 1 Comparative analysis between the existing and proposed system

Parameters	Existing system	Proposed system
Decentralized	No	Yes
Profitable	To organization	To general people
Secure	No	Yes
Employability opportunity	None	To car owners
User verification	Centralized	Distributed

any other car in future. Here, anyone who own a car can earn from their car by giving it on rent for a specific period of time, and any other person who is in need to get a car for approximately the same period of time can come and rent the car. So from the customer point of view, their problem of renting a car from home to airport and airport to hotel is gone, and from the owner point of view, the centralized hold on cars is gone.

For increasing the security, consensus and proof of work algorithm are applied to the blocks [5]. The proof of work increases the computational time of the hash and hence makes the task of changing data of blockchain a big challenge. To change the blockchain's data after the application of proof of work needs a group of miner that can control 50% of the network mining rates, this is known as 51 attack and is not easy to implement it on a well-developed blockchain. The proposed approach starts with gathering user's information and adding it to the specified block, computing hash for that required block by applying proof of work and storing hash of previous block in it (Table 1).

4 Conclusion

Current car hiring system is completely dependent on a single organization, centralized system. In case of zoom car or instance of yolo cycles all work on same framework and business model. Here, a completely different model is proposed for car hiring system. The proposed system with blockchain technology has provided security and hence has acted as a helping hand in gaining customer trust. The proposed model aims to remove organizational hold on car hiring system, and this system would not only benefit car renters but also will provide employment opportunity to many. Its user interface could be further enhanced by associating a web application or website with it.

References

1. Patidar, K., Jain, S.: Decentralized E-voting portal using blockchain. In:10th ICCCNT 2019, 6–8 July 2019, IIT-Kanpur, Kanpur, India (2019)
2. Shrestha, A.K., Joshi, S., Vassileva, J.: Customer data sharing platform: a blockchainbased shopping cart. In: IEEE International Conference on Blockchain and Cryptocurrency (ICBC) (2020)
3. Talreja, R., Chouksey, R., Verma, S.: A study of blockchain technology in farmer's portal. In: Second International Conference on Inventive Research in Computing Applications (ICIRCA-2020). IEEE Xplore (2020)
4. Khan, N.D., Chrysostomou, C., Nazir, B.: Smart FIR: securing e-FIR data through blockchain within smart cities. In: IEEE 91st Vehicular Technology Conference (VTC2020-Spring) (2020)
5. Jain, R., Dogra, A.: Solar energy distribution using blockchain and IoT integration. In: IECC'19, Okinawa, Japan (2019)
6. Memon, M., Hussain, S.S., Bajwa, U.A., Ikhlas, A.: Blockchain beyond bitcoin: blockchain technology challenges and real-world applications. In: 2018 International Conference on Computing, Electronics & Communications Engineering (iCCECE) (2018)
7. Vujičić, D., Jagodić, D., Ranđić, S.: Blockchain technology, bitcoin, and ethereum: a brief overview. Ministry of Education, Science, and Technological Development of the Republic of Serbia (2018)
8. Zorzo, A.F., Nunes, H.C., Lunardi, R.C., Michelin, R.A., Kanhere, S.S.: Dependable iot using blockchain-based technology. In: Eighth Latin-American Symposium on Dependable Computing (LADC) (2018)
9. Choi, S.S., Burm, J.W., Sung, W., Jang, J.W., Heo, Y.J.: A blockchain-based secure iot control scheme. In: 2018 International Conference on Advances in Computing and Communication Engineering (ICACCE-2018) Paris, France (2018)
10. Syed, T.A., Alzahrani, A., Jan, S., Siddiqui, M.S., Nadeem, A., Alghamdi, T.: A comparative analysis of blockchain architecture and its applications: problems and recommendations. Deanship of Research, Islamic University of Madinah, KSA (2019)
11. Gemeliarana, I.G.A.K., Sari, R.F.: Evaluation of proof of work (pow) blockchains security network on selfish mining. In: International Seminar on Research of Information Technology and Intelligent systems (ISRITI) (2018)
12. Gervais, A., Karame, G.O., Wüst, K., Glykantzis, V., Ritzdorf, H., Capkun, S.: On the security and performance of proof of work blockchains (2016)

A Correlation Blockchain Matrix Factorization to Enhance the Disease Prediction Accuracy and Security in IoT Medical Data

P. Renuka and B. Booba

Abstract An IoT software product's reliability is the probability of the product working "correctly" under or over a given time. New opportunities are the result of expansion in the fast-paced Internet of Things (IoT) space. IoT technologies on the collected datasets improve disease progression technology, disease prediction, patient self-management and clinical intervention. To propose, the IoT with cipher block chaining in the traditional cryptographic operation mode will be used for cryptographic processing. Developing models for the supervised learning classification and security of imbalanced datasets is challenging, especially in the medical field. However, most real-time IoT datasets present most traditional machine learning algorithms challenging unbalanced datasets. Proposed a new framework for the Correlation Blockchain Matrix Factorization Classifier (CBMFC) related to comprehensive medical records. CBMFC uses a multiple class label machine learning that represents an independent population model based on disease meta functions such as profile age, group, or cognitive function keys. The Pairwise Coupling Multi-Class Classifier (PCMC) is used to prove the model's correctness. This produces more comprehensive data in various machine learning environments, such as predictive classification, similar to real data performance. For the results of security analysis confirmation, the proposed IoT application model's effectiveness can withstand various attacks, such as selected cryptographic attacks. In this proposed CBMFC system, classification accuracy, precision, recall, execution time and security matrix are used to evaluate performance.

Keywords Internet of Things · Correlation Blockchain Matrix · Pairwise Coupling Multi-Class Classifier · Cryptography

---

P. Renuka (✉) · B. Booba
VISTAS, Vels University, Chennai, India

K. Khanna et al. (eds.), *Cyber Security and Digital Forensics*,
Lecture Notes on Data Engineering and Communications Technologies 73,
https://doi.org/10.1007/978-981-16-3961-6_30

1 Introduction

The use of the Internet of Things (IoT) in healthcare is sharply increased among a variety of specific Internet-use cases. The Medical Internet of Things takes more effort to improve care itself, with remote monitoring as the broader primary application of remote medicine. The synergies between medicine and technology have taken great strides around the world. For example, the Internet of Things (IoT) data analytics is gaining popularity, providing the next generation of electronic healthcare and mobile healthcare services. They are transforming traditional institutions based on the management of large data with a blockchain solution. The chain's evolution is in favor and permitted by technical means to maintain the support of strategic applications needed for its potential growth. In the cloud-enabled network blockchain, there are trading and mining nodes both in the cloud and on-premises. According to embodiments, the node may be an enterprise-level server. The final level of overall collaboration on evolution is based on cloud and blockchain applications as the basis for distributed chains' operation. The devices configured to use public blockchain services and private blockchain nodes clouds to communicate securely via APIs. The IoT devices combining blockchain technology as a security framework IoT system using secure distributed key management techniques make it possible to discover each other and transaction encryption machine-to-machine.

In this, Fig. 1 represents the cloud blockchain network communication. Machine learning techniques have been widely used in the medical field. Most of the medical data resources can use to transfer valuable knowledge to assistive scientific decision making. It is stored in each hospital or other medical institution separately, which poses a major issue to the medical data applied to the constructed prediction model, its quality and efficiency. For medical professionals, researchers, there are several machine learning techniques to be used throughout disease research. Medical data keeps a history of patient records and will be unused in the future. This can be analyzed and considered for future research. This huge database is analyzed to identify machine learning techniques used by healthcare staff to predict patient risk.

The most commonly used machine learning technology, which may be registered in the category. Occupies a set of pre-classified patterns to create a demographic model classification. Learning and classification are involved in a process called data classification. The training data is being learned and analyzed by a classification algorithm. The test data is used for classification to approximate the classification rules. The rules can apply to the new data tuple with acceptable accuracy. The pre-classification example has been used in the classifier training algorithm to verify the group that requires the appropriate identification parameters. This section discusses the process of classification of disease analysis using IoT clinical data and machine learning methods.

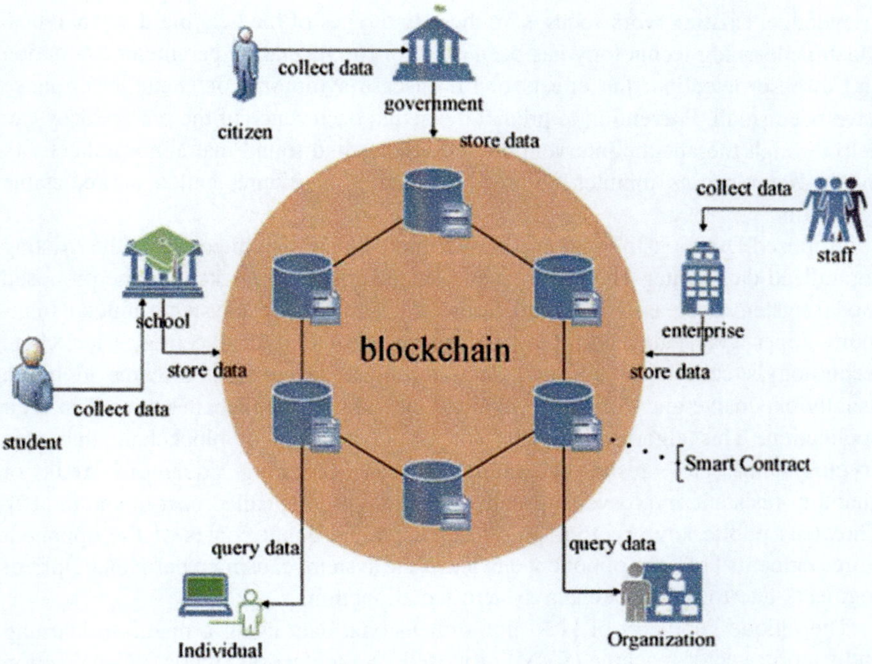

Fig. 1 Cloud blockchain network model

2 Related Work

The literature survey for various IoT blockchain security methods and disease classification strategies used to predict the result is discussed below.

Co-clustering is used simultaneously to group clinical features and patients to characterize block-wise data loss patterns [1] (A) group-based feature selection and data imputation for specific patient subgroups. (B) Miss the predictive model to consider data availability. The machine learning (ML) method of decision making and the medical field's data has proven highly predictive and supportive. ML technology's latest developments in the Internet of Things (IoT) [2] are being used. The task of breath estimation task [3] compares the performance of different machine learning techniques. Problems can be divided into two categories: high and low breathing work based on information extracted from pressure, volume and flow through signals recorded by mechanical ventilation.

Select the feature to solve the problem, and the proposed method is novel, rapid provision of mutual information to select the feature [4]. The feature selection algorithm improves the classification accuracy and is used to select the function to reduce the classification system's execution time. The neural network modification uses structured data and unstructured data and recommends a conventional neural network (CNN) basic peak disease risk prediction algorithm [5]. To the best of our knowledge, existing work focuses on these two types of medical big data analytics. Machine learning technology has been used for [6] vegetation parameter estimation and disease detection; the effects of the disease symptoms on their performance have been small. Prevention to predict the actual occurrence of the pre epilepsy can help through therapeutic intervention [7]. Studies had found that abnormal activity in the brain begins minutes before the onset of a seizure, called a predictable condition.

Proposed DualFog-IoT is compared to the IoT-based architecture of the existing centralized data center. Having a genetic characteristic of blockchain, the proposed model system decreases rates and furthers existing IoT ecosystems unload minimum upper grayscale and [8, 9] to reduce the cloud data center. Blockchain technology's complexity is maintained for most developers or teams to build, usually expensive and difficult to monitor the support blockchain network in their application. This algorithm thereby forces the redesign of blockchain that uses cryptosystems to resist quantum attacks, creating quantum-fixed or quantum-resistant cryptosystems, called quantum proofs, called post-quantum [10]. Threatens public key cryptographic hash functions. If all copies of the opponent store segments [11], the opponent can leave the system, causing a permanent loss of segments due to the blockchain system's malfunction.

The unique geometry of [12] high-dimensional data using a manifold learning and support vector machine (SVM) proposed PVC detection and data visualization method. [13, 14] has proposed some of the blockchain-based storage systems in recent years. In most cases, the blockchain acts as a "witness of the agreement" with publishers. This method is, so far, because it is not managed to reduce the size of the blockchain itself, and it will not be able to avoid the storage model of Bitcoin. Maintaining this also a complete record of blockchain implementation in the IoT environment helps to insufficient storage capacity on edge devices. At the same time, the system does not require any significant transactions per second [15]. Even segment blockchain is used to improve blockchain sharing by separating transaction storage from transaction validation.

Our main contribution is to link the probability of failure as a group to each epoch by using the probability limit as the sum of the upper limit hypergeometric [16] and the two types of distribution. It is a reliable data management scheme based on blockchain in edge computing (BlockTDM called) [17] that has been proposed to solve the above problems. The flexible configuration of blockchain architecture, mutual authentication protocol, flexible consensus, smart contract management module and transaction data and blockchain node management and deployment are among them. Algorithms included artificial neural networks (ANN) [18], support vector machine (SVM), naïve Bayesian classifier (NBC), boosted decision trees (BDT) and multivariable logistic regression (MLR).

However, one of the important problems of fog computing and blockchain [19] integration is scalability. The group chain has proposed a new type of scalable public blockchain for the double-stranded structure of IoT services computing fog of computing. Despite its potential, there are some urgent problems to be solved to make the IoT services are widely used. Various loosely coupled distributed intelligence [20] to adjust the connection operation requires an IoT device for managing the system. Developers' point of view analyzed blockchain from [21] blockchain, a larger software system, highlighting data storage and combination key concepts and considerations.

Two methods are currently introduced based on the heterogeneous network, protein interactions, genotype—is built using the phenotype correlations and phenotypic similarity. In HeteSim_MultiPath (HSMP) [22], HeteSim scores different routes contributing to the longer path and the damping constant. Therefore, a non-invasive diagnostic system based on machine learning (ML) has been developed to solve these problems. The decision-making system of experts based on the application of machine learning classifiers and artificial fuzzy logic is an effective diagnostic result; the mortality rate has decreased [23]. Therefore, no clear requirements should be treated as a technique to identify the most appropriate parameters during predictive analysis.

Instead, the feature extraction time-series EEG is converted to a time-frequency distribution (TFD) [24]. Gradient boosters use a set of programs aside to train the entire TFD directly. This paper proposes a machine learning method for predicting particulate matter concentrations from wind and precipitation levels, based on [25] two-year weather and air pollution data.

This paper [26] has been used to record valid composite minorities oversampling to generate new composites. Baldwinian learning and PSO (BLPSO) [27] are based on a novel hybrid algorithm to increase particle diversity and prevent premature convergence of PSO. Ability to compare calculated heart rate variability (HRV) baselines before the start of daytime antibiotics (LOS group) or during a randomly selected period (control group) during the calibration period [28].

In the future, Moderate Resolution Imaging Spectrum radix statistical model for predicting fire activity for 1–5 days to use satellite fire counts and meteorological data from temporary reanalysis [29] to develop. The component contains an apnea event [30], which automatically proposes the first use of EIT boundary voltage data from infants to obtain the main function of research apnea detection using machine learning. Factorial switching in linear dynamic systems is a common framework for addressing this issue [31].

This analysis of the previous method has a low classification accuracy and less security performance to introduce a new method in the next section.

3 Implementation of the Proposed Method

This IoT application model's main motivation is to provide a computationally secure key generation for protected data through encryption with blockchain technology. A master key is a secret key that is accepted between communicating parties before a communication protocol begins. The essential characteristic of machine learning is the design of heterogeneous data applications with different dimensionalities. Various IoT application consists of different information collected by recording the data for producing diverse data representations. The collection of data is carried out by distributed and decentralized control with autonomous data sources.

Correlation Blockchain Matrix Factorization Classifier (CBMFC) algorithm is used to classify the given data into several columns and provide security in medical data information. Blockchain HMAC encryption is the enforcement of access control mechanisms, digital signatures, routing controls, notarization, etc., to provide data security services against attacks and prevention. The digest is calculated for ciphertext messages using the HMAC encryption algorithm. HMAC stands for the Hash-based Message Authentication Code. This authentication uses a key to implement the hash function product along with the content of the message. From Fig. 2, first, preprocess the data using Co-relational Matrix Data Preprocessing to remove the noise. The model is mainly executed in the preprocessing task. It presents a preprocessing task to remove noise and inconsistent data obtained from various sources. Interval-based measures are the discretization factors from large data applications, and the measured values reach the closest values. Data preprocessing is usually done to see if there is redundant information in the dataset. All attributes are initially considered separate subsets, with the final combination of attributes and features is marked as the highest subset of features. The hash value is stored on each block; if any new record enters the storage, it will update into the previous block. This proposed framework reduces the clustering task; a grouping of the same attributes occurs by using our framework for predicting the disease using a given disease dataset.

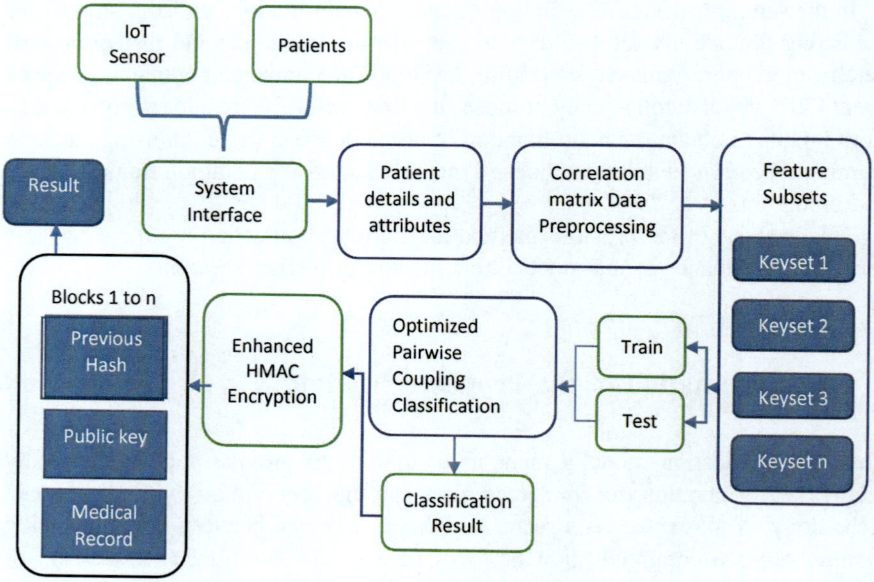

Fig. 2 Proposed Method CBMFC block diagram

3.1 Correlation Matrix Data Preprocessing

The initial stage in Correlation Matrix Data Preprocessing is finding the unrelated data and normalized datasets. While processing the data, it is essential to calculate the univariate stats like the mean value, standard error, and rate of recurrences to inspect the volume of missing data. In complete information, maximum likelihood the population criteria have approximated that would probably generate the estimate values from the sample set which is analyzed. The classification accuracy depends upon the accuracy of data, so data should be non-ambiguous, correct and complete. Data collection methods are loosely controlled, resulting in inappropriate values like out-of-range missing values. Correlation matrices are useful for showing the correlation coefficient (or degree of relevance) between variables. The correlation matrix is symmetric, just as the correlation between *V*1 and *V*2 is the same as the correlation between *V*2 and *V*1.

Let the data matrix *V* composed to *n*-dimensional observation dataset, and individual variable values (*V*) size of $[R \times n]$. To assume the row of *R* has centered, observe all the row values *i* to *n*. The correlational $\sum$ of each row *V* data.

$$\sum = cor(V) = \mathrm{E}\left[V^T\right] \tag{1}$$

To estimate the correlation data matrix.

$$\mathrm{E}\left[V^T\right] \approx \frac{V^T}{n} \tag{2}$$

Form the Eq. (2), matrices as the product of two simpler matrices E and L, using a procedure known as Eigenvalue Decomposition.

$$\sum = EL^{-1} \tag{3}$$

The data matrix *E* is resized $[R \times C]$ matrix, where each column to apply the eigenvector.

From Eqs. (2, 3), derivation of the correlation matrix is as follows,

$$\rho(R, C) = \mathrm{corr}(R, C) = \frac{\mathrm{cov}((R, C)V)}{\sigma_R \sigma_C} = \frac{E[\mathrm{cov}(((R - \mu_R)(C - \mu_C))V)]^{-1}}{\sigma_R \sigma_C} \tag{4}$$

All such data discrepancies can lead to wrong research results; thus, data is processed before applying an algorithmic technique for better and improved results. Data needs to be preprocessed to ease the entire data process to improve machine techniques' efficiency. The main steps used for preprocessing the data include data cleaning, data integration, data conversion and data reduction. In this, fill attribute values with the correlation value for unknown instances and convert all disease databases in a single file format.

3.2 Medical Feature Selection

A feature is that a subset selection which is a preprocessing step used in machine learning. It aims to increase learning accuracy to reduce and eliminate invaluable and irrelevant data dimensions. It indicates specific problems and their functions and the type of prediction that will be useful.

The input dataset is fed into the feature selection method block, where the feature selection is made according to the given dataset. It will take this way to reduce the number of attributes selected for a given number of dependent attributes.

The medical algorithm feature selected by randomly sampling instances from the training data and the selection process is shown in Fig. 3. It has been found that the nearest value class (adjacent value) of the same class is opposite each time to select the best models. The characteristic weight is based on its value to distinguish the example for detecting the model or has been updated from the latest hit and latest features.

$$\text{Weight } (W_m) = \sum_{i,j=0}^{w} f_m - \frac{(U,\ V,\ S\ )^{2}(i)}{N} \tag{5}$$

$$\text{Gain } (G_m) = \sum_{i=o}^{n} S_i log_2 \tag{6}$$

where f_m, the weight for attributes U, V, which are randomly sampled instances, S is the latest hit and N is the number of randomly sampled instances. The function diff calculates the difference between two instances of a given attribute.

f_m is an attribute of U, V, if they are an approximate model example, S is the recent success and N is the approximate number of sample events. The functional differences are calculated as the difference between the two events of a given

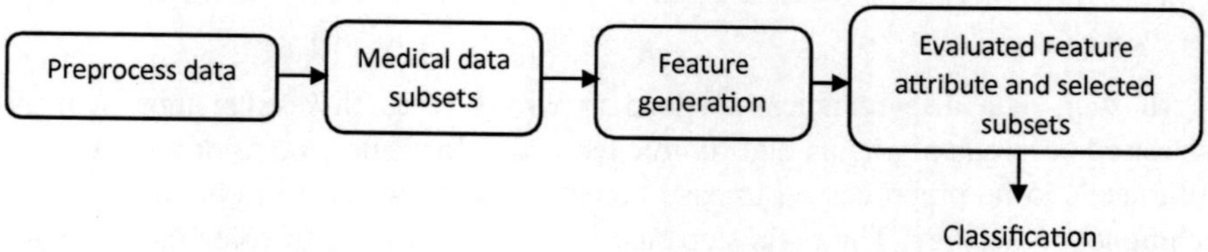

Fig. 3 Best feature selection process

attribute. The difference between the continuous attribute is the actual difference normalized to the interval [0, 1].

Algorithm steps

Input: training data D and feature subsets f_{n-1}

Input: training data D and feature subsets f_{n-1}

Step 1: Initialize the Attribute $S = S_1$

Step 2: Calculate the feature weight and gain using above eqs. (5) and (6)

Step 3: $\alpha_f = \max(S_i, G, W)$ to evaluate the best feature (α_f)

Step 4: For each current iteration $\leq N$

If $\alpha_f > th$, then // where th is represented a threshold value.

$$f'(D) = S \cup \left(\frac{G(n)+W(n)}{\alpha_f}\right) \quad (7)$$

Feature ⊛ f';

End

If available in the presence and corresponding class distribution of the feature, feature selection measures the amount of information about the class prediction bit.

3.3 *Optimized Pairwise Coupling Classification (OPCC)*

It also serves as a classification mark supervised learning method and a statistical method of classification. It considers a basic natural model and assigns it to us by determining the probability of imprisonment with uncertainty in the moral model. This feature is characterized by information gain, and then, the best ranking features are selected as the best attributes to use in the classification.

Algorithm steps:

Step 1: Read data Ds= {{P_i, Q_i}/i=1, 2, 3…n}} be the set of training data.

Step 2: Initialize the random weight support values, $W(0)$.

Step 4: For each training data (P_i, Q_i) ∈, Ds.

To analysis the predicted output Y_i^ (k)

For each weight, we do.

Update the weight (k+1) = W_j (k) + (y_i − y_i^(k)) x_{ij}.

End for.

End for.

Step 5: Until disease predicts the output.

The attributes are determined for each attribute's information to gain to classify a set of data segments. Then the information gain must be selected maximized attributes. After the classification result was stored in the database using HMAC encryption with blockchain.

3.4 HMAC Encryption and Blockchain Security Model

The Hash-based Message Authentication Code (HMAC) algorithm is implemented using binary operations and hash functions. HMAC is calculated with any cryptographic hash function; the resulting MAC algorithm is called HMAC-MD5. The security strength provided by the HMAC algorithm depends on the HMAC key, the most basic hash algorithm and the security features of the MAC Tag length. MAC is calculated using the data on the HMAC function, and the following operation is performed:

$$\text{HMAC}\ (k_1, k_2,\ \text{data}) = \ \text{hash}\ ((k_1,\ \text{inner})\ ||\ (\ k_2,\ \text{outer}))\ \text{t} \tag{8}$$

where t = time, k_1 *and* k_2 = a pair of keys (encryption and authentication).

Algorithm steps

Input: document, MK- master key, HMAC encryption key (k1, k2).

Step 1: To initialize the MK, $k1$, $k2$.

Step 2: compute the HMAC tag for authentication to create a block.

Step 3: To encrypt the document using the XOR function

Step 4: read the data hash function to generate the MK, $k1$, $k2$

XOR⊗generate 64{MK, $k1$, $k2$}

Step 5: If the tag T equal, then

to connect the blockchain network and authenticate the user.

Else

return

Step 6: Store the document SD

Step 7: mapping the authentication key to the array and remove the special characters, spaces from SD.

Step 8: for n to string

$K_{SD} = H_{k1'}(A_i)$// where, $H_{k1(x)}$is hash key 1 and A_i authentication key

Convert to K_{SD} to integer list and remaining truncate bits

End

Step 9: Decryption process and verification key

for n to the number of character

If (SD is an integer), then.

$P(C_i - K_{SD(i)})$ mod 10 // where p is integer array and C_i, character array.

Else

$P(C_i - K_{SD(i)}) \bmod 26$

Return

End

End

Step 10: Mapping and validate the p array data to x.

It is considered an easy way to compute the MAC value $h(x)$ for a given k and an arbitrary input x (Fig. 4).

$H_{k1(x)}$ is mapped to a message x of arbitrary length value having n number of fixed bits.

It is considered impossible to calculate the MAC value $hk(m)$ of a new message m if the key k is unknown, though we get MAC values of other messages.

Protecting HMAC is used to demonstrate the correct data relationship between the designer's embedded strength and the HMAC hash function. In terms of an embedded hash function, the HMAC function's strength for basic hash function encryption within security depends in a certain way. HMAC function is usually a safe and predetermined number of successes based on the probability of fraud. Created based on the time spent using the same key to create a MAC message.

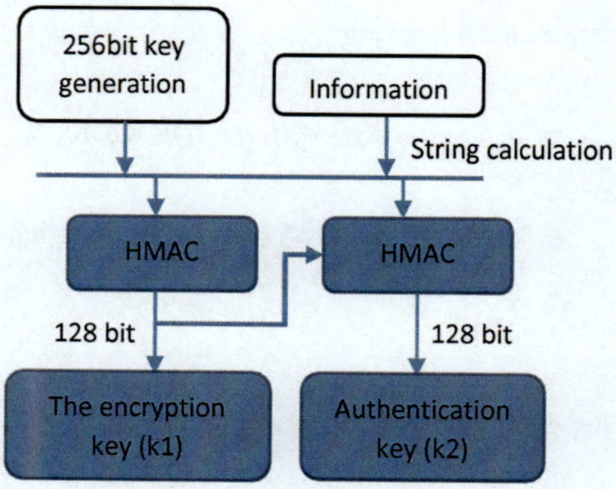

Fig. 4 Process of Key Generation

3.5 *Blockchain Network Construction Algorithm:*

Step 1: To initialize the index of the variable (IN), Timestamp (Ts), Node Information List (NI), Previous Hash (PH), Data Identity (DI).

Step 2: Create a new block.

IN = Chain.Count // in this chain count is the total number of the node to enter the network

TS = DateTime.UTCNow

NI = {user request, request ID, ssLocation}.to list

PH = Get hash (Chain.Last)

DI = current data-id

Chain.add(IN, TS, NI, PH, DI)

Step 3: To create a user transaction data-id (DI)

DI =0

While (Is not valid (last DI, Current DI, PH))

DI++

Return DI

Step 4: To register the users in blocks.

For each user in users

URL_{node} =$"http// {user id}"

Register user (URL_{user})

Insert (user.count()) //new user is added.

End

Step 5: Check the user information and database block

If (block is empty == 0)

Go to step 2

Else

Verify the block attacker or not

Record = data size

IN = block.IN, TS = block.TS, NI = block.NI, PH = block.PH

For the user in u //. u=users

PH (k) = HMAC [k] // k is keys

To evaluate the user (Li) using.

$$U(i) = IN_{n(i)}.\text{getvalue} + TS_{n(i)}.\text{getvalue} + NI_{n(i)}.\text{getvalue} + PH_{n(i)}.\text{getvalue}$$

End for

End if

From the above algorithm step to form the network group, the user can upload their document to a centralized server with encrypted format help of HMAC encryption. The HMAC authenticates the user using the master key for help to encrypt and decrypt the documents. This blockchain network incorporates patient medical data to diagnose and predict disease, and the resultant data is stored securely.

4 Result and Discussion

Statistics provide a strong basic background for quantifying and assessing results. However, it needs to be modified and tweaked for statistics-based algorithms before being applied to the IoT blockchain method. This section presents the results of a work that proposed a technique for predicting disease using machine learning.

Table 1 Simulation parameters

Parameters	Values
Language	C#
Dataset name	UCI Machine learning repository
Diseases	Cardiology
Total number of data	1500
Train data	1000
Test data	500

Section analysis to take cardiology disease data and some parameters (Temperature, heart rate and blood pressure) are considered to predict the disease level. Table 1 represents the simulation parameter of the proposed method to use.

The number of correctly classified files with patient data according to the total number of files is defined as classification accuracy. This proposed method evaluates the following Eqs. (9, 10, 11, 12) for classification accuracy, precision, recall, F1 score, security and authentication time analysis. The Correlation Blockchain Matrix Factorization Classifier (CBMFC) method prediction accuracy is compared to the random forest, Bayesian classifier and SVM methods. Similarly, the proposed method's security analysis, Enhanced HMAC Blockchain (EHMACB) security, compares to existing method Blowfish, MD5 and MidChain methods (Fig. 5).

Different bytes of data are taken to estimate the method proposed in this execution time analysis. In this proposed method, EHMAC is compared to existing methods Blowfish, MD5 and MidChain. The proposed method is to authenticate users and store the data to the network within 270 ms less execution time than the HMAC blockchain method.

Figure 6 represents the comparison of the proposed and existing method graph. In this analysis of security result, the proposed method EHMACB provides a 93.5% security compare to existing methods MidChain has 91.3%, MD5 has 85.6% and blowfish has 83.4% security in the medical blockchain network. The machine

Fig. 5 Execution Time Analysis

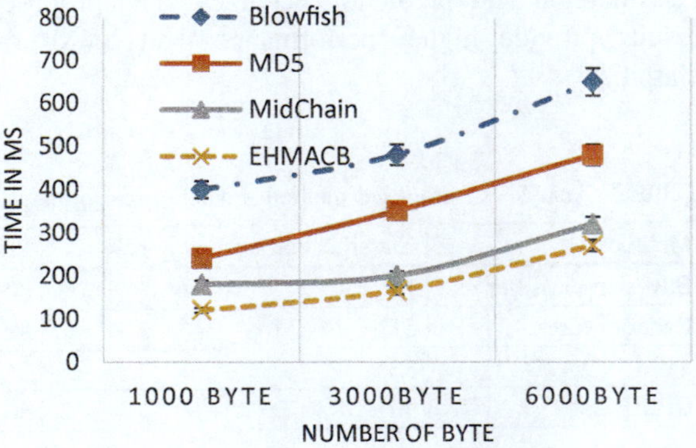

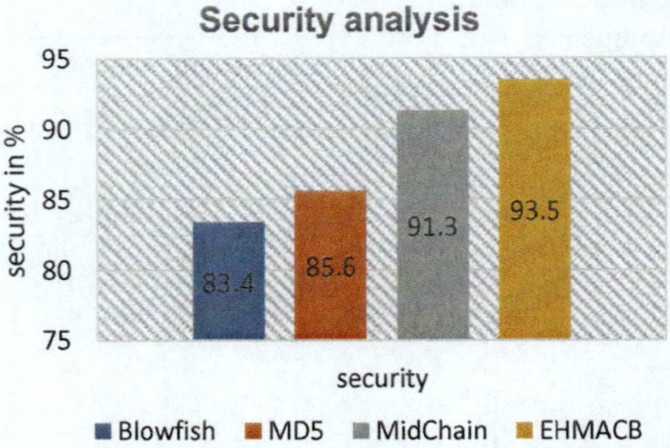

Fig. 6 Comparison of Security Analysis

learning performance of the clinical dataset classification after the analysis is evaluated using the equation below.

$$CA = \frac{\text{Number of classified files}}{\text{number of files}} * 100 \quad (9)$$

$$\text{Precision} = \frac{\text{true positive}}{\text{true positive} + \text{false positive}} * 100 \quad (10)$$

$$\text{Recall} = \frac{\text{true positive}}{\text{true positive} + \text{false negative}} * 100 \quad (11)$$

$$F1 = \frac{\text{Precision} * \text{Recall}}{\text{Precision} + \text{Recall}} \quad (12)$$

Table 2 shows a comparison of the existing and proposed method disease prediction performance. Table 2 shows the classification accuracy, precision, recall, proposed method CBMFC, existing methods SVM, random forest and Bayesian classifier (Fig. 7).

Using a pairwise coupling method to evaluate a two-pair feature matrix improves classification and prediction accuracy. The proposed machine learning method results provide higher performance than SVM, random forest and Bayesian classifier.

Table 2 Analysis of proposed method prediction performance

Methods	Classification accuracy in %	Precision in %	Recall in %
Bayesian classifier	79	80	78
Random forest	81	82	85
SVM	82	84	81
CBMFC	90	86	87

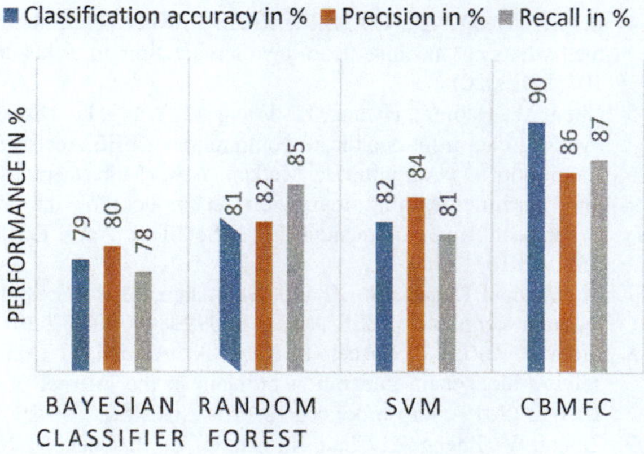

Fig. 7 Proposed method performance of prediction analysis

5 Conclusion

A given patient's cardiology disease needs to be diagnosed accurately and in time. The proposed Correlation Blockchain Matrix Factorization Classifier (CBMFC) analysis the IoT data disease prediction. First, to apply the co-relational matrix for preprocessing to remove the noise from the IoT dataset. Finally, a Correlation Blockchain Matrix Factorization Classifier (CBMFC) method uses the train data and predicts it. The Blockchain Enhanced Hash-based Message Authentication Code (EHMAC) Encryption is used to provide security, it encrypts the user request, and records are stored on blocks. This analysis of the proposed method simulation result has a 90% of classification accuracy, 86% of precision, 87% of recall values and 93.5% of security with 270 ms execution time more efficiently compared to the existing method. This proposed method to implement into hospitals for analysis disease and secure the data form unknown person.

References

1. Wang, H., Huang, Z., Zhang, D., Arief, J., Lyu, T., Tian, J.: Integrating co-clustering and interpretable machine learning for the prediction of intravenous immunoglobulin resistance in kawasaki disease. IEEE Access **8**, 97064–97071 (2020)
2. Mohan, S., Thirumalai, C., Srivastava, G.: Effective heart disease prediction using hybrid machine learning techniques. IEEE Access, 1–1 (2019)
3. Castro, L.F.B., Santacruz, L.F.E., Sánchez, M.B.S.: Work of breathing estimation during spontaneous breathing test using machine learning techniques. In: 2020 IEEE Colombian Conference on Applications of Computational Intelligence (IEEE ColCACI 2020), Cali, Colombia, pp. 1–6 (2020). https://doi.org/10.1109/ColCACI50549.2020.9247855

4. Li, J.P., Haq, A.U., Din, S.U., Khan, J., Khan, A., Saboor, A.: Heart disease identification method using machine learning classification in E-healthcare. IEEE Access **8**, 107562–107582 (2020)
5. Chen, M., Hao, Y., Hwang, K., Wang, L., Wang, L.: Disease prediction by machine learning over big data from healthcare communities. IEEE Access **5**, 8869–8879 (2017)
6. Ashourloo, D., Aghighi, H., Matkan, A.A., Mobasheri, M.R., Rad, A.M.: An investigation into machine learning regression techniques for the leaf rust disease detection using hyperspectral measurement. IEEE J. Sel. Top. Appl. Earth Obs. Remote Sens. **9**(9), 4344–4351 (2016)
7. Muhammad Usman, S., Khalid, S., Aslam, M.H.: Epileptic seizures prediction using deep learning techniques. IEEE Access **8**, 39998–40007 (2020)
8. Memon, R., Li, J., Nazeer, I., Khan, A., Ahmed, J.: DualFog-IoT: additional fog layer for solving blockchain integration problem in the internet of things. IEEE Access **7**, 169073–169093 (2019). https://doi.org/10.1109/ACCESS.2019.2952472
9. Zhang, W., Zheng, Z., Chen, X., Dai, K., Li, P., Chen, R.: NutBaaS: a blockchain-as-a-service platform. Comput. Sci. IEEE Access **7**, 134422–134433 (2019)
10. Fernandez-Carames, T.M., Fraga-Lamas, P.: Towards post-quantum blockchain: a review on blockchain cryptography resistant to quantum computing attacks. IEEE Access, 1–1 (2020)
11. Xu, Y., Huang, Y.: Segment blockchain: a size reduced storage mechanism for blockchain. IEEE Access, 1–1 (2020)
12. Ribeiro, B.R., Henirques, J.H., Marques, A.M., Antunes, M.A.:Manifold learning for premature ventricular contraction detection. In: 2008 Computers in Cardiology, Bologna, Italy, pp. 917–920 (2008). https://doi.org/10.1109/CIC.2008.4749192
13. Xu, Y.: Section-blockchain: a storage reduced blockchain protocol, the foundation of autotrophic decentralized storage architecture. In: 2018 23rd International Conference on Engineering of Complex Computer Systems (ICECCS), pp. 115–125. IEEE (2018)
14. Xu, Q., Aung, K.M.M., Zhu, Y., Yong, K.L.: A blockchain-based storage system for data analytics in the internet of things. In: New Advances in the Internet of Things, pp. 119–138. Springer (2018)
15. Ren, Y.J., Leng, Y., Cheng, Y.P., Wang, J.: Secure data storage based on blockchain and coding in edge computing. Math. Biosci. Eng. **16**, 1874–1892 (2019)
16. Hafid, A., Hafid, A.S., Samih, M.: New mathematical model to analyze security of sharding-based blockchain protocols. IEEE Access **7**, 185447–185457 (2019)
17. Zhaofeng, M., Xiaochang, W., Jain, D.K., Khan, H., Hongmin, G., Zhen, W.: A blockchain-based trusted data management scheme in edge computing. IEEE Trans. Ind. Inf. 1–1 (2019)
18. Mueller, M., Wagner, C.C., Stanislaus, R., Almeida, J.S.: Machine learning to predict extubation outcome in premature infants. In: The 2013 International Joint Conference on Neural Networks (IJCNN), Dallas, TX, USA, pp. 1–6 (2013). https://doi.org/10.1109/IJCNN.2013.6707058
19. Lei, K., Du, M., Huang, J., Jin, T.: Group chain: towards a scalable public blockchain in fog computing of IoT services computing. IEEE Trans. Serv. Comput. **13**(2), 252–262 (2020)
20. Li, Z., Kang, J., Yu, R., Ye, D., Deng, Q., Zhang, Y.: Consortium blockchain for secure energy trading in the industrial Internet of Things. IEEE Trans. Ind. Informat. **14**(8), 3690–3700 (2018)
21. Paik, H.-Y., Xu, X., Bandara, H.M.N.D., Lee, S.U., Lo, S.K.: Analysis of data management in blockchain-based systems: from architecture to governance. IEEE Access **7**, 186091–186107 (2019)
22. Zeng, X., Liao, Y., Liu, Y., Zou, Q.: Prediction and validation of disease genes using HeteSim scores. IEEE/ACM Trans. Comput. Biol. Bioinf. **14**(3), 687–695 (2017)
23. Ansarullah, S.I., Kumar, P.: A systematic literature review on cardiovascular disorder identification using knowledge mining and machine learning method. Int. J. Recent Technol. Eng. **7**(6S), 1009–1015 (2019)

24. Murphy, B.M., Goulding, R.M., O'Toole, J.M.:Detection of transient bursts in the EEG of preterm infants using time–frequency distributions and machine learning. In: 2020 42nd Annual International Conference of the IEEE Engineering in Medicine & Biology Society (EMBC), Montreal, QC, Canada, pp. 1023–1026 (2020). https://doi.org/10.1109/EMBC44109.2020.9175154
25. Mbarak, A., Yetis, Y., Jamshidi, M.:Data—based pollution forecasting via machine learning: case of Northwest Texas. In: 2018 World Automation Congress (WAC), Stevenson, WA, USA, pp. 1–6 (2018). https://doi.org/10.23919/WAC.2018.8430438
26. Davagdorj, K., Lee, J.S., Park, K.H., Ryu, K.H.: A machine-learning approach for predicting success in smoking cessation intervention. In: 2019 IEEE 10th International Conference on Awareness Science and Technology (iCAST), Morioka, Japan, pp. 1–6 (2019). https://doi.org/10.1109/ICAwST.2019.8923252
27. Leon, C., Carrault, G., Pladys, P., Beuchee, A.: Early detection of late onset sepsis in premature infants using visibility graph analysis of heart rate variability. IEEE J. Biomed. Health Inform. https://doi.org/10.1109/JBHI.2020.3021662
28. Wang, W., Chen, L., Jie, J., Wang, H., Xu, X.: A novel hybrid algorithm based on baldwinian learning and PSO. In: 2010 International Conference on Computational Aspects of Social Networks, Taiyuan, China, pp. 299–302 (2010). https://doi.org/10.1109/CASoN.2010.73
29. Graff, C.A., Coffield, S.R., Chen, Y., Foufoula-Georgiou, E., Randerson, J.T., Smyth, P.: Forecasting daily wildfire activity using poisson regression. IEEE Trans. Geosci. Remote Sens. **58**(7), 4837–4851 (2020). https://doi.org/10.1109/TGRS.2020.2968029
30. Vahabi, N., Yerworth, R., Miedema, M., van Kaam, A., Bayford, R., Demosthenous, A.: Deep analysis of EIT dataset to classify apnea and non-apnea cases in neonatal patients. IEEE Access **9**, 25131–25139 (2021). https://doi.org/10.1109/ACCESS.2021.3056558
31. Quinn, J.A., Williams, C.K.I., McIntosh, N.: Factorial switching linear dynamical systems applied to physiological condition monitoring. IEEE Trans. Pattern Anal. Mach. Intell. **31**(9), 1537–1551 (2009). https://doi.org/10.1109/TPAMI.2008.191

A Self-Sovereign Identity Management System Using Blockchain

Tripti Rathee and Parvinder Singh

Abstract Blockchain is emerging as a functional technology for remodeling current technologies and also for creating new applications which in practical was not possible earlier. In this paper, we are building a self-sovereign identity management system using blockchain *(BSelSovID)*. The application will serve as an interface to all the entities involved, viz. user, authority, and verifier. The use of blockchain is to dodge the centralized identity manager. The user gets digital identity after getting authority's verifiable claim and the identity is stored on interplanetary file system (IPFS) and the content address of these is stored on the blockchain. This makes it easy for the user to interact with different services without having identity for each one of them. The security of user data is ensured by encrypting the data and giving user the complete control over his data stored in the IPFS. The blockchain cannot extract the user's data but can only request the user to provide some data attribute or some verification as the verifier demands. Thus, the system provides security and privacy to users' data as well as user has control over his data.

Keywords Blockchain · Cryptography · Digital identity · Privacy · Security

1 Introduction

Blockchain technology gained its importance after the release of a white paper [1] in which Satoshi Nakamoto (which is an alias) introduced the trustless concept. If we utter the word blockchain to a group of people, most of them will relate it to bitcoin or cryptocurrency. But the application of blockchain is far wide than just

T. Rathee (✉)
Maharaja Surajmal Institute of Technology, New Delhi, India

T. Rathee · P. Singh
Deenbandhu Chhotu Ram University of Science and Technology, Murthal, Haryana, India

K. Khanna et al. (eds.), *Cyber Security and Digital Forensics*,
Lecture Notes on Data Engineering and Communications Technologies 73,
https://doi.org/10.1007/978-981-16-3961-6_31

cryptocurrency [2–6]. Since the world in which we are living is data-driven, therefore protecting the user's data and identity is the need of hour. After the rise of Internet at global level, we are fighting the identity management (IM) challenges which include security and privacy. In real world, we have Aadhaar card, driver's license and even passport to create personal identity. In digital world, there are systems to create digital identities; however, these systems seem to be imperfect. There is no system for online authentication of identities or the digital entity's identity. Moreover, the users do not have control over their personal data. They do not face any national or international boundaries. Blockchain technology offers a solution to the above problem by delivering a system which can be used to create a digital identity on the blockchain. The system further can help the individual to manage and create their own digital identity, allowing them to have tremendous control over their personal data. Unfortunately, there is a possibility of Sybil attack if the whole system is distributed [7]. The inherent distributed principle of blockchain when combined with identity verification can be used to create a digital identity which can be used to perform any transaction over the Internet. It can encrypt and store the encrypted digital identity, so as to allow the users to share and manage their personal data on their own terms. The execution of smart contracts on blockchain can help everyone in the network to establish decentralized trust.

1.1 Our Contribution

In this paper, we propose *BSelSovID*; a blockchain-based self-sovereign identity management system that uses blockchain to eliminate the problems faced by centralized systems. The architecture deals with mainly three actors namely—user, verifier, and authority. A user uses the system to create and store identity, the authority checks users' data from their database which allows the user to create their identity and verifier/third party uses our system to check/verify users' identity. All the users' information is stored on IPFS while blockchain only stores the content address of the users' data and their public key which is a unique identifier of their information. All the verification is done by smart contracts that run on the blockchain. In this way, smart contracts can send a yes or no message to the verifier while verifying instead of revealing complete users' information.

The rest of the paper is organized as follows: Sect. 2 describes the background of blockchain technology and needs the need for the new type of self-sovereign identity management system. Section 3 describes the proposed *BSelSovID*, Sect. 4 concludes this paper, and finally, Sect. 5 presents the limitations and future work.

2 Preliminaries

2.1 Blockchains

A blockchain is composed of some logical elements called as *nodes*, a distributed ledger consensus algorithms with cryptographic integrity [8]. Blockchain mainly consists of three layers, i.e., point-to-point network (P2P) [9] consisting of nodes, data storage, and asymmetric key cryptography [10]. The distributed P2P network allows the non-trusted members to interact with each other without the involvement of central authority [11]. It consists of transactions which are signed by the peers. The nodes in the network are responsible for grouping the transactions in the *blocks* and to determine the validity of the transactions. A valid transaction means, for example, User A receives one bitcoin from User B. However, User B may have tried to send the same bitcoin to user C. Therefore, it is now the responsibility of existing nodes to decide which transaction should be added into the blockchain so as to avoid any fraud entry in the blockchain [11, 12]. The *consensus* mechanism here plays a vital role in determining whether a transaction is valid or not. There are various consensus mechanisms available in the literature [13–16]. Asymmetric key cryptography is to ensure privacy and security of the data. The inherent property of blockchain provides immutability to the data stored in the blocks. Figure 1 describes the overall blockchain structure.

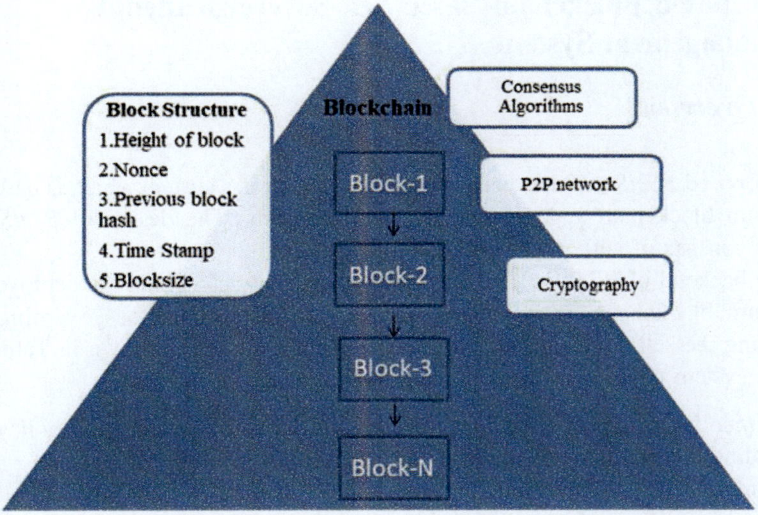

Fig. 1 Blockchain structure

2.2 Related Work Summary

The decentralized identity management systems are built using bitcoin blockchain in [17, 18] which provides a reliable and public identity. An identity system based on biometric is built in [19]. There has been a lot of research going on in the field of identity management [20–23]. But identity management using blockchain technology needs more attention. The privacy has been decentralized using blockchain technology in [24]. Blockchain has been used to access medical data records in [25]. The authors in [26] have proposed the utilization of auditable contacts using blockchain for fair data access. Similarly, the authors in [27] have proposed a framework to provide online rating based on user behavior. The authors in [28] have surveyed and analyzed three tools that use blockchain for identity management. The authors in [29] have produces a system for storing the identities of users on blockchain based on their credentials. The authors in [30] have concluded in their studies that in Canada, IdM system based on blockchain can enhance online services. The authors in [31] have proposed a framework for blockchain-based ID as service. The authors in [32] have proposed a framework for identity management based on blockchain in business to enable different parties to exchange identities. However, the above-mentioned models do not solve the identity management challenges defined by [33]. Therefore, it is a necessity to build strong IAM models.

3 Proposed Blockchain-Based Self-Sovereign Identity Management System

3.1 Overview

The proposed *BSelSovID* has been shown in Fig. 2. It is a system which is based on Ethereum blockchain and guarded by smart contacts. The design *of BSelSovID* mainly consists of authority, verifier, and end user.

The backend of BSelSovID is based on smart contracts which are deployed on Ethereum blockchain. These smart contracts are responsible for governing and managing the self-sovereign identity management modules as shown in Table 1.

The system consists of the following modules.

1. *Verification of data by Authority* deals with the whole verification policy and sending the verifiable claims back to the user.
2. *Validation* deals with validating the digital signature of authority and storing digital identifier (DID) on blockchain.
3. *Interaction with verifier* deals with the user interaction and verification did using third-party policy.

These modules and the smart contracts operation are discussed in the following sub-section

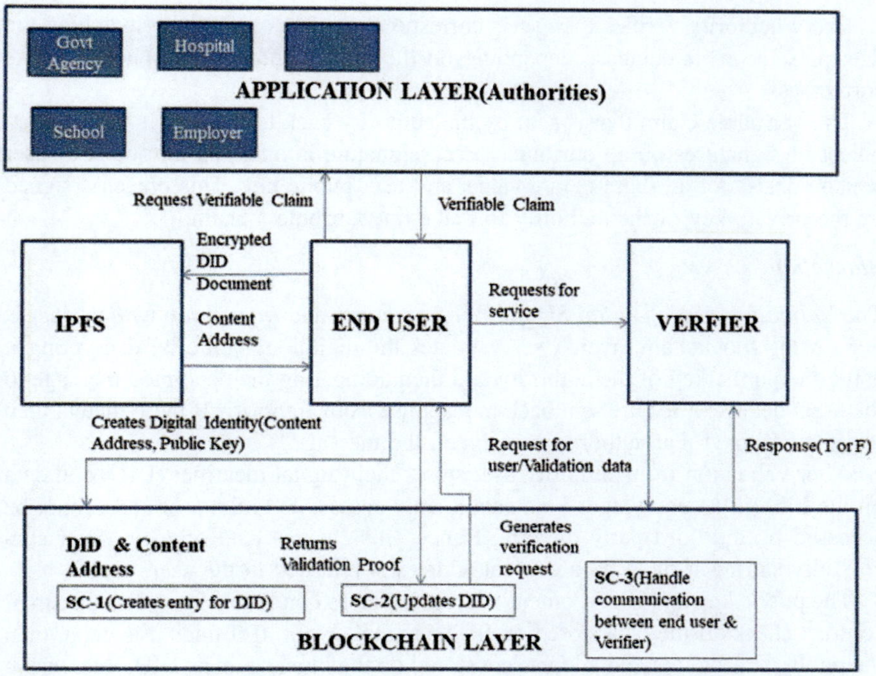

Fig. 2 Proposed self-sovereign identity management system

Table 1 Smart contract per module

Module	Smart contracts	Purpose
Verification of data by Authority	– createIdentity – getIdentity – editIdentity	• Creation of Digital Identity • Getting Digital Identity • Editing Digital Identity
Validation	– triggerDataRequest – triggerDataResponse	• Trigger Data Request • Trigger Data Response
Interaction with verifier	– triggerVerifyRequest – triggerVerifyResponse	• Trigger Verification Request • Trigger Verification Response

Application Layer:

The verification of data by authority is performed at application layer. The user is given a list of authorities to choose from. He chooses a particular authority to sign on a particular data attribute. This list of attributes along with their values and the specified authority whose sign is needed on that attribute by the user is sent on a backend API.

Every authority verifies the user's corresponding data attribute by matching the data present on its database depending on the implementation and checks for its correctness.

The verifiable claim that is sent by the authority back to the user is basically an object of signatures on all attributes. Each signature is made on an object of user which consists of attribute name, value, and user public key. This object is signed by the private key of the authority to make one attribute's claim.

Blockchain Layer:

The *Validation of the Digital Signature of Authority* and *Interaction with verifier* is done using blockchain layer. User validates the digital signature by decrypting it using the public key of the authority and then comparing the decrypted message to the users details which are sent back in response from authority. If these match, then it gets confirmed that authority has signed the message.

After validation from authority user stores their digital identifier (DID) and data on the IPFS in an encrypted. This data is only accessible to the user and cannot be accessed by the third party directly. Hence, this ensures data privacy. After successfully storing it on IPFS a content address is returned to the user.

The public key and IPFS content address are stored on blockchain after the smart contract checks if the data stored on IPFS is valid or not. If it finds it is valid, then the public key and content address are stored on the blockchain as DID otherwise it shows an error message.

User interacts with Verifier:

A user tries to login into a service provided by a third party; in order to prove their identity, the user provides their public key and data to the verifier. The verifier upon coming across the user public key and the data which needs to be true for the user in order to provide services emits an event through a blockchain smart contract which asks for the user to verify an attribute. Whenever the user reads an event asking for verification of data attribute, the user sends back a signed response. This is done by the user by emitting an event through a blockchain smart contract. This sign is verified at the smart contract to check whether the sign is made by a legitimate authority, and also whether the data on which sign is made matches with the value available with the authority. The verifier now constantly reads for the response event emitted by the user. It receives a true or false for the attribute, which depicts whether the attribute's value was successfully verified by the user through the smart contract event or not. This method ensures a complete safety to the user's data, as it verifies the data without actually disclosing the user details. This procedure ensures that the data provided by the user to the verifier is completely consensual and is provided by the user itself; hence, it ensures privacy to the user.

Table 2 Comparison of proposed system with existing systems

Laws of identity	uPort	Sovrin	ShoCard	BSelSovID
User control and consent	X	√	√	√
Minimal disclosure for a constrained use	√	√	X	√
Justifiable parties	√	√	X	√
Directed identity	√	√	√	√
Design for a pluralism of operators and technology	√	√	X	√
Human integration	X	X	X	X
Consistent experience across contexts	√	X	√	√

Source This table is inspired from [37]

3.2 Key Generation

The authenticity of each transaction should be verified using blockchain technology. The use of digital signature helps to guarantee the authenticity and non-repudiation of the data. RSA has been one of the very popular algorithms for key generation [34]. However, there is a continuous effort in improving the security of digital signatures. The algorithm used for key generation in our proposed system is elliptical key cryptography (ECC) [35].

3.3 System Implementation Details

The system is implemented on Intel(R) Core(TM) i5, 3.30 GHz CPU, 4 cores, and 8 GB RAM. Table 2 shows the comparison of proposed system with the existing identity management solution described in [28]. It has been observed that our system satisfies the laws of identity defined by [36] as compared to the other IdM solutions.

4 Conclusion

This paper has introduced the concept of *BSelSovID* which is based on the core concept of blockchain and self-sovereign identity. The architecture aims to replace the conventional centralized systems. It is a generic framework which can be applied to any use case. It makes use of smart contract to perform the verification process and hence speeds up the processing. The user's data is encrypted before it is stored on IPFS, which adds more security to the data. Moreover, the data control is in the hands of the user which adds privacy. The system is feasible and easy to implement.

5 Limitations and Future Work

Although the system is easy to implement, it has certain limitations. Firstly, there is no provision to control the further distribution of user's data by the verifiers. Secondly, in case the smart contract fails, the system will lack reliability. As of now, we have implemented the system on local blockchain only. In the future work, a framework for public permissionless blockchain can be developed.

References

1. Nakamoto, S.: Bitcoin whitepaper. https://bitcoin.org/bitcoin.pdf (Дата обращения: 17.07. 2019) (2008)
2. Ali, M., Nelson, J., Shea, R., Freedman, M.J.: Blockstack: A global naming and storage system secured by blockchains. In: 2016 ${$USENIX$}$ annual technical conference (${$USENIX$}$${ATC}$ 16), pp. 181–194 (2016)
3. Fromknecht, C., Velicanu, D., Yakoubov, S.: A decentralized public key infrastructure with identity retention. IACR Cryptol. ePrint Arch. **2014**, 803 (2014)
4. Lee, J.-H., Pilkington, M.: How the blockchain revolution will reshape the consumer electronics industry [future directions]. IEEE Consum. Electr. Mag. **6**, 19–23 (2017)
5. Lee, B., Lee, J.-H.: Blockchain-based secure firmware update for embedded devices in an Internet of Things environment. J. Supercomput. **73**, 1152–1167 (2017)
6. Pilkington, M., Crudu, R., Grant, L.G.: Blockchain and bitcoin as a way to lift a country out of poverty-tourism 2.0 and e-governance in the Republic of Moldova. Int. J. Internet Technol. Secur. Trans. **7**, 115–143 (2017)
7. Douceur, J.R.: The sybil attack. In: International workshop on peer-to-peer systems, pp. 251–260 (2002)
8. Vitenberg, R.: Debunking blockchain myths. NISK J. **11** (2018)
9. Park, H., Ratzin, R.I., van der Schaar, M.: Peer-to-peer networks: Protocols, cooperation and competition. In: Streaming Media Architectures, Techniques, and Applications: Recent Advances, pp. 262–294. IGI Global (2011)
10. Thomsen, S.S., Knudsen, L.R.: Cryptographic hash functions (2005)
11. Christidis, K., Devetsikiotis, M.: Blockchains and smart contracts for the internet of things. IEEE Access. **4**, 2292–2303 (2016)
12. Vukolić, M.: The quest for scalable blockchain fabric: Proof-of-work vs. BFT replication. In: International workshop on open problems in network security, pp. 112–125 (2015)
13. Antonopoulos, A.M.: Mastering Bitcoin: unlocking digital cryptocurrencies. O'Reilly Media, Inc. (2014)
14. Dannen, C.: Introducing Ethereum and solidity. Springer (2017)
15. Pilkington, M.: Blockchain technology: principles and applications. In: Research handbook on digital transformations. Edward Elgar Publishing (2016)
16. Castro, M., Liskov, B.: Practical Byzantine fault tolerance and proactive recovery. ACM Trans. Comput. Syst. (TOCS) **20**, 398–461 (2002)
17. Augot, D., Chabanne, H., Clémot, O., George, W.: Transforming face-to-face identity proofing into anonymous digital identity using the Bitcoin blockchain. In: 2017 15th Annual Conference on Privacy, Security and Trust (PST), pp. 25–2509 (2017)
18. Augot, D., Chabanne, H., Chenevier, T., George, W., Lambert, L.: A user-centric system for verified identities on the bitcoin blockchain. In: Garcia-Alfaro, J., Navarro-Arribas, G., Hartenstein, H., Herrera-Joancomartí, J. (eds.) Data Privacy Management, Cryptocurrencies and Blockchain Technology, pp. 390–407. Springer International Publishing, Cham (2017)

19. Othman, A., Callahan, J.: The Horcrux protocol: a method for decentralized biometric-based self-sovereign identity. In: 2018 International Joint Conference on Neural Networks (IJCNN), pp. 1–7 (2018)
20. Halperin, R., Backhouse, J.: A roadmap for research on identity in the information society. Identity Inf. Soc. **1**, 71–87 (2008)
21. Pfitzmann, A., Hansen, M.: A terminology for Talking About Privacy by Data Minimization: Anonymity, Unlinkability, Undetectability, Unobservability, Pseudonymity, and Identity Management (2010)
22. Jensen, J.: Federated identity management challenges. In: 2012 Seventh International Conference on Availability, Reliability and Security, pp. 230–235 (2012)
23. Jensen, J., Jaatun, M.G.: Federated identity management-we built it; why won't they come? IEEE Secur. Priv. **11**, 34–41 (2012)
24. Zyskind, G., Nathan, O., others: Decentralizing privacy: Using blockchain to protect personal data. In: 2015 IEEE Security and Privacy Workshops, pp. 180–184 (2015)
25. Azaria, A., Ekblaw, A., Vieira, T., Lippman, A.: Medrec: Using blockchain for medical data access and permission management. In: 2016 2nd International Conference on Open and Big Data (OBD), pp. 25–30 (2016)
26. Kaaniche, N., Laurent, M.: A blockchain-based data usage auditing architecture with enhanced privacy and availability. In: 2017 IEEE 16th International Symposium on Network Computing and Applications (NCA), pp. 1–5 (2017)
27. Yasin, A., Liu, L.: An online identity and smart contract management system. In: 2016 IEEE 40th Annual Computer Software and Applications Conference (COMPSAC), pp. 192–198 (2016)
28. Dunphy, P., Petitcolas, F.A.P.: A first look at identity management schemes on the blockchain. IEEE Secur. Priv. **16**, 20–29 (2018)
29. FAIDELLA, D.C., Schukai, R.J., Manuel, S.R., Pierleoni, M., Thomas, J.A.: Methods and systems for identity creation, verification and management (2020)
30. Wolfond, G.: A blockchain ecosystem for digital identity: improving service delivery in Canada's public and private sectors. Technol. Innov. Manag. Rev. **7** (2017)
31. Lee, J., Member, S.: BIDaaS: Blockchain based ID as a service. IEEE Access. **6**, 2274–2278 (2018). https://doi.org/10.1109/ACCESS.2017.2782733
32. Ebrahimi, A.: Identity management service using a blockchain providing certifying transactions between devices (2017)
33. Torres, J., Nogueira, M., Pujolle, G.: A survey on identity management for the future network. IEEE Commun. Surv. Tutor. **15**, 787–802 (2012)
34. Rivest, R.L., Shamir, A., Adleman, L.: A method for obtaining digital signatures and public-key cryptosystems. Commun. ACM **21**, 120–126 (1978)
35. Johnson, D., Menezes, A., Vanstone, S.: The elliptic curve digital signature algorithm (ECDSA). Int. J. Inf. Secur. **1**, 36–63 (2001)
36. Cameron, K.: The laws of identity. Microsoft Corp. **12**, 8–11 (2005)
37. El Haddouti, S., El Kettani, M.D.E.-C.: Analysis of identity management systems using blockchain technology. In: CommNet, pp. 1–7 (2019)

Blockchain and IoT for Auto Leak Unearthing

Pooja Sapra, Vaishali Kalra, and Simran Sejwal

Abstract Blockchain and Internet of Things (IoT) are important constituents of Internet-enabled era of information technology. Both technologies are distributed, autonomous, and decentralized systems. IoT devices require the strengthening of its security features, and security is an intrinsic aspect of blockchain due to cryptographic mechanisms. On the other hand, blockchain needs contribution from the distributed nodes and IoT includes within its architecture. So, blockchain can aid in the settlement of major security requirements in IoT. Blockchain features such as decentralization, immutability and transparency (DIT), auditability, and data encryption help to solve various IoT architectural problems. The main goal of water supply sector is to provide a solution to get shielded, authentic, and cost-effective water supply through well-regulated arrangements. It is very hard to achieve these goals. This paper introduces an algorithm for implementing a smart water management system that identifies and quantifies the water requirement by an individual consumer within a given locus and also identifies leaks (if any) in the plumbing system. The system proposed monitors both water quality and water scarcity aspects within the supplied vicinity. The smart water management system is collateral to a decentralized system implemented using smart tanks that uses the Internet of Things (IoT) for implementation and blockchain technology for providing a more rooted mechanism.

Keywords Blockchain · Internet of Things · Water management · Security

P. Sapra (✉) · V. Kalra
The NorthCap University, Gurugram, India
e-mail: poojasapra@ncuindia.edu

V. Kalra
e-mail: arya.vaishali17@gmail.com

S. Sejwal
Netaji Subhas University of Technology, Dwarka, India
e-mail: simran.cs19@nsut.ac.in

K. Khanna et al. (eds.), *Cyber Security and Digital Forensics*,
Lecture Notes on Data Engineering and Communications Technologies 73,
https://doi.org/10.1007/978-981-16-3961-6_32

1 Introduction

The significance of the Internet of things has progressed on account of the mix of different advancements, real-time analytics, AI, ware sensors, and embedded systems. Internet of Things has been a blessing across each domain of modus and numerous business ventures, and some of the emerging fields in which IoT is largely used includes client engagement platforms, optimization of technology, waste reduction, enabling smart homes, elderly care, transportation, buildings and home automation, infrastructure application, health care, and many more. A major drawback of IoT in terms of water management is dispensing a sense of security within the system that has been lacking in the traditional methods for the same. This void can be replenished using the advantages of blockchain technology.

In a perfect world, water resource management planning has regard to all the contending requests for water and tries to designate water on an impartial premise to fulfill all uses and requests [1]. Probably the greatest worry for our water-based assets later on is the manageability of the current and future water asset allocation. As water turns out to be rarer, the significance of how it is overseen develops vastly. Finding a harmony between what is required by people and what is required in the earth is a significant advance in the maintainability of water assets. Non-revenue water (NRW) real losses include leaks and bursts from the underground pipeline network. About 20% of the water produced is simply wasted because of underground water leakage which goes undetected, and some old or worn-out pipes might end up wasting 50% of water in supply [2]. Most leaks remain unidentified being underground. Water leakage detection program identifies and reports leakage problem and reduce previously undetected leaks.

The system is required because leakage in pipes lead to:

- Health issues of people due to contaminated groundwater.
- Ineffective supply management.
- Ineffective demand management.
- Weakening of the physical infrastructure.
- Customer dissatisfaction due to unreliable service of poor quality.

This system is explicit, anchored, ensures to an extent that the condition of water scarcity does not arise and also encourages individuals to conserve water along with earning some reward. Also, the proposed system utilizes sophisticated methodologies of the buzzing technology—Internet of Things (IoT) [3, 4], blockchain, and smart contracts for implementing such systems [5].

Blockchain Technology: A chain of blocks, having information about each transaction, is called blockchain. Each block has the address of its previous block that makes data provenance easy and difficult to tamper. Further, if data received from IoT sensors is placed in blockchain, it would reduce the possible attacks of leakage.

Smart Contracts: Smart contracts are the logical piece of codes, recorded in the blockchain and are triggered when an event takes place. Ethereum virtual machine

(EVM) is deployed to avoid the hacking of the smart contracts. There are many languages and IDEs available to develop the smart contracts like serpent, solidity, Go, etc. In this paper, we have considered solidity language to develop the smart contracts on Remix IDE. Smart contracts are deployed using Ganache tool.

2 Literature Survey

Various methods have been considered and implemented toward providing a secure management plan, and in this section, we would discuss the related work toward a secure water management plan.

Author in [6] proposed a system that takes into account the acoustic emission techniques to identify leak in the pipeline. This method uses acoustic (sound) sensors which work on the phenomena of sound propagation and uses sound waves to detect leakage. Sound waves would travel back and forth in a pipe with a certain wavelength, and the wavelength noted on a leakage would differ from the original ones, and hence, the leakage is detected.

The proposed system has a few drawbacks associates to it which includes: (i) Relying on the concept of sound propagation might not be accurate enough to detect the exact location of the leak. (ii) Acoustic sensors do not yield accurate results which might cause errors or failure in detection.

Authors in [7–9] incorporated methods that uses global system for mobile communication (GSM) to monitor leaks. Once the leak is identified, GSM technology is used to deliver an alerting message to the users designated mobile. Major disadvantage of this method is that GSM technology only allows producing an altering system and does not cater to provide the exact location of the leak within the plumbing system. Some other orthodox methods for management and for detection of leakage are using water sensors. This mechanism is very feeble and of low precision as (i) Water sensor would fail to identify leakage in underground pipes, (ii) Water sensor would detect minor leakages or underground water, (iii) The degree of accuracy decreases drastically. Such methods would be beneficial at small scale such as home automation or finding minor faults in the water supply.

Author in [4] caters to the issue in a sophisticated format using GSM technology for altering system on leakage, liquid crystal display (LCD) to remotely monitor inflow and outflow rate of water and flow sensors to determine the mass flow rate of inflow and outflow of water used in the pipelines. The system lacks a sense of security that could be provided taking into account the advancements in blockchain and cloud computing as blockchain has been seen as a cut above the traditional ways to secure our IoT devices.

The software-based solution for water management is provided in [1], wherein the author installed the smart water meters, and based on the data collected from the meter, the predictions are made using AI forecasting techniques. Based on the obtained predicted data, the meta-heuristics algorithms have been applied for smart management of non-renewable resource water. This one technique would also help in reducing the pumping activity. In order to save the water usage, the author [10] brings an android-based solution for small to medium size garden that measure the soil moisture level, air condition, and humidity level using sensors and schedule the watering of the plants. The proposed framework is built on fuzzy rules and introduced block chains to provide access to trusted devices only. In 2018, IBM [11] also proposed a blockchain model with IoT integration to establish clarity in the use of water in the asymmetry industry and to illustrate its importance in sustainable water management. This blockchain solution also help in providing the transparent trading of water contracts which further saved our resources [12]. This blockchain with IoT implementation is also proposed for creating the future smart cities [13, 14].

3 Proposed System

Water being a non-renewable resource is critical for human sustainability, and proper methods are to be incorporated to ensure more longevity and better management for future needs; hence, a fool proof method is required for water management, and detection of underground leaks to eradicate the issues that are associated with it. In our proposed system, we have tried to eradicate some of the issues related to the same, which includes the following:

- Hydrostatic pressure sensors as they are more accurate sensors as compared to existing models
- Using GPS technology allows fetching the exact geographical location of the leak
- Introducing blockchain to secure the system
- Chances of error or failure decrease drastically
- Higher accuracy and exact location detection
- The proposed system can function underground as well.

The work flow of the proposed model given in Fig. 1:

i. **Connection and Establishment of Smart Tanks**: For identification of leaks, the proposed model will include implementation of a smart tank, using hydrostatic pressure sensor and pH level indicator for water quantity and

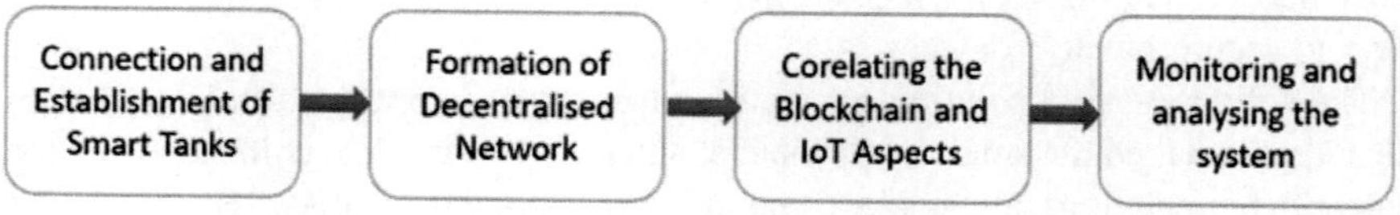

Fig. 1 Workflow diagram for creating an IoT-based Blockchain solution for water leakage management

quality management and lastly identifying leaks in underground pipeline. The devices required are:

- **Arduino board**: The monitoring of the sensors would be done by a concept of Internet of Things (IoT) circuit Arduino. Arduino is an open-source platform which relies on easy and convenient manner hardware and software programs. The hardware reads potential input, input from any connected sensor, a click on a button, or messages (via the internet platforms such as Twitter) and converts it into the required output. The user can send a set of instructions to the micro-controller on the board. The Arduino programming language which is used for writing syntax is the Arduino Software (IDE).
- **Pressure sensor**: The system would require an industry-based pressure sensor. A pressure sensor is a device used for measuring pressure value for gases and liquid. Example for applications for pressure sensor would be in the measuring of combustion pressure measured inside an engine cylinder or in a gas turbine. Pressure sensors are commonly manufactured from piezoelectric materials like as quartz.
- **GPS sensor**: Global positioning systems that can help detect leakage in underwater pipelines without any signal disparity. The global positioning system (GPS) is a satellite-based radio navigation sensor. GPS is a global navigation satellite sensor that gives geolocation and time information to a GPS receiver which provides output anywhere on or near the Earth surface where there is an unobstructed object or line which is in sight to four or more GPS satellite system. Obstructions such as mountains and buildings block the weak GPS signals.
- **pH Level sensor**: To determine the quality of water supplied, we would require a pH level indicator. A pH sensor is categorized as one of the most essential tools that is currently used for water quality measurements. This sort of sensor can gauge the measure of alkalinity and acidity in water and different solutions. The standard pH scale is from 0 to 14. When the pH value is seven, it is considered to be neutral. If pH value is above seven, it represents alkaline substances and substances with pH value less than seven are more acidic.

  For determining the quality and quantity factor of the water supplied within the vicinity, the idea is to equip every water consuming unit (a house) within the decentralized network (a colony or sector) with a smart tank which automatically detects the water level in the tank using hydrostatic pressure level sensors and water quality using the pH sensor. If the water level in the smart tank is below a certain threshold, it fetches water from the main source and thereby checking it quality set by the administrator, and if the water stored in the tank is more than the average requirement, it can be shared within the network among another consuming unit.

ii. **Formation of Decentralized Network**: Here, the entire process is decentralized to make it peer-to-peer and eliminates the need for a middle man. A transparent ledger is then generated to document the specifics of water use and delivery. Wherever the consumer drinks or distributes water, the entry for the same is registered in the decentralized ledger of the network.

iii. **Correlating the Blockchain and IoT**: A pool of water token is distributed to each water consuming unit, and water token are deducted if water is consumed, while water token is earned if water is distributed by the unit.

iv. **Monitoring and Analyzing the System**: The spent and unspent water token details help in data analysis and monitoring the water requirements.

Step by Step Procedure:

The step by step process can be seen in Fig. 2, where the pressure sensor first notes the difference in pressure between the two points and plots the same graph showing the peak value, the precise longitude, and latitude details of the leakage (if any) obtained after identification of the peak pressure value, and the GPS can include the geographical position of the leakage by which we can identify the location of the leakage. A decentralized ledger/record is created where the data (water tokens) of each water consumption unit are stored and can be used for review in a crisis situation; the water level indicator in each tank will assist with the data interpretation of each unit offering a deeper view into the locality's water requirements (Fig. 3 and 4).

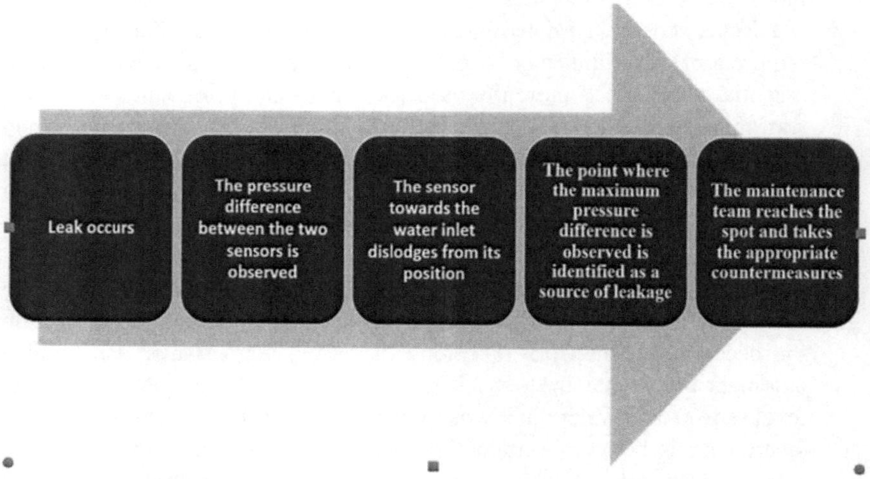

Fig. 2 Identification of the leaks

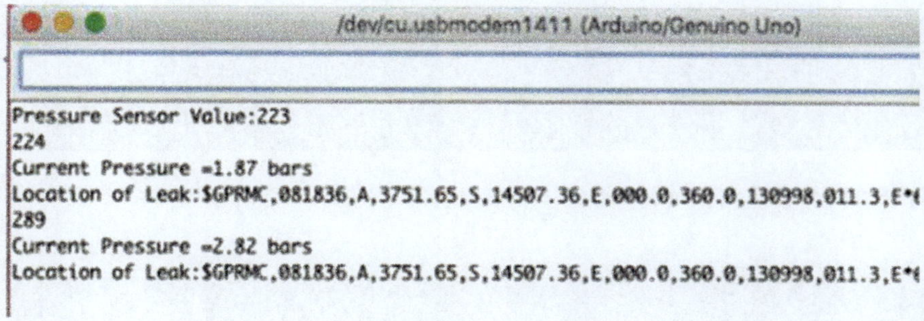

Fig. 3 GPS location and pressure value

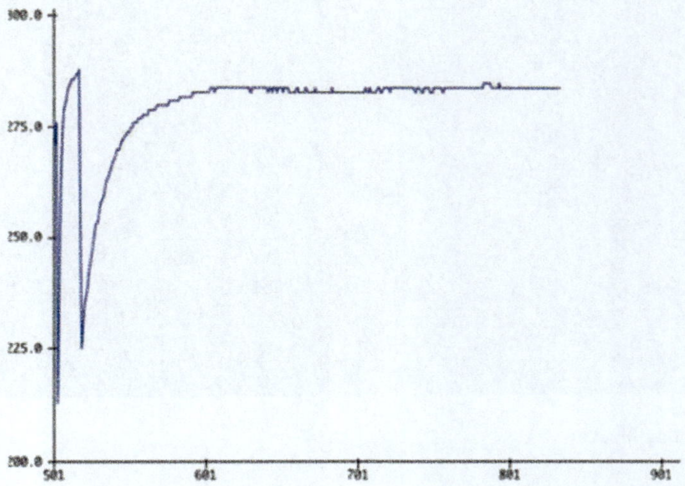

Fig. 4 Pressure sensor graph

4 Implementation and Results

For implementing the smart water management system, a blockchain is created.

By creation of blockchain, we ensure:

- Water requirement for each unit (house) can be recorded and monitored
- Addition and deduction of water tokens
- To provide a valid and secure blockchain mechanism

In Figs. 5 and 6, we showed that how the block chain is created, and the chain of blocks is formed to make the data provenance easy and difficult to tamper, and in Fig. 7, the transactions are shown. Each transaction is hash code encrypted which is not easy to temper.

```
Block's Hash: 99999eeae168b0aef3cb1ee98540246957b78be27937687f889aef8430b86a40
Block added
Simran has 20 water_tokens on her account
The Chain is valid and secure
```

Fig. 5 Creating blockchain

```
Simran started minning
Previous Block's Hash: 999997e5ac38b6ae5b720cd4e325d9ef583502428b8becfe27e1e57f98e896de
['{\n'
 '    "fromWallet": "Zining",\n'
 '    "toWallet": "Alex",\n'
 '    "amount": 0.01\n'
 '}',
 '{\n'
 '    "fromWallet": "Tom",\n'
 '    "toWallet": "Ankit",\n'
 '    "amount": 100\n'
 '}',
 '{\n'
 '    "fromWallet": "Raymond",\n'
 '    "toWallet": "Ankit",\n'
 '    "amount": 1e-07\n'
 '}']
Block's Hash: 99999ea8e9f36c128622bb966115613a7a66aa0802b2d975cf8e3a91e358d84b
Block added
Simran has 20 water_tokens on her account
The Chain is valid and secure
```

Fig. 6 Creating blocks

```
Simran started minning
Previous Block's Hash: 83247c122a3e29320a4e93412c3f76f11d9d3373a471fd183c8cfd61b84fa894
['{\n'
 '    "fromWallet": "Ankit",\n'
 '    "toWallet": "Alex",\n'
 '    "amount": 3.2\n'
 '}',
 '{\n'
 '    "fromWallet": "Ankit",\n'
 '    "toWallet": "Raymond",\n'
 '    "amount": 1\n'
 '}',
 '{\n'
 '    "fromWallet": "Alex",\n'
 '    "toWallet": "Raymond",\n'
 '    "amount": 5.12\n'
 '}']
Block's Hash: 999997e5ac38b6ae5b720cd4e325d9ef583502428b8becfe27e1e57f98e896de
Block added
```

Fig. 7 Transaction

5 Challenges

- The following issues can be encountered during the system implementation and run and the following countermeasures or suggestions can be carried out for better accuracy in the system.
- The longevity of the sensors might vary for every smart tank installed.
- Longevity of the sensors depends on various external factors and quality of the sensor (average sensor life is between 3 to 4 years)
- The working of sensors can be affected due to various factors (water, pressure of water current)
- Need to monitor pressure and other factors on regular basis.
- Installation and retrieval of the system can be tedious.
- It would require a team of 5–6 members for the same and can use additional features of an in-built camera for better surveillance.
- Data collection and analysis of raw data.
- Blockchain technology and its advantages can be well utilized.
- External factors that may vary data.
- Need regular monitoring regarding external factors such as corrosion, pipe structure, and tanks.

6 Conclusions

This proposed paper contemplates a highly accurate and full proof method toward management of water. It assists ways to check quality and quantify water requirements within a sector. The system also identifies leakage in underground pipes along with furnishing the exact geographical location of the leak which helps in taking the required measures appropriately and within the least possible time frame to avoid unnecessary wastage of water. Key features the system showcases are—The pressure difference can be classified into categories to report the severity of the leak. The electronic mechanism shall be enclosed in an air-tight ball-like structure, To reduce response time in sealing, leaks shall provide a lot of direct and indirect benefits. The sensors shall be mounted on the inner walls of the pipes with an electronically controlled switch to dislodge. To identify potential leak sites, the recorded measurements can be timely sent to a central server. A similar device can be used for oil and gas pipeline. Blockchain helps in providing a decentralized ledge and security measures blockchain technology and Internet of Things (IoT) aids the proposed system to function in a coherent manner and provide it with the required contrivance and tools.

References

1. Grigoras, G., Bizon, N., Enescu, F.M., Guede, J.M. L., Salado, G.F., Brennan, R., Alalm, M.G.: ICT based smart management solution to realize water and energy savings through energy efficiency measures in water distribution systems. In: 2018 10th International Conference on Electronics, Computers and Artificial Intelligence (ECAI), pp. 1–4. IEEE (2018)
2. Maouriyan, N., Krishna, A.A.: Aquachain-water supply-chain management using distributed ledger technology. In: 2019 3rd International Conference on Computing and Communications Technologies (ICCCT), (pp. 204–207). IEEE (2019)
3. Robles, T., Alcarria, R., de Andrés, D.M., de la Cruz, M.N., Calero, R., Iglesias, S., Lopez, M.: An IoT based reference architecture for smart water management processes. JoWUA **6**(1), 4–23 (2015)
4. Ntuli, N., Abu-Mahfouz, A.: A simple security architecture for smart water management system. Proc. Comput. Sci. **83**, 1164–1169 (2016)
5. Dogo, E.M., Salami, A.F., Nwulu, N.I., Aigbavboa, C.O.: Blockchain and internet of things-based technologies for intelligent water management system. In: Artificial Intelligence in IoT, pp. 129–150. Springer, Cham (2019)
6. Nicola, M., Nicola, C., Vintilă, A., Hurezeanu, I., Duţă, M.: Pipeline leakage detection by means of acoustic emission technique using cross-correlation function. J. Mech. Eng. Auto **8**, 59–67 (2018)
7. Lee, S.W., Sarp, S., Jeon, D.J., Kim, J.H.: Smart water grid: the future water management platform. Desalin. Water Treat. **55**(2), 339–346 (2015)
8. Zhang, D., Gersberg, R.M., Wilhelm, C., Voigt, M.: Decentralized water management: rainwater harvesting and greywater reuse in an urban area of Beijing, China. Urban Water J. **6** (5), 375–385 (2009)
9. Bordel, B., Martín, D., Alcarria, R., Robles, T.: A blockchain-based water control system for the automatic management of irrigation communities. In: 2019 IEEE International Conference on Consumer Electronics (ICCE), pp. 1–2. IEEE (2019)
10. Munir, M.S., Bajwa, I.S., Cheema, S.M.: An intelligent and secure smart watering system using fuzzy logic and blockchain. Comput. Electr. Eng. **77**, 109–119 (2019)
11. Chohan, U.W.: Blockchain and environmental sustainability: case of IBM's blockchain water management. Notes on the 21st Century (CBRI) (2019)
12. Pee, S.J., Nans, J.H., Jans, J.W.: A Simple blockchainbased peer-to-peer water trading system leveraging smart contracts. In: Proceedings on the International Conference on Internet Computing (ICOMP), pp. 63–68. The Steering Committee of The World Congress in Computer Science, Computer Engineering and Applied Computing (WorldComp) (2018)
13. Lazaroiu, C., Roscia, M.: Smart district through Iot and blockchain. In: 2017 IEEE 6th International Conference on Renewable Energy Research and Applications (ICRERA), pp. 454–461. IEEE (2017)
14. Ye, Z., Yin, M., Tang, L., Jiang, H.: Cup-of-Water theory: a review on the interaction of BIM, IoT and blockchain during the whole building lifecycle SSS. In: Proceedings of the International Symposium on Automation and Robotics in Construction, vol. 35, pp. 1–9. ISARC Publications (2018)

Coin Drop—A Decentralised Exchange Platform

Vanita Jain, Akanshu Raj, Abhishek Tanwar, Mridul Khurana, and Achin Jain

Abstract In today's world, cryptocurrency has seen a boom in users' number, and the numbers are increasing day by day. There are multiple cryptocurrencies, so there must be a platform to provide an exchange of cryptocurrencies. These days, many platforms provide users with the service, but they lack speed and are limited to some cryptocurrencies. Thus, we have proposed and developed a system that will increase the transactions rate by performing it off-chain to increase the transactions' speed and perform exchange between any cryptocurrencies. The proposed system combines the best practices of both the decentralised and centralised exchange platforms.

Keywords Cryptocurrency · Centralised cryptoexchanges · Decentralised exchange · Blockchain · Ethereum · Smart contract

1 Introduction

A cryptocurrency [1] is a form of currency available only virtually and is protected using various cryptography techniques. Researchers and developers proposed it in the early 1980's giving hints of the systems having digital money. Cryptocurrency is a term that refers to a digital asset of a network that does all the money-related work using that cryptocurrency, and the network is distributed across a large number of computers. Thus, the network is considered to have a decentralised structure with no single authority having all the power.

We can also consider cryptocurrencies as the systems that allow users to make online payments securely [2]. These payments are denominated in terms of virtual

V. Jain (✉) · A. Raj · A. Tanwar · M. Khurana · A. Jain
Bharati Vidyapeeth's College of Engineering, New Delhi, India
e-mail: vanita.jain@bharatividyapeeth.edu

K. Khanna et al. (eds.), *Cyber Security and Digital Forensics*,
Lecture Notes on Data Engineering and Communications Technologies 73,
https://doi.org/10.1007/978-981-16-3961-6_33

"tokens", which are represented by ledger entries internal to the system. "Crypto" refers to the various encryption algorithms and cryptographic techniques that safeguard these entries, such as elliptical curve encryption, public–private key pairs and hashing functions [3].

In today's world, there are thousands of cryptocurrencies with various functions and specifications. Some of these cryptocurrencies are clones or forks of popular cryptocurrencies like bitcoin and Ethereum, and also, some new currencies are built from scratch. Some popular cryptocurrencies apart from bitcoin [4] and Ether [5] include Litecoin [6], Peercoin, Namecoin [7] and Cardano [8]. The total estimated value of the cryptocurrencies in existence can be around $214 billion [9].

However, there are several risks involved while using cryptocurrencies. The actual value as currency is effective $0, so the only store of value is in other utility for a distributed trustless public append-only ledger [10]. Since the government does not monitor cryptocurrency, this creates a massive security risk as the criminals could use unmonitored money [11]. The private key protects the cryptocurrencies wallet, and once someone gets hold of the private key of the user, he can move the money from one account to another, and since the cryptocurrency is not monitored by the government or any central authority (in the case of decentralised exchange platforms), the owner cannot even file a report [12].

We already know that there are multiple cryptocurrencies globally, then there must be a system to allow trading between these currencies; this leads to the introduction of decentralised exchange platforms. Decentralised exchange (DEX) platforms provide the users of cryptocurrencies with a platform to perform cryptocurrency exchange. It allows users to have direct peer-to-peer cryptocurrency transactions online securely and without the need for any intermediary authority. Some of the famous DEX platforms are AirSwap, Atomex, Bancor and dex. blue [13].

However, these systems have some drawbacks in terms of speed, and they are also limited to the number of cryptocurrencies they can take into consideration in their respective platforms; these issues are discussed later in this paper, and a solution to overcome all those shortcomings is proposed.

The proposed system (coin drop) is comparatively much faster than the already present decentralised systems and can work with almost all kinds of cryptocurrencies available. In Section II of this paper, we have discussed cryptographic networks, their working, and the different types of wallets. In Section III, we have discussed the cryptocurrencies exchange platforms. Section IV discusses the working and architecture of "coin drop". Section V holds the results obtained by comparing various cryptographic exchange platforms against the proposed system. In Section VI, we have concluded the paper.

2 Cryptonetworks and Wallets in Today's World

2.1 *Cryptographic Networks*

To create a network that uses cryptocurrency for online payments, we need to work with blockchain [14]. This technology can be used to keep track of all the transactions that have taken place ever been in a ledger. Blockchain provides the developers with a way to create a structure of the ledger that is to be stored. It is made sure that the structure is secure. This structure is then shared with the entire network, individual nodes or computer maintaining a ledger copy. When a majority of nodes in the network are in favour of the structure, then only the structure is established in the network. A set of transactions is forged into a new block verified by each node of the network. If the nodes confirm the block, then only the block is added to the chain. Thus, this process makes it almost impossible for the mishappenings of the transaction histories. The main goals of such a network are to allow members of the network to synchronise their view of the system state and then to disseminate peer information to allow peers to reenter the system after a disconnection [15].

One such network is the bitcoin network [16]. Bitcoin is a cryptocurrency created in January 2009 after the housing market crash. Bitcoin follows the ideas that are set out in a whitepaper by the mysterious and pseudonymous Satoshi Nakamoto. Bitcoin uses the blockchain network that keeps balances and maintains a public ledger that everyone has transparent access to; all bitcoin transactions are verified by a massive amount of computing power [17].

2.2 *Hot and Cold Wallets*

These cryptocurrencies are stored in digital wallets. There are two types of digital wallets: hot wallet and cold wallet [18].

A wallet that is connected to the network is referred to as hot wallet. They are generally easy to set up, access and be given more tokens. Hot wallets can hold any cryptocurrencies. These wallets are mainly for everyday cryptocurrency users. Nevertheless, they are open to hackers and other technical vulnerabilities because they are connected to the Internet. Some examples of hot wallets are Coinbase and Blockchain.info.

Any wallet which is not connected to the cryptographic network is termed as cold wallet. They are more secure than hot wallets, but they do not accept as many cryptocurrencies when compared to hot wallets. Cold wallets are mainly designed devices that are designated physical cryptocurrency storage. One of the main advantages these wallets provide is that we can have our cryptos beside us. They are expensive to buy, while hot wallets are free of cost and easy to access. Some examples of cold storage devices are Trezor and Ledger [19, 20].

3 Cryptocurrency Exchange Platforms

3.1 Centralised Cryptoexchanges

They are trading platforms like the stock market, where a company or a third party has total control over all the transactions made by both parties. In centralised cryptoexchanges [21], the user has to trust the third party because they do not have access to the private keys of exchange account wallets. Trust is the main factor in centralised exchange platforms. All the transactions are controlled by third parties, which could lead to hacks in these platforms or malicious activities by the service providers. Centralised exchange platforms have their system's off-chain. The transactions are not handled by the blockchain, due to which it is prone to security breaches and other attacks on the systems. Today most of the platforms are centralised, which run on high regulatory risk. Some famous CCEs are Binance, Bittrex, Bitfinex, Coinbase and Kraken.

None of the centralised cryptoplatforms is immune to hacks. More than 30 cryptocurrency exchange hacks have occurred in the last nine years, for example, MTGox, BitGrail, and Coincheck. Some government bodies have also banned CCEs in recent years. China, South Korea and Russia are among them. Using CCEs often comes with a large amount of risk, the user has to provide sensitive information about themselves, including bank details, address and govt. issued ID as well [22].

3.2 Decentralised Cryptocurrency Exchanges

Decentralised cryptocurrency exchanges overcome many disadvantages of the centralised structure as it operates on a peer-to-peer marketplace directly on the blockchain. Using this way, the traders do not have to reveal their sensitive information to a third party. Since it works on blockchain, it is less immune to security attacks, unlike centralised cryptoexchanges. Still, most of the platforms have a centralised structure because building decentralised exchange platforms is complicated and too costly. Currently, most of the platforms do not have the liquidity to compete with centralised cryptoexchanges.

Decentralised cryptocurrency exchanges operate on a peer-to-peer network because their nodes are distributed; they experience a lower risk of attacks compared to centralised cryptoexchanges [13]. There is no involvement of a third party in DEX which will have control over all the transactions. They are less prone to security attacks and malicious activities by service providers. In DEX, all the payments use cryptocurrencies. DEX enables users to remain in control of their funds by operating essential functions over the blockchain. It overcomes the centralised structure's main limitation since there is no point of failure in this exchange, making blockchain a powerful technology. There are many through backs and inefficiencies in a centralised structure, due to which there is an introduction of many semi-decentralised exchanges which are coming into action. These are models that operate between the centralised and decentralised marketplace, which are highly efficient than centralised ones. CCEs lack security, transparency and efficiency, due to which demand for decentralised exchanges has increased. DEX promises two significant advantages: security control and the global marketplace. EtherDelta is one of the oldest projects in this field. It has a simple user interface and basic trading features. It has already gained sufficient attraction in the market. Some of the popular DEX are Wavesdex, Bancor protocol, Kyber Network, EtherDelta and Airswap.

4 Proposed System and Working

We have proposed a system that provides decentralised cryptocurrency exchange facilities using blockchain, which is merged with an interactive, user-friendly interface created using ReactJS. The proposed system is much faster than the traditional methods used and combines the best features from both the centralised and the decentralised exchange platforms. The working of the whole system is explained in great detail. Figure 1 shows the workflow of the system.

The trader interacts with the interface (web app in the diagram). The smart contract that we have to build is deployed on Ethereum blockchain and connected with the front end. For testing purpose, we have deployed a contract on the Rinkeby network and used injected web3 environment.

First of all, the user deposits the quote currency in the wallet provided in the exchange and any other token listed on the exchange. Figure 2 shows the wallet that a user will use.

Coin drop has two types of orders: one is market orders and the other one is limit orders. Market orders are filled immediately, irrespective of the price of the token.

To create a market order, there is a function in the smart contract called createMarketOrder that takes the ticker, amount of the token, and the side of the order book. For selling tokens, we can identify if the seller has enough tokens, but while buying market order, we cannot determine the price at which the market order will be fulfilled because one order can have several trades under it, i.e. market orders can be fulfilled partially as per availability in the order book. It is also checked whether the token for which the order is being created exists or not.

In the case of limit orders, there is a function called createLimitOrder that is used to create a new order in the order book. It takes ticker, amount, price of the token and side to place the order. It is checked that the token exists for which the order is being placed, and the token is not quoted currency. The cost to buy/sell one token is specified and is only filled if matching order is found in the counter order book, i.e. if a user wants to sell tokens using limit order, the order can only be filled if there is a matching order to buy.

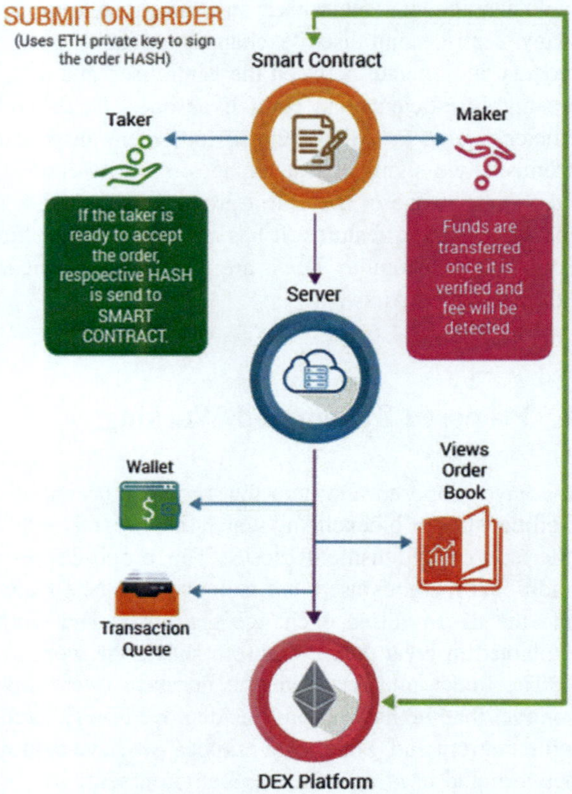

Fig. 1 Flow chart of the system

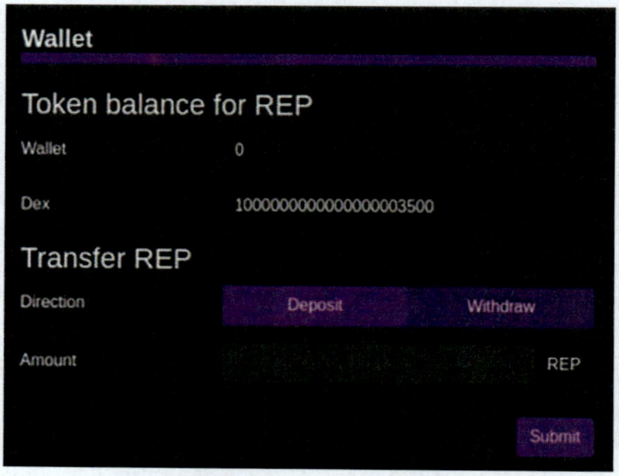

Fig. 2 User wallet in the coin drop

The getOrders function in the smart contract is used to list all the current orders from both sides of the order book, i.e. from the sell-side and buy-side. A getTokens function is used to list all the tokens purchased or sold on the exchange. It returns an array of tokens. The addToken process in the smart contract is used to record a

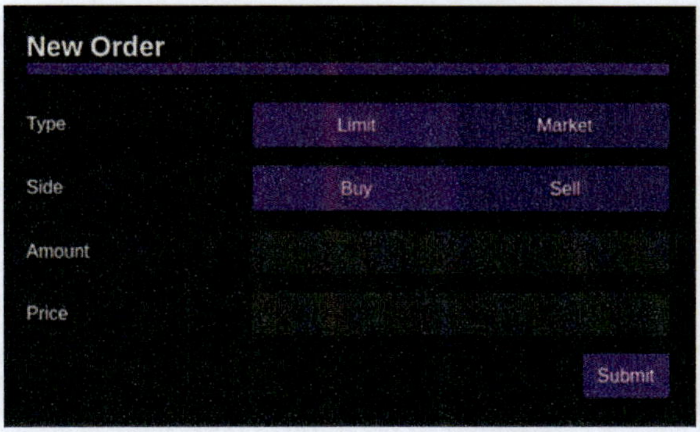

Fig. 3 New order function

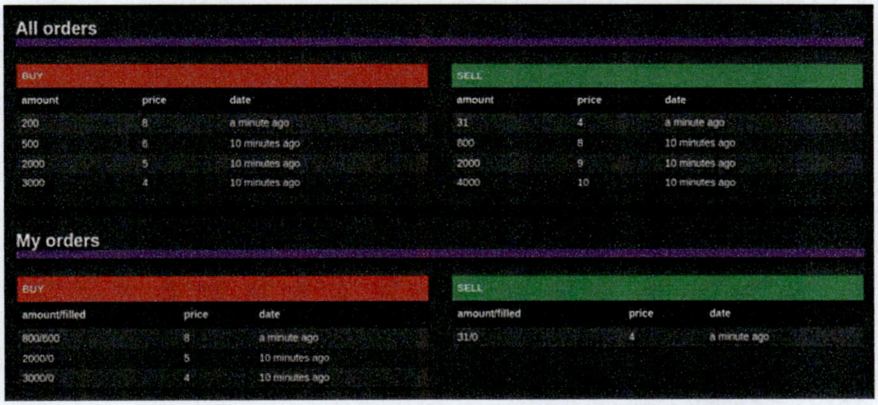

Fig. 4 Orders placed on the coin drop platform

new token on the exchange. It takes token address and ticker (symbol) as an argument. The admin of exchange can only call this function. The Deposit function is used to deposit tokens into the exchange, and the Withdraw function is used to withdraw the tokens from the exchange into external wallets.

DAI token is used as quote currency using which any listed token can be purchased or sold for DAI. DAI is used because it is a stable coin and its value always remains close to $1.

Figure 3 shows the function where a user can place new orders.

Figure 4 shows all the orders placed on the platform by the users worldwide and the orders that a particular user places on the coin drop platform in my orders' section.

Characteristics	Centralized Exchange Platforms	Decentralized Exchange Platforms	Coin Drop (Proposed System)
Control	platform has the most control	User has the most control	User has the most control
Security	Risk of Hackers	No chance of hacking	No chance of hacking
Transaction Fees	Charges fees for using the platform	Charges zero or very minimal fees	Charges zero fees
Type of Transactions	On-Chain	On-chain	Off-chain
Number of Currencies	2-5	2-5	Almost all
Liquidity	High Liquidity	Low Liquidity	High Liquidity
Speed	Executes orders in milliseconds	Can take up to an hour to execute orders	Executes orders in milliseconds

Fig. 5 Comparison of centralised exchange [13–15], decentralised exchange [13, 14, 22] and coin drop

5 Result

From the proposed system stated in the paper, we found that the decentralised platform that is proposed and created has the control in the hands of users and is less prone to attack by the malicious users as compared to the centralised exchanges as the amounts in the personal wallets are comparatively less when compared to the centralised banks. The proposed system aims to take no amount to create or validate a transaction, and the transactions are performed off-chain, which increases the speed of transactions. The proposed system can handle any cryptocurrencies using DAI token, a stable currency form, as an intermediatory currency when exchanging between two currencies. Thus, all these points were taken into consideration, and a comparison of current centralised and decentralised exchange platforms with the system proposed in the paper is shown in Fig. 5.

6 Conclusion

The cryptocurrency users have increased, and so the worth of cryptocurrencies. Many platforms provide cryptoexchanges services, but they have certain limitations, and they work with a few cryptocurrencies. We have developed and proposed a system that increases the cryptographic transactions' speed as they are done in the off-chain. The coin drop system can handle exchange between any cryptocurrencies. Thus, the developed system provides a better approach and takes the right parts of both the centralised and decentralised systems.

References

1. Farell, R.: An analysis of the cryptocurrency industry (2015)
2. Peters, G., Panayi, E., Chapelle, A.: Trends in cryptocurrencies and blockchain technologies: a monetary theory and regulation perspective. J. Financ. Perspect. **3**(3) (2015)
3. Bucko, J., Pal'ova, D., Vejacka, M.: Security and trust in ´ cryptocurrencies. In: Central European Conference in Finance and Economics, pp. 14–24 (2015)
4. Bohme, R., Christin, N., Edelman, B., Moore, T.: Bit- ¨ coin: Economics, technology, and governance. J Econ. Perspect. **29**(2), 213–238 (2015)
5. Bouoiyour, J., Selmi, R.: Ether: Bitcoin's competitor or ally? arXiv preprint arXiv:1707.07977 (2017)
6. Reed, J.: Litecoin: an introduction to litecoin cryptocurrency and litecoin mining (2017)
7. Gkillas, K., Bekiros, S., Siriopoulos, C.: Extreme correlation in cryptocurrency markets. Available at SSRN 3180934 (2018)
8. Guides, T.S.: Why cardano ada deserves your attention—cardano cryptocurrency strategy (2018)
9. Wei, W.C.: Liquidity and market efficiency in cryptocurrencies. Econ. Lett. **168**, 21–24 (2018)
10. Liu, Y., Tsyvinski, A., Wu, X.: Common risk factors in cryptocurrency. National Bureau of Economic Research, Technical Report (2019)
11. Barone, R., Masciandaro, D.: Cryptocurrency or usury? Crime and alternative money laundering techniques. Eur. J. Law Econ. **47**(2), 233–254 (2019)
12. Twomey, D., Mann, A.: Fraud and manipulation within cryptocurrency markets. In: Alexander, C., Cumming, D. (eds.) Corruption and Fraud in Financial Markets: Malpractice, Misconduct and Manipulation, pp. 205–250 (2020)
13. Lin, L.X.: Deconstructing decentralised exchanges. Stanf J Blockchain Law Policy, **2** (2019)
14. Scott, B.: How can cryptocurrency and blockchain technology play a role in building social and solidarity finance? UNRISD Working Paper, Technical Report (2016)
15. Bashir, I.: Mastering Blockchain: Distributed Ledger Technology, Decentralization, and Smart Contracts Explained. Packt Publishing Ltd (2018)
16. Velde, F., et al.: Bitcoin: A primer (2013)
17. Nakamoto, S.: Bitcoin whitepaper (2008). https://bitcoin.org/bitcoin.pdf (17.07. 2019)
18. Khan, A.G., Zahid, A.H., Hussain, M., Riaz, U.: Security of cryptocurrency using hardware wallet and qr code. In: 2019 International Conference on Innovative Computing (ICIC). IEEE, pp. 1–10 (2019)
19. Das, P., Faust, S., Loss, J.: A formal treatment of deterministic wallets. In: Proceedings of the 2019 ACM SIGSAC Conference on Computer and Communications Security, pp. 651–668 (2019)
20. Jasem, F.M., Sagheer, A.M., Awad, A.M.: Enhancement of digital signature algorithm in bitcoin wallet. Bull. Electr. Eng. Inf. **10**(1), 449–457 (2021)
21. Bacon, J., Michels, J.D., Millard, C., Singh, J.: Blockchain demystified: a technical and legal introduction to distributed and centralised ledgers. Rich. JL Tech. **25**, 1 (2018)
22. Pop, C., Pop, C., Marcel, A., Vesa, A., Petrican, T., Cioara, T., Anghel, I., Salomie, I.: Decentralizing the stock exchange using blockchain an ethereum-based implementation of the bucharest stock exchange. In: 2018 IEEE 14th International Conference on Intelligent Computer Communication and Processing (ICCP). IEEE, pp. 459–466 (2018)

Smart Contracts and NFTs: Non-Fungible Tokens as a Core Component of Blockchain to Be Used as Collectibles

Akash Arora, Kanisk, and Shailender Kumar

Abstract Non-fungible tokens are one of the most important future application domains for smart contracts. Ethereum is the pioneer of a blockchain-based decentralized computing platform that has ultimately standardized these types of tokens into a well-defined interface, now known as ERC-721. Blockchain-based cryptocurrencies have received extensive attention recently. Massive data has been stored on permissionless blockchains. This paper aims to analyze blockchain and cryptocurrencies' technical underpinnings, specifically non-fungible tokens or "crypto-collectibles," with the help of a blockchain-based image matching game. While outlining the theoretical implications and use cases of NFTs, this paper also gives a glimpse into their possible use in the domain of human user verification to prevent misuse of public data by automated scripts. This demonstrates the interaction of the ERC-721 token with the Ethereum-based decentralized application. Further, we aim to reach a definitive conclusion on the benefits and challenges of NFTs and thus reach a solution that would be beneficial to both researchers and practitioners.

Keywords Blockchain · Dapp · Decentralized · Ethereum · NFT · ERC-721 · Smart contract · Truffle

A. Arora (✉) · Kanisk · S. Kumar
Department of Computer Science Engineering, Delhi Technological University, Shahbad Daulatpur, Bawana, New Delhi, India
e-mail: akasharora_2k17se13@dtu.ac.in

Kanisk
e-mail: Kanisk_2k17se13@dtu.ac.in

S. Kumar
e-mail: shailenderkumar@dce.ac.in

K. Khanna et al. (eds.), *Cyber Security and Digital Forensics*,
Lecture Notes on Data Engineering and Communications Technologies 73,
https://doi.org/10.1007/978-981-16-3961-6_34

1 Introduction

The concept of blockchain was initially envisaged as a decentralized network that could store records of transactions that happened on it and thus act as a source of trust. "The data stored on the blockchain is immutable and updated by the peer-to-peer network" [1]. By design, entities called "blocks" constitute a blockchain. These blocks have the capability by design to store transactions that have been broadcasted. Only such transactions are deemed to be valid that have been broadcasted and hence been verified. To put it quite simply in layman terms, a blockchain is a database [1]. To grasp this concept clearly, we have to understand what a database is. A database points to collecting stored information either electronically or otherwise on either a computer system or a ledger. In databases, information or data is typically structured in table format to allow for easier searching and filtering specific details. While a blockchain ultimately aims to serve a similar purpose as a database, it offers many advantages over the latter. The primary difference arises in the way that the data is stored in a blockchain. While in a database, data is often stored in a tabular format, a blockchain stores data in small chunks or blocks (and hence the name blockchain) that hold sets of information that are all connected together.

The entities which constitute a blockchain and are known as "Blocks" which are designed to have storage capacities. When the storage capacity of one such block is completely full, it is then attached to the previously filled block. In this manner, a chain of data known as the "blockchain" is created. New information is written after creating that freshly added block is added into a newly created block that will also be added to the chain once it is completely full (Fig. 1).

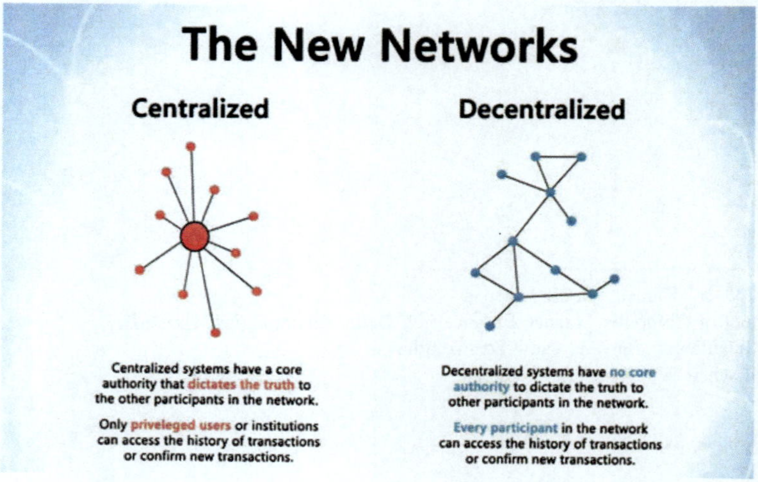

Fig. 1 Centralized versus decentralized system. *Source* https://blockgeeks.com/blockchain-infographics/

One of the central tenets of blockchain technology is "Decentralization" [2, 3]. The concept of decentralization can be best understood by contrasting it with a centralized network. The basic idea behind any centralized network is straightforward, "any centralized entity will anyways store all the data, and hence, an individual will always have to interact solely with this particular centralized entity to get whatever information or data that one requires" [2]. The traditional Client–Server network model is a perfect example of a centralized system. However, centralized systems have their drawbacks. The primary disadvantages include security vulnerabilities, lack of redundancy, and excessive dependency on the central node, which increases the chances of complete failure. These are the drawbacks that a decentralized system aims to address. By definition, in a decentralized system, "the information will not be stored by any one single entity. Everyone in the network owns the information" [2, 3].

Every node in a blockchain is designed to have a complete record of all the data that has been stored on the blockchain since its inception. The error in one node's data can be corrected by referencing the correct data in thousands of other nodes. This way, the creators of blockchain have ensured by a design that a single malicious node cannot inject information on the entire chain. This also means that the transaction history present in each block that constitutes the Bitcoin blockchain cannot be altered or modified. Another advantage of this decentralized nature of blockchain is transparency. A user can independently verify any transaction transparently, either by a personal node or by using automated scripts that crawl through the blockchain. This would enable interested individuals to witness ongoing transactions. Moreover, every constituting node contains a private version of the blockchain that gets refreshed as new blocks are verified and included.

Non-fungible tokens are an integral part of this paper. Non-fungible tokens are dissimilar tokens that do not conform to the concept of fungibility. The idea of fungibility deals with the currency's ability to maintain a standard value and uniform acceptance. Fungibility implies the immunity of the currency's value from its precedents; this ensures that each piece of that currency is equal in value to every other piece [4]. This means that one ₹100 note in my pocket is equal in value, identical, and replaceable with any other ₹100 note in any other individual's pocket. Hence, a ₹100 note is a fungible asset.

On the contrary, non-fungible tokens are blockchain assets that are designed not to be equal. Non-fungibility is the USP of such investments. This characteristic is of enormous importance when we are talking about rare and unique collectibles. The value of such items is derived from their non-fungibility (Fig. 2).

Another essential part of a blockchain network is smart contracts. Strictly speaking, a smart contract can be classified as a self-executing contract with the terms of the agreement between the buyer and seller being written from the get-go into the lines of code that make it up. The script and the contract contained within the smart contract are present across a distributed ledger and a decentralized network. The code controls the execution, and transactions are trackable and are hence irreversible.

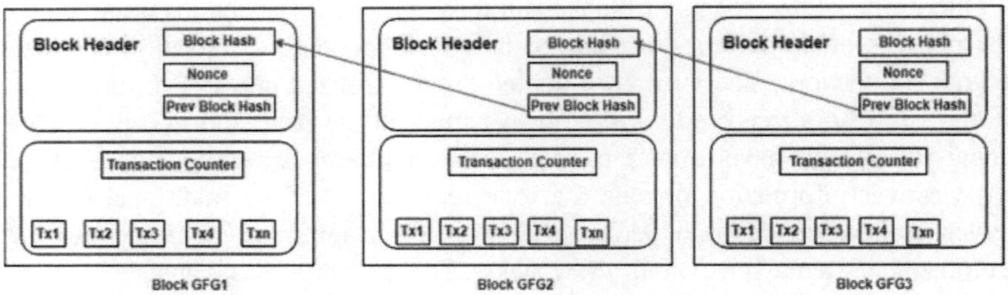

Fig. 2 Blockchain structure. *Source* https://www.geeksforgeeks.org/proof-of-work-pow-consensus/

"The biggest advantage of smart contracts is that they permit trusted transactions and agreements to be carried out among disparate, anonymous parties without the need for a central authority, legal system, or external enforcement mechanism" [5]. Smart contracts were originally proposed in the late twentieth century by Nick Szabo. Nick was a western computer scientist who had invented a virtual currency called "Bit Gold." His original intention behind creating smart contracts was to scale up the capabilities of contemporary transaction methods, such as point of sale (POS) to the digital realm.

Having glossed through this paper's fundamental building blocks, it would be prudent to revert to this paper's intention. We intend to showcase the interaction of NFTs built on the ERC-721 standard with smart contracts on the Ethereum blockchain using an Ethereum Dapp. To achieve this, we will be mining NFTs using an image tile-matching game. In this game, we will be collecting tokens in the form of collectibles. Furthermore, this collection is driven by smart contracts, and hence, the interaction between smart contracts and NFTs is established on the Ethereum network.

2 Related Work

2.1 Blockchain

A blockchain can be referred to as a decentralized network that can keep transaction records and thus get the capability to act as a bank of reliability [6]. The electronic details cached on the blockchain are perceived to be unchangeable, and it is designed to be continually updated by the peer-to-peer network. These blocks have the capability by design to store transactions that have been broadcasted. Only such transactions are deemed to be valid that have been broadcasted and hence been verified.

Blockchain technology can be said to have evolved from the development of Bitcoin as a distributed, immutable ledger that is maintained and verified on a network of peers. Since then, several industries have sought to explore the fundamental peer-to-peer technology and its myriad other applications, including creating extremely cost-effective and decentralized business network models or architecture.

Blockchain can also be a working specimen of a distributed computing system with high secure fault tolerance. Satoshi Nakamoto pioneered blockchain in 2008, who initially conceptualized it and then implemented it in the following year as a core component of the digital currency Bitcoin. Blockchain forms the public ledger for all transactions that are done using Bitcoin. Blockchain databases are designed to be managed without any human intervention, i.e., autonomously using a peer-to-peer network and a distributed timestamping server. Blockchain helped Bitcoin to attain the status of the first digital currency to solve the double-spending problem.

Blockchains, based on ownership and exclusivity, can be classified into private and public. The primary difference between the two is that anyone can join and contribute to a public blockchain. In contrast, one requires permission from a central authority to do so on a private blockchain. An advantage of public blockchains is that they are pseudo-anonymous by design as all transactions are public and hence hard to link them to identity.

2.2 Ethereum

Ethereum was initially conceptualized in 2013 by Vitalik Buterin, a 19-year-old Canadian developer. He sought to expand on the Nakamoto whitepaper and published a paper which was titled "A Next-Generation Smart Contract and Decentralized Application Platform." Buterin worked on all the primary strengths of Bitcoin, which included an incentive scheme for miners, proof of work, and hashing, to name a few, and further capitalized it to create a new Blockchain known as Ethereum. A significant advantage possessed by Ethereum can be said to be the implementation of a turing-complete language. What that meant was that the Ethereum blockchain could handle complex code to exploit the immense computing power that was being harnessed (but ultimately wasted) by the incentive scheme, something which could not be done successfully or optimally by the Bitcoin blockchain due to inherent defects of designs. It can be said that while Bitcoin was envisaged to be an unhackable store of value with few capabilities other than this one objective. Ethereum was designed and intended from the start to expand and ultimately realize the full scope of capabilities that could potentially be offered by blockchain technology and hence, theoretically, create a decentralized computer that could operate on a global scale.

It would be worth noting that Ethereum is a blockchain designed to build on the abstractions introduced by Bitcoin. Hence, Ethereum does not have an overarching outfit to handle the transactions leading to all recorded events on a blockchain. A consensus algorithm called proof of work is used to secure this open and decentralized network. The general public has complete access to all events

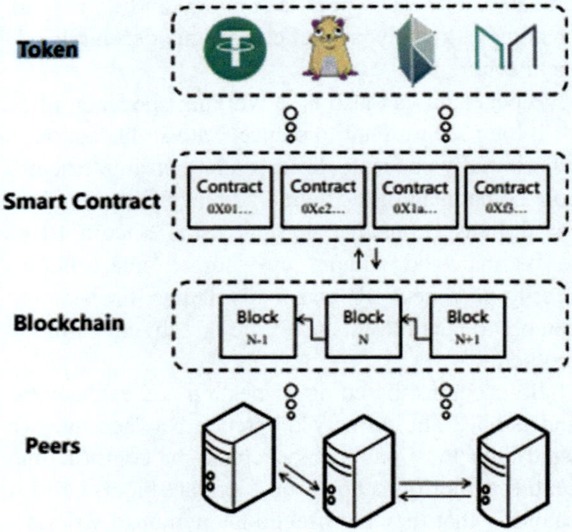

Fig. 3 Overview of ethereum blockchain. *Source* Zheng et al. [7]

happening on the blockchain. Ethereum is supposed to tackle use cases related to Bitcoin, with error avalanche, underlining and dictating the creation of coins on the network and arriving at a network-wide consensus. A spinoff of this capability is that Ethereum has become one of the most popular platforms for developing new blockchain-based applications (Fig. 3).

2.3 *Nodes*

Nodes can be described as the participants in the blockchain. All the nodes are connected on a peer-to-peer network. Every node has complete access to the blockchain and can thus verify any incoming transactions. The operational requirement of a full node is a copy of the blockchain and ample storage space. The basic requirement for a user to become a miner is to run a full node. There is one other type of node, which is known as a lightweight node that stores only the hashes of the blocks. It occupies significantly less space as it receives additional information from full nodes.

2.4 *Mining*

Mining is the method through which new transactions are added to the blockchain. Hence, mining can be described as the process through which a block of transactions can be created and added to the Ethereum blockchain. Ethereum currently uses a consensus mechanism known as proof of work (PoW) [8]. Mining is said to be the lifeblood of proof of work. Ethereum miners are generally computers running specific software which use time and high computation power to process transactions and produce blocks. Further, miners are greatly incentivized to compete with each other because finding a reward in the form of newly minted coins and transaction fees is guaranteed on finding the right solution. When compared to Bitcoin, Ethereum does not have a limit on the number of possible Ether that can be mined. Subsequent to the successful creation of the target block and its verification by the network, “miners start competing for the next block” [1].

2.5 *Proof of Work*

It is a type of consensus algorithm which is also used in Ethereum blockchain. Proof of work can be best described as the algorithm that has to be solved by miners so as to find an appropriate hash for the next block.

Hash rate denotes the rate of mining activity on the network. We can suitably adjust the difficulty of this algorithm if the hash rate increases. This algorithm is said to be innovative because not only does it allow unrestricted entry to anyone who is running a full node but it also accredits the network to frame a protocol that can be used to amend the blockchain. Furthermore, it has error avalanche handling capability, which means that it can positively handle setbacks and malicious users that disrupt the network (Fig. 4).

2.6 *Addresses and Wallet*

Asymmetric cryptography is a fundamental concept to blockchain. Any user can spawn a random digital signature in the form of a private key. This key can be used to spawn a public key. The user address can be spawned using the public key only. This is the address where the number of funds can be stored, and then, the user can use his private key to sign transactions from his address. The public key shall be used to check the origin of those transactions and get them verified by the network.

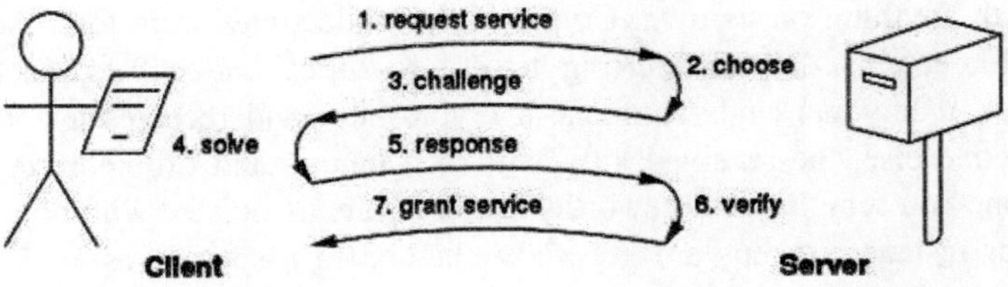

Fig. 4 PoW consensus. *Source* https://en.wikipedia.org/wiki/Proof_of_work

This means that access to funds would be lost if the private key is lost. Wallets help users to keep track of their financial assets and addresses. In the case of Ethereum, the wallet also can communicate internally with smart contracts and Dapp.

2.7 Transactions

"Messages that are signed by the address that has triggered them and can thus change the state of the network are known as transactions" [1, 9]. The fundamental concepts of the proprietary algorithm used are as follows:

Proof of authority: Unique signature of the owner of the private key.

Non-repudiation: The above is undeniable.

Unchangeability of transaction data: The transaction contents are unchangeable, and its integrity is not compromised.

The transaction structure on the Ethereum blockchain is the following.

Nonce: Numerical quantity of transactions that have been sent from an address.

Gas price: The financial cost to be levied on the initiator of the transaction for its execution.

Gas limit: The maximal gas permitted.

Recipient: The smart contract address or the payable address for the transaction.

Value: The financial weight in terms of Ether in the transaction.

Data: Quantity of binary input.

v, r, s: Worth of transaction signature.

2.8 Non-Fungible Tokens (NFTs)

Non-fungible tokens possess a specific digital signature that is contained within their smart contract. This identifying information is what makes each NFT different from another, and hence, an individual cannot swap them for another token. What this means is that they cannot be supplanted one for another, considering no two are undifferentiated. To put things in perspective, banknotes can be simply exchanged one for another if they possess the same value, and thus, it makes no difference to the holder. Non-fungible tokens are indivisible, in the same way as it is not possible to send someone part of a movie ticket. Part of a movie ticket is not redeemable and is not worth anything on its own. Crypto-kitties collectibles were the first famous non-fungible tokens. When referring to crypto-kitties collectibles, each cat is unique; i.e., if anyone sends someone a crypto-kitty and receives a crypto-kitty from someone else, the received kitty will be a unique and different crypto-kitty from the one you sent. Some noteworthy attributes are associated with non-fungible tokens. For instance, tokens are generally chained to a specific asset. Moreover, they can also prove the right of possession over items like vinyl's used in games and even the ownership of physical assets. Other tokens can be purported to possess

Fig. 5 Crypto-kitties collectibles. https://www.investopedia.com/news/cryptokitties-are-still-thing-heres-why

the characteristic of fungibility in the same way as any fiat currency is said to be fungible. Fungible tokens are different from NFTs as they are identical, and thus, they have the same denomination when exchanged.

NFTs are as follows:

(a) Rare (b) Unique (c) Indivisible (d) Transferable (Fig. 5).

2.9 Smart Contracts

For a layman, "smart contracts can be described as lines of code that are stored on a blockchain and are automatically executed when predetermined or predefined terms and conditions are met" [10]. Technically speaking, smart contracts were defined in 1994 by Nick Sczabo as "a computerized transaction protocol which can be used to execute the terms of any contract." As is evident from the description, business collaborations can demonstrate most of the smart contracts' upsides. "Smart contracts can be used to enforce some agreement so that all participants can be certain of the outcome without an intermediary's involvement in such business collaborations" [10]. Ethereum pioneered the use of smart contracts among blockchain, which led to the development of high-utility applications in various sectors, including the finance sector. "The code can only be deployed on the network after its high-level code is compiled to low-level code and is sent to a 0×0 address" [1]. Subsequently, "an address is allotted to it which can subsequently be used to utilize the contract's functionalities by calling its functions. Although the high-level

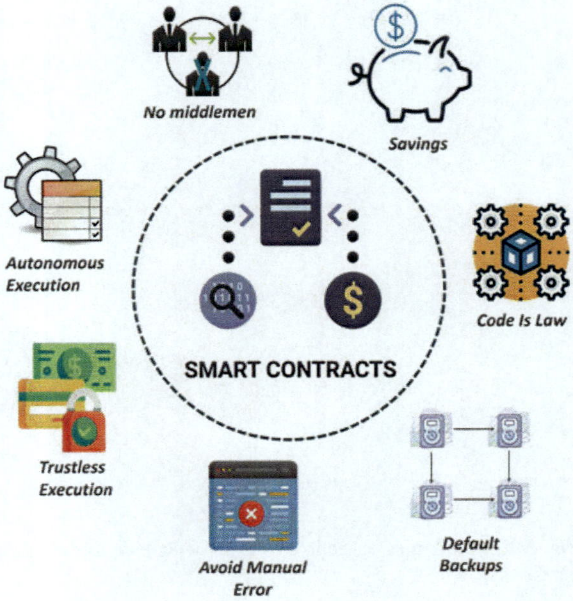

Fig. 6 Representation of smart contracts. *Source* https://www.edureka.co/blog/smart-contracts/

language of choice for writing Ethereum code in solidity, there are various other high-level languages which can be used for writing Ethereum code" [1] (Fig. 6).

3 Literature Review

After assessing a substantial number of research papers on the subject matter at hand, we learned the following points. Our learnings can be categorized under the following sub-headings.

3.1 Decentralized Finance (DeFi)

The YoY growth of the market for NFTs is expected to be above 60% in this FY. Adopting an iterative research approach to the use-case about ownership of NFTs, a novel solution that proposes to increase market penetration of NFTs has been reached. This solution will provide a platform for developing increasingly complex financial instruments, enabling NFT users to leverage and lease their valuable assets linked/stored in tokens [1].

Some proposed use cases include:

1. Leasing tokens to generate income for passive users.
2. Lending/leveraging tokens generate quick liquidity, thereby eliminating the overhead costs associated with selling the asset.

3. Creating financial tools on the lines of futures will help to stabilize the market.
4. Creating a legal framework for the tokenization of real estate assets.

3.2 Blockchain Development Platforms

There are multiple platforms that users can utilize for the development and implementation of smart contracts based on the level of advancement attained by the user and the specific use cases that the user is targeting. Truffle suite is best suited for developers who are just starting and are nascent to this technology owing to the limited capacity of truffle suite to implement multi-layered smart contracts. Developers who are looking for a robust compiler and development environment would be better suited to utilizing remix. An ideal combination would be using truffle suite for testing while remix would be utilized for compiling the smart contracts, and the deployment would be handled by Mist and Geth [10].

3.3 Recent Advancements and Future Trends in Smart Contracts

Smart contracts are fundamental to the optimal exploitation of the immense potential offered by blockchain and associated technologies. Due to this, a large amount of research is going to be done on smart contracts in the academic field, and the resultant interest generated in the industry will be immense. The trump card of smart contracts is their immutable and irreversible characteristic. These characteristics facilitate the exchange of money and other assets in a way which avoids the involvement of a third party. Smart contracts face certain challenges which include reentrancy vulnerability, transaction ordering dependence, time-stamp dependence, untrustworthy data feed, and privacy issues. The future of society is a transition into a cyber physical social system (CPSS). Blockchain and smart contracts will play an instrumental role in aiding this transition [5].

3.4 Trends on Crypto-Collectibles

The meteoric rise of crypto-collectibles on the blockchain network can be best understood by taking a glance at crypto-kitties. Crypto-kitties can be said to be the pioneer among collectibles in terms of raising financial interest. An initial amount of \$12 million followed by another \$15 million was successfully raised by VC method. Thus, it can be argued that crypto-games, just like crypto-Kitties, are financially viable. These games would serve to amplify the movement toward blockchain-based gaming. Coming back to crypto-kitties, the game works on the principle of ownership of tokens, which are very valuable due to a limited supply by algorithmic design, being proven by blockchain. This lends to value addition of collectibles. Moreover, cryptocurrencies also reflect a kind of "digital materiality" due to this design [9].

Hence, it can be concluded that the major factors which would determine the worth of crypto-game tokens can be refined to three most influential ones. The first factor can be said to be the limited technical infrastructure which impedes the capabilities of the network. Limited technical infrastructure can refer to the lack of scalability of the blockchain itself or the steep learning curve involved in learning how to operate crypto-wallet. The second limiting factor is the unequal playing field for blockchain users. By an unequal playing field, we are referring to the gas fees required to be paid by the users to cryptocurrency miners for the purpose of infrastructure maintenance which gives an unfair advantage to users with greater financial resources. Finally, the third limiting factor is the unclear or unvalidated legal ownership of tokens due to the anonymity accorded to the users by design on a blockchain. Any attempt to validate or prove legal ownership of tokens can only be done after the user de-anonymizes himself which would then defeat the purpose of cryptocurrencies based on a blockchain network.

3.5 Blockchain Digital Art

One of the major use cases of blockchain technology lies in the creative industry. The creative industry relies heavily on the revenue generated by the trio of rights management, licensing, and data management of digitized/digitizable products. Blockchain technology could be leveraged to manage the aforementioned in a manner which is automated and decentralized. Block technology will increase the power in the hands of the original content creators by offering a shared data layer on the application level rather than on the data level which is offered by existing Internet-based solutions. A business model which is based on token issuance and management by the creator can be built on the framework provided by blockchain technology [6].

4 Research Background and Methodology

4.1 Decentralized Applications (DApps)

Dapps were created to operate as applications compatible with smart contracts enabled, which had the capacity to work on distributed ledgers. Users have the capability to interact with a wide range of Dapps on Ethereum.

4.2 How Do You Make a Token?

The definition of tokens is contained within smart contracts. A token smart contract, once deployed, keeps track of the number of tokens owned by any address. Addresses also have the capability to transfer tokens attributed to them to other addresses. The smart contract keeps track of the number of tokens owned by each individual and not the underlying blockchain itself in contrast to ETH [11]. Hence, we would need to query the smart contract in order to find out how many tokens an address has. This is a distinction that everyone must recognize. When one is querying how many ETH an address has, one is querying the blockchain. When user query the amount of ETH (Ethreum currency) linked to an address, blockchain is queried. In contrast, when user query the amount (number) of tokens linked to an address,smart contract ae queried. Hence, to know how many ETH and tokens a wallet has, an individual needs to know all the addresses of where the token smart contracts are deployed on the blockchain [1].

4.3 ERC-721

ERC-721 has been established as the basest standard that a smart contract will have to adhere to so that it is allowed to own and trade unique tokens. ERC-721 has not decreed a caliber for token metadata and neither does it restrict adding supplemental functions. Thus, ERC-721 can be described as a protocol that is subject to free peer assessment, which advises how to build NFTs on the Ethereum decentralized network of nodes. A whole lot of tokens are fungible (fungibility implies that every token is exactly the same to and changeable with any other token), ERC-721 tokens have to be all one of a kind. ERC-721 tokens are non-fungible in nature. "Their lack of fungibility connotes that each token has a set of characteristics and standards associated with it which are exclusionary to it" [1]. The "uniqueness of such tokens makes ownership more desirable especially in the case of collectibles and other such highly sought-after tokens" [1].

4.4 Open Zeppelin

Open Zeppelin is a library that was created for the purpose of safe contract development. It is built on a solid foundation of community-vetted code. It enables the implementation of standards like ERC20 and ERC721. It boasts of a flexible role-based per missioning scheme. Further, components of solidity, which are used to build custom contracts and complex decentralized systems, are reusable, which enhances its utility. Moreover, top-grade interoperability with the gas station network for systems with no gas fees also contributes to its varying uses. It has been audited by leading security firms and is, thus, perfectly safe.

4.5 Truffle Suite

"Truffle is a development environment, testing framework, and asset pipeline for Ethereum, which aims to make the life of an Ethereum developer much easier" [12]. Truffle offers the following advantages to an Ethereum developer:

Deployment and binary management linked using built-in smart contract compilation.

Computer robotized testing of contract.

Custom build processes supported by configurable build pipeline.

Scriptable deployment and migrations framework.

Public and private networks deployed using network management.

Scripts executed within a truffle environment by an external script runner.

4.6 Web3.js

Web3.js enables an Ethereum developer to fulfill his/her second responsibility, i.e., creating modules that communicate with the Ethereum decentralized network. Web3.js is basically an amalgamation of non-volatile computer code that permits the developer to perform actions like the transfer of Ether and modify smart contracts. The diagram demonstrates how a client does tête-à-tête with Ethereum.

Web3.js talks to the Ethereum blockchain using a JSON RPC, which stands for the "Remote Procedure Call" protocol. As Ethereum is based on a peer-to-peer network of nodes that needs to save an additional version of important information on the blockchain, Web3.js allows a developer or a user to make requests to an individual Ethereum node with JSON RPC in order to modify or access the data in the network. An appropriate would be the cohesion of jQuery with a JSON API to modify a web server.

4.7 Metamask

"MetaMask is a browser plugin that also serves as an Ethereum wallet and is installed like any regular plugin. It allows users to store Ether and other ERC-20 tokens and thus enables them to consummate transactions with any Ethereum address" [13].

4.8 Ganache

Ganache is used for setting up an individual Ethereum decentralized network of nodes for validating solidity contracts. Moreover, it provides more functionalities when compared to other similar software.

5 Design and Implementation Setup

See Figs. 7, 8.

A. **Installing the Following Dependencies**:

```
{
``name'': ``blockchain-game'',
``version'': ``0.1.0'',
``description'': ``token collection game'',
``author'': ``akash and kanisk'',
``dependencies'': {
``@openzeppelin/contracts'': ``^2.3.0'',
``babel-polyfill'': ``6.26.0'',
``babel-preset-env'': ``1.7.0'',
``babel-preset-es2015'': ``6.24.1'',
```

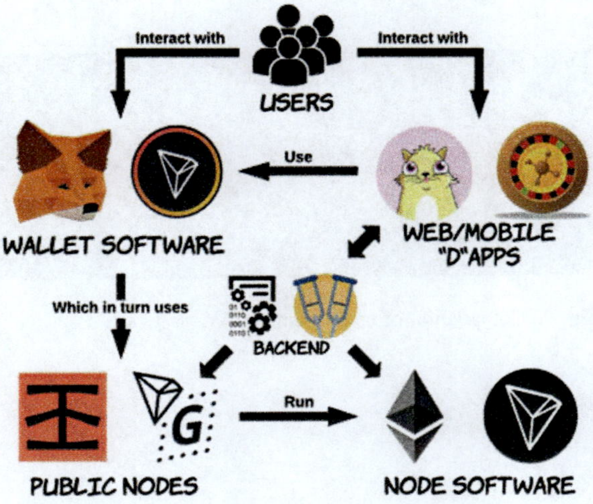

Fig. 7 Design workflow: https://medium.com/free-code-camp/how-to-design-a-secure-backend-for-your-decentralized-application-9541b5d8bddb

Fig. 8 Setting up environment

```
``babel-preset-stage-2'': ``6.24.1'',
``babel-preset-stage-3'': ``6.24.1'',
``babel-register'': ``6.26.0'',
``bootstrap'': ``4.3.1'',
``chai'': ``4.2.0'',
``chai-as-promised'': ``7.1.1'',
``chai-bignumber'': ``3.0.0'',
``react'': ``16.8.4'',
``react-bootstrap'': ``1.0.0-beta.5'',
``react-dom'': ``16.8.4'',
``react-particles-js'': ``^3.4.1'',
``react-scripts'': ``2.1.3'',
``truffle'': ``5.0.5'',
``truffle-flattener'': ``^1.4.2'',
``web3'': ``1.0.0-beta.55''
n}
```

See Figs. 9, 10.

6 Application. Conclusions

See Figs. 11, 12, 13, 14, 15, 16, 17.

Fig. 9 Compilation of smart contract

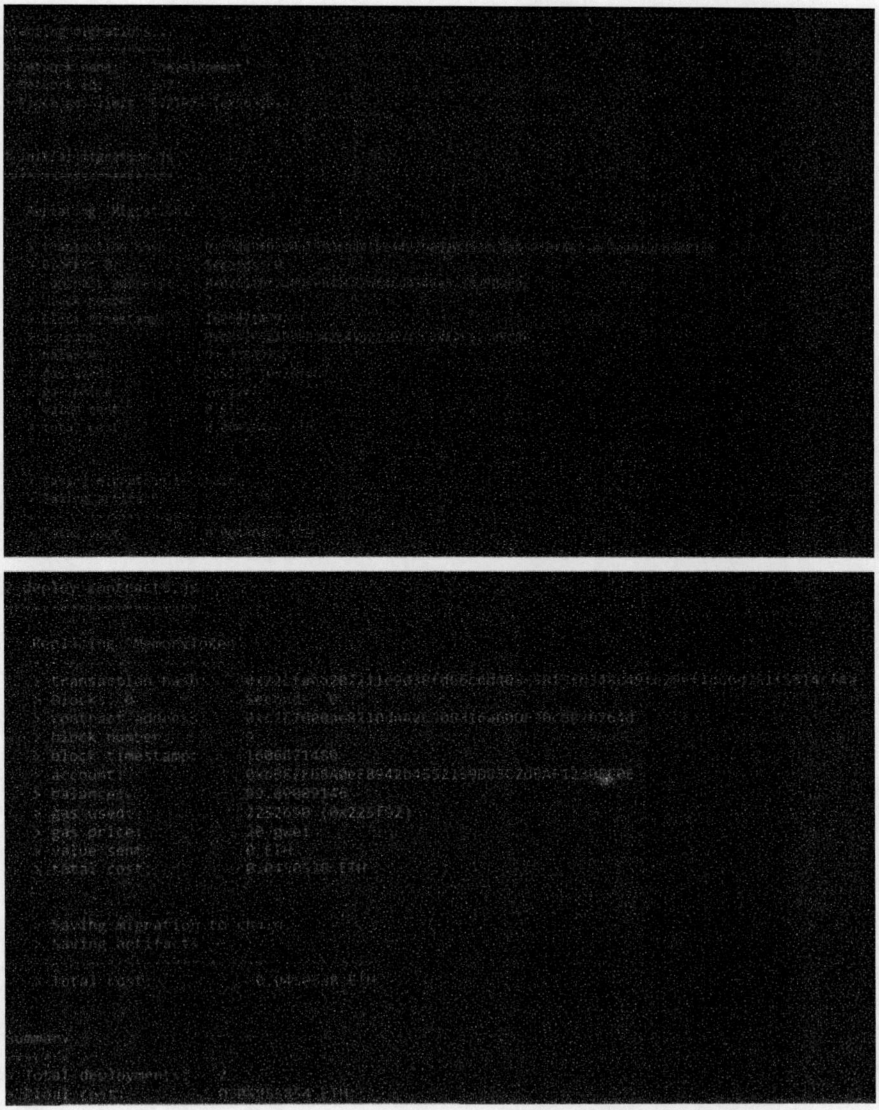

Fig. 10 Deploying the smart contract

Fig. 11 Smart contract address

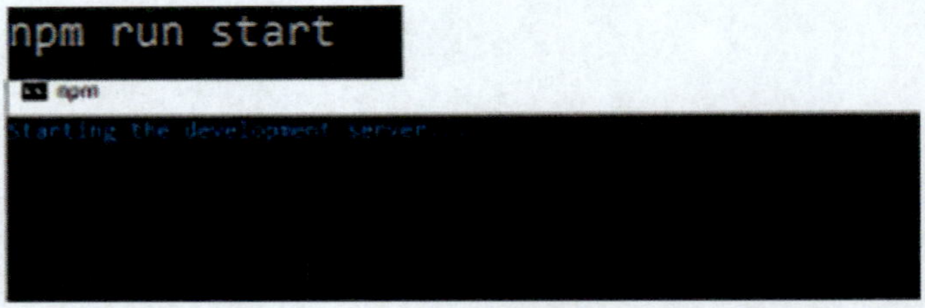

Fig. 12 For starting the server

Fig. 13 Application home page

Fig. 14 Flow 1: tiles did not match (unsuccessful attempt)

Fig. 15 Flow 2: tiles match (successful attempt)

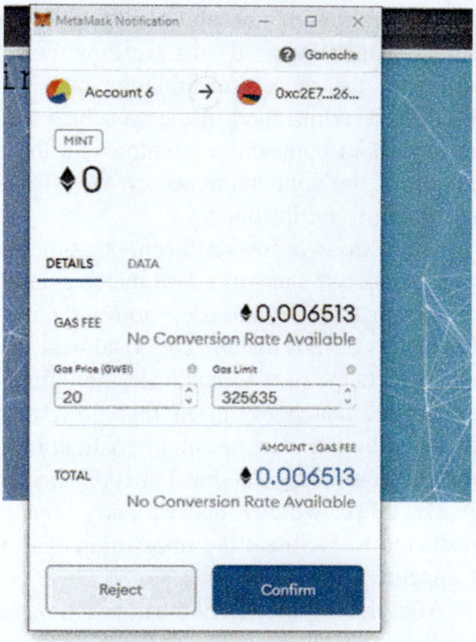

Fig. 16 Transaction notification by MetaMask on successful matching and collection of tokens initiated

Fig. 17 Token collected

7 Results

The application environment is initially set up by installing the required dependencies. Consequent to the successful installation of the required dependencies, we then proceed to initiate the Ethereum node on the blockchain network using Ganache. Pursuant to the successful initiation of the Ethereum node, we then move on to smart contracts. The custom-tailored smart contract, written in solidity, has to be compiled by using truffle suite. Upon the successful compilation of the smart contract, we then deploy the smart contract. Once the smart contract is successfully deployed, a unique address is provided to the smart contract using which all further transactions can be accounted for and tracked [14]. This completes the back-end setup of our Ethereum decentralized application (Dapp).

The front-end part of the application is initiated by starting the application server, which is based on a React.js framework. It should be noted that a prerequisite for the successful initiation of the application server is that the plugin for MetaMask is installed in the web browser. The absence of which will lead to an unsuccessful connection attempt with the back-end server. On the successful initiation of the application server, we will see an image tile-matching game hosted successfully on localhost.

There are two flows of events that may take place henceforth. The first flow can be labeled as the unsuccessful match event. In this flow, as two similar-looking tiles are not matched by the user, no transaction is initiated by the smart contract as the token IDs do not match. The local host displays an unsuccessful attempt message and directs the user to make another attempt. The second flow can be labeled as a successful match event. In this flow, as two similar-looking tiles are correctly matched by the user, the smart contract initiates the transaction as the token IDs are matched correctly. The local host displays a successful attempt message to the user. A MetaMask window subsequently pops up to inform the user of the gas charge deducted to facilitate the transaction. The transaction details can be tracked using Ganache.

After the completion of the transaction, the token is collected and displayed on the application window. Hence, we have successfully gained a collectible NFT. By gaining this collectible NFT, the user can establish the provenance of the collectible as that collectible now possesses the uniqueness and rareness of the NFT it is linked to.

8 Conclusion and Future Scope

To conclude, we can state that the transaction of non-fungible tokens using deployable smart contracts has immense commercial and academic implications. The ERC-721 standards have given us immense power to accomplish next-generation advancements in fields as varied as art, culture, online gaming, rare collectibles, and more. This whole process hence confirms the non-fungibility of the tokens created on the ERC-721 standard and their interaction with the smart contract as shown with the help of an Ethereum Dapp.

In the future, this work can be extended to the domains of authentication and human user verification so as to prevent misuse of cyber facilities and public data by automated scripts. These non-fungible tokens can also be used to validate the authenticity and pedigree of high value, rare, vintage collectibles, which could otherwise be faked so as to commit fraud. This will give the power of proof of originality/ownership to the owner or creator in the digital world. Hence, advanced applications can be made along similar lines of thoughts, which would give us a deeper insight into this domain and hence help us in verifying the antecedents of the product or any other item of value.

References

1. Musan, D.I.: NFT . finance leveraging non-fungible tokens. Imperial College London, Department of Computing (2020)
2. Kumar, A., Kumar, S.: A systematic review of the research on disruptive technology—blockchain. In: 2020 5th International Conference on Communication and Electronics Systems (ICCES), Coimbatore, India, pp. 900–905 (2020). https://doi.org/10.1109/ICCES48766.2020.9138055
3. Kumar, A., Kumar, S.: Implementation of decentralized electronic polling system using ethereum blockchain. Int. J. Adv. Sci. Technol. **29**(4), 10717–10728. (SCOPUS Indexed) ISSN: 2005–4238 (2020)
4. Uribe, D., Waters, G.A.: Privacy laws, genomic data and non-fungible tokens. J. Br. Blockchain Assoc. **3**(2) (2020) [Online]. Available: https://jbba.scholasticahq.com/article/13164-privacy-laws-genomic-data-and-non-fungible-tokens
5. Wang, S., Yuan, Y., Wang, X., Li, J., Qin, R., Wang, F.Y.: An overview of smart contract: architecture, applications, and future trends. IEEE Intell. Veh. Symp. Proc. 2018-June, no. Iv, pp. 108–113 (2018). https://doi.org/10.1109/IVS.2018.8500488
6. Chevet, S.: Blockchain technology and non-fungible tokens: reshaping value chains in creative industries. Sylve CHEVET Under the supervision of Alain BUSSON, pp. 1–73 (2017)
7. Zheng, P., Zheng, Z., Wu, J., Dai, H.: XBlock-ETH: Extracting and exploring blockchain data from ethereum. IEEE Open Journal of the ComputerSociety, pp. 1–1 (2020). https://doi.org/10.1109/OJCS.2020.2990458
8. Buterin, V.: A next-generation smart contract and decentralized application platform. Etherum, no. January, pp. 1–36 (2014) [Online]. Available: http://buyxpr.com/build/pdfs/EthereumWhitePaper.pdf
9. Bal, M., Ner, C. arXiv: NFTracer: A Non-Fungible Token Tracking Proof-of-Concept Using Hyperledger Fabric, pp. 1–9 (2019)
10. Chirtoaca, D., Ellul, J., Azzopardi, G.: A framework for creating deployable smart contracts for non-fungible tokens on the ethereum blockchain. In: 2020 IEEE International Conference on Decentralized Applications and Infrastructures (DAPPS), Oxford, United Kingdom, pp. 100–105 (2020). https://doi.org/10.1109/DAPPS49028.2020.00012
11. Serada, A., Sihvonen, T., Harviainen, J.T.: CryptoKitties and the New Ludic Economy: How Blockchain Introduces Value, Ownership, and Scarcity in Digital Gaming. Games Cult. (2020). https://doi.org/10.1177/1555412019898305

12. Anilkumar, V., Joji, J.A., Afzal, A., Sheik, R.: Blockchain simulation and development platforms: survey, issues, and challenges. 2019 International Conference on Intelligent Computing and Control Systems ICCS 2019, no. Iciccs, pp. 935–939 (2019). https://doi.org/10.1109/ICCS45141.2019.9065421
13. Dagher, G.G., Marella, P.B., Milojkovic, M., Mohler, J.: Bron covote: secure voting system using ethereum's blockchain. ICISSP 2018—Proceedings of 4th International Conference on Information System Security and Privacy, vol. 2018-Janua, pp. 96–107 (2018). https://doi.org/10.5220/0006609700960107
14. Mofokeng, N.E.M., Matima, T.K.: Future tourism trends: Utilizing non-fungible tokens to aid wildlife conservation. Afr. J. Hosp. Tour. Leis. **7**(4) (2018)

Blockchain in Health Care: A Review

Sanya Bindlish, Sargam Chhabra, Kshitij Mehta, and Pooja Sapra

Abstract Blockchain is taken to be a system of ledger which handles data and manages their transactions with the help of their time stamps through cryptographic hashing and serves in a decentralized technique over the computer networks. Despite the fact that blockchain concept is used on a large scale for cryptocurrency, this paper explores some potential applications of the blockchain concept in the healthcare industry as well. It discusses the use of blockchain technology in the healthcare sector and how it can help in overcoming the present challenges like interoperability, security, and cost of maintenance in the traditional healthcare system. This paper highlights the applications of blockchain in detail and why it is hard to implement in the healthcare sector.

Keywords Blockchain · Internet of things · Security · Health care

1 Introduction

Blockchain [1] is a type of database that consists of information in the form of blocks connected to each other using cryptographic methods, with each block containing the hash of the previous block, the records, and their time stamps. These blocks are consistently growing in number with the present value at 672,425 (recorded at 05:00 PM February 27, 2021).

S. Bindlish (✉) · S. Chhabra · K. Mehta · P. Sapra
The NorthCap University, Gurugram, India
e-mail: sanyabindlish@gmail.com

S. Chhabra
e-mail: sargamchhabra15@gmail.com

K. Mehta
e-mail: kshitij.mehta28@gmail.com

P. Sapra
e-mail: poojasapra@ncuindia.edu

K. Khanna et al. (eds.), *Cyber Security and Digital Forensics*,
Lecture Notes on Data Engineering and Communications Technologies 73,
https://doi.org/10.1007/978-981-16-3961-6_35

The features of blockchain are:

1.1 ***Immutability***: Blockchain does not allow any modification or deletion of the data once it has been recorded. This is because each block contains the hash of the previous one and if we modify the data then the hash of every following block will have to be altered.

1.2 ***Decentralization***: Blockchain does not have a central entity that manages or controls the system. Instead, it works on the concept of a distributed network in which data is shared with all the computers on the network. This is called a ledger. Each time a new block is added, all these computers display the addition as every connection has the same copy of the ledger.

1.3 ***Transparency***: This feature of blockchain allows anyone to join the network and validate the transactions of their peers so they can be added to a block. A history of all the transactions is also available.

1.4 ***Trust and Consensus***: The blockchain system is established on trust as data is secured with the help of cryptographic methods. Consensus is a way of validating transactions on a distributed ledger by establishing an agreement between the peers. It eliminates the need of a central authority.

In this paper, Sect. 2 discusses the literature available for the applications of blockchain in health care, and Sect. 3 focusses on challenges in the healthcare sector and various domains where blockchain can be implemented. Section 4 concludes the paper.

2 Literature Survey

In [2], authors have presented a structured literature survey of research on blockchain applications such in the healthcare area. This analysis includes 42 article write-ups presenting the most up-to-date knowledge on present-day implications and loopholes in the blockchain execution for improvement in the healthcare sector. The findings in the sheet show that blockchain is being utilized to build advanced and up-to-the-minute interventions to upgrade the universal standards of handling, processing, and sharing of medical information/data and health records of the patients. This technology is going through an evolution in the healthcare sector where it has added a positive impact through improved efficiency, technological advancements, access control, security of data handling, privacy protection, etc. The sheet also presents some discoveries that suggest that some limitations primarily prevail to model the performance as well as have some effect on the cost of execution.

This research paper [3] addresses the potential areas where the future researchers or experimenters could contribute some value in the areas like system architecture, data security, etc. Concluding the research, the findings suggest that the researches in the future will contribute in the deployment of this technology on a larger scale and also in the critical issues related to legal consents, security measures, medical diagnostics, and improving patient care in some cases like remote monitoring or emergency situations.

In this research paper [4], the researchers have surveyed a lot of interesting implementations of blockchain technology. Apart from the financial sector, a lot of sectors have been implementing the blockchain concept where its beneficial features add innovations and solutions to some compromised areas. These sectors which are mentioned in this paper are—health care, electronic voting, etc. For each of these topics, they have given a detailed study including the problems and solutions given by the blockchain concept.

In this research paper [5], authors reviewed important use cases of blockchain in the healthcare industry: patient record management, research, and billing claims management alongside the related projects. The authors have also discussed the challenges in the implementation of blockchain technology in the healthcare industry. Most of the blockchain projects and concepts are still in the form of prototypes and white papers. However, the use of blockchain technology is expanding.

In this research paper [6], authors gave an overview about the application of blockchain in the healthcare sector and how it helps in automation of healthcare services. The authors have not only focussed on the different applications of the technology but also identified the limitations of existing research and surveys. The authors conclude with stating the need of research for better understanding and development of this technology.

This research paper [7] gives us knowledge of how blockchain is gaining popularity in the market due to its concept of digitally shared ledger and consensus algo removing all the risks and threats of the mediators and intercessors. It also shows that the usage of blockchain has now extended to all the major sectors of research including banking, governance, IoT, health care, education, etc. The study reviews the existing literature about this field so as to list the major issues in health care and inspect the features of blockchain that could rectify the identified issues. However, a few challenges and limitations of this technology still need more research.

In this research paper [8], authors gave an overview of benefits and drawbacks related to blockchain technology. This paper also discusses the existing challenges related to crypto-currencies. Authors have discussed the advantage of using blockchain in the healthcare industry. There are various potential challenges in implementing blockchain in the healthcare sector; authors have also provided solutions to some of the challenges. Blockchain can help in the advancement of the medical sector in various ways and authors expect many more applications of blockchain in the near future.

In [9], the authors proposed a medical data management system using blockchain technology, which is much more secure than the traditional model and also allows patients ownership over their records while granting hospitals a temporary and easy access to the data. This patient-centered blockchain model is based on Ethereum and it stores the actual medical records on a decentralized cloud system, Ethereum Swarm. Each medical record having its unique swarm hash is combined with the decryption key to form the root chunk and only the authorized personnel have access to the content in the root chunk.

In this research paper [10], authors have presented a basic introduction of blockchain technology and its structure. The authors have discussed the implementation of blockchain in the healthcare sector. Blockchain is an emerging technology but has some major challenges including usability and authentication of patients. The paper also briefly mentions the impact of blockchain in the near future. Future of blockchain in the healthcare sector is promising and will authorize the patients to take control of their medical records so that they can keep track of their health condition.

This research paper [11] discusses the management of confidential medical information of patients, sharing of medical records, image sharing, and log management. The authors have also reviewed studies that intersect with other areas and have performed SWOT analysis on the basis of these to analyze their positive and negative aspects. The authors provide guidance to future researchers by assessing the benefits and limitations of the medical concepts of blockchain. The paper concludes by summarizing the techniques used in health care, their applications, and pros and cons.

3 Blockchain and Health care

With the concept of blockchain becoming increasingly popular, it is believed that this technology will also evolve the healthcare sector in the near future. It can grant a new framework for health information exchanges (HIE) by listing medical records of patients more systematically keeping privacy, security, and interoperability in check.

3.1 Blockchain Versus Current Healthcare Challenges

This section provides an insight on how blockchain technology can help in overcoming some of the existing challenges in the healthcare field. Some of these are:

(i) ***Interoperability***: It refers to the characteristic of health information systems to operate together with other such organizations to deliver quality healthcare services to people more efficiently.
Blockchain overcomes this challenge by retrieving data with the help of APIs, which helps in achieving data format standardization.

(ii) ***Security and Integrity***: It refers to the set of guidelines and standards to protect the confidentiality and integrity of personal health information. Data tampering and breach is a matter of concern in the healthcare sector.
Blockchain provides security as it is immutable and data is stored in encrypted form so it is not possible to modify information.

(iii) ***Access Rights***: Managing health records across the organization is a challenge where blockchain technology can come in use.
It makes sure that records are updated and available at every node. Also, access rights can be defined using the concept of smart contracts.

(iv) ***Cost of Maintenance***: Maintenance of healthcare systems includes various actions such as backup and recovery mechanisms.

In the case of blockchain, there is an open-distributed network where data is shared with all the nodes in the network. Therefore, failure of one node does not affect the other nodes. This reduces the need of a backup mechanism.

3.2 Applications of Blockchain in Health care

In this section, several important use cases and applications of blockchain in health care are discussed below:

(i) ***Blockchain for Electronic Health Record (EHR)***:
Ideally, it is required that EHR should maintain a lifetime medical record which is accessible to the entire medical staff. In the present application of EHR, patient information is stored in multiple organizations throughout his lifetime. The physician conducts the diagnosis, specifies the problem, medication, and this record remains with him even after the patient is treated.
In the blockchain implementation of EHR, only the individual will have access to his medical record and doctors will be granted access through smart contracts when required.
At present, interoperability and data fragmentation is a huge challenge among different organizations as they are reluctant to share data with each other. Using blockchain, the owner can set viewership permissions on the data shared with another entity. This is supported by smart contracts which are self-executing conditional contracts in which the agreement is coded. New records can be updated regularly on the patient's profile and he can decide which entity has access to which content, hence providing authorization.

(ii) ***Blockchain in Clinical Research***:
This also centralizes around the issue that an individual's medical data is scattered across the databases and records of multiple organizations which causes hindrance in sharing of data. Not only does it leads to inefficient treatment of patients, but also causes a setback in the domain of clinical research.

Clinical trials require real-time medical information of patients to conduct a sound hypothesis and research which is backed up by a set of realistic data. Gathering all these records from the scratch can be a very hard, time consuming, and costly process. Surveys show that people are ready to share their medical records for research provided that their data is not compromised.
Blockchain supports this due to its features of transparency and trust. The data is public and anyone can write to it but no one has the permission to store it because of encryption. Patients can manage their data and also have the rights to deny access to it anytime. This framework would be beneficial for research entities to get hold of secure and accurate data.

(iii) ***Blockchain in Drug Supply Chain Management*:**
Transparency and security of the supply chain are major challenges in the pharmaceutical sector. These services of this sector directly impact the lives of their consumers which is why safety and efficiency of the product are necessary. The product reaches the consumer after passing multiple stages: manufacturing, transportation, handling, storage, redistribution, and retail. Anything can go wrong in these stages from a simple human error to a malicious fraud as in the traditional system, identifying the cause of the issue can be very difficult. This is because records are maintained on a personal level and can be tampered with very easily. Such an unorganized set of data is not available to the company itself, let alone the government and consumers.
Blockchain solves this problem by providing a distributed ledger shared among all concerned members in the supply chain. The records stored in the form of blocks are immutable, permanent, and transparent. This reduces the possibility of errors and frauds significantly and also provides organized data visible even to the consumers.

(iv) ***Blockchain in Claim and Billing Management*:**
Fraudulent claims and billing is one of the issues in the healthcare sector that needs to be eliminated. Some of the challenges faced in the claim and billing process are overcharging, performing unnecessary medical tests, and claiming charge for non-performed services. There are a lot of third parties involved in resolving the claim process, to ensure faster processing of claims and cost reduction for suppliers and clients.
Blockchain technology can help in automation of claim processing with the help of a smart contract between the involved parties. Also, the challenges faced in this process can be reduced to a great extent because of the transparent and distributed network in blockchain.

(v) ***Blockchain Application in Health Information Exchanges (HIE)*:**
HIE's goal is to offer secure delivery of healthcare data beyond geographical boundaries. To achieve this goal, data privacy and security should be given priority.

Security: Securing patient data is important and blockchain can be useful in doing so. On a blockchain, data is in encrypted form and can be accessed by using the shared public key of a patient. Use of private and public keys defines access rights and ensures data privacy.

Decentralization: Centralized network is a traditional setup for HIE in which data from patients is stored in a central storage system where failing of the central node will lead to loss of medical records. Blockchain is an open-distributed ledger where each node will have the updated records and patients can define the access rights of their medical records, thereby scaling the infrastructure to a decentralized network from a traditional centralized network.

Interoperability: It refers to the characteristic of health information systems to operate together with other such organizations to deliver quality healthcare services to people more efficiently.

Blockchain offers access to patient's data with regular updates across all healthcare organizations. Itss distributed-ledger feature removes the chances of duplicity as the same copy of data is available to all the organizations.

3.3 Blockchain Technology—Challenges in Health Care

Applying blockchain in the domain of health care has numerous advantages but the practical implementation of this technology is not as easy as it seems. This portion of the paper will discuss the challenges which blockchain technology has yet to overcome.

(i) Blockchain has already been widely accepted in the field of banking but the awareness regarding this technology in the healthcare sector is yet to come by. Organizations are not aware of this technology and even if they are, the crucial challenge is understanding how it works and the benefits of adopting this technology.
(ii) Organizations are comfortable with working in a centralized environment and do not prefer to shift to a decentralized system as it incurs cost for ensuring speed and efficiency.
(iii) Another reason why healthcare entities are resistant to change is that shifting from their current system would require a lot of time and effort.

4 Conclusion

Blockchain is an emerging technology and has the capability to change the traditional business model to a brand-new advanced framework. Some organizations have started to adopt this technology in the healthcare sector, and it is expected to take over at least 55% of the global healthcare system by the year 2025. Blockchain technology offers promising solutions to the healthcare industry, and at this rate, it will soon become an integral part of the data management and supply chain systems. There are various advantages of this technology in the healthcare sector; however, there are some challenges related to the implementation of blockchain such as lack of awareness regarding the working and implementation of technology that have yet to be tackled. Another challenge in implementation of blockchain in the healthcare sector is the amount of time and effort that will be required in shifting from the current centralized system to the decentralized system. This paper presents a review of numerous studies that deals with the use of blockchain technology in the healthcare sector.

On the basis of all the studies reviewed, it is evident that blockchain technology has a lot of scope in the future to develop and take over most of the traditional healthcare models. With this advancement in the health care, patients will have the assurance that their private data will not fall in the wrong hands and be misused.

References

1. Swan, M.: Blockchain: "Blueprint for a New Economy"; O'Reilly Media, Inc., Sebastopol, CA, USA (2015)
2. Nakamoto, S.: Bitcoin: "A Peer-to-Peer Electronic Cash System" (2008). Available online: www.bitcoin.org (accessed on 12 March 2019)
3. Hölbl, M., Kompara, M., Kamišali´c, A., Zlatolas, L.N.: A systematic review of the use of blockchain in healthcare. Symmetry **10**, 470 (2018)
4. Patel, V.: A framework for secure and decentralized sharing of medical imaging data via blockchain consensus. Health Inform. J. (2018)
5. Katuwal, G.J., Pandey, S., Hennessey, M.: Bishal Lamichhane: Applications of Blockchain in Healthcare: Current Landscape & Challenges (2018)
6. Fekih, R.B., Lahami, M.: Application of blockchain technology in healthcare: a comprehensive study. In: ICOST 2020: The Impact of Digital Technologies on Public Health in Developed and Developing Countries, 23, June 2020 (2020)
7. Yaqoob, S., Khan, M.M., Talib, R., Butt, A.D., Saleem, S., Arif, F., Nadeem, A.: Use of blockchain in healthcare: a systematic literature review. Int. J. Adv. Comput. Sci. Appl. (IJACSA), **10**(5) (2019)
8. Kuo, T.-T., Kim, H.-E., Ohno-Machado, L.: Blockchain distributed ledger technologies for biomedical and health care applications. J. Am. Med. Inform. Assoc. **24**(6), November 2017 (2017)
9. Chen, H.S., Jarrell, J.T., Carpenter, K.A., Cohen, D.S., Huang, X.: Blockchain in healthcare: a patient-centered model. Biomed. J. Sci. Tech. Res. (2019)
10. Revanna, N.: Blockchain in healthcare. Int. Res. J. Eng. Technol. (IRJET) **07**(04) (2020)
11. De Aguiar, E.J., Faiçal, B.S., Krishnamachari, B., Ueyama, J.: A survey of blockchain-based strategies for healthcare. ACM Comput. Surv. **53**(2), Article 27 (2020)

Section–C

A Comparative Study of the Energy-Efficient Advanced LEACH (ADV-LEACH1) Clustering Protocols in Heterogeneous and Homogeneous Wireless Sensor Networks

Nitin Kumar, Vinod Kumar, and Pawan Kumar Verma

Abstract Wireless sensor network (WSN) is a promising technology for monitoring the physical world. The energy limitation of WSNs makes it an energy saver technology. A number of diverse routing protocols can be used to enhance the network lifetime of WSNs. There exist two types of clustering-based approaches, namely homogeneous and heterogeneous. In the former case, all nodes have the same technical characteristics (bandwidth, processor, initial energy, etc.). In contrast, in the latter case, nodes have different technical characteristics, i.e., some of the nodes have higher ability than others in terms of the parameters listed above. In this paper, we analyze the ADV-LEACH1 algorithm for the homogenous WSNs. The mathematical modeling and simulation results achieved by MATLAB-2017b show a comparative analysis of heterogeneous and homogeneous WSNs in terms of the energy consumption, alive/dead nodes, and network lifetime. The ADV-LEACH1 with heterogeneous network performs better than that with the homogenous network because of advanced node presence in the network with higher energy level.

Keywords Node deployment · Heterogeneous · Homogeneous · WSN

N. Kumar · V. Kumar (✉)
Department of Electronics and Communication Engineering, SRM Institute of Science and Technology, Ghaziabad, UP, India
e-mail: vkchaudhary.rs.ece@iitbhu.ac.in

P. K. Verma
Department of Electronics and Communication Engineering, Dr. B. R. Ambedkar National Institute of Technology, Jalandhar, PB, India
e-mail: vermapk@nitj.ac.in

K. Khanna et al. (eds.), *Cyber Security and Digital Forensics*,
Lecture Notes on Data Engineering and Communications Technologies 73,
https://doi.org/10.1007/978-981-16-3961-6_36

1 Introduction

The productive enhancement of WSNs has grown into an evolving subject matter of concern in communication. WSNs have erratically dispersed sensors that convey the data to base station (BS) nodes over the network wherever information can be retrieved and analyzed, as shown in Fig. 1. A BS is essentially considered the same as an interface to interrelate among an operator and a system. Utilizing radio signals, a sensor node (SN) can interact with another node. Afterward, arranging nodes above, sensors generate an appropriate infrastructure with multihop as well as single-hop communication. It also consists of a shortage of power because of the slight handling speed and storage capacity. To confirm a long-lasting network lifetime, lengthy detachment communication must be remained separately [1, 2]. Wireless Sensor Networks are different from Ad-hoc networks such as Vehicular Ad-hoc Network (VANET) [3] and Mobile Ad-hoc Network (MANET) [4] based on the mobility models, node mobility, higher node density, power consumption and network lifetime.

A clustering method is organizing the sensor nodes so that advantageous energy effectiveness and productivity are attained. Nodes in a cluster can convey that information to the cluster-head (CH) intended for the BS as exhibited. CH assists in exchanging data amid BS and SN [5].

Clustering can be categorized as homogeneous as well as heterogeneous. The SNs in homogeneous WSNs have equal energy, memory space, handling ability, etc. On the other hand, SNs in heterogeneous WSNs are different in one or additional characteristics like initial energy, memory space, handling ability, etc. [6].

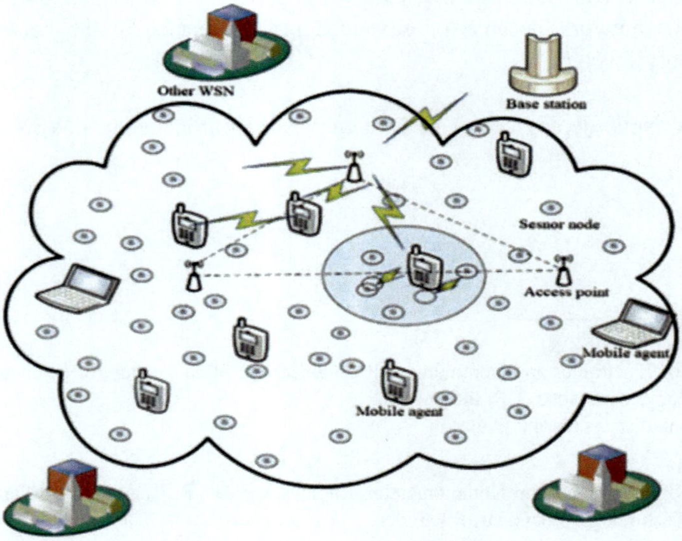

Fig. 1 A wireless sensor network

This paper proposes and analyzes the homogenous and two-tier heterogeneous network model as per the ADV-LEACH1 [7]. In this paper, we implement and analyze the clustering algorithm of ADV-LEACH1 [7]. This algorithm is a threshold-based cluster algorithm and is implemented on the heterogonous wireless sensor networks (HWSNs). We propose the same clustering algorithm on homogenous wireless networks and analyze the model performance. Then, we evaluate the network performance as per ADV-LEACH1 for our proposed model by considering the homogeneous network model. Therefore, the homogenous ADV-LEACH1 (HOM-ADV-LEACH1) assumes all SNs in this model having the same amount of energy. As per ADV-LEACH1 [7], the protocol is present with two-tier heterogeneous WSNs.

The rest of the work is structured as follows. Section 1 presents the layout and basic information about WSNs. In Sect. 2, literature of clustering algorithm is mentioned. Section 3 gives the clustering algorithm parameter for heterogeneous and homogeneous WSNs model. Section 4 discusses the WSN model and radio energy dissipation model. The experimental results of the proposed layout are given in Sect. 5 and finally, conclusion in Sect. 6.

2 Literature Review

In general, the power supply of SNs in WSNs is provided by a battery that is not rechargeable or replaceable. Improving energy efficiency and optimizing the lifetime of WSNs are key challenges. In recent years, various protocols have been deliberately centered explicitly on energy use to prolong the network's existence. In this context, the low-energy adaptive clustering hierarchy (LEACH) approach reduces energy consumption to extend network lifetime [8]. It chooses few SNs as CH is dynamically determined based on the remaining energy by every round in the WSN. The CHs fetch data from the SNs within the clusters, collect it, and then forward the collected information to the BS LEACH-C [9], and enhancement through the LEACH protocol is managed by spreading CHs across the network to achieve better output. Stable election protocol (SEP) [10], an extension of LEACH, having bi-level heterogeneity in the heterogeneous WSNs provides a longer constant region due to efficient SNs generating extra energy; heterogeneity cannot be prolonged to multilevel heterogeneous WSNs. Distributed energy-efficient cluster (DEEC) [11] approach considers two-level and multilevel heterogeneity of energy.

HEED's primary objectives are to enhance the network's lifetime by absorbing energy supplies and the clustering run within the number of rounds [12, 13]. It minimizes overhead control and creates well-distributed CHs and compact clusters. HEED regularly selects CHs as a primary parameter, allowing two clustering constraints to be combined, namely the remaining energy of every SN and the cost of intracluster interaction as a method of adjacent closeness or cluster density as a secondary parameter. The initial parameter is utilized to elect the CH in a probabilistic manner. HEED shows the best load balancing in the network. HEED clustering technique extends the network lifetime as compared to the LEACH approach. LEACH algorithms choose CHs randomly, which may reduce the network performance and make some nodes die more easily. In HEED, the final CHs chosen are well spread across the network, and connectivity costs are minimized. Another energy-efficient clustering approach [14] suggests the existence of a WSN extended by dividing the network into clusters. To reduce the cluster's size, the CHs are selected in every round based on the cluster's number of SNs.

3 Network Model

We consider two different types of the sensor networks, i.e., heterogeneous and homogeneous network (in Fig. 2) having N SNs, which are randomly distributed over the monitoring area. A radio energy model is used for energy dissipation. There are some assumptions that are considered about the network and its SNs:

(1) The SNs are inside the communication range. The SNs communicate with each other. CH node directly communicates with the BS.
(2) All nodes have heterogeneous and homogeneous sensing, communication capabilities, and cost computing.
(3) SNs are randomly distributed in the monitoring area.
(4) BS is positioned in the inside and outside of the area, and BS has a source with infinite energy.
(5) All SNs in the homogeneous network have equal initial energy, and heterogeneous networks have different initial energy for both types of nodes as advanced node has higher energy than the intermediate node.

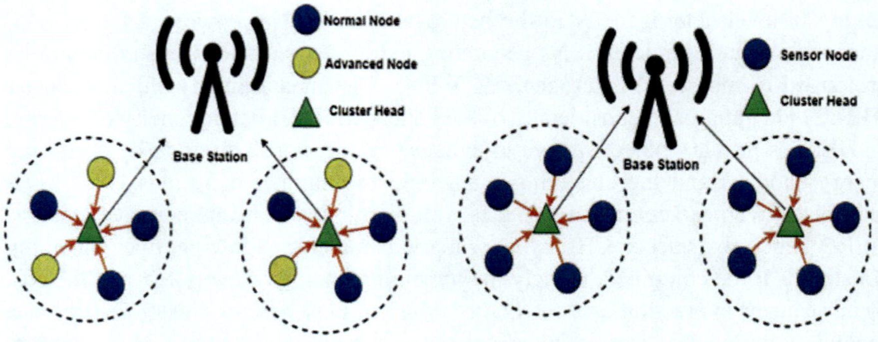

Fig. 2 Heterogeneous and homogeneous wireless sensor network

3.1 Energy Consumption Model

To dissipate communication energy, simplified model is used as presented in [9]. According to the distance among the sender and receiver, as shown in Fig. 3, free space (d^2) and the multipath fading (d^4) power loss channel systems are utilized. The l-bit packet is used for the transmission over distance d. The energy model is:

$$E_{TX}(k,d) = E_{Tx-\,\mathrm{elec}}(k) + E_{Tx-\mathrm{amp}}(k,\ d) = \begin{cases} E_{\mathrm{elec}} * k + \ \epsilon_f \ * k * \ d^2 & , d < d_o \\ E_{\mathrm{elec}} * k + \ \epsilon_m * k * \ d^4 & , d > d_o \end{cases} \quad (1)$$

$$E_{Rx}(k) = \ E_{\mathrm{elec}} * k \quad (2)$$

where $d_o = \sqrt{\frac{\epsilon_f}{\epsilon_m}}$, E_{elec} denotes the energy depletion of radio dissipation, ϵ_f and ϵ_m denote the energy depletion for amplifying radio, which is used depending upon the communication distance.

The sensory information is believed to be highly correlated. In contrast, the degree of correlation of sensory data from various clusters is comparatively low. The CH can still aggregate the information collected from its cluster members (CM) into a only fixed-length packet, and the energy is absorbed as E_{DA} by the process of data aggregation.

4 Proposed Clustering Method

In our proposed model, four stages consist of cluster-based routing protocols: CH election, cluster creation, aggregation of data, and data communication. As shown in Fig. 4, the setup state begins with selecting the CH and continues with the

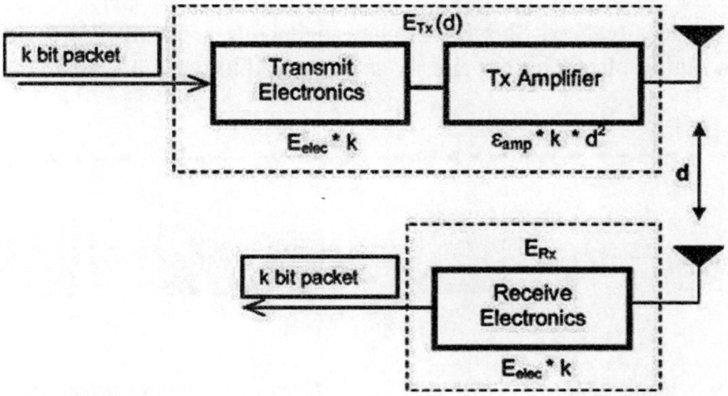

Fig. 3 Radio dissipation energy model

creation of clusters. The state of setup is monitored by the steady-state data communication that is partitioned into phases of data aggregation and transmission. Every round of running a cluster approach, which iterates over running the protocol or the network lifetime, forms the setup and steady data communication. SNs in clustering approach can be clustered into four categories, depending on the role:

Cluster head (CH): The key duties of a CH are the organization of a community of SNs located within the transmission range, the aggregation of sensed information by the cluster members (CM), and the transfer of aggregated information to the next hop.

Base station (BS): Because of the high computing capabilities and the infinite energy source, BS can be the network coordinator and the BS where entirely the aggregated information is processed based on the type of application and the end user's demands.

General node (GN): The common of network nodes that provide sensed data only based on the type of use.

The LEACH [9] is the first and the most common self-organizing clustering protocol for WSNs. Every node chooses whether to be a CH or not for the setup process in LEACH. CH's choice is based on a node decision with the choice of arbitrary numbers among 0 and 1. If the value is not greater than the default *T* (*n*) threshold, then the SNs for the current round is considered to be CH. We may define *T*(*n*) as the threshold as:

$$T(n) = \begin{bmatrix} \frac{P_{CH}}{1-P_{CH}*\left(r_c \bmod \left(\frac{1}{P_{CH}}\right)\right)}, & \text{if } n \in G \\ 0, & \text{otherwise} \end{bmatrix} \tag{3}$$

Here, r_c is a random number lies among 0 and 1; probability P_{CH}. The proportion of SNs is selected as a CH while data transmission whereas G is the SNs group that was not CHs in current rounds.

The proposed algorithm is improved by the remaining energy stored in the SNs and distance (*d*) between the CH and SNs. The SNs are having high energy for advance SN and low energy for normal SNs. The remaining energy is considered along with the distance of the SNs for the election of the new CH. SNs are divided into normal and advanced SNs for the heterogeneous network [7, 10, 15]; α is a component that implies a higher energy as compared to normal SNs; *m* denotes the

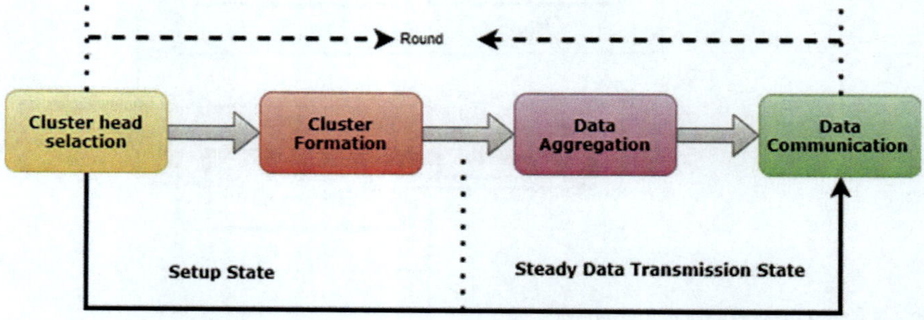

Fig. 4 Categories clustering algorithm model

ratio of advanced SNs. P_{ad} and P_{erm} stand for the probability of advanced as well as normal SNs nominated to be CH, respectively, in Eqs. (4 and 5).

$$P_{nrm} = \frac{P}{1 + \alpha^{*} m} \tag{4}$$

$$P_{adv} = \frac{P * (1 + \alpha)}{1 + \alpha * m} \tag{5}$$

The combination of current energy ($E_{current}$) and initial energy factor ($E_{initial}$), as well as the distance of the current SN ($D_{current}$), is used to design a new probability formulation. The (D_{max}) is the highest value of SNs to BS distance that is utilized to design an enhanced formulation of *T*(*n*) for CH election. The main purpose to design a new formulation is to decrease the CH SNs energy consumption. Equations (6)–(8) are taking to deploy the SNs of WSNs. The new formulation of *T* (*n*) for normal and advanced nodes by [7].

For normal SNs (heterogeneous WSN) [7]:

$$T(n)_{nrm} = \begin{bmatrix} \frac{P_{bnrm} * \left(u\left(\frac{E_{current}}{E_{start}}\right) + v\left(\frac{d_{current}}{d_{max}}\right)\right)}{1 - P_{nrm} * \left(\text{rmod}\left(\frac{1}{P_{nrm}}\right)\right)}, & n \in G \\ 0, & \text{otherwise} \end{bmatrix} \tag{6}$$

For advanced SNs (heterogeneous WSN) [7]

$$T(n)_{adv} = \begin{bmatrix} \frac{P_{badv} * \left(u\left(\frac{E_{current}}{E_{start}}\right) + v\left(\frac{d_{current}}{d_{max}}\right)\right)}{1 - P_{adv} * \left(\text{rmod}\left(\frac{1}{P_{adv}}\right)\right)}, & n \in G \\ 0, & \text{otherwise} \end{bmatrix} \tag{7}$$

According to Eqs. (6), (7), $P_{bnrm} = b * P_{nrm}$, $P_{badv} = a * P_{adv}$ is the weight of P_{nrm}, P_{adv}, respectively. The value of *b* is a proportional constraint according to network size. *u*, *v* is the ratio coefficient, and having the value between 0 and 1 and $u + v = 1$.

For general SNs (homogeneous WSN):

$$T(n)_{GN} = \begin{bmatrix} \frac{P_{CH} * \left(u\left(\frac{E_{current}}{E_{start}}\right) + v\left(\frac{d_{current}}{d_{max}}\right)\right)}{1 - P_{CH} * \left(\text{rmod}\left(\frac{1}{P_{CH}}\right)\right)}, & n \in G \\ 0, & \text{otherwise} \end{bmatrix} \tag{8}$$

5 Result and Simulation

The network terminology with their value is mentioned in Table 1. In the network 20% of advance and 80% of normal sensor node of the total nodes. The BS is located at (50, 50) and (50, 150). The initial parameters are mentioned in Table 1.

Both the algorithms are implemented using MATLAB (R2017b) on the Windows 10 platform. The line plot is also designed on MATLAB with the respective color code. The outcome shows that ADV-LEACH1 (HETRO) performs better based on these alive/dead node analysis metrics, throughput, network life time, residual energy, and optimized CH selection of network. ADV-LEACH1 (HETRO) further improves efficiency; however, ADV-LEACH1 (HETRO) performs best as compared to the ADV-LEACH1 (HOMO).

Figure 5 shows that after only 1600 rounds, the average energy of the ADV-LEACH1 (HOMO) approach comes to 0, wherein ADV-LEACH1 (HETRO) continues till $\sim$2000 rounds. This becomes possible since energy has been saved for those SNs with higher energy in ADV-LEACH1 (HETRO); that node can choose CH for a few more. ADV-LEACH1 (HOMO) did not have any higher energy nodes; all nodes have equal energy, so as SNs get selected as CHs for a few rounds, then their energy gets depleted faster, and a SN becomes a dead node. As per Fig. 5 ADV-LEACH1 works better in the heterogeneous sensor network. ADV-LEACH1 (HETRO) works better than the ADV-LEACH1 (HOMO).

The proposed method for alive nodes up to 2000 rounds is shown in both Fig. 6. We notice that the number of alive nodes in ADV-LEACH1 (HOMO) drops significantly around 1600th round and continues to decrease as compared to other ADV-LEACH1 (HETRO) approaches. In contrast, the number of alive nodes in ADV-LEACH1 (HETRO) is much higher than ADV-LEACH1 (HOMO).

Figure 6 (left) compares the number of dead nodes in every round when both the ADV-LEACH1 (HETRO) and ADV-LEACH1 (HETRO) approaches are run on the network with similar parameters. The results demonstrate that the SN could not remain alive after 2000 rounds in the case of ADV-LEACH1 (HOMO). However, the last node died after 2000 rounds in ADV-LEACH1 (HOMO), but ADV-LEACH1 (HETRO) approach still has them alive nodes after 2500 round. The ADV-LEACH1 (HETRO) approach has better results in the dead/alive node analysis than the ADV-LEACH1 (HOMO) approach.

In addition to network lifetime, its performance is another metric to determine an approach's effectiveness. The efficiency of the proposed solution is verified by a CH having more data packets. In a sense, but not always, throughput relies on network life. Figure 7 clearly demonstrates the throughput of ADV-LEACH1 (HETRO) that

Table 1 Initial parameter

Parameters	Value
No. of round®	2500
P	0.2
E_{TX} or E_{RX}	50 nj/bit
$\in_f$	10 pj/bit/m^2
E_{elec}	5 nj/bit/message
$\in_m$	0.0013 pj/bit/4
No. of sensor (N)	200
Packet size	4000 bits

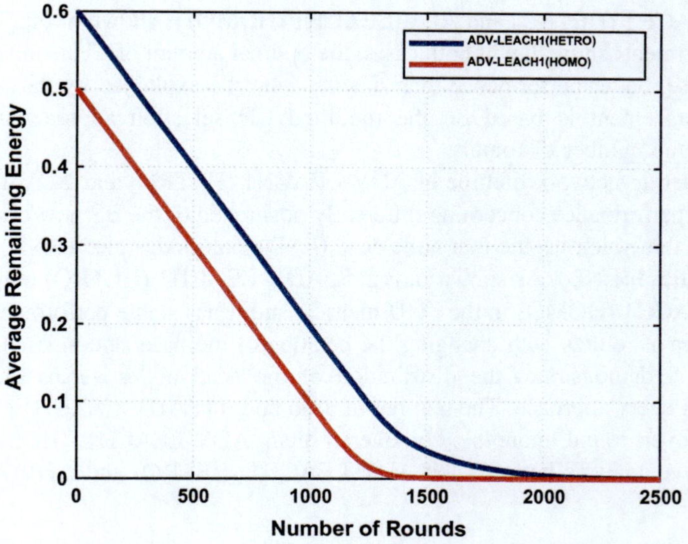

Fig. 5 Comparative analysis of average remaining energy for homogeneous and heterogeneous sensor network versus the number of nodes

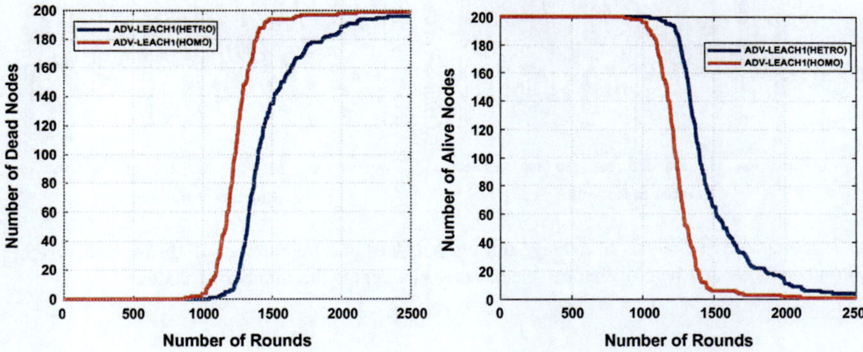

Fig. 6 Comparative analysis of number of dead (left) and alive nodes (right) for homogeneous and heterogeneous sensor network versus number of nodes

is quite similar to the approach of ADV-LEACH1 (HOMO). The performance increase is caused by limiting the amount of data transmission obtained on the CH side. In ADV-LEACH1 (HETRO), the number of packets received at CH is equivalent to 500 nodes.

The energy efficiency of WSNs is significantly influenced by a selection of CHs. Thus, these CHs are going to die sooner. Thus, the stability of the CH numbers around an optimum number is needed in successive rounds to obtain balanced energy consumption. The number of CHs in each round as opposed to

ADV-LEACH1 (HETRO) and ADV-LEACH1 (HOMO) is shown in Fig. 7 (right). The experiments show that in both cases, the optimal amount of CH is mismatched. In this case, better performance than others cannot be explained by the approach. This enhancement is based on the modified CH selection approach that also increases the number of rounds.

Considering network lifetime of ADV-LEACH1 (HETRO) and ADV-LEACH1 (HOMO) performance concerning differently positioned of the BS network lifetime is used in three metrics, the first node dead (FND), half nodes died (HND), and the last node dead (LND). As shown in Fig. 8, ADV-LEACH1 (HETRO) outperforms ADV-LEACH1 (HOMO) in the FND metrics, and it has stable performance based on number of rounds with changing the position of the base station (BS).

Figure 8 demonstrates the distribution of the dead nodes versus number of rounds for every approach. The number of dead nodes in ADV-LEACH1 (HETRO) changes over round more slowly overall than ADV-LEACH1 (HOMO). The comparative analysis between the ADV-LEACH1 (HETRO) and ADV-LEACH1

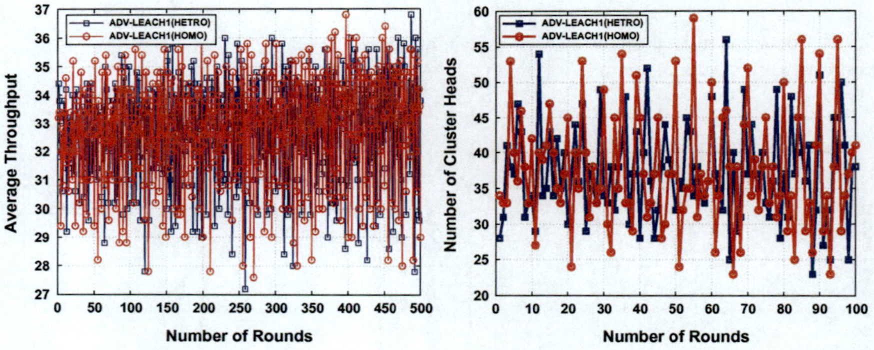

Fig. 7 Comparative analysis of average throughput (left) and the number of cluster heads (right) for homogeneous and heterogeneous sensor networks versus the number of nodes

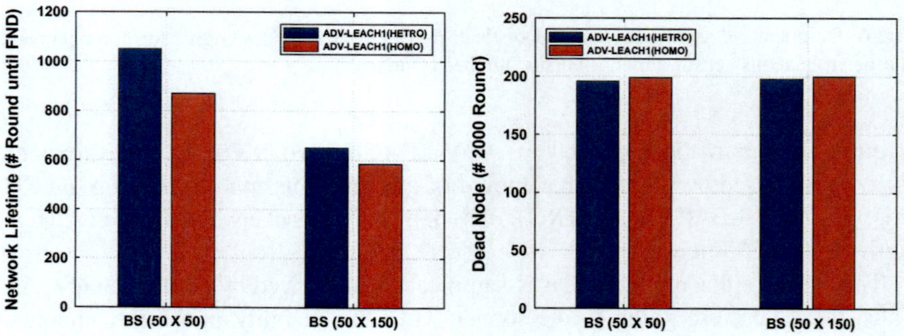

Fig. 8 Comparative analysis of network lifetime (left) and dead node (right) for homogeneous and heterogeneous sensor network based on the base station (BS) position

Table 2 Comparative analysis

Algorithms	BS Position (50 × 50)		BS Position (50 × 150)	
	Network LifeTime (#FND)	Dead node (# 2500 rounds)	Network LifeTime (#FND)	Dead node (# 2500 rounds)
ADV-LEACH1 (HETRO)	1051	196	650	198
ADV-LEACH1 (HOMO)	870	199	586	200

(HOMO) based on the network lifetime and dead nodes till 2000 nodes, as mentioned in Table 2.

6 Conclusion

In this paper, we have analyzed the ADV-LEACH1 for the heterogeneous and homogeneous networks based on the analysis of the latest clustering algorithm ADV-LEACH1 with two different types of network and different base station positions (BS). The SNs are randomly deployed in the square area and base station in the center of the area. Also, a solution is proposed in the form of ADV-LEACH1 (HETRO) and ADV-LEACH1 (HOMO) approaches that uses the modified threshold probability of clustering for the CH selection. The performance is evaluated based on the various performance parameters. The simulation results show that ADV-LEACH1 (HETRO) performed better in terms of residual energy and network life time compared to ADV-LEACH1 (HOMO).

References

1. Ngangbam, R., Hossain, A., Shukla, A.: Lifetime improvement for hierarchical routing with distance and energy based threshold. In: International Conference on Intelligent Data Communication Technologies and Internet of Things, pp. 77–83. Springer, Cham (2019)
2. Sara, G.S., Sridharan, D.: Routing in mobile wireless sensor network: a survey. Telecommun. Syst. **57**(1), 51–79 (2014)
3. Hussain, N., Rani, P.: Comparative studied based on attack resilient and efficient protocol with intrusion detection system based on deep neural network for vehicular system security. In: Distributed Artificial Intelligence, pp. 217–236. CRC Press (2020)
4. Jawandhiya, P.M., Ghonge, M.M., Ali, M.S., Deshpande, J.S.: A survey of mobile ad hoc network attacks. Int. J. Eng. Sci. Technol. **2**(9), 4063–4071 (2010)
5. Jan, B., Farman, H., Javed, H., Montrucchio, B., Khan, M., Ali, S.: Energy efficient hierarchical clustering approaches in wireless sensor networks: a survey. Wirel. Commun. Mob. Comput. (2017)

6. Singh, S.: Energy efficient multilevel network model for heterogeneous WSNs. Eng. Sci. Technol. Int. J. **20**(1), 105–115 (2017)
7. Kumar, N., Kumar, V., Ali, T., et al.: Prolong network lifetime in the wireless sensor networks: an improved approach. Arab. J. Sci. Eng. (2021). https://doi.org/10.1007/s13369-020-05254-3
8. Heinzelman, W.R., Chandrakasan, A., Balakrishnan, H.: Energy-efficient communication protocol for wireless microsensor networks. In: Proceedings of the 33rd annual Hawaii international conference on system sciences, p. 10. IEEE (2000)
9. Heinzelman, W.B., Chandrakasan, A.P., Balakrishnan, H.: An application-specific protocol architecture for wireless microsensor networks. IEEE Trans. Wirel. Commun. **1**(4), 660–670 (2002)
10. Smaragdakis, G., Matta, I., Bestavros, A. SEP: A stable election protocol for clustered heterogeneous wireless sensor networks. Boston University Computer Science Department (2004)
11. Qing, L., Zhu, Q., Wang, M.: Design of a distributed energy-efficient clustering algorithm for heterogeneous wireless sensor networks. Comput. Commun. **29**(12), 2230–2237 (2006)
12. Younis, O., Fahmy, S.: Distributed clustering in ad-hoc sensor networks: A hybrid, energy-efficient approach. In: IEEE INFOCOM 2004, vol. 1. IEEE (2004)
13. Younis, O., Fahmy, S.: HEED: a hybrid, energy-efficient, distributed clustering approach for ad hoc sensor networks. IEEE Trans. Mob. Comput. **3**(4), 366–379 (2004)
14. Singh, M.P., Gore, M.M.: A new energy-efficient clustering protocol for wireless sensor networks. In: 2005 International Conference on Intelligent Sensors, Sensor Networks and Information Processing, pp. 25–30. IEEE (2005)
15. Chen, X., Wang, Z., Wu, J.: The improved wireless sensor network routing algorithm based on LEACH. Chin. J. Sens. Actuators **1** (2013)

Cognition of Driver Drowsiness to Inculcate Predictive Analysis

Abra Shafiq Siddiqi, Md. Afshar Alam, Sherin Zafar, Samia Khan, and Nida Iftekhar

Abstract Life being the most valuable assets is lost myriads of time due to dreadful vehicular accidents. However, a real-time drowsiness detector used on a car can significantly obviate these accidents and save valuable lives worldwide. The main reason being inattention of driver mainly referred to as driver's drowsiness. The driver drowsiness monitoring system works with a high frequency detection system using an incoming video stream of a driver focusing on the natural visual changes in driver such as the steady closure of eyes and slow changing facial expression using artificial intelligence and scientific drowsiness detection measure. The proposed work focuses on the various monitoring systems used in the drowsiness detection as well as the process of the detection system. The driver drowsiness detector attached to crucial predictive analysis system is proposed to be utilized. This research paper focuses on the analysis of various drowsiness system so that better predictive analysis could be done. Machine learning is a very important paradigm for predictive analysis, so this paper focuses on the various machine learning techniques and their efficiency for detection of drowsiness in various systems.

Keywords Drowsiness detection · Driver drowsiness detector · Facial expression · Predictive analysis · Machine learning algorithms

1 Introduction

AI The available statistical data prescribes over 20% people die and 76,000 people suffering fatal injuries caused by road accidents [1]. The statistics further emphasis that majority of road accidents are caused by driver drowsiness. Driver when falls asleep during a journey ultimately there exists a loss of control over the driving vehicle that majority times causes a serious crash either with another running

A. S. Siddiqi · Md. A. Alam · S. Zafar · S. Khan · N. Iftekhar (✉)
Jamia Hamdard, New Delhi, India
e-mail: nida.iftekhar@jamiahamdard.ac.in

K. Khanna et al. (eds.), *Cyber Security and Digital Forensics*,
Lecture Notes on Data Engineering and Communications Technologies 73,
https://doi.org/10.1007/978-981-16-3961-6_37

vehicle or any standing objects. The term "drowsy" literally means an inclination to fall asleep. Drowsiness is synonymous to sleepy. Further the stages of sleep can be divided as awake (not sleepy), non-rapid eye movement sleep (NREM), and rapid eye movement sleep (REM). NREM can be further subdivided into the following given three stages below [2].

Stage I: change from awake to asleep (drowsy)
Stage II: light sleep
Stages III: deep sleep

In order to determine driver drowsiness, researchers mainly have evaluated Stage I (change from awake to asleep) that is the drowsiness stage. Various crashes mainly occurring due to drowsiness of driver have varied characteristics features [3].

- Occurs mainly at late night
- Driver is frequently alone
- Involving an individual vehicle running off the road
- Mainly occurs mainly on high-speed roadways
- Driver usually a young male
- Blood alcohol level that are actually below the legal driving limit
- Absolutely zero skid figures or braking signal in relation to these features.

However, the police databases use the following given indices to determine accidents caused due to drowsiness [4].

- if vehicle has any mechanical failure or defect
- no brakes applied
- weather conditions and clear visibility
- lack of sleep
- if vehicle ran off the appropriate road or onto the back of another moving or standing vehicle
- excluding "driving too close to the vehicle in front" and "speeding" as potential causes

The statistical data obtained using these methods cannot usually ascertain fully for road accidents that are caused by drowsiness in drivers because of the various number of complexities involved. Thus, the accidents recognized due to driver drowsiness could be more ugly, devastating, and fatal than these statistics data reveal. Hence, to obviate accidents, it becomes out most significant to effectively measures and detect driver drowsiness such that the drivers are left alert. The later parts of the research paper focus on the drowsiness detection process and the advancements in the field of the driver drowsiness detection systems and the future advancement scopes in this field.

In the research paper, further Sect. 2 defines literature review. Section 2 talks about the gaps in literature. Further, Sects. 3, 4, 5 discuss methodology, objective of study, the process of driver drowsiness detection, and conclusion, respectively.

The research paper lastly discusses the conclusion under Sect. 6. References and the resources utilized during the research paper construction are enlisted under title references.

2 Literature Review and Gaps

This literature reviewed the head movement-based change for detection of driver drowsiness [5]. The research paper emphasizes on general measures and methods utilized to determine the drowsiness in drivers and also provides a well distinguished analysis of different systems detection drowsiness utilized throughout the globe. In addition to this, a well distinguished system for determining the drowsiness fatigue caused in drivers is signified [6]. The research paper also focused on the works that used varied methods to obviate catastrophic road havocs and accidents caused by drowsiness [7]. However, a great number of literature have been gathered around the detection of drowsiness systems, this field actually draws an immense advancement, and this time demands great need for a review literature for machine learning techniques and approaches which can be applied to drowsiness detection. This research paper now basically attempts to address and evaluate various cardinal needs by ascertaining behavioral techniques specifically focusing on the machine learning technologies for distinguishing the various drowsiness stages.

The research paper has extensively undergone the evaluation and consultation of various works done on the drowsiness detection systems which are accepted worldwide and are turning into cardinal norms in detection and measuring of drowsiness in vehicles to obviate the fatal accidents and save precious lives. Various works and methodologies have been adopted to present and increase the efficiency of detection systems which are broadly discussed in the research paper. However, there also exists certain areas which definitely need to filled and worked upon further to overcome the gaps and improve the efficiency of the detection system which are roughly presented below:

- While using the subjective measures for detection, drowsiness levels are frequently measured every 5 min approximately. However, rapid changes cannot be determined or measured using these subjective measures
- An additional gap identified using various subjective ratings suggests that the self-introspection majorly alerts the driver, therefore dropping their levels of drowsiness. In addition to this, it is also very hard to evaluate feedbacks related to drowsiness from a driver in a real driving situation. Since, subjective ratings are very useful in detecting drowsiness in a replicated environment; therefore, the remaining measures and indices may be better obtained for the drowsiness detection in a real environment
- The vehicle-based method used can reliably function only at specific surrounding which are also dependent on varied geometric individuality as well as

the matrices of road but to lesser extent on the dynamic features of the vehicle which needs to be amended such that more focus is on the characteristic features of the vehicle

- Another limitation identified by this study is a vision-based approach that uses lighting since usual cameras cannot perform efficiently during night hours. Researches should use active and efficient illumination methods such as a light emitting diode (LED) of infrared nature. However, LEDs work fairly significant during night hours but less significant during day time.

Therefore, it is very essential to use a high frequency hybrid technology to overcome the above-mentioned gaps and increase the efficiency of the detection system through the predictive analysis techniques.

3 Objective of Study

Analysis the research paper mainly focuses on the driver drowsiness detection and identification of its various technologies used. Driver drowsiness detection system being a safety technology which helps saving the life of driver by obviating accidents, while the driver is drowsy. The main objective of the research paper is initially designing a system to detect the drowsiness by:

- continuously examining the eye retina
- The proposed system works despite the fact driver wears spectacles and also considers various lighting conditions
- Alerting the driver of drowsiness utilizing an buzzer or alarm system through analytics
- Reducing the speed of the vehicle after predictive analysis
- It can efficiently reduce accidents by maintaining the traffic management
- Fetching authenticated data of driver drowsiness and performing predictive analysis for getting best possible results through artificial intelligence/machine learning/ deep learning algorithms.

4 Methodology

The behavioral measurement techniques accurately monitor the levels of drowsiness by using high definition mounted cameras fixed in a car such that the change in facial expressions including yawning, eye state, blinking rate, slow eye closure, head movement, etc. The scientific approaches mostly practice specific procedure to process driver features of face recorded from the provided camera as depicted in Fig. 1. After gathering the above-mentioned features, the further work is counted in order to monitor drowsiness levels. This is specifically obtained by using

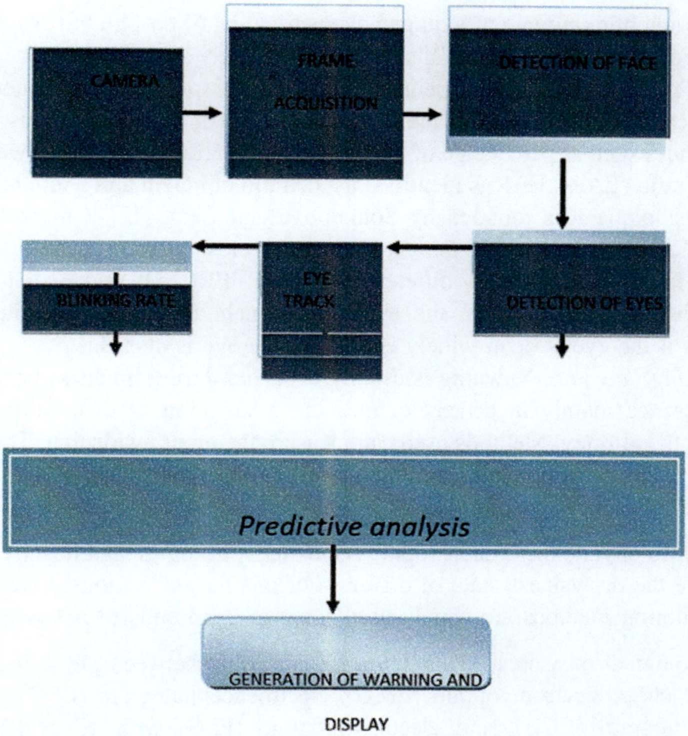

Fig. 1 Drowsiness detection process

technologies of machine learning which frequently include convolutional neural networks or hidden Markov models support vector machines (SVM) [8]. These procedures, however, are applied by specific characteristics as well as labeled data matrices in order to construct systems detecting drowsiness and obviate fatal hazards. However, the most challenging part in detection process is to find huge dataset that contains featured variations throughout different course of race as well as skin pigments. This being a significant concern because of myriad of security as well as discretionary issues that can surely arise during the whole process of publishing these datasets for academic as well as commercial purpose [9].

The following expressions mentioned below are marked from the facial features which are specifically gained from a face of driver:

1. *Changing facial expression analysis*: Here the technique combines more than one facial expression such that the driver drowsiness detected. This encompasses expressions such as wrinkles found specially on the forehead also the head poses that are sometimes extreme [10].
2. *Rate of eye blink*: This technique detects the rate of blinking by considering the frequency with which the eye-blinks such that the drowsiness is measured. Since

the normal blinking rate of a human eye is roughly 10 per [5], the drowsiness is detected if the blinking rate falls the given number.

3. *Eye closure analysis*: This being an important expression is frequently and specifically used to measure the drowsiness level in the driver. This includes techniques such as percentage of the eye closure (PERCLOS) [11] as well as eye aspect ratio (EAR). EAR is identified as the ratio of height and width of the eye. This technique was founded by Soukupova and Cech [12]. On the contrary, PERCLOS is calculated as the percentage eye closure over a certain period of time. However, the primary difference between PERCLOS and EAR is that the later distinguishes between the eye ratio, which descends while the former detects if the eye is open widely enough or the eye is closed.
4. *Analyzing Yawning*: Yawning is usually observed during tiredness or boredom. Its presence mainly in drivers defines the predication of their falling asleep during the journey. Methods used can measure the mouth widening. This can be traced using yawning features in the driver on observing shape of the mouth as well as lip corners position [13].

In order to obviate these catastrophic accidents, it becomes significantly cardinal to measure the drowsiness state of drivers. For this reason, various driver drowsiness monitoring methods are widely used some of them enlisted below:

(a) Physiological measures—This defines relationship between physiological signals such as electromyogram (EMG), electroencephalogram (EEG) [2], electrocardiogram (ECG), and electrooculogram (EoG) with driver [5]. These measurements yield a highly efficient and accurate measurement data but are not globally prevalent mainly due to certain practical downfalls.

(b) Vehicle-based measures—This includes various metrics such as pressure on the acceleration pedal, braking patterns, movement of the steering wheel, deviations from lane position, etc., which are constantly evaluated including any changes in these metrics that crosses a particular threshold. Thus, indicating an essential increased probability of driver drowsiness [14]. Among all the methods in vehicle-based measures, steering wheel yields the most accurate results. However, vehicle-based methodology being non-invasive, this may in the long run not be a reliable and efficient determining technique mainly as they are solely reliant on the road nature as well as driving skills of drivers.

(c) Behavioral or computer vision measures—this method is much steadfast than the vehicle-based method as it focus mainly on the person itself than the monitoring system based on vehicle. This focuses on driver's natural responses gathering information using cameras for detecting minute change in facial features of the drivers which includes yawning of the person, closure of eye, blinking of eye, change in head pose, etc., is monitored through a camera, and the driver is alerted if any of these drowsiness symptoms are detected [15]. This method being non-invasive is gaining immense popularity in drowsiness detection determination. The rate of positive detection reduced to a significant amount when the experimental work when conducted in actual atmosphere.

Table 1 enlists the work of various authors referred during this work who had used behavioral measures for driver drowsiness detection.

There exist numerous procedures and techniques for detection of face as well as to draw and determine features collected from the video data. However, majority of these works conducted approach various databases which could follow the individual algorithms. This exists mainly because of insufficiently systematic datasets which could be mainly applied as a point of references in the study. Thus, it is

Table 1 Previous works list on driver drowsiness detection using behavioral measures

Sensor used	Drowsiness measure	Detection techniques	Feature extraction	Classification	Positive detection rate (%)
CCD camera	Yawning	Grey projection and gravity-center template	Gabor wavelets	LDA	91.97
Simple camera	Eye blinking	Algorithm diamond searching for tracing the face and cascaded classifiers algorithm to detects face	Frequency of continuous blinks, frequency of eye blink, duration of eyelid closure	Region mark algorithm	98
CCD micro camera with infrared illuminator	Pupil	AdaBoost	Red eye effect, detection method for texture	Eye-height, eye-width ratio	92
Camera	Multi-scale dynamic featuring	Gabor filter	LBP	AdaBoost	98.33
Digital video camera	Facial action	Gabor filter	Decomposing wavelet	SVM	96
IR camera	Eye state	Gabor filter	Condensation algorithm	SVM	93
Fire wire camera and webcam	Eye closure frequency and duration of eye closure	Hough transform	Transforming discrete wavelet	Neural classifier	95
IR Illuminator attached to camera	PERCLOS	Detecting face by Haar algorithm	Kalman filter algorithm unscented	SVM	99

difficult to distinguish methods by merely determining the reported data efficiencies. The techniques of machine learning studies are mainly used to determine the different stages of drowsiness that are now listed along with their discussions below. Also, including the analysis of various measures and methods that sufficiently form a drowsiness detection system for drivers.

(a) *Hidden Markov Model (HMM)*

This is basically a statistical data models and methods applied to predict and determine about the hidden states based on the actual perceived states which are defined by the probability methods. HMMs were extracted during late 1960s period and early 1970s period [1]. In this era, these methods have a global application in various technologies such as modeling sequence errors, DNA gene annotation, and face expression recognition. Table 2 shows the varied range of features as well as perspectives which are used by hidden Markov model-based driver drowsiness detectors. However, [10, 16] have excluded some of the cardinal information perspective for the formation of the differentiation which are not actually used for the meta-analysis step formation. The authors have beautifully enumerated various fresh facial characteristics using changing wrinkles that are differentiated by calculating the edge potencies from the driver's facial features by using a camera of infrared nature to exclude the variations in illumination such that it permits the proper functioning both in day and night conditions. Howbeit, the deeper and strong wrinkles formed on faces of the older people may provide the false results generated by this system. In contrast, the HMM techniques implied for the eye tracking movements which are based on the varied geometrical features and color of eyes. For eliminating the illumination, authors have used a 2-stage Lloyd-max quantization method which is calculated to be sufficient enough to eliminate changes in illumination. Unfortunately, the system suggested is drafted for the inside conditions which does not succeed to display the facial features when the driver is actually not forward facing. Table 2 enlists the HMM technique below used for driver drowsiness detection.

Table 2 Driver drowsiness detection based on HMM technique

Author list	Year	Metric	Classifiers	Frame per second (fps)	Accuracy %
Choi et al.	2016	State of eye and position of head	HMM	16–20	N/A[a]
Aboalayon et al.	2019	State of eye	HMM	N/A[a]	95.9
Deng et al.	2019	Blink of eye	SVM and HMM	61	90.99
LaRocco et al.	2020	Eye state	HMM	3	99.7
Tadesse et al.	2014	Closure of eye and other varied features	HMM and SVM	20	97
Muhammad et al.	2019	Blinking eye	HMM	25	95.7

[a]Dependent variable

(b) *Convolutional Neural Network (CNN)*

This technique is very similar to a neural network commonly used that is also compiled of the neurons which further consists of the learnable weights. CNNs use varied spatial convolutions layers that are convenient enough for drawing images used to display the strong spatial relations. The methods have shown success in certain areas of image recognition, classification, and video analysis [1]. First to describe CNNs for vision of computer, however, this was manifested in 2012 only, when fabulous results were found using in object recognition deep convolutional neural networks. Table 3 provides a brief CNN-based techniques used for the determination drowsiness in driver. Dwivedi et al. [17] provide an database algorithm for detection of drowsiness by using a demonstration of this learning. However, this stage witnessed algorithm by Jones, and for recognition of faces, Viola was used. The images were trimmed to the dimensions of (48 * 48 square) which were incorporated into the initial layer of the network that comprises basically of 20 filters. The overall network consists of 2-layer systems. CNN provided a data which preceded to one layer softmax for the purpose of classification. CNN system, however, could not contemplate the change in the head pose, thus, can result in a fail. Nevertheless, authors [14] have applied a featured 3D deep neural network for attaining more efficient results. In this system, the face feature followed by a combined filter predicts for correlation encompassed with a Kalman filter system to track face robustly. The regions of face are drawn out and passed directly to a 3D-CNN system further to a gradient boosting machine for classification in an optimum manner. The advantage added to this system is that it also works well with the change in head position.

(c) *Support Vector Machines (SVM)*

SVMs are specialized, supervised, and efficient learning techniques used as regression and classification system. The methods were initially utilized in [15]. These specialized methods focused to find a hyperplane to discontinue data of training into pre-defined classes. To analyze the varied states of the driver drowsiness obtained from the labeled data, SVMs are initially used. A significant

Table 3 Driver drowsiness detection using the CNN technique

Author	Year	Metric	Methods	Classifiers	Accuracy %
George and Routray	2016	Gaze of eye	Algorithm of Jones and Viola	CNN	98.32
Zhang et al.	2017	State of eye	PERCLOS, AdaBoost and LBF	CNN	95.18
Reddy et al.	2017	State of eye	Eye state and mouth	MTCNN and DDDN	91.6
Dwivedi et al.	2014	Visual features	Algorithm of Jones and Viola	Softmax layer CNN	78

Table 4 Detection of driver drowsiness using SVMS

Author	Year	Measure	Classifiers	Frame per second (fps)	Accuracy %
Pauly and Sankar	2015	State of eye	HOG and SVM	5	91.6
Manu	2016	Eye closure and Yawning	Binary SVM including linear kernel	15	94.58
ALAnizy et al.	2015	Closure of eye	Haar features with SVM	60	99.74
Ren et al.	2014	State of eye	SVM	25	98.4
Punitha et al.	2014	State of eye	SVM	15	93.5

work on SVM done to mark its capabilities for drowsiness detection. Innumerous measures and techniques have been put forward to determine a significant amount of drowsiness in driver's using SVMs. The contrast obtained in these techniques presented in Table 4 with their perspectives embraced are reported in the table [18] have submitted a total mechanized system with the capacity of driver drowsiness detection. The detection of face and extraction of eye measurements was calculated using Haar featured algorithm. In the work further, classification of SVM was demonstrated to signify open or closed eyes such that an alarm is activated. Similarly, the proposed approaches have provided a system to detect distractions caused by the driver drowsiness. Here also using Jones algorithm and Viola for face detection in addition to this also a color-based histograms including local binary pattern system (LBP), data was obtained for smooth tracking of face over the frames. This technique further attained a 100% accuracy for detection of face howbeit, and a potential downfall still exists in the approach that is the minute rate of mount attainable which possibly could end in insufficient and inappropriate expressions obtained from facial. Table 4 represents the list of authors who worked on driver drowsiness detection using SVMS technology including their measures and data.

(d) *META-ANALYSIS*

Even though a considerable amount of work has been amalgamated to gather the data, however, a notable room exists to make up more accurate and efficient drowsiness detection systems. The significant concern observed according to this system analysis is that varied datasets have been used to obtain the goals but are not actually contrasted. However, the datasets which are reviewed so that the bound orders are maintained within a guided surroundings which could easily conduct an authentic world scenario surveillance. For a reasonable differentiation, 25% of the data collected under this work has been mainly due to the use of meta-analysis [1]. These collections of papers formally use sorting accuracy and efficiency to aggregate the systems production. Approximated performance observed that more accurate results were submitted by the CNNs that too when actually differentiated with HMMs and SVMs [19]. Skillings-Mack test actually non-parametric in nature was used that provided a 6.66 Chi-square estimate which was notable $p = 0.035709$ [20]. This stipulates an actual contrast between the procedures conducted in presentation. Further the tests were conducted acquiring exactness using Eye-Chimera, ZJU Eye blink Database ULg Multimodality Drowsiness Database (DROZY), and the NTHU-drowsy driver detection video dataset Yawn Detection Dataset (YawnDD).

However, all the literatures conducted showed significant majority classifiers used were SVMs than by HMMs and later followed by CNN. Therefore, it is also been observed that there is a significant increase in the use of CNN for driver drowsiness detection since 2012, including the growing increase in use of deep learning.

Further, the machine learning aspects like SVM, HMM, and CNN were also bought up into the consideration during this paper. However, it is significantly difficult to sum up these perspectives, due to their restricted number of datasets systematized that actually exist. Meta-analysis was executed in an attempt to obtain these results. This procedure enshrined CNNs performance which excelled other existing approaches, but it also observed a huge need for datasets with quality specification features for detection of drowsiness.

5 Conclusion and Future Scope

As described in the research paper, different technologies and methods do exist to detect and determine the driver drowsiness fatigue. The research paper tried to evaluate various evolving techniques such that the best possible approaches are designed to obviate fatality caused by drowsiness crashes. This research paper has reviewed and analyzed varied methodologies that are accessible to detect the state of driver drowsiness. However, no universally accepted definition for drowsiness exists but this research paper certainly enshrines various systems and ways in which drowsiness can be obtained in a simulated environment. The various methods used to detect drowsiness mainly consists of subjective, behavioral measures, and vehicle-based as well as physiological, which were also discussed in detail including their advantages and disadvantages. The accuracy and efficiency rate to detect drowsiness while using physiological measures is high being highly intrusive. Howbeit, these intrusive nature of physiological measures can be resolved utilizing contactless electrode placement. Thus, it would be worth developing hybrid system using vehicle-based measures as well as physiological measures such as ECG and behavioral measures such that an appropriate and efficient drowsiness detection system is developed. Also, it becomes very essential for maintaining and focusing on the driving environment to obtain optimal adequate results. Authors further purposed to find out dataset and apply better machine learning algorithms to get better results which are not considered in this research study and will be our further work of implementation. Random forest and decision trees are the proposed algorithms that authors propose to an authenticated data set for building much accurate analytical models.

References

1. Deng, W., Wu, R.: Real-time driver-drowsiness detection system using facial features. IEEE Access **7**, 118727–118738 (2019). https://doi.org/10.1109/ACCESS.2019.2936663
2. LaRocco, J., Le, M.D., Paeng, D.G.: A systemic review of available low-cost eeg headsets used for drowsiness detection. Front. Neuroinform. **14**, 553352 (2020). https://doi.org/10.3389/fninf.2020.553352
3. Parkhi, O.M., Vedaldi, A., Zisserman, A.: Deep face recognition. BMVC **1**, 6 (2015)
4. Manu, N.: Facial features monitoring for real time drowsiness detection. In: Proceedings of the 12th International Conference on Innovative and Information Technology, IIT 2016, pp. 78–81 (2017)
5. Shakeel, M., Bajwa, N.A.: Detecting driver drowsiness in real time through deep learning based object detection. In: Advances in Computational Intelligence, pp. 283–296 (2019)
6. Nair, I.R., Ebrahimkutty, N., Priyanka, B., Sreeja, M., Gopu, D.: A survey on driver fatigue-drowsiness detection system. Int. J. Eng. Comput. Sci. **5**(11), 19237–19240 (2016)
7. Gill, Chisty: A Review : Driver Drowsiness Detection System, vol. 3, no. 4, pp. 243–252 (2015)
8. Reddy, B., Kim, Y., Yun, S., Seo, C., Jang, J.: Real-time driver drowsiness detection for embedded system using model compression of deep neural networks. Comput. Vis. Pattern Recognit. Work. (2017)
9. Han, W., Yang, Y., Bin Huang, G., Sourina, O., Klanner, F., Denk, C.: Driver drowsiness detection based on novel eye openness recognition method and unsupervised feature learning. In: Proceedings of the IEEE International Conference on System Man, Cybernetics, SMC 2015, pp. 1470–1475 (2016)
10. Choi, I.H., Jeong, C.H., Kim, Y.G.: Tracking a driver's face against extreme head poses and inference of drowsiness using a hidden Markov model. Appl. Sci. **6**(5) (2016)
11. Provisional Registrations or Sales of New Vehicles (2018) [online] Available: http://www.oica.net/wp-content/uploads/
12. Cech, J., Soukupova, T.: Real-time eye blink detection using facial landmarks. In: 21st Computer Vision Winter Working (2016)
13. Bin Zainal, M.S., Khan, I., Abdullah, H.: Efficient drowsiness detection by facial features monitoring. Res. J. Appl. Sci. Eng. Technol. **7**(11), 2376–2380 (2014)
14. Huynh, P., Kim, Y.G.: Detection of Driver Drowsiness Using 3D Deep Neural Network and Semi-Supervised Gradient Boosting Machine, vol. 10116 (2017)
15. Ren, S., Cao, X., Wei, Y., Sun, J.: Face alignment at 3000 FPS via regressing local binary features. In: Proceedings of the IEEE Conference on Computer Vision Pattern Recognition, pp. 1685–1692 (2014)
16. Zhang, B., Wang, W., Cheng, B.: Driver eye state classification based on co-occurrence matrix of oriented gradients. Adv. Mech. Eng. **7**(2) (2015)
17. Dwivedi, K., Biswaranjan, K., Sethi, A.: "Drowsy driver detection using representation learning" Souvenir 2014. In: IEEE International Advance Computer Conference IACC 2014, pp. 995–999 (2014)
18. AL-Anizy, J., Nordin, M.J., Razooq, M.M.: Automatic driver drowsiness detection using Harr algorithm and support vector machine techniques. Asian J. Appl. Sci. (2015)
19. Nandy, G.T., Manna, N.: Real time eye detection and tracking method for driver assistance system. Adv. Med. Electron. (2015)

20. Ngxande, M., Tapamo, J., Burke, M.:Driver drowsiness detection using behavioral measures and machine learning techniques: a review of state-of-art techniques. In: 2017 Pattern Recognition Association of South Africa and Robotics and Mechatronics (PRASA-RobMech), Bloemfontein, pp. 156–161 (2017). https://doi.org/10.1109/RoboMech.2017.8261140

Correlation Between K-means Clustering and Topic Modeling Methods on Twitter Datasets

Poonam Vijay Tijare and Jhansi Rani Prathuri

Abstract Twitter is a popular platform for people to express their feelings on any subject or topic irrespective of place and time in the world. The view expressed by the Twitter community gives enormous information about them and the trend going on. To identify these trends and patterns, many data science techniques are used. However, using appropriate clustering technique always remains a challenge for researchers. The research concentrates on using hard clustering approaches like K-means and soft clustering approaches like Latent Dirichlet Allocation (LDA) and Latent Semantic Indexing (LSI). The proposed methodology uses K-means on numerical attributes, LDA, and LSI on textual attributes. The experiments are done using different Twitter datasets and tested on Sabarimala temple tweet dataset. The first time experimental study shows the promising correlation of the K-means cluster sentiments with topic sentiments. This helps in understanding the stance of the topics formed. The paper concludes by highlighting the relevance between the results of K-means clusters with the topics formed using LDA and LSI techniques.

Keywords LDA · LSI · Clustering · Sentiment · K-means · Twitter dataset

1 Introduction

Users around the world with diverse backgrounds express their opinion on any matter using various social media platforms. Social media platforms generate a lot of data. Diving into the world of huge data and drawing important inference out of that seems a most difficult task. This task is possible with the help of recent trends and techniques like clustering and classification.

P. V. Tijare (✉) · J. R. Prathuri
Department of Computer Science and Engineering (VTU RC), CMR Institute of Technology, Bengaluru, India

K. Khanna et al. (eds.), *Cyber Security and Digital Forensics*,
Lecture Notes on Data Engineering and Communications Technologies 73,
https://doi.org/10.1007/978-981-16-3961-6_38

Twitter data analysis involves dealing with multiple Twitter data attributes including different formats like date, numeric fields such as latitude and longitude values, and textual data such as tweets. Though these attributes hold different data formats, nevertheless they are highly coherent and correlated in terms of analyzing subject dominance, a stance of the subject or its sentiments. For example, tweets holding the negative sentiment tweeted from the border region of India holds more sensitivity. Border region can be identified by latitude and longitude values.

As Twitter data have a broad spectrum of opinions and topics, finding their sentiments will help in understanding the emotions related to a particular subject. Tweets on behavioral analysis make the messages tweeted by user as one of the most important attributes compared to other attributes discussed above to perform such analysis and to draw out inferences. Sentiment analysis of the tweets is one of the important sources to capture the emotions or opinions of people who tweeted. Grouping the tweets based on sentiments will help to understand the sensitivity of that subject in public, and it can be further used for many applications [1].

This paper proposes a novel method to perform clustering analysis of Twitter datasets. The proposed method concludes the coupling between numerical attributes and textual attributes. Numerical attributes considered as sentiment score of the tweets and textual attribute as 'Tweet' which are tweets twitted by Twitter uses.

Numerical values are used to cluster the tweets based on their sentiment score. The textual attribute is processed further to identify important topics. The correlation between the above two factors is derived by calculating the common tweets in both approaches. This helps to analyze the important topics and sentiments associated with those topics. This work focuses on the application of the K-means algorithm to cluster the data on numerical fields and topic modeling algorithms such as LSI and LDA to find clusters based on textual data such as tweets. Figure 1 displays the clustering techniques applied based on the type of attributes [2].

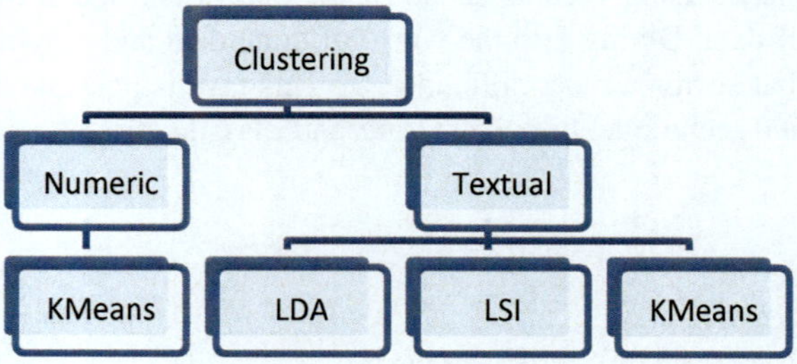

Fig. 1 Choosing clustering technique based on type of data

2 Related Work

Clustering is an important task in data analysis. The approaches used for clustering depend on the data. In textual clustering, the methodology uses sentiments, rule-based methods, lexicons, etc. When the clustering is applied on short texts like Twitter, the traditional methodologies give unpredictable results. Singh et al. highlighted the textual similarity in words used by social media users. Strength matrix is created for each user based on words used. WorldNet, K-means, and spectral K-means are used. Authors found that spectral K-means gives good results on sparse data at the expense of computational cost [6]. Asif et al. used the lexicon dictionary with the word intensity weight created to detect multilingual sentiment and found that the repeated words in the lexicon reduce the accuracy [7]. Han et al. build the model with word embeddings to create lexicons to interpret personality [8].

The rule-based approach for sentiment analysis is also popular for short text analysis. Singh and Sachan present the system SentiVerb using rule-based classifiers. The dictionary on opinion is created. The novel concept of negative parameter threshold is introduced [9]. Vashishtha and Susan proposed the fuzzy rule-based approach for sentiment analysis. The nine rules are formulated to find the sentiment of each tweet. The approach is tested on online datasets using SentiWordNet, AFINN and VADER lexicons [10]. Researchers also presented work on the anomalies with textual analysis on social media along with sentiment diffusion and sentiment reversal techniques. It is observed that self-supervised learning algorithms give better accuracy than traditional supervised algorithms [11, 12].

Various clustering methods used for short text involve hybrid clustering. The approach involves clustering on different levels, such as first on location and then on the text. The author used DBSCAN and K-means algorithms to compare results [13–15]. Self-learning-based max-margin clustering, the Silhouette score, and Calinski–Harabasz score and Tf/Idf-based word embeddings with topic modeling algorithm are used to improve the precision of clustering [16, 17]. Gopal et al. proposed method by tagging keyword to perform topic modeling using LDA. This method gives better result at the expense of time taken to tagging [18].

Multiple methods and approaches were proposed by the research community, each having its pros and cons. The approach used in this paper is based on error value, sentiment score, and its correlation with the topics generated on the different dataset is presented in Sects. 4 and 5.

3 Datasets

As mentioned in Sect. 1, there are heterogeneous data fields available in Twitter data and choosing only one type of dataset would result in biased implementation. To overcome this issue, a variety of Twitter-based datasets are considered. The two benchmark datasets used are ‘Twitter US Airline sentiments’ and ‘Social media disaster tweets' [4, 5]. The third dataset is on Sabarimala temple tweets. The Twitter Airline sentiments are the tweets expressed on US Airlines in the month of February 2015. The dataset has 26 fields. Social media disaster tweets dataset is collected from 'Data for everyone website.' This dataset classifies the tweets into two categories that are relevant and irrelevant to disaster.

The third dataset is generated from Sabarimala temple tweets from January 2019 to May 2019. The dataset tries to collect data of patriarchy culture about Kerala temple, India. Earlier, the temple was open only for men to offer prayers inside the premises of the temple. When the Supreme Court of India decided that the temple will also be open for women, it created huge discussions and repulsions from its devotees and became the topic of discussion. This dataset contains the tweets based on this event. Some of the important fields are datetime, tweets, and then sentiment and subjectivity are calculated for implementation. Variety of Twitter datasets will give better understanding on clustering output using different techniques [3].

Sabarimala temple tweet datasets subjectivity score for each tweet computed. This will help to understand that the tweets in the dataset are related to the stated subject. Figure 2 shows the subjectivity score for each tweet. The *x*-axis is an individual tweet and the y-axis is the score of the tweet ranging from 0 to 1, where 0 indicates not relevant or objective sentence and 1 indicates highly relevant to the subject. Almost all of the tweets collected from Twitter on Sabarimala temple topic shows high subjectivity score which proves that the dataset is relevant to the subject Sabarimala temple.

4 Methodology

The Twitter datasets have various types of data such as text, numeric, and spatial. The approach used for the analysis also changes based on the characteristics of data attributes presenting the approach used for analysis of datasets in Fig. 3. The figure describes the different clustering techniques used based on the type of data. The results obtained are then compared and correlated.

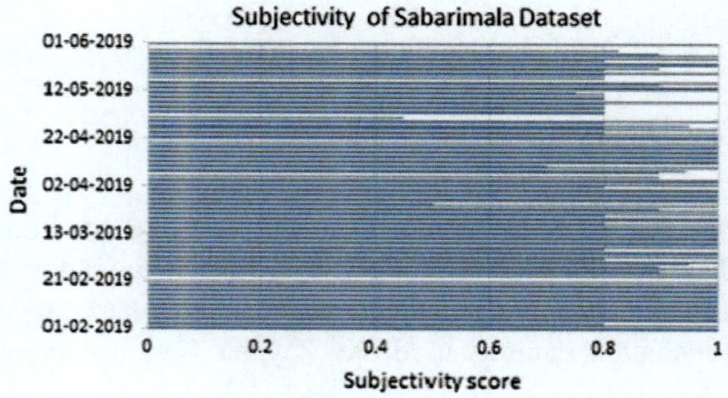

Fig. 2 Subjectivity score on Sabarimala temple dataset

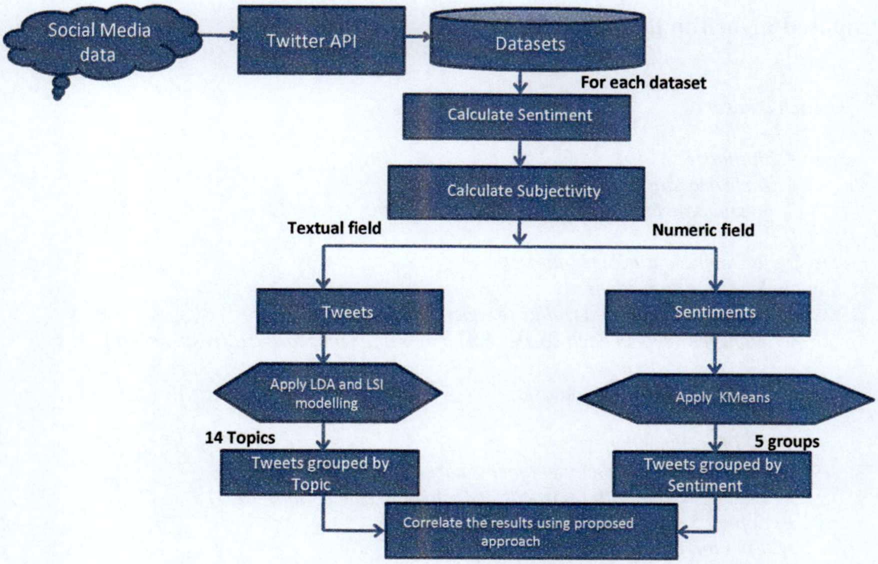

Fig. 3 Process to determine the correlation on Sabarimala temple verdict dataset

4.1 Proposed Algorithm

This section describes the algorithmic approach used for correlating clusters formed using techniques shown in Fig. 3. The parameters for correlation are sentiment score and topics formed using keywords in tweets. The sentiment score of the tweet is very important parameter which gives insight into emotions of the user. The proposed approach uses the density of the clusters formed using K-means based on sentiment scores and correlates with the results of topic modeling techniques applied.

Proposed algorithmic approach

```
for each dataset ds

step 1:   initialize:
          numeric_attribute
          textual attribute

step 2:  for numeric attribute_dataset:
    apply K-means algorithm:
          calculate center and radius for each cluster
          identify tweets in each cluster based on sentiment score (say resultset R1)

step 3:  for textual attribute_dataset:

    A) apply LDA algorithm
          Find topics and their sentiment score
          identify tweets with keywords in each topic (say resultset R21)
       B) apply LSI algorithm
          Find topics and their sentiment score
          identify tweets with keywords in each topic (say resultset R22)

step 4 :  Novel method to compare the clustering results to identify overlapping between
above results :

    for each cluster in resultset R1:            //K-means
       for each topic in resultset R21:    //LDA
          if sentiment_cluster_center ~ sentiment_topic:
             find common tweets and
             consider that as "overlapping_factor"

       for each topic in resultset R22:     //LSI
          if sentiment_cluster_center ~ sentiment_topic:
             find common tweets and
             consider that as "overlapping_factor"
```

For each dataset mentioned in Sect. 3, apply the K-means method on sentiment scores of each tweet. Form clusters and find the centers of each cluster. Identify the tweets in each cluster density. The tweets identified are considered as a result set R1. This process is needed to be done for all the datasets considered as input. Apply LDA and LSI topic modeling techniques for textual attribute column Tweet in each dataset. This will give topics with 15 words in each topic forming that topic. These topics with top words are considered as result set R21 and R22. Find the topic sentiment. Topic sentiment is average sentiment of all the tweets containing the words in that topic. Compare the topic sentiment score with the cluster center.

If the values are almost close, then find common tweets and the overlapping tweets in topics formed in step 3 and clusters formed in step 2. Repeat this process for all topics formed using LDA and LSI method. The stated approach successfully shows the correlation between the clusters formed using numerical attribute and textual attribute.

4.2 *Implementation and Results*

The algorithm described in Sect. 4.1 is applied to the datasets stated in Sect. 3. This section shows the implementation results on Twitter datasets.

4.2.1 Clustering Analysis Using K-means

As explained earlier about the dataset, the experiment is done by considering the text which involves the public tweets. After cleaning these tweets, the K-means algorithm applied on tweets of the datasets. Figure 4 shows the result of K-means applied on tweets of Sabarimala temple tweet dataset.

Figure 4 shows large number of clusters formed. These clusters are plotted using centroids. Majority of clusters are concentrated at the center of the graph. Clusters are not well separated. The clustering results on tweets were not worth considering, and the next analysis is done on sentiment scores. The sentiment and subjectivity scores found and added as an attribute.

Fig. 4 K-means labels on tweets

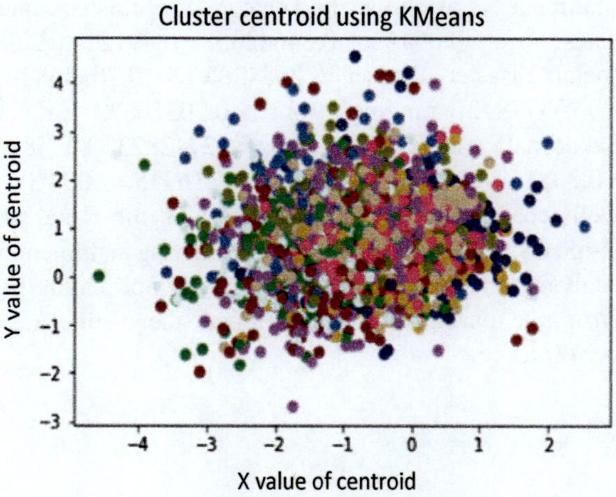

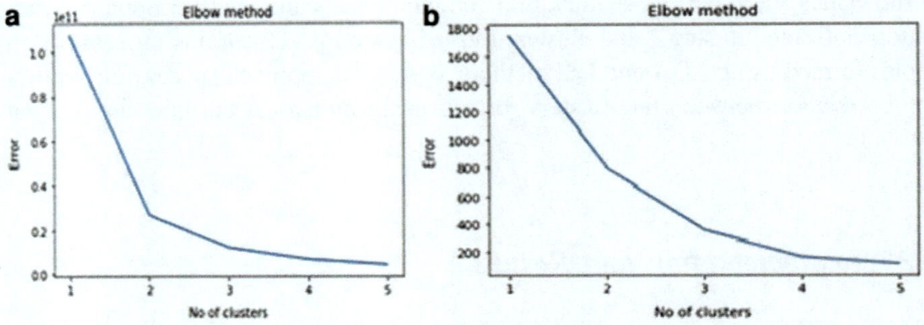

Fig. 5 K-means error rate **a** Benchmark datasets, **b** Sabarimala temple dataset

Usually, sentiments are grouped into three categories: positive, negative, and neutral. K-means is applied on sentiment with three clusters and five clusters. Analysis of cluster quality is done by plotting graphs with the error value. The curve in Fig. 5a, b shows the error curve formed using Elbow method with the K-means on benchmark and Sabarimala temple tweet datasets. The *x*-axis is number of clusters starting from 1 to 5, and *y*-axis is an error value for the number of clusters. Elbow method is used to find the appropriate number of clusters. This method is applied for clusters ranging from 1to 5, and average error is calculated. The distortion is computed using mean squared error method between cluster centers and the point. It is seen from Fig. 5a, b that the error value is nearing to zero after crossing number of clusters as three, stabilizes to almost zero with cluster size 5. The results obtained on the Airline and social disaster dataset are verified on the Sabarimala temple dataset. Based on the above analysis, it is concluded to use the number of clusters as five.

The centers for each cluster using the K-means method with five clusters of sentiments is as shown in Table 1. The cluster centers for US Airline Sentiment dataset are −0.004226, 0.6464205, −0.89121, 0.2706736, −0.34945168 and for social disaster dataset 0.26843858, −0.70356829, 0.59311648, 0.00351392, −0.23977812, ranging from −0.70356829 to 0.6464205. The centers for Sabarimala temple dataset are −0.12742823, 0.0358207, 0.53544311, 0.21365713, −0.39216915, ranging from −0.39216915 to 0.59311648. The centers formed are well separated on all the datasets using the K-means method on sentiments. The K-means analysis is also done considering sentiment and subjectivity scores. In the analysis, it is found that subjectivity is not adding much value with the increased error rate during clustering. Hence, the results are not considered due to poor performance.

Table 1 Center of five clusters found using K-means

Dataset	Cluster center for 5 clusters				
Airline dataset	−0.00422636	0.6464205	−0.89121	0.2706736	−0.34945168
Social disaster dataset	0.26843858	−0.70356829	0.59311648	0.00351392	−0.23977812
Sabarimala temple dataset	−0.12742823	0.0358207	0.53544311	0.21365713	−0.39216915

4.2.2 Clustering Analysis of Text Data Using LDA and LSI

The result of K-means was not up to the mark on tweets (textual data). The topic modeling algorithms LDA and LSI are used to perform micro topic modeling on the Airline, social disaster dataset, and Sabarimala temple tweet dataset. These methods processes cleaned tweets and state the number of micro-topics. The decision on the number of micro-topics can be done by performing LDA GridSearchCV method with the number of topics as [5, 10, 15, 19, 20] and 'learning_decay' as [0.5, 0.7, 0.9]. This method takes combinations of both the parameters. The combination which gives best score is taken for optimal LDA model.

The log-likelihood score help in understanding the quality of fit. The perplexity is a measure used to decide the prediction ability of the probability model on a given sample. The log-likelihood score and perplexity help in deciding the ideal model. Higher log-likelihood score is always preferable. Table 2a shows the log-likelihood and perplexity scores of LDA. Table 2b shows the dominance of each topic and the number of tweets forming each topic. The 'Topic' in Table 2b indicates the clusters formed on tweets. Table 2b shows that topic 0 in US Airline sentiment dataset and topic 1 in the social disaster dataset is the most dominant topic formed using a maximum number of tweets compared to other topics.

Figures 6 and 7 show a log-likelihood curve for the three learning tests for choosing optimal LDA model for US Airline sentiment and social disaster dataset. The curve is formed using three learning rates as 0.5 (Blue), 0.7 (Orange), and 0.9 (Green). The best learning rate for dataset US Airline sentiment and social disaster dataset has come as 0.9 and the number of clusters as 5. The final LDA model is formed, and the resulting clusters are visualized using pyLDAvis Python package.

Table 2 **a** Best scores on datasets. **b** Dominance of topics on tweets on datasets

(a)

Dataset	Best log-likelihood score	Model perplexity
US Airline	−28,862.72	1183.2674
Social disaster	−24,552.78	887.6521

(b)

Topic	Airline tweets	Social tweets
0	7400	1932
1	6496	2150
2	7233	1504
3	7297	1420
4	6063	1198

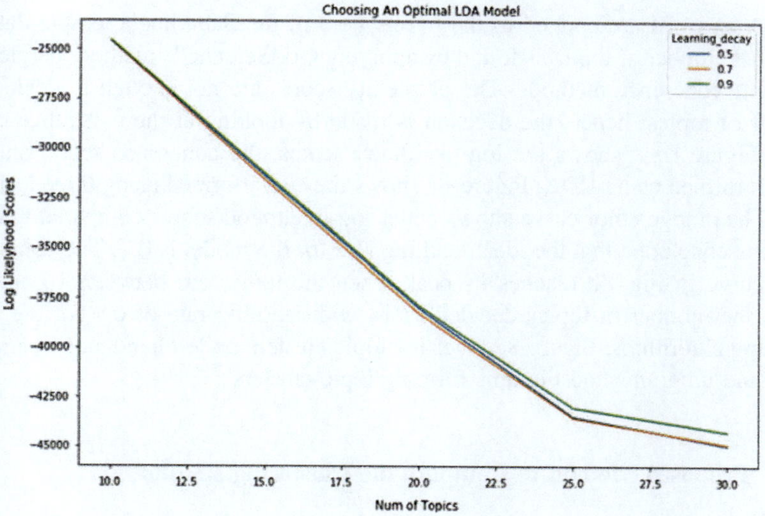

Fig. 6 Log-likelihood score to choose optimal LDA model

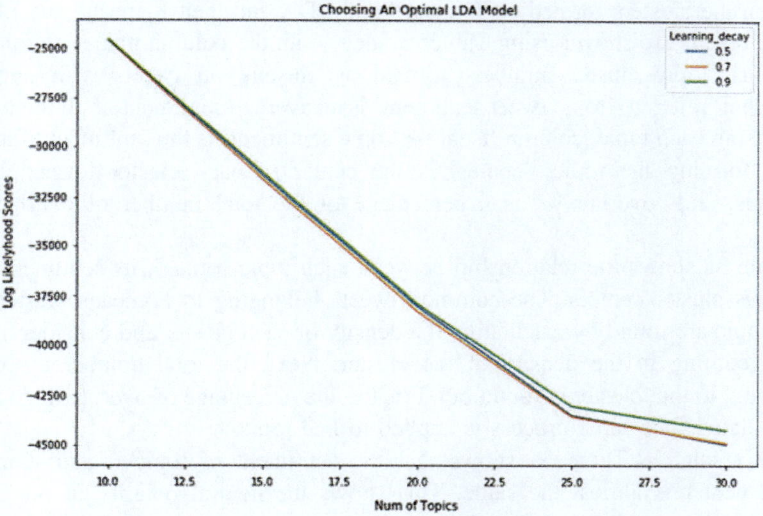

Fig. 7 Log-likelihood score to choose optimal LDA model

The topics are formed using the LSI method with the number of topics as 5 and log-likelihood score as in Table 2a. The topics formed using LDA and LSI models are further analyzed by finding their sentiment scores. This analysis will help in understanding the topic wise sentiment flow.

LDA and LSI are applied to the tweets field of the Sabarimala temple dataset. The ideal number of topics is found by applying GridSearchCV method, perplexity, and topic coherence methods. The perplexity scores are not enough to decide the number of topics; hence, the decision is made by looking at the coherence curve score. Figure 8a–c shows the log-likelihood scores, the coherence score, and the clusters formed using LDA. Figure 8a shows the curve formed using three learning rates. The orange color curve shows better log-likelihood scores compared to other two. It is concluded that the ideal learning rate for the model is 0.7. The coherence score curve in Fig. 8b reaches its peak when the topics are between 10 and 15; hence, the number of topics decided as 14 with learning rate of 0.7 for the topic modeling algorithms. Figure 8c gives the topic clusters on left hand side in circular shape and unigrams and bigrams forming topic clusters.

4.2.3 Proposed Method to Compare the Clustering Results

The results are compared and correlated using the algorithmic approach proposed in Sect. 4.1. This method is applied on the clusters generated by K-means and topic modeling methods. The clusters are compared by finding the percentage of overlap between the clusters formed by K-means and LDA and then K-means and LSI.

The results are shown using Tables 3 and 4 with the column names as follows: 'Tno' as 'topic cluster number' formed as clusters on tweets with number, 'tweetsent' refers to 'topic tweet sentiment' is an average sentiment of all the tweets comprising each topic, column 'tsent' is 'topic sentiment' is the sentiment of all the words forming the topic, 'center' is the center of each cluster formed using K-means, and 'overlap %' is a percentage of the total number of overlapping tweets.

Table 3a shows the relationship between each topic formed, its sentiment, and K-means cluster centers. The common tweets belonging to K-means cluster and LDA topic are found by calculating the density of each cluster and considering the tweets coming in the density of the cluster. Next, the total number of tweets belonging to that cluster is found out. Finally, the percentage of overlapping tweets is calculated. The same process is applied to LSI topics.

The results in Table 3a shows that the sentiment of topic 0 and K-means cluster center is almost the same. This shows the neutral polarity of that topic with more than 65% similarity. Topic 1 has sentiment −0.2 with K-means center as −0.2397 with 43% of overlapping tweets. Similarly, Table 3b shows the result of US Airline sentiment dataset using LSI and K-means. The 'topic tweet sentiment' for topic 0 is −0.5 and % of overlapping tweets more than 56% of rows 1, 3, and 4 of topic 0 in Table 3b. The remaining topics are also showing a high overlapping percentage.

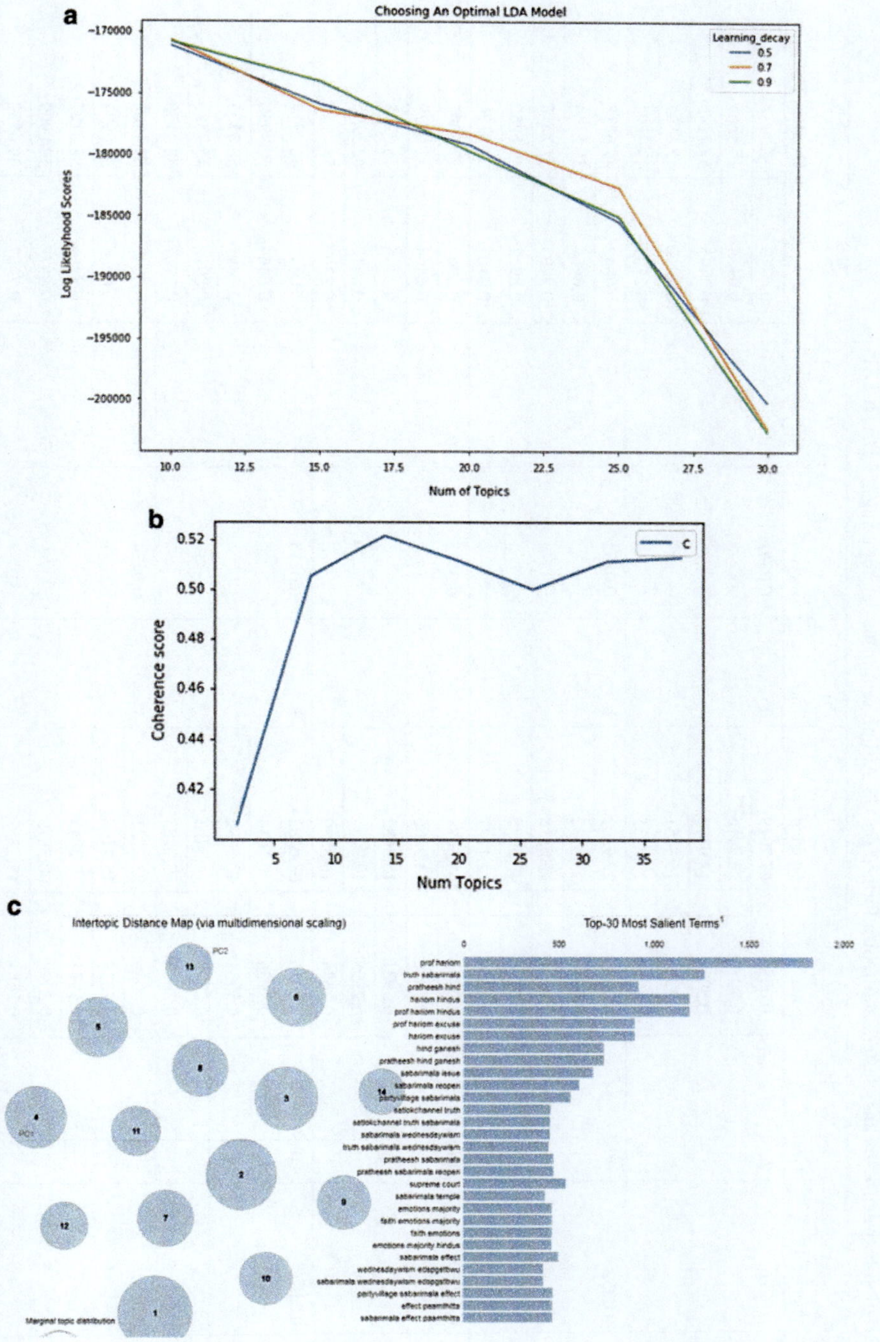

Fig. 8 **a** Log-likelihood curve, **b** coherence score on number of topics, **c** clusters using LDA on Sabarimala temple tweet dataset

Table 3 Relation between K-means clusters and topics formed on US Airline sentiment dataset using **a** LDA, **b** LSI

(a)					(b)				
Tno	tweetsent	tsent	center	Overlap %	Tno	tweetsent	tsent	center	Overlap %
0	0.0226	0	0.0035	67.7019	0	*−0.0381*	−0.5	−0.0042	*57.7432*
			−0.7036	23.2416				0.6464	46.5365
			0.5931	13.9773				−0.3495	*57.1145*
			−0.2398	15.8451				−0.8912	*69.8864*
			0.2684	17.2535				0.2707	49.4204
1	0.0253	*−0.2*	0.0035	22.9302	1	0.0382	*0.1*	−0.0042	49.6213
			−0.7036	23.8532				0.6464	35.5164
			0.5931	15.6818				−0.3495	*51.7538*
			−0.2398	*43.5035*				−0.8912	40.6250
			0.2684	22.1127				0.2707	53.8176
2	0.0600	0.1667	0.0035	63.5638	2	0.0415	*0.2*	−0.0042	47.8092
			−0.7036	11.9266				0.6464	51.0579
			0.5931	15.5682				−0.3495	46.3203
			−0.2398	12.3239				−0.8912	42.3295
			0.2684	16.3380				*0.2707*	58.4930
3	0.0182	−0.0633	0.0035	68.0282	3	0.0373	*−0.3*	−0.0042	46.5724
			−0.7036	10.7034				0.6464	40.9950
			0.5931	10.5682				*−0.3495*	58.5705
			−0.2398	16.5493				−0.8912	45.7386
			0.2684	9.7183				0.2707	38.4158
4	0.0321	−0.35	0.0035	23.2721	4	0.0224	*−0.5333*	−0.0042	57.3808
			−0.7036	16.8196				0.6464	34.6977
			0.5931	9.2045				*−0.3495*	52.8127
			−0.2398	60.3873				*−0.8912*	59.0909
			0.2684	15.5259				0.2707	39.6832

Table 4 Relation between K-means clusters and topics formed on social disaster dataset using **a** LDA, **b** LSI

(a)					(b)				
Tno	tweetsent	tsent	center	Overlap %	Tno	tweetsent	tsent	center	Overlap %
0	0.0226	*0*	*0.0035*	67.7019	0	0.0388	*0*	*0.0035*	63.8129
			−0.7036	23.2416				0.5931	24.5455
			0.5931	13.9773				−0.7036	22.9358
			−0.2398	15.8451				−0.2398	25.6162
			0.2684	17.2535				0.2684	23.3099
1	0.0253	*−0.2*	0.0035	22.9302	1	0.0476	*0.1*	*0.0035*	63.3317
			−0.7036	23.8532				0.5931	21.5909
			0.5931	15.6818				−0.7036	21.4067
			−0.2398	43.5035				−0.2398	18.3099
			0.2684	22.1127				0.2684	20.9859
2	0.0600	*0.1667*	*0.0035*	63.5638	2	0.0153	*0*	*0.0035*	68.2550
			−0.7036	11.9266				0.5931	9.2045
			0.5931	15.5682				−0.7036	13.4557
			−0.2398	12.3239				−0.2398	15.7570
			0.2684	16.3380				0.2684	13.3099
3	0.0182	*−0.0633*	*0.0035*	68.0282	3	0.0376	*0*	*0.0035*	70.8231
			−0.7036	10.7034				0.5931	10.2273
			0.5931	10.5682				−0.7036	6.1162
			−0.2398	16.5493				−0.2398	10.7394
			0.2684	9.7183				0.2684	10.2689
4	0.0321	−0.35	0.0035	23.2721	4	0.0324	*0*	*0.0035*	64.8714
			−0.7036	16.8196				0.5931	21.3636
			0.5931	9.2045				−0.7036	21.4067
			−0.2398	60.3873				−0.2398	26.6725
			0.2684	15.5259				0.2684	22.0423

Table 5 Relation between K-means clusters and topics formed on Sabarimala temple dataset using **a** LDA, **b** LSI

(a)						(b)					
Tno	K-means cluster center	Total tweets	K-means	Common tweets (3 and 4)	Topic sentiment	Tno	K-means cluster center	Total tweets	K-means	Common tweets (3 and 4)	Topic sentiment
1	0.0035	124	7111	80	*−0.6*	1	*0.1940*	124	*65*	57	0.3167
	0.5931		880	11			−0.1375		29	25	
	−0.7036		*327*	*94*			0.5354		6	6	
	−0.2398		1136	15			−0.3922		6	6	
	0.2684		1420	14			0.0205		40	30	
5	*0.0035*	10	*7111*	*6*	*−0.125*	5	0.1940	10	40	3	−0.05
	0.5931		880	1			−0.1375		29	0	
	−0.7036		327	0			0.5354		6	0	
	−0.2398		1136	2			−0.3922		6	0	
	0.2684		1420	1			0.0205		65	7	
6	0.0035	29	7111	3	*0.2571*	6	0.1940	29	40	8	0
	0.5931		880	2			−0.1375		29	5	
	−0.7036		327	1			0.5354		6	1	
	−0.2398		1136	3			−0.3922		6	0	
	0.2684		*1420*	*20*			0.0205		65	15	
10	*0.0035*	17	*7111*	*10*	*0.3167*	10	0.1940	17	40	3	0
	0.5931		880	2			−0.1375		29	3	
	−0.7036		327	0			0.5354		6	1	
	−0.2398		1136	1			−0.3922		6	0	
	0.2684		1420	13			0.0205		65	10	
11	*0.0035*	6	*7111*	1	*0.33*	11	0.1940	6	40	2	0
	0.5931		880	3			−0.1375		29	1	
	−0.7036		327	0			0.5354		6	0	
	−0.2398		1136	1			−0.3922		6	0	
	0.2684		1420	1			0.0205		65	3	

Table 4a shows the result of LDA compared with K-means cluster centers. The overall sentiment value for 'Topic 0' is 0, and K-means cluster center value is 0.0035. Topic 1 has −0.2 topic sentiment and K-means cluster center value for 4th cluster is −0.2398. Similarly for other topics also values are very close. Table 4b shows the LSI topics of social disaster dataset. The topic sentiment for all the topics is the same as the first cluster center using K-means. The analysis is done on the results from the US Airline sentiment dataset, and the social disaster dataset shows that the topic sentiment matches with K-means cluster center, and the results show more than 60% overlapping tweets.

Table 5a, b shows the result for the Sabarimala temple tweet dataset using approach mentioned in Sect. 4.1. Topic 1 of Table 5a with cluster center −0.7036 has 94 tweets common with the average sentiment of −0.6 which is quite close to the cluster center value. Topic 5 has an average sentiment value of −0.125, and cluster center at 0.0035 has 6 tweets in common. Topic 6 has K-means center of 0.2684 and topic sentiment as 0.2572. Topic 10 with the last cluster values as 0.2685 is close to the topic average sentiment value 0.3166. Similarly, the results in Table 5b also show a lot of closeness in sentiment and cluster center values with the topics formed using LSI.

Considering the results shown in Tables 3, 4, and 5, it is observed that the topics formed using topic modeling algorithm on tweets has more than 60% similarity with the clusters formed using sentiment.

5 Conclusion and Future Work

Twitter datasets on US Airline sentiment and social disaster are evaluated and tested on Sabarimala temple event dataset created with the help of LDA, LSI, and K-means clustering approach. The clustering is done on numerical values like the sentiment of each tweet and textual values as tweets tweeted by users. The analysis done using proposed approach shows significant correlation between the clusters formed using K-means and the topics formed using techniques LDA and LSI for each dataset. The average sentiment of each topic and the cluster centers is showing matching polarity at many places. It is observed that there is a maximum overlapping of tweets for the most relative centers and topics. Further, this work can be extended to find an improved correlation between each topic and its associated sentiment. Analysis on blog can be done after finding important topic using above work to find opinion and perform stance analysis. Results can be more precisely validated by comparing events from the blogs by performing analysis on blogs.

References

1. Kursuncu, U., Gaur, M., Lokala, U., Thirunarayan, K., Sheth, A., Arpinar, I.B.: Predictive analysis on Twitter: techniques and applications. In: Emerging Research Challenges and Opportunities in Computational Social Network Analysis and Mining, pp. 67–104. Springer, Cham (2019)
2. Crockett, K.A., Mclean, D., Latham, A., Alnajran, N.: Cluster analysis of Twitter data: a review of algorithms. In: Proceedings of the 9th International Conference on Agents and Artificial Intelligence, vol. 2, pp. 239–249. Science and Technology Publications (SCITEPRESS)/Springer Books (2017)
3. Chronology of events on entry of women into Sabarimala temple: New Delhi, November 14, 2019 IST 17:11 Updated: November 14, 18:29 IST. https://www.thehindu.com/news/national/chronology-of-events-on-entry-of-women-into-sabarimala-temple/article29972784.ece (2019)
4. Twitter US Airline Sentiment: Analyze how travelers in February 2015 expressed their feelings on Twitter. Updated: Wed October 16, 2019 IST 5:34:05. https://www.kaggle.com/crowdflower/Twitter-airline-sentiment (2015)
5. Disasters on social media: Which tweets are relevant news and which are just banter? Updated: Sun 15 April 2018 IST 12:34:59. https://www.kaggle.com/jannesklaas/disasters-on-social-media (2018)
6. Singh, K., Shakya, H.K., Biswas, B.: Clustering of people in social network based on textual similarity. Perspect Sci. **8**, 570–573. https://doi.org/10.1016/j.pisc.2016.06.023 (2016)
7. Asif, M., Ishtiaq, A., Ahmad, H., Aljuaid, H., Shah, J.: Sentiment analysis of extremism in social media from textual information. Telematics Inform. **48**, 101345. https://doi.org/10.1016/j.tele.2020.101345 (2020)
8. Han, S., Huang, H., Tang, Y.: Knowledge of words: an interpretable approach for personality recognition from social media. Knowl. Based Syst, 105550. https://doi.org/10.1016/j.knosys.2020.105550 (2020)
9. Singh, S.K., Sachan, M.K.: SentiVerb system: classification of social media text using sentiment analysis. Multimedia Tools Appl. **78**(22):32109–32136. https://doi.org/10.1007/s11042-019-07995-2 (2019)
10. Vashishtha, S., Susan, S.: Fuzzy rule based unsupervised sentiment analysis from social media posts. Expert Syst. Appl. **138**, 112834. https://doi.org/10.1016/j.eswa.2019.112834 (2019)
11. Kokatnoor, S.A., Krishnan, B.: Self-Supervised Learning Based Anomaly Detection in Online Social Media. https://doi.org/10.22266/ijies2020.0630.40
12. Wang, L., Niu, J., Yu, S.: SentiDiff: combining textual information and sentiment diffusion patterns for Twitter sentiment analysis. In: IEEE Transactions on Knowledge and Data Engineering. https://doi.org/10.1109/TKDE.2019.2913641 (2019)
13. Cresci, S.: Detecting malicious social bots: story of a never-ending clash. In: Multidisciplinary International Symposium on Disinformation in Open Online Media, pp. 77–88. Springer, Cham (2019)
14. Jiang, Y., Li, Z., Ye, X.: Understanding demographic and socioeconomic biases of geotagged Twitter users at the county level. Cartogr. Geogr. Inf. Sci. **46**(3), 228–242 (2019)
15. Bamakan, S.M., Nurgaliev, I., Qu, Q.: Opinion leader detection: a methodological review. Expert Syst. Appl. **115**, 200–222 (2019)
16. Gupta, S., Banerjee, B.: Unsupervised event detection using self-learning-based max-margin clustering: analysis on streaming tweets. IETE J. Res, 1–0 (2019)

17. Curiskis, S.A., Drake, B., Osborn, T.R. and Kennedy, P.J.: An evaluation of document clustering and topic modelling in two online social networks: Twitter and Reddit. Inf. Process. Manag. **57**(2), 102034 (2020)
18. Gopal, G.N., Kovoor, B.C. and Mini, U.: Keyword template based semi-supervised topic modelling in tweets. In: International Conference on Innovative Computing and Communications, pp. 659–666. Springer, Singapore (2021)

Design and Analysis of 2 × 4 Microstrip Patch Antenna Array with Defected Ground Structure for 5G Mobile Communication

Sameena Zafar, Vineeta Saxena, and R. K. Baghel

Abstract Wireless communication application is not possible without using the antenna. Nowadays, microstrip patch antenna (MPA) is using broadly due to its major advantages than others. Most of the electronics device is using MPA. There are various shapes, and pattern is proposed by antenna researchers. Microstrip patch antenna provides better performance in wireless communication applications. This paper proposed a novel design of microstrip patch antenna array for 5G Wi-Fi, wi-max applications. The CST microwave studio software is used to make a proposed antenna design and simulation. The resonant frequency of this antenna is 6.9 GHz. Overall bandwidth achieved by proposed antenna is 903 GHz. The large bandwidth is required for 5G communication applications.

Keywords Microstrip · Patch · Antenna · CST · FR4 · VSWR · Return loss

1 Introduction

It is generally utilized in versatile remote gadgets in light of the simplicity of creating it on printed circuit sheets. Numerous fix reception apparatuses on a similar substrate called microstrip radio wires can be utilized to make high pick-up cluster receiving wires and staged exhibits in which the shaft can be electronically directed. With the headway of distant correspondence propels, it has gotten continuously appealing for present day specific contraptions to facilitate diverse correspondence rules, for instance, 2G/3G/4G/5G. Consequently, radio wires with broadband execution are well known for multi-standard incorporation. To achieve broadband movement, various plans of microstrip gathering mechanical assemblies have been

S. Zafar (✉) · V. Saxena
UIT RGPV, Bhopal, India

R. K. Baghel
MANIT, Bhopal, India

K. Khanna et al. (eds.), *Cyber Security and Digital Forensics*,
Lecture Notes on Data Engineering and Communications Technologies 73,
https://doi.org/10.1007/978-981-16-3961-6_39

represented, for instance, adjusting the condition of the microstrip arms, improving dealing with methodologies, stacking parasitic radiators, and using magneto-electric complementary structures [1].

Besides, radiation execution is furthermore needed for some distant correspondence applications, for instance, indoor sign consideration, far off sections, and little scope base stations. Considering the multimode PFDA, a clear and incredible structure to achieve radiation execution is made by setting two of the proposed PFDAs successive. The ensuing gathering contraption shows extraordinary radiation plans in the even plane with level increment assortment of under 1.27 dB. The presented traditionalist broadband radio wire is a fair competitor for indoor sign incorporation. Three reverberating modes are gotten by using a changed planar imploded microstrip and its coupled dealing with structure. Combining the shorting pins and parasitic patches, various resonating modes in the reception apparatus are controlled, moved, and a while later joined to fabricate the impedance information transmission. Using this thought, a model of a multimode fell microstrip is arranged, made, and assessed [1]. Normal coupling between two microstrip reception apparatus with different estimations put at self-assured equivalent positions is bankrupt down using simultaneous key conditions with unequivocal parts and restricted opening feeds [2]. A story plan procedure for a wideband twofold delighted gathering device is presented by using shorted microstrips, consolidated baluns, and crossed feed lines. Reenactment and proportionate circuit examination of the reception apparatus are given.

To endorse the arrangement procedure, a receiving wire model is arranged, updated, produced, and assessed as depicted in Fig. 1 [3]. An epic super wide band immovably coupled microstrip reflects exhibit radio wire is presented in this work. This reflects cluster radio wire involves a wideband feed and a wideband reflecting surface. By joining the advantages of reflecting cluster gathering contraptions and those of solidly coupled show radio wires, the proposed TCDR reception apparatus achieves ultra-wide exchange speed with diminished capriciousness and creation cost [4].

Microstrip radio wire is especially notable considering the way that the information transmission of more modest scope strip microstrip receiving wire is high when appeared differently in relation to the scaled down scale strip fix gathering contraption. Little scope strip microstrip gathering mechanical assembly will be the

Fig. 1 Microstrip antenna

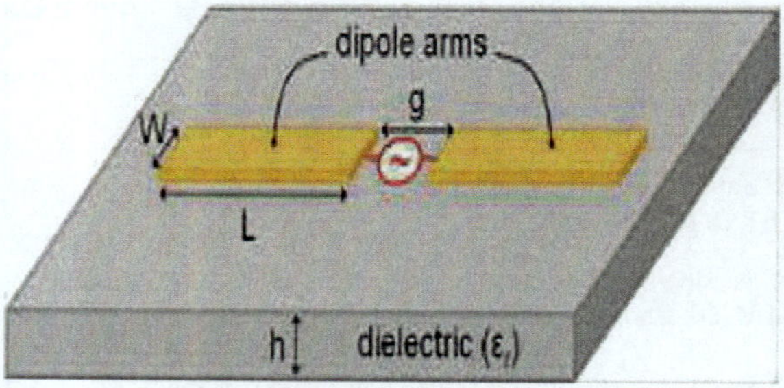

rule point of convergence of this work. A scaled down scale strip microstrip is a utilization of the standard microstrip radio wire on a dielectric piece, which can be easily made with existing PCB systems. Rather than standard little scope strip gathering device, that use one side of the piece as the ground plane, the more modest scope strip microstrip basically uses the dielectric segment as the host material. This radio wire is picked considering the way that it is direct, however, then has potential for future improvement. A traditionalist wideband twofold entranced reception apparatus with improved upper out-of-band camouflage is presented in this work. The proposed gathering mechanical assembly is a corresponding course of action of two electric microstrips and two appealing microstrips using the crossed shunt circles. The breaker of the electric and alluring microstrips enables the radio wire to achieve wide impedance bandwidth and a moderate radiator size. Likewise, four parasitic strips are inserted near the inside edge of the four circles to overhaul the radio wire upper out-of-band disguise at 3.5 GHz with improved impedance information move limit.

In the upcoming sections in this paper, authors will focus on background study of design and analysis of 4 × 4 microstrip patch antenna array with defected ground structure for 5G mobile communication in Sect. 2. Proposed design and results are presented in Sects. 3 and 4 that will highlight on the model building for predictive analysis, followed by conclusion and references.

2 Background

A. Singh et al. present a ten-part multi-gathering mechanical assembly terminal for huge different data diverse yield (MIMO) in the fifth-age (5G) sub-6 GHz with wide impedance bandwidth (IBW) is proposed. The most outrageous reproduced envelope relationship coefficient is viewed as 0.21 and remembering that the base receiving wire adequacy is 78.4%. Each element has a structure factor of 17.2 $\times 3.8\text{mm}^2$ and engraved on a thick Rogers 4003 substrate. The proposed multi-antenna terminal is suitable for 5G sub-6 GHz (3.2–6.1 GHz) correspondence including long-term evolution bands 42, 43 and 46 [1].

Ojaroudi Parchin et al. present another various information/different yield (MIMO) gathering mechanical assembly setup is introduced for future PDAs. The proposed structure contains four arrangements of twofold dealt with round ring resonators arranged at different edges of the mobile phone printed circuit board (PCB) with a FR-4 substrate and a segment of 75 × 150 mm^2. A respectable repeat information move limit (S 11 ≤ -10 dB) of 3.3–3.9 GHz has been gotten for the mobile phone receiving wire group. Regardless, for S 11 ≤ -6 dB, this value is 3.1–4.3 GHz. More than 3 dB recognized increment and 80% complete capability are refined for the single-segment radiator. The structure gives not simply satisfactory radiation consideration supporting different sides of the mainboard yet also the polarization grouped assortment [2].

Feng et al. present twofold straightforwardly charmed (LP) high-demand mode radio wire with high increment, and high separation is proposed for 5G millimeter-wave (mm-wave) applications. To get high increment and wide information move limit, a 2 × 2 space dealt with magneto-electric (ME)-group receiving wire segments are stimulated by a high-demand mode TM 430 cavity. Assessment results show that a covered repeat move speed of 14.6% (36.8–42.6 GHz) with a zenith increment of 25.8 dBi and a withdrawal of more significant than 45 dB for the 8 × 8 twofold LP gathering mechanical assembly bunch are refined, which guarantee strong fast data transmission and against impedance limit regarding 5G exchanges [3]. Hussain et al. present multiplication and assessment outcomes of a wideband planar organized bunch for 5G. The ideal wideband action is cultivated using an immovably coupled group display (TCDA). The group has a height of simply 0.144λ l, and it shows high expands, high profitability >71% (greatest yield point), and extraordinary cross-polarization. The reenactments are endorsed by production and assessment of a display model, which shows agree to reproduced regards. The proposed twofold delighted radio wire bunch can be sent later on bar checking applications for 5G [4].

Serhsouh et al. present novel reconfigurable opened flawed wave gathering contraption (LWA) considering a substrate joined waveguide (SIW) with a fixed—repeat bar controlling capacity is presented in this paper. The pin diodes are tunable by changing the DC inclination voltage, which achieves column checking at a fixed repeat of around 27 GHz. In this manner, an electronically controlled steerable SIW radio wire has been organized and likely watched that the radiation edge shifts from −33° to +33 [5]. Pérez et al. present concerning the structure and masterminding of new radio interfaces for the fifth-age (5G) systems, and this paper presents a supportive obligation to the depiction of the wideband indoor radio redirect in the 3–4-GHz repeat band. An assessment campaign has been done in two unmistakable indoor circumstances to look at indisputably the main wideband limits of the expansion channel, including a cautious assessment of its direct to meet the new radio advancement challenges [6].

3 Proposed Antenna Design

Figure 2, showing top perspective on proposed microstrip antenna array, one side of a dielectric substrate goes about as an emanating microstrip and opposite side of substrate goes about as ground plane. As in Fig. 2 demonstrated as follows, top perspectives on a rectangular microstrip radio wire with coaxial feed have microstrip, and ground plane together makes bordering fields, and this field is answerable for making the radiation from the receiving wire. Microstrip reception apparatus is proposed because of little size and better reference. Resonant frequency of proposed reception apparatus is approx. 6.9 GHz that implies it work under 5G C-band. Thus, proposed radio wire ought to be valuable for all C-band application.

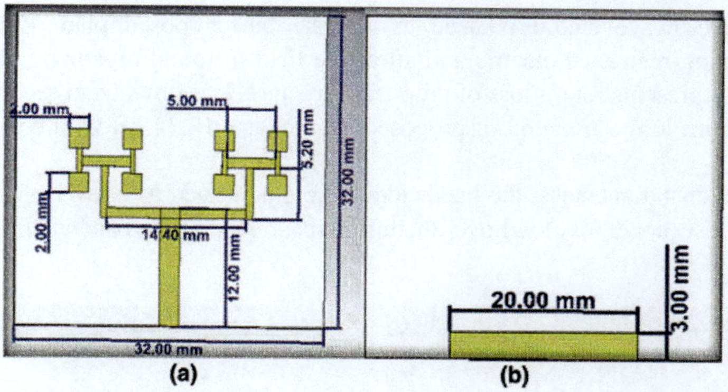

Fig. 2 **a** Top view, **b** defected ground structure of proposed microstrip antenna array

Above figure is indicating proposed microstrip antenna of plan. The top and ground layer is made by lossy copper material, and substrate is made by FR4 material which having 4.4 dielectric steady worth.

4 Result Analysis

The geometry of the proposed structure of microstrip for C-band applications is appeared in Fig. 2. The general size of the structure is 32 mm × 32 mm × 1.64 mm (L × W × H) and imprinted on fire resistant 4 (FR4), with an overall permittivity of 4.4, and a loss digression of 0.024. Table1 records the element of the proposed antenna. The antenna is taken care of by 50-Ω and 0.5 W coaxial link or straightforward. The antenna utilizes the microstrip structure with one opening for C-band applications.

Table 1 Design parameters for proposed antenna

S.No.	Parameter	Value
1	Lower frequency (f_L)	4 GHz
2	Higher frequency (f_H)	8 GHz
3	Dielectric constant (ε_r)	4.4/FR4
4	Ground ($L \times W$)	3 mm × 120 mm
5	Ground height	0.0.35 mm
6	Substrate ($L \times W$)	32 mm × 32 mm
7	Substrate height (h)	1.57 mm
8	Line impedance	50 Ω
9	Tangent loss	0.06
10	Input watt	0.5 W

CST microwave studio is used to recreate the proposed plan. Figure 3 is demonstrating reenacted electric and attractive field in round organize framework.

Figure 4 presents return loss of proposed structure. It is obvious to see this chart, and the return loss estimation of proposed antenna is −18.34 dB with 6.9 resonant frequency.

For broadband antennas, the bandwidth is communicated as a level of the recurrence contrast (upper less low) over the inside recurrence of the bandwidth as shown

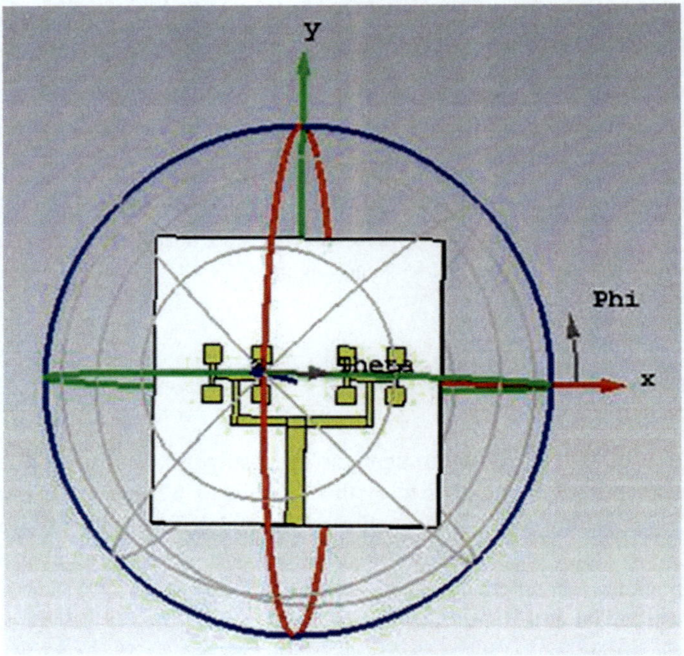

Fig. 3 Simulation and fields of proposed antenna

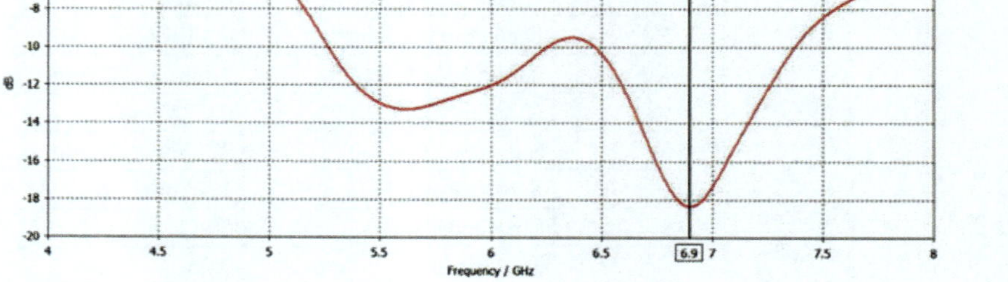

Fig. 4 Return loss

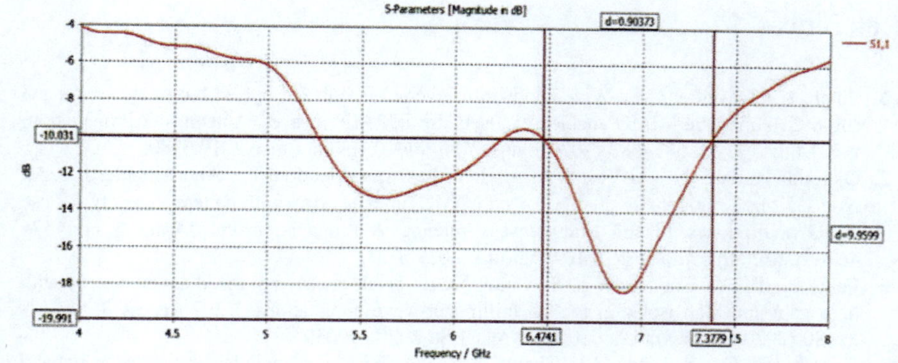

Fig. 5 Bandwidth

Table 2 Simulated results of proposed antenna

Sr. No.	Parameter	Value
1	Return loss or S11	−18.34 dB
2	Bandwidth	903.73 MHz
4	Resonant frequency	6.9 GHz

in Fig. 5. The bandwidth of proposed antenna is 903.73 MHz, (7.3779–6.4741 GHz), for optimized band. It is a variety of the force transmitted by an antenna as an element of the heading ceaselessly from the antenna.

Table2 shows about terms of every vital parameter like return loss, bandwidth, VSWR, and resounding recurrence. It is clear by observing reenacted values from Table 2, and proposed antenna accomplishes significant improved outcome.

5 Conclusion and Future Scope

A microstrip antenna array is planned and recreated utilizing CST reenactment programming. The reproduction results are introduced and examined. Structure of proposed antenna is basic and reduced in size of 32 × 32 × 1.6 mm^3. Reenacted results exhibit that the antenna bandwidth covers S-band and C-band, at full recurrence 6.9 GHz. The bandwidth is huge accomplished better than microstrip fix antenna structure. Microstrip antenna array configuration is recently exploring theme among specialists. The general bandwidth is 903.73 MHz. Subsequently, proposed antenna is reasonable and meets to current correspondence requests. In the future, above design can be implemented for higher number of elements thereby increasing the array size, and thus, better results can be expected using slots or fractal design.

References

1. Singh, A., Saavedra, C.E.: Wide-bandwidth inverted-F stub fed hybrid loop antenna for 5G sub-6 GHz massive MIMO enabled handsets. In: IET Microwaves, Antennas & Propagation, vol. 14, no. 7, pp. 677–683 (2020). https://doi.org/10.1049/iet-map.2019.0980
2. Ojaroudi Parchin, N., Al-Yasir, Y.I.A., Jahanbakhsh Basherlou, H., Abd-Alhameed, R.A., Noras, J.M.: Orthogonally dual-polarised MIMO antenna array with pattern diversity for use in 5G smartphones. In: IET Microwaves, Antennas & Propagation, vol. 14, no. 6, pp. 457–467 (2020). https://doi.org/10.1049/iet-map.2019.0328
3. Feng, B., Tu, Y., Chen, J., Yin, S., Chung, K.L.: Dual linearly-polarized antenna array with high gain and high isolation for 5G millimeter-wave applications. IEEE Access **8**, 82471–82480 (2020). https://doi.org/10.1109/ACCESS.2020.2990494
4. Hussain, S., Qu, S., Zhou, W., Zhang, P., Yang, S.: Design and fabrication of wideband dual-polarized array for 5G wireless systems. IEEE Access **8**, 65155–65163 (2020). https://doi.org/10.1109/ACCESS.2020.2984613
5. Serhsouh, I., Himdi, M., Lebbar, H., Vettikalladi, H.: Reconfigurable SIW antenna for fixed frequency beam scanning and 5G applications. IEEE Access **8**, 60084–60089 (2020). https://doi.org/10.1109/ACCESS.2020.2983001
6. Pérez, J.R., et al.: Empirical characterization of the indoor radio channel for array antenna systems in the 3 to 4 GHz frequency band. IEEE Access **7**, 94725–94736 (2019). https://doi.org/10.1109/ACCESS.2019.2928421
7. Shen, X., Liu, Y., Zhao, L., Huang, G., Shi, X., Huang, Q.: A miniaturized microstrip antenna array at 5G millimeter-wave band. IEEE Antennas Wirel. Propag. Lett. **18**(8), 1671–1675 (2019). https://doi.org/10.1109/LAWP.2019.2927460
8. Kim, Y., Kim, H., Yoon, I., Oh, J.: 4 × 8 Patch array-fed FR4-based transmit array antennas for affordable and reliable 5G beam steering. IEEE Access **7**, 88881–88893 (2019). https://doi.org/10.1109/ACCESS.2019.2926379
9. Chattha, H.T.: 4-Port 2-element MIMO antenna for 5G portable applications. IEEE Access **7**, 96516–96520 (2019). https://doi.org/10.1109/ACCESS.2019.2925351
10. Li, A., Luk, K., Li, Y.: A dual linearly polarized end-fire antenna array for the 5G applications. IEEE Access **6**, 78276–78285 (2018). https://doi.org/10.1109/ACCESS.2018.2884946
11. Khan, R., Al-Hadi, A.A., Soh, P.J., Kamarudin, M.R., Ali, M.T., Owais: User influence on mobile terminal antennas: a review of challenges and potential solution for 5G antennas. IEEE Access **6**, 77695–77715 (2018). https://doi.org/10.1109/ACCESS.2018.2883788
12. Mukhopadhyay, A.: J. C. Bose's scientific inventions confirmed the truth of consciousness. IJOHMN **4**(6), 1–20 (2018). https://doi.org/10.24113/ijohmn.v4i6.72
13. Naqvi, S.H.R., Ho, P.H., Jabeen, S.: A novel distributed antenna access architecture for 5g indoor service provisioning. IEEE J. Sel. Areas Commun. **36**(11), 2518–2527 (2018). https://doi.org/10.1109/JSAC.2018.2874144
14. Bengtsson, E.L., Rusek, F., Malkowsky, S., Tufvesson, F., Karlsson, P.C., Edfors, O.: A simulation framework for multiple-antenna terminals in 5G massive MIMO systems. IEEE Access **5**, 26819–26831 (2017). https://doi.org/10.1109/ACCESS.2017.2775210
15. Li, M., Xu, Z., Ban, Y., Sim, C., Yu, Z.: Eight-port orthogonally dual-polarised MIMO antennas using loop structures for 5G smartphone. In: IET Microwaves, Antennas & Propagation, vol. 11, no. 12, pp. 1810–1816 (2017). https://doi.org/10.1049/iet-map.2017.0230
16. Neil, C.T., Shafi, M., Smith, P.J., Dmochowski, P.A., Zhang, J.: IEEE Transactions on Impact of Microwave and Wave Channel Models on 5G Systems Performance

Efficiency Analyzing on Vehicle Tracking Systems

L. Rahunathan, D. Harish, A. Antony Samson, and D. Sivabalaselvamani

Abstract The vehicle tracking system is the fundamental part in our everyday life. Yet, the vehicle tracking system regularly utilized GSM/GPS procedure. Presently what the issue with is, we should need a web association with doing this kind of vehicle following, but a portion of the cases and in no organization regions this sort of vehicle following won't work. Along these lines, in this paper, we investigate the demonstration of GPS/GSM tracking system and the chance of LoRa-based vehicle tracking system. Results uncovered that all measurements measured with in re-enacted experiments were comparable to the results of a genuine examination; notwithstanding, we thought that by looking for a more realistic relationship with real inquiries, the model could be strengthened.

Keywords Arduino UNO · LoRa · GSM · GPS · Vehicle tracking system

1 Introduction

LoRa stands for "Low Power Wide Area Network" and requests decreased organization framework to cover an enormous region, with low force utilization. With regards to vehicle following, LoRa does have capacity to help with observing and appropriate path applications, whereby company can share small volumes of information non-concurrently. By and by, Lora's assessments are needed in setting up vehicle communications. LoRa (short for long reach) is a spread range adjustment strategy got from peep spread range (CSS) innovation. Utilizing of this gadget we can impart the GPS co-ordinates starting with one LoRa then onto the next. By sharing the co-ordinates, we can undoubtedly follow that LoRa area without utilizing web. Also, just as we would disk about the GSM/GPS vehicle global positioning framework for better comprehension about vehicle following.

L. Rahunathan (✉) · D. Harish · A. A. Samson · D. Sivabalaselvamani
Department of Computer Applications, Kongu Engineering College, Perundurai, Erode, Tamilnadu, India

K. Khanna et al. (eds.), *Cyber Security and Digital Forensics*,
Lecture Notes on Data Engineering and Communications Technologies 73,
https://doi.org/10.1007/978-981-16-3961-6_40

1.1 IoT

An IoT, or Internet of Things, helps billions of actual web-related devices across the globe to exchange and capture knowledge. It is imaginable to transform something, from anything as small as a pill for anything as large as a mountain to an IoT entity, because of its appearance of insanely the pervasiveness of remote organizations and central processors, Associating and attaching sensors to all these different articles brings a layer of specialized expertise to devices that would normally be idiotic, allowing them to relay continuous data without requiring a human. Through consolidating the advanced and real worlds, the Internet of Things renders the design of our general world more sensitive and more brilliant.

1.2 GPS

The GPS, or the Global Positioning System, world broad route satellite network that provides synchronization of location, time and speed. The Global Positioning System (GPS) is a free assistance that worked and claimed by the U.S. Government and is consistently accessible. GPS is all over. You can discover GPS frameworks in your vehicle, your cell phone and your watch (Fig. 1).

GPS causes you get where you are going, from point A to point B. What is GPS? Peruse this article to become familiar with how it functions, its set of experiences and future advancements. Firstly, At a given point, a sign of timing is transmitted from the GPS satellite. Thusly, a time distinction between the purpose of time clock and GPS time which a GPS beneficiary receives the time sign could be determined to generate the distinction to the satellite from the receiver. With three other open

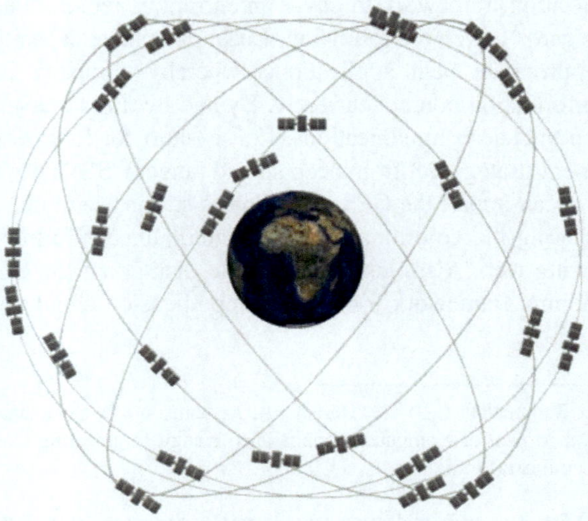

Fig. 1 GPS satellites. Copyright 2017 by NASA

satellites, a similar period will be completed. Determining the condition of the GPS receiver from a distance from the GPS collector is determined for three satellites. In any event, the location created by the methods for such a technique is not accurate, since there is an error in the distance calculated between the satellites and the GPS receiver, which is the result of a time error on the clock fused into the GPS collector. A nuclear clock is consolidated for a satellite to obtain data on the spot time, but the time generated by tickers entering GPS beneficiaries is not as precise as the time created on satellites by nuclear timekeepers. Hence, a fourth satellite assume its task and it is possible to use the isolation form of the satellite to a collector to work out the situation in relation to the location information generated by the distance between the 3 satellite and the receiver, thus reducing the space for the precision of give and take on the position.

1.3 Vehicle Tracking System

With its most straightforward definition, a vehicle global positioning framework is the framework that permits following and controlling of vehicles through an online PC, advanced mobile phone, tablet, and so forth on a day in and day out premise on account of GPS satellites. Vehicle global positioning frameworks make it conceivable to have a prompt and history following of vehicle speeds, the courses they followed, halting focuses, sitting occasions on guides furnishing a vault and checkpoint with over a wide span of time reports. Essentially, the vehicle global positioning frameworks work in a circle of GPS, GSM/GPRS, computerized guides and uncommon programming. Versatile information gadgets mounted on vehicles send two snippets of data they get from satellites—constant when the satellite data was sent and position of the satellite in circle at the hour of transmission and communicate telemetric data, for example, temperature to control and correspondence focuses by means of the GSM/GPRS network. Data got as such are then gathered gratitude to unique programming and recorded in a databank on workers. On the client side, vehicles can be followed on their prompt and history records through a PC or a cell phone/tablet by utilizing exceptional programming making it conceivable to picture all data from vehicles and to alter alerts and program statuses of vehicles. This structure shapes the reason for working of the vehicle global positioning frameworks.

2 Literature Review

The LoRa is a Semtech's physical layer (PHY) assurance [1], low power associations and planned for long reach. The development enables the accessibility of shrewd things at range of the solicitation for km, fundamental for Internet of Things networks, with low-energy use. To do the same, LoRa introduces a ludicrous prohibitive way of distributing guidelines, which should be a variety of chirp spread range (CSS) [2], which changes repeat peep beats without altering the stage between touching images [3], encoded the data [1, 4].

Therefore, LoRa material obtained the resultant sign impenetrable to interference block or near-frequency signals. In addition, LoRa is represented by a decrease in the unpredictability of gear, a decrease in multifaceted existence header size in relation to hops and a propensity to participate in critical two-way correspondence using an unrelated low-energy method. LoRa physical layer switches, Repeat gatherings with unauthorized repetition of Industrial Research and Medical Band in sub GHz radio signals (ISM), To Brazil, a synchronized repeat band for Industrial Science and Medical band (ISM) is somewhere in the 928 and 915 MHz (AU 915 to 928 Megahertz) range, as per the National-Telecommunications-Agency (ANATEL) [5]. Also, a 433 MHz repeat could be executed for channel with very some devices [6]. A couple of limits for plan the LoRa PHY are according to the accompanying: Bandwidth, Spreading Factor, Code-Rate and Carrier repeat. For the transmitting band, the carrier repeat illustrates the middle repeat. It is defined by the area of action of the substance. Thus, as seen by the submission, this cap is usually not portable. The volume of the repeat distance used is defined by information transmission, with 3 programmable characteristics: 125, 250, and 500 kHz. To address a picture, SF chooses the quantity of peeps used [4, 7]. It thus determines the magnitude between the spot rate and also the twitter price. 6 extraordinary characteristics for SF limit are listed in the LoRa detail: Spread factor 7, 8, 9, 10, 11 and 12 [2], which enables even channel to be molded, so that there are no incidents between them along with different SFs. High SF raises the level of the social occasion by restricting the degree of sign power, by extending the relaxing time on air (ToA) and by reducing the association's transmitting movement [8].

Vangelista et al. [9] examine the various technologies of the less Power Large location Network (LoRa) and provide the detailed depiction of the success of LoRa. 2 studies are carried out: (i) The LoRa personal association is formed to monitor the humidity and temperature components of such a structure (19 stories) and (ii) an inspection is carried out in Italy, Padova, causing approximately 2 km length to be covered. A manufacture believes that LoRa-subject radio associations will interface equipment organized by several kilometers.

Gomez et al. [10] consider joining the creation of LoRa in a proposal for vehicular communication. They analyze two situations: I the topography of a vehicle-to-infrastructure and the geology of a vehicle-to-vehicle. The manufacturers research the RSSI levels for SF-7 and SF-12 in the critical circumstances, assuming that SF-12 had a highest level of RSSI, having a wider 10 km radius, higher than SF-7 (6 km). They survey the area covered by a left system inside of the vehicles for the resulting situations, submitting informed warnings. SF-12 appeared at a distance of 6 km, with much more distinct shadow areas and SF-7, which appeared at a distance of 2.5 km. The manufacturers believe that the refined presentations in terms of elimination are preferable to any other cell growth and that the findings of vehicle-to-vehicle limited the spectrum of integration by a stature of an access point collection unit in comparison to the vehicle-to-infrastructure situation.

Mikhaylov et al. [11] offer an examination of the application of the production with LoRa and examine its kindness toward the way of Doppler's. Roedig et al. [12] are using the SimPy system to create a discrete event diversion instrument called LoRa-Sim. This gadget makes a 2-dimensional space for dispensing *N* centers and *M* passages. In order to survey the most extraordinary requirements of the end devices in a LoRa interface, each center point is portrayed by transmitting limits unambiguous to LoRa growth.

Haxhibeqiri et al. [13] are operating an NS3 LoRa library that joins a slip-up model used to pick spectrum, analogous to the impedance between separate simultaneous transmissions. The device also enables bi-directional correspondence in addition. The manufacturers claim that it is important to coordinate implementation of the complex limit selection on endpoint contraptions. Similarly, they conclude that the restricted downlink movement amazingly debases the degree of bundle transport; the test could be alleviated by increasing the number of passages. In the LoRa physical layer executed [14] and [13], all SF and the all code values defined there in particular layer are retained.

3 Method

Simulations are done by the simulation model NS3 tool. In [14] that simulates a LoRa framework and physical layers, we add the LoRa publicly available libraries. In such, libraries are tailored for such a research to a specifications of the testing situations suggested. For all this, we reproduce a simulation model where the nodes interact in actual experiments at speeds and distances specified. In the NS3 simulation software, we choose a LoRa physical layer method to replicate a vehicular scenario, with such a prediction method for wide areas. A normal distribution variable is measured to evaluate faded loss of signal, an acceleration attribute of a different setting, in order to incorporate a probabilistic method.

4 GSM/GPS-Based Vehicle Tracking System

4.1 Arduino Uno

An Arduino is an open source, programming association and PC gear, attempt, produce microcontroller packs and customer bunch that plans for building modified gadgets and wise article that can identify and control inquiries in reality. The initiation of an Arduino Broadens began at the Institute of Interaction Design in Ivrea, Italy. A Creative Commons Attribution Share appropriates the hardware comparison plans. The device works from 1.8 to 5.5 V. The ATmega 328 is a solitary—microchip in the mega AVR family made by ATmega. The device achieves bandwidth shifting toward one MIPS per each Megahertz. B. Force Supply AC connectors are utilized with electrical gadgets that require control yet don't contain inside parts to infer the necessary voltage and force from mains power. The inner hardware of an outer force supply is fundamentally the same as the plan that would be utilized for an inherent or interior stock. Outside force supplies are utilized both with hardware with no other wellspring of intensity and with battery-fueled gear, where the stockpile, when connected, can some of the time charge the battery notwithstanding driving the hardware. Utilization of an outside force supply permits transportability of gear fueled either by mains or battery without the additional heft of inner force parts and makes it pointless to deliver hardware for utilizing just with a predefined power source; a similar gadget can be controlled from 120 VAC or 230 VAC mains, vehicle or airplane battery by utilizing an alternate connector. Another bit of leeway of these plans can be expanded wellbeing; since the dangerous 120- or 240-V mains power is changed to a lower, more secure voltage at the divider source and the machine is dealt with by the client.

4.2 GPS Module

The GPS stands for Global Positioning System, is a global route satellite architecture focused on space that offers solid region and time data in all temperatures and continuously and everywhere on or near the Planet, when and when at least four GPS satellites are viewed unhindered.

4.3 GSM Module

As the world's most commonly used phone invention, GSM, this describes the Worldwide Framework for Phone Lines, regulations. By searching for PDA tower in the surrounding area, PDA uses a wireless assistance transporter GSM network. The GSM module is a SIM 900 Quad-band/SIM 900A Dual-band GPRS/GSM module breakout board and minimum arrangement. By way of AT instructions, it will speak with regulators (GSM 07.07 SIMCOM and 07.05 and improved at a Commands). This module maintains the on and resets power of programming. It has an 850/900/1800/1900 Megahertz quad-band and a 900/1900 Megahertz double band. It has power by way of AT instructions, an exceptionally poor 1.5 mA use of force (rest mode).

4.4 Relay

The Relay is an electronic switch that closes and opens heavily influenced by another electronic circuit. In the first structure, that switch is operated by an electromagnet to close or open one or several arrangements of contacts. The Relay can monitor a yield circuit of higher force than the information circuit; it very well may be viewed as, from a wide perspective, a type of an electric enhancer.

4.5 Working Principle

The principle expectation of this venture is to discover the specific area of the taken vehicle by burglaries and advise to the concerned authority through a SMS. This GSM-based vehicle burglary control framework recovers the specific area of a vehicle as far as its longitude and scope. This information is taken care of to the Arduino, that is interfaced to a GSM modem. The Arduino recovers the specific area subtleties from the GPS and sends a SMS to the concerned authority over GSM modem. An LCD show is associated with the Arduino for intersection the information got prior to being sent over GSM. This venture will be helpful to individuals to monitor their vehicles. Further, this venture can likewise have the option to stop the start of the vehicle by the proprietor distantly by sending a SMS in robbery circumstances. A bit of leeway of this undertaking is that the proprietor of the vehicle can likewise send back the SMS, which will deactivate the start of the vehicle (Fig. 2).

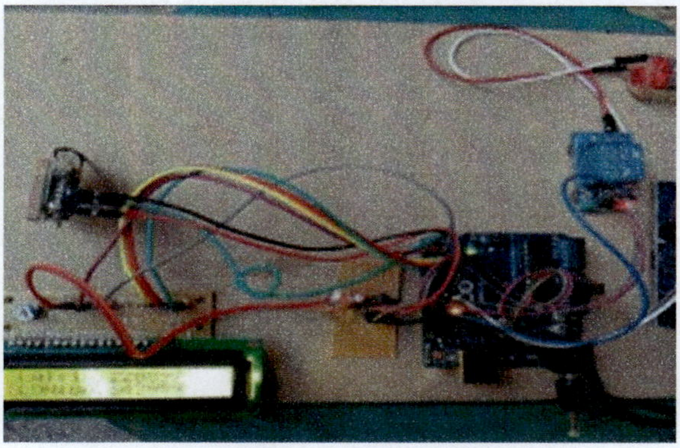

Fig. 2 Circuit setup from vehicle tracking system using GPS and GSM

5 LoRa-Based Tracking System

5.1 Lora

LoRa (short for long reach) is spreads arrive at rule procedure got from tremble spread reach (CSS) progression. Semtech's LoRa contraptions and removed radio recurrent headway is a long reach, low force distant stage that has become the recognized advancement for Internet of Things (IoT) networks all over. LoRa contraptions and the open LoRa show connect sharp IoT applications that settle no doubt the best inconveniences confronting earth: energy the board, brand name asset decay, contamination control, foundation productivity, disaster avoidance and the sky is the limit starting there. A few hundred recognized use cases for sharp metropolitan organizations, sharp homes and structures, mind-blowing growth, wise metering, impressive stock association and joint efforts have been amassed by Semtech's LoRa contraptions and the LoRa display, and the sky is the limit from there more than 167 million gadgets related with networks in 99 nations and making, LoRa gadgets are making a Smarter Planet, Which consolidated super low force utilization with a powerful long-range. Semtech's utilized on the intensity of the network through a consortium and changed the innovation into one of the critical drivers of the ebb and flow wave of computerized change being initiated by IoT. Utilizing LoRa innovation, gadgets can commonly convey over a scope of 13–20 km with the capacity to go similarly as 80 km in certain Line-of-sight arrangements, and to the extent a few 100 s of KM from space as shown by FOSSAT and Lacuna. This reach is accomplished at an extremely low force which makes LoRa more reasonable than other correspondence conventions, for distant, battery-controlled IoT gadgets that are relied upon to run for quite a long time (or years) on a solitary battery charge. Peer to Peer communication permits two gadgets with LoRa radios to converse with one another in a way like how two Bluetooth gadgets impart, with the significant distinction being the way that the reach increments enormously and less force is burned through (Table 1).

To identify a symbol length [16], which is really the time required to submit two SF chips at a chip value. It is distinguished by:

$$\boldsymbol{Ts} = 2\frac{SF}{BW}, \quad \text{with } SF \in \{7 - 12\}. \tag{1}$$

The length of a preamble is calculated by:

$$\boldsymbol{Tp} = (\boldsymbol{np} + 4.25) \times \boldsymbol{Ts}. \tag{2}$$

Table 1 Comparison with related simulation versus experimental paper works

Reference papers	Analysis types		Scenarios	Metrics
	Experiment	Simulation		
[9]	Yes	No	Wide area	Coverage
[10]	Yes	No	Vehicular	To the end packet rate of loss
[15]	Yes	No	Vehicular indoor walking	To the end delay packet rate of rate
[12]	Yes	Yes	Large size	Package quality of receipt Level of extracting data
[11]	Yes	No	Vehicular boat analytical rotation analysis	Success ratio for the packet Coverage doppler impact of the RSSI standard
[14]	No	Yes	City sides	The ratio of packet distribution Limitation of service period Success of packet Likelihood Coverage
[13]	No	Yes	Fixed location	Analysis of scalability Ratio for packet distribution

The count of signs in the header and payload is calculated by:

$$\text{payloadSymb}\,Nb = 8 + \max\left(\frac{8PL - 4SF + 28 + 16 - 20H}{4(SF - 2DE)}(CR + 4),\ 0\right). \quad (3)$$

The amount of payload signs multiplies by the time interval equals the payload length. It is distinguished by:

$$\text{Tpayload} = \text{payloadSymb}\,Nb \times Ts. \quad (4)$$

The packet length is calculated by:

$$\text{Tpacket} = \text{Tpreamble} + \text{Tpayload}\,. \quad (5)$$

For a Band Width = 500 kHz and Coding Rate = 4/5, Table 2 displays the possible Time on air values of different SFs. These estimated results are being used to evaluate the receiving time period between successive packets, allowing the vehicle's situational perception and environment to be measured. In an attempt to create the conditions for programs that use LoRa and communications technology, we equate these factors to experimentally collected values.

Table 2 Theoretical data rate of Lora and time on air

Spread factor	Time on air (ms)
SpreadFactor 7	24.104
SpreadFactor 8	42.087
SpreadFactor 9	78.056
SpreadFactor 10	134.632
SpreadFactor 11	268.264
SpreadFactor 12	494.568

5.2 NS3 Simulator

The simulations were performed using version 3.28 of the NS3 simulator software. We use the LoRa open source libraries, which replicate the LoRa network and physical frameworks and can be found in [14]. This library has been updated to satisfy the needs of the theoretical conditions suggested in this report. To accomplish this, we created a simulation environment in which the modules interact at predetermined speeds and distances based on experimental findings. In order to explain the actions of the vehicular touch interaction in a virtual environment, it is critical to choose a transmission model that closely matches the real-world scenario. It shows how electromagnetic waves behave and travel from a transmitter to a receiver. Long-distance transmission is the transmission type that best suits the condition selected for research. This technique is commonly used in large situations and communications to sub-gigahertz radio frequency networks. Finally, Eq. (6) predicts signal attenuation that is equivalent to real-world results using signal strength and distance as variables.

$$PL(d) = PL(do) + 10nLog(ddo). \tag{6}$$

We used an irregular standard normal attribute to classify the damages suffered by the vanishing problem because Log Distance is a typical problem as a result, and Eq. (7) is used in the simulation analysis for going to compare simulated and real experimentation:

$$PL(d) = PL(do) + 10nLog(ddo) + X\sigma. \tag{7}$$

6 Experimentation Results

6.1 Efficiency Analyzing with Experimental Results

A RSSI amount is seen in Fig. 3. When the vehicle is moving, signal intensity with all spread factors is comparable with just an RSSI amount ranging from 120

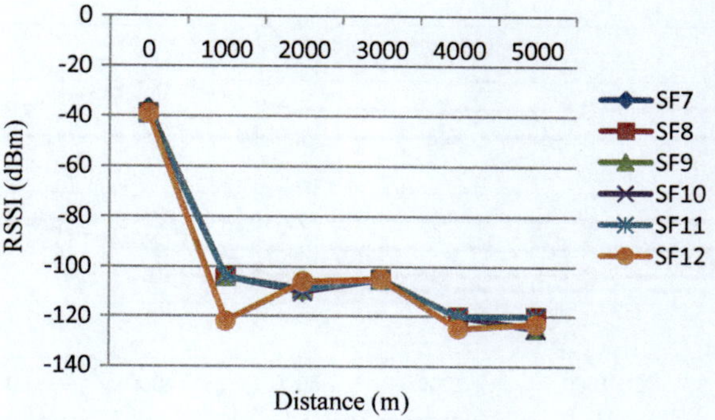

Fig. 3 The grade of RSSI acquired in the receiving unit using various SF for the highest communication range scenario

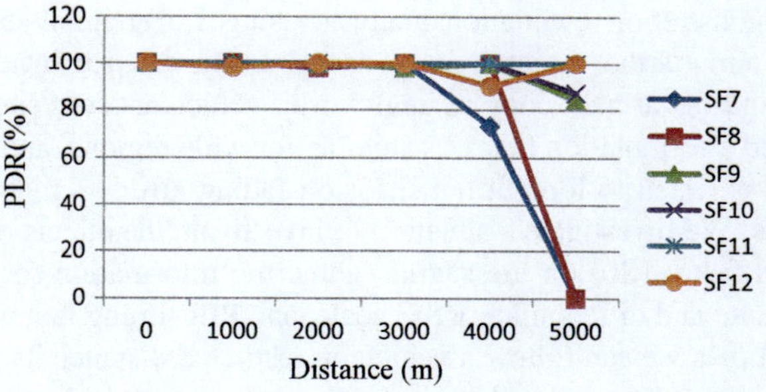

Fig. 4 The PDR acquired in the receiving unit using various SF for the highest communication range scenario

decibel-milliwatts for the actual demonstration to 110 decibel-milliwatts for the virtual demonstration at 5000 m.

A PIR period as in versatility case is depicted in Fig. 4. We described few minor time differences among frequencies that aren't important; however, this time reaction could be useful in defining contact connection outages.

A PDR measured at each stage recorded mostly by Geo location as in system impedance is shown in Fig. 5. This is clear to see that spread factor 7 to spread factor 12 have a similar behavior and after template, with such a PDR of more than 90% with both rates.

7 Conclusion

In general, cellular and Internet-based vehicle tracking is used. However, vehicle tracking allowed by the Internet and cellular does not fall into this category in off-limits areas. In this work, in these non-cellular and non-Internet environments,

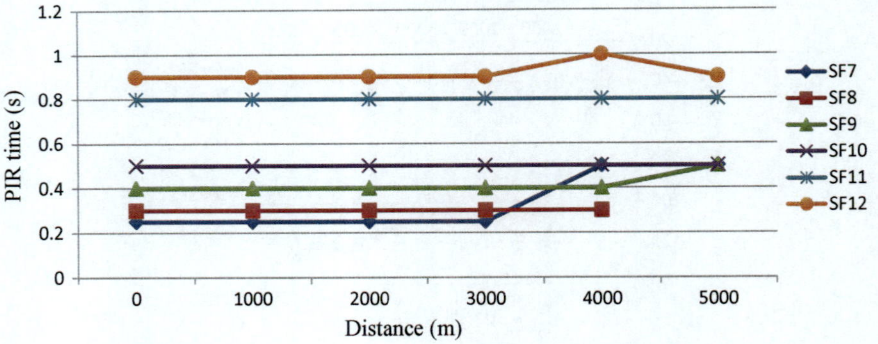

Fig. 5 The PIR time acquired in the receiving unit using various SF for the highest communication range scenario

we lead a demonstration evaluation analysis using LoRa innovation in vehicle tracking. The aim of the research is to see if LoRa can replicate a real-world environment and if it can copy a real-world vehicular correspondence. This experiment used a formulation that was suitable for wide regions, and then a regular odd variable was tested to look at transmission fading errors, which are typical in LoRa platforms. We investigate a situation to investigate the effects of range, speed rate and spread factor-(SF) on messaging, gathering information for the simulated studies to evaluate and differentiate RSSI scale, and PIR timing factors. Be that as it may, utilizing LoRa we can follow a vehicle in restricted distance itself. Finally, we conclude that in the future the distance will be increased with the assistance of forthcoming innovations.

8 Future Work

To make that product function as a real-world monitoring system, certain modifications must be made. According to interference from local structures such as homes, trees and hills, devices may not always accept the exact direction of a signal. The transmitter throughput determines its capacity to implement this concept. The performance is best when the throughput is high, and vice-versa. To perform simulation work, the LoRa physical model applied in NS3 emulator has been used. We prove that the LoRa-based vehicle monitoring system is effectively executed without the Internet with the restricted range region by the outcome of our analysis report. Although this is not appropriate for the condition and challenges that are coming up. We would try to expand the range of this approach with the use of future technologies in functionality.

References

1. LoRaalliance technology, LoRa Alliance: Available in online: https://www.lora-alliance.org/technology. Last accessed by November 2020 (2016)
2. Lora modulation basics by Semtech Corp.: Available in online: https://www.semtech.com/uploads/documents. Last accessed by November 2020 (2015)
3. Gorce, J.-M., Goursaud, C.: Dedicated networks for IoT: PHY/MAC state of the art and challenges. EAI Endorsed Trans. Internet Things (2015)
4. Roedig, U., Bor, M., Vidler, J.: LoRa for the internet of things. In: International-Conference-on-Embedded-Wireless-Systems-and-Networks (EWSN) (2016)
5. Radio-frequency-bands-usable by restricted-radiation-equipment in ANATEL: http://www.anatel.gov.br/legislacao/resolucoes/2018. Last accessed by November 2020 (2018)
6. Frequency-assignment, attribution and destination plan on Brazil by ANATEL: http://www.anatel.gov.br/institucional/acervo-documental. Last accessed by November 2020 (2016)
7. LoRa modem designer's guide by Semtech Corp.: Availability oninternet: https://www.semtech.com/uploads/documents. Last accessed by November 2020 (2013)
8. Sooriyabandara, M., Raza, U., Kulkarni, P.: Low power wide area networks: an overview. IEEE Commun. Surv. Tutor. **19**(2) (2017)
9. Vangelista, L., Centenaro, M., Zorzi, M., Zanella, A.: Long-range communications in unlicensed bands: the rising stars in the IoT and smart city scenarios. IEEE Wirel. Commun. (2016)
10. Gomez, J.S., Gomez-Skarmeta, A.F., Sanchez-Iborra, R., Santa, J., Fernandez, P.J.: Integrating LP WAN Communications Within the Vehicular Ecosystem (2017)
11. Mikhaylov, K., Petajajarvi, J., Pettissalo, M., Iinatti, J., Janhunen, J.: Performance of a low-power wide-area network based on LoRa technology. Doppler Robustness, Scalability, and Coverage (2017)
12. Roedig, U., Alonso, J.M., Bor, M.C., Voigt, T.: Do LoRa low-power wide-area networks scale? In: Analysis and Simulation of Wireless and Mobile Systems of ACM International Conference on Modeling, (MSWiM) (2016)
13. Haxhibeqiri, J., Van den Abeele, F., Hoebeke, J., Moerman, I.: Scalability analysis of large-scale LoRaWAN networks in NS-3. IEEE Internet Things (2017)
14. Centenaro, M., Magrin, D., Vangelista, L.: Performance evaluation of LoRa networks in a smart city scenario. IEEE International Conference on Communications (ICC) (2017)
15. Won, M., Patel, D.: Experimental study on low power wide area networks (LPWAN) for mobile internet of things. In: IEEE 85th Vehicular-Technology-Conference (2017)
16. Antoine-Santoni, T., Gualtieri, J.-S., Manicacci, F.-M., Aiello, A.: AMBLoRa: a wireless tracking and sensor system using long range communication to monitor animal behavior. In: SMART: The Seventh International Conference on Smart Cities, Systems, Devices and Technologies (2018)

Evaluation and Transformation Analysis of the Mithi River

Saumya Deshmukh, Shrishti Karkera, Prachi Rawale, and Chhaya Narvekar

Abstract Excessive enrichment of nutrients can cause any water body to undergo eutrophication, thus damaging the water body and sometimes making it completely unusable and futile. Further, anthropogenic activities lead to cultural/accelerated eutrophication that becomes native to several species of algae including cyanobacteria and cyanotoxins causing algal blooms, organic pollution caused due to organic and inorganic waste, etc. Monitoring of the river and lakes for such phenomena using satellite images has become increasingly popular to understand the trends and patterns in order to further protect the environment. The Mithi River in Mumbai that travels 17.84 km from the Vihar Lake to the Mahim Bay has been studied in this paper using multispectral satellite images and remote sensing to understand the different processes that take place due to human and biological factors which degrade the river. Chlorophyll-a (vegetation index), chemical oxygen demand (COD) and biological oxygen demand (BOD) are the parameters that we have taken into consideration to measure the water quality index at eleven different predefined points (Table 1). Sentinel 2A multispectral images have been used to study the parameters along the stretch of the river by applying cloud masks to eliminate the cloud and the cloud shadow effects on the image collection.

Keywords Eutrophication · Biological oxygen demand · Chemical oxygen · Demand · Remote sensing · Sentinel 2 MSI

1 Introduction

Water is the necessity for all life on earth. Hence, not just water, but clean water should be available for all and it is our responsibility to keep such resources clean and safe. Mithi river situated on the island of Mumbai city is formed by the convergence of the tail water discharges from the Powai and Vihar lakes and is a

S. Deshmukh (✉) · S. Karkera · P. Rawale · C. Narvekar
Xavier Institute of Engineering, Mumbai, India
e-mail: chhaya.n@xavier.ac.in

K. Khanna et al. (eds.), *Cyber Security and Digital Forensics*,
Lecture Notes on Data Engineering and Communications Technologies 73,
https://doi.org/10.1007/978-981-16-3961-6_41

seasonal river by nature which rises and overflows during monsoon [1]. It is a 17.84 kms stretch which is subjected to various illegal activities such as washing of cattle shed, vessels, oil/chemical drums, dumping of solid waste by the slums living along the stretches of the river, several drainage systems released into the river [2]. The pollution caused due to this is to a scale where the river can no longer be used which makes it futile. Many restoration projects are established to revive the river. The paper focuses on analyzing and studying the vegetation index (Chlorophyll a), the BOD and the COD index across the stretch using sentinel-2 satellite data.

2 Background

Remote sensing is the procurement of information about a geographical phenomenon or object without physically attending it or making any physical contact with it, remotely and thus, sensing it remotely by detecting and monitoring its physical characteristics by using satellites for measuring the reflected and emitted radiation from a distance [3, 4]. The image collection bands that are obtained by the remote sensing mechanism vary from satellite to satellite which depends on the sensor system (hyperspectral, multispectral, radar, etc.). These bands are used to obtain different perspectives of the same image and can be convoluted with the help of algorithms, formulas, expressions in order to analyze and study the area under observation. Sentinel 2 MSI image dataset which has a multispectral sensor system comprising of 13 bands is used for the purpose [5]. The different resolutions available are 10, 20 and 60 m according to the band. The river can be analyzed on various parameters like TSS, pH, etc. The parameters that are focused and analyzed in this paper are Chlorophyll-a (vegetation index), BOD and COD.

3 Data and Research Area

Since the river originates at an altitude of 246 m above sea level in the hills located in the east of the Sanjay Gandhi National Park, it gathers water from the streams and spillway discharges of the Tulsi, Vihar and Powai Lakes to the Mahim bay (Figs. 1 and 2).

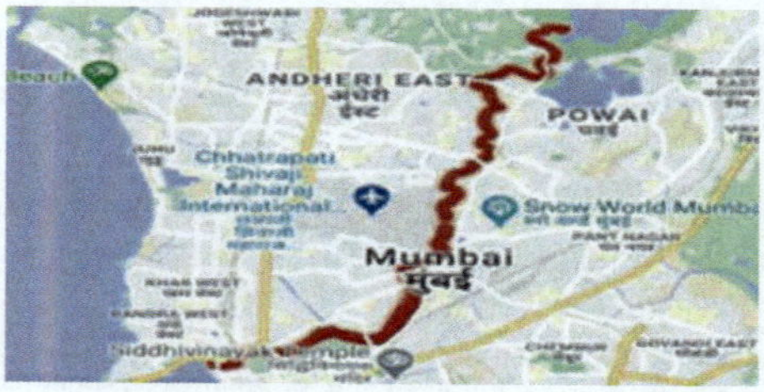

Fig. 1 Mithi River flow across Mumbai

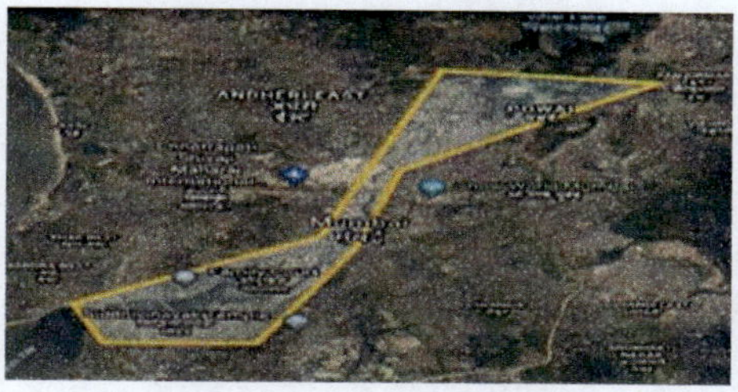

Fig. 2 Mithi River vicinity

The table demonstrates the division of the study area into eleven zones or regions [6] which helps in analyzing the stretch and study individual areas (Fig. 3 and Table 1).

Algal blooms are essential in photosynthesis; hence, it is directly related to Chl-a concentration [7]. Trophic state of the water body is predicted by Chl-a. Chl-a takes up most of its energy from wavelengths of violet-blue and orange-red light due to which the reflectance of Chlorophyll appears to be green. The spectrum absorption was then extended by the addition of Chl-b. Chl-a's dispersion-absorption characteristics include heavy absorption between 450–475 nm (blue) and 670 nm (red) and reflectance peaks at 550 nm (green) and near 700 nm (NIR). A variety of algorithms

Fig. 3 Blue placemarks denote zones corresponding to the ones in Table 1

Table 1 Division of Mithi River into 11 zones

Zone	Description
1	Mithi River meeting the sea (Estuary)
2	Convergence of River Mithi with Vakola nalla
3	Slums and zopad
4	Small scale commercial units and few illegal residents
5	Airport area
6	Commercial area with industries and hotels
7	Resident area with few industries
8	Residential area with both structured and unstructured houses
9	Open area with natural vegetation
10	Slum area with few industries
11	Natural vegetation

to recover Chl-a in turbid waters were developed using the reflectance peak near 700 nm and its ratio to reflectance at 670 nm [8, 9].

The biological oxygen demand can be defined as the amount of oxygen consumed by micro-organisms like bacteria, to decompose the organic matter at a specific temperature under aerobic (rich in oxygen) conditions [10]. BOD is not an optically active component, and hence, an accurate and precise system is required to calculate BOD using satellite imagery [11, 12]. The paper [13] proposes a system which uses Landsat satellite images using National Sanitation Foundation Water Quality Index (NSFWQI) to analyze the pollution status of water body. The NSFWQI equation is

$$\text{NSFWQI} = \sum_{i=1}^{P} w_i I_i \quad (1)$$

wherein I_i is the sub-index for the ith water quality parameter, w_i is the weight associated with the ith water quality parameter p is the number of water quality parameters. The sub-index equations used for BOD are

$$I = 96.67 - 7 * \text{BOD} \quad (\text{for the range 0-10}) \quad (2)$$

$$I = 38.9 - 1.23 * \text{BOD} \quad (\text{for the range 10-30}) \quad (3)$$

The regression equations used for the prediction of NSFWQI, pH, DO, BOD and FC from radiance values from green, red, NIR and SWIR bands of satellite imagery were developed. For the regression model, radiance of satellite imagery is used as independent variables, and one of the above parameters is used as dependent variable. SPSS software is used to form different linear regression equations to find the most accurate one for each parameter. The best equation for BOD is the multiple regression equation formed by the radiance in red and SWIR bands. The equation is as follows

$$\text{BOD} = 32.465 - 11.923 * \text{Rr} + 3.799 * \text{Rm} \quad (4)$$

The conc. of BOD in mg/l, Rr—Radiance in red band, Rm—Radiance in MIR band.

The chemical oxygen demand (COD) index is used to understand the organic pollution of a water body. The measurement of COD can be done using CODCr (potassium dichromate oxidation) and CODMn (potassium permanganate index). CODCr is mainly for the wastewater from the industries and sewage monitoring which is a very time-consuming process. Hence, CODMn is used in order to measure the organic pollution of rivers, lakes, surface water, drinking water and reservoirs [14, 15]. Huang et al. [16] establish a method to measure the COD index with Landsat TM bands 1, 2, 3 and 4. Paper [17] uses the method in Wang and Ma [18] to estimate the CODMn index of Lake Dongting with Landsat TM bands 2 and 3 as shown below:

$$\text{CODMn} = e(0.3671 + 1.2454 * \ln(\text{T2}/\text{TM3})) \tag{5}$$

where CODMn is the potassium permanganate index, TM2 and TM3 are the Landsat multispectral bands 2 and 3. The formula is further modified with respect to sentinel 2 bands and is used to estimate the COD index in the Mithi River.

4 Methodology

Chlorophyll-a is one of important components of the algae. Chlorophyll-a is an important parameter for reflecting the eutrophication degree. Due to this, it can determine spectral reflectance properties of water. If the concentration of Chlorophyll-a in an area is lower, it also has a lower value of trophic state index. If the concentration of Chlorophyll-a in an area is higher, it also has a higher value of trophic state index [19].

$$\text{Chlorophyll-a}\ \mu\text{g/L} = ((\text{B5} + \text{B6})/\text{B4}) \tag{6}$$

where B4, B5, B6 are bands in Sentinel-2A [20].

The BOD test involves the standard oxidation (or incubation) for a period of 5 days at 20 °C. At times where time isn't a constraint and highest accuracy is preferred, the test can go up to 20 days. BOD testing includes two stages—the first is carbonaceous stage where the oxygen demand is involved in the conversion of carbon-to-carbon dioxide and the second is nitrogenous stage where organic nitrogen, ammonia and nitrite are converted to nitrate [21]. BOD of water using remote sensing is calculated using the formula from [13] with suitable modifications, as shown below

$$\text{BOD} = (32.465 - 11.923 * \text{Rr} + 3.799 * \text{Rs})/100) \tag{7}$$

The bands selected Rr and Rs are the bands B4 (664.5 nm, 10 m) and B12 (2202.4 nm, 20 m), respectively, of sentinel 2. The original formula is derived for Landsat satellite. The required modification involves reflectance instead of spectral radiance and normalizing the data by 1000 to get the appropriate results.

The COD test involves potassium permanganate in this case since it is a strong chemical oxidizing agent. It is then used to oxidize the sample taken under the conditions of heat and strong acid. As mentioned above, the formula being used to determine the COD index is the same [17] with a slight variation since the equation should correspond to the Sentinel-2 bands:

$$\text{CODMn} = (e(0.3671 + 1.2454 * \ln(\text{B3}/\text{B4})) * 100) \tag{8}$$

where CODMn is the potassium permanganate index (Mg/L), B3 and B4 are the bands in Sentinel-2 MSI, i.e., green Band (560 nm-S2A) and the red Band (664.5 nm-S2A), respectively.

5 Results and Analysis

5.1 Preprocessing of Satellite Images

Function to mask clouds using sentinel-2A QA band: The image shows the GUI of earth engine code editor using cloud mask in order to eliminate the effects produced by the cloud [22, 23]. The filters applied are the area bound and time series. The image collection shows the 54 images that are valid as per the parameters (Figs. 4 and 5).

```
function maskS2clouds(image) {
  var qa = image.select('QA60');

  // Bits 10 and 11 are clouds and cirrus, respectively.
  var cloudBitMask = 1 << 10;
  var cirrusBitMask = 1 << 11;

  // Both flags should be set to zero, indicating clear conditions.
  var mask = qa.bitwiseAnd(cloudBitMask).eq(0)
      .and(qa.bitwiseAnd(cirrusBitMask).eq(0));

  return image.updateMask(mask).divide(10000);
}

var dataset = ee.ImageCollection('COPERNICUS/S2_SR')
                  .filterDate('2020-01-01', '2020-01-30')
                  // Pre-filter to get less cloudy granules.
                  .filter(ee.Filter.lt('CLOUDY_PIXEL_PERCENTAGE',20))
                  .map(maskS2clouds);

var visualization = {
  min: 0.0,
  max: 0.3,
  bands: ['B1', 'B3', 'B8'],
};

Map.setCenter(83.277, 17.7009, 12);
```

Fig. 4 Snip of code for preprocessing

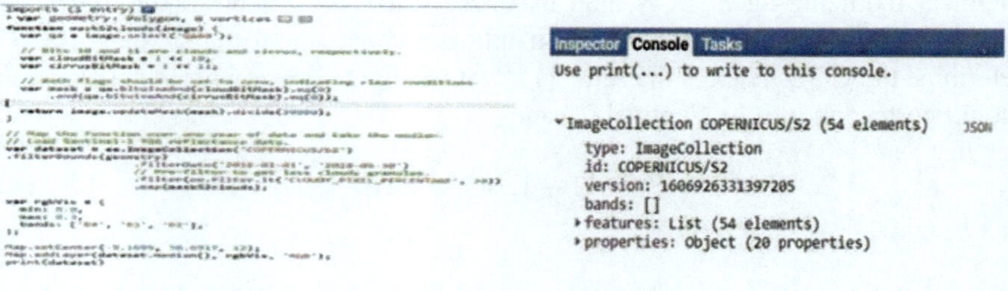

Fig. 5 Image collection with band, features and properties of the image

5.2 *Chlorophyll-a*

Figure 7 shows us the zone wise distribution of Chlorophyll-a during year 2018 to year 2020. In zone 1 and 10, the value is increasing. In zone 2, 4, 8, the value has increased from 2018 to 2019 then decreased in 2020. In zone 3, the value is almost equal in 2018 and 2020 but decreased in 2019. In zone 6, the value is decreasing. In zone 5 and 11, the value is decreasing in 2018 to 2019 then increasing again in 2020. In zone 7, there is a sudden increase in value from 2019 to 2020. In zone 9, there is a sudden decrease in value from 2019 to 2020 (Figs. 6 and 8).

Fig. 6 Chl-a observed in December 2018, 2019 and 2020

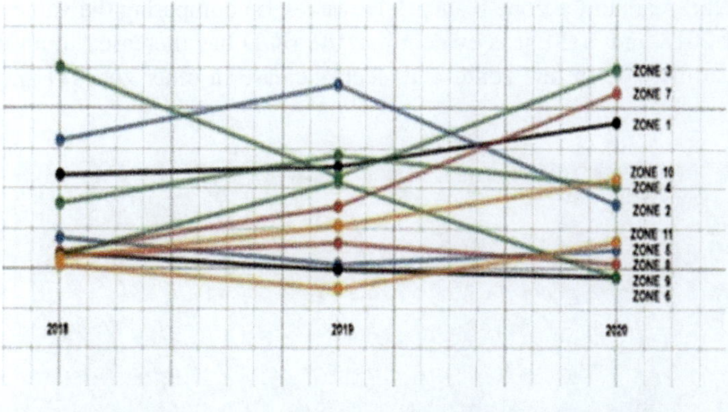

Fig. 7 A graph showing Chlorophyll-a trends from year 2018 to 2020 for all 11 zones

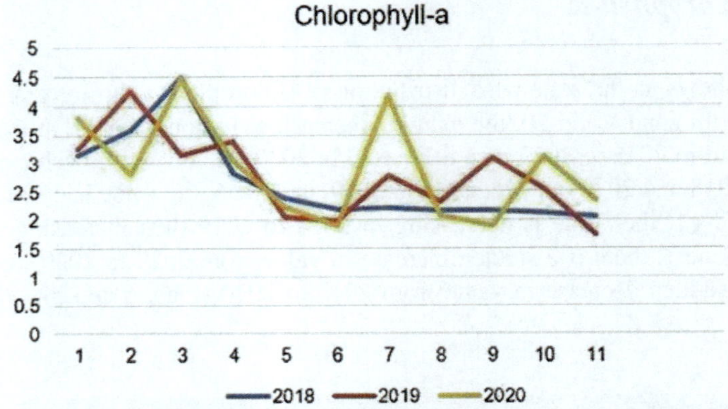

Fig. 8 A graph showing Chlorophyll-a trends from year 2018 to 2020

5.3 *Biological Oxygen Demand*

Equation (4) is originally derived for Landsat satellite. It is converted for sentinel 2 bands which use reflectance without atmospheric correction Eq. (7) instead of spectral radiance. Thus, the value obtained is negative. The negative sign of the value is irrelevant for the study. After normalization, the BOD obtained is approximately 1.5–8 mg/l which concludes that the water in River Mithi is highly contaminated and unsuitable for living organisms. The concentration of BOD also varies in different zones. BOD is more where toward the lower part and less near the origin (Fig. 9).

Figure 10 resembles a 'Inverted V' structure, indicating that the BOD values were high in the year 2019 and lower in the year 2018 and 2020. This trend can be seen clearly in zone 1, 2, 3, 4, 5, 6, 7, 9 and 11. There is a regular increment in the value of BOD over the years in the zone 8 and10. However, on comparing the values of BOD in the year 2018 and 2020, it is evident that the BOD has increased in all the zones with gradual increase in few zones and steep increase in other zones (Fig. 11).

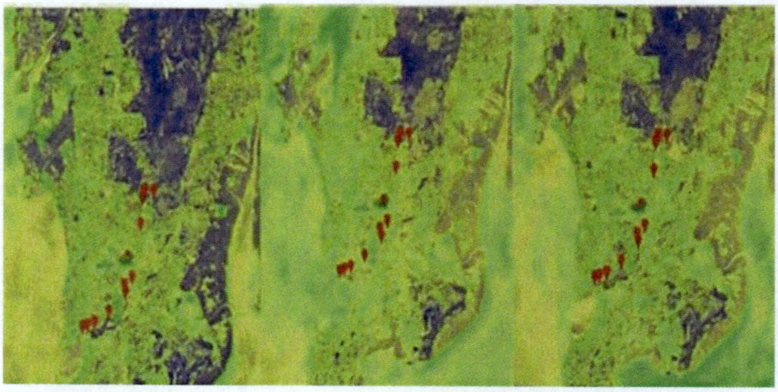

Fig. 9 BOD of 11 zones in December 2018, 2019 and 2020

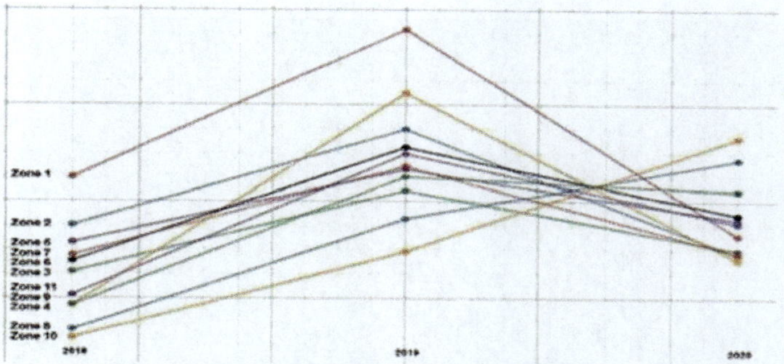

Fig. 10 A graph showing BOD trends from year 2018 to 2020 for all 11 zones

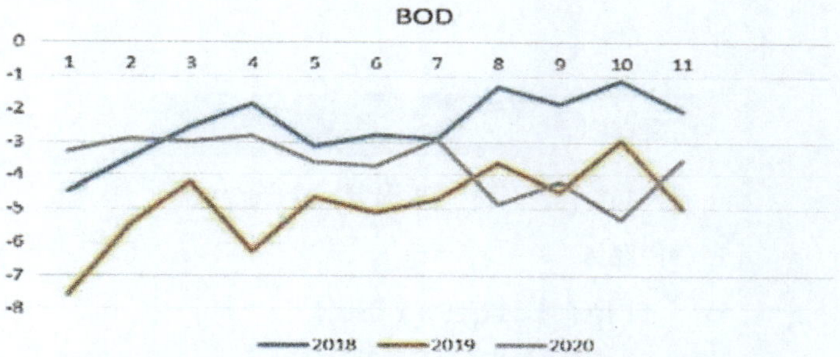

Fig. 11 Zonal trends in BOD for year 2018, 2019 and 2020

5.4 *Chemical Oxygen Demand*

Equation (5) was originally used in [17] for Landsat ETM and is converted for Sentinel-2 MSI surface reflectance in Eq. (8) in the paper to produce the COD index of Dec 2018, 2019 and 2020, respectively, as shown below. In the Inspector window, the COD index value in each position can be found out. A collection of such values will be helpful to further find patterns and to estimate future values of COD index using machine learning algorithms (Figs. 12 and 13).

The COD values are further plotted to analyze from 2018 to 2020. The trends help us understand and further study the real reasons behind the food or bad practices which would make us aware of the consequences (Fig. 14).

Figure 15 is a zone wise distribution of COD index from 2018 to 2020. There is a clear trend in zones 1, 2, 4, 5, 7, 9, 10, 11. The COD values in these zones were lower in 2019 than in 2018 and rose again in the year 2020 due to which there is a valley shape in the graph of these zones. Zones 6 and 8 have their COD values higher in 2018 as compared to 2019 and 2020. The values decreased in 2019 and further in 2020 as well. Zone 3 has a very unique trend where in the COD value in the year 2018 was lower than in 2019, which further increased in 2020. In majority

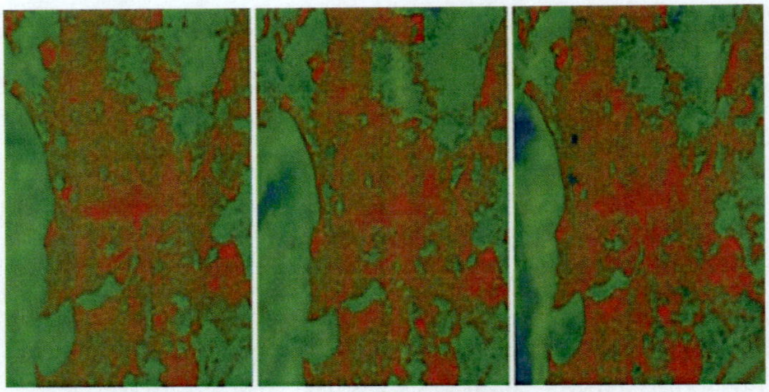

Fig. 12 COD observed as per sentinel-2 in Dec 2018, 2019 and 2020

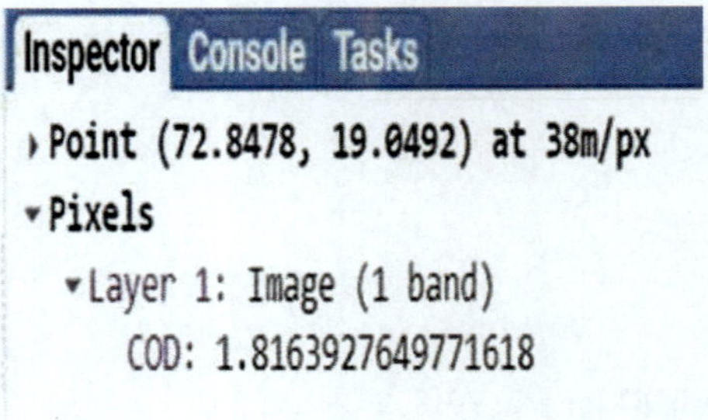

Fig. 13 COD between the zones 2 and 3

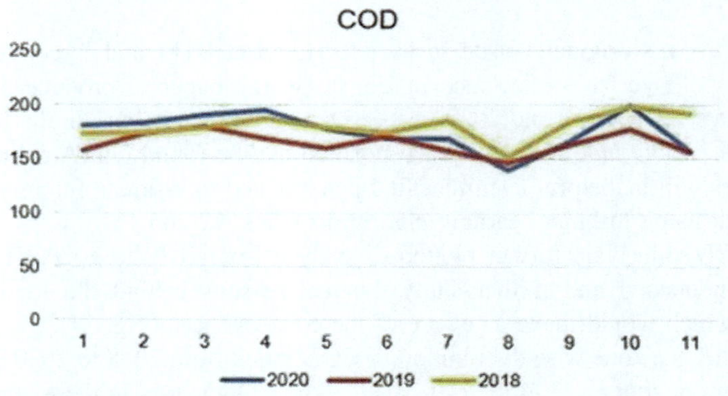

Fig. 14 COD variation of 3 years

of the cases, COD values have increased dramatically in the year 2020. Zonal analysis and surveys let us the different possible reasons due to which this has happened, which further lets us understand the problem and amend our practices.

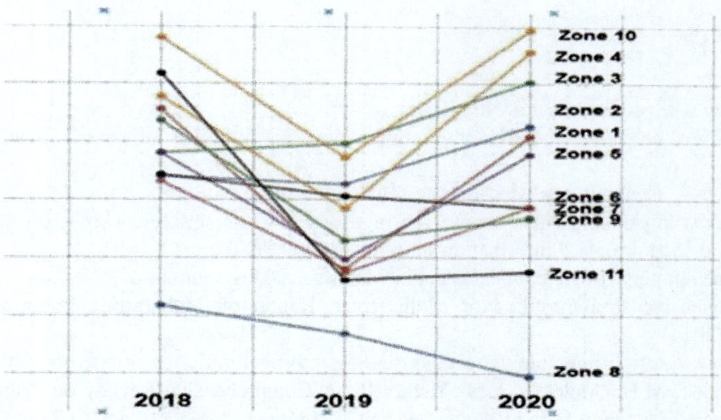

Fig. 15 Zone wise distribution of COD

6 Conclusion

From the graph, we can say that there are different trends in different zones for Chlorophyll-a value. In most of the areas, value is increasing from 2018 to 2019. In some areas, there is a sudden drop in 2020. However, this is not same in the case of BOD. From the analysis done, we can say that the trends in BOD show an overall increase. The BOD was lower in 2018 as compared to 2019 but has decreased again in 2020. Based on the data obtained and algorithms applied, COD index has mostly decreased from 2018 to 2019 and then increased again from 2019 to 2020.

7 Future Scope

Data from the eleven zones will be divided into training data and testing data. Using training data multiple regression analysis will be carried out to determine the model coefficient [24].

$$\text{WQI} = a + b * x1 + c * x2 + d * x3 \quad (9)$$

where $x1$-BOD, $x2$-COD, $x3$-Chlorophyll-a.

Further, parameters can be added like TSS [25], SDT, turbidity, etc., which will be combined with the current parameters to improve the WQI [24]. Then, based on the final WQI results, specific regions can be analyzed- the industries, the various activities can be carefully placated for the betterment of that region.

References

1. https://en.wikipedia.org/wiki/Mithi_River
2. https://www.thehindu.com/news/cities/mumbai/how-the-mithi-was-destroyed/article26298525.ece
3. https://www.usgs.gov/faqs/what-remote-sensing-and-what-it-used
4. https://en.wikipedia.org/wiki/Remote_sensing#:~:text=Remote%20sensing%20is%20the%20acquisition,acquiring%20information%20about%20the%20Earth
5. https://developers.google.com/earth-engine/datasets/catalog/sentinel-2
6. Comprehensive study/profiling of Mithi River: Report by Maharashtra Pollution Control Board (2014)
7. https://www.soft.farm/en/blog/vegetation-indices-ndvi-evi-gndvi-cvi-true-color-140
8. Gholizadeh, M.H., Melesse, A.M., Reddi, L.: A Comprehensive Review on Water Quality Parameters Estimation Using Remote Sensing Techniques. https://doi.org/10.3390/s16081298
9. Gitelson, A.: The peak near 700 nm on radiance spectra of algae and water: relationships of its magnitude and position with chlorophyll concentration. Int J Remote Sens **13**, 3367–3373 (1992). https://doi.org/10.1080/01431169208904125
10. https://www.usgs.gov/special-topic/water-science-school/science/biological-oxygen-demand-bod-and-water?qt-science_center_objects=0#qt-science_center_objects
11. Ansper, A., Alikas, K.: Retrieval of Chlorophyll a from Sentinel-2 MSI Data for the European Union Water Framework Directive Reporting Purposes. https://doi.org/10.3390/rs11010064
12. https://www.researchgate.net/post/What_is_the_relationship_between_the_COD_and_BOD_values_in_Waste_water#:~:text=COD%20or%20Chemical%20Oxygen%20Demand,present%20in%20water%20%2F%20waste%20water
13. Sheela, A.M., Letha, J., Joseph, S., Ramachandran, K.K., Justus, J.: Assessment of pollution status of a Coastal Lake system using satellite imagery. J. Remote Sens. GIS. https://doi.org/10.4172/2169-0049.1000110
14. Lv, J.J., Yang, H., Chen, J., Gao, L., Xu, X.F., Liu, Z.Q.:China. Environ. Sci. **24**, 307–310 (2004)
15. Yan, Z., Xie, Z.W., Zhou, N., Feng, L., Dan, D.Z.: Chem. Res. Appl. **18**, 455–458 (2006)
16. Huang, M.F., Qi, X.P., Yu, W.Y., Zhang, Y.M.: Arid. Land. Geog. **29**, 885–893 (2006)
17. Yang, B., Liu, Y., Ou, F., Yuan, M.: Temporal and spatial analysis of COD concentration in East Dongting Lake by using of remotely sensed data. In: 3rd International Conference on Environmental Science and Information Application Technology (2011). https://doi.org/10.1016/j.proenv.2011.09.420
18. Wang, X.J.: T. Ma. Environ. Sci **11**, 65–68 (2000)
19. Zeng, N., Liu, Z., Miao, Z., Wei, Y.: Design and implementation of Chlorophyll a and eutrophication remote sensing monitoring system based on ArcGIS Engine. https://doi.org/10.1109/ETTandGRS.2008.254
20. Kaymaz, Ş.M., Ates, E.: Estimating Chlorophyll-A Concentration using Remote Sensing Techniques. https://doi.org/10.19080/ARR.2018.04.555633
21. https://www.envexp.com/labmatters/236-understanding-biochemical-oxygen-demand-bod
22. https://developers.google.com/earth-engine/tutorials/community/sentinel-2-s2cloudless
23. https://developers.google.com/earth-engine/tutorials/tutorial_global_surface_water_01
24. Erhan Alparslan, H., Coskun, G., Alganci, U.: Water Quality Determination of Küçükçekmece Lake, Turkey by Using Multispectral Satellite Data. https://doi.org/10.1100/tsw.2009.135
25. Topliss, B.J., Almos, C.L.: Int. J. Remote Sens. **11**, 947–966 (1990)

Human-Sensing Technologies for Business Solutions

Rajeev Tiwari, Kamal Kumar, Satyam Kumar, and Shelly

Abstract Detection of human presence can be used for intelligently switching on and off devices. This will save a lot of electricity as well as help in building an intuitive experience for smart homeowners. In the future, technologies like artificial intelligence may also benefit from this information by taking smart decisions and providing more contextual responses. One can find many research proposals and commercial products for efficiently counting humans. However, these systems are very costly, extremely difficult to install, and many a time obstructs people. It is the reason why they have still not made it to residential homes yet. In this paper, we have discussed some of the most cutting-edge technologies that can be used for sensing humans. We analyze them based on six factors and determine how economic and easy to deploy they are. Some of the solutions deliver promising results and have the potential of becoming a mass-market product.

Keywords Human counting · Person detection · Sensing · Noise · Environment

1 Introduction

The smart home market is expected to be doubled by the year 2024 [1]. Detection of human presence is something that is still missing from the category of smart electronic devices being sold. There have been studies that recorded a 25% reduction in energy consumption [2] when HVAC systems are efficiently managed. It must be noted that the calculation of the number of people in individual rooms and commercial offices is a crucial factor for efficiently managing HVAC systems. Another study in the United States of America found that around 40% of energy is

R. Tiwari · Shelly (✉)
Department of Virtualization, University of Petroleum and Energy Studies, Dehradun, India
e-mail: shelly@ddn.upes.ac.in

K. Kumar · S. Kumar
National Institute of Technology, Srinagar, Uttarakhand, India
e-mail: kamalkumar@nituk.ac.in

K. Khanna et al. (eds.), *Cyber Security and Digital Forensics*,
Lecture Notes on Data Engineering and Communications Technologies 73,
https://doi.org/10.1007/978-981-16-3961-6_42

wasted by residential homes and commercial buildings [3]. Room occupancy can be used for safety purposes. It will open new opportunities for building futuristic experiences by making artificial intelligence-powered agents more aware of the context.

Many researchers have tried to use various detecting techniques to get an accurate count of the number of persons in a room. The most widely used and very cheap solution is a pyroelectric infrared sensor [4]. It can be easily found in modern washrooms, being used to automatically switch the lights when a person enters or exits the room. Detecting techniques commonly used by researchers as well as industry professionals include RGB cameras, laser beams (when combined in form of a linear array at the doorway), infrared cameras, ultrasonic sensors, light sensors, vibration sensors, piezoelectric crystals, etc. Multiple authors [5, 6] have tried to accumulate works and compare these methodologies based on scalability, cost, counting, functionality, location, and identity detection capabilities. In recent years, there has been a surge in the number of research articles and patents based on technologies that use disturbances in radio signals emitted by Wi-fi routers [7–11], changes in CO_2 levels, and air pressure sensing [12], etc., that have not been discussed by the authors. New researches have been made possible due to the increased penetration of smart gadgets and the internet.

This paper accumulates the various technologies available and discusses on drawbacks and unique offerings of each technology. We also compare them based on counting capability, accuracy, or reliability, ease of installation, and economic feasibility. Our goal is to investigate the new solutions being proposed by researchers as well as examine the results of combining multiple technologies for greater accuracy and reliability.

2 Challenges

Common challenges that are faced when detecting humans are listed below.

2.1 Environment

Designing a system that automatically adapts to any environment or works irrespective of surroundings is somewhat difficult to build. In real-world conditions, any sudden or unexpected changes in the surroundings contribute the most to dropping the accuracy rate of any system. For example, PIR sensors might be affected by the presence of hot objects inside the room/building, or by heat currents flowing in centralized air conditioners, or even by hot weather. Image processing-based systems can be easily affected by lighting conditions moving objects and shadows.

2.2 *Noise*

Depending on how well the sensor hardware is, the sensor data obtained is usually affected by noise. Most sensors make use of small entities (like photons are used by camera sensors). This makes them prone to errors like statistical fluctuations in the speed with which particles arrive. A good sensor design configuration can easily remove noises like pink noise and thermal noises.

2.3 *Similarity*

There can be similarities in detecting background signals and humans. The primary component of most human detection techniques is to differentiate between the background environment with humans. Since the real-world environment can be very complex, it becomes difficult to get things working outside laboratory conditions.

To sensors, different people seem similar. It becomes very difficult to identify and track people. This makes it challenging to count the number of people in the room. Image processing systems find it very challenging when all the people in the room are wearing similar dresses for example in a gathering with a strict dress code.

3 Factors Considered

We analyze different human-sensing techniques based on the following six factors.

3.1 *Presence*

This criterion is used to check whether there is any person in the room. Presence is the most trivial criteria that any human-sensing technology must have. PIR sensors are the most commonly deployed sensors that are used to detect human presence and can be easily found in modern-day washrooms. In places where humans can be provided with wearable devices, solutions like radio frequency identification, i.e., RFID can be easily deployed, for example, incorporate buildings. Detection of presence can then be used for applications life automatically switching on and off lights, air conditioners, fans, televisions, projectors and other electric appliances and devices. It will certainly help Smart home manufacturers create intuitive experiences for their consumers as well.

3.2 Counting

This factor determines if we can accurately count the number of people present inside a particular room. Counting the number of people in an environment can be done by either the sensors that observe the entire room or by deploying special sensors that only count people when they are entering and exiting the room. The person count can be used to smartly control electronic devices like air conditioners depending on the number of people that are present in the room. It can also have several commercial applications like determining the rush, counting the exact number of people which is not possible manually in a crowded place, etc.

3.3 Tracking

This factor helps us determine if any particular human tracking system can distinguish and efficiently track different humans. It also includes reporting the entry and exit times of a particular person. The person, however, can be labeled with temporary person IDs and tracking may be anonymous. The temporary identification can be lost when a person leaves the room and re-enters the scene later. This will help futuristic artificial intelligence-powered agents to provide much more contextual information.

3.4 Adaptability

This criterion is used to check if the system can be easily adapted to any new environment without any need for customization. It plays an important role in checking if the system is versatile enough to perform well outside laboratory conditions and handle complex environmental uncertainty. As we will discuss, some systems require automatic training before they can start functioning. In this study, we will still consider such systems as adaptable because they do not require any customization by the installation engineer and can be used easily by a normal person. It is a key factor to consider when we are looking at the commercial scalability of any particular system.

3.5 Cost

This factor determines how much resources and manpower does a particular system required to be built installed. If the system is costly, it could only be used by premium segment or by commercial owners. On the other hand, the system is cheaper it can be easily used by the mass public like PIR sensors for example. The economic feasibility of any technology determines if it has the potential to become widely popular and used as a consumer device.

3.6 Ease of Installation

This criterion determines how easy is it to install a human tracking system at a location without any prior experience or expertise. Any system which is easy to install must also be adaptable at the same time. However, vice versa may not be true. If any system cannot be easily installed by a non-technical person, then it attracts additional installation charges which automatically adds to the overall cost of the system. This factor plays an important role in deciding scalability and feasibility of the solution is.

An interesting property to note is that if factor 3 is satisfied, then factor 2 gets automatically satisfied. This is because if we can individually track humans, then trivially we will also be able to count the number of people in a room. Similarly, if factor 2 is satisfied, then factor 1 automatically gets satisfied. Factor 2 can be seen as a more generalized measurement of factor 1 for count >1.

4 Survey of Technologies

We classify the human-sensing technologies in to natural characteristic-based and motion-based sensing methods.

4.1 Natural Characteristic-Based Sensing

These technologies detect the presence of one or more humans based on natural characteristics of human body like emission of CO_2 or absorption of radio signals by human body.

Wi-Fi. Multiple researchers [7–11] have proposed ways to count the number of people by using Wi-Fi signals. The basic idea behind these types of approaches is that humans create disturbances in received Wi-Fi signals. These disturbances are then processed by either advanced algorithms (based on some complex mathematical calculations [7, 8, 10] or neural networks [9–11]) which then determine the number of people inside the room. It must be noted that the receiver must be stationary, and hence, a device like a smartphone cannot be used for the same purpose. This microcontroller or computer runs the algorithms or neural networks and yields the count of people as output.

Light Sensing. It works on the principle that the presence of humans reduces the intensity of light in the room. Photodetectors are used to convert light signals into electrical signals, and then, a device like ADC is used to convert electric signal to digital signal which can be easily processed by a microcontroller [13–15]. There are two light-based sensing techniques for recognizing any human activity which are as follows:

Active sensing. The user wears a device/or a simple LED that emits light of a specific frequency. This emitted light is then used to track the position of a person using a camera or a photodiode [13, 14].

Passive sensing. It doesn't require people to wear any special device. Photodetectors installed at some places in the room are used to sense ambient light and any flickering frequencies for the reconstruction of human blueprint and indoor location [15].

CO_2 levels. Humans produce carbon dioxide due to the natural respiration process. This increases the percentage of carbon dioxide in the atmosphere which can be used to identify the presence of humans in a room. Some of the systems require machine learning-based model training for each environment before it can start producing results [12]. Due to the slow speed of diffusion of carbon dioxide in the air, it takes some time before the carbon dioxide released by human beings reaches the designated sensor. Hence, it is not suitable for applications that need instant feedback on whether a human is present or not.

4.2 Motion-Based Sensing

These technologies detect the motion of people and use it to detect the presence or count and track the persons inside a room.

Beam Break Sensor. A beam break sensor is used to detect the entrance and exit of humans at doorways. They work by having an emitter on one side that emits light and a receiver on the other side which measure the amount of light incident on it. These types of sensors are also commonly found in lifts, metro token pathways, etc. However, they have very strict installation requirements and cannot detect when multiple people pass by simultaneously. We also need at least two beam break sensors installed perpendicular to the direction pathway for identifying the direction of movement. Under properly managed situations, these do a wonderful job of counting the number of people present inside the room. However, they have very strict installation requirements and cannot detect when multiple people pass by simultaneously.

Sensing Air Pressure. This system work on the principle that when human move between rooms because of disturbances in airflow at the ductwork of air conditioning (HVAC) systems found in many buildings and houses. Human movement changes the static pressure in the HVAC air handling unit. A significant change in pressure is noticed when doors are opened or closed and when people go from one room to another. The pressure variation is noted by sensors installed at air filters which are then used to classify the movement events [16]. Ongoing research is going on for testing similar systems at various places. The accuracy level is lower compared to the technologies discussed above.

Array of IR Cameras. This system works on the principle that human presence causes a sudden change in temperature patterns recorded by infrared sensors. In [17], the authors used an array of very low-cost IR imaging devices at the doorway for counting entry and exit events. As with break beam sensors, this method also fails to identify events when multiple people enter and exit simultaneously and pose similar installation challenges.

Image Processing. RGB cameras are widely used by machine learning and image processing enthusiasts [18, 19] for counting and tracking people. This method can achieve high levels of accuracy at places where a complete field of view is available. In environments where it works, it can track different persons very effectively. It, however, fails to work under dark lighting conditions and poses some serious privacy issues at some obvious places.

Ultrasonic Sensors. Multiple research groups [20–23] have investigated the use of ultrasonic waves for solutions that count the number of people. The underlying idea is that the human body actively reflects any ultrasonic waves incident on it. It is recommended to use sensors [24, 25] that are capable of generating higher frequencies because they tend to be more accurate than cheaper ones that do not produce such high frequencies. This method can count the number of people; however, tracking is not possible. Data generated can become big data, so analytical techniques can also be useful for such big data [26–28].

5 Discussion

We have investigated the different human-sensing technologies available for precise estimation of people count. Unlike many researchers that have shared their opinions on the technologies for measuring the human presence, we have taken into account the economic, technological, and operational feasibility. There has been very little or no discussion on newer technologies like differential air pressure sensing, and CO_2 reading in existing researches. We have taken into consideration the perspective of smart device manufacturers and business owners (Table 1).

In this paper, we have given the reader understanding of how these technologies work and discussed the potential of them becoming a reality. Talking in terms of cost, break beam sensors, IR array and PIR sensors remain to be the most cost-effective solutions to date. Technologies like Wi-Fi signals demonstrate the

Table 1 Summary of discussed sensing technologies

Technology	Presence	Count	Track	Adaptable	Cost	Ease of deployment
Wi-Fi	Yes	Yes	No	No	Med	Easy
CO_2	Yes	Yes	No	Yes	High	Easy
Active light sensing	Yes	Yes	Yes	No	High	Hard
Passive light sensing	Yes	No	No	Yes	Low	Easy
Beam break sensor	Yes	Yes	No	Yes	Med	Hard
Air pressure sensing	Yes	No	–	–	High	Hard
IR camera array	Yes	Yes	No	Yes	Med	Hard
Image pro-cessing	Yes	Yes	Yes	Yes	Med	Easy
Ultrasonic sensors	Yes	Yes	No	Yes	Med	Hard

huge potential for becoming a mass-market technology provided further research makes it adaptable to real-world environments outside laboratories.

6 Conclusion

Similar to how computing devices transitioned from desks to people's pockets. In the world we live in, there will be an increasing demand for more powerful artificially intelligent agents. In this paper, we examined and discussed the pros and cons of the existing methods to acquire such information. We anticipate that future human presence detecting systems will likely consist of a mixture of one or more technologies discussed in the paper. With the current advancements in technology, we are more hopeful than ever to see people counting devices making their way to residential homes and commercial buildings.

References

1. MARKETSANDMARKETS: Smart Home Market by Product (Lighting Control, Security & Access Control, HVAC, Entertainment, Smart Speaker, Home Healthcare, Smart Kitchen, Home Appliances, and Smart Furniture), Software & Services, and Region-Global Forecast to 2024 https://www.marketsandmarkets.com/Market-Reports/smart-homes-and-assisted-living advanced-technologie-and-global-market-121.html (2018)

2. Beltran, A., Erickson, V.L., Cerpa, A.E.: Thermosense: occupancy thermal based sensing for HVAC control. In: Proceedings of the 5th ACM Workshop on Embedded Systems for Energy-Efficient Buildings, pp. 1–8. ACM (2013)
3. Buildings Energy Data Book. 2011: Energy Efficiency and Renewable Energy. US Department of Energy (2011)
4. Moghavvemi, M., Seng, L.C.: November. Pyroelectric infrared sensor for intruder detection. In: 2004 IEEE Region 10 Conference TENCON 2004, vol. 500, pp. 656–659. IEEE (2004)
5. Teixeira, T., Dublon, G., Savvides, A.: A survey of human-sensing: methods for detecting presence, count, location, track, and identity. ACM Comput. Surv. **5**(1), 59–69 (2010)
6. Mohammadmoradi, H., Yin, S., Gnawali, O.: Room occupancy estimation through wifi, UWB, and light sensors mounted on doorways. In: Proceedings of the 2017 International Conference on Smart Digital Environment, pp. 27–34 (2017)
7. Depatla, S., Muralidharan, A., Mostofi, Y.: Occupancy estimation using only WiFi power measurements. IEEE J. Sel. Areas Commun. **33**(7), 1381–1393 (2015)
8. Depatla, S., Mostofi, Y.: Crowd counting through walls using WiFi. In: IEEE International Conference on Pervasive Computing and Communications (PerCom) 2018 Mar 19, pp. 1–10. IEEE
9. Tiwari, R., Kumar, N.: Minimizing query delay using cooperation in IVANET. Procedia Computer Science **57**, 84–90 (2015)
10. Yang, Y., Cao, J., Liu, X., Liu, X.: Wi-Count: Passing People Counting with COTS WiFi Devices (2018). https://doi.org/10.1109/ICCCN.2018.8487420
11. Sobron, I., Del Ser, J., Eizmendi, I., Velez, M.: A Deep Learning Approach to DeviceFree People Counting from WiFi Signals. In: Del Ser, J., Osaba, E., Bilbao, M., SanchezMedina, J., Vecchio, M., Yang, X.S. (eds.) Intelligent Distributed Computing XII. IDC 2018. Studies in Computational Intelligence, vol 798. Springer (2018)
12. Arief-Ang, I.B., Salim, F.D., Hamilton, M.: DA-HOC: semisupervised domain adaptation for room occupancy prediction using CO_2 sensor data. In: Proceedings of the 4th ACM International Conference on Systems for Energy-Efficient Built Environments (BuildSys '17), pp. 1–10. Association for Computing Machinery, New York, NY, USA, Article 1 (2017)
13. Kuo, Y.-S., Pannuto, P., Hsiao, K.-J., Dutta, P.: Luxapose: Indoor positioning with mobile phones and visible light. In: MobiCom'14, pp. 447–458. ACM (2014)
14. Li, L., Hu, P., Peng, C., Shen, G., Zhao, F.: Epsilon: A Visible Light Based Positioning System. In: NSDI'14, pp. 331–343. USENIX Association, Seattle, WA. https://www.usenix.org/conference/nsdi14/technical-sessions/presentation/li (2014)
15. Zhang, C., Zhang, X.: LiTell: robust indoor localization using unmodified light fixtures. In: MobiCom'16, pp. 230–242. ACM (2016)
16. Sharma, I., Tiwari, R., Anand, A.: Open source big data analytics technique. In: Proceedings of the International Conference on Data Engineering and Communication Technology. Springer, Singapore (2017)
17. Mohammadmoradi, H., Munir, S., Gnawali, O., Shelton, C.: Measuring people-flow through doorways using easy-to-install ir array sensors. In: 2017 13th International Conference on Distributed Computing in Sensor Systems (DCOSS), pp. 35–43. IEEE (2017)
18. Barandiaran, J., Murguia, B., Boto, F.: Real-time people counting using multiple lines. In: 2008 Ninth International Workshop on Image Analysis for Multimedia Interactive Services, pp. 159–162. IEEE (2008)
19. Zhao, X., Delleandrea, E., Chen, L.: A people counting system based on face detection and tracking in a video. In: Sixth IEEE International Conference on Advanced Video and Signal Based Surveillance, 2009 (AVSS'09), pp. 67–72. IEEE (2009)
20. Hnat, T.W., Griffiths, E., Dawson, R., Whitehouse, K.: Doorjamb: unobtrusive room-level tracking of people in homes using doorway sensors. In: SenSys'12, pp. 309–322. ACM (2012)
21. Nasir, N., Palani, K., Chugh, A., Prakash, V.C., Arote, U., Krishnan, A.P., Ramamritham, K.: Fusing sensors for occupancy sensing in smart buildings. In: ICDCIT'15, pp. 73–92. Springer (2015)

22. Shih, O., Rowe, A.: Occupancy estimation using ultrasonic chirps. In: ICCPS'15, pp. 149–158. ACM (2015)
23. Mishra, D., Khan, A., Tiwari, R., Upadhyay, S.: Automated irrigation system-IoT based approach. In: 2018 3rd International Conference on Internet of Things: Smart Innovation and Usages (IoT-SIU), pp. 1–4. IEEE (2018)
24. Tiwari, R., Kumar, N.: A novel hybrid approach for web caching. In: 2012 Sixth International Conference on Innovative Mobile and Internet Services in Ubiquitous Computing. IEEE (2012)
25. Sharma, I., Tiwari, R., Rana, H.S., Anand, A.: Analysis of mahout big data clustering algorithms. In: Intelligent Communication, Control and Devices, pp. 999–1008. Springer, Singapore (2018)
26. Anand, A., et al.: Comparative analysis between proprietary software vs. open-source software vs. free software. In: 2018 Fifth International Conference on Parallel, Distributed and Grid Computing (PDGC). IEEE (2018)
27. Kumar, S., Tiwari, R.: Optimized content centric networking for future internet: dynamic popularity window based caching scheme. Comput. Netw. **179**, 107434 (2020)
28. Khan, E., et al.: Automated toll tax collection system using cloud database. In: 2018 3rd International Conference on Internet of Things: Smart Innovation and Usages (IoT-SIU). IEEE (2018)

Identification and Minimization of Churn Rate Through Analysing Financial Routines Using Machine Learning

Rahul Pahuja, Niket Dheeryan, Lovish Sethi, Preeti Nagrath, and Rachna Jain

Abstract In recent years, source of revenue for companies across all industries is subscription products. Hence, it is crucial to know the user area of interest so that customer churn from the company is minimized. Machine learning is gaining scope day by day. Its application has also found a way to predict and analyse the churn rate. The model is build using logistic regression and support vector machine (SVM) algorithms. Objective of the model is to predict which user is likely to cancel product subscription. Accuracy obtained in the model using logistic regression was found to be 61.2 and using SVM was found to be 60.7%.

Keywords Churn · Prediction · Accuracy · Logistic regression · SVM

1 Introduction

This research paper with the churn analysis of subscription products. This involves identifying user's area of interest and behavioural patterns that acts as rapid fire in disengagement of a particular product. Every company wants that subscription to the products should not be cancelled and for that they first need to identify which services provided by them is fulfilling the need of customer or not. Efforts are made accordingly by providing new features to the user like low interest rate for loans. Machine learning has recently advanced with a great pace. Logistic regression one of the algorithms is applied in this model and prediction of churning is obtained for all the users using the features. Logistic regression is used to identify statistical probability in form of binary variables whether churn will take place or not.

R. Pahuja · N. Dheeryan · P. Nagrath · R. Jain
Department of C.S.E., Bharati Vidyapeeth College of Engineering, New Delhi 110063, India
e-mail: preeti.nagrath@bharatividyapeeth.edu

R. Jain
e-mail: rachna.jain@bharatividyapeeth.edu

L. Sethi (✉)
Department of IT, Bharati Vidyapeeth College of Engineering, New Delhi 110063, India

K. Khanna et al. (eds.), *Cyber Security and Digital Forensics*,
Lecture Notes on Data Engineering and Communications Technologies 73,
https://doi.org/10.1007/978-981-16-3961-6_43

Features which are important to predict the response variable were analysed in this model building process so that amendment can be made in those services and company minimize the cancelling of subscription and hence retain the customers. Churning of customers depends on quality of services provided to them. Many churn predicting models based on machine learning has been designed in past few years, but all models cannot work efficiently. If best services are provided to the customers at proper time, then chances of retaining the customers to a product increase with great pace. Machine learning models can easily predict the parameters which are forcing the user to leave the product. Although we can see that machine learning is advancing and used in almost every area, it suffers from few following limitations:

1. Problem of overfitting arises when generally small datasets are used.
2. Large possibility of errors.
3. Training data that is used for fitting the model must be unbiased and should be of best quality.

Machine learning has proved to be very helpful to the big companies and businessmen as they get to know which factor is making competitions high among the markets.

Support vector machine (SVM)—It is supervised machine learning algorithm that is used mainly for classification purposes. In this algorithm, every data point of dataset is plotted in n dimensional space where n is number of features in dataset with each feature's value being the value of a particular coordinate. It consists of a hyperplane to differentiate and perform classification.

Various classification techniques serve the purpose of determining the chances of churning. Banking systems are also applying random forest algorithms along with hyper tuning of dataset to classify the churn possibility of the users. Standardization and Normalization of data has to be done to get effective results of prediction. Feature extraction is necessary to determine which features have to be used for prediction in machine learning models. In machine learning, most of the dataset can be read through pre-existing pandas and Sklearn library.

Logistic regression is one of classification algorithm that uses a logistic sigmoid function to return probability value that maps to discrete classes.

1.1 Motivation

Determination of churning of user from the company or from subscription products is the major demand of every company in order to increase the profit. By knowing the areas in which users are churning the most, they can introduce new features to the products.

Various authors have made efforts in this area in past few years. Different methodologies have been adopted for making prediction. Authors in [1] have designed a hybrid model (a two-stage model). Other works include the use of different algorithms like decision tree, SVM, fuzzy *c* means algorithm. Hybrid approaches gave much better result as compared to single algorithm. Machine learning is a growing field of science and it contains various algorithms which need to be explored further.

1.2 Contribution

In this paper two machine learning algorithms, i.e. Logistic regression and support vector machine (SVM) are proposed for predicting the churning of user. The proposed approach uses various techniques in model building like pre-processing, plotting various plots for getting distribution of features, One hot encoding, Feature scaling and balancing, Feature selection using RFE function. Accuracy obtained with this approach was found to be 61.2% on Fintech dataset for 27,000 users using logistic regression while SVM algorithm has provided accuracy of 60.7%.

Following are the contributions of paper:

1. Histograms and pie charts are plotted to get distribution of various features of dataset.
2. Effect of various features on churning of user is determined using correlation plot.
3. Efficiency of algorithms is enhanced by extracting few features which are necessary for accomplishing objective of prediction.

1.3 Uniqueness of the Paper

1. Relationship between different features and churning of user is studied using correlation plots. This gives a brief overview that whether churning is positively related to that feature or negatively related.
2. Feature selection technique is applied to achieve top best feature required for making prediction which increases the efficiency of algorithms.

1.4 Organisation

The remaining part of the paper is organized as follows. Section 2 contains the related works that have been done for churn prediction in different areas using various approaches. Section 3 describes the proposed approach. Performance evaluation of the proposed approach is indicated in Sect. 4. Section 5 is the concluding part of the paper.

2 Related Works

Weighted voting method was used to build a final decision tree from several small decision trees in the work by Wei et al. [2] in 2002. Final classification of churning users was obtained using the decision tree obtained. Decision tree's ability to classify better even in presence outliers has made it a recursively used algorithm.

Hadden et al. [3] had published a work in 2006 in which they are identifying the most important variables to identify churn possibility. Artificial neural network, decision trees and regression were used to model the data. Churn risk plot made in artificial neural network predicted the result to 72% accuracy. Highest accuracy found for the model was from decision trees.

Qi et al. [4] in 2006 had built Tree Logit model consist of logistic regression and A D-tree (decision tree with boosting algorithm). Logistic model was built for every decision tree that is obtained after categorization to make customer churning prediction.

In 2007, probability of churning is predicted using self-organizing map (an artificial neural network) by Chu et al. [5]. Clustering of subscribers according to important characteristics was done. SOM layers size was fixed so to build an effective model better version of SOM was used, i.e. GHSOM with 85% accuracy.

In another work done by Zhao and Dang [1] in 2008, that used SVM with 4 kernels for prediction of customers churning from bank and then compared it with different classifiers like logistic regression, Naïve Bayes, ANN. In the analysis, SVM gave highest prediction accuracy of 59.74%.

Various algorithm was analysed like regression, Naïve Bias, decision tree, support vector machine (SVM) was applied by Coussement and Van den Poel [6] for classification of most important feature in churn minimization. To handle the nonlinear relationships between independent fields of the dataset, RBF kernel was used for SVM model implementation. This model worked well with newspaper subscription dataset and prediction was accurate to great extent. PCC metric was used to determine the probability of most likely churning user to the least likely churning user, hence making subscription analysis effective in that model.

Churn prediction in retail banking was determined using Fuzzy c means clustering by Popovic and Dalbelo Basic [7] in 2009. It was the more accurate algorithm as compared to hierarchical methods and k means clustering. It was operated well on the outliers existing in dataset.

Hybrid model was built in 2009 by Bose and Chen [8]. This model involved two stages. In first stage, various clustering algorithms like KMD, KM, SOM, BIRCH, FCM were applied. Second stage was the final prediction stage which used decision tree (C 5.0) along with boosting procedure. Decision tree was used because when compared to other algorithms as it is faster approach.

Neural network based on genetic algorithm was proposed by Pendharkar [9] in 2009. Managing hidden nodes effectively played an important role in making churn predictions. Efficient feature selection in GANN had enabled it to make more accurate predictions.

Another work by Bangzhu [10] in 2010 includes the use of Least Square Support Vector Machines, Rough sets in addition to SMC for predicting Churn rate of customers. Accuracy of the model was somewhat higher in this model. LSSVM worked well with inseparable data and was able to make prediction better than SVM.

Data mining techniques were applied to telecom dataset by Umayaparvathi and Iyakutti [11] in 2012. Performance of decision tree and a neural network was analysed and decision tree was found to be more efficient in predicting users more likely to churn. Neural network and decision tree were trained using rules defined for the features of dataset.

Qureshi et al. [12] in 2013, had analysed and classified active telecom customers and churn customers using various approaches. Linear regression, Artificial neural networks, 4 types of decision trees, i.e. CART (Classification and regression tree), CHAID (Chi-squared automatic interaction detector), Exhaustive CHAID and QUEST (Quick unbiased and efficient statistical tree) were used. Exhaustive CHAID was the most efficient version of decision tree with 70% accuracy.

In 2014 another work done by Zhao [13] was able to predict the ecommerce customer churning rate using Artificial Neural Network and SVM. Genetic algorithm was used which means result of SVM prediction and artificial neural network were combined and passed to SVM model again to obtain better results. Since ecommerce data was highly nonlinear, model using single algorithm could not give efficient result in that scenario and time consumed was also higher.

Data mining was used for predicting churn of telecom user in the work presented by Dahiya and Talwar [14] in 2015.

Khan et al. [15] had obtained 90% accuracy in 2015. Subscribers churning from a telecom industry using different algorithms like Linear Regression, SVM, KNN, random forest. Trees that were trained using K fold cross-validation-based approach were used for feature selection.

Predicting the next activity and its timestamp can be accomplish using LSTM (long short-term memory). LSTM was used by Tax et al. [16] in their project to predict remaining cycle time and next process on the basis of past activity in 2016. It worked well for prediction as LSTM is special RNN which has long-term dependencies power. LSTM may have very long training time which is the limitation for its working.

Churn prediction in telecom industry was performed by Dalvi et al. [17] which incorporated the use of two algorithms, i.e. decision tree and logistic regression in 2016 with a speciality that the system can be used with any type of data. System consists of 3 modules for providing different functionalities. First module was for providing interface, second for extracting features and third for making predictions.

Role of attributes selection in making prediction of customers churn from telecom industry was analysed with data mining techniques by Umayaparvathi and Iyakutti [18] in 2016. Gradient boost algorithm has given highest accuracy of 73 and 95% on two datasets out of 7 classifiers used for prediction.

Jain et al. [19] had made an effort in 2017 to predict churn using decision tree along with classification approach. Decision tree proves to be an efficient approach while making prediction as it studies every node and considers all possible outcomes. RMD metric was used to predict, how much accurate the model was. Pruning technique was used to improve the results of decision tree. Hence, number of churning user was determined by the model. Data needs to be properly managed as changes make decision tree unstable.

K means clustering was combined with many classifiers like Naïve Bayes, random forest, decision tree, SVM in 2017 by Rajamohamed and Manokaran [20]. Rough k means clustering algorithm with SVM was able to achieve highest accuracy of 96%. Single classification algorithms were less accurate as compared to hybrid models.

Improvement in FCM algorithm was provided by Cui et al. [21] in 2017 to overcome the previous inefficiencies in the algorithm like inability to predict number of clusters for making final customer churn prediction. The prediction accuracy was obtained to be 80.3% by enhanced algorithm.

Four algorithms, i.e. logistic regression, SVM, gradient boost and random forest for prediction of churn were used in the work suggested by Gaur and Dubey [22] in 2018. Analysis of area under curve (AUC) values has proved gradient boost to be most accurate method for prediction with its value being 84.59.

Artificial neural network can also be used for classification of churn. The most popular work was done in the paper by Xia and He [23] in which the use of BP neural network was incorporated with SVM to predict the churn. Various plots were plotted between different commodities and churn. Results of combined algorithms proved to be efficient with accuracy of 93%. But on the other side rate of convergence of BP neural network is quite slow and also numerical stability of predicted results is poor for this algorithm.

In 2018 Sahu et al. [24] had made an effort to predict churn of telecom customers using data mining techniques. Pattern among the already churned customers was found. LASSO was used to penalize the regression coefficient size. Then Lasso was fitted to logistic regression to evaluate the result. Decision tree used in the model was more effective than logistic regression.

Factors responsible for churning were determined, and churning chances of users were determined for telecom using neural network with an 80% accuracy by Agrawal et al. [25] in 2018.

Logistic regression was incorporated for making prediction in the model developed by Sai and Sasikala [26] in 2019. ROC was used for model evaluation and value of Area under curve was determined to be 0.83 while making prediction with value of threshold being 0.5.

In 2019, behaviour of ecommerce customers was analysed to predict the churn rate using K Nearest neighbour and decision trees. Hybrid approach was followed in this model by Kareena [27]. In the beginning, relationship was determined between response and features using decision tree (Base Classifier). Then final classification of customers into different categories was made using KNN (Meta Classifier). This is known as hybrid approach.

Table 1 Comparative analysis of various works

Reference no.	Year	Algorithm used	Accuracy (%)
[3]	2006	1. Decision tree 2. Regression 3. Artificial Neural Network	82 81 74
[5]	2007	1. Self-organizing map	85
[1]	2008	1. SVM 2. Logistic regression 3. Naïve Bayes 4. ANN	59.74 58.90 55.49 54.79
[11]	2012	1. Decision tree 2. Neural network	98.88 98.43
[15]	2015	1. SVM 2. Random forest 3. KNN 4. AdaBoost 5. Logistic regression	89.4 88.4 88.2 89.2 89.3
[20]	2017	1. KNN 2. DT 3. SVM 4. RF 5. NB	76.05 83.7 93.2 85.6 77.9
[27]	2019	1. Naïve Bayes 2. Hybrid model (Decision tree and KNN)	74.11 90.75

Random forest was used in the work suggested by Zhou and Yang [28] in which review helpfulness was predicted using textual and numerical features. In this analysis, review length was found to be crucial criteria for prediction (Table 1).

3 Proposed Approach

Machine learning is a subfield of artificial intelligence. It provides computer the ability to learn and practise without being explicitely programmed. Machine learning contains many algorithms with which it passes data and trains from it. One of the machine learning algorithms known as logistic regression is used in this model to predict whether a user will churn or not.

3.1 Reading Dataset

Dataset is read using Pandas library. Pandas is an open source library whose key feature is the data frames (manipulation is done in form of rows and columns) read function is used to read the file.

Description of dataset: Dataset used for model building is a Fintech Company dataset that provides financial services to the customers. Dataset contains information of the 27,000 users using different services from the company. Number of features in the dataset is 31 out of which some contains binary data and some contain numerical data. Dataset contains user identification number for identification of the services used by the user. Source: https://www.kaggle.com/niketdheeryan/fintech-users-data.

3.2 Cleaning of Data

Data cleaning is very crucial step in implementing machine learning algorithms. Null values in the given dataset are checked using prebuilt ISNA () function. If a particular column contains large number of null values, then that particular column is dropped using drop () function in order to accomplish model building process.

3.3 Plotting Histograms and Pie Chart

Histograms and pie chart are plotted to get the distribution of various features of dataset.

Histogram plots the data by dividing into interval called bins as shown in Fig. 1. Distribution obtained from histograms is less informative as binary variables are not

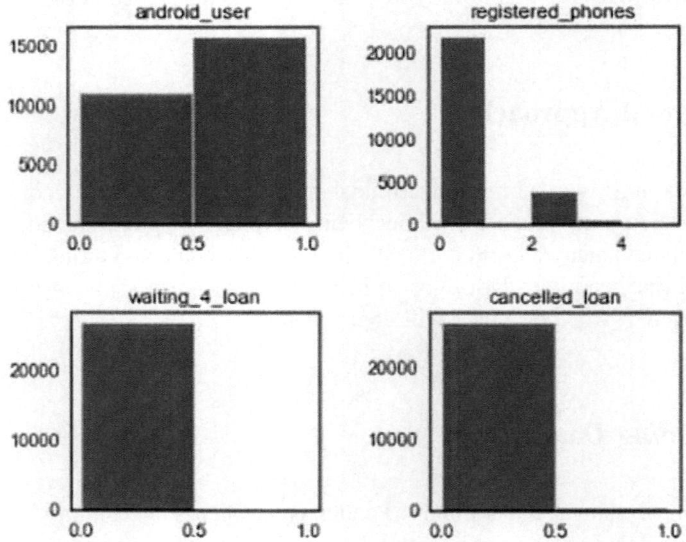

Fig. 1 Histogram for Fintech users' dataset

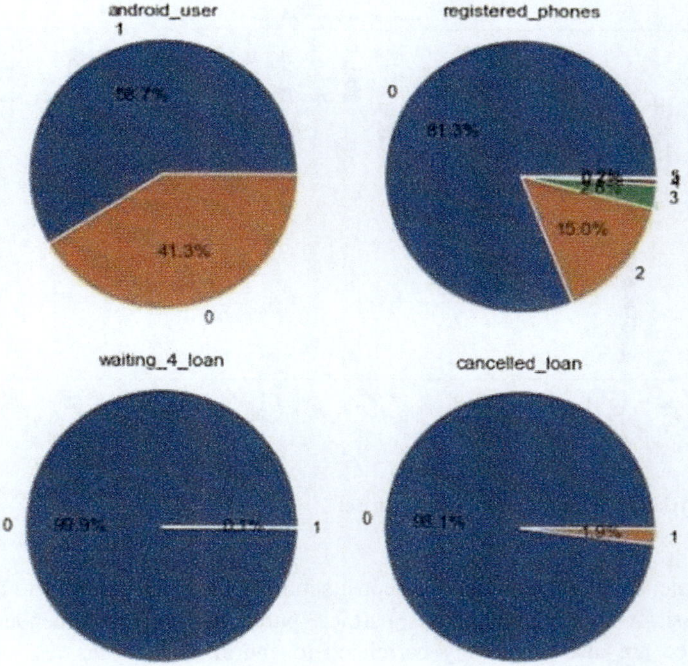

Fig. 2 Pie Chart for Fintech user's dataset

properly examined in histograms. So, pie chart is plotted in order to obtain better visualization in form of percentages.

In Fig. 2, depicting pie chart for Fintech user's dataset, fields which are evenly distributed (i.e. 50%–50% or 40%–60%) are not of much concern. Small subsets have to be explored further in order to know whether they are useful for prediction or not. For example, android_user plot is evenly distributed and there is nothing to worry about those distributions. On the other hand waiting_4_loan, registered_phones, cancelled_loan have subsets which are very small have to be taken care off because there are not enough results for those small subsets to include both rows where churn variable is 0 and 1.

3.4 *Correlation Plot*

This plot is used to gaze the correlation between independent and dependent variables of dataset. Categorical variables are removed and binary variables are kept because this correlation plot requires numeric value fields. In Fig. 3 showing the correlation plot among various fields and churning possibility, the fields whose bar plot is below origin line are negatively correlated with the churn chance.

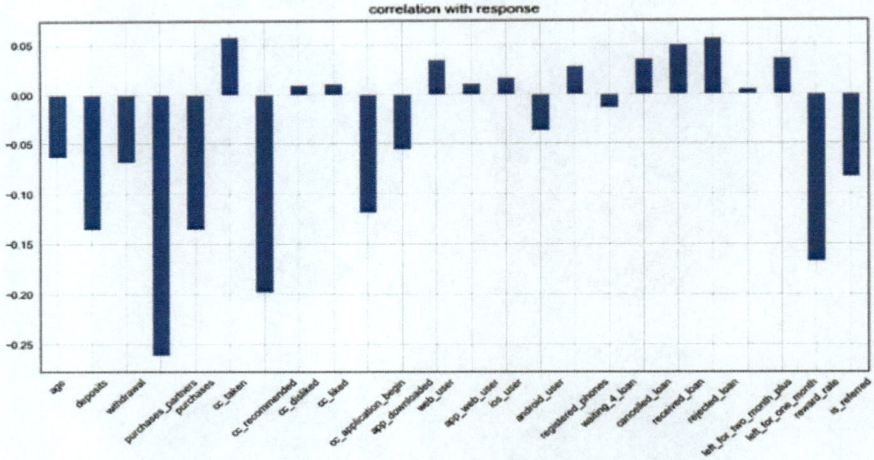

Fig. 3 Correlation plot for Fintech user's dataset

For example, smaller the number of deposits more is the chance of user to be churn. age, deposit, withdrawl, purchases_partner, purchases, cc_recommended, app_-downloaded are also negatively correlated to churn. This means less activity a person is having in particular field; he is more likely to churn. In case of cc_recommended, if more recommendations are provided, more likely the user will stay with the company. In a similar manner, cc_taken, web_user, app_web_user, ios_user, registered_phones are positively correlated to response variable, i.e. churn. It is giving an insight that people are not satisfied with these features and enhancement has to be done to prevent churn of users.

3.5 *One Hot Encoding*

It converts all categorical variables into their own independent binary variables because python models cannot read categorical variables. One hot encoding is done by getting dummies () function present in panda library. Correlation among the one hot encoded column is dropped using drop () function.

3.6 *Splitting the Dataset into Train Set and Test Set*

train_test_split () is a function in Sklearn model selection for splitting data arrays into two subsets: train set and test set. Test set is used to predict the accuracy of the designed model while train set is used to train the model.

3.7 Feature Scaling and Balancing

Balancing the amount of zeroes and ones that training set has in its response variable guarantees that whatever accuracy we are getting is result of model itself rather than random permutation of how many zeroes or ones are present in training set.

Feature scaling is a pre-processing step applied to features of dataset. It normalizes all numerical fields in such a way that its weight is not too large in equation of logistic regression standard scalar present in sklearn pre-processing standardize features by removing the mean and scaling to unit variance.

3.8 Feature Selection Process

Top best features were selected from the dataset. Useful attributes can be extracted to minimize the time taken for machine learning algorithm. RFE function present in sklearn feature selection is used for that purpose. RFE support used in the project returns list of Boolean values meaning that these values are actually mapping which columns are to be included in final result.

3.9 Fitting Model to Training Set

Logistic regression (LR) is a statistical method similar to linear regression since LR finds an equation that predicts an outcome for a binary variable, Y from one or more response variables X.

Modelling of model is done in such a way that probability maps to either 0 or 1.

Now let us consider a generalized model having parameter θ then we have function,

$$h_\theta(X) = 1/\left(1 + \mathrm{e}^{-\theta TX}\right) = \Pr(Y = 1|X;\theta) \tag{1}$$

exponent is used in modelling logistic regression because probability is always greater than 0.

Hence,

$$\Pr(Y = 0|X;\theta) = 1 - h_\theta(X) \tag{2}$$

$$\text{And since } Y \in \{0, 1\}, \text{therefore } \Pr(y|X;\theta) = h_\theta(X)^y(1 - h_\theta(X))^{1-y} \tag{3}$$

Now assumption is made that all observations are Bernoulli distributed, likelihood function is given as:

$$L(\theta|y;x) = \Pr(Y|X;\theta) \tag{4}$$

$$= \prod_i \Pr(yi|xi;\theta) \tag{5}$$

$$= \prod_i h_\theta(x_i)^{yi}(1 - h_\theta(x_i))^{1-yi} \tag{6}$$

log likelihood is maximized,

$$N^{-1} \log L(\theta|y;x) = N - 1\sum_{i=1}^{N} \log \Pr(yi|xi;\theta) \tag{7}$$

Gradient descent is used to maximize Eq. (7).
Now (*x*, *y*) pair are drawn in uniform manner from distribution,
Applying limit to *N*,

$$\operatorname*{Lim}_{n\to\infty} \sum_{i=1}^{N} \log \Pr(yi|xi;\theta) \tag{8}$$

$$= \sum_{x\in X}\sum_{y\in Y} \Pr(X = x; Y = y) \log \Pr(Y = y|X = x;\theta) \tag{9}$$

$$= \sum_{x\in X}\sum_{y\in Y} \Pr(X = x, Y = y)\left(-\log\frac{\Pr(Y = y|X = x)}{\Pr(Y = y|X = x;\theta)} + \log \Pr(Y = y|X = x)\right) \tag{10}$$

$$= -D_{\mathrm{KL}}(Y||Y_\theta) - H(Y|X) \tag{11}$$

where D_{KL} is the Kullback Leiber divergence and $H(X|Y)$ is conditional entropy.
The odds of any event in logistic regression are given by:

$$p/1 - p = e^{w_0 + w_1x_1 + \cdots + w_nx_n} \tag{12}$$

which is further log transformed.
Now to define log likelihood we take log of Eq. (12)

$$\log(p/1 - p) = w_0 + w_1x_1 + \cdots + w_nx_n \tag{13}$$

3.10 *Predicting Output for Test Set and Evaluation of Accuracy*

Test data is fed to predict function of the classifier and predicted values are determined. Accuracy of model is an important parameter for evaluation of any model. Various accuracy scores are imported using sklearn metrics.

Accuracy score is defined as:

$$\text{Accuracy Score} = \frac{\text{TRUE POSITIVE} + \text{TRUE NEGATIVE}}{\text{TRUE POSITIVE} + \text{TRUE NEGATIVE} + \text{FALSE POSITIVE} + \text{FALSE NEGATIVE}}$$

Precision score is defined as:

$$\text{Precision Score} = \frac{\text{TRUE POSITIVE}}{\text{TRUE POSITIVE} + \text{FALSE POSITIVE}}$$

Precision score is the ability of classifier not to label as positive if sample is negative.

Recall score is defined as how many relevant items are selected.

$$\text{Recall Score} = \frac{\text{TRUE POSITIVE}}{\text{TRUE POSITIVE} + \text{FALSE NEGATIVE}}$$

Diagrammatic representation of model building process.

Figure 4 depicts the flow of steps involved in model building. Initially, dataset is read using read function in step 1. Then in step 2 pre-processing (cleaning) of dataset is done. Distribution of various features isplotted using various plots in steps 3 and 4. In step 5, One hot encoding is applied. In step 6, we split data into two sets —train and test. Then feature scaling and balancing is applied for normalising the fields in step 7. Best features are then selected in step 8. After that model was fitted to training set. In last step, evaluation of model is done.

4 Performance Evaluation

4.1 *Discussion on Results*

For building the project standard libraries like Sklearn, Pandas, NumPy, seaborn, matplotlib were used.

Sklearn is used because it supports python numerical libraries like NumPy and is freely available machine learning library for python. matplotlib pyplot is used for plotting various plots to get distribution of various fields in the dataset.

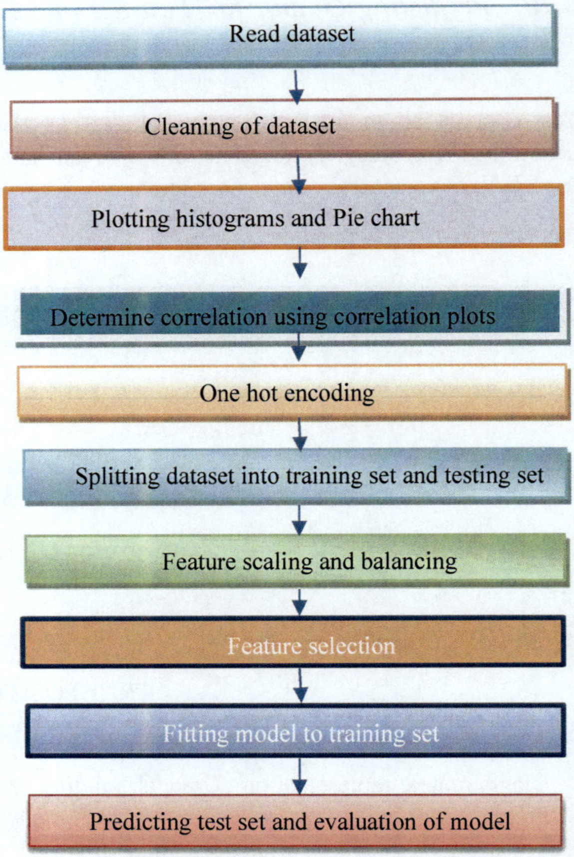

Fig. 4 Flow chart for model building process

Pandas library is most popular library and is used because its key feature includes Dataframe and helps to manipulate data in form of rows and columns.

Dataset imported was splitted into two parts in order to train the model and then evaluate on test data. Model was fitted to logistic regression and accuracy score was calculated. Accuracy Score is the important parameter for evaluation of model (Table 2).

The heat map is the representation of matrix data in a graphical form. It is used to represent the confusion matrix or any two-dimensional data in form of colour.

Table 2 Various accuracy scores obtained on testing

Algorithm used	Accuracy score	Precision score	Recall score	F1 score
Logistic regression	61.2	55.4	74.8	61.4
SVM	60.75	51.6	75.4	61.5

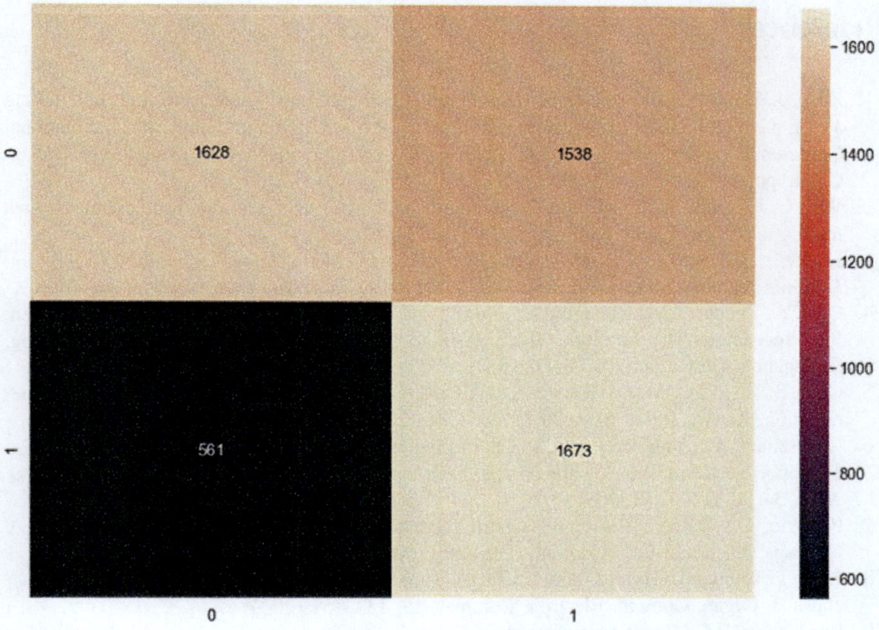

Fig. 5 Heat map for confusion matrix

In Fig. 5, Heat map representing the confusion matrix is shown. It has successfully able to identify 1665 true positives, 1541 false positives, 1625 true negatives and 569 false negatives.

The accuracy can be further improved in future by designing hybrid models which involve combining two or more algorithms together to form an efficient model and also by applying neural networks and deep learning algorithms. Hybrid approach gives better result as compared to single algorithm.

5 Conclusion

This paper has justified the need of churn analysis in order to minimize the customer loss from the subscription to the particular product. Simple model using logistic regression, SVM algorithm was built to identify the areas in which churning from the company has occurred. This is the need of every company to know area of interest of user using the product so as to retain their customers. The current model build using machine learning has involved data cleaning and other pre-processing steps like feature scaling which need to be minimized further. In future, this process may be implemented through other models like artificial neural networks, deep learning. Models should be built in such a manner that pre-processing steps like feature scaling, balancing steps are minimized to greater extent in order to ensure effectiveness of the model.

References

1. Zhao, J., Dang, X.: Bank customer churn prediction based on support vector machine: taking a commercial bank's VIP customer Churn as the xxample. In: 2008 4th International Conference on Wireless Communications, Networking and Mobile Computing, Dalian, China, pp. 1–4 (2008). https://doi.org/10.1109/WiCom.2008.2509
2. Wei, C.-P., Chiu, I.-T.: Turning telecommunications call details to churn prediction: a data mining approach. Expert Syst. Appl. **23**(2), 103–112 (2002)
3. Hadden, J., Tiwari, A., Roy, R., Ruta, D.: Churn prediction: does technology matter. Int. J. Intell. Technol. **1**(1), 104–110 (2006)
4. Qi, J.Y., Zhang, Y.M., Zhang, Y.Y., Shi, S: TreeLogit model for customer churn prediction. In: Proceedings of the 2006 IEEE Asia-Pacific Conference on Services Computing, Guangzhou, China, pp. 70–75, 12–15 Dec. 2006
5. Chu, B.-H., Tsai, M.-S., Ho, C.-S.: Toward a hybrid data mining model for customer retention. Knowl. Based Syst. **20**(8), 703–718 (2007)
6. Coussement, K., Van den Poel, D.: Churn prediction in subscription services: an application of support vector machines while comparing two parameter-selection techniques. Expert Syst. Appl. **34**(1), 313–327 (2008)
7. Popovic, D., Dalbelo Basic, B.: Churn prediction model in retail banking using Fuzzy C-means clustering. Eng. Comput. 036–130 (2009)
8. Bose, I., Chen, Xi.: Hybrid models using unsupervised clustering for prediction of customer churn. J. Organ. Comput. Electron. Commer. **19**, 133–151 (2009)
9. Pendharkar, P.C.: Genetic algorithm based neural network approaches for predicting churn in cellular wireless network services. Expert Syst. Appl. **36**(3), 6714–6720 (2009)
10. Bangzhu, Z.: The prediction of e-business customer churn based on smc-rs-lssvm model. Syst. Eng. Theor. Pract. **11**, 1960–1967 (2010)
11. Umayaparvathi, V., Iyakutti, K.: Applications of data mining techniques in telecom churn prediction. Int. J. Comput. Appl. (2012)
12. Qureshi, S.A., Rehman, A.S., Qamar, A.M., Kamal, A., Rehman, A.: Telecommunication subscribers churn prediction model using machine learning. In: Digital Information Management (ICDIM), 2013 Eighth International Conference, pp. 131–136 (2013)
13. Zhao, X.: Research on E-commerce customer churning modeling and prediction. Open Cybern. Syst. J. **8**, 800–804 (2014)
14. Dahiya, K., Talwar, K.: Customer Churn prediction in telecommunication industries using data mining techniques—a review. In: International Journal of Advanced Research in Computer Science and Software Engineering (2015)
15. Khan, M.R., Manoj, J., Singh, A., Blumenstock, J.: Behavioral modeling for churn prediction: early indicators and accurate predictors of custom defection and loyalty. In: Proceedings of the IEEE International Congress on Big Data, BigData Congress 2015, pp. 677–680 (2015)
16. Tax, N., Verenich, I., La Rosa, M., Dumas, M.: Predictive Business Process Monitoring with LSTM Neural Networks (2016)
17. Dalvi, P.K., et al.: Analysis of customer churn prediction in telecom industry using decision trees and logistic regression. In: 2016 Symposium on Colossal Data Analysis and Networking (CDAN), Indore, India, pp. 1–4 (2016). https://doi.org/10.1109/CDAN.2016.7570883
18. Umayaparvathi, V., Iyakutti, K.: Attribute selection and customer Churn prediction in telecom industry. In: 2016 International Conference on Data Mining and Advanced Computing (SAPIENCE), Ernakulam, India, pp. 84–90 (2016) https://doi.org/10.1109/SAPIENCE.2016.7684171
19. Jain, R., Rajpoot, V., Richhariya, V.: Predicting Churn in Telecom Sector Using Classification & Decision Tree, vol. 3(5) (2017). ISSN: 2395-3853
20. Rajamohamed, R., Manokaran, J.: Improved credit card churn prediction based on rough clustering and supervised learning techniques. Cluster Comput. **21**, 65–77 (2017)

21. Cui, S., Ding, N.: Customer Churn prediction using improved FCM algorithm. In: 2017 3rd International Conference on Information Management (ICIM), Chengdu, pp. 112–117 (2017). https://doi.org/10.1109/INFOMAN.2017.7950357
22. Gaur, A., Dubey, R.: Predicting customer Churn prediction in telecom sector using various machine learning techniques. In: International Conference on Advanced Computation and Telecommunication (ICACAT), Bhopal, India, pp. 1–5 (2018). https://doi.org/10.1109/ICACAT.2018.8933783
23. Xia, G., He, Q.: The research of online shopping customer Churn prediction based on integrated learning. In: 2nd International Conference on Mechanical, Electronic, Control and Automation Engineering (MECAE 2018), China
24. Sahu, M.K., Pandey, R., Silakari, S.: Analysis of customer Churn prediction in telecom sector using logistic regression and decision tree. J. Appl. Sci. Comput. **5** (2018)
25. Agrawal, S., Das, A., Gaikwad, A., Dhage, S.: Customer Churn prediction modelling based on behavioural patterns analysis using deep learning. In: 2018 International Conference on Smart Computing and Electronic Enterprise (ICSCEE), Shah Alam, Malaysia, pp. 1–6 (2018). https://doi.org/10.1109/ICSCEE.2018.8538420
26. Sai, B.N.K., Sasikala, T.: Predictive analysis and modeling of customer Churn in telecom using machine learning technique. In: 2019 3rd International Conference on Trends in Electronics and Informatics (ICOEI), Tirunelveli, India, pp. 6–11 (2019). https://doi.org/10.1109/ICOEI.2019.8862625
27. Kareena, R.K.: A consumer behavior prediction method for E-Commerce application. Int. J. Recent Technol. Eng. (IJRTE). **8**(2S6) (2019). ISSN: 2277-3878
28. Zhou, Y., Yang, S.: Roles of review numerical and textual characteristics on review helpfulness across three different types of reviews. IEEE Access **7**, 27769–27780 (2019)

Machine Learning-Based Predictive Analysis to Abet Climatic Change Preparedness

Abra Shafiq Siddiqi, Md. Afshar Alam, Deepa Mehta, and Sherin Zafar

Abstract Global warming and the corresponding climatic changes have affected the world adversely. Climatic changes encompass the soaring temperatures, extremity in weather phenomenon, disrupting habitats, rising water levels in seas, and plenty of other impacts. As these changes emerge, the humans attempt to reduce the carbon emissions. This research paper aims to study the changing temperatures as a result of the industrial activities and greenhouse effect over the last 5 decades. The analysis utilizes data analytic tools to analyze the percentage at which the decades are affected as we moved into technologically advanced era. After studying the effects, the research paper also aims to predict the changes in the mean temperatures for the next decade using the time series prediction model with the help of machine learning algorithms. The dataset includes monthly temperatures for about 150 countries for a period of 58 Years, and machine learning algorithms aim to predict the rise and fall of temperatures for the next decade successfully.

Keywords Global warming · Prediction analysis · Time series · Machine learning

1 Introduction

AI and man-made consciousness are required to give huge new experiences into understanding environmental change and how to battle it, as indicated by an investigation distributed today in Environmental Research Letters. Studies give a convincing contention that AI (ML) and man-made reasoning (AI) can fill a portion

Present Address:
A. S. Siddiqi · Md.A. Alam · S. Zafar (✉)
Jamia Hamdard, New Delhi, India
e-mail: sherin.zafar@jamiahamdard.ac.in

Md.A. Alam
e-mail: aalam@jamiahamdard.ac.in

D. Mehta
GD Goenka University, Sohna, Haryana, India

K. Khanna et al. (eds.), *Cyber Security and Digital Forensics*,
Lecture Notes on Data Engineering and Communications Technologies 73,
https://doi.org/10.1007/978-981-16-3961-6_44

of the holes that exist in environment science. Educator Chris Huntingford, from the Center for Ecology and Hydrology, is the investigation's lead creator. He said: "In spite of the fact that environment research is by and large viewed as a cycle drove movement, it is likewise incredibly complex, thus the use of measurably based AI calculations will clarify new climate examples and associations."

Co-creator Professor Mike Bonsall said: "ML calculations have progressed significantly lately, and have empowered momentous discoveries in other examination sec-peaks. There is each motivation to expect they can be utilized to help environment analysis." Co-creator Hannah Christensen added: "There is a wide scope of potential uses of ML in the environment sciences. While a few people stay wary of ML, truth be told climate researchers have been utilizing attempted and tried ML strategies for a long time with-out acknowledging it. We feature this in our paper, just as point out new zones where ML could change our field." Co-creator Dr. Hui Yang remarked: "Planet Earth is observed at extraordinary levels, and particularly by satellites gathering tremendous quantities of environment-related information. Specialists need further developed calculations and techniques to utilize such enormous measures of information, to describe patterns, practices and interconnections."

To represent explicit capability of ML and AI, the analysts inspected three test plea where they could be utilized to acquire knowledge into environment occasions —the UK summer 2018 dry season; the environment "break"; and earthbound biological system condition fabricating—for instance how plant nourishment interfaces with environment and impacts the measure of carbon that can be put away on the land surface. Favorable to Professor Huntingford said: "regarding understanding the 'rest', it is fundamental that for an environment highlight so unmistakable, we join all strands of proof to create a complete clarification. As the rest is likely an element of concurrent cooperation's in the environment framework, ML can help describe these, and highlight any environment model lacks [1–3]."

Co-creator Dr. Elizabeth Jeffers said: "In biological system condition building, ML can help distinguish which plant synthetic characteristics and natural conditions are driving input to the nitrogen and carbon cycles, empowering the improvement of interaction-based conditions for use in demonstrating supplement restriction." Co-writer Thomas Lees said: "Our examination sums up where explicit environment framework segments have just been researched with ML". Our call is to go a lot further and utilize ML techniques to the whole the Earth framework, with an accentuation on surveying its inner associations. Supportive of Professor Huntingford summed up: "Environment demonstrating needs a stage change to lessen vulnerability in projections—ML probably has a significant job in accomplishing that" [4–6].

In the upcoming sections in this paper, authors will focus on machine learning-based predictive analysis to abet climatic change preparedness, various research issues and questions are discussed in Sect. 2. Methods and results are presented in Sects. 3 and 4 that will highlight on the model building for predictive analysis, followed by conclusion and references.

2 Research Problem and Research Questions

2.1 Research Problem

In the utilization of AI-based examination, is there a methodology that can be effortlessly utilized by amateurs to make information investigation and the understanding of results workable for individuals not familiar with programming? Moreover, is there a doable route for fledglings to play with explicit factors in different PC reenactment situations utilizing easy to use AI programming? How might the bits of knowledge from the investigated information be passed on in natural non-specialized terms that others, e.g., strategy producers, who are more worried about the "generally speaking greater picture," can plainly get it? Creators propose that there is an answer that can be utilized to determine these previously mentioned issues. The easy to use AI-empowered methodology that will be used in this paper depends on Bayesian network (BN) probabilistic thinking [7, 8]. It very well may be applied utilizing a product bundle that is appropriate for fledglings, as it doesn't need any product programming abilities. Utilizing this methodology, people can start to lead the pack and consider over the predictive models produced by AI. Probabilistic thinking utilizing BN models is more intuitive for speaking with numerous partners across multidisciplinary area verticals, as deduction as far as likelihood is nearer to normal human idea, compared to utilizing old style frequentist connections or p-values [9–12]. The models in the current paper will likewise encourage openings for exploring different avenues regarding, e.g., how the change of ecological factors could foresee various results of climatic changes [13–15].

2.2 Research Questions

The qualities of the unique worldwide climatic changes framework are investigated utilizing these examination questions:

RQ1: What are the elements that add to the current conditions that characterize the worldwide climatic changes?

RQ2: What are the prescient effects of the direst outcome imaginable in the versatility of the normal assets on the result of climatic changes, with the goal that strategy producers can be educated and endeavor to stay away from them?

3 Methods

Analysis involves the publicly available dataset from National Aeronautics and Space Administration Goddard Institute for Space Studies (NASA-GISS). The dataset describes the surface temperature changes across various countries.

The FAOSTAT temperature change domain includes the temperature change pertaining to each country, with updates provided annually. The available dissemination involves details from the year 1961–2019.

Data analysis is performed using pandas library in python. Analysis studies the effect of global warming over the three decades. The yearly data available from 1961 to 2019 are divided into six decades. The temperature trends are studied for India during the summer, winter, fall and spring. After, studying the effects on temperatures in the Indian subcontinent. The analysis further investigates the effect on the 190 countries.

The countries around the world are investigated for the changes in temperatures over the similar 6 decades. After the analysis, the final section involves the use of time series prediction algorithm to predict the temperature changes in the coming decade. The involvement of Time series prediction is due to the fact that it:

Time series permits understanding and comparison while assuring that none of the extremely relevant information is lost with respect to the background of "time." Time series also enables forecasting.

The time series prediction includes the use of Facebook's open-source library prophet. Prophet renders the ability to permit time series predictions with great accuracy with the use of simple yet intuitive parameters. The following section discusses the results achieved.

4 Result Analysis

Global warming has been on the rise particularly during the last decade, and this research investigates the effect of this on the Indian subcontinent during the last decade compared to the previous five decades. In Fig. 1, the graph indicates that the summer temperatures have risen considerably in the last decade. Figure 2 indicates a sharp decline in the winter temperatures during the last two decades. While spring also has similar fluctuation to winter, fall has a similar trend to summer as depicted by Figs. 3 and 4. Results are indicative of the last decade that is 2010–2019 has been among the ten hottest years on record in terms of mean annual temperatures.

Figure 5 depicts the pair plot analysis of the six decades and their correlation. A pair plot is indicative of the relationships between the decades.

The animation map in Fig. 6 demonstrates the seriousness of the rising temperatures year after year. From our above analysis result, I came up winter season getting hotter. The results are a proof of global warming that is severely affecting the global temperatures across the countries.

Figure 7 enables the visualization of the temperature trends across the six decades within the 190 countries across the world. The darker areas indicate rising temperatures across last two decades as a result of global warming.

Figures 8 and 9 predicts the temperatures expected in the next two decades from 1970 to 2040. The predictions indicate that temperatures will rise further drastically in case of absence of significant effects to abet the climate changes. The reason for

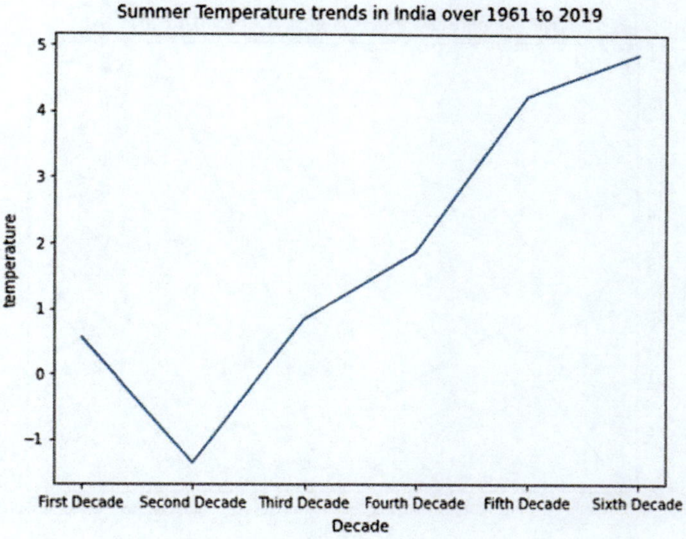

Fig. 1 Summer temperature trends in India over 1961–2019

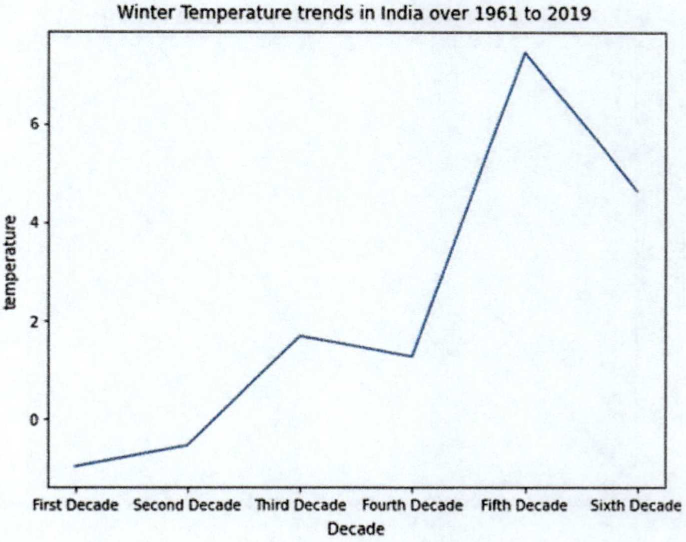

Fig. 2 Winter temperature trends in India over 1961–2019

the drastic changes Earth's climate during the last decade is contributed by human activities which are increasing the heat trapped and therefore, increased greenhouse gas levels in Earth's atmosphere, increasing Earth's average surface temperature.

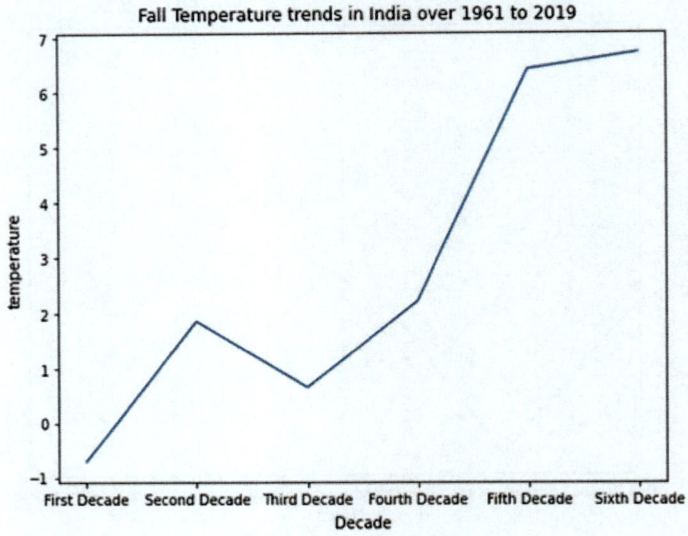

Fig. 3 Fall temperature trends in India over 1961–2019

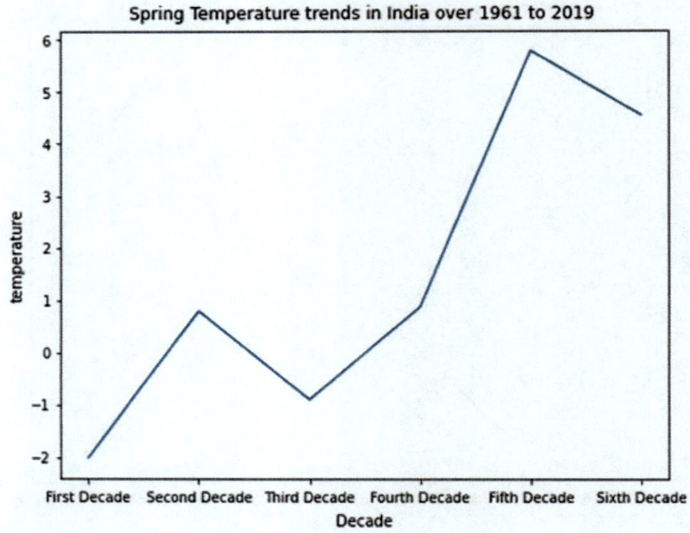

Fig. 4 Spring temperature trends in India over 1961–2019

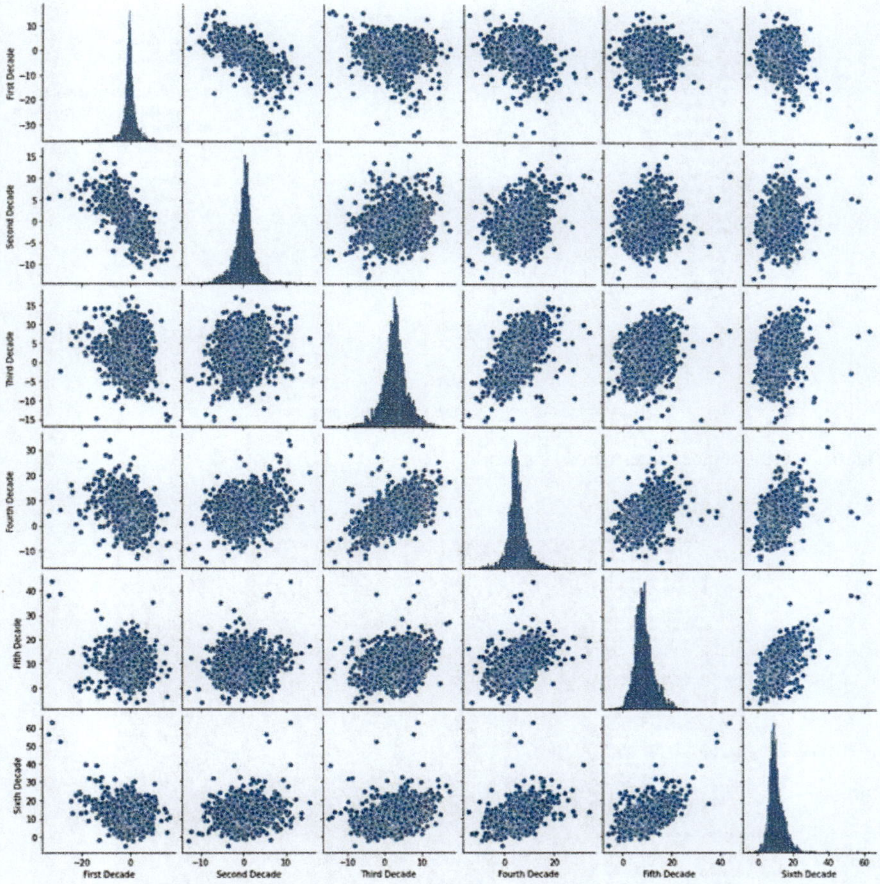

Fig. 5 Pair plot analysis of all decades

5 Conclusion and Future Scope

This study proposes a machine learning (ML) model to abet climatic change preparedness. In this study, data analysis is performed using pandas library in python. Analysis studies the effect of global warming over the three decades. The yearly data available from 1961 to 2019 are divided into six decades. The temperature trends are studied for India during the summer, winter, fall and spring. After, studying the effects on temperatures in the Indian subcontinent. The analysis further investigates the effect on the 190 countries. The countries around the world are investigated for the changes in temperatures over the similar 6 decades. After the analysis, the final section involves the use of time series prediction algorithm to predict the temperature changes in the coming decade. The accuracy of the model was validated using the historical records, and the results produced predictive

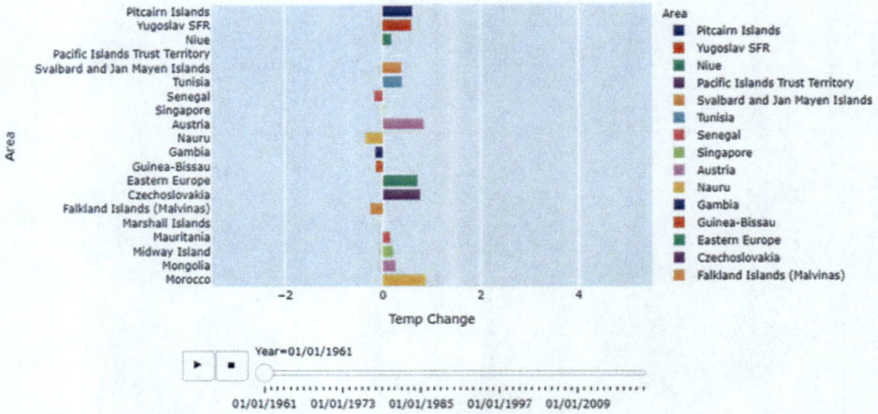

Fig. 6 Varying temperatures across the world

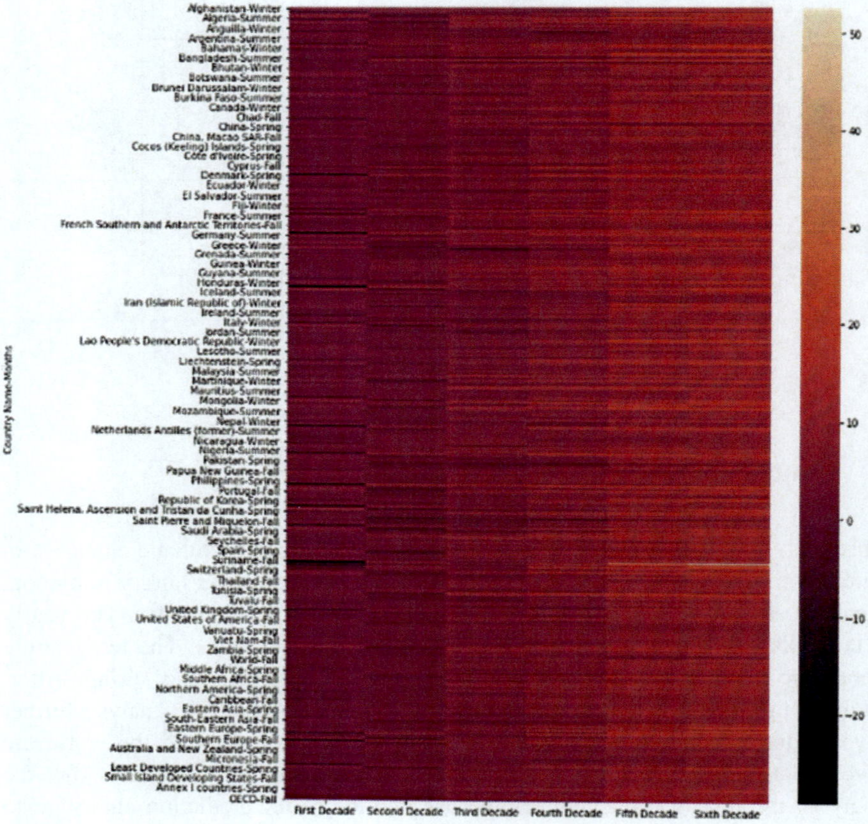

Fig. 7 Rising temperatures over the six decades across the world

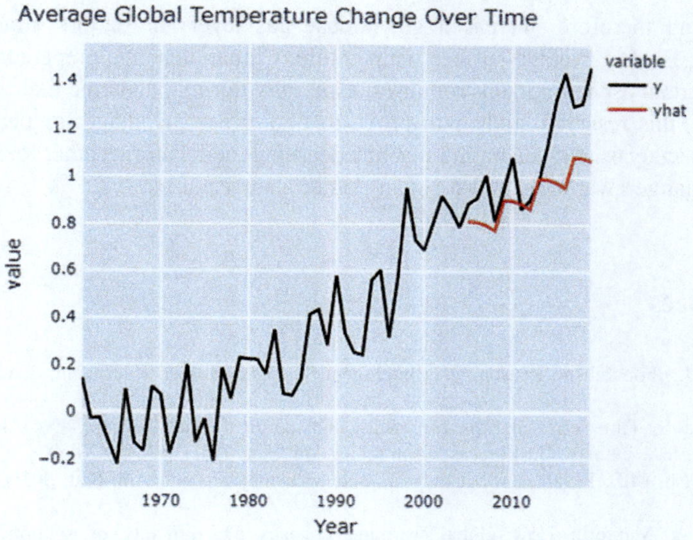

Fig. 8 Average global temperature change over time (1970–2010)

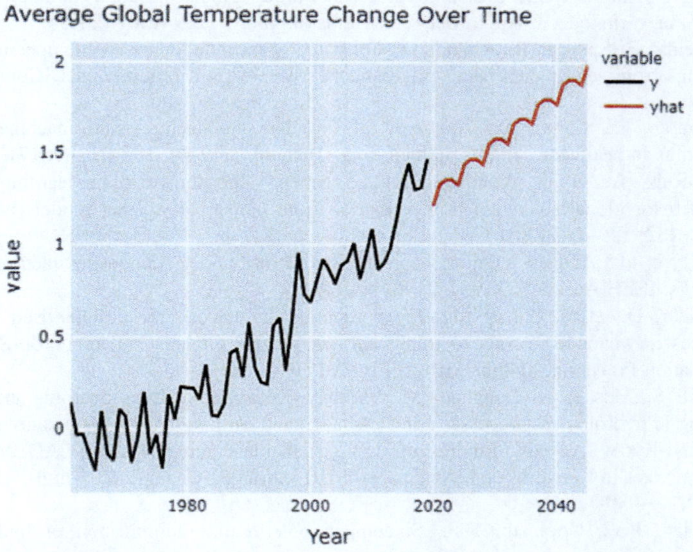

Fig. 9 Average global temperature change over time (1980–2040)

analysis model for optimized model generation. The predictions indicate that temperatures will rise further drastically in case of absence of significant effects to abet the climate changes. The reason for the drastic changes Earth's climate during the last decade is contributed by human activities which are increasing the heat trapped and therefore, increased greenhouse gas levels in Earth's atmosphere, increasing Earth's average surface temperature. Climatic change preparedness is a very important aspect in many countries, especially during pandemic like covid-19. Therefore, this research study will prove to be a great advantage for performing predictive analysis and generating machine learning models for weather forecasting, climate change awareness and for many other changes also.

References

1. Pearl, J.: Probabilistic Reasoning in Intelligent Systems. Morgan Kaufmann, San Francisco (1988)
2. Wulff, S.S.: Time series analysis: forecasting and control. J. Qual. Technol. **49**(4), 418 (2017)
3. Taylor, S.J., Letham, B.: Forecasting at scale. Am. Stat. **72**(1), 37–45 (2018)
4. Schervish, M.J.: P values: what they are and what they are not. Am. Stat. **50**(3), 203–206 (1996)
5. Scher & Messori.: How global warming changes the difficulty of synoptic weather forecasting. Geophys. Res. Lett. **46**(5), 2931–2939 (2019)
6. FAOSTAT.: Temperature change [Online]. http://www.fao.org/faostat/en/#data/ET, last accessed 2021/02/10
7. Hwang, Y., Carbone, G.J.: Ensemble forecasts of drought indices using a conditional residual resampling technique. J. Appl. Meteorol. Climatol. **48**(7), 1289–1301 (2009)
8. Al-Obeidat, F., Spencer, B., Alfandi, O.: Consistently accurate forecasts of temperature within buildings from sensor data using ridge and lasso regression. Future Gener. Comput. Syst. (2018)
9. Manogaran, G., Lopez, D.: A survey of big data architectures and machine learning algorithms in healthcare. Int. J. Biomed. Eng. Technol. **25**(2–4), 182–211 (2017)
10. Aybar-Ruiz, A., et al.: A novel grouping genetic algorithm–extreme learning machine approach for global solar radiation prediction from numerical weather models inputs. Sol. Energy **132**, 129–142 (2016)
11. Yin, C., et al.: A deep learning approach for intrusion detection using recurrent neural networks. IEEE Access **5**, 21954–21961 (2017)
12. Borthakur, D., et al.: Smart fog: fog computing framework for unsupervised clustering analytics in wearable internet of things. In: 2017 IEEE Global Conference on Signal and Information Processing (GlobalSIP), IEEE (2017)
13. Ardabili, S., Mosavi, A., Dehghani, M., Várkonyi-Kóczy, A.R.: Deep Learning and machine learning in hydrological processes climate change and earth systems a systematic review. In: Várkonyi-Kóczy, A. (eds.) Engineering for Sustainable Future. INTER-ACADEMIA 2019. Lecture Notes in Networks and Systems, vol. 101. Springer, Cham (2020). https://doi.org/10.1007/978-3-030-36841-8_5
14. Gopirajan, P.V., Gopinath, K.P., Sivaranjani, G., et al.: Optimization of hydrothermal liquefaction process through machine learning approach: process conditions and oil yield. Biomass Conv. Bioref. (2021). https://doi.org/10.1007/s13399-020-01233-8
15. Anuj, K., Kumar, V.: Big data in climate: opportunities and challenges for machine learning. In: Proceedings of the 23rd ACM SIGKDD International Conference on Knowledge Discovery and Data Mining (KDD '17). Association for Computing Machinery, New York, NY, USA, 21–22 (2017). https://doi.org/10.1145/3097983.3105810

Deep Learning Approaches for Anomaly and Intrusion Detection in Computer Network: A Review

Khushnaseeb Roshan and Aasim Zafar

Abstract In recent years, a great deal of attention has been given to deep learning in the field of network and information security. Any intrusion and anomaly in the network can significantly impact many areas, such as security of the private and social data, national security, social and financial concerns, etc. Therefore, network and information security are a broad research domain for which researchers are actively utilizing the functionally improved, emerging deep learning technique and report the improved result. In this review paper, we have analysed several deep learning methods in the area of network anomaly, intrusion detection, network traffic analysis and its classification. We have presented a comprehensive review of widely known deep learning approaches. And then, we conclude with open research challenges and unresolved issue for further study. This review paper provides an overall background for the researchers interested in network anomaly and intrusion detection based on deep learning methods.

Keywords Deep learning · Intrusion detection · Network anomaly · Network attack · Network security

1 Introduction

Network and communication security is an important part of our daily life, and its protection has continuously emerged as a major concern. Also, the global accessibility of the network exposes individuals and organizations to network attack. Antivirus software, firewalls, Intrusion detection, and anomaly detection methods are various approaches to protect network and individual from attack [1]. Network traffic analysis and classification are also important for network anomaly detection, which try to identify the signs of intrusion and abnormal behaviour using profiling method (IDS anomaly based) [2].

K. Roshan (✉) · A. Zafar
Aligarh Muslim University, Aligarh, Uttar Pradesh 202002, India
e-mail: azafar.cs@amu.ac.in

K. Khanna et al. (eds.), *Cyber Security and Digital Forensics*,
Lecture Notes on Data Engineering and Communications Technologies 73,
https://doi.org/10.1007/978-981-16-3961-6_45

Moreover, outsiders can cause security instances and network anomalies, such as any suspicious activity aiming to disable the services, steal private data and information. Other reasons like server crashes, power outages, link congestion are known as inside factors. And these factors significantly impact on network and communication security [3].

The recent rise of interest in deep learning (DL) algorithms resulted in major advancements in network security and anomaly detection methods. Also, the availability of the extensive computational power encourages researchers to adopt deep learning techniques for attack detection, network intrusion and anomalies detection [4–6].

This review paper has mainly used journal and conference articles in network anomaly detection (attack, intrusion and abnormal behaviour) and network traffic analysis. This review article organizes as follows: Section 1 presents a brief introduction. Section 2 provides a detailed description of various deep learning methods applied in network anomaly and intrusion detection. Section 3 discusses open research challenges and issue for future work with suggestions. Section 4 concludes this review paper.

2 Deep Learning for Network Anomaly and Intrusion Detection

Deep learning methods are very effective to discover the relationship in highly complex data without human intervention. Feature learning and data classification are the main task of DL methods and hence very useful in handling complex network traffic data. This review analysis provides a complete overview of DL methods with their advantages and challenges for future exploration.

As shown in Fig. 1, deep learning architecture broadly classified into three categories, i.e. generative, discriminative and hybrid. Each category has its working advantages and disadvantage that we have included in Table 1.

2.1 Autoencoder (AE)

Autoencoder networks were initially developed as compression techniques that aim to covertly give input to output with the least amount of distortion. This compression achieves by using less number of nodes than the nodes used in the hidden layer. Autoencoders are theoretically simple and widely used for unsupervised anomaly detection. In high dimensional data, these networks are used for feature extraction (reduction in dimensionality) by the hidden layer between the encoder and decoder [7].

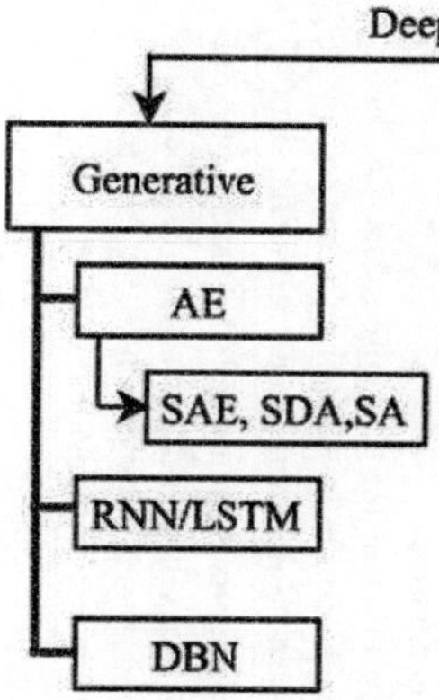

Fig. 1 Deep learning architecture

The authors of [8] proposed IDS model using autoencoder for cybersecurity. The model was capable enough to detect not only malicious packets but also classify specific types of attacks with 81.31% recognition accuracy. In the study of [9], a new IDS model based on autoencoder proposed, this model has been trained and tested on two popular benchmark network datasets, i.e. KDD99 and NSL-KDD. The model successfully achieved a recall of 98.11%. The authors also used the memetic algorithm for final classification.

In the latter study, the authors of [10] proposed a deep learning framework named SU-IDS. They used both semi-supervised and unsupervised technique to develop the Intrusion Detection System using autoencoder. The experimental results on NSL-KDD and CICIDS2017 dataset were quite acceptable. The model achieved approximately 99.65% accuracy for intrusion detection. Then other effective deep learning approach proposed in [11], authors used sparse autoencoder mechanism for feature selection and SVM for final classification, which improved its detection capability for intrusion and classification accuracy. Moreover, the results of the proposed approach are also compared with other classification methods, such as SVM, naive Bayesian, random forest and J48.

2.2 *Convolutional Neural Networks (CNNs)*

Convolutional neural networks are the advanced version of ANNs and were initially developed for pattern recognition task within complex images. CNNs are formed by stacking three layers, namely convolutional layers, pooling layers and fully connected layers. These networks extract higher resolution features and then convert them into more complex features at a coarser resolution in image classification task [12]. However, in some recent research, they are also performed well in classifying non-image data (like network traffic data, anomaly detection).

Table 1 Single DL techniques

Detection category	DL approach	Datasets used	Measur ment metrics used in papers	Performance	Workings and purpose of the methods
IDS, Cyber Security (2017) [8]	AE	KDD-CUP99, NSL-KDD	ACC	90.12% ACC	AEs are generative or unsupervised deep learning methods, mainly used for unlabelled datasets, also used for dimensionality reduction in highly complex data like network traffic datasets
IDS (2017) [9]	AE	KDD-CUP99, NSL-KDD	DR, CR	0.98 DR	
IDS (2018) [31]	AE	NSL-KDD, CICIDS2017	ACC, DR, FAR	0.99DR	
IDS (2018) [11]	AE	NSL-KDD	ACC, P,R, F	99.396% ACC	
NTC (2017) [13]	CNN	USTC-TFC2016 (Self-collected)	ACC, P, R, F	99.41% ACC	CNNs are discriminative or supervised deep learning methods used for labelled datasets in prediction and classification purposes, very efficient in image classification
IDS (2018) [14]	CNN	NSL-KDD	ACC, P, R, F	81.84%ACC	
IDS (2018) [15]	CNN	ISCX2012	ACC, DR, FAR	99.21%	
IDS (2017) [17]	DBN	KDD-CUP99, NSL-KDD	ACC, P, R, F	99.7% ACC	DBNs are unsupervised methods based on RBMs, can be used for dimensionality reduction, can also be used for classification purpose with an additional discriminative layer
IDS (2017) [18]	DBN	KDD-CUP99	ACC, DR, FAR	99.04% ACC	
IDS (2017) [20]	LSTM	KDD-CUP99	ACC, P, DR, FAR	97.54% ACC	LSTMs are generative methods, very efficient for time-series data, LSTM uses a feedback loop in its hidden cells, makes it capable of learning timestamps and sequential data
NTC (2020) [21]	LSTM	NSL-KDD	ACC, DR, FPR	84.25% ACC	
IDS (2020) [22]	LSTM	NSL-KDD	ACC, P, R, F, FPR	83.85%ACC	
NTC (2017) [23]	SAE	AWID	DR, FPR	92.7% DR	SAEs are unsupervised methods and a variant of AEs, composed of many hidden layers, performed better in terms of accuracy with raw data like network traffic, can also reduce high dimensional data with few features
NTC (2017) [24]	SAE	AWID	ACC, DR,F, FAR	99.91 ACC%	
NTC (2017) [25]	SAE	AWID + Emulated	ACC	98.66% ACC	
NTC (2016) [26]	SAE	Simulated Environment	ACC, TPR, FPR	99.99% ACC	
IDS (2019) [27]	SAE	KDD-CUP99, UNSW-NB15	ACC, P, R, F, FAR	99.99% ACC	

(continued)

Table 1 (continued)

Detection category	DL approach	Datasets used	Measur ment metrics used in papers	Performance	Workings and purpose of the methods
IDS (2017) [28]	SDA	CTU-13 + UNB ISCX IDS 2012 dataset	ACC, P, R, F	99.48% ACC	Another variant of AEs, used to produce refined data from corrupted and noisy input data
IDS (2016) [29]	SA	NSL-KDD	ACC, P, R, F	95.95% DR	Another variant of AEs, have only one hidden layer in its architecture
IDS (2020) [30]	GAN + AE	NSL-KDD, UNSW-NB15	ACC, P, R, F	95.19% ACC	GANs are hybrid methods, i.e. composed of both generative and discriminative techniques

GAN—Generative Adversarial Network; **CNN**—Convolutional Neural Network; **RNN**—Recurrent Neural Network; **LSTM**—Long Short-Term Memory; **AE**—Autoencoder; **SAE**—Stacked AE; **DBN**—Deep Belief Network; **SVM**—Support Vector Machine; **SDA**—Stacked Denoising Autoencoder; **SA**—Sparse AE; **RBM**—Restricted Boltzmann Machine; **DBM**—Deep Boltzmann Machine; **DNN**—Deep Neural Network; **DL**—Deep learning; **DT**—Decision Tree; **RF**—Random Forest; **ACC**—Accuracy; **FAR**—False alarm rate; **DR**—Detection Rate; **R**—Recall; **P**—Precision, **F**—F-measure; **FPR**—False positive rate; **TPR**—True positive rate; **NTC**—Network traffic classification

The authors of [13] presented a novel representation learning approach using CNNs and applied it on network traffic data to detect malware. The model has trained on raw network traffic data by converting it into images and achieved an average accuracy of 99.41%. However, the authors used only spatial feature in the training process and ignored the temporal feature completely, which is the major drawback of the method. In the latter study [14], the same approach is applied to IDSs, where the authors used NSL-KDD dataset and graphics conversion technique to train the model on transformed data. The model used the full feature set as input and performed well as compared to standard classifier but did not improve the state-of-the-art method. Future work can improve the graphical conversion technique for better result.

The authors of [15] presented a new model named TR-IDS to detect network intrusion. The model utilizes both statistical and payload data for better result. Word embedding and text-convolutional neural are two modern NLP techniques adopted by the model to extract feature from payloads, and ISCX2012 dataset was used for training. Furthermore, the RF algorithm was used in the final classification, and the detection rate achieved by the model was 99.26%.

2.3 Deep Belief Networks (DBNs)

A deep belief network is generated by stacking restricted Boltzmann machine (RBM), and the input of an RBM is the result of earlier RBMs. These networks contain a layer of visible nodes and layers of hidden nodes like other neural network architecture. All layers are connected with a unidirectional connection, except for the top two layers, which provide a two-way connection [16].

The authors of [17] suggested a new method based on DBN and PNN (probabilistic neural network) for intrusion detection, which also addressed the problem of falling into local optimal. This method somehow shortens the model's training and testing time by converting raw network data into a low dimension. The benchmark dataset KDD CUP-1999 is used for testing the performance of the method, and results were compared with traditional methods like PNN, PCA-PNN and un-optimized DBN-PNN. However, the actual network environment is more complex than the dataset used; therefore, further, improvement can be applied in real network traffic as an enhancement.

The latter study [18] compares deep and shallow network architecture in network anomaly identification. Authors distinguished between the normal and attack connection. Additionally, model was capable of categorizing an attack to its corresponding category. Experimental results showed that deep networks are essential for real-time anomaly identification (Figs. 2 and 3).

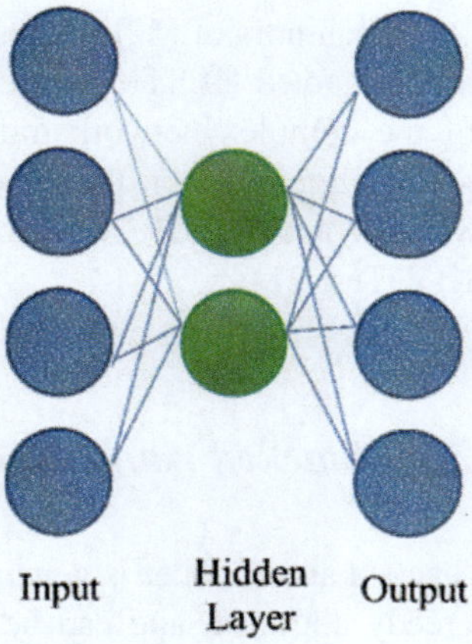

Fig. 2 Autoencoder

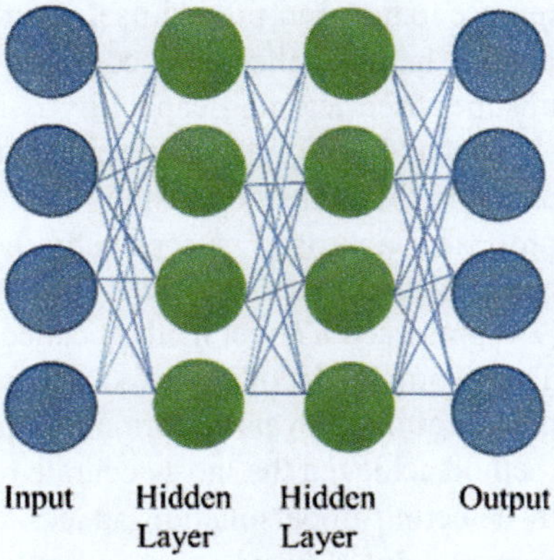

Fig. 3 Deep belief network

2.4 Recurrent Neural Networks (RNNs)

These networks contain at least one cycle in its connection graph, and these nodes are also known as the recurrent cell (or nodes). This network architecture can adopt the internal memory to process random sequences of inputs. This architecture is required to perform a complex task. This technique is good for handling real-time learning and processing time-series data. Presently, RNN based on Long Short-Term Memory (LSTM) is a widely known architecture [19].

The authors of [20] proposed a classification model for IDS. This model is based on LSTM-RNN and nadam optimizer as optimizing technique. The experimental results successfully achieved an accuracy of 97.54%. The false alarm rate of 9.98% is also quite acceptable, having a detection rate of 98.95%. Recently, a novel model named BAT-MC was proposed by combining bidirectional Long Short-Term memory (BLSTM) and attention mechanism [21]. The model well-described network traffic behaviour and improved the detection rate and accuracy. The model was tested on KDDTest+ and KDDTest-21 dataset, including NSL-KDD.

The authors of [22] suggest another approach for cyber-attack detection, and the model named HLSTM was capable enough to learn temporal hierarchical features of the complex network traffic data. The detection rate of the system has been greatly improved on the benchmark datasets. The multi-classification accuracy of 83.85% and 69.73% was achieved on both the dataset, namely KDDTest+ and KDDTest-21.

2.5 *Stacked Autoencoder (SAE)*

Stacked autoencoder is generated by adding (stacking) more hidden layer using the greedy approach and can be trained using the same methods for each additional layer. SAE is ideal for transforming the original feature set into more meaningful representation and proved useful for unsupervised learning. SAE efficiently extract feature in unlabelled and complex data, i.e. raw network traffic data. And this unsupervised learning capability is important to detect unknown attacks.

The authors of [23] presented an optimized method to detect impersonation attack in Wi-Fi network using ANN and SAE by selecting the abstract features. The proposed approach successfully achieved a quite acceptable result. But this approach is limited to detect specific attack only. In a latter study, the authors of [24] presented a novel method called deep feature extraction and selection (D-FES). They combined both stacked and weighted feature selection approach to collect meaningful, relevant information from raw network traffic data. The proposed method achieved the most accurate results in detection accuracy and false alarm rate in detecting impersonation attacks in Wi-Fi networks, i.e. 99.918% and 0.012%, respectively. However, future work can enhance the model to detect more attack classes and not limit it for impersonation attacks.

The method presented by [25] is not just limited to impersonation, which was the drawback of the previously mentioned approaches. Here, they presented a solution based on anomaly detection and considered a multi-class problem, i.e. flooding, injection, normal traffic and impersonation attack. The proposed frameworks were developed using the SAE architecture with two and three hidden layers. The model achieved an accuracy of 98.67%, which is quite acceptable as compared to other state-of-the-art methods. A new approach based on anomaly detection to identify DoS attacks in encrypted network traffics was presented [26]. The authors tested this technique on a realistic cyber environment that generated realistic traffic patterns. However, the study can be improved in terms of accuracy and can be tested to perform better in the bigger dataset, i.e. by capturing the network traffic data for several days.

Recently, a model is proposed named TSDL based on deep stacked autoencoder. This model contains two stages and two hidden layers at each stage with softmax classifier. The model is trained in a semi-supervised manner [27]. Results were quite impressive; model achieved good recognition rates of 99.99% and 89.13% for both KDD99 and UNSW-NB15 dataset, respectively, with a low false alarm rate.

Furthermore, future work can improve its performance by combining the proposed approach with novel reinforcement learning.

2.6 *Stacked Denoising Autoencoder (SDA)*

A new approach in session-based Intrusion Detection System model was presented by the authors of [28]. They implemented an SDA-based deep architecture, and used botnet traffics dataset for evaluation. Researchers obtained relatively impressive results, but the future study can use the network application layer's complete payload data for further improvement.

2.7 *Sparse Autoencoder*

A Network Intrusion Detection System (NIDS) is suggested using the concept of self-taught learning (STL) and a deep learning-based architecture, i.e. sparse autoencoder [29]. The benchmark dataset, NSL-KDD were used for evaluation. The model performed well, but it has been observed that the proposed technique can be improved by using other deep learning architecture. Additionally, instead of using the derived feature, future work can be done to learn feature from raw network traffic headers on the go.

2.8 *Generative Adversarial Network*

GANs are composed of two neural networks, i.e. Generator and Discriminator. Generators generate the sample data and discriminators distinguish it from actual data. GANs are based on hybrid deep learning method which utilizes generative features at initial layers and discriminative data at later stages. The authors of [30] proposed intrusion detection model based on GAN and AE named as SAVAER-DNN. The issue of imbalance nature of network traffic data is also addressed by generating new attack samples.

3 Open Issues, Findings and Future Directions

Although, a number of deep learning methods have been presented in this review paper for network anomaly and intrusion detection, still so many unresolved issues and challenges are open for future work.

We found that single deep learning methods explored earliest, but now day's researchers are exploring hybrid deep learning methods. We would also like to encourage researchers to leverage hybrid deep learning methods such as GANs to move towards effective and efficient solutions. Further exploration of hybrid deep learning methods would be worthwhile as they have shown quite acceptable and promising results [30, 32–34].

3.1 Datasets Challenges

There are several publically available network traffic and intrusion detection datasets as shown in Table 1. But most of them are problematic and do not even simulate the real network environment. Also, the attacking scenario is constantly advancing due to which feature selection for one attack situation might not work effectively for others. The other issue is the lack of availability of labelled dataset. And creating the labelled dataset through testbed or simulated environment gathered over a long period is also troublesome.

The Imbalance and the diverse nature of the network traffic datasets are another obstacle for anomaly detection. It will change the actual perspective of the data and might increase false alarm rate. However, the majority of the researchers used KDD99 and its modified version NSL-KDD dataset for intrusion and anomaly detection. Still, both are unreliable and less application in a real network environment. Furthermore, the benchmarks datasets are very old, do not have real-time scenarios and recent traffic behaviour.

As a solution, we would like to encourage researchers to use network traffic from the simulated environment. Real-world and big datasets are also recommended to enhance the performance and detection accuracy of the model. The authors also developed a method based on deep learning and tested it on real-world environment [33].

3.2 Security Challenges

Security for every device existing over the internet is crucial as they are prone to be hacked or attacked at any time. To secure the network, deep learning model should be capable enough to detect all attack, whether successful or unsuccessful immediately. It is critical to secure the network (NIDS) to protect all devices such as hosts, various resources and servers. Furthermore, a powerful threat or attack detection can be achieved through better detection techniques and hence an improved IDS and anomaly detection method based on deep learning would be a key solution to improve the security of the network.

Anomaly-based hybrid deep learning methods can overcome some security issues as they are capable enough to reduce false alarm rate and have a good detection rate for both known and unknown attacks.

3.3 Real-Time Performance

Real-time implementation of anomaly detection model is an important issue to be addressed. Also, the nature of the network anomalies keeps changing with time. Hence, whenever unknown attacks are detected, it is important to update the profile in real-time without compromising the performance.

Deep learning methods are highly computation intensive and required high-performance CPUs and GPUs. Therefore, the improved performance methods can achieve real-time detection of unknown attacks. There are various deep learning techniques like SAE, DBN, etc. that proved efficient for unsupervised feature extraction of network traffic data. The authors of [35] suggested using multi-core CPU and GPU, with highly complex data, like network traffic data for intrusion detection system and anomaly detection methods.

3.4 Complexity of Network Traffic Data

The network traffic data's scale and density are very complex and rapidly growing over the years. To handle such large traffic volume and highly diverse data distribution is very challenging. To reduce network traffic data complexity, developing the appropriate and fast feature selection deep learning techniques would be a key solution.

This survey aimed to provide the complete understating of network anomaly, network traffic classification, intrusion detection problems, and its various aspects. It also includes multiple deep learning methods proposed in this domain with its advantage, drawbacks and improvements for further research.

4 Conclusion

This review paper has explained the importance of computer network security from anomalies, intrusion and various network attacks (like User to Root (U2R), Remote to Local (R2L), Trojan, Probe, etc.). Then we have reviewed and analysed several deep learning methods, techniques in network anomaly and intrusion detection. Two broad categories were identified, i.e. single and hybrid deep learning methods. In this review paper, we have included only single DL approaches. We will also review hybrid deep learning methods in our further study. Finally, we have discussed open issues and recommendations for researchers and practitioners who are willing to work for network anomaly detection and its security. Even though several open issues and challenges need to be addressed, in contrast, researchers should also keep in mind the quite promising guidelines for further investigation in this domain.

References

1. Choo, K.K.R.: The cyber threat landscape: Challenges and future research directions. Comput. Secur. **30**(8), 719–731 (2011)
2. Iglesias, F., Zseby, T.: Analysis of network traffic features for anomaly detection. Mach. Learn. **101**(1–3), 59–84 (2015)
3. Löf, A., Nelson, R.: Annotating network trace data for anomaly detection research. In: Proc. Conf. Local Comput. Networks, LCN, vol. 2014-Novem, no. November, pp. 679–684 (2014)
4. Saied, A., Overill, R.E., Radzik, T.: Detection of known and unknown DDoS attacks using Artificial Neural Networks. Neurocomputing **172**, 385–393 (2016)
5. Kang, M.-J., Kang, J.-W.: Intrusion detection system using deep neural network for in-vehicle network security. PLoS One, **11**(6), e0155781 (2016)
6. Naseer, S., et al.: Enhanced network anomaly detection based on deep neural networks. IEEE Access **6**(8), 48231–48246 (2018)
7. Yu, Y., Long, J., Cai, Z.: Network intrusion detection through stacking dilated convolutional autoencoders. Secur. Commun. Netw. **2017** (2017)
8. Alom, M.Z., Taha, T.M.: Network intrusion detection for cyber security on neuromorphic computing system. In: Proc. Int. Jt. Conf. Neural Networks, vol. 2017-May, pp. 3830–3837 (2017)
9. Mohammadi, S., Namadchian, A.: A new deep learning approach for anomaly base IDS using memetic classifier. Int. J. Comput. Commun. Control **12**(5), 677–688 (2017)
10. Long, E.M.B.J., Liu, Q., Cui, J., Cai, Z.: SU-IDS : A Semi-supervised and Unsupervised Framework for Network Intrusion Detection, vol. 1. Springer International Publishing (2018)
11. Al-qatf, M., Lasheng, Y.U., Al-habib, M., Al-sabahi, K.: Deep Learning Approach Combining Sparse Autoencoder with SVM for Network Intrusion Detection, vol. 6 (2018)
12. O'Shea, K., Nash, R.: An Introduction to Convolutional Neural Networks, pp. 1–11 (2015)
13. Wang, W., Zhu, M., Zeng, X., Ye, X., Sheng, Y.: Malware Traffic Classification Using Convolutional Neural Network for Representation Learning (2017)
14. Li, Z., Qin, Z., Huang, K., Yang, X., Ye, S.: Intrusion detection using convolutional neural networks for representation learning. In: Lect. Notes Comput. Sci. (including Subser. Lect. Notes Artif. Intell. Lect. Notes Bioinformatics), vol. 10638 LNCS, pp. 858–866 (2017)
15. Min, E., Long, J., Liu, Q., Cui, J., Chen, W.: TR-IDS : anomaly-based intrusion detection through text-convolutional neural network and random forest. Secur. Commun. Netw. **2018** (2018)
16. Ferrag, M.A., Maglaras, L., Janicke, H., Smith, R.: Deep Learning Techniques for Cyber Security Intrusion Detection : A Detailed Analysis, pp. 126–136 (2019)
17. Zhao, g., Zhang, c., Zheng, l.: Intrusion detection using deep belief network and probabilistic neural network. In: Proc. 2017 IEEE Int. Conf. Comput. Sci. Eng. IEEE/IFIP Int. Conf. Embed. Ubiquitous Comput. CSE EUC 2017, vol. 1, pp. 639–642 (2017)
18. Vinayakumar, R., Soman, K.P., Poornachandran, P.: Evaluating effectiveness of shallow and deep networks to intrusion detection system. In: 2017 Int. Conf. Adv. Comput. Commun. Informatics, ICACCI 2017, vol. 2017-Janua, pp. 1282–1289 (2017)

19. Ferrag, M.A., Maglaras, L., Janicke, H., Smith, R.: Deep learning techniques for cyber security intrusion detection: a detailed analysis. In: 6th International Symposium for ICS & SCADA Cyber Security Research 2019 6, pp. 126–136 (2019)
20. Le, T.T.H., Kim, J., Kim, H.: An effective intrusion detection classifier using long short-term memory with gradient descent optimization. In: 2017 Int. Conf. Platf. Technol. Serv. PlatCon 2017 - Proc., pp. 0–5 (2017)
21. Su, T., Sun, H., Zhu, J., Wang, S., Li, Y.: BAT: deep learning methods on network intrusion detection using NSL-KDD dataset. IEEE Access **8**, 29575–29585 (2020)
22. Hou, H., et al.: Hierarchical long short-term memory network for cyberattack detection. IEEE Access **8**, 1–1 (2020)
23. Oh, E., Kim, T., Cho, T.: Detecting impersonation attack in WiFi networks using deep learning approach. Wisa **1**, 186–197 (2017)
24. Aminanto, M.E., Choi, R., Tanuwidjaja, H.C., Yoo, P.D., Kim, K.: Deep abstraction and weighted feature selection for Wi-Fi impersonation detection. IEEE Trans. Inf. Forensics Secur. **13**(3), 621–636 (2017)
25. Thing, V.L.L.: IEEE 802.11 network anomaly detection and attack classification: a deep learning approach. In: 2017 IEEE Wirel. Commun. Netw. Conf., pp. 1–6 (2017)
26. Zolotukhin, M., Hamalainen, T., Kokkonen, T., Siltanen, J.: Increasing web service availability by detecting application-layer DDoS attacks in encrypted traffic. In: 2016 23rd Int. Conf. Telecommun. ICT 2016 (2016)
27. Khan, F.A., Gumaei, A., Derhab, A., Hussain, A.: TSDL: a two-stage deep learning model for efficient network intrusion detection. IEEE Access **7**, 30373–30385 (2019)
28. Yu, Y., Long, J., Cai, Z.: Session-Based Network Intrusion Detection Using a Deep Learning Architecture, no. January, pp. 144–155 (2017)
29. Niyaz, Q., Sun, W., Javaid, A.Y., Alam, M.: A deep learning approach for network intrusion detection system. In: EAI Int. Conf. Bio-inspired Inf. Commun. Technol. (2015)
30. Yang, Y., Zheng, K., Wu, B., Yang, Y., Wang, X.: Network intrusion detection based on supervised adversarial variational auto-encoder with regularization. IEEE Access **8**, 42169–42184 (2020)
31. Min, E., Long, J., Liu, Q., Cui, J., Cai, Z., Ma, J.: Su-ids: a semi-supervised and unsupervised framework for network intrusion detection. In: International Conference on Cloud Computing and Security, pp. 322–334 (2018)
32. Zhang, H., Li, Y., Lv, Z., Sangaiah, A.K., Huang, T.: A real-time and ubiquitous network attack detection based on deep belief network and support vector machine. IEEE/CAA J. Autom. Sin. **7**(3), 790–799 (2020)
33. Kim, A., Park, M., Lee, D.H.: AI-IDS: application of deep learning to real-time web intrusion detection. IEEE Access **8**, 70245–70261 (2020)
34. Zhong, Y., et al.: HELAD: a novel network anomaly detection model based on heterogeneous ensemble learning. Comput. Netw. **169**, 107049 (2020)
35. Potluri, S., Diedrich, C.: Accelerated deep neural networks for enhanced intrusion detection system. In: 2016 IEEE 21st International Conference on Emerging Technologies and Factory Automation (ETFA), pp. 1–8 (2016)

Forest Cover Change Detection Using Satellite Images

Achal Kalwar, Rohan Mathur, Shubham Chavan, and Chhaya Narvekar

Abstract Deforestation, which has contributed to adverse effects on the natural environment, is one of the challenges to reducing biodiversity and global climate change. Thus, early detection of deforestation is of utmost importance. Inspired by the above situation, this work provides an examination of the automated deforestation detection method. Change detection is used to figure out whether or not the changes occurred using remote sensing images at two different times. This work proposes an idea of effective method for detecting relevant changes in the equivalent scene between two temporally different images. This research analyzes image data from a remote sensing satellite called Landsat-8 in order to track changes in forest cover over a period of time. The findings of such a study will lead to taking steps to conserve the environment.

Keywords Remote sensing · Multi-temporal · Landsat-8 · QGIS · Change detection

1 Introduction

Forests provide us with essential ecological and economic services like clean water and air, soil conservation, climate modulation, timber, food, and shelter for the animals. Forest cover changes are dynamic, expedite, and extensive process [1]. Owing to climate change and man-made factors, the forest cover regions in many parts of the world have been forced to degrade. Development in a metropolitan city like Mumbai comes at the cost of degradation of forest covers. Many forest areas of Mumbai have been forced to deteriorate due to the development of infrastructure. Mumbai's economic development poses a danger to the ecological balance of the forests. Many scientists have predicted that in the coming few decades, and the city will lose its green lungs.

A. Kalwar (✉) · R. Mathur · S. Chavan · C. Narvekar
Xavier Institute of Engineering, Mumbai, India
e-mail: chhaya.n@xavier.ac.in

K. Khanna et al. (eds.), *Cyber Security and Digital Forensics*,
Lecture Notes on Data Engineering and Communications Technologies 73,
https://doi.org/10.1007/978-981-16-3961-6_46

Remote sensing is acquiring information about objects or areas without making physical contact with the object [2]. One of the commonly used applications of remote sensing is tracking of the forest covers. Remote sensing forms to be an economic tool for forest mapping as it is cheaper and also faster compared to other methods of surveying. Change detection using remote sensing technique is based on a series of multi-temporal satellite images which uses various classification algorithms along with geographic information system (GIS) tool that provides a suitable platform for data analysis. The paper mainly focuses on studying and monitoring the pattern of forest cover change using temporal satellite data from different time periods and deep learning image classification techniques.

2 Problem Definition

Around 3.4 billion hectares of the world is covered by forest. For development purposes, 15,000 km^2 of forests have been diverted in the past three decades [3]. There is not any continuous monitoring system for forest cover change which includes forest inventory or geographical information system which keeps track of land use which can give actual figures. This lack of the system and great concern involved in the increased destruction of forest cover change has led to the need to propose a system for the proper forest, management, and decision improvement [4].

Due to this lack of the system and great concern involved in the increased destruction of forest cover change has led to the need to propose a system for proper forest management and decision improvement. From the above context, the following problems are identified:

- Decline in the area under forest.
- Loss of biodiversity and habitat.
- Transformation of the reserved and protected forest areas into developed and agricultural areas [5].

The proposed system is designed to tackle the above-mentioned problems by studying the change cover detection of forest areas over the past few years. Remote sensing (RS)-based tools along with deep learning techniques are used to classify the multi-temporal Landsat 8 images.

3 Data and Study Area

The research area is Sanjay Gandhi National Park, a protected area spread over approximately 87 km^2 in Mumbai. It covers a geographical location between 19° 8′ N 72° 53′ E and 19° 21′ N 72° 58′ E. The park occupies the majority of the northern suburbs of Mumbai. The suburbs Goregaon, Malad, Kandivali, Borivali,

Table 1 Satellite data details

Date	Satellite/Sensor
November 19, 2013	Landsat 8 OLI
November 01, 2018	Landsat 8 OLI

Fig. 1 Study area

and Dahisar lie to the west. Bhandup and the Mulund suburbs lie to the east [4] and to the south lie Aarey Milk Colony. The forest enters Thane City in the north. The Sanjay Gandhi National Park is home to more than 270 bird species, 35 mammal species (including leopards), and 1300 plant species [6].

Landsat 8 OLI multi-temporal remote sensing data for November 2013 and November 2018 has been acquired. The remote sensing data was gathered from the geological survey of the United States (USGS) archives [7, 8]. Table 1 provides the details of satellite data used (Figs. 1, 2 and 3).

4 Literature Survey

Usually, current change detection techniques adopt one of two ways, using either post classification analysis or analysis of difference image. Owing to the high-resolution nature of satellite images, these approaches are also resource-intensive and time-intensive. The post classification method would first classify the two temporally distinct images of the equivalent scene and then compare them to determine the changes. The second approach is comparative study, which produces a difference image (DI). In order to determine the extent of the changes, further DI research is then undertaken [9, 10].

Deep learning (DL) methods have recently been successfully applied to the image processing of remote sensing (RS). It is possible to learn several layers of data representation using deep neural networks (DNNs) and extract more robust and abstract features, which typically offer more useful knowledge than hand-crafted

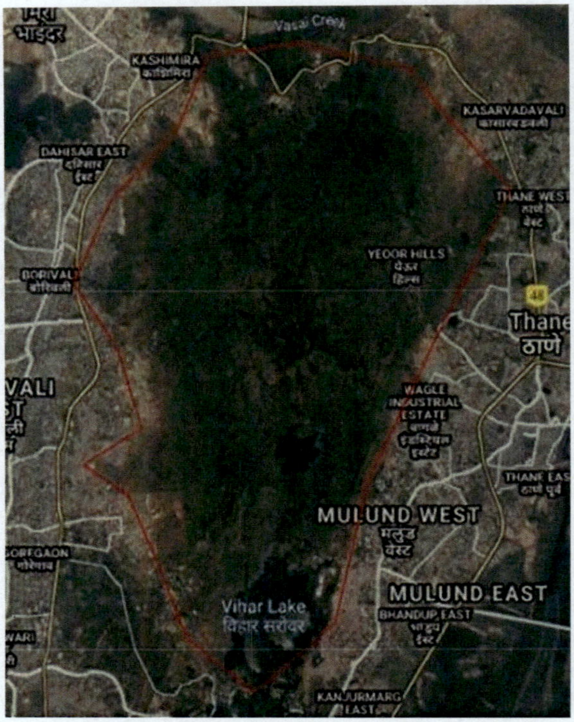

Fig. 2 Satellite view of Sanjay Gandhi National Park

Fig. 3 Raw satellite image downloaded from USGS

ones. In this context, variants of DNNs are possible candidates for automatic deforestation detection, such as convolutional neural networks (CNNs) and Siamese networks [11, 12].

In remotely sensed data, the prevalent approaches to change detection can be classified into two main classes—low-level local approaches and object-based approaches. Low-level approaches allow use of statistical indexes extracted from spectral image pixel values. Object-based methods take qualitative information into account by focusing on homogeneous pixels, which are typically clustered together on the basis of their appearance, position, and/or temporal properties [13].

Before detecting the change, the feature extraction stage is necessary when using machine learning algorithms. The feature extraction step is used to improve multispectral image precision. But, when using deep learning algorithms, no separate feature extraction process is necessary. The accuracy of the classification is dependent on the algorithms used to classify the changes, as well as the resolution of the images. We also concluded that deep learning algorithms have provided higher accuracy than techniques for machine learning [14].

Image processing techniques, image fusion, fuzzy clustering, and difference image (DI) are the earlier methods of change detection. The traditional methods are not very accurate in detecting the changes from the satellite images. Change detection methods can be classified as either supervised or unsupervised. To derive an appropriate training set, the supervised approach requires a ground truth. The unsupervised method, by allowing a simple comparison of multi-temporal images without adding any additional detail, performs change detection [3, 15].

In general, relative to other machine learning approaches, the drawbacks of deep learning are the need for large and high-quality training data, as well as hardware restrictions related to GPU computing capacity. The most notable benefit of deep learning is the degree of automation and the high ability to generalize by using large quantities of representative training data, which, however, may not always be available, particularly with regard to ground-truth labels that might be scarce or not exist at all [16, 17].

5 Methodology

5.1 Extraction of Study Area

The initially downloaded multi-temporal satellite images were clipped, and the area of Sanjay Gandhi National Park was extracted with the help of QGIS clipper tool [18].

QGIS is a free and open-source cross-platform desktop geographic information system (GIS) program that facilitates accessing, editing, and analyzing geospatial data [19].

The clipped images were further processed to make several atmospheric corrections. Clipped images were the input source for the QGIS semi-automatic classification plugin tool, which generated reflectance images as its output [18]. Reflectance images typically contain spectral details of the specified area and are

Fig. 4 Clipped image

used for image classification or image processing. Figure 4 shows the raw satellite image of the study area which is clipped using the QGIS clipper tool.

5.2 Study of Spectral Bands

A spectral band is a matrix of points identified by three dimensions, their coordinates, and their radiance-related strength [20]. In different combinations, the multispectral image bands were combined and applied to the obtained reflectance images. To get the final processed image, the band combination 5-4-3 was used. The 5-4-3 band combination uses the near-infrared band (NIR), which is more useful for showing and distinguishing land cover from urban and agricultural areas [21, 22].

6 Conclusion

The dataset has been acquired successfully from the USGS website. We have performed preprocessing on the data and have successfully extracted the area of study by clipping the initially acquired images for the year 2013 and 2018 and also applied a band combination 5-4-3 using QGIS. It is possible to monitor forest cover changes using satellite images [23]. This study tests the technological potential of satellite imagery to quantify and monitor forest cover along with the use of deep learning techniques to classify multi-temporal satellite images to assess if there has been some change in the forest cover being analyzed. This research will help to take precautions to conserve forests. Government agencies should take adequate steps to conserve the environment in Mumbai and not allow it to be sacrificed at the cost of urban development (Figs. 5 and 6).

Fig. 5 Satellite image of November 19, 2013

Fig. 6 Satellite image of November 01, 2018

7 Future Scope

The satellite images obtained after performing preprocessing steps, extracting the area of interest, applying atmospheric the correction, and using a suitable band combination are the final images that will be used as an input to the convolutional neural network (CNN) model. Further, a CNN model will be developed using a suitable activation function and optimization algorithms [24]. To learn the sequence of information compared to the global pattern of a conventional neural network, CNN uses filters on the raw pixel of an image [25]. This model will extract features from each input image and will compare them for image classification. The image classification will result in an output image that will display the pattern of forest cover change over the years.

References

1. http://www.fao.org/3/XII/0586-C1.htm, last accessed 2020/09/23
2. https://en.wikipedia.org/wiki/Remote_sensing, last accessed 2020/02/11
3. https://www.encyclopedia.com/environment/energy-government-and-defense-magazines/forest-resources, last accessed 2020/03/13
4. https://en.wikipedia.org/wiki/Sanjay_Gandhi_National_Park, last accessed 2020/04/25
5. Yismaw, A., Gedif, B., Addisu, S., Zewudu, F.: Forest cover change detection using remote sensing and GIS in Banja district, Amhara region, Ethiopi. Int. J. Environ. Monit. Anal., 23 Dec 2014
6. https://mumbaimirror.indiatimes.com/mumbai/cover-story/park-in-peril/articleshow/62773396.cms, last accessed 2020/08/11
7. https://earthexplorer.usgs.gov/, last accessed 2020/04/11
8. https://www.usgs.gov/core-science-systems/nli/landsat/landsat-8-data-users-handbook, last accessed 2020/12/11
9. de Jong, K.L., Bosman, A.S.: Unsupervised Change Detection in Satellite Images Using Convolutional Neural Networks, 21 Mar 2019
10. Boriah, S., Mithal, V., Garg, A., Steinbach, M., Kumar, V.: Automated Detection of Forest Cover Changes. Igarss (2010)
11. Ortega, M.X., Bermudez, J.D., Happ, P.N., Gomes, A., Feitosal, R.Q.: Evaluation of deep learning techniques for deforestation detection in the amazon forest. ISPRS Ann. Photogram. Remote Sens. Spat. Inf. Sci. **IV**-2/W7 (2019)
12. Bragilevsky, L., Bajic´, I.V.: Deep Learning for Amazon Satellite Image Analysis. IEEE (2017)
13. Khan, S.H., He, X., Porikli, F., Bennamoun, M.: Forest change detection in incomplete satellite images with deep neural networks. IEEE Trans. Geosci. Remote Sens. (August 2018)
14. Vignesh, T., Thyagharajan, Ramya, K.: Change detection using deep learning and machine learning techniques for multispectral satellite images. UJITEE, **9**(Issue-IS) (November 2019). ISSN: 2278–3075
15. Henderson, J., Piwowar, J.: Analysis of Changes in Vegetation Condition in Grasslands National Park Using Remote Sensing. IEEE (2006)
16. Hamdi, Z.M., Brandmeier, M., Straub, C.: Forest Damage Assessment Using Deep Learning on High Resolution Remote Sensing Data, MDPI, 22 Aug 2019
17. Bhatt, A., Ghosh, S.K., Kumar, A.: Automated Change Detection in Satellite Images Using Machine Learning Algorithms for Delhi, India. IGARSS (2015)

18. https://plugins.qgis.org/plugins/SemiAutomaticClassificationPlugin/, last accessed 2020/11/06
19. https://en.wikipedia.org/wiki/QGIS, last accessed 2020/05/16
20. https://www.sciencedirect.com/topics/earth-and-planetary-sciences/spectral-band, last accessed 2020/06/21
21. https://openweather.co.uk/blog/post/satellite-imagery-landsat-8-and-its-band-combinations, last accessed 2020/06/01
22. Dsouza, H., Gupta, S.: A study in change in vegetation cover in an urban environment: a multi-spectral, multitemporal analysis of Mumbai suburban district using remote sensing. J. Agroecol. Nat. Resour. Manage. **3**(2) (July-September 2016)
23. Coppin, P.R., Bauer, M.E.: Change Detection in Forest Ecosystems with Remote Sensing Digital Imagery (1996)
24. Hu, F., Xia, G.-S., Hu, J., Zhang, L.: Transferring Deep Convolutional Neural Networks for the Scene Classification of High-Resolution Remote Sensing Imagery", MDPI, November 2015
25. https://www.guru99.com/convnet-tensorflow-image-classification.html, last accessed 2020/10/19

FPGA-Based Design Architecture for Fast LWE Fully Homomorphic Encryption

Sagarika Behera and Jhansi Rani Prathuri

Abstract A high-speed field-programmable gate array (FPGA) implementation architecture is purposed to implement fast learning with error (LWE) fully homomorphic encryption. Currently, there are many security issues with conventional cryptosystems. In addition, encrypting and decrypting a large volume of data consume enormous computing time which makes the conventional cryptosystems ineffective. In this work, a novel fully homomorphic encryption algorithm, LWE, has been analyzed using linear algebraic equations. The same has been simulated in Python. In addition, to map the LWE scheme in the FPGA, digital circuits are conceptualized to implement mathematical operations such as modulo adder, multiplier, and noise generator.

Keywords Field programmable gate array (FPGA) · Learning with error (LWE) · Modulo adder · Multiplier · Noise generator

1 Introduction

In this new digital era of transmission, communication, and storing of data in the cloud, cryptography plays a major role. Various industries produce massive quantities of data every day, which is then processed in the cloud. Industries such as financial sectors, health industries, stock markets, defense sectors, IoT applications, IT industries, and power sectors generate large volume of data every day. Various cloud service providers (CSP) are used by various industries to save their information in the cloud. As a result, cloud computing has become increasingly important in the digital age. Since data are stored in some other location that is not in control of the client, the CSP is responsible to secure the client information. User encrypts his or her data before uploading it to the cloud server. CSP maintains privacy of the user and the secrecy of the data. Cryptography is required for all

S. Behera (✉) · J. R. Prathuri
CMR Institute of Technology, Bengaluru, India
e-mail: sagarika.b@cmrit.ac.in

K. Khanna et al. (eds.), *Cyber Security and Digital Forensics*,
Lecture Notes on Data Engineering and Communications Technologies 73,
https://doi.org/10.1007/978-981-16-3961-6_47

these operations. Cloud computing provides different types of services such as PaaS, SaaS, IaaS, and XaaS with less price. With these services, cloud computing also provides a powerful computing model with low cost. Cloud computing is facing various security issues such as data leakage, the privacy of sensible data, cyber-attacks, and account hijacking. If the client wants to do some computing on the encrypted data saved on the server in the cloud, then the client has to share the private key with the server to decrypt the encrypted data. This may lead to different types of data security issues. To avoid this type of situation, many prefer smart computation on encrypted data. In this computation method, the client will not share the secret key with the third party. Computation on encrypted data will be carried out, and the encrypted form will be the output of the operation. Homomorphic encryption (HE) enables encrypted data to work in such a way. This enables the cloud server to compute the encrypted data without accessing the customer's secret key. This technology came to the picture just 12 years back. Since then, more researchers started working in this area. In the cloud computing and big data environment, customers are accepting this technology, since they are not going to lose the confidentiality and secrecy of their critical and sensitive data. Fully homomorphic encryption (FHE) is the most common type of homomorphic encryption technique as it allows various operations without decrypting the encrypted data. It enables arbitrary circuits of unbounded depth to be evaluated. The basic concept of FHE is shown in Fig. 1 and explained below.

Let $M1$ and $M2$ are two messages, *op* is an operator and R is the result one gets after performing the operation, and $R1$ is holding the encrypted form of the result.

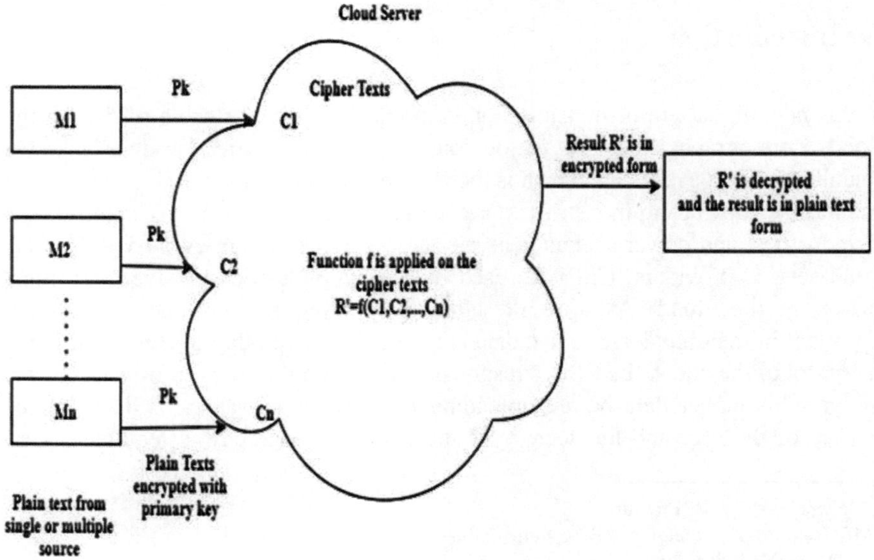

Fig. 1 FHE processing on the cloud data

$$R = M1\ op\ M2 \tag{1}$$

$$R1 = \text{Encrypted }(R) = \text{Encrypted }(M1\ op\ M2) \tag{2}$$

Let $C1$ and $C2$ be the encrypted versions of $M1$ and $M2$, respectively. Same operation has been carried out on the encrypted form. The result is denoted by R'.

$$R\prime = C1\ op\ C2 \tag{3}$$

As per the homomorphic encryption scheme, Eq. (2) = Eq. (3)

$$\text{Encrypted }(M1\ op\ M2) = \text{ Encrypted }(M1)\ op \text{ Encrypted }(M2) = C1\ op\ C2 \tag{4}$$

After decryption the result R' will be equivalent to R.

$$\text{Decr }(C1\ op\ C2) = M1\ op\ M2 \tag{5}$$

There are four steps involved in the FHE scheme. So the FHE scheme can be represented as the quadruplet of algorithms (KeyGen, Encr, Eval, Decr).

- KeyGen(r): It is an algorithm for generating keys. It takes a random number r as input and returns Pk as a public key and Sk as a secret key as output.
- Encr (M, Pk): Encryption algorithm takes two inputs: a plain text message M and a public key Pk. As an output, it generates ciphertext C.
- Eval (f, C1, C2, …, Cn): It is an evaluation algorithm. This is a homomorphic-specific operation, which takes cipher texts C1, C2, …, Cn as input and applies the function f on those ciphertexts. The output of the operation will be in encrypted form. The most important point is that here, the format of the output after the evaluation process should be preserved in order to decrypt it correctly.
- Decr(C, Sk): Decryption algorithm accepts the C as ciphertext and the secret key Sk as inputs and plain text M is retrieved.

2 Related Work

In the year 2009, Craig Gentry [1] gave the most powerful smart computation FHE concept to the research community. Since then, researchers are working in this field to make it more efficient. Before that, homomorphic encryption concept was there, but all of them were partial homomorphic encryption. Partial homomorphic encryption method supports only one type of operation (multiplication or addition).

Some cryptographic algorithms which support partial homomorphic encryption are RSA, ElGamal, and Paillier encryption methods. RSA and ElGamal encryption methods support the multiplicative homomorphic property. Paillier encryption method supports the additive homomorphic property. Gentry [1] gave an FHE scheme which is lattice-based homomorphic encryption. However, as the depth of the circuit grows, so does the noise. He then modified his scheme and made it bootstrapping. In bootstrapping scheme, a class of circuit also contains its own decryption circuit. Van Dijk et al. [2] proposed an FHE scheme which is based on the concept of partially homomorphic encryption over integers. Two types of the gate (addition and multiplication) can be evaluated by somewhat homomorphic encryption scheme. The complete working model of FHE scheme was proposed by Gentry and Halevi [3]. Halevi [4] explained all the concepts of different types of the homomorphic encryption schemes.

Without bootstrapping, Brakerski et al. [5] proposed a fully homomorphic concept. For constructing leveled fully homomorphic encryption schemes, this scheme uses learning with error (LWE) and ring-LWE (RLWE). Brakerski and Vaikuntanathan [6] improved the scheme presented in [5]. They stated that the ciphertext generated by their proposed scheme is shorter and can be used for the single-server private data retrieval (PIR) protocol based on LWE. Chatterjee and Aung [7] explained the different types of homomorphic encryption schemes with circuit and their application in the real world. Yi et al. [8] briefed about the homomorphic encryption schemes and their applications. Su et al. [9] proposed FPGA-based architecture for ring-LWE (RLWE) problem.

Luo et al. [10] made a learning with rounding (LWR)-based FHE scheme to eliminate the tangly modulus problem using approximate eigenvector method. They extended their scheme to construct multi-key FHE which is an alternative to LWE-based multi-key FHE. They claimed that their scheme can be applied to multiparty computation with high efficiency. LWE-based FHE scheme presented in [5, 6] is more secure and efficient compared to the scheme presented in [1, 2, 11, 12]. Kim et al. [13] performed a secure search operation on encrypted genomes which were encrypted using homomorphic encryption. Applications of homomorphic encryption in different fields such as privacy preserving data mining, private data retrieval, privacy preserving prediction, multiparty computation, and training neural networks over encrypted data are given in [14–17]. Since 2009, after Gentry's FHE scheme [1], many researchers are working in this area to improve the performance. Fan et al. [18] compared LWE scheme with RLWE scheme. They have analyzed various homomorphic operations such as multiplication and bootstrapping and derived worst case bounds on the noise caused by these operations. In the paper [19], the authors claimed that their modified LWE scheme which is based on approximate eigenvector method is easier to understand and asymptotically faster than the previous LWE scheme. Ducas et al. [20] claimed that their bootstrapping method for homomorphic encryption takes about half a second to run on a personal computer. Fast fully homomorphic encryption (TFHE) [21] is the improvement of FHE scheme which is based on GSW and its ring variants.

The authors of the paper [21] demonstrated a new bootstrapping circuit that converts LWE ciphertexts to low-noise RingGSW ciphertexts in few milliseconds.

LWE, a modern fully homomorphic encryption algorithm, is analyzed using linear algebraic equations in this paper. The same has been simulated in Python. Further, this LWE has been converted to binary strings, and mathematical circuits such as modulo adder, multiplier, and a noise generator are conceptualized to map the LWE scheme in the FPGA architecture.

In Sect. 3, LWE-based homomorphic encryption scheme is explained. In Sect. 4, the design architecture for the FPGA-based LWE scheme is covered. Section 5 explains the implementation and simulation result of the LWE scheme. Finally, the paper is concluded in Sect. 6.

3 LWE-Based Homomorphic Encryption Scheme

Oded Regev first proposed the LWE problem in 2005. After that, it is put to use in a public key cryptosystem. Brakerski et al. [6] proposed a LWE-based fully homomorphic encryption scheme. First, we will describe the LWE problem, and then, we will describe how to do encryption with LWE.

This is how the LWE problem is described. Let $\mathbb{Z}_q$ be the ring of modulo q integers and $\mathbb{Z}_q^n$ be the set of n number of vectors over $\mathbb{Z}_q$. There is an unknown linear function *f*, defined as f: $\mathbb{Z}_q^n \rightarrow \mathbb{Z}_q$. The LWE problem takes a pair $(\mathbf{x}, \mathbf{y})$ as input, $\mathbf{x} \in \mathbb{Z}_q^n$ and $\mathbf{y} \in \mathbb{Z}_q$. As a result $y = f(x)$ having high probability. Due to some minor error, some deviation will be there from the equality.

Now we will describe how to create a FHE scheme using this LWE concept. Let $\vec{S}$ is a secret vector and $\vec{S} \in \mathbb{Z}_q^n$, **A** is a uniform matrix and $A \in \mathbb{Z}_q^{n \times M}$, $\vec{\eta}$ is a small noise vector where $\vec{\eta} \in \mathbb{Z}_q^M$ and the elements of this noise vector are very small, where $|\eta_i| \leq \alpha q$ and α are very tiny. $\vec{S}$ and $\vec{\eta}$ are selected randomly. The secret vector is multiplied with the uniform matrix and noise is added to it. We got another vector $\vec{b}$ and $\vec{b} \in \mathbb{Z}_q^M$.

$$\vec{b} = \vec{S}.A + \vec{\eta} \tag{6}$$

So Eq. (6) tells that if we have a random noisy linear equation with mod integer (q), then it is computationally indistinguishable from the uniform equation.

Now we will consider a new matrix ($-A$) where $A \in \mathbb{Z}_q^{(n+1) \times M}$ and 1 is concatenated with $\vec{S}$. So the new secret vector $\vec{S} \in \mathbb{Z}_q^{n+1}$.

$$\vec{\eta} = \vec{b} - \vec{S}.A \tag{7}$$

So small noise vector can be written as $\vec{S}.A = \vec{\eta}$. Now we will discuss how the encryption of the message can be performed from LWE.

Here matrix A is the public key and $\vec{S}$ is the secret key. A uniform binary vector $\vec{r}$ is multiplied with A and encoded message $\vec{y}$ is added to it, and we got the cipher text $\vec{C}_y$. So the cipher text is

$$\vec{C}_y = A.\,\vec{r} + \vec{y} \tag{8}$$

So the attacker can see the public key and the cipher text. It will be difficult for the attacker to find the message since it cannot find out the secret key. If we want to encrypt "0," then $\vec{y}$ will be 0. If we want to encrypt "1," then $\vec{y}$ will be a random vector. Now we will discuss the decryption process. We will multiply the secret key with the cipher text.

$$\vec{S}\,\vec{C}_y = \vec{S}\mathrm{A}.\vec{r} + \vec{S}\vec{y} = \vec{\eta}\vec{r} + \vec{S}\,\vec{y} \tag{9}$$

In Eqs. (8) and (9), we have shown the encryption and decryption processes for the vector. Now it can be generalized to matrices where A, R, and Y are the matrices.

So the encryption process will be $A\ R + Y = \vec{C}_y$ where R is the binary matrix consists of $\{0.1\}$ of size $M \times N$, and Y is the encoded message of size $(n + 1) \times N$. Decryption process will be $\vec{S}\vec{C}_y = \vec{\eta}R + \vec{S}Y$.

Now we will discuss the eigenvector method and approximate eigenvector, and then, how to perform encryption and decryption on approximate eigenvector. It was observed that if C_1 and C_2 are two matrices with same eigenvector $\vec{S}$ and m_1, m_2 are respective eigenvalues with respect to same eigenvector $\vec{S}$, then it holds the following properties.

1. $C_1 + C_2$ has eigenvalue $(m_1 + m_2)$ with respect to same eigenvector $\overrightarrow{S}$.
2. $C_1.C_2$ has eigenvalue m_1m_2 with respect to same eigenvector $\vec{S}$.

It is satisfying the fully homomorphic properties, but it is not secure. Eigen vector can be easily calculated. So we will take approximate eigenvector where secret key $\vec{S}$ is the eigenvector and C is the cipher text, m is the encoded message and $\vec{e}$ is small noise or error. The corresponding linear equation is given below.

$$\vec{S}.C = m\vec{S} + \vec{e} \tag{10}$$

It is decryptable if $|\vec{e}| < q$. Fully homomorphic properties of approximate eigenvector are given below. Let C_1 and C_2 are two cipher texts. Applying Eq. (10) on C_1 and C_2, we got the following two equations.

$$\vec{S}C_1 = m_1\vec{S} + \overrightarrow{e1} \quad \text{and} \quad \vec{S}.C_2 = m_2\vec{S} + \overrightarrow{e2}$$

After performing addition operation on these two cipher texts

$$\begin{aligned} \text{Cadd} &= C_1 1 + C_2, \Rightarrow \vec{S}\,.\,(C_1 + C_2) = \vec{S}.C_1 + \vec{S}.C_2 \Rightarrow m_1\vec{S} + \overrightarrow{e1} + m_2\vec{S} + \overrightarrow{e2} \\ &\Rightarrow (m_1 + m_2)\vec{S} + (\overrightarrow{e1} + \overrightarrow{e2}) \end{aligned} \quad (11)$$

After addition of two cipher texts, noise grows a little. Now we will perform the multiplication operation on these two cipher texts.

$$\begin{aligned} C_{\text{mult}} = C_1.C_2 &\Rightarrow \vec{S}\,.\,(C_1.C_2) \Rightarrow \left(\mathrm{m}_1\vec{S} + \overrightarrow{e_1}\right)C_2 \Rightarrow m_1 C_2\vec{S} + C_2\overrightarrow{e_1} \Rightarrow m_1\left(m_2\vec{S} + \overrightarrow{e_2}\right) + C_2\overrightarrow{e_1} \\ &\Rightarrow m_1 m_2\vec{S} + \left(m_1\overrightarrow{e_2} + C_2\overrightarrow{e_1}\right) \end{aligned} \quad (12)$$

Compared to addition operation, noise grows more in multiplication operation. If we will generalize the growth of noise during homomorphic circuit evaluation, then it can be represented as below.

The noise increases by the factor $(N + 1)$ at each step. At depth "*d*," noise will grow by $(N + 1)^d$, where N is the size of polynomial. For decryption, we need the noise to be small.

4 FPGA-Based Design Architecture for LWE Scheme

FPGA-based design architecture for LWE scheme contains three blocks as shown in Fig. 2. The components and functions of three blocks are explained below.

- **Key Generator**: It contains true randomness generator (TRG), which generates a symbol series that resembles a uniformly random sequence. It generates the public key matrix *A*. Polynomial multiplier multiplies A with the secret key *S*. The modular operation is performed on the output of the multiplier, and it is added to the noise e. So the output of the polynomial adder is the public key *b*. The output of the key generator is two public keys *A* and *b*.
- **Encryption**: This is the second module which takes input message m and gives the encrypted message in the form (*u*,*v*). Public key *A* is multiplied with *r* (binary uniform vector), and a modular operation is performed on it to get *u*. Message *m* and *q*/2 are multiplied by scalar multiplier, and the result is added to public key *b* and modular reduction is applied on it to get the value v.
- **Decryption:** In this module, *u* and *S* are multiplied, and the result is subtracted from *v*. If the result is less than *q*/2, it is 0 otherwise 1.

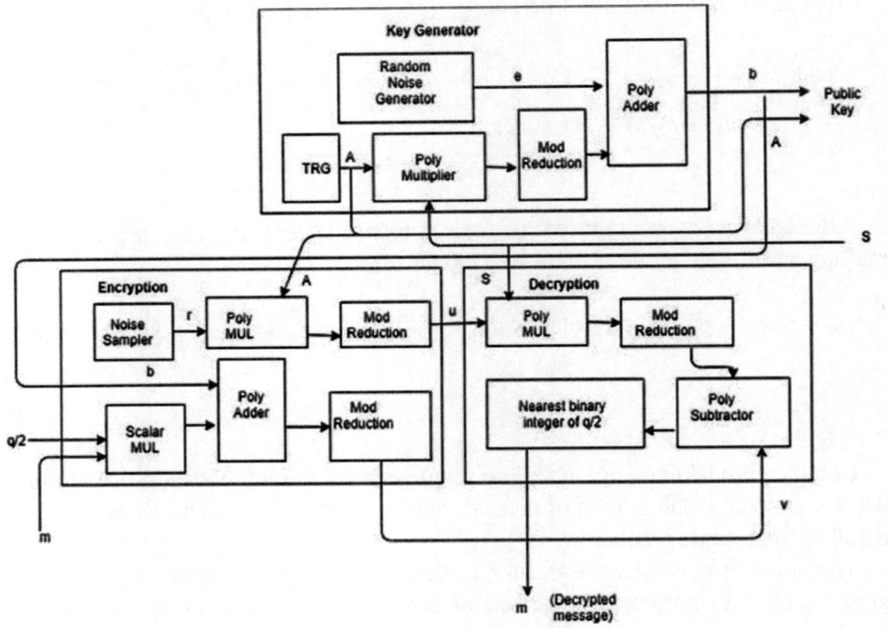

Fig. 2 FPGA-based design architecture for LWE scheme

5 Result and Implementation

Implementation of LWE and the result are discussed here. First, a random series of values are selected for our public key A. We have selected 20 random values in the range of 0–137.

Public key (A): [4, 46, 82, 132, 117, 10, 28, 47, 75, 7, 125, 45, 22, 104, 16, 57, 65, 18, 48, 122]

Prime number q is 137, the secret key S is 43, and e is small error values.

Error vector (e): [2, 2, 2, 1, 1, 2, 2, 2, 3, 1, 2, 1, 2, 1, 1, 2, 3, 2, 2, 1]

List B is generated using the method $Bi = A_i S + e_i \pmod q$

List (B): [37, 62, 103, 60, 100, 21, 110, 105, 77, 28, 34, 18, 126, 89, 4, 124, 58, 91, 11, 41]

We took two message values 9 and 5 for encryption.

Bits to be ciphered: [1, 0, 0, 1, 0, 0, 0, 0], [1, 0, 1, 0, 0, 0, 0, 0]

We calculated two values u and v using the following method. Where $u = \Sigma A_i$ mod q and $v = \Sigma B_i + q/2.M \pmod q$. The encrypted message sample (u,v) is given below.

[11, 9], [3, 5], [14, 8], [11, 4], [0, 9], [17, 1], [12, 6], [16, 6]
For decryption, we have used the following calculation.

$$\text{Decr} = v - Su(\text{mod} q)$$

If Decr < $q/2$, then the message bit is 0 else it is 1.
Result bit0 is: 0, result bit1 is: 0, result bit2 is: 1, result bit3 is: 1

6 Conclusion and Future Work

Mathematical analysis and simulation of a fully homomorphic LWE algorithm have been carried out in order to map all the mathematical expressions into FPGA. It is concluded that the mathematical operations namely polynomial adder, subtractor, multiplier, modulo reduction, and random number generator for a large length of binary strings can easily be implemented in modern FPGA. FPGA-based architecture also implements a large amount of mathematical operations which are typical for LWE fully homomorphic encryption. The computation is also faster compared to central processing unit (CPU), graphics processing unit (GPU), and general purpose computing on graphics processing unit (GPU). Our future work will be the simulation of LWE in VHDL and real-time implementation in FPGA board to quantify the claim of this work.

References

1. Gentry, C.: Fully Homomorphic Encryption Scheme. Stanford University (2009)
2. Van Dijk, M., Craig Gentry, Halevi, S., Vaikuntanathan V.: Fully homomorphic encryption over the integers. In: Proceedings of Advances in Cryptology, pp. 24–43, EUROCRYPT'10 (2010)
3. Gentry, C., Halevi, S.: Implementing Gentry fully homomorphic scheme. In: Proceedings of Advances in Cryptology, EUROCRYPT'11, pp. 129–148 (2011)
4. Halevi, S.: Homomorphic Encryption, In Tutorials on the Foundations of Cryptography, pp. 219–276. Springer Cham (2017)
5. Brakerski, Z., Craig Gentry, Vaikuntanathan, V.: Fully homomorphic encryption without bootstrapping, In: Innovations in Theoretical Computer Science (ITCS'12) (2012)
6. Brakerski, Z., Vaikuntanathan, V.: Efficient fully homomorphic encryption from (Standard) LWE. SIAM J. Comput. **43**(2), 831–871 (2014)
7. Chatterjee, A., Aung, K.M.M.: Fully Homomorphic Encryption in Real World Applications. http://www.springer.com/series/15213
8. Yi, X., Paulet, R., Bertino, E.: Homomorphic Encryption and Applications. https://www.springer.com/series/10028
9. Su, Y., Yang, B., Yang, C., Tian, L.: FPGA-based hardware accelerator for leveled ring-LWE fully homomorphic encryption, 168008–168025. IEEE Access 8 (2020)
10. Luo, F., Wang, F., Wang, K., Li, J., Chen, K.: LWR-Based Fully Homomorphic Encryption, Revisited, Security and Communication Networks (2018). https://doi.org/10.1155/2018/5967635

11. Brakerski, Z., Vaikuntanathan, V.: Fully homomorphic encryption from ring-LWE and security for key dependent messages. In: Annual Cryptography Conference, vol. 6841 of Lecture Notes in Computer Science, pp. 505–524, Springer, Heidelberg, Berlin, Germany (2011)
12. Smart, N.P., Vercauteren, F.: Fully homomorphic encryption with relatively small key and ciphertext sizes, Public Key Cryptography, vol. 6056 of Lecture Notes in Computer Science, pp. 420–443, Springer, Berlin, Germany (2010)
13. Kim, M., Song, Y., Cheon, J.H.: Secure searching of biomarkers through hybrid homorphic encryption scheme. BMC Med. Genomics, **10**(2), 42 (2017)
14. Yi, X., Kaosar, M.G., Paulet, R., Bertino, E.: Single database private information retrieval from fully homomorphic encryption. IEEE Trans. Knowl. Data Eng. **25**(5), 1125–1134 (2012)
15. Bos, J.W., Castryck, W., Iliashenko, I., Vercauteren, F.: Privacy-friendly forecasting for the smart grid using homomorphic encryption and the group method of data handling. In: International Conference on Cryptology in Africa, Springer, pp. 184–201 (2017)
16. Gilad-Bachrach, R., Dowlin, N., Laine, K., Lauter, K.,Naehrig, M., Wemsing, J.: Cryptonets: applying neural networks to encrypted data with high throughput and accuracy. In: International Conference on Machine Learning, pp. 201–210 (2016)
17. Xu, R., Joshi, J.B., Li, C.: CryptoNN: training neural networks over encrypted data. In: IEEE 39th International Conference on Distributed Computing Systems (ICDCS), pp. 1199–1209 (2019)
18. Fan, J., Vercauteren, F.: Somewhat practical fully homomorphic encryption, IACR Cryptol. 144 (2012)
19. Gentry, C., Sahai, A., Waters, B.: Homomorphic encryption from learning with errors: conceptually-simpler, asymptotically-faster, attribute-based. In: Annual Cryptology Conference, Springer, pp. 75–92 (2013)
20. Ducas, L., Micciancio, D.: FHEW: bootstrapping homomorphic encryption in less than a second. In: Annual International Conference on the Theory and Applications of Cryptographic Techniques, Springer, pp. 617–640 (2015)
21. Chillotti, I., Gama, N., Georgieva, M., Izabachene, M.: TFHE: fast fully homomorphic encryption over the torus. J. Cryptol. **33**(1), 34–91 (2020)
22. Clear, M., McGoldrick, C.: Multi-identity and multi-key leveled FHE from learning with errors. In: Proceedings of the Annual Cryptography Conference, pp. 630–656, Springer, Heidelberg, Berlin, Germany (2015)

Hierarchal Communication Architecture for Multi-level Energy Harvesting Support in Underwater Sensor Network

Anuradha, Amit Kumar Bindal, Devendra Prasad, and Afshan Hassan

Abstract Communication by using an energy-saving approach is a vital requirement of an energy-constrained underwater sensor network. The underwater acoustic communication technique ordinarily governs the usage of clout. The network's capability to communicate the collected facts is affected due to minimal renewing capacity in case of the underwater wireless sensor network. The proposed work describes the reliable resources available in underwater for the sensors. The proposed scheme provides the mechanism for energy harvesting from tidal energy into electrical energy to charge the Li-ion cells used in sensors. To cover the maximum area of the ocean sensor, the data aggregators are deployed at three different levels (i.e. bottom, middle and top levels). The sensors deployed in oceans at the top level are static, whereas they are mobile at middle and bottom levels.

Keywords Underwater sensor network (UWSN) · Energy harvesting · Li-ion · Underwater acoustic sensor networks (UW-ASNs)

Anuradha
ECE Department, M. M. Engineering College, M. M. (Deemed To Be University) Mullana, Ambala, Haryana, India

A. K. Bindal (✉)
Department of Computer Science and Engineering, M. M. Engineering College, M. M. (Deemed to be University) Mullana, Ambala, Haryana, India
e-mail: amitbindal@mmumullana.org

D. Prasad · A. Hassan
Institute of Engineering and Technology, Chitkara University, Chandigarh, Punjab, India
e-mail: devendra.prasad@chitkara.edu.in

A. Hassan
e-mail: afshan.hassan@chitkara.edu.in

K. Khanna et al. (eds.), *Cyber Security and Digital Forensics*,
Lecture Notes on Data Engineering and Communications Technologies 73,
https://doi.org/10.1007/978-981-16-3961-6_48

1 Introduction

The current ear of learning of UWSNs has appealed speedily mounting interests [1–3]. UWSNs are the arising and inspiring frameworks that entitle an extensive range of energetic short-term and long-term applications like distributed tactical surveillance, disaster preclusion, ecological monitoring, oceanographic statistics and abetted navigation [4]. Untethered sensor nodes are used in the disposition in such challenging to reach localities. The major constraint related to unleashed nodes is restricted battery volume which means nodes will work for a predetermined period. Numerous practices have been introduced to increase the lifetime of SNs powered by the battery. To lengthen the network life, essential energy is kept in the hardware in any form. If the category of battery is preferred, the Li-ion battery is the utmost capable technology for UWSNs because of their greater energy yield and power density in comparison to nickel–cadmium and nickel–metal hydride [5]. An alternative way of solving energy problems is the energy harvesting approach, which generates energy by itself. Harvesting energy means the restoration of power by absorbing it from the atmosphere or using other energy sources, such as the heat of a body, stroke of a finger and tidal energy [6–8].

2 Literature Review

Communication Architecture

Following communication, architectures are considered for UWSNs, i.e. 2D and a 3D architecture [9].

(i) ***Two-Dimensional (2-D) UW-ASNs***

As shown in Fig. 1, the 2D architecture is affixed in the lowermost of the marine. UW-SNs are associated with one or more UW-gateways by acoustic transmission media.

The surface station, fortified by an acoustic transceiver, which might be capable of handling bar scores of corresponding communication with the UW-gateways and by a long-range radio or satellite transmitter, desires communication with onshore or surface sink [9].

(ii) ***Three-Dimensional (3-D) UW-ASNs***

Figure 2 shows 3D architecture; in this structural design, sensors sink at dissimilar depths [9]. In the design of 3D architecture, the problem of coverage area is a little bit resolved. We cover more areas by deploying the sensors in

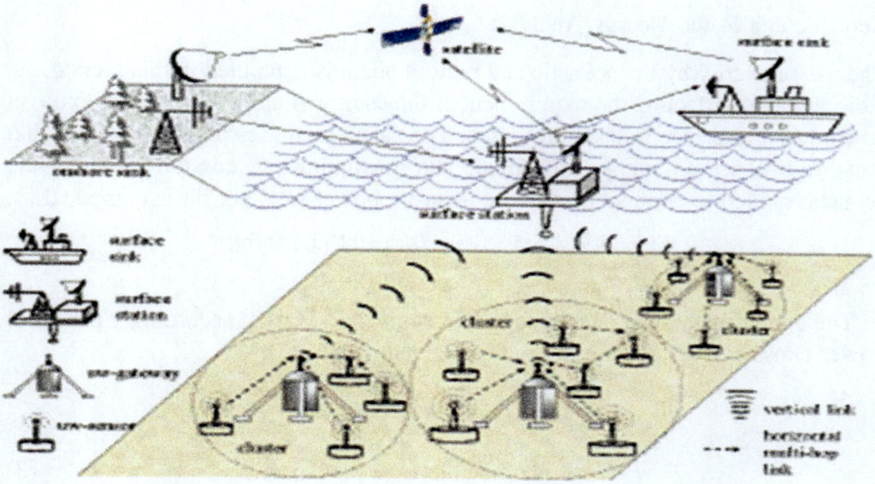

Fig. 1 2-Dim UW-ASNs architecture [9]

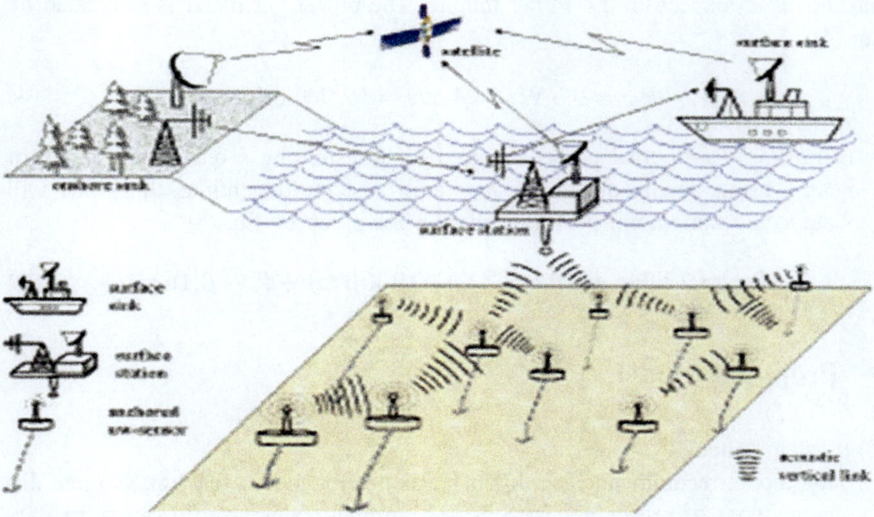

Fig. 2 3-Dim UW-ASNs architecture [9]

depth of the ocean, but the energy issue is there. This problem arises due to more energy consumption because each node is working as a gateway in 3 D architecture.

Acoustic and Radio Energy Analysis

The use of energy by the acoustic and radio modem is computed in this section. As radio requires surfacing, the requirement of the energy of any node in order to travel upward to the surface is also considered. The calculation uses pragmatic inputs that match our observed energy utilization. The maximum transmission power of the acoustic modem is around 10 W. The acoustic power (*Pa*) per bit is computed as:

$$P_a = 5\,\mathrm{W}/(5.5 * 8\,\mathrm{bits/s}) = 113.6\,\mathrm{mJ/bit}. \tag{1}$$

The approximate data rate for the radio is given in Eq. (2) accordingly per bit of power consumed (P_r) using the radio is given by Eq. (3).

$$76{,}800/6/2\,\mathrm{bits/s} = 6400\,\mathrm{bits/s} \tag{2}$$

$$P_r = 1\,\mathrm{W}/6400\,\mathrm{bits/s} = 0.16\,\mathrm{mJ/bit} \tag{3}$$

However, the SN must go up to the surface first, with the help of a depth adjustment system, to transmit data. This arrangement consumes around 0.6 watts and moves at a speed of 2.4 m per minute. The power per meter is calculated by Eq. (4).

$$P_w = 0.6\,\mathrm{W}/\ 0.04\,\mathrm{m/s} = 15{,}000\,\mathrm{mJ/m} \tag{4}$$

In Eq. (5), P_{rw} is total power required for transmitting *k* bits from a depth of *d* meters, using depth adjustment and radio system (pretentious that node will relocate to the same point post transmission), is given by Eq. (5).

$$P_{rw} = 2\,d\mathrm{Pw}\ + k\mathrm{Pr} = 2\,d \leftarrow 15000\,mJ + k \leftarrow 0.16\,mJ \tag{5}$$

3 Proposed Model

(C) System Model
The model recommended in this paper is heterogeneous that entails three different types of sensor nodes (SNs), i.e. ordinary SNs used to sense the phenomenon that occurred as aggregators accumulate the data from ordinary SNs and BS receives the data from aggregators and use for scrutiny.

(D) **Network architecture**
In the proposed model, SNs are deployed at 3 layers, i.e. top, middle and bottom layer. Individual layer comprises of data aggregator accompanied by ordinary SNs, as shown in Fig. 3. The base station (BS) is floating on the surface of the water. Sensors are fitted out with the depth acclimation system, which strengthens the network.

Fig. 3 Network architecture of the proposed model

(E) Data dissemination mechanism
To deliver the data from the various layers to the BS, the data aggregator deployed at bottom level transfers the data to the data aggregator at middle level, middle-level data aggregator transfers the data to the top-level data aggregator, and top-level data aggregator node transfers the data to the floating BS. Finally, the BS delivers data to the data centre through satellite as shown in Fig. 3. In the process, data is delivered to the BS from various layers using shortest path (computed through any standard algorithm).

(F) Energy Model
The sensors deployed at various levels keep on watching its energy level. If the remaining energy (E_r) goes below some threshold energy (E_{th}), sensors open their charging circuitry and start charging without interrupting their communication. Once its battery is full, the sensor closes its charging circuitry. In the proposed model, tidal energy is converted into electrical energy and provide to the sensor's circuitry for charging purpose.

(i) **Acoustic Energy Consumed**
As TDMA slot is capable of sending as well as receiving data packet of 20 bytes, containing 14 bytes of payload, in every 4 s; hence, the throughput is 7 bytes/s. The acoustic power per bit is computed as in Eq. (6).

$$P_{pb} = 5\ \text{W}/\left(7 * \frac{8\ \text{bits}}{\text{sec}}\right) = 89.29.4\ \text{mJ/Bit} \tag{6}$$

(ii) **Acoustic Energy Gain through Tidal Energy**
The primitive equation of tidal energy is given by Eq. (7).

$$e = \frac{1}{2} * a_{\text{layer}} * \rho_{sw} * g * h_t^2 \tag{7}$$

where e is the energy, a_{layer} is an area of the basin, ρ_{sw} is water density (1025 kg/m^3), g is the acceleration due to gravity (9.81 m/s^2), and h_t is the height of the tide in the ocean. The equation is implemented only if it is dimensionally correct.

$$e = m^2 * \frac{Kg}{m^3} * \frac{m}{s^2} * m^2$$
$$e = Kg\left(\frac{m^2}{s^2}\right) \tag{8}$$

$$K.E. = 1/2\ m\,v^2 \tag{9}$$

Similarly, by putting the units of various parameters in Eq. (9), the dimensionality comes out to be as given in Eq. (10).

$$K.E. = Kg\left(\frac{m^2}{s^2}\right) \tag{10}$$

After comparing the dimensionality of tidal energy and kinetic energy, we can conclude that Eq. (7) used in the proposed model is correct. As only 30% of the total energy can be utilized, the net energy available can be calculated as below:

$$e = 0.15 * a_{\text{layer}} * \rho_{sw} * g * h_t^2 \tag{11}$$

After substituting the value of ρ_{sw} and g in Eq. (11), the tidal energy is coming out to be as given in Eq. (12).

$$e = 0.15 * a_{\text{layer}} * 1025 * 9.81 * h_t^2$$
$$e = 1508.29 * a_{\text{layer}} * h_t^2 \tag{12}$$

From Eq. (12), we conclude that the tidal energy is directly proportional to the area of the basin and the square of the height/depth.

(i) **Computation of energy at various levels**: Eq. (12) can be used to find the energy produced by tide at any layers.

- ***Top Level***

The height of the tide at top level is 1 m. The energy available at top level (E_{top}) for sensors can be computed by putting the value of the height of tide at the top level and basin area in Eq. (12).

$$E_{top} = 1508.29 * 2.5 * 10^5 * 1^2\,\mathrm{J} = 3.8 * 10^{11}\,\mathrm{mJ} \tag{13}$$

Equation (13) gives the energy available for all sensors deployed in the entire area. So the energy available for a single sensor

$$E_{single\_top} = 3.8 * 10^{11} \,/\, 500\ \mathrm{mJ} = 7.6 * 10^8\ \mathrm{mJ} \tag{14}$$

According to the principle of energy conservation, the energy supplied by the tidal waves will be related to the electrical energy produced by the transducers. So electrical energy generated by a single transducer at the top level will be equal to $7.6 * 10^8$ mJ.

To charge a 5 V Li-ion cell, 1500 mA of current is required. The total power (T_{ptl}) produced for Li-ion cell, at top level, is computed by Eq. (15).

$$T_{ptl} = \mathrm{E/C} = 0.6 * 10^8 \,/\, 1 = 7.6 * 10^8\ \mathrm{V} \tag{15}$$

- ***Middle Level***

Similarly, for the middle level, the value of E_{mid}, E_{single_mid} and T_{pml} are computed as above and given by Eq. (16) through Eq. (18).

$$\begin{aligned} E_{mid} &= 1508.29 * 2.5 * 10^5 * 3^2\,\mathrm{J} = 3770.7 * 10^5 * 3^2\,\mathrm{J} = 33.9 * 10^8\,\mathrm{J} \\ &= 33.9 * 10^{11}\ \mathrm{mJ} \end{aligned} \tag{16}$$

$$E_{single\_mid} = 33.9 * 10^{11} \,/\, 500\ \mathrm{J} = 6.78 * 10^9\ \mathrm{mJ} \tag{17}$$

$$T_{pml} = \mathrm{E/C} = 6.78 * 10^9/1 = 6.78 * 10^9\ \mathrm{V} \tag{18}$$

- ***Bottom Level***

By putting the value of the basin area and height of the tide at the bottom level, the value of E_{bot}, E_{single_bot} and T_{pbl} are computed as earlier and given by Eq. (19) through Eq. (21).

$$E_{\mathrm{bot}} = 1508.29 * 2.5 * 10^5 * 5^2 = 94.3 * 10^8\mathrm{J} = 94.3 * 10^{11}\mathrm{mJ} \tag{19}$$

$$E_{\mathrm{single\_bot}} = 94.3 * 10^{11}/500 = 18.86 * 10^9\mathrm{mJ} \tag{20}$$

$$T_{\mathrm{pbl}} = \mathrm{E/C} = 18.86 * 10^9/1 = 18.86 * 10^9\ \mathrm{V} \tag{21}$$

4 Simulation Result and Discussion

Figures 4, 5 and 6 represent the energy consumption in various slots under proposed work. From various figures, we can observe that the energy consumption in the proposed work is very less as compared to the existing schemes for all levels, i.e. at the top, middle and bottom level.

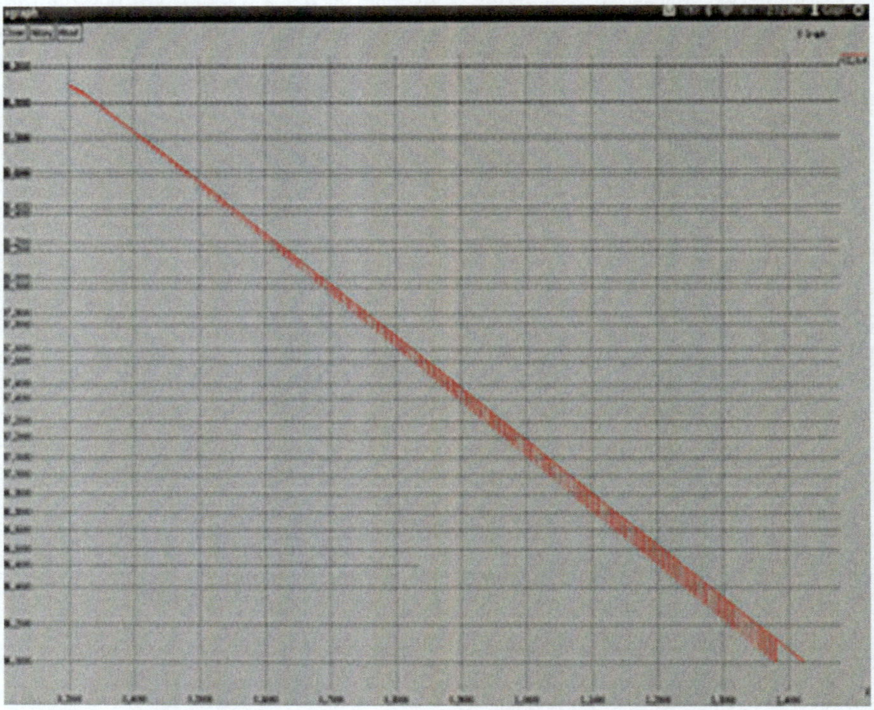

Fig. 4 Energy consumed at top level

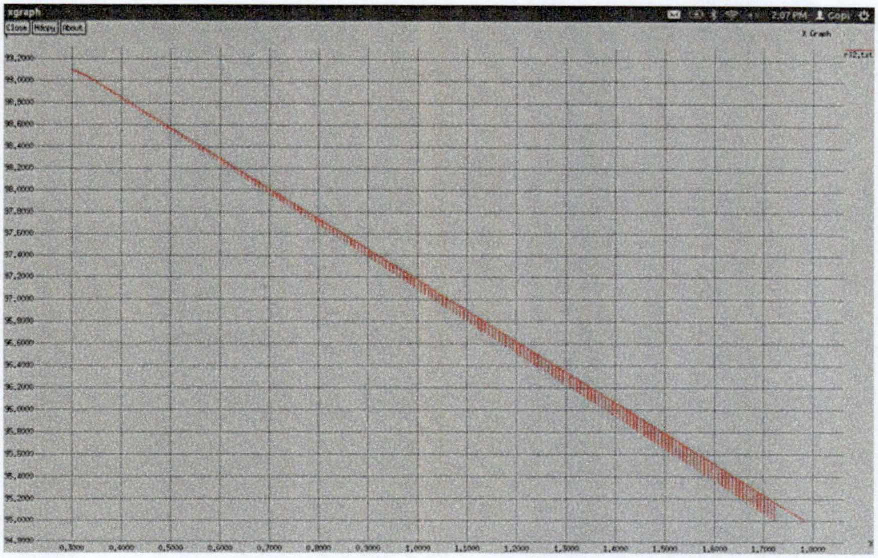

Fig. 5 Energy consumed at middle level

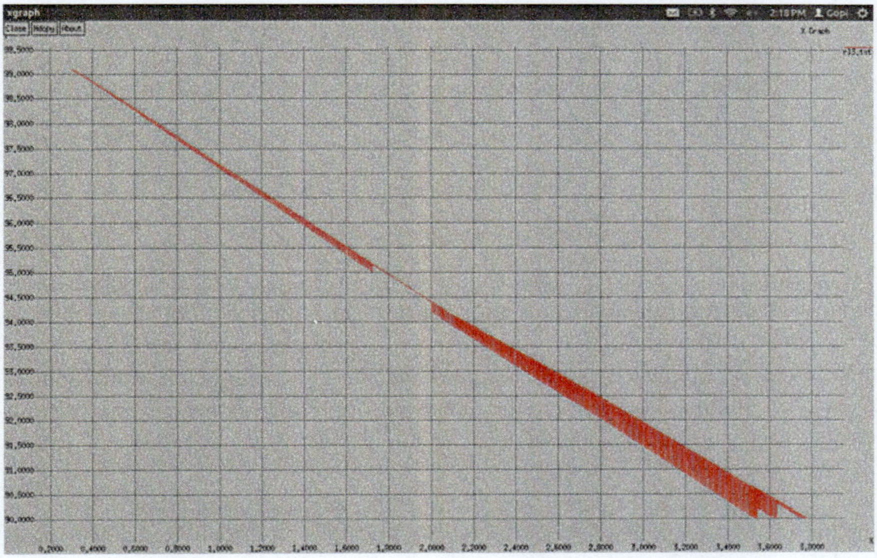

Fig. 6 Energy consumed at bottom level

As in the proposed work, we are providing the concept of energy harvesting to resolve the reliable resource of energy for charging the battery of sensors at various levels. We observed from the simulation that there is a huge amount of energy gain at bottom level as compared to the middle and top levels; the amount of gain in

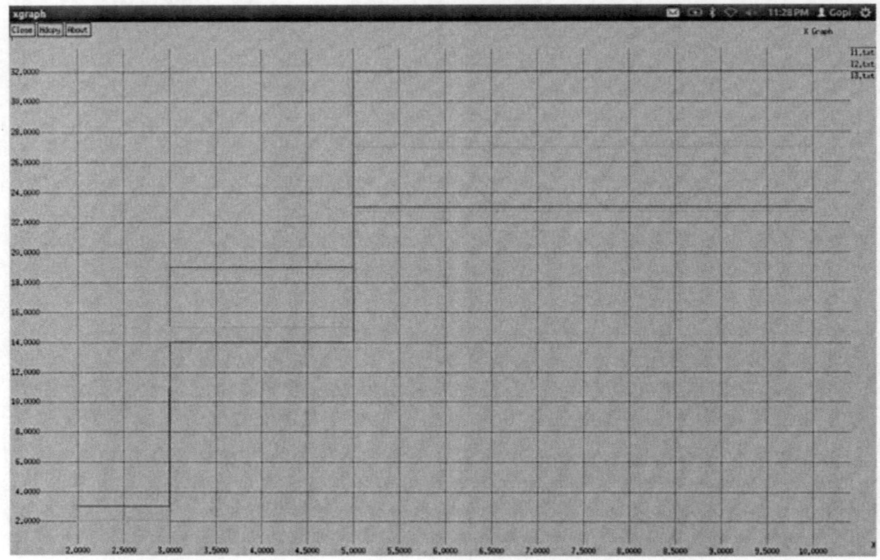

Fig. 7 Energy gain at top level, middle level and bottom level

energy at middle level is more as compared to the top level. In Fig. 7, energy gain at the bottom level is presented by the blue line, at the middle level presented by the green line and the top level is presented by the red line.

5 Conclusion

This work proposed a layered architecture to monitor the activities underwater. In the presented work, data of the various levels are collected through aggregator deployed at various levels. In the process of data dissemination, energy consumption at various levels is observed. As the energy consumption at the bottom levels is very high as compared to the top level, for prolonging the network life, the density of sensors at the bottom level should be very less as compared to the top-level.

As future work, researchers may analyse the effect of sensor density on the amount of tidal energy required for recharging the Li-ion cell so that network should work continuously without any disaster occurs. One can also investigate the various parameters that must be considered in the conversion of tidal energy into electrical energy.

References

1. Goel, K., Bindal, A.: Energy issues in underwater wireless sensor network: a survey report. IJRSR, **3**(2), 1–4 (2014)
2. Yang, H., Ren, F., Lin, C., Liu, B.: Energy efficient cooperation in underwater sensor networks. In: IEEE 18th International Workshop on Quality of Service (IWQoS) (2010). https://doi.org/10.1109/IWQoS.2010.5542761
3. Pompili, D., Akyildiz, I.: Overview of networking protocols for underwater wireless communications. Commun. Mag. IEEE **47**(1), 97–102 (2009)
4. Xu, J., Li, K., Min, G., Lin, K., Qu, W.: Energy-efficient tree-based multipath power control for underwater sensor networks. IEEE, **23**(11), 2107–2116 (2012)
5. Ovaliadis, K., Savage, N., Kanakaris, V.: Energy efficiency in underwater sensor networks: a research review. JESTR, **3**(1), 151–156 (2010)
6. Sudevalayam, S., Kulkarni, P.: Energy harvesting sensor nodes: survey and implications. IEEE, **13**(3), 443–461 (2011)
7. Pompili, D., Melodia, T., Akyildiz, I.F.: Deployment analysis in underwater acoustic wireless sensor networks. In: 1st ACM International Workshop on Underwater Networks, Los Angeles, California, USA, pp. 48–55 (2006)
8. Bindal, A.: 3-tier architecture for sustainable underwater wireless sensor networks. Adv. Math.: Sci. J. **9**(3), 1205–1212 (2020)
9. Detweiler, C., O'Rourke, M., Basha, E.: Multi-model communications in under water sensor networks using depth adjustment. ACM 978–1–4503–1773–3–3/12/11, WUWNet'12, Los Angeles, California, USA, pp. 1–5 (2012)

Review of Evolutionary Algorithms for Energy Efficient and Secure Wireless Sensor Networks

Rajiv Yadav, S. Indu, and Daya Gupta

Abstract Wireless sensor network (WSN) finds vast real-world applications in the field of energy control, security, health care, defense, and environment monitoring. WSNs are subdued by limited power with a specific battery backup. Due to the large distance between sensor nodes and sink, more consumption of power takes place in the sensors. Limited energy of sensor nodes is a major drawback to empower a large network coverage area. Therefore, the battery life and location of cluster heads play an important role in increasing the efficiency and lifetime of sensor nodes for long-term operation in WSNs. While there are many algorithms leading to the optimization of performance using convergence, comparison of such algorithms and their advantages and challenges is addressed. Different types of attacks and security goals are described for high-level security and privacy in WSNs. This paper tabulates a systematic survey of the evolutionary algorithms of WSNs based on nature. This paper also intends to reflect on the security challenges of WSN and proposes effective techniques to address them.

Keywords GA · PSO · ACO · BBO · GWO · Energy utilization · Wireless sensor network · IoT

1 Introduction

Wireless sensor networks (WSNs) having a group of sensor nodes through which these nodes communicate with each other to identify useful information and data. The primary target of WSN is to detect the behavior of the current event and transmits its information to the sink. This phenomenon of distribution protects the loss of data in a specific coverage area. Due to the large distance, more consumption

R. Yadav (✉) · S. Indu
Department of ECE, Delhi Technological University, Delhi, India
e-mail: s.indu@dce.ac.in

D. Gupta
Department of CSE, Delhi Technological University, Delhi, India

K. Khanna et al. (eds.), *Cyber Security and Digital Forensics*,
Lecture Notes on Data Engineering and Communications Technologies 73,
https://doi.org/10.1007/978-981-16-3961-6_49

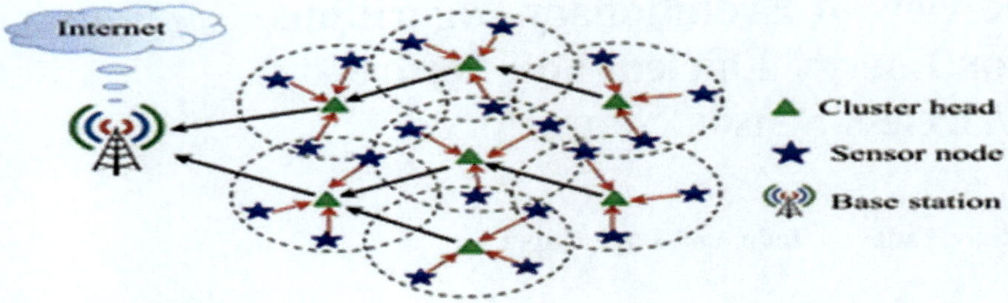

Fig.1 Wireless sensor network topology

of power takes place in the sensors. Limited energy of sensor nodes is a major drawback to empower a large network coverage area. The lifetime of a WSN depends on the CH lifetime, which makes optimum use of CH battery an important parameter for an energy-efficient WSN.

In designing a sensor network, the focus must be on utilizing energy consumption and battery lifetime. The existing WSN model is presented in Fig. 1 [1]. Previous researches on WSN gave different ideas to extend network lifetime by utilizing the present resources. Clustering and routing algorithms also have a specific contribution to the process. There are many performance matrices associated with the efficient working of a WSN. For an energy-efficient network, optimum use of the CH battery is essential, load balancing is important for the stability of the WSN, and the hole-free deployment of sensor nodes is essential for the uninterrupted coverage of the surveillance region.

The introduction of IoT-based applications demands a new algorithm to ensure the efficient working of WSN. In WSN, different objects at different places are empowered to transmit and receive the signals generated by them. High security and privacy are required in WSN through the environment. Information storage, authentication, and management are the main issues in security concerns. Different types of layers can be designed for the safety of WSN, in which the first layer includes many sensors and actuators for taking the information from different functionalities. The main security concerns that the network layer has faced are access attacks, data transit attacks, routing attacks, and unlawful attacks.

2 State of Art

2.1 Energy Efficiency in WSN

The selection of cluster heads randomly from sensor nodes with less residual energy is defined by LEACH [2]. There have been other differences in LEACH with respect to WSN performance enhancement. The fuzzy logic-based clustering algorithm in [3] contains three: fuzzy node energy, centrality, and concentration variables. But a big problem faced by this logic is the involvement of a single cluster in the process. Authors in [4] worked on energy dissipation of the network, but the distance parameter is not mentioned. This avoidance of the distance parameter creates unbalancing in the network. Authors in [5] considered only one parameter node centrality to minimize energy consumption. Other important parameters like load balancing, transmission distance, and uniformity in deployment are not considered. A novel algorithm in [6] is presented to minimize the distance between sink and CH. The transmission distance is specified in this algorithm, but the communication distance between sensor nodes and CHs is not considered.

To enhance the lifetime of a WSN, a partial swarm optimization (PSO) in [7] is used. In PSO, they consider the fitness function parameters like network lifetime and cluster head to sink distance to achieve an energy-efficient network. But the sensor node to CH distance is not considered in PSO. Energy description during the process is also not considered. Authors in [8] proposed a significant algorithm to tackle optimization problems. This algorithm is highly flexible and more desirable for automatically configured techniques. In complex applications, ACO can be used. Changes such as new distances may be adjusted. It can also search among a population in parallel. The challenges faced are probability distribution changes for each iteration. ACO has a difficult theoretical analysis and uncertain time to convergence.

A new nature influenced by the authors in [9] suggested a technique of optimization based on biogeography (BBO-C) in WSN for energy-efficient clustering. To maximize the performance of the network, a complex and effective fitness feature is used. Authors in [10] proposed a clustering and routing algorithm (EBBO-CR) for extending the lifetime of WSN. It also minimizes the distance from its respective CHs and the number of hopes required to move data from CH to sink. But in exploiting the solutions, BBO and EBBO are bad. They have no provision for selecting each generation's best members and the resulting fitness solution.

A novel version of BBO was proposed by the authors in [11]. The implementation of the age-structured BBO (ASBBO) is the impact of people's age on the migration method. Individuals are implemented to capture age-related trends of birth rates and survival rates. Age systems enable information to be shared not only on the basis of the present goodness. In [12], the writers mimic the leadership structure and hunting method of gray wolves. Four forms of the gray wolves were used to determine their fitness, i.e., alpha, beta, delta, and omega. For the energy-efficient WSNs and load balancing of the cluster heads, they suggested DBSCDS-GWO and DBSC-GWO. Yeah, but the author does not take WSN's accuracy and convergence.

2.2 Load Balancing in WSN

Recent researches described many algorithms to enhance the performance of a network. In [13], the authors described a breadth-first tree using a clustering algorithm to find the gateway which is at low load. The drawback of this algorithm is the time of execution while calculating large computations. Authors in [14] build a load-balanced CDS using GA and load-balanced allocation of a sensor node to cluster heads. Before this, a limited study has been done to achieve a load-balanced CDS WSN. The problem that is to be observed in this algorithm is high energy dissipation and slow convergence rate. LEACH-C with GA has been proposed later to minimize energy dissipation for the entire network. For congestion management of wireless sensor networks, a load-balancing protocol inspired by GA [15] has been introduced. A fuzzy-based PSO routing algorithm was proposed by the authors in [16] to balance the load and scalability of the network. The authors improved three important factors like density, energy consumption, and transmission distance by using FBPSO. They also balanced the cluster heads' energy dissipation and increased the lifespan of WSNs. The choice of cluster head and network stability was not considered by them.

Authors in [17] proposed a data dissemination strategy using nature-inspired ant colony optimization (ACO) named TMLBS. This paper identified three load balancing systems, which are the system of load decentralization, the system of maintenance, and the system of load diversion. The authors tried to create multiple transmission paths for the entire network at different places. In these schemes, they implemented multipath subtrees to avoid excessive load, pheromone update mechanism, and paths with low data load, respectively. But they did not focus on the lifetime of the network and convergence speed. New nature-inspired techniques have been proposed by authors in [18] biogeography-based optimization (BBO-C and BBO-CR) for WSN energy production. This leads to better sensor distribution and a well-balanced method of clustering. But there is no provision for the BBO-C and BBO-CR to pick the best members from each generation. The resulting fitness solution could not be concluded when too many alternatives were produced.

The authors in [19], proposed DBSCDS-GWO and DBSC-GWO to calculate the distance of transmission and balance the load on the network. A selection of dominator nodes is deterministically chosen by DBSCDS-GWO, and the meta-heuristic method GWO is used to find the optimal location. Authors have also tried to minimize the effective transmission distance (ETD) parameter, as follows:

$$\text{ETD} = \Phi \text{ or } \eta(\text{whichever is shorter}) \tag{1}$$

Authors achieved stability and balanced the load of the network up to an extent, but GWO has a bad local searching ability, low solving precision, and low speed of convergence.

2.3 Security Enhancement in WSN

In the past years, a large increment has been seen in the field of WSNs devices. These devices are connected wirelessly with each other and communicate wirelessly without any human intervention. During transmission of data or information from sensor node to sink, security can be compromised. WSN protection is very important. Otherwise, it can lead to significant losses of property and life as well. Data or information can be modified or removed by the attackers to break the security of WSNs. Attackers try some specific attacks (active or passive) to alter the specifications of the network. These attacks are classified according to the layers of WSNs. Each wireless environment consists of the following four layers on which different types of attacks are imposed in [20–23].

To protect these layers from attacks and alterations, there is a need to follow some security protocols. So, three main factors need to describe the security of WSNs, which are:

- **Attacks in WSNs**: There are primarily two kinds of attacks: active (aggressive) attacks and passive attacks. Active attacks are DOS attack, physical attack, routing attack, node replication attack, node outage attack, node malfunction attack, false node, etc. Passive attacks are traffic monitoring, traffic analysis, and eavesdropping.
- **Security goals of WSNs**: WSN security goals are divided into two parts, i.e., primary and secondary goals. Primary goals are mainly confidentiality, authentication, integrity, and availability. Secondary goals are secure localization and resilience to attacks.
- **Security protocols for WSNs**: Authors have identified various types of protection in [24, 25]. WSN protocols based on SPIN, line-selected multicast (LSM) [26], randomized efficient and distributed (RED) [25], CRS-A [27], logical key hierarchy-based model (LKHM) [28], localized encryption and authentication protocol (LEAP), certificate-less effective key management (CL-EKM) [29], low-storage clone detection (LSCD) [30].

2.3.1 Security in IoT

IoT is the upcoming era of communication. According to recent reports, there will be around 9.5 billion connected devices in 2020. IoT applications are rapidly increasing all over the world but are majorly used in North America, Western Europe, and China. To expand the digital economy, IoT is the major upcoming market, according to data declares in reports. Recent and existing applications of IoT are highly promising and more efficient for the users, but high-level privacy and authentication for these applications are required. Existing solutions using different technologies like machine learning, fog computing, edge computing, and blockchain are implemented for high-level security and privacy in IoT, as discussed in Table 1.

Table 1 Comparison of security techniques in IoT

Blockchain technique-based solutions	Fog computing-based solutions	Machine learning-based solutions	Edge computing-based solutions
1. WSNs are vulnerable to exposing private data. Then, by using permission blockchain, WSNs can be secured 2. Moving toward decentralization using blockchain, exponential growth can be handled in WSNs 3. Data available on different nodes of the network by blockchain technique. Hence, no distinct failed node is observed	1. Verifiable computation is proposed by fog computing 2. Server-aided exchange and data analytics provided by fog computing 3. Sharing data securely and designing protocols are also provided by fog computing 4. Identity recognition and access management are also done by using fog computing	1. Machine learning techniques help to secure the network from attacks and in increasing accuracy 2. Machine learning helps us to prevent the devices from eavesdropping 3. Machine learning helps us to increase detection and classification accuracy 4. The techniques of machine learning also help reduce the false alarm rate and the average error rate	1. Edge computing eliminates the transit of data and thus avoids the possibility of data theft and data breaches 2. Organizations can manage the data within their boundaries using edge computing to ensure data enforcement 3. Using edge computing, surveillance cameras can also be empowered to achieve faster response time

3 Comparison of Evolutionary Algorithms

3.1 Genetic Algorithm (GA)

The authors in [31] suggested GA for distance-conscious routing. They used this algorithm to choose the head of the cluster and to achieve the calculation of fitness. Authors in [32] used GA for node deployment. They found efficient positions for sensor nodes by using this algorithm. The authors mainly focused on three factors in this paper which are to minimize no. of sensor nodes, maximize connectivity and maximize coverage region. Authors in [33, 34] proposed GA for routing of WSN. In choosing the cluster head, they considered three weaknesses and achieved a prolonged network lifetime. Authors in [35] used GA for load balancing of the network. They also identified a prolonged lifetime of the network after simulation. Authors in [36, 37] GA is used to schedule the set cover problem and to minimize the exposure path. They scheduled the sensors for better conservation of energy and approached redundant trend scheduling. They also implemented a cross-over operator for exploitation purposes and an upside-down operator to avoid premature convergence. In order to find a new generation with better fitness appeal, GA effectively explores the new combinations with the information available.

3.2 *Particle Swarm Optimization (PSO)*

Since its introduction in 1995, particle swarm optimization (PSO) [38] has undergone many improvements. They have derived new models, developed new implementations, and published theoretical studies on the effects of different parameters and aspects of the algorithm as researchers have learned about the technique. Authors in [7, 39] proposed PSO for routing of the wireless sensor network. They used an unsupervised learning technique to achieve efficient routing between sensor nodes. They also used PSO to find routing paths for data transfer and to balance the load of WSN. Using PSO, the authors also found optimal solutions by implementing double-layer encoding. Authors in [40] proposed discrete PSO to overcome the scheduling problem in PSO. The proposed model of discrete PSO improved task allocation and utilization of available resources. Authors in [41, 42] proposed a new methodology to reduce the overhead of CH named cluster assistant node (CAN). They also focused on coverage area and node positioning of sensor nodes under CHs to minimize energy dissipation. In recent work, it is observed that PSO has few parameters to adjust and short computational time.

3.3 *Ant Colony Optimization (ACO)*

Authors in [1] proposed ACO for the deployment of sensor nodes. They achieved maximum coverage by allocating the nodes using a probability distribution. Another technique, greedy migration, was also implemented by authors for fast convergence. Authors in [43] proposed ACO in terms of QoS-based routing. They considered energy consumption and network performance for finding a highly efficient route from sensor nodes to sink. They achieved a better ratio of packet delivery and low consumption of energy by using this algorithm. Authors in [44] proposed ACO for routing based on trust values of sensor nodes. The authors also considered number of hopes and re-laid nodes for balance the network and secure transfer of data. In recent research, it is found that ACO can be used in complex applications. Changes such as new distances may be adjusted. It can also perform simultaneous searches among a population.

3.4 *Biogeography-Based Optimization (BBO)*

BBO is an evolutionary algorithm that is applied to preserve population diversity through the transfer of species from one habitat to another habitat. A person is represented by habitat in the BBO. There can be multiple environments in a group, such as chromosomes in genetic algorithms. Habitat suitability index (HSI) values are determined for every habitat. In this way, several changed vectors/habitats allocated to sensor nodes will be obtained. For all these resulting ecosystems, measure HSI values again. Retain the best HSI value for the habitat. If after adjustment, the HSI value is higher than the HSI value before modification, the output of the batter network is achieved. The entire situation is then simulated and optimal efficiency is achieved. In [45], the authors suggested the algorithm (MSDR-BBO) to pick the optimal routing CHs in MSWSN according to the data transfer requirements of SNs and CHs.

3.5 Gray Wolf Optimization (GWO)

Authors in [46] proposed recent work (2014–2017) based on GWO. They described GWO based on wolf hunting behavior which shows their domination while hunting their food. The key phases of hunting, which track, chase, and approach the prey, were also explained by the authors. They found problems such as cost estimation, parameter tuning, economic dispatch problems, and classification of genes in this research. GWO is also capable of addressing both single and multi-objective questions. To solve the distinct optimization problem, improved and hybrid GWO are increasing. A hybrid protocol MLHP was proposed by the authors in [47] to centralize selection processes for the selection of suitable CH, and deterministic selection at levels two and three. The algorithm used in this paper performed 500 times better as compare to LEACH in terms of the lifetime of WSN.

Authors in [49, 50] focused on localization problems and energy utilization of WSNs. They also achieved the fitness criteria by which nodes consumed desired energy. Network lifetime, stability, residual energy, and performance are considered during the entire implementation of the methodology. In a recent study, it is observed that GWO provides automatic configuration of the set of parameters in a manner to balance the load. It also provides more accuracy instability and energy efficiency as compared to previously used algorithms. The comparison and contribution of previously used evolutionary algorithms are described in Table 2.

4 Conclusion and Future Directions

The main purpose of this paper is to review various algorithms and clustering routing protocols used to improve the energy efficiency of WSNs. Recently, many algorithms are described in WSN in order to improve the stability and lifetime of WSN. Now, a better methodology or hybrid approach is required to increase energy efficiency. Hence, this paper elaborates various evolutionary algorithms and clustering routing protocols to enhance the energy efficiency in WSN. This survey also describes the security techniques of IoT in different scenarios. An introduction on energy efficiency, load balancing, and security of WSNs has been described. Some

Table 2 Comparison of evolutionary algorithms [48]

Algorithms	Contribution
GA	1. Used to find the efficient location of sensor nodes and CHs 2. Used for whole area coverage by using less number of sensor nodes 3. Used to prolong the network lifetime and exploitation purposes
PSO	1. Used for exploration and perform better than GA 2. Used routing to minimize the number of hops and efficient data transfer 3. Used double-layer encoding to find the optimal solution
ACO	1. Used probability distribution to achieve maximum coverage for the network 2. Used multiple constraints for multi-hop routing 3. Achieved better packet delivery ratio and energy consumption
BBO	1. The introduced phenomenon of extinction and evolution of species 2. Used nonlinear mapping used to enhance the stability of the network 3. Achieved gateways optimum location faster to reduce energy dissipation
GWO	1. Achieved a more stable and energy-efficient network using DBSCDS-GWO and DBSC-GWO 2. Used to reduce effective transmission distance (ETD) 3. Overall performance of GWO is better than previously used algorithms

advantages and challenges for evolutionary algorithms are explained in the paper to optimize future problems. Table 1 shows a comparison of security techniques-based solutions for IoT. The review paper mainly comprises of the research articles published in the past few years and the details of many other surveys are also described here. In the future, the hybridization technique in the above-discussed algorithms can be implemented to obtain better results. This paper also can be expanded in the future for the hybridization of evolutionary algorithms.

References

1. Liu, X., He, D.: Ant colony optimization with greedy migration mechanism for node deployment in wireless sensor networks. J. Netw. Comput. Appl. **39**, 310–318 (2014). https://doi.org/10.1016/j.jnca.2013.07.010
2. Heinzelman, W.R., Chandrakasan, A., Balakrishnan, H.: Energy-efficient communication protocol for wireless microsensor networks. In: Proc. Annu. Hawaii Int. Conf. Syst. Sci. 2000-Janua, 1–10 (2000)
3. Gupta, I., Riordan, D., Sampalli, S.: Cluster-head election using fuzzy logic for wireless sensor networks. In: Proc. 3rd Annu. Commun. Networks Serv. Res. Conf. 2005, pp. 255–260 (2005). https://doi.org/10.1109/CNSR.2005.27
4. Tamandani, Y.K., Bokhari, M.U.: SEPFL routing protocol based on fuzzy logic control to extend the lifetime and throughput of the wireless sensor network. Wirel. Netw. **22**, 647–653 (2016). https://doi.org/10.1007/s11276-015-0997-x
5. AbdulAlim, M.A., Wu, Y.C., Wang, W.: A fuzzy based clustering protocol for energy-efficient wireless sensor networks. Adv. Mater. Res. 760–762, 685–690 (2013). https://doi.org/10.4028/www.scientific.net/AMR.760-762.685

6. Gupta, S.K., Kuila, P., Jana, P.K.: GAR: an energy efficient GA-based routing for wireless sensor networks. In: Lect. Notes Comput. Sci. (including Subser. Lect. Notes Artif. Intell. Lect. Notes Bioinformatics). 7753 LNCS, pp. 267–277 (2013). https://doi.org/10.1007/978-3-642-36071-8_21
7. Kuila, P., Jana, P.K.: Energy efficient clustering and routing algorithms for wireless sensor networks: particle swarm optimization approach (2014)
8. Blum, C., López-Ibáñez, M.: Ant colony optimization. Intell. Syst. (2016). https://doi.org/10.4249/scholarpedia.1461
9. Kaushik, A., Indu, S., Gupta, D.: Optimizing and enhancing the lifetime of a wireless sensor network using biogeography based optimization. Commun. Comput. Inf. Sci. **899**, 260–272 (2019). https://doi.org/10.1007/978-981-13-2035-4_23
10. Zhang, S., Xu, S., Zhang, W., Yu, D., Chen, K.: A hybrid approach combining an extended BBO algorithm with an intuitionistic fuzzy entropy weight method for QoS-aware manufacturing service supply chain optimization. Neurocomputing **272**, 439–452 (2018). https://doi.org/10.1016/j.neucom.2017.07.011
11. Shukla, K., Verma, M., Gupta, D.: Age-Structured Biogeography-based Optimization, pp. 339–346 (2020)
12. Kaushik, A., Indu, S., Gupta, D.: A grey wolf optimization approach for improving the performance of wireless sensor networks. Wirel. Pers. Commun. 1429–1449 (2019). https://doi.org/10.1007/s11277-019-06223-2
13. Low, C.P., Fang, C., Ng, J.M., Ang, Y.H.: Efficient load-balanced clustering algorithms for wireless sensor networks. Comput. Commun. **31**, 750–759 (2008). https://doi.org/10.1016/j.comcom.2007.10.020
14. He, J., Ji, S., Yan, M., Pan, Y., Li, Y.: Load-balanced CDS construction in wireless sensor networks via genetic algorithm. Int. J. Sens. Netw. **11**, 166–178 (2012). https://doi.org/10.1504/IJSNET.2012.046331
15. Raha, A., Kanti Naskar, M., Paul, A., Chakraborty, A., Karmakar, A.: A genetic algorithm inspired load balancing protocol for congestion control in wireless sensor networks using trust based routing framework (GACCTR). Int. J. Comput. Netw. Inf. Secur. **5**, 9–20 (2013). https://doi.org/10.5815/ijcnis.2013.09.02
16. Balaji, S., Julie, E.G., Rajaram, M., Robinson, Y.H.: Fuzzy based particle swarm optimization routing technique for load balancing in wireless sensor networks. Int. J. Comput. Inf. Eng. **10**, 1418–1427 (2016)
17. Liu, X., Qiu, T., Wang, T.: Load-balanced data dissemination for wireless sensor networks: a nature-inspired approach. IEEE Internet Things J. **6**, 9256–9265 (2019). https://doi.org/10.1109/JIOT.2019.2900763
18. Kaushik, A., Indu, S., Gupta, D.: A novel load balanced energy conservation approach in WSN using biogeography based optimization. AIP Conf. Proc. **1884**, 1–4 (2017). https://doi.org/10.1063/1.5002507
19. Bozorg-Haddad, O.: Studies in Computational Intelligence—Advanced Optimization by Nature-Inspired Algorithms (2018)
20. Lghd, V., Wkh, D., Fkdoohqjhv, I., Dqg, J., Glvdvwhu, F., Lqirupdwlrq, E., Dqg, F., Lpsruwdqfh, P., Wkh, V.R., Fkdudfwhuv, X., Wkhvh, R.I., Wkh, D.Q.G., Sxusrvhv, D., Xvhg, D.U.H., Surylgh, W.R., Iru, W., Dqg, L., Dwwdfnv, R., Wkhuh, Q.W., Whupv, D.U.H., Whup, Q., Xvhg, L. V, Wkh, D.L., Wkh, S.: $ 6xuyh\ rq 6hfxulw\ $wwdfnv lq :luhohvv 6hqvru 1hwzrunv. 536–539 (2016)
21. Wang, Q., Zhang, T.: A survey on security in wireless sensor networks. Secur. RFID Sens. Networks. 293–320 (2016). https://doi.org/10.5121/ijnsa.2017.9103
22. Karakaya, A., Akleylek, S.: A survey on security threats and authentication approaches in wireless sensor networks. In: 6th Int. Symp. Digit. Forensic Secur. ISDFS 2018—Proceeding. 2018-Janua, 1–4 (2018). https://doi.org/10.1109/ISDFS.2018.8355381
23. Teymourzadeh, M., Vahed, R., Alibeygi, S., Dastanpor, N.: Security in Wireless Sensor Networks: Issues and Challenges. arXiv (2020)

24. Perrig, A., Szewczyk, R., Tygar, J.D., Wen, V., Culler, D.E.: SPINS: security protocols for sensor networks. Wirel. Netw. **8**, 521–534 (2002). https://doi.org/10.1023/A:1016598314198
25. Conti, M., Di Pietro, R., Mancini, L., Mei, A.: Distributed detection of clone attacks in wireless sensor networks. IEEE Trans. Dependable Secur. Comput. **8**, 685–698 (2011). https://doi.org/10.1109/TDSC.2010.25
26. Gligor, V.D.: Emergent properties in ad-hoc networks: a security perspective. In: Proc. 2006 ACM Symp. Information, Comput. Commun. Secur. ASIACCS '06. 2006, 1 (2006). https://doi.org/10.1145/1128817.1128819
27. Ren, J., Zhang, Y., Zhang, K., Shen, X.: Adaptive and channel-aware detection of selective forwarding attacks in wireless sensor networks. IEEE Trans. Wirel. Commun. **15**, 3718–3731 (2016). https://doi.org/10.1109/TWC.2016.2526601
28. Di Pietro, R., Mancini, L. V., Law, Y.W., Etalle, S., Havinga, P.: LKHW: a directed diffusion-based secure multicast scheme for wireless sensor networks. In: Proc. Int. Conf. Parallel Process. Work. 2003-Janua, pp. 397–406 (2003). https://doi.org/10.1109/ICPPW.2003.1240395
29. Seo, S.H., Won, J., Sultana, S., Bertino, E.: Effective key management in dynamic wireless sensor networks. IEEE Trans. Inf. Forensics Secur. **10**, 371–383 (2015). https://doi.org/10.1109/TIFS.2014.2375555
30. Dong, M., Ota, K., Yang, L.T., Liu, A., Guo, M.: LSCD: a low-storage clone detection protocol for cyber-physical systems. IEEE Trans. Comput. Des. Integr. Circuits Syst. **35**, 712–723 (2016). https://doi.org/10.1109/TCAD.2016.2539327
31. Bhatia, T., Kansal, S., Goel, S., Verma, A.K.: A genetic algorithm based distance-aware routing protocol for wireless sensor networks. Comput. Electr. Eng. **56**, 441–455 (2016). https://doi.org/10.1016/j.compeleceng.2016.09.016
32. Gupta, S.K., Kuila, P., Jana, P.K.: Genetic algorithm approach for k-coverage and m-connected node placement in target based wireless sensor networks. Comput. Electr. Eng. **56**, 544–556 (2016). https://doi.org/10.1016/j.compeleceng.2015.11.009
33. Shokouhifar, M., Jalali, A.: A new evolutionary based application specific routing protocol for clustered wireless sensor networks. AEU Int. J. Electron. Commun. **69**, 432–441 (2015). https://doi.org/10.1016/j.aeue.2014.10.023
34. Khalesian, M., Delavar, M.R.: Wireless sensors deployment optimization using a constrained Pareto-based multi-objective evolutionary approach. Eng. Appl. Artif. Intell. **53**, 126–139 (2016). https://doi.org/10.1016/j.engappai.2016.03.004
35. Pal, V., Yogita, Singh, G., Yadav, R.P.: Cluster head selection optimization based on genetic algorithm to prolong lifetime of wireless sensor networks. Procedia Comput. Sci. **57**, 1417–1423 (2015). https://doi.org/10.1016/j.procs.2015.07.461
36. Zhang, X.Y., Zhang, J., Gong, Y.J., Zhan, Z.H., Chen, W.N., Li, Y.: Kuhn-Munkres parallel genetic algorithm for the set cover problem and its application to large-scale wireless sensor networks. IEEE Trans. Evol. Comput. **20**, 695–710 (2016). https://doi.org/10.1109/TEVC.2015.2511142
37. Ye, M., Wang, Y., Dai, C., Wang, X.: A hybrid genetic algorithm for the minimum exposure path problem of wireless sensor networks based on a numerical functional extreme model. IEEE Trans. Veh. Technol. **65**, 8644–8657 (2016). https://doi.org/10.1109/TVT.2015.2508504
38. Kiranyaz, S.: Particle swarm optimization. Adapt. Learn. Optim. **15**, 45–82 (2014). https://doi.org/10.1007/978-3-642-37846-1_3
39. Taherian, M., Karimi, H., Kashkooli, A.M., Esfahanimehr, A., Jafta, T., Jafarabad, M.: The design of an optimal and secure routing model in wireless sensor networks by using PSO algorithm. Procedia Comput. Sci. **73**, 468–473 (2015). https://doi.org/10.1016/j.procs.2015.12.028
40. Guo, W., Li, J., Chen, G., Niu, Y., Chen, C.: A PSO-optimized real-time fault-tolerant task allocation algorithm in wireless sensor networks. IEEE Trans. Parallel Distrib. Syst. **26**, 3236–3249 (2015). https://doi.org/10.1109/TPDS.2014.2386343

41. Rejinaparvin, J., Vasanthanayaki, C.: Particle swarm optimization-based clustering by preventing residual nodes in wireless sensor networks. IEEE Sens. J. **15**, 4264–4274 (2015). https://doi.org/10.1109/JSEN.2015.2416208
42. Yan, Z., Goswami, P., Mukherjee, A., Yang, L., Routray, S., Palai, G.: Low-energy PSO-based node positioning in optical wireless sensor networks. Optik (Stuttg). **181**, 378–382 (2019). https://doi.org/10.1016/j.ijleo.2018.12.055
43. Wang, Y.L., Song, M., Wei, Y.F., Wang, Y.H., Wang, X.J.: Improved ant colony-based multi-constrained QoS energy-saving routing and throughput optimization in wireless Ad-hoc networks. J. China Univ. Posts Telecommun. **21**, 43–53 (2014). https://doi.org/10.1016/S1005-8885(14)60267-3
44. Sun, Z., Wei, M., Zhang, Z., Qu, G.: Secure routing protocol based on multi-objective ant-colony-optimization for wireless sensor networks. Appl. Soft Comput. J. **77**, 366–375 (2019). https://doi.org/10.1016/j.asoc.2019.01.034
45. Kaushik, A., Indu, S., Gupta, D.: Adaptive mobile sink for energy efficient WSN using biogeography-based optimization. Int. J. Mob. Comput. Multimed. Commun. **10**, 1–22 (2019). https://doi.org/10.4018/IJMCMC.2019070101
46. Hatta, N.M., Zain, A.M., Sallehuddin, R., Shayfull, Z., Yusoff, Y.: Recent studies on optimisation method of Grey Wolf Optimiser (GWO): a review (2014–2017). Artif. Intell. Rev. **52**, 2651–2683 (2019). https://doi.org/10.1007/s10462-018-9634-2
47. Al-Aboody, N.A., Al-Raweshidy, H.S.: Grey Wolf optimization-based energy-efficient routing protocol for heterogeneous wireless sensor networks. In: 2016 4th Int. Symp. Comput. Bus. Intell. ISCBI 2016, pp. 101–107 (2016). https://doi.org/10.1109/ISCBI.2016.7743266
48. Balasubramanian, D., Govindasamy, V.: Study on evolutionary approaches for improving the energy efficiency of wireless sensor networks applications. EAI Endorsed Trans. Internet Things. **5**, 164856 (2020). https://doi.org/10.4108/eai.13-7-2018.164856
49. Rajakumar, R., Amudhavel, J., Dhavachelvan, P., Vengattaraman, T.: GWO-LPWSN: grey wolf optimization algorithm for node localization problem in wireless sensor networks. J. Comput. Networks Commun. **2017** (2017). https://doi.org/10.1155/2017/7348141
50. Sharawi, M., Emary, E.: Impact of grey Wolf optimization on WSN cluster formation and lifetime expansion. In: 9th Int. Conf. Adv. Comput. Intell. ICACI 2017, pp. 157–162 (2017). https://doi.org/10.1109/ICACI.2017.7974501

Utilization and Energy Consumption Optimization for Cloud Computing Environment

Rajeev Tiwari, Roohi Sille, Nilima Salankar, and Pardeep Singh

Abstract In a cloud environment, the workload that has to be maintained using visualization is limited by the available hardware resources of virtual machines (VMs). So utilization of VMs becomes significant to do more work with lesser infrastructure. Thus, in recent times, major thrust was shown by researchers in the field of task allocation algorithms on VMs. There are many techniques discussed in the literature, which uses different allocation methods, which can improve the performance by changing the working of cloud environment. In this research work, analysis, implementation and performance comparison of the existing allocation techniques have been performed using CloudSim. So performance tuning is being done analytically and practically for the task allocation algorithm. VMs and cloudlets are configured for experimental purposes and parameter results are obtained. Parameters recorded are execution time, makespan, utilization ratio and power consumption. A new algorithm is proposed for task allocation algorithm (Tiwari et al in Int J Adv Intell Syst Comput, 2016 (Tiwari and Kumar in Telecommun. Syst. 62:149–165, 2016)). These parameters are calculated for FCFS, SJF, Hungarian and the proposed algorithm. Then, result analysis is done and majorly got a speedup in utilization ratio of the proposed algorithm *w.r.t.* to FCFS as 53.20%, 18.08% *w.r.t.* to SJF and 10.52% *w.r.t.* to Hungarian. For power consumption, the algorithm has shown a significant decrease in power consumption from 37.21, 16.52 and 10.52% *w.r.t.* to FCFS, SJF and Hungarian algorithms.

Keywords Cloud environment · Optimization · Utilization rate · Energy consumption · Virtualization · Speedup percentages

---

R. Tiwari (✉) · R. Sille · N. Salankar · P. Singh (✉)
School of Computer Science, UPES, Dehradun, India
e-mail: rajeev.tiwari@ddn.upes.ac.in

P. Singh
e-mail: pardeep.singh@ddn.upes.ac.in

K. Khanna et al. (eds.), *Cyber Security and Digital Forensics*,
Lecture Notes on Data Engineering and Communications Technologies 73,
https://doi.org/10.1007/978-981-16-3961-6_50

1 Introduction

Cloud computing [1–3] is enabling any infrastructure, development environment and software as a services on-demand basis. These are provisioned rapidly through web enablement using a shared pool of resources like compute, storage, network etc. with minimum manual intervention. This is considered as resource management, which is a current research topic in cloud [4]. Any application can be deployed in cloud environment, and this environment is a complex system that works with a number of configurations requirements. For researches and testing purposes, a tool is created called CloudSim [5] where any workload that needs to be executed is termed as cloudlet. It needs to be allocated on virtualized hardware through virtual machines [6]. Any application service that needs to be executed is to be allocated on virtual machines in such a way that load can be managed dynamically and provisioned rapidly. This application allocation policy of cloudlet on virtual machine is termed as task allocation policy (TAP) [7, 8]. There are many number of TAPs available for different metrics like work load, utilization, energy consumption, more throughput, etc.

Using TAP in a cloud environment, executions can be done fast, and throughput can be increased significantly [8], with proper utilization of resources. So most of the researches are going in the direction of optimization of performance parameters for cloud environment.

Other most common and effective method of performance tuning for such environments is in network caching, where caches are deployed on nodes for prefetching and forwarding required data. So optimization in terms of power consumption and utilization is also given by authors in ACIT [9] and ECCI [10] techniques.

2 Literature Review

There are a number of scheduling algorithms being proposed by researchers in history. A few are being discussed in the coming sections. First scheduling algorithm is famous for its simplicity, i.e. first come first serve scheduling algorithm (FCFS), shortest job fastest resource (SJFR), longest job fastest resource (LJFR), min–min algorithm and max–min algorithms are evolved as per their need. As these algorithms consider all processes of same priority to execute at same time, while this is not the case in in real scenarios. So different cloudlets with workloads may have different significance and priority, so another pool of algorithms is designed. Some algorithm [11] offers first preference to a high-complexity job, and other has given high priority to level of parallelism. However, due to preference of high priority jobs by these algorithms, there is a limitation of increase in waiting time for low priority jobs.

First come first serve (FCFS) [12] is the simplest version of task allocation algorithms. Processes are allocated to VMs as per their order of arrival in queue. It is a non-preemptive algorithm. FCFS is the default technique for task allocation scheduling. But due to its simplicity, waiting time of processes may increase rapidly. It considers all processes as equal, but process or workload may be of different significance, so it may exhibit losses in business.

In Shortest Job First (SJF) algorithm [11, 13] or Shortest-Process-Next (SPN) algorithm, out of number of work processes waiting for execution, a smallest length cloudlet process is selected to be allocated to available virtual machine. It is a non-preemptive algorithm. The SJF algorithm is suitable for batch jobs, with predefined execution times. Since this type of scheduling results in the lower average time, it is optimal in nature. However, the algorithm should have a prior information of time required by each cloudlet for executing completely. This prediction about execution time of each cloudlet prior to its actual execution is a major task and is an overhead to system. Then at times, estimation can be erroneous if some new type of workload is to be executed whose history is not well known.

The Hungarian algorithm [14] uses combinatorial optimization for allocating cloudlets on VMs. This can optimize the whole task execution, proposed by Harold Kuhn. Another advancements done in the same field by James Munkres which found that it is strongly polynomial. So using combinations, it can solve any non-polynomial problem in polynomial times, so this was an optimization achieved in utilization of resources, and their efficiency in execution. In CloudSim, the Hungarian method is used for having least cost for whole cloudlet execution for an objective to attain with whole VMs to run effectively [15]. Then, some recent research works are also done which optimize the cloud performance on time constraints as given in [16]. Another optimization approach that is suitable for clouds is using Cournot and Bertrand games utilized for resource management in the federated cloud as proposed in [17].

3 Problem Formulation

In a cloud environment, hardware is associated with virtual machines (VM) which are the working units in cloud environment. Task allocation algorithms are mechanisms for allocating workloads on these VMs. A number of algorithms and techniques are discussed in previous section. For the main theme of cloud computing, resource utilization must be maximized for better worth of infrastructure of VMs, as per their capabilities. Optimization of servers to lessen the energy resources and proper tuning of energy usage in these active resources is the prime concern nowadays. So this work has proposed an algorithm for tuning of energy consumed and the resource usage in task allocation algorithm. Optimization through this work may result in good performance for a cloud environment. It can be done by making modifications in parameters like execution time of the cloudlets, which in turn will affect makespan and may change throughput, energy consumed and utilization ratio of the hardware resources [18, 19].

3.1 Open Issues

According to research literature discussed in section 2 above, a number of research gaps are identified, in which direction future researches can be perused. Following are the areas of interest in this field.

- Power consumption is the main area in allocation algorithms for work; system should consume power efficiently.
- Cloud demands a dynamic and rapid provisioning of services through infrastructure. So faster response time is always a requirement for good cloud environment.
- Explored techniques have achieved lesser makespan but have utilized more power consumption.
- Makespan and execution time must be approaching to each other, lesser the gap better is the algorithm.
- The price/performance ratio helps estimate the cost and further design the service catalogue.

3.2 Contributions

1. As per problem statement, an objective function is defined with all assumptions and constraint fixed for scenario.
2. Designed algorithm is implemented on CloudSim.
3. Execution time of cloudlets is reduced.
4. Reduction in energy consumption is attained.
5. Then, performance of proposed work with respect to peer techniques is shown for execution time, makespan, utilization ratio and power consumption parameters.

3.3 Objective Function

For the optimization of the task scheduling technique, in the given cloud environment, the resources allotted need to be used to their maximum efficiency. For maximizing the utilization ratio (UR), we have proposed an objective function given below in equation no. 1, subject to constraints followed. Table 1 describes various symbols and their meaning used in this paper.

Table 1 Utilization ratio per VM in each technique

	FCFS	SJF	Hungarian	proposed
VM0	1	1	0.84	1
VM1	0.42	0.95	1	0.95
VM2	0.42	0.45	0.60	0.89

$$\text{Maximize}\{UR\}\text{Subject to:}\begin{cases} \text{ET: minimized} \\ MS - ET \to 0 \\ ET \propto \frac{1}{IT} \\ P_{\text{cons}}\text{: minimized} \\ S_{\text{up}}\text{: increased} \\ \text{POW}_{\text{rate}} \propto UR \end{cases} \tag{1}$$

where UR → utilization ratio, ET → execution time, MS → makespan, Pc → power consumed, Sup → speedup percentage, POW_{rate} → rate of power consumed, and IT → ideal time.

For maximization of utilization ratio, following constraints are being followed like: execution time needs to be minimized, difference between makespan and execution time should limit towards zero, i.e. makespan and execution times should be same. Execution time is inversely proportional to ideal time. P_{cons} should be minimized and S_{up} should be more for a suitable viable algorithm.

3.4 Problem Statement

For the main theme of cloud computing, resource utilization must be maximized for better worth of infrastructure of VMs, as per their capabilities. Due to the depletion of energy from the servers, proper energy tuning approach is required in server optimization, this is a scope of research in this area. So this work has implemented an algorithm for tuning of energy consumed and the resource usage in task allocation algorithm. Optimization of these techniques results in performance tuning of the cloud environment by affecting parameters such as execution time of the cloudlets, which in turn affects makespan and may change throughput, energy consumed and utilization ratio of the hardware resources.

3.5 Methodology

For having tuning of parameters for a task allocation algorithm, we have to design a new algorithm for task allocation. Our proposed algorithm was given in [20]. For implementing and testing the performance of this proposed algorithm with its peer algorithms, CloudSim tool [5, 21] is used. Its performance parameters are computed, noted and calculated for peer comparison on same platform. Techniques for comparison taken are FCFS, SJF and Hungarian matching.

4 Implementation and Output

In this section, implementation description of our algorithm and its peers is discussed. Then, its output is measured and given.

4.1 Proposed Solution

The proposed technique aims to tune the performance of the cloud environment by reducing the execution time of the cloudlets on VMs and the energy consumption of the system. For the implementation of our goal, we have considered the following performance parameters like execution time of the cloudlets, makespan, throughput of the system, utilization ratio of the hardware resources, rate of energy consumption of the system and total energy consumed by the system.

Overview Task allocation should be done in a cloud environment such that high performance can be achieved. For this requirement, a new algorithm is proposed. The proposed algorithm is a hybrid algorithm with a function for weights to allocate workload with VMs capabilities. It allocates the appropriately biased application with suitable VM. Like VMs with high memory are being offered with cloudlets of high memory requirements. VMs with more IOs are offered with cloudlets with more IOs requirements. So workload nature-wise biasing is being done by heuristic function. So proper utilization of VMs resources can be done.

Flowchart The proposed algorithm can be depicted using flowchart as shown in Fig. 1.

Output The output screen on implementing the proposed technique is shown in Fig. 2.

As a result utilization ratio is more and power consumption would be lesser. As a result, utilization ratio is more and power consumption would be less. Same can be evident from implementation and performance parameter values.

5 Performance Testing

The following performance parameters are taken into account:

- Execution time of the cloudlets
- Makespan

Fig. 1 Working flowchart

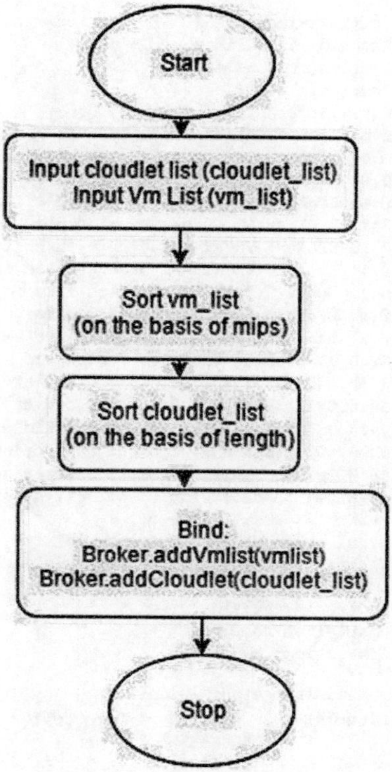

- Throughput of the system
- Utilization ratio of the hardware resources
- Rate of energy consumption of the system
- Total energy consumed by the system

All the techniques have been tested with the same cloudlet and VM configurations and compared for performance, and the results have been represented graphically [22].

5.1 Configurations of VMs and Cloudlets

The virtual machines and the cloudlets that have been used for experimentation have the configurations [22].

VMs are configured with MIPS of 250, 150 and 200 with a size of 10,000, RAM is 512, bandwidth is 1000, processing elements are 1000, and virtual machine manager is Xen hypervisor in each instance.

While **Cloudlets** are configured with one processing element, length is 250,000, 350,000 and 200,000, respectively. File size and output size parameters are both set to 300.

```
Initialising...
Starting CloudSim version 3.0
Datacenter_0 is starting...
Broker is starting...
Entities started.
0.0: Broker: Cloud Resource List received with 1 resource(s)
0.0: Broker: Trying to Create VM #0 in Datacenter_0
0.0: Broker: Trying to Create VM #2 in Datacenter_0
0.0: Broker: Trying to Create VM #1 in Datacenter_0
0.1: Broker: VM #0 has been created in Datacenter #2, Host #0
0.1: Broker: VM #2 has been created in Datacenter #2, Host #0
0.1: Broker: VM #1 has been created in Datacenter #2, Host #0
0.1: Broker: Sending cloudlet 2 to VM #0
0.1: Broker: Sending cloudlet 0 to VM #2
0.1: Broker: Sending cloudlet 1 to VM #1
1250.1: Broker: Cloudlet 0 received
1333.4333333333332: Broker: Cloudlet 1 received
1400.0973333333332: Broker: Cloudlet 2 received
1400.0973333333332: Broker: All Cloudlets executed. Finishing...
1400.0973333333332: Broker: Destroying VM #0
1400.0973333333332: Broker: Destroying VM #2
1400.0973333333332: Broker: Destroying VM #1
Broker is shutting down...
Simulation: No more future events
CloudInformationService: Notify all CloudSim entities for shutting down.
Datacenter_0 is shutting down...
Broker is shutting down...
Simulation completed.
Simulation completed.

========== OUTPUT ==========
Cloudlet ID    STATUS    Data center ID    VM ID    Time       Start Time    Finish Time
    0          SUCCESS        2              2      1250          0.1         1250.1
    1          SUCCESS        2              1      1333.33         0.1         1333.43
    2          SUCCESS        2              0      1400          0.1         1400.1
Proposed Algorithm  Executed!
```

Fig. 2 Output results from CloudSim

5.2 *Utilization Ratio*

The ratio of time taken by a cloudlet to execute completely to the total time a VM runs is known as the utilization ratio of that VM. The utilization ratio of each VM in the techniques implemented has been shown in Table 1.

5.3 *Power Consumption*

Studies [23, 24] have shown that on average, when an idle server is fully utilized, it consumes almost 70% of the total power consumed. Further, these studies additionally demonstrate that this power consumption can be accurately described by a linear relationship [25] between the CPU utilization (utilization ratio) and power consumption. Therefore, for our experimental studies, we define the power consumption as a function of the CPU utilization (Pow_{rate}):

$$\text{Pow}_{\text{rate}}(t) = i * P_{\max} + (1 + i) * P_{\max} * UR \; \text{Pow}_{\text{rate}}(t) = P_{\max} * (0.7 + 0.3 * UR)$$

where

I = power consumed by an idle server in fraction; UR = CPU utilization (utilization ratio) where $P_{\max}$ is assumed as:

250 mW/ms for VM0

200 mW/ms for VM1

230 mW/ms for VM2

Thus, total power consumed is shown in Fig. 3.

The rate of power consumption per VM for each technique considered can be seen in Table 2.

Due to variations in the workload, the utilization ratio may change over time and is, thus, defined as a function of the time: UR(t). Therefore, to define the total energy consumption (P_{consumed}) by a server we use [26]:

$$P_{\text{consumed}} \int_0^t \text{Pow}_{\text{rate}}(t)\text{d}t$$

Thus, to reduce the total energy consumption, we improve the utilization ratio.

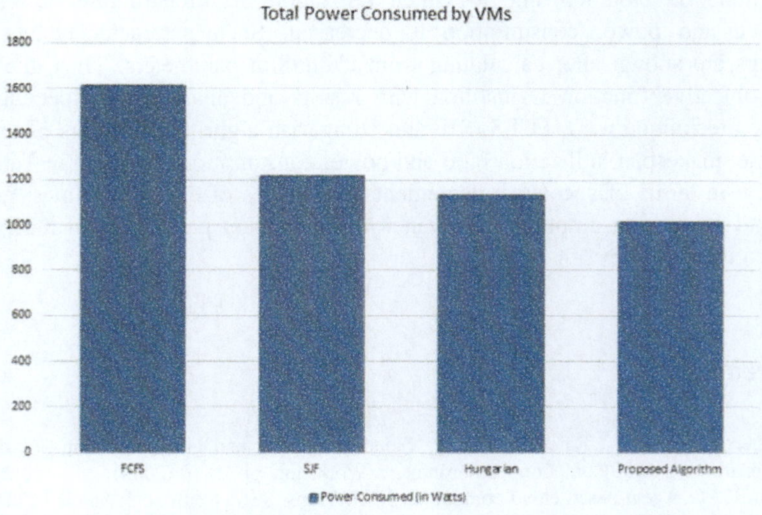

Fig. 3 Total power consumed by each technique

Table 2 Rate of power consumption per VM for each technique

FCFS SJF Hungarian proposed				
VM0	250	250	237.99	250
VM0	250	250	237.99	250
VM1	165.71	197.14	200	197.14
VM2	190.57	192.54	202.39	222.60

Table 3 Speedup of proposed algorithm versus FCFS, SJF and Hungarian for various parameters

FCFS			SJF Hungarian
Execution time (%)	8.07	5.53	2.04
Makespan (%)	39.99	20	16
Utilization ratio (%)	53.20	18.08	16.60
Power consumption (%)	37.21	16.52	10.52

6 Conclusion

In this paper, various task-scheduling algorithms are studied, mainly FCFS, SJF and Hungarian in cloud environment. A new task allocation algorithm is proposed for achieving higher utilization and lesser energy consumption. Due to the hybrid allocation function of proposed algorithm, appropriate allocations of workloads are being done on cloudlets due to which utilization of infrastructure of VM is increased and power consumption is decreased. Such parameter results and speedups are shown after calculating from CloudSim parameters. Thus, the proposed objective function is justified with results and analytically. Speedups of proposed technique *w.r.t.* FCFS, SJF and Hungarian algorithm in terms of execution time, makespan, utilization ratio and power consumption are given in Table 3.

While in terms of speedups, decrement percentages of execution times, makespan and power consumption are given while increment percentage of utilization ratio is given.

References

1. Foster, I., Zhao, Y., Raicu, I., Lu, S.: Cloud computing and grid computing 360-degree compared. In: Grid Computing Environments Workshop, pp. 1–10 (2008)
2. Sim, K.M.: Agent-based cloud computing. IEEE Trans. Serv. Comput. **5**, 564577 (2012)
3. Mell, P.M., Grance, T.: The NIST Definition of Cloud Computing. NIST special publication 800-145 (2011)
4. Wang, X., et al.: Online cloud resource prediction via scalable window waveform sampling on classified workloads. Future Gener. Comput. Syst. **117**, 338–358 (2021)
5. Beloglazov, A., Buyya, R.: CloudSim: a toolkit for modeling and simulation of cloud computing environments and evaluation of resource provisioning algorithms. Wiley Online Library (wileyonlinelibrary.com) (2010)

6. Theja, P.R., Khadar Babu, S.K.: Resource optimization for dynamic cloud computing environment: a survey. Int. J. Appl. Eng. Res. **9**(24) (2014). ISSN 0973-4562
7. Lei, J., et al.: Cloud task and virtual machine allocation strategy in cloud computing environment. In: Network Computing and Information Security, Second International Conference, NCIS Proceedings (2012)
8. Tayal, S.: Tasks scheduling optimization for the cloud computing systems. (IJAEST) Int. J. Adv. Eng. Sci. Technol. **5**(2) (2011)
9. Tiwari, R., Kumar, N.: An adaptive cache invalidation technique for wireless environments. Telecommun. Syst. **62**(1), 149–165 (2016)
10. Tiwari, R., Kumar, N.: Cooperative gateway cache invalidation scheme for internet-based vehicular adhoc networks. Wireless Pers. Commun. **85**(4), 1789–1814 (2015). https://doi.org/10.1007/s11277-015-2867-3
11. Rekha, S., Santhosh Kumar, R.: MJHP—job scheduling algorithm for cloud environment. Int. J. Tech. Res. Appl. (2014). e-ISSN: 2320-8163
12. Yahyapour, R., Schwiegelshohn, U.: Analysis of first-come-first-serve parallel job scheduling. In: Proceedings of the Annual ACM-SIAM Symposium on Discrete Algorithms (1998)
13. Hong, C., Caesar, M., Godfre, P.: Finishing flows quickly with preemptive scheduling. SIGCOMM Comput. Commun. Rev. (2012)
14. Kunh, H.W.: The Hungarian method for the assignment problem. Naval Res. Logistics Q. **1**, 83–97 (1955)
15. Kushang, P., et al.: Virtual machine allocation policy in cloud computing using CloudSim in Java. Int. J. Grid Distrib. Comput. (2015)
16. Halima, R.B., Kallel, S., Gaaloul, W., Maamar, Z., Jmaiel, M.: Toward a correct and optimal time-aware cloud resource allocation to business processes. Future Gener. Comput. Syst. **112**, 751–766 (2020)
17. Khorasani, N., Abrishami, S., Feizi, M., Esfahani, M.A., Ramezani, F.: Resource management in the federated cloud environment using Cournot and Bertrand competitions. Futur. Gener. Comput. Syst. **113**, 391–406 (2020)
18. Vijindra, Shenai, S.: Survey on scheduling issues in cloud computing. Procedia Eng. **38** (2012)
19. Endo, P.T.: Resource allocation for distributed cloud: concept and research challenges. IEE, 42–46
20. Tiwari, R., Lal, G, Goel, T.: Published paper "Performance tuning approach for cloud environment". Int. J. Adv. Intell. Syst. Comput. (2016)
21. Ranbhise, S.M., Joshi, K.K.: Simulation and analysis of cloud environment. Int. J. Adv. Res. Comput. Sci. Technol. (2014)
22. Majumdar, S.: Resource management on cloud: handling uncertainties in parameters and policies. CSI Commun. 16–19 (2011)
23. Fan, X., et al.: Power provisioning for a warehouse-sized computer. In: Proc. of the 34th Annual International Symposium on Computer Architecture, pp. 13–23 (2007)
24. Kusic, D., et al.: Power and performance management of virtualized computing environments via look ahead control. Clust. Comput. **12**(1), 1–15 (2009)
25. Beloglazov, A., Buyya, R.: Adaptive threshold-based approach for energy efficient consolidation of virtual machines in cloud data centers. In: Proceedings of the 8th International Workshop on Middleware for Grids, Clouds and E-Science—MGC '10 (2010)
26. Beloglazov, A., Buyya, R.: Energy efficient resource management in virtualized cloud data centers. In: 2010 10th IEEE/ACM International Conference on Cluster, Cloud and Grid Computing (2010)

Author Index

K. Khanna et al. (eds.), *Cyber Security and Digital Forensics*,
Lecture Notes on Data Engineering and Communications Technologies 73,
https://doi.org/10.1007/978-981-16-3961-6

Lecture Notes in Networks and Systems

Volume 291

The series "Lecture Notes in Networks and Systems" publishes the latest developments in Networks and Systems—quickly, informally and with high quality. Original research reported in proceedings and post-proceedings represents the core of LNNS.

Volumes published in LNNS embrace all aspects and subfields of, as well as new challenges in, Networks and Systems.

The series contains proceedings and edited volumes in systems and networks, spanning the areas of Cyber-Physical Systems, Autonomous Systems, Sensor Networks, Control Systems, Energy Systems, Automotive Systems, Biological Systems, Vehicular Networking and Connected Vehicles, Aerospace Systems, Automation, Manufacturing, Smart Grids, Nonlinear Systems, Power Systems, Robotics, Social Systems, Economic Systems and other. Of particular value to both the contributors and the readership are the short publication timeframe and the world-wide distribution and exposure which enable both a wide and rapid dissemination of research output.

The series covers the theory, applications, and perspectives on the state of the art and future developments relevant to systems and networks, decision making, control, complex processes and related areas, as embedded in the fields of interdisciplinary and applied sciences, engineering, computer science, physics, economics, social, and life sciences, as well as the paradigms and methodologies behind them.

Indexed by SCOPUS, INSPEC, WTI Frankfurt eG, zbMATH, SCImago.

All books published in the series are submitted for consideration in Web of Science.

More information about this series at http://www.springer.com/series/15179

João Manuel R. S. Tavares · Paramartha Dutta ·
Soumi Dutta · Debabrata Samanta
Editors

Cyber Intelligence and Information Retrieval

Proceedings of CIIR 2021

Springer

Editors
João Manuel R. S. Tavares
Faculdade de Engenharia
Universidade do Porto
Porto, Portugal

Paramartha Dutta
Department of Computer and System Sciences
Visva-Bharati University
Santiniketan, India

Soumi Dutta
Institute of Engineering and Management
Kolkata, India

Debabrata Samanta
Department of Computer Science
CHRIST (Deemed to be University)
Bengaluru, India

ISSN 2367-3370 ISSN 2367-3389 (electronic)
Lecture Notes in Networks and Systems
ISBN 978-981-16-4283-8 ISBN 978-981-16-4284-5 (eBook)
https://doi.org/10.1007/978-981-16-4284-5

This Springer imprint is published by the registered company Springer Nature Singapore Pte Ltd.
The registered company address is: 152 Beach Road, #21-01/04 Gateway East, Singapore 189721, Singapore

Preface

Since the most recent couple of decades, there has been a critical need of cyber intelligence and intelligent computing in various fields of science and innovation because of the impeccable addition in information. Improvement in science and innovation is legitimate to raise the inward quality of the nation. Be that as it may, exhibiting the flaws and procedures of progress with an unprejudiced post advantages to refresh the current information and advances. In the present circumstance, every one of the regions of related fields of information retrieval strategies presents versatile frameworks that approve the comprehension of hypothesis just as information in some troublesome and evolving situations.

The first international conference entitled "Computational Intelligence and Information Retrieval" (CIIR 2021) is organized by the Institute of Engineering and Management (IEM), Kolkata, West Bengal, India, on 20 and 21 May 2021. The conference is directed towards the knowledge and structure of positive research in different applications of cyber intelligence and information retrieval; those are leading and governing the technological domain. CIIR 2021 has attracted more than 170 submissions based on the thematic areas. With a thorough rigorous peer review process, the editors have chosen just 56 great papers by sophisticated knowledgeable domain expertise evaluators preferred from the country and abroad. The papers are reviewed with respect to the commitment, specialized substance, clearness and creativity of some most recent discoveries and research. The entire system (includes the submission of paper, review and acceptance process) is done electronically. We had a great time in collaboration with the advisory, programme and technical committees for call for papers, review and confirm papers for the proceedings of CIIR 2021.

The international conference CIIR 2021 goals at encircling new type of educators, technologists and researchers for global success. The conference features various keynote addresses in the scope of advanced information regarding CIIR 2021 topics. The sessions are largely categorized according to the significance and interdependency of the papers pertaining to the key concept of the conference, and they have established plentiful prospects for presentations.

The accepted papers (both research and review papers) have been estranged to highlight the modern spotlight of computational intelligence and information retrieval techniques. We anticipate the author's individual study, and estimations include value to it. Principal and foremost are the authors, columns and editorial whose workings have computed the conference an immense success. We require an enormous time putting together this proceedings. The CIIR 2021 conference is an acknowledgment to a vast collection of people, and everybody should feel proud for the outcome. We enlarge our deep sense of thankfulness to all those for this affectionate encouragement, motivation and support for building it attainable. Last but not least, the editorial members from Springer have helped us in an enormous manner, our significant legitimate appreciation to them is not just for assembly, and our apparition comes true in the form of proceeding, but likewise for its brilliant in-time and get-up productions in an agile manner.

Thanking all the authors who have chosen CIIR 2021 as the publication platform for their research, we would like to express our hope that their papers will help in further developments in the design and analysis of engineering aspects of complex systems, being a valuable source material for scientists, researchers, practitioners and students who work in these areas.

Porto, Portugal João Manuel R. S. Tavares
Santiniketan, India Paramartha Dutta
Kolkata, India Soumi Dutta
Bengaluru, India Debabrata Samanta

Contents

Editors and Contributors

About the Editors

Dr. João Manuel R. S. Tavares graduated in Mechanical Engineering at the Universidade do Porto, Portugal in 1992. He also earned his M.Sc. degree and Ph.D. degree in Electrical and Computer Engineering from the Universidade do Porto in 1995 and 2001, and attained his Habilitation in Mechanical Engineering in 2015. He is a senior researcher at the Instituto de Ciência e Inovação em Engenharia Mecânica e Engenharia Industrial (INEGI) and Full Professor at the Department of Mechanical Engineering (DEMec) of the Faculdade de Engenharia da Universidade do Porto (FEUP). João Tavares is co-editor of more than 60 books, co-author of more than 50 book chapters, 650 articles in international and national journals and conferences, and 3 international and 3 national patents. He has been a committee member of several international and national journals and conferences, is co-founder and co-editor of the book series "Lecture Notes in Computational Vision and Biomechanics" published by Springer, founder and Editor-in-Chief of the journal "Computer Methods in Biomechanics and Biomedical Engineering: Imaging & Visualization" published by Taylor & Francis, Editor-in-Chief of the journal "Computer Methods in Biomechanics and Biomedical Engineering" published by Taylor & Francis, and co-founder and co-chair of the international conference series: CompIMAGE, ECCOMAS VipIMAGE, ICCEBS and BioDental. Additionally, he has been (co-)supervisor of several MSc and PhD thesis and supervisor of several post-doc projects, and has participated in many scientific projects both as researcher and as scientific coordinator.

Dr. Paramartha Dutta is presently Professor in the Department of Computer and System Sciences of Visva Bharati University, West Bengal, India and did his Bachelors and Masters in Statistics from the Indian Statistical Institute, Calcutta, in the years 1988 and 1990, respectively. After that, he completed his Masters in Technology in Computer Science in 1993 from the same Institute. Prof. Dutta served in various research projects funded by the Government of India as Research Scientist in Indian Statistical Institute, Calcutta, and Bengal Engineering and Science University, Shibpur. Prof. Dutta has about 240 research papers published to his credit in different peer-reviewed international and national journals as well as proceedings of reputed international and national conferences along with a few book chapters. He has authored eight books and has thirteen edited books published. In recognition to his vibrant research capability, he has more than fifteen invited lectures/talks to his credit.

Dr. Soumi Dutta is Associate Professor at Institute of Engineering and Management, India. She has completed her Ph.D. (CST), IIEST, Shibpur. She received her B.Tech. (IT) and M.Tech. (CSE) securing 1st position (Gold medalist), from MAKAUT. She is certified as Publons Academy Peer Reviewer, 2020, and Certified Microsoft Innovative Educator, 2020. Her research interests are data mining, information retrieval, online social media analysis, and image processing. She has published many conferences and journal papers in Springer, IEEE, IGI Global, Taylor & Francis, etc. She has 2 Book Chapters in Taylor & Francis Group and IGI Global. She is a peer reviewer and TPC member in different international journals. She was the editor in CIPR2019, IEMIS2018 for 3 volumes, SI 2 volumes in IJWLTT, IEMIS 2020, CIPR 2020, and CIIR 2021. She is the member of several technical functional bodies such as IEEE, IFERP, MACUL, SDIWC, Internet Society, ICSES, ASR, AIDASCO, USERN, IRAN, and IAENG. She has published 3 patents.

Dr. Debabrata Samanta is presently working as Assistant Professor, Department of Computer Science, CHRIST (Deemed to be University), Bengaluru, India. He obtained his Ph.D. in Computer Science and Engineering from National Institute of Technology, Durgapur, India, in the area of SAR Image Processing. He is keenly interested in interdisciplinary research and development and has experience spanning fields of SAR image analysis, video surveillance, heuristic algorithm for image classification, deep learning framework for detection and classification, blockchain, statistical modelling, wireless ad hoc network, natural language processing, and V2I communication. He has published 9 books. He has edited 1 book available on GOOGLE BOOK server. He has authored and co-authored of 18 book chapters. He also serves as an acquisition editor for Springer, Wiley, CRC, Scrivener Publishing LLC, Beverly, USA, and Elsevier. He is Professional IEEE Member, Life Member of Computer Society of India (CSI) and Indian Society for Technical Education (ISTE).

Contributors

V. S. Abhijith Department of Information Technology, Thiagarajar College of Engineering, Madurai, Tamil Nadu, India

Ambuj Kumar Agarwal Chitkara University Institute of Engineering and Technology, Chitkara University, Punjab, India

Prashant Ahlawat Manipal University Jaipur, Jaipur, India

Md. Ezaz Ahmed Department of Computer Science, Saudi Electronic University, Al Madina, Kingdom of Saudi Arabia

Mostaq Ahmed Department of Computer Science and Engineering, Comilla University, Cumilla, Bangladesh

Saniya Zubair Ahmed Ansari Bachelor of Information Technology, Amity University, Dubai, UAE

Sabyasachi Bala Department of Engineering and Architecture, Amity University, Dubai, UAE

Hritam Banerjee Computer Science Engineering Department, University of Engineering and Management Kolkata, Kolkata, India

Rupayan Banerjee Institute of Engineering and Management, Calcutta, India

Barnali Gupta Banik Department of Computer Science and Engineering, Koneru Lakshmaiah Education Foundation, Deemed to be University, Hyderabad, Telangana, India

Swarnali Banik Department of Computer Science and Engineering, St. Mary's Technical Campus, Kolkata, India

Somnath Bera Department of Computer Science, Vidyasagar University, Midnapore, West Bengal, India

Abishek Bhattacharya Institute of Engineering and Management, Kolkata, India

Sujoy Bhattacharya Institute of Engineering and Management, Calcutta, India

Prajakta Bhosale Pimpri Chinchwad College of Engineering, Nigdi, Pune, Maharashtra, India

Debapriya Bhowmick Department of Computer Application and Science, Institute of Engineering and Management, Kolkata, India

Srimoyee Bhowmick Computer Science Engineering Department, University of Engineering and Management Kolkata, Kolkata, India

Sandipan Biswas Department of Computer Science and Engineering, Brainware University, West Bengal, India

Shreya Bose Computer Science Engineering Department, University of Engineering and Management, Kolkata, India

Partha Chakraborty Department of Computer Science and Engineering, Comilla University, Cumilla, Bangladesh

Arjama Chatterjee Institute of Engineering and Management, Kolkata, West Bengal, India

Siddhartha Chatterjee Department of Computer Science and Engineering, Gargi Memorial Institute of Technology, Kolkata, India

Anjuli Chaubey Dr. Harisingh Gour University, Sagar, India

Tanupriya Choudhury Department of Informatics, School of Computer Science, University of Petroleum and Energy Studies (UPES), Energy Acres, Dehradun, Uttarakhand, India

Abhinandan Das Department of Computer Application and Science, Institute of Engineering and Management, Kolkata, India

Prasenjit Das Chitkara University School of Computer Applications, Chitkara University, Solan, Himachal Pradesh, India

Prithwidip Das Institute of Engineering and Management, Kolkata, West Bengal, India

Soumik Das Institute of Engineering and Management, Kolkata, India

Debashri Das Adhikary Department of Computer Science, Vidyasagar University, Midnapore, West Bengal, India

Digbijoy Dasgupta Computer Science Engineering Department, University of Engineering and Management, Kolkata, India

Aditya Dawda Department of Computer Science and Engineering, MIT School of Engineering, MIT Arts Design and Technology University, Pune, India

Anasuya Dev Department of Computer Application and Science, Institute of Engineering and Management, Kolkata, India

Aditya Devchakke Department of Computer Science and Engineering, MIT School of Engineering, MIT Arts Design and Technology University, Pune, India

Tanushree Dey Department of Computer Science, Vidyasagar University, Midnapore, West Bengal, India

Namrata Dhanda Department of Computer Science and Engineering, Amity University, Lucknow Campus, India

Souvik Dihidar Department of Computer Science, Eminent College of Management and Technology, Barasat, West Bengal, India

Ashwini Doke Department of Information Science and Engineering, CMR Institute of Technology, Bengaluru, India

Soumi Dutta Department of Computer Application and Science, Institute of Engineering and Management (IEM), Kolkata, India

Rajiv Kumar Dwivedi University Department of Mathematics, Vinoba Bhave University, Hazaribag, Jharkhand, India

Naived George Eapen Department of Computer Science, CHRIST (Deemed to be University), Bengaluru, India

Bethoven Filomeno Department of Humanities Arts and Applied Sciences, Amity University, Dubai, UAE

Gagandeep Department of Computer Science, Punjabi University, Patiala, India

Mohammad Gouse Galety College of Engineering and Technology, Catholic University in Erbil, Erbil, Kurdistan Region, Iraq

Vinay Gautam Chitkara University Institute of Engineering and Technology, Chitkara University, Punjab, India

R. B. Geeta KLE Technological University, Hubli, Karnataka, India

Ashley George Pimpri Chinchwad College of Engineering, Nigdi, Pune, Maharashtra, India

Sujata Ghatak Institute of Engineering and Management, Kolkata, India

Ahona Ghosh Department of Computer Science and Engineering, Maulana Abul Kalam Azad University of Technology, West Bengal, India

Goldina Ghosh Department of Computer Application and Science, Institute of Engineering and Management, Kolkata, West Bengal, India

Polok Ghosh Computer Science Engineering Department, University of Engineering and Management Kolkata, Kolkata, India

Ritusree Ghosh Institute of Engineering and Management, Calcutta, India

Shivnath Ghosh Department of Computer Science and Engineering, Brainware University, West Bengal, India

Surajit Goon Department of Computer Science and Engineering, Brainware University, Barasat, West Bengal, India

Sonal Gore Symbiosis International (Deemed University), Lavale, Pune, Maharashtra, India;
Pimpri Chinchwad College of Engineering, Nigdi, Pune, Maharashtra, India

Sanchita Guchhait Department of Computer Science, Vidyasagar University, Midnapore, West Bengal, India

Divya Gupta Chitkara University Institute of Engineering and Technology, Chitkara University, Punjab, India

Deepika S. Hegde Department of Computer Science, CHRIST (Deemed to be University), Bengaluru, India

Jayant Jagtap Symbiosis Institute of Technology, Symbiosis International (Deemed University), Lavale, Pune, Maharashtra, India

Syeda Surma Jahanapi Department of Computer Science and Engineering, Comilla University, Cumilla, Bangladesh

Shaily Jain Chitkara University Institute of Engineering and Technology, Chitkara University, Solan, Himachal Pradesh, India

Afsha Jaweed Department of Computer Science and Engineering, New Delhi, India

Prajakta Joshi Pimpri Chinchwad College of Engineering, Nigdi, Pune, Maharashtra, India

Parul Kalra Amity University Uttar Pradesh, Noida, India

Adarsh Kandwal School of Computer Science, University of Petroleum and Energy Studies (UPES), Energy Acres, Dehradun, Uttarakhand, India

Raktim Karmakar Computer Science Engineering Department, University of Engineering and Management, Kolkata, India

Rohan Karmakar Computer Science Engineering Department, University of Engineering and Management Kolkata, Kolkata, India

Nivedita Kasturi KLE Technological University, Hubli, Karnataka, India

Mohammad Nadeem Khalid Department of Engineering and Architecture, Amity University, Dubai, UAE

Vikas Khullar Chitkara University Institute of Engineering and Technology, Chitkara University, Punjab, India

Anuj Kinge Department of Computer Science and Engineering, MIT School of Engineering, MIT Arts Design and Technology University, Pune, India

Ramkumar Krishnamoorthy Vishwakarma University, Pune, India

Varad Kshemkalyani Department of Computer Engineering, Vivekanand Education Societys Institute of Technology, Mumbai, India

Heer Kukreja Department of Computer Engineering, Vivekanand Education Societys Institute of Technology, Mumbai, India

Vinay Kukreja Chitkara University Institute of Engineering and Technology, Chitkara University, Rajpura, Punjab, India

Nilima Kulkarni Department of Computer Science and Engineering, MIT School of Engineering, MIT Arts Design and Technology University, Pune, India

Hardeep Kumar Department of Engineering and Architecture, Amity University, Dubai, UAE

Surya Bhushan Kumar University Department of Computer Applications, Vinoba Bhave University, Hazaribag, Jharkhand, India

Pabitra Kundu Computer Science Engineering Department, University of Engineering and Management Kolkata, Kolkata, India

Prerana Kundu Computer Science Engineering Department, University of Engineering and Management Kolkata, Kolkata, India

S. Kuppa Department of Computer Science, Kuvempu University, Karnataka, India

Gururaj Harinahalli Lokesh Vidyavardhaka College of Engineering, Mysuru, India

Rooprekha Baksi Maiti Amity University Kolkata, Kolkata, India

Snehomoy Maitra Computer Science Engineering Department, University of Engineering and Management Kolkata, Kolkata, India

D. Malathi SRM Institute of Science and Technology, Kattankulathur, Chennai, India

Sohini Mallik Computer Science Engineering Department, University of Engineering and Management Kolkata, Kolkata, India

Ankita Mandal Institute of Engineering and Management, Kolkata, India

Pratim Mandal Computer Science Engineering Department, University of Engineering and Management Kolkata, Kolkata, India

Ranjan Kumar Mandal Department of Computer Science and Engineering, Ranchi University, Ranchi, Jharkhand, India

Ratna Mandal Institute of Engineering and Management, Kolkata, India

Rabin Mazumder Institute of Engineering and Management, Kolkata, India

Deepti Mehrotra Amity University Uttar Pradesh, Noida, India

Pratheek Menon Department of Computer Engineering, Vivekanand Education Societys Institute of Technology, Mumbai, India

Udayan Misra Computer Science Engineering Department, University of Engineering and Management, Kolkata, India

Uzayruddin Siddiqui Mohammed Bachelor of Information Technology, Amity University, Dubai, UAE

Ashwin Mohan Pimpri Chinchwad College of Engineering, Nigdi, Pune, Maharashtra, India

Mahsa Monfared Bachelor of Information Technology, Amity University, Dubai, UAE

Anurag Mor School of Computer Science, University of Petroleum and Energy Studies (UPES), Energy Acres, Dehradun, Uttarakhand, India

Aditi Moudgil Chitkara University Institute of Engineering and Technology, Chitkara University, Punjab, India

Kuntal Mukherjee Department of Computer Science and Engineering, BIT Mesra, Ranchi, Jharkhand, India

Sabyasachi Mukherjee Asansol Engineering College, Asansol, India

Ankit Mukhopadhyay Department of Computer Science and Engineering, MIT School of Engineering, MIT Arts Design and Technology University, Pune, India

Vishal Nagar Department of Computer Science and Engineering, Pranveer Singh Institute of Technology, Kanpur, India

Mauparna Nandan Department of Computational Science, Brainware University, Kolkata, India

V. Narendra Department of Computer Science and Engineering, Shri Madhwa Vadiraja Institute of Technology and Management Bantakal, Udupi, Karnataka, India

Ishita Pahari Computer Science Engineering Department, University of Engineering and Management, Kolkata, India

Deepthi G. Pai Shri Madhwa Vadiraja Institute of Technology and Management, Bantakal, Udupi, India

Yogeesha Pai Manipal Institute of Technology, Manipal, India

Debdutta Pal Department of Computer Science and Engineering, Brainware University, Barasat, West Bengal, India

Shayan Pal Institute of Engineering and Management, Kolkata, West Bengal, India

Sudipta Basu Pal Computer Science Engineering Department, University of Engineering and Management Kolkata, Kolkata, India

Jeebananda Panda Delhi Technological University, Delhi, India

Digesh Pandey RA (CST-UP), Dr. Shakuntala Mishra National Rehabilitation University, Lucknow, India

Ravi Prakash Pandey IET, Dr. Rammanohar Lohia Avadh University, Ayodhya, India

M. Shahina Parveen Department of Information Science and Engineering, CMR Institute of Technology, Bengaluru, India

Ashwini P. Patil Department of Computer Applications, CMR Institute of Technology, Bengaluru, India

Jagdish Chandra Patni School of Computer Science, University of Petroleum and Energy Studies, EnergyAcres, Dehradun, Uttarakhand, India

Bachchu Paul Department of Computer Science, Vidyasagar University, Midnapore, West Bengal, India

Sabyasachi Pramanik Haldia Institute of Technology, Haldia, India

Mathur Pratistha Manipal University Jaipur, Jaipur, India

A. Priyadharshini Department of Information Science and Engineering, CMR Institute of Technology, Bengaluru, India

D. S. Raghukumar Department of Computer Science, Kuvempu University, Karnataka, India

V. Rahul Chiranjeevi SRM Institute of Science and Technology, Kattankulathur, Chennai, India

Sanskar Rai Computer Science Engineering Department, University of Engineering and Management Kolkata, Kolkata, India

Shalli Rani Chitkara University Institute of Engineering and Technology, Chitkara University, Punjab, India

A. Raghavendra Rao Department of Data Science, CHRIST (Deemed to be University), Bengaluru, India

Nismon Rio Robert Department of Computer Science, CHRIST (Deemed to be University), Bengaluru, India

Birol Roy Department of Computer Application and Science, Institute of Engineering and Management, Kolkata, India

Moumita Roy Department of Computer Science, Christ Deemed To Be University, Bengaluru, India;
Department of Computer Applications, CMR Institute of Technology, Bengaluru, India

Manjima Saha Institute of Engineering and Management, Kolkata, West Bengal, India

Prasenjit Saha Department of Computer Application and Science, Institute of Engineering and Management, Kolkata, India

Tiyasa Saha Institute of Engineering and Management, Calcutta, India

Sahana Department of Computer Science and Engineering, Shri Madhwa Vadiraja Institute of Technology and Management Bantakal, Udupi, Karnataka, India

Sakshi Chitkara University Institute of Engineering and Technology, Chitkara University, Rajpura, Punjab, India

Debabrata Samanta Department of Computer Science, CHRIST (Deemed to be University), Bengaluru, India

Sougata Samanta Institute of Engineering and Management, Calcutta, India

Shamindra Nath Sanyal Bennett University, Greater Noida, India

A. R. Sathya Department of Computer Science and Engineering, Koneru Lakshmaiah Education Foundation, Deemed to be University, Hyderabad, Telangana, India

Komal Saxena Amity Institute of Information Technology, Amity University, Noida, India

Yogesh Shahare MGMCET, Kamothe Navi Mumbai, India

Abhishek Sharma Computer Science Engineering Department, University of Engineering and Management, Kolkata, India

Chetan Sharma Chitkara University, Solan, Himachal Pradesh, India

Hitesh Kumar Sharma School of Computer Science, University of Petroleum and Energy Studies (UPES), Energy Acres, Dehradun, Uttarakhand, India

Preeti Sharma School of Computer Science, University of Petroleum and Energy Studies (UPES), Energy Acres, Dehradun, Uttarakhand, India

Robin Sharma Department of Humanities Arts and Applied Sciences, Amity University, Dubai, UAE

Shivani Sharma School of Computer Science, University of Petroleum and Energy Studies, EnergyAcres, Dehradun, Uttarakhand, India

Kumar Shaswat Amity University Uttar Pradesh, Noida, India

Raj Shree CST-UP, Babasaheb Bhimrao Ambedkar University, Lucknow, India

Ashwani Kant Shukla DCS, Babasaheb Bhimrao Ambedkar University, Lucknow, India

Vinod Kumar Shukla Department of Engineering and Architecture, Amity University, Dubai, Dubai, UAE

Vivek Shukla DIT, Babasaheb Bhimrao Ambedkar University, Lucknow, India

Farheen Siddiqui Department of Computer Science and Engineering, New Delhi, India

Avantika Singh Amity University Uttar Pradesh, Noida, India

Jaiteg Singh Chitkara University Institute of Engineering and Technology, Chitkara University, Rajpura, Punjab, India

Monisha Singh Department of Computer Applications, CMR Institute of Technology, Bengaluru, India

Saravjeet Singh Chitkara University Institute of Engineering and Technology, Chitkara University, Rajpura, Punjab, India

Rohit Singhee Computer Science Engineering Department, University of Engineering and Management Kolkata, Kolkata, India

Ashish Sinha Delhi Technological University, Delhi, India

B. Sowmiya Department of Information Technology, Thiagarajar College of Engineering, Madurai, Tamil Nadu, India

S. Sowmya Department of Computer Science and Engineering, Shri Madhwa Vadiraja Institute of Technology and Management Bantakal, Udupi, Karnataka, India

R. K. Srivastava CST-UP, Dr. Shakuntala Mishra National Rehabilitation University, Lucknow, India

S. Sudersan Department of Information Technology, Thiagarajar College of Engineering, Madurai, Tamil Nadu, India

Mariyam Sultana Department of Computer Science and Engineering, Comilla University, Cumilla, Bangladesh

M. Suresha Department of Computer Science, Kuvempu University, Karnataka, India

Y. Swathi.Y Department of Information Science and Engineering, CMR Institute of Technology, Bengaluru, India

Mathur Swati Manipal University Jaipur, Jaipur, India;
CMR Institute of Technology, Banglore, India

Abha Tewari Department of Computer Engineering, Vivekanand Education Societys Institute of Technology, Mumbai, India

M. Thangavel Department of Computer Science and Engineering, Siksha 'O' Anusandhan (Deemed to be University), Bhubaneswar, Odisha, India

Reuben Thomas Department of Computer Engineering, Vivekanand Education Societys Institute of Technology, Mumbai, India

Raj Gaurang Tiwari Chitkara University Institute of Engineering and Technology, Chitkara University, Punjab, India

S. G. Totad KLE Technological University, Hubli, Karnataka, India

P. Varalakshmi Department of Computer Technology, Madras Institute of Technology, Anna University, Chennai, Tamil Nadu, India

Amit Verma HTMi Switzerland, Dubai, UAE

Rajat Verma Department of Computer Science and Engineering, Amity University, Lucknow Campus, India

Shivani Wadhwa Department of Computer Science, Punjabi University, Patiala, India;
Chitkara University Institute of Engineering and Technology, Chitkara University, Punjab, India

Cyber Intelligence

DTNMA: Identifying Routing Attacks in Delay-Tolerant Network

Siddhartha Chatterjee, Mauparna Nandan, Ahona Ghosh, and Swarnali Banik

Abstract The delay-tolerant networks (DTNs) are mainly characterized by decentralized and self-managed structures possessing unique features and characteristics. However, limited network resources and buffer space, bandwidth issues, high mobility nodes and disconnected paths are the characteristics that lead to network vulnerabilities and potential threats. These malicious attacks by compromised nodes in DTN systems can interrupt and degrade the normal functioning of the routing protocols in the network, thereby resulting in delivery interruption due to the possibility of dropping the delivered messages by the selfish or malicious nodes which can further lead to performance degradation, compromising with message confidentiality and integrity, denial of service (DoS) attacks, etc. A variety of security attacks, namely packet dropping attack, flooding attack, gray hole attack, worm hole attack, black hole attack, etc., affect the network connectivity, thereby increasing the delay in the message delivery rate. Thus, the effective mechanisms need to be identified and implemented to eliminate the different types of vulnerability attacks before implementing anything in DTN. This paper mainly addresses the areas of identifying the different types of routing attacks as well as some open issues in order to establish a reliable and secure DTN in the future.

Keywords Delay-tolerant networks · Routing attacks · Security · Routing protocols

S. Chatterjee (✉)
Department of Computer Science and Engineering, Gargi Memorial Institute of Technology, Kolkata 700144, India

M. Nandan
Department of Computational Science, Brainware University, Kolkata 700125, India

A. Ghosh
Department of Computer Science and Engineering, Maulana Abul Kalam Azad University of Technology, West Bengal 741249, India

S. Banik
Department of Computer Science and Engineering, St. Mary's Technical Campus, Kolkata 700126, India

J. M. R. S. Tavares et al. (eds.), *Cyber Intelligence and Information Retrieval*, Lecture Notes in Networks and Systems 291,
https://doi.org/10.1007/978-981-16-4284-5_1

1 Introduction

Delay-tolerant networks (DTNs) [1, 8] are a specific type of networks that are designed to operate effectively over an extreme distance as they can sustain long delays and are able to store (or buffer) the data packets in the intermediate nodes and forwarding them to other nodes unless their target is reached in order to address the challenges of delivering an end-to-end service where an end-to-end communication route may not be established. DTN is a newly emerging topic in the field of scientific research and has gained significant importance due to its wide applications in extreme challenging environments, namely space communications [2], low-cost Internet connection in isolated areas and high-speed vehicular ad hoc networks [3], military operations [4], underwater sensor networking [5], relief rescues in remote and disaster-affected areas, etc. DTNs are becoming more significant and prevalent with the advancement of communication networks in the current scenario. DTNs are heterogeneous networks and its self-organizing nature along with other malicious attacks are posing a threat to its security challenges. DTNs are at high risks for different types of flood attacks. Flood attacks occur in the network when the buffer of the server is overloaded with data packets, thereby causing the network to shut down its service for which the outcome maybe the denial of service (DoS).

Thus, DTNs are prone to numerous security threats and also present various new security challenges. Various other types of insider attacks are also present, namely packet dropping attack [8], jelly fish attack [18], gray hole attack [16], worm hole attack [24], black hole attack [10], tailgating attack [19], packet misrouting attack [9], routing information falsification attack [8], acknowledgment falsification attack [8], Sleep deprivation attack [20], identity impersonation attack [11], lack of cooperation attack [9], byzantine attack, flooding attack [8, 12], etc., have a significant impact on the network, thereby compromising with the information integrity, authenticity, privacy and the overall system performance.

In this paper, we provide an in-depth illustration of the key challenges and the identification of various types of malicious activities in delay-tolerant systems. The impact of these attacks on the network is also highlighted. Figure 1 depicts the diagram of store and forward message switching technique, where data is incrementally moved and stored throughout the network in hops or nodes, that will eventually reach its destination.

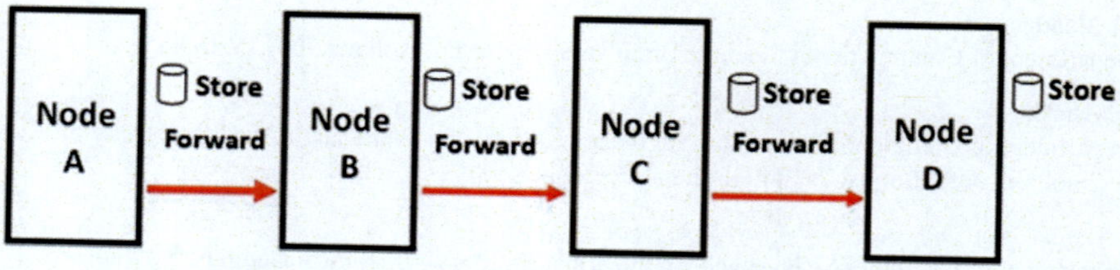

Fig. 1 Sample diagram of a store and forward message switching technique in DTN network

2 Key Challenges of Routing Attacks

In DTN environment, during the dissemination of information, it was observed that it frequently suffered from connectivity link cutoffs leading to delivery interruption due to the presence of some misbehaving nodes, commonly categorized as malicious node or selfish node that mainly hampers the overall or the partial network routing process. Here, the malicious node creates an abnormal situation in a network routing system as a consequence of which the overall the network routing system degrades as well as the message delivery rate also diminishes. These unwanted malicious activities can lead to illegal access and usage of DTN supplies, thereby causing a serious threat to security issues and also hampering the normal functioning of the DTNs.

We have studied different papers on attacks and their challenges in DTN networks but none of these papers have discussed every possible security attacks. Our paper here is more precise in this respect. Here, Fig. 2 clearly specifies this concept.

3 Identification of Routing Attacks

Delay-tolerant networks (DTNs) had been developed in the past decade mainly to address the technological issues of heterogeneous and challenging networks characterized by intermittent network connectivity, long delays, frequent disruption in message delivery, etc. Although DTN system employs hop-to-hop routing and the store and forward platform for resolving the end-to-end connectivity, still security and privacy issues pose a major threat to DTN networks.

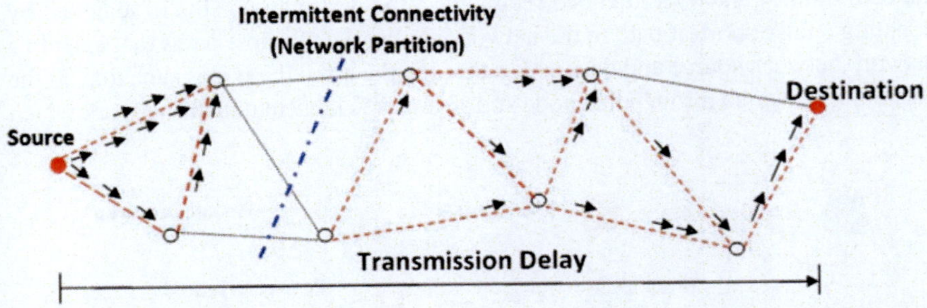

Fig. 2 Schematic diagram of a delay-tolerant network (DTN) where the malicious nodes are denoted by red circles and the regular nodes by blue circles

3.1 Packet Dropping Attack

This is the most commonly occurring attack in delay-tolerant networks [13] and is addressed as one category of denial of service (DoS) attack. The attacks are induced by the compromised nodes when they misbehave by dropping packets even when sufficient buffer space is available. This leads to routing misbehavior [14] which is another major concern in DTN networks and is caused by selfish nodes that are reluctant to transmit the packets to other nodes, and they consume resources like power and buffer space. This type of attack can also be caused by malicious nodes that drop the packets being forwarded to them by employing random, selective or periodic method for launching denial of service attack in the network. Here, the compromised node acts as an attacker and drops the packets continuously in the place of forwarding it to the receiver node and, thereby, disrupting the message forwarding service in the network. Since these attacks are created within the network and are executed by the participating malicious nodes that usually act normally until they are infected, and therefore, detecting them becomes very difficult. Figure 3 represents the packet dropping scenario due to the presence of malicious nodes.

3.2 Black Hole Attack

Another form of denial of service (DoS) attack that is quite eminent in a DTN network is the black hole attack [15, 16] where the malicious node projects itself as a perfect node to each destination node but then drops all the packets received from other nodes. Black holes attacks in the network silently discard (or drops) both the incoming and the outgoing messages without updating the sender node that the data did not reach its intended recipient in the destination. This is achieved by injecting a false shortest path in the network among the multiple paths that exist in a network between source and destination through the affecting nodes such that all the packets are diverted through that node and an attack is launched in the network which

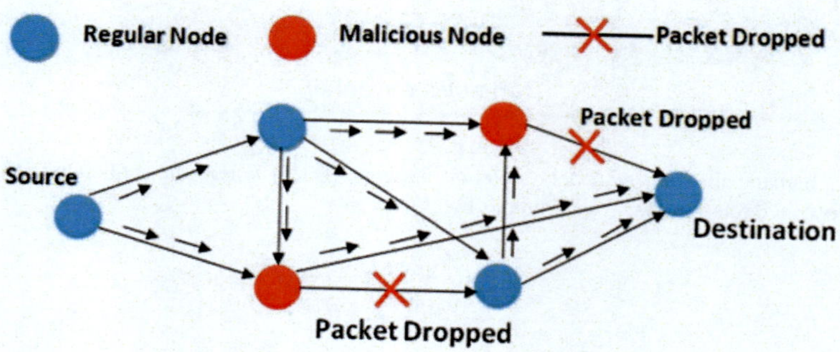

Fig. 3 Packet dropping attack

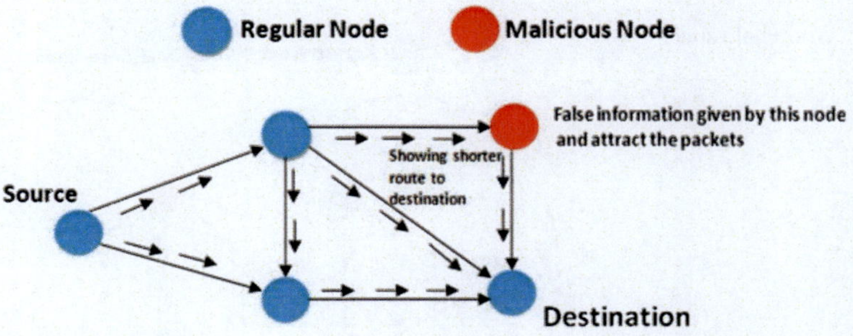

Fig. 4 Black hole attack

further results in data loss. The malicious node publicizes its obtainability of fresh routes regardless of the contents in the routing table. Here, the attacker exploits this advertising strategy of route request and always respond to the path request issued by the sender node even prior the actual answer from the destination node is received and then initiates to forward packets through that specific malicious node. In this manner, a malicious and forged path is generated and introduced in the network. The impact of this attack leads to failure of particular applications, degrades network communications, isolates nodes and creates routing loops, etc. Figure 4 shows the false shortest route in the network by malicious nodes.

3.3 Worm Hole Attack

Worm hole attack is a very devastating category of attack and is commonly observed in wireless networks like DTN [24]. It is one category of denial of service (DoS) attack where the attacker creates two or more malicious nodes at different powerful strategic positions in the network with the help of low latency links. By utilizing their locations, these malicious nodes create a fake route which comprises the shortest route compared to other original routes within the network. This shortest route is nothing but a virtual tunnel is established between two or more malicious nodes which can confuse the routing mechanism as it solely relies upon the knowledge about distance between the several couples of nodes in the network. The worm hole attackers create a virtual tunnel for recording the packet transmission at one position of network and channelize to other points of network. This tunnel is either a wired connection or a high-frequency connection which is commonly known as the wormhole link. This creates the impression that the channel's two end points are close to each other. The attackers then advertise their routes to other nodes in the network for transmitting their data. Then the attackers keep on listening to the channel to record the transmitted information such that the packets can be captured from one location and transmitted to a distant node which is then distributed locally. When the new route is created and the network traffic starts to use this shortcut wormhole link,

Fig. 5 Worm hole attack

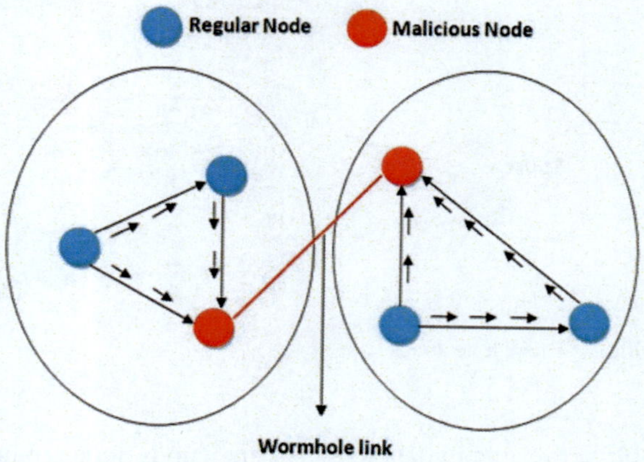

the malicious node selectively starts dropping data packets to disrupt communication. Routing mechanisms get confused which causes severe damage to the network. By analyzing the data packets obtained, the attacker may execute several other forms of more violent attacks like man-in-the-middle, cipher breaking, protocol reverse engineering, etc. The wormhole attack can be still carried out without acquiring any previous concept of the network or even if the attacker has not compromised with any legitimate nodes with the aid of cryptographic keys even though the network communication provides network confidentiality and authenticity. Figure 5 clearly explains the existence of virtual tunnel between two malicious nodes.

3.4 *Gray Hole Attack*

It is one of the most popular attacks in DTN networks [16, 17] and is essentially a packet dropping attack where a affecting or a gray hole node decides to contribute in the transmission of data packets to other network nodes but then drops the packets selectively to mislead the network. Because of the autonomous, dynamic nature of the ad hoc network and want of any static architecture makes the network malicious to different attacks. The gray hole attack is implemented by advertising a false route as a valid one having the intention to intercept data packets. This false path is publicized as the shortest path toward the destination so that the other nodes agree to forward packets through this path, and the malicious nodes can then take advantage of this route request advertising strategy of the sender node to launch packet dropping attacks in the system. In comparison with the black hole attack, this attack is much more tough for identification because here, the malicious nodes drop the data packets very carefully and selectively to ignore arousing suspicion and detection from other regular nodes. Another category of gray hole attackers may also be present where the attacker nodes behave maliciously for some duration of time unless the packets are dropped and then immediately return to their usual activity mode, thereby making it much

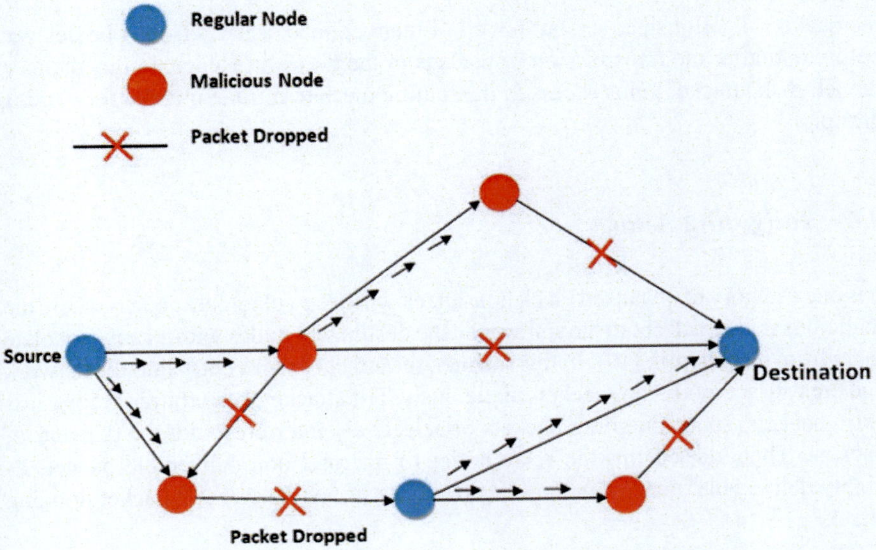

Fig. 6 Gray hole attack

more difficult to identify them. The gray hole attack (known also as node misbehaving attack) is a category of denial of service (DoS) attack. Figure 6 demonstrates the selective packet dropping and false route injection in a network with the intervention of malicious nodes.

3.5 *Jelly Fish Attack*

It is basically a passive attack [18] where the malicious node utilizes the protocol maliciousness and sometimes drop, delay and/or reorder packets. It is a kind of denial of service (DoS) attack in which the attackers tend to raise the packet end-to-end delay and introduces network jitters. This type of attack is mainly targeted in closed-loop flows in which the attacker node strictly follows the protocol rules so that it cannot be easily detected. This attack is due to transmission control protocol (TCP) vulnerability. Ad hoc on-demand distance vector (AODV), for example, is a routing protocol that is mainly affected by the jellyfish attack. The primary objective of the jellyfish node is to decrease the network throughput by delaying or dropping some packets. A jellyfish attacker participates in route discovery and packet forwarding process in order to remain undetected. It is similar to blackhole attack where the jellyfish attacker first captures all the packets, but unlike the blackhole attack where the attacker node drops the data packets, the attacker may choose to drop some of the packets in jellyfish attack but not all of them at once or packets may be delayed before and after the transmission of packets in the network. Hence, in this manner, the jellyfish attacker gains access to the routing mesh to intrude into the forwarding group.

3.6 *Tailgating Attack*

It is one category of advanced black hole attack when the vulnerable node is where the malicious node tracks both the source and the destination nodes with an aim to isolate them from the network [19]. In this manner, the attacker seeks entry into the network and then attracts and drops packets maliciously. The attacker then arbitrarily forwards extra packets, spoof outgoing packets or selectively interferes with the delivery of packets. This attack helps the relay nodes to interpret non-deliverable packets as highly deliverable, thus extending the effectivity of non-deliverable packet flooding attack.

3.7 *Packet Misrouting Attack*

It is a kind of packet mistreatment [9] or packet losing attack in which the packets are disrupted from reaching their destination nodes by an middle compromising node. However, the compromising node forwards the packet to give the impression as a legitimate node to its neighbors in the system. This form of attack is achieved by changing the data packet's final destination address or forwarding a data packet to a wrong next hop on the path to reach the destination. When the packets are unable to find its destination, the packets are dropped by the network. Thus, in this manner, a legitimate node is treated as a malicious node as it drops packets. This attack is mainly applicable to unacknowledged packets and can cause network congestion and minimizing the throughput to launch DoS attacks in the network.

3.8 *Routing Information Falsification Attack*

In a route falsification attack [8], the routing information of a network stored in a routing table is attacked by a malicious node, thereby causing all the packets to get delivered to wrong destinations in the network due to the generation of false routing table updates. In this attack, the malicious can operate in both the directions and falsifies the existence of the shortest path to both the path request and the acknowledging packets to attract a wide part of the network traffic. In this manner, the source node is forced to select this falsified route so that the malicious nodes can drop the packets silently. Thus, a denial of service (DoS) may be initiated by continuously transmitting route error messages identifying a broken connection on the path for the prevention of the sender from interacting with the receiver. This decreases the packet delivery predictability which further degrades the routing services in the total network. Secure routing information protocol (SRIP) is designed to completely remove routing information falsification attack.

3.9 *Acknowledgment Falsification Attack*

Acknowledgment is a very effective mechanism for packet delivery since it requires less overhead. In this attack [8], the regular nodes either receive packet acknowledgment before packet delivery or receive fake acknowledgments propagated form other nodes. This, in fact, misleads the sender node to understand that whether the original data packets transmitted by it have reached the specific destination or not. In most cases, if a node receives an original packet acknowledgment before the packet, it describes a malicious acknowledgment. In DTN networks, whenever a regular node encounters a malicious acknowledgment before the receipt of the original acknowledgment, uncertainty arises regarding the packet delivery among different network nodes, and the entire network is left in a state of confusion. As a result of this attack, the routing services in the network get disrupted, thereby leading to performance degradation of the entire network and primarily timing delay sets in between the different nodes forwarding packets at different time interval.

3.10 *Sleep Deprivation Attack*

It is one category of flooding attack in which either a particular node or a set of nodes are aimed to keep them awake by exhausting their resources. This attack is implemented when a vulnerable node transmits huge amount of data packets to a targeted node with an aim to exhaust its network bandwidth, battery and computation power of the latter by sending false requests to existing or non-existing destination nodes. As a result, the targeted node is unable to process the requests coming from genuine nodes. The primary objective of the malicious node is to minimize the lifetime of the targeted node by consuming its valuable resources. As a result, the target node is unable to engage in routing schemes as it becomes inaccessible to other network nodes. Thus, with the aid of unnecessary flooding of the routing packets, the malicious node launches the sleep deprivation attack [20, 21] in the network.

3.11 *Identity Impersonation Attack*

An impersonation attack commonly known as spoofing attack [22] is an attack where the opponent successfully assumes the identity of any legitimate node and injects fake messages into the network. This is the most popular direct attack against a routing protocol primarily targeting the routing information shared between two nodes by spoofing, modifying or replaying routing information. This leads to the disruption of traffic in the network which includes creation of routing loops, extending and curtailing of source routes, attracting or repelling network traffic from selected nodes, network partitioning, generating fake error messages and improving end-to-end latency. The standard approach for this attack is the authentication in which routers only accept valid routing information.

3.12 Lack of Cooperation Attack

In this attack [9], the regular nodes are discouraged to cooperate in the network operations. Misbehaving regular nodes (malicious nodes) employ different strategies to save their limited resources. They might refuse to forward any packets to other nodes in the network as well as in their devices when they are not transmitting any packets in the network. In DTN networks, the malicious node discourages packet forwarding to the regular nodes, and they are compelled to work according to the malicious node specification. The main intention of these malicious nodes is to save their battery power by cutting off packet forwarding service which in turn disrupts the routing services in the network.

3.13 Byzantine Attack

Byzantine attack [23] also named as spectrum sensing data falsification (SSDF) attack is a kind of insider attack and is commonly observed in cooperative spectrum sensing (CSS) where an intermediate malicious node or a group of intermediate malicious nodes work collectively to launch this attack. In this attack, an attacker node creates routing loops, forward packets through non-optimal paths/routes or may selectively drop data packets which results in the performance degradation of the entire network. Secure message transmission (SMT) is the secured data communication protocol invented to prevent this attack.

3.14 Flooding Attack

Flooding attacks [24, 25] are the most dangerous threats against different computer networks. It is a type of active attack wherein the primary aim of the attacker is to exhaust the network resources, namely the bandwidth utilization, consumption of battery power and reduction of network lifetime with an intention to disrupt the routing operations which eventually leads to severe network degradation in network performance. The attack commences by flooding the network with fake data packets as a consequence of which the network is unable to process legitimate requests since it becomes overloaded with invalid requests. Such malicious flows in DTN networks cause most of the legitimate packets to be dropped at the routers without reaching their destination. This is because it exhausts the memory buffer of the server and connection paths are closed, thereby reducing the probability of data transmission which further leads to the blocking of the entire network.

Name of the attack	Area of identification	Nature & Activities	Impact on network
Packet dropping	Wireless sensor network	Drops packets frequently	Reduce network efficiency
Black hole	Mobile ad hoc network	DOS attack	Reduce network performance
Wormhole	Delay-tolerant network	DOS attack	Reduce network efficiency
Gray hole	Delay-tolerant network	DOS attack	Reduce network efficiency
Jellyfish	Mobile ad hoc network	Kind of DOS attack	Reduce network throughput
Tailgating	Delay-tolerant network	Kind of DOS attack	Reduce network throughput
Packet misrouting	Delay-tolerant network	Kind of DOS attack	Reduce network throughput
Routing information	Delay-tolerant network	Kind of DOS attack	Reduce network throughput
Falsification	Delay-tolerant network	Kind of DOS attack	Reduce network throughput
Acknowledgment falsification	Delay-tolerant network	Send fake acknowledgment	Disrupted network
Sleep deprivation	Wireless sensor network	Kind of flooding attack	Collapse network bandwidth
Identity impersonation	Identity impersonation	Spoofing attack	Traffic jam
Lack of cooperation	Wireless network	Discourages packet transmission	Wastage of network battery power
Byzantine	Delay-tolerant network	Insider attack	Performance degradation
Flooding	Delay-tolerant network	Disrupts routing operations	Network block

4 Conclusion and Future Work

We are all well acquainted regarding the connectivity background of delay-tolerant networks (DTN) and its extensive applicability in various fields. This paper presented a brief explanation about the different types of malicious activities, their outcome on the networks and and the various challenges faced during transmission of message in DTN networks. We have also focused on the misbehavior of malicious nodes that is very harmful for any kind of wireless networks. Thus, network security is one of the most challenging and exciting research areas in the field of computer networks including DTN where certain procedures and algorithms need to be designed to secure the network from illegal access, misuse or modification of network resources. In the future work, the authors would like to address proper detection strategies, mitigation approaches for avoiding various attacks in DTN. DTN routing protocol security and optimum message delivery among various nodes are also earmarked as future attention.

References

1. Fall K (2003) A delay tolerant networking architecture for challenged internet. In: Proceedings of the 2003 conference on applications, technologies, architectures, and protocols for computer communications, SIGCOMM' 03, pp 27–34
2. Kate A, Zaverucha G, Hengartner U (2007) Anonymity and security in delay tolerant networks. In: The 3rd international conference on security and privacy in communications networks and the workshops, pp 504–513
3. Lin X, Lu R, Zhang C, Zhu H, Ho P-H, Shen X (2008) Security in vehicular Ad Hoc networks. IEEE Commun Mag 46(4):88–95
4. Pathak P, Shrivastava A, Gupta S (2015) A survey on various security issues in delay tolerant networks. J Adv Shell Programming 2(2):12–18
5. Lu R, Lin X, Zhu H, Ho P-H, Shen X (2008) ECPP: efficient conditional privacy preservation protocol for secure vehicular communications. In: The 27th IEEE International Conference on Computer Communications, INFOCOM
6. Jia Z, Raffay MA, Li L et al (2012) A secure message delivery scheme with path tracking for delay tolerant networks. In: 1st IEEE International Conference on Communications in China (ICCC), pp 161–166
7. Nagrath P, Aneja S, Purohit GN (2019) Attacks in delay tolerant networks: classification and analysis'. In: 11th International Conference on Communication Systems and Networks (COMSNETS)
8. Burgess J, Bissias GD, Corner MD, Levine BN (2007) Surviving attacks on disruption-tolerant networks without authentication. In: Proceedings of the 8th ACM International Symposium on Mobile Ad Hoc Networking and Computing, MobiHoc, pp 61–70
9. Jayaprakash N, Balaji BS, Manikandan J, Gayathri VM (2014) Identification of misrouting in wireless Adhoc network. Int. J. Eng. Res. Technol. (IJERT) 3(3)
10. Ren Y, Chuah MC, Yang J, Chen Y (2010) Detecting blackhole attacks in disruption-tolerant networks through packt exchange recording. In: IEEE international symposium on "A World of Wireless, Mobile and Multimedia Networks" (WoWMoM)
11. Uddin MYS, Khurshid A, Jung HD, Gunter C, Caesar M, Abdelzaher Z (2011) Making DTNs robst against spoofing attacks with localized countermeasures. In: 8th annual IEEE communications society conference on sensor, mesh and ad hoc communications and networks
12. Hu Y-H, Choi H, Choi H-A (2004) Packet filtering to defend flooding-based DDoS attacks. In: IEEE Sarnoff symposium on advances in wired and wireless communications
13. Sangeetha R, Vijayabhasker R (2018) Analysis of packet dropping, storage issue, security issue and routing misbehavior in disruption tolerant networks. Int J Eng Adv Technol 8(2)
14. Li Q, Cao G (2012) Mitigating routing misbehavior in disruption tolerant networks. IEEE Trans Information Forensics Security 7(2)
15. Singh S, Elayaraja K (2015) A survey of misbehaviors of node and routing attack in delay tolerant network. Int J Sci Eng Technol Res 4(2)
16. Dixit S, Joshi KK, Joshi N (2015) A review: black hole & gray hole attack in MANET. Int J Future Generation Netw 8(4):287–294

17. Arya M, Jain YK (2011) Grey hole attack and prevention in Mobile Adhoc Network. IJCA 27(10)
18. Kaur S, Kaur R, Verma AK (2015) Jellyfish attack in MANETs: a review. In: IEEE international conference on electrical, computer and communication technologies
19. Li F, Wu J, Srinivasan A (2009) Thwarting blackhole attacks in disruption-tolerant networks using encounter tickets. In: IEEE INFOCOM, pp 2428–2436
20. Bhattasali T, Chaki R, Sanyal S (2012) Sleep deprivation attack detection in wireless sensor network. Int J Comput Appl 40(15)
21. Jain S (2014) Security threats in MANETS: a review. Int J Information Theor 3(2)
22. Alam S, De D (2014) Analysis of security threats in wireless sensor network. Int J Wireless Mobile Netw 6(2)
23. Vempaty A, Tong L, Varshney PK (2013) Distributed interference with Byzantine data. IEEE Signal Process Mag 30(5):65–75
24. Hu YC, Perrig A, Johnson DB (2003) Packet leashes: a defense against wormhole attacks in wireless networks. In: Proceedings IEEE-INFOCOM, vol. 3 (1976–1986)
25. Meher R, Ladhe S (2014) Review paper on flooding attack in MANET. Int J Eng Res Appl 4(1):39–46

Security Aspects for Mutation Testing in Mobile Applications

Naived George Eapen, A. Raghavendra Rao, Debabrata Samanta, Nismon Rio Robert, Ramkumar Krishnamoorthy, and Gururaj Harinahalli Lokesh

Abstract Due to the increase in the number of Android Platform Devices, there are more and more applications being developed across various domains. It is interesting to see the involvement of bugs/crashes even in the deployed applications even though it has been through various test phases. Unit tests are essential in a well-trusted testing environment; however, it does not guarantee that the range of test caries every component of the application. This writes up discusses the overview of mutation testing method concerning Android Applications. Even though mutation testing is found out to be very effective in other applications, it is not that easy to implement the same for an Android Developed Application because of additional resources it would hold. Further, various measures for mutation testing are discussed with types of mutant operators, tools etc. The current studies of mutation analysis mainly focus on testing all the functionalities irrespective of the resource usage. However, the target of the future mutation tests must be also to evaluate the efficiency of the applications under the same test cases.

Keywords Android · Mobile apps · Mutation testing · GeneTic

N. G. Eapen · D. Samanta (✉) · N. R. Robert
Department of Computer Science, CHRIST (Deemed to be University), Bengaluru, India

N. R. Robert
e-mail: nismon.rio@christuniversity.in

A. R. Rao
Department of Data Science, CHRIST (Deemed to be University), Bengaluru, India
e-mail: RAlthar@firstam.com

R. Krishnamoorthy
Vishwakarma University, Pune, India

G. H. Lokesh
Vidyavardhaka College of Engineering, Mysuru, India
e-mail: gururaj1711@vvce.ac.in

J. M. R. S. Tavares et al. (eds.), *Cyber Intelligence and Information Retrieval*,
Lecture Notes in Networks and Systems 291,
https://doi.org/10.1007/978-981-16-4284-5_2

1 Introduction

Mobile applications (mobile apps) are software agendas which are connected and implemented on mobile tools, for example, smart phones, tablets, smart TVs, and smart watches. As the costs of mobile devices are progressively becoming inexpensive, and the mobile networks are speedily increasing to virtually every country, more people around the world are procuring mobile tools. Every now and then, one individual may own manifold mobile tools: he or she may employ a Smartphone for calling household members and friends, may have a smart TV for viewing news and movies, and may use a smart watch for stalking whereabouts and bodily hustle [1]. Mobile devices have emerged from moderate tools with a microscopic, less firmness, and black and white display which are only accomplished with voice calling, texting, and surfing news deprived of any images or animations, to potent convenient computers and amusement stations. Mobile tools are civilizing people's life, comprising the technique we market, amuse, navigate, connect, and make friends [2, 3].

The number of mobile applications in the market is growing because of several rea-sons such as price drop, user-friendliness, marketing strategy, and a high number of devices. Android, being the dominating mobile operating system in the world, can run any compatible application for various functional applications. When the "world of applications" is growing more and more, on the other side, there are various test me-thods made to ensure the quality of the application. It is a very common problem that many applications reach the market with errors (or otherwise known as bugs) leading into application crashes during the use [4]. The ineffective testing methods even for a language with comparatively new features are one of the main causes of such issues.

2 Mobile Applications and Testing

Mobile applications could be viewed as the software programs that are installed a mobile device to execute various functionalities. Any mobile user has access to the very large set of applications that are developed across various domains. In today's scenario, a life without mobile applications (smart phone-based) is almost unimaginable. From ordering food to transacting money, games to educational applications, forums to social media, the impact of such applications in communication, shopping, entertainment, infotainment etc. is extremely high [5, 6]. There are millions of applications available for users in various platforms such as android, iOS etc. We restrict the scope of this write-up to Android applications alone. The increasing number of applications that are open to users results in need of better user experience and error-free implementations because of the competition that is happening between the developing companies [7]. Thus, the demand for high-quality applications is very much going high. In the industry, any developed software is tested through multiple phases to ensure good quality. For any static software, the testing becomes easy as the scope of usage is predictable. However, for mobile applications, there are a number of limitations [8] that do not make normal software testing methods feasible. There a number of unique characteristics of Android Applications that has to be taken care of during the testing phase [9].

1. Unique lifecycle: Every application must go through the same set of life-cycle and any improper handling of an error would result in error of the entire package.
2. Use of XML files: Apart from other programming languages, Android depends highly on XML files for configuration, temporary data storage etc.
3. Content-awareness: Android devices collect data through sensors and these are not directly fed by the user. Thus, even minor changes in these factors must be taken care of in the testing environments.
4. Dynamic Screen Orientation: Unlike other devices, mobile phones have portrait/landscape modes.
5. Event-based programs: The events can be considered as touch, drag, release etc.
6. Connections/Signals: The devices are equipped with multiple forms of connections that need to be taken care of in the testing phase.
7. Limited battery life.

3 Mutation Testing

Mutation Testing affirms to be competent in recognizing sample test data which can be employed to discover actual errors. However, the number of such likely errors for a specified program is giant; it is difficult to produce variants indicating all of them. Thus, customary Mutation Testing marks only a subgroup of these errors, those which are adjacent to the precise form of the program, with the hope that these will be adequate to prompt all errors [10]. Mutation testing modifies a software component to generate its tweaked versions called mutants that are expected to be faulty. These changes are not random but are created by applying rules, otherwise called as mutation operators. When a mutant and its corresponding original file are available, the tester could create tests that cause them to exhibit different behaviors [11]. If found that the testing will give different outcomes for mutant and original software, the change in the mutant is discarded to be put in the next version of the original testing software; and this process is called killing the mutants. Sometimes, the mutation operators force the testers to change test designs to stronger ones that are likely to produce invalid outputs in the mutants and proper outputs in the original file. With this property in hand, we could say that Mutation Testing is nothing but a test for test cases. The fundamental idea of mutation testing is that if a mutant is found to be alive after the testing, it either means that the mutant is alternative, or, that block of code where the mutant must exhibit a different approach was never been executed and the test case must be modified. Thus, they can be seen as tests for individual unit test designs (overall test suite) to provide more integrity for the application [12].

It means that proper implementation of mutation testing can make existing test suites stronger. This is because of the fact that mutation tests require not only require the original software to pass the test, but also have to ensure that the mutant is failing the test. It is possible that some programs may give the same outputs as mutant programs and these mutants are called equivalents. The 'stillborn' mutants are avoided from testing as this type gives syntax errors under compilation [13]. Mutation operators can be defined using either of the below approaches. When fault models are already available, mutation operators can be defined from them where a corresponding mutation operator is designed to force the testing of those fault instances. Alternatively, the syntactic analysis of the programming language can be also considered to design mutation operators such that mutants involve mistakes similar to typical syntax errors that usually programmers might produce [14]. Fundamentally, there are three types of mutation testing:

1. Value Mutation: Value of constants or parameters are changed
2. Decision Mutation: The condition of a decision-making block is changed
3. Statement Mutation: The occurrences of statements are modified by changing the order of lines/duplicating a line/deleting a line etc.

3.1 Mutation Analysis of Android Applications

The method that is used for traditional Java programs cannot be applied in Android Application Mutation analysis. It is because of the need to analyze the XML files that define the Android UI design along with configurations. Furthermore, Android applications may require additional processing before the package deployment. In Java mutation of either byte codes or source code is allowed. When executed, byte code files for the code are dynamically linked by the language system. Even though this step is needed for any Android program as it involves the usage of Java programs, it is important to note that even the mutants could be deployed or compiled as an APK to evaluate in either emulators or mobile devices for testing. The above requirements clearly show that the design of mutation analysis tools for Android becomes extremely hard.

3.2 Crashes/Bugs in Android

Android applications are constructed with respect to a pre-defined structure with a mandatory XML manifest file and four types of components (Activities, Services, Broadcast Receivers, and Content Providers). These XML file will have information about the application including individual component and configuration information. Therefore, when one is going to find all the possible errors in an application, all these components have to be taken care [15]. There are no large-scale studies describing the taxonomy of Android Application crashes or bugs currently. Moreover, there is no mobile-specific fault detection testing services. The bugs and crashed that are faced by current programs define the set of future mutant programs. In the study of [16], they have manually analyzed six different sources of bugs/crashes. They are as follows:

- *Bug reports:* Bug reports of applications are the widest sources to identify the crashes, especially in order to observe the common bugs affecting an application. Several applications will provide a "send bug report to developer" option along with the application itself so that the users itself could submit crash data to them.
- *Bug-fixing commits*: It is possible that some bugs are not identified in open source development at initial stages. However, the community developers' discussion will lead into direct fixation. These are very much true for small and directly fixable bugs in cloned repositories.
- *Android-related Stack Overflow discussions*: The Stack Overflow discussions are one of the most widely used techniques to quickly resolve any bugs that need additional help, especially when they experience a difficulty in fixing the application. There are thousands of Android bug-related discussions in such forums.
- *The hierarchy of exceptions in Android application program interfaces*: unhandled exceptions and risky statements are sources of faults in android applications to a great extent.
- *Crashes/bugs described in testing studies*: During the test phase of application development, it is possible that new bugs may show up. The analysis of test-documents with application version, bug status etc. could give information about the bugs.
- *Reviews posted in App Stores*: A reduced set of reviews are very much helpful for developers in useful for developers that points to very prominent application bugs.

4 Mutation Operators in Android

The mutation operators in Android must be defined in such a way that it would be able to incorporate occurrences of any bugs to a great extent.

4.1 Based on Bug Taxonomy

Figure 1 borrowed from [17] represents a very large set of Defined Android Taxonomy. It defines various faults that could result as part of various test methods. These include false positive results, activity/intent/data-based bugs, both android and java specific bugs and other functional and non-functional requirements along with their types and tags etc.

The analysis from the studies of shows that:

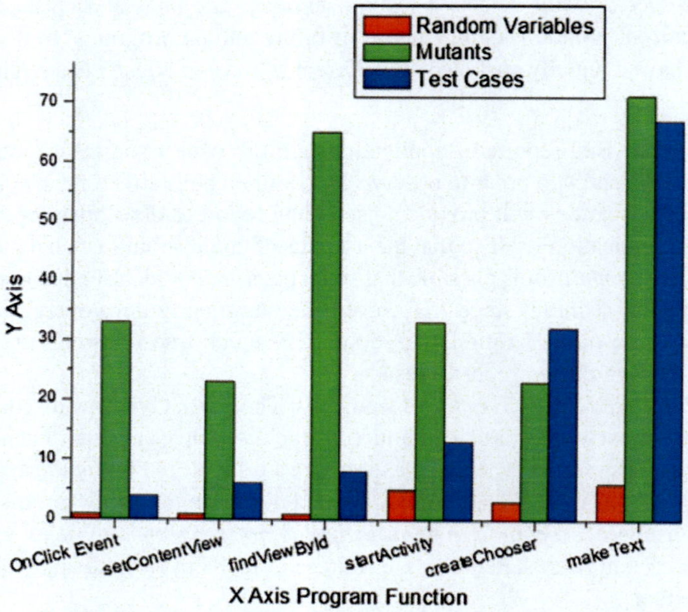

Fig. 1 Outcome of various mutant test-cases

- Architecture-dependency of bugs is 26%
- Android-specific bugs are 35% of the entire identified set.

This embosses the importance of using Android-specific testing instruments tailored for each one of the possible bugs. On the other hand, even if we develop Andro-id-specific mutation operators, the other 65% of the bugs are dependent on JAVA platform itself. Hence, the set of Android-specific mutation operators must be differ-ent from the Java mutation operators, able to make test suites in such a way that it identifies faults from applications, minimize the number of unwanted mutants (still-born, trivial etc.), and able to simulate faults by modification of statements and ele-ments in the application resources [18, 19]. For instance, two of the most common bugs are:

4.2 Fail-On-Null

Null Pointer Exception is one of the most shared thrown exemptions in the testing as the elements are randomly checked and one is usually not aware that whether one is accessing a null object. The defined mutants for such cases, called "fail-on-null" mutants add a statement before each object is accessed. The mutated statements are inserted to the locations before accessing any object members [20].

4.3 Orientation Lock

Mobile devices have the unusual characteristic of being able to adjust the orientation of the phone. Thus, as the screen orientation toggles, several apps modify the configuration of the graphical user interface. For example, when the orientation is shifted from portrait to landscape, video playing apps automatically turn to the full screen [21, 22]. Nevertheless, it is not true that different Android devices have different requirements in screen sizes and resolutions for all applications. When switching to landscape mode, it is very common for applications that function perfectly in portrait mode to become very sloppy. Orientation lock mutant's adds a special locking declaration into the software to freeze an activity's screen orientation to be in one of the modes only. Therefore, in order to destroy these mutants, the test cases during the testing process of the application must specifically alter the screen orientation [23].

4.4 Based on APP-ELEMENTS

Some types of mutation operators have been defined in [2], four of which are based on Android Application elements and the last is based on crash-es/bugs programs. Because in the above sub-section the mutant's dependent on software bugs were addressed, we do not consider that scenario here. Purpose, Operation Lifecycle, Event Handler and XML are the basis of four types of mutation operators. It was possible to identify the mutation operators [24].

(1) Intent Payload Replacement (IPR)

Intents can carry different parameters as key-value pairs. For example, in the given example below, The sendMessage() method takes the name of the key as the first parameter, and the value as the second. The mutation method replaces the default parameter with the second parameter, values (0 for numeric data types, null for strings and arrays etc.) depending on data type.

```
i.sendMessage(Parameter1, message); //Original
i.sendMessage(Parameter1, “”); //Mutated
```

(2) Intent Target Replacement (ITR)
The destination intent is explicitly specified by the developer in such cases. This forces the tester to enterprise test-cases that check not only parameters, but the receiving element also.

(3) Activity Lifecycle Mutation Operators
This type of operator modifies the methods that are used to fulfill transitions be-tween different states int the application lifecycle, by means of overriding or replacing such methods.

(4) Lifecycle Method Deletion

The lifecycle method deletion operator's delete the overriding methods created by the programmer. The test suites must ensure that the app is in the proper expected state throughout execution.

```
//ORIGINAL code:
public voidActivityA (View v){
Intent intent = new Intent (Example1.this, Example2.class);
startActivity (intent);
}
//MUTATED code:
public voidActivityA (View v){
Intent intent = new Intent (Example1.this, Example3.class);
startActivity (intent);
}
```

(5) Event Handler Mutation Operators

The fundamental property of Android Applications is that they are event-based. Therefore, event handlers are used to identify and respond to events. For user activities such as tapping and dragging, events are generated. In such mutation operations, corresponding event handlers are replaced with other compatible handlers. The two types are:

(6) OnClick() Event Replacement.

After it searches and identifies all the occurrances, each handler of OnClick() event is replaced by a compatible alternative in this method.

```
GeneTic Algorithm for mutant generation
NAME    Mutants_genetic_algo{
Generate first value of Mutants Population
Evaluate population
while(stopping Mutants_genetic_algo criteria not acceptable)
for 1 to (number of mutants)
Execute any of the mutation operator
        (Select/Find/Remove/Create …)
Generate mutant program
Comparison with Original Program
Accept/Delete mutants based on comparison
End for loop
            End while loop
Energy Efficiency Evaluation
Finish
```

Currently, mutation analysis is a very energy costly process as it involves generation of multiple mutant programs that are almost as big as the original application itself. Therefore, the energy efficiency evaluation must be able to give inferences for developing energy efficient mutation testing methods in the future [25, 26]. Table 1 express number of mutants and test-cases generated for different program functions. Figure 1 shows Outcome of various mutant test-cases.

Table 1 Number of mutants and test-cases generated for different program functions

Program function	Random variables	Mutants	Test cases
OnClick Event	1	33	4
setContentView	1	23	6
findViewById	1	65	8
startActivity	5	33	13
createChooser	3	23	32
makeText	6	71	67
Total	17	248	130

5 Mutation Test Results for Sample Program

The following table describes the test-results for a sample program.

6 Conclusion

Mutation testing is the most comprehensive method to evaluate the test cases of any application. However, the process is time consuming with respect to Android even by means of using frameworks; and a large number of mutants must be created from an existing program to effectively check an application. This leads to compilation and execution of a very large number of copies of the program. By keeping this disadvantage apart, mutation tests can be used to identify better test cases and learn about error propagation in the program. The tools and frameworks that are currently available are not capable of giving a complete testing satisfaction with respect to all factors. The fault injection methods are capable of identifying most of the functional outcomes of an application; but on the other hand, the growing complexity of android applications and their domains requires new modules in mutation analysis which are able to reduce energy usage for maximum efficiency and speed in the testing phase.

References

1. Linares-Vásquez M, Bavota G, Tufano M, Moran K, Di Penta M, Vendome C, Bernal-Cárdenas C, Poshyvanyk D (2017) Enabling mutation testing for android apps. In: ESEC/FSE 2017 proceedings of the 2017 11th joint meeting on foundations of software engineering, pp 233–244. https://doi.org/10.1145/3106237.3106275
2. Jabbarv R, Malek S (2017) μDroid: an energy-aware mutation testing frame-work for android. ESEC/FSE'17, Septe 4–8, 2017, Paderborn, Germany, https://doi.org/10.1145/3106237.3106244
3. Deng L, Offutt J, Samudio D (2017) Is mutation analysis effective at testing an-droidApps? In: 2017 IEEE international conference on software quality, reliability and security. https://doi.org/10.1109/QRS.2017.19

4. Gurunath R, Agarwal M, Nandi A, Samanta D (2018) An overview: security issue in IoT network. In: Proceeding of IEEE—2nd international conference on I-SMAC (IoT in Social, mobile, analytics and cloud) (I-SMAC 2018), 30–31 Aug 2018 by SCAD Institute of Technology at Coimbatore, India
5. Kumar A, Samanta D, Chakrabarti S (2015) Principles of mobile computing. In: Proceeding of national conference on convergence in operational and computational technology, 18 Mar 2015 in Bangalore.
6. Bhatia V, Singhal A (2016) Design of a fuzzy model to detect equivalent mutants for weak and strong mutation testing. In:2016 International conference on information technology (InCITe)—the next generation it summit on the theme—internet of things: connect your worlds, Noida, 2016, pp. 1–6. https://doi.org/10.1109/INCITE.2016.7857578
7. Watanabe H, Tokuoka H, Wu W, Saeki M (1998) A technique for analysing and testing object-oriented software using coloured petri nets. In: Software engineering conference
8. Guha A, Samanta D (2021) Hybrid approach to document anomaly detection: an application to facilitate RPA in title insurance. Int J Autom Comput 18:55–72. https://doi.org/10.1007/s11633-020-1247-y
9. Samanta D, Sivaram M, Rashed A, Boopathi CS, Amiri IS, Yupapin P (2020) Distributed feedback laser (DFB) for signal power amplitude level improvement in long spectral band. J Opt Commun https://doi.org/10.1515/joc-2019-0252
10. Pramanik S, Samanta D, Bandyopadhyay S, Ghosh R, A new combinational technique in image steganography. Int J Inf Secur Privacy (IJISP) 15(3). ISSN: 1930–1650|EISSN: 1930–1669
11. AK Biswal, Singh D, Pattanayak BK, Samanta D, Yang M-H, IoT-based smart alert system for drowsy driver detection, Wirel Commun Mobile Comput. ISSN: 1530–8677 (Online)
12. Gurunath R, Samanta D, A novel approach for semantic web application in online education based on steganography. Int J Web-Based Learn Teach Technol (IJWLTT). ISSN: 1548–1093
13. Arnomo SA, Ibrahim NB (2019) Priority path for mutant repairs on mutation testing.2019 2nd International conference on applied information technology and innovation (ICAITI), Denpasar, Bali, Indonesia, 2019, pp 71–76. https://doi.org/10.1109/ICAITI48442.2019.8982158
14. Kintis M, Papadakis M, Papadopoulos A, Valvis E, Malevris N, Le Traon Y (2017) How effective are mutation testing tools$f$ An empirical analysis of Java mutation testing tools with manual analysis and real faults. Empir Softw Eng 1–38
15. Rojas JM, White TD, Clegg BS, Fraser G (2017) Code defenders: crowdsourcing effective tests and subtle mutants with a mutation testing game. In: 2017 IEEE/ACM 39th international conference on software engineering (ICSE) 2017, pp 677–688
16. Bowes D, Hall T, Harman M, Jia Y, Sarro F, Wu F (2016) Mutation-Aware fault prediction mutation testing and mutation metrics. In: ISSTA 2016—proceedings of the 25th international symposium on software testing and analysis, pp 330–341
17. Loise T, Devroey X, Perrouin G, Papadakis M, Heymans P (2017) Towards security-aware mutation testing. In: Proceedings—10th IEEE international conference on software testing, verification and validation workshops. ICSTW 2017, pp 97–102
18. Iida C, Takada S (2017) Reducing mutants with mutant killable precondition (2017) In: 2017 IEEE international conference on software testing, verification and validation workshops (ICSTW), Tokyo, 2017, pp 128–133. https://doi.org/10.1109/ICSTW.2017.29
19. Gopinath R, Alipour MA, Ahmed I, Jensen C, Groce A (2016) On The limits of mutation reduction strategies. In: Proceedings of 38th international conference on software engineering (ICSE 2016), pp 14–22
20. Jia Y, Harman M (2008) Constructing subtle faults using higher order mutation testing. In: Proceedings of 8th IEEE international working conference on source code analysis and manipulation (SCAM 2008), pp 249–258
21. Syed KAK, Samanta D, Paul D (2014) Novel approach for message security. Int J Inf Sci Intell Syst (IJISIS) 3(1):47–52
22. Maheswari M, Geetha S, Selva kumar S, Karuppiah M, Samanta D, Park Y, PEVRM: probabilistic evolution based version recommendation model for mobile applications. IEEE Access https://doi.org/10.1109/ACCESS.2021.3053583

23. Gomathy V, Padhy N, Samanta D et al (2020) Malicious node detection using heterogeneous cluster based secure routing protocol (HCBS) in wireless adhoc sensor networks. J Ambient Intell Human Comput 11:4995–5001. https://doi.org/10.1007/s12652-020-01797-3
24. Sivakumar P, Nagaraju R, Samanta D et al (2020) A novel free space communication system using nonlinear InGaAsP microsystem resonators for enabling power-control toward smart cities. Wireless Netw 26:2317–2328. https://doi.org/10.1007/s11276-019-02075-7
25. Khamparia, A, Singh, PK, Rani, P, Samanta, D, Khanna, A, Bhushan, B (2020) An internet of health things-driven deep learning framework for detection and classification of skin cancer using transfer learning. Trans Emerging Tel Tech e3963. https://doi.org/10.1002/ett.3963
26. Althar RR, Samanta D (2021) The realist approach for evaluation of computational intelligence in software engineering. Innov Syst Softw Eng. https://doi.org/10.1007/s11334-020-00383-2

Classification Framework for Fraud Detection Using Hidden Markov Model

Deepika S. Hegde, Debabrata Samanta, and Soumi Dutta

Abstract Machine learning is described as a computer program that learns from experience E with regard to some task T and some performance measure P, if its performance on T improves with E as measured by P. Suppose we have a credit card fraud detection which watches which transactions we mark as fraud or not, and on the basis, it knows how to filter better fraudulent transactions then, E is watching your transactions is fraud or not, T is classifying your transactions as fraud or not, P is number of transactions correctly differentiated as spam or not spam. Machine learning has two types: supervised learning and unsupervised learning. Supervised learning is the type of machine learning where machine is provided with input mapped with its output, and these inputs and outputs are used to make a machine learn a particular function from the trained dataset. There are two branches of supervised learning, i.e., classification and regression. In unsupervised learning, we do not supervise model instead we allow machine to work on its own to discover information. Clustering is type of unsupervised learning.

Keywords Hidden Markov method · Emission probability · Initial probability · Machine learning · Supervised learning · Credit card · Fraud detection

1 Introduction

In our everyday life, we use our credit card for our most of our financial transactions. We may use our credit card manually, or we may also use it for purchasing something online. When we use our card physically, there is less chance of it being fraudulent as the fraudulent transaction may require the actual credit card in order to steal

D. S. Hegde · D. Samanta (✉)
Department of Computer Science, CHRIST (Deemed to be University), Bengaluru, India

D. S. Hegde
e-mail: deepika.hegde@cs.christuniversity.in

S. Dutta
Institute of Engineering and Management, Kolkata, India

J. M. R. S. Tavares et al. (eds.), *Cyber Intelligence and Information Retrieval*, Lecture Notes in Networks and Systems 291,
https://doi.org/10.1007/978-981-16-4284-5_3

Fig. 1 Model representation

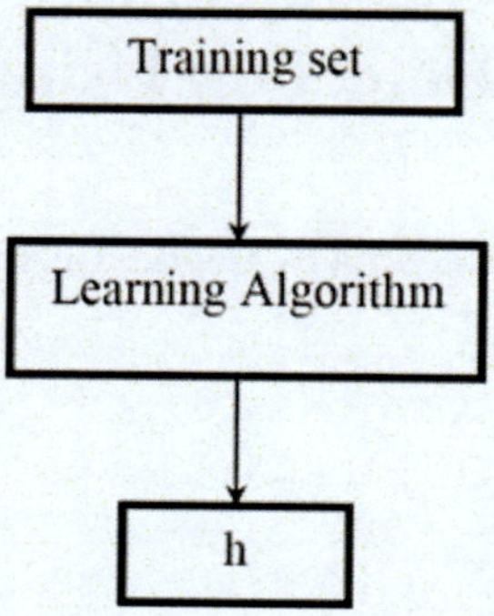

money [1, 2]. But the chances of transactions being Fraudulent increase in online purchase as compared to manual transactions as in online transactions only card information are required for any purchase (like card number, expiry date, cvv) [3–5]. We can define credit card fraud as the transactions made by the person illegally and also where the owner or issuer of card is unaware of the transactions made. The fraud detection algorithms are usually used on the task of classification [14–16]. Classification can be defined as the method of separating datasets by adding labels into various categories or classes [6, 7]. We can define classification as the process of taking dataset, analyzing it, and then dividing the dataset on the basis of different conditions. Here, his hypothesis links between the input variables/features to output variables/target variables [17–20]. Figure 1 expresses model representation.

Records:

m: Count of examples for the training.

x: input variable/function data.

y: output variables/target variables.

A much more formal definition of the supervised learning is to learn a function given a training seth: $x -> y$ to that $h(x)$ is a good prediction for classification of y. That is to classify if the transaction is fraudulent or not.

2 Literature Survey

Credit card fraud detection is one of the major issues which can be solved using machine learning techniques like classification, regression, and clustering [8–10]. There are several ways of using these techniques, some of which include neural networks, random forest, decision tree, linear regression method, logistic regression method, fuzzy logic, K-nearest neighbor, Bayesian network, etc. [21–23, 27]. According to "a comparative analysis of various credit card fraud detection techniques [11], artificial neural network is considered to be one of the best techniques used. According to Geeks for Geeks [12, 13, 24], random forest gives more accurate results. But whatever technique we use, we will get false positive values. Our task is to minimize these false positive values and give more accurate results.

3 Proposed System

The proposed system uses hidden Markov model for the classification of fraudulent transactions. In this system, we have used our own example dataset to show how the model works. In this model, we are predicting that the transaction is fraudulent or not using the location of the transactions based on the previous transactions of the person. Here, we take the information of the particular person using the credit card

Table 1 Transactions made by a particular person

Number of transactions	Location	Fradulant/Not fradulant
1	Koramangala, Bangalore, Karnataka	Not fradulant
2	Adugodi, Bangalore, Karnataka	Fradulant
3	Koramangala, Bangalore, Karnataka	Not fradulant
4	Koramangala, Bangalore, Karnataka	Fradulant
5	Koramangala, Bangalore, Karnataka	Fradulant
6	Adugodi, Bangalore, Karnataka	Not fradulant
7	Adugodi, Bangalore, Karnataka	Not fradulant
8	Koramangala, Bangalore, Karnataka	Not fradulant
9	Koramangala, Bangalore, Karnataka	Not fradulant
10	Koramangala, Bangalore, Karnataka	Not fradulant

and also the information of the locations where these cards has been used. Table 1 denotes transactions made by a particular person.

The above table shows the transactions made by the particular person. So for now, we have considered only ten transactions made by that person. Let us consider second, fourth, and fifth transactions to be fraudulent, and all other transactions are genuine. Out of ten transactions, seven transactions are made in Koramangala and three transactions are made in Adugodi [25, 26].

Therefore, the probability of transactions made in Adugodi is 0.3, and the probability of the transactions made in Koramangala is 0.7. The probability of fraudulent transactions made in koramangala is 0.14, and the probability of transactions are not fraudulent is 0.86. The probability of fraudulent transactions made in Adugodi is 0.33, and the probability of transactions are not fraudulent is 0.67. These probabilities are called emission probabilities in hidden Markov method. Table 2 shows emission probabilities. Figure 2 defines emission probabilities.

Therefore, emission probabilities of the above transactions are transactions in both Koramangala and Adugodi. The fraudulent transactions are less when compared to genuine transactions. Hidden Markov method uses another probability, i.e., transition probability. This probability tells about the transactions in the place, i.e., if the present transaction is not fraudulent, then it tells about the next transaction. Therefore, transition probabilities of the above transactions is that now let us consider another five transactions made by the same person which our model is expected predict if the transactions are fraudulent or not.

Table 2 Emission probabilities

Location	Fradulant	Not fradulant
Koramangala	0.14	0.86
Adugodi	0.33	0.67

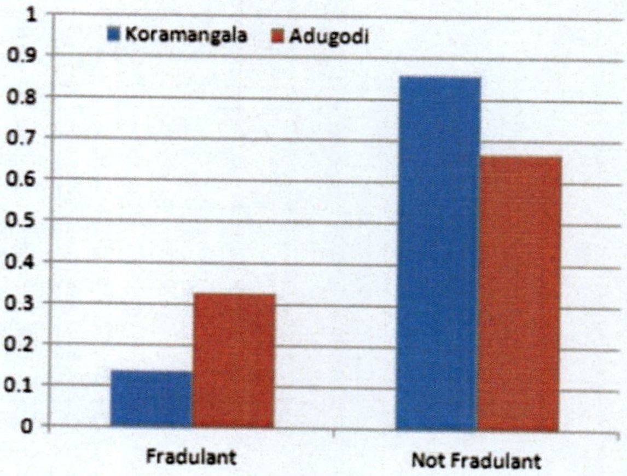

Fig. 2 Emission probabilities

4 Algorithms

So the algorithm to be used to predict if the transactions given above or fraudulent or not: Using the above algorithm, we can find if the transaction was can send a personal question to the registered number of the user if the transaction turns out to be fraudulent [14–16]. The question can be anything like his DOB, Aadhar number, PAN card number, etc. If the answer given by the user is correct, then the transaction must be allowed [17, 18]. Else transaction must not be allowed. Figure 3 shows model representation.

1.Note the transition probability

2.Note initial probabilities

3.Note the emission probabilities

4.Next transactions whose transactions are to be predicted has to be given as input

5.For all the transactions

1. Check if transaction is made in 'koramangala'

*A. Genuine_transaction=max(last_transaction_genuine*transition_probability_genuine_to_genuine *emission_probability_koramangala_genine, last_transaction_fradulant* transition_probability_fradulant_to_genuine * emission_probability_koramangala_genine).*

B. Fradulant_transaction=max(last_transaction_genuine transition_probability_genuine_to_fradulant*emission_probability_adugodi_geni*

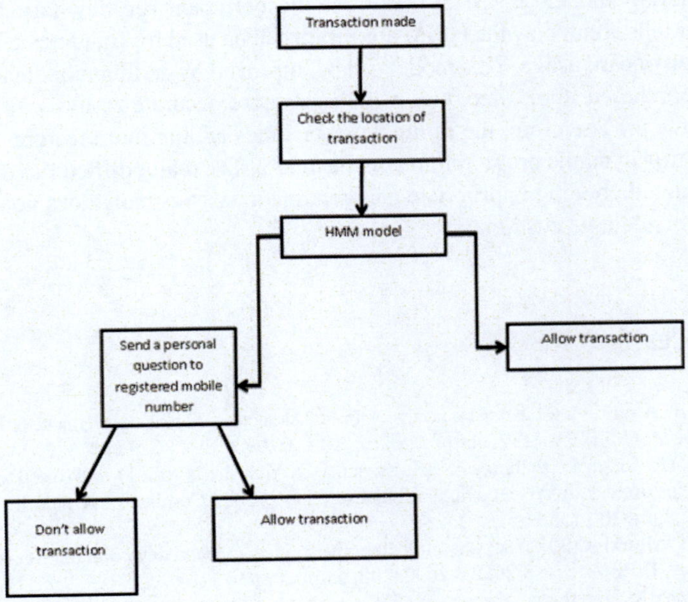

Fig. 3 Model representation

*ne,last_transaction_fradulant*transition_probability_fradulant_to_fradulant*emissi on_probability_adugodi_genine)*
else if transaction is made in 'adugodi'
*A. Genuine_transaction=max(last_transaction_genuine*transition_probability_gen uine_to_genuine *emission_probability_koramangala_fradulant, last_transaction_fradulant* transition_probability_fradulant_to_genuine * emission_probability_koramangala_fradulant)*
*B. Fradulant_transaction=max(last_transaction_genuine * transition_probability_genuine_to_fradulant*emission_probability_adugodi_frad ulant, last_transaction_fradulant * transition_probability_fradulant_to_fradulant* emission_probability_adugodi_fradulant)*
3.else
1.Fraudulent
6. ifprobability_genuine>probanility_fradulant
1. genuine
7. else
1. Fraudulent

5 Conclusion

Hidden Markov model (HMM) is one of the most efficient very few false positives. This paper tells about how the HMM algorithm can be used for fraud detection using the location of card usage. This model can be improved by adding more features like amount, purchased item, time, etc., which get more accurate results. We can also use dynamic pr. Therefore, the future work includes adding more features and also implementing dynamic programming on the model. The major difficulties during the research include finding appropriate and genuine dataset as bank does not give any information about its customer.

References

1. Omair B, Alturki A (2020) Taxonomy of Fraud Detection Metrics for Business Processes. IEEE Access 8:71364–71377. https://doi.org/10.1109/ACCESS.2020.2987337
2. Biswal AK, Singh D, Pattanayak BK, Samanta D, Yang M-H (2021) IoT-Based smart alert system for drowsy driver detection. Wirel Commun Mobile Comput 13 Article ID 6627217. https://doi.org/10.1155/2021/6627217
3. Omair B, Alturki A (2020) A systematic literature review of fraud detection metrics in business processes. IEEE Access 8:26893–26903. https://doi.org/10.1109/ACCESS.2020.2971604
4. Maheswari M, Geetha S, Kumar SS, Karuppiah M, Samanta D, Park Y (2021) PEVRM: probabilistic evolution based version recommendation model for mobile applications. IEEE Access. https://doi.org/10.1109/ACCESS.2021.3053583

5. Sun C, Yan Z, Li Q, Zheng Y, Lu X, Cui L (2019) Abnormal group-based joint medical fraud detection. IEEE Access 7:13589–13596. https://doi.org/10.1109/ACCESS.2018.288711
6. Gomathy V, Padhy N, Samanta D et al (2020) Malicious node detection using heterogeneous cluster based secure routing protocol (HCBS) in wireless adhoc sensor networks. J Ambient Intell Human Comput 11:4995–5001. https://doi.org/10.1007/s12652-020-01797-3
7. Makki S, Assaghir Z, Taher Y, Haque R, Hacid M, Zeineddine H (2019) An Experimental study with imbalanced classification approaches for credit card fraud detection. IEEE Access 7:93010–93022. https://doi.org/10.1109/ACCESS.2019.2927266
8. Sivakumar P, Nagaraju R, Samanta D et al (2020) A novel free space communication system using nonlinear InGaAsP microsystem resonators for enabling power-control toward smart cities. Wirel Netw 26:2317–2328. https://doi.org/10.1007/s11276-019-02075-7
9. Carrasco RSM, Sicilia-Urbán M-Á (2020) Evaluation of deep neural networks for reduction of credit card fraud alerts. IEEE Access 8:186421–186432. https://doi.org/10.1109/ACCESS.2020.3026222.
10. Khamparia A, Singh PK, Rani P, Samanta D, Khanna A, Bhushan B (2020) An internet of health things-driven deep learning framework for detection and classification of skin cancer using transfer learning. Trans Emerging Tel Tech e3963. https://doi.org/10.1002/ett.3963
11. Zanetti M, Jamhour E, Pellenz M, Penna M, Zambenedetti V, Chueiri I (2019) A tunable fraud detection system for advanced metering infrastructure using short-lived patterns. IEEE Trans Smart Grid 10(1):830–840. https://doi.org/10.1109/TSG.2017.2753738
12. Althar RR, Samanta D (2021) The realist approach for evaluation of computational intelligence in software engineering. Innov Syst Softw Eng. https://doi.org/10.1007/s11334-020-00383-2
13. Fang W, Li X, Zhou P, Yan J, Jiang D, Zhou T (2021) Deep learning anti-fraud model for internet loan: where we are going. IEEE Access 9:9777–9784. https://doi.org/10.1109/ACCESS.2021.3051079
14. Li Z, Liu G, Jiang C (2020) Deep representation learning with full center loss for credit card fraud detection. IEEE Transactions on Computational Social Systems 7(2):569–579. https://doi.org/10.1109/TCSS.2020.2970805
15. Guha A, Samanta D (2021) Hybrid approach to document anomaly detection: an application to facilitate RPA in title insurance. Int J Autom Comput 18:55–72. https://doi.org/10.1007/s11633-020-1247-y
16. Zhang Z, Chen L, Liu Q, Wang P (2020) A fraud detection method for low-frequency transaction. IEEE Access 8:25210–25220. https://doi.org/10.1109/ACCESS.2020.2970614
17. Samanta D, Sivaram M, Rashed A, Boopathi CS, Amiri, IS, Yupapin P (2020) Distributed feedback laser (DFB) for signal power amplitude level improvement in long spectral band. J Opt Commun https://doi.org/10.1515/joc-2019-0252
18. Hu T, Guo Q, Shen X, Sun H, Wu R, Xi H (2019) Utilizing Unlabeled data to detect electricity fraud in AMI: a semisupervised deep learning approach. IEEE Trans Neural Netw Learn Syst 30(11):3287–3299. https://doi.org/10.1109/TNNLS.2018.2890663
19. Samanta D, Podder SK (2019) Level of green computing based management practices for digital revolution and new India, Int J Eng Adv Technol (IJEAT) 8(3S). ISSN: 2249–8958
20. P. Sivakumar, R Nagaraju, Samanta D, M. Sivaram, Nour HindiaIraj, S Amiri, A novel free space communication system using nonlinear InGaAsP microsystem resonators for enabling power-control toward smart cities, Wirel Netw J Mobile Commun Comput Inf ISSN: 1022–0038 (Print) 1572–8196 (Online)
21. Mahua B, Podder SK, Shalini R, Samanta D (2019) Factors that influence sustainable education with respect to innovation and statistical science. Int J Recent Technol Eng 8(3). ISSN:2277–3878
22. Praveen B, Umarani N, Anand T, Samanta D, Cardinal digital image data fortification expending steganography. Int J Recent Technol Eng 8(3). ISSN:2277–3878
23. Dhanush V, Mahendra AR, Kumudavalli MV, Samanta D (2017) Application of deep learning technique for automatic data exchange with air-gapped systems and its security concerns. In: Proceedings of IEEE international conference on computing methodologies and communication [ICCMC 2017] @IEEE 18–19 July 2017, Erode

24. Kumar R, Rishabh K, Samanta D, Paul M, Vijaya Kumar CM (2017) A combining approach using DFT and FIR filter to enhance Impulse response. In: Proceedings of ieee international conference on computing methodologies and communication [ICCMC 2017] @IEEE , 18–19 July 2017, Erode
25. Ghosh G, Samanta D, Paul M, Janghel NK (2017) Hiding based message communication techniques depends on divide and conquer approach. In: Proceedings of IEEE international conference on computing methodologies and communication [ICCMC 2017] @IEEE, 18–19 July 2017, Erode
26. Singh RK, Begum T, Borah L, Samanta D (2017) Text encryption: character jumbling. In: Proceedings of IEEE international conference on inventive systems and control [ICISC 2017] @IEEE, 19–20 Jan 2017, Coimbatore. 978-1-5090-4715-4/17/$31.00 @ 2017 IEEE
27. Dutta S, Ghatak S, Dey R et al (2018) Attribute selection for improving spam classification in online social networks: a rough set theory-based approach. Soc Netw Anal Min 8:7. https://doi.org/10.1007/s13278-017-0484-8

A Review on Security Issues in Healthcare Cyber-Physical Systems

V. S. Abhijith, B. Sowmiya, S. Sudersan, M. Thangavel, and P. Varalakshmi

Abstract The rapid advancements in computation methodologies and cloud computing along with the rapid expansion of the Internet of things have resulted in tremendous advancements in Cyber-Physical Systems (CPS) especially in health care systems. The competencies for adaptation of advanced healthcare monitoring systems have been raised. In healthcare industries, today CPS ranges from providing smart feedback systems, automated updating of the patient data digitally, continuous passive monitoring of patients using biosensors, and much more. Cyber-Physical Systems are changing into life-saving decision-making systems. Moreover, CPS has developed an unbreakable bonding with Wireless Sensor Networks abbreviated as WSN. The main motive has been to consolidate the major works in the areas of data privacy and security in CPS and healthcare in the survey. In this paper, researchers' contributions related to the security of Healthcare CPS have been discussed in general and included the latest updated papers in the domain, later, has extended the focus to Wireless Sensor Networks and the vulnerabilities faced along with their defense as an overview. The concept of Sensor Only Networks, an advancement to WSN based on the security policies has been overviewed.

Keywords Cyber-Physical systems · Data privacyy · Healthcare · Security in MCPS · Wireless sensor networks

---

V. S. Abhijith · B. Sowmiya · S. Sudersan
Department of Information Technology, Thiagarajar College of Engineering, Madurai, Tamil Nadu, India

M. Thangavel (✉)
Department of Computer Science and Engineering, Siksha 'O' Anusandhan (Deemed to be University), Bhubaneswar, Odisha, India

P. Varalakshmi
Department of Computer Technology, Madras Institute of Technology, Anna University, Chennai, Tamil Nadu, India

J. M. R. S. Tavares et al. (eds.), *Cyber Intelligence and Information Retrieval*, Lecture Notes in Networks and Systems 291,
https://doi.org/10.1007/978-981-16-4284-5_4

1 Introduction

In recent years, the rise in the demand for quality products and services along with systems that could integrate the control mechanisms with high computational capabilities lead to the evolution of Cyber-Physical Systems (CPS). CPS integrates with Wireless Sensor Networks (WSN) to provide better integration with the cyber world. The transformations in various fields have also affected healthcare to have a demand for new and innovative healthcare devices that included the concept of Cyber-Physical Systems. Healthcare is an important factor in improving the quality of life since ancient times. But due to the current changes in the lifestyle and various other factors that occur around us, there is a high chance of people getting affected by diseases. And during 1999, the first attempt at the development of eHealth came into existence. But the first evolution occurred at the beginning of the twentieth century where doctors and other health specialists moved towards technology-based assistance for the diagnosis and consultation of patients. [10, 11] Thus, evolved telemedicine from telecommunication and information technology. Various methods and techniques were proposed for securing patient data online after the development of cloud technologies for storing the data. Gradually, the world faced the need for more sophisticated medical devices that could moreover act as life-saving devices. The great urge led to the evolution of the Cyber-Physical System in the area of healthcare. CPS initially acted as the paradigm that monitors and controls various biosensors and integrated systems to provide personalized assessment over the medical records. Thus, the healthcare network through CPS developed as a distributed system with the sharing of information. The use of sensors and actuators which was the signature of CPS followed certain processes and procedures:

1. data collection from various sensor data,
2. sending it to the cloud via a gateway for processing,
3. querying and real-time for processing of the data for further processing,
4. decision making from the processed data
5. providing the actual results and recommendations.

2 Why Does Privacy Stand as a Major Factor in Healthcare?

The main vulnerabilities and areas of risk assessment include data leakage, improper disposal of data, unauthorized access, data loss, and hacking IT incidents which have become the major part. According to stats, over 13% of the cloud services used in the healthcare industry are at high risk and, about 77% of them are at medium risk. (Source: Blog at appknox.com). Cyber-attacks that have become so popular today stand as a perfect threat to the present industry. Wireless Sensor Networks (WSN) as it suggests consist of many and many nodes and small devices connected to it each with sensing, processing, communicating, and retrieving of millions of bytes of data every moment. But due to the lack of processing power, limited memory they had to be accustomed to traditional security methodologies to defend all these vulnerabilities. Since they have a larger processing time and comparatively higher overhead, various methodologies and algorithms have been proposed to deal with them.

3 MCPS and Vulnerabilities Explained

Medical Cyber-Physical Systems: MCPS is the integration and incorporation of medical implementations through Cyber-Physical Systems. Though they are complex systems they face more challenges related to interoperability, privacy, the security of data. These are used as implantable devices and as automated devices in hospitals and clinics. Ensuring security should begin from the design phase concentrating on the security aspect while developing every module and functionality. A Cyber-attack is an intentional attack on system architecture with the intention of either exploiting the resources or gaining sensitive information from the victim. Even though CPS is a complex system they are still vulnerable to these attacks if not maintained and secured. These systems being an integration of the physical system to the cyber world could act as the best measure that could lead to the compromising of a very large amount of sensitive data since they connect all those physical systems with the internet world where the data has a lot of exposure to attacks. Some of the common threats that the Cyber-Physical Systems are vulnerable to are: *Attack from the outside (external environment):* The attack usually occurs from an external environment where the attackers focus on exploiting the victim environment. *Attack from the inside (internal attacks):* In this type of attacks, the employee in the organization or an employee who has left the organization in the near past but still having access to the organizational environment may be acting as the compromising factor which could act as a backdoor for the attackers to enter into the secured environment and exploit the data. *Worms and Viruses* (Neither external nor internal): These are also critical types of exploitation methodology that may occur through carelessness and insecure scanning of the network or connection through a public wireless network.

Concentrating on deeper attacks, Denial of service stands at the top of the list. Considering healthcare services as the main area of risk assessment Denial of Service is critical. Any denial of service in the healthcare environment or network can produce catastrophic results. Though the vulnerability could be restored even downtime of a few seconds could even cause a high and threatening danger to the network and the device connected to it. Man in the Middle Attack, which occurs when the hacker represents himself as an authorized person between the client and the server to steal sensitive information. Another important risk that could affect every field of technology is the Phishing attack. The emails or links sent would always look casual and secure enough and once clicked may execute a code that could even change the password of the system. The fundamental requirement of security in Medical CPS would be robustness and response to unpredictable conditions. CPS majorly depends on the communication networks that may not be secured always Though cryptography techniques seem to be comparatively providing security to the systems, these methods are now facing increasing risks with the increase in the development of cloud and quantum computing methodologies that help attackers to extract information with much faster rates than that was available to date. The concept of Physical Layer Security (PLS) introduced achieved the next level of security in the cryptosystems as they didn't require pre-shared keys for the transfer of information.

The security and privacy in MCPS systems, providing security, and high confidence in the transfer of data from the sensors especially biosensors to the cloud gateway and after processing to the physical systems that provide personalized medical suggestions still stand a challenge for researchers. In this paper, various contributions related to the field of security in CPS healthcare particularly MCPS have been elaborated. In the later section, the gaps in the area have been identified with proposed solutions.

4 CPS in Health Care

The technological revolution oversaw considerable progress in the field of healthcare technologies. However, these new technologies have also made healthcare data not only much bigger but also much more arduous to handle, process, and have contributed to a higher response time. These issues can be overcome by the use of Cyber-physical systems (Fig. 1).

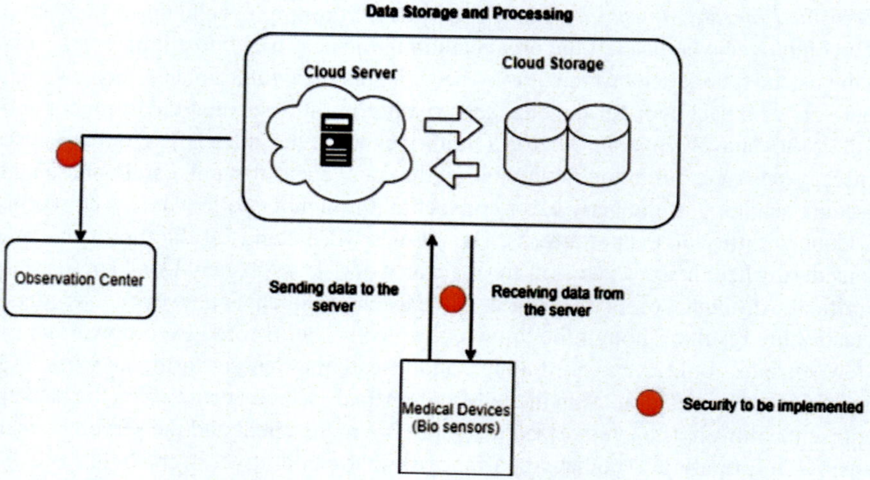

Fig. 1 Block diagram on the communication and controls in MCPS

4.1 Security Requirements in Case of Sensor Networks

Data has been the main focus and this section describes the more detailed part of Confidentiality, Integrity, and Availability in the context of Sensor Networks.

Confidentiality.: A node should not be able to access the data or values or readings of any other nodes until and unless they are provided authorization and access to. The mechanism should be robust and secure at the same time ensuring privacy. Not only the data, but the information of the nodes such as the session ids, the keys, and other node-based information shall be encrypted.

Integrity: The sensor information should be transmitted and received by the nodes and the server ensuring that the data received is the same as sent by the sender node and has not been modified. Also, to ensure that the packet has not been captured in an earlier time and has been replayed.

Availability: As an extension to the internal, external, and phishing attacks, DoS (Denial of Service) has been a major issue in securing WSN. Many methodologies have been proposed to prevent this attack.

Authentication: The main motive is to ensure that the data or information received is from the authorized node.

Though several methodologies have been proposed to ensure all these policies and services, extending the question in the case of Wireless Sensor Networks still stands uncovered to the fullest (Fig. 2).

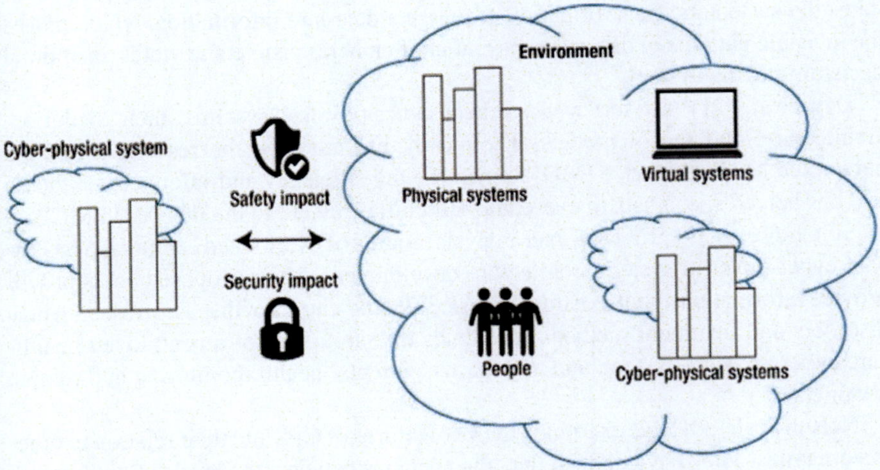

Fig. 2 Representation of CPS and security impact. *Source* IEEE Computer Society

5 State of Art

A large number of papers have been taken into consideration not specifically focusing on WSN but to the at most methodologies in the domain of healthcare in CPS and has focused on portraying them to WSN in providing an additional layer of security techniques and methodologies.

Nair et al. [19] detailed the range of attacks that are possible against data privacy breaches in Medical Cyber-Physical Systems and contrasted the characteristics of Intrusion Detection Systems (IDS) in MCPS especially devices. The article has been concluded with the need of improving efficiency and several crucial issues in it.

Khan et al. [25] investigates the methodology of an attacker developing a custom tool in the Man in the Middle (MITM) attack against synchrophasor devices. The authors have detailed the Anatomy of MITM Attack against a Substation and demonstrated the Distributed Intrusion Detection System and to protect the synchrophasor device against those attacks. Have further put forth the lack of authentication, encryption, and cryptographic signatures.

Nithya et al. [21] have put forth the crucial behavior of Medical devices, the architecture of Medical Cyber-Physical Systems (MCPS), and the issues and challenges in MCPS. The authors have also described a described survey on cryptography, access control, anomaly detection, authentication, and many other solutions in the field.

Rahaman et al. [23] focuses on existing trends in the development of big data analytics and the architecture for CPS (cyber-physical systems and wireless sensor networks (WSNs) for cloud computing for healthcare applications (such as life support) and its application in monitoring and decision support systems. CPS facilitates the systematic transformation of massive data into information, which makes the invisible patterns of degradations and inefficiencies visible and yields an optimal decision-making system.

Dey et al. [21] has surveyed medical cyber-physical systems, their evolution, architecture, and sensor models. Their work has enriched the researchers of the networked Medical Device (MD) to increase the efficiency and safety of healthcare and has helped specialists to overcome difficulties related to the devices in MCPS.

Sultanov et al. [9] provide a data analysis module of overlooked centralized healthcare cyber-physical system architecture development. Their proposed concept will provide information about existing diseases, identify causes of diseases, research new disorders and treatment methods for clinics that are a part of a centralized healthcare cyber-physical system, and automatize patients' health monitoring and remote treatment.

Nelson et al. [20] have examined the key features of CPS and their relation to other system types. They have defined the dependencies between the levels of automation and human roles in CPS from a systems engineering viewpoint. They have applied system thinking to describe a multi-layered diagrammatic representation of CPS for combined safety and security risk analysis and have demonstrated an application in the maritime sector to analyze an autonomous surface vehicle.

Gu et al. [18] have integrated fog computation and MCPS to build fog computing supported MCPS (FC-MCPS) and have jointly investigated base station association, task distribution, and virtual machine placement towards cost-efficient FC-MCPS. They have formulated the problem into a mixed-integer nonlinear program (MINLP) and then linearized it into mixed-integer linear programming (MILP).

Xu et al. [24] enhanced privacy communication by physical layer security enhancement methodologies. Specifically, they have used an amplify and forward (AF) relay network wherein all devices are equipped with a single antenna. Numerical results have also been presented to show the effectiveness of their proposed algorithms.

Jiang et al. [30] have conducted the first study on applying execution time verification to cooperate with current DSS based on real-time data. Within the proposed technique, a user-friendly domain-specific language, named DRTV, was designed to specify important real-time data sampled by medical devices and temporal properties originating from clinical guidelines. In terms of warning detection, it outperforms human inspection and improves the quality of clinical health care of the hospital.

Dogaru et al. [8] neatly presents the role of Cyber-physical Systems in healthcare networks. The authors proposed a general framework for the medical and related devices and services and have put forth the various security issues in it. The author has also thrown light on the lack of authentication for medical devices and equipment.

Seifert et al. [7] surveys the cyber-physical system architecture that is currently available. Clearly described the architectural qualities of health care CPS in which the scalability, adaptability, reliability, unique challenges, architectural quality, reactivity have been discussed. A complete case study on the Infusion Pump has been put forth to explore the security issues in the area.

Rao et al. [26] proposes an effective mechanism for handling data security through a novel security key exchange model. The results after simulation expressed have been claimed to provide increased throughput with the least overhead and end to end delay. A Data Privacy Mechanism has been proposed that could target both the attacks from inside and outside. Which utilizes a TTP certificate.

Giraldo et al. [13] has given a detailed survey on surveys on the topics of CPS Domains, attacks, research trends, network security, security level implementations. Explained from the basics of CPS Security and privacy along with the taxonomy. The author has given deeper insights into the Security of medical devices, ICS, Manufacturing, intelligent transportation, aircraft drones, and other important CPS domains. The author has further extended the survey with research directions for the future.

Almohri et al. [12] has focused on creating a thorough understanding of the threat model in Medical Cyber-Physical Systems (MCPS The author has also given a clear relevance to the Trust and the Threat models in detail along with threat analysis. The security remedies have also been detailed into Cryptographic Techniques, System Hardening techniques, and so on.

Bolbot et al. [27] has described the sources of complexities in CPS concerning their variation in time, the sources of structural complexity, sources of dynamic complexity, and the safety assurance methods of CPS. The author has also pointed out the hazards and provided verification and validation through model checking, fault injection, and so on along with testing along with a pictorial representation of the result and analysis.

Saleem et al. [16] has explained the Common threats to Wireless Body Area Network in eHealthcare systems. The author has divided the area of assessment of the security into different levels like Administrative level, Network level, MAC Level, and so on.. The article gives a clear-cut explanation of the Cloud Security for Healthcare and also the development of Healthcare with 5G networks and the security to be implemented.

Zao et al. [33] have proposed proxy-oriented identity-based outsourcing in cloud-based MCPS with public auditing scheme using elliptic curve cryptography. The proposed scheme enables a user to authorize the proxy to generate and upload the signatures of medical data and corresponding encrypted data to cloud-based MCPS. The medical data can be audited efficiently by any third-party auditor (TPA), without retrieving the entire medical data set. The proposed scheme is designed on identity-based systems, which can avoid complex certificate management.

Mohammed et al. [34] define the field of CPS forensics and its dimensions viz. Technical, Organisational, and Legal and reviews examples of current research efforts in the field and the types of tools and methods they propose for CPS forensics The authors then use the review outcomes to discuss future research directions to address challenges and create more effective, efficient, and safe forensics tools and for CPS.

6 Possible Cyber Measures and Breaches in Healthcare

See Table 1.

7 Comparative Study

See Table 2.

8 Security Advancements Proposed

Since the vulnerabilities keep evolving always even when it is felt to be secured, providing maximum security and identification of all the vulnerabilities is important. In the case of Complex Systems like CPS, it is not possible to make the entire system secure by a single methodology. There are a lot of security methodologies either technical and non-technical to be framed to ensure safety and security in terms of data and other aspects. A few of the identified security implements are:

Table 1 Possibility of breaches and cyber structure available

S. No.	Area of assessment	Cyber structure available	Breach possibility techniques	Physical structure
1	Confidentiality	Crypt algorithms and protocols	Sniffing, MITM	Stealing documents and immediate information-
2	Integrity	Advanced hashing anti-malware	False data injection phishing data manipulation	Agreement breach social engineering
3	Availability	Spread spectrum backup devices firewalls	DDoS, replay attack	Physical damage/natural destruction
4	Authentication	Strong password multi-factor authentication	Password recovery vulnerabilities, cracking using quantum computing	Privilege escalation and unauthorized access, Insider attacks
5	Privacy	Anti-malware, antivirus protection	Virus, trojan, backdoor	Phishing, carelessness

Table 2 Methodologies proposed so far and their inference

S. No.	Methodologies/solution	Inference
1	Intrusion detection system (IDS)	Difficult to connect to actuator systems esp. in the case of implantable medical devices
2	ARP poisoning and real synchro phaser systems	Poor observability in terms of threats. inability in a real-time environment
3	Network attack by using One-class SVM, approached by Ming Zhang et al.	The new upcoming event behavior cannot be recognized and be declared as not safe or vulnerable network
4	Implementation of the HSM module	Security in terms of network and other factors to be analyzed
5	Crypt DB encryption	Better protection but can be cracked through quantum computing if not strong enough
6	Implementation of software defined networking (SDN) and network function virtualization (NFV)	Still vulnerable to an extend by Hypervisor vulnerabilities, security service insertion, and Improper network isolation (to some extend possibility)

8.1 Strong Security Policies

Most of the physical attacks mentioned in Table 1 may be controlled by maintaining proper and secure security policies. Maintaining a strong security policy among the employees and administrators on non-compromising of the security and authenticity should be maintained. The Inside attacks could be prevented by revoking the credentials within a period.

8.2 Sensor Only Network

The researcher introduces the concept of Sensor Only Network (SON), the advancement of the form of Wireless Sensor Network, wherein the following security policies are to be strictly maintained and followed:

The sensor or the actuator connected to the network may have very little access control over the other nodes in the network that too on authorization by the node. Robust and secure key-based encryption methodology wherein even the public keys are encrypted thereby preventing the chances of MITM and network traffic based attacks. The nodes shall be self-organized and distributed. Each transmission made would be encrypted with time as a factor and would be dropped by the receiver node unless received within the time scheduled, thereby preventing the chances of replay attacks. SON would mean very little interpretation of other devices in the network secured via Universal 2nd Factor. The devices in the network may be authenticated thereby preventing MITM to the fullest extent possible. A cryptographic addition is made where a flag is affixed as a header with the transmitted packet which is programmed to be changed if the encoding changes in the scope of the transmission thereby preventing modification. Though the methodologies proposed to appear to be complex and require a large amount of overhead, the Sensor Network proves to be made available especially in the domain of health care which in contrast prove to cause real-world damage to the life and safety of the people.

8.3 Future Scope

The proposed series of methodologies is a conceptual approach for preventing threats and attacks in the world of Sensor networks. The methodologies have not yet been experimentally analyzed and implemented. The concept of advancement in Wireless Sensor Networks so-called Sensor Only Networks may prove to be the best security when implemented. The nearest future scope may include the analysis of the feasibility of the methodologies proposed.

9 Conclusion

Security being the major concern of the Wireless Sensor Networks, the paper would be a great approach for consolidation of the advancements in the field along with the advancements that could be proposed in the future. The concept of sensor only networks (SoN) introduced appears to address the complexity and the overhead that the CPS systems are now facing especially in the area of healthcare systems that involves sensors and cloud. Moreover, the security aspects have been studied considering various aspects of complexity, privacy related to both hardware and software and the combination of all the best methodologies proposed in the field. The main aim has been to introduce a technology that could provide maximum security thereby reducing overheads. Thus the survey paper stands a very good reference filling the gaps in the research areas concerned with the security and privacy aspects of Cyber Physical Systems in Healthcare. The paper also proposed the advancements that could be made in terms of administrative security, and the concept of Sensor Only Networks, an advancement to Wireless Sensor Networks.

References

1. Rao A, Carreón N, Lysecky R, Rozenblit J (2018) Probabilistic threat detection for risk management in cyber-physical medical systems. IEEE Softw 35(1):38–43
2. Humayed A, Lin J, Li F, Luo B (2017) Cyber-Physical systems security—a survey. IEEE Internet Things J 4(6):1802–1831
3. Pérez AH, Clemente MG, Perez FJM (2018) Sustainable securing of medical cyber-physical systems for the healthcare of the future. Sustain Comput Inf Syst 19:138–146
4. Gatouillat A, Badr Y, Massot B, Sejdić E (2018) Internet of medical things: a review of recent contributions dealing with cyber-physical systems in medicine. IEEE Internet Things J 5(5):3810–3822
5. Bhatnagar V, Jain A (2017) Concoction of ambient intelligence and big data for better patient ministration services. Int. J. Ambient Comput. Intell. 8(4):19–30
6. Konstantinou C, Maniatakos M, Saqib F, Hu S, Plusquellic J, Jin Y (2015) Cyber-physical systems: a security perspective. In: Proceedings - 2015 20th IEEE european test symposium, ETS 2014. Institute of Electrical and Electronics Engineers Inc.
7. Seifert D, Reza H (2016) A security analysis of cyber-physical systems architecture for healthcare. Computers 5(4):27
8. Dogaru DI, Dumitrache, I (2015) Cyber-physical systems in healthcare networks. In: 2015 E-Health and bioengineering conference (EHB), Iasi, pp 1–4
9. Sultanovs E, Romānovs A (2016) Centralized healthcare cyber-physical system's data analysis module development. In: 2016 IEEE 4th workshop on advances in information, electronic and electrical engineering (AIEEE), Vilnius, pp 1–4
10. Grispos G, Glisson WB, Choo KR (2017) Medical cyber-physical systems development: a forensics-driven approach. In: 2017 IEEE/ACM international conference on connected health: applications, systems and engineering technologies (CHASE), Philadelphia, PA, pp 108–113
11. Qiu H, Qiu M, Liu M, Memmi G (2020) Secure health data sharing for medical cyber-physical systems for the healthcare 4.0. IEEE J Biomed Health Inform 24(9):2499–2505
12. Almohri H, Cheng L, Yao D, Alemzadeh H (2017) On threat modeling and mitigation of medical cyber-physical systems. In: Proceedings of the second IEEE/ACM international conference on connected health: applications, systems and engineering technologies (CHASE '17). IEEE Press, pp 114–119
13. Giraldo J, Sarkar E, Cardenas AA, Maniatakos M, Kantarcioglu M (2017) Security and privacy in cyber-physical systems: a survey of surveys. IEEE Design Test 34(4):7–17
14. Yaacoub JPA, Salman O, Noura HN, Kaaniche N, Chehab A, Mallia M (2020) Cyber-physical systems security: limitations, issues and future trends. Micro Process Microsyst 77:103201

15. Stankovic JA (2017) Research directions for cyber physical systems in wireless and mobile healthcare. ACM Trans Cyber-Phys Syst 1(1):12. Article 1
16. Saleem K, Tan Z, William B (2017) Security for cyber-physical systems in healthcare. https://doi.org/10.1007/978-3-319-47617-9_12
17. Choo KKR, Kermani MM, Azarderakhsh R, Govindarasu M (2017) Emerging embedded and cyber physical system security challenges and innovations. IEEE Trans Dependable Secur Comput 14(3):235–236
18. Gu L, Zeng D, Guo S, Barnawi A, Xiang Y (2017) Cost efficient resource management in Fog computing supported medical cyber-physical system. IEEE Trans Emerg Top Comput 5(1):108–119
19. Nair MM, Tyagi AK, Goyal R (2019) Medical cyber physical systems and its issues. Procedia Comput Sci 165:647–655
20. Carreras Guzman NH, Wied M, Kozine I, Lundteigen M (2020) Conceptualizing the key features of cyber-physical systems in a multi-layered representation for safety and security analysis", Systems Engineering. 189–210, 2020.
21. Dey N, Ashour AS, Shi F et al (2018) Medical cyber-physical systems: a survey. J Med Syst 42:74
22. Nithya S, Sangeetha M, Apinaya Prethi KN (2018) Role of cyber-physical systems in health care and survey on security of medical data. Int J Pharma Res Health Sci 6:2075–2080
23. Rahaman MO, Shuvo A, Abul Kashem M, Cyber physical systems for healthcare. Int J Adv Res 6(Nov):36–46
24. Xu Q, Ren P, Song H, Du Q (2016) Security-Aware waveforms for enhancing wireless communications privacy in cyber-physical systems via multipath receptions. IEEE Internet Things J 2327–4662
25. Khan R, McLaughlin K, Laverty JHD, David H, Sezer S (2018) Demonstrating cyber-physical attacks and defense for Synchrophasor technology in smart grid. In: 2018 16th Annual conference on privacy, security and trust (PST), Belfast, pp 1–10
26. Rao KR, Rao SNT, Reddy C (2018) An effective data privacy mechanism through secure session key exchange model for MANET. Int J Electr Comput Eng (IJECE) 8(5):3267–3277
27. Bolbot V, Theotokatos G, Bujorianu ML, Boulougouris E, Vassalos D (2019) Vulnerabilities and safety assurance methods in cyber-physical systems: a comprehensive review. Reliab Eng Syst Saf 182:179–193
28. Devi PV, Kalaichelvi V (2017) Security issues in medical cyber-physical systems (MCPS)—a survey. Int J Pure and Appl Math 117(20):319–324
29. Li W, Meng W, Su C, Kwok LF (2018) Towards false alarm reduction using fuzzy if-then rules for medical cyber physical systems. IEEE Access 6:6530–6539
30. Jiang Y, Song H, Wang R, Gu M, Sun J, Sha L (2017) Data-Centered runtime verification of wireless medical cyber-physical system. IEEE Trans Industr Inf 13(4):1900–1909
31. Fu Z, Guo C, Ren S, Jiang Y, Sha L (2017) Modeling and integrating physical environment assumptions in medical cyber-physical system design. In: Proceedings of the Conference on design, automation & test in europe (DATE '17). European design and automation association, Leuven, BEL, pp 1619–1622
32. Sen J (2009) A survey on wireless sensor network security. Int J Commun Netw Inf Secur 1:59–82
33. Zhang X, Zhao J, Mu L, Tang Y, Chunxiang X (2019) Identity-based proxy-oriented outsourcing with public auditing in cloud-based medical cyber-physical systems. Pervasive Mobile Comput 56
34. Mohamed N, Al-Jaroodi J, Jawhar I (2020) Cyber–Physical systems forensics: today and tomorrow. J Sens Actuator Netw 9(3):37

Analysis of the Beaufort Cipher Expansion Technique and Its Usage in Providing Data Security in Cloud

Deepthi G. Pai and Yogeesha Pai

Abstract Cloud computing usually refers to the usage of computational resources that is delivered as a service over the internet. Virtualization can be considered as the main technology behind cloud computing. There is a need for providing the security of the data in the cloud. Several encryption techniques have been used for providing the data security in the cloud. In this paper, analysis of the Beaufort expansion technique is carried out, and it is used for providing the security for the cloud data. Beaufort expansion technique provides better security against crypt analysis and pattern prediction compared to the original Beaufort cipher.

Keywords Beaufort · Cloud · CloudSim · Encryption · Cryptography · Plaintext · Cipher text · Decryption · Key

1 Introduction

Cloud cryptography usually refers to the technique that is used for encrypting the data that is used or stored in the cloud. Cloud cryptography allows the customer to securely access the data since encryption techniques are used for protecting the data. Cloud encryption transforms the cloud data into the cipher text.

1.1 Cloud Computing

Cloud computing is the delivery of computing services—servers, storage, databases, networking, software, analytics, and more—over the Internet (the cloud). Cloud

D. G. Pai (✉)
Shri Madhwa Vadiraja Institute of Technology and Management, Bantakal, Udupi, India

Y. Pai
Manipal Institute of Technology, Manipal, India
e-mail: yogish.pai@manipal.edu

J. M. R. S. Tavares et al. (eds.), *Cyber Intelligence and Information Retrieval*, Lecture Notes in Networks and Systems 291,
https://doi.org/10.1007/978-981-16-4284-5_5

computing usually provides the customer with virtual cloud infrastructure where they can run their applications Marinescu [1].

Section 2 refers to Beaufort cipher, Sect. 3 includes the literature survey, Sect. 4 depicts the implementation and simulation, and Section V discusses conclusion.

2 Beaufort Cipher

Beaufort cipher is an example of the poly alphabetic substitution cipher. Beaufort cipher is a substitution cipher made by Sir Francis Beaufort which is similar to Vigenere cipher. The Beaufort cipher component depends on the Beaufort square which is the same as that of Vigenere square. Beaufort ciphers are additionally called as the reciprocal cipher [2].

2.1 Beaufort Cipher Algorithm

Beaufort cipher algorithm makes use of the tabula recta which is represented in Fig. 1. Both the sender and receiver share the same secret key, which contains one or more words.

The steps include.

Fig. 1 Tabula recta [3]

	A	B	C	D	E	F	G	H	I	J	K	L	M	N	O	P	Q	R	S	T	U	V	W	X	Y	Z
A	A	B	C	D	E	F	G	H	I	J	K	L	M	N	O	P	Q	R	S	T	U	V	W	X	Y	Z
B	B	C	D	E	F	G	H	I	J	K	L	M	N	O	P	Q	R	S	T	U	V	W	X	Y	Z	A
C	C	D	E	F	G	H	I	J	K	L	M	N	O	P	Q	R	S	T	U	V	W	X	Y	Z	A	B
D	D	E	F	G	H	I	J	K	L	M	N	O	P	Q	R	S	T	U	V	W	X	Y	Z	A	B	C
E	E	F	G	H	I	J	K	L	M	N	O	P	Q	R	S	T	U	V	W	X	Y	Z	A	B	C	D
F	F	G	H	I	J	K	L	M	N	O	P	Q	R	S	T	U	V	W	X	Y	Z	A	B	C	D	E
G	G	H	I	J	K	L	M	N	O	P	Q	R	S	T	U	V	W	X	Y	Z	A	B	C	D	E	F
H	H	I	J	K	L	M	N	O	P	Q	R	S	T	U	V	W	X	Y	Z	A	B	C	D	E	F	G
I	I	J	K	L	M	N	O	P	Q	R	S	T	U	V	W	X	Y	Z	A	B	C	D	E	F	G	H
J	J	K	L	M	N	O	P	Q	R	S	T	U	V	W	X	Y	Z	A	B	C	D	E	F	G	H	I
K	K	L	M	N	O	P	Q	R	S	T	U	V	W	X	Y	Z	A	B	C	D	E	F	G	H	I	J
L	L	M	N	O	P	Q	R	S	T	U	V	W	X	Y	Z	A	B	C	D	E	F	G	H	I	J	K
M	M	N	O	P	Q	R	S	T	U	V	W	X	Y	Z	A	B	C	D	E	F	G	H	I	J	K	L
N	N	O	P	Q	R	S	T	U	V	W	X	Y	Z	A	B	C	D	E	F	G	H	I	J	K	L	M
O	O	P	Q	R	S	T	U	V	W	X	Y	Z	A	B	C	D	E	F	G	H	I	J	K	L	M	N
P	P	Q	R	S	T	U	V	W	X	Y	Z	A	B	C	D	E	F	G	H	I	J	K	L	M	N	O
Q	Q	R	S	T	U	V	W	X	Y	Z	A	B	C	D	E	F	G	H	I	J	K	L	M	N	O	P
R	R	S	T	U	V	W	X	Y	Z	A	B	C	D	E	F	G	H	I	J	K	L	M	N	O	P	Q
S	S	T	U	V	W	X	Y	Z	A	B	C	D	E	F	G	H	I	J	K	L	M	N	O	P	Q	R
T	T	U	V	W	X	Y	Z	A	B	C	D	E	F	G	H	I	J	K	L	M	N	O	P	Q	R	S
U	U	V	W	X	Y	Z	A	B	C	D	E	F	G	H	I	J	K	L	M	N	O	P	Q	R	S	T
V	V	W	X	Y	Z	A	B	C	D	E	F	G	H	I	J	K	L	M	N	O	P	Q	R	S	T	U
W	W	X	Y	Z	A	B	C	D	E	F	G	H	I	J	K	L	M	N	O	P	Q	R	S	T	U	V
X	X	Y	Z	A	B	C	D	E	F	G	H	I	J	K	L	M	N	O	P	Q	R	S	T	U	V	W
Y	Y	Z	A	B	C	D	E	F	G	H	I	J	K	L	M	N	O	P	Q	R	S	T	U	V	W	X
Z	Z	A	B	C	D	E	F	G	H	I	J	K	L	M	N	O	P	Q	R	S	T	U	V	W	X	Y

1. Identify the current plaintext letter in the top horizontal row.
2. Traverse down the column to find the current key letter.
3. The leftmost letter in the current row gives the next letter of the cipher text.

The same algorithm can be used for decryption also. That is, encrypting the cipher text for the second time outputs the original text or plaintext.

2.2 *Demerits of Beaufort Cipher*

Beaufort cipher can be easily broken using the brute force attack. Here, the intruder tries to decrypt the cipher text by trying all the words that are found in the dictionary as Beaufort keys.

Decryption of Beaufort cipher text can also be done by using the frequency analysis since the Beaufort encryption algorithm does not mask the frequently occurring symbols in the cipher text.

Breaking the Beaufort cipher:

Keyword: HAI Plaintext: HELLO *Cipher*: AWXWM.

Steps include:

- Map all the letters in the plaintext to numbers 0–25. HELLO to 7 4 11 11 14
- Map your keyword letter to numbers 0–25. HAI to 7 0 8
- Subtract plaintext from the key in following manner: (7 0 8 7 0) – (7 4 11 11 14) resulting in 0 −4 −3 −4 −14.
- Take each resulting number mod 26 resulting in 0 22 23 22 12
- Map each number back to alphabet to get the final cipher text

A W X W M.

The decryption process can be carried out just by reversing the above steps. Kasiski test [4] can be used to break the Beaufort cipher without using the keyword.

Beaufort cipher expansion technique

The Beaufort cipher expansion method makes use of some transformation techniques on the plaintext to create the cipher text which masks the connection between the both thus making the crypt analysis difficult.

Steps in the encryption process:

Figure 2 explains the different steps in the Beaufort cipher expansion algorithm. The expansion algorithm takes the plaintext and the key as input and outputs an intermediate cipher. This intermediate cipher undergoes a series of transformation techniques to yield the final cipher text.

Comparing Beaufort cipher technique and Beaufort cipher expansion technique

Extended Beaufort cipher algorithm provides more security for the data compared to the normal Beaufort cipher algorithm since the extended algorithm undergoes the double encryption process which increases the security.

Encryption algorithm calculation:

Step 1: Start

Step 2: Read the plain text

Step 3: Read the key for the Beaufort cipher

Step 4: Calculate the intermediate cipher based on the basic Beaufort encryption algorithm making use of the tabula recta.

Step 5: Calculate the ASCII values of the key and intermediate cipher.

Step 6: Calculate the binary value of key and intermediate cipher based on its ASCII value calculated.

Step 7: Perform the XOR operation between the binary of key and binary of the intermediate cipher.

Step 8:Calculate the decimal value of the XOR result.

Step 9: Stop

Fig. 2 Beaufort cipher expansion algorithm

- **Brute force attack**:

Brute attack is a method to break the Beaufort cipher. In this method, the intruder tries to decrypt the cipher text by trying all the words in the dictionary file as Beaufort keys. Cipher text is more vulnerable to attacks if the Beaufort key is an English word which can be found in a dictionary. This attack can be reduced to a certain extent if some random words are used for the Beaufort key. But certain techniques have been developed to break the Beaufort cipher text with random Beaufort keys. In the extended Beaufort cipher algorithm, Beaufort keys are used twice for the encryption process making the brute force attack troublesome.

- **Frequency analysis**:

For example, using the keyword CRYPTO, the translation from the plaintext to the cipher text becomes:

Plain text symbols: A B C D E F G H I J K L M N O P Q R S T U V W X Y Z.

Cipher text symbols: C Q W M P J W K Q G J D Q E K A D X K Y E U X R E S.

Based on the monogram frequency distribution which is represented in Fig. 3, it can be concluded that the letter 'e,' 'k,' and 'q' are the most common, followed by the letters 'd,' 'j,' 'w,' and 'x.' If the encryption algorithm does not mask the frequently occurring symbols in the cipher text, then crypt analysis can be easily carried out just by looking at the cipher text.

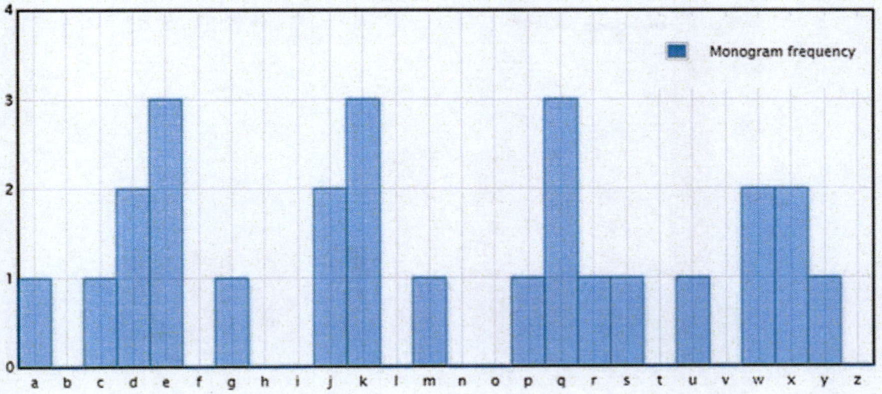

Fig. 3 Monogram counts

2.3 Frequency Analysis of the Extended Beaufort Cipher Algorithm

For example,

Plaintext symbols: N E T W O R K.

Key: S E C U R I T Y.

Intermediate Cipher text symbols: F A J Y D R J.

So the monogram frequency distribution of the intermediate cipher text looks like.

Cipher text: 132,875,039,162,901.

Monogram frequency distribution of the intermediate cipher text and cipher text is represented in Figs. 4 and 5. Here, analysis of letter frequency does not provide an

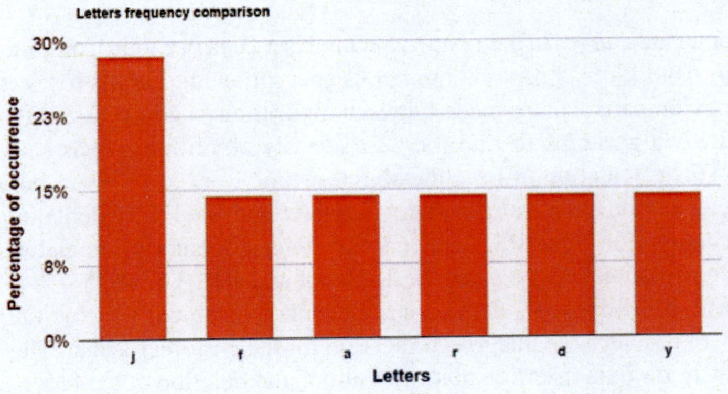

Fig. 4 Monogram counts of intermediate cipher text

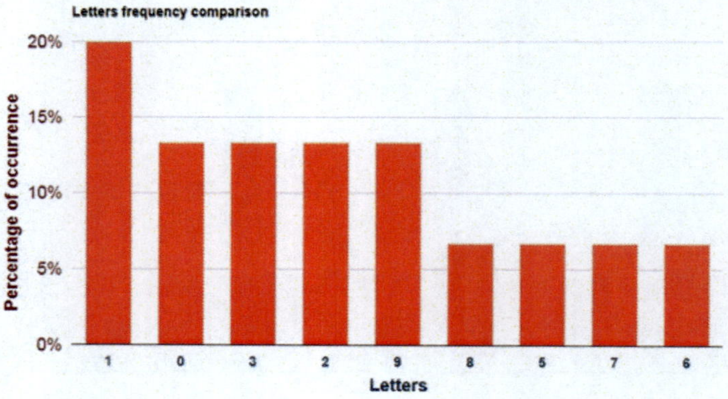

Fig. 5 Monogram counts of final cipher text

idea about the plaintext since the cipher text is yielded in digits, masking the relation between the plaintext and cipher text.

3 Literature Survey

This portion discusses the different strategies which were utilized to provide security for the cloud data.

The author Singh et al. [5] proposed a scheme where security is provided for the data through the authentication and cryptographic procedures. Here, password authentication technique is used rather than the normal authentication mechanism. For implementation, elliptic curve digital signature algorithm (ECDSA) is used. Company authentication and modified ECDSA schemes use very short time and are efficient in both key generation, key signing, and signature verifying operation. This enhances the efficiency of storage and retrieval of data or information in cloud computing.

Arora and Parashar [6] in their paper, Secure User Data in Cloud Computing Using Encryption Algorithms, proposed numerous encryption algorithms for securing the cloud data. Comparisons are made between algorithms such as AES, DES, Blowfish, and RSA algorithms to find the best security algorithm. It has been discovered that AES calculation utilizes the slightest time to execute cloud information. Blowfish calculation has the slightest memory prerequisite. DES calculation devours minimum encryption time. RSA expends the longest memory size and encryption time. Implementation is done using the IDE tool and JDK 1.7.

Raju et al. [7] proposed a data integrity checking protocol for cloud repository. This protocol provides the integrity protection for the customer's data. The protocol paves the way for data insertion, data alteration, and deletion at the block level and additionally supports public verifiability. The theoretical analysis and experimental analysis were carried out on the protocol, and it demonstrated that the protocol has good strength within the communication, computation, and storage price aspects.

Handa and Singh [8] in their paper, Data Security in Cloud Computing using Encryption and Steganography, proposed a model which works on the principle of securing data both during transmission and while data-at rest at servers. The basic processes of encryption and steganography are done in this model. This model also provides data security at the two levels, that is when data is moved to the remote servers and when the data is being saved and guarded by the cloud service provider officials.

Sidhu et al. [9] proposed a scheme to provide improved security in cloud infrastructure. This scheme delivers encryption in the form of a hashing function to the messages which have been exchanged from various customer machines to server machines in the cloud framework. Usage of the AES and MD5 is explained here as the hybrid approach. The proposed algorithm begins with plain text which is later transformed to a whitened text.

Singla and Singh [10] in their paper, Cloud Data Security using Authentication and Encryption Technique, proposed a scheme where the data security for the cloud is provided using two different approaches such as Extensible Authentication protocol CHAP and Rijndael encryption algorithm. The proposed model uses challenge handshake authentication protocol CHAP for the authentication purposes and Rijndael algorithm (standard symmetric key encryption algorithm) for encrypting the sensitive information.

4 Implementation and Simulation

4.1 Importance of Simulation in the Cloud Environment

To gain admittance to cloud benefits, the cloud service providers usually demand price based on the service space provided to the customer. In innovative work, it is not continuously conceivable to have the genuine cloud foundation for playing out the examinations. For any academician or researcher, it is not possible to hire the cloud services each time for conducting the experiments.

With the end goal of research, advancement, and testing, the open source libraries are accessible which can be utilized to perform the executions and experience the cloud services. Nowadays, cloud simulators are broadly utilized by the researchers and specialists without paying any add up to any cloud service provider.

The cloud developers can use the software based library of the cloud simulator and can evaluate the outcomes in various parameters such as integrity, confidentiality, bandwidth, and power.

4.2 CloudSim

CloudSim can be defined as the toolkit for modeling and simulating the cloud environments. This simulation toolkit allows the developers to execute their provisioning strategies in a controllable environment, free of cost. Since CloudSim is a simulator, it does not run any actual software [11].

4.3 Implementation

The CloudSim platform is used to implement the Beaufort cipher expansion technique algorithm. The simulation using the predefined classes is given by CloudSim for the cloud environment. CloudSim is a Java-based tool kit that can be implemented using either the Eclipse IDE or the Netbeans IDE. The Netbeans IDE is used here for implementation.

We first need to install the Netbeans to run the CloudSim on Netbeans, and after installing the Netbeans IDE, download, extract, and import the CloudSim package into the Netbeans.

In proposed work, we have illustrated our expansion algorithm in the form of Java code, and for this purpose, we have to create new classes. Figures 6 and 7 represent screenshots that depict the working of our algorithm on the CloudSim framework.

The screenshots show the execution of the encryption process. Predefined functions such as createCloudlet(), createBroker(), and createDatacenter() are used for creating various entities such as Cloudlets, Virtual Machines, Broker, and Datacenter [12].

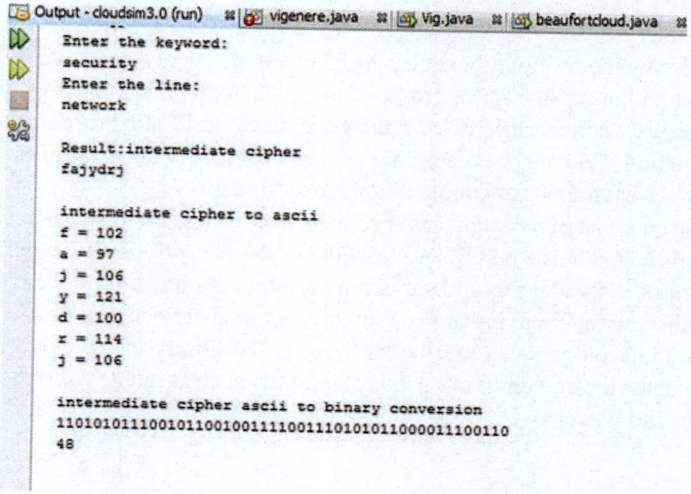

Fig. 6 Simulation: Encryption process

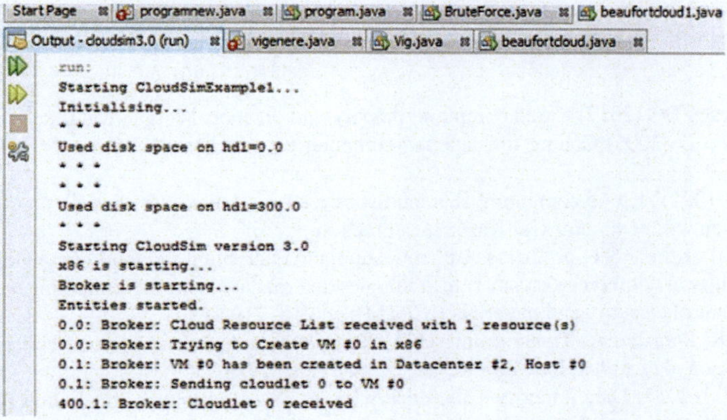

Fig. 7 Simulation: Encryption process

5 Comparison with Other Methods

This section gives a brief comparison between the other cipher methods and the proposed technique. The Caesar cipher provides only a minimum security for the data that is to be transmitted. Decryption of the Caesar cipher can be easily carried out by just identifying the frequency of the letter pattern. The demerit of the Vigenere cipher is the repeating nature of its key. If the crypt analyst correctly guesses the key's length, then the cipher text can be easily broken by treating it as an interwoven Caesar ciphers. But the proposed Beaufort cipher expansion technique yields the cipher text in digits which makes the frequency analysis difficult thus enhancing the security of the cipher.

6 Conclusion

Cloud computing provides the computational resources to the customers over the internet. Providing the security for the data in the cloud can be considered as a major concern in the cloud computing environment. Numerous algorithms have been developed for providing data security in the cloud. Beaufort cipher expansion algorithm overcomes the demerits of the existing Beaufort cipher and provides the security for the data in cloud by using a double encryption process where the relation between the plaintext and cipher text is masked and makes the crypt analysis process more troublesome.

References

1. Marinescu DC (2017) Cloud computing: theory and practice. Morgan Kaufmann
2. Salomon D (2005) Coding for data and computer communications. Springer. ISBN 0–387–21245–0
3. Olive C(2011) Cloud computing characteristics are key. General Physics Corporation
4. https://en.wikipedia.org/wiki/Kasiskiexamination.
5. Singh JP, Kumar S et al (2015) Authentication and encryption in cloud computing. In: 2015 International conference on smart technologies and management for computing, communication, controls, energy and materials (ICSTM), pp 216–219. IEEE
6. Arora R, Parashar A, Transforming CCI (2013) Secure user data in cloud computing using encryption algorithms. Int J Eng Res Appl 3(4):1922–1926
7. Raju RV (2013) Data integrity using encryption in cloud computing. Int J Glob Res Comput Sci (UGC Approved Journal) 4(5):40–43
8. Handa K, Singh U (2015) Data security in cloud computing using en- cryption and steganography. Int J Comput Sci Mob Comput 4(5):786–791
9. Sidhu A, Mahajan R (2014) Research article enhancing security in cloud computing structure by hybrid encryption. Int J Recent Sci Res 5:128–132
10. Singla S, Singh J (2013) Cloud data security using authentication and encryption technique. Glob J Comput Sci Technol
11. Alnazir MKAM, Mustafa ABAN, Ali HA, Yousif AAO (2017) Performance analysis of cloud computing for distributed data centers using cloud-sim. In: 2017 International conference on communication, control, computing and electronics engineering (ICCCCEE), pp 1–6. IEEE
12. Arora A, Khanna A, Rastogi A, Agarwal A (2017) Cloud security ecosystem for data security and privacy. In: 2017 7th International conference on cloud computing, data science & engineering-confluence, pp 288–292. IEEE

Distribution of Internet Banking Credentials Using Visual Cryptography and Watermarking Techniques

Surajit Goon, Debdutta Pal, and Souvik Dihidar

Abstract Distribution of Internet banking credentials like user ID, login password, and transactional password is the most critical step and prone to forgery and hacking. We proposed a methodology by which banks can distribute the same in a much safer way. This methodology uses visual cryptography and digital watermarking techniques to encrypt the user login password and transactional password to be embedded into the user login ID. One part of the secret is being digitally transmitted and another trough the traditional postal services. This system also shows that reconstructed image quality is very good in respect to other visual cryptography schemes.

Keywords Visual cryptography · Halftone image · Error diffusion · Meaningful shares · Digital watermarking

1 Introduction

Most of the recent banking systems are secure in their underlying structure. But the distribution of the login credentials may encounter some interventions by the intruders. In net banking systems, most of the bank still sends the login credentials by traditional post and that can produce a serious threat to the whole system. The username and password may be misused by the hackers. In this paper, we propose a more secure way to distribute the login credentials along with the transactional password. Here, we used visual cryptography schemes with digital image watermarking techniques to achieve our goal. Visual cryptography (VC) is a simple technique that

S. Goon (✉) · D. Pal
Department of Computer Science and Engineering, Brainware University, Barasat, West Bengal, India

D. Pal
e-mail: ddp.cs@brainwareuniversity.ac.in

S. Dihidar
Department of Computer Science, Eminent College of Management and Technology, Barasat, West Bengal, India
e-mail: souvik@ecmt.in

J. M. R. S. Tavares et al. (eds.), *Cyber Intelligence and Information Retrieval*, Lecture Notes in Networks and Systems 291,
https://doi.org/10.1007/978-981-16-4284-5_6

generates two or more shares at the encryption stage from any hidden image or code, and at the decryption end, it reveals the secret message with our visuals by only stacking those shares one after another.

Naor and Shamir [1] first introduced the visual cryptography scheme (VCS) by giving a model where a secret image generates two or more binary transparent share images in such a way that until we stack a certain number of those shares one after another and we cannot get the original secret image. Though it was simple, but very difficult to break. Most of all it requires zero computational power; at the decryption end, only human visual system is enough to reveal the secret. The basic concept is to translate a hidden code into an image and then generate n number of shadow images from it, and at the time of decryption, it will require minimum k number of shadows to reveal the secret. A digital image watermark [2] embeds a secret code or signature into an image in such a way that cannot be noticed easily by the intruders. The entire system includes two basic components: encoder and decoder. Encoding function can be defined as $C_w = E(C_o,W,K)$ where C_w: watermarked image, E:encoding algorithm, C_o: carrier image, W: watermark, and K: security key.

In our proposed model, we use the above-mentioned two techniques to distribute net banking user name, user login password, and transactional password to the user. In Sect. 2, we did a literature review of existing systems. Then, in Sect. 3, we introduced our model. In the next Sect. 4, we discussed about the implementation of our model. Here, we take user login password as the secret image input of a VCS and generates two meaningful shares using one natural image and user ID as the cover images of these two shares. Now again insert the transactional password into the natural image share through digital watermarking technique and transmit it through Internet. The other share which has user ID as its cover image was transmitted to the user through traditional post. After getting both the meaningful shares, user simply decodes the transactional password using the decryption algorithm and then finally get the user password by simply stacking these two shares one after another. In the next Sect. 5, we show the result of our experiments and compare them with the existing VCSs output. In the last section VI, we conclude with some future expansions possibilities.

2 Literature Review

VCS is a cryptographic technique for encrypting visual data, so that it can be decrypted only by the human visual system. The k-out-of-n VCS, abbreviated as (k,n) VCS [3], is the basic scheme. The encryption is set up in such a way that it takes k or more of the n generated images to reconstruct the original image. In the (2, 2) VCS, each pixel P in the original image is encrypted into two sub pixels called shares. Figure 1 depicts the shares of a white pixel and a black pixel. Ordered dithering [4] is a technique which can be used to quickly and in parallel transform a gray-level image into an equal-sized binary image. After transforming a gray-level image into halftone image, Zhou [5] used binary VC schemes to generate gray-scale shares. Extended visual cryptography (EVC) is a method developed by Ateniese

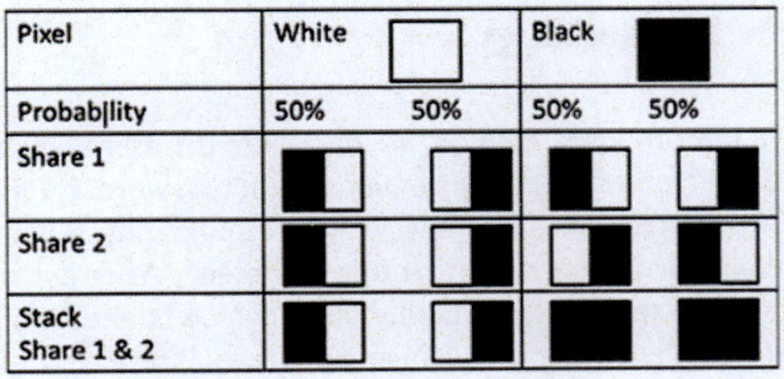

Fig. 1 Shares of a white pixel and a black pixel

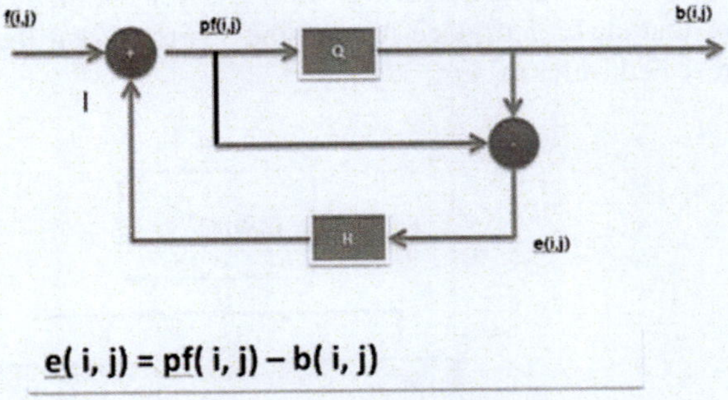

Fig. 2 Floyd–Steinberg's error diffusion

[6], in which, shares contain not only confidential information but also meaningful images. Halftoning is a form of noise that is used to randomize quantization error. It is also known as "blue noise." The idea of keeping track of the error came from Floyd–Steinberg's error diffusion algorithm [7]. The Floyd–Steinberg algorithm is depicted in Fig. 2.

Hegde et al. [8] proposed a method for parsing a customer's signature and then splitting it into shares. Depending on the bank's chosen system, the total number of shares to be issued can vary. When two people have a mutual interest, one is stored in the bank database, and the other is generated. The consumer retains ownership of the product. The customer must present his or her share throughout all of his business dealings.

To build a safe framework, Roy and Venkateswaran [9] proposed combining text-based steganography and visual cryptography. It reduces the amount of knowledge shared between the user and the online world merchant; they must, however, be able to move funds from the consumer's account to the merchant's account while ensuring customer's security.

Pathak et al. [10] used a variant of visual cryptography to generate encryption share by combining the authentication parameter as a hidden image with a color image. Unlike other cryptography techniques, they do not encrypt the secret image, instead creating a new image that can be linked to the secret image if the keys (color image and passkey) are identified.

3 Methodology

In our proposed method, we distribute the user name as one of the cover images of the secret image which contains login password. Login password actually splits into two shares using halftone visual cryptography scheme. Then one of the shares is embedded using watermarking technique. After embedding, it is transmitted using Internet to the user. Another share which is already hidden under the user name is transmitted to the user using manual postal services. After receiving the digital share, user will have to extract the watermark message that is the transactional password using the extracting algorithm already given to him by the bank. Then, take a printout of that share and place it upon the share, so that the user login password can be revealed (refer to) Fig. 3.

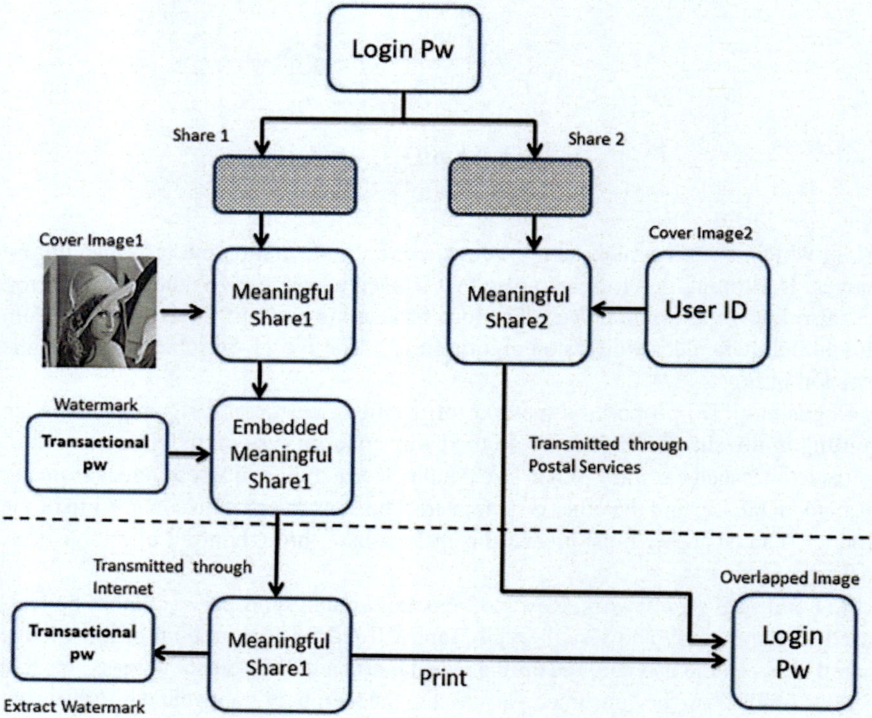

Fig. 3 Proposed method

4 Implementation

Implementation of our proposed model has two phases 1. Encryption at the bank side and 2. Decryption at the user end.

4.1 Encryption

Step 1 User login password is chosen as the secret image.
Step 2 Convert it to halftone image using Floyd–Steinberg technique.
Step 3 Generate shares using (2,2) VCS.
Step 4 Share 1 is embedded with a natural image producing meaningful share 1.
Step 5 Share 2 is embedded with an image which represents the user ID and produces meaningful share 2.
Step 6 Insert transactional password as digital watermark into the meaningful share 1.

4.2 Decryption

Step 1 After receiving the watermarked meaningful share 1, extract the transactional password.
Step 2 To make a hardcopy printout in transparent paper. technique.
Step 3 Stack both meaningful shares to reveal the login password.

Figure 4 shows the outputs we got in each step. Here, we used discrete cosine transform (DCT) watermarking into the meaningful share 1 which is a natural image.

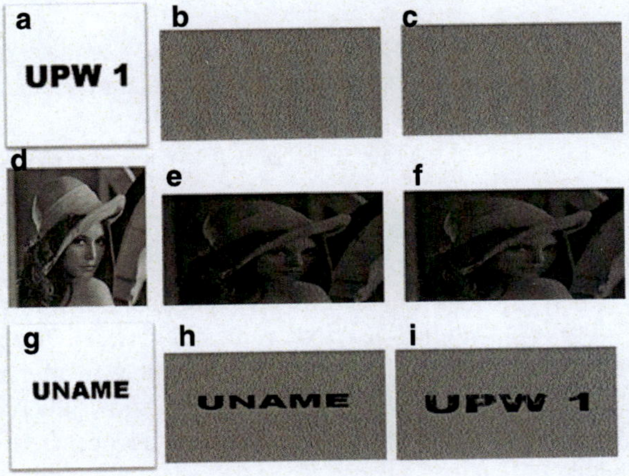

Fig. 4 From top left to right **a** secret image, **b** share1, **c** share 2, **d** cover image 1, **e** meaningful share 1, **f** watermarked share 1, **g** cover image 2, **h** meaningful share 2, **i** decrypted image

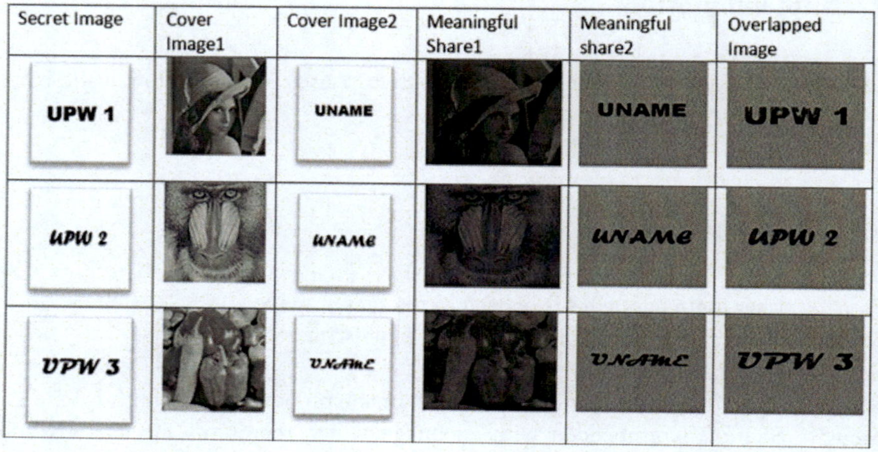

Fig. 5 Outputs of various image samples

We give the result of various samples in Fig. 5,where in each sample we took separate secret image and separate cover images.

5 Result

We compared the decrypted images which were produced by traditional (2,2) VCS with the decrypted images of our system using the values of MSE, PSNR, and SSIM. Peak signal–noise ratio (PSNR) finds the correlation between two images and is represented by

$$\mathrm{PSNR} = 10 * \log\left(255^2/\mathrm{MSE}\right) \tag{1}$$

And mean squared error (MSE) is relative to the zero (lower the better), i.e., mean of the squared difference between original and recreated one

$$\mathrm{MSE} = 1/(M * N)\sum\sum\left[I(i,j) - I^{\sim}(i,j)\right]^2 \tag{2}$$

Structural similarity index (SSIM) also measures the degradation of original and reconstructed image.

In our proposed model, we compare both the decrypted images and got the MSE values less than 0.005, PSNR values in between 70 to 75 dB, and SSIM values in between 0.95 to 1.0. This result shows that even after attack, our final outcomes are good in respect to the other VCS. Table 1 shows the results. Figures 6, 7, and 8 also prove that there are very less information loss occurred, and the reconstructed images quality are high.

Table 1 MSE, PSNR, and SSIM values of reconstructed images without attack against the images with attack of each samples

Secret image	Reconstructed image without attack	Reconstructed image with attack	MSE	PSNR	SSIM
UPW 1	UPW 1	UPW 1	0.0044	71.6492	0.9719
UPW 2	UPW 2	UPW 2	0.0028	73.5888	0.9777
UPW 3	UPW 3	UPW 3	0.0033	72.9310	0.9760

Fig. 6 MSE values of reconstructed images

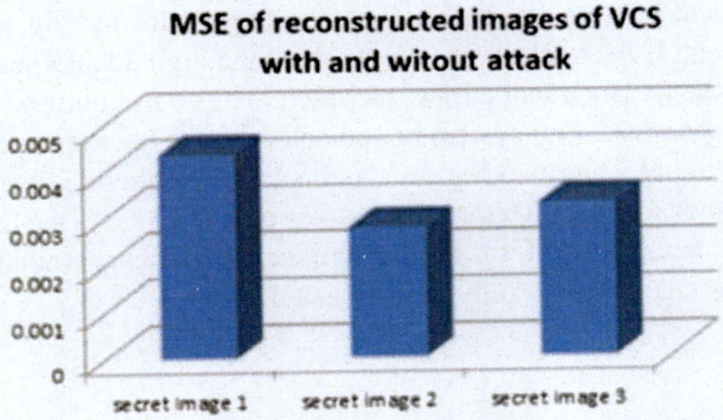

Fig. 7 PSNR values of reconstructed images

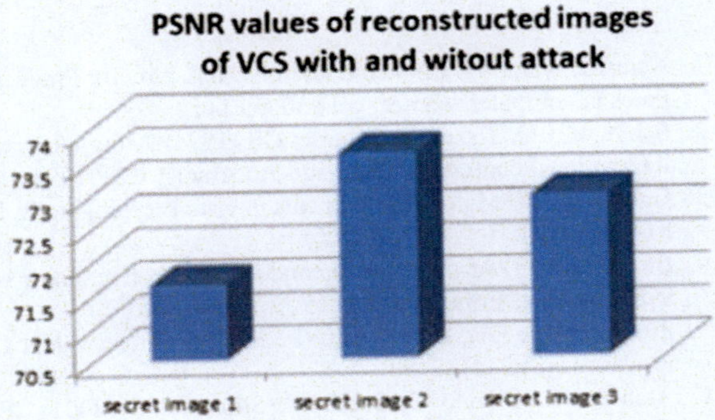

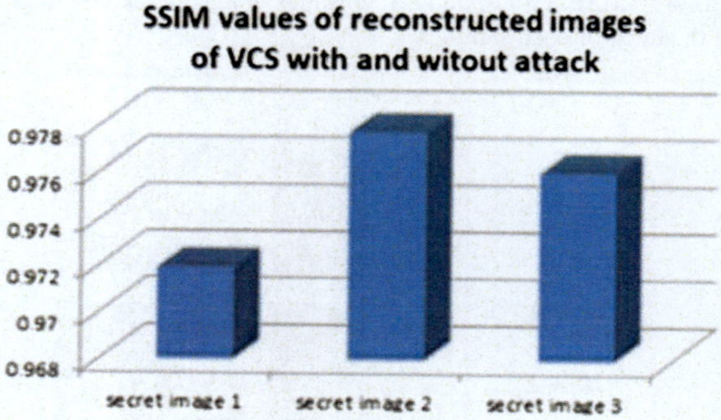

Fig. 8 SSIM values of reconstructed images

6 Conclusion

We have implemented a model where bank can encrypt the user login password along with the transactional password using visual cryptography and digital watermarking techniques in such a way that this information are carried with an image of user ID and a natural image. So, if at the time of transmission, any attempt of forgery occurs, and if the hackers be able to crack the digital watermark, then also he will only get the transactional password which is of no use until he knows the user ID and login password. Our system has some limitation because users have to pre-equip with the digital watermarking decryption software provided by the bank. Also users have to own a printer. This system losses some data due to the implementation of meaningful shares, but still it is a successful one, and there should be no problem to decrypt the image by using only human visual system.

References

1. Naor M, Shamir A (1994) Visual cryptography. In: Proceedings of Euro crypt 1994, lecture notes in computer science, vol 950, pp 1–12
2. Schyndel RG, Tirkel A, Osborne CF (1994) A digital watermark. In: Proceedings of IEEE international conference on image processing, ICIP-1994, pp 86–90
3. Goon S (2019) Major developments in visual cryptography. Int J Eng Adv Technol, 9(1S6), pp 81–88
4. Bayer B (1973) An optimum method for two-level rendition of continuous-tone pictures (PDF). IEEE Int Conf Commun 1:11–15
5. Zhou Z, Arce GR, DiCresenzo G (2006) Halftone visual Cryptography. IEEE Trans Image Process 15(8):2441–2453
6. Ateniese G, Blundo C, Santis A, Stinson DR (2001) "Extended Capabilities for Visual Cryptography", Acm Theor. Comput Sci 250:143–161
7. Floyd RW, Steinberg L (1976) An adaptive algorithm for spatial grey scale. Proceedings of the Society of Information Display 17:75–77
8. Hegde C, Manu S, Shenoy PD, Venugopal KR, Patnaik LM (2008) Secure authentication using image processing and visual cryptography for banking applications. In: IEEE transaction, 16th international conference on advance computing and communications, pp 65–72

9. Roy S, Venkateswaran P (2014) Online payment system using steganography and visual cryptography. In: IEEE students' conference on electrical, electronics and computer science
10. Pathak B, Pondkule D, Shaha R, Surve A (2020) Visual cryptography and image processing based approach for bank security applications. In: ICCNCT 2019, LNDECT vol 44, pp 292–298

Pattern Recognition

Virtual Keyboard Using Image Processing and Computer Vision

Polok Ghosh, Rohit Singhee, Rohan Karmakar, Snehomoy Maitra, Sanskar Rai, and Sudipta Basu Pal

Abstract Today, with this fast-changing world, portability is one of the critical factors for consumers when choosing a personal computer (PC). Still, the conventional hardware keyboard is a significant limitation in desktops and notebooks. Virtual keyboard prototype solves hardware limitations of the keyboard by presenting an easy and economical alternative using minimum external hardware. In this paper, we have tried to design a virtual keyboard model, which is based on image processing using an inbuilt camera. After that, the basic framework of this can be modified to serve various types of augmented reality (AR), paper, and air keyboard purposes.

Keywords Masking · Threshold binarization · Contour · Finger pointer · Image processing · OpenCV

1 Introduction

Computers are an essential part of our lives nowadays. We tend to use them nearly every day to work out our day-to-day tasks. However, it is not invariably attainable to take computers or laptops with us whenever we tend to depart. This can be one of the explanations why we came up with the concept of virtual keyboards that ultimately takes care of the user's comfort. It is merely another example of the expansion within the field of image processing or computer technology as a whole. Virtual keyboards create use of the image processing algorithms, and therefore, the main camera of a device allows users to type on a plane surface or in the air using gestures.

Virtual keyboards, to the present day, are given varied forms that use the external software package applications and external 3D cameras to evaluate the user's finger distance from the camera. This technology that we are presenting requires new human–computer interaction processes that are enforced to supply multiple interactions between users and machines, so that the device does not face any drawback to discover the human's gesture. Image processing or commonly computer

P. Ghosh · R. Singhee · R. Karmakar · S. Maitra · S. Rai · S. B. Pal (✉)
Computer Science Engineering Department, University of Engineering and Management Kolkata, Kolkata, India

J. M. R. S. Tavares et al. (eds.), *Cyber Intelligence and Information Retrieval*, Lecture Notes in Networks and Systems 291,
https://doi.org/10.1007/978-981-16-4284-5_7

vision researchers have put importance in the development of image vision-based devices. These are primarily seen in high-security areas with CCTV cameras and large company cars. The researchers are also seen to be using infrared cameras to detect the interaction, but these techniques make these technologies unavailable to the common people.

This paper describes a system that utilizes both the grayscale and contour masking to comprehend the user input. The tracking of the fingers is not based on skin color detection; however, we used color detection algorithms to separate the wearable pointer from their surroundings. More about the process, camera calibration, and algorithms will be discussed in the later sections.

2 Review Work

The new technological enhancements are stunning us each day. Computers and technologies have made our daily tasks easier and efficient. As technology advances, transformation takes place in virtually every field to require care of user's comfort and their wants, similar is the case with computers and keyboards.

ENIAC, the primary laptop which came into being in 1946, was large in proportions. Then, the researchers and engineers came up with microchips. The evolution of microchips opened new opportunities to form smaller devices like CPUs, integrated circuits, and microcontrollers.

The next part is keyboards, initially all exhausting switches were used as keys. Then QWERTY keyboards came into action which frequently led to lost or broken keys. These days, dupe keypads are well-liked. These keypads are largely used for specific applications primarily in ATMs.

The virtual keyboard has been implemented throughout a spread of assorted forms, one of those supported 3D optical and CCTV cameras, as they are primarily based on image processing. Also, engineers have tried to increase the use of virtual keyboards by making a holographic image of the keyboard.

The virtual keyboard that we are focusing on in the building should have a camera, keyboard layout sketched on a sheet of paper with no additional hardware necessities, and should make use of hand-driven gestures and basic tools of image processing like threshold, segmentation, grayscale conversions, edge detection, etc. [1].

Researchers are also seen to use the process of tip recognition and a custom keyboard. In the tip recognition system, a filter is employed to sleek the image. This filtered image is divided into RGB channels. The B/W-MORPH rule is employed to extract a user's hand from its surroundings and then to extract the tip from the finger. Then, the character recognition should use a feature extraction system and try to remember the exact location of the paper keyboard. The system monitors the console and also the tip perpetually. Users will naturally use their fingers to type out commands. Moreover, the system can even give 3D data, to regulate 3D virtual objects by three-dimensional positioning and inclination of the visual panel. For detection of hands and finger joints, contour analysis is performed. Experimental analysis results encourage the buildup of errorless vision-based virtual keyboards in not only desktops but also mobile devices additionally to existing little keypads for ease of use for the users.

3 Result Analysis and Discussion

There are various steps involved in the successful implementation of the keyboard project. Starting from a raw camera feed seen in Fig. 1 to CV (computer vision) computations and finger-pointer detection for a flawless and hassle-free typing experience. As sequentially mentioned below.

3.1 Masking

Masking is a procedure to extract the parts of an image that are required for further processing. This is based on color or movement. Here, we are using a supportive program to first mask the preferred colored object to be used as a pointer using the manual calibration of the HSV (hue, saturation, value) parameters and thus saving

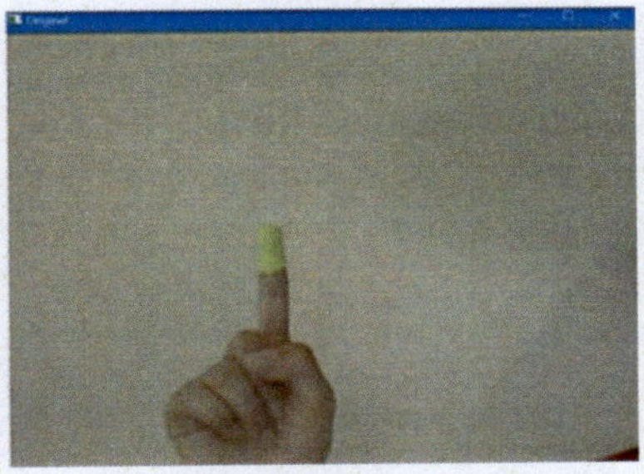

Fig. 1 Original image

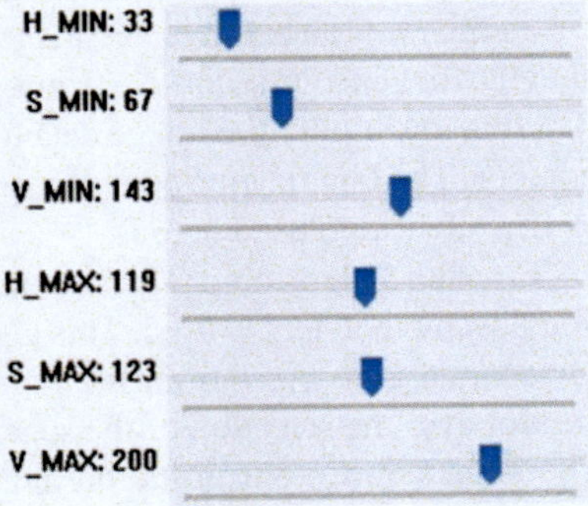

Fig. 2 Image calibration trackbar

the values for final reference displayed in Fig. 2. This is done with the help of the inRange function [2].

```
inRange()
void cv::inRange(InputArray src,
                 InputArray lowerb,
                 InputArray upperb,
                 OutputArray dst
                )
```

3.2 Image Feed Preprocessing

The raw camera feed contains unwanted areas and noise which might be a disturbance in final calibration and object detection. This is solved by first cropping the image to the required area and then softening the image by little using the inbuilt Gaussian blur and median blur functions of OpenCV. This causes a little loss of data but serves our purpose in detecting the pointer more accurately. Algorithms used here are:

Grayscale algorithm: This algorithm converts the original red, green, and blue values into gray values. The values of the pixel of the color digital image are retrieved, and each pixel is converted into a single gray value [3]. Converting the color image into a binary image helps in faster recognition of the specific object removing the color information, leaving the luminance of each pixel. In a single image, there are several regions present. To choose a blob as the important one, several filtering techniques are used among which the thresholding algorithm is being used.

3.3 Threshold Binarization

The threshold is a function to bar unnecessary information and keep only the required. Using this, we can binarize the image; thus, the function output would be much more suitable for generating contour. Here, in our project, we have used the binary threshold function to simplify the image data to get our required object as the largest rectangle contour. The binarization function output can be seen in Fig. 3.

The algorithm used here is:

Threshold algorithm: Thresholding is a part of image segmentation and is the simplest method that is used. This algorithm is used to apply on the grayscale image to separate the object from the regions in the background. The user has manual control over the selection of the specific object which is the input given by the user. The object shines out in the digital image when the setting of the object is done based on the threshold value on the pixel intensity of the original image creating a

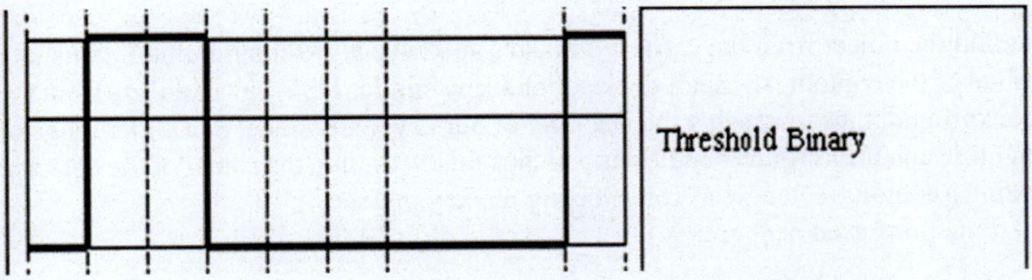

Fig. 3 Threshold binarization graphical representation

Fig. 4 Original versus masked threshold image

binary image [4]. After masking, the threshold binarization of detected color pointer is clearly visible in Fig. 4.

cv.THRESH_BINARY.

The first argument is the source image, which should be a grayscale image.

The second argument is the threshold value, which is used to classify the pixel values.

The third argument is the maximum value, which is assigned to pixel values exceeding the threshold.

```
THRESH_BINARY
Python: cv.THRESH_BINARY
dst(x,y)={maxval    if src(x,y)>thresh
            0     otherwise    }
```

3.4 *Contour Generation*

Contours can be explained simply as a curve joining all the continuous points (along the boundary), having the same color or intensity [5]. The contours are a useful tool for shape analysis and object detection and recognition. Here, we have used contour to find the object from the extracted images and have also set a minimum threshold value of the contour, so that it does not pick any similar HSV value objects from the background, thus increasing the usability of our keyboard model in a wider range of lighting and background conditions and thus finally finding the center of the detected pointer contour rectangle as for selecting the key pressed.

Functions used here are:

```
cv.FindContours
cv.drawContours
```

3.5 Finger-Pointer Detection

After detection of the edges with the contour function, the last and the most important part is the fingertip detection. For the finger-pointer detection part, the edges which were detected by contour processed to the 2D plane for computer vision accessibility purposes. This is done by applying a linear fractional transformation specified by a matrix of the position of each pixel from the image itself. We as humans, when we see any scene with our eyes, the objects far away from our eyes seem smaller than objects that are closer, and this is called perspective. So, in the projection part (orthographic projection), do not consider this effect for measuring the correct and perfect perspective and portray a real viewing experience. The processing to the 2D plane does not happen automatically, this happens with the process called image perspective transformation algorithm which takes [image, a size] as the arguments (where a is the matrix referred to earlier, image is referred to as the input array src, and size is the referred as the size of the specified image size). As the value of the contour area increases, then it detects the keypress and starts its computational work. The function used here is [6]:

```
do_keypress(img, center, row_keys_points):
```

3.6 Keyboard Output and Workflow Model

One can move his/her finger while wearing the pointer with the help of the visual keyboard visible on screen toward the desired letter and flick(moving the tip toward the camera in a sudden burst of motion) the finger to make a keypress. Thus, the final layout of virtual keyboard is clearly shown in Fig. 5. The detailed architectural workflow model of this paper is explained in Fig. 6. Also, an example of the user typed word "hello" in virtual keyboard can be seen in Fig. 7.

Fig. 5 Final keyboard layout and keypress

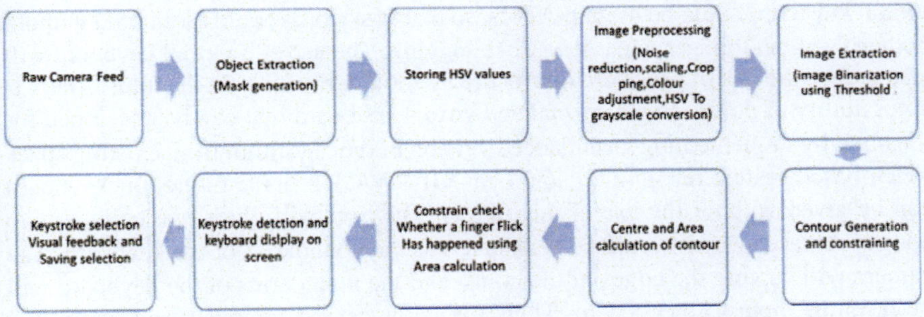

Fig. 6 Workflow model

Fig. 7 Typed word "hello"

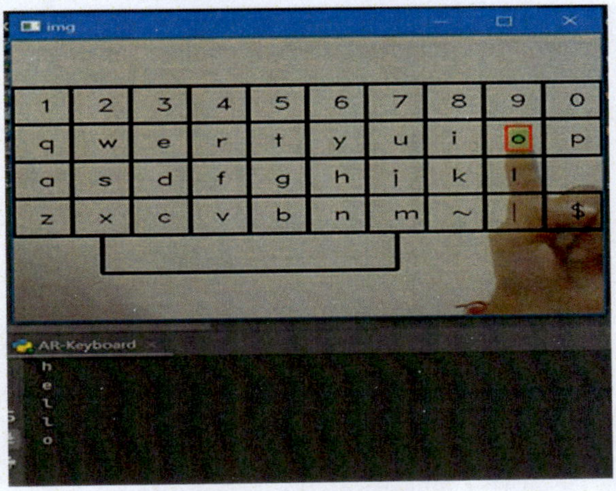

4 Future Scope

In the future, we will take a piece of paper and print a keyboard on it and color the keyboard keys with different colors. A mobile selfie camera or a laptop camera or a webcam will be able to detect the different colors (which will be the keys) and the estimated fingertip positions which in result will give us the correct output. Future development of the product can be done using the hard connection between the two cameras or using only one camera after appropriate installation and calibration.—ontour processing and curvature calculation.—Multiple camera integration. The main achievement of this project is fingertip detection. The method used to detect the fingertip (R-SoG) was implemented by us, with no reference. We could not find cases using this method for this purpose. The virtual keyboard can be used to perform all basic computer work. This keyboard could be used for noise-free typing. This keyboard could also be used to enter passwords and other private data. The virtual keyboard could be used in ATMs, so that passwords could be entered without bothering about things such as keystroke logging. The use of a virtual keyboard will also reduce the risk of password disclosure by shoulder surfing. In the future, there is a possibility of developing a specialized virtual keyboard that can be developed for the blind by implementing an enhanced version of the algorithm from existing space touch-based gesture recognition, and LeapKII-SW/GUI. In the future, the keyboard can be given without the use of any object. The user will place two coins at two ends of surfaces which would be considered as the boundaries of the keyboard. The camera will capture the edges of the coins, and the boundaries of the keyboard will be taken as input to the system. Then, the coins can be removed, and the virtual keyboard with the desired size will be displayed on the device. Thus, a keyboard can be used for typing and other applications without the use of any object or surface material [7]. In future, this keyboard can be implemented on an AR specific technology also. In doing so, we have to use some image processing algorithms, such as edge detection, Hough transforms, morphological operations, and image filtering, homographic transform, and correlation.

5 Conclusion

With the recent rise in the popularity of small screen mobile devices, there is a requirement of a convenient and easy to use data unit that will be designed for a user who will be trying to put information into the device but does not want to carry an extra hardware around, which would affect the mobility and portability of mobile devices. There are already many data input devices and different virtual keyboards available in the market. But these devices lack accuracy and are less precise. To eliminate some of the above issues, a new projection keyboard called "reconfigurable virtual keyboard," which provides basic touch typing style with the use of a webcam and a paper keyboard layout. The reconfigurable virtual keyboard is more convenient in applications like gaming, but as its processing is slow, it takes more time to compare input characters with the characters in the database. But as computers are getting faster day by day, we can create a fast processing keyboard in future [8]. We can create a keyboard using the system, in any layout as per our need or convenience. In a country like India which has a rich variety of languages, a augmented keyboard can be used to create a multi-language keyboard, such that the user can create a convenient keyboard by changing the inside mapping of characters. No change in hardware is required, it will make a multi-language keyboard economically reasonable. And also the cost of a virtual keyboard is less than that of the current keyboard. A virtual keyboard is also extremely easy to use, and for security purposes, it will be far better than the current keyboard.

References

1. https://en.wikipedia.org/wiki/Virtual_keyboard
2. https://docs.opencv.org/3.4/da/d97/tutorial_threshold_inRange.html
3. https://www.ijert.org/reconfigurable-virtual-keyboard-based-on-image-processing (Year of Publication: 2015, vol 3, no 6, p 3, Conference-International Journal Of Engineering Research And Technology ,ISSN 2278–0181)
4. https://docs.opencv.org/master/d7/d4d/tutorial_py_thresholding.html
5. https://docs.opencv.org/3.4/d4/d73/tutorial_py_contours_begin.html
6. Virtual Keyboard. Int J Sci Eng Technol Res (IJSETR) 3(3). http://ijsetr.org/wp-content/uploads/2014/03/IJSETR-VOL-3-ISSUE-3-518-522.pdf
7. Priyanka I, Kogila M, Mohanapriya, Kumar V (2014) A survey on portable virtual keyboard based on threshold and perspective transform. IJERT 3(2); Naykodi T, Patil A, Sase K, Singh S (2015) Reconfigurable virtual keyboard based on image processing. In: ICONECT-2015 conference proceedings, special issue vol 3, no 6. https://www.ijert.org/a-survey-on-portable-virtual-keyboard-based-on-threshold-and-perspective-transform
8. Zhang Y, Yan W, Narayanan A (2017) A virtual keyboard implementation based on finger recognition. In: 2017 International conference on image and vision computing New Zealand (IVCNZ), Christchurch, 2017, pp 1–6. https://doi.org/10.1109/IVCNZ.2017.840245

Facial Expression Recognition Using Convoluted Neural Network (CNN)

Prerana Kundu, Pabitra Kundu, Sohini Mallik, Srimoyee Bhowmick, Pratim Mandal, Hritam Banerjee, and Sudipta Basu Pal

Abstract This paper aims to fabricate a facial expression recognition system using a convolution neural network (CNN). Since human evolution, human emotion interpretation has been an effective method for non-verbal communication. It helps to contend human perspectives or to reciprocate feelings and his or her mental state. With the development in science and technology, face detection and face recognition from sources have become relatively more straightforward. This project targets the fabrication of a real-time recognition system that traces the very mood of human beings through popular computer vision libraries and convolution neural networks. Facial action coding system (FACS), which uses action units (A.U.s), has been used as the base toward the integration of this research. This research has predominantly endeavored to detect specific subtle changes in facial components (A.U.s). Thus, relating those changes with the corresponding combinations has finally successfully seen the underlying emotion. "OpenCV," an open-source computer vision library, has been used in the classification of the seven primary human emotions: fear, sadness, happiness, anger, disgust, surprise, and neutral. After the face is detected, image processing based on the feature point extraction method is used to extract a set of feature points. A considerable amount of significant research has been contributed toward the advancement of human–computer interaction (HCI).This paper tries to focus on the integration and implementation of a successful facial expression recognition system using CNN consists of a detailed work review and result from the analysis of automatic facial expression recognition systems.

Keywords CNN · Facial action coding system (FACS) · Open-CV · Computer vision · FER2013 database · HCI · MaxPoolong2

P. Kundu · P. Kundu · S. Mallik · S. Bhowmick · P. Mandal · H. Banerjee · S. B. Pal (✉)
Computer Science Engineering Department, University of Engineering and Management Kolkata, Kolkata, India

J. M. R. S. Tavares et al. (eds.), *Cyber Intelligence and Information Retrieval*, Lecture Notes in Networks and Systems 291,
https://doi.org/10.1007/978-981-16-4284-5_8

1 Introduction

In today's world, real-time facial expression recognition based on CNN [1–3] is a very renounced research area. Emotions play a crucial role in human life. Facial expressions help us to communicate psychologically, through emotions and allows us to peep deep into one's conscience. In the "Expression of the Emotions in Man and Animals", Charles Darwin believed that emotions to be species-specific rather than culture-specific [4]. Darwin's research worked in 1872, facial expression detection and emotion analysis has been a popular topic of research for several science enthusiasts. Suwa et al. in 1978 made his attempt to comprehend emotions from facial expressions automatically. This helped lay the stepping stone for further advancements in facial expression recognition systems in the later years.

Data is an integral part of the existing approaches. User has to put big data, so that the algorithm automatically extracts features and their motion information, subtly discriminates the face emotion, and gives the desired result. The identification of facial expressions plays a crucial role in identifying patterns and image processing. Facial recognition technology (FER) is based upon the intricacies hidden among the subtle changes in facial components.

Facial action coding system (FACS) [5] analyzes the facial movement of 44 areas and refers to them as the action units (A.U.s). The scrutiny of severity and combinations of these A.U.s can help determine human emotions [6]. OpenCV is being used here for successful and more comfortable face recognition from sources like images or videos. By the application of machine learning, the computer can read and compare human expressions. Four necessary steps are as follows:

1. Facial detection,
2. Preprocessing of data,
3. Data extraction,
4. Data testing,
5. Facial expression analysis and recognition.

This part of computer vision has been praised widely around the globe for its useful applications in the fields of cyber security, social studies, market management, and also in psychiatric and medical examinations. One of the significant challenges faced by this system, which is an unbiased analysis, regardless of culture, gender, and context, remains unsolved.

1.1 Background

The two popular models of studying FER are mainly action units (A.U.s) [7] and the valence–arousal space (V–A space) [8]. The V–A space is a universal model widely used in continuous emotion recognition tasks of audio, visual, and physiological signals. Facial expression recognition (FER)-based study has received extensive attention in psychology, computer vision, and pattern recognition. This study has broad applications in multiple domains, including human–computer interaction [9, 10], virtual reality, augmented reality [11], advanced driver assistance systems, education, and entertainment. The A.U.s encode essential facial muscles' movements, and the combination of A.U.s could be utilized for FER. In [12], a framework is proposed to apply A.U.s to estimate the V–A intensity. In this paper, mainly, the analysis is entirely based on the FER concept.

2 Problem Statement

There are seven raw emotions that human facial expressions can easily depict: happiness, sadness, surprise, anger, fear, neutral, and disgust [13]. These subtle expressions help us sometimes to understand the inner motives of an individual. The effects that content and services have on their consumers can be measured through an easy and low-cost procedure by facial emotion recognition. In Fig. 1, the basic facial recognition system is depicted clearly.

For example,

- To provide better service by analyzing patients' mental states during treatment,
- To monitor the audience's engagement in events and shows,
- To assess customer interest.

Over the past few years, different research has been conducted on facial expression recognition and emotion detection. It has grown out to be one of the successful applications of image analysis. Due to its diverse applications, computer vision has also created a remarkable base for several psychiatric and neurosurgical studies, e.g., in child psychiatry, criminal psychology, etc.

PROBLEMS IN DATASET FER2013:

1. Imbalance problem sol: data augmentation,
2. Intra-class variation of FER sol: avoid overfitting,
3. Occlusion (mostly with hands),
4. Contrast variation,

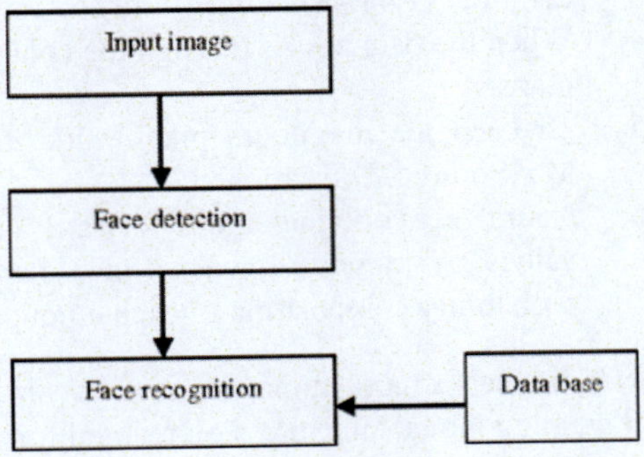

Fig. 1 Diagrammatic representation of facial expression recognition system

5. Eyeglasses,
6. Outliners.

3 Tools and Libraries

Convolutional neural networks (ConvNets or CNNs) are a type of neural networks that are very useful in areas such as image recognition and classification.

(1) This research focused on creating a CNN, which was able to classify the basic human emotions with a high degree of accuracy.
(2) Python 3.6 was used to write the programs
(3) Jupyter Notebook was used as the IDE.
(4) The CNN for facial expression detection was built using Keras
(5) FER 2013 was used for the datasets of images. These images have been registered, so that the face covers large parts in the center of the image and fits in the same amount of space in each image (Saransh & Dr. Muthamil Selvan T, 2019).
(6) OpenCV: OpenCV has the advantage of being a multi-platform framework; it supports both Windows and Linux, and more recently, Mac OS X (EMAMI & SUCIU, 2015).
(7) **Keras**: This is one of the libraries which is used to code deep learning models. In its backend, it uses Tensorflow.

Algorithm

1. A dataset of images is obtained. In this study, we are using the FER2013 dataset. It has prerecorded data of 48-by-48-pixel, 35887 cropped, grayscale images of facial expressions, each depicting different types of raw emotions. (Sransh & Dr. MuthamilSelvan T, 2019)
2. Before the images' preprocessing, facial detection is applied to the photographs, after which the photographs are cropped and converted to gray scale. They are hence giving us a more normalized form of the testing data.
3. Grayscale images, converted from cropped images, were fed into the convolution 2D array (1,48,48) NumPy array.
4. When the data passes through the convolutional layer, the layer creates feature maps.
5. To keep the maximum pixel value in the feature map, a pooling method, MaxPooling2D, is applied.
6. Neural network undergoes forward and backward propagation on the pixel values. A model named softmax function is being used, from which the probability composition of each emotion in the face was being gathered.

Now, the six basic human expressions are shown in Fig. 2, and the six major steps of creating the facial expression recognition system are depicted in Fig. 3.

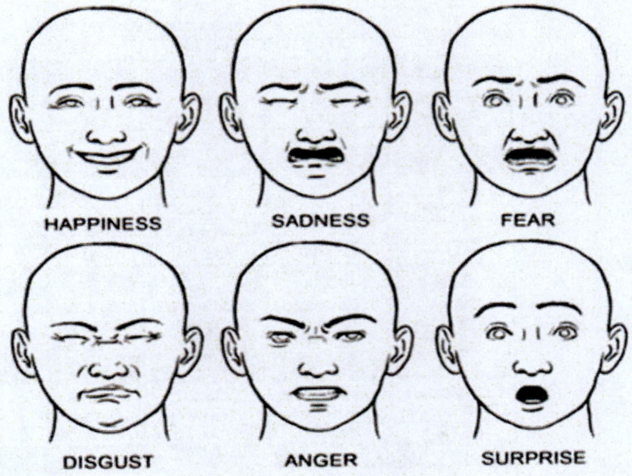

Fig. 2 Six basic human expressions [14]

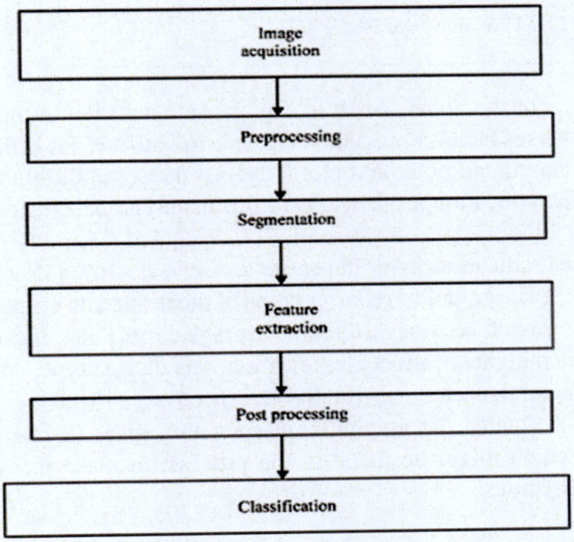

Fig. 3 Facial expression recognition process [16]

4 Result Analysis and Discussion

The facial expression analysis in most cases is done with the help of FACS action units. In the later years, the researchers Ekman and Friesen and Izard et brought at the table the theory which lies behind the several facial expressions, accompanying a certain human emotion such as lowering the lips in sadness or raising the eyebrows as a gesture to greet, etc. Figure 4 shows the FACS data.

FACS, which stands for facial action coding system, is known for its usefulness toward expression analysis and emotion detection. It uses a technology-based upon human inspection. FACS codes and records facial expressions, either separately or in combination, and are popularly referred to as action units [5].

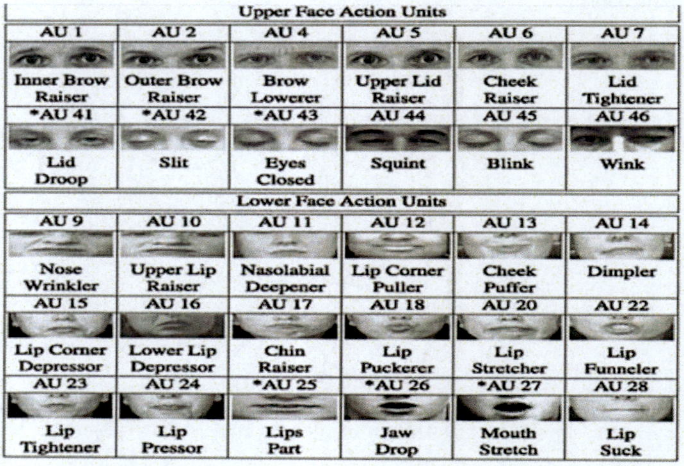

Fig. 4 Examples of FACS AUs

In the above graph in Fig. 5, we have tried to plot the correct classification % versus facial expressions. We have tried to take the different measurements of correct classification %, and plot it against the seven different basic emotions. The correct classification % of anger is 75.00; disgust is 65.38%; fear is 72.22%; happy is 98.18%; sad is 94.44%; surprised is 100.00%; and neutral is 100.00%. The average correct classification % for the entire testing is 86.46% [15].

The graphical representation of the result data gives us a clear representation of the research accuracy. Here, the average correct classification % is 86.46%. The lowest and highest correct classification % is disgust type, 65.38%, and both surprised and neutral, each being 100.00%, respectively [15] .

Finally, the graphical interpretation gives us clarity about the further changes which might be made to the process, to make it, even more, data-efficient, and accurate.

Fig. 5 Correct classification versus expression

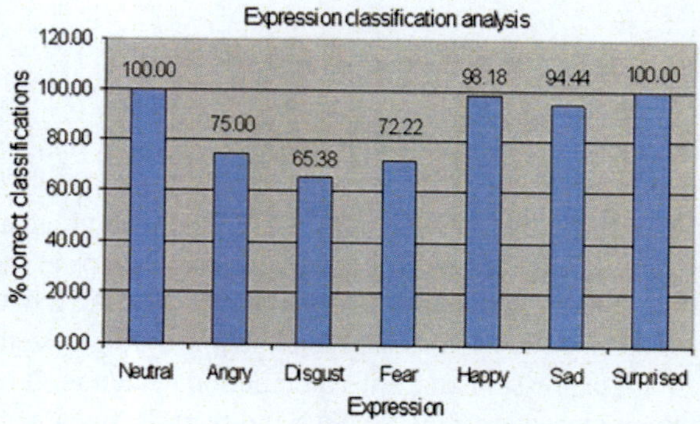

5 Conclusion

The study was concentrated on the application of a convoluted neural network to determine basic emotions from facial expression, the accuracy of which depends on its database. Its strength lies in the fact that convoluted neural networks are faster, but it requires more memory which makes it expensive. So, it can be concluded that in the future, the proposed technique would improve to consider wide ranges of emotion with accuracy.

References

1. Lopes T, de Aguiar E, Oliveira-Santos T (2015) A facial expression recognition system using convolutional networks. In: 2015 28th SIBGRAPI conference on graphics, patterns and images, Salvador
2. Halder R, Sengupta S, Pal A, Ghosh S, Kundu D (2016) "After the face is" detected, image processing basedon feature point extraction method isused to extract a set of feature points. Int J Comput Appl 0975–8887.
3. Liu K, Hsu C, Wang W, Chiang H (2019) Real-Time facial expression recognition based on CNN. In:2019 International conference on system science and engineering (ICSSE), Dong Hoi, Vietnam
4. Darwin C, The expression of the emotions in man and animals. Create Space Independent Publishing Platform.
5. Ekman P, Friesen WV (1978) The facial action coding system: a technique for the measurement of facial movement. Consulting Psychologists Press; San Francisco. [Google Scholar] [Ref list]
6. Scherer K, Ekman P (1982) Handbook of methods in nonverbal behavior research. Cambridge Univ. Press; Cambridge, UK. [Google Scholar] [Ref list]
7. Tian YI, Kanade T, Cohn JF (2001) Recognizing action units for facial expression analysis. IEEE Trans. Pattern Anal Mach Intell 23:97–115. [Google Scholar] [CrossRef] [PubMed]
8. Russell, J.A. A circumplex model of affect. J. Pers. Soc. Psychol. 1980, 39, 1161. [Google Scholar] [CrossRef]
9. Cowie R, Douglas-Cowie E, Tsapatsoulis N, Votsis G, Kollias S, Fellenz W, Taylor JG (2001) Emotion recognition in human-computer interaction. IEEE Signal Process Mag 18:32–80. [Google Scholar] [CrossRef]
10. Bartlett MS, Littlewort G, Fasel I, Movellan JR (2003) Real time face detection and facial expression recognition: development and applications to human computer interaction. In: Proceedings of the conference on computer vision and pattern recognition workshop, Madison, WI, USA, 16–22 June 2003; vol 5, p 53. [Google Scholar]
11. Chen CH, Lee IJ, Lin LY (2015) Augmented reality-based self-facial modeling to promote the emotional expression and social skills of adolescents with autism spectrum disorders. Res Dev Disabil 36:396–403. [Google Scholar] [CrossRef] [PubMed]
12. Hamm J, Kohler CG, Gur RC, Verma R (2011) Automated facial action coding system for dynamic analysis of facial expressions in neuropsychiatric disorders. J Neurosci Methods 200:237–256. https://doi.org/10.1016/j.jneumeth.2011.06.023. [PMC free article] [PubMed] [CrossRef] [Google Scholar]
13. Kaulard K, Cunningham DW, Bülthoff HH, Wallraven C (2012) The MPI facial expression database—a validated database of emotional and conversational facial expressions. PLoS ONE 7:e32321. https://doi.org/10.1371/journal.pone.0032321. [PMC free article] [PubMed] [CrossRef] [Google Scholar]

14. Dubey M, Singh L, Automatic emotion recognition using facial expression: a review. IRJET 03(02). e-ISSN: 2395 -0056. p-ISSN: 2395–0072
15. Liliana DY (2019) Emotion recognition from facial expression using deep convolutional neural network. J Phys Conf Ser
16. Premaladha J, Sujitha S, Priya ML, Ravichandran KS (2014) A Survey on melanoma diagnosis using image processing and soft computing techniques. Res J Inf Technol 6:65–80

Next Step to the Future of Restaurants Through Artificial Intelligence and Facial Recognition

Uzayruddin Siddiqui Mohammed, Vinod Kumar Shukla, Robin Sharma, and Amit Verma

Abstract In the previous generation, the resources we had for our technology were limited. The development was not enough. There were big dreams to be fulfilled in technology by only imagining our favorite books and watching our favorite sci-fi movies. Through the years, as our resources expanded, so did our technology. Artificial Intelligence technology has developed a lot through the years and is still developing. As our resources grow, we all are surrounded by technology in this age and time. It helps to make our lives a little easier than before, and Artificial Intelligence (AI) is a technology used to an industrial level to make things easier. Restaurants are commonly visited places by any middle-class man; let it be for lunch or Birthday Party. It is very common to visit a restaurant. There are issues regarding the restaurants' hygiene, from the kitchen countertop to the waiter's hygiene serving you food. During the period of normalizing going out in the Covid-19 pandemic, there were many issues regarding this? In this paper, the core point is to avoid physical contact and enhance restaurants' future with AI and its upcoming potential to deliver a new and safer experience to the customer. This will help prevent contact with the waiter, create a unique customer experience, and create a hygiene environment while implementing newer things.

Keywords Artificial intelligence · Facial recognition · Future restaurant · Machine learning · Big data

---

U. S. Mohammed
Bachelor of Information Technology, Amity University, Dubai, UAE

V. K. Shukla (✉)
Department of Engineering and Architecture, Amity University, Dubai, UAE

R. Sharma
Department of Humanities Arts and Applied Sciences, Amity University, Dubai, UAE
e-mail: rsharma@amityuniversity.ae

A. Verma
HTMi Switzerland, Dubai, UAE
e-mail: amit.verma@htmidubai.com

J. M. R. S. Tavares et al. (eds.), *Cyber Intelligence and Information Retrieval*,
Lecture Notes in Networks and Systems 291,
https://doi.org/10.1007/978-981-16-4284-5_9

1 Introduction

People visit restaurants regularly. Many restaurants worldwide live up to the hygiene standards, and some provide an experience like never before, but what if there was a way to improve the standard of the whole term restaurants? It only makes sense that as time passes by and the technology in the world increases, we utilize it for a more technological tomorrow. This can be made possible by using Artificial Intelligence. Artificial Intelligence has been used in many things for a good time now. Let's say this on a busy day, you choose a table, and you go through the menu have your order ready to be given to the waiter. Due to the heavy rush on the weekend, you have to wait for the waiter to finish with the other customers and then come and take your order. Here is another case, you are ready to order, but it's the pandemic. You are afraid that the waiter is visiting every table and then coming to you over there, there is a risk and a high possibility to catch the Covid-19 virus. Even if it's not the virus, it can cause a possible health hazard or any hygiene issue. Now all this can be avoided by the use of simple Artificial Intelligence. Imagine you come to a restaurant you choose a table you have a sit you decide what you want. Instead of waiting for the waiter, you immediately tell it to the Artificial Intelligence ready and installed by the table and avoid any hygiene issue or deadly flu in the pandemic.

2 Literature Review

Food is a necessity of life. Without it, there is no survival for any living creature. Some food consumption is needed to survive. According to the studies of the United States Department of Health, an average adult female needs about 1600–2400 cal a day, whereas an average adult male needs roughly 2000–3000 cal a day to maintain his physical health and stay fit, it varies from person to person and even children according to their rate of physical growth [1].

Table 1 shows a brief chart of how many calories needs to be consumed by a Child, a Female, and a Male according to their age and their activeness varying from Sedentary, Moderately Active, and Active [2].

In 2014 just the United Arab Emirates average daily calorie consumption was 9.9 billion calories a day which meant that an average United Arab Emirate consumer bought about 1080 kcal of miscellanies food items like soft drinks and packaged foods, biscuits, etc. [3]. When it comes to calorie consumption in the Middle East, Saudi Arabia has the most calorie consumers; their packaged food consumption is 1463.22 kcal per consumer. When it comes to consuming the most significant number of calories, Mexico has the highest number of calories consumed to the ratio of per consumer consuming 1.927.51 kcal just from packaged food items, followed by Germany consuming 1.867.94 kcal [3].

Table 1 Calories consumption chart

Gender	Age (Years)	Sedentary (kcal)	Moderately active (kcal)	Active (kcal)
Child	2–3	1000	1000–1400	1000–1400
Female	4–8	1200	1400–1600	1400–1800
	9–13	1600	1600–2000	1800–2200
	14–18	1800	2000	2400
	19–30	2000	2000–2200	2400
	31–50	1800	2000	2200
	51+	1600	1800	2000–2200
Male	4–8	1400	1400–1600	1600–2000
	9–13	1800	1800–2200	2000–2600
	14–18	2200	2400–2800	2800–3200
	19–30	2400	2600–2800	3000
	31–50	2200	2400–2600	2800–3000
	51+	2000	2200–2400	2400–2800

3 Artificial Intelligence

Everyone has their own opinion on what Artificial Intelligence is. Some think it's a revolution, and some believe it's something that will take over human beings. According to Minsky and McCarthy, back in the 1950s, they described Artificial Intelligence to be something that can perform tasks by a type of machinery that would have required human intelligence for the task to be achieved.

The simulation of human intelligence processed by machinery such as computer systems is known as artificial intelligence. Artificial intelligence can learn from its own experience. They are not any normal instrument that follows the command. Artificial intelligence learns to deal with new inputs to perform tasks capable of being done by human intelligence, for example, a self-driving car. The idea of Self-driving cars is futuristic, and it has been made possible by companies like Tesla and BMW, etc. A lot of natural learning is processed into a self-driving car, and they also rely very heavily on deep learning [4–6] (Fig. 1).

In the above Fig. 2, we can see how many accidents were caused before and after Tesla's Autosteer Artificial Intelligence was implemented in their cars. There was a 40% drop in the casualties because the Autosteer Artificial Intelligence identifies a potential accident and steers the vehicle away from it, avoiding the accident. This is only possible by Artificial Intelligence's heavy deep learning [7].

There is a lot put into the workings of artificial intelligence. Large quantities of data are combined with really quick processing and many intelligent algorithms into software that then learns on its own by various things like data provided, the input patterns, or even the artificial intelligence pattern. Artificial Intelligence is a broad field that requires many other subfields such as deep learning, Natural language processing, machine learning, neural network, cognitive computing, and computer

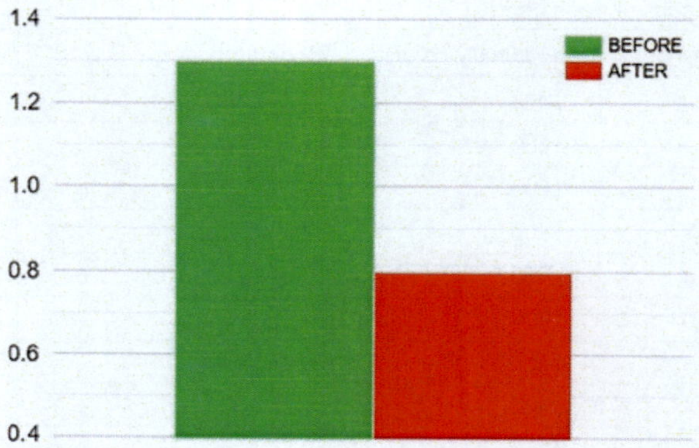

Fig. 1 Before and after Tesla Autosteer [7]

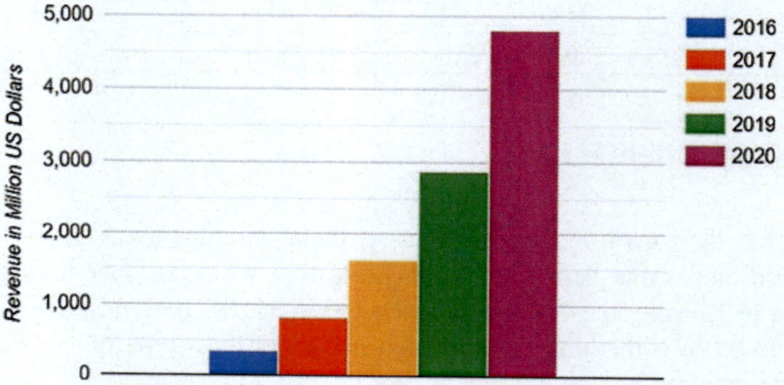

Fig. 2 Growth of artificial intelligence 2016–2020

vision, etc. [8]. Artificial intelligence is broad, but there is not only one type of artificial intelligence. There are *four types of artificial intelligence: theory of mind, reactive machines, self awareness, and limited memory.*

3.1 Theory of Mind

Theory of mind is the artificial intelligence understanding that living things such as human beings, animals, etc. have their way of thinking that their thoughts and the way they feel, their emotions is what brings an effect to one's behavior.

3.2 Reactive Machines

This type of artificial intelligence acts on what it sees; it is the most basic form of artificial intelligence as it is just reactive. Reactive machines have no sense of memory it does not need any experience to make decisions. It predicts what it has to do it works with an algorithm and chooses the best possible among the other various possibilities. The most famous reactive machine is the Deep Blue by IBM. It is a chess-playing artificial intelligence that does not rely on any past experiences but instead focuses on predicting what possibly could be the opponent's next move.

3.3 Self Awareness

Self Awareness artificial intelligence, in a way is the extension to the theory of mind. This type of artificial intelligence needs to have consciousness. It should be able to understand and predict feelings. It's the same way we humans learned it but instead of human intelligence, its artificial intelligence.

3.4 Limited Memory

Limited Memory is an artificial intelligence that is already being used in self-driving cars like tesla's. This type of artificial intelligence has a sense of memory of past and experience they learn from their pasts, learn new patterns and add them up with old habits and learn from the experiences [9].

4 Growth of Artificial Intelligence

There is no doubt to say that Artificial intelligence took a giant leap in the past ten years, a lot of significant business cooperation who have already implemented artificial intelligence in their company or products and they have come far with it too. If you take voice assistant artificial intelligence, then Apple's iPhones, Siri or Amazon Alexa are ruling the market. If someone had even thought back then ten years ago that we would be asking an artificial intelligence to set our alarms, then they would laugh.

In the above Fig. 2 we can see Artificial Intelligence's growth in the past five years worldwide. The market revenue rose from 357.89 million dollars in 2016 to 4806.3 million dollars as of now in 2020 [10]. There is no saying how far artificial intelligence can go. It can be the face of the future in this decade. The major companies in the race of artificial intelligence are shown in Fig. 3 [11].

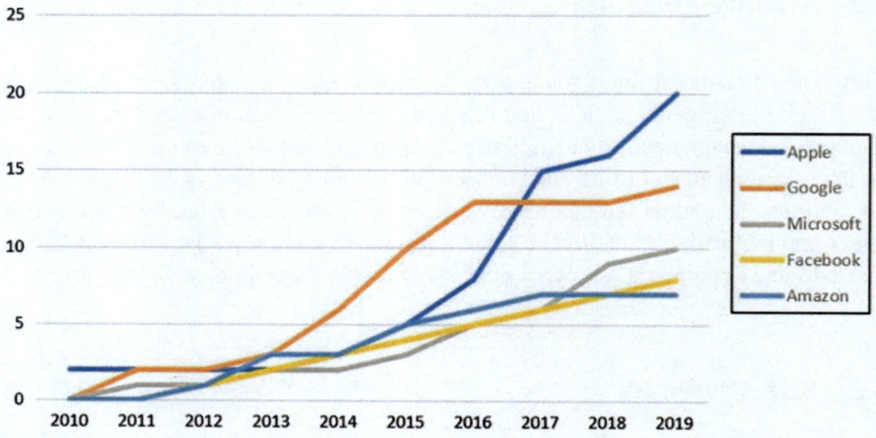

Fig. 3 Race of artificial intelligence

In the above Fig. 3, we can see that as of 2019, Apple is the leading company in the Race of Artificial Intelligence. According to the figure, apple was the only company among the other companies that started implementing artificial intelligence before any other race. There are many other enterprises, small and big, using artificial intelligence in one way or the other to make their work easier [11].

There is more to come to artificial intelligence in the coming time. Artificial intelligence is a technology that is growing day by day. According to Forbes business, 84% of business enterprises believe in artificial intelligence. They think that having artificial intelligence being used in their business enterprise will give their business a greater competitive advantage against other business enterprises in their fields. Artificial intelligence can possibly bring more new business to the market. According to Forbes business, 75% of business enterprises expect others to start an enterprise in the artificial intelligence field, which gives the existing business new methods and ways to gain a higher place in the artificial intelligence market [10].

In the above Fig. 4, we can see a significant leap expected to be taken by the field in the next five years. By 2025 the artificial intelligence field is expected to have 31,236.92 million dollars in revenue [10]. As the technology develops more and more, then in the future there is a possibility that due to the increase in the artificial intelligence market and the fluctuation of the demand and the availability of the technology at that time may cause a drop in the price of the technology due to the heavy possible competition in the market.

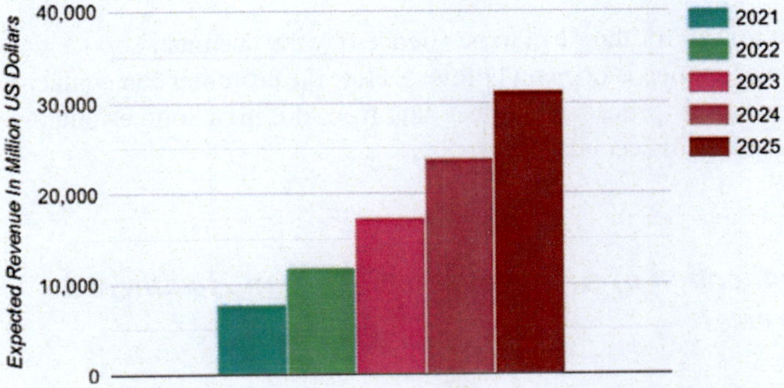

Fig. 4 Growth of artificial intelligence in the future (2021–2025)

5 Applications of Artificial Intelligence

Artificial intelligence has been implemented in a few industries already. In the paper, there are mentions of prevalent artificial intelligence technology being used worldwide; the following are a few other artificial intelligence applications.

5.1 Application of Artificial Intelligence in Finance

There are many brilliant business tycoons who know how the trading market works and its patterns. In the trading market, there is profit only if you have the talent to predict which stock will fall or rise in the near future, and this is all done by analyzing large amounts of data. Machines can also process patterns and understand them and learn them they also have an incredible speed of reading and understanding the data in a very short amount of time which gives the machines a plus point compared to humans as they are more accurate when they give the possibility percentage and are able to process a large amount of data faster than any human can which allows trading companies an opportunity to improve their overall performance and boost their chances of earning profits [12].

5.2 Application of Artificial Intelligence in Banking

The technology of artificial intelligence is already being implemented in many banks worldwide to detect anomalies caused to their bank users and provide customer service to their customers regularly. Many banks use chatbots to assist their customers with their queries. HDFC Bank of India started attending to their customer's queries with a chatbot called Electronic Virtual Assistant, in short EVA. Now the customers don't have to wait for their call to be attended by the customer service or make sure that customer service is open on holidays. Now the customer can contact EVA 24/7, and artificial intelligence will collect data from different sources and answer their query in a matter of seconds [12].

5.3 Application of Artificial Intelligence in Healthcare Services

Artificial Intelligence hasn't developed much in the healthcare industry but it has a lot of rising potential. It has already begun its journey in the industry by safeguarding hospital documents to many government confidential records. Artificial intelligence is also being implemented in robots to get assisted surgeries done. This is still not in mass use as this technology is still not ready to help save lives. Other than that, artificial intelligence has worked its way in assisting patients from their homes. They don't have to make calls or come to the hospital to book appointments. Instead, they can go online and book an appointment or with artificial intelligence enable virtual assistants. This virtual assistant also helps the nurses save 20% of their time from assisting with just information. Pharmaceutical companies are also using artificial intelligence to collect data worldwide and research and develop medicines and medical supplies that can save lives [13].

5.4 Application of Artificial Intelligence in Travel

Travelling is something we do regularly. Sometimes we go to places without knowing the way when we are in such a situation. We take the help of online navigation such as google maps or Waze. For example, let's take google maps; when we put a destination, it shows us the route, how long it will take to reach there, Kilometers or Miles, and the traffic. What happens to give us all the data is that an artificial intelligence scans all the possible routes to your destination and displays the fastest way possible to the destination? The artificial intelligence also shows you how much traffic you have on the road by showing yellow or red on the route line depending on the traffic; this is information is given to the user even if they are 10 or 100 km away from the traffic. The artificial intelligence gathers this information by recognizing the number of users available there at the traffic spot and at what speed is there vehicle is moving. If the vehicle speed is low and the number of users in that spot is a lot, it indicates red, moderate traffic, then yellow, and no traffic then blue [13].

5.5 *Artificial Intelligence in Social Media*

Social Media is a platform being used by billions of people globally, but many that billion people don't know the social media uses artificial intelligence. Let it be suggesting you add your mutual friend or even auto-update your phone. A lot of decisions are made by artificial intelligence than we think it does. For example, you are looking to buy something on an online shopping website, let's say a mobile phone, you spend some time on a particular product. You don't buy it then; later on, you are on your preferred social media platform, and you are scrolling to your feed. Suddenly you see an ad promoting the mobile phone you were searching to buy now this is made possible by the artificial intelligence Analyzing your web data. It does analyze way more than your shopping websites. It sees what you are thinking of purchasing. If that particular website is sponsoring that specific social media, amazon, as it is a well-established company, only the advertisement is done. These social media platforms only analyze website data with some of the other links to the medium [14].

5.6 *Security in Artificial Intelligence*

The term security in artificial intelligence refers to security being provided with taking artificial intelligence into leverage; cybersecurity companies use artificial intelligence to provide a better service to their clients. They use Artificial intelligence on heavy algorithms to scan the system and detect harmful viruses like trojans or malware. The artificial intelligence also runs pattern recognitions in the software to recognize and hold data on new viruses and malware and be up to date. Artificial Intelligence is widely used on social media platforms also such as Instagram. Instagram is a central social media platform used by 1 billion people every month, and it's Instagram's responsibility to maintain a safe social experience and they do it by implementing artificial intelligence to prevent cyberbullying. They have an on/off setting to hide offensive comments. If anyone decides to comment in an offensive on your post, then the artificial intelligence detects it and filters it out, and it does not display. Social media is an open platform with the freedom of speech, and there is a lot of offensive and fake posts which the artificial intelligence recognizes it and shows a prompt like it can be fake news, offensive post, or graphical imagery either way it can be recognized by the artificial intelligence or can be reported [15–18].

6 Framework for Restaurant Using Artificial Intelligence and Facial Recognition

Let's consider this, in a restaurant, a tablet is artificial intelligence enabled on the table's side attached to the wall, easily accessible by the restaurant's customer. To that tablet, a facial recognition device connected the tablet to the table. Hence, when a registered customer walks into the table booth. They immediately get recognized by the artificial intelligence using facial recognition, which then the tablet displays the most frequent or favorite food order by that person. Then artificial intelligence enabled chatbot asks for the customer's order and takes it verbally using the tablet's microphone; this gives the customer a futuristic experience. From there, when the order is decided, it gets sent to the kitchen. The customers don't have to wait for the waiter to come and take your order. All the waiters have to do is deliver you the food, which also maintains a sanitary environment in the surrounding, which can be a little assurance for the customer that it is safe to be there in situations like the COVID-19 pandemic (Fig. 5).

6.1 *Tablet Device*

A tablet device is which can be used as a portable computing device. It uses its touchscreen panel as an input and it has a microphone. Tablets in the market depend on different companies offering different prices depending on the tablet specifications; some tablets can be used by electronic pens designed for that tablet [19]. In above Fig. 4 shows a diagrammatic representation of the framework; the diagram shows how the data is passed from the table to the kitchen; the kitchen will have a screen

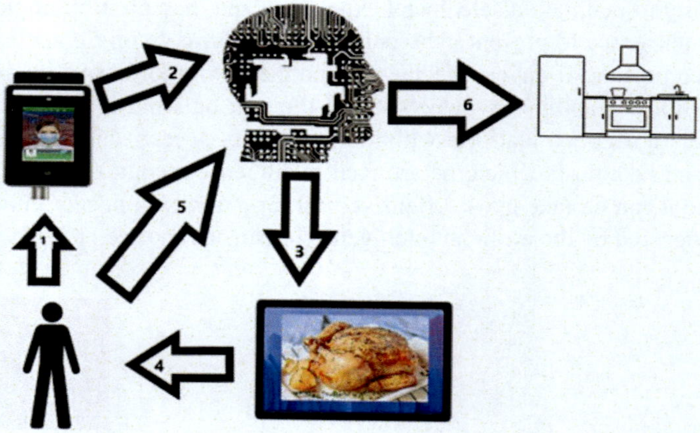

Fig. 5 Diagram of the proposed framework

panel that displays all the orders given with their respective table number. This technology can be used in any restaurant willing to provide a futuristic experience to the customer and stay one step ahead of their competition.

6.2 Face Recognition

Face recognition is the recognition of a human face by an electronic device. A facial recognition device maps your face structure using biometrics. This might be a privacy concern to many people. There is a lot of complications when it comes to a machine understanding a human face structure. When you meet someone for the first time, it is pretty hard to remember every exact feature of someone's face. Still, a face recognition device can scan the face in a few seconds and store that as data in its system, and this storing of faceprints is done with the help of various algorithms, the data stored can be accessed. A facial recognition device's working can be explained in four steps; let's take the Federal Bureau of Investigation as an example [20].

Step 1: *Your face is captured through a video or picture source; depending on the facial recognition device, you may or may not have to have a straight picture for it to run.*

Step 2: *Then the software runs that photograph or video and scans and reads the geometry of your facial structure, which includes your key features such as the color of your eye, the distance between your features, and beauty spots when the process is done the result is to be known as your facial signature.*

Step 3: *The facial signature just made then is compared to millions or billions of faces in the respective databases, in this case, the Federal Bureau of Investigation.*

Step 4: *After all these processes are done, your face print may or may not match with the Federal Bureau of Investigations database* [20].

There are other ways a facial recognition device can work. It depends from device to device when it comes to different applications, and facial recognition is pretty mainstream now since mobile devices are using facial recognition to access their mobile devices. Artificial Intelligence is getting popular in many domains such as banking [21], healthcare [22] and tourism [23]. Soon there will be a time when AI will be an integral part of any business around the world as today we take the help of internet and office automation technologies.

7 Conclusion

The paper discusses many technologies that have come so far in the market; by bringing them all together, we can take restaurants' future to the next level. The technology is not perfected it still has issues that can eventually be fixed by developing and testing more and more mathematical algorithms for sanitary safety. This technology can come handy than it seems to be, it is avoiding the waiter being near you.

At the same time, you decide and give the order if you make contact with the waiter just two times on a regular basis and with this technology just once then the sanitary measure is being increased by 50%. This is a one-time investment for a restaurant, and the main focus is for it to be cost-effective while not compromising the quality of the hardware.

References

1. Calorie Calculator (online). https://www.calculator.net/calorie-calculator.html#:~:text=Though%20it%20differs%20depending%20on,the%20U.S%20Department%20of%20Health
2. Albuquerque DD, Shukla VK, Verma A, Tyagi SK, Sharma P (2020) Enhancing sustainable customer dining experience through QR code and geo-fencing. In: 2020 International Conference on Computation, Automation and Knowledge Management (ICCAKM), Dubai, United Arab Emirates, pp 190–196. https://doi.org/10.1109/ICCAKM46823.2020.9051470
3. Cleofe Maceda. UAE consumers eating more calories than ever (online) https://gulfnews.com/uae/health/uae-consumers-eating-more-calories-than-ever-1.1469548#:~:text=As%20of%202014%2C%20the%20average,donut%20per%20person%20per%20day
4. Artificial Intelligence (online) https://builtin.com/artificial-intelligence
5. Linda Tucci. Artificial Intelligence (online) https://searchenterpriseai.techtarget.com/definition/AI-Artificial-Intelligence
6. Artificial Intelligence (online) https://www.sas.com/en_ae/insights/analytics/what-is-artificial-intelligence.html
7. Galeon D. Report: Tesla crashes dropped by 40% after autopilot was installed (online). https://futurism.com/report-tesla-crashes-dropped-by-40-after-autopilot-was-installed
8. Grobbelaar W et al (2021) J Phys Conf Ser 1714 012032
9. Hintze A. Understanding the four types of artificial intelligence (online) https://www.govtech.com/computing/Understanding-the-Four-Types-of-Artificial-Intelligence.html
10. Columbus L. 10 Charts that will change your perspective on artificial intelligence's growth (online). https://www.forbes.com/sites/louiscolumbus/2018/01/12/10-charts-that-will-change-your-perspective-on-artificial-intelligences-growth/?sh=5e1cd65b4758
11. The race for AI: here are the tech giants rushing to snap up artificial intelligence startups (online). https://www.cbinsights.com/research/top-acquirers-ai-startups-ma-timeline/
12. Johari A. AI applications: top 10 real world artificial intelligence applications (online). https://www.edureka.co/blog/artificial-intelligence-applications/
13. Daley S. 19 Examples of artificial intelligence shaking up business as usual (online) https://builtin.com/artificial-intelligence/examples-ai-in-industry
14. Applications of AI—real life use cases in different sectors (online). https://data-flair.training/blogs/applications-of-artificial-intelligence/
15. AI Security (online). https://awakesecurity.com/glossary/ai-security/#:~:text=AI%20security%20refers%20to%20tools,on%20similar%20or%20previous%20activity
16. Laurence A. The impact of artificial intelligence on cyber security (online). https://www.cpomagazine.com/cyber-security/the-impact-of-artificial-intelligence-on-cyber-security/
17. Costa CD. Instagram explains how it uses AI to choose content for your explore tab? (online). https://uxplanet.org/instagram-explains-how-it-uses-ai-to-choose-content-for-your-explore-tab-f98803e0b114#:~:text=Recently%2C%20Instagram%20revealed%20that%20it,like%2C%20unlike%20an%20individual%20post
18. Constine J. Instagram now uses machine learning to detect bullying within photos (online). https://techcrunch.com/2018/10/09/instagram-bullying-photos/
19. Tablet (online). https://techterms.com/definition/tablet

20. Symanovich S. How does facial recognition work? (online). https://us.norton.com/internetsecurity-iot-how-facial-recognition-software-works.html
21. Suhel SF, Shukla VK, Vyas S, Mishra VP (2020) Conversation to automation in banking through Chatbot using artificial machine intelligence language. In: 2020 8th International Conference on Reliability, Infocom Technologies and Optimization (Trends and Future Directions) (ICRITO), Noida, India, pp 611–618. https://doi.org/10.1109/ICRITO48877.2020.9197825
22. Athota L, Shukla VK, Pandey N, Rana A (2020) Chatbot for healthcare system using artificial intelligence. In: 2020 8th International Conference on Reliability, Infocom Technologies and Optimization (Trends and Future Directions) (ICRITO), Noida, India, pp 619–622. https://doi.org/10.1109/ICRITO48877.2020.9197833
23. Verma A et al (2021) J Phys Conf Ser 1714 012037

Extensive Feature Analysis and Baseline Model for Stance Detection Task

Kumar Shaswat, Avantika Singh, Parul Kalra, and Deepti Mehrotra

Abstract Identifying and curtailing the spread of fake news is a complex and challenging task. Automated stance detection can be an important first step in alleviating the tedious process of fact checking. In this paper, we try to evaluate various language representation, and statistical similarity technique to discern the best possible mathematical modeling of the sentences for the task of stance evaluation. Our work is based on the dataset of the fake news challenge 1. The paper expounds our implementation details, the features, and models that work best for the given task.

Keywords Fake news challenge (FNC) · Term frequency-inverse document frequency (TF-IDF) · Natural language processing (NLP) · Gradient boosted decision tree (GBDT)

1 Introduction

The primary objective of the stance detection task is to categorize the tone of a given statement against a topic as 'agreeing', 'disagreeing', or 'neutral'. Although, it does not completely solve the problem of fake news, and stance detection is an effective and time saving automation for the human fact checkers to flag any false claims on a topic. Therefore, to promote the development of robust models for this task, the first stage of FNC was organized [1]. The task stipulated the development of a system that takes in a user statement and compares it with a news article to classify their relationship into a fixed set of four labels. There has been stance detection system previously that detected stance of a single sentence, however, the models developed for FNC-1 dataset had to classify the stance of an entire document which created new conceptual and algorithmic challenges [1].

There have been models for the FNC-1 that are state of the art and very tough to beat [2–4]. However, the aim of the paper is not to beat these models but rather focus on the language representation and feature engineering. Here, we are trying work on

K. Shaswat · A. Singh (✉) · P. Kalra · D. Mehrotra
Amity University Uttar Pradesh, Noida, India

J. M. R. S. Tavares et al. (eds.), *Cyber Intelligence and Information Retrieval*,
Lecture Notes in Networks and Systems 291,
https://doi.org/10.1007/978-981-16-4284-5_10

various ways to represent textual data and extract and learn about the feature that are the most important for the stance detection task. Our hypothesis is that there are set of vector representation and features that are universally more important than others in this task. But, the results of our implementation contradict this statement.

Rest of Section is as follows: Sect. 2 presents Related Work, Sect. 3 presents Methodology, Sect. 4 presents Result, Sect. 5 presents Conclusion.

2 Related Work

The stance detection task is not a novel idea, although it was popularized the FNC-1 conducted by fakenewschallenge.org. A paper written by Riedel et.al describing the stance detection model that came in third in the FNC-1 competition, the researcher used bag-of-words and TF-IDF representation for the both headline and body. Using the multi-layer perceptron model, they were able to achieve FNC score of 81.72% [2]. Another paper by Throne et.al, working on the same data, used a five model ensemble technique. In all their models, they used either the Word2Vec or the TF-IDF representations [5]. They got an upper limit score or 90.89 by using logistic regression, and various forms of MLP [4]. Augenstein et.al, and Zarrella et.al worked on the same task, but on the SemEval-2016 dataset. The task was focused on Twitter stance detection corpus [6]. Augenstein et.al compared five baseline models that contained n-grams, Bag of words, and LSTMs encodings to avail an f-1 score of 0.58. Zeralla et.al used the word2vec skip-gram method for text embedding and transfer learning to create the stance detection model. The achieved an f1-score of 0.67 on the test set [7].

The work most similar to ours is the one done by Henselowski et.al. In their paper, they compared all the systems that were submitted in the FNC-1 challenge. They also compared many of the text encoding techniques such as the COOC, LSI-300 etc. [3]. However, they did not include the vector representations that are widely used in another NLP task which are TF-IDF and Word2Vec. Except this paper, all the other papers were their own isolated research works, and only focused on building the stance model. Our work on the other hand focuses more on feature analysis.

3 Methodology

3.1 Data Preparation

The entire dataset is divided into four csv files that contain the heading and body of news articles separated, one for training and one for testing. The nltk text processing library and regex in python are used to clean the data of irregularities. Firstly, non-English words and special words are removed from the dataset as they will only generate noise. Next, the document is tokenized using the word tokenizer feature readily available in the nltk package, then the WordNetLemmatizer is used to lemmatized all adjective, adverbs, verbs, and nouns [1]. Finally, all stop words are removed according to the stopwords list in nltk corpus. After the cleaning process, datasets are stored in a dataframe for further use.

In the training set, we founded 1669 distinctive news article bodies and 1648 distinctive headlines by observing the datasets and using the unique method for the dataframe. The 3317 selected documents are then put in a new variable. Further, the dataframes containing the headlines and the news body are merged for the better and efficient processing.

3.2 Vector Representation

For this task, the Word2Vec and TF-IDF vector representation techniques are used. These are the most popularly used representation methods used in the domain of NLP. The significance of a word to a document in a group can be reflected by the TF-IDF value, which is the term frequency of a word in a document to the inverse document frequency [8]. Word2Vec is multi-dimensional vector representation of words in which the coordinates of a word are logically associated with its similarity to other words [9]. We only use the training set to build this representation. The representations built on tf-idf represent each document as 1844-dimensional vector and 3467-dimensional vector, respectively. For the td-idf calculation, the unique collection mentioned above is used.

A total of 18,201 distinct words were collected as the model vocabulary from the training set, and index ranging from 0 to 18,200 was assigned to them. For computational efficiency, the inverse frequency document of every word was computed and stored beforehand. And then, the TF-IDF of every word was easily calculated using the inputs calculated above. A vector of 18,201 dimension was received as an output, with each dimension signifying the TF-IDF value of the lexicon. Now, the TF-IDF vector that we received has too many dimensions, and the information density is also low as not every word would be important. Therefore, the ease the computational pressure we will threshold the vector and keep only key selected vectors. Top 3 and 30 TF-IDF of headings and bodies will be treated as key features and the rest are dumped. The new lexicon now contains of only 3298 words. The words are then re-indexed and stored in a new vector.

The Genism python library is used for the representation based on Word2Vec. The headline was merged with the correlated news body to create a single document. The Word2Vec model was trained by the total of 48,753 merged documents created above. The function parameters namely size, and min_count is set to 200 and 1. Finally, to produce the vectorized document, all the vectorized word present in the merged document is added. As different documents may have different ranges, the documents are normalized by dividing all the words in the document by their average.

Table 1 Instances and their Cosine Similarity attributes

	Headline	Body Id	Stance	articleBody	Cos_sim	Cos_sim_key	Cos_sim_w2v
0	[police, find, mass, graf, least, body, near,...	712	Unrelated	[danny, boyle,direct, Untitled,film,seth,r...	0.000000	0.000000	−0.137995
1	[hundreds, Palestinians, flee, flood, gaza, is.	158	Agree	[hundreds, Palestinians, evacuate, home, sunda...	0.525937	0.658249	0.775353

$$\cos(\theta) = \frac{A \cdot B}{|A||B|} \tag{1}$$

Now that the dataset is represented in the appropriate mathematical form, we need to use a type of similarity coefficient that can quantify the relationship between a headline and its corresponding news article. For this purpose, we use the Cosine similarity as shown in Eq. (1). It calculates the cosine of the angle created between two vectors [10]. Table 1 displays two varying values of cosine similarities. The first instance has a very low similarity value in all three attributes which show a weak, and even a negative, correlation between the headline and the body. On the other hand, the second instance show a high similarity values which indicates their higher correlation.

3.3 *Language Model Based Representations*

For this subtask, language models are created which is based on the interpolation and n-gram language model. In the probabilistic language model, the n-gram model is broadly used to predict the succeeding word according to the n-1 words before [11].

To resolve the data sparsity issue which is common in the NLP area, a smoothing technique is needed, while building the n-gram language model. In this project, we have used interpolation smoothing which is an effective and simple smoothing technique to interpolate between n-gram and n-1-g models. In this project, two types of language models are created which are the unigram language model and the interpolation model based on the unigram and bigram model.

The Ngram_Im permits us to a built instance of an n-gram language model based on training data and a stated n-order. For this given class, the probability function can return the appearance probability of a certain word assigned a certain history. These two types of language models are created for each pair of headlines and bodies. Further to predict the probability of words in the headlines, both headline model and body model are used, and the result is returned as vectors. For the unigram model,

Table 2 KL-divergence attribute

Stance	Kl_head_uni	Kl_head_bi
0 unrelated	12.274821	107.867233
1 agree	3.443514	20.116824

vectors whose length is similar to the length of history are returned. Since the length of history is one, the vector length of interpolation models equals the length of the headline [12]. To predict bodies to get a vector, we tried to use this model but this method is abandoned to poor performance.

$$D_{\mathrm{KL}}(P||Q) = -\sum_{i} P(i) \ln \frac{Q(i)}{P(i)} \tag{2}$$

Equation (2) shows KL-divergence can be estimated as we have two pairs of vectors for each pair of headlines and bodies. The divergence across two probability distribution can be measured by KL divergence. To estimate the difference between the headline's and body's language model, KL can be used effectively. We calculated two KLs for each pair of headlines and bodies based on the unigram and bigram interpolation model, respectively, as shown in Table 2.

3.4 Propose and Implement Alternative Features

For this subtask, three kinds of features are presented and implemented which are spearman correlation, Euclidean distance, and the number of same words. Equation (3) displays the Euclidean distance formula, it calculates the straight-line distance across two points in Euclidean n-dimensional space where the n-dimensional vector correlates to a point.

$$d(p,q) = \sqrt{(p_1 - q_1)^2 + (p_2 - q_2)^2 + \cdots + (p_i - q_i)^2 + \cdots + (p_n - q_n)^2} \tag{3}$$

The Spearman correlation is a non-parametric measurement of non-linear correlation that relates to the association between two variables. The spearman correlation as shown in Eq. (4) where x and y refer to the rank of original data. The number of similar words is a self-defined feature which means the number of unique words present in the headlines and the bodies exhibits similar if there are more common words.

$$p = \frac{\sum_i (x_i - \overline{x})(y_i - \overline{y})}{\sqrt{\sum_i (x_i - \overline{x})^2 \sum_i (y_i - \overline{y})^2}} \tag{4}$$

	Stance	euc_dist	euc_dist_key	euc_dist_w2v	spear_corr	spear_corr_key	spear_corr_w2v	same_word
0	unrelated	1.110189	0.956684	1.505255	-0.001715	-0.005329	-0.148394	0.0
1	agree	1.508805	1.388314	0.685887	0.193569	0.368627	0.733944	7.0

Fig. 1 Instance sample with all features

Figure 1 displays all features generated by us from the previous step. As we used three types of vector representation and calculated their Euclidean distance, and Spearman correlation, we were able to produce a total of seven features. We can observe from the 'agree' stance example below that its similar stances score higher in the same word, and Spearman correlation, and lower on metrics such as Euclidean distance and Word2Vec. This is definitely an indication that these features are good indicators of stance dissimilarity. However, Euclidean distance of TF-IDF metrics gets an unusually high score which in not in line with our assumptions.

3.5 Analysing Key Features

We have plotted the frequency distribution of all the features, all four stances are represented with different colors. From these plots, we can observe a few important characteristics. We discern that all the features that are derived by Word2Vec show homographic distribution. Similarly, both types of TF-IDF features that are based on Spearman correlation and cosine similarity are analogous. However, the same word and unigram models with KL-divergence are comparatively unique. Another important observation is that most of the features that we generated are able to differentiate unrelated and rest of the stances. But, KL-divergence derived bigram interpolation and TF-IDF based Euclidean distance models are the not able to separate the four kinds of stances effectively.

Now, as the key-TF-IDF based cosine similarity feature has proved to be marginally superior than other similar features, we choose it as there representative, and the unigram-KL is selected as distinct well performing attribute.

From Figs. 2 and 3, it is evident that the frequency distributions of 'unrelated' label instances significantly different as compared to other three types of stance instances. However, both of these features are not able to distinguish between the remaining three stance labels namely 'discuss', 'agree', and 'disagree'. Therefore, we combine them as they are generated by dissimilar methods.

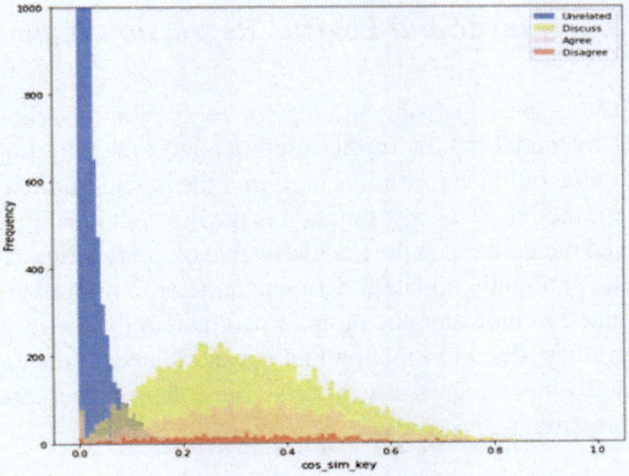

Fig. 2 Frequency distribution of cosine similarity TF-IDF keys and KL-divergence Unigram

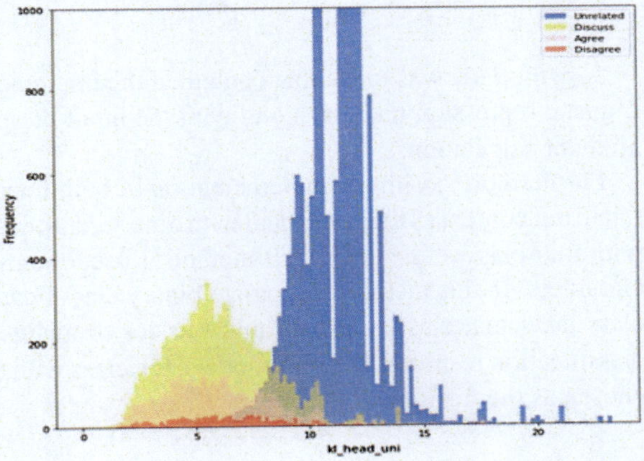

Fig. 3 Frequency distribution of cosine similarity TF-IDF keys and KL-divergence Unigram

3.6 *Train-Validate Split*

Before building the stance classification model, we split the training set in training and validation set. We use a 9:1 train-validate ratio and shuffle the dataset before the split to make the ratio of the four classes similar as shown in Table 3.

Table 3 Split ratio table

	Agree	Disagree	Discuss	Unrelated
Training set	0.073601211	0.016809412	0.1782798367	0.73130953
Training subset	0.073218459	0.016987219	0.1788104502	0.73098367
Validation subset	0.077046227	0.015209125	0.1735041024	0.73424053

3.7 Linear and Logistic Regression Implementation

The linear regressing and logistic regression models are used for a baseline model. This model will be further used to analyse feature importance. The gradient decent optimization algorithm is used find the global minima of the cost function related to our models. The cost function is the mathematical function of the model being used and the gradient is the first derivative of the function. The gradient descent is applied to continually update function parameter to minimize the error. The minimized cost function indicates the optimal parameters for the model. Equation (5) displays the gradient descent function that is used to update the parameter ‘θ’ found in Logistic and Linear regression models. Here, h(x) corresponds to the linear and sigmoid function, respectively.

$$\theta_j := \theta_j - \alpha \frac{1}{m} \sum_{i=1}^{m} \left(h_\theta\left(x^{(i)}\right) - y^{(i)}\right) x_j^{(i)} \quad (5)$$

A python file was made that contained the implementation of both the linear and Logistic regression models, along with the input pre-processing, cost function, and gradient calculation.

Furthermore, a simple implementation of both linear and logistic regression will fail in our context as they only deal with binary classification. Since our data is labeled with four classes, the one-vs-all method is used the make these models multi-label classifiers. In this method, a separate binary classification model is defined for each class label that classify any instance as for or against that label. Aggregating the classification results of the four models, the class with the highest probability can be chosen as the final prediction.

3.8 Model Evaluation

Accuracy, precision, recall and f1-score are used as evaluation metrics for the models. Accuracy is the ratio of number of correctly predicted instances against the total number of instances. It is a useful score that can provide heuristic intuition about model performance, however in cases of class asymmetric data it is not a trustworthy metric. Hence, precision, recall and F1-score metrics are used to curb the above inconsistency. Precision is the ratio of the true positives against the total number of selected instances, while recall in the ratio of true positives against all relevant elements. And lastly, the f1-score in the combination of the above two metrics, and can show a more comprehensive snippet of model performance.

The overall accuracy of the linear regression model was measured as 0.73 and of logistic regression was 0.80, when tested on the test set. The complete metrics evaluation is displayed in Tables 4 and 5.

Table 4 Score table of logistic

	Precision	Recall	F1-score	Support
Agree	0.36	0.01	0.01	1903
Disagree	0.06	0.45	0.11	697
Discuss	0.43	0.12	0.19	4464
Unrelated	0.94	0.96	0.95	18,349
Accuracy			0.73	25,413
Macro avg	0.45	0.39	0.31	25,413
Weighted avg	0.78	0.73	0.72	25,413

Table 5 Score table linear regression

	Precision	Recall	F1-score	Support
Agree	0	0	0	1903
Disagree	0.07	0.62	0.12	697
Discuss	0	0	0	4464
Unrelated	0.93	0.97	0.95	18,349
Accuracy			0.72	25,413
Macro avg	0.25	0.4	0.27	25,413
Weighted avg	0.67	0.72	0.69	25,413

Although, the linear regression model is able to achieve a marginally higher accuracy score, the logistic regression model performs substantially better on f1-score, recall, and precision. As discussed earlier, in case of asymmetric class distribution f1-score and recall are more reliable indicators of classification model performance. Therefore, we can conclude that the logistic regression model beat the linear model in this classification task. When the learning rate or ‘α’ is set in between 0.1 and 0.5, both models work appropriately. Even increasing the value further than 0.5 does not affect the models, only the number of iterations to optimize the model decrease. However, gradient descent optimizer starts to falter if ‘α’ is set to an aggressively high value like 40. The algorithm jumps around the global minima, and even move away in some cases to an arbitrarily large number.

3.9 *Model Tuning and Enhancement*

Class imbalance is a major issue in our dataset. The class label with a disproportionately low-sample count is difficult to classify. Skewed thresholding can be used for each model to alleviate the imbalance. For instance, the threshold value of 0.34 can be set for classifying sample as positive for the logistic regression model. However, instead of guessing the optimal threshold, we have found the best value by testing a range of thresholds on the validation set. The thresholding process when tested on the test set improve the f1-score. However, the accuracy on the set sees a marginal dip.

Another model deployed to handle the class imbalance problem is the down-sampling. It is a technique in which we train the majority class with disproportionally fewer instances to prevent it from the overpowering other classes in prediction. After testing for the optimal down-sampling on validation, rate of 0.5 for 'disagree' and 0.3 for 'agree' was selected. The improved classification ability, with respect to minority classes, of the linear regression model in evident for table below.

4 Results

4.1 Importance of Features

The model parameters of all 12 features that were generate using both the vector, and language model representation are plotted in Figs. 4 and 5.

The theta value or weights indicate the importance of each feature for the models. The similarity of weight distribution is clearly seen between the two models used. Features 10, 7, 2 which are Word2Vec Spearman correlation, Word2Vec Euclidean distance, and Word2Vec cosine similarity have highest weight for both the models.

Fig. 4 Theta values of all features in linear regression model and logistic regression model, respectively

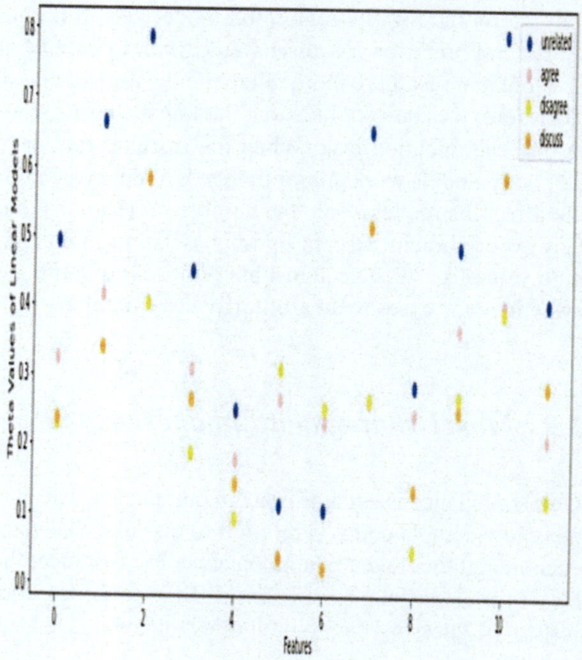

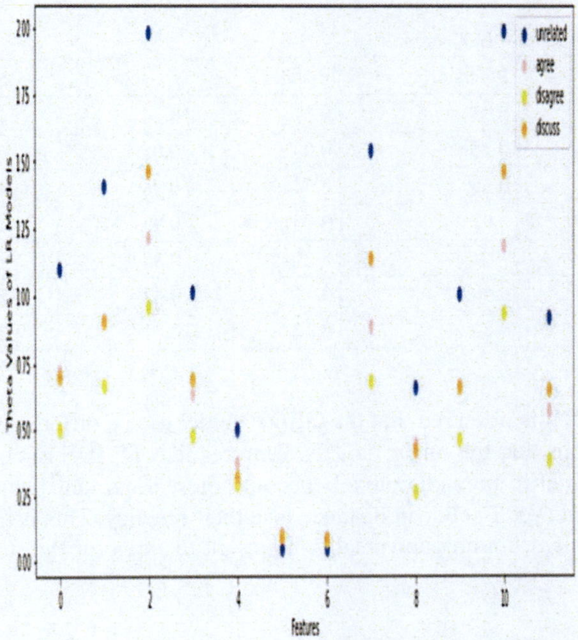

Fig. 5 Theta values of all features in linear regression model and logistic regression model, respectively

And, key TF-IDF Euclidean distance, TF-IDF Euclidean distance, and bigram KL-divergence which correspond to features 6, 5, 4, respectively, have to low weightage. Also, the frequency distribution graphs agree with these results. We can conclude that the Word2Vec representations perform better than TF-IDF representations in the context of stance detection.

4.2 *Final Baseline*

For the final base line model, GBDT model is used. The implementation is done using the XGBoost python gradient boosting library. Boosting techniques are required to minimize train data overfitting as normal decision tree have a very high variance. The main advantage, however, of using decision trees is that they are able to learn non-linear functions, unlike to linear functions used previously. The raw XGBoost model performance was to unsatisfactory, however, applying 0.2 down-sampling rate on 'unrelated' class helped the model improve significantly. The GBDT model shows a comprehensive overall increase in performance than the linear and logistic regression model and supported by the performance table as shown in Table 6.

Table 6 Score table for GBDT model

	Precision	Recall	F1-score	Support
Agree	0.31	0.37	0.34	1903
Disagree	0.15	0.03	0.05	697
Discuss	0.59	0.54	0.56	4464
Unrelated	0.93	0.96	0.95	18,349
Accuracy			0.82	25,413
Macro avg	0.49	0.47	0.47	25,413
Weighted avg	0.8	0.82	0.81	25,413

Furthermore, it is observed that the GBDT model used a different set of features for its predictions that the linear models. Features like TF-IDF Euclidean distance that were nominal in previous models become most important features here, and others like Word2Vec Euclidean distance lose their potency. This is a contradiction to our hypothesis of finding universally important features for the stance detection task.

5 Conclusion

This paper showed that both Word2Vec and TF-IDF are effective methods for text encoding. However, they are not universally significant for all types of classification models. Models with high linear bias showed a higher inclination toward Word2Vec based metrics, whereas models with higher degrees of freedom like the decision tree deemed TF-IDF based features as more important. Furthermore, a simple model like GBDT after dealing with class imbalance and, hyperparameter tuning delivers an acceptable performance. The stance detection task that we performed was no were near the best possible our insight in feature generation selection can help future research on the same task.

References

1. Bhatt G, Sharma A, Sharma S, Nagpal A, Raman B, Mittal A (2018). Combining neural, statistical and external features for fake news stance identification. In: Companion proceedings of the the web conference 2018, pp 1353–1357
2. Riedel B, Augenstein I, Spithourakis GP, Riedel S (2017) A simple but tough-to-beat baseline for the fake news challenge stance detection task. arXiv preprint arXiv:1707.03264
3. Hanselowski A, PVS A, Schiller B, Caspelherr F, Chaudhuri D, Meyer CM, Gurevych I (2018) A retrospective analysis of the fake news challenge stance detection task. arXiv preprint arXiv:1806.05180

4. Thorne J, Chen M, Myrianthous G, Pu J, Wang X, Vlachos A (2017) Fake news stance detection using stacked ensemble of classifiers. In: Proceedings of the 2017 EMNLP workshop: natural language processing meets journalism, pp 80–83
5. Christian H, Agus MP, Suhartono D (2016) Single document automatic text summarization using term frequency-inverse document frequency (TF-IDF). ComTech: Comput Math Eng Appl 7(4):285–294
6. Augenstein I, Rocktäschel T, Vlachos A, Bontcheva K (2016) Stance detection with bidirectional conditional encoding. arXiv preprint arXiv:1606.05464
7. Zarrella G, Marsh A (2016) Mitre at semeval-2016 task 6: transfer learning for stance detection. arXiv preprint arXiv:1606.03784
8. Hiemstra D (2000) A probabilistic justification for using tf× idf term weighting in information retrieval. Int J Digit Libr 3(2):131–139
9. Adewumi TP, Liwicki F, Liwicki M (2020) Word2vec: optimal hyper-parameters and their impact on nlp downstream tasks. arXiv preprint arXiv:2003.11645
10. Sitikhu P, Pahi K, Thapa P, Shakya S (2019) A comparison of semantic similarity methods for maximum human interpretability. In: 2019 artificial intelligence for transforming business and society (AITB), IEEE, vol 1, pp 1–4
11. Lopez-Gazpio I, Maritxalar M, Lapata M, Agirre E (2019) Word n-gram attention models for sentence similarity and inference. Expert Syst Appl 132:1–11
12. Mikolov T, Grave E, Bojanowski P, Puhrsch C, Joulin A (2017) Advances in pre-training distributed word representations. arXiv preprint arXiv:1712.09405

A Comparative Study into Stock Market Prediction Through Various Sentiment Analysis Algorithms

Sandipan Biswas and Shivnath Ghosh

Abstract From past times, there is strong link between finances, currencies and economic growth the future and progress of humankind. Now a days, the future of economic development is dictated by fortunes and vagaries prevalent in share stock markets. Researchers have found that it is possible to make forecasts with large historical data on stock market and up-down in prices of share values. So, the fact that stock markets play a really vital role in national and global economy, which is today undeniable. Stock markets can be profitable by speculations provided, of course though the future behavior can be forecast with a constant degree of accuracy. In this study, the authors propose a model which help to guess stock market trends consistently and with minimal of error value. The model discussed here makes the use of sentiment analyses based on financial news and also historical patterns of stocks in share markets and can offer more accurate results to analyses data from multiple news sources and historical price movement of individual stocks. By using a two-step process, the model offers a minimum prediction accuracy value of 72%. In the first step, Naïve Bayes algorithm is used to evaluate text polarity to obtain a fix on public sentiment based on news feeds collected and received. In the second step, the future stock prices are forecasted by combining the evaluation results on text polarity with historical data on stock value price up-down.

Keywords Stock Market · Sentiment Analysis · Naïve Bayes Algorithm · Logistic Regression · Support Vector Machine

1 Introduction

In the decade of globalization and proliferation of the Internet on a scale unprecedented and, possibly, never even have been imagined even a time ago, even the

S. Biswas (✉) · S. Ghosh
Department of Computer Science and Engineering, Brainware University, West Bengal, India

S. Ghosh
e-mail: dsg.cs@brainwareuniversity.ac.in

J. M. R. S. Tavares et al. (eds.), *Cyber Intelligence and Information Retrieval*, Lecture Notes in Networks and Systems 291,
https://doi.org/10.1007/978-981-16-4284-5_11

smallest of waves affecting economy and money matters can change stock prices shooting upward and plummeting downhill. The reason for which stock market values are so strongly watched and observed with surgical accuracy of the fact that global economy depends on stock markets and are highly sensitive of changes in share prices across major share trading in this world. The first origins of stock market and share trading can be traced back to The Wall Street, New York City, USA [1]. On that time, and as still is now to certain extent, holding and trading of stocks meant a stake in ownership of actual companies. A stock was a piece of paper that signified that a holder of a stock was, and so, a shareholder actually owned a part of that company.

The problem which are related to stock market analyses comply public sentiments and politics at play [2]. We can see sentiments of investors are critical and important factor [3]. Although, in certain regions of the global investments which may be more susceptible to taking risks, most traders in buying or selling decisions based on sentiments that are influenced by news data circulated in the market [4]. With this advent and explosion of the Internet and digital social media, news that results in sentiments of investors can be conveyed and circulated faster than ever before [5]. Weblogs and Twitter feeds have been more fruitful by researchers to cause significant impact insofar as noise trading is concerned. Researchers and market observers can predict share prices which holds the key to maximizing profits from trading in stock markets [6]. However, by computing the direction of share prices which are expected to move is serious for research for both of academic interests and for economies of nations across the globe [7]. Here in Table 1, we have described different source and data tools.

2 Related Work

Sentiment analysis (SA) approaches the problem of assessing an author's view in relation to any subject or topic. SA involves exploring data available on the Web and categorizing the text within the data. The objective of conducting SA is to be able to learn about the predilection of an author in respect of any topic—be it economical, literary, political, and social, sports or such like. The result of sentiment analysis can either be positive, negative or neutral [15]. Researchers have used several techniques to mine data and observe patterns through this SA [16]. The prominent techniques are Naïve Bayes classification, Logistic Regression, Support Vector Machine and k-nearest-neighbor classification. Sentiment analysis can take the form of either machine learning or lexicon-based or hybrid approach and it can be further subdivided into supervised and unsupervised learning. Naïve Bayes classification and Support Vector Machine (SVM) are examples of supervised learning techniques which are broadly used in research involving opinion mining, data mining or SA [16]. Here in Table 2, we have described different supporting algorithms and tools.

Table 1 Description of different source and data tools

Topic	Description
A. Twitter as a Source for Data	The rise of Twitter among digital social media platforms makes Twitter an interesting choice for researchers to conduct their studies is the fact that the tweets/messages posted by users not only can have images, but also geographical locational data with timestamps embedded [8]. In the opinion of many researchers, it can out-perform any syndromic surveillance tool or approach and, to offer a macro level overview of human behavior [9]. As such, the potential to mine Twitter data to conduct research and social experiments is significant enough to warrant a deeper understanding on how Twitter feeds or messages can be mined to extract and extrapolate user sentiments [10]. In a study [11] conducted by researchers, experiments using Python programming language coupled with Twitter's API (or application programming interface), have focused/revealed that sentiment analysis in respect of stock market movements is a practical approach in understanding how current affairs can shape investor action
B. Analytical Approach to Evaluating Twitter Feeds	Using Twitter data as a basis of evaluating sentiments, while not exactly new, has been effectively used by researchers to analyse sentiments [12] on evaluating sentiments based on structured and unstructured data, researchers have revealed which can be mapped and modeled using unsupervised classification algorithm
C. Data Mining	In a more recent study [8], researchers have proposed a model based on Naïve Bayes classifier to conduct analyses of sentiments based on Twitter data that approaches to predict fluctuations in stock prices based on Twitter feeds. The approach to making predictions on stock market movements relies on data mining. Data mining has also been termed as knowledge discovery from data (KDD) [13], and with good reason. Patterns from data can be extracted and turned into knowledge for use in wide variety of applications for data mining classified into two sections: descriptive (serve to identify, and segregate properties of data being analyzed) and predictive(use existing data for analyses to come up with results that help in identifying future trends;).for example, predicting stock market prices based on available data [14]
D. Using the Predictive Approach	For our forecasting in stock market trends through predictive tasks and classification analysis we have considered Naïve Bayes classification, Logistic Regression, Support Vector Machine and k-nearest-neighbor classification

Table 2 Description of different supporting algorithms and tools

Tools	Description
A. Support Vector Machine and Particle Swarm Optimization Technique	If we consider accuracy factor while making predictions, researchers are aware of the need to devise a classification technique suitable for classifying data. In a study [17], research has shown that Support Vector Machines (SVMs), a machine learning tool, can be used to obtain better prediction results. Researchers [18] have observed that SVM is an innovative form of machine learning approach that can help in arrangement of sentiments. However, that is easier said than done as this approach relies heavily on availability of appropriate parameters. The researchers used Particle Swarm Optimization (PSO) technique together with SVM to classify sentiments based on movie reviews posted on Twitter [19]. PSO is a kind of searching algorithm that was designed by the researchers, Kennedy and Eberhart, drawing inspiration from the way birds flock and fish school. The cooperative nature of birds to find nests or food was the basis of the keystones of PSO [20]. In achieving high prediction results involving analyses based on social media and financial news, researchers [21] have also detected that machine learning algorithms are able to regulate whether public opinion and views can change investor sentiments and views when capitalizing in stock markets. The problem of manually sifting through huge volumes of text posted as news feeds and tweets in social media is imposing. In assessing a best solution to automating data mining, researchers [22] have observed that this is possible at aspect, document and at sentence level. The authors of the paper [22] with their model that relies on aspect level opinion mining combined with a syntactic method. Their experiments discovered that their proposed solution yielded 78.04% accuracy which was a development over equivalent Part-of-Speech method on explained test set by around 6%
B. The Hybrid Approach using SVM and PSO	Application of supervised data mining techniques is a well-trodden path among researchers. And with good reason, as researchers [23] have demonstrates in their paper wherein the authors propose an algorithm based on Naïve Bayes classifier to detect sentiments of movie goers based on reviews posted online. Since, a sizeable chunk of data stored in online database systems is in text, information can be freely exchanged and viewed for people to read and contribute. Under such circumstances, text mining of web pages, or opinion mining leading to SA, grew to be immensely significant. Realizing the need to refine SA techniques considering that text mining is of relevance and significance in the context of machine learning, the authors [24] have proposed Support Vector Machine algorithm combined with Particle Swarm Optimization technique According to the research submitted by the authors, the precision and accuracy of results achieved using their proposed approach for SA were significantly better than that of SVM approach alone. In a similar study conducted on text mining techniques, the authors [25] expounded that a hybrid approach using SVM and PSO certainly has its merits by using their proposed model reveal that a hybrid approach has the potential to classify two classes of sentiments—either positive or negative

(continued)

Table 2 (continued)

Tools	Description
C. Natural Language Processing	In the opinion of a section of researchers, however, the very many systems that depend on SA utilize, what the authors [26] term as "bag-of-words" principle to mine data from social media sites and web sites. This, it can lead to fallacious and often poor results owing to dependency on traditional machine learning algorithms such as Naïve Bayes classifier, Support Vector Machines, etc. The reason, the authors offer, is that these traditional ML algorithms tend to be biased. So, the authors propose a model based on Natural Language Processing (NLP) and addition of semantics in their model
D. In Support of SVM by Tuning of Parameters	In the view of another group of researchers, results yielded following application and use of SVM are hopeful and predictions obtained were found to be reliable under most situations. However, the authors [27] contend that performance in classifying data using SVM is largely dependent on tuning certain sets of parameters. Parameter tuning is also referred to as model selection though, linear SVMs can be used in case of all linearly separate problems, and the same does not hold true when non-linearly separable problems are involved. A grid-search method during cross-validation to enable identification and selection of a parameter set that best matches the requirements is then applied to training data set in order to derive the classifier. The generated classifier can then be used on a given testing data set to obtain generalization accuracy. Arguing on similar lines, researchers [28] have posited that though SVM is a classifier that has seen widespread use, its primary disadvantage lies in the fact the user is required to supply regular cost parameters and kernel parameters. To progress performance of SVMs, researchers [29] have offered a method concerning active learning. In their proposed approach an idea referred to as "version space" is used to enhancing performance of SVM in inductive and transudative ML settings, the authors present, which enables an SVM to choose instances from a pool of unlabeled instances and to later affix labels to some or all such cases. Although researchers in the paper [30] presented by in support of SVM by current standards, to yield meaningful results, which are considered important even in the present context of the research paper acquiesced by the present authors. The researchers [30] clearly outline and offer clarifications with examples as a technique how SVM can be refined and put to good use by following the proposed measures presented in their paper

3 Proposed Work

Here, we have described various algorithms that we have used in our proposed research to evaluate the efficient ones among them. We have also shown how induction of the better performing algorithms can aid us in building our proposed model aimed to achieve higher accuracy figures than any single one of the algorithms we have discussed.

3.1 Naïve Bayes Classifier

Research [31] has shown that the Naïve Bayes classifier remains relevant in almost all use cases where classifier may return correct classification decisions despite corresponding probability estimates being incorrect or even wildly inaccurate. This paper also highlights that the Naïve Bayes classifier is good at solving problems in scenarios that involve either independent or dependent features where the former is expected in the view of the authors, the latter is not consistent with general expectations of this classifier. In paper [32] also highlights the efficacy of Naïve Bayes classifier over similar other ML algorithms in the context of social media. In paper [33] the authors have observed that there was a marked improvement in predicting stock market movements by involving Naïve Bayes classifier to analyze factors influencing markets.

3.2 Logistic Regression

Logistic regression (LR) model is frequently used where responses can generally have only two outcomes and it is possible for LR models to be applied in cases where there may be more than two events which are usually less common [34]. The main advantage to using LR is that this algorithm can reveal for the factors those are under assessment and has a high probability. It can also enable in measuring the magnitude of the possible influence [35].

3.3 Support Vector Machine

Support Vector Machines also referred to as Support Vector Networks [36, 37] are a class of supervised machine learning models where type of classifier is predominantly used in cases where classification and regression analyses are involved. Researchers have categorized SVM as a non-probabilistic binary type of linear classifier.

3.4 k-Nearest Neighbor Algorithm (k-NN)

The k-Nearest Neighbor algorithm is a non-parametric algorithm that is employed during classification and regression exercises [38] where inputs for this algorithm consist of k closest training examples constituted in feature space.

4 The Proposed Model

In this section, we describe in details the working of our proposed model. Before we go deep into its features, we present an easy-to-understand flowchart as depicted in Fig. 1.

4.1 Data Description

For performing experiments, we have collected data from Reddit—a US social news aggregation website, and news from Yahoo! Finance web site. The data sets are of two sources constitute the entire bulk of text (in the form of news) and numerical

Fig. 1 Flowchart of proposed model

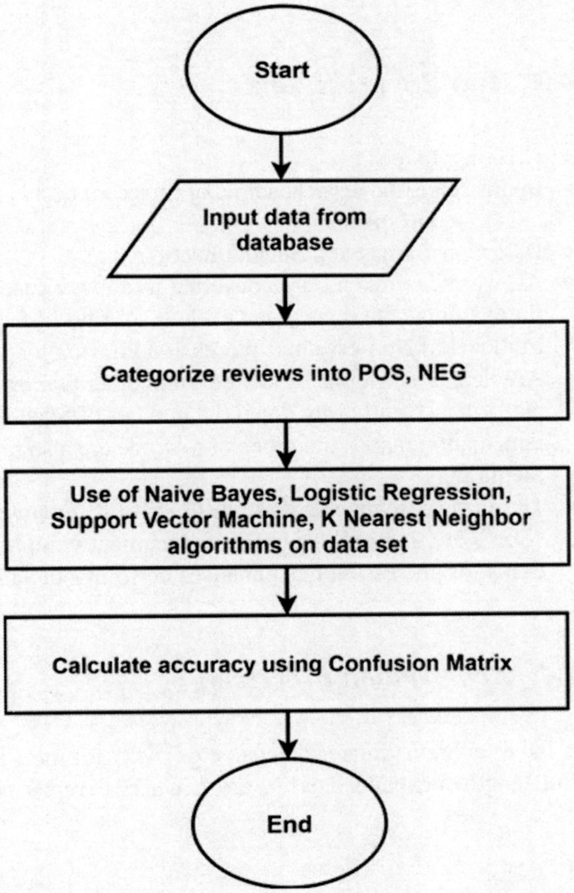

information. Articles which are posted on news pages on company and stock information was collected each day and stock prices at the opening bell, highest price and lowest price, along with closing values were also collected to facilitate in making predictions on future stock market rates.

4.2 Sentiment Analysis Component

At this stage of our proposed model, we have performed an analysis on biasness and partiality of stock news which is rendered as such: in case of news headlines, the aim is to segregate news as containing either positive sentiments or negative ones by conducting pre-processing of news headlines. The following section details the steps involved in the process:

4.3 Text Pre-processing

- Creating Tokens.
- In this stage, the news headline or finance report is first splitted into words which are termed as tokens.
- Data Consistency and Standardization.
- All words, or tokens, are converted into lower case and compiled in a document for enabling consistency and evenness for proposed data.
- Removal of Non-essential Words and Punctuation Marks.
- Articles, like "a", "an" and "the", and other prepositions are removed to enhance performance and bring down the number of features that need to be tracked and punctuation marks and other stop-words are also removed at this step.
- Stemming.
- The Porter Stemmer algorithm [39, 40] to remove suffixes, e.g., "-ed", "-ing", "-ion", etc. is used to minimize document complexity by retaining only single terms for processing that enhances performance of our proposed model.

4.4 Abbreviation Processing

A list of abbreviations is created, e.g., "US" for the United States of America, etc., (full length or expanded terms) are then used to replace corresponding abbreviations.

5 Programming Language and Simulator

We have used Python (high-level general-purpose programming language), as an interpreted programming language developed by Guido van Rossum in 1991. Python has come a long way in scientific research work, software domain, and as an academic tool that emphasizes readability of code which endears it to research scholars and software engineers in various form or size of project.

In our experiments based on a random selection of stocks that are generally traded in high volumes, we have extracted comments and reviews posted by people regarding the stocks selected among them. If reviews are found positive, we proceed to "buy" the stock; otherwise, we "sell" it. In our research, which are not actual buying or selling of stocks in the market is rather a virtual representation. We have used the Naïve Bayes classifier, Logistic Regression, Support Vector Machine and k-Nearest Neighbor algorithms on the data presented in tabular form in order of dates on *Confusion Matrix*, for building accuracy score. We use *Confusion Matrix* to compare predicted values with actual class values in the matrix (Table 3).

The labels TP (True Positive), TN (True Negative), FP (False Positive) and FN (False Negative) are described.

- Naïve Bayes Algorithm

Table 4 shows the confusion matrix by using Naïve Bayes algorithm.

- Logistic Regression Algorithm

The confusion matrix of the used Logistic Regression algorithm is presented in Table 5.

- Support Vector Machine Algorithm

Table 3 Confusion matrix for TP, FP, FN, and TN

	Negative	Positive
Negative	FN	FP
Positive	TN	TP

Table 4 Confusion matrix for the Naïve Bayes Algorithm

	Negative	Positive
Negative	158	1
Positive	6	0

Table 5 Confusion matrix for the logistic regression algorithm

	Negative	Positive
Negative	159	0
Positive	6	0

Table 6 Confusion matrix for the support vector machine algorithm

	Negative	Positive
Negative	156	3
Positive	6	0

Table 7 Confusion matrix for the K-Nearest Neighbor Algorithm

	Negative	Positive
Negative	159	0
Positive	6	0

Table 8 Comparison between different algorithms

Algorithm	Accuracy score (%)
Naïve Bayes Algorithm	95.76
Logistic Regression Algorithm	96.36
Support Vector Machine Algorithm	94.54
K-Nearest Neighbor Algorithm	96.36

Table 6 shows the confusion matrix for the support vector machine.

- K-Nearest Neighbor Algorithm

The confusion matrix using K-NN algorithm in is presented in Table 7.

Table 8 represents the comparison of the four algorithms and their test results to compare individual accuracy. Based on that analysis, it becomes evident that not all are equally suited for implementation insofar as stock markets and share trading are concerned and we are of the view that Logistic Regression and k-Nearest Neighbor algorithms have rank (NN = 3). In our view, further research work may be required with other optimization techniques for comparable viability in this research direction.

6 Conclusion and Future Scope

In our proposed model, we have discussed and analysed the performance of four prominent machine learning algorithms and also explored how these algorithms are effective in predicting stock market trend movements based on user reviews and comments on social media. The comparative study conducted during research also indicates that Logistic Regression and k-NN approaches are significantly more effective at returning higher rates of accuracy insofar for predicting stock market trends.

We have used Confusion Matrix [41, 42] to identify classifier performance based on our sample data and test results have shown that our proposed model can improve prediction accuracy. Further research is recommended to improve the accuracy over time and across multiple sources of online news.

References

1. Wyss B (2001) Fundamentals of the stock market. McGraw Hill, pp 1–245
2. Khan W, Ghazanfar MA, Azam MA, Karami A, Alyoubi KH, Alfakeeh AS (2020) Stock market prediction using machine learning classifiers and social media, news. Jo Ambient Intelligence Humanized Comput
3. See-To EWK, Yang Y (2017) Market sentiment dispersion and its effects on stock return and volatility. Electron Mark 27:283–296
4. Xindan LI, Bing Z (2017) Stock market behavior and Investor sentiment: evidence from China. Front Bus Res China 2(2):277–282
5. Smailovic J, Grcar M, Lavrac N, Znidarsic M (2013) Predictive sentiment analysis of tweets: a stock market application. Human-computer interaction and knowledge discovery in complex, unstructured, big data. HCI-KDD 2013. Lect Notes Comput Sci 7947:77–88
6. Wang H, Ou P (2009) Prediction of stock market index movement by ten data mining techniques. Mod Appl Sci 3(12):28–42
7. Shu-e Y, Qiang Z (2009) Noise trading, investor sentiment volatility, and stock returns. Syst Eng Theor Pract 29(3):40–47
8. Pal R, Pawar U, Zambare K, Hole V (2020) Predicting stock market movement based on twitter data and news articles using sentiment analysis and fuzzy logic. In: Second international conference on computer networks and communication technologies. Lecture Notes on Data Engineering and Communications Technologies, vol 44, pp 561–571
9. Bose R, Dey RK, Roy S, Sarddar D (2018) Analyzing political sentiment using twitter data. Information and communication technology for intelligent systems. Smart Innov Syst Technol 107:427–436
10. Federer LM, Belter CW, Joubert DJ, Livinski A, Lu Y-L, Snyders LN, Thompson H (2018) Data sharing in PLOS ONE: an analysis of data availability statements. PLoS ONE 13(5):1–12
11. Biswas S, Sarkar I, Das P, Bose R, Roy S (2020) Examining the effects of pandemics on stock market trends through sentiment analysis. J Xidian Univ 6:1163–1176
12. Biswas S, Ghosh AK, Chakraborty S, Roy S, Bose R (2020) Scope of sentiment analysis on news articles regarding stock market and GDP in struggling economic condition. Int J Emerging Trends Eng Res 8(7):3594–3609
13. Han J, Kamber M, Pei J (2011) Data mining concepts and techniques. Morgan Kaufmann 3:1–744
14. Namugera F, Wesonga R, Jehopio P (2019) Text mining and determinants of sentiments: twitter social media usage by traditional media houses in Uganda. Comput Soc Netw 6(3):1–21
15. Aqlan AAQ, Manjula B, Lakshman Naik R (2019) A study of sentiment analysis: concepts, techniques, and challenges. In: International conference on computational intelligence and data engineering. Lecture Notes on Data Engineering and Communications Technologies, vol 28, pp 14–162
16. Sharma D, Sabharwal M, Goyal V, Vij M (2020) Sentiment analysis techniques for social media data: a review. In: First international conference on sustainable technologies for computational intelligence 1045:75–90
17. Kothari AA, Patel WD (2015) A novel approach towards context sensitive recommendations based on machine learning methodology. In: Fifth international conference on communication systems and network technologies 2015:1114–1118
18. Basari ASH, Hussin B, Ananta GP (2012) Opinion mining of movie review using hybrid method of support vector machine and particle swarm optimization. In: Malaysian Technical Universities Conference on Engineering and Technology (MUCET) 4:545–552
19. Gopal A, Sultani MM, Bansal JC (2019) On stability analysis of particle swarm optimization algorithm. Arab J Sci Eng 1–10
20. Kennedy J, Eberhart R (1995) Particle swarm optimization. In: International conference on neural networks 4:1942–1948

21. Khan W, Ghazanfar MA, Azam MA, Karami A, Alyoubi KH, Alfakeeh AS (2020) Stock market prediction using machine learning classifiers and social media news. J Ambient Intelligence Humanized Comput
22. More P, Ghotkar A (2016) A study of different approaches to aspect-based opinion mining. Int J Comput Appl 145(6):11–15
23. Smeureanu I, Bucur C (2012) Applying supervised opinion mining techniques on online user reviews. Informatica Economica 16(2):81–91
24. Sharma D, Sabharwal M (2019) Sentiment analysis for social media using SVM classifier of machine learning. Int J Innov Technol Exploring Eng (IJITEE) 8(9):39–47
25. Umamaheswari K, Rajamohana SP, Aishwaryalakshmi G. Opinion mining using hybrid methods. Int J Comput Appl 18–21
26. Sarddar D, Dey RK, Bose R, Roy S. Topic modeling as a tool to gauge political sentiments from twitter feeds. Int J Natural Comput Res (IJNCR) 9(2):1–22
27. Srivastava DK, Bhambhu L (2009) Data classification using support vector machine. J Theor Appl Information Technol 1–7
28. Hastie T, Rosset S, Tibshirani R, Zhu J (2004) The entire regularization path for the support vector machine. J Mach Learn Res 5:1391–1415
29. Tong S, Koller D (2001) Support vector machine active learning with applications to text classification. J Mach Learn Res 45–66
30. Chang C-C, Lin C-J (2011) LIBSVM: a library for support vector. ACM Trans Intelligent Syst Technol 2(3):1–27
31. Watson TJ (2001) An empirical study of the naive Bayes classifier, pp 1–6
32. Ahmad S, Asghar MZ, Alotaibi FM, Awan I (2019) Detection and classification of social media-based extremist affiliations using sentiment analysis techniques. Human Centric Comput Information Sci 9:24
33. Ding G, Qin L (2019) Study on the prediction of stock price based on the associated network model of LSTM. Int J Mach Learn Cybern 1–11
34. Wilson JR, Lorenz KA (2015) Standard binary logistic regression model. In: Modeling binary correlated responses using SAS, SPSS and R. ICSA Book Series in Statistics, vol 9, pp 25–54
35. Tolles J, Meurer WJ (2016) Logistic regression relating patient characteristics to outcomes. JAMA Guide Statistics Methods 316(5):533–534
36. Phienthrakul T, Kijsirikul B, Takamura H, Okumura M (2009) Sentiment classification with support vector machines and multiple kernel functions. In: Leung CS, Lee M, Chan JH (eds) Neural information processing. ICONIP 2009. Lecture Notes in Computer Science, vol 5864, pp 583–592
37. Cai L, Hofmann T (2004) Hierarchical document categorization with support vector machines. In: CIKM'04, ACMI, pp 78–87
38. Altman NS (1992) An introduction to Kernel and nearest-neighbor nonparametric regression. Am Stat 46(3):175–185
39. Verma P, Om H (2019) A novel approach for text summarization using optimal combination of sentence scoring methods. Sadhana 44(110)
40. Porter M (1980) An algorithm for suffix stripping. In: Program: electronic library and information systems, vol 14, issue 3, pp 130–137
41. Ting KM (2017) Confusion matrix. Encyclopedia of machine learning and data mining
42. Singh R, Baidya D (2019) Usage of data science to predict string integrity failures. Society of Petroleum Engineers

The Survey on Handwritten Mathematical Expressions Recognition

Sakshi, Chetan Sharma, and Vinay Kukreja

Abstract Recognition of handwritten mathematical expressions is one of the challenging problems that has been studied and researched for a long. Many techniques have been researched to build a system that produces high-performance mathematical expression recognition systems. Plentiful research has been done on handwritten mathematical text though it is symbols and or two-dimensional mathematical expressions. The amount of research witness the origins and trends of several techniques and displays the shift of trendline of recognition techniques from grammar-driven or parsing based methods to machine learning and deep learning models. Remarkable research has been done in the last few years in this domain of handwritten mathematical scripts. In this paper, a survey of various research efforts has been compiled to develop and enhance handwritten mathematical expressions. Recognition systems of handwritten math text are presented. Comparative analysis in terms of classification technique used, datasets, and their accuracies have been performed.

Keywords Handwritten math recognition · Survey · Recognition techniques · Handwritten mathematical expressions

1 Introduction

Automation of the task of mathematical expression recognition is genuine of great practical as well as scientific importance. In case the technology evolves and becomes mature enough, the printed documents or handwritten equations could be directly

Sakshi · V. Kukreja
Institute of Engineering and Technology, Chitkara University, Punjab, India
e-mail: sakshi@chitkara.edu.in

V. Kukreja
e-mail: vinay.kukreja@chitkara.edu.in

C. Sharma (✉)
Chitkara University, Himachal Pradesh, India
e-mail: chetan.sharma@chitkarauniversity.edu.in

J. M. R. S. Tavares et al. (eds.), *Cyber Intelligence and Information Retrieval*,
Lecture Notes in Networks and Systems 291,
https://doi.org/10.1007/978-981-16-4284-5_12

fed to the systems for access and retrieval [1]. The following tasks after this evolution are going to be a lot more advantageous if the task of recognition is achieved a considerable accuracy. Recent times is evidential of machine learning and deep learning-based recognition models [2] which are capable of achieving competent accuracy rates. Recognition of handwritten expressions can be useful with a data tablet, which can be an alternative and convenient way of writing rather than writing a LATEX expression [3]. It is a vital recognition problem as mathematical expressions constitute to be an essential part of the notation in most of the scientific disciplines. Moreover, the search and retrieval operations on the digital mathematical text are more challenging than other handwritten script data.

In this paper, the authors are endeavouring to summarize the significant works of the associated topic and prepare a condensed report of comparison highlighting the performances of the different recognition models chosen.

2 Motivation

Almost all of the scientific information makes the use of mathematical data and notations. So, digitalizing these mathematical equations and expressions will make a lot of difference and contribute generously to the growth and development of search and access methods for mathematical and scientific data extraction. For the present time, there exists a bulk of mathematical and scientific literature, and this literature if digitized, could prove to be advantageous to the growth of knowledge and literature. Also, if the math content and formula once introduced into the computer machine, it can be used against myriad applications. It also has academic motivation as it can help the students and researchers search and access any scientific, mathematical data easily. For evaluation of students' math answer sheets too, this digitalization and recognition has also of practical motivation and has sufficient potential of social motivation as these recognition techniques can also be used of ease of access of data for the disabled person [4]. For instance, a printed or handwritten mathematical expression can be translated into Braille notation using text-to-speech tools.

3 Overview of Mathematical Symbols and Expression

By definition, mathematical expressions mean a finite set of symbols arranged according to rule or experimental study related to some context, mostly science, and research. These mathematical expressions include symbols, which can be digits, operators, constants, variables, functions, parenthesis, memorable characters, and letters (Greek, Latin) in a well-formed order in accordance with some formal propositional rules. Thus, the mathematical expression uses symbols, characters, and notations to purposefully represent some mathematical, scientific, and technical rules or formulas. Moreover, a mathematical expression is not just random symbols.

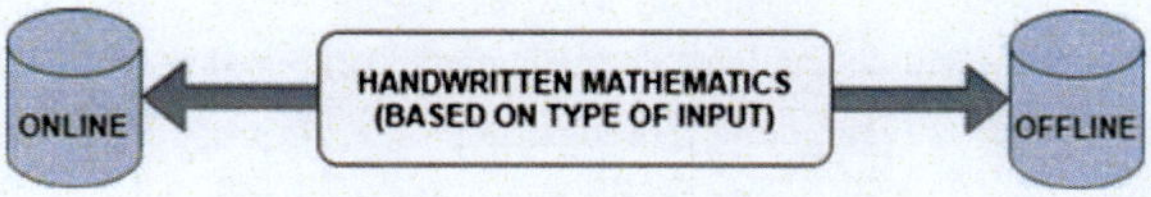

Fig. 1 Types of handwritten mathematics

They have a well-organized structure that is subject to the rules of the system of mathematical notation.

Furthermore, it is essential to understand that the input of a mathematical text from a handwritten device like tablet or touchscreens and input obtained by scanning the handwritten paper content imposes to produce different recognition accuracies [5]. The scanned images of handwritten data constitute the offline form of the dataset. In contrast, the direct input of feeding the mathematical data on tactile devices like touchscreens and tablets constitutes the online form of a dataset.

Both the two types of input modes classify the handwritten math characters in two categories represented in Fig. 1.

Online handwritten text: It majorly concentrates on mathematical data directly drawn or written on the contact-sensitive surface like the screens of tablets and touch-pads. The online handwritten math expression depicts the dynamic representation of data, and it stimulates simultaneous interactions with the connecting system and displays the results on screens and monitors as illustrated by Fig. 2. Every single character analysis is depended on the stroke analysis that implies the handwriting is recognized after each stroke for forming every symbol in the expression is correctly recognized. Stroke defines itself along the trajectory of motion of a symbol drawn; that is, one stroke is a series of points drawn from pen down to pen lift.

Offline handwritten mathematical text: It deals with mathematical expressions existing in printed forms or handwritten form but on a piece of paper. This kind of data needs to be scanned carefully for input as depicted by Fig. 2. Unlike online handwritten data, this offline mathematical text depicts the static representation of data, which are once written on paper and then scanned for output. The mathematical inputs include scanned images and bitmaps of handwritten mathematical data [6].

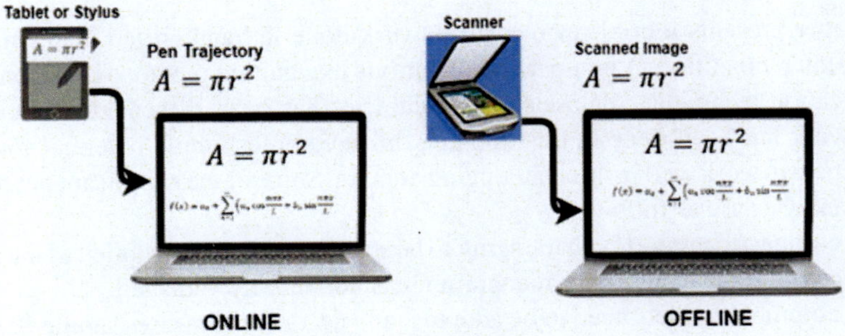

Fig. 2 Online versus Offline handwritten mathematical expressions

Table 1 Offline handwritten mathematical expressions

Authors	Technique used	Recognition method	Dataset	Dataset frequency	Accuracy rate (%)
[7]	Machine Learning	Extreme Learning Machine and SVM	Self created	9200 Symbols	Symbol classification = 98
[8]	Machine Learning	Recursive Cortical Network	MNIST	60,103 Symbols	Segmentation = 80
[9]	Machine Learning	CNN and Fractional Order Darwinian Particle Swarm Optimization	Self created	264,780 images	99
[10]	Machine Learning	Simple Linear Iterative Clustering	MNIST	6000 Images	SqueezeNet = 90
[11]	Machine Learning	Back Propagation Neural Network	Self created	50 Math equations	97.11
[12]	Machine Learning	CNN	Self created	6300 Formula symbol images	94.25

4 Survey on Online and Offline Handwritten Mathematical Expressions

Survey of offline and online handwritten mathematical expression is given in Tables 1 and 2, respectively.

5 Conclusion

This paper presents a brief review of the work done in handwritten mathematical expression recognition. A comparison of various recognition methods based on their classification techniques, datasets, and accuracies has been presented. Although a lot of work has been done in this domain, this area holds ample potential for new researchers to work well in this challenging domain. Some of the significant outcomes of our analysis are as follows:

The comparison report is challenging to be standardized and compiled as the trend of self-creating the datasets for recognition has not entirely vanished.

The common datasets need to be adopted, and accuracy measures should be standardized and uniformly used in the recognition process. Presented work is sure to help the researchers for developing new practical techniques for recognition, especially the works associated with deep learning-based recognition system needs to be enhanced.

Table 2 Online handwritten mathematical expressions

Authors	Technique used	Recognition method	Dataset	Dataset frequency	Accuracy rate (%)
[13]	Machine Learning	Paired Adversarial Learning	CROHME 2014 and CROHME 2016	8835 Handwritten MEs	CROHME 2014 = 54.87% CROHME 2016 = 57.89
[14]	Machine Learning	CNN for Clustering	CROHME 2014 and CROHME 2019	8834 for CROHME 2014 and 9993 for CROHME 2019	Clustering purity = 0.99
[15]	Machine Learning	CNN	CROHME 2014 and CROHME 2016	9822 for CROHME 2014 and 10,969 for CROHME 2016	Expression Recognition 49.46 for CROHME 2014 and 46.06 for CROHME 2016
[16]	Non-Machine Learning	Graph Grammar	CROHME 2016	10,969 Handwritten MEs	Recall and Precision = 40.80
[17]	Machine Learning	CNN	CROHME 2014 and CROHME 2016	Not specified	Expression recognition rate CROHME 2014 = 48.78 and CROHME 2016 = 45.60
[18]	Machine Learning	Neural Network	CROHME 2013, 2014, 2016	CROHME 2013 is used for Validation, 9822 for CROHME 2014 and 10,969 for CROHME 2016	CROHME 2014 = 61.16 and CROHME 2016 = 57.02
[19]	Machine Learning	DenseNet	CROHME 2014 and CROHME 2016	9822 for CROHME 2014 and 10,969 for CROHME 2016	Expression Recognition CROHME 2014 = 52.8 and CROHME 2016 = 50.1
[20]	Machine Learning	CNN BLSTM	CROHME 2014	9999 Symbols	Stroke Order = 92

References

1. Zanibbi R, Blostein D (2012) Recognition and retrieval of mathematical expressions. Int J Doc Anal Recognit 15(4):331–357. https://doi.org/10.1007/s10032-011-0174-4

2. Kukreja V, Dhiman P (2020) A Deep Neural Network based disease detection scheme for Citrus fruits. In: International Conference on Smart Electronics and Communication (ICOSEC) 2020:97–101
3. Zhang J, Ming W, Liu P (2020) A two-stage framework for mathematical expression recognition. ICLR 2020:1–11
4. Lin H, Tan J (2020) Application of Deep Learning in handwritten mathematical expressions recognition. Lect Notes Comput Sci (including Subser Lect Notes Artif Intell Lect Notes Bioinformatics) 12068 LNCS, 137–147. https://doi.org/10.1007/978-3-030-59830-3_12
5. Kukreja V, Kumar D, Kaur A et al. GAN-based synthetic data augmentation for increased CNN performance in Vehicle Number Plate Recognition. In: 2020 4th International Conference on Electronics, Communication and Aerospace Technology (ICECA), pp 1190–1195
6. Álvaro F, Sánchez JA, Benedí JM (2016) An integrated grammar-based approach for mathematical expression recognition. Pattern Recognit 51:135–147. https://doi.org/10.1016/j.patcog.2015.09.013
7. Abirami M, Jaganathan S (2019) Handwritten mathematical recognition tool. In: ICCIDS 2019—2nd Int Conf Comput Intell Data Sci Proc, pp 1–4, 2019. https://doi.org/10.1109/ICCIDS.2019.8862155
8. Dai J, Sun Y, Su G, Ye S, Sun Y (2019) Recognizing offline handwritten mathematical expressions efficiently. In: Proceedings of the 10th international conference on E-Education, E-Business, E-Management and E-Learning, pp 198–204. https://doi.org/10.1145/3306500.3306543
9. Fontenele Marques Jr FDC et al. (2019) Recognition of simple handwritten polynomials using segmentation with fractional calculus and convolutional neural networks.In: Proc—2019 Brazilian Conf Intell Syst BRACIS 2019, pp 245–250. https://doi.org/10.1109/BRACIS.2019.00051
10. Nazemi A, Tavakolian N, Fitzpatrick D, Suen CY et al. (2019) Offline handwritten mathematical symbol recognition utilising deep learning. arXiv Prepr. arXiv1910.07395
11. Shinde S, Waghulade RB, Bormane DS (2018) A new neural network based algorithm for identifying handwritten mathematical equations. In: Proceedings—international conference on trends in electronics and informatics, ICEI 2017, vol 2018, January, pp 204–209. https://doi.org/10.1109/ICOEI.2017.8300916
12. Wan Z, Fan K, Wang Q, Zhang S (2019) Recognition of printed mathematical formula symbols based on convolutional neural network. In: DEStech Trans Comput Sci Eng no. ica, pp 80–85. https://doi.org/10.12783/dtcse/ica2019/30711
13. Wu JW, Yin F, Zhang YM, Zhang XY, Liu CL (2020) Handwritten mathematical expression recognition via paired adversarial learning. Int J Comput Vis. https://doi.org/10.1007/s11263-020-01291-5
14. Nguyen CT, Khuong VTM, Nguyen HT, Nakagawa M (2020) CNN based spatial classification features for clustering offline handwritten mathematical expressions. Pattern Recognit Lett 131:113–120. https://doi.org/10.1016/j.patrec.2019.12.015
15. Wang H, Shan G (2020) Recognizing handwritten mathematical expressions as LaTex sequences using a multiscale robust neural network
16. Julca-Aguilar F, Mouchère H, Viard-Gaudin C, Hirata NST (2020) A general framework for the recognition of online handwritten graphics. Int J Doc Anal Recognit 1–18. https://doi.org/10.1007/s10032-019-00349-6
17. Le AD, Indurkhya B, Nakagawa M (2019) Pattern generation strategies for improving recognition of Handwritten Mathematical Expressions. Pattern Recognit Lett 128:255–262. https://doi.org/10.1016/j.patrec.2019.09.002
18. Zhang J, Du J, Dai L (2018) Track, Attend, and Parse (TAP): an end-to-end framework for online handwritten mathematical expression recognition. IEEE Trans Multimed 21(1):221–233. https://doi.org/10.1109/TMM.2018.2844689
19. Zhang J, Du J, Dai L (2018) Multi-scale attention with dense encoder for handwritten mathematical expression recognition. In: Proc—Int Conf Pattern Recognit, vol 2018, August, pp 2245–2250. https://doi.org/10.1109/ICPR.2018.8546031

20. Le AD, Nguyen HD, Indurkhya B, Nakagawa M (2019) Stroke order normalization for improving recognition of online handwritten mathematical expressions. Int J Doc Anal Recognit 22(1):29–39. https://doi.org/10.1007/s10032-019-00315-2

Bangla Document Categorization Using Deep RNN Model with Attention Mechanism

Mostaq Ahmed, Partha Chakraborty, and Tanupriya Choudhury

Abstract Text classification is a strategy where the document categories are decided. Report classification is the assignment of relegating archives to a predefined space category. In this paper, a new method has classified Bangla text documents which consist of two deep learning models in terms of Bangla text categorization. The main purpose of this research is to text classification from Bangla text documents. This system can easily categorize from a large number of text documents. Firstly, the collected data was processed for the Bangla text document. Secondly, by designing the model architecture training data it was fitted out into the model. Finally, by calculating the accuracy and F1-score on the testing dataset, the model performance was evaluated. The deep learning recurrent neural network attention layer and recurrent neural network with BiLSTM attained 97.72% and 86.56% accuracy, respectively, which is compared to the other well-known classification algorithms in Bangla text classification.

Keywords Document classification · Deep learning · NLP · RNN · RNN with Attention Layer · LSTM · BiLSTM

1 Introduction

Document classification or document categorization becomes a citable area in the field of natural language processing in recent years. The main task of document classification is to assign a document to one or more classes or categories. The documents to be classified may be text, images, videos, etc. Each kind of docu-

M. Ahmed · P. Chakraborty (✉)
Department of Computer Science and Engineering, Comilla University, Cumilla 3506, Bangladesh
e-mail: partha.chak@cou.ac.bd

T. Choudhury
Department of Informatics, School of Computer Science, University of Petroleum and Energy Studies, Dehradun, India

J. M. R. S. Tavares et al. (eds.), *Cyber Intelligence and Information Retrieval*, Lecture Notes in Networks and Systems 291,
https://doi.org/10.1007/978-981-16-4284-5_13

ments possesses its special classification problems. Document classification covers of great importance in real-life applications such as sentiment analysis, topic labeling, online document management, spam detection, a recommendation system, and so on. It can also be categorized with hierarchically structured datasets [1] and multiple labels containing datasets [2] The rapid growth of internet activity and exponentially rising social media usage has not only necessitates but also implied the task of great importance. Algorithms utilized for report classification point to reveal the covered-up connections, patterns, and designs which are a strong base for commerce choice making. The commonly used unsupervised algorithm is K-means clustering. Moreover, a Naïve Bayes classifier based on the so-called Bayesian Theorem is one of the best text classification techniques. K-Nearest neighbor (KNN), support vector machine (SVM), neural networks (NN), latent dirichlet allocation (LDA) are also some popular techniques for document classification [3].

Bengali is one of the foremost widely utilized dialects within the world. A huge number of people express their words using Bengali as their native language. It is used in almost all types of text documents such as newspaper articles, social media posts, blog posts, etc. So, it becomes an important task to provide some contribution in this sphere.

Different strategies for Bangla report classification have been proposed in later considers. In this paper, two systems consisting of recurrent neural network (RNN) with attention mechanism and recurrent neural network with BiLSTM as feature selection techniques were used for classifying the documents. After classifying, we have tested this system with a great number of datasets collected from some renowned newspaper articles. After evaluation, we have shown a comparative study among those datasets.

We collected an existing dataset of 40 K Bangla newspaper articles from Prothom Alo as our dataset. To increase our accuracy, we used preprocessing algorithms like one hot encoder, label encoder, word embedding, etc. Finally, we work with two models 'RNN' and 'RNN with attention layer'. The accuracy of RNN is 86.56% and RNN with attention layer is 97.72%. The main key goal of this research is: To do Bangla text classification using two deep learning models which are RNN and RNN attention layer. Then, compare the two different deep learning models for identifying which model gives the best performance.

The leftover portion of the paper is organized as takes after: Segment 2 depicts the related works; Area 3 appears that the complete technique. The result investigation and comparison are examined in Segment 4. At last, the conclusion, limitations, and future works are discussed in Sect. 5.

2 Literature Review

Text categorization of content records completely different dialects could be a well-studied range is teaching a few supervisors to classify content records algorithms utilized. Researchers have adopted different types of methods for text categorization.

Text classification is not new for the English language. A large number of researcher's works have worked in this field. Kim [4] and Kalchbrenner et al. [5] both used the model of a single channel convolutionary neural network (CNN) to verify CNN's efficiency in the task of classifying sentences. The former tested his model with the help of six English datasets, but the latter tested their model with four separate data sets. When both studies consolidated the same national CNN structures, a study [4] showed better performance than the study [5]. Also, word2vec embedded with SVM is used [6] to classify English documents. Trstenjak et al. [7] term frequency-inverse document frequency (TF-IDF) and K-Nearest neighbor (KNN) have created a model. They tested their model with 500 online documents. The efficiency of that classification was based on different categories because the same type of document was more than appropriate. They further identify that with a larger number of documents. Classification performance has declined. A deep attention-based recurrent neural network was involved [8] by Changshun Du, Lei Huang for text classification which is an approach that achieved F-values by 88.5% and 51.8% on the two datasets Chines and English, respectively. Here, they first use the text preprocessing technique to remove low-frequency words and stop words, then use the recursive neural network with the attention mechanism to extract the feature vector of the article, and finally pass the feature vector to the softmax classifier.

Compared to other languages, there are very few works to classify Bangla articles that have been addressed. In [9], the author analyzes the ability to classify n-gram based texts for the Bengali language with the help of the one-year annual news corpus of Prothom Alo. The text classification of the Bengali language has not yet found better accuracy than that of English. However, many approaches have already been developed using a variety of machine learning algorithms. Islam et al. [10], used and Chi-square feature selection strategies, including two separate TF-IDF and Stochastic Gradient Descent, Naive Bayes, Vector Machine Supported algorithms, found higher accuracy in TF-IDF-based models. Kaur et al. [11] using the Stochastic Gradient Descent algorithm, the appropriate value was 0.9386, the recall value was 0.9388, and the F1 score was 0.9385 with comparatively more accuracy than other algorithms. A deep convolutional neural network was involved by M. R. Hossain et al. for text classification. This approach depicted 94.96% accuracy [12]. A Bangla document classification using deep recurrent neural network with BiLSTM by Saifur Rahman, and Partha Chakraborty et al. for text classification. This approach achieved 98.33% accuracy and an F1-score of 0.98 [13]. The collected data were processed for the Bangla text document and designing the model architecture training data was fitted out into the model. A Bangla reports classification utilizing transformer-based profound learning models by Md Mahbubur Rahman, Md. Aktaruzzaman Pramanik, Rifat Sadik, Monikrishna Roy, and Partha Chakraborty et al. for text classification [14]. They have applied the BERT and ELECTRA models for Bangla text classification and have used three different Bangla text datasets for their experiment. Their ELECTRA model achieved higher accuracy and the F1 score implies that it performs better than the BERT model and on the other hand, by applying the concept of convolutional neural network (CNN) for recognizing Bengali Sign Language gesture images for digits only in real-time. Bengali sign language images for digits are analyzed by the paper [15]. An efficient automated design to generate UML diagram from natural language specifications by Gulia et al. [16] the main aim of their paper is to focus on the production of activity diagram and sequence diagram through natural language specifications and on the other hand, by applying a conceptual design for making a mobile edge computing environment including the appropriateness of deep learning algorithms. Human-Robot Interaction System: A deep learning-based approach by the paper [17]. Detection of fake profile in online social networks using machine learning by Singh et al. [18] Researchers are studying these online social networks which see how they affect people some malicious accounts are used for misinformation and agenda-setting. The main goal of their paper is to identify fake profiles and eliminate the fake accounts on online social networks.

3 Methodology

The first step in a series of studies proposed in this demonstration was to collect data and clear the raw data. Then, clean text data preprocessing and total dataset distribution to training sets and test sets. Next, design and apply the RNN model and RNN with attention layer model on the train set. Finally, test data and model evaluation forecast. The successive steps of the proposed study appeared in Fig. 1.

3.1 Data Collection

Datasets are rare to find in Bangla. However, was able to collect a dataset of 40,000 Bangla news articles [19] for the proposed study. The dataset contains 3 columns, sections, content, and titles. 'Section' is a news class; the title is the name of the news article and the content of the news article in the news. The dataset was cleaned

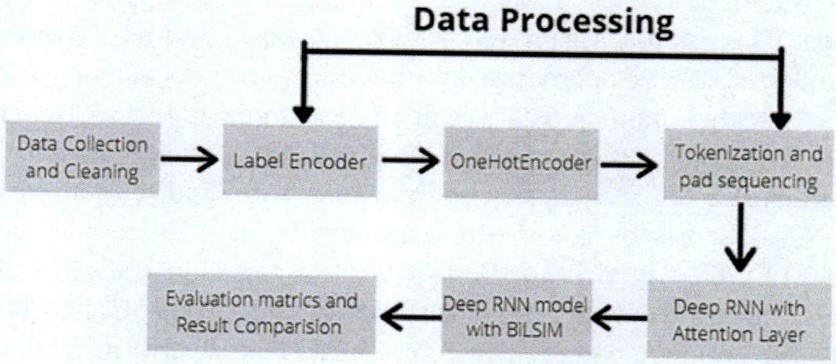

Fig. 1 The step-by-step method of the proposed study

by deleting English characters, mail addresses, HTML tags, unnecessary spaces, etc. Initially, the dataset contained news from 13 sections. An item was removed because the item had an unlabeled section.

3.2 Document Classification Dataset

Figure 2 shows the data of 10 sections. Most of the news articles collected are in the Bangladesh section, the number 12238 is correct. Information on other statistics about the dataset is shown in Table 1.

3.3 Data Preprocessing

The dataset was preprocessed to ensure a good fitting and good accuracy before fitting into the RNN model. Scrapped data was not in the form to fit the model. Classified data can be encoded in a variety of ways. First, label encoder was to convert the department's data into numbers. Then, one hot encoder was used to encode our sections for more accuracy.

Each word in the sentence is tokenized, and these tokenized words are saved in the python list for further processing. The main goal of this tokenization is to present each word in such a way that the machine can create a map with a numerical value [20].

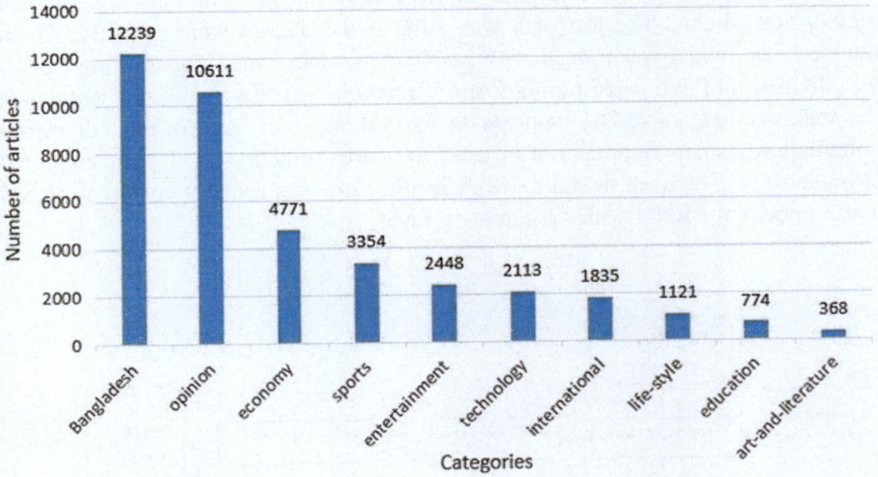

Fig. 2 Number of articles by category

Table 1 Statistical information about datasets

Total Characters (count)	108,970,652
Total Words	16,889,327
Total Sentences	13,810
Vocabulary	413,542
Most occurred word	ও, এই, এ, করে, থেকে, করা, জন্য
Most occurred word (count)	197,848
Maximum sentences in an article	437 (গদ্যরচনায় নজরুল মানস)
Maximum words in an article	5968 (গদ্যরচনায় নজরুল মানস)

3.4 Classification Model

News corpus was ready for training after preprocessing. The main goal of this study was to find a way to guess which news is a part of which class. This way each news can be categorized into their correct sections. The neural network model was made up of an embedding layer, two bidirectional layers, and a dense layer. The embedding layer was used to create word vectors and simultaneously perform classification tasks. The bidirectional layers were LSTM units with an input level of 128. Figures 3 and 4 show the RNN architecture and RNN attention architecture of the model, respectively.

Attention is a mechanism combined with RNN that enables easy learning and high quality when predicting specific parts of the output sequence. The combination of attention mechanisms has enabled improved performance in many activities as it is an integral part of modern RNN networks. The RNN Attention layer model is performed better than the RNN model because its works with specific parts of a sentence. So, its accuracy is higher than the RNN model. The dataset contained 39,999 news items. The test data was 30% of the dataset which was 12,000 in number. The model was trained with 31,707 news data items. The training dataset was divided into two parts, training, and validation. There were 25,366 samples in the training part and 6,341 samples in the validity part. At the time of model compilation, Adam optimizer was used as a loss function [21] with classified cross-entropy. The deep neural network architecture was tested 9 epochs for RNN and 5 epochs for RNN with the attention layer.

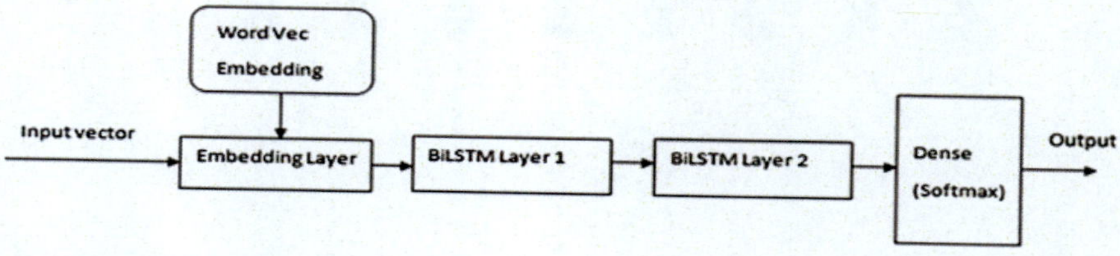

Fig. 3 RNN architecture

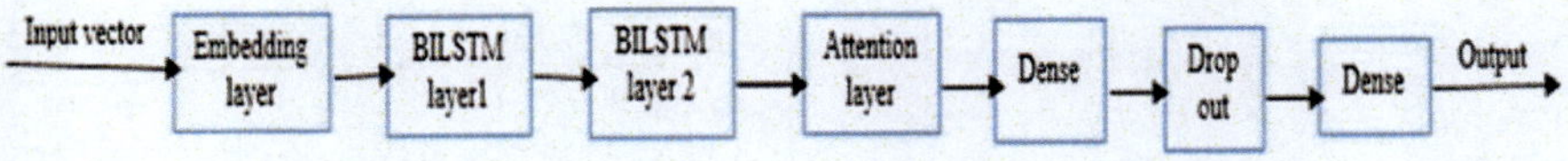

Fig. 4 RNN attention architecture

This section describes the training and architecture of the proposed models for the classification of Bengali documents. The proposed model used recurrent neural network (RNN) with attention mechanism and recurrent neural network with BiLSTM unit for text classification.

4 Experimental Setup and Result Analysis

4.1 Result Analysis

News corpus was ready for training after preprocessing. The main goal of this study The RNN model has achieved 86.56% which is well-known among others the classification is compared with the algorithm. Figure 5 shows the training and validation accuracy of the model. Here, we can see that the accuracy of the training increased a lot between 1 and 4 epochs, and remained the same from 4 to 9 epochs. On the other hand, the appropriateness of legitimacy increases slowly and stays within the same limits.

Figure 5 also shows the training and validity loss of the model. Here, we can see that after a certain epoch the loss of training using BLSTM and deep RNN with a reduction in validity started to decrease even more than Bangla document classification. On the other hand, the decrease in validity first decreases at a slower rate than remains the same for the rest of the training period compared to the rate of the training loss.

The RNN with attention layer model has achieved 97.72% which is higher accuracy than recent works in Bangla document classification. Figure 6 shows the

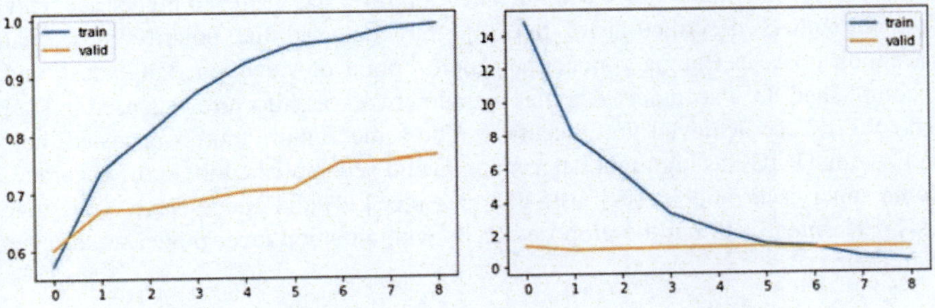

Fig. 5 Training and validation accuracy and loss of RNN

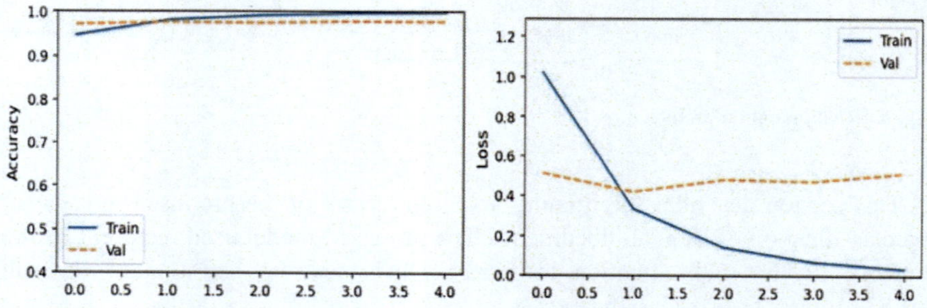

Fig. 6 Training and validation accuracy and loss of RNN attention layer

training and validation accuracy of the model. Here, we can see that the accuracy of the training increased between 1 and 5 epochs.

Figure 6 also shows the training and validity loss of the model. Here, we can see that after a certain epoch the loss of training using RNN with attention layer model a reduction in validity started to decrease even more than Bangla document classification. On the other hand, the decrease in validity first decreases at a slower rate than remains the same for the rest of the training period compared to the rate of the training loss. Here, four evaluation matrices precision, recall, f1-score, and accuracy were used for performance evaluation. A confusion matrix was formed to calculate these matrices. From the confusion matrix, we have calculated precision, recall, and f1-score for our model. In Table 2, full classification report is given. The accuracy of our model RNN and RNN with attention layer is 86.56% and 97.72%, respectively.

4.2 *Result Comparison*

In this task, we used two models RNN and RNN with an attention layer. The accuracy of RNN is 86.56% and the accuracy of RNN with the attention layer is 97.22%. From the comparison, we can see that from comparison Table 3 it is clear that the proposed model RNN with the attention layer has achieved higher accuracy than the others described here. In [22], they detected the polarity of textual Facebook posts in Bangla containing people's point of views on Bangladesh and accomplished 83% accuracy. Another neural network architecture was used in [23] with Word2Vec achieved 96% accuracy. The same authors moreover tested their show with TF-IDF as highlight determination and accomplished the same accuracy. Some supervised models, such as SVM classifier [24] and Naive Bayes Classifier [24] performed well, but the proposed RNN with attention layer model was higher than them.

Table 2 Classification report of RNN and RNN with attention layer

Class name	Precision		Recall		F1-score	
	RNN	RNN*	RNN	RNN*	RNN	RNN*
Bangladesh	0.82	0.31	0.83	0.43	0.82	0.36
Opinion	0.82	0.88	0.77	0.92	0.79	0.90
Economy	0.79	0.85	0.71	0.87	0.75	0.86
Sports	0.92	0.64	0.91	0.65	0.92	0.65
Entertainment	0.79	0.93	0.73	0.87	0.76	0.90
Technology	0.65	0.76	0.68	0.67	0.66	0.71
International	0.58	0.68	0.60	0.71	0.59	0.70
Life-style	0.34	0.95	0.46	0.93	0.39	0.94
Education	0.39	0.99	0.57	0.93	0.47	0.96
Art-and-literature	0.21	0.86	0.25	0.83	0.85	0.85
Weighted average	0.77	0.89	0.76	0.88	0.76	0.88

Here, RNN* = RNN with Attention Mechanism

Table 3 Result Comparison with recent works in Bangla text classification

Model	Precision	Recall	F1-score	Accuracy (%)
Mandal, A. K. et al. SVM [24]	N/A	N/A	0.89	89.14
Mandal, A. K. et al. Naive Bayes [24]	N/A	N/A	0.85	85.22
M., Jubayer et al. TF-IDF (SGD) [10]	0.92	0.92	0.92	92
M., Jubayer et al. TF-IDF (SVM) [10]	0.92	0.92	0.92	92
Chakraborty M. et al. DenseNN2 [22]	0.85	0.85	0.85	85.208
Alam, M. T. et al. TF-IDF* (NN) [23]	0.96	0.96	0.96	96
Hossain, M. R. et al. (CNN) [12]	N/A	N/A	0.94	94.96
S. Rahman and P. Chakraborty et al. RNN with BiLSTM [13]	0.98	0.98	0.98	98.33
Mostaq et al. Proposed Model RNN	**0.86**	**0.86**	**0.86**	**86.56**
Mostaq et al. Proposed Model RNN with an attention layer	**0.97**	**0.97**	**0.97**	**97.72**

5 Conclusion

In this paper, deep learning-based Bangla document classification was implemented. Two deep learning-based models, namely RNN and RNN with attention layer used for this classification task. The accuracy of RNN is 86.85% and RNN with attention layer is 97.72%. These studies have shown that the neural network architecture achieves higher accuracy than the supervised learning models according to KNN, SVM, Naive Bayes, Decision tree, etc. The models based on the neural network work well in the classification of Bangla documents.

Compared to other studies of the feature in the study, it was also found that selection methods work better with neural networks, according to word2vec and TF-IDF. Besides the two deep learning models, namely, BERT and ELECTRA work well in Bangla text classification. This work used 40 K datasets. Moreover, the dataset was not enough for our purpose. In the future, we can use large datasets and our results might improve.

Despite the many satisfactory results, we have achieved, there are opportunities for future improvement in our work. So, in the future, we will use better models which will work well with neural network for Bangla text documents such as Bert, transformer-based deep learning models.

References

1. Krendzelak M, Jakab F (1990) Text categorization with Machine Learning and Hierarchical Structure In: The Technical University of Kosice, pp 396–402. https://www.researchgate.net/publication/307854016 Text categorization with machine learning and hierarchical structures (Book download)
2. McCallum A (1999) Multi-label text classification with a mixture model trained by them. In: AAAI workshop on text learning, pp 1–7
3. Lewis DD, Ringuette M (1994) A comparison of two learning algorithms for text categorization. In Third annual symposium on document analysis and information retrieval, vol 33, pp 81–93
4. Kim Y (2014) Convolutional neural networks for sentence classification. In: Conference on empirical methods in natural language processing
5. Kalchbrenner N, Grefenstette E, Blunsom P (2014) A convolutional neural network for modelling sentences. In: Proceedings of the 52nd annual meeting of the association for computational linguistics, pp 655–665
6. Lilleberg J, Zhu Y, Zhang Y (2015) Support vector machines and word2vec for text classification with semantic features. In: ICCI*CC, Ge N, Lu J, Wang Y, Howard N, Chen P, Tao X, Zhang B, Zadeh LA (eds) IEEE computer society, pp 136–140
7. Trstenjak B, Mikac S, Donko D (2014) KNN with TF-IDF based framework for text categorization. Proc Eng 69:1356–1364
8. Du C, Huang L (2018) A deep attention-based recurrent neural networks. Int J Comput Commun Control (IJCCC)
9. Mansur M (2006) Analysis of n-gram based text categorization for Bangla in a newspaper corpus. In: Proceedings of international conference on computer and information technology
10. Islam M, Jubayer FEMS, Ahmed I (2017) A comparative study on different types of approaches to Bengali document categorization. In: Proceedings of the international conference on engineering research, innovation and education 2017
11. Kaur J, Saini JR (2015) A study of text classification natural language processing algorithms for Indian languages. VNSGU J Sci Technol 162–167
12. Hossain MR, Hoque MM (2019) Automatic Bengali document categorization based on deep convolution nets. In: Emerging research in computing, information, communication, and applications. Springer, Berlin, pp 513–525
13. Rahman S, Chakraborty P (2020) Bangla document classification using deep recurrent neural network with BiLSTM. In: Proceedings of international conference on machine intelligence and data science applications

14. Rahman MM, Pramanik MA, Sadik R, Roy M, Chakraborty P (2002) Bangla documents classification using transformer based deep learning models. In: 2nd international conference on Sustainable Technologies for Industry 4.0 (STI), 19–20 December, Dhaka, Bangladesh
15. Ahammad K, Shawon JAB, Chakraborty P, Islam MJ, Islam S (2021) Recognizing Bengali sign language gestures for digits in real time using convolutional neural network. Int J Comput Sci Information Security (IJCSIS) 19(1) September
16. Gulia S, Choudhury T (2016) An efficient automated design to generate UML diagram from natural language specifications. In: 6th international conference-cloud system and big data engineering (confluence), pp 641–648
17. Chakraborty P, Ahmed S, Yousuf MA, Azad A, Alyami SA, Moni MA (2021) A Human-Robot interaction system calculating visual focus of Human's attention Level. IEEE Access 9:93409–93421
18. Singh N, Sharma T, Thakral A, Choudhury T (2018) Detection of fake profile in online social networks using machine learning. In: International Conference on Advances in Computing and Communication Engineering (ICACCE), pp 231–234
19. Zakir Hossan M (2018) 40k Bangla Newspaper Article. September 2018. https://www.kaggle.com/zshujon/40k-bangla-newspaper-article
20. Gupta M. Tokenization algorithms in Natural Language Processing (NLP). https://medium.com/data-science-in-your-pocket/tokenization-algorithms-in-natural-language-processing-NLP-1fceab8454af
21. Kingma DP, Ba J (2014) Adam: a method for stochastic optimization. In: 3rd international conference for learning representations
22. Faruque MA, Rahman S, Chakraborty P, Choudhury T, Um JS, Singh TP (2021) Ascertaining polarity of public opinions on Bangladesh cricket using machine learning techniques. Spatial Information Research
23. Alam MT, Islam MM (2018) BARD: Bangla article classification using a new comprehensive dataset. In: International conference on bangla speech and language processing, August 2018
24. Mandal AK, Sen R (2014) Supervised learning methods for Bangla web document categorization. Int J Artif Intelligence Appl (IJAIA) 5(5), September 2014

Bangla Handwritten Digit Recognition

Partha Chakraborty, Syeda Surma Jahanapi, and Tanupriya Choudhury

Abstract Due to the wide range of shapes, sizes, and writing styles, handwritten digit recognition has always been difficult. For all of the recognition algorithms built in this thesis, the rewarded method is used. NumtaDB is the largest dataset. It is a collection for Bengali handwritten digits. This is a massive dataset with over 85,000 images in it. This dataset, however, is incredibly difficult to work with due to its complexity. The aim of this paper is to preprocess images that can be used to train deep learning models with high accuracy. The reason for this is that, unlike the MNIST for English digits, there is no preprocessed data to deal with when it comes to Bengali digit recognition. This unbiased dataset, NumtaDB, is used in this paper for Bangla digit recognition. In this paper, various preprocessing techniques are developed for image processing, with a deep convolutional neural network (CNN) functioning as the classification algorithm. On the NumtaDB image dataset, the performance is systematically evaluated of this process. Finally, 93% accuracy is obtained for NumtaDB dataset, and 92% accuracy rate is obtained for our own dataset using same of the proposed method in experiments. In this paper, the accuracy for every training dataset in NumtaDB is separately calculated which has the maximum accuracy 98%.

Keywords NumtaDB dataset · Deep CNN · Deep learning · Bangla digit recognition

P. Chakraborty (✉) · S. S. Jahanapi
Department of Computer Science and Engineering, Comilla University, Cumilla 3506, Bangladesh
e-mail: partha.chak@cou.ac.bd

T. Choudhury
Department of Informatics, School of Computer Science, University of Petroleum and Energy Studies, Dehradun, India

J. M. R. S. Tavares et al. (eds.), *Cyber Intelligence and Information Retrieval*,
Lecture Notes in Networks and Systems 291,
https://doi.org/10.1007/978-981-16-4284-5_14

1 Introduction

Academic and commercial sectors also interest automated handwriting character or numeral recognition. There are currently available algorithms for learning to recognize handwritten digits. The most difficult aspect of handwritten digit classification is dealing with the vast variety of handwriting types used by different authors in various languages. In certain languages, digits are written separately from one another in some situations, depending on the language (such as Japanese, Thai, and Laos). In certain cases, they are written in cursive. The task of identifying individual digits is also complicated by the various differences in writing styles. Deep learning, especially the convolutional neural network (CNN) is now more effective than other machine learning methods in classifying these types of recognition work. The Bangla language has ten numerical digits. Depending on the handwritten, the accuracy of digit recognition algorithms can differ significantly. Bangla digit recognition has a wide range of applications, including Bangla OCR, automatic license plate recognition for vehicles, National ID number recognition, post office automation, parking lot management, post office automation, online banking, and many others. Everybody has their own unique writing style. The difficult task is to achieve stable output and high accuracy for the NumtaDB [1] dataset, which is massive, unbiased, unprocessed, and highly augmented. The entire manuscript is organized in such a way that the literature review is in Sect. 2, the system methodology is in Sect. 3, the experimental details are in Sect. 4, the result analysis is in Sect. 5, and the final Sect. 6 is conclusion.

2 Literature Review

Since NumtaDB dataset was not available at the time, most paper studies relied on skewed datasets such as CMATERDB 3.1.1. Bangla digits were also identified using the local binary pattern [2]. The convolutional neural network, deep learning approach [3, 4], was used to improve supervised learning and performance. Handwritten digit recognition was proposed by EMNIST [5] for both balanced and imbalanced datasets. Researchers used a local binary pattern to identify Bangla digits, achieving more accuracy [2]. Yann LeCun created the newer ConvNet architecture LeNet5 in 1998 [6]. The Le-Net5 architecture was able to successfully recognize digits on bank checks as well as recognize handwritten digits. As a result of the recent success of deep learning, especially convolutional neural network (CNN) for computer vision, many researchers have been encouraged to use the CNN to recognize handwritten characters and digits as a computer vision challenge [7]. Subhadip Basu et al. proposed handwritten Bangla digit recognition using classification combinations by Dempster–Shafer (DS) technique [8]. They used the DS technique and MLP classifier for classification on the training dataset of 6000 handwritten samples, as well as threefold cross-validation. Their technique resulted in 95.1% test accuracy.

One of the most important tasks in the field of character recognition is local area identification, which involves extracting the best unambiguous features. N. Das et al. use this concept in their genetic algorithm-based handwritten digit recognition technique [9]. M.S. Islam and his colleagues suggested Bayanno-Net, in which they used CNN to understand Bangla handwritten numerals. Their model had a 97% accuracy rate. A review article [10] by M.N. Hoq et al. offered a comparative analysis of existing classification algorithms for recognizing Bengali handwritten numerals. Bengali sign language images for digits are analyzed by the paper [11]. A model based on handwritten digits was proposed by Shawon et al. [12]. An efficient automated design to generate UML diagram from natural language specifications by Gulia et al. [13]; the main aim of their paper is to focus on the production of activity diagram and sequence diagram through natural language specifications and on the other hand by applying a conceptual architecture for creating a human computing environment involving the applicability of deep learning algorithms. Intelligent human computing: A Deep Learning-Based Approach by the paper [14]. A Bangla document classification using deep recurrent neural network with BiLSTM by Saifur Rahman and Partha Chakraborty et al. for text classification. This approach achieved 98.33% accuracy and an F1-score of 0.98 [15]. The collected data were processed for the Bangla text document and designing the model architecture training data; it was fitted out into the model. A Bangla documents classification using transformer-based deep learning models by Md Mahbubur Rahman, Md. Aktaruzzaman Pramanik, Rifat Sadik, Monikrishna Roy, and Partha Chakraborty et al. for text classification [16]. They have applied the BERT and ELECTRA models for Bangla text classification. In [17], researchers are studying these online social networks which see how they affect people some malicious accounts are used for misinformation and agenda setting. The main goal of their paper is to identify fake profiles and eliminate the fake accounts on online social networks.

3 Methodology

Three main steps are applied to recognize NumtaDB handwritten digits. Among them, feature extraction step and classification step are combined in one step. So the number of main steps of our digit recognition is two. One is preprocessing of images, and another is deep CNN model. These steps are shown in Fig. 1.

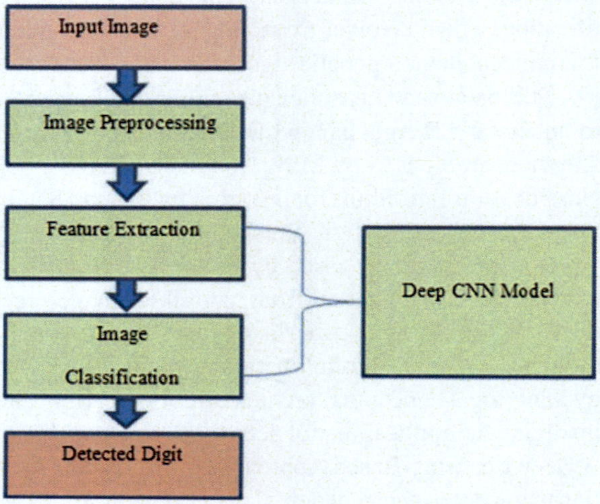

Fig. 1 Block diagram of proposed Bangla digit recognition

4 Experimental Details

4.1 Data Collection

A very challenging dataset NumtaDB is used here for training and testing model. The NumtaDB dataset includes more than 85,000 handwritten digital images from Bangla. NumtaDB is a large, standard, unprocessed, and reviewed dataset. Each dataset image is about 180 pixels to 180 pixels. Sample images are in Fig. 2

We have our own dataset consisting of 1500 handwritten digits. These handwritten images are collected from different people with different handwriting style. The training and testing set have different images. There is no repeat of images in both of training and testing images. The written digits are checked with the camera on the handset. After that, each image in the dataset is resized to 180 × 180 pixels. Training and testing data are split into 75–25 percent. That is 1125 data presents in training folder and 375 data in testing folder. Sample images are shown in Fig. 3.

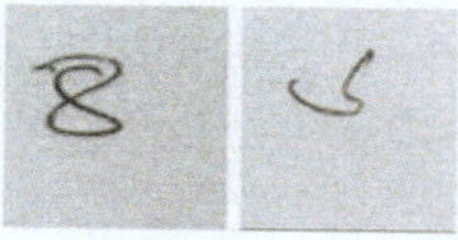

Fig. 2 Images of Bangla handwritten digits from NumtaDB dataset

Fig. 3 Bangla handwritten digit images from our dataset

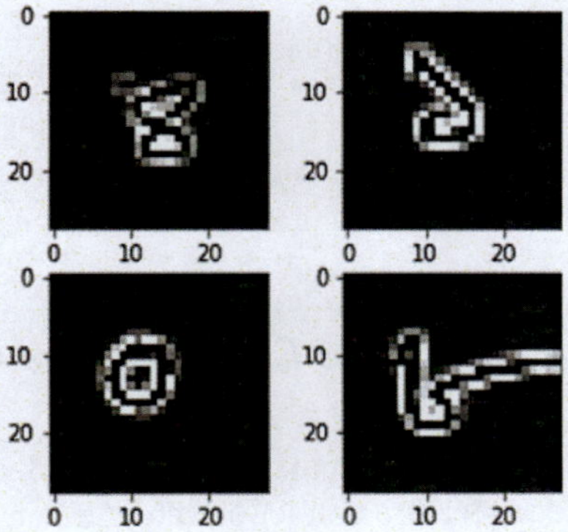

Fig. 4 Images after preprocessing

4.2 Dataset Preparation for Overall NumtaDB and Own Dataset

Image resizing: Each image in the NumtaDB dataset is 180 × 180 pixels. Each image's dimension is resized to 28 × 28. This is called photograph resizing.

Image grayscaling: This RGB format image has to be converted to a grayscale format. This means that the color channel will be transformed from three channels to one channel.

Image interpolation: After resizing images, we employ inter-area interpolation. Smoothing images: Gaussian blur smooths uneven pixel values in an image. If images are taken in low light, Gaussian blur can mute noise from pictures. In order to get a de-blurred image, a weighted portion of that mask is added.

Sharpening images: Laplacian filter is used here. Here, filer is 3 × 3 matrix.

Noise removal from images: From NumtaDB images, we suppress noise. For noise removal, median filter is used. The images become smooth, noise free, and sharp after preprocessing. After preprocessing, images are shown in Fig. 4

4.3 Dataset Preparation for Every Single Training Dataset

Image resizing: Each image in every training dataset is 180 × 180 pixels. Each image's dimension is resized to 28 × 28.

Image grayscaling: This RGB format image has to be converted to a grayscale format. As a result, all RGB images must be converted to grayscale.

Image interpolation: The method by which a small picture is rendered larger is interpolation. After resizing images, inter-area interpolation is employed. The experiment begins by training the model with the above-mentioned preprocessed dataset.

Table 1 Model parameters for training

Parameter name	Value
Learning rate	0.001
Epoch	30
Batch size	64

4.4 Feature Extraction and Classification Steps

Deep convolutional neural network automatically extracts and classifies features. It does feature extraction through convolution and maxpooling layer, and it does the classification through fully connected layer.

Here, proposed complete dataset architecture consists of total six convolutional layers for feature extractions and also two dense layers that are fully connected for classification. Input image is 28 × 28 with channel 1 size in CNN. There are 32 filters in the first two layers, 128 filters in the second two layers, and 256 filters in the third two convolutional layers. As activation function for all of convolutional layers, the rectified linear unit (ReLu) is used. After every two layers, Maxpooling layers and Batch Normalization are used. Here, maxpooling layer's pool size is 2 × 2. To input it to the next layer, flattening transforms the data into a one-dimensional array. Machine Learning techniques [18] are applied to improve accuracy. Here, we used dropout 20% that means only 20% neuron dropout in each of the epoch. The first fully connected layer has 64 filters, and the last fully connected layer has 10 filters for 10 digits. In our neural network, softmax transforms the output of the last layer into what is basically a distribution of probability. To update weights, we use the Adam [19] optimizer. Trainable programs are 1,359,978 after applying the model among 1,360,810 programs, and 8322 are the nontrainable programs. Model parameters are given in Table 1

4.5 Architecture of Model for Every Training Single Dataset

Proposed architecture for a single training dataset consists of six convolutional layers. Our architecture consists of three fully connected dense layers. Here, input image size is 28 × 28, and each image is with channel 1. The first two convolutional layers have total 32 filters, the size of first filter is 5 × 5, and the size of second filter is 4 × 4. The next two convolutional layers consist of 64 filters. The size of each filter is 3 × 3. The last two convolutional layers have 64 filters, and the size of each filter for these layers is 3 × 3. As an activation function, rectified linear unit (ReLu) is used for all layers. Same padding is used here. Maxpooling layers are present here. The size of pool of our maxpooling layer is 2 × 2 with stride number 2. Dropout (30%) is added after each maxpooling layers. After completing these steps, features are extracted from our input images. From first convolutional layer, low level features are extracted, and from next convolutional layers, high level features are extracted.

Table 2 Model summary of deep CNN architecture for every single training dataset

Layer	Output shape
Conv2D 1	(None, 28, 28, 32)
Conv2D 2	(None, 28, 28, 32)
MaxPooling2D 1	(None, 14, 14, 32)
Dropout 1	(None, 14, 14, 32)
Conv2D 3	(None, 14, 14, 64)
Conv2D 4	(None, 14, 14, 64)
MaxPooling2D 2	(None, 7, 7, 64)
Dropout 2	(None, 7, 7, 64)
Conv2D 5	(None, 7, 7, 64)
Conv2D 6	(None, 7, 7, 64)
MaxPooling2D 3	(None, 3, 3, 64)
Dropout 3	(None, 3, 3, 64)
Flatten 1	(None, 576)
Dense 1	(None, 256)
Dense 2	(None, 256)
Dropout 4	(None, 256)
Dense 3	(None, 10)

Dropout (50%) is used after first and second fully connected layers. The first two fully connected layers each have filters, where the number of filter is 64. The third layer has 10 filters. Here, we used activation function, softmax function for ten-digit classification at last. Model summary is given in Table 2.

Here, we use RMSprop optimizer in our model. Here, the learning rate is changing over time. We use 1000 batch size and 100 epochs for training our model. With learning rate reduction (from 0.001 to 0.00001), we can achieve higher accuracy on the tested for Bangla digit. After applying model among 362,602 programs, our trainable programs are 362,602 and non-trainable programs are 0.

5 Result Analysis

Of the 85000+ images, the NumtaDB dataset's train and test split ratio is 85–15% [1]. For every training dataset A to E, accuracy is calculated. Here, train and test split ratio is 80–20%. The train and test split ratio for our dataset is 75–25%.

Three outcomes are obtained after 30 epochs for NumtaDB, and for own dataset, after 100 epochs. Results are given in Table 3.

It also tests the processing time of training. Eight seconds are required per epoch. The total training time is 240 s.

Table 3 Training, validation, and testing accuracy for NumtaDB dataset

NumtaDB dataset	Accuracy (%)
Training	99
Validation	98
Testing	93

Table 4 Training, validation, and testing accuracy for own dataset

Own dataset	Accuracy (%)
Training	99
Validation	84
Testing	92

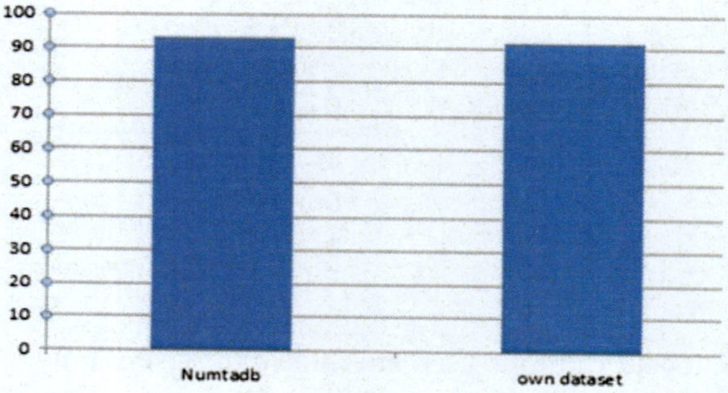

Fig. 5 Comparison of NumtaDB testing and own dataset testing accuracy

A result is observed from our own dataset. Model is tested accuracy from that dataset. Training, validation results, and test results after 100 epochs for our dataset are given in Table 4.

Testing accuracy is compared between all testing datasets of NumtaDB and our own dataset in Fig. 5.

At first the training first began, the level of training failure or error was extremely very high, which is 3.91. The amount of training accuracy was very low, and the value of training accuracy is 0.11. However, as the training epochs increased, the value of this loss became lower dramatically, and the value of training accuracy increased. The error rate dropped to almost 0.0091 after 100 epochs, and training accuracy becomes 0.99.

When the training first began, the level of validation failure was extremely high, with a value of 2.30. The amount of validation accuracy was very low, and the value of validation accuracy was 0.15. However, as the number of training epochs grew, the value of validation failure decreased dramatically, then slowed, while the value of validation accuracy increased. After 100 epochs, the error rate was nearly 0.46, and the validation accuracy was 0.90. These graphs are shown in Figs. 6 and 7.

Accuracy for each of the training dataset (a, b, c, d, and e) are computed, and scikit model is used to calculate the accuracy. But we do not get good accuracy by using scikit. Keras sequential model is also used for calculating accuracy. By using CNN

Fig. 6 Training accuracy and validation accuracy

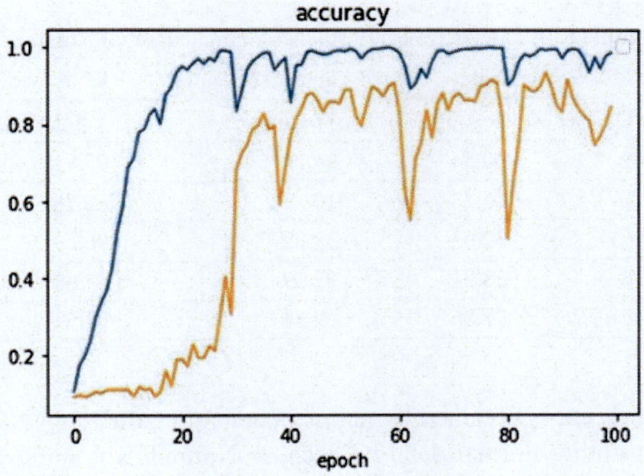

Fig. 7 Training loss and validation loss

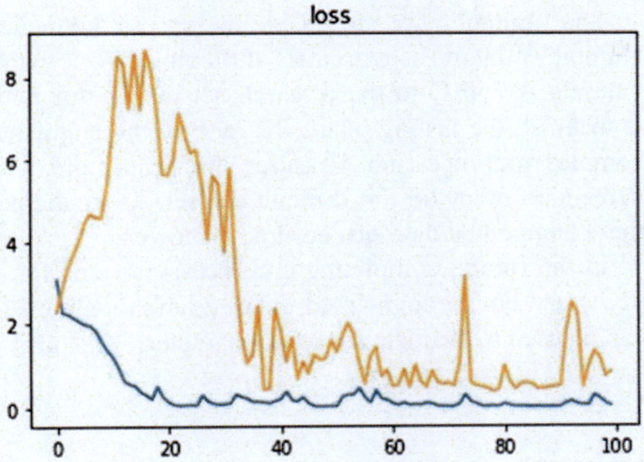

model, higher accuracy is obtained. Because in CNN, our model is trained with more and more convolutional layers, so that it can extract higher level features from the images. These higher level features are used to classify our digits more accurately. Here, we split our training and testing dataset 80% to 20%. Now accuracy calculated for this two models will be compared. Applying keras model, accuracy for every dataset is above 90 %. Model CNN is best for classification. More convolutional layers are used for extracting high-level features which is essential for classification. Table 5 shows the comparison of accuracy for every dataset by applying two models.

6 Conclusion

In this paper, Bangla digit recognition method based on deep convolutional neural networks is presented for a normal and unbiased dataset. Testing accuracy of 93

Table 5 Accuracy for every single training dataset of NumtaDB by applying two models

Model	scikit	keras (sequential)	keras (sequential)	keras (sequential)
Dataset	Testing (%)	Training (%)	Validation (%)	Testing (%)
A	60.54	99.29	97.91	97.83
B	15.28	100	82.76	77.78
C	32.02	99.22	96.63	96.87
D	38.91	92.09	97.63	97.75
E	71.06	99.54	97.52	98.02

percent is obtained, which is better for challenging NumtaDB dataset as compared to other normal datasets because NumtaDB is unbiased. On our own dataset, we achieve 92% accuracy. For different training datasets of NumtaDB, 90% up accuracy has got.

The limitation we are unable to process augmented datasets at this time. The NumtaDB dataset is extremely difficult to work with. The addition of augmented datasets A and C to the research set makes this dataset more difficult. Our aim is to finish the testing phase for each of the augmented datasets separately, using complex preprocessing measures. By proper preprocessing, we can achieve 100 percent accuracy for this difficult dataset. As we did not complete preprocessing for these augmented datasets, accuracy is lower.

In the future, completing preprocessing steps for each augmented datasets are necessary. For the augmented, datasets have to achieve 100% accuracy. The algorithm can be used to evaluate for other characters. This is also left for future.

References

1. Alam S, Reasat T, Doha RM, Humayun AI (2018) NumtaDB Assembled Bengali handwritten digits. arXiv:1806.02452 [cs.CV]
2. Hassan T, Khan AH (2015) Handwritten Bangla numeral recognition using Local Binary Pattern. In: 2015 international conference on Electrical Engineering and Information Communication Technology (ICEEICT), pp 1–4. IEEE
3. Akhand MAH, Ahmed M, Rahman MMH (2016) Convolutional neural network training with artificial pattern for Bangla handwritten numeral recognition. 5th International conference on informatics, electronics and vision (ICIEV) pp 625–630. https://doi.org/10.1109/ICIEV.2016.7760077
4. Alom M, Sidike P, Taha T, Asari V (2017) Handwritten Bangla digit recognition using deep learning. arXiv:1705.02680v1 [cs.CV]
5. Cohen G, Afshar S, Tapson J, Schaik A (2017) EMNIST: An extension of MNIST to handwritten letters. arXiv:1702.05373[cs.CV]
6. LeCun Y (2015) LeNet-5, convolutional neural networks (Online). Available: http://yann.lecun.com/exdb/lenet. Accessed 20 Oct 2016
7. Ghosh A, Sufian A, Sultana F, Chakrabarti A, De D (2019) Fundamental concepts of convolutional neural network. In: VEB et al. (eds), Recent trends and advances in artificial intelligence and Internet of Things

8. Basu S, Sarkar R, Das N, Kundu M, Nasipuri M, Basu DK (2005) Handwritten Bangla digit recognition using classifier combination through ds technique. In: Pattern recognition and machine intelligence, pp 236–241
9. Das N, Sarkar R, Basu S, Kundu M, Nasipuri M, Basu DK (2012) A genetic algorithm based region sampling for selection of local features in handwritten digit recognition application. Appl Soft Comput 12(5):1592–1606. https://doi.org/10.1016/j.asoc.2011.11.030
10. Hoq MN, Islam MM, Nipa NA, Akbar MM (2020) A comparative overview of classification algorithm for Bangla handwritten digit recognition. In: Proceedings of international joint conference on computational intelligence, pp 265–277
11. Ahammad K, Shawon JAB, Chakraborty P, Islam MJ, Islam S (2021) Recognizing Bengali sign language gestures for digits in real time using convolutional neural network. Int J Comput Sci Information Security (IJCSIS) 19(1), September 2021
12. Shawon A, Rahman MJ, Mahmud F, Zaman MMA (2018) Bangla handwritten digit recognition using deep CNN for large and unbiased dataset. In: ICBSLP, 21–22 September 2018
13. Gulia S, Choudhury T (2016) An efficient automated design to generate UML diagram from Natural Language Specifications. In: 6th international conference-cloud system and big data engineering (confluence), pp 641–648
14. Chakraborty P, hmed S, Yousuf MA, Azad A, Alyami SA, Moni MA (2021) A human-robot interaction system calculating visual focus of human's attention level. IEEE Access 9:93409–93421. https://doi.org/10.1109/ACCESS.2021.3091642.
15. Rahman S, Chakraborty P (2020) Bangla document classification using deep recurrent neural network with BiLSTM. In: Proceedings of international conference on machine intelligence and data science applications
16. Rahman MM, Pramanik MA, Sadik R, Roy M, Chakraborty P (2020) Bangla documents classification using transformer based deep learning models. In: 2nd international conference on sustainable technologies for industry 4.0(STI), 19–20 December, Dhaka, Bangladesh
17. Singh N, Sharma T, Thakral A, Choudhury T (2018) Detection of fake profile in online social networks using machine learning. In: International Conference on Advances in Computing and Communication Engineering (ICACCE), pp 231–234
18. Chakraborty P, Yousuf MA, Rahman S (2021) Predicting level of visual focus of human's attention using machine learning approaches. In: Kaiser MS, Bandyopadhyay A, Mahmud M, Ray K (eds) Proceedings of international conference on trends in computational and cognitive engineering. Advances in Intelligent Systems and Computing, vol 1309. Springer, Singapore. https://doi.org/10.1007/978-981-33-4673-4_56
19. Kingma D, Ba J (2015) Adam: a method of stochastic optimization. arXiv:1412.6980[cs.LG]

Audio Watermarking in Linear Canonical Transform Domain Using Frequency-Dependent Clustering

Ashish Sinha and Jeebananda Panda

Abstract Audio watermarking is considered as a challenging domain owing to the requirement of imperceptibility post watermarking, while keeping the robustness intact (Dhar PK, Robust FFT based watermarking scheme for copyright protection of digital audio data, IEEE, China, pp. 181–184, 2011). This paper proposes an effective methodology for watermarking of the audio signal using linear canonical transform (LCT), which has found its use in a multitude of signal processing applications, due to its flexibility (Xiang et al., The linear canonical transform and its application to time-frequency signal analysis, IEEE, Wuhan, China, pp 1–4, 2009). The proposed work uses a watermark sequence of complex numbers which is suitably embedded into the transformed audio signal using frequency-dependent clustering. The watermark detection is blind and performed reversibly. It is found that the watermark is extremely resilient against a wide range of signal processing and feature distortive attacks. The technique is also secure, imperceptible, stable and offers sufficient payload capacity of nearly 15% of the audio signal length.

Keywords Audio watermarking · Transform domain · Frequency-dependent clustering · Linear canonical transform

1 Introduction

Recent strides made in the field of digital media and related technologies have spurred an unprecedented growth and have also led to misutilization of data and copyright infringements, thus imperiling intellectual property rights and causing monumental financial losses [3, 4]. Copyright infringement to exploit the work of artists and make

A. Sinha (✉) · J. Panda
Delhi Technological University, Delhi, India
e-mail: ashishsinha_2k20spd02@dtu.ac.in

J. Panda
e-mail: jpanda@dce.ac.in

J. M. R. S. Tavares et al. (eds.), *Cyber Intelligence and Information Retrieval*, Lecture Notes in Networks and Systems 291,
https://doi.org/10.1007/978-981-16-4284-5_15

undue profits has been recognized as a grave concern in the music industry [5]. As a counteractive measure, audio watermarking has been broached upon in this paper.

The scheme chiefly entails hiding the watermark information (owner/distributor identity, file manipulation/ transactions information, etc.) into the original audio signal, in an imperceptible manner and without affecting the fundamental characteristics of the audio signal [6].

An effective and efficient watermarking scheme, in general, is characterized by robustness, security, imperceptibility, and high payload capacity [6]. Imperceptibility is vital as it has been observed that embedding the watermark often leads to perceptible distortions. Further, watermarking schemes with blind extraction are preferred for their higher efficiency and security. Techniques for watermarking of audio signals are primarily classified into frequency domain, spread spectrum-based, time domain, echo hiding, and patch work-based techniques [7, 8].

Transform domain watermarking provides a simple yet robust approach for watermarking [9, 10] as it is derived from indirect modification of spectral coefficients instead of the time domain samples. Further, the techniques also provide the requisite degree of imperceptibility, due to prudent selection of the watermarking locations. In contrast, the methodologies based on echo coding and the patch work scheme are accompanied by lack of robustness and are unsuitable for signals with sparse frequency distribution [6].

This manuscript puts forth a secure watermarking approach for audio signals in linear canonical transform domain. The key feature of the proposed technique is the watermarking of mid-frequency transform coefficients, which has a twofold advantage of stability and robustness. Security is ensured by the three degrees of freedom afforded via the LCT. Extraction of the watermark is done in a blind and reversible manner, and the watermark is detected adaptively. The technique also exhibits desirable imperceptibility and adequate payload capacity.

2 Related Work on Transform Domain Audio Watermarking

Dhar et al. [1] used the fast Fourier transform (FFT) to formulate a watermarking scheme for audio signals. Therein, the magnitude spectrum was segmented, and watermark was embedded in the significant peaks of the energy segments in the high energy band. The scheme exhibited desired imperceptibility and robustness.

Ali Al-Haj [3] developed an audio watermarking methodology of semi-blind nature by cascading the DWT and SVD. This resulted in complementary but enhanced levels of imperceptibility, robustness, and payload capacity.

Wu and Wu [4] proposed a blind and robust watermarking algorithm of audio data by exploiting the DWT's multi-resolution nature and DCT's energy concentration capability. Experiments confirmed the technique's imperceptibility, high payload capacity, and robustness to attacks.

Megias et al. [6] performed a tuning analysis of the frequency domain audio watermarking scheme. The compiled results showed intrinsic relations between the data transmission parameters and key watermarking requirements.

Dhar et al. [9] proposed a scheme for watermarking of audio signal using DCT and SVD. The audio cover work was segregated into non-overlapping frames followed by DCT computation on a per frame basis. Experimental results revealed the scheme's effectiveness in terms of blind detection, low error rates, and high robustness.

Attari et al. [11] leveraged the spread spectrum in DWT domain for the development of a blind audio watermarking principle. The watermarks were embedded in the coefficients marked by low frequency, and a psychoacoustic model was proposed for enhancing imperceptibility and robustness, along with Savitsky–Golay filter, for better extraction accuracy.

Liu et al. [12] developed a watermarking algorithm for audio signals using DCT, based on vector quantization (VQ). The middle frequency coefficients vector was quantized in accordance with the watermark information, which also ensured robustness of the scheme.

Panda et al. [13] used empirical mode decomposition (EMD) for watermarking of audio signal. The blind scheme was found to be robust against some of the non-blind methodologies and also offered adaptive threshold during watermark extraction. Further, the watermark was resilient to filtering attacks as the extraction could be done from a non-masked intrinsic mode function (IMF).

3 The Linear Canonical Transform: A Primer

3.1 Continuous Domain Transformation

The one-dimensional continuous linear canonical transform (LCT) [14] of a signal $s(t)$, computed using the real-valued parameter matrix $X = \begin{pmatrix} X_{11} & X_{12} \\ X_{21} & X_{22} \end{pmatrix}$, is given by Eq. 1

$$T_X(\mu) = L^X(s(t)) = \begin{cases} \int_{-\infty}^{\infty} s(t) R_X(t, \mu) \mathrm{d}t, & X_{12} \neq 0 \\ \sqrt{\lambda/\tau}\, e^{\frac{j\left(-\lambda + \frac{\lambda^2 \zeta^2}{\tau^2}\right)}{2\mu^2}} s(\mathrm{d}\mu), & X_{12} = 0 \end{cases} \quad (1)$$

where the parameter matrix elements satisfy $X_{11}X_{22} - X_{12}X_{21} = 1$ and $X_{11} = \frac{\zeta}{\tau}$, $X_{12} = \frac{1}{\tau}$, $X_{21} = -\tau + \frac{\lambda\zeta}{\tau}$ and $X_{22} = \frac{\lambda}{\tau}$, λ, τ and ζ are real and t and μ are continuous time and frequency domain variables, respectively. The LCT transformation kernel, denoted by $R_X(t, \mu)$, is expressed as in Eq. 2

$$R_X(t,\mu) = \sqrt{\frac{\tau}{j2\pi}} e^{j\left(\frac{\lambda\mu^2}{2} - \mu t\tau + \frac{\zeta t^2}{2}\right)} \tag{2}$$

A list of important relations and properties of the LCT along with necessary proofs can be found at [2, 14, 15].

3.2 Discrete Domain Transformation

By setting $\delta_t = \delta_\mu = (N|\tau|)^{-1/2}$, $t = v\delta_t$, $\mu = w\delta_\mu$,, in the equations for discrete LCT as discussed in [14], with $v, w = 0, 1, \ldots, N-1$, the N—size discrete linear canonical transform (DLCT) of $s(v)$ can be expressed as in Eq. 3

$$s_X(w) = \sum_{v=0}^{N-1} s(v) R_X(w, v) \tag{3}$$

with the discrete LCT kernel therein being given by Eq. 4

$$R_X(w, v) = \frac{\sqrt{\tau} e^{-\frac{j\pi}{4}}}{\sqrt{N|\tau|}} \exp\left[\frac{j\pi}{N|\tau|}(\lambda w^2 - 2\tau v w + \zeta v^2)\right] \tag{4}$$

Here, δ_t and δ_μ pertain to the sampling intervals in the time and LCT domains, respectively, N is the number of samples, τ is real, and v and w are discrete time and frequency domain variables, respectively.

4 Proposed Methodology

4.1 Watermark Embedding Stage

a. The DLCT of the audio signal is computed using Eqs. 3 and 4, for a set of LCT matrix parameters, and the coefficients thus obtained are reordered in decreasing (non-monotonic) series. Let the transform coefficients be represented by $K_i = L_i + jM_i$, $|K_i| < |K_{i+1}|$, $i = 1, 2, \ldots, N$.
b. Watermark is a set of K normally distributed complex numbers [16], with zero mean and $\frac{\sigma^2}{2}$ variance and represented as $k_i = l_i + jm_i$, $i = 1, 2, \ldots, K$.
c. If the position of the middle coefficient is denoted by N_{median}, then the DLCT coefficients K_i with position in the range $i = N_{\text{median}} + 1, N_{\text{median}} + 2, \ldots, N_{\text{median}} + K$ are clustered for insertion of the watermark, in adherence with the requirements of robustness and imperceptibility.
d. The watermark is embedded by direct modulation in the following manner

$$K_i^w = K_i + l_i|L_i| + jm_i|M_i|, i = N_{\text{median}} + 1, \ldots, N_{\text{median}} + K$$

e. The DLCT coefficients in the reordered series obtained in a are substituted with the watermarked DLCT coefficients calculated in d, at the designated positions.
f. The coefficients are reordered as per the originally obtained transform series.
g. The inverse DLCT (IDLCT) of the transform coefficients obtained in f is computed using Eqs. 3 and 4, with the inverse of the LCT matrix used in a, to yield the watermarked audio cover work.

4.2 Watermark Detection Stage

The detection parameter η is used as a metric for distinguishing and isolating the originally embedded watermark. It is computed as in Eq. 5

$$\eta = \sum_{j=N_{\text{median}}+1}^{N_{\text{median}}+K} (l_i - jm_i)K_i^{\text{attack}} \quad (5)$$

where K_i^{attack} is the set of DLCT transform coefficients used for embedding the watermark. A dynamic scheme is proposed in [16] which involves the computation of η for a large set of random watermarks, calculation of their mean (say ρ) and standard deviation (say ϕ), and setting threshold as $\omega = \rho + c\phi$, where c is found heuristically.

Based on the foregoing, the watermark detection technique is given below.

a. Computation of the DLCT of the received audio signal using Eqs. 3 and 4, with the same LCT parameter matrix used for watermark embedding.
b. Reordering the DLCT coefficients in decreasing (non-monotonic) order and selection of the watermarked transform coefficients K_i^{attack}.
c. Creation of a large set (e.g., 1000) of complex random watermark sequences, all generated using the technique described in the embedding stage, step b. The originally embedded watermark is included at the median location (i.e., 501st position) in this set.
d. Calculation of the detection parameter η using Eq. 5, for the set of watermarks obtained in c.
e. Calculation of the threshold ω for the detection parameters adaptively.
f. Detection of the originally embedded watermark by comparing the values of the detection parameter with the threshold obtained in e.

5 Experimental Configuration and Simulation Parameters

The proposed technique was implemented on MATLAB®R2018a running on Intel® Core™ i3 7th generation processor clocked at 3.90 GHz. Key parameter settings are given below.

a. The audio signal is recorded on a single channel, for 2 s, at a sampling frequency of 4 kHz with 16-bit encoding.
b. For watermark embedding, the LCT parameter matrix is $X = \begin{pmatrix} \frac{1}{6} & \frac{5}{6} \\ -\frac{17}{30} & \frac{1}{3} \end{pmatrix}$. Equivalently, $(\lambda = 0.2, \tau = 0.6, \zeta = 0.1)$.
c. Standard deviation of the normally distributed watermark sequence is 50.
d. The embedding location of the watermark is selected from the median sample position (i.e., 50th percentile location) to nearly up to 65th percentile sample position.
e. Number of normally distributed watermark sequences is 1000.
f. Constant c in the threshold ω is set as 3, based on experimentations for $c \in [1, 10]$.

6 Results

The proposed method is assessed on the basis of the following figures of merit, given by Eqs. 6, 7, and 8

$$\text{Normalized Cross} - \text{Correlation}(NC) = \frac{\sum_{q=1}^{K} \psi(q) \cdot \psi^*(q)}{\sqrt{\sum_{q=1}^{K} \psi(q)^2} \sqrt{\sum_{q=1}^{K} \psi^*(q)^2}} \tag{6}$$

$$\text{Similarity Index(SIM)} = \frac{\sum_{q=1}^{K} \psi(q) \cdot \psi^*(q)}{\sqrt{\sum_{q=1}^{K} \psi^*(q)^2}} \tag{7}$$

where ψ and ψ^* denote the embedded and extracted of watermarks of length K.

$$\text{Signal to Noise Ratio (SNR)} = 10 \log_{10} \left(\frac{\sum_{q=1}^{N} s(q)^2}{\sum_{q=1}^{N} (s(q) - s^*(q))^2} \right) \tag{8}$$

where s and s^* denote the original and retrieved audio signals of length N.

The values of the above figures of merit for different signal processing and feature distortive attacks of various strengths/degree are given in Table 1.

Table 1 Figures of merit for different strength/degree of attacks

Type of attack	Strength/Degree	Similarity index (SIM)	Normalized cross-correlation (NC)	Signal-to-noise ratio (SNR) dB
AWGN (Strength of noise in dB)	25	0.8781	0.6332	95.3002
	50	0.8628	0.7705	95.8859
	75	0.9313	0.6243	95.9028
	100	0.8659	0.6978	95.5513
LPF (Passband cutoff frequency in Hz)	10	0.5594	0.9936	99.8594
	20	0.8761	0.9919	99.5708
	30	0.9448	0.9672	99.5756
	50	0.8919	0.9915	99.3354
HPF (Passband cutoff frequency in Hz)	100	0.6822	0.6853	96.5952
	150	0.8267	0.7212	97.1797
	200	0.7750	0.7335	97.7530
	300	0.7355	0.6584	97.1428
Decimation (Percentage of samples cropped out)	2	0.7204	0.9973	83.3639
	5	0.8265	0.9977	73.2754
	7	0.5489	0.9983	67.0179
	10	0.9025	0.9990	60.7395
MP3 compression (Sampling rate in KHz)	4	0.7199	0.7280	94.3109
	8	0.7419	0.5350	94.2596
	16	0.9094	0.6000	94.2612
	44.1	0.8221	0.7531	96.1731

Test of Imperceptibility

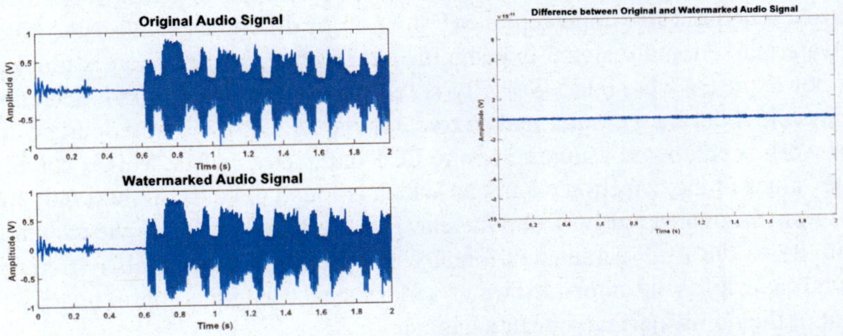

Comparative Plots of Original and Watermarked Audio in Time Domain

Contrariety between the Original and Watermarked Audio (10^{-10}~10^{-11})

Fig. 1 Comparison between original audio cover work and watermarked audio signal

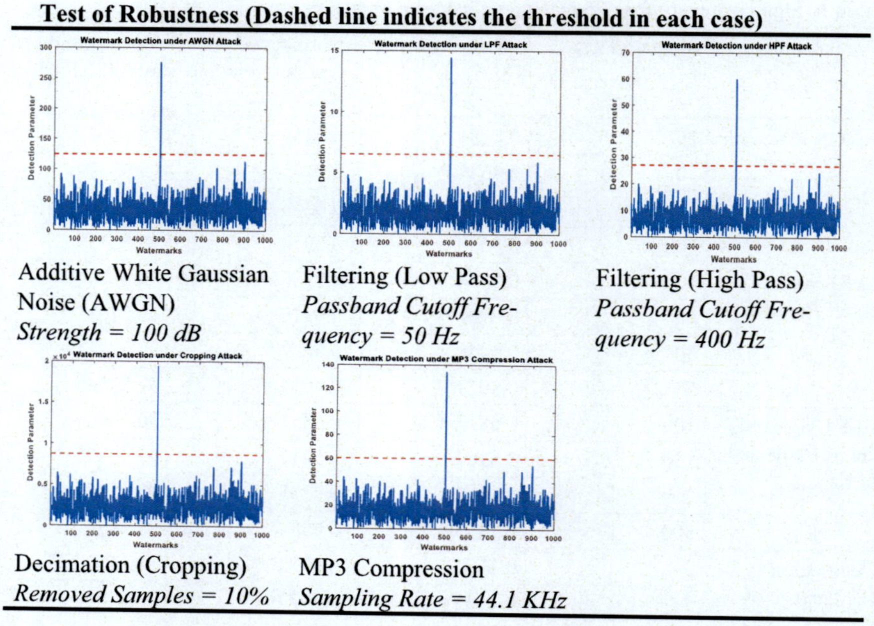

Fig. 2 Type of attacks at their maximum strength/degree *(Italicized)*

7 Conclusion

The scheme's robustness against various signal processing and feature distortive attacks is verified from Table 1 and Fig. 2. The stability is also evinced by the observation that the values of the detection parameters for the random watermarks are much lower than the stipulated threshold in each type of attack (Fig. 2). The proposed scheme is also imperceptible (Fig. 1), as the difference between the original and watermarked audio signal is insignificant. Further, the detection is blind as it does not require cover signal. Security is reinforced as the watermark detection is not possible if during the watermark extraction, the DLCT of the watermarked audio cover work is computed using the wrong LCT parameter matrix. This is confirmed by the values of the detection parameter which is found to be the highest only in the case of the original set of LCT parameters (Fig. 3), thus reaffirming the technique's sensitivity to the LCT parameters. Finally, the proposed scheme also offers good payload capacity as the embedded watermark sequence length is approximately 15% of that of the non-watermarked audio signal.

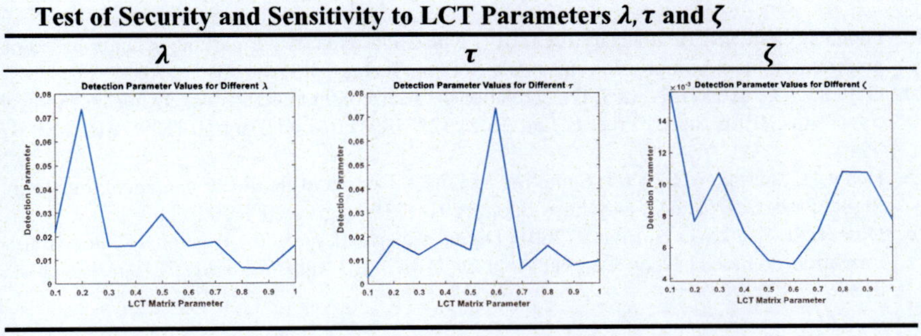

Fig. 3 Detection parameter values for $(\lambda, \tau, \zeta) \in [0.1, 1]$

References

1. Dhar PK (2011) Robust FFT based watermarking scheme for copyright protection of digital audio data. In: Seventh international conference on intelligent information hiding and multimedia signal processing journal. IEEE, China, pp 181–184. https://doi.org/10.1109/IIHMSP.2011.85
2. Xiang Q, Qin K (2009) The linear canonical transform and its application to time-frequency signal analysis. In: 2009 international conference on information engineering and computer science. IEEE, Wuhan, China, pp 1–4. https://doi.org/10.1109/ICIECS.2009.5365117
3. Al-Haj Ali (2014) An imperceptible and robust audio watermarking algorithm. EURASIP J Audio Speech Music Process 2014(1). https://doi.org/10.1186/s13636-014-0037-2
4. Wu Q, Wu M (2018) A novel robust audio watermarking algorithm by modifying the average amplitude in transform domain. Appl Sci 8(5):723. https://doi.org/10.3390/app8050723
5. Global Music Report 2021. International Federation of Phonographic Industry (IFPI). https://www.ifpi.org/wpcontent/uploads/2020/03/GMR2021_STATE_OF_THE_INDUSTRY.pdf. Last accessed 2021/04/29
6. Megías D, Herrera J, Minguillón J (2005) Robust frequency domain audio watermarking: a tuning analysis. In: Cox IJ, Kalker T, Lee HK (eds) Digital watermarking. IWDW 2004. Lecture notes in computer science, vol 3304, pp 244–258. Springer, Heidelberg. https://doi.org/10.1007/978-3-540-31805-7_21
7. Cvejic N, Seppanen T (2007) Digital audio watermarking techniques and technologies: applications and benchmarks, 1st edn. IGI Global, Hershey PA, USA
8. Guang H, Huang J, Shi Y, Goh J, Thing V (2016) Twenty years of digital audio watermarking-a comprehensive review. Signal Process 128:222–242. https://doi.org/10.1016/j.sigpro.2016.04.005
9. Dhar PK, Shimamura T (2011) Blind audio watermarking in transform domain based on singular value decomposition and exponential-log operations. Radioengineering 26(2):552–561. https://doi.org/10.13164/re.2017.0552
10. A time-frequency perspective on audio watermarking. https://arxiv.org/abs/2002.03156. Last accessed 29 Apr 21
11. Attari A, Shirazi A (2019) Robust and transparent audio watermarking based on spread spectrum in wavelet domain. In: 2019 international joint conference on electrical engineering and information technology (JEEIT). IEEE, Amman, Jordan, pp 366–370. https://doi.org/10.1109/JEEIT.2019.8717415
12. Liu J, Lu Z, Pan J (2008) A robust audio watermarking algorithm based on DCT and vector quantization. In: 2008 eighth international conference on intelligent systems design and applications. IEEE, Kaohsuing, Taiwan, pp. 541–544. https://doi.org/10.1109/ISDA.2008.31

13. Panda J, Gera KR, Bhattacharya A (2014) Non-blind audio watermarking scheme based on empirical mode decomposition. Int J Adv Sci Eng Technol 3(3):38–43
14. Oktem FS, Ozaktas HM (2009) Exact relation between continuous and discrete linear canonical transforms. IEEE Signal Process Lett 16(8):727–730. https://doi.org/10.1109/LSP.2009.2023940
15. Healy JJ, Sheridian JT (2009) Sampling and discretization of the linear canonical transform. Signal Process 89(4):641–648. https://doi.org/10.1016/j.sigpro.2008.10.011
16. Djurovic I, Stankovic H, Pitas I (2001) Digital watermarking in the fractional Fourier transformation domain. J Netw Comput Appl 24(2):167–173. https://doi.org/10.1006/jnca.2000.0128

Secure Cloud Services Using Quantum-Blockchain Technology

Surya Bhushan Kumar, Ranjan Kumar Mandal, Kuntal Mukherjee, and Rajiv Kumar Dwivedi

Abstract Data privacy and data security are basic agenda prior to outsource data to cloud environment. It is observed that cloud technology is centralized so any attacker can easily tamper with data store in it. On the other hand, blockchain technology is decentralized. Cloud provides on demand computing, storage of data, etc. Blockchain technology provides a very strong security mechanism. In this manuscript, blockchain technology has been used to provide data security at cloud environments. Here, a novel framework of cloud computing has been proposed along with algorithms based on quantum mechanics. The quantum superimposition principle has been used to detect data tampering. Experimental results have been proposed to lay bare the effectiveness of proposed approach.

Keywords Blockchain and Quantum in Cloud · Blockchain in Cloud Security · Quantum Computing in Cloud · DDoS Attack in Cloud

1 Introduction

Authentication and authorization are the two basic security features of cloud data security. For these features, a Cloud service provider and their users believe in trusted third party. However, data leakage and data tampering often happened in recent

S. B. Kumar
University Department of Computer Applications, Vinoba Bhave University, Hazaribag, Jharkhand 825301, India

R. K. Mandal
Department of Computer Science and Engineering, Ranchi University, Ranchi, Jharkhand 834001, India

K. Mukherjee (✉)
Department of Computer Science and Engineering, BIT Mesra, Ranchi, Jharkhand 834001, India
e-mail: kmukherjee@bitmesra.ac.in

R. K. Dwivedi
University Department of Mathematics, Vinoba Bhave University, Hazaribag, Jharkhand 825301, India

J. M. R. S. Tavares et al. (eds.), *Cyber Intelligence and Information Retrieval*, Lecture Notes in Networks and Systems 291,
https://doi.org/10.1007/978-981-16-4284-5_16

years by some third party for economy benefit. Another cutting edge technology is blockchain technology. The World Economic Forum in [1] reported that investment in blockchain is growing day by day. In recent times, more than 20 countries are involved into it [1]. Nowadays blockchain, distributed ledger-related projects blockchain-based security scheme, etc., are major interest area by different banks across the globe. According to the worldwide financial trading, the marketplace for security-based-blockchain technology has reached up to USD 20 billion in 2020 [1]. Hence, the latest trend of adopting data security in cloud environment using blockchain technology is observed [2].

Blockchain (Block chain) is an open, shared, decentralized to distributed databases ledger. A blockchain comprises a series of chronological blocks that hold an entire list of all legitimate transactions based upon peer-to-peer (point-to-point) information transfer in the network technology and disabled arbitrary tampering [3–5]. Blockchain functions the same way as cloud data storage but there is no third-party provider is involved [3, 4]. Additionally, a user builds their own database and stores the data in blocks on distributed and shared storage. Each blockchain data block consists of a block header and a block body. In the block, detailed transaction data and a transaction counter reside in a block body. Timestamp, Merkle root hash, parent hash value and random number are stored by a block header. The maximum number of transactions depends on block size and size of each transaction. The information is recorded in a Merkle tree [6, 7]. The major disadvantage of cloud technology is security and its downtimes, network dependency, limited control and no legal liability for vendors [6–8]. These disadvantages can be overcome by adopting the decentralized and shared blockchain technology in the cloud environment. The blockchain technology infrastructure in [2] is based on seven-layer model. It comprises of data layer, network layer, consensus layer, incentive layer, contract layer and application layer, as shown in Fig. 1 [2].

In research paper [2], the authors have described the blockchain-based cloud data integrity verification scheme. In this security framework, virtual machine proxy model is designed to check information change under smart contract. This data trust verification happens through the unique hash value generation by the Merkel hash tree. Trustworthy data storage in cloud environment, verification and monitoring are carried out by virtual machine agent mechanism. The framework generates a warning message for data modification or data tampering; in addition, a "block-and-response." Furthermore, authors in [3] have mentioned the wide adoption of blockchain technology by different sectors across the globe due to its high demand. Blockchain is distributed public ledger-based on the concept of cryptography, which can provide security of data in cloud environment using pay-per-last-N shares (PPLNS) scheme [3].

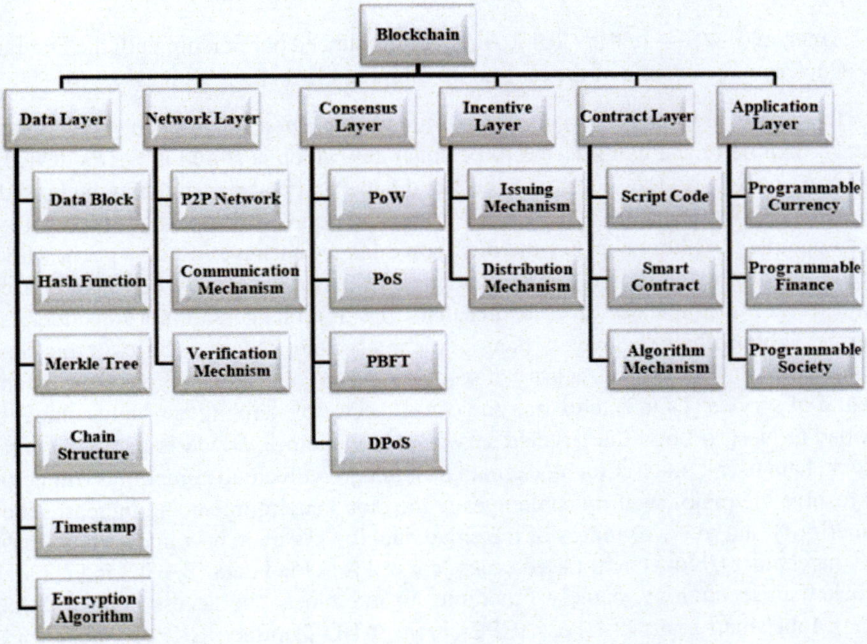

Fig. 1 Blockchain infrastructure [2]

2 State-Of-The-Art

It is observed that distributed ledger technology (DLT), well known as blockchain has evolved as a consumer-focused technology (form Bitcoin) [2]. The prime agenda of cloud storage provider is to ensure the security of user's data [3]. Furthermore, it is observed that blockchain technology may play a significant role in order to provide security of data in cloud environment [3, 4]. That is why, the giant corporations like IBM (Hyperledger Fabric), Amazon (Digital Currency Group), Microsoft (partnership with Ethereum) and JP Morgan (Quorum) are involved in blockchain as a service (BaaS) for cloud environment to enhance its security architecture [4]. Nowadays, blockchain as a service (BaaS) is the latest trend of providing security services to the users of cloud computing. The authors in [5] have pointed out the major advantages of Blockchain as a service (BaaS) in cloud environment. Some giant corporations, who have adopted block chain technology in order to provide security of data in cloud environment are [6]:

- Microsoft (Azure)—It is a cloud-based service architecture. In November 2016, Microsoft publicized a joint venture with ConsenSys to provide Blockchain Ethereum as a Service (EBaaS) in the Azure cloud environment [6].
- IBM (BlueMix)—In February 2017, IBM's blockchain as a service using Hyperledger [6].

- Amazon (AWS)—In May 2017, Amazon announced partnership with the Digital Currency Group, one of the biggest investors in blockchain businesses [6].

To tamper blockchain (for example, Bitcoin) [6], first, an attacker does reuse attack then he or she gains access to personal key which is stored in peer's device. If he or she does so the security of the cloud data can be at risk and may be leaked or damaged. On the other hand, one of the strong points of Blockchain is that it is very complex to alter or falsify the distributed ledger because so many peers work on decentralized and shared transaction ledgers [7]. So, it is almost not possible for any attacker to gain access over greater than 50% of all ledgers. Some considerable points for blockchain security in [6] are settlement of blockchain, security of transaction, security of wallet and security of software. Also, some attackers use distributed denial of service (DDoS) to disrupt the cloud service by sending enormous amount of bad request to flood the targeted servers [6]. Therefore, DDoS is dangerous for blockchain users. Integrating blockchain technology with cloud computing will help to resolve the major security challenges in the cloud environment and increases the superiority and trustworthiness of the cloud data [6]. Again, a new interesting security mechanism, blockchain-based access control for cloud data (BACC) in [7] uses blockchain technology, namely Ethereum, to implement the access control tasks using four smart contracts (i.e., ACPContract, AUD-Contract, LKP-Contract, and CLP-Contract). In this mechanism, the authors used AES-128 encryption algorithm to protect confidentiality. In addition to secure cloud storage audit framework in [8], the authors have demonstrated that an Ethereum-based blockchain may play significant role. In order to protect cloud data, all useful information of the user is stored in the form of a block in blockchain technology [9]. The malicious or fake information of the cloud user is examined by a series of block to guarantee the security of auditing data source. To audit cloud data, in this framework, two Merkle trees are also used to save user behavior records and metadata information in the block body [9]. To verify block information, an error could be generated and the block would be rejected if the block data is tampered. Moreover, the authors in [9] have introduced the concept of blockchain-as-a-cloud, which is a "thin cloud" and is more suited to carry out the new types of "thin" programs, specifically called "smart contracts". In order to compare with a traditional cloud computing infrastructure, smart contracts are the business logic such as cloud applications (for example, Microsoft Azure or Amazon Web Services) on blockchain virtual machines (for example, on Ethereum) but actually runs on physical servers [9, 10]. The authors in [10] have proposed a blockchain-based access control framework with privacy protection in cloud environment called AuthPrivacyChain. In this method, identity is referred as the account address of the node in blockchain. Simultaneously, in blockchain, access control permission of cloud data is encrypted and stored [10]. This feature not only enhances the security of data in cloud environment but also takes care of authorized privacy of the user [10, 11].

2.1 Research Gap

Thus, a block in blockchain technology comprises of data, digital signature (called hash value) and hash value of previous block. From the above literature review, it is observed that there is a significant lack in research of strong security mechanism for hash value based on the principles of quantum mechanism. Hence, in this endeavor, the superposition state of quantum mechanism has been used to detect the tampering with data in block of block chain technology. The quantum superposition state can easily identify the tampering with data in block. It is a superior technique as comrade to traditional approach.

3 Proposed Framework

A specific security mechanism for data storage in cloud environment has been proposed in this endeavor, which is shown in Fig. 2.

The proposed security mechanism provides security of outsourced data in cloud environment using blockchain technology. Apart from the existing blockchain technology, the proposed technology has used quantum phenomena. The data block in proposed framework 2 comprises of data, hash function and previous hash function based on quantum phenomena. The block is shown in Fig. 3.

If anyone tampers with the data then the superposition state described in the hash function would be collapsed and the state would be either 0 (denoted by black) or 1(denoted by white). Based on the quantum mechanics principle, the proposed technique has been discussed below.

Fig. 2 Proposed framework

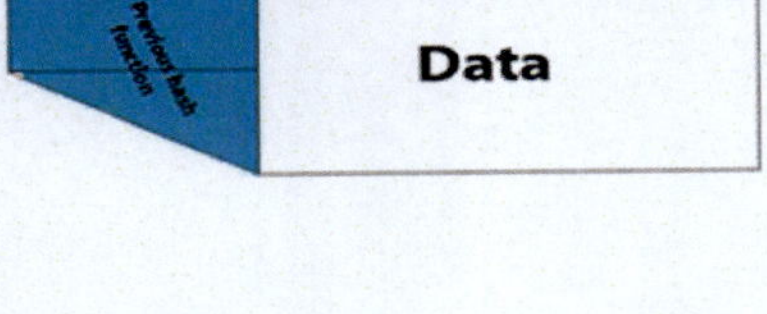

Fig. 3 Block

Table 1 User Details

Form	Unicode equivalent
Name	161
Aadhar card no	250
Address	340
Gender	21

3.1 *Significance of the Proposed Approach*

The proposed approach in this endeavor can contain a large number of information due to the superposition behavior of quantum mechanics. Each and every superposition state can map each and every point on block sphere and hence can accommodate huge data. Furthermore, if anyhow data is tampered with, then the superposition state will collapse. Hence, this security mechanism makes this approach unique as compared to traditional approach.

3.2 *Proposed Algorithm*

For the sake of simplicity, only four fields are considered, say, Name, Aadhar-Card Number, Address and Gender. Let the Unicode values for a specific example, are 161, 250, 340, and 21, as shown in Table 1.

Now, let us propose a certain scheme as follows:

Rule 1: Consider a complex number, say, Z_1 as:

$Z_1 = (-1)^x$ (Unicode equivalent of Name) $+ (-1)^x$ (Unicode Equivalent of Aadhar No) i.

where $i = \sqrt{-1}$.

Rule 2: Consider a complex number, say, Z_2 as:

$Z_2 = (-1)^x$ (Unicode equivalent of Address) $+ (-1)^x$ (Unicode equivalent of Gender) i.

where $i = \sqrt{-1}$.

Rule 3: If Unicode value is even, then $x = 2$, otherwise $x = 1$.

For the above example, shown in Table 1, the proposed rules can be written as:

$$Z_1 = -161 + 250i$$
$$Z_2 = 340\text{--}21i$$

Rule 4: Obtain $K1 = \sqrt{|Z_1|^2 + |Z_2|^2}$.

In above example, $K1 = \sqrt{}(|-161 + 250\ i|^2 + |340 - 21\ i|^2)$.

Rule 5: Obtain probability amplitude *a*, *b* as:

$$a = |Z_1|^2/K_1^2$$
$$b = |Z_2|^2/K_1^2$$

For above example, it is written as $a = (|-161 + 250i|^2)/{K_1}^2$ and $b = (|340 - 21i|^2)/{K_1}^2$.

Rule 6: Let the above rules are implemented for a particle whose Eigen values are $|0\rangle$ and $|1\rangle$. Here, Eigen values are described in terms of energy. Therefore, general state can be described as:

$$|\psi\rangle = a|0\rangle + b|1\rangle$$

This superposition state is considered as hash value of the proposed block. Based on the above discussions, the proposed algorithms of creating block of blockchain and checking its integrity are shown in Figs. 4 and 5, respectively.

Based on the above algorithms, the proposed block is shown in Fig. 6.

4 Experimental Setup and Result

This manuscript has used blockchain for cloud security services. Here, implementation of proposed algorithms has been done using Python programming language. Flask library has been for creating the Web and for displaying the blockchain. JSON has been used to store the data in proposed blockchain. Python library "haslib" has been used for calculating the hash in order to add digital fingerprints to the blocks.

Hash value ($|\psi\rangle$) has been implemented using quantum computing and its corresponding proposed quantum circuit is shown in Fig. 7.

5 Conclusion

To resolve security issues, this manuscript introduces a novel security framework with proposed algorithm using quantum-blockchain in cloud computing. Once the data is secured and immutable using quantum-blockchain framework, it will not be maliciously tampered. An experimental result has been shown to lay bare the effectiveness of the proposed approach.

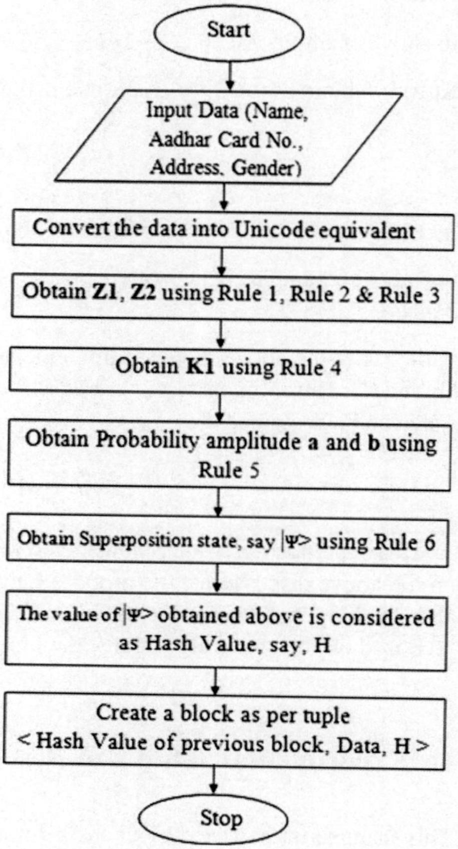

Fig. 4 Proposed block of blockchain

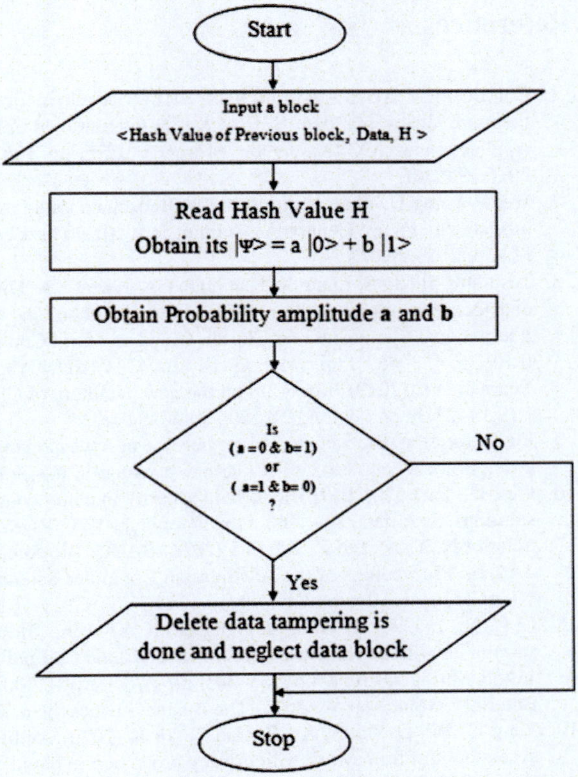

Fig. 5 Proposed flowchart for checking integrity of data in block

Fig. 6 Block of blockchain

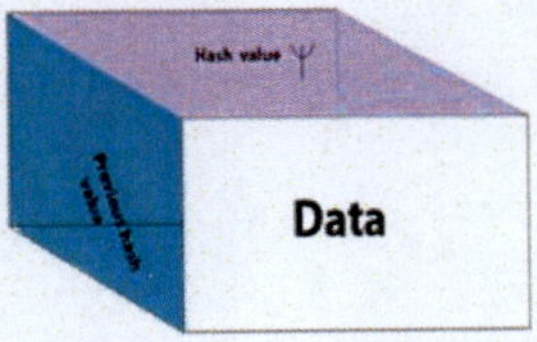

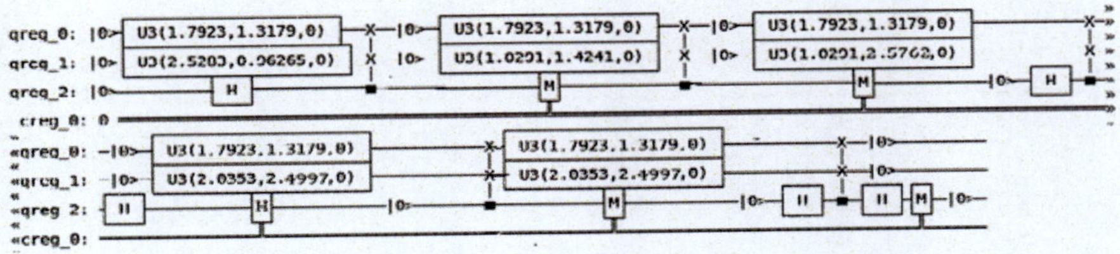

Fig. 7 Quantum circuit of proposed algorithm

References

1. Conoscenti M, Vetrò A, Martin JCD (2016) Blockchain for the Internet of Things: a systematic literature review. In: 2016 IEEE/ACS 13th International Conference of Computer Systems and Applications (AICCSA), Agadir, Morocco. IEEE, pp. 1–6. https://doi.org/10.1109/AICCSA.2016.7945805
2. Wei P, Wang D, Zhao Y et al (2020) Blockchain data-based cloud data integrity protection mechanism. Future Generation Comput Syst 102:902–911 (Elsevier). https://doi.org/10.1016/j.future.2019.09.028
3. Tosh DK, Shetty S, Liang X, Kamhoua CA, Kwiat KA, Njilla L (2017) Security implications of Blockchain cloud with analysis of block withholding attack. In: 2017 17th IEEE/ACM International Symposium on Cluster, Cloud and Grid Computing (CCGRID), Madrid, Spain, IEEE, pp 458–467. https://doi.org/10.1109/CCGRID.2017.111
4. Cearnau D-C (2019) Block-Cloud: the new paradigm of Cloud Computing. Econ Informatics J 19:14–22. https://doi.org/10.12948/ei2019.01.02
5. Ed Featherston. Blockchain as a service: the new weapon in the cloud wars. https://dzone.com/articles/blockchain-asa-service-the-new-weapon-in-the-clou. Last accessed 2021/1/21
6. Park JH, Park JH (2017) Blockchain security in cloud computing: use cases, challenges, and solutions. Symmetry 9(8):164. https://doi.org/10.3390/sym9080164
7. Sohrabi N, Yi X, Tari Z, Khalil I (2020) BACC: Blockchain-based access control for cloud data. In: Proceedings of the Australasian Computer Science Week (ACSW 2020), USA, vol 10. ACM, pp 1–10 https://doi.org/10.1145/3373017.3373027
8. Li C, Hu J, Zhou K, Wang Y, Deng H (2018) Using Blockchain for data auditing in cloud storage. In: 4th International Conference on Cloud Computing and Security, ICCCS 2018, vol 11065. Springer, China, pp. 335–345 https://doi.org/10.1007/978-3-030-00012-7_31
9. Buterin V, Mougayar W (2016) The Business Blockchain. Wiley
10. Yang C, Tan L, Shi N, Xu B, Cao Y, Yu K (2020) AuthPrivacyChain: a blockchain-based access control framework with privacy protection in cloud. In: IEEE Access, vol 8. IEEE, pp 70604–70615 https://doi.org/10.1109/ACCESS.2020.2985762
11. https://en.wikipedia.org/wiki/Blockchain. Last accessed 2021/2/11

Information Retrieval

Multi-label Classification: Detailed Analysis

Mathur Swati and Mathur Pratistha

Abstract Multi-label classification is a classification problem wherein output domain multiple labels are assigned to each instance. Because of this reason importance of multi-label classification has increased with time. It has applications in almost all fields. To summarize the research work under multi-label classification an attempt has been made to put forward the state-of-the-art of MLC techniques with their benefits and drawbacks. Evaluation of MLC is more complicated than single-label classification, thus based on the study all evaluation metrics are summarized in one place. Extending it to outline various online tools and resources available for research in MLC. As a conclusion, various challenges and scope present in MLC are discussed so that further research direction can be explored.

Keywords Multi label · Metrics · Label · Correlation · Imbalance

1 Introduction

Data classification is a technique to classify the instances into one of the predetermined set of labels. Our aim is to predict the label for the new instance. So, by designing a general model we can achieve this goal. Model is designed by following two steps: First phase is a training phase in which a model is designed by making the model learn using the training set data. Various algorithms are proposed to design the model. Second phase is a testing phase. In this phase designed model is used to predict class labels and also compare this value with the target value present in the data set.

M. Swati (✉) · M. Pratistha
Manipal University Jaipur, Jaipur, India

M. Pratistha
e-mail: pratistha.mathur@jaipur.manipal.edu

M. Swati
CMR Institute of Technology, Banglore, India

J. M. R. S. Tavares et al. (eds.), *Cyber Intelligence and Information Retrieval*,
Lecture Notes in Networks and Systems 291,
https://doi.org/10.1007/978-981-16-4284-5_17

Table 1 Binary classification example

Example	Output-1	Output-2
Citizenship	Yes	No
Sentiment	Positive	Negative

Various classification algorithms are designed depending on the various parameters like how many outputs is there, nature of data, etc.

(i) Binary Classification: In this technique, class labels can take only two values. It is in the form of 0/1, True/False, Positive/Negative, Yes/No. Each instance of the dataset can only hold a single value from the available set. For this, traditional classification techniques can be applied. Table 1 highlights few applications of binary classification.

(ii) Multi-class Classification: It is an extension of the previous approach. Here, class-label set contains multiple values. Single value is allocated to the testing instance among all possible available values. Table 2 presents few of the applications of multi-class classification.

(iii) Multi-label Classification: In this technique set of relevant labels are assigned to the instance set. In other words, multiple labels can be allocated to a specific testing data set. Label values are hand-pick from label space. It has achieved paramount significance in the research domain and has application in diversified fields like sentiment analysis, Fake news identification, etc. Table 3 presents few of the applications.

Now, we can define Multi-label classification as: "Let U represents n-dimensional input space of numerical or categorical features. Let $L = \{l_1, l_2 \dots l_p\}$ is output space having p number of labels where $p > 1$. Then set (u, Z) where $u = (u_1, \dots, u_n) \in U$ is an n-dimensional instance which has a set of labels associated $Z \subseteq L$. Label associations can be also represented as a p dimensional binary vector $z = (z_1, z_2, \dots, z_p) = \{0,1\}^p$, where each element is 1 if the label is relevant and 0 otherwise "[1].

Table 2 Multi-class classification

Example	Output set
Types of soil	Sandy, Loam, Chalk, Clay
Crops	Food, Cash, Horticulture, Plantation
Mood disorder	Bipolar I, Bipolar II, Cyclothymic, Major Depressive

Table 3 Multi-label classification

Example	Output set
Movie genre	Comedy, Romance
Fashion shopping site	Multiple red dress options
Object detection in an image	Cat, Car, lady, truck

Our work has been arranged in the following sections: Sect. 1 deal with introduction, Sect. 2 explains the various MLC domains and techniques proposed, Sect. 3 about the various evaluation parameters for MLC. Section 4 highlights the extensive comparison among all methods. Section 5 discussed about the various available tools. Section 6 is focused about the various challenges for MLC. In the last section, the conclusion is discussed.

2 Multi-label Classification Techniques

We can assort multi-label classification by two ways:

First way, three levels are defined: In first order approach, data set is mutually exclusive, thus ignore dependencies among labels. Second order approach focuses on pair-wise association among labels. Higher order approach is more complicated because it tries to establish association among all labels.

Second way, various techniques are designed for multi-label classification. All techniques are proposing different methods for different set of application or data. These techniques are: Problem Transformation, Algorithm adaption, Ensemble based, Correlation based, Semi-Supervised Learning based, Online Multi label, Input Encoding based, Generative Modeling based, Large Scale Data processing, Hierarchical Based, Label Transformation based methods, shared subspace and Label Subset Selection based. Detailed diagram of all the methods proposed by different techniques is summarized in Fig. 1 which is presented in the form of mind map of multi-label classification techniques.

a. Problem Transformation: Herein, MLC problem is transformed into single-label classification. Then by applying traditional classification method these problems can be solved. After that, obtained result is further transferred into multi-label format.
b. Algorithm adaption: It modifies the single-label classifier to make multi-label classification predictions.
c. Ensemble based method: It trains multiple hypotheses all together to solve the problem. Thus it gives more precise and accurate results.
d. Correlation-based methods: This method looks for the probability of labels occurring together.
e. Semi-Supervised Learning based methods: Labeling of data is very time consuming and expensive also. Semi supervised provides a way to exploits unlabeled datasets to make prediction in MLC.
f. Online Multi-label methods: When the nature of data is in the form of streams, then complexity is a major concern. So for such application, this method is suitable.
g. Input Encoding based methods: To improve the efficiency, ambiguity in labels in the input domain needs to be control rather than output domain.

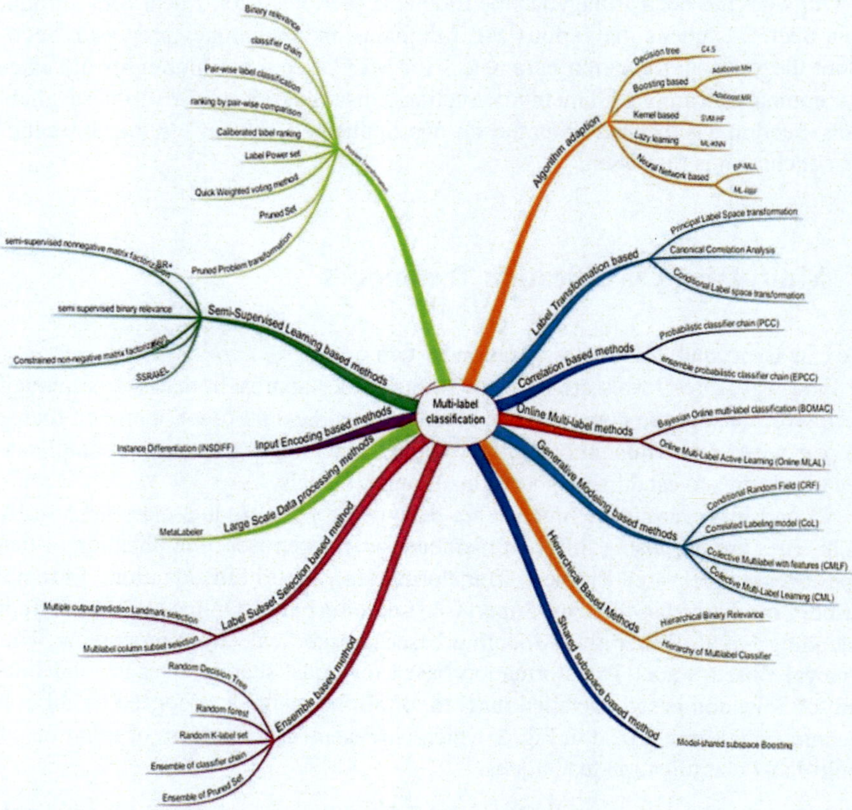

Fig. 1 Mind map of multi-label classification techniques

h. Generative Modeling based methods: General assumption is all class labels are independent. This technique offers a framework to leverage the dependencies between all classes.
i. Large Scale Data processing methods: When you have massive online content, then this method is designed to handle such applications.
j. Hierarchical Based Methods: This type of method is suitable for data where can create sub-classes and create hierarchical form.
k. Label subset selection based method: As a count of labels rises progressively, challenge of influencing the outcome will increase. Thus, only label subset is trained in place of the entire label set.
l. Shared subspace based method: To control the data redundancy and to look after for label correlation, all labels share a common subspace.
m. Label Transformation based methods: This method also minimizes the label set size.

3 Evaluation Metrics and Statistics

Evaluation of MLC is more complicated which can't be performed by traditional methods. There are three categories for evaluation: Bipartition, ranking, and label based metrics. [2]

3.1 Bipartition Based Metrics

Further two categories are formed which are discussed below

(i) Example based measures: It is measured case by case.
(a) Hamming Loss: "It is the fraction of wrong labels to the total number of labels."

$$HL = \frac{\sum_{1}^{m} |Uj \Delta Vj|}{ml} \tag{1}$$

(b) Accuracy: "It is the ratio of the number of instances classified correctly to the total number of instances."

$$\text{Accuracy} = \frac{1}{m} \sum\nolimits_{j=1}^{M} \frac{|Uj \cap Vj|}{|Uj \cup Vj|} \tag{2}$$

(c) Precision: "It is the fraction of positive prediction to the actual correct number."

$$\text{Precision} = \frac{1}{m} \sum_{j=1}^{m} \frac{|Uj \cap Vj|}{Vj} \tag{3}$$

(d) Recall: Recall is "the fraction of actual positive prediction to the total number of retrieved instances."

$$\text{Recall} = \frac{1}{m} \sum_{j=1}^{m} \frac{|Uj \cap Vj|}{|Uj|} \tag{4}$$

(e) F1-score(or F-measure):

$$F - \text{measure} = \frac{2 * \text{Precision} * \text{recall}}{\text{precision} + \text{recall}} \tag{5}$$

(f) Subset accuracy (Exact match or classification accuracy): "It indicates the percentage of samples that have all their labels classified correctly."

$$\text{subsetaccuracy} = \frac{1}{M}\sum\frac{|Uj \cap Vj|}{|Uj| + |Vj|} \tag{6}$$

(ii) Label based measures: In this measure, each label is to be calculated differently. "Let we have l number of labels and we have true-positive (tp_i), true-negative (tn_i), false-positive (fp_i) and false-negative (fn_i)."

a. Micro-averaged precision:

$$\text{MIP} = \frac{\sum_{i=1}^{l} tpi}{\sum_{i=1}^{l} tpi + \sum_{i=1}^{l} fpi} \tag{7}$$

b. Macro-averaged precision:

$$\text{MAP} = \frac{1}{l}\sum_{i=1}^{l}\frac{tpi}{tpi + fpi} \tag{8}$$

c. Micro-averaged recall:

$$\text{MIR} = \frac{\sum_{i=1}^{l} tpi}{\sum_{i=1}^{l} tpi + \sum_{i=1}^{l} fni} \tag{9}$$

d. Macro-averaged recall:

$$\text{MAR} = \frac{1}{l}\sum_{i=1}^{l}\frac{tpi}{tpi + fni} \tag{10}$$

e. Micro-averaged F-measure:

$$\text{MIF} = \frac{2^{*}\text{MIR}^{*}\text{MIP}}{\text{MIR} + \text{MIP}} \tag{11}$$

f. Macro-averaged F-measure:

$$\text{MAF} = \frac{1}{l}\sum_{i=1}^{l}\frac{2 * tpi}{2 * tpi + fpi + fni} \tag{12}$$

3.2 *Ranking-Based Metrics*

a. Ranking Loss: "It imposes penalty when a pair of label is incorrectly classified."
b. One-Error: "It assesses the number of times top ranked labels are not available in the label set."

c. Average Precision: "It calculates the proportion of labels ranked ahead of a certain relevant label."
d. Coverage: "It provides us the number of labels to be incorporated to consider all relevant labels."

3.3 *Label Based Metrics*

Hierarchical Loss: It ignores the subtree which is rooted at a wrong predicted node.

There are three measure statistical approaches to analyze the statistical properties of labels in the dataset.

a. Label density: "It is calculated by taking ratio of a number of labels per sample to the total number of labels."

$$LD = \frac{1}{N}\sum_{j=1}^{N}\frac{|Uj|}{l} \tag{13}$$

b. Label cardinality: "It is calculated by taking the average number of labels per instance in the set."

$$LC = \frac{1}{N}\sum_{j=1}^{N}Uj \tag{14}$$

c. Distinct Label sets: "It gives the count of distinct label sets in the dataset."

4 Comparative Analysis of MLC Algorithms

After analyzing all the methods presented in Fig. 1, all the methods proposed under different techniques have their own benefits and some drawbacks. Below in Table 4, detailed comparison is performed in all the methods which are as follows:

5 Progress of Multi-Label Classification Techniques

Over the years various multi-label classification methods were introduced by various researchers. Each technique is addressing various challenges and handling different types of datasets. The progress of multi-label classification techniques can be seen in the following Table 5. The seed laid for multi-label classification is in year 2000. Progressively every year researchers are coming up with new techniques. With the

Table 4 Comparative analysis of multi-label classification methods

Methods	Benefits	Drawbacks
Binary relevance [3]	• First-order approach • Simplest method • Running time is in linear order • Takes less memory space • Quickest method • It handles unlabeled data	• Relationship between class label is not considered • Problem of class-imbalance
Classifier chain [4, 5]	• High-order approach • It considers the relationship between class labels • Support all base classifier • Consumes low memory	• Comparable to BR running time of CC is insignificantly high • For unlabeled dataset,this technique can't be used • Suffers from overfitting of classification model • With different arrangement we will get different result • It can't be parallelized
Label Power set [6, 6]	• It considers the relationship between class labels	• High computational complexity • Suffers from overfitting of training data
Pair-wise label classification [8]	• Simple method	• Running time is Quadratic • It does not handle labels overlapping
ranking by pair-wise comparison [9]	• Flexible • It is completely in line with preference modeling	• Require large Storage space • Takes more time in prediction
Calibrated label ranking [10]	• Handles ranking also • Generalization is possible	• Computation is expensive • For unlabeled dataset, this technique can't be used
Pruned Set [6, 6]	• Fast execution • During classification it can handle new label set • It handles irregular labeling • Label correlation is considered	• High complexity • Label combinations that are available in the training set can only be predicted • Mapping to lower dimensional label space is not clear
Quick Weighted voting method [8]	• In comparison with pair-wise approach, prediction performance has improved	• Only quadratic numbers of classifiers are stored
ML-C4.5 [12]	• Easy learnability • This technique helps to select the best attribute among all to generate the tree	• Relationship between class label is not considered • For unlabeled dataset, this technique can't be used

(continued)

Table 4 (continued)

Methods	Benefits	Drawbacks
ML-KNN [13]	• Performance has improved	• For unlabeled dataset, we cannot utilize this technique • Not suitable for high dimensionality data
Adaboost.MH [14]	• Finest accuracy • Reduced hamming loss	• For unlabeled dataset, we cannot utilize this technique • Performance has reduced due to generalization
Adaboost.MR [14]	• Improved accuracy • Reduced ranking loss	• For unlabeled dataset, we cannot utilize this technique • Performance has reduced due to generalization
BP-MLL [15]	• Better ranking loss • Time complexity of prediction is slight	• Computational complexity in training phase is high • For unlabeled dataset, this technique can't be used
ML-RBF [16]	• It addresses correlation between labels	• It is not a fast learner
SVM-HF [17]	• Correlation among label is estimated • Better accuracy	• With unlabeled data accuracy gets reduced
Random K-label set (RAkEL) [18]	• Computationally simple • More capable in prediction • Label correlation is taken into consideration	• High running time complexity • For unlabeled dataset, this technique can't be used • For specific parameters this method generates good result
Random decision tree [19]	• Number of class label doesn't affect the training complexity • It guarantees improved performance	• Relationship between class label is not considered • For unlabeled dataset, this technique can't be used
Ensemble of pruned set [20]	• Scalable • Prediction performance has enhanced significantly • Allow parallelism	• For unlabeled dataset, this technique can't be used
Ensemble of classifier chain [5]	• Overfitting problem of classifier chain is resolved	• For unlabeled dataset, this technique can't be used • Base classifiers are learned using random subsets of data with random label ordering By averaging over the base model final conclusion is drawn
Random forest [21]	• No problem of overfitting	• Interpretation is complicated

(continued)

Table 4 (continued)

Methods	Benefits	Drawbacks
Hierarchical binary relevance [22]	• Handles dependencies among label	• For unlabeled dataset, this technique can't be used
Hierarchy of multi-label classifier (HOMER) [23]	• Handles dependencies among label • Reduced operational cost	• For unlabeled dataset, this technique can't be used
Online multi-label active learning (Online MLAL) [24]	• Multi-label classification methods can be applied	• Relationship between class label is not considered • For unlabeled dataset, this technique can't be used
Bayesian online multi-label classification (BOMAC) [25]	• Suitable for large scale online data	• For unlabeled dataset, this technique can't be used
Probabilistic classifier chain [26]	• Optimal prediction • It does not propagate the inference error down the chain	• Computational complexity at prediction time is intractable
MetaLabeler [27]	• For large dataset this method suitable • No cross-validation is required	• Less flexible • For unlabeled dataset, this technique can't be used
Instance differentiation [28]	• Relationship between input and output ambiguities is exploited	• Space and time complexity has increased
Model-shared subspace boosting (MSSBoost) [29]	Information redundancy during the learning process is reduced	• Generalization can be done • For unlabeled dataset, this technique can't be used
Correlated Labeling model[30]	Achieved best F-score performance Successfully established the correlation between classes	The inference process is expensive for large number of labels
Conditional random field [31]	• Capture the relations among the class labels Directly models the conditional joint distribution	• Computationally very demanding
Collective multi-label learning [31]	• Label correlation is taken into account	• Restricted to limited number of labels • For unlabeled dataset,this technique can't be used
Collective multi label with features [31]	• Reduced error in subset accuracy, macro, and micro averages It considers the relationship between class labels	• Restricted to limited number of labels • For unlabeled dataset,this technique can't be used

(continued)

Table 4 (continued)

Methods	Benefits	Drawbacks
Constrained nonnegative matrix factorization [32]	• Input documents representation is very effective • This method is suitable for large scale datasets	Parameter selection is vital
Semi-supervised nonnegative matrix factorization [33]	It can extract more discriminative features	Complexity is very high
BR+	It handles unlabeled data	• For labeled and unlabeled data separate feature space is allotted • This method is not efficient for high label space
SSLP [34]	It handles unlabeled data	• Selection of unlabeled data is very costly because of randomness • Relationship between class label is not considered
SSRAkEL [34]	It handles unlabeled data	• Selection of unlabeled data is very costly because of randomness • Relationship between class label is not considered
Multiple output prediction landmark selection [35]	• Suitable for large scale of labels	• More encoding time • Expensive optimization problem
Multi-label column subset selection problem	More adaptive to change in label as compared to other label selection method	• Slow method
Principal label space transformation [36]	• Fast method • Improved space complexity	• For unlabeled dataset,this technique can't be used
Conditional label space transformation [37]	• Improved space complexity	• It is slow as compared to PLST
Canonical correlation analysis [38]	• Improved space complexity	• For unlabeled dataset,this technique can't be used

past few years scope of multi-label classification has amplified as multiple domains are working and solving their problems with Multi-label classification.

6 Challenges

(i) High dimensionality of label space: This is the most challenging issue with MLC. As the number of label increases, dimension complexity will get

Table 5 Year wise progress of various multi-label classification techniques

Year	Algorithm proposed
2000	Boosting method-adaboost.MH, Adaboost.MR [14]
2004	SVM-HF [17]
2005	Collective multi label with features [31], Collective multi-label learning [31]
2006	BP-MLL[15], Hierarchical binary relevance [22]
2007	Quick weighted voting method [8], ML-KNN [13], Instance differentiation [28], Model-shared subspace boosting (MSSBoost) [29]
2008	Ranking by pair-wise comparison [9], Caliberated label ranking [10], Pruned set [11], Ensemble of Pruned set [20], Online Multi-Label Active Learning [24], Correlated labeling model [31], Hierarchy of multi-label classifier[23]
2009	Classifier chain [5], ML-RBF [16], MetaLabeler [27]
2010	Random decision tree [19], Probabilistic classifier chain [26], semi-supervised nonnegative matrix factorization [33], Bayesian online multi-label classification [25]
2011	Label power set [7], RAkEL [18], Canonical correlation analysis [38]
2012	Principal label space transformation [36], Conditional label space transformation [37], Multiple output prediction Landmark selection [35]
2013	SSLP [34], SSRAkEL [34]
2015	Random forest [21]

increased. So some mechanism is required to reduce the dimension to achieve high accuracy.

(ii) Correlation among labels needs to be studied deeply for multi-label learning: Many algorithms have been suggested that are based on the correlation among labels, but they are very complex processes. So the challenge is to develop an algorithm with linear complexity.

(iii) Hierarchical Multi-Label Classification (H- MLC): To handle Hierarchical relationship, tree, and DAG approaches are there. But still, much research is required to improve its performance.

(iv) Data Imbalance: This is the case where some labels are more relevant than the labels which are occurring frequently. This problem will lead to data imbalance problem.

(v) Exploits stratification methods which are suitable for the nature of the multi label data: Stratification is a sampling technique which creates disjoint groups based on the information from the population. Handling stratification is complex in the case of multi label.

(vi) Formulate new features selection based methods that are appropriate for multi-label datasets: Feature selection is a very basic step. With complexity of multiple labels, selection of suitable features to handle multi-label data is an alarming concern.

Despite progressive work done in the Research of MLC, it still leaves many challenges unexplored [7].

7 Conclusion

In this study herein, we have described MLC in detail from different perspectives. In our study, we have presented various MLC techniques with their detailed comparison. Correlation between labels is addressed by methods named classifier chain, Label power set, pruned set, pruned problem transformation, ML-RBF, RAkEL, HBR, collective Multi-label learning, HOMER, Conditional random field, Correlated Labeling model, and collective multi label with features. Among all classifier chain and label power set suffers from overfitting problem. However, adaboost and SVM-HF are giving better accuracy. QWVM and ensemble of pruned set both methods are giving improved prediction performance. However, if we analyze computation complexity then Label power set, CLR, pruned set, BP-MLL, RAkEL, probabilistic classifier chain, instance differentiation, CRF and semi-supervised nonnegative matrix factorization is giving high computational complexity. Were as, BR, BR+ , SSLP, SSRAkEL are the methods that handle the unlabeled data.

Further, we also explored performance evaluation metrics. The metric explains the performance of a model. However, it is very much important to choose the right metric to evaluate the model.

Then we discussed about the various tools available in the market. In the end, we proposed the trending challenges in the research.

With this, we can say that multi-label classification has a large scope for researchers to research in this domain and explore more domains where we can apply it. Empirical comparison of various methods is needed to be further explored.

References

1. Gibaja E, Ventura S (2015) A tutorial on multilabel learning. ACM Comput Surv 47(3). https://doi.org/10.1145/2716262
2. Wu XZ, Zhou ZH (2017) A unified view of multi-label performance measures. In: 34th International conference machanics learning ICML 2017, vol 8, pp 5778–5791
3. Zhang ML, Li YK, Liu XY, Geng X (2018) Binary relevance for multi-label learning: an overview. Front Comput Sci 12(2):191–202. https://doi.org/10.1007/s11704-017-7031-7
4. Read J, Pfahringer B, Holmes G, Frank E (2019) Classifier chains: a review and perspectives. pp 1–19. [Online] Available http://arxiv.org/abs/1912.13405
5. Read J, Pfahringer B, Holmes G, Frank E (2009) Classifier chains for multi-label classification. Springer, vol 5782, pp 254–269. https://doi.org/10.1007/978-3-642-04174-7_17
6. Dharmadhikari SC (2011) A comparative analysis of supervised multi-label text classification methods. Int J Eng Res Appl 1(4):1952–1961
7. Sharma S, Mehrotra D (2018) Comparative analysis of multi-label classification algorithms. In: ICSCCC 2018—1st International conference on secure cyber computing and communications, pp 35–38. https://doi.org/10.1109/ICSCCC.2018.8703285
8. Park SH, Fürnkranz J (2007) Efficient pairwise classification. In: Machine learning: ECML 2007, vol 4701 LNAI, pp 658–665. https://doi.org/10.1007/978-3-540-74958-5_65
9. Johannes F (2008) [2006] Hullermeller—Label ranking by learning pairwise preferences. ACM 172:16–17. https://doi.org/10.1016/j.artint.2008.08.002

10. Fürnkranz J, Hüllermeier E, Loza Mencía E, Brinker K (2008) Multilabel classification via calibrated label ranking. Mach Learning 73(2):133–153. https://doi.org/10.1007/s10994-008-5064-8
11. Read J (2008) A pruned problem transformation method for multi-label classification. In: New Zeal. Computing science research student conference NZCSRSC 2008—proceedings, pp 143–150
12. Clare A, King RD (2001) Knowledge discovery in multi-label phenotype data. In: Lecture notes computer science (including Subseries Lecture Notes Artificial Intelligence Lecture Notes Bioinformatics), vol 2168, pp 42–53. https://doi.org/10.1007/3-540-44794-6_4
13. Zhang M, Zhou Z (2007) Ml-knn: a lazy learning approach to multi-label learning,.40(7). https://doi.org/10.1016/j.patcog.2006.12.019
14. Schapire RE, Singer Y (2000) BoosTexter: a boosting-based system for text categorization. Mach Learn 39(2):135–168. https://doi.org/10.1023/A:1007649029923
15. Zhang ML, Zhou ZH (2006) Multilabel neural networks with applications to functional genomics and text categorization. IEEE Trans Knowl Data Eng 18(10):1338–1351. https://doi.org/10.1109/TKDE.2006.162
16. Zhang ML (2009) Ml-rbf: RBF Neural networks for multi-label learning. Neural Process Lett 29(2):61–74. https://doi.org/10.1007/s11063-009-9095-3
17. Godbole S, Sarawagi S (2004) Discriminative methods for multi-labeled classification. In: Lecture notes computer science (including Subseries Lecture Notes Artificial Intelligence Lecture Notes Bioinformatics), vol 3056, pp 22–30. https://doi.org/10.1007/978-3-540-24775-3_5
18. Tsoumakas G, Katakis I, Vlahavas I (2011) Random k-labelsets for multilabel classification. IEEE Trans Knowl Data Eng 23(7):1079–1089. https://doi.org/10.1109/TKDE.2010.164
19. Zhang X, Yuan Q, Zhao S, Fan W, Zheng W, Wang Z (2010) Multi-label classification without the multi-label cost. In: Proceedings 10th SIAM international conference data mining, SDM 2010, pp 778–789. https://doi.org/10.1137/1.9781611972801.68
20. Read J, Pfahringer B, Holmes G (2008) Multi-label classification using ensembles of pruned sets. In: Proceedings—IEEE international conference data mining, ICDM. pp 995–1000. https://doi.org/10.1109/ICDM.2008.74
21. Liu F, Zhang X, Ye Y, Zhao Y, Li Y (2015) MLRF: Multi-label classification through random forest with label-set partition. 1(Cc):275–281. https://doi.org/10.1007/978-3-319-22053-6_44
22. Cesa-Bianchi N, Gentile C, Zaniboni L (2006) Incremental algorithms for hierarchical classification. J Mach Learn Res 7:31–54
23. Tsoumakas G, Katakis I, Vlahavas I (2008) Effective and efficient multilabel classification in domains with large number of labels. In: Proceedings ECML/PKDD 2008 workshop mining multidimensional data, pp 30–44. [Online] Available http://lpis.csd.auth.gr/publications/tsoumakas-mmd08.pdf
24. Hua XS, Qi GJ (2008) Online multi-label active annotation: towards large-scale content-based video search. In: MM'08—proceedings 2008 ACM international conference multimedia, with co-located symposium workshop, pp 141–150. https://doi.org/10.1145/1459359.1459379
25. Zhang X, Graepel T, Herbrich R (2010) Bayesian online learning for multi-label and multi-variate performance measures. J Mach Learn Res 9:956–963
26. Dembczyński K, Cheng W, Hüllermeier E (2010) Bayes optimal multilabel classification via probabilistic classifier chains. In: ICML 2010—proceedings 27th international conference machanics learning, pp 279–286
27. Tang L, Rajan S, Narayanan VK (2009) Large scale multi-label classification via MetaLabeler. In: WWW'09—Proceedings 18th international world wide web conference, pp 211–220. https://doi.org/10.1145/1526709.1526738
28. Zhang ML, Zhou ZH (2007) Multi-label learning by instance differentiation. Proc Natl Conf Artif Intell 1:669–674
29. Yan R, Tesic J, Smith JR (2007) Model-shared subspace boosting for multi-label classification. In: Proceedings ACM SIGKDD international conference knowledge discovery data mining, pp 834–843. https://doi.org/10.1145/1281192.1281281

30. Wang H, Huang M, Zhu X (2008) A generative probabilistic model for multi-label classification. In: Proceedings-IEEE international conference data mining, ICDM, pp 628–637. https://doi.org/10.1109/ICDM.2008.8
31. Ghamrawi N, Mccallum A (2005) Collective multi-label classification. ACM 195–200
32. Liu H, Wu Z, Member S, Li X, Member S (2012) Constrained nonegative matrix factorization. 34(7):1299–1311
33. Lee H, Yoo J, Choi S (2010) Semi-supervised nonnegative matrix factorization. 17(1):4–7
34. Rodrigues FM, De M Santos A, Canuto AMP (2013) Using confidence values in multi-label classification problems with semi-supervised learning. Proceed Int J Conf Neural Netw 707–714. https://doi.org/10.1109/IJCNN.2013.6706806
35. Balasubramanian K, Lebanon G (2012) The landmark selection method for multiple output prediction. In: Proceedings 29th International conference machanics learning ICML 2012, vol 1, pp 983–990
36. Tai F, Lin HT (2012) Multilabel classification with principal label space transformation. Neural Comput 24(9):2508–2542. https://doi.org/10.1162/NECO_a_00320
37. Chen Y-N, Lin H-T (2012) Feature-aware label space dimension reduction for multi-label classification. ACM 1:1529–1537
38. Sun L, Ji S, Ye J (2011) Canonical correlation analysis for multilabel classification: a least-squares formulation, extensions, and analysis. IEEE Trans Pattern Anal Mach Intell 33(1):194–200. https://doi.org/10.1109/TPAMI.2010.160

A Comparative Study on Sentiment Analysis Influencing Word Embedding Using SVM and KNN

Bachchu Paul, Sanchita Guchhait, Tanushree Dey, Debashri Das Adhikary, and Somnath Bera

Abstract Development of sentiment analysis is one of the most active research areas that relates natural language and social networks. In our proposed work, we have done sentiment analysis on an annotated list of positive and negative sentiment words from dataset opinion-lexicon-English. Here to perform our task, we used pretrained word embedding that converts words into numeric vectors and forms the basis for a classifier. Word2vec, a commonly used algorithm, includes CBOW and Skip-gram model in learning word embedding which is basically used for calculating the word vector. Finally, feature vectors are used to train SVM and KNN classifier. Here, we got testing accuracy of 96.2% for SVM classifier and 93.4% for KNN classifier.

Keywords NLP · Sentiment analysis · Word embedding · Word2vec · SVM · KNN

1 Introduction

Sentiment is a thought influenced by feelings and emotions. The thought or feeling intended to be conveyed by words, acts, and gestures. With increasing popularities of various social networking sites, the way of expressing views, ideas have changed [1]. In recent trends, sentiment analysis is becoming a necessary tool to monitor and recognize the sentiment to upgrade customer's satisfaction and decision making. Sentiment analysis is essential to understand different opinions of the customers in business purpose which helps to take correct decisions [2]. Sentiment analysis is the interpretation and classification of emotions within text data using word embedding techniques. Sentiment analysis or opinion mining is a type of natural language processing for tracking the writer's feelings, and its aim to examine the opinions of the writer about specific entities is in positive and negative view [3].

B. Paul (✉) · S. Guchhait · T. Dey · D. Das Adhikary · S. Bera
Department of Computer Science, Vidyasagar University, Midnapore, West Bengal 721102, India

J. M. R. S. Tavares et al. (eds.), *Cyber Intelligence and Information Retrieval*, Lecture Notes in Networks and Systems 291,
https://doi.org/10.1007/978-981-16-4284-5_18

Everyday a huge amount of data is generated through e-mails, social media conversations, chats, surveys, articles, etc. It is very difficult to handle these data manually. So, to make this process automatic and cost effective, various techniques of sentimental analysis are used. There are two types of approach that are normally used to develop sentiment analysis model, which are symbolic technique or knowledge-based approach and machine learning approach. In knowledge-based approach, a complete database which contains predefined emotion information is required and a well-organized representation of knowledge is needed for identifying sentiments. In machine learning approach, for identifying the emotions of each word, we need both training and testing datasets. Like knowledge-based approach, machine learning approach does not require the database of words, and therefore, machine learning approach is better than knowledge-based approach. Word embedding can collect semantic details of the words so that similar words have similar vectors.

Word2vec is commonly used in learning word embedding that consists with two popular models where the first one is continuous bag-of-words model and second one Skip-gram model. In our proposed method, we worked with two machine learning algorithms SVM and KNN which are used to predict the sentiment of the words using their corresponding vector representation.

2 Literature Review

In previous so many researchers share their thoughts and ideology through their research work about sentiment analysis. Among them someone completed their research by user comment labeling to positive or negative, someone uses the techniques of sentiment detection from the given text. Both symbolic and machine learning techniques are used for analyzing the sentiments. Sentiments of people are analyzed from their tweets where basic sentiment score is computed. After calculating the sentiment score, it is classified as positive or negative. Knowledge-based approach and machine learning approach both are used here. Sanders analytic datasets are used for analyzing the tweets. Several parameters, like emotions, hashtags, etc., are included in the feature vectors and used to evaluate the sentiments by using the machine learning algorithm. By using the mixture of above two approaches, i.e., hybrid approach, they achieved 100% accuracy [1]. SVM is used for classifying the features extracted from input comments and tweets of different candidates involved in politics. With the help of TF-IDF and SVM, information is retrieved and preprocessed here. SVM is used to categorize the texts for finding the polarity of comment used in text. Recall and precision are used here for evaluating the performance for SVM and comparing with ANN to categorize the texts [3]. People's opinions are expressed in different languages through sentiment detection where a fast and flexible method is used. In this ML approach, vectors represent the textual document and used to train the polarity classification model. Online review of user-based four dataset on English and Greek languages is used here. Minimal computational resources are required here. The hybrid approach is used to predict the sentiment where lexicon-based approach and word embedding-based approach both are combined to achieve the high accuracy [4]. The techniques like SVM, semantic analysis (WordNet), Maximum Entropy and Naïve Bayes are used for classifying the sentence level sentiment analysis from twitter dataset. A comparative study is made up in this research work where it has been seen that 88.2% accuracy is achieved for Naïve Bayes, 83.8% accuracy for maximum entropy, 85.5% for SVM and finally 89.9% accuracy for semantic analysis. From this observation, it can also say that a better result is achieved by using Naïve Bayes technique than maximum entropy [5]. Sentence-level sentiment analysis based on word embedding is proposed here where word weight is considered for sentence vector calculation based on their importance. Trained data is used to calculate the word importance, and its proper utilization is done to calculate the sentence vector. XGBoost algorithm is used to classify the sentence vector which are unlabeled. Finally, it is seen that higher accuracy is achieved by using the effect of word importance [6].

The rest of the paper organized as Sect. 3 briefly describes our proposed methodology, Section 4 discusses result and analysis, and finally Sect. 5 gives the conclusion of the paper.

3 Methodology

The schematic diagram of our proposed work is shown in Fig. 1.

Here Fig. 1 shows the methodological way to express the text-based opinion into two polarities, i.e., positive and negative using word embedding approach.

3.1 Dataset

For our proposed work, we use opinion-lexicon-English dataset [7]. Here, total 6789 words are used, among which 2006 words are positive data and 4783 are negative data. We used a pretrained English word embedding corpus of 16 billion token

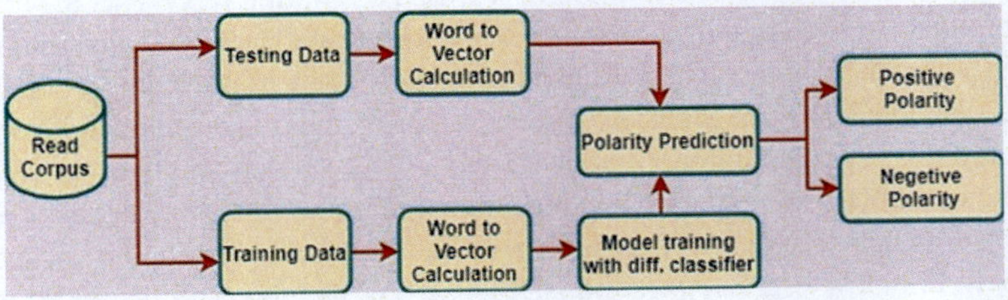

Fig. 1 Schematic diagram of the proposed method

words from word embedding support package namely “fastTextWordEmbedding”. Here, 90% data is used for training purpose and 10% data is used for testing purpose.

3.2 *Feature Extraction*

Feature extraction is done to obtain the feature vector of the word to make the classifier to work properly. In our work, we use word2vec that obtains word vectors.

3.2.1 Word Embedding

Word embedding is a type of word representation that helps to communicate between human and machine. Word embedding is attempt to taking syntactic and semantic meaning of words from the text collection. Here, text is represented in n-dimensional space where similar words have similar representation, i.e., two similar words are represented by similar vectors which are nearly placed in a vector space [8].

Word2vec Method

Word2vec is an unsupervised prediction-based word embedding approach, and the concept of this method comes from the scattered representation of word. There are so many algorithms from which two word2vec algorithms are used here:

(a) The continuous bag-of-words algorithm (CBOW) (b) Skip-gram algorithm.

(a) **Continuous bag-of-words algorithm (CBOW)**

CBOW algorithm is applicable to predict a single word among the surrounding words. It is used to understand the word context which is taken as input. The basic function of this model is to predict the contextually accurate word. Vector representation of word is needed to present the CBOW model where a predefined sliding window is used to prepare the vector presentation of words. The column represents the dimension of total no. of words and the row represents the vocabulary size. Some minimum frequency threshold is used here to use the word sequence in training the CBOW model. CBOW model has generally three layers, i.e., input layer, hidden layer, and output layer, where the number of nodes in the input layer and output layer depends upon the vocabulary size we used and the number of nodes in the hidden layer represents the dimensionality and it may vary from one word to the maximum vocabulary size. Here in below Fig. 2 shows the simple CBOW method with one word in the context.

The task of the system expressed in Fig. 2 is to identify the missing word which is missing from the context. Now from the above diagram, it is seen that the value of x_1,

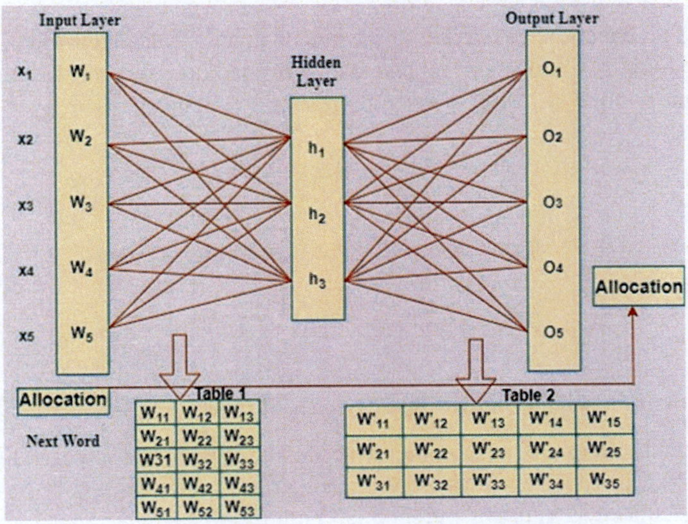

Fig. 2 Simple CBOW method with one word in the context [10]

Table 1 Weighted representation from input to hidden layer

*W*11	*W*12	*W*13
*W*21	*W*22	*W*23
*W*31	*W*32	*W*33
*W*41	*W*42	*W*43
*W*51	*W*52	*W*53

Table 2 Weighted representation from hidden to output layer

W′11	*W*′12	*W*′13	*W*′14	*W*′15
W′21	*W*′22	*W*′23	*W*′24	*W*′25
W′31	*W*′32	*W*′33	*W*′34	*W*′35

$x_2, \ldots, x_5$ in the input layer will be 0 or 1 according to the context. If the context has two words then the value of x_1 and x_2 will be set to 1 and the rest will be set to 0. In the hidden layer, the hidden nodes are defined as h_1, h_2, h_3 where h_1 is the minimum node. Now Table 1 shows the matrix form of the weighted representation from input to hidden layer, and Table 2 shows the matrix form of the weighted representation from hidden to output layer. After getting the optimal weight of all the neuron, the trained CBOW model is used to predict the next word for the sequence what we used. In CBOW model, first we have to train the system by three following steps which are: (i) forward propagation, (ii) error calculation, and (iii) backpropagation.

(i) **Forward Propagation**

In forward propagation, the weight of the hidden layer is calculated. Here, different from traditional ANN, no sigmoid calculation is used here but only Σ phase used. So the value at node h_1, h_2, h_3 is given in Eqs. 1– 3 [9, 10]

$$h_1 = (w_{11}x_1 + w_{21}x_2 + w_{31}x_3 + w_{41}x_4 + w_{51}x_5) \tag{1}$$

$$h_2 = (w_{12}x_1 + w_{22}x_2 + w_{32}x_3 + w_{42}x_4 + w_{52}x_5) \tag{2}$$

$$h_3 = (w_{13}x_1 + w_{23}x_2 + w_{33}x_3 + w_{43}x_4 + w_{53}x_5) \tag{3}$$

Weight calculation from hidden layer to output layer

To calculate the weight of each node, two different calculations are used here. First one is association represented as Σ, and second one is softmax output. Association phase is calculated as a sum of weight of all numbers that connected to output layer O_1 and the weight of associated hidden layer nodes. The equations of O_1 to O_5 are represented in Eqs. 4–8 here [9, 10].

$$\text{Net}(O_1) = u_1 = \left(W'_{11}h_1 + W'_{21}h_2 + W'_{31}h_3\right) \tag{4}$$

$$\text{Net}(O_2) = u_2 = \left(W'_{12}h_1 + W'_{22}h_2 + W'_{32}h_3\right) \tag{5}$$

$$\text{Net}(O_3) = u_3 = \left(W'_{13}h_1 + W'_{23}h_2 + W'_{33}h_3\right) \tag{6}$$

$$\text{Net}(O_4) = u_4 = \left(W'_{14}h_1 + W'_{24}h_2 + W'_{34}h_3\right) \tag{7}$$

$$\text{Net}(O_5) = u_5 = \left(W'_{15}h_1 + W'_{25}h_2 + W'_{35}h_3\right) \tag{8}$$

Finally, the association score is generalized in Eq. 9 [9, 10]

$$\text{Net(O)} = W'_{11}h = \begin{vmatrix} W'_{11}\,W'_{21}\,W'_{31} \\ W'_{12}\,W'_{22}\,W'_{32} \\ W'_{13}\,W'_{23}\,W'_{33} \\ W'_{14}\,W'_{24}\,W'_{34} \\ W'_{15}\,W'_{25}\,W'_{35} \end{vmatrix} \begin{vmatrix} h_1 \\ h_2 \\ h_3 \end{vmatrix} \tag{9}$$

The softmax output for any node, i.e., just like as Out (O_1) represented as y_1 is given in Eq. 10 [10].

$$\text{Out}(O_1) = y_1 = \frac{e^{\text{Net}(O_1)}}{e^{\text{Net}(O_1)} + e^{\text{Net}(O_2)} + e^{\text{Net}(O_3)} + e^{\text{Net}(O_4)} + e^{\text{Net}(O_5)}}$$

$$= \frac{e^{u_1}}{e^{u_1} + e^{u_2} + e^{u_3} + e^{u_4} + e^{u_5}} \tag{10}$$

Similarly, the values of Out (O_2), Out (O_3), Out (O_4), Out (O_5) are also calculated here.

The softmax output for the word $\mathrm{W_j}$ and the given input context words $\mathrm{W_I}$ can be given in Eq. 11 by conditional probability in terms of exponentials [10], i.e.,

$$p(w_j|w_I) = y_j = \frac{e^{\mathrm{Net}(O_j)}}{\sum_{j=1}^{v} e^{\mathrm{Net}(O_j)}} = \frac{e^{(u_j)}}{\sum_{j=1}^{v} e^{(u_j)}} \tag{11}$$

(ii) **Error Calculation**

Let W_0 is the actual output word and W_1 is the given input contexts and the size of input context is V. So the loss function can be defined in Eq. 12 [10].

$$\max P(W_0|W_I) = \max(y_{j^*}) = \max(\log(y_{j^*})) \tag{12}$$

So for minimizing the error, E is represented in Eq. 13 [10].

$$E = -\log P(W_0|W_1) \tag{13}$$

And finally the generalized form of E is represented in Eq. 14 [10].

$$E = \log \sum_{j=1}^{v} e^{u_j - u_{j^*}} \tag{14}$$

where j^* is the index of actual output word in the output layer.

Loss function also be understood as the special of cross-entropy between two probabilistic distributions also.

The derivation is generalized in Eq. 15 [10].

$$\frac{dE}{du_j} = \frac{d\left(\log \sum_{j=1}^{v} e^{u_j} - u_{j^*}\right)}{du_j} = \frac{e^{u_j}}{\sum_{j=1}^{v} e^{u_j}} = \frac{d(u_{j^*})}{du_j} = y_j - t_j = e_j \tag{15}$$

Note : $t_j = 1,\ \mathrm{if}(t_j = t_{j^*})\ \mathrm{else} t_j = 0$

(i) **Backpropagation**

Backpropagation is calculated from output layer to hidden layer. The main aim of this technique is to update the weight of all neurons from hidden to output layer, i.e., W_{11}, W_{12}, ….., W_{35}. The main target is to update the weight of entire matrix which is shown in Table 1.

In the first phase, gradient of E has to be taken. Then the weights of hidden to output layer are to be updated [9, 10] which are shown in Table 2.

(b) Skip-gram algorithm

Skip-gram algorithm uses the central word to predict the surrounding word. The main concept behind this algorithm is that, we randomly initialize the vector for every word in the vocabulary. This algorithm aims to maximize average value of the log probability for the given set of training words $w_1, w_2, \ldots, w_T$ [11, 12].

Now for predicting the context word from the given center word, we will compute the probability of our model using the Formula 16 [11, 12].

$$\frac{1}{T}\sum_{t=1}^{T} \sum_{\substack{-m \leq j \leq m \\ j \neq 0}} \log p\left(w_{t+j} | w_t\right) \tag{16}$$

Here m is the window size. For getting better accuracy, the size of m will be large that contains more training context. w_t = Target word or center word and w_{t+j} = Context word within a window size m. $p(w_{t+j}|w_{t)}$ = the probability where w_{t+j} occur within the neighbor of w_t and the basic Skip-gram algorithm defines $p(w_{t+j}|w_{t)}$ using the softmax function 17 [11, 12]

$$P(w_O|w_I) = \frac{\exp(v'_{wo}\, v_{wI}^{\top})}{\sum_{w=1}^{W} \exp(v'_{w} v_{wI}^{\top})} \tag{17}$$

where w is the total vocabulary set. v_w is the input and v'_w is the output representation of w. Here, we compute the exponential of dot product of all our words. But it will be very time consuming when we have large vocabulary size. Therefore, a new method called negative sampling is used which will minimize the probability of random words appearing around the center word and maximizes the probability of real context word. Negative sampling is defined by the Formula 18 [12]

$$\log_{\sigma}(v'_{w} v_{wI}^{\top}) + \textstyle\sum_{i=1}^{k} \mathbb{E}_{w_{i} \sim P_n(w)} \left[\log_{\sigma}(-v'_{w} v_{wI}^{\top})\right] \tag{18}$$

Which is accustomed to substitute each log $p(w_O|w_I)$ term of the Skip-gram algorithm where k is the negative samples for every data sample and $p_n(w)$ is noise distribution parameter.

3.3 *Classifier*

After extracting feature vector, classification is done using SVM and KNN separately to classify the data into binary class such as positive class and negative class. Here, vector space is used as input for the specific classifier and gets output as positive or negative.

3.3.1 Working of Support Vector Machine (SVM)

SVM is one of the learning techniques that performs well on sentiment classification. It is used for dividing the datasets into classes for finding the maximum marginal hyperplane. The goal of SVM is done by following two steps which are:

Step 1: The hyperplanes are initiated iteratively by SVM which divides the classes in best way.

Step 2: Next it chooses that hyperplane which are used for separating the classes properly.

3.3.2 Working of KNN

KNN is used to classify the new data points based on distance function. Classification is done by a maximum vote to its neighbors. Here, the Euclidean distance is measured to find the distance between the data points of the training set and the test data point. Euclidean distance is calculated between two point $A_1\ (X_1, Y_1)$ and $B_2\ (X_2, Y_2)$ which is given in Eq. 19 [13].

$$\sqrt{(X_2 - X_1)^2 + (Y_2 - Y_1)^2} \tag{19}$$

4 Result and Analysis

We developed our model with two machine learning algorithms: one is SVM classifier and other is KNN classifier. We performed experiment on the dataset Opinion-Lexicon-English so that, the model works more effectively and efficiently that depends on true positive(TP), false positive (FP), true negative (TN), and false negative (FN).

The confusion matrix for measuring testing accuracy using SVM and KNN of our proposed work is given in Tables 3 and 4, respectively.

Using the true positive (TP), false positive (FP), true negative (TN), and false negative (FN) on testing data, we get precision, recall, F-score, FPR, and FNR as for measuring performance of SNM and KNN classifier which is expressed in Table 5.

Table 3 Confusion matrix for SVM

Output class \ Target class	Positive class	Negative class	
	190	11	Positive class
	14	437	Negative class

Table 4 Confusion matrix for KNN

Output class \ Target class	Positive class	Negative class	
	177	16	Positive class
	27	432	Negative class

Table 5 Performance management

Classifier	Precision		Recall		F-score		FPR	FNR
	Positive	Negative	Positive	Negative	Positive	Negative		
SVM	0.945	0.968	0.931	0.975	0.938	0.971	0.025	0.069
KNN	0.917	0.94	0.868	0.964	0.891	0.952	0.0357	0.132

Comparative chart of precision, recall measurement for positive and negative polarity, and performance analysis of accuracy according to SVM and KNN classifier are shown in Figs. 3, 4, and 5, respectively.

Visualization of plotting the positive and negative sentiments according to the prediction scores in word cloud is given in the following figures where Fig. 6 shows the visualization of predicted positive class for SVM classifier and Fig. 7 shows the visualization of predicted positive class for KNN classifier.

5 Comparative Study

After studying some previous work related to sentiment analysis, it has been clearly analyzed that based on different dataset and different classifiers, several research work achieved more accuracy, some achieved less accuracy than our proposed work.

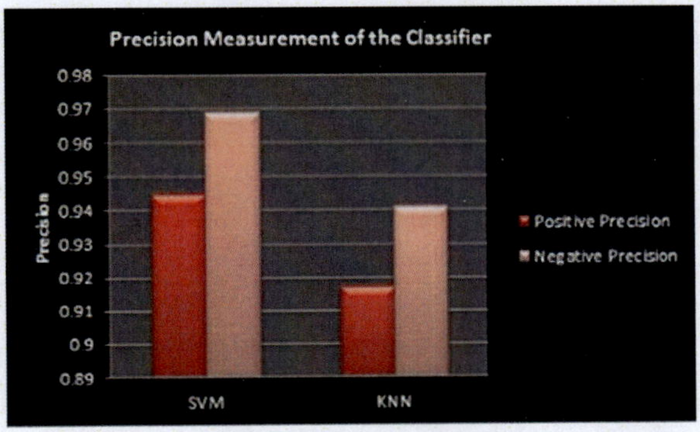

Fig. 3 Comparative chart on precision measurement

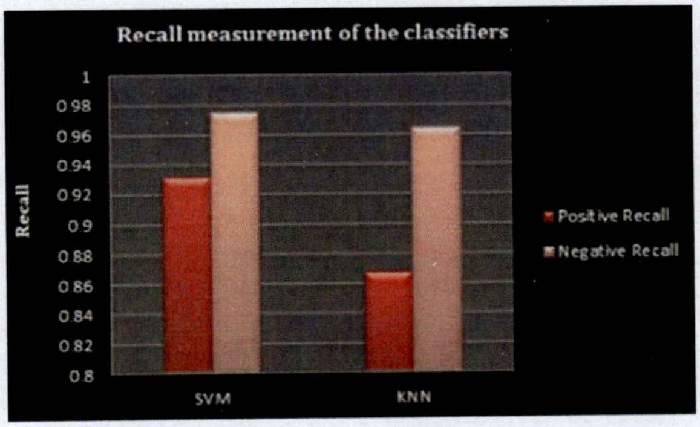

Fig. 4 Comparative chart on recall measurement

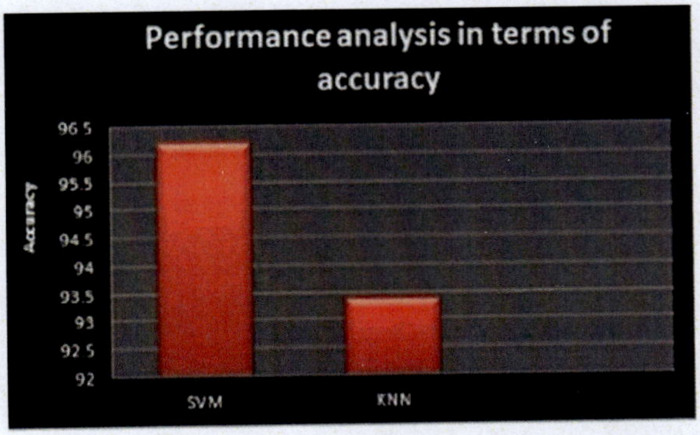

Fig. 5 Comparative chart on accuracy

But enough methods are not available for comparing with our implemented method except Gautam et al. [5] which uses twitter dataset contain 19,340 data labeled by unigram where 18,340, i.e., 94.83% for training and 1000, i.e., 05.17% for testing purpose and uses SVM, semantic analysis, maximum entropy, and Naïve Bayes for

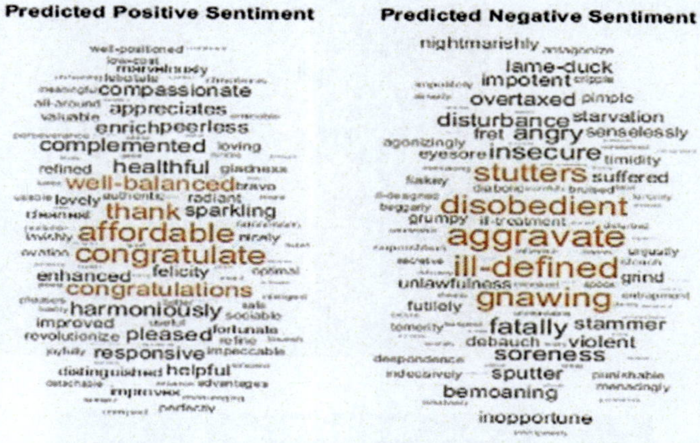

Fig. 6 Visualization of predicted positive class for SVM

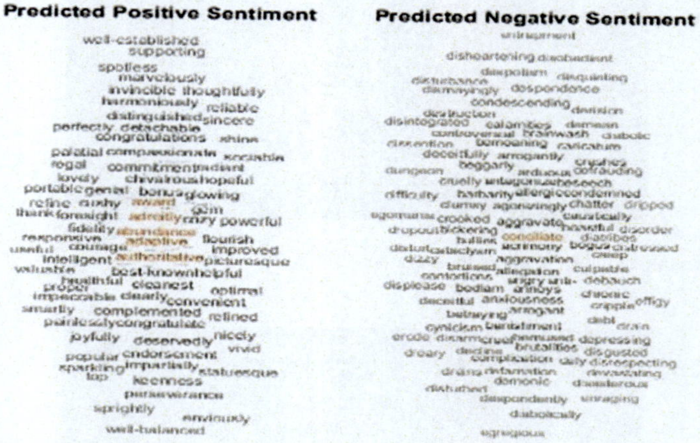

Fig. 7 Visualization of predicted positive class for KNN

classifying the sentence. Among of different classifier, by using SVM on 19340 dataset, they achieved 85.5% accuracy whereas in our proposed work, SVM and KNN are used on Opinion-Lexicon-English dataset contain 6789 dataset extracted by word2vec where 90% data for training and 10% for testing purpose and achieved 96.2% accuracy by using SVM. In our implemented work, two classes are used and it can be clearly says that the model is less complex than other model.

6 Conclusion

In this research work, our experiment achieved 96.2% accuracy by SVM classifier and 93.4% accuracy by KNN classifier. Our proposed work is based on CBOW and skip-gram model which is better than other word embedding techniques as both of them are probabilistic in nature. Here also, we used two simple machine learning algorithms which can get better accuracy and make our research work more flexible. But after analyzing the result, we can conclude that SVM gives better sentiment polarity than KNN. The drawback of our proposed technique is sentiment analysis which is not possible for sentence vector and predict only two polarities (positive and negative). In future, in order to use efficiency and flexibility of our methodology, we will apply sentiment analysis on Bengali words. Further, we will focus for multiple polarity and work on sentence level. We will apply our methodology for large datasets of various popular social network like chat, tweets of twitter, comments used in e-commerce sites, etc.

References

1. Suchdev R, Kotkar P, Ravindran R, Swamy S (2014) Twitter sentiment analysis using machine learning and knowledge-based approach. Int J Comput Appl 103(4)
2. Preety SD, Dahiya S (2015) Sentiment analysis using SVM and Naive Bayes algorithm. Int J Comput Sci Mob Comput 4(9):212–219
3. Patil G, Galande V, Kekan V, Dange K (2014) Sentiment analysis using support vector machine. Int J Innov Res Comput Commun Eng 2(1):2607–2612
4. Giatsoglou M, Vozalis MG, Diamantaras K, Vakali A, Sarigiannidis G, Chatzisavvas KC (2017) Sentiment analysis leveraging emotions and word embeddings. Expert Syst Appl 69:214–224
5. Gautam G, Yadav D (2014) Sentiment analysis of twitter data using machine learning approaches and semantic analysis. In: 2014 seventh international conference on contemporary computing (IC3) August, IEEE, pp 437–442
6. Hayashi T, Fujita H (2019) Word embeddings-based sentence-level sentiment analysis considering word importance. Acta Polytechnica Hungarica 16(7):152–252
7. https://www.cs.uic.edu/~liub/FBS/sentiment-analysis.html#lexicon
8. Deho BO, Agangiba AW, Aryeh LF, Ansah AJ (2018) Sentiment analysis with word embedding. In: 2018 IEEE 7th International conference on adaptive science and technology (ICAST) August, IEEE, pp 1–4
9. Mikolov T, Chen K, Corrado G, Dean J (2013) Efficient estimation of word representations in vector space. arXiv preprint arXiv:1301.3781
10. Rong X (2014) word2vec parameter learning explained. arXiv preprint arXiv:1411.2738
11. Mikolov T, Sutskever I, Chen K, Corrado G, Dean J (2013) Distributed representations of words and phrases and their compositionality. arXiv preprint arXiv:1310.4546
12. Kaji N, Kobayashi H (2017) Incremental skip-gram model with negative sampling. P. *Machine Learning, 1e.* Pearson Education India. arXiv preprint arXiv:1704.03956.Education
13. Huq MR, Ali A, Rahman A (2017) Sentiment analysis on Twitter data using KNN and SVM. (IJACSA) Int J Adv Comput Sci Appl 8(6):19–25

ML and GIS-Based Approaches to Flood Prediction: A Comparative Study

Abha Tewari, Varad Kshemkalyani, Heer Kukreja, Pratheek Menon, and Reuben Thomas

Abstract Floods are one of the most devastating natural disasters and cause damage worth millions of dollars every year. The research on models predicting floods and advances in the methods applied have significantly reduced risk and loss of life and allowed better suggestions for policies to reduce harm to property due to floods. Empirical approaches use historical data to predict floods that might occur on the basis of statistics. Although this approach has received tons of attention in recent years thanks to advances in supporting technology, particularly within the areas of massive data and machine learning, it lacks in modeling generalized systems. The spatial distribution of the input features like meteorological conditions and other physical parameters can be captured by using physical models. Since the physical models are dynamic, they may overcome the empirical models which are less accurate and also lack generalizability. However, these models are data intensive, difficult to build, and require expertise in the domains. This paper reviews various hydrological modeling methods and predictive models and proposes a hybrid approach which combines the two approaches to predict floods in the vicinity of the river basins.

Keywords Flood prediction · Machine learning · Natural disasters · Artificial neural networks · LSTM · Remote sensing · GIS

1 Introduction

With the drastic climate change over the past few years, the degree and extent of floods have increased massively with extensive flash floods bringing fast-moving and rapidly-rising water with force, resulting in huge loss in life and property as well as disruption of services worldwide. Government authorities are failing to develop accurate flood risk maps for flood-prone areas and planning for flood risk management that specialize in prevention, protection, and preparedness. In hazard assessment

A. Tewari · V. Kshemkalyani (✉) · H. Kukreja · P. Menon · R. Thomas
Department of Computer Engineering, Vivekanand Education Societys Institute of Technology, Mumbai, India
e-mail: 2017.varad.kshemkalyani@ves.ac.in

J. M. R. S. Tavares et al. (eds.), *Cyber Intelligence and Information Retrieval*, Lecture Notes in Networks and Systems 291,
https://doi.org/10.1007/978-981-16-4284-5_19

and extreme event management, flood prediction algorithms hold utmost importance. Advancement in the methods applied will contribute to the reduction of risk and loss of life as well as property damage related to floods and evacuation modeling.

Thus, flood prediction and flood control are important issues for policymakers and designers [1]. During the past two decades, machine learning (ML) methods and Artificial intelligence (AI) have contributed highly to the enhancement of prediction systems resulting in better performance and cost-effective solutions. Many ML algorithms, e.g., artificial neural networks (ANNs), neuro-fuzzy, support vector machines (SVM), and support vector regression (SVR); also, Bayesian forecasting models which use probability distribution from historical streamflow data can effectively predict floods with some accuracy. Although these methods are fairly simple to build and understand, they fail to capture the characteristics of the watersheds which physical approaches like VIC (Variable Infiltration Capacity) do.

2 Hydrological Modeling

Hydrological models are rudimentary implementations of real-world systems which are used for simulating and predicting flood events and determining what anthropogenic protection measures may be necessary in the near future. In general, hydrological forecasting models can be broken down into two categories: empirical models and physically-based models [2]. Empirical models are purely observation-based models which use mathematical or statistical functions to link the observed input to the output. These models are further divided into two categories on the basis of their approach: the artificial intelligence approaches such as ANNs and statistical approaches such as regression and support vector machines [3]. They do not consider properties of the hydrological system, instead are focusing entirely on existing data and hence are also called data-driven models. The second group is of theoretical models, which aims to represent the physical environment in terms of equations, processes, and physical interactions. These models are usually very complex but give more accurate results. The focus of this paper will be a comparative analysis of the two traditional models based on data-driven machine learning like SVM, ANN, and Random Forest and physical models that consider the geographic characteristics of the target area as well.

3 Related Work

3.1 Predicting Flood Susceptibility Using Long Short-Term Memory (LSTM) Neural Network Model

In this paper, Fang et al. [4] proposed a flood susceptibility mapping technique that uses a local spatial sequential (LSS-LSTM) neural network for the Shangyou County in China. According to the authors, the LSS-LSTM method can get both the effects of flood conditioning factors and also has the capability to manage regional spatial relationships using sequential modeling techniques. The authors used feature engineering to transform raw flood data into one-dimensional feature vectors. The entire study area was converted into raster form with each pixel being classified to determine whether it will be susceptible to flooding or not. A total of 13 flood conditioning factors including slope, altitude, aspect, NDVI, distance from river, land use, wetness index, and annual average precipitation were chosen to build the model. Each of these factors were converted into 30 m resolution rasters over the entire study area and stacked together to form a multi-band image. A 3×3 window is iteratively applied across the raster. At each step, a central pixel and its eight neighboring cells are extracted to form a $3 \times 3 \times 13$ image patch composed of 9 vectors. Each of these vectors contains flood conditioning factors at the corresponding grid cell. [4] These vectors are sorted according to their spatial continuity to form sequential data of size 9×13. The sorted vectors are progressively sent to the LSTM architecture and the results are output only at the final time step. According to the authors, this process allows the model to capture not only the morphological, geological, and hydrological conditions but also the spatial relationships to the neighboring environment [4]. The authors used Information Gain Ratio (IGR) to understand the relationship between the flood conditioning factors and flood occurrences and found that slope, distance from river, land use, and NDVI are the major contributing factors to flooding. Their model was further evaluated using Receiver Operator Curves (ROC) and Area under Curve (AUC) method. The LSS-LSTM method achieves an AUC value of 0.965 and a test set prediction accuracy of 93.75% which the authors feel can be further improved by the use of multi-temporal data.

3.2 Flood Prediction Using Support Vector Machines (SVM)

In this study, the authors' Yu Shi et al. [5] have put forth a Support Vector Machine (SVM) based approach to forecasting river flood events on the basis of a blend of rain gauge and river water discharge measurement recorded from the River Don catchment, their research was aimed at showcasing the potential of SVMs for predicting future Their study showed that SVMs are extremely accurate forecasters of both difference in the river flow as well as peak values of flow over lesser lead times (less than 48 h). The authors evaluated the results for multiple lead times based on the

RMSE of the prediction. They found that the rainfall is most correlated to the river discharge for a lead time of 1 but progressively deteriorates with an increase in lead time. The lowest RMSE for the validation set was 0.450 when the lead time was 1. This method was capable of providing warnings for imminent floods on a short notice. But beyond a 48- hour window, predictions concerning peak flow were not satisfactorily accurate to denote occurrences of floods, and the model decomposed over longer prediction horizons. However, the authors state that long-term predictive accuracy of the method could improve by including soil moisture and water as additional explanatory variables. The downside to this approach is that arriving at confidence intervals surrounding the forecasts is impossible. The accuracy of the predictions can be verified only by means of correlation with factual data.

3.3 Artificial Neural Networks for Flood Susceptibility Mapping in Data-Scarce Urban Areas

Falah et al. [6] in their study mapped flood inundated areas of Emam-Ali township in Iran by applying the Artificial Neural Network (ANN) model for delineating flood inundated areas. They prepared a flood inventory map of the study area which illustrates the thematic layers of factors affecting floods, including elevation, distance from drainage, slope angle, drainage density, slope aspect, and land use [6]. The authors used receiver operating characteristic (ROC) curves to check for the accuracy of the generated inundation map. According to the authors, flood risk maps are invaluable tools for predicting the magnitude and probable damage due to floods. The evaluation resulted in having a ROC curve (AUC) having 94.6% accuracy for training data and 92% for the validation set which is good enough to detect the inundated areas [6]. The authors state that by taking into account the duration of precipitation we can further improve the accuracy of the model.

3.4 Height Above the Nearest Drainage, a Hydrologically Relevant New Terrain Model

Nobre et al. [7] proposed a new terrain model called the HAND—Height above nearest drainage which normalizes topography according to the local relative heights along the drainage network and thus gives us a picture of the gravitational potential and the local drainage potentials [7]. This model shows a correlation between the water table and the surrounding environments of soil and water. The Relative vertical flow path distances to the nearby drainage were used to classify the normalized draining potentials. These classified values have a comparable hydrological significance in the studied catchment and for the surrounding ungauged catchment.

The Height Above Nearest Drainage model is a digital elevation model that has been drainage normalized. The normalized local height in the HAND model, its z-axis variable, indicates the vertical distance from a surface cell of a hillslope to the nearest outlet drainage cell [7]. This study thereby revealed strong correlations between solid water and the segmented classes which were explained using the local draining potential. This model can hence be used as a baseline for classifying and mapping the terrain under study. According to the authors, HAND classes were validated with verified terrains and its correspondence with landforms. Field-optimization thresholds were used to classify the HAND model into classes. The points obtained were matching with the field environments with only a few unavoidable localization errors due to resolution mismatch rather than errors in classification.

3.5 *Flood Modeling of Mangalam River Using GIS and HEC-RAS*

Sunilkumar et al. [8] proposed a solution in which a flood hydrograph of the target region was developed by using the time distribution coefficient method with the Hydrologic Engineering Center's Hydrologic Modeling System (HEC-HMS) tool and then the flood risk map modeled using the Hydrologic Engineering Center's River Analysis System (HEC-RAS) tool. The inundation maps were created using GIS. The study area is the Mangalam river, which is 30 km long spread over the Palakkad district in Kerala.

The first step in the proposed methodology is the flood estimation wherein precipitation data along with other data was used to compute runoff volume with the use of the HEC-HMS software. The next step is the hydraulic analysis wherein the HEC-RAS software was used to prepare a flood inundation map by computing water surface elevations which made use of two inputs for flow computation: geometric data and flow data. HEC GeoRAS was used for pre-processing of geometric data and post-processing of outputs. Flood inundation map was created with HEC GeoRAS which uses functions associated with the Spatial Analyst and 3D analyst tools in ArcGIS and only makes use of elevation data. In the pre-processing step, the river network was established using HEC-RAS wherein the river, along with the river banks and flow paths were digitized, and then the were cross-sections created, which are the main input to HEC-RAS. Next, the river reach, bank lines, and cross section lines from the previous step were imported into HEC-RAS, and unsteady flow analysis was performed on them and downstream boundary conditions defined. Finally, in the post-processing step, necessary file format conversions were performed at the end of which a bounding polygon defining the extent analysis for inundation mapping, and then the inundation extent was mapped with the profile with the maximum flow followed by floodplain delineation.

The proposed solution is extremely helpful in flood mitigation plans, especially in the regions which are prone to flash floods. Obtained values can be used to design new structures such as dams in the region and they also provide one with an estimation of the highest probable flood discharge which can come in very handy in mitigation plans.

Owing to the absence of a pre-existing historical flood map, the results were validated with on-field cross section data. The model did not consider tributaries of the river due to low spatial resolution of the DEM. One can extend the study area until the sea at which point the effects of the flood get neutralized.

3.6 Application of ASTER DEM in Watershed Management as Flood Zonation Mapping in Koyna River of the Western Ghats

Shinde et al. [9] have made use of the Advanced Spaceborne Thermal Emission and Reflection Radiometer (ASTER) and Global Digital Elevation Model (GDEM) based GIS layers to carry out flood level mapping of the Koyna river basin using the ILWIS software, and then the output of the ASTER DEM analysis was checked with flood risk data obtained from other remote sensing and GIS sources. It makes use of a 1 m DEM of the study area, in order to minimize errors. Highest Flood Level (HFL) values (in cusecs) were overlaid on the ASTER DEM after digitizing in the value domain. Keeping this in mind the ASTER DEM was subjected to multiple rounds of slicing and then the fill sink procedure to generate the continuous flow direction and flow accumulation maps [9]. df/dx and df/dy filters were applied and a slope percentage map, slope map in degrees, slope aspect map, and slope shape map were prepared [9]. From the Landsat 7 TM data, the required submaps were extracted and a band ratioing method was used for performing supervised classification of Landsat data. Additional discharge was calculated from each confluence point of the tributaries with the Koyna mainstream by using flow accumulation map, thus readying the flood level segment map for further analysis. This flood level map was interpolated and a digital flood discharge model was created. On overlaying high-resolution satellite data on this model, a click on any point of the imagery displays the flood discharge value in cusecs required for inundation of that point on ground. In this study, the degree of floods along the banks of the Koyna river was assessed by splitting the flood discharge model into 8 flood hazard zones.

As stated by the authors, ASTER GDEM is an effective tool for planning in flood management and mitigation since it has an accuracy level higher than topographic map-generated DEM. The output of the model was correlated with flood hazard related data from IRS 1D PAN data to verify the results.

3.7 Terrain Analysis Enhancements to the Height Above Nearest Drainage Flood Inundation Mapping Method

Garousi-Nejad et al. [10] have proposed a terrain enhancement to the HAND model [7] in this paper. The authors suggest that empirical models for flood inundation mapping could be better alternatives than comprehensive hydraulic models as they are more practical and have lower data demands.

The authors have used the Bear River flood event of 2017 as a case study to evaluate and develop improvements in empirical methods for Continental-Scale Flood Inundation Mapping (CFIM) based on the Height above Nearest Drainage (HAND). Comparing the flood map generated by the HAND model and high-resolution Planet RapidEye Satellite imagery of the flood event, the authors found that there were quite a few differences in the areas which were forecasted to be flooded and actual flooded areas. They suggest that the shortcomings of the HAND derived from National Elevation Dataset as well as the dissimilarity between the forecasted and observed discharges could be a reason for the differences. The authors then propose a set of improvements to address the shortcomings of the model to reduce the differences seen in the inundation areas. These methods include using high-resolution hydrography to condition the topography, dispersing nodes that subdivide the river into reaches and catchments, and using a high-resolution digital elevation model [10].

Another approach suggested by the authors involves obtaining a reach specific Manning's constant (n) from the observed inundation.

3.8 Flood Risk Mapping Using GIS and Multi-criteria Analysis: A Greater Toronto Area Case

Rincón et al. [11] have developed flood risk maps by making use of analytical hierarchy process (AHP) for assigning weights to those factors that can play a part in the flood in conjunction with GIS and multi-criteria analytics. Flood risk maps for different criteria were created based on four different scenarios. The factors considered included distance to stream (DS), slope(S), height above nearest drainage (HAND), curve number (CN), and also included social and economic vulnerabilities. These factors helped them get a better understanding of the impacts and also aided in the decision making of flood prevention planning. This approach was flexible, low cost, and was easily extensible to areas where data is scarce. Depending upon the importance, the planners can adjust the weights of the vulnerabilities because of the AHP weighting method which allowed the planners to focus on minimizing human and economic losses. In order to improve the accuracy of the maps, the study suggests using effective precipitation in place of total rainfall. Thus, the factors that were considered in the study are sufficient to obtain maps that are accurate. Different scenarios were compared to a base map which gave different percentages of likeliness to the base map.

Table 1 Analysis and comments

Related work	Results and analysis
3.1	**Approach used**: Long short-term memory (LSTM) neural network **Hardware/Software used**: Python **Datasets used**: 30 m ASTER GDEM, LANDSAT 7, Lithology, Soil, and Precipitation **Results**: Test set prediction accuracy of 93.75% **Limitations**: The model cannot capture temporal changes. Use of multi-temporal data might improve model accuracy
3.2	**Approach used**: Support Vector Machine (SVM) **Hardware/Software used**: Python **Datasets used**: Rainfall and river flow over 30 years for river Don **Results**: Lowest RMSE of 0.45 for lead time of 1 **Limitations**: While results are pretty good for shorter lead times, the model starts decomposing for longer horizons
3.3	**Approach used**: Artificial Neural Network (ANN) **Hardware/Software used**: Python **Datasets used**: ASTER DEM, LANDSAT 7, Flood inventory maps **Results**: Accuracy of 94.6% and 92% for training and validation sets respectively **Limitations**: The model considers neither temporal nor local spatial dependencies in the watershed
3.4	**Approach used**: Height Above Nearest Drainage (HAND) **Hardware/Software used**: ArcGIS **Datasets used**: SRTM-DEM **Results**: Strong and robust relation between the results of HAND model with the field collected data with few misclassified points largely due to resolution mismatch **Limitations**: Results depend strongly on the availability of high resolution of DEM's
3.5	**Approach used**: Geographical Information System (GIS) **Hardware/Software used**: ArcGIS, HEC-RAS, HEC-HMS **Datasets Used**: Precipitation data, DEM/TIN elevation data **Results**: Useful for flash flood situations and aids in the design of new hydraulic structures **Limitations**: In absence of historical flood maps, on-field data will be needed for result validation. Low DEM resolution results in tributaries not being taken into account
3.6	**Approach used**: Geographical Information System (GIS) **Hardware/Software used**: ILWIS software **Datasets Used**: ASTER GDEM, IRS 1D PAN data **Results**: Higher accuracy than approach using topographically generated DEM **Limitations**: Useful in regions having steep relief inaccessible from a traditional survey point of view

(continued)

Table 1 (continued)

Related work	Results and analysis
3.7	**Approach used**: Hydrologically enhanced HAND **Hardware/Software used**: ArcGIS, TauDEM **Datasets Used**: 10 m DEM, streamlines, historical streamflow data, 5 m satellite images **Results**: Overcame the discrepancies of the HAND by reducing the differences seen in the inundation areas using Manning's constant **Limitations**: Requires manual calibration of Manning's constant to model accurate rating curve
3.8	**Approach used**: Geographical Information System (GIS), Multi-Criteria Decision Making **Hardware/Software used**: HEC-RAS, MAUT **Datasets Used**: HAND, slope, soil type, stream layer, precipitation, floodplain, census **Results**: Flood risk maps based on different scenarios, compared with base maps gave accurate results for flood risk assessments **Limitations**: The use of effective precipitation instead of total precipitation might improve model accuracy

4 Analysis

Table 1 summarizes some of the existing systems in the domain of flood prediction, both machine learning based and GIS-based approaches, and highlights their results and limitations, setting the foundation for our proposed system.

5 The Proposed System

The proposed system is a hybrid model which is novel in the way that it combines the power of both machine learning as well as geospatial analysis. The system will consist of two modules, one which forecasts the river stage height based on current weather phenomena and the other which will generate a flood risk map for the given stage height.

The stage height will be predicted for a moderate horizon using ensemble models for multivariate time series forecasting based upon meteorological conditions. Once the stage has been predicted, the flood inundation map will be generated using the HAND model [9]. This way the proposed model will make use of both meteorological as well as physical conditions to give accurate predictions of the occurrence and extent of flooding in a watershed.

LSTM (Long Short-Term Memory) networks were used for modeling the complex relationships between rainfall and river stage, which does away with several complexities involved and leaves the task of learning to the computers, making the implementation and deployment of such crucial systems much easier. The accuracy of the

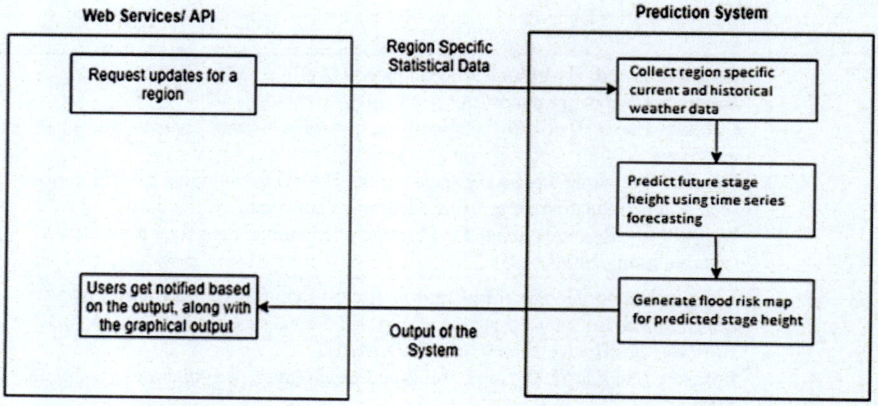

Fig. 1 System block diagram

HAND generated map can be further improved by the use of DEM's of higher resolutions and calibration of the flow accumulation thresholds. Our proposed system gives priority to quick development and real-time predictions without any significant compromise on the accuracy (Fig. 1).

6 Conclusions

Traditional ML-based flood prediction models are not as accurate or dynamic or are unable to prove both long-term and short-term predictions. Furthermore, data scarcity and inaccurate data can be a hindrance to the use of such models. Table 2 demonstrates how our proposed system builds upon the existing systems by coupling GIS-based systems with ML-based data-driven models, which can be a viable solution to these problems and can provide better results. The use of hydrographic modeling systems by taking into consideration the geographic conditions can provide a better way to

Table 2 Comparison of the proposed system with existing systems

Existing systems	Proposed system
Lack of long-term flood prediction capabilities	Will be capable of both short term as well as long-term prediction
Does not consider all relevant and crucial parameters	Will consider parameters which are region specific and which have a higher effect on flooding
Data intensive models	Usable for data-scarce regions
Cannot determine the extent of floods in a region	Will give extent of flooding in the region

identify flood risk zones in the catchment and predict the occurrence and extent of floods.

References

1. Fi-John C, Kuolin H, Li-Chiu C, Yinghao Y, Hongbo Z, Vijay S, Jianzhong Z, Mohammad E , Faridah O, Zaher Y , Haitham A, Mohammed A, Gokmen T, Saeed F, Hojat K, Nazanin F , Ozgur K , Aida J , Deg-Hyo B, Amir M (2019) Flood forecasting using machine learning methods. https://doi.org/10.3390/books978-3-03897-549-6
2. Jajarmizadeh M (2012) A review on theoretical consideration and types of models in hydrology. J Environ Sci Technol 5:249–261
3. An D, Kim NH, Choi J-H (2015) Practical options for selecting data-driven or physics-based prognostics algorithms with reviews. Reliabil Eng Syst Safety 133. https://doi.org/10.1016/j.ress.2014.09.014
4. Fang Z, Wang Y, Peng L, Hong H (2020) Predicting flood susceptibility using long short-term memory (LSTM) neural network model. J Hydrol. https://doi.org/10.1016/j.jhydrol.2020.125734
5. Shi Y1, Cheng T1, Taalab KP1 (2016) Flood prediction using support vector machines (SVM) 1 SpaceTimeLab, Department of Civil, Environmental and Geomatic Engineering, University College London, Chadwick Building, Gower Street, London WC1E 6BT, United Kingdom January 5
6. Falah F, Rahmati O, Rostami M, Ahmadisharaf E, Daliakopoulos IN, Pourghasemi HR (2019) Artificial neural networks for flood susceptibility mapping in data-scarce urban areas. Spatial Model in GIS and R for Earth Environ Sci
7. Nobre AD, Cuartas LA, Hodnett M, Rennó CD, Rodrigues G, Silveira A, Waterloo M, Saleska S (2011) Height above the Nearest Drainage, a hydrologically relevant new terrain model. J Hydrol
8. Sunilkumar P, Vargheese KO (2017) Flood modelling of mangalam river using GIS and HEC-RAS. Department of Civil Engineering, Government Engineering College, Thrissur, India
9. Tarboton D, Zheng X, Maidment D, Liu Y, Wang S (2018) Using digital elevation model derived height above the nearest drainage for flood inundation mapping and determining river hydraulic geometry. HydroShare http://www.hydroshare.org/resource/8ffaac4118db485badbe48bed96633be
10. Garousi-Nejad I, Tarboton DG, Aboutalebi M, Torres-Rua AF (2019) Terrain analysis enhancements to the height above nearest drainage flood inundation mapping method
11. Rincón D, Khan UT, Armenakis C (2018) Flood risk mapping using gis and multi-criteria analysis: a greater toronto area case study

Implementation of Machine Learning in Lung Cancer Prediction and Prognosis: A Review

Afsha Jaweed and Farheen Siddiqui

Abstract The machine learning is essentially the computer assessment inevitably developed over practice which is a part within the sphere of artificial intelligence. Machine learning algorithms are based upon models that are data driven which is the training set provided to the algorithm resulting in making prediction. Statistical, probabilistic knowledge is used to site the patterns from past examples which contain data sets which are huge, defeating, or hard. This skill is explicitly acceptable for medical applications especially for cancer detection and diagnosis as they rely on complex measurements of proteomics and genomics. The application of machine learning in cancer prediction is around 20 years now. The accuracy of the algorithms has gradually increased from past till now. The intention of this paper is to evaluate various machine learning algorithms particularly for the lung cancer detection to look for a void in the future improvement of lung cancer detection. Each technique is analyzed, and the overall disadvantages are pointed out. Various types of machine learning algorithms like Naive Bayes, support vector machine (SVM), logistic regression, and artificial neural network (ANN) have been applied in the healthcare sector for analysis and prognosis of lung cancer.

Keywords Machine learning · Training data · Genomics · Naïve Bayes · Support vector machine (SVM) · Artificial neural networks (ANN) · Logistic regression · Lung cancer

1 Introduction

The field of science in which machines can understand and behave like humans is machine learning. In this, machines are fed with facts and knowledge in the form of evidence and real-world experiences, and they develop their learning over time

A. Jaweed (✉) · F. Siddiqui
Department of Computer Science and Engineering, Jamia Hamdard, New Delhi, India

F. Siddiqui
e-mail: fsiddiqui@jamiahamdard.ac.in

J. M. R. S. Tavares et al. (eds.), *Cyber Intelligence and Information Retrieval*,
Lecture Notes in Networks and Systems 291,
https://doi.org/10.1007/978-981-16-4284-5_20

Fig. 1 PubMed publications on lung cancer from India

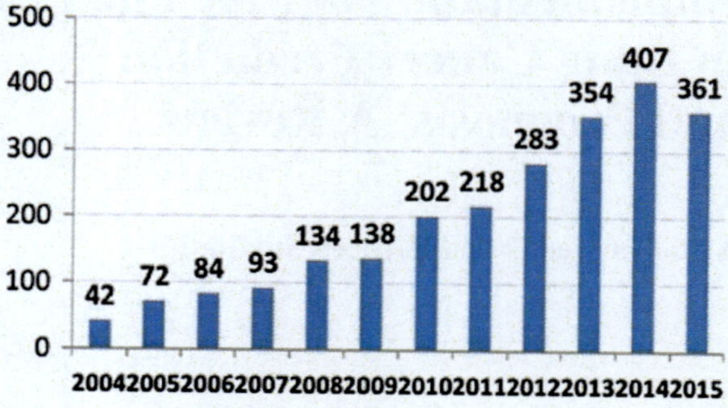

independently. Machine learning is not new to cancer science. Machine learning algorithms like artificial neural networks, decision trees have been used for almost 20 years. Besides cancer detection, these are used for CRT image identification classification of tumors and X-ray [1]. Cancer is considered as a most critical illness that results in death. Among 14.1 million cases of cancer worldwide in 2012, lung cancer stands at 13 percent that means 8.2 million deaths due to this. Cancer is transmitted through lymph exit or the bloodstream into the lungs. Small cell lung malignancies (SCLC) and non-small cells (NSCLC) are the two pulmonic diseases that circulates unusually in our body. In general, about 13% of all lung cancers are SCLC, and 84% are NSCLC. If diagnosed early, the survival rate will rise to 55% approximately. Consequently, it is challenging for doctors to diagnose adenocarcinoma. It can be detected using medical images such as evaluated images, chest X-ray, and CT scans. The predictions of lung cancer were mentioned by the American cancer society for 2021, and according to them, in USA approximately 235, 760 new lung cancer cases were detected (119,100 men and 116,660 in women). And about 131,800 lung cancer deaths were reported (69,410 men and 62,470 women). It is a serious problem in India also as it is clearly visible from Fig. 1 depicts the articles mentioned in PubMed from India on lung cancer [2].

Machine learning methods help us to understand the key features of complex lung cancer [3]. Decision trees, linear regression, random forest, SVM, Naïve Bayes, K-nearest neighbors, etc., are some machine learning algorithms that have deep impact on medical aid.

The foremost objective of this research work is to propose a brief work on lung cancer using machine learning algorithms. There are four stages of lung cancer. Stage 1: confined to lungs. Stage 2: confined to chest. Stage 3: confined to chest but reveal the most belligerent tumors. Stage 4: spread to different parts of the body. Cough, chest pain, dyspnea, shoulder pain, paralysis of vocal cords, weight loss, weakness, fatigue, hemoptysis, or coughing of blood are some of the common symptoms faced by the patients who suffer from lung carcinoma [27].

In this paper, Sect. 2 discusses about the growth of lung cancer in the past decade worldwide as well as in India. Section 3 describes the literature review of this paper. Section 4 compares the various papers, machine learning algorithms, and the accuracy obtained. Section 5 proposes the future work and concludes the paper.

2 Growth of Lung Cancer in Past Decade

Due to the beginning of tobacco epidemic in the world, fatalities due to lung cancer are rising tremendously in many countries. Cigarette smoking is the leading cause of this disease. The cases have increased from 347 in 1912 to 1.8 million cases in 2012 and 1.6 million deaths in the same year. In 2008, 1.6 million cases were diagnosed, and 1.4 million deaths have taken place. In 2018, USA declared around 234,030 cases which is about 641 cases a day. [4]. According to the data collected from ICMR cancer registry, about 41,000 lung cancer cases were diagnosed as on July 1, 2002, with the M: F ratio of 5:1 depending on the age and the smoking status of the person[5].

3 Related Works

Cruz et al. [1] calculated the success rate of cancer prediction and prognosis by surveying different kinds of machine learning methods and the incorporated data. Former machine learning methods like artificial neural networks are proved to be more reliable in case of prostate and breast cancer as compared to the recently developed machine learning methods, and it is also proved that these algorithms are more accurate over the regular statistical or expert-based systems. Pradhan and Chawla [3] evaluated the lung cancer using different machine learning algorithms associated with IoT devices, and for the betterment in this field, a gap was searched toward the future improvement in medical IoT. Benefits as well as challenges were noted. In this paper, detection of lung cancer at an early stage was given. Numerous data sets were observed be it standard or manually gathered, and the performance metrics were specified with its simulation platforms Raoof et al. [6] applied several machine learning algorithms like Naïve Bayes, logistic regression, support vector machine, and artificial neural networks is given in reference to lung cancer with their respective advantages and disadvantages. The most accurate result was produced using supervised learning methods, and accordingly, a cancer prediction system was introduced. Shanthi and Rajkumar [7] examined the need for an early diagnosis of lung cancer resulting in a greater chance of successful treatment. Along with the Gabor filter extraction technique, this focuses on grey level co-occurrence matrix. Stochastic diffusion search (SDS), an algorithm of feature selection is proposed. SDS has better execution in contrast to current approaches like minimum redundancy, maximum relevance, and co-relation-based feature selection. Xie et al. [8] proved that the patients with lung cancer can survive if diagnosed early. Here an attempt was made, and Chinese plasma metabolites were discovered which were the biological indicators of/for lung cancer and were developed using machine learning models. They recommended Naïve Bayes as an extraordinary tool for the detection of lung cancer which was not limited to lung cancer only but other kind of cancers as well. Kadir and Gleeson [9] described the main interventions used for nodule recognition and the prediction of lung cancer using Ct scans. Various data sets have been used like if the patient is a smoker or a non-smoker or if the patient is suffering from any disease or the patient has a history of any malignancy. Alam et al. [10] radiologists can use an intelligent computerized recognition system for prediction and diagnosis of lung cancer. Here support vector machine (SVM) is used for cancer prognosis and chances of cancer. Hazra et al. [11] observed Support vector machine (SVM) and logistic regression algorithms, the subsistence of cancer patients by forecasting these methods and recording their accuracy, precision, recall, F1 score, and confusion matrix helping the doctors to make correct decisions. Adetiba and Olugbara [12] compared artificial neural network (ANN) with support vector machine (SVM) for predicting lung cancer. Here, a protein known as epidermal growth receptor is altered which are collected by samples received from patient's nucleotides. Kirsten rat sarcoma viral oncogene and tumor suppressor p53 genomes collected as biomarkers from the IGDB. Nasser et al. [13] recorded various symptoms of lung cancer like jaundice, anxiety, chronic disease, exhaustion, allergic diseases, acute infection in bronchitis, coughing, dyspnea, esophagitis, and pleuritis, and accordingly, an artificial neural network(ANN) was designed which detected if the cancer was present or not and accuracy was measured.

4 Comparative Analysis

The inferences from some of the papers studied are collected in Table 1, and the conclusions are compared based on the type of validation conducted during survey.

The accuracies of different classifiers are presented in Fig. 2. It can be clearly seen that support vector machine and Naïve Bayes show maximum accuracy of 97 and 100%, respectively, as compared to other classifiers, followed by LFDA-SVM with an accuracy of 96.77%, ANN with an accuracy of 96.67%, ensemble MLP-ANN with an accuracy of 95.90%, DNN with an accuracy of 94.56%, SDS-NN with an accuracy of 89.63%, and logistic regression with the minimum accuracy of 77.40%.

5 Conclusion and Future Work

'LUNG CARCINOMA' which is another name for lung cancer are prime cause of cancer deaths in the past few years. It is a deadly tumor that has uncontrolled cellular development. To prevent metastasis, its early detection is very critical. The predominant smoking of tobacco for a sustained period is the major cause which leads to 85% of lung cancer. Here, various papers are reviewed were different machine learning algorithms like SVM, logistic regression, Naïve Bayes, decision tree, and artificial neural network are used for prediction and prognosis of lung cancer. The accuracies of these are compared and mentioned in Table 3. From the comparison table, we observe that SVM and ANN give the most accurate result. The major

Table 1 Comparative analysis

Paper	Data set	Algorithms used	Accuracy obtained (%)
Kanchan Pradhan et al. [3]	UCI repository data set and National lung screening trial (NLST)	LFDA-SVM ANN Decision tree Logistic regression	96.77 95 93.6 82.2
Syed Saba Raoof et al. [6]	Primary data set: Chest X-rays, MRI scans, CT scans, bone scan LUNA 16, lung 1, TCIA data set	KNN NN DNN SVM	94.56
Shanthi et al. [7]	Primary data set: 140 usual and 130 unusual images were used from TCGA CT, ultrasound, PET, MRIs were used	SDS-NN MRMR-DT CFS DT SDS DT MRMR-NB CFS-NB SDS-NB MRMR-NB	89.63 83.33 82.22 87.41 81.85 81.48 88.52
Xie et al. [8]	Primary data of 110 patients (43 healthy)	K-nearest Naïve Bayes Ada boost SVM Random forest Neural network	93.6 100.0 88.5 97.4 97.4 100.0
Kadir et al. [9]	Input CT image, LIDC-IDRI data set	SVM regression CNN	97
Alam et al. [10]	Multiclass SVM classifiers		
Animesh et al. [11]	Primary data set: Cancer imaging archieve which has 422 observations and 10 variables Clinical genomic data, e.g., patient id, age, etc.)	SVM Logistic regression	76.20 77.40
Adetiba et al. [12]	CCDS (collaborative consensus coding sequence) Data from IGDB-NSCLS (integrated genomic database of non-small cell lung cancer	Multi-layered perceptron artificial neural network(MLP-ANN) Non-ensemble MLP-ANN Ensemble MLP-ANN Non-ensemble SVM Ensemble SVM	87.6 95.9 86.5 91.9

(continued)

Table 1 (continued)

Paper	Data set	Algorithms used	Accuracy obtained (%)
Ibrahim et al. [13]	Primary data of people who may or may not have lung cancer, e.g., gender, age, etc.	ANN anxiety	96.67

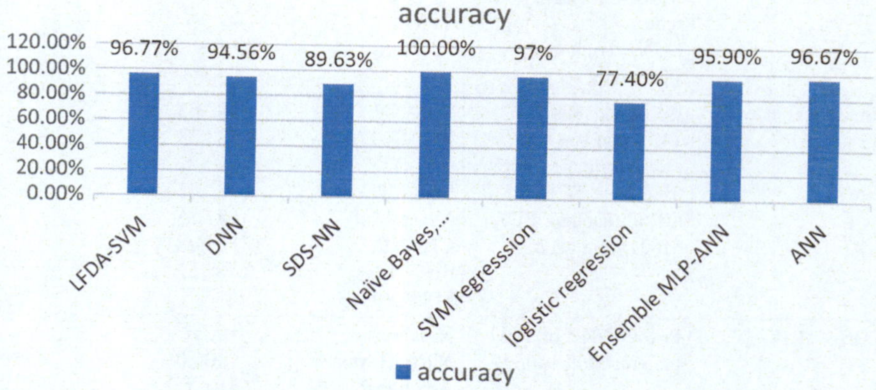

Fig. 2 Highest accuracies achieved in each paper

intention of this study is to provide a detailed overview of the current developments and trends in the field of machine learning w.r.t lung cancer so that the researchers can understand what has already been achieved and what are the expectations. It will also act as a base for future researchers and will help them to choose best method by simply modifying the parameters of the existing ones.

While reviewing the papers, we came across various machine learning algorithms like SVM, logistic regression, artificial neural network, decision tree, Naïve Bayes, recurrent neural networks, etc., that have been used in detection and prediction of lung cancer. The purpose of this research is to develop a machine learning algorithm in future, which would be useful in detection of lung cancer and will give us more accurate results. The data set we will use for designing such algorithms can be obtained from clinics like MRI report of the patient, X-ray of the patient and the primary data which includes age, name, sex, symptoms, etc., of the patient. The machine learning algorithm should be cost effective and should give us the accurate results in optimum time at an early stage. The algorithm can be visualized through confusion matrix which determines the performance of the algorithm.

References

1. Cruz DSWJA (2006) Applications of machine learning in cancer prediction. Cancer Info 2006: 2 59–78 2:59–78
2. A. A. R. B. G. N. G. R. S. R. B. G. S. B. S. K. M. S. A. V. A. V. a. S. H. P. M. Parikh (2016) Lung cancer in India: Current status and promising strategies. South Asian J Cancer 93–95
3. Chawla KPAP (2020) Medical internet of things using machine learning algorithms for lung. J Manage Anal 7(4):591–623
4. de Groot CCWBWCARFMPM (2018) The epidemiology of lung cancer. Translational Lung Cancer Res. 220–233
5. Behera BTD (2004) Lung cancer in India. Indian J Chest Dis Allied Sci 81:46–269
6. Raoof MAJASAFSS (2020) Lung cancer prediction using machine learning: a comprehensive approach. IEEE 108–115
7. Rajkumar SSNA (2020) Lung cancer prediction using stochastic diffusion search (SDS) based feature selection and machine learning methods. Springer US
8. Xie W-YMR-ZLEAY (2020) Early lung cancer diagnostic biomarker discovery by machine learning. Elsevier 1936–5233
9. Gleeson TKAF (2018) Lung cancer prediction using machine learning and advanced imaging techniques. Transl Lung Cancer Res 7:304–312
10. Janee Alam SAH (2018) Multi-stage lung cancer detection and prediction using multi-class SVM classifier. IEEE pp 1–4
11. Animesh Hazra NBM (2017) Predicting lung cancer survivability using SVM and logistic regression algorithm. Int J Comput Appl 174:19–24
12. Olugbara EAAOO (2015) Lung cancer prediction using neural network ensemble with histogram of oriented gradient genomic features. Sci World J. 1–17
13. Nasser SSA-NIM (2019) Lung cancer detection using artificial neural network. Int J Eng Info Syst (IJEAIS) 3(3):17–23

Fetching Information Through Crowdsourcing Within a Social Networking Site

Goldina Ghosh, Soumi Dutta, Abhinandan Das, Anasuya Dev, Birol Roy, Debapriya Bhowmick, and Prasenjit Saha

Abstract A collaborative activity leads to a formation of a lot of information. These activities are usually in a group in any social networking sites. Different social networking sites can form different groups, and different types of information generated from these groups are usually unique. When a lot of people start communicating and sharing ideas in social networking site, it forms a crowd and this crowd lead to some source of information. Here in this paper, the concept is framed. ResearchGate social networking sites for researches are considered. We conducted an experimental analysis on a dataset comprising over ten types of research domains performed by 50 researchers in different institutes. A number of researchers from different institutes working on numerous domains are assembled together for the analysis. The analysis is done through crowdsourcing in a social networking site to fetch information.

Keywords Social network analysis · Crowdsourcing · Collaborative activity · Fetching of information · Grasshopper algorithm

1 Introduction

ResearchGate is an online social networking site that provides a platform for sharing all kinds of information related to research. All researches on different fields are available. Different research works take place in collaboration with different institutes or in same institute. Once the paper is published, all data regarding the journal, conference, and book chapter along with the author details like name, designation, and institute name are all available from the site. From all these records, a good analysis can be performed related to collaboration, diffusion of knowledge through crowdsourcing

G. Ghosh (✉) · S. Dutta · A. Das · A. Dev · B. Roy · D. Bhowmick · P. Saha
Department of Computer Application and Science, Institute of Engineering and Management, Salt Lake, Kolkata, India
e-mail: goldina.ghosh@iemcal.com

S. Dutta
e-mail: soumi.dutta@iemcal.com

J. M. R. S. Tavares et al. (eds.), *Cyber Intelligence and Information Retrieval*, Lecture Notes in Networks and Systems 291,
https://doi.org/10.1007/978-981-16-4284-5_21

in a social networking site. To justify the concept in this paper, ResearchGate site data has been used to analyze how crowdsourcing is helping in fetching information. There can be novice researchers or there can be experienced researchers working in a same research domain under a particular team. The researches can work together from same institute or in collaboration from different institutes in a same domain or different domain. Hence, keeping all these conditions under consideration a set of experimental work had been planned, and they are as follows:

Researchers with common research domain
Researchers belonging in the same institute with common research domain
Researchers belonging in the different institute with common research domain
From RG score select, the best or experienced researchers, so that new researchers can collaborate with them

While doing a research in a group, there can be both experienced or novice researchers. Both the levels are important in a team work as the experienced researcher can deliver innovative ideas and novice researchers can explore these ideas in details and give their inputs. Based on these inputs, the experienced researchers can further carry on the research which leads to a success. Such a kind of active is also visible in nature where different species work in collaboration. One such group is the grasshoppers. Based on their behavior of searching for food has led to the development of grasshopper algorithm. There are two stages of the grasshoppers, they are, nymph and the adult. In the nymph stage they are too young to fly so they hop around the ground to research the destination of the food. Where as in the adult stage they can fly and can explore a wider range to research the destination. The concept of exploration and exploitation of the nymph and the adult is observed in grasshopper algorithm [1, 2]. Here in this paper a demonstration is given about the concept of exploration and exploitation among the researchers based on the grasshopper algorithm. The two levels of researcher, that is novice and experienced, can work together and can get great success in there publication work. Researchers from different research domain can work in collaboration from same institute or from different institutes.
To perform the analysis different survey related to crowdsourcing, social networking and grasshopper algorithm technique had to be done.

2 Related Work

To perform the work, certain related papers were studied related to social network analysis and crowdsourcing. Information filtering and retrieval is one of the most challenging area in online social networking data analysis. Several different types of researches [3–13] are ongoing in this domain An instrumental achievement in achieving a better consideration of human–environment interactions based on the analysis of data from social media and social networking sites was performed in [14]. A quantitative review was done on 169 studies which discovered that most studies concentrates on the examination of people's behavior and observations of the environment along with environmental monitoring and applications in environmental planning and governance. The authors in [15] has used a theoretical framework to understand and explain the behaviors of users to give an idea on crowdsourcing platforms for supporting organizational innovation. They had named it as social interdependence theory. The study had 19,516 data for analyzing both behavioral and social network of registered users of a crowdsourcing platform. Social network being an Internet-based collaboration platform that provides vital role in information spread, opinion-forming, trend-setting, and hence keeping everyone connected. Based on this concept, the paper in [16] discusses on how effectively social networks can provide information propagation in the context of viral marketing. They gave a detailed discussion on how effectively social networks can be used for information propagation in viral marketing context. Data is the only valuable element with wide capability to create business value which is not seen in the traditionally executed business. Based on this context, the research paper [17] provides an attempt to give a theoretical framework and its empirical analysis to apply analytics effectively. This helps in removing uncertainties in business for its stability. Social network-based crowdsourcing idea was applied for recruiting participants [18]. The reliability and effectiveness of multimedia intricate behavioral testing was integrated. Crowdsourcing-based research for a music recommendation problem was incorporated. A novel task assignment method was proposed by Zhiyong Zhang et.al to enable crowd assessment. This improves the security and trustworthiness of social networking platforms. It also behaves as a task assignment algorithm which is a social-domain associated situational analytics based on SocialSitu [19, 22]. The increase in collaborative work in interdisciplinary field has been justified by the concept of graph theory and genetic algorithm. The analysis was done on ResearchGate social networking site. Each researcher's performance was evaluated by calculating their grade score, and this value was passed as a fitness function value in genetic algorithm [21]. This work was further extended by using PSO in [20]. Exploitation was done using genetic algorithm in [21], and exploration was done with PSO using the same grade score value to the objective function [20, 22].

3 Dataset

The dataset was prepared according to the ResearchGate social networking site. The essential attributes are considered for the site. They are researcher's name, RG score, number of articles published by that researcher (journal, conference, book chapter), research domain, and the respective institute names. The corresponding data related to these fields have been developed. The dataset was prepared for 50 records. Two tables were prepared. Table 1 consists of "Researcher's name," "RG score," "Number of articles," "Research domain," and "Institute name." Table 2 consists of "Researcher's name," "Co-authors," "Research domain," "Institute name of researcher," and "Institute name of co-authors." Table 1 and Table 2 records are available from the corresponding author on request.

4 Problem Statement

To analyze the crowdsourcing, the data are collected from Table 1. To accomplish the research, four different algorithms were developed.
Algorithm 1 demonstrates about the number of the researchers who are doing research on common domain.

Algorithm 1: Researchers with common research domain

Input: Table 1 DATA.csv, Preferred domain
Variables: Two empty list z[], r[]
Calculations:
1. Enter the Domain name from user (to check for the Researchers researching on it)
2. Use "While" loop to search for the element (researcher) and "If-Else" to check for the match
3. Display Researcher(s) name(s)

Output: The list of Researchers researching on that Domain

In order to find a team work existing in the same institute, doing research on some common domain is expressed in Algorithm 2.

Algorithm 2: Researchers belonging in the same institute with common research domain

Input: Table 1 DATA.csv, Preferred domain, Institute name
Variables: Two empty list z[], r[]
Calculations:
1. Enter the domain name and institute name for which we are searching the researchers from user
2. Use "While" loop to search for the elements (researchers) from the table and "If-Else" to check for the match

Output: The list containing the names of all the researchers working in the given domain under that institute

Again to find a team working in different institutes and doing research on some common domain is expressed in Algorithm 3.

Table 1 Researcher's details

Researcher's name	RG score	Number of articles	Research domain	Institute name
R1	3.66	5	BIOINFORMATICS	I1
R2	10.01	21	CLOUD COMPUTING	I2
R3	17.33	25	ARTIFICIAL INTELLIGENCE	I3
R4	14.11	36	CYBER SECURITY	I4
R5	7.10	12	CLOUD COMPUTING	I2
R6	9.20	19	INTERNET OF THINGS	I6
R7	3.44	9	BIOINFORMATICS	I7
R8	10.44	27	ARTIFICIAL INTELLIGENCE	I3
R9	12.36	23	GRAPHICS AND VISUALIZATION	I9
R10	2.59	6	CYBER SECURITY	I10
R11	13.71	28	INTERNET OF THINGS	I11
R12	4.02	8	CLOUD COMPUTING	I12
R13	4.51	11	BIOINFORMATICS	I1
R14	3.99	7	ROBOTICS	I10
R15	10.42	17	ARTIFICIAL INTELLIGENCE	I15
R16	7.27	14	CLOUD COMPUTING	I2
R17	6.25	15	CYBER SECURITY	I4
R18	9.76	19	BIG DATA	I7
R19	14.10	31	ARTIFICIAL INTELLIGENCE	I3
R20	13.12	25	ROBOTICS	I20
R21	2.21	6	BIG DATA	I7
R22	5.52	14	BLOCKCHAIN TECHNOLOGY	I22
R23	4.81	9	CYBER SECURITY	I23
R24	6.12	19	ARTIFICIAL INTELLIGENCE	I24
R25	13.12	27	BLOCKCHAIN TECHNOLOGY	I25

(continued)

Table 1 (continued)

Researcher's name	RG score	Number of articles	Research domain	Institute name
R26	6.65	11	GRAPHICS AND VISUALIZATION	I9
R27	15.25	31	COMPUTER VISION	I27
R28	3.25	9	CYBER SECURITY	I4
R29	7.22	19	CLOUD COMPUTING	I29
R30	2.25	6	CYBER SECURITY	I12
R31	5.02	12	BIOINFORMATICS	I1
R32	4.03	10	ARTIFICIAL INTELLIGENCE	I3
R33	7.42	19	COMPUTER VISION	I27
R34	4.12	8	BLOCKCHAIN TECHNOLOGY	I25
R35	14.18	28	BIOINFORMATICS	I1
R36	3.96	6	GRAPHICS AND VISUALIZATION	I15
R37	7.36	20	CYBER SECURITY	I37
R38	10.45	22	GRAPHICS AND VISUALIZATION	I9
R39	6.27	15	COMPUTER VISION	I27
R40	9.31	23	INTERNET OF THINGS	I6
R41	14.42	38	ARTIFICIAL INTELLIGENCE	I41
R42	7.44	16	BLOCKCHAIN TECHNOLOGY	I42
R43	7.66	18	INTERNET OF THINGS	I22
R44	10.15	24	ROBOTICS	I20
R45	5.31	11	BIOINFORMATICS	I1
R46	11.36	21	CLOUD COMPUTING	I37
R47	8.12	16	BLOCKCHAIN TECHNOLOGY	I29
R48	5.56	10	ARTIFICIAL INTELLIGENCE	I48
R49	7.32	14	INTERNET OF THINGS	I6
R50	13.25	27	BIG DATA	I7

Table 2 Researcher's with their research details

Researcher's Name	Co Authors	Research Domain	Institute Name of Researcher	Institute Name Co-Authors
R1	R7	BIOINFORMATICS	I1	I7
	R13			I1
	R31			
	R35			
	R45			
R2	R5	CLOUD COMPUTING	I2	I2
	R12			I12
	R16			I2
	R29			I29
	R46			I37
R3	R8	ARTIFICIAL INTELLIGENCE	I3	I3
	R15			I15
	R19			I3
	R24			I24
	R32			I3
	R41			I41
	R48			I48
R4	R10	CYBER SECURITY	I4	I10
	R17			I4
	R23			I23
	R28			I4
	R30			I12
	R37			I37
R5	R2	CLOUD COMPUTING	I2	I2
	R12			I12
	R16			I2
	R29			I29
	R46			I37
R6	R11	INTERNET OF THINGS	I6	I11
	R40			I6
	R43			I22
	R49			I6
R7	R1	BIOINFORMATICS	I7	I1
	R13			
	R31			
	R35			
	R45			

(continued)

Table 2 (continued)

Researcher's name	Co authors	Research domain	Institute name of researcher	Institute name co-authors
R8	R3	ARTIFICIAL INTELLIGENCE	I3	I3
	R15			I15
	R19			I3
	R24			I24
	R32			I3
	R41			I41
	R48			I48
R9	R26	GRAPHICS AND VISUALIZATION	I9	I9
	R36			I15
	R38			I9
R10	R4	CYBER SECURITY	I10	I4
	R17			I4
	R23			I23
	R28			I4
	R30			I12
	R37			I37
R11	R6	INTERNET OF THINGS	I11	I6
	R40			I6
	R43			I22
	R49			I6
R12	R2	CLOUD COMPUTING	I12	I2
	R5			I2
	R16			I2
	R29			I29
	R46			I37
R13	R1	BIOINFORMATICS	I1	I1
	R31			
	R35			
	R45			
R14	R20	ROBOTICS	I10	I20
	R44			

(continued)

Table 2 (continued)

Researcher's name	Co authors	Research domain	Institute name of researcher	Institute name co-authors
R15	R3	ARTIFICIAL INTELLIGENCE	I15	I3
	R8			I3
	R19			I3
	R24			I24
	R32			I3
	R41			I41
	R48			I48
R16	R2	CLOUD COMPUTING	I2	I2
	R5			I2
	R12			I12
	R29			I29
	R46			I37
R17	R4	CYBER SECURITY	I4	I4
	R10			I10
	R23			I23
	R28			I4
	R30			I12
	R37			I37
R18	R21	BIG DATA	I7	I7
	R50			
R19	R3	ARTIFICIAL INTELLIGENCE	I3	I3
	R8			I3
	R15			I15
	R24			I24
	R32			I3
	R41			I41
	R48			I48
R20	R14	ROBOTICS	I20	I10
	R44			I20
R21	R18	BIG DATA	I7	I7
	R50			
R22	R25	BLOCKCHAIN TECHNOLOGY	I22	I25
	R34			I25
	R42			I42
	R47			I29

(continued)

Table 2 (continued)

Researcher's name	Co authors	Research domain	Institute name of researcher	Institute name co-authors
R23	R4	CYBER SECURITY	I23	I4
	R10			I10
	R17			I4
	R28			I4
	R30			I12
	R37			I37
R24	R3	ARTIFICIAL INTELLIGENCE	I24	I3
	R8			I3
	R15			I15
	R19			I3
	R32			I3
	R41			I41
	R48			I48
R25	R22	BLOCKCHAIN TECHNOLOGY	I25	I22
	R34			I25
	R42			I42
	R47			I29
R26	R9	GRAPHICS AND VISUALIZATION	I9	I9
	R36			I15
	R38			I9
R27	R33	COMPUTER VISION	I27	I27
	R39			I27
R28	R4	CYBER SECURITY	I4	I4
	R10			I10
	R17			I4
	R23			I23
	R30			I12
	R37			I37
R29	R2	CLOUD COMPUTING	I29	I2
	R5			I2
	R12			I12
	R16			I2
	R46			I37

(continued)

Table 2 (continued)

Researcher's name	Co authors	Research domain	Institute name of researcher	Institute name co-authors
R30	R4	CYBER SECURITY	I12	I4
	R10			I10
	R17			I4
	R23			I23
	R28			I4
	R37			I37
R31	R1	BIOINFORMATICS	I1	I1
	R13			
	R35			
	R45			
R32	R3	ARTIFICIAL INTELLIGENCE	I3	I3
	R8			I3
	R15			I15
	R19			I3
	R24			I24
	R41			I41
	R48			I48
R33	R27	COMPUTER VISION	I27	I27
	R39			
R34	R22	BLOCKCHAIN TECHNOLOGY	I25	I22
	R25			I25
	R42			I42
	R47			I29
R35	R1	BIOINFORMATICS	I1	I1
	R13			
	R31			
	R45			
R36	R9	GRAPHICS AND-VISUALIZATION	I15	I9
	R26			I9
	R38			I9
R37	R4	CYBER SECURITY	I37	I4
	R10			I10
	R17			I4
	R23			I23
	R28			I4
	R30			I12

(continued)

Table 2 (continued)

Researcher's name	Co authors	Research domain	Institute name of researcher	Institute name co-authors
R38	R9	GRAPHICS AND-VISUALIZATION	I9	I9
	R26			I9
	R36			I15
R39	R27	COMPUTER VISION	I27	I27
	R33			
R40	R6	INTERNET OF THINGS	I6	I6
	R11			I11
	R43			I22
	R49			I6
R41	R3	ARTIFICIAL INTELLIGENCE	I41	I3
	R8			I3
	R15			I15
	R19			I3
	R24			I24
	R32			I3
	R48			I48
R42	R22	BLOCKCHAIN TECHNOLOGY	I42	I22
	R25			I25
	R34			I25
	R47			I29
R43	R6	INTERNET OF THINGS	I22	I6
	R11			I11
	R40			I6
	R49			I6
R44	R20	ROBOTICS	I20	I20
	R44			

(continued)

Table 2 (continued)

Researcher's name	Co authors	Research domain	Institute name of researcher	Institute name co-authors
R45	R1	BIOINFORMATICS	I1	I1
	R13			
	R31			
	R35			
R46	R2	CLOUD COMPUTING	I37	I2
	R5			I2
	R12			I12
	R16			I2
	R29			I29
R47	R22	BLOCKCHAIN TECHNOLOGY	I29	I22
	R25			I25
	R34			I25
	R42			I42
R48	R3	ARTIFICIAL INTELLIGENCE	I48	I3
	R8			I3
	R15			I15
	R19			I3
	R24			I24
	R32			I3
	R41			I41
49	R6	INTERNET OF THINGS	I6	I6
	R11			I11
	R40			I6
	R43			I22
R50	R18	BIG DATA	I7	I7
	R21			

(continued)

Algorithm 3: Researchers belonging in the different institute with common research domain.

Input: Table 1 DATA.csv, Preferred domain, Institute name
Variables: Two empty list z[], r[]
Calculations:
1. Searching the researchers with common research domain
2. Use "While" loop to search for the elements (researchers) from the table and "If-Else" to check for the match

Output: The list containing the names of all the researchers working in the given domain in different (other) institutes

To form a better way of collaboration, the best researcher in a particular domain needs to be detected. Hence, the best or experienced researchers are denoted in Algorithm 4 using Table 2 record set.

Algorithm 4: From RG score select the best or experienced researchers so that new researchers can collaborate with them.

Input: Table 1DATA.csv, Preferred domain
Variables: Two empty list z[], r[], Max
Calculations:
1.Declare first researcher's RG score as Max from a preferred domain based sorted list
2. Check the Max with other researcher's RG score using "While" loop and "If-Else" statement

Output: Name of best Researcher with RG score and domain name

From the algorithms, it was analyzed that if different levels of researchers working in a group the concept of crowdsourcing takes place. Based on the grasshoppers algorithm if two different levels that is nymph and adult is considered, it can be correlated with novice and experienced researchers. To perform a further analysis, Table 3 has been designed from Table 1. Table 3 gives the detail about research's name, their RG score, and the team they are working on. Based on the RG score, a distinct analysis has been done on broadly dividing the two levels of researcher's that is novice and experienced.

Based on the RG score, the different levels of researchers can be identified depending on the two conditions. The conditions are as follows:

Condition-1

IF RG score ≥ 6 THEN
Researcher = experienced researcher.
Hence, when compared with grasshopper algorithm, it is the adult stage.

Condition-2

IF RG score ≤ 5 THEN
Researcher = novice researcher.
Hence, when compared with grasshopper algorithm, it is the Nymph stage.

Table 3 Researcher's with their research team

Research's name	Research team	RG score
R7	A	3.66
R13		3.44
R31		4.51
R35		5.02
R45		14.18
R1		5.31
R5	B	10.01
R2		7.1
R12		4.02
R16		7.27
R29		7.22
R46		11.36
R3	C	17.33
R8		10.44
R15		10.42
R19		14.1
R24		6.12
R32		4.03
R41		14.42
R48		5.56
R4	D	14.11
R10		2.59
R17		6.25
R23		4.81
R28		3.25
R30		2.25
R37		7.36

Thus, if both levels of researchers work in a same team, then different forms of the research works can be completed through crowdsourcing. Novice can learn from experienced researchers, whereas the experienced researchers can give some experimental work to the novice who can help in accomplishing the research completely. Here lies the concept of exploration and exploitation. Grasshoppers search globally in large-scale to find an optimal solution of their destination by exploration. They tend to move locally when they are in the final steps allowing them to exploit the search space. There exists a comfort zone where the grasshoppers gradually balance exploration and exploitation. Here, this comfort zone is the section both grades of researchers can work together successfully.

5 Results and Discussion

For performing the analysis, the four different algorithms have been implemented using IDLE and Matplotlib module of Python to find out the number of researchers with common domain and Microsoft Power BI to show tabular data in graphical form. Figure 1 has the result of Algorithm 1 when applied on Table 1 and Fig. 2 gives the graphical output.

A combined representation of Algorithm 2 and 3 is given as a graphical representation in Fig. 3.

Algorithm 4 implemented in Table 2 for finding experienced researchers from RG score, so that new researchers can collaborate with them is explained in Fig. 4.

Both exploration and exploitation can take place in both these stages for doing a better research work. Experienced researchers can work for finding new techniques, and new researchers can do an intensive study to know more about the technique and share it with the experienced researchers. With this phenomenon, new ideas can be formed through exploration and exploitation. As per Table 3 following results were observed:

From Research Team A, it is observed that there is only one experienced researcher and all others are novice. But everyone's contribution leads to a successful research work. From Research Team B, it is observed that there is only one novice researcher and all others are experienced researchers. Research Team C has a major number of experienced researchers, and Research Team D has a major number of novice researchers. From the four different teams, it is observed how the crowdsourcing is taking place. An experienced researcher can train the novice researchers which is observed in Team A, whereas in Team B, it is observed that all experts have also

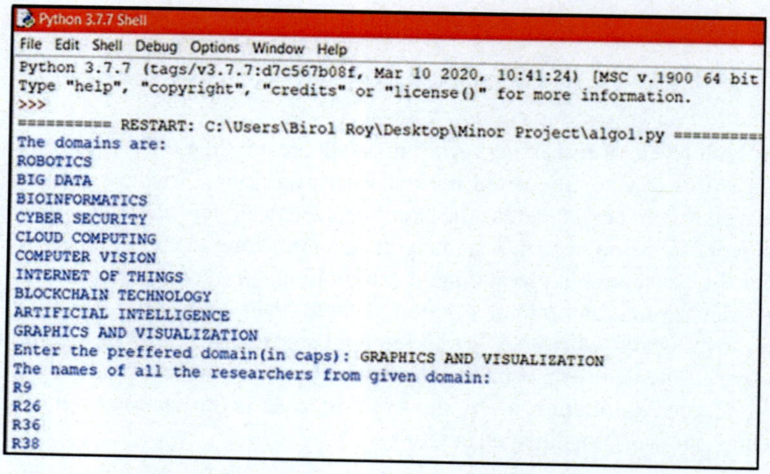

Fig. 1 Result related to number of researchers with common domain

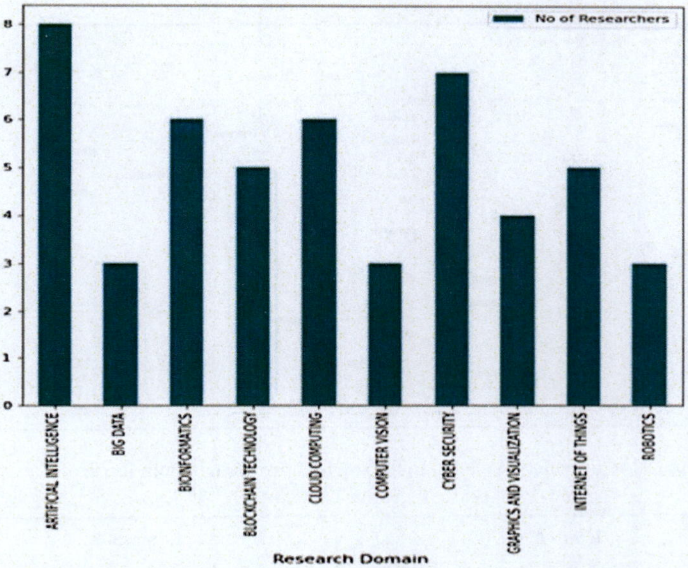

Fig. 2 Graphical representation related to number of researchers with common domain

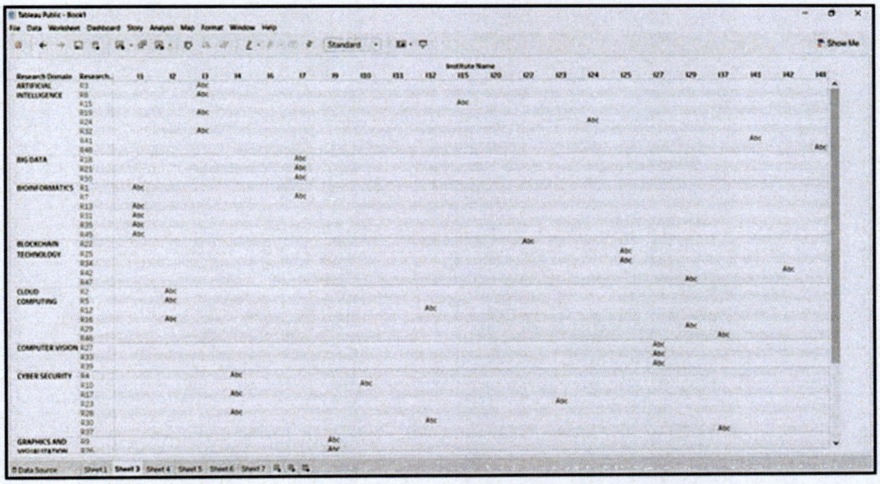

Fig. 3 Graphical representation related to researchers belonging in the same institute/ different institute

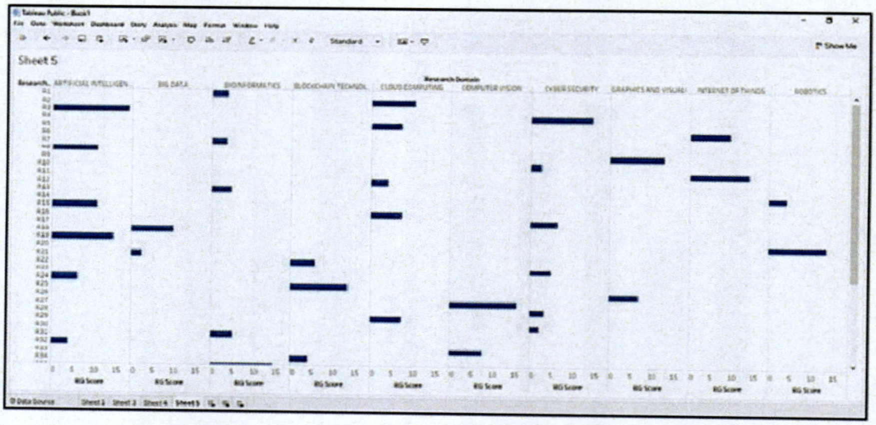

Fig. 4 Graphical representation related to finding best researchers from RG score

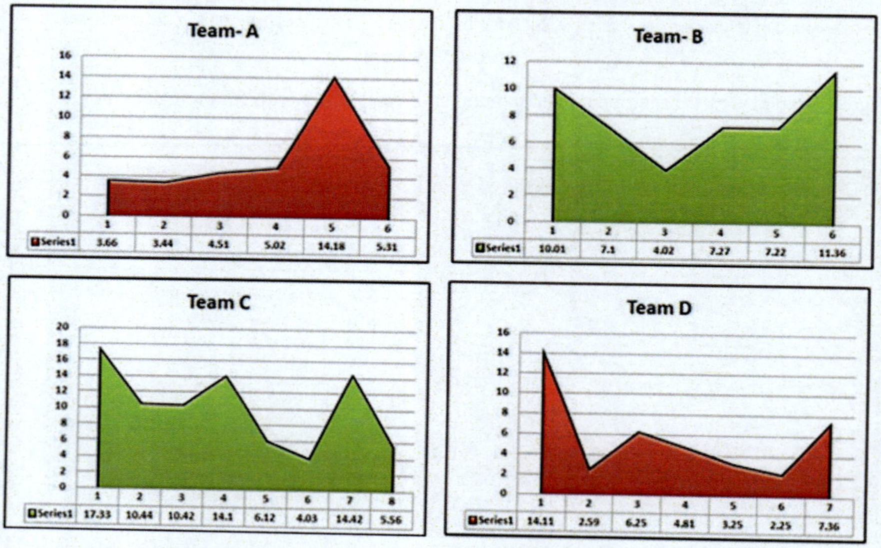

Fig. 5 Analytical report of exploration and exploration concept with crowdsourcing in social network

included a novice to encourage the work flow. The amalgamation of work with the novice and experienced researchers has been observed in both Team C and D. In Team C, there are a major number of experienced researchers, and in Team D, there are a major number of novice researchers. Figure 5 describes each of the teams with the researcher's RG score.

6 Conclusion

Social media and social networking services have become an area for doing different analytical research work. This analysis when applied with crowdsourcing helps in gaining a number of perceptions. When applying crowdsourcing, it is essential to describe about the goals and expectations in order to help the society for some improvement and finding a possibility of innovation. Here in this paper, this has been proved by considering the ResearchGate social networking site. Through crowdsourcing, it was possible to select the best researches and researchers from different institutes. It was observed how experienced and novice researchers work in a group successfully. Even a group having a major number of novice researchers can have a successful research work. Four different algorithms were developed to find "Researchers with common research domain," "Researchers belonging in the same institute with common research domain," "Researchers belonging in the different institute with common research domain," and "From RG score, select the best or experienced researchers, so that new researchers can collaborate with them." A basic idea to grasshopper behavior in grasshopper algorithm is also provided.

Future Scope The work can be extended in future by trying to implement the grasshopper algorithm for finding experienced researches in a particular domain or working in some interdisciplinary field in an optimized manner.

References

1. Saremi S, Mirjalili S, Lewis A (2017) Grasshopper Optimisation Algorithm: Theory and application. Advances in Engineering Software 105:30–47
2. Saxena A, Shekhawat S, Kumar R (2018) Application and development of enhanced chaotic grasshopper optimization algorithms. Hindawi Modell Simul Eng 2018. Article ID 4945157. https://doi.org/10.1155/2018/4945157
3. Dutta S, Chandra V, Mehra K, Das AK, Chakraborty T, Ghosh S (2018) Ensemble algorithms for microblog summarization. IEEE Intell Syst 33(3):4–14. https://doi.org/10.1109/MIS.2018.033001411.
4. Dutta S, Ghatak S, Dey R, Das AK, Ghosh S (2018) Attribute selection for improving spam classification in online social networks: a rough set theory based approach. In: Social network analysis and mining. Springer
5. Sarkar A, Bhattacharya A, Dutta S, Parikh KK (2018) A survey on recent trend in cloud data mining. In: Proceedings of IEEE and Springer sponsored international conference on emerging technologies in data mining and information security (Springer's AISC Series), 2018. vol 813, pp 565–578
6. Dutta S, Das AK, Dutta G, Parikh Parikh KK, Das AD, Ganguly D (2018) Community detection based tweet summarization. In: International conference on emerging technologies in data mining and information security (IEMIS2018)
7. Dutta S, Chandra V, Mehra K, Ghatak S, Das AK, Ghosh S (2018) Summarizing microblogs during emergency events: a comparison of extractive summarization algorithms. In: International conference on emerging technologies in data mining and information security (IEMIS2018)

8. Dutta S, Das AK, Dutta G, Gupta M (2018) A comparative study on cluster analysis of microblogging data. In: International conference on emerging technologies in data mining and information security (IEMIS2018)
9. Dutta S, Ghatak S, Das A, Gupta M, Dasgupta S (2017) Feature selection based clustering on micro-blogging data. In: International conference on computational intelligence in data mining (ICCIDM-2017)
10. Dutta S, Ghatak S, Ghosh S, Das AK (2015) A genetic algorithm based Tweet clustering technique series of IEEE Xplore digital library indexed. In: International conference on computer communication and informatics (ICCCI 2017)
11. Dutta S, Ghatak S, Roy M, Ghosh S, Das AK (2015) A graph based clustering technique for Tweet summarization series of IEEE Xplore digital library indexed. In: 4th international conference on reliability, infocom technologies and optimization (ICRITO 2015) in SCOPUS. Google Scholar, DBLP, EI-Compendex, OCLC, SCImago, Summon by serial solutions, pp 441–446
12. Das A, Roy M, Dutta S, Ghosh S, Das AK (2014) Predicting trends in the Twitter social network: a machine learning approach. In: The 5th joint international conference on swarm, evolutionary and memetic computing (SEMCCO 2014) and fuzzy and neural computing (FANCCO 2014), in Swarm, Evolutionary, and Memetic Computing, volume 8947 of the series lecture notes in computer science. Dec 2014, pp 570–581
13. Dutta S, Roy M, Das AK, Ghosh S (2014) Sentiment detection in online content: a wordNet based approach. In: Paper presented in the 5th joint international conference on swarm, evolutionary and memetic computing (SEMCCO 2014) and fuzzy and neural computing (FANCCO 2014), in Swarm, evolutionary, and memetic computing, volume 8947 of the series lecture Notes in computer science pp 409–420. Dec 2014
14. Ghermandi A, Sinclair M (2019) Passive crowdsourcing of social media in environmental research: A systematic map. Global Environmental Change. 55:36–47
15. Renard D, Davis JS (2019) Social interdependence on crowdsourcing platforms. Journal of Business Research. 103:186–194
16. Sumith N, Annappa B, Bhattacharya S (2017) A holistic approach to influence maximization. In: Banati H et al (eds) Hybrid intelligence for social networks. Springer. https://doi.org/10.1007/978-3-319-65139-2-6
17. Garg A, Goyal DP (2019) Sustained business competitive advantage with data analytics. Int J Bus Data Anal 1(1)
18. Choi JH, Lee JS (1991) Online social networks for crowdsourced multimedia-involved behavioral testing: an empirical study. Front Psychol 6. https://doi.org/10.3389/fpsyg.2015.01991
19. Zhang Z, Jing J, Wang X, Choo KKR, Gupta BB (2020) A crowdsourcing method for online social networks security assessment based on human-centric computing. Hum. Cent. Comput. Inf. Sci. 10:23. https://doi.org/10.1186/s13673-020-00230-0
20. Ghosh G, Akki CB, Kasturi N (2019) A PSO based investigation of research fields of researchers. J Kybernetes 49(6):1767–1782. Publisher: Emerald Publishing Limited. Publication date: 2019/9/14
21. Ghosh G, Kasturi N (2017) Expansion of social connectivity: a concept of big data analysis and genetic algorithm modeling. Inf Commun Technol (CICT) pp. 1–6. (Publisher: IEEE, Publication date: 2017/11/3)
22. Dutta S, Saha N, Ghosh S, Das AK (2019) Clustering model for microblogging sites using dimension reduction techniques. IGI Glob Int J Inf Syst Model Des (IJISMD)

Soil Nutrient Assessment and Crop Estimation with Machine Learning Method: A Survey

Yogesh Shahare and Vinay Gautam

Abstract The field of agricultural research is extremely important, as India or every farmer worldwide faces numerous problems in increasing crop production. In the context of this situation, many people worked in agriculture, such as soil quality, crop management, irrigation of agriculture, cultivation, harvesting, and pesticide activities. The machine learning approach is a crucial decision-making tool to estimate crop yield and to understand which crops are to be cultivated and what processes to be carried out while growing seasonal crops. This paper focuses on a literature review of different approaches to machine learning to estimate the crop yield and various soil properties. Three categories of soils include physical soil, chemicals, and biological characteristics. A selected agricultural soil and method study and analysis of different soil nutrients for plant prevention, for instance, primary and secondary nutrients. Such an approach will help farmers resolve their crop production problem.

Keywords Soil fertility · Machine learning Algorithm · Soil nutrients · Agricultural crop

1 Introduction

Agriculture is a very significant and crucial area of research, even though, to communicate with farmers, all civilians are helping farmers to seek a suitable approach in the existing system, research, and technology in real agriculture and the Government is also offering some funding for different innovative agriculture development projects, which will effectively help to the farmers. More work has been done in the computer science domain in the agriculture application using the internet of things and artificial technology. This research examined the various kinds of machine learning

Y. Shahare (✉)
MGMCET, Kamothe Navi Mumbai, India

V. Gautam
Chitkara University, Punjab, India
e-mail: vinay.gautam@chitkara.edu.in

J. M. R. S. Tavares et al. (eds.), *Cyber Intelligence and Information Retrieval*,
Lecture Notes in Networks and Systems 291,
https://doi.org/10.1007/978-981-16-4284-5_22

approaches presented and proposed by different authors through the past several years and observing different chemical characteristics of soil nutrients such as organic carbon, electrical conductivity, potassium, pH, nitrogen, phosphorus, secondary soil characteristics (S), and other parameters (Cu, Zn, Fe, Mn and B), etc.

Machine learning (ML) techniques can be used in many domains, differing from convenience stores to user behavior appraisal to customer phone usage prediction. For several years, machine learning was also used in agriculture. Crop yield forecasting is one of the most complex and difficult challenges in crop improvement, and so far many approaches have been developed and reviewed Machine learning is a technology which provides an instant desire for learning and learning. It includes many well-developed models, gathers and utilizes complex algorithms to achieve the desired outcomes. Different machine learning algorithms have been used in the application of the agriculture domain for growing crop productivity and find better quality cultivation. The machine learning algorithm will work to find specific crops based on agricultural factors and their situation. Soil is the key to the plant's cultivation. Liquids and solids (minerals and organic matter) (water) are made up of living organisms and gases (primarily carbon dioxide and oxygen). All the physical and chemical properties are all components required to sustain fertility, better yield, and environmental protection; necessary for the soil to be properly nourished. Nutrients from soil analysis are very helpful in determining the nutrient type [29]. This paper studied soil fertility analysis and predict the suitable crop using different approaches of the machine learning algorithm. This classification can be used to produce a report of soil fertility index, and fertility decisions should be made using the intelligence system, and it will make it easier to compare the levels of soil fertility, henceforth the importance of research to identify soil nutrients fertility indices such as Nitrogen (N), Phosphorus (P), Potassium (K), Electrical conductivity (EC), and organic carbon(OC) [28].

Various meta-parameters of both methods are specified for a better classification of soil nutrient analysis. Machine learning methods have been available in recent years applied in the field of agriculture by different researchers. The following analysis using various machine learning practices during the last few years is presented in different parts. The remaining part of this paper is ordered as surveys: Part 2: describe contextual of the literature review. Part 3: discuss and describe the different problem approaches. Part 4: conclude the paper.

2 Related Work

The information about agricultural production is an essential task to make accurate decisions at the country level. Assess the model of adequate crop yield helps farmers for which crops are needed and how the particular crops are grown. There are different ways of predicting crops. This study focused on what was done in crops using machine learning. One of the results measures to analyze the publications reviewed is that they are a review or a traditional review paper. Below are some research reviews.

Akpot et al. [14] authors have reviewed the agricultural farmland suitability was produced through online data in a systematic view. This review articles used factors like atmospheric conditions, nutrients and produced soils, root zone water availability, and behavioral and analytical requirements; these variables were further categorized, and also, to get a complete view of the ecosystems of crops and inherent local behavioral factors should be considered, few factors can be explained and enhance the particular yield.

Colezea et al. [17] discussed a combined web service forum aimed at increasing the quality of farm-grown products and promoting business development in the field of agriculture. Also, this platform offers more services and intercommunication between them, presents the framework stream design and execution test results that show the effectiveness of the stage.

Bünemann et al. [18] reviewed the quality of the soil and conceptual frameworks in terms of its scope, research methods, and accurate measure collection and explanation. They have identified the most commonly used soil performance metrics for farmland use. The suggested concept is associated with soil estimation of soil fertility, nutrient availability, soil ability, growing conditions as well as improves soil and also mentioned the Associating soil quality to soil utilities and ecosystem facilities.

Jeong et al. [19] this paper authors have described, Using supervised learning methods and spatial distribution of various features such as topsoil carbon, nitrogen and available phosphorus in elevation regions were established, defining the cognitive fertilizer application form in South Korea watershed of the particular lake, functional landscape analysis was carried out. Examine the various algorithms like generalized additive model algorithm (GAM), random forest algorithm (RF), and support vector regression algorithm (SVR).

Sirsat et al. [20] this paper, they have portrayed compound soil estimations for groupings more proper soil boundaries like explicit town insightful soil nutrients highlights like natural carbon (OC), Phosphorus pentoxide (P_2O5), Manganese (Mn), and Iron (Fe), soil pH worth and types, soil supplements nitrous oxide (N_2O), P_2O5 and potassium oxide (K_2O). Based on these parameters they have recommended suitable amounts of fertilizers as well as preferable crops.

Rahman et al. [21] the model suggested in this paper can indicate the required crops and estimate the soil sequence including various land types in accordance with the soil evaluations. Supervised algorithm of K-nearest neighbor (k-NN), the Bagged Trees, and the Gaussian Kernel-based weighted aid vector machines (SVM). The experimental findings and suggested SVM-based process is better than other processes.

3 Methodology

Studied various literature review papers such as Google Scholar, Web of Science, Science Direct, Scopus, and springer to find the research outcome is examined as shown in Fig. 1 and Conceptual proposed research framework as shown in Fig. 2. This survey paper differs from previous literature research based on some research questions.

1. What type of machine learning algorithm has been used for soil fertility prediction and suggested appropriate crop?
2. Which type of soil, chemical features, climatic features have been used for the prediction of soil fertility and suggested appropriate crop?
3. Which evaluation techniques have been used for soil fertility prediction and crop?
4. What are the challenges of soil fertility prediction and suggested the crop

3.1 *Soil Fertility*

In the field of agriculture, soil fertility is a very important aspect which is the full strength of the soil and the supply of the necessary fertilizer of the crop and water in soil presence in appropriate quantities and amounts for progressing the crop and reflection without the presence of toxic substances that may prevent crop grown. Soil consists of several components such as inorganic units, organic materials, and soil water contents, soil air content, and living organisms including bacteria that help crop decomposition [23].

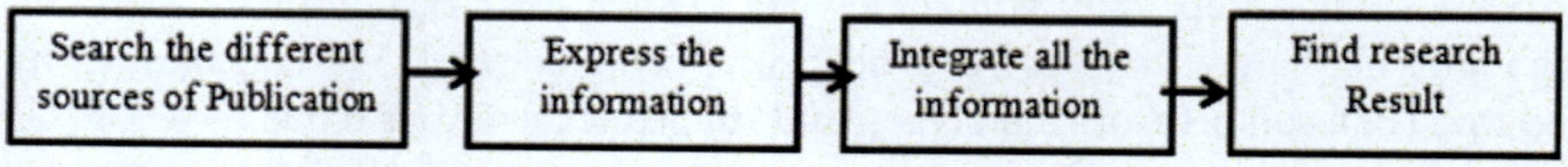

Fig. 1 Details of survey report steps

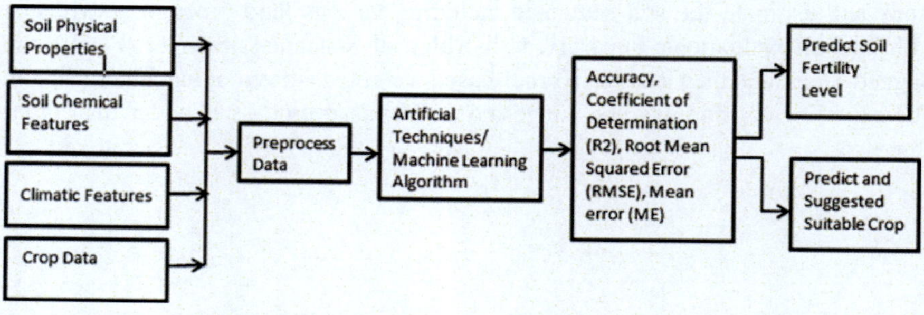

Fig. 2 Conceptual framework of review

3.1.1 Soil Features

Soil consists of three categories of features such as chemical feature, physical feature, and biological feature. This paper describes the chemical features of the soil. There are several chemical features are available such as primary soil elements of Nitrogen (N), various components of Phosphorous (P), available Potassium (K), and secondary soil components are given as a specific value of Iron (Fe), value of manganese (Mn), value of Zinc(Zn), value of copper (Cu), value of Sulfur(S), value of Boron(B), value of Electrical conductivity (EC, and) value of Organic Carbon (OC) [26].

3.1.2 Soil PH

Soil PH value is a very useful content for growing the crop. Soil PH value is derived in different levels such as Acidity, Neutral, and Alkaline, and all these levels should depend on a specific range for identifying the present amount of PH content in the soil. PH < 4.5 then very powerfully acidic, 4.6–5.5 is powerfully acidic, 5.6–6.5 then medium acidic, 6.6–6.9 then little acidic, 7.0 then neutral, 7.1–8.0 then little alkaline, 8.1–9.0 then medium alkaline, 9.1–10.0 then powerfully alkaline, 10.1–11.0 then very powerfully alkaline [32].

3.2 Machine Learning Algorithm

3.2.1 Decision Tree Classification and Regression

Decision tree learning divides the entire region of the samples into a repeated, smaller space that is sufficient for a simple model to formulate. There is a whole sample space in the tree of the root node (first node). Splitting sample space into smaller subsamples means bifurcating a root node in children's nodes, in which each node can be divided recurrently into leaf nodes (a node that cannot be further divided). The nodes are in the tree but the leaf node, Split sample space, based on the input attribute collection of condition(s), and the leaf node assigns the input attributes on the root-leaf path in the tree to the output value. The nodes consider a parameter as an impurity (e.g., standard ID3 deviation; Gini Index used in C4.5) and node (number of information present in a node) Node based on the criteria calculation of impurity [24].

3.2.2 Random Forest Classification and Regression

Random forest is a most common and strongly supervised machine training algorithm able to perform the regression tasks of both classifications and building several decision-making bodies during training and the class output that is the class model (classification) or the average prediction (regression) of the individual trees. The benefit of the algorithm of random forest is that it is less efficient than the problem of the Random forest, unlike algorithms for learning decision tree machines. No cutting is important to forest random algorithms that can be built in parallel with random forest machine learning algorithms [22].

3.2.3 Support Vector Machine

Support vector machine is an algorithm that works on the basis of a concept of decision-making plans that define boundaries for decision making. A decision limits the objects of a class separate from that of a class decision limits Vectors of support are the closest to the hyperplane data points. Input is transformed in a higher dimension space with a kernel function that is used to separate nonlinear data. This survey proposed an SVM method that the Gaussian radial functional kernel is used [20].

3.2.4 Principal Component Regression (PCR) and Partial Least Square Regression (PLSR)

When comparing the Principal component algorithm which is a unique model known as multiple linear regression model, the X-variables is consist of principal components that are not correlated by spectral data acquisition is filtered. Partial least squares regression is a method that is quite similar to PCR. The PLSR method is distinguished by the projection of both spectroscopy data from soil samples called X-variables and various soil fertility properties such as Y-variables. The PLSR consider a different approach to constructs and design an orthogonal of various factors are oriented along with the different directions of maximum covariance between the spectral matrix and the response vector [31].

4 Results

This survey paper studied more literature on soil fertility using machine learning algorithms, and other technologies. Figure 3 has shown as more distribution of all machine learning algorithms and other algorithms.

5 Discussion

Discussed various machine learning for analysis of soil chemical features such as soil primary component of Nitrogen (N), content of Phosphorous (P), content of Potassium (K), secondary component of soil value is Iron (Fe), value is manganese

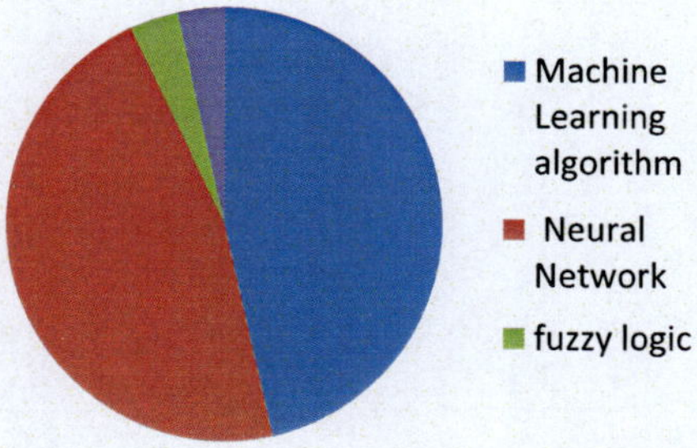

Fig. 3 Shows distribution of Machine Learning Algorithm and others

(Mn), value is Zinc(Zn), value is copper (Cu), value is Sulfur(S), value is Boron(B), value is Electrical conductivity (EC), value is Organic Carbon (OC), and also some climatic parameters like temperature, humidity, rainfall. Different soil types were studied including clay, silt, and loamy. Analyzing all the parameters and predict soil fertility and suitable crops using a machine learning algorithm. Table1 is shown a literature survey of soil chemical components.

6 Conclusion

In this paper survey soil fertility and predict the suitable crops. Analysis of soil fertility including soil chemical features such as thr primary feature of Nitrogen (N), Phosphorous (P), Potassium (K), and secondary features of value is Iron (Fe), value is manganese (Mn), value is Zinc(Zn), value is copper (Cu), value is Sulfur(S), value is Boron(B), value is Electrical conductivity and Organic Carbon (EC, OC), and also other climatic parameters like temperature, humidity, rainfall. Different soil types were studied including clay, silt, and loamy. Based on all parameters can predict the crops by using a different machine learning algorithm. Find a better performance model with more features, and different machine learning techniques. Most of the papers are shown better model performance by evaluating different techniques. So it is very clearly stated that based on this survey paper, find the better performance method used in an artificial neural network like PLSR and PCR and decision tree and random forest algorithm.

In future work, based on this survey outcome of this study and develop a better performance model for the prediction of the specific crop by analyzing a large amount of soil dataset.

Table 1 Shows soil nutrient features and analysis using different artificial techniques

Reference no and year	Existing work	Technology	Resources / Dataset location	parameter	Accuracy
Sonal Agarwal et al. [3]	Prediction of crop performance with machine and deep learning	SVM, LSTM, RNN	Kaggle.com	temperature, rainfall, pH value, relative humidity, and area, and Crop dataset	97%
Stephen Gbenga Fashoto et al. [4]	Use multiple regression and backward elimination to estimate the maize yield to implement and develop the model	MLR, backword elimination algorithm	United States Department of agriculture library and Eswatini Minister of agriculture	Water input like rainfall and irrigation, Nitrogen added with fertilizer, Crop evapotranspiration, Average air, temperature, vapor pressure, Solar radiation	R^2, RMSE, Adjusted R^2
Dongxue Zhao et al. [2]	Identify particular models and developed relevant calibrations using multi-depth data	Bagging-PLSR, Cubist	vis–NIR spectra Australian cotton area dataset	Physical properties of Soil depth: Topsoil, Subsurface, Subsoil, multi-depth Chemical soil properties: clay, silt, sand CEC, ESP, PH	Calibration: R^2, LCCC Validation: LCCC, RPD
Gilson Augusto Helfer et al. [6]	Proposed system framework evaluates the soil's based on context history using Partial Least Squares Regression algorithm	Partial Least Squares Regression algorithm	Central Analitica soil laboratory,	Clay, pH, electrical conductivity, luminosity, rainfall, pressure, temperature, and humidity,	R^2: 0.8725, RMSEC: 0.25 T/ha and RMSECV: 0.88 T/ha.-

(continued)

Table 1 (continued)

Reference no and year	Existing work	Technology	Resources / Dataset location	parameter	Accuracy
Sonal Jain et al. [7]	Produce the maximum amount of crop yield based on some climatic feature and other components of soil, and suggests the suitable crop sowing period time	RNN, RF	Weather data from last five years	Minimum Temperature (°C), Maximum Temperature (°C) Rainfall (mm) different Soil classification, and pH, Capacity of Water Soil, Fertility Crop	RMSE
Suchithra et al [10]	Analysis and prediction of village wise soil chemical elements and improves soil health and environmental quality	methods such as Gaussian radial basis, sine-squared, hyperbolic tangent, triangular basis, and hard limit	Kerala state soil Dataset	Phosphorus (P), Potassium (K), Organic Carbon (OC), and Boron (B), Soil Reaction (pH)	90% for P, 87% for p 88% for p, 86% for P 86% for P and B
Rajat Chaudhari et al. [9]	Soil data to predict the fertility of soil considering various chemical elements	Naive Bays, KNN, SVM, Decision Tree	Sample dataset of Soil fertility	(pH, EC, OC), (N, P,K), and (Fe, Zn, Mn, Cu)	43%, 45%, 60% 56%
Agus Arip et al. [1]	Discuss spectroscopic dataset and calibration models database of near infrared spectroscopy (NIRS)	Principal component regression (PCR) and partial least square regression (PLSR)	Department of Agricultural Engineering and Department of Soil Science, Syiah Kuala University, Banda Aceh, Indonesia	soil nitrogen content (N), phosphorus (P), potassium (K), soil pH (pH), magnesium (Mg), and calcium (Ca)	R^2, r, RMSEC, RPD

(continued)

Table 1 (continued)

Reference no and year	Existing work	Technology	Resources / Dataset location	parameter	Accuracy
Vaishali Pandith et al. [11]	Identify Performance and Evaluation of Machine Learning Techniques for estimating the Mustard Crop Yield according to soil features	KNN, NB, LR ANN, RF	Dataset collected from the laboratory of soil testing in the Department of Agriculture, Talab tillo, Jammu	pH, EC, OC, N, P, K, S, Cu, Fe, Zn, Mn	86%, 72% 80%, 76% 94%
Saranya et al [5]	Developed and implement a model for classifying the various soil according to the certain amount of fertilizers and predicting the crop	KNN, SVM, and logistic regression	Soil dataset and crop dataset	pH value, soil salinity, Organic matter, Potassium, Sulfur, Zinc, Boron, Calcium, Magnesium, Copper, Iron, Manganese	96%
Rajeswari et al. [12]	Predicts the level of soil fertility and provides soil information for recommendations in crop selection and sowing using mobile application	Soil Fertility level Prediction C4.5, C5.0, C5.0: ADT Suggestion of crop selection Prediction C4.5 90% 10% C5.0 C5.0: ADT classifier	Soil data for Virudhunagar District, Tamilnadu	Soil pH value, (EC) Electrical conductivity, (OC) Organic Carbon, (N) Nitrogen, (P) Phosphorus, (K) Potassium, (S) Sulfur, (Zn) Zinc, (Fe) Iron, (Cu) Copper, (Mn) Manganese, (Ca) Calcium, (B) Boron, Fertility Level	Accuracy and Error rate 87% and 13 90% and 10 92% and 8 90% and 10% 92% and 8% 95% and 5

(continued)

Table 1 (continued)

Reference no and year	Existing work	Technology	Resources / Dataset location	parameter	Accuracy
Ramiro Recena et al. [27]	Vis–NIR spectroscopy soil fertility evaluation: soil prediction instead of availability indexes	Modified Partial Least Squares Regression (MPLSR)	agricultural lands of Spain	Nutrient availability, P, K, Fe, NIR	Standard error of calibration(SEC), Standard error of cross validation (SECV), R^2
Felipe B. de Santana et al. [13]	Green method for organic soil analysis in combination with a learning machine using the national spectral library	Random forest regression model	Around (43,000) samples of soils National Soil Spectral Library	Soil sample (SOM)	The values of RMSE and R^2
Tatiane Melo et al. [30]	Determine the total elementary soil sample composition and anticipate soil fertility with simple or multiple regression as a proxy with pXRF data	Linear regression model	Agricultural soil from the Brazilian Cerrado based on pXRF data	pH, Clay, Silt, Sand, Ca2, Mg2, K, Al3, CEC1, CEC2, SOM, P, S-SO4, Zn, Cu, Fe, Mn	r, R^2, RMSE
Renuka, Sujata Terdal et al. [25]	Estimates of crop yield include estimates for crop yields from historical data	Decision tree KNN, SVM_rbf, SVM_linear	Kaggle.com and data.gov.in	Saline, alkaline, sodic, non-alkaline, Soil composition (PH, N, P, K, EC, OC, Zn, F)	Accuracy and MSE, 99 and 0.01, 58.078 and 0.603 58.952 and 0.559123 53.712 and 6.20

(continued)

Table 1 (continued)

Reference no and year	Existing work	Technology	Resources / Dataset location	parameter	Accuracy
M.S. Sirsat et.al [16]	Automatically predicting the soil fertility of indices will be important in terms of reduceing the quantity of soil nutrients	NN, DL SVM regressor, RF, bagging and boosting lasso, ridge regression, Bayesian models, and more	Marathwada region, State Government of Maharashtra (India)	P_2O_5, Fe, Mn, Zn, EC, OC, B, Cu, Nitrous oxide	R^2, Regressor
Louis Kouadio et.al [15]	Using Artificial approach for estimating coffee production and analyze the several soil factors	Extreme Learning Machine (ELM), Multiple Linear Regression (MLR), and Random Forest (RF)	Dataset collected from the Institute of Agriculture and Forestry Science, Vietnam's Western	SOM, available K, B, S, Zn, N, exchangeable calcium, Mn, and pH	RMSE, MAE, RRMSE, RMAE
Suparna Biswas et al. [8]	Classification of different Soil data collected from Various Geotechnical Properties by using various Data Mining Techniques to analyze the soil fertility	Fully connected neural network (FCNN), Generalized Regression Neural Network(GRNN), Linear Regression (LR), Support Vector Regression (SVR)	8 different types of soil have been classified using SPT and depth method	soil type in depth: SPT, sufficient sand, silty clay, clayey silt with sufficient sand, sufficient sand with silt, clayey silt, silty sufficient sand, sandy silt, and rubbish	R2 value, MSE MAE

References

1. Munawar AA, Yunus Y, Devianti , Satriyo P (2020) Calibration models database of near infrared spectroscopy to predict agricultural soil fertility properties. Data in Brief
2. Zhao D, Arshad M, Li N, Triantafilis J (2021) Predicting soil physical and chemical properties using vis-NIR in Australian cotton areas. Catena
3. Agarwal S, Tarar S (2021) A hybrid approach for crop yield prediction using machine learning and deep learning algorithms. J Phy
4. Fashoto SG, Mbunge E, Ogunleye G, Burg JVD (2021) Implementation of machine learning for predicting maize crop yields using multiple linear regression and backward elimination. Malaysian J Comput 6(1):679-697
5. Saranya N, Mythili A (2020) Classification of soil and crop suggestion using machine learning techniques. IJERT 9(02)
6. Helfer GA, Barbosa JLV, dos Santos R (2020) A computational model for soil fertility prediction in ubiquitous agriculture. Comput Electron Agri 175
7. Jain S, Ramesh D (2020) Machine learning convergence for weather based crop selection. In: Conference on electrical, electronics and computer science
8. Biswas S, Faysal TI, Promiti AS, Hossain MS, Bazlul L, Sarwar AM, Shaiban SM, Rahman RM (2020) Analysis of soil and various geo-technical properties using data mining techniques. In: International conference on intelligent systems
9. Chaudhari R, Chaudhari S, Shaikh A, Chiloba R, Khadtare TD (2020) Soil fertility prediction using data mining techniques. Mukt Shabd J 9(6). ISSN NO: 2347–315
10. Suchithra MS, Pai ML (2020) Improving the prediction accuracy of soil nutrient classification by optimizing extreme learning machine parameters. Info Process Agri 7:72–82
11. Pandith V, Kour H, Singh S, Manhas J, Sharma V (2020) Performance evaluation of machine learning techniques for mustard crop yield prediction from soil analysis. J Sci Res 64(2)
12. Rajeswari S, Suthendran K (2019) C5.0: Advanced decision tree (ADT) classification model for agricultural data analysis on cloud. Comput Electron Agri 156:530–539
13. de Santana FB, de Souza AM, Popi RJ (2019) Green methodology for soil organic matter analysis using a national near infrared spectral library in tandem with learning machine. Sci Total Environ 658:895–900
14. Akpoti K, Kabo-bah AT, Zwart SJ (2019) Agricultural land suitability analysis: State-of-the-art and outlooks for integration of climate change analysis, Agri Syst 172–208
15. Kouadio L, Deo RC, Byrareddy V, Adamowski JF, Mushtaq S (2018) Van Phuong Nguyen, Artificial intelligence approach for the prediction of Robusta coffee yield using soil fertility properties. Comput Electron Agric 155:324–338
16. Sirsat MS, Cernadas E, Fernández-Delgado M, Barro S (2018) Automatic prediction of village-wise soil fertility for several nutrients in India using a wide range of regression methods. Comput Electron Agric 154:120–133
17. Colezea M, Musat G, Pop F, Negru C, Dumitrascu A, Mocanu M (2018) CLUeFARM: integrated web-service platform for smart farms. Comput. Electron. Agri 154
18. Bünemann EK, Bongiorno G, Bai Z, Creamer R, Deyn GD, de Goede R, Fleskens L, Geissen V, Kuyper TW, Mäder P, Pulleman M, Sukkel W, van Groenigen JW, Brussaard L (2018) Soil quality–a critical review. Soil Biol Biochem 120
19. Jeong G, Oeverdieck H (2017) Soo Jin Park, Bernd Huwe, Mareike Ließ, Spatial soil nutrients prediction using three supervised learning methods for assessment of land potentials in complex terrain. CATENA 154:73–84
20. Sirsat MS, Cernadas E, Fernández-Delgado M, Khan R (2017) Classification of agricultural soil parameters in India. Comput Electron Agric 135:269–279
21. Rahman SAZ, Mitra KC, Mohidul Islam SM (2017) Soil classification using machine learning methods and crops suggestion based on soil series. In: International conference of computer and information technology (ICCIT), 21–23 December
22. Donges N (2020) A complete guide to the random forest algorithm. from https://builtin.com/datasciencerandomforestalgorithm.html

23. Jayalakshmi R, Devi MS (2019) Relevance of machine learning algorithms on soil fertility prediction using R. Int J Comput Intell Info 8(4):193–199
24. Singh V, Sarwar A (2017) Analysis of soil and prediction of crop yield (Rice) using machine learning approach. Int J Adv Res Comput Sci 8(5)
25. Renuka, Terdal S (2019) Evaluation of machine learning algorithms for crop yield prediction. Int J Eng Adv Technol
26. Supriya DM (2017) Analysis of soil behavior and prediction of crop yield using data mining approach. Int J Innov Res Comput Commun Eng
27. Recena R, Fernández-Cabanás VM, Delgado A (2019) Soil fertility assessment by Vis-NIR spectroscopy: predicting soil functioning rather than availability indices. Geoderma 337
28. Gholap J, Ingole A, Gohil J, Gargade S, Attar V (2012) Soil data analysis using classification techniques and soil attribute prediction. IJCSI
29. Baskar SS, Arockiam L, Charles S (2013) Applying data mining techniques on soil fertility prediction. Int J Comput Appl Technol Res ISSN: 2319–8656
30. de Lima TM, Weindorf DC, Curi N, Guilherme LRG, Lanaa RMQ, Ribeiro BT (2019) Elemental analysis of Cerrado agricultural soils via portable X-ray fluorescence spectrometry: inferences for soil fertility assessment. Geoderma 353:264–272
31. Zhang H, Wu P, Yin A, Yang X, Zhang M, Gao C (2017) Prediction of soil organic carbon in an intensively managed reclamation zone of eastern China: a comparison of multiple linear regressions and the random forest model. Sci Total Environ
32. Paul M, Vishwakarma SK, Verma A (2015) Analysis of soil behaviour and prediction of crop yield using data mining approach. In: International conference on computational intelligence and communication networks, IEEE 978–1–5090–0076–2015

COVID-19 Pandemic Diagnosis and Analysis Using Clinical Decision Support Systems

Jagdish Chandra Patni, Hitesh Kumar Sharma, Shivani Sharma, Tanupriya Choudhury, Anurag Mor, Md. Ezaz Ahmed, and Prashant Ahlawat

Abstract Since the pandemic began, the attention of the entire world has turned to the digital health environment to provide potential health options to minimize the effects of this pandemic in this moment of unparalleled medical crisis. Paper presents the survey on the role of Artificial Intelligence in decision support systems to diagnose COVID-19 and we are presenting the Artificial Intelligence (AI) model for Decision Support System (DSS), Challenges faced during earlier pandemic due to lack of faster and efficient methods of diagnosis and Opportunities and Challenges of Clinical Decision Support System (CDSS). A three-dimensional deep learning system, called COVID-19 Neural Network detection (COVNet), is developed to detect COVID-19 based on CT images of the chest. We have seen that Decision support system using AI in Healthcare that is termed as Clinical decision support system is a very helpful, fast and advanced technique because during pandemics like COVID-19 the cases are increasing exponentially day by day so it's not possible to diagnose each patient manually.

Keywords Artificial Intelligence · Clinical support system · Decision support system · Severe acute respiratory syndrome · Intelligent decision support system · Machine learning · Artificial neural networks · Intelligent agents

J. C. Patni · H. K. Sharma (✉) · S. Sharma · T. Choudhury · A. Mor
School of Computer Science, University of Petroleum and Energy Studies, EnergyAcres, Bidholi, Dehradun, Uttarakhand 248007, India

S. Sharma
e-mail: 500074240@stu.upes.ac.in

Md. E. Ahmed
CS Department, Saudi Electronic University, Al Madina KSA, Saudi Arabia
e-mail: m.ezaz@seu.edu.sa

P. Ahlawat
Manipal University, Jaipur, India

J. M. R. S. Tavares et al. (eds.), *Cyber Intelligence and Information Retrieval*, Lecture Notes in Networks and Systems 291,
https://doi.org/10.1007/978-981-16-4284-5_23

1 Introduction

On March 11, 2020, the World Health Organization (WHO) declared coronavirus as a pandemic [11]. This disease is caused by a virus which is known as Severe Acute Respiratory Syndrome Coronavirus-2 (SARS-CoV-2) or Novel Coronavirus [3, 4]. This global health crisis sparked a hunt for the medical industry and technologies to track and manage the spread of pandemic COVID-19 [5, 6]. This is the time when correct planning and decision-making matters to prevent it from spreading, its diagnosis and its vaccine which can prevent thousands from this virus. Some of the countries implemented pandemic lockdown to save their people under this crucial situation, many courageous citizens are contributing their services in this situation to avoid it. But several countries suffered from the dramatic rise in the number of coronaviruses. Yet there's no vaccine available yet. Researchers, Scientists, technologists, etc. are working to support the world in their fields to find possible methods to cure it, either forecasting the number of COVID-19 cases, finding the vaccine, helping people to save themselves from the virus. The main focus of this paper is on using a Decision support system in Healthcare sector that needs the help of emerging technology such as Artificial Intelligence (AI), Machine learning, IoT to fight against the new disease.

2 Decision Support System in Healthcare

Human beings have the power to take decisions. It can be structured, semi-structured or unstructured. Structured decision issues have been found to be an ideal solution and thus do not need support for the decision. Whereas Unstructured decision–making problems have no defined criteria or solution and depend on the decision-maker's preferences. There are wide variety of semi-structured issues they usually have some accepted parameters and still requires human input or preferences for decision with a limited set of criteria [1, 7]. Decision Support System (DSS) is the domain of Information Systems (IS) that focuses on supporting and enhancing decision-making in the management. DSS is an interactive computer-based system particularly designed to help decision-makers to use data and models to define, solve and determine problems. It is helpful in many sectors like Business, manage inventory, sales optimizations, Farming, Health care, Education, etc.

The benefits of using DSS are that they enhance the performance and speed of decision-making activities, Promote learning, saves time, etc. There is a type of Decision support system called Clinical Decision Support System (CDSS) [8, 9]. This system is expected to enhance human services conveyance by upgrading medical decisions with focused clinical information, persistent data and other well-being information. Typically CDSS consist of software which is developed to provide aid to CDSSs today are essentially utilized for the purpose of care, for the clinician to join their insight with data or recommendations gave by the CDSS. Progressively, be that as it may, there are CDSS being created with the ability to use information and perceptions, in any case, ridiculous or uninterpretable by people [1].

Artificial intelligence can be used to create the alternatives techniques that are called an intelligent decision Support system (IDSS) [10–12].

AI methods like Fuzzy Logic, Machine Learning, Artificial Neural Networks and Intelligent Agents, when used with DSS then it provides aid in solving complicated problems, either they are real-time problems which includes huge amount of structured data. Like, Car alarms, Fire detectors, house alarms, surveillance systems. In figure given below shows the steps used by intelligent support system.

3 Literature Review

HIV/AIDS was declared as a pandemic in 2006 by WHO. HIV is a retrovirus which leads to AIDS, a disease which affects human immune system and leads to life threatening infections. AIDS has infected more than 25 million people, making it one of the most devastating pandemics in history. HIV is expected to kill 90 million people in Africa according to latest figures, resulting in a total estimation of 18 million orphans [9, 11, 12]. This paper explored the use of decision support systems as a computer simulation model to investigate tuberculosis and HIV in the Russian Federation. A simulation model was built to portray the dynamics of tuberculosis and HIV transmission that is known as System Dynamics (SD) [13–15]. This approach was selected because infectious disease transmission is complex, including the multiple interacting variables, intertwined feedback loops and non-linear causal-effect relationships. Their behaviour is usually hard to predict considering the dynamic complexity of such systems. Such processes can be defined by structured system-based modelling methods, such as System Dynamics, simulates their behaviour and predicting the effect of health system interventions to assist policymakers in determining the most effective steps to enhance health outcomes [16, 17]. This SD model was used to estimate the effect of implementing the globally endorsed policies on the spread of such two diseases in an area of the Russian Federation and the subsequent deaths within the next decade. The simulation model was structured to be simple and easy to use so that decision-makers in public health and medical personnel, with limited computer skills, can easily use it. The simulation model served as a DSS to assist decision-makers in various parts of the national health program in performing studies, evaluating outcomes, learning about the dynamics of these two diseases and gaining a common understanding of the reason for adopting simulation model policies [2, 18, 19]. The finding of jungle fever follows a progression of reasoning and explanatory procedures that start with acquiring patient's narratives, leading physical assessments and research centre tests, trailed by the assessment of the general significance of the distinctive introducing indications and their classification as per their criticalness. Differential analysis is made by successfully precluding some other conditions, by barring one ailment after another so as to show up at a positive analysis of jungle fever. This procedure is not an exact science and

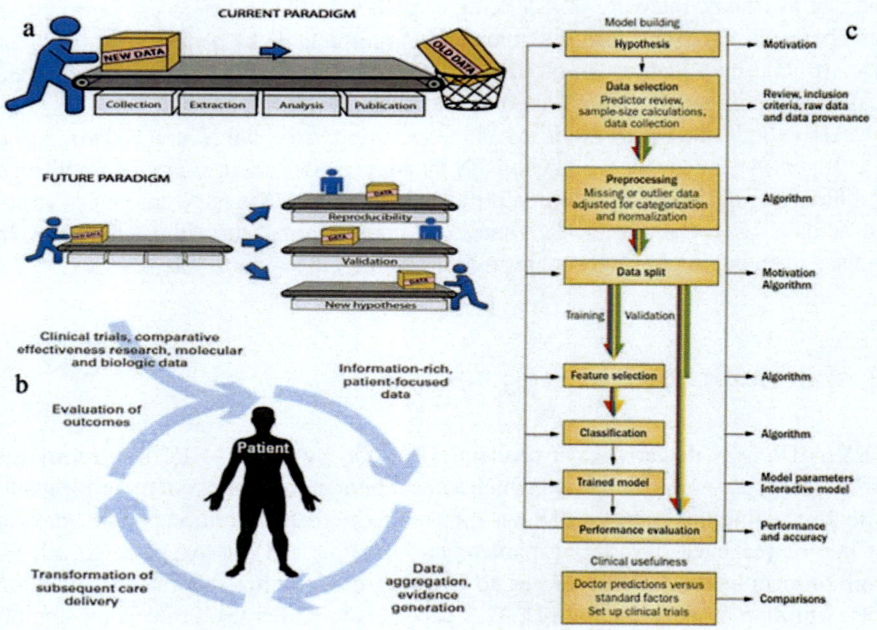

Fig. 1 Clinical Decision Support System

could demonstrate truly confounding and oppressive for an unpractised doctor. The finding of intestinal sickness follows the ordinary procedure of clinical determination demonstrated in Fig. 1). Initial social affair of abstract data which incorporate past history, history of presentation to mosquito nibbles, family ancestry and other emotional side effects. 2). Generation of plausible indicative rundown, regularly up to five presumed pathologies are explored and the doctor at this point may start treatment dependent on speculated determination, what's more, emotional data gathered. Confirmation of analysis through a social event of target proof like estimating the patients' fundamental signs, running blood tests and acquiring radiological assessments [20]. The doctor at that point assesses the overall significance of the distinctive introducing manifestations. Sort them as indicated by their relative significance and in the thought of target research centre examinations. Last determination is shown up by adequately precluding other related conditions utilizing some combinatorial investigation [21–23].

4 Challenges Faced During Pandemics

Diagnostic tests are a crucial part of an effective outbreak management technique that is used in each step of the outbreak, from preliminary diagnosis to potential resolution. Development of diagnostic tests are useful for the prevention and control of epidemic is technically difficult and process for development, testing and implementation is complex and time consuming. In this paper, they analyzed common issues during diagnosis faced in many outbreaks like Ebola and Marburg, Lassa fever, MERS-CoV, SARS, Nipah and henipaviral diseases, etc. Listed below are some of the common problems faced in outbreaks [5].

1. Lack of diagnosis in a field use adapted type
2. Low test market feasibility over non-outbreak cycles
3. Restricted access to samples leading to further delays in production of the diagnosis
4. Restricted expert / laboratory collaboration with pathogen-specific knowledge
5. Delays in the sharing of diagnostic data which affect response times and containment
6. Weak diagnostic and supervisory capability in several nations at national scale.

5 Decision Support System for COVID-19 Diagnosis

1st phase—Early Detection system

Computer-based intelligence can rapidly break down unpredictable side effects and other 'warnings' and accordingly alert the patients and the human services specialists. It assists with giving a quicker dynamic, which is practical. It assists with building up another conclusion and the executive's framework for the COVID 19 cases, through helpful calculations.

A bunch of patients were conceded with fever, cough, shortness of breath, etc. If these symptoms are noticed then patients were checked using Computed tomography (CT), which results in varied opacities in contrast with pictures of sound lungs. This discovering prompted the underlying finding of pneumonia. Extra nucleic acid analysis using multiplex real-time polymerase chain reaction (PCR). In Fig. 2 there are two methodologies that are utilized to Diagnose at the fundamental advance and take a choice on its premise [6].

B. 2nd Phase—Diagnosis System

Data Collection

This review study was affirmed by the morals councils of the taking part medical clinics. Further assent was postponed with endorsement. It incorporates 4536 3D volumetric CT tests of chest. Total 3506 patients were obtained at 6 clinical from 16 of Aug 2016 and 17 of Feb 2020. This model included the following criteria: (1) differentiate CT test; (2) tests with cut thickness > 3 mm. In the wake of applying the rejection standards, 4356 3D chest CT tests from 3322 patients were chosen in this investigation. The normal age is in between 30 to 60 years and it included more number of male patients than female patients (1838 versus 1484 where p-value was 0.29). CT tests with different recreation portions at a similar imaging meeting or gained at various time focuses were incorporated. The last dataset comprises 30% tests for Corona Virus Disease, 40% for CAP and 30% for not having pneumonia. All these Corona Virus Disease patients were found CORONA VIRUS DISEASE

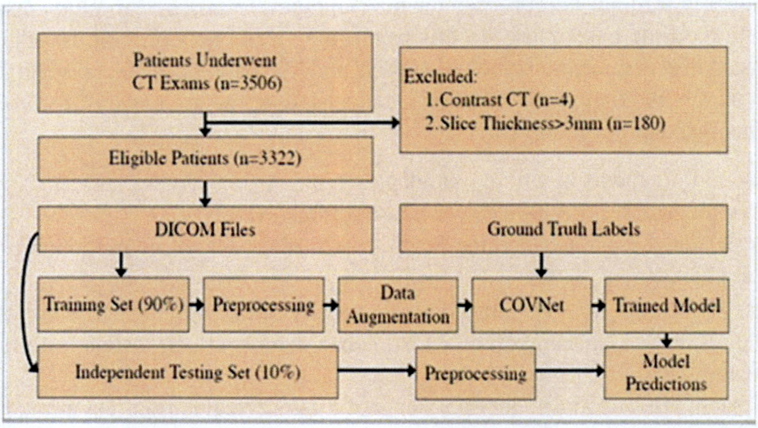

Fig. 2 DSS Model for COVID-19 diagnosis

positive by RT-PCR test which contained the data of patients from 31 of Dec 2019 to 17 of Feb 2020.

A simple flow of CNN layer diagram is shown in Fig. 3. The most widely recognized indications were hack (66%) and fever (81%). For every patient, one or numerous CT tests at a few time focuses over the span of the malady were gained (on average CT tests per persistent were 1.8 and the range from 1 to 6 CAP and having no pneumonia dataset of patients was randomly taken from different hospitals. The ration of Training and Testing Dataset was 9:1. The training dataset was divided further for internal validation and model training. Statistics of patients who were used are given below in Table 1.

Deep Learning Model

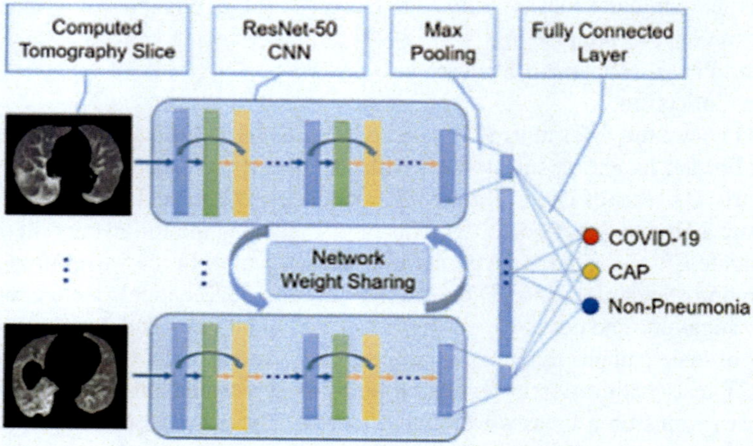

Fig. 3 Detection neural network

Table 1 Summery of training and individual testing datasets

Training					Independent testing			
	COVID-19	CAP	Non-Pneumonia	P value	COVID-19	CAP	Non-Pneumonia	P value
Patients	400(14%)	1396 (47%)	1173 (40%)	–	68 (19%)	155 (44%)	130 (37%)	
Exams	1165 (30%)	1560 (40%)	1193(31%)	–	131 (30%)	175 (40%)	132 (30%)	
Male patients	212 (53%)	782 (56%)	645 (55%)	0.55	30 (52%)	96 (62%)	68 (52%)	0.17
Age (Year)	5 + −15	5 + −19	41 + −13	< 0.001	52 + −17	5 + −20	4 + −13	< 0.001

Note Values in parenthesis are the percentage, ages are reported as moon + - standard deviation, COVID-19, CAP-community acquired pneumonia

Clinical specialists can figure out how to identify CORONA VIRUS DISEASE cases that can be diagnosed using chest CT assessments, but their tasks are manual and time consuming, particularly when required to look at a lot of cases. So AI strategies, particularly Deep learning, have been utilized to process and analyze clinical imaging information to help radiologists and specialists to improve performance. Moreover, the current CORONA VIRUS DISEASE pandemic has seen various examinations concentrating on programmed identification of CORONA VIRUS DISEASE using Deep learning.

A three-dimensional deep learning system, called CORONA VIRUS DISEASE Neural Network detection (COVNet), is developed to detect CORONA VIRUS DISEASEbased on CT images of the chest. Three types of CT images, including CORONA VIRUS DISEASE, Community Acquired Pneumonia (CAP) and few other cases of non-pneumonia, are mixed to test the robustness of the proposed model, COVNet is a convolutionary ResNet-50 model. The COVNet system consists of a backbone RestNet50, in Which small slices of CT Image as input and creates a set of attributes for the respective. Then these extracted characteristics from all these small slices are put together via a Max-Pool layer process. To produce a probability for each form (CAP, Corona Virus Disease and No pneumonia), the final feature map is given as an input to a completely softmax function and fully connected layer. The images should be in standard measurements, each volumetric data should consist 51 to 1094 CT slices having slice-thickness from 0.5 mm to 3 mm that can vary. The reconstruction matrix is 512×512 pixels with spatial resolution of in-plane pixels from 0.29×0.29 mm2 to 0.98×0.98 mm2.

Analysis of statistics

The efficiency of the model was calculated by using the independent training dataset. This analysis was performed using R programming language. To compare the difference between groups for dichotomous variables and continuous variables we used chi-square and ANOVA table. The value of p was considered as 0.05 which was less than the two-sided p-value. Then were estimated. After all processes, AUC (Area Under Curve) is calculated using a 94% of level of confidence and an ROC curve was generated to prove accuracy of the model.

6 Results

The results can be obtained in 4.51 s by using high quality system specifications [7] after developed the deep learning model it will give the following results shown in Fig. 4 and Table 2.

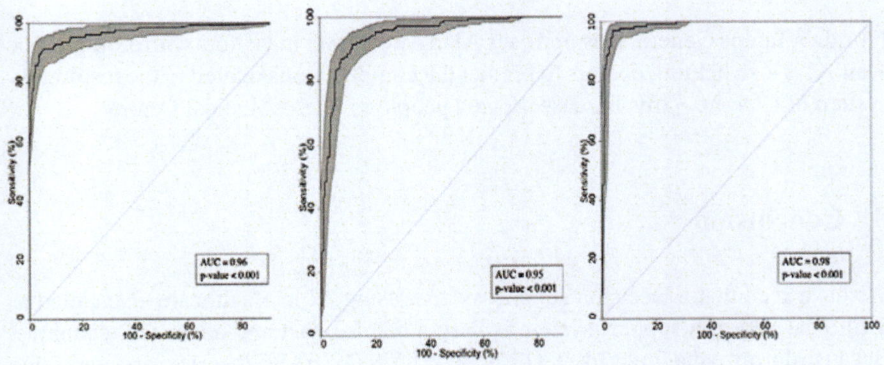

Fig. 4 Result Analysis

Table 2 Deep learning framework on the independent testing set

	Sensitivity%	Specificity%	AUC	P-value
CORONA VIRUS DISEASE	90 (114 of 127) [83,94]	96(294 of 307) [93,98]	0.96 [0.94,0.99]	< 0.001
CAP	87(152of175) [81,91]	92(239of259) [88,95]	0.95 [0.93,0.97]	< 0.001
Non-Pneumonia	94(124of132) [88,97]	96(291of302) [94,98]	0.98 [0.97,0.99]	< 0.001

Note Values in parentheses are the numbers for the percentage calculation

7 Challenges and Opportunities of CDSS

If we will talk about challenges and opportunities in a pandemic is that how accurate the model is, what is the data quality, quantity of the data, speed of the model is and ease of the system. Many companies are competing in this pandemic to find out optimal and accurate solutions for diagnosis where some are successful than others. So on the basic hand, we need a good decision support system and the correct choice of algorithms for the best diagnosis system [8]. Let's understand it by taking an example of two models, one is of chest CT and another is of RT-PCR. If we will talk about sensitivity Chest CT has higher sensitivity than RT-PCR (98% vs. 71% resp., p < 0.001). The reasons for nucleic acid detection's low efficiency could include: (1). Nucleic acid detection technology is not developed in a correct way. (2) Variations in detection levels from manufacturers to manufacturers (3) improper clinical sampling [9].

DSS using AI has a large scope in future. It can help in other pandemics Diagnoses too. Recently, we have also seen moving photos of healthcare workers working relentlessly across the globe to treat CORONA VIRUS DISEASEpatients, sometimes endangering their own lives. AI may play a crucial role in lightening their burden while ensuring that they do not suffer from the standard of treatment. For example,

Florida's Tampa General Hospital uses AI to detect fever in visitors with a simple face scan.AI is also helping doctors to predict the complications related to the respiratory system of COrona VIrus Diseaseaffected people in Sheba Medical Centre.

8 Conclusion

We have seen that a Decision support system using AI in Healthcare also called as a Clinical decision support system is very a helpful, fast and advanced technique. Because during pandemics like CORONA VIRUS DISEASE the cases are increasing exponentially day by day so it's not possible to diagnose each patient manually. It is more time consuming and we are not sure that during this kind of situation when the number of patients is increasing in mass, we can get 100% correct results there could be some human errors we can face. So here the machines play an efficient role. We need to train machines under a perfect model so that they can help fight against the pandemic. These are very useful in areas which are still lacking in a good medical staff. Variety of AI techniques such as fuzzy logic, case-based reasoning, Artificial Neural Networks, rough sets to allow decision support systems to perform better under uncertain conditions.As AI came later in trend so it wasn't used in earlier pandemics efficiently. If we have an already preformed DSS for a pandemic then we can reframe it by making some small changes for a new pandemic with some similarities.

References

1. Sutton RT, Pincock D, Baumgart DC et al (2020) An overview of clinical decision support systems: benefits, risks, and strategies for success. npj Digit Med 3:17
2. Lebcir RM1, Choudrie J, Atun RA, Coker RJ (2009) Using a decision support systems computer simulation model to examine HIV and tuberculosis: the Russian Federation. Int J Electron Healthc 5(1):14–32
3. Uzoka F-ME, Osuji J, Obot OU (2011) Clinical decision support system (DSS) in the diagnosis of malaria: a case comparison of two soft computing methodologies. Expert Syst Appl 38(3):1537–1553
4. Patni JC, Sharma HK (2019) Air quality prediction using artificial neural networks. In: International conference on automation computational and technology management (ICACTM)
5. Khanchi I, Ahmed E (2019) Automated framework for real-time sentiment analysis. In: International conference on next generation computing technologies (NGCT-2019)
6. Sharma HK, Singh T, Kshitiz K, Singh H, Kukreja P (2017) Detecting hate speech and insults on social commentary using nlp and machine learning. Int J Eng Technol Sci Res 4(12):279–285
7. Taneja S, Ahmed E, Patni JC (2019) I-Doctor: An IoT based self patient's health monitoring system. In: 2019 International conference on innovative sustainable computational technologies, CISCT 2019
8. Thakkar BA, Hasan MI, Desai MA (2010) Health care decision support system for swine flu prediction using naïve bayes classifier. In: 2010 International conference on advances in recent technologies in communication and computing, Kottayam, pp 101–105. https://doi.org/10.1109/ARTCom.2010.98

9. KellyCirino CD, Nkengasong J, Kettler H et al (2019) Importance of diagnostics in epidemic and pandemic preparedness. BMJ Global Health 4:e001179
10. Vaishya R, Javaid M, Khan AH, Haleem A (2020) Artificial intelligence (AI) applications for CORONA VIRUS DISEASE pandemic. Diabetes Metabolic Syndrome: Clinical Res Revie 14(4):337–339
11. Li L, Qin L, Xu Z et al (2020) Artificial intelligence distinguishes CORONA VIRUS DISEASE from community acquired pneumonia on chest CT, RSNA
12. Kumar K (2020) The world wasn't prepared for CORONA VIRUS DISEASE : In: future AI will curb the next pandemic , AI can play a huge role in predicting an outbreak and also minimising or stalling its spread, economictimes
13. Khanchi I, Agarwal N, Seth P, Ahlawat P (2019) Real time activity logger: a user activity detection system. Int J Eng Adv Technol 9(1):1991–1994
14. Fang Y, Zhang H, Xie J et al (2020) Sensitivity of chest CT for CORONA VIRUS DISEASE : comparison to RT-PCR, RSNA
15. Smeulders AW, Van Ginneken AM (1989) An analysis of pathology knowledge and decision making for the development of artificial intelligence-based consulting systems. Anal Quant Cytol Histol 11(3):154–165
16. Gupta R, Ghosh A, Singh AK, Misra A (2020) Clinical considerations for patients with diabetes in times of CORONA VIRUS DISEASE epidemic. Diabetes Metabolic Syndrome Clin Res Rev 14 (3):211–212
17. Sohrabi C, Alsafi Z,'Neill NO, Khan M, Kerwan A, Al-Jabir A, Iosifidis C, Agha R (2019) World Health Organization declares global emergency: a review of the 2019 novel coronavirus (CORONA VIRUS DISEASE). Int J Surg
18. Sharma HK, Shastri A, Biswas R (2012) A framework for automated database tuning using dynamic SGA parameters and basic operating system utilities. Database Syst J 3(4)
19. Sharma HK (2013) E-COCOMO: the extended cost constructive model for cleanroom software engineering. Database Syst J 4(4):3–11
20. Kshitiz K, Shailendra (2018) NLP and machine learning techniques for detecting insulting comments on social networking platforms. In: International conference on advances in computing and communication engineering (ICACCE) 2018, IEEE, pp 265–272
21. Joshi A, Sabitha AS et al. (2017) Crime analysis using K-means clustering. In: 2017 3rd International conference on computational intelligence and networks, Odisha, pp 33–39
22. Mittal V, Gupta S et al. (2018) Comparative analysis of authentication and access control protocols against malicious attacks in wireless sensor networks. In: Smart computing and informatics, Springer, Singapore, pp 255–262
23. Jayaraman S, Kumar P et al (2017) Analysis of classification models based on cuisine prediction using machine learning. In: 2017 International conference on smart technologies for smart nation (SmartTechCon), IEEE, pp 1485–1490

Bengali Abstractive News Summarization Using Seq2Seq Learning with Attention

Mariyam Sultana, Partha Chakraborty, and Tanupriya Choudhury

Abstract Text summarization is the technique for generating short and succinct summaries from long texts that focuses on the most important information but keeps the overall exhaustive signification of the whole text. This paper presents a method of generating short abstractive summaries from long Bengali news articles using some basic NLP approaches with various Recurrent Neural Network (RNN) architectures, such as Bidirectional RNN, Encoder-Decoder RNN, Sequence to Sequence (Seq2Seq) Learning with Attention mechanism, Longest Short-term Memory (LSTM), etc. Dataset collected by the authors from different Bengali online newspapers is used here. Then, the dataset is preprocessed. After that, word embedding and vocabulary counts are done. Finally, the deep learning approaches are applied to generate an abstractive summary for each news article. The model used the Seq2Seq algorithm with an attention mechanism which reduced the training loss up to 0.001 successfully. Another existing dataset is used to evaluate the results and found satisfactory results for own dataset than the other.

Keywords Seq2Seq · Abstractive · Text summarization · NLP · Attention

1 Introduction

People are thirsty for acquiring knowledge by nature. A huge number of people continuously look for information from anywhere at every moment. Nowadays, it becomes easier to gather information because of the widespread usage of modern technologies. When it comes to large texts, generally they don't want to read their whole contents spending a lot of time. Rather, they want to focus on the important

M. Sultana · P. Chakraborty (✉)
Department of Computer Science and Engineering, Comilla University, Cumilla 3506, Bangladesh
e-mail: partha.chak@cou.ac.bd

T. Choudhury
Department of Informatics, School of Computer Science, University of Petroleum and Energy Studies, Dehradun, India

J. M. R. S. Tavares et al. (eds.), *Cyber Intelligence and Information Retrieval*,
Lecture Notes in Networks and Systems 291,
https://doi.org/10.1007/978-981-16-4284-5_24

words from the text that are enough to describe the main theme of a large text. That's why text summarization is needed.

Text summarization can be separated into two parts. One is extractive text summarization that generally picks the critical words from the whole text. Another is abstractive which usually focuses on the understanding of the texts rather than just picking up the important words. It also generates a short, concise summary from a long text, but not by just copying the most salient sentences from the text.

In present days, non-structured data is almost everywhere because of digitization. So, it becomes almost impossible to summarize all of the texts manually from the vast repository of information. Most of them are redundant and inappropriate. Hence, the need to develop an automatic text summarization system arises. It let the people get the main theme from them easily. Automatic text summarizers are capable of extracting useful information that removes unnecessary things from the texts. It increases the readability of documents and lessens the time spent looking for information. Thus, a huge number of real-life problems can come to a greater solution with automatic text summarization.

The increased popularity of internet usage among the people who use Bengali daily has led to the expansion of Bengali text documents on the web. But there is very little work done in Bengali language regarding this topic compared to other languages.

In this paper, a method for summarizing automatic abstractive news articles that are collected from two popular Bengali newspapers is described. For that, some preprocessing techniques such as lexical analysis, adding contractions, removing stop words, lemmatization, etc. are applied to produce cleaned and purified texts and summaries. A pre-trained Bengali word vector file is used for word embedding and count vocabulary.

The most popular method for text summarization named Seq2Seq model with multilayered bidirectional RNN for the input text is applied. Two layers of RNN with LSTM are used to produce more efficient summaries. Bahdanau attention model [4] is used to make the output more efficient. In Encoder-Decoder architecture, the encoder encodes all of the input news articles into a fixed-length vector and the decoder generates an output sequence with the help of the attention [4] mechanism. Though this model was created for Neural Machine Translation (NMT), it also works well for automatic text summarization.

2 Literature Review

Since 1950, many significant works have been done in order to address the challenge of automatic text summarization. At the beginning of the twenty-first century, with the huge popularity of artificial intelligence and machine learning technologies, many remarkable works have been done regarding text summarization. Wang et al. [1] introduced a system named "Domain-Independent Abstract Generation for Focused Meeting Summarization" in 2013. By applying Multiple-Sequence

Alignment, they addressed the challenge of generating natural language abstractive summaries for spoken meetings in a domain-independent fashion. A technique named overgenerated-rank technique was used to create and rank postulant abstracts. Then Li et al. [2] proposed an extractive document summarization using bigram-based supervised method in the ILP framework with a regression model for each bigram. They showed that their system consistently outperformed the previous ILP method on different TAC data sets. Again, Qian et al. [3] described "Fast Joint Compression and Summarization via Graph Cuts" on TAC2008 dataset in the same year.

Bahdanau et al. [4] introduced an approach named "Neural Machine Translation by Jointly Learning to Align and Translate" to increase the performance of traditional encoder and decoder methods. They extended basic encoder-decoder architecture by permitting a model to consequently explore parts of a source sentence relevant to anticipating a target word. Luong et al. [5] examined two effective categories of attention mechanism, the global approach continually attends to all source words and the local one only focuses on a subset of source words at a time. Sutskever et al. [6] described an end-to-end approach for Seq2Seq learning that makes the least assumptions on the sequence structure. They used a multilayered LSTM to map the input sequence to a vector of fixed dimensionality, and another deep LSTM to decode the target sequence from the vector. Liu et al. [7] used decoder-only architecture that can scalably attend to very long sequences to generate fluent Wikipedia paragraphs and whole articles. Summary from long text sequence and multi-document summary from a large and paralleled dataset. Shang et al. [8] proposed a neural network-based response generator named Neural Responding Machine (NRM) for Short-Text Conversation which takes the traditional encoder-decoder frameworks.

In Bengali, very little works have been done regarding this topic. Sheikh Abujar et al. [9] created a new sort of sentence scoring process for Bengali extractive text summarization. To create understandable and meaningful summaries of Bengali text documents, Talukder et al. [10] developed an abstractive text summarization using BiLSTM. Rahman et al. [11] have classified Bengali text documents using transformer-based models named BERT and ELECTRA. S. Rahman and P. Chakraborty [12] proposed a new way of classifying Bengali news documents using a Deep Recurrent Neural Network with BiLSTM.

Islam et al. [13] used Word Movers Distance (WMV) Sentence Similarity Technique for Summarizing Online Product Reviews in Bengali to validate the outcome by text similarity. In Bengali, there was no enriched dataset for abstractive text summarization before. In recent days, Bhattacharjee et al. [14] created BANS dataset containing 19,096 news articles which is the biggest dataset for Bengali abstractive text summarization technique so far. But it contains so short articles that summarization of them is not needed for them.

Choudhury et al. [15] proposed a paper titled "An Efficient Automated Design to Generate UML Diagram from Natural Language Specifications" to focus on the production of an activity diagram and sequence diagram through Natural Language Specifications. They also discussed the fundamentals of Artificial Neural Network (ANN) and Recurrent Neural Network-Long Short Term Memory (LSTM) based Deep learning (DL) architectures measuring attention in another paper [16].

3 System Architecture

In this research, a model for text summarization is used following some deep learning methods. RNN is used to solve text-related problems. Before applying these methods, a dataset was created by the authors. Then all of the preprocessing steps such as lexical analysis, adding contractions, removing unnecessary stop words, whitespaces, characters punctuations, lemmatization, etc. are applied to get purified texts. Then Word2Vec is applied for counting vocabulary and word embedding. After that, the proposed models named RNN Encoder-Decoder and Seq2Seq with Bahdanau attention [4] mechanism are applied. The outputs from the system are compared to another dataset to show the difference with the own dataset. A short overview of the proposed system architecture is shown in Fig. 1.

4 Experimental Details

4.1 Dataset Creation

The dataset created by the authors is used in this model. Texts are collected from two popular Bengali newspapers named "The Daily Prothom Alo" and "The Daily Ittefaq". 500 Long articles containing 250 from "Prothom Alo" and 250 from "Ittefaq" with articles of all categories are collected. After collecting the articles, a short-length summary for each article has been produced manually. BANS [14] dataset from "kaggle.com" containing 19 k short articles and 19 k short summaries is also collected and observed for the evaluation of the outputs.

4.2 Data Preprocessing

After creating the dataset, it needs to be cleaned and preprocessed to produce purified news and articles for the model. Data preprocessing begins with the lexical analysis step. Firstly, it takes the altered source code from language preprocessors written in sentences and breaks the syntax into a series of tokens by expelling all whitespace or comments. Each language including Bengali has contractions of some words which the machine can't understand. So, the full meanings of the contractions are added. Stop Words are removed to filter out redundant data. Punctuations and characters which can influence the text summarization processing are removed also. After that, the lemmatization process is applied to find the roots of words. Lemmatization algorithm collects all inflected forms of a word to break them down to their root dictionary form. Words are broken down into a part of speech (POS) by using the rules of grammar. After all of these preprocessing steps, the purified texts are found where no articles or summaries have any punctuations, characters, stop words

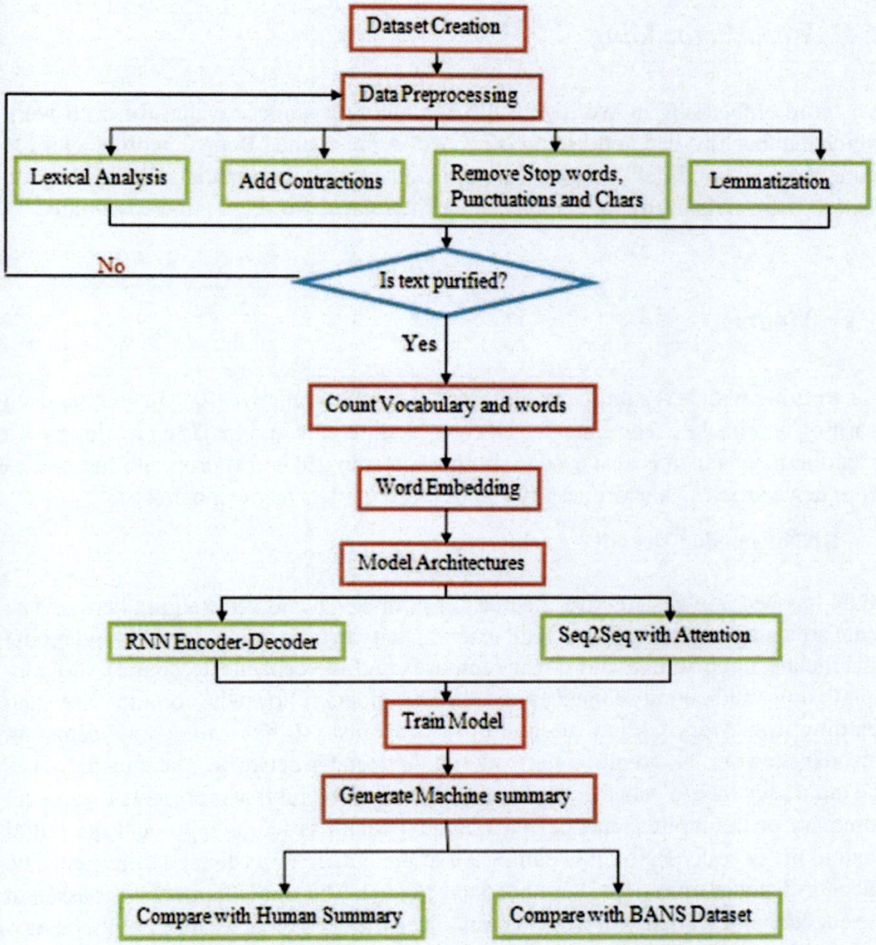

Fig. 1 Proposed system architecture

or extra whitespaces. Both Purified texts and summaries are used as input sequences of the Bengali automatic news summarization model.

4.3 Count Vocabulary

Before word embedding, Vocabulary is counted from own dataset. For news articles, there are 12,676 unique words in this dataset. Among them, 11,712 words are used in the model. For summaries, the total number of words is 61,163 with 1176 unknown words.

4.4 Word Embedding

For word embedding, a Word2Vec file containing a numeric value for each word which the machine can understand is needed. A pre-trained Bengali word vector file named "bn w2v model" collected online is used for this method. The vectors are used as the inputs of the model to produce the related words as targeted outputs.

4.5 Models

For working with text summarization, LSTM is more effective than most other deep learning approaches. The Encoder-Decoder architecture and Seq2Seq learning with attention mechanism is also used to produce meaningful and appropriate summaries from news articles. A short description of these models is given below:

i. **RNN Encoder-Decoder Architecture**

RNN has become an effective method for both NMT and seq2seq prediction. The main advantages of this approach lie in its capacity to train a single end-to-end model specifically upon source and target sentences and the capacity to control variable-length input and output sequences of text. This model follows the common Seq2Seq learning framework with an attention [4] mechanism. It has three components: an encoder network, an attention network, and a decoder network. The encoder takes the input sequence to generate the fixed-size context vector that represents a semantic summary of the input sequence. This context vector is fed as input and the initial state to the decoder RNN. It is connected to the hidden units at each time step. The encoder-decoder time steps don't need to be equal. The single input text is passed in the encoder as a sequence of word vectors. Then the decoder takes this input sequence and produces an output sequence from the relevant input sequence. A graphical view of this architecture is given in Fig. 2.

If a is the target sequence of sentences, the maximum probability of the word vector sequence will be a. Here, b is the source sequence of the sentences. The probability will be, $\arg(\max_b p(a|b))$.

If A is the long input sequence like Eq. (1),

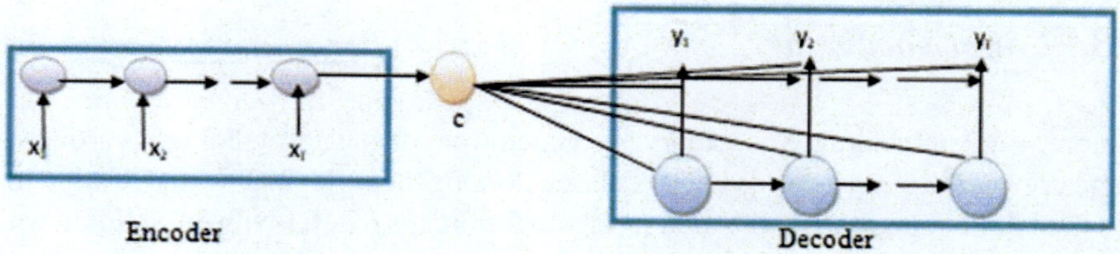

Fig. 2 RNN encoder-decoder architecture

$$A = (a_1, a_2, a_3, \ldots, a_n) \tag{1}$$

c is the context vector and f, q are non-linear functions, then the sequence will be like Eqs. (2) and (3),

$$h_t = f(a_t - h_{t-1}) \tag{2}$$

and,

$$c = q(h_1, \ldots, h_{Ta}) \tag{3}$$

ii. Sequence to Sequence (Seq2Seq) Learning with Bahdanau [4] Attention

The main purpose of a Seq2Seq model is to map a fixed-length input with a fixed-length output. The length of the input and output may vary. Here, a regular LSTM network can't be used to map each word from the input sequence to the output sequence. A token is used by each sequence to identify the first and last point of the sequence. To manage the sequence in the encoder and decoder, some special tokens such as <GO>, <EOS>, <PAD>, etc. are used in this model. For the encoder, the hidden state h_t is given in Eq. (4),

$$h_t = f(W_{hh}h_{t-1} + W_{hx}x_t) \tag{4}$$

Here, the hidden states are calculated using the weights from the previous states and the input vector, x_t. For the decoder, the hidden state h_t is given in Eq. (5),

$$h_t = f(W^{hh}h_t - 1) \tag{5}$$

The output at time step t is calculated using Eq. (6)

$$y_t = \mathrm{softmax}\left(W^s h_t\right) \tag{6}$$

A graphical view of this Seq2Seq Model with Bahdanau [4] is given in Fig. 3.

5 Result Analysis

In this paper, the methods discussed in the previous section are used to produce nearly the actual summary from news articles. For that, this model is trained for more than 3 h using the usual training parameters such as epochs = 100, RNN size = 256, batch size = 2, learning rate = 0.001, and probability rate = 0.70 to 0.75. "Adam" optimizer is applied to minimize training loss. It minimized the average training loss up to 0.001. To check the output further, the model is saved in

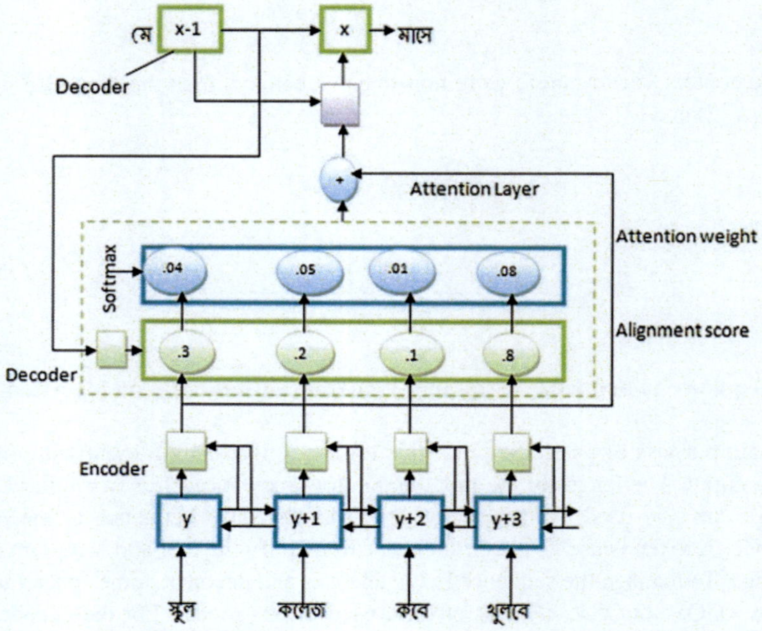

Fig. 3 Seq2Seq model with attention

"modell.ckpt" file. Then TensorFlow session with version 1.15.0 is created to reload the saved graph. The frame of data for news and summaries is also defined randomly. Actually, this is an input–output system. This model gives nearly the actual outputs for most of the articles. But sometimes it outputs less similar summaries corresponding to actual summaries. Two sample outputs from the results using this dataset are given in Table 1.

To evaluate the results, another dataset named Bengali Abstractive News Summarization (BANS) [14] is collected which is the largest publicly published dataset for Bengali text summarization model. But their text size is too short to summarize. In reality, people don't want so short texts to summarize. No newspaper publishes so short articles. During the training process, the same training parameters such as epochs, batch size, RNN size, learning rate with the same attributes as own dataset is kept. Using these parameters, the average loss found for the final update is 0.004. To compare the results of these two datasets, the number of similar words between the actual and response summary is counted for the first 3 sentences from both datasets. The comparison is given in the following Fig. 4. Besides, outputs of two other recent works related to this topic are considered to evaluate this work. Talukder et al. [10] reduced the average training loss up to 0.008. In 2020, Islam et el. [13] improved this by reducing the average loss up to 0.0034. Comparing with all of these works, it can be seen clearly that, our model gives better results among all.

Table 1 Output from the model using own dataset

Input Text	Summary
গত ২৪ ঘণ্টায় দেশে করোনা ভাইরাসে আক্রান্ত হয়ে আরও ৭ জনের মৃত্যু হয়েছে। এ নিয়ে দেশে মোট মৃতের সংখ্যা দাঁড়ালো ৮ হাজার ৩৫৬ জন। নতুন করে ৩৬৬ জন শনাক্তসহ দেশে মোট শনাক্তের সংখ্যা দাঁড়িয়েছে ৫ লাখ ৪৩ হাজার ৭১৭ জন। গত ২৪ ঘণ্টায় মারা যাওয়া সাতজনের চারজন পুরুষ ও তিনজন নারী। দেশে গত ২৪ ঘণ্টায় ১১ হাজার ১০৩ জনের নমুনা পরীক্ষা হয়েছে। সোমবার দুপুরে করোনা ভাইরাসের বিজ্ঞপ্তিতে স্বাস্থ্য অধিদফতরের প্রেস □িজ্ঞপ্তিতে এ তথ্য জানানো হয়। প্রেস বিজ্ঞপ্তিতে জানানো হয়, এ সময়ের মধ্যে সুস্থ হয়েছেন ৬৯২ জন। এ নিয়ে দেশে মোট সুস্থ হয়েছেন ৪ লাখ ৯২ হাজার ৫৯ জন।	**Actual:** গত ২৪ ঘণ্টায় মারা যাওয়া সাতজনের চারজন পুরুষ ও তিনজন নারী **Response:** গত ২৪ ঘণ্টায় মারা যাওয়া সাতজনের চারজন পুরুষ
২৫ থেকে ৪০ বছর বয়সী তরুণদের নিয়ে গঠিত, উন্নয়ন ও গঠনমূলক কাজে নিয়োজিত বিশ্বব্যাপী প্রতিষ্ঠান 'জেসিআই' ঢাকা ইস্ট জোন শুক্রবার (১২ ফেব্রুয়ারি) □িকেলে রাজধানীর সারিনা হোটেলে সফলভাবে পালন করল □ছরের প্রথম সাধারণ সদস্য সভা। সভায় ২০২১ সালের হিসা□ □ি□রণ অনুমোদন ও বছরের কর্মপরিকল্পনা নির্ধারিত হয়। আসন্ন জাতীয় □ক্ত□্য ও বিতর্ক প্রতিযোগিতা কার্□ক্রম সম্পর্কে সদস্যদের অভিহিত করা হয়। উপস্থিতির মাধ্যমে অনুষ্ঠানটির শোভাবর্ধন করেন জেসিআই বাংলাদেশের কর্ণধার ও সিনেটর, তুরস্কে নিযুক্ত বাংলাদেশের রাষ্ট্রদূত মাসুদ মান্নান।	**Actual:** চাটমোহরে মশার উপদ্রব □েড়েছে। **Response:** চাটমোহরে মশার উপদ্রব □েড়েছে

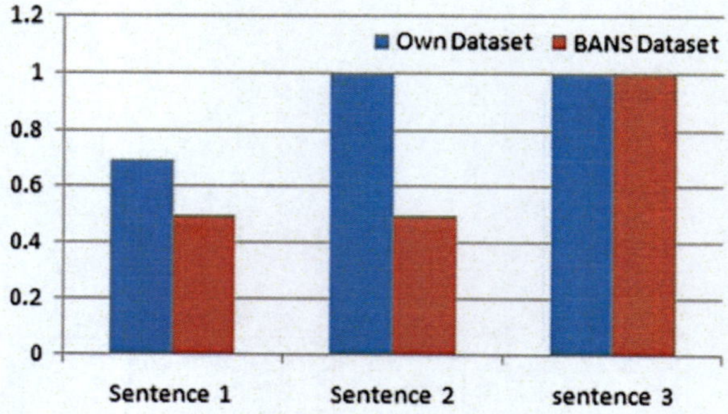

Fig. 4 Comparison of the outputs of BANS dataset with own dataset

6 Discussion

Working with a large number of texts is very challenging. And when it comes to abstractive text summarization, it becomes much harder. Despite having so many difficulties, this model gives greater results according to the inputs. Before using Bengali texts, this model is applied for English texts for reviewing Amazon products. It gives amazing results for English texts also. So, it is clear that this model works well for any language. But, in this work, the Bengali language is focused on abstractive text summarization, so the results for Bengali texts are shown. The dataset with long texts created by the authors gives comparatively better results than all of the existing Bengali abstractive text summarization models. The outcomes of this model are fully satisfactory.

7 Conclusion

This work mainly aims to create and analyze an automatic Bengali abstractive news summarization approach. Here, both news articles are used as the inputs and the generated summaries of those sequences are of Bengali language. Though own dataset is too small for this model, the system gives a greater response. There is no rich dataset with long texts available in Bengali for generating summaries. This model is built for generating short summaries from long texts. Here, sequence length and summary length are defined for generating the output. For variable-length-based sequences, this model doesn't show better outputs. Moreover, this model gives comparatively better results for short articles than long articles. It cannot output the numerical data used in the texts. Resolving the limitation issues can be helpful to generate a more efficient architecture. Increasing the datasets of the Bengali texts for summarization, preparing a working model for any kind of text length, improving the model for better accuracy, using performance metrics like ROUGE, BLEU measures for evaluation can be the future implications of this work.

References

1. Wang L, Cardie C (2013) Domain-independent abstract generation for focused meeting summarization. In: Proceedings of the 51st annual meeting of the association for computational linguistics. Association for Computational Linguistics, Sofia, Bulgaria, pp. 1395–1405
2. Li C, Qian X, Liu Y (2013) Using supervised bigram-based ILP for extractive summarization. In: Proceedings of the 51st annual meeting of the association for computational linguistics. ACL, Sofia, Bulgaria, pp. 1004–1013
3. Qian X, Liu Y (2013) Fast joint compression and summarization via graph cuts. In: Proceedings of the conference on empirical methods in natural language processing. ACL, Seattle, Washington, USA, pp. 1492–1502
4. Bahdanau D, Cho K, Bengio Y (2014) Neural machine translation by jointly learning to align and translate. arXiv preprint arXiv:1409.0473
5. Luong MT, Pham H, Manning CD (2015) Effective approaches to attention-based neural machine translation. arXiv preprint arXiv:1508.04025
6. Sutskever I, Vinyals O, Le QV (2014) Sequence to sequence learning with neural networks. In: Advances in neural information processing systems. NIPS, pp. 3104–3112
7. Liu PJ et al (2018) Generating wikipedia by summarizing long sequences. In: International conference on learning representation. arXiv:1801.10198
8. Shang L, Lu Z, Li H (2015) Neural responding machine for short-text conversation. In: Association for computational linguistics, vol 1. ACL, pp. 1577–1586
9. Abujar S, Hasan M, Shahin M, Hossain SA (2017) A heuristic approach of text summarization for Bengali documentation. In: 8th international conference on computing, communication and networking technologies. IEEE, pp. 1–8
10. Talukder MAI, Abujar S, Masum AKM, Faisal F, Hossain SA (2019) Bengali abstractive text summarization using sequence to sequence RNNs. In: 10th international conference on computing, communication and networking technologies. IEEE, pp. 1–5
11. Rahman MM, Pramanik MA, Sadik R, Roy M, Chakraborty P (2020) Bangla documents classification using transformer based deep learning models. In: 2nd international conference on sustainable technologies for industry 4.0. IEEE, Dhaka, Bangladesh, p. 15
12. Rahman S, Chakraborty P (2020) Bangla document classification using deep recurrent neural network with BiLSTM. In: Proceedings of international conference on machine intelligence and data science applications
13. Islam MJ, Ahmed K, Chowdhury MKH (2020) Summarizing online product reviews in Bengali based on similarity using sequence to sequence RNNs. In: 11th international conference on electrical and computer engineering. IEEE, Dhaka, Bangladesh
14. Bhattacharjee P, Mallick A, Islam MS, Jannat M (2020) Bengali abstractive news summarization (BANS): a neural attention approach. In: 2nd international conference on trends in computational and cognitive engineering. arXiv:2012.01747
15. Gulia S, Choudhury T (2016) An efficient automated design to generate UML diagram from natural language specifications. In: 6th international conference-cloud system and big data engineering (Confluence). IEEE, pp. 641–648
16. Chakraborty P, Ahmed S, Yousuf MA, Azad A, Alyami SA, Moni MA (2021) A human-robot interaction system calculating visual focus of human's attention level. In: IEEE Access, vol. 9, pp. 93409–93421. https://doi.org/10.1109/ACCESS.2021.3091642

An Approach to Improve Searching Textual Data in Email Dataset

Manjima Saha, Arjama Chatterjee, Prithwidip Das, and Shayan Pal

Abstract In current times it is very hard to imagine a day without the use of a computer. Communication among people has increased drastically with the help of the internet using social media platforms. Among these various modes of communication, email plays a vital role in connecting people globally using a unique ID called email ID which is used to identify different users, there's also instant messaging and other modes of communication over the internet via a social media service such as Facebook, WhatsApp, etc. Using these services users can communicate by sending messages and data in the form of attachments to one another. In this paper, we try to implement feature selection techniques such that we can find the strongest relationships between users which are low cost and can be easily implemented to recognize users writing patterns and also help in searching and providing human-like search results and responses. In this case, we have used the Enron email corpus which is made public for research by the CALO project.

1 Introduction

With the rapid pace of technological advances, email, instant messaging and other forms of interactions on social media have gained a lot of popularity, being the go-to mode of communication for connecting with one another. The growth and ease of use of these technologies have led many to transform traditional modes of communication and adopt these in a positive aspect so much so, that they have become the official mode of communication in offices, educational institutions, social interactions and others.

This has led to an increase in the amount of data that is generated. Such data can be used to understand consumer behaviour, perform prediction based analysis, communication networks, pattern recognition and many more implementations. But

M. Saha (✉) · A. Chatterjee · P. Das · S. Pal
Institute of Engineering and Management, GN-34/2, Sector–V, Saltlake Electronics Complex, Kolkata, West Bengal 700091, India
e-mail: manjima.saha@iemcal.com

J. M. R. S. Tavares et al. (eds.), *Cyber Intelligence and Information Retrieval*, Lecture Notes in Networks and Systems 291,
https://doi.org/10.1007/978-981-16-4284-5_25

it is difficult to process through countless such data. The reason for such a problem is the nature of the data as most of it is unlabelled and unstructured, i.e. the data is very text heavy and has no definite manner or format of order, hence it makes it difficult to process, information extraction and analysis, leading to uncertain outcomes.

The advent of machine learning and data mining techniques has led us to accomplish a huge feat of analyzing large datasets. Text mining can be described as a technology that helps to extract information from textual data such as emails, messages, documents, tweets and others, through various methods such as Natural Language Processing (NLP) to gain valuable insights. It is very essential as the raw data present is very unstructured, but with the help of text mining, the data can be structured and then used for analytics or knowledge. Using the data mining techniques like classification, it labels the attributes related to the numerous types of data present and then arranges them in categories to identify and understand the structure better. For example, a simple classification in a dataset containing emails would be to categorize them between "spam" and "not spam" emails. Clustering is another learning technique that is implemented to understand the relationships, i.e. similarities and dissimilarities among the different types of data. The points are then grouped into specific clusters containing similar types of data points. To understand it better, a graphical approach is always employed such that the distribution of data can be portrayed.

Feature selection is mostly used to evaluate and find the strongest connections between each input variable and the target variables. These are mostly used to reduce the computation cost and complexity of a model and also enhancing it which makes these methods very quick and effective. The significance of using feature selection is to find a subset of features by evaluating them based on certain predefined criteria.

We used feature selection to analyze the patterns in the author's style of writing such that we can find the frequently used words by the author and suggest words accordingly. We have used the Enron email dataset and analyzed the different words used in the data set and how frequently they have been used. It is believed that the behaviour and style of writing of an author can be analyzed and even designed by using certain key features taken from their writings. This helped us to tag certain writing patterns to different users in our dataset.

2 Methods and Algorithms

In this chapter, we will describe the different methods and algorithms we have used for our research.

2.1 K-Means Clustering Algorithm

Clustering can be defined as a set of techniques used to split given data into clusters where the data in the cluster share some relationship or are similar to one another. The K-Means algorithm (shown in Algorithm 1) is used to minimize the sum of distances between the data points and their respective clusters. It is an unsupervised machine learning technique that is used to associate clusters of data objects present in a dataset. An unlabeled dataset can be given as input and the algorithm categorizes the data objects into K number of centroids, with a predetermined value of K and then assigns each data object to the nearest cluster as well as keep the centroid as small as possible. The K points in the group are called means because they are the mean values of the objects categorized in them.

Algorithm 1 K-Means Clustering Algorithm

```
1: State number of clusters (k) to allocate
2: randomly select and initialize k centroids
3: repeat
4:     expectation: Each data point is allocated to the closest centroid
5:     maximization: Centroid(mean) of each cluster is calculated
6: until Position of the centroids do not change
```

2.2 Text Vectorization

Word or Text Vectorization can be defined as a procedure to convert text into its numerical form. One can achieve this using NLP to map words present in a text to a vector of real numbers which can be used to find word predictions, semantics, etc. Using Cosine Similarity we calculate the cosine of the angle between two vectors in an N-dimensional vector space to match similar texts.

2.3 TF-IDF Scores

Term Frequency—Inverse Document Frequency or TF—IDF is a method used to know the value of a word in a document by assigning a weight to each word to signify its importance. The Term Frequency metric provides an estimate of the frequency of occurrence of a word in the text or document. We normalize the frequency value by dividing it by the total number of words in the text to eliminate any biases.

$$\text{tf}\ (t, d) = \frac{\text{count of terms } (t) \text{ in document } (d)}{\text{number of words in document } (d)} \tag{1}$$

The count of the number of documents in which a term has occurred gives us the Document Frequency (DF) metric.

Inverse Document Frequency (IDF) measures the knowledge imparted or importance of a term. This helps us remove StopWords since they are most occurring and are less informative so they have very low IDF scores. But for a large dataset, the IDF scores are hard to calculate so we take the log of IDF. When querying, if a word occurs which is not in the set then the df value will be 0, so to avoid such a case, we add 1 to the denominator.

$$\text{idf} = \log\left\{\frac{\text{number of document } (N)}{\text{number of document in which the word appears (df)} + 1}\right\} \tag{2}$$

Now, by multiplying Eq. (1) and (2), we obtain the TF-IDF scores.

$$\text{tf - idf } (t, d) = \text{tf}(t, d) \times \log\left\{\frac{N}{(\text{df} + 1)}\right\} \tag{3}$$

After obtaining the TF-IDF scores we get to understand the significance or importance of that word in the text.

3 Related Works

The Enron email dataset contains over millions of data that needed to be cleaned and structured so that it can be used for text mining, analysis and other purposes. In [1], the authors tried to implement some strategies to clean the dataset such as identifying the entities, extracting and generalizing the formats, remove duplicates and understand key aspects of these emails in large corporations where the cleaned dataset can be used for future work.

On comparing various text categorization methods in [2] the authors found that out of the numerous classification methods, KNN is found to be having the best performance even when it is used in very noisy and vast classification problems.

Reference [3] provides a survey on the text mining techniques, classification techniques and clustering algorithms currently in demand.

Using WEKA tool [4] different features from the email dataset are classified using Neural network, SVM, Naive Bayes and Decision Tree into suspicious or regular emails. Out of these, simple ID3 classifier showed promising results and higher detection rates.

To reduce the training time, a parallel SVM based on MapReduce algorithm is used [5] for email classification to overcome the difficulties from email folder and document classification. But this reduces the accuracy of the set which can be improved using ontology based semantics.

In [6], various classification algorithms such as SVM, NB, J48 and others were used by the authors to gain higher accuracy of fraudulent email detection. In this pursuit, they observed that accuracy is directly related to the choice of efficient features and frequency of them and is not entirely dependent on the choice of the classification algorithm used because similar feature sets give similar results for most classification methods.

Since regression-based classifiers contain information about annotator judgements so they offer better performance than binary classifiers so [7] a machine learning approach based on regression is used for summarizing email threads. It is found that Gaussian and Bagging processes have the highest weighted recall after an extensive comparison of various regression-based classifiers.

Feature selection is one of the core aspects required to find an understanding of the relationship between the data by removing the redundant data and help in reducing the training time in most machine learning techniques. A comparison of feature selection methods has been studied in-depth [8] where the authors conducted tests by exposing it to different datasets and rate the performance of various classification algorithms and two feature selection algorithms Boruta and Chi-Square were used. The authors reached a conclusion that Boruta Algorithm has better performance for datasets that contained more attributes but Chi-Square performs better when fewer attributes are present. When applied with different classification algorithms such as Random Forest and Gradient Boosting, Boruta feature selection algorithm had higher performance and greater accuracy overall.

Using writing-style features [9] such as syntactic, lexical, structural and content-specific features, a learning algorithm is built to identify authorship. With structural features and content-specific features, a clear demarcation is made between the authors. For a given training set SVM had 83% accuracy which further concluded that the writing-style features analysis is generally effective for a large number of authors.

Applying feature selection in training phase [10] to select the best feature subset and the attributes of the categories using Bayesian model of selection then a generative probabilistic model is used where the categories are described by distributions and using a binary exclusion/inclusion latent vector. This enhances and reduces the dimensionality and improves accuracy of text classification with the selected features.

In [11], for social network analysis and identifying organizational structures, an adjacency matrix of the communication graph is used and graph theoretical analysis is performed by calculating numerous graph metrics. But dealing with large datasets can be very challenging and to overcome such problems, randomized algorithms have been proposed [12] which show significant improvement in performing low-rank matrix approximation using random sampling by finding a subspace that can perform most of the actions of the matrix. A low-rank approximation of the email communication data and the pre-processing of data has a high impact on the communication graph.

Another approach for social network analysis using a new model called Author-Recipient-Topic (ART) is proposed in [13] based on Latent Dirichlet Allocation (LDA) and the AuthorTopic (AT) model. The model learns the distribution of topics based on the direction-sensitive messages between two nodes, by adding key attributes on the distribution between the sender and receiver to predict the roles better and gives low perplexity.

To find the communication patterns in a social network through text analysis, a theoretical approach [14] is proposed by the authors where they found that people communicate more with other people with whom they share similar value patterns but that does not imply that these people can be all put under a single value type.

Apart from feature selections and the methods of text classification, [15] email formality is also an important subject in communication. The Brown and Levinson's model of politeness is used to conduct the research by the authors to test how power, imposition and social distance can affect the formality and choice of words.

In [16] for classifying texts, Deep Learning models are used after pre-processing, particularly, Long Short-Term Memory (LSTM) networks are found to be having high efficiency for binary and multi-label text classification problems. Training the model on text and common bigrams proved to be highly effective in identifying malicious texts.

4 Experiment Methodology

First, we took the Enron dataset and converted it into data frames using the pandas library in python. The dataset contained a lot of data objects which were not necessary for our research so we created a new frame where the emails are divided into lines as per some predetermined selection criteria, then the body or the content of the email and the index numbers are extracted. The extracted data is converted from Map to Lists with 3 categories, 'body', 'to' and 'from'. Here, the unique email addresses are retrieved and the duplicate, redundant and blank emails are dropped from the dataset.

The extracted data is then tokenized and converted to a document-term matrix. Terms appearing frequently or common in the dataset are included in English stop-words like ['ect', 'hou', 'com', 'recipient']. The dataset is then vectorized using "TfidfVectorizer()".

We selected a frequency range between 0.3 and 2.0 where 0.3 is set to ignore all terms that appear in more than 30% of the document and 2.0 is set so that all the terms that appear in less than 2 documents are ignored.

To determine vocabulary and IDF, a matrix-type variable is taken which returns document-term matrix using fit transform() method.

We used text vectorization technique by using TF-IDF, we found the vectorization or tf-idf score using Eq. (3) for every word, i.e. frequency score by which we can build an understanding of which words has been used mostly or how many times the word has appeared.

For feature extraction, we use an array mapping method to map feature integer indices to their corresponding feature names. On the basis of vectorization/tf-idf scores of words, we printed the top few terms.

We built clusters based on the previous outputs but since there's no knowledge about the email content and due to large number of emails an unsupervised machine learning technique is used called K-Means Clustering algorithm. We divided the dataset into training and testing parts and used the K-Means Clustering technique using algorithm 1, then 3 clusters are fixed according to the previously generated word's tf-idf scores.

The output of K-Means Cluster is stored in a sparse matrix. The cluster centres are computed and cluster indexes are predicted for each sample.

For our research, a smaller batch size is considered so 'MiniBatchKMeans' is not implemented. Else, for large batch sizes the use of 'MiniBatchKMeans' is recommended. The number of clusters can be increased depending on the use case but for our research, we fixed it at 3 clusters. To visualize the 3 clusters, the sparse matrix is used to build a graph with 2-axes using Principal Component Analysis (PCA).

To reduce the linear dimensionality to a lower dimensional space so that the data is oriented and scaled can be implemented depending on the number of components to extract and the form of the input data using the LAPACK execution of the full SVD, the ARPACK application of the truncated SVD or using the method of [5] randomized truncated SVD.

Figure 1 represents the graph generated for the 3 clusters after calculating the word scores using TF-IDF scores for each word consisting of the words having maximum TF-IDF scores in that cluster, those are the top used words.

The output gave us insights on which words are used the most and which are used the least and even, some employee names of the company also are shown such as "John" which appear very frequently.

Figure 2 shows the 3 clusters in different colours and the TD-IDF scores of the terms are plotted. Cluster—2 has the highest number of word scores so it is

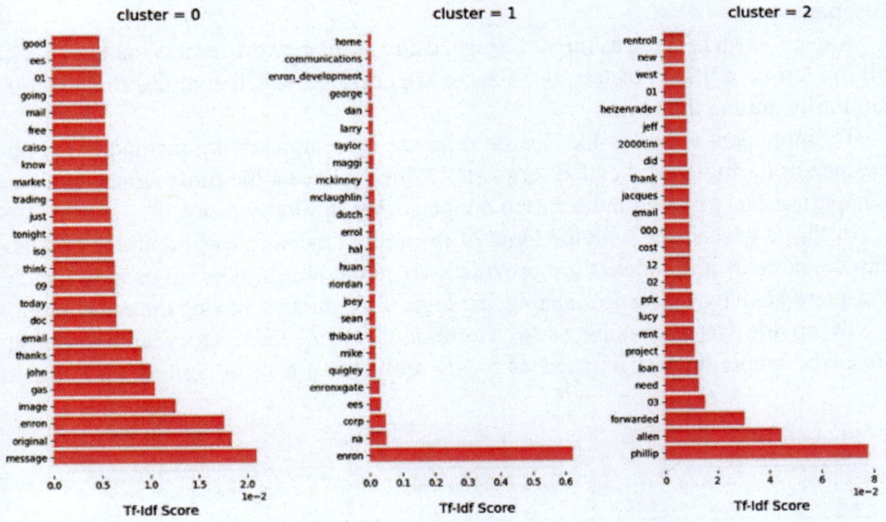

Fig. 1 Represents the top terms per cluster of the 3 clusters

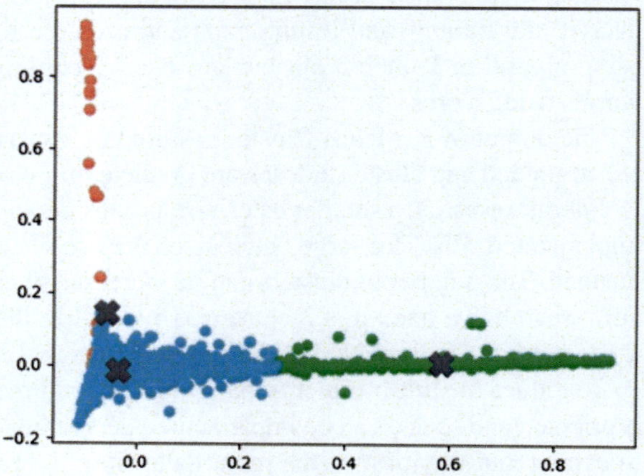

Fig. 2 Represents the 3 clusters in a scatter plot

represented towards the top with higher values. The cross markings on the graphs represent the represents the centroids or the centres of the clusters.

From the above research, it helped us build a knowledge base on which words are similar and which words are used the most so we implemented a query system. A system where an input can be given and all the terms exactly matching with the query as well as all the related terms with the text associated with them and the index numbers of the emails will be provided giving it a more of a human-like search and response mechanism.

We used cosine distance to find out that if there's any email that has been repeated or closely related to any other email then what is the amount of similarity between the mails.

A query term is taken as input. Using Cosine Similarity the term is matched with all the terms in the 3 clusters and a score is generated which signifies the level of similarity among the terms.

To implement a human-like search response system, a sorting method based on cosine similarity is used called 'argsort()' which gives us the most related indices comparing every related index based on the cosine similarity score.

In Fig. 3 we can see with the input of the query "meeting" again, all the emails, associated texts and indexes are provided where the term might not exactly match the query but have the same meaning, are logically similar or having the same intent.

To provide exactly matching search terms as the query and reduce search response time, the whole dataset is taken as a 2-D array consisting of 'index' and 'email

```
[1. 0. 0. ... 0. 0. 0.]

[ 139 2422  722 1869 8038 2195 2138 2299  186 2186 2581  294  896 3068 5771 6946 4122 8218 8649 8549 3984 3337 7232 2158 8542 8719 8384 9445 8448 8914 8679 8525 8352
8774 4746 4179 8526 8773 8354 1360 4505 4727 5962 3792 2256 4771 4274 4752 3809 6581 6234 8232 8486 8637 6281 2219 7352 3473 4577 37 2283 8716 8874 8486 9098 2198 8499
8601 8272 81 2366  661 3698 3759 9024 4300 5725 6022 6884 3006 4061 1440 1946 4367 4764 8516 8214 8653 4929  176 2298 3667 7545 3748 3820 4751 8393 8547 8582]

['I would look at properties in San Antonio or Dallas.']
['Alan,You should have received updated numbers from Keith Holst.  Call me if youdid not receive them.Phillip']
```

Fig. 3 Relevant search results

```
[   1  825 1197  271 4120 1170  725 4237 4126]

["Traveling to have a business meeting takes the fun out of the trip.  Especially if you have to prepare a presentation.  I would suggest holding the business plan
meetings here then take a trip without any formal business meetings.  I would even try and get some honest opinions on whether a trip is even desired or necessary.As
far as the business meetings, I think it would be more productive to try and stimulate discussions across the different groups about what is working and what is not.
Too often the presenter speaks and the others are quiet just waiting for their turn.   The meetings might be better if held in a round table discussion format.My
suggestion for where to go is Austin.  Play golf and rent a ski boat and jet ski's.  Flying somewhere takes too much time."]
```

Fig. 4 Exactly matching search results

content'. We retrieve the email indices of the terms that exactly match the query term and are shown with their indexes as the output.

In Fig. 4 we can see with the input of the query "meeting", all the emails, associated texts and indexes are provided where the term exact term "meeting" is present.

5 Conclusions

In this research work, we have tried to provide a different approach to searching through large email datasets. In traditional approaches, the emails are often tagged or labelled either manually by users or just placed randomly under a designated label. In our approach, with the use of unsupervised machine learning algorithms, we can cluster the data into groups that are found to be closely related by the machine thus automatically categorizing them. The TF-IDF scores are used to find the frequently used terms and can be used as keywords by the algorithm while searching. Thus, this would effectively reduce the number of email data to process by making the results appear faster. Since the authors writing patterns and behaviours are analyzed so, over time the machine can be trained and be able to make future predictions of frequently used terms and can provide the results more quickly and with higher accuracy.

This helps us to provide an approach to a simple, low cost and easy to implement solution where one can easily be able to learn the frequently used words, provide better text prediction and also meaningful search results which would include human-like related search responses.

6 Applications and Areas of Further Research

During our research work only worked with textual data, but this work can be used to expand and incorporate the email attachments such as documents, media, externals links, etc. which are of similar context in future. One can further extend this work and implement it on crime data for counter terrorism measures through social media posts, blogs and using characteristic trait analysis, text analysis and also email data analysis provide more appropriate results or responses with higher accuracy. In scenarios, where more human-like understanding and responses are necessary such as voice assistants, chatbots, speech recognition software, etc. can use this approach to easily implement such features in the system.

References

1. Zhou Y, Goldberg M, Magdon-Ismail M, Wallace W (2007) Strategies for cleaning organizational emails with an application to Enron email dataset
2. Yang Y (1999) An evaluation of statistical approaches to text categorization. Inf Retrieval 1:69–90
3. Allahyari M, Pouriyeh S, Assefi M, Safaei S, Trippe ED, Gutierrez JB, Kochut K (2017) A brief survey of text mining: classification, clustering and extraction techniques. ArXiv abs/1707.02919 (2017): n. pag.
4. Balamurugan S, Rajaram R, Govindaraj A, Muthupandian M (2021) Data mining techniques for suspicious email detection: a comparative study
5. Xu K, Wen C, Yuan Q, He X, Tie J (2014) A MapReduce based parallel SVM for email classification. J Netw 9. https://doi.org/10.4304/jnw.9.6.1640-1647
6. Nizamani S, Memon N, Glasdam M, Nguyen DD (2014) Detection of fraudulent emails by employing advanced feature abundance. Egypt Inf J 15(3):169–174. ISSN 1110-8665
7. Ulrich J, Carenini G, Murray G, Ng R (2009) Regression-based summarization of email conversations. In: Proceedings of the international AAAI conference on web and social media 3(1)
8. Bhalaji N, Sundhara Kumar KB, Selvaraj C (2018) Empirical study of feature selection methods over classification algorithms. Int J Intell Syst Technol Appl 17(1/2):98–108
9. Zheng R, Li J, Chen H-C, Huang Z (2006) A framework for authorship identification of online messages: writing-style features and classification techniques. JASIST. 57:378–393. https://doi.org/10.1002/asi.20316
10. Feng G, Guo J, Jing B-Y, Hao L (2012) ABayesian feature selection paradigm for text classification. Inf Proc Manage IPM 48. https://doi.org/10.1016/j.ipm.2011.08.002
11. Chapanond A, Krishnamoorthy M, Yener B (2005) Graph theoretic and spectral analysis of Enron email data. Comput Math Organ Theory 11:265–281. https://doi.org/10.1007/s10588-005-5381-4
12. Halko N, Martinsson P-G, Tropp J (2011) Finding structure with randomness: probabilistic algorithms for constructing approximate matrix decompositions. SIAM Rev 53:217–288. https://doi.org/10.1137/090771806
13. McCallum A, Wang X, Corrada-Emmanuel A (2007) Topic and Role Discovery in social networks with experiments on Enron and academic email. J Artif Intell Res 30:249–272
14. Zhou Y, Wallace W, Fleischmann K (2010) Automatic text analysis of values in the Enron email dataset: clustering a social network using the value patterns of actors. In: 2014 47th Hawaii international conference on system sciences, Koloa, Kauai, Hawaii, pp 1–10. https://doi.org/10.1109/HICSS.2010.77
15. Peterson K, Hohensee M, Xia F (2011) Email formality in the workplace: a case study on the Enron corpus, pp 86–95
16. Baccouche A, Ahmed S, Sierra-Sosa D, Elmaghraby A (2020) Malicious text identification: deep learning from public comments and emails. Information 11(6):312. https://doi.org/10.3390/info11060312

Analyze Cardiotocograph to Classify the Fetal Status Using Various Machine Learning Algorithms

Mathur Swati

Abstract During pregnancy, study of fetus inside the womb is required to know. Cardiotocography (CTG) is a technique with the help of which we can monitor baby's heart rate and mother's contraction simultaneously. Fetal state can be categorized into normal, suspicious, and pathological. Various machine learning algorithms are used to study their performance to analyze and compare using various evaluation metrices.

Keywords Cardiotocograph · Fetal status · Svm · Decision tree · Adaboost · Xgboost

1 Introduction

During pregnancy to know the status of fetus, manual examination of womb is now not in practice as well as gives error also. Clinical study is needed can be done for reducing morbidity and mortality. Cardiotocography (CTG) exam is a technique with the help of which we can capture uterine contractions (UC) and fetal heart rate (FHR) simultaneously. FHR is based on ultrasound concept and UC is based on pressure transducers.

Initially, gynecologists used to review CTG recording through visual interpretation which is not a reliable approach to fetal morbidity. Thus there is a need for an automated tool which helps to analyze the CTG recording. Machine learning provides various classification techniques which help us to detect the status of fetal and make decisions. Fetal state can be categorized into normal, suspicious, and pathological.

In this paper, for automatic diagnosis of CTG various machine learning algorithms are used. SVM, decision tree, adaboost, and xgboost algorithms are implemented.

M. Swati (✉)
CMR Institute of Technology, Banglore, India

J. M. R. S. Tavares et al. (eds.), *Cyber Intelligence and Information Retrieval*, Lecture Notes in Networks and Systems 291,
https://doi.org/10.1007/978-981-16-4284-5_26

2 Literature Review

In [1], comparing feedforward artificial neural network and extreme learning machine for Cardiotocography signals. Accuracy for extreme learning is better than ANN.

In Research paper [2], studying various machine learning algorithms like RF, GLM, KNN, ANN, and SVM. By applying 10-repeated cross validation, SVM algorithm is giving better results.

V. Subha et al. [5] compared various classification techniques. It compared naïve bayes, decision tree, MLP, and radial basis function. Performance of these classifiers is evaluated using accuracy, sensitivity, specificity, and F-score. Decision tree is giving better results as compared to other techniques.

WEKA is a powerful tool. Many researchers are using this tool to compare various classification algorithms [6, 7, 9] to know the status of fetal. Hakan Sahin [9], is comparing the various classification algorithms using WEKA. ANN, SVM, SL, RBF, C4.5, CART and RF. Each algorithm is tested on 10-fold cross validation. Random forest is giving the best result. In [6], comparing different algorithms like neural network, decision tree, bagging classifier, KNN, SVM. This paper [6] is proposing rough neural network which is giving better results than others.

Following research papers [3, 4, 8, 11–14] are comparing various algorithms of machine learning. Recall, accuracy, precision, F-Score are the common parameters are used for performance evaluation.

3 Experimental Setup

3.1 Dataset

In our study, Cardiotocography dataset [15] available publicly at UCI Machine learning repository. Dataset contains 2126 observations. It is using 21 attributes. Classification is done with respect to both FHR class [1–10] and NSP [P = pathologic, S = suspect, N = normal]. Details of attributes are given in Table 1. Dataset is divided into training and testing datasets. Size of the training dataset is 1594 and size of the testing dataset is 532.

3.2 Classification Techniques

As per our literature study, we found that SVM and decision trees are giving good results. So in my research work, comparing these algorithms to check which algorithm is performing better. Xgboost always gives better results than random forest. Thus, predictive analytics techniques used for analysis are discussed below:

Table 1 Attribute information

Feature	Description
LB	FHR baseline (beats per minute)
UC	No. of uterine shrinking per second
AC	No. of accelerations per second
DL	No. of light decelerations per second
FM	No. of fetal movements per second
ASTV	Percentage of time with abnormal short term variability
DS	No. of severe decelerations per second
MSTV	Mean value of short term variability
DP	No. of prolonged decelerations per second
MLTV	Mean value of long term variability
Width	HR histogram width
ALTV	Percentage of time with abnormal long term variability
Max	Maximum of FHR histogram
Min	Minimum of FHR histogram
Nzeros	No. of histogram zeros
Nmax	No. of histogram peaks
Mean	Mean in histogram
Mode	Mode in histogram
Median	Median in histogram
Tendency	Tendency in histogram
Variance	Variance in histogram

a. Support Vector Machine: It belongs to a family of supervised machine learning algorithm which takes data points and separates the data classes into hyperplanes. SVM is used for both classification and regression. Accuracy of kernel function depends on the training data. Rbf kernel is used for implementing SVM.
b. Decision Tree algorithm: It is a supervised learning algorithm. Using training data decision rules are prepared which is used to predict the class of variable.
c. AdaBoost algorithm: It is used with weak learners. Adaboost is used to improve the performance by combining the results of various weak learners and combining their predictions to form one strong rule.
d. XGBoost algorithm: It uses gradient boosting framework. During implementation, learning rate is set as 0.001.

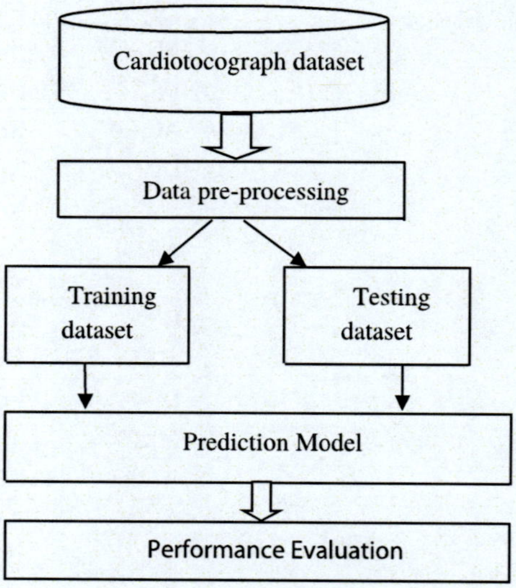

Fig. 1 System architecture for prediction of fetal status

3.3 *Performance Measures*

Performance measures considered in this study are Precision, accuracy, F1-score, and accuracy to measure various classification techniques.

a. Precision: It is defined as ratio of relevant instances out of retrieved instances.
b. Recall (sensitivity): It is defined as the ratio of relevant instances that are retrieved.
c. F1-score: It is defined as the harmonic mean of recall and precision.
d. Accuracy: It specifies the percentage of data predicted correctly.

3.4 *Implementation*

Various classification algorithms like SVM, decision tree, Adaboost, xgboost are used to classify the fetal status. System architecture for prediction is given in Fig. 1. Python tool is used for the implementation of all classification algorithms.

4 Results

In this experiment, various classification algorithms are used to evaluate the result of fetal state. Classification algorithms used are support vector machine (SVM), decision tree, Adaboost, XGBoost algorithm. After performing the test on Cardiotocograph dataset, performance of algorithms are evaluated on precision, recall, f1-score, support, and accuracy (Tables 2, 3, 4 and 5).

Table 2 Precision score of various classification algorithms

Algorithm	Precision
SVM	0.84
Decision Tree	0.78
Adaboost	0.72
XGBoost	0.86

Table 3 Recall score of various classification algorithms

Algorithm	Recall
SVM	0.63
Decision Tree	0.70
Adaboost	0.70
XGBoost	0.59

Table 4 f1-score score of various classification algorithms

Algorithm	f1-score
SVM	0.69
Decision Tree	0.73
Adaboost	0.71
XGBoost	0.66

Table 5 Accuracy score of various classification algorithms

Algorithm	Accuracy
SVM	85.3
Decision Tree	86.8
Adaboost	86.2
XGBoost	84.9

In the experiment, XGBoost is performing good for precision but not showing results on Recall, accuracy, and F1-score. Decision Tree is giving good results for F1-score and accuracy. Adaboost and Decision tree both are performing well for recall. However, performance of SVM is average for all performance measures. In order to visualize the performance of various algorithms is represented in pictorial

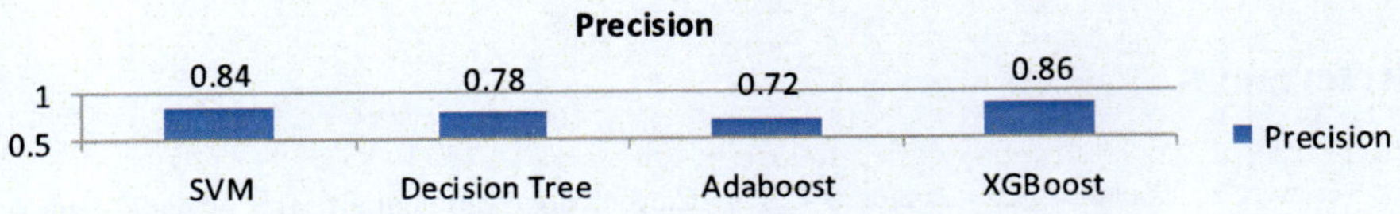

Fig. 2 Precision of various classification algorithms

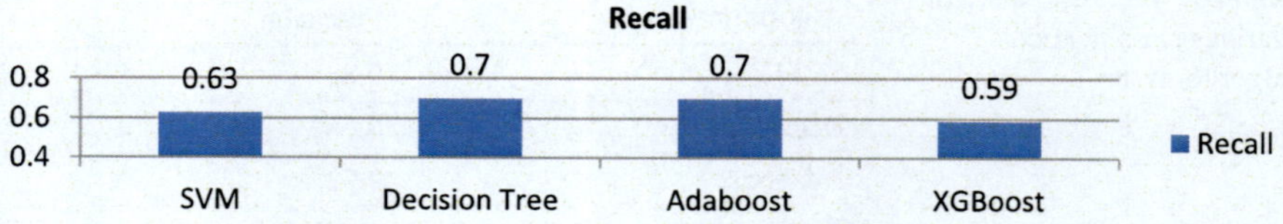

Fig. 3 Recall of various classification algorithms

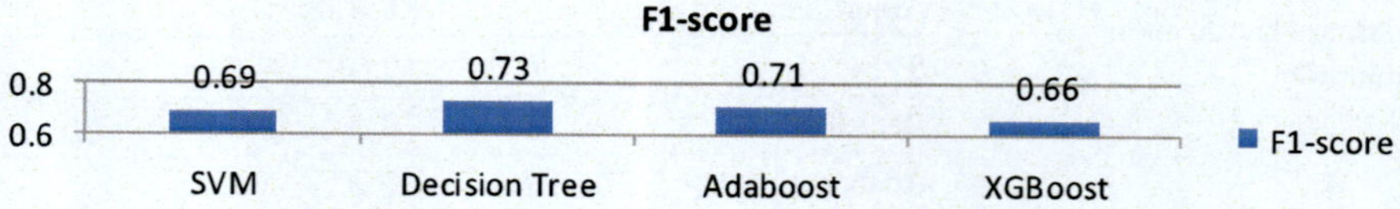

Fig. 4 F1-score of various classification algorithms

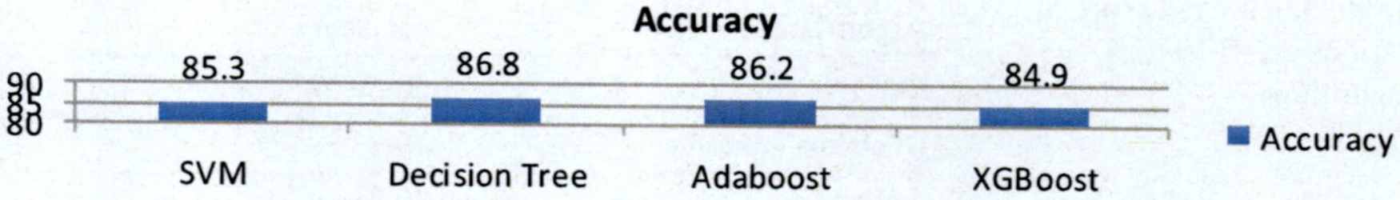

Fig. 5 Accuracy of various classification algorithms

representations. Figure 2 explains about the precision of various classification algorithms. Figure 3 demonstrate recall of various classification algorithms. Analysis of F1-score on various classification algorithms can be seen in Fig. 4. Figure 5 gives the analytical view of accuracy on various classification algorithms.

5 Conclusion

In this work, we used various classification algorithms to study the CTG results. Fetal state can be normal, suspect, or pathologic. Performance is analyzed using various metrics like precision, recall, accuracy, and f1-score. To study the status of fetal SVM, Decision tree, Adaboost, and XGBoost algorithms have been implemented.

References

1. Comert Z, Kocamaz AF, Gungor S (2016) Cardiotocography signals with artificial neural network and extreme learning machine. In: IEEE, 2016 24th signal processing and communication application conference (SIU)
2. Sontakke SA et al (2019) Classification of cardiotocography signals using machine learning. Springer

3. Comerta Z, Kocamazb AF (2016) Comparison of machine learning techniques for fetal heart rate classification. In: ICCESEN 2016
4. Fergus P et al (2016) Prediction of intrapartum hypoxia from cardiotocography data using machine learning. Appl Comput Med Health
5. Subha V et al (2013) Comparative analysis of classification techniques using cardiotocography dataset. IJRIT 1(12):274–280
6. Amin B (2019) Classifying cardiotocography data based on rough neural network. IJACSA 10(8)
7. Bhatnagar D, Maheshwari P (2016) Classification of cardiotocography data with WEKA. IJCSN 5(2)
8. Zhang Y, Zhao Z (2017) Fetal state assessment based on cardiotocography parameters using PCA and AdaBoost. In: 2017 10th international congress on CISP-BMEI
9. Sahin H, Subasi A (2015) Classification of the cardiotocogram data for anticipation of fetal risks using machine learning techniques. 33:231–238
10. Tang H et al (2018) The design and implementation of cardiotocography signals classification algorithm based on neural network. Comput Math Methods Med
11. Hoodbhoy Z et al (2019) Use of machine learning algorithms for prediction of fetal risk using cardiotocographic data. Int J Appl Basic Med Res 9(4):226–230
12. Batra A (2017) Cardiotocography analysis using conjunction of machine learning algorithms. In: CMVIT
13. Agrawal K, Mohan H (2019) Cardiotocograph analysis for fetal state classification using machine learning algorithms. In: ICCCI-2019, 23–25 January 2019, Coimbatore, India. IEEE
14. Sahin H, Subasi A (2012) Classification of fetal state from the cardiotocogram recordings using ANN and simple logistic. In: ISSD, May 2012

Dataset

15. Bernardes J et al (2010) Cardiotocography data set. In: UCI Machine learning repository, 07-09-2010

Predictive Analysis of the Recovery Rate from Coronavirus (COVID-19)

Abishek Bhattacharya, Goldina Ghosh, Ratna Mandal, Sujata Ghatak, Debabrata Samanta, Vinod Kumar Shukla, Sabyasachi Mukherjee, Soumi Dutta, and Ankita Mandal

Abstract Estimation of recovery rate of COVID-19 positive persons is significant to measure the severity of the disease for mankind. In this work, prediction of the recovery rate is estimated based on machine learning technology. Standard data set of Kaggle has been used for experimental purpose, and the data sets of COVID cases in Italy, China and India for these countries are considered. Based on that data set and the present scenario, the proposed technique predicts the recovery rate.

Keywords COVID-19 · Predictive analysis · Classification · Situational analysis · Summarization

---

A. Bhattacharya · G. Ghosh · R. Mandal · S. Ghatak · S. Dutta (✉) · A. Mandal
Institute of Engineering and Management, Kolkata, India

A. Bhattacharya
e-mail: abhishek.bhattacharya@iemcal.com

G. Ghosh
e-mail: goldina.ghosh@iemcal.com

R. Mandal
e-mail: ratna.mandal@iemcal.com

S. Ghatak
e-mail: sujata.ghatak@iemcal.com

A. Mandal
e-mail: ankita.mandal@iemcal.com

D. Samanta
CHRIST (Deemed To Be University), Bengaluru, India

V. K. Shukla
Amity University Dubai, Dubai, UAE
e-mail: vshukla@amityuniversity.ae

S. Mukherjee
Asansol Engineering College, Asansol, India

J. M. R. S. Tavares et al. (eds.), *Cyber Intelligence and Information Retrieval*, Lecture Notes in Networks and Systems 291,
https://doi.org/10.1007/978-981-16-4284-5_27

1 Introduction

Coronaviruses are a genus of viruses that mostly infect the respiratory tract and gut of humans. Infections including pneumonia, the common cold and disorders like severe acute respiratory syndrome (SARS) are all linked to the viruses. During the COVID-19 pandemic, human-to-human transmission of SARS-CoV-2 was verified on January 20, 2020. Within a range of roughly 1.8 m, transmission is predominantly via respiratory droplets from coughs and sneezes (6 ft). Infection can also be spread through indirect contact with contaminated surfaces. According to the preliminary study, the virus may survive for up to three days on plastic and steel, but not for more than one day on cardboard or for more than four hours on copper; the virus is inactivated by soap, which destabilizes its lipid bilayer. Infected people's stool samples have also been discovered to contain viral RNA. As has been widely reported, the world passed 3 million confirmed cases, with the USA responsible for over a third of them and the overall number of deaths. European countries, on the other hand, have the highest number of cases and deaths per 1000 people. However, because the number of confirmed cases is a function of the level of testing, it is not the same as the actual total number of cases. The COVID-19 outbreak, which was caused by the severe acute respiratory syndrome coronavirus 2 (SARS-CoV-2), has overburdened healthcare systems, necessitating the fast establishment of medication, at the very least to reduce the prevalence of COVID-19. Over the last two decades, various strains of coronavirus have been reported from all over the world causing considerable number of deaths. The transmission of the first strain of coronavirus to humans was first recorded in December 2019 which mainly originated from the wholesale seafood market in Huanan city, located in the Hubei Province of China [1, 2]. The earliest known cases had link with the wholesale food market. The rapid increase in positive cases all over the world led the World Health Organization to declare the outbreak as a pandemic on March 11, 2020 [1]. According to the reports as of on March 5, 2021, more than 2.5 million deaths have been recorded all over the world due to COVID-19.

The virus responsible for this outbreak is a single-stranded RNA virus having zoonotic origin [3]. It spreads vary rapidly, from patients being asymptomatic to having life-threatening health issues leading to casualties. The human-to-human transmission is primarily via respiratory droplets which are released into air as an infected person breathes, speak, sneeze or cough. People remain infectious for up to two weeks and can spread the virus even if they are asymptomatic [2].

In this work, the objective is to predict the rate of the recovered people and the number of the people of confirmed cases, affected by coronavirus (COVID-19) and the number of people who passed away due to COVID-19.

2 Related Work

There has been a lot of research going on about today's pandemic which has taken the world down to its knees. Thousands of research groups from all around the world are collaborating to collect data and develop solutions. Machine learning algorithms [4] have proven to be quite useful in anticipating dangers across a wide range of fields. In the case of medical risk, machine learning could be useful in three ways: infection risk, severity risk and outcome risk. Machine learning may be able to assist in the prediction of all of these dangers [5]. Machine learning was utilized by DeCapprio et al. [6] to create an initial vulnerability index for COVID-19. Wearing masks, washing hands and social separation are all likely to have an impact on total risk. Because of an undeniable degree of vulnerability and absence of fundamental information, standard models have shown low accuracy for long-term expectation. Different literature incorporates a few endeavors to address this issue, the fundamental speculation and vigor capacities of existing models should be improved. Authors [7] present a similar investigation of AI and machine learning models to foresee the COVID-19 flare-up as an option in contrast to susceptible-exposed infectious-removed (SEIR) models and susceptible-infected-recovered (SIR) models. Among a wide scope of AI models explored, two models showed promising outcomes (i.e., multilayered perceptron, MLP, ANFIS). In view of the outcomes announced here, and because of the exceptionally perplexing nature of the COVID-19 flare-up and variety in its conduct across countries, this investigation proposes machine learning as an effective technique. Motivationally, inside a brief time frame since COVID-19 episode, AI-based methods have been utilized in the classifications of COVID-19 genomes [8], survival forecast of extreme COVID-19 patients [9] and finding potential medication applicants against COVID-19 [10], CRISPR-based COVID-19 identification test [11]. Authors in [12] have proposed deep learning system to screen coronavirus disease 2019 pneumonia.

3 Experimental Data set

The data set utilized in this investigation is described in this section. The data set contains the number of effected people, i.e., the confirmed cases, the number of deaths due to COVID-19 and the rate of recovery, i.e., the cases which have been recovered from the adverse effect of the virus. This data set is collected and updated from different countries of the world. Especially, data set of China, Italy and India considered to compare the affection list of India than that of the other countries such as China and Italy. Table 1 is generated from cases in China for a limited time period only, and an interval is considered for one week base. Table 2 is created from cases in Italy for a limited time period only, and an interval is considered for one week base. Table 3 is derived from cases in India for a limited time period only, and an interval is considered for one week base.

Table 1 Summary of the cases of China

Date	Country	Confirmed	Deaths	Recovered
22/1/2020	China	5509	131	101
29/1/2020	China	23,707	491	843
4/2/2020	China	44,386	1112	4636
11/2/2020	China	74,211	2003	14,206
18/2/2020	China	77,754	2665	27,676
25/2/2020	China	80,261	2947	47,450
3/3/2020	China	80,887	3139	60,181
10/3/2020	China	81,058	3230	68,798
17/3/2020	China	81,591	3281	73,280
24/3/2020	China	82,198	3308	73,923
31/3/2020	China	82,279	3309	73,973
7/4/2020	China	82,718	3335	75,110
14/4/2020	China	83,306	3345	75,833
21/4/2020	China	83,853	4636	75,862

Table 2 Summary of the cases of Italy

Date	Country	Confirmed	Deaths	Recovered
22/1/2020	Italy	0	0	0
29/1/2020	Italy	0	0	0
4/2/2020	Italy	3	0	0
11/2/2020	Italy	3	0	0
18/2/2020	Italy	322	10	1
25/2/2020	Italy	2502	79	160
3/3/2020	Italy	10,149	631	724
10/3/2020	Italy	31,506	2503	2941
17/3/2020	Italy	69,176	6820	8326
24/3/2020	Italy	101,739	11,591	14,620
31/3/2020	Italy	105,792	12,428	15,729
7/4/2020	Italy	135,586	17,127	24,392
14/4/2020	Italy	162,488	21,067	37,130
21/4/2020	Italy	183,957	24,648	51,600

Table 3 Final summary data for the cases of India

Date	Country	Confirmed	Deaths	Recovered
22/1/2020	India	0	0	0
29/1/2020	India	3	0	0
4/2/2020	India	3	0	0
11/2/2020	India	3	0	3
18/2/2020	India	3	0	3
25/02/2020	India	5	0	3
3/3/2020	India	56	0	4
10/3/2020	India	142	3	14
17/3/2020	India	536	10	40
24/3/2020	India	1251	32	102
31/3/2020	India	1397	35	123
7/4/2020	India	5311	150	421
14/4/2020	India	11,487	393	1359
21/4/2020	India	20,080	645	3975

4 Proposed Methodologies

4.1 Predictive Analysis

The goal of the study is to predict the number of confirmed cases and number of recovery cases to control the growing rate of COVID cases, that is needed to get rid of the coronavirus, so that, again people of our country can roam around just like earlier days, and we will also analyze in how many days, the number of recovery case should be maximum, and the number of confirmed case could be minimized. In the previous section, a detailed guideline has been given regarding the experimental data set. Data has been collected for different countries, so that the predictive outcome can be compared to the other countries. India, in the first wave to some extent, has lesser case than that of the other two countries, i.e., discussed. But, there are number of cases that have been gradually increased day by day, and in the second wave there is a huge number of cases with respect to other countries. However, in this work, we have tried to predict the future reference, that, how many confirmed cases can be visualized or how much recovery cases can be achieved and by which point of time. Also, we have tried to enhance the present scenario through our analysis. For predictive analytical model, linear regression and time series algorithm (ARIMA model) are used mainly.

Linear Regression: Linear regression can be used to find a link between two continuous variables. The predictor or independent variable is one, while the response or dependent variable is the other. It searches for statistical relationships rather than deterministic ones. If one variable can be precisely described by the other, the relationship between the two variables is said to be deterministic. It is possible to precisely forecast Fahrenheit using temperature in degrees Celsius, for example. In determining the relationship between two variables, statistical relationships are not correct. For instance, consider the link between height and weight.

Time Series Analysis: Time series analysis, often known as trend analysis, is a statistical methodology that works with time series data. Data that is in a sequence of specific time periods or intervals is referred to as time series data. ARIMA model which we used for our model stands for auto regressive integrated moving average model. Auto regressive parameter is denoted by p; when p is 0, no autocorrelation is there, but when p is 1 the series autocorrelation is lagging by 1. Integrated is denoted by d, for the analysis we have to make our series stationary so if d is 1 or more than that it is not stationary and we have to make it stationary, i.e., we have to take the difference so as to predict the further value in the series (more than two time difference is not reliable). Moving average denoted by q means that it is an error term.

5 Result and Analysis

Confirmed cases can be visualized or how much recovery cases can be achieved and by which point of time. Hence, a comparative study on the confirmed cases with India and China has been made in Fig. 1a. It is also obtained that compared to China, confirmed cases are 7% lesser (approximately) in India. This is based on training data set of two countries. Also, another statistical review of the confirmed cases on test data sets has been made as shown in Fig. 1b. However, not only with China, we have also compared our data set with the data set of Italy as well. So, the comparative study seems to be in Fig. 2a.

Also, just like before, the rate of confirmed cases has been calculated, and it has revealed that 79% (approximately) confirmed cases are there more in Italy than that of India. This prediction is based on the training data set. Just like previous comparison,

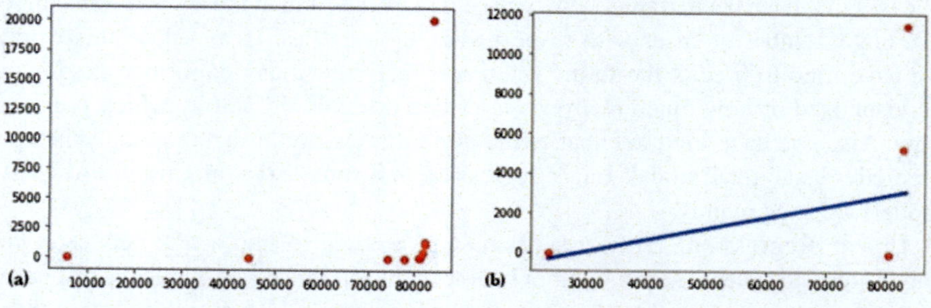

Fig. 1 **a** Prediction of comparative study of confirmed cases of India and China **b** comparison based on the test data set of two countries

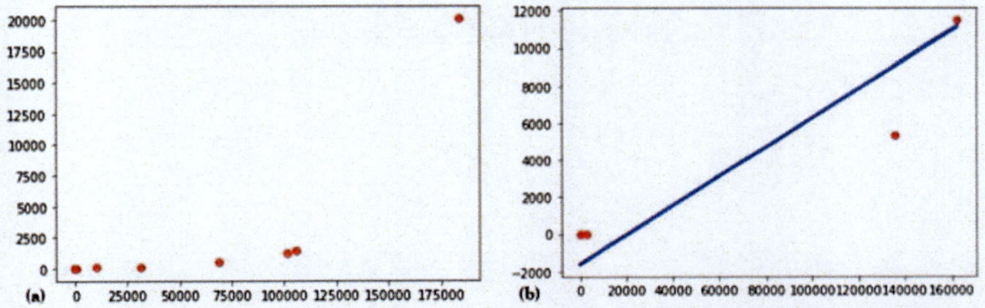

Fig. 2 **a** Graph depicting the confirmed cases of Italy and India **b** Comparative study of India and Italy (confirmed cases)

another graphical review of the confirmed cases of India versus Italy, based on the test data set, is made as shown in Fig. 2b. Hence, this is how predictions have been initialized, by measuring the present situation, and attempts are made to visualize the current situation and made a comparative study between two countries.

5.1 *Situational Analysis*

As we have discussed in Sect. 4, we have tried to visualize the present scenario, comparing to the other countries. However, as we all know, that, some of the states in India, such as Delhi, Maharashtra, Gujarat, Tamil Nadu and Andhra Pradesh, are highly affected of the COVID-19. However, we can have the study of the active cases as follows: From Fig. 2, we can conclude the confirmed cases of Italy and India. Figures 3a and 4 have been obtained from the data set we are having, and the same is depicted on the confirmed and active cases all over the country. Hence, if we

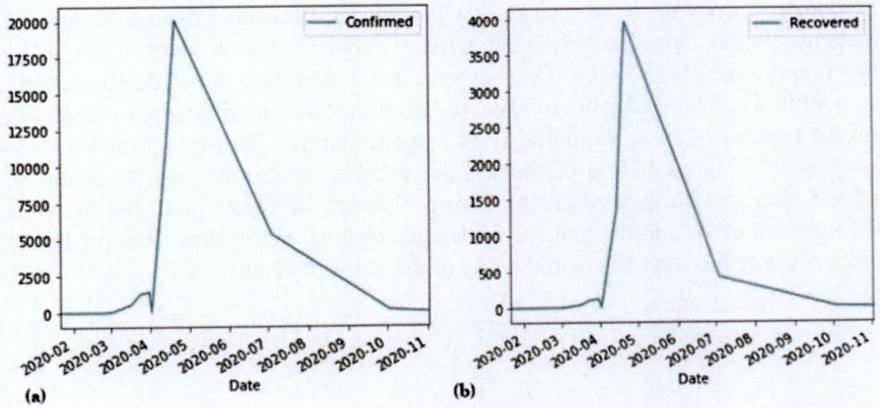

Fig. 3 Study of India **a** confirmed cases and **b** recovered cases

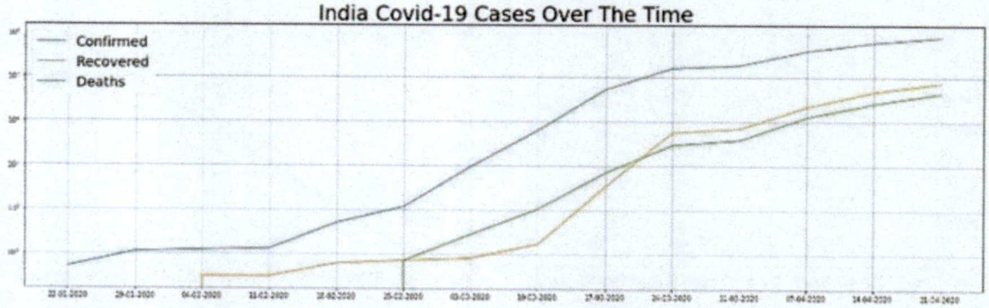

Fig. 4 Graph of the recovered cases, confirmed and the deaths in India

consider the present situation, the active cases are mostly reduced, due to the effect of the lockdown, i.e., announced by our Honorable Prime Minister.

Now, if we consider the recovered cases, we may have the following figure, such as, this is how we have settled up the present situation model. Finally, we can have a look on the current scenario through the following diagram with the details of the deaths, recovery and confirmed, i.e., active cases.

5.2 Future Predictive Analysis

In the previous section, we have discussed how we model the problem of predicting the future, based on the present situation. This section describes the properties of the future prediction analysis by taking the help of the present scenario, based on the features extracted for the recovery cases and the death cases. However, the future confirmed cases are said to be as follows: Hence, as per our prediction, the confirmed cases would be 28,293 (approximately). N.B.: The rate of confirmed cases, as per the prediction we are getting like 130% (approximately).

Also, we are having the recovery rate with respect to the data analysis regarding the predictive methodology, i.e., as per our predictive analysis, the recovery rate would be 83% (approximately). However, we have predicted the future rate of deaths as well, i.e., within the discussed point of time, the death rate will be 67% (approximately), and the number of cases would be 1034 (approximately). Figures 4, 5 and 6 show the graphical representation of the recovered cases, confirmed and the deaths in India, China and Italy, respectively. Figure 7 shows the future predicted outcome of confirmed cases and the number of deaths, whereas Fig. 8 shows the predictive analysis of the recovery rate and the rate of the cases of deaths.

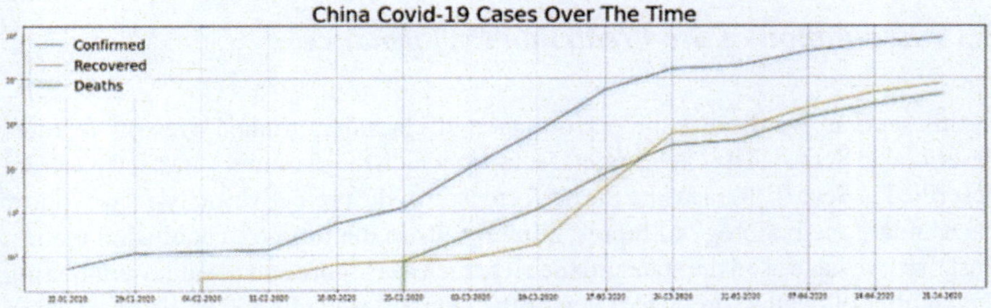

Fig. 5 Graph of the recovered cases, confirmed and the deaths in China

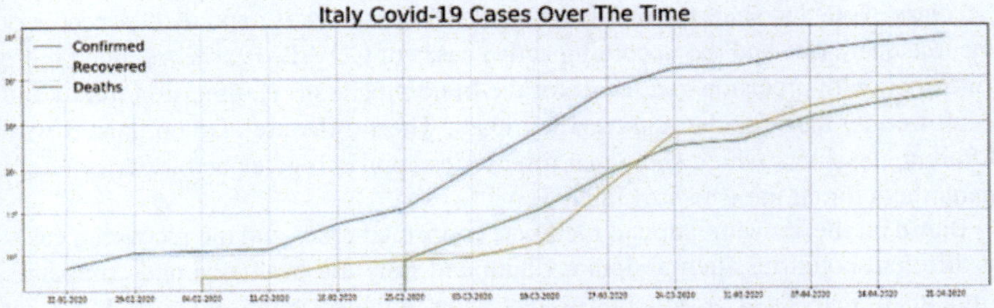

Fig. 6 Graph of the recovered cases, confirmed and the deaths in Italy

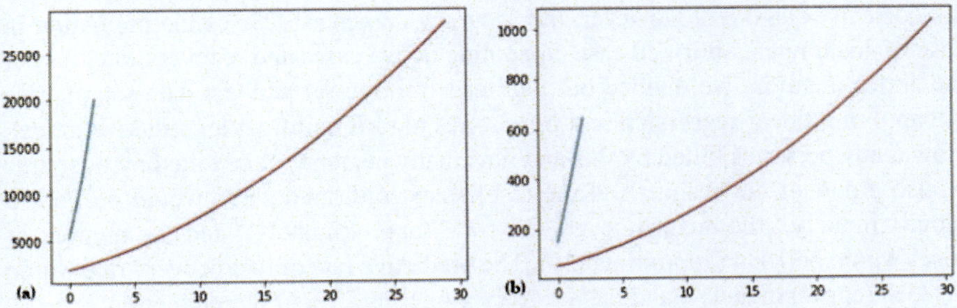

Fig. 7 **a** Confirmed cases in the future reference **b** the number of deaths in future reference

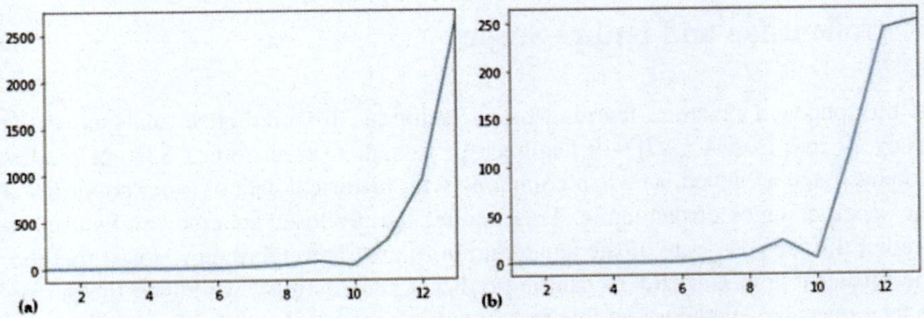

Fig. 8 **a** Predictive analysis of the recovery rate **b** the rate of the cases of deaths

5.3 Evaluation of the Prediction Performance

In this section, we discuss the performance of classifiers trained over the features described in Sect. 5. The experimental setup in which the classifiers are used has been described in Sect. 4. To measure the performance of the predictive analysis, we follow the standard methodology of representing results in the form of a confusion matrix. Post that, we have used the standard metrics, such as F-score and accuracy comparing the number of deaths (predicted) versus the number of confirmed cases (predicted). The precision with respect to a certain class is the ratio of the number of items that are correctly predicted to be in that class, to the total number of items predicted in that class. Note that since the ultimate objective of the work is to predict the recovery rate, the death rate and the upcoming active cases of COVID-19, we are much more concerned with precision and recall for the trending class (i.e., precision trends and recall trends) than for the non-trending class. The results are also compared with different classifiers which show that time series approach gives better performance than others for all the data set (Table 4).

Based on the abovementioned methods, confirmed cases and the recovered cases of different countries such as India, China and Italy are evaluated and compared. Situational analysis is also demonstrated using graphical representation. In Result section, we will get a detailed explanation of the predicative analysis. Finally, the accuracy of classification (A) is the proportion of the total number of items that are correctly classified. However, the accuracy comes as 92% of the prediction in case of death rate, confirmed case, upcoming active cases and recovery rate. As per the above mention, we divided our data into training set and test data set, and we are applying linear regression and time series algorithm to get a predictive model. How many person is killed by this and how many person will be killed by this virus is also a part of our result. And the results are confirmed cases would be 28,293 (approximately), the death rate will be 67% (approximately), and the number of cases would be 1034 (approximately). The predictive rate of the recovery rate would be 83% (approximately), and the accuracy comes as 75% of the prediction.

6 Conclusion and Future Scope

In this paper, a machine learning-based technique for predictive analysis of the recovery rate from COVID-19 has been proposed. A precision of 83% in prediction has been achieved, as when compared with historical data of other countries at the specific stages of pandemic. This implies the proposed scheme can be used to predict the recovery rate of the pandemic in future. Through the proposed method, the further rise of COVID-19 can be predicted successfully. Moreover, the performance measures evaluated in this research work are highly useful for checking the

Table 4 Final classification outcome for the different cases of India, Italy and China

Methods	Confirmed	Cases	Recovered	cases	Predictive rate		Death	Rate
	Accuracy	F-Score	Accuracy	F-Score	Accuracy	F-Score	Accuracy	F-Score
Data set 1	**India**							
Time series	93.4468	0.9340	93.4468	0.9340	92.9399	0.9290	92.9399	0.9290
Linear regression	84.5692	0.8460	84.5692	0.8460	88.4340	0.8710	86.1434	0.8610
naïve Bayes	89.8914	0.8990			89.4931	0.8950	89.9493	0.8950
RBF network	82.8434	0.8170	82.8434	0.8170	81.6056	0.8110	83.9881	0.8350
Logistic	81.2182	0.8030	81.2182	0.8030	75.4085	0.7570	83.9881	0.8030
LogitBoost	81.7270	0.8170	81.7270	0.8170	81.7840	0.8190	83.6772	0.8370
Decision Table	88.2530	0.8786	88.2326	0.8783	86.7788	0.8666	88.6153	0.8791
Data set 2	**China**							
Time series	85.5090	0.8370	87.5597	0.8730	86.5947	0.8690	84.5597	0.8480
Linear regression	82.6297	0.8230	81.3185	0.8100	83.9000	0.8150	83.2328	0.8300
naïve Bayes	79.4041	0.7370	78.5247	0.7850	84.6454	0.8420	82.3691	0.8230
RBF network	74.0772	0.7290	83.9566	0.8280	83.9566	0.8210	82.5090	0.8110
Logistic	81.2135	0.8170	82.0929	0.8160	83.2135	0.8290	80.6610	0.8030
LogitBoost	81.5766	0.8110	81.5247	0.8151	83.8359	0.8300	82.0422	0.8140
Decision table	81.0772	0.8140	84.0772	0.8340	84.0772	0.8340	82.0772	0.8240
Data set 3	**Italy**							
Time series	89.6000	0.8860	87.1596	0.8840	87.6620	0.8637	87.6620	0.8639
Linear regression	87.2864	0.8200	83.6620	0.8590	84.0376	0.8300	84.0376	0.8300
naïve Bayes	84.6948	0.8370	83.2864	0.8280	85.8216	0.8550	85.8216	0.8550
RBF network	86.9484	0.8670	84.6948	0.8370	83.9437	0.8280	83.9437	0.8280
Logistic	87.6056	0.8720	82.2300	0.8700	86.9484	0.8670	86.9484	0.8670
LogitBoost	81.4085	0.8070	84.6056	0.8720	82.6056	0.8210	81.6056	0.8710
Decision table	81.4085	0.8070	81.4085	0.8070	81.4085	0.8070	81.4085	0.8070

correctness of the systems models developed. However, based on the paper, calculations can be made about how many cases could be aroused and how much people can be affected in future, as the prediction is limited to a certain point of time period.

References

1. WHO Director-General's opening remarks at the media briefing on COVID-19–11 March 2020
2. Sharma N India's swiftness in dealing with Covid-19 will decide the world's future, says WHO
3. China's first confirmed Covid-19 case traced back to November 17, March 2020
4. Alimadadi A, Aryal S, Manandhar I, Munroe PB, Joe B, Cheng Xi (2020) Artificial intelligence and machine learning to fight covid-19. Phys Genom 52(4):200–202 PMID: 32216577
5. Stadnytskyi V, Bax CE, Bax A, Anfinrud P (2020) The air-borne lifetime of small speech droplets and their potential importance in SARS-CoV-2 transmission. Proc Natl Acad Sci USA 117(22):11875–11877
6. DeCapprio D, Gartner J, McCall CJ, Burgess T, Kothari S, Sayed S (2020) Building a covid-19 vulnerability index. medRxiv, 2020
7. Ardabili SF, Mosavi A, Ghamisi P, Ferdinand F, Varkonyi-Koczy AR Reuter U, Rabczuk T, Atkinson PM (2020) Covid-19 out- break prediction with machine learning. Algorithms 13(10)
8. Machine learning using intrinsic genomic signatures for rapid classification of novel pathogens: Covid-19 case study. PLOS ONE 15(4):1–24 (2020)
9. Yan L, Zhang H-T, Xiao Y, Wang M, Sun C, Liang J, Li S, Zhang M, Guo Y, Xiao Y, Tang X, Cao H, Tan X, Huang N, Jiao B, Luo A, Cao Z, Xu H, Yuan Y (2020) Prediction of survival for severe covid-19 patients with three clinical features: development of a machine learning-based prognostic model with clinical data in wuhan. medRxiv
10. Ge Y, Tian T, Huang S, Wn F, Li J, Li S, Yang H, Hong L, Wu N, Yuan E, Cheng L, Lei Y, Shu H, Feng X, Jiang Z, Chi Y, Guo X, Cui L, Xiao L, Li Z, Yang C, Miao Z, Tang H, Chen L, Zeng H, Zhao D, Zhu F, Shen X, Zeng J (2020) A data-driven drug repositioning framework discovered a potential therapeutic agent targeting covid-19. bioRxiv
11. Al-Emra M, Al-Kabi MN, Marques G (2021) A srvey of using machine learning algorithms during the COVID-19 pandemic. Springer International Publishing, Cham, pp 1–8
12. Butt C, Gill J, Chun D, Babu BA (2020) Retracted article: deep learning system to screen coronavirus disease 2019 pneumonia. Appl Intell

Data Science and Data Analytics

Deep Learning: An Application Perspective

Sakshi, Prasenjit Das, Shaily Jain, Chetan Sharma, and Vinay Kukreja

Abstract Nowadays, machine learning and deep learning have proven to be an encouraging field after artificial intelligence. Deep learning being operative, supervised, time and cost-effective technique evolving has its myriad applications not just in the field of pattern recognition and prediction but also can be utilized for addressing some important issues in the field of data science and data analytics. Deep learning came into existence with the availability of the exabyte of data collected with time. This subset technique of machine learning acquires the instructive, differential, and distinct features of data. Deep learning is a subset of machine learning that has made a noteworthy breakthrough as it is inspired by the working of the brain and its cells called neurons now coming up with significant performance, efficient evaluation and remarkable outcomes in a huge range of applications, and deep learning techniques have set their way with useful security tools and are now widely used in the field of statistics, neuroscience, data analytics, pattern recognition, image sensing, and the list goes on. This paper covers the basics of deep learning, its several commonly used algorithms and techniques, their architectures along with several application areas

Sakshi · V. Kukreja
Chitkara University Institute of Engineering and Technology, Chitkara University, Rajpura, Punjab, India
e-mail: sakshi@chitkara.edu.in

V. Kukreja
e-mail: vinay.kukeja@chitkara.edu.in

P. Das
Chitkara University School of Computer Applications, Chitkara University, Solan, Himachal Pradesh, India
e-mail: prasenjit.das@chitkarauniversity.edu.in

S. Jain
Chitkara University Institute of Engineering and Technology, Chitkara University, Solan, Himachal Pradesh, India
e-mail: shaily.jain@chitkarauniversity.edu.in

C. Sharma (✉)
Chitkara University, Solan, Himachal Pradesh, India
e-mail: chetan.sharma@chitkarauniversity.edu.in

J. M. R. S. Tavares et al. (eds.), *Cyber Intelligence and Information Retrieval*, Lecture Notes in Networks and Systems 291,
https://doi.org/10.1007/978-981-16-4284-5_28

where these techniques have shown prominent differences and revolutionary change with the time. We have also tried to generalize the characteristics of different used methods of deep learning. At last, we present the challenges for the growth of these methods, concluding with our perspectives and future scope.

Keywords Deep learning · Neural network · Applications · AI · Machine learning

1 Introduction

Artificial intelligence (AI) is a broader term in which the machines are trained through the program to perform tasks more smartly and show the same intelligence as humans learning and reasoning. How computers could perform Human cognitive reasoning was proposed and found satisfactory through the "The Turing Test" in 1950 [1]. Numerous research fields are evolved when AI comes in existence and subfields of AI became the need of society to solve real-life problems. Now, the subfield of AI, i.e., machine learning has constantly evolved. Every time a new problem appears, machine learning makes some advancement in its techniques to efficiently solve that problem. Deep learning is the outcome of this strategy. Deep learning is considered as a subpart of AI neural network architecture, in the term deep popularly known as depth that exists between input and output layer. Machine learning is mainly divided into three types, named as supervised, unsupervised, and reinforced learning. It is highly believed that with time, deep learning methods will reach a high level of accuracy and provide better results enhancing the overall future and scope of machine learning and artificial intelligent systems. The general deep learning architecture consists of the following: -

- An input layer.
- A large number of hidden layers and
- An output layer.

The input layer takes the input from the environment or the user, and this layer results turn out to be the input of the first hidden layer. Each hidden layer adds an abstraction to the features, and these features then become the input to the next higher layer and this process continues. The hidden layer results act as an input for the output layer which is known as the last layer, and the results produced by the output layer are based upon the calculation, classification, and feature extraction performed in the previously hidden layers of the model. The first deep learning model was a multilayer perceptron with backpropagation thereafter deep learning has been implemented in many ways like deep belief networks, convolutional neural networks, recurrent neural networks, etc. Deep learning gives a better performance when applied to larger datasets. It has been extensively used in recognizing images from larger datasets, voice recognition, self-driving cars, etc. Google uses deep learning in its voice image recognition algorithms, whereas Netflix and Amazon use deep learning in predicting the next purchasable item by the customer. Research subfields of AI

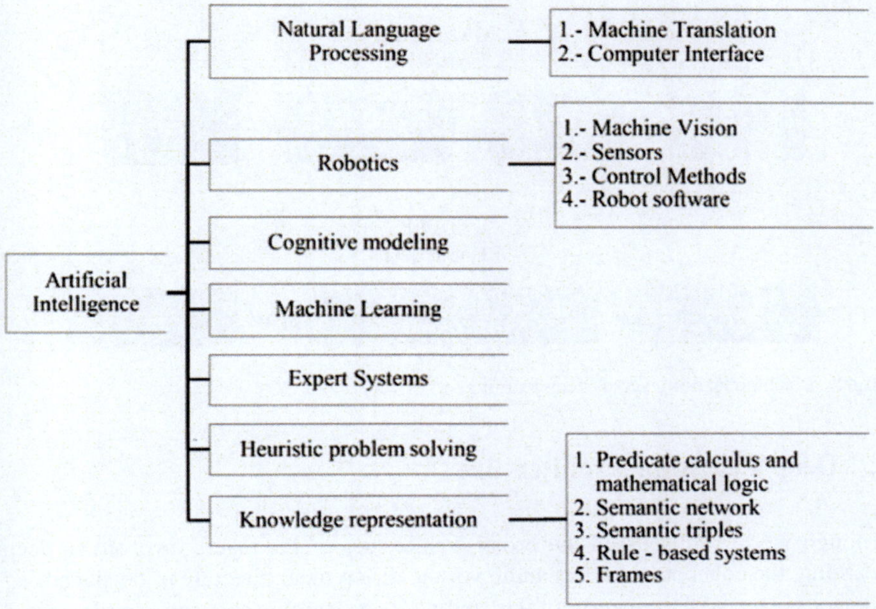

Fig. 1 Research subfield of AI [2]

on which researchers have started to work upon are shown in Fig. 1 [2]. With the evolution of time and technology, AI is replaced by machine learning in around 1960s and become the popular topic of research among the researchers. Artificial intelligence (AI) that gives drastic subsection to the world is machine learning which benefits to learn automatically from the training and experience. Machine learning has constantly evolved. Every time a new problem appears, machine learning makes some advancement in its techniques to efficiently solve that problem. Deep learning is the outcome of this strategy. Over the years' data is collected and stored in databases [3] for static purposes to conclude the scenario of past and present to predict the future. Statically wanted and unwanted data is increasing day by day in huge amounts which we have to manage, for that data mining is introduced by the researchers and combined field of machine learning and data mining become the revolution to the world.

Previously in the earlier decade, only the standard technique, algorithms, and processes were used to process the data, and optimization of those applied techniques gives very effective self-learning [4]. Undoubtedly, deep learning follows the procedures and algorithms of machine learning, but it works primarily on the model of artificial neural networks and the significant efficiency of the algorithms is depending upon the volume of the data. The difference between these two has been shown in Fig. 2. The general deep learning architecture consists of the three layers namely an input layer, a large number of hidden layers, and an output layer.

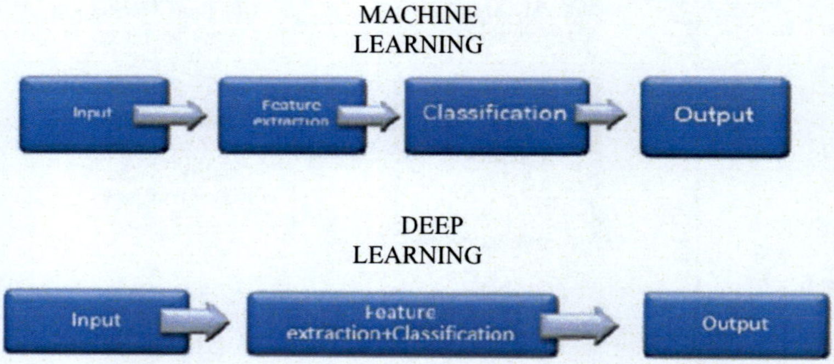

Fig. 2 Machine learning versus deep learning

2 Deep Learning Architecture

Though we have discussed the broad architecture of the layers involved in deep learning, the conceptual understanding of the different architecture of deep learning methods is one of the concerns of this study. A deep learning concept is implemented through a wide variety of architectures and algorithms [5]. The year-wise distribution of the different architecture of deep learning is shown in Table 1, and we will discuss the different architecture as mentioned in Fig. 3 later in this section.

2.1 Recurrent Neural Networks (RNN)

It is like basic building blocks for all the other architectures available while using deep learning. The major and core concept involved is the concept of "hidden layer" which is a kind of memory that stores information that is input to it from the previous layers. This information is stored in the memory which already has the accumulated data and previous results. It remembers the parameters for every type of input as it is likely to perform the same operations on all the inputs to generate the output accordingly. Unlike the traditional neural networks and architectures where the inputs and outputs to the layers were independent of each other, this layer not only works in a feed-forward manner but also has connections that feedback to the prior and previous layer connected in the entire architecture. For summarizing, it can be concluded that this architecture can memorize the previous outputs before producing new outputs

Table 1 Deep learning architecture usage [6]

Architecture	Recurrent neural network	Convolutional neural network	Long short-term memory	Deep belief network	Deep stacked network	Gated recurrent unit
Year	1990–95	1995–2005	1995–2005	2005–10	2010–17	2010–17

for the given inputs. This feature of remembering the features inputs is called the long short-term memory. Further, this RNN can also be paired up with CNN and deep learning architectures.

2.2 Convolutional Neural Networks

Convolutional neural network (CNN) is one of the most popular deep learning architectures which is responsible for the exponential rise of interest in the field of deep learning [7]. Being biologically inspired by the visual cortex of the human eye, this architecture is not only used for the field of image processing but also created a hype while it comes to the processing of natural languages, handwriting recognition, recommender systems, and the list goes on. The main exciting fact that is responsible for such a huge popularity of this architecture is that it is trained to automatically extract the features without any human intervention or external supervision. Broadly summarizing the former layers of the architecture extracts the features from the given set of inputs and the later layers recombine the extracted features into characteristics of the higher order which is used to classify the sets of inputs fed to the layers. CNN has two major tasks of extraction and classification which are performed with the help of four basic layers of the architecture. These layers include convolution, ReLU, pooling, and fully connected layers. Each layer with specific purpose works in sequence to produce the outputs, classified according to the extracted characteristics, the motivation behind working on deep learning methods.

Deep learning is the technology in today's era which makes life easier by enhancing the performance of different applications. We are trying to making our machine intelligent humans from a very earlier time and we achieved up to some extent but deep learning makes a revolution in this field to make the machine behave like humans. A deep learning network is very deep as compared to previous technology which makes the result more effective and good with the support of higher graphical processing units.

2.3 Characteristics of Deep Learning

1. **Self-Connectivity**: Being multilayered, the neurons in a layer are connected either fully or partially. This connectivity pattern between the neurons is inspired by the structural organization of the visual cortex present in the eye of humans and animals. This special region of the animal eye has small regions of cells that are sensitive to specific regions of the visual sight and field. It is highly significant to note that these visual neurons in the brain recognize and stimulate the presence of certain edges of specific orientation like those of horizontal and vertical nature. All these are possible due to the interconnectivity of the neurons in the layers.

2. **Multilayered Network**: Most of the deep learning algorithms are known to have a multilayer architecture and network. This multilayer architecture is what that the deep learning a really "deep" thing to understand. As already discussed, the layered architecture of all basic algorithms of deep learning has been observed that the multilayer feature of the model makes the automated feature extraction possible. Though it consists of multiple layers, still every layer can produce a nonlinear output. Having multiple layers, we can use the hidden layer to store inputs that can be forwarded to the next layer in the network. This is how features can be extracted using multilayer architecture.
3. **Cognitive Skills**: Deep learning focusing on the concept and feature extraction aims to imitate the cognitive and learning skills of the human brain. Acquiring basic mind skills is one of the important targets of deep learning. Deep learning on its onset has been successful to recognize the visual objects and patterns. This remarkable application of deep learning insight and vision has lighted hopes of its growth in the other aspects of cognitive thinking of a human brain. After years of research, it has been proven that the emergence of deep learning can perform well the higher cognitive functions more or like a human brain.
4. **Functional Derivative Nature**: Hereby considering deep learning as a mathematical algorithm, let us discuss the derivative nature of its algorithms. It has been observed that almost the entire deep learning algorithm has a predictive nature and forms that can be expressed in terms of statistical interplay and derivative forms. This functional derivative characteristic of deep learning takes forms of comparative analytics, regression values, discrete and continuous patterns, Bayesian networks, and so on.

3 Deep Learning Applications Review

Detailed literature has been conducted to know the different applications of deep learning models. And, it has been found that it is usually used in many applications and these applications are elaborated by many researchers. The survey has been explained in Table 2.

4 Challenges Involved in Using Deep Learning

Deep learning models and methods are achieving the best results par our expectations and are capable of solving complex problems without human intervention with the dint of its power to extract features and perform extensive classification using the several layers of its architecture. A definite fact which is proven and acknowledged is about the accuracy of the outcomes by deep learning tools. The accuracy of the systems is highly appreciable yet there are some challenges which this evolving technology has to admit and overcome. Some of the challenges are the requirement

Table 2 Deep learning applications literature review

Application	Description	References
Acoustic modeling	Using deep learning algorithms for phone recognition. Paper presents unique way to represent the acoustic modeling for speech recognition in the presence of noise using deep learning methods	[8, 9]
Automotive industry	Applications and tools are developing to implement in the automotive industry using deep learning methods using a dataset proposed in the paper	[10]
Big data	Using deep learning methods big data can be converted into useful data and that data can be used to predict the information required in different sectors of the research	[11, 12]
Document analysis and recognition	Detect or observe the different handwritten words or digits from the different types of images or documents. Different datasets are used by the researchers to perform the experiments on the same	[13, 14]
Human activity recognition	A model to monitor the continuous behavior of the human in a different environment was proposed and for that Skoda, PAMAP2, WISDM, UCI datasets are used for experiments	[15, 16]
Health care	The researchers endeavored to analyze the model which helped in medical diagnosis used by healthcare professionals, and researchers in this field to discover the hidden patterns in medical data collected over a time, which in turn will be helpful to serve the healthcare industry	[17, 18]
Image recognition and classification	CIFAR dataset is used for the experiment and achieved higher accuracy rate and decreases the complexity by changing the filters in deep learning algorithms	[19, 20]
Object detection	Model for object recognition and detection is analyzed using multiple CNN and SVM deep learning methods. Caltech-101 and Caltech Pedestrian Datasets are used for the experiment	[21]
Radio wireless networks	The proposed model is based on the long short-term memory (LSTM) recurrent neural network for boosting the forecasting accuracies in spectrum channels, which is used by the licensed users	[22, 23]

(continued)

Table 2 (continued)

Application	Description	References
Semantic image segmentation	Semantic image segmentation is done by associating each pixel of an input with a class label. These labels are linked to specific class names such as cat and dog, and it involves the classification of image pixels; combined computer vision and CNN model of deep learning can be applied to perform semantic analysis on the image	[24]
Smart city	Deep learning techniques for video analytics for the smart cities. Audio and video databases required automatic processing and analysis for that author's recommended technique. GPUs were used by the author to reduce the training time	[25]
Speech recognition	Different deep learning approaches like deep neural network (DNN) and hidden Markov model (HMM) are used for speech recognition by different researchers for differencing the audio from noise and when the speaker changes his speaking style	[26]
Stock market analysis	Recommended a method for the stock market target, time frame, and stop loss. Used three unsupervised learning methods for this purpose	[27, 28]
Synthetic aperture radar	For computation and reconstruction of the image, RNN techniques are used which are best suited for image formulation	[29]
Word spotting	They identified the method to extract the keywords from the handwritten document. Using different deep learning algorithms, they spot the different keywords from the handwritten document from the different databases like RIME	[30, 31]
Writer identification	Every person has a unique writing style. Many features are defined and proposed by different researchers about the writing style of different persons as the writing style varies from person to person and also depends upon age, skills, and other factors	[32, 33]

of a large dataset for training and testing. Undoubtedly the deep learning tools and machines work well when provided large input data, but in many cases, the data acquisition and availability becomes a challenge in itself. Moreover, the hardware requirements of the functioning of deep learning-based systems and machines are all about installations of graphical processing units of high power and performance.

These GPU requirements are highly recommended. The abstraction of deep learning methods is also an advantage and a disadvantage. The high level of abstraction involved bypass human intelligence at many stances. Adding to these challenges, the call for maintenance of the input store of data frequently is also one of the challenges involved. It also involves excessive high power computational tools and hardware which again is a challenging task.

5 Conclusion and Future Scope

In this paper, we have discussed the preliminaries of the deep learning technique, and a basic discussion covers the brief study of the several architectures involved in this deep learning methods. Being a highly stimulating an emerging field, the basics, architectures, and its myriad applications need to be studied highly proficiently and is a daunting task. We have also tried to add another dimension to our study by analyzing the characteristics of methods that come in this domain of deep learning. This brief study of analyzing the key factors of this theme explains the prerequisite knowledge needed to understand this technology. Our future scope includes delving into deeper details of this emerging technology and focus on the primary architecture under this umbrella including some implementation tasks involved to accrue to the conceptual clarity about the topic.

References

1. Moor J (2003) The turing test: the elusive standard of artificial intelligence, 30th ed. Springer Science & Business Media, Berlin, Germany
2. Adeli H (1988) Expert systems in construction and structural engineering. Chapman & Hall, United Kingdom
3. Kim W, Oh JM (2017) Deep learning: from chemoinformatics to precision medicine. J Pharm Investig 47(4):317–323. https://doi.org/10.1007/s40005-017-0332-x
4. Safdar S, Zafar S, Zafar N, Khan NF (2018) Machine learning based decision support systems (DSS) for heart disease diagnosis: a review. Artif Intell Rev 50(4):597–623. https://doi.org/10.1007/s10462-017-9552-8
5. Kukreja V, Dhiman P (2020) A deep neural network based disease detection scheme for citrus fruits. In: 2020 international conference on smart electronics and communication (ICOSEC), pp. 97–101
6. Dargan S, Kumar M, Ayyagari MR, Kumar G (2019) A survey of deep learning and its applications: a new paradigm to machine learning. Arch Comput Methods Eng 1–22. https://doi.org/10.1007/s11831-019-09344-w
7. Kukreja V, Kumar D, Kaur A et al (2020) GAN-based synthetic data augmentation for increased CNN performance in vehicle number plate recognition. In: 2020 4th international conference on electronics, communication and aerospace technology (ICECA), pp 1190–1195
8. Mohamed A-R, Dahl G, Hinton G (2009) Deep belief networks for phone recognition . Nips Work Deep Learn Speech Recognit Relat Appl 1(9):31–39. https://doi.org/10.4249/scholarpedia.5947

9. Santana D, Queiroz LM, Santos RM, Matos LN, Macedo HT (2018) Deep neural networks for acoustic modeling in the presence of noise. IEEE Lat Am Trans 16(3):918–925. https://doi.org/10.1109/msp.2007.4317458
10. Luckow, Cook M, Ashcraft N, Weill E, Djerekarov E, Vorster B (2016) Deep learning in the automotive industry: applications and tools. In Proceedings—IEEE international conference on big data, Big Data 2016, pp 3759–3768. https://doi.org/10.1109/BigData.2016.7841045.
11. Chen XW, Lin X (2014) Big data deep learning: challenges and perspectives. IEEE Access 2:514–525. https://doi.org/10.1109/ACCESS.2014.2325029
12. Gheisari M, Wang G, Bhuiyan MZA (2017) A survey on deep learning in big data. In: Proceedings—IEEE international conference on computational science and engineering (CSE) and IEEE/IFIP international conference on embedded and ubiquitous computing EUC) pp 173–180. https://doi.org/10.1109/CSE-EUC.2017.215
13. Wu Z, Swietojanski P, Veaux C, Renals S, King S (2015) A study of speaker adaptation for DNN-based speech synthesis. In: Proceedings of the annual conference of the international speech communication association, INTERSPEECH, pp 879–883
14. Ashiquzzaman A, Tushar AK (2017) Handwritten Arabic numeral recognition using deep learning neural networks. In: 2017 IEEE international conference on imaging, vision and pattern recognition, icIVPR, pp 3–6. https://doi.org/10.1109/ICIVPR.2017.7890866
15. Ignatov A Real-time human activity recognition from accelerometer data using convolutional neural networks. Appl Soft Comput J 62:915–922. https://doi.org/10.1016/j.asoc.2017.09.027
16. Nweke HF, Teh YW, Al-garadi MA, Alo UR (2018) Deep learning algorithms for human activity recognition using mobile and wearable sensor networks: State of the art and research challenges. Expert Syst Appl 105:233–261. https://doi.org/10.1016/j.eswa.2018.03.056
17. Loh CS, Then PHH (2017) Deep learning for cardiac computer-aided diagnosis: benefits, issues & solutions. MHealth 3:45–45. https://doi.org/10.21037/mhealth.2017.09.01
18. Dai Y, Wang G (2018) A deep inference learning framework for healthcare. Pattern Recognit Lett 1–12. https://doi.org/10.1016/j.patrec.2018.02.009
19. Ciregan UM, Schmidhuber J (2012) Multi-column deep neural networks for image classification. In: Proceedings of the IEEE computer society conference on computer vision and pattern recognition, pp 3642–3649. https://doi.org/10.1109/CVPR.2012.6248110
20. Liu PH, Su SF, Chen MC, Hsiao CC (2015) Deep learning and its application to general image classification. In: Proceedings international conference on informative and cybernetics for computational social systems (ICCSS), pp 7–10. https://doi.org/10.1109/ICCSS.2015.7281139
21. Zhou X, Gong W, Fu W, Du F (F) Application of deep learning in object detection. In: Proceedings 16th IEEE/ACIS international conference on computer and information science (ICIS), pp 631–634. https://doi.org/10.1109/ICIS.2017.7960069
22. López ER, Gualdron O (2019) Primary user characterization for cognitive radio wireless networks using a neural system based on deep learning. Artif Intell Rev 7(1):1–27. https://doi.org/10.1177/1550147718811828
23. Yu X, Wu X, Luo C, Ren P (2017) Deep learning in remote sensing scene classification: a data augmentation enhanced convolutional neural network framework. GIScience Remote Sens 54(5):741–758. https://doi.org/10.1080/15481603.2017.1323377
24. Chen LC, Papandreou G, Kokkinos I, Murphy K, Yuille AL (2018) DeepLab: semantic image segmentation with deep convolutional nets, Atrous convolution, and fully connected CRFs. IEEE Trans Pattern Anal Mach Intell 40(4):834–848. https://doi.org/10.1109/TPAMI.2017.2699184
25. Wang L, Sng D (2015) Deep learning algorithms with applications to video analytics for a smart city: a survey. ArXiv Prepr. arXiv1512.03131, pp 1–8
26. Bhardwaj V, Kukreja V (2021) Effect of pitch enhancement in Punjabi children's speech recognition system under disparate acoustic conditions. Appl Acoust 177:107918
27. Arevalo A, Nino J, Hernandez G, Sandoval J (2016) High-frequency trading strategy based on deep neural networks. In: Proceedings of the international conference on intelligent computing, pp 424–436. https://doi.org/10.1007/978-3-319-42297-8

28. Chong E, Han C, Park F (2017) Deep learning networks for stock market analysis and prediction. Expert Syst Appl 83:187–205. https://doi.org/10.1016/j.eswa.2017.04.030
29. Yonel EM, Yazici B (2018) Deep learning for passive synthetic aperture radar. IEEE J Sel Top Signal Process 12(1):90–103. https://doi.org/10.1109/JSTSP.2017.2784181
30. Thomas S, Chatelain C, Heutte L, Paquet T, Kessentini Y (2015) A deep HMM model for multiple keywords spotting in handwritten documents. Pattern Anal Appl 18(4):1003–1015. https://doi.org/10.1007/s10044-014-0433-3
31. Wicht B, Fischer A, Hennebert J (2016) Deep learning features for handwritten keyword spotting. In: IEEE proceedings of international conference on pattern recognition (ICPR), pp 3434–3439. https://doi.org/10.1109/ICPR.2016.7900165
32. Chu J, Srihari S (2014) Writer identification using a deep neural network. In Proceedings of the 2014 Indian conference on computer vision graphics and image processing, pp 1–7. https://doi.org/10.1145/2683483.2683514
33. Dhieb T, Ouarda W, Boubaker H, Alimi AM (1870) Deep neural network for online writer identification using Beta-elliptic model. In: Proceedings of the international joint conference on neural networks, pp 1863–1870. https://doi.org/10.1109/IJCNN.2016.7727426

Detection of Weapons in Surveillance Scenes Using Masked R-CNN

V. Rahul Chiranjeevi and D. Malathi

Abstract Weapon detection in surveillance videos is one of the major research topics in computer vision and machine learning. Current techniques require human intervention to identify the weapons in the scenes by monitoring the videos continuously. To reduce the human monitoring, an automated weapon detection framework is defined. In the proposed method, Masked R-CNN is implemented to detect the weapons in the surveillance scenes. A background network called Resnet50 is combined with feature pyramid network to extract the features. For every feature map, region proposals are created using region proposal network (RPN). Proposed method is experimented on dataset containing guns and knives. Experimental results in real-time environment show that Masked R-CNN technique performs well for weapon detection without human intervention.

Keywords Masked RCNN · Weapon detection · Feature pyramid network · Regional proposal network

1 Introduction

In modern times, use of surveillance cameras has increased rapidly. Closed-circuit television (CCTV) is being installed in public places like shopping malls, theatres, railway stations and airports, etc. to monitor the public safety. These systems record huge number of videos which makes cumbersome for the human operator to identify the abnormal activities and monitoring capability is not efficient. Anomaly detection is one of the major tasks in video surveillance. Anomalous events like accidents, mishaps, illegal activities, intruder detection, etc. hardly occur so automatic surveillance systems can be utilized to detect the occurrence of abnormal activities. In recent years, development of various algorithms in computer vision, deep learning

V. Rahul Chiranjeevi (✉) · D. Malathi
SRM Institute of Science and Technology, Kattankulathur, Chennai 603203, India

D. Malathi
e-mail: malathid@srmist.edu.in

J. M. R. S. Tavares et al. (eds.), *Cyber Intelligence and Information Retrieval*, Lecture Notes in Networks and Systems 291,
https://doi.org/10.1007/978-981-16-4284-5_29

and machine learning has gained an increased attention. These algorithms can be used to model the development of intelligent surveillance systems. In 2018, sultani et al. developed a large dataset and a solution for anomaly detection using multiple instance learning. Anomalies are learnt at the video level instead of clip level which reduces the complexity of the model. Various other models like Landi et al. [1] and Xu et al. [2] used localized detection method rather processing the full video. In many instances, algorithms are developed for addressing a particular anomaly like accident detection, theft identification, intruder detection etc. These methods cannot be implemented for other anomalous events as they are developed for particular anomaly. Various challenges occur during detection of anomalies like illumination change, occlusion, blurred scenes, environmental conditions, etc. High rate of false alarms increases since the environment in the scenes changes rapidly. Abnormal events in the real world are diverse and difficult to detect every anomalous event. To increase the anomaly detection rate, algorithms need to be trained with large number of data collected in various environmental conditions.

2 Related Works

Detecting anomalies in the surveillance videos is one of the challenging problems in the field of computer vision and machine learning. Various approaches have been developed for detecting particular anomalies. One of the approaches is designed by Nasaruddin et al. [3], who developed a model to learn the anomaly behaviour using attention regions. They used a background subtraction method to extract the motion in the videos followed by three-dimensional neural network. To differentiate the normal and abnormal events, deep convolutional neural Nntwork is utilized. Kooij et al. [4] developed a surveillance model called CASSANDRA to detect the abnormal behaviour in public places. Persons are tracked using 3D cameras and the features are extracted from the scenes. Both audio and visual cues are extracted to reduce the uncertainty in the scenes. Audio cues like screaming, shouting and throwing an object aggressively are extracted. They used Dynamic Bayesian Network (DBN) to identify the abnormal behaviour. Using Lagrangian particle dynamics approach, Shandong Wu et al. [5] developed a model to detect the anomalies in crowded places. Optical flow technique has been used for detecting the crowd flow and the trajectories are clustered to obtain representative trajectories. In 2015, Xu et al. [2] proposed an image fusion approach for detecting concealed weapons. Images are obtained from sensors and differentiated into low frequency and high frequency bands using Double-Density Dual Tree Complex Wavelet Transform (DDDTCWT). At the end of processing, they applied an inverse of DDDTCWT to obtain the fused image. Different datasets are used to perform experiments and achieved higher accuracy compared to other state of art methods. A new approach using fuzzy classifier [6] is to detect fire arms and knives in CCTV videos. Sliding window technique is used to detect the knife in the image and classification is done using SVM classifier. For gun detection, PCA and neural network architectures are used. A new dataset is created containing many CCTV recordings. Experimental results show high specificity and sensitivity of detection.

In 2017, Kibra et al. [7] performed a comparison between bag of words (BoW), speed up robust features (SURF), histogram of gradients (HoG) and Alex net, to identify knifes in surveillance scenes. Alex net architecture achieved higher accuracy compared to other methods. Upadhyay [8] developed an image fusion technique to detect weapons under the clothes in surveillance scenes. Multi-focus and multi-exposure fusion methods are used to combine the images. Various colour models have been used for better understanding. Similarly, a deep auto-encoder method is developed by Vallez et al. [9] to detect guns in a particular scenarios. Faster R-CNN is adopted and experiments are performed on public dataset which produced state of art results. Mask R-CNN has been used in many applications [10–14] for instance object segmentation. In 2019, Malbog et al. [10] developed a R-CNN model for pedestrian detection. Resnet 101 architecture is utilized as a backbone. A new dataset has been created for experimentation. A low-cost weapon detection mechanism is developed for identifying weapons in school environments [15]. Acoustic signals are used to detect the acoustic signatures for identifying the weapons. High pitch sound is projected at various frequencies to obtain the signatures and a fast Fourier transform is applied to extract the features. Random forest algorithm also utilized for detecting the weapons.

3 Weapon Detection Framework

Training procedure of Mask R-CNN for detecting weapons is shown in Fig. 1. It consist of Resnet50 as a backbone network, regional proposal network (RPN) [16] for generating bounding box proposals, Masked R-CNN for regression. For instance, segmentation mask branch is used and RoI alignment is replaced by RoI pooling.

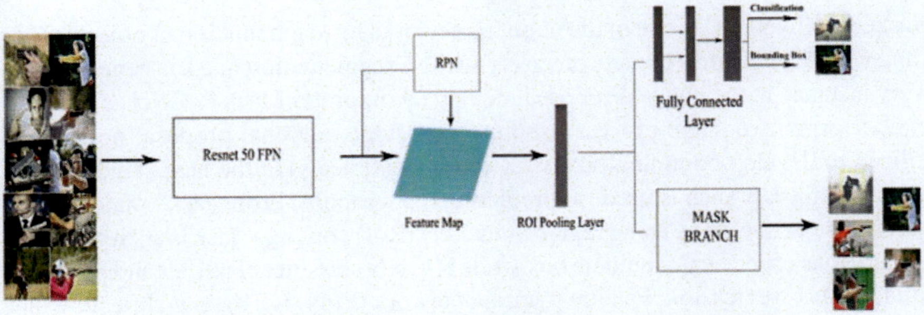

Fig. 1 Training procedure of Masked R-CNN for weapon detection in surveillance scenes

3.1 Backbone Network

Backbone network of Mask R-CNN is Resnet 101 where the number of network layers is 10. In the proposed method, Resnet 50 is used as a backbone since the weapon detection is relatively simple and does not require too many layers. By using a smaller number of layers, speed of the model is increased. Since the images are of different sizes, single neural network will not be able to extract all the features in the image; therefore, Resnet50 and a feature pyramid network (FPN) [17] are also utilized. FPN is a top-down model with connections from lateral which uses single input to develop feature pyramid by solving the problem of extracting objects from the images.

3.2 Regional Proposal Network

In the proposed framework, FPN network is adopted and the image features are extracted in different sizes. Large targets are distinguished by narrow features and the small targets are distinguished based on deep features. Feature maps generate by FPN network are fed as input to the regional proposal network. RPN extracts the different size of RoI features from the feature pyramid; therefore, detection accuracy improved without increasing the computational time. RPN is similar to the sliding window technique. The sliding window produces frame anchors of various sizes to covers many images of different sizes. For detecting the weapon in the image algorithm, adjust the anchors to various sizes.

3.3 Masked R-CNN

Masked R-CNN [18] is one of the main frameworks for segmentation of objects. This framework can identify objects precisely while a segmentation mask is generated for every instance in the image. Proposed method incorporates Mask R-CNN as baseline and performs two stage process. In the initial stage, regional proposal network is utilized to develop bounding boxes for candidate objects. In the next stage, fast R-CNN and mask branch is used for predicting the weapons. From each bounding box features are extracted by using region of interest (RoI) pooling. Mask branch produces a binary mask for every candidate box while R-CNN classifier classifies and performs bounding box regression. Feature pyramid network (FPN) is a main architecture that plays important role in extracting the features. For detecting various type of weapons, FPN architecture uses a top-down approach to develop feature maps of high level at different scales. A feature pyramid with various scale features is developed by taking weapon image as input. Regional proposal network is used to generate region of interest (RoI) in batches for obtained features and fed to RoI pooling layer. These features are fed as input to Mask R-CNN classifier which produces bounding boxes for the input images. Features from RoI are also fed to mask branch which consist of a neural network layer for predicting the weapons in the images.

4 Results and Discussion

Experiments are performed using TensorFlow and Keras deep learning framework. For GPU acceleration, nvidia1080 with Intel i7 CPU with 16 GB memory is used. For training and testing, a new dataset is created containing 6000 gun and knife images. (80% of training and 20% for validation). Reliability of the model is tested using 300-gun and 200 knife images. Images with gun and knife are detected, marked with scores with segmentation masks. Results are obtained by performing the experiments based on two aspects, performance of detection and instance segmentation is improved.

Precision and recall rate values are used to determine the performance of the proposed method. Segmentation performance and computational time of the model are evaluated using Mean Intersection over Union (MIoU).

$$\text{Precision} = \frac{\text{True Positive}}{\text{True Positive} + \text{False Positive}}$$

$$\text{Recall} = \frac{\text{True Positive}}{\text{True Positive} + \text{False Negative}}$$

$$\text{IoU} = \frac{\text{TP}}{(\text{TP} + \text{FP} + \text{FN})}$$

True positive rate is the total number of samples classified by the model correctly. False positive rate defines the number of negative of positive sample that are marked incorrectly. False negative rate defines number of correct images that are marked negatively. Precision is the accuracy rate of the model. Average accuracy of model is achieved from the P-R graph which is drawn in Fig. 2 based on the predicted values of the test data. Using the below P-R graph, accuracy of the model can be seen in (Table 1).

In Fig. 3, weapons in the images are detected and labelled using bounding boxes followed by the output of mask branch. Weapons are identified and masked by eliminating the background in the images.

5 Conclusion

Using the proposed system, weapons in the surveillance scenes can be detected. A prelabelled dataset is created containing images of guns and knives. Proposed Masked

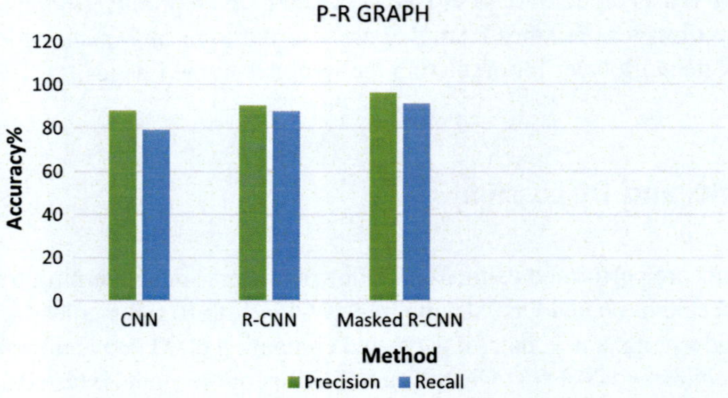

Fig. 2 P-R graph for weapon detection

Table 1 Precision and recall rates for weapon detection model

Parameters	Gun	Knives	Overall
Precision	96.22	91.98	94.33
Recall	95.41	91.26	93.47
MIoU accuracy	82.45	81.77	80.34

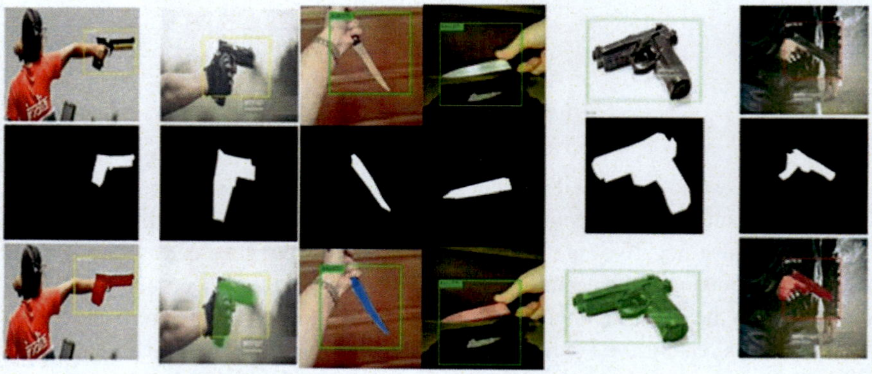

Fig. 3 Detection of weapons and mask branch output

R-CNN method achieved an accuracy of 94.33% which is better compared to other methods and performed successfully in recognizing various types of weapons. At present, computational time of the model is slightly slow due to huge amount of processing the data and more calculations in between the deep layers. As a future enhancement, more productive techniques and large datasets can be utilized to recognize and track the weapons in surveillance scenes by reducing the computational time.

References

1. Landi F, Snoek CGM, Cucchiara R (2019) Anomaly locality in video surveillance. arXiv, 2019
2. Xu T, Wu QMJ (2015) Multisensor concealed weapon detection using the image fusion approach. IET Semin. Dig. 5:2015. https://doi.org/10.1049/ic.2015.0108
3. Nasaruddin N, Muchtar K, Afdhal A, Dwiyantoro APJ (2020) Deep anomaly detection through visual attention in surveillance videos. J Big Data 7(1). https://doi.org/10.1186/s40537-020-00365-y
4. Kooij JFP, Liem MC, Krijnders JD, Andringa TC, Gavrila DM (2016) Multi-modal human aggression detection. Comput Vis Image Underst 144:106–120. https://doi.org/10.1016/j.cviu.2015.06.009
5. Wu B, Moore E, Shah M (2010) Chaotic invariants of lagrangian particle trajectories for anomaly detection in crowded scenes. In: IEEE computer society conference on computer vision and pattern recognition, pp. 2054–2060. https://doi.org/10.1109/CVPR.2010.5539882
6. Grega M, Matiolański A, Guzik P, Leszczuk M (2016) Automated detection of firearms and knives in a CCTV image. Sensors (Switzerland) 16(1). https://doi.org/10.3390/s16010047
7. Kibria SB (2017) An analysis of feature extraction and classification algorithms for dangerous object detection. pp 1–4
8. Upadhyay EM, Rana NK (2014) Exposure fusion for concealed weapon detection. In Proceeding IEEE international caracas conference devices, circuits Systems ICCDCS, pp 1–6. https://doi.org/10.1109/ICDCSyst.2014.6926141
9. Noelia Vallez(B) OD, Velasco-Mata A, Jose Corroto J (2013) Weapon detection for particular scenarios using deep learning. Uncertain. In: Artificial Intelligence - Proceedings 29th Conference UAI 2013, vol 15, no 50, pp 371–386
10. Malbog MA (2019) MASK R-CNN for pedestrian crosswalk detection and instance segmentation. In: 2019 6th IEEE International Conference on Engineering Technologies and Applied Sciences (ICETAS), pp 2–6. https://doi.org/10.1109/ICETAS48360.2019.9117217
11. Songhui M, Mingming S, Chufeng H (2019) Objects detection and location based on mask RCNN and stereo vision. In: 2019 14th IEEE international conference on conference electronic measurement & instruments ICEMI 2019, pp 369–373. https://doi.org/10.1109/ICEMI46757.2019.9101563
12. Zhang Q, Chang X, Bian SB (2020) Vehicle-damage-detection segmentation algorithm based on improved mask RCNN. IEEE Access 8:6997–7004. https://doi.org/10.1109/ACCESS.2020.2964055
13. Jeong DK, Kang HS, Kim DE, Lee JM (2019) Mask-RCNN based object segmentation and distance measurement for robot grasping. Int Conf Control Autom Systv 671–674. https://doi.org/10.23919/ICCAS47443.2019.8971673
14. He P, Zuo L, Zhang C, Zhang Z (2019) A value recognition algorithm for pointer meter based on improved mask-RCNN 2019. In: 9th International conference on information science and technology (ICIST), pp 108–113.https://doi.org/10.1109/ICIST.2019.8836852
15. Je B, Ceh D, Ceh I, Eh I, Ki DJ, Dej I (2019) Low-cost concealed weapon detection for school environments using acoustic Signatures, pp 5–9
16. He K, Zhang X, Ren S, Sun J (2016) Deep residual learning for image recognition. In: Proceedings of the IEEE conference on computer vision and pattern recognition, pp 770–778. https://doi.org/10.1109/CVPR.2016.90
17. Li X, Lai T, Wang S, Chen Q, Yang C, Chen R (2019) Weighted feature pyramid networks for object detection. In: 2019 IEEE international conference on parallel & distributed processing with applications, big data & cloud computing, sustainable computing & communications, social computing & networking (ISPA/BDCloud/SocialCom/SustainCom), pp 1500–1504. https://doi.org/10.1109/ISPA-BDCloud-SustainCom-SocialCom48970.2019.00217
18. He K, Gkioxari G, Dollár P, Girshick R (2020) Mask R-CNN. IEEE Trans Pattern Anal Mach Intell 42(2):386–397. https://doi.org/10.1109/TPAMI.2018.2844175

Deep Learning Approaches for Spatio-Temporal Clues Modelling

M. Suresha, S. Kuppa, and D. S. Raghukumar

Abstract Multimedia domain, understanding of video content requires vibrant semantic features, which are static and temporal information, static information contains spatial clues, and temporal information contains short-term motions clues, as well as long-term motions clues. Exploring these visual features is a challenging task. Conventional handcrafted features extraction methods are inefficient to analyse these features. Inspired by the significant progress of deep learning models in the image domain, this research extends to the design and exploration of deep neural models and in video domain. We addressed the evolution of spatio-temporal features in two ways in this paper. First, we used various "convolutional neural networks (CNN)" and their pipelines to investigate video features, fusion strategies and their results. Second, the drawbacks of convolutional neural networks for long-term motion clues are overcome by using sequential learning models such as the long short-term memory (LSTM) network in conjunction with multimodel fusion techniques. The main objective of this study is to explore spatial-temporal clues, different models available for the problem, and determine the most promising approaches.

1 Introduction

Today's rapidly expanding digital contents include multimedia such as images, text, audio, video and so on. Video, in particular, is a modern mode of communication that makes use of rich multimedia devices. Understanding video content has recently become a heavily researched subject in the research community, and substantial progress has been made in recent days in developing an effective function for understanding video content, with several classes of variations.

M. Suresha (✉) · S. Kuppa · D. S. Raghukumar
Department of Computer Science, Kuvempu University, Karnataka 577451, India
e-mail: suresham@kuvempu.ac.in

J. M. R. S. Tavares et al. (eds.), *Cyber Intelligence and Information Retrieval*,
Lecture Notes in Networks and Systems 291,
https://doi.org/10.1007/978-981-16-4284-5_30

Traditional video content understanding methods are primarily focused on the image domain; however, it can be expanded to use temporal dimension video data, and it [1, 2] focuses on exploring robust spatio-temporal features using classical vision approaches, which are then categorized using different classifiers. Recent developments in "convolutional neural network (CNN)" [3], in the area of image classification [4], extend into temporal dimension [5] by stacking the visual frames. Furthermore, attempts to action prediction, video content classification and motion recognition, etc., [2, 6] by stacked spatial and optical flow frames to capture the spatial and motion information. Evolution of this information named as spatio-temporal features makes a new trend line and provides more substantial clues for video understanding from natural or raw video data.

Deep learning models that are cutting-edge are investigating deep features for understanding video content, using different neural models to evaluate these features, for example, two-stream models [7] with fusion-based pooling techniques and hybrid models [8–10]. Two-stream networks commonly used two CNN pipeline for modelling spatial and temporal patterns with optical flow frames. Furthermore, extension of this two-stream model for evaluate long-term motion patterns introduced long short-term memory (LSTM) [11, 12] layer.

This article provides spatial and temporal clues modelling approaches and problems as a result of these overall observations. To the best of our ability, we investigate different approaches to modelling spatial and temporal patterns here. Moreover, it focuses on various deep learning models to extract deep features using special neural models. A significant contribution can be summarized as follows:

1. The investigation of various classical and machine vision procedures for the extraction of spatio-temporal clues. In addition, investigate cutting-edge deep learning two-stream and hybrid multimodels capable of extracting spatio-temporal patterns for video understanding.
2. In addition, investigate cutting-edge deep learning two-stream and multimodels that can derive spatio-temporal patterns for video understanding.

This study's reminder is structured as follows. Section 1.1 examines related works. The classical and machine vision techniques for extracting deep video features are defined in Sect. 2. State-of-the-art models for evaluating deep features with different modality are discussed in Sect. 3, and Sect. 4 discusses the results and interpretation. Finally, in Sect. 5, the end of the section is followed by a conclusion.

1.1 Literature Review

In the multimedia domain, understanding of video content has been longstanding active research; it addresses multiple problems such as captioning and video classification, action recognition, activity classification and salient activity recognition. Most successful systems majorly depend on the effective and accurately extracted video clues, and most recent works mainly concentrated on designing robust feature sets. The advances in great success from the image domain motivate so many video representation methods: spatial, which can extend to utilize the time video data, and temporal. Understanding video content requires the modelling of this spatio-temporal clue. Lowe et al. [13] and [14] use classical features such as scale-invariant feature transform (SIFT) [14] to capture local still image spatial clues in video. In addition to local spatial clues about visual features, global visual clues like motion are very significant. A common feature extraction method implemented for this is the dense trajectory [15], which evaluates a feature densely sampled with pooled local temporal patches over time and evaluates multiple features based on the trajectories.

Conventional feature extraction mechanisms are constrained in their capacity to [16] process raw natural data. Growth of learning with robust features from natural data is possible by deep neural networks (DNN). A basic convolutional neural network (CNN) model is used to learn to explore optical flow features directly from raw data. Deep feature extraction models, convolutional neural networks (CNN) and others have achieved strong results on a variety of problems such as object localization [17], speech recognition [18] and image classification [19]. Upgrading normal CNN to spatial-temporal feature called two-stream network but it modeling only short-term motion patterns because model learning rate decreased [7, 9, 11, 20, 21]. To address these limitations, [10] proposes multimodel deep learning networks; in video level, it can learn interfeature relationship by combining "regularized feature fusion network", and it models both short-term motions patterns as well as long-term motions patterns.

From this review we discussed, evolution of spatial–temporal clues from image to temporal domain, various deep learning models used to extract spatio-temporal clues, the importance of spatio-temporal clues and different fusion techniques for understanding video content.

2 Spatio-Temporal Feature

Understanding video content requires the assessment of spatial–temporal features. The spatial–temporal features design was inspired by the two segment theory principle in neurology, which states that the human brain visual cortex has two paths: the ventral stream (which performs spatial object recognition) and the dorsal stream (which perceive the motion: temporal) [22, 23]. Handcrafted representation methods have investigated some common feature sets such as HOG and 3D Histogram of Gradient [24], which created spatial clues such as texture, while others such as "Histogram of Optical Flow (HOF)," spatio-temporal SIFT [25], Space-Time Interest Points (STIP) [26], SIFT-3D [27], dense trajectory [15], and its revised version [28]. Creating the best feature set is a critical topic in the multimedia domain. In this section, we categorize two methods for extracting spatio-temporal clues. First, traditional and machine vision techniques for extracting spatial cues are discussed. Second, traditional and machine vision strategies for extracting temporal optical flow features.

2.1 Spatial Domain

The assessment of spatial features from general static video frames. It investigates the exact spatial field of similarity between objects as well as the semantic relationship between the objects in video frames. It is extremely useful to distinguish between static and dynamic contexts. Moving with an object, performing the activity concerning the view and static context are some dynamic context activities where moving object performing the activity at constant position. Spatial semantics provides the location of movement in each object, allowing us to distinguish between static and dynamic relationships. In this paper, we look at two different approaches to spatial feature modelling.

2.1.1 Classical Vision

These methods explore spatial clues using space-time interest point (STIP) detectors or various sampling methods for each frame. In classical feature extraction, [25–27] and [24] are defined as handcrafted spatial feature representation techniques.

2.1.2 Machine Vision

In machine vision, deep neural network (DNN) [16] has been served in the area of object detection, recognition and classification, and it provides better results compared to classical vision techniques. Especially the most promising results are from convolutional neural networks (CNN). For example, stacked frames as inputs to CNN for spatial feature extraction or use pretrained networks and explore to annotated pretrained images in the spatial network like ImageNet challenge dataset [4] and more [29–31]. Using above mentioned image-net trained parameters by using transfer learning techniques we can easily solve spatial domain problems.

2.2 Temporal Domain

The temporal features that are used to recognize movement or classify a moving object from motion in the context of optical flow [32]. Optical flow is a motion pattern of an object in a series of video frames, and its primary application is structural motion evaluation. It is important to model the temporal optical flow clues such as short term and long term. For exploring long-term optical flow functions, the multimedia community derives several different classical descriptors such as HOF [33] and its enhanced variant "Histogram Oriented Optical Flow (HOOF)" [34], dense trajectory, hidden Markov models (HMM) [25], conditional random fields (CRF) [27] and Bayesian networks (BN) [26].

Deep learning robust models, such as CNN, RNN and others, are used to investigate optical flow features using machine vision. In DNN convolutional neural networks, a promising achievement for optical flow estimation has been developed. Various optical flow extraction networks, such as Lite FlowNets, FlowNet2 and EV-FlowNets, are also available to achieve good accuracy.

2.2.1 Classical Vision

The method of extracting temporal features is divided into two types: optical flow and deep optical flow. Optical flow is based on two fundamental phenomena. First, the pixel intensities of an object remain constant through stacked consecutive frames. Second, the motion of neighbouring pixels is identical.

For example, in the first frame, a pixel *I (x,y,t)* moves by distance $(\mathrm{d}x, \mathrm{d}y)$ in the next frame taken after *dt* time. So, since these pixels have the same strength, we may assume that there is no shift.

$$I(x, y, t) = I(x + \mathrm{d}_x, y + \mathrm{d}_y, t + \mathrm{d}t) \tag{1}$$

Taylor series approximation in RHS eliminates related terms and divides by $\mathrm{d}t$, as seen in Eq. 2.

$$f_x u + f_y v + f_t = 0 \tag{2}$$

where:

$$f_x = \frac{\partial f}{\partial x};\ f_y = \frac{\partial f}{\partial y} \tag{3}$$

$$u = \frac{\mathrm{d}x}{\mathrm{d}t};\ v = \frac{\mathrm{d}y}{\mathrm{d}t} \tag{4}$$

The above-mentioned equation is known as the equation for calculating optical flow. Where will I find *fx* and *fy* image gradients, *ft* is the gradient with time in the same way. However, *(u, v)* is uncertain. It is difficult to solve a single equation with two enigmatic variables. So various methods are proposed to solve this problem, one of which is the Lucas-Kanade method (LK). Based on the second premise, the LK approach takes a $3X3$ chunk around the point where the motions are identical so that it can locate,

$$(f_x, f_y, f_t) \tag{5}$$

Then solve,

$$\begin{bmatrix} u \\ v \end{bmatrix} = \begin{bmatrix} \sum_i {f_{x_i}}^2 & \sum_i f_{x_i} f_{y_i} \\ \sum_i f_{x_i} f_{y_i} & \sum_i {f_{y_i}}^2 \end{bmatrix}^{-1} \begin{bmatrix} -\sum_i f_{x_i} f_{t_i} \\ -\sum_i f_{y_i} f_{t_i} \end{bmatrix} \quad (6)$$

Using Eq. 6, we can obtain optical flow as well as scale. It only calculates optical flow for a sparse feature set. Dense optical flow provides optical flow for entire frame regions. Inspired by this evaluate a dense trajectory [15, 28], it computes dense points from each frame as well as a motion-based optical flow field from the displacement vector.

2.2.2 Machine Vision

Various FlowNets for optical flow estimation were demonstrated in machine vision. However, the most recent conventional approach focuses on small displacement and natural data. FlowNets do not engage in vector procedures. They primarily work on large temporal data sets, using cutting-edge deep learning architecture; in general, convolutional neural networks (CNN) have been used to test optical flows. FlowNets 2.0, introduced by Ilg et al. [35], learns optical from natural data. Figure 1 depicts an overview of the FlowNets networks, which are comprised of four stages, each of which can simulate large and small displacement fields. It yields three big results: During preparation, it first focuses on natural training data and preprocessed presenting data. Second, for warp of the next consecutive image with transitional optical flow, staked architecture "Motion Stacked Difference Image (MSDI)" [36]. Third, entangle a small collection of displacements to another abstract paradigm for exploring small set motions. As a result, when running positive reiterative frame rates, it produces more improvement in image quality and efficiency than the previous methods. Hui et al. suggested a modified version of FlowNets called Lite FlowNets [37], which is completely optimized by a large reduction in network size and implements a hierarchical structure to prevent losses in features, pixel-to-pixel level to the subpixel level.

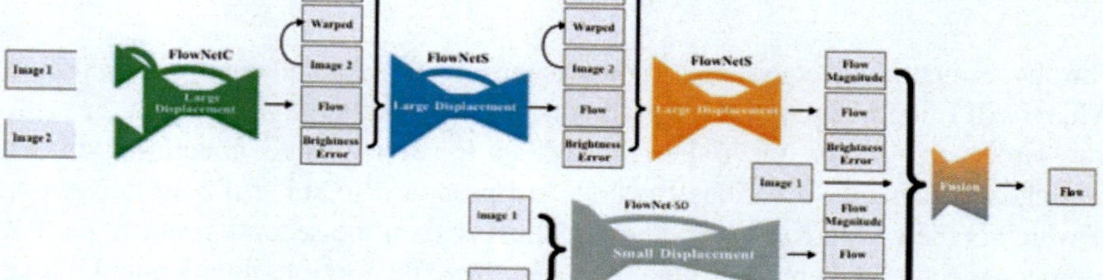

Fig. 1 Overview of FlowNet architecture

3 Feature Extracting Proposal Models and Methods

Many neural designs, such as two-stream systems and multimodel systems, model spatial–temporal clues. Two-stream networks that measure spatio-temporal feature from layered optical flow and still frame using a range of techniques. Multilevel abstraction in video frame loops, such as long-term motion clues, is the subject of hybrid multimodels.

3.1 *Two-Stream Approach*

The two-stream convolutional neural network [7] was constructed using spatial–temporal features, according to the neurology [22, 23]. Figure 2 depicts an overview of the two-stream CNN with two convolutional pipelines for static frames and optical flow frames with score fusion. In [7], two-stream CNN for dense optical flow motion estimation. The proposed network may have two distinct CNN stages, with one step extracting spatial features from still video frames and the second extracting motion clues from optical frames. They both provide good results when modelling the spatio-temporal clues. Each stream's features are fused to perform additional operations.

Modelling the temporal features only considers short-term motion patterns computed in short sequences. It was strengthened further by modelling long-term temporal clues to add long short-term memory (LSTM) [38, 39] layers, modelling temporal clues within and between the sequences using bidirectional LSTM followed by CNN. Figure 3 depicts the CNN and LSTM designs for pooling spatial and temporal long-term relationships by incorporating the LSTM network by recurrent neural network. Exploring long-term hints maps spatial patterns with temporal relationships using fusion features and CNN and LSTM outputs.It makes use of various spatial feature fusions and LSTM for temporal dynamics patterns.

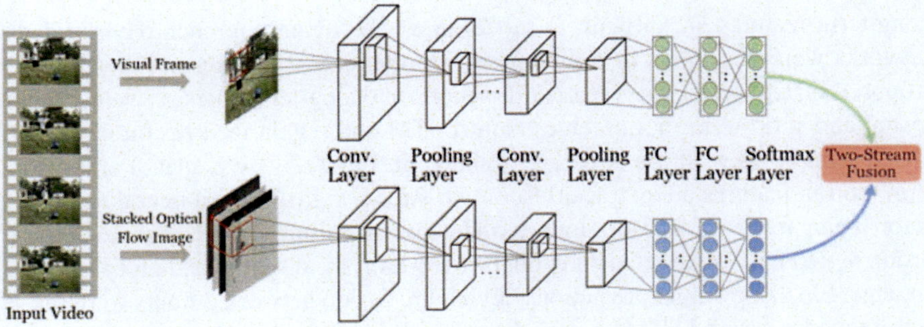

Fig. 2 Pipeline architecture of the two-stream CNN

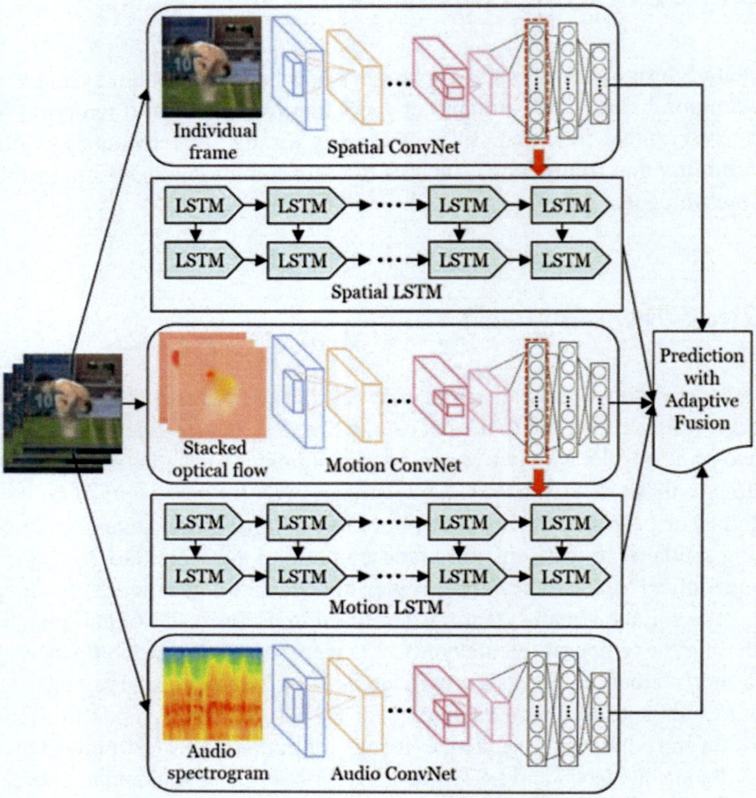

Fig. 3 Long short-term memory (LSTM) with two-stream

3.2 Hybrid Multimodel Approach

Long-term features are difficult to model in a two-stream approach. Hybrid deep networks were developed to overcome this limitation. According to [10, 40], this proposed system investigates spatial, short-term and long-term clues, as well as fuses spatial and short-term motion clues from LSTM and regularized fusion networks. The schematic view of the structure is shown in Figure 4. Both spatial and short-term motion features are extracted from two-stream convolutional neural networks before being fed into LSTM for long-term feature modelling. The regularized feature fusion network [40] extracted and polled video-level local spatial and global motion features. Finally, for final predictions, all feature fusion network outputs were fused with sequence-based LSTM.

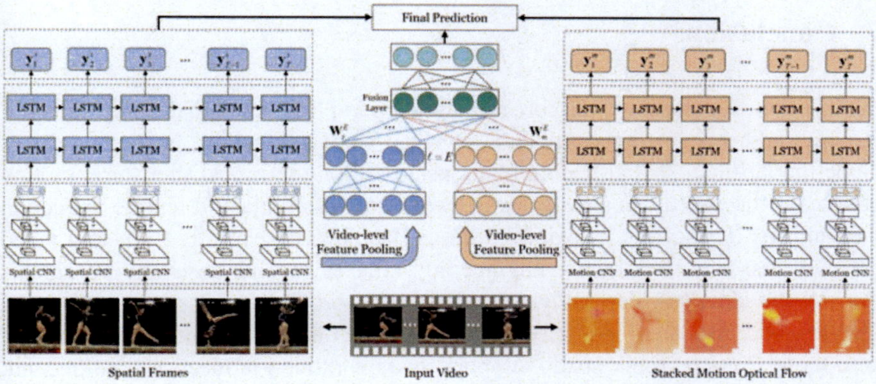

Fig. 4 Long short-term memory (LSTM) with two-stream

4 Results and Analysis

A different form of understanding problems such as actions recognition, classification and captioning is resolved through the modelled spatial and temporal clues. Table 1 shows the spatio-temporal models results on video understanding applications by using the HMDB51 and UCF101 data sets. In classification and recognition, accuracy improves with the usage of the multiple abstraction in pyramid network without loss of long-term motion patterns. The definitions of the two data sets COCO CS5 and COCO 201-A are used generate the captions, as a result of word arrangements with movement patterns gives huge variations compare to classification and recognition.

Table 1 Spatio-temporal clues modelling results on video understanding problems

Models	Action recognition		Video classification		Video captioning	
	UCF101	HMDB51	UCF101	HMDB51	COCO CS5	COCO 201-A
Two-stream CNN	87.9	66.8	88.0	59.4	55.1	56
Two-stream LSTM	94.6	69.0	93.5	69.2	62.8	68.2
Hybrid Muliti-model	93.1	72.1	92.6	71.2	85	94.6
Pyramid network	94.6	68.9	94.2	69.4	62	68.2

5 Conclusion

In this paper, we discussed the core spatial–temporal features with deep neural models, also addresses various problems in video analysis domain. In the context of modelling abstract spatial feature relationship on long-term and short-term motion patterns is discussed using hybrid frameworks with self-attained feature extraction mechanisms .Using this knowledge, we can achieve best results from raw video data with less feature extraction engineering skills. Hence, our observation from this overall study, major video analysis problems are sort out by using spatio-temporal features, and also there is a need of standard deep learning framework contained with these features, First: spatio-temporal feature extraction mechanisms, Second: exploit a fusion score among the selected feature frames, Third: explore abstract inter class feature relationship between different video content with local and global level. Therefore, our future works are concentrate to develop a standard the deep learning framework contained with above utilities.

References

1. Bobick AF, Davis JW (2001) The recognition of human movement using temporal templates BT—pattern analysis and machine intelligence. IEEE Trans 23(3):257–267
2. Laptev I et al (2008) Learning realistic human actions from movies. In: 26th IEEE conference computer vision pattern recognition, CVPR. https://doi.org/10.1109/CVPR.2008.4587756
3. Sermanet P, Chintala S, LeCun Y (2012) Convolutional neural networks applied to house numbers digit classification. In: Proceedings of the 21st international conference on pattern recognition (ICPR2012), pp 3288–3291 (IEEE, 2012)
4. Gonzalez TF (ed) (2007) Handbook of approximation algorithms and metaheuristics. CRC Press. https://doi.org/10.1201/9781420010749
5. Karpathy A et al (2014) Large-scale video classification with convolutional neural networks. In: Proceedings of the IEEE computer society conference on computer vision and pattern recognition, pp 1725–1732. https://doi.org/10.1109/CVPR.2014.223
6. Burghouts GJ, Schutte K (2013) Spatio-temporal layout of human actions for improved bag-of-words action detection. Pattern Recognit Lett 34(15):1861–1869. https://doi.org/10.1016/j.patrec.2013.01.024
7. Simonyan K, Zisserman A (2004) Andrew: two-stream convolutional networks for action recognition. In: NIPS'14 proceedings of the 27th international conference on neural information processing systems, vol 1 pp 568–576. https://doi.org/10.1016/0006-2952(83)90587-7
8. Jiang YG et al (2018) Modeling multimodal clues in a hybrid deep learning framework for video classification. IEEE Trans Multimed 20(11):3137–3147. https://doi.org/10.1109/TMM.2018.2823900
9. Wang Y et al (2017) Spatiotemporal pyramid network for video action recognition. In: Proceedings - 30th IEEE conference on computer vision and pattern recognition, CVPR 2017. pp 2097–2106. https://doi.org/10.1109/CVPR.2017.226
10. Wu Z et al (2015) Modeling spatial-temporal clues in a hybrid deep learning framework for video classification. In: MM'15 proceedings of the 23rd ACM international conference on multimedia, pp 461–470
11. Peng Y et al (2018) Two-stream collaborative learning with spatial-temporal attention for video classification. IEEE Trans Circ Syst Video Technol. https://doi.org/10.1109/TCSVT.2018.2808685

12. Singh B et al (2016) A Multi-stream Bi-directional recurrent neural network for fine-grained action detection. In: Proceedings of the IEEE computer society conference on computer vision and pattern recognition. https://doi.org/10.1109/CVPR.2016.216
13. Lowe DG (2004) Distinctive image features from scale-invariant Keypoints David. Int J Comput Vis 1–28
14. Yao L (2016) Extract the relational information of static features and motion features for human activities recognition in videos. Comput Intell Nerosci 2016:1–7. https://doi.org/10.1155/2016/1760172
15. Wang H et al (2011) Action recognition by dense trajectories To cite this version: Cvpr'11 (2011)
16. Lecun Y et al (2015) Deep learning. https://doi.org/10.1038/nature14539
17. Ouadiay, F.Z. et al.: Simultaneous object detection and localization using convolutional neural networks. 2018 Int. Conf. Intell. Syst. Comput. Vision, ISCV 2018. 2018-May, 1-8 (2018). https://doi.org/10.1109/ISACV.2018.8354045
18. Guiming D et al (2017) Speech recognition based on convolutional neural networks. In: 2016 IEEE international conference on signal and image processing, ICSIP, pp 708–711. https://doi.org/10.1109/SIPROCESS.2016.7888355
19. Russakovsky O et al (2015) ImageNet large scale visual recognition challenge. Int J Comput Vis 211–252. https://doi.org/10.1007/s11263-015-0816-y
20. Feichtenhofer C et al (2016) Convolutional two-stream network fusion for video action recognition. In: Proceedings of the IEEE computer society conference on computer vision and pattern recognition, pp 1933–1941. https://doi.org/10.1109/CVPR.2016.213
21. Ye H et al (2015) Evaluating two-stream CNN for video classification. https://doi.org/10.1145/2671188.2749406
22. Goodale MAM, Melvyn DM (2004) Separate visual pathways for perception and action. Lit Theol 11(1):80–92. https://doi.org/10.1093/litthe/11.1.80
23. Kruger N et al (2013) Deep hierarchies in the primate visual cortex: what can we learn for computer vision? IEEE Trans Pattern Anal Mach Intell 35(8):1847–1871. https://doi.org/10.1109/TPAMI.2012.272
24. Palmer R et al (2012) Scale proportionate histograms of oriented gradients for object detection in co-registered visual and range data. In: 2012 international conference on digital image computing techniques and applications (DICTA 2012). https://doi.org/10.1109/DICTA.2012.6411699
25. Li F, Du J (2012) Local spatio-temporal interest point detection for human action recognition. In: 2012 IEEE 5th international conference on advanced computational intelligence ICACI 2012, pp 579–582. https://doi.org/10.1109/ICACI.2012.6463231
26. Laptev I, Lindeberg T (2003) Space-time Interest Points Ivan. In: Ninth IEEE international conference on computer vision (ICCV'03) 0-7695-1950-4/03:0-7
27. Peng K et al (2009) 3D reconstruction based on SIFT and Harris feature points. 2009 IEEE International Conference on Robotics and Biomimetics (ROBIO), ROBIO 2009. 960, 1, 960–964 (2009). https://doi.org/10.1109/ROBIO.2009.5420735
28. Papadopoulos K et al (2019) Localized trajectories for 2D and 3D action recognition
29. He K et al (2016) Deep residual learning for image recognition. In: Proceedings of the IEEE computer society conference on computer vision and pattern recognition. pp 770–778. https://doi.org/10.1109/CVPR.2016.90
30. Simonyan K, Zisserman A (2014) Very deep convolutional networks for large-scale image recognition, pp 1–14
31. Szegedy C et al Rethinking the inception architecture for computer vision. In: Proceeding IEEE computer society conference on computer vision and pattern recognition. 2016-Decem, pp. 2818–2826. https://doi.org/10.1109/CVPR.2016.308
32. Baker S et al (2011) A database and evaluation methodology for optical flow. Int J Comput Vis 92(1):1–31. https://doi.org/10.1007/s11263-010-0390-2
33. Wang T, Delaunay-lms IC (2012) Histograms of optical flow orientation for visual abnormal events detection. In: IEEE Ninth international conference on advanced video signal-based surveill, pp 13–18. https://doi.org/10.1109/AVSS.2012.39

34. Chaudhry R et al (2009) Histograms of oriented optical flow and Binet-Cauchy kernels on nonlinear dynamical systems for the recognition of human actions. In: 2009 IEEE conference on computer vision and pattern Recognition (CVPR) Work. CVPR Work. 2009 IEEE, pp 1932–1939 (2009). https://doi.org/10.1109/CVPRW.2009.5206821
35. Ilg E et al (2017) FlowNet 2.0: evolution of optical flow estimation with deep networks. In: Proceeding - 30th IEEE conference computer vision pattern recognition, CVPR 2017. pp 1647–1655. https://doi.org/10.1109/CVPR.2017.179
36. Wang L et al (2017) Three-stream CNNs for action recognition. Pattern Recognit Lett 92:33–40. https://doi.org/10.1016/j.patrec.2017.04.004
37. Hui TW et al (2018) LiteFlowNet: a lightweight convolutional neural network for optical flow estimation. In: IEEE computer society conference on computer vision and pattern recognition, pp 8981–8989. https://doi.org/10.1109/CVPR.2018.00936
38. Sun C et al (2015) Temporal localization of fine-grained actions in videos by domain transfer from web images. https://doi.org/10.1145/2733373.2806226
39. Zhao R et al (2017) Two-stream RNN/CNN for action recognition in 3D videos. IEEE Int Conf Intell Rob Syst 2017-Septe, 4260–4267 (2017). https://doi.org/10.1109/IROS.2017.8206288
40. Wu Z et al (2018) Modeling multimodal clues in a hybrid deep learning framework for video classification. IEEE Trans Multimed 20(11):3137–3147. https://doi.org/10.1109/TMM.2018.2823900

Covid-19—Analysis and Prediction—A Case Study Using Machine Learning

Abhishek Sharma, Digbijoy Dasgupta, Shreya Bose, Udayan Misra, Ishita Pahari, Raktim Karmakar, and Sudipta Basu Pal

Abstract In this research paper, machine learning-based models are used for predicting the number of countries that have been affected by the COVID-19 virus and what was the situation of them. Then, the analysis is performed whether the "Lockdown" is the savior or not, and how India faced the whole problem. After that the prediction model for India was also prepared. These are the selected ones that are comprised in this paper. The COVID virus has spread globally, causing thousands of deaths and having an enormous impact on our health systems and economies. As soon as the Covid-19 pandemic was hit hard worldwide, lots of researchers were trying to figure out lot more things, among them when and from where did the virus make its an outbreak. The whole analysis part is based on the data science and machine learning algorithms and utterly dependent on the data collected from various trusted sources. In this paper, decision tree-based linear (DTBL) model and Random Forest Regression (RFR) models are applied to ensure the accuracy of the models. The Auto-Regressive Integrated Moving Average (ARIMA) and Prophet Algorithm are used for the prediction model. Preventive measures to reduce the spread of the COVID virus in respective zones are also suggested. We hope that this research work will help in the understanding and eradication of the threatening disease. This study can be applied by other countries for predicting COVID-19 cases at the state or national level.

Keywords Covid-19 · Corona virus · Machine learning · Data analysis · Prediction model

1 Introduction

The whole world is struggling through the Covid-19 pandemic situation due to the break out of the virus named Corona Virus or officially called the Covid-19. Most of the people who were affected by Covid-19, were passed through simple and casual

A. Sharma · D. Dasgupta · S. Bose · U. Misra · I. Pahari · R. Karmakar · S. B. Pal (✉)
Computer Science Engineering Department, University of Engineering and Management, Kolkata, India

J. M. R. S. Tavares et al. (eds.), *Cyber Intelligence and Information Retrieval*, Lecture Notes in Networks and Systems 291,
https://doi.org/10.1007/978-981-16-4284-5_31

symptoms of fever, dry cough and had recovered from the virus attack without having any kind of special treatment unless and until the virus majorly affected his/her body. A Covid-19 affected patient's common symptoms are fever, dry cough and fatigue [1].

On these following topics, we have done our research work.

Where did this virus outbreak? WHO (World Health Organization) named the virus as Novel Corona Virus, which made its first outbreak at Wuhan city of the Hubei Province of China. As per the WHO, 30th January is the date when they declared the outbreak of the virus worldwide due to the public health safety and emergency of international concern [2].

HOW IT SPREADS? As per the research work of John Hopkins University [3], they have distinguished and simulated the whole situation of the transmission of the virus from one person to another. The virus was spreading in an exponential manner as soon as the infected person sneezes or coughs in the presence of another disinfected person(s). The droplets of the cough and sneezes of the infected person fall on the surfaces as they were very heavy to be hang in the air, and as soon as they came down to the surfaces, if the disinfected persons touch those surfaces then he/she will be affected. The two main ways by which the virus can enter someone's body are by touching nose or, eyes after touching the infected area or, surface.

How it affected the major countries along with the whole world? The outbreak of the virus was in Wuhan, China but it became the world wide pandemic situation as it affected 246 countries among them European countries are the major victims, after that USA was affected in a large scale and then the virus came to India [4]. All the major countries who were affected by this virus, all of them were in the uncertainties like when this will be over or, how long it will take to heal the whole world against the virus, as it was taking millions of lives continuously in each day.

LOCKDOWN: Whether it was savior or, not? One thing that every affected country have done during the pandemic as they have imposed lockdown in the countrywide and the international borders were closed as every country was in an isolated situation. But imposing lockdown is not the only thing that will remove the threat of the virus, planning of the regular testing procedure, social awareness and increasing health facility system are the major improvements that will ensure the mortality rate will be under control. Some of the countries have strictly follow these, but the majority of the countries have failed to do these and became the worst victims of this deadly virus[5].

INDIA: Current situation and analysis. As per the data collected up to 15th November, 2020, we have seen that India is the third worst victim after USA and Brazil. India's Government and Ministry of Health and Welfare of India have imposed the unplanned the lockdown in the early stages of the outbreak in India, when the number of confirmed cases were 3–4, which shows the awareness of India, but unfortunately the unplanned behavior of the Government was not even rectified over the period of the whole lockdown, as the testing procedure was not done properly, health facility wards were not improved to fight against this virus and the Government was failed to increase the social awareness within the citizens of India [6].

Whether the health facility of India is up against Covid-19. During this research work, we have seen majority faults were done by the health facility systems, as the hospitals and health facility wards were not improved although the Government imposed the lockdown in the early stage of the break out of the virus. The Health Facility of India [7] was too poor that if the second wave of the virus hit along with the first wave as it did in the European countries. No one in India will be safe from this virus, as India is already running out of beds, and that will be the nightmare for the country.

The Prediction Model. India has not reached the peak yet, once it reaches the peak the prediction model can be built to show that how much time it will take to get things back to the normal. India has never experienced such pandemic in last 100 years so what do people think about this pandemic, lockdown, government approach/policies, etc., can be studied to have sentiment insight of this pandemic. The descriptive analysis on the above-mentioned points is given in this project.

2 Review Work

Covid-19 is such a pandemic situation that no one has estimated that earlier. In this emergency situation, data science and machine learning play an important role to find out the prediction of this pandemic. Also data is wealth for detecting any kind of situation and it also shows the right path how the situation will go on and how the parameters can affect the situation and also how the situation can affect the parameters too. Various works have done during the pandemic trying to figure out the correct prediction model based on the raw data. Such an example is John Hopkins University has done a research work where they showed that how the transmission of the virus was going on and how it affected the people exponentially. Based on that studies we have determined few factors.

First of all we have found that from where this virus was make it's outbreak, then how many major countries have been affected by this virus and how they have implemented the precautionary lockdown period and also whether it was successful or not (based on the data up to 15th November, 2020).

After that the whole study was centralized toward India. Previous reports of India, published by the Ministry of Health and Welfare, Government of India, showed that India's confirmed cases count are increasing and also the recovery rate is quite satisfactory. Based on that reports, we have predicted the upcoming situation of India, keeping constant that no vaccination is being started. And also how much the health facility of India is capable of fighting against the virus is also analyzed using the data analysis and machine learning models.

3 Methods and Algorithms Used

While working with huge datasets, machine learning provides us lots of luxury and algorithms through which we can determine the correct analytics and can have a good accuracy score or, rather a good cross validation score. The three main factors by which we can implement the model are, (a) Total no. of Confirmed cases, (b) Total no. of Recovered cases, (c) Total no. of Death cases. We have think of the factors like, the model has to be effective, the model should consume less time and will provide the results in an interesting manner so, that it can help the future researchers for their works. For that nonlinear regression is the best way to implement the models for the individual countries so that we can visualize the parameters clearly and it will provide the better result than any other models. Apart from this, we have used decision tree-based linear (DTBL) model and Random Forest Regression (RFR) models to ensure the accuracy of the models. Multiple linear models are used here to overcome the defects of one particular model and it will help to solve all the complex parameters as well as the nonlinear parameters and the accuracy was 92%.

Talking about the implementation of the prediction model, the best ways are Auto-Regressive Integrated Moving Average (ARIMA) [8] and Prophet Algorithm. The prediction models use more or less same algorithm of training the dataset and testing the dataset and then implement the future predictions. Similarly, ARIMA model is deployed here to find the prediction of upcoming 25 days, for that 2/3 of the data is used to train the model and 1/3 data is used to test the model. For ARIMA model implementation, the accuracy score is 0.84, which shows the efficacy of the model.

Facebook developed the Prophet [9] Algorithm for machine learning and data analysis purposes where we can forecast the model and predict the model in a larger scale. In this algorithm too, we have to train the dataset and then test the dataset and then have to implement the model. While implementing the model the cross validation score is 0.96, which shows the efficacy of the model. Prophet model also uses the nonlinear regression and random forest regression model.

4 Result Analysis and Discussions

This research paper includes the following problem statements,

- What is the origin of the virus and what are data is talking about the origin of the virus?
- How it affected the whole world?
- Lockdown procedure whether it was a blessing or a nightmare?
- How India has faced the situation and what are the health facility conditions of India.

By this paper, we can conclude these problem statements and will answer a lot more queries by the help of the data analysis and machine learning models and algorithms. Machine learning is the domain where we can distinguish the whole situation based on the raw data collected from the various sources and can implement a model through which we can understand the situation and also can do future predictions from the model.

4.1 Origin of the Virus

According to the research work and data analysis of the given data, it was found that the Covid-19 virus was first found in Wuhan, China in the month of December, 2019. In the month of December and January 2019–20, the total no. of cases of Wuhan reached the upper limit of 70,000, 90% of the total cases of the China in a span of 60 days [10]. From this data and analysis, we can easily conclude that the outbreak was in Wuhan, China. After that the Pathogen identified that the virus as Novel Corona Virus.

4.2 Spread Out of the Virus

As per the data analysis, Covid-19 has effected 246 countries up to 15th November, 2020 and 57% of the total world population was affected by this, which means the no. of confirmed cases in the world was 54,283,840. Approximately, 21% of the total confirmed cases of the world is found at USA, 16.27% from India and 10.8% from Brazil. And the values are increasing exponentially.

4.3 Lockdown: The Savior or, not!

To face this pandemic, lockdown is the only that was chosen by the major countries. Some of them have successfully overcome the situation by imposing it and some of them were not. As per the data Italy, Japan, Germany, China have successfully overcome and controlled the situation, while USA, India have failed miserably. And the situation became worst when the Government revoked the lockdown and the cases were breaking records every now and then.

4.4 Analysis of India

As per the data updated till 15th November, 2020, India reached at the mark of 9,227,557 confirmed cases with 134,804 casualties. Among all the negatives coming from the health facility of India, there was just one positive figure for India, that is, India was having the lowest mortality rate of 3.14% and the recovery rate was above 60%. Among all the states, Maharashtra became the worst victim and caused more than 50% of the total confirmed cases of India. Mumbai named as the worst affected city of India with confirmed cases above 100,000.

According to the data analysis and studies, the distribution of males and females in case of confirmed cases are 66.8 and 33.2%, respectively. As Fig. 1 shows the following data in a more simplified manner. Now as per the data analysis, we have visualized the distribution of confirmed cases with respect to the age. The following Table 1 shows the distribution.

So, obviously from this we can determine that the age group of above 80 years is the most endangered one and the age group of 0–9 years is the least endangered one. While in between the ages of 40 to 79 years they hold the distribution of all total 57.78% which is moderately endangered area of age group.

The outbreak of Covid-19 made an impact over the health facilities of the tech and money giants like USA and China. But it is difficult for the third world countries like India, where the health facility conditions are not so good to face off the pandemic situation. Still India is trying to make over the situation with the health facilities by taking emergency actions on various fields.

In this pandemic situation, the health facility is the only way which can save the country. But unfortunately for Maharashtra, the state is running out beds since the

Fig. 1 Percentage of gender

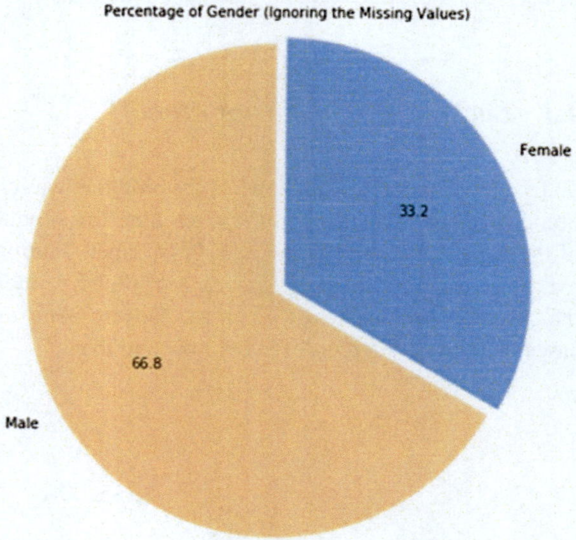

Table 1 Distribution of confirmed cases with respect to age group

Age group (years)	Distribution (%)
0–9	2.22
10–19	4.44
20–29	6.67
30–39	8.89
40–49	11.11
50–59	13.33
60–69	15.56
70–79	17.78
$> = 80$	20

month of October, 2020. States like Tamil Nadu and West Bengal are well ahead in the number of public beds. Same goes to Uttar Pradesh, Madhya Pradesh and also Gujarat. As per the logistic growth, Maharashtra needed 50,000 more beds within 14 days so that they can control the situation. Despite of Maharashtra, other states are in a good situation to fight against the virus till now.

4.5 Prediction Model for India

As per the ARIMA model, we have obtained that up to 8th December, 2020 the total no. of active cases for India will be 1,000,000. (It is a prediction of 25 days) Now in case of next 6 months (up to April 2021) we have determined the model using the Prophet Library. Using this library, we have implemented the model for next 6 months, where it shows that to flatten the curve based on the above-mentioned factors, it will take time to January 2021. After January 2021 the curve will be flattened and at the month of April 2021 the curve will be flattened completely. Based on the studies, it is really difficult to predict the exact model of any prediction but we have implemented this model with the accuracy score of 92.8% and the cross validation score is 89%. At the month of January 2021, the number of active cases will be near about 1,000,000 and at the end of the April 2021 it will be 1,400,000. The predicted curve of the Active cases are given below for the reference in Fig. 2.

5 Conclusion

In this paper, we have analyzed the whole dataset and predicted the future condition of India and how the health facility should encounter the problem.

- As per our research and studies, the virus has affected 246 countries, with USA, India and Brazil were the worst victims.

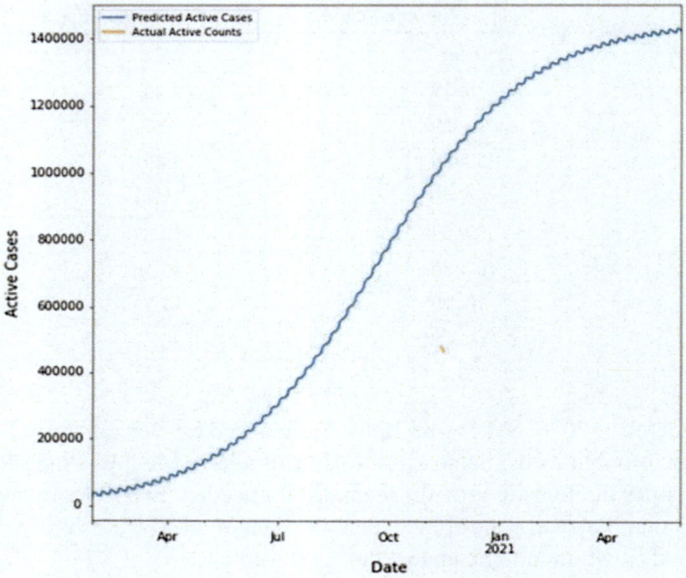

Fig. 2 Prediction model for India upto April ‘21

- Imposing of lockdown did not prevent the pandemic situation of the major countries like USA, India.
- For India, Maharashtra is the worst affected state with more 50% of the total confirmed cases of India.
- For India, males are doubly affected than the females. Also the age group of more than 70 years is the endangered age group and 10–20 years age group has the highest resistance capacity.
- We can also conclude the health facility situations of the states like, Bihar, Gujarat, Tamil Nadu, West Bengal, Uttar Pradesh, Karnataka and Maharashtra needed to be improved in terms of hospital beds otherwise if the second wave of the virus hit these states then nothing will save these states.
- According to the research, if the vaccination is not being started in the month of January and proper testing have done then the curve will be at its peak in the month of January, 2021 and will be sloping down toward the month of April, 2021.
- We do not have any kind of conclusive evidence when the second wave of the virus will hit India, because it came to various countries at the various stages of the pandemic, so, it is hard to predict this, but we have to take precautionary measurements and have to be safe enough to fight against this virus even after the vaccination starts.

References

1. Symptoms of Covid-19 updated on Dec 22 (2020). https://www.cdc.gov/coronavirus/2019-ncov/symptoms-testing/symptoms.html
2. WHO Report. https://www.who.int/docs/default-source/coronaviruse/who-china-joint-mission-on-covid-19-final-report.pdf
3. John Hopkins University Data Analysis. https://www.bostonherald.com/2020/11/27/johns-hopkins-tips-to-make-the-air-in-your-home-safer-from-coronavirus/
4. Data Analysis and detailed algorithm. https://github.com/abhisheks008/Covid-19-Analysis-and-Prediction
5. WHO (2020) WHO Corona virus Disease (COVID-19) Dashboard. Retrieved from https://covid19.who.int/?gclid=Cj0KCQjwrIf3BRD1ARIsAMuugNuI-pJWSxU9q1fy7kJCoEgMh-_CXUU7sy1IMeHw_bSxN4L4PQlRaHYaAoArEALw_wcB
6. The Economic Times (2020) India Covid count, April 29: Death toll crosses 1000 mark, total cases over 31,000. Retrieved from https://economictimes.indiatimes.com/news/politics-and-nation/india-covid-count-april-29-death-toll-crosses-1000-mark-total-cases-over-31000/articleshow/75441195.cms
7. Ministry of Health and Welfare : Government of India . https://www.mohfw.gov.in/
8. ARIMA Model definition and details. https://www.machinelearningplus.com/time-series/arima-model-time-series-forecasting-python/#:~:text=ARIMA%2C%20short%20for%20'Auto%20Regressive,used%20to%20forecast%20future%20values
9. Prophet Model in Python and machine learning . https://machinelearningmastery.com/time-series-forecasting-with-prophet-in-python/
10. Roosa K, Lee Y, Luo R, Kirpich A, Rothenberg R, Hyman JM, Chowell G (2020) Real-time forecasts of the COVID-19 epidemic in China from February 5th to February 24th. Infectious Disease Model 5:256–263

A Feature Based Classification and Analysis of Hidden Markov Model in Speech Recognition

R. K. Srivastava, Raj Shree, Ashwani Kant Shukla, Ravi Prakash Pandey, Vivek Shukla, and Digesh Pandey

Abstract Speech recognition is to change over the acoustical sign got from a spokesman or a phone which produces a lot of words. Speech recognition can also regardless called computer speech cognizance which means making the digital device understand what we are talking about. It helps users direct their systems for work, and avoids typing their work because a system can write words faster than a human being. There are several Hmm based models have been developed by the researchers for speech recognition but due to daily advancement of the technology landscape still need a robust techniques in the field of speech recognition. Due to its significant ability, it get classified and with the help of them there are several speech recognition techniques has been developed. The comparative analysis of the various Hmm model shows their efficiency and proposed the effective model in field of speech recognition.

Keywords Speech recognition · Hidden Markov Model (HMM) · Speech-to-text · Soft computing

R. K. Srivastava
CST-UP, Dr. Shakuntala Mishra National Rehabilitation University, Lucknow, India

R. Shree (✉)
CST-UP, Babasaheb Bhimrao Ambedkar University, Lucknow, India

A. K. Shukla
DCS, Babasaheb Bhimrao Ambedkar University, Lucknow, India

R. P. Pandey
IET, Dr. Rammanohar Lohia Avadh University, Ayodhya, India

V. Shukla
DIT, Babasaheb Bhimrao Ambedkar University, Lucknow, India

D. Pandey
RA (CST-UP), Dr. Shakuntala Mishra National Rehabilitation University, Lucknow, India

J. M. R. S. Tavares et al. (eds.), *Cyber Intelligence and Information Retrieval*,
Lecture Notes in Networks and Systems 291,
https://doi.org/10.1007/978-981-16-4284-5_32

1 Introduction

In the current technological perspective, the behavioral sequence of the human characteristics for communicating such as human-to-human (H2H), human-to-machine (H2M), and machine-to-machine (M2M) which is one of the fundamental approaches in different ideas shaped in the way of the speaker are imparted by diction as words, expressions, and sentences by implementing relevant linguistic rules. Diction is an essential method of interaction among a person and the most common and proficient type of exchanging information among people is speech. Speech is the main significant essential needs of people and the most helpful method for communication between individuals. Language contributes to give plans as standard interfaces so that computerized element can reach the majority and encourage the exchange of data across various individuals communicating in various languages. These advancements assume an indispensable part in multi-lingual social directives. Voice to text change takes contribution from the speaker as voice input and then it is changed over into text structure [1]. Verbal exchange management is the analysis of the pitch signals, and the different techniques that are utilized to deal with them. In this cycle, different applications, such as speech coding, synthesis, recognition, and speaker recognition technologies, speech processing is employed. Among the above mentioned, speech recognition [2] is the most significant one. The primary purpose of speech recognition is to change over the acoustical sign got from a mouthpiece or a phone, which produce a lot of words. Speech recognition can also regardless be called computer speech cognizance which means that digital devices can understand what we are talking about. All things are considered to be a program, a computer with the ultimate objective that it can hear us and contact us. By cognize, it means that change over voice into an appropriate process for instance text. These functions can be divided into two classes, depending on the type of information signals, to be specific single-channel and multi-channel. In single-channel multi-speaker communication isolation and confirmation tasks, some strategies have been proposed, which accomplish significant progress. One such method is deep clustering (DPCL). In DPCL, a neural network is designed to plan the vector for each time-iteration unit, which is used by the clustering algorithm [2, 3] to assign each unit to a source [4–8].

Hidden Markov Model (HMM) is an enormous and accommodating class of stochastic measures. It is portrayed by Markov property which suggests that the future state of the cycle depends simply on the present status, not on the gathering of events that went before it [9]. Hidden Markov Model can be classified into various types which shows in Fig. 2, comparative and featured based study of various Hidden Markov model are shown in Tables 1, 2, 3, 4, 5, 6, 7, 8, 9 and 10.

Three primarily problems of the HMM (shows in Fig. 1) model have been discussed below:

i. **The Evaluation Problem:** Given the model, process the probability that a specific output arrangement was delivered by that model.
ii. **The Decoding Problem:** Given the model, locate the most probable sequence of hidden states which might have produced a given output structure.

Table 1 Pros and cons of HMM models

S.N	Name of methods	Pros	Cons	References
1	First order HMM	Better recognition rate	It has stability issue	[19, 21]
2	Higher order HMM	Utilizing this technique one can tackle the three hmm issues for any HOHMM by utilizing a first order HMM plan	Changing the model boundary	[19, 22]
3	Hidden semi Markov model	New limits, include saliencies, are familiar with the model and used to includes that perceive states	Algorithm doesn't have a convergence investigation	[15, 18. 19]
4	Factorial HMM	Permits quite a few states in the Markov chain and created two Markov chain Monte Carlo algorithms	FHMM doesn't hold the Weighted entirety model because of resonation and clamor segment	[19, 23, 24]
5	Second order HMM	Reduces time complexity by log N	Their assurance accuracy's really remained to rise, and their examination types should to be extended	[19, 25]
6	Layered HMM	LHMM gives 10% upgrades in word mistake rates	LHMM has the second most likely bigger state	[11, 19]
7	Autoregressive HMM	It required less training time	ARHMM should have better performance for idle states	[19, 26]
8	Non-stationary HMM	NSHMM Method tackle the inalienable issue of the ARHMM	Execution is better just in non-fixed environment	[2, 19]
9	Hierarchal HMM	Accuracy and speed is of high quality	For heterogeneous data sources, it requires an effective preparing model	[19, 27]

iii. **The Learning Problem:** Given a set of the output structure, locate the most probable arrangement of state change and output probabilities.

2 Literature Review

Karita et al. [1], the proposed ASR model for diction and auto-text encryption that imparts encryption and decoding to programmed discourse acknowledgment (ASR). In order to improve the execution of ASR with tremendous speech and text-just

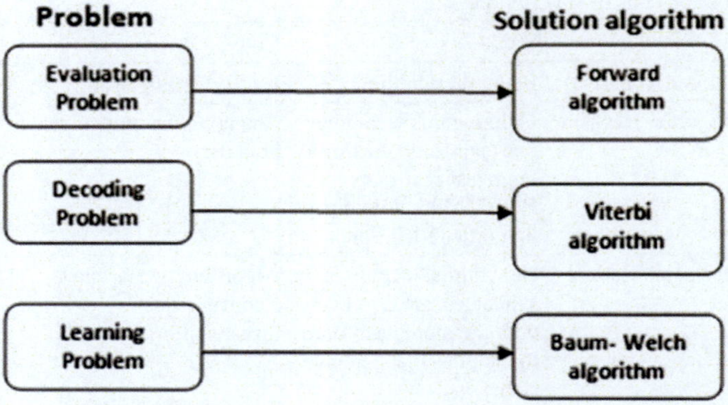

Fig. 1 Fundamental problems and solution algorithm of HMM models [10]

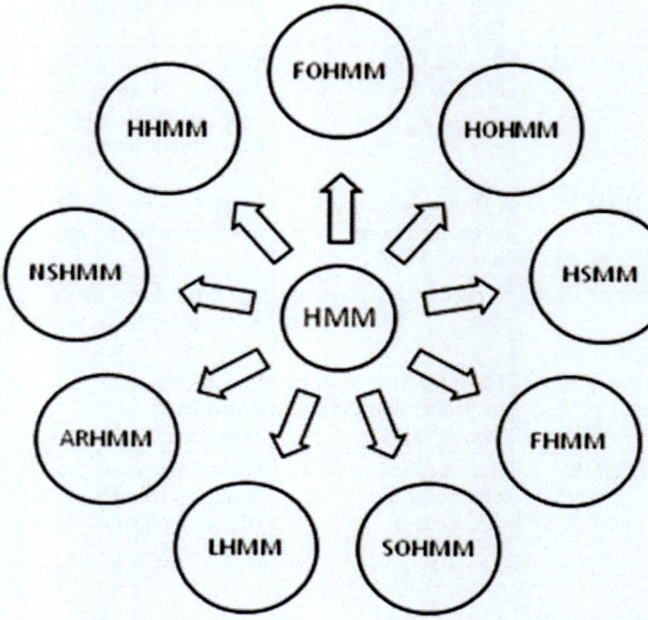

Fig. 2 Classifications of HMM models

planning datasets. It is an augmentation of semi-directed end-to-end programmed speech cognizance. It can utilize the TTS model, discourse and text auto-encryption, with as long as 100 h of matched diction or text, just 300 h of speech, and 460 h of text-just information. This semi-directed model with MMD-based between area misfortune diminishes CER from 10.4 to 8.4 and WER from 20.6 to 18.0%.

Zeineldeen et al. [11], typical LN LSTM variations applied to the LSTM the RNN model for speech recognition suggests that these models are more steady during speech recognition and give an overall introduction to deviations from the LNM model without the LTM model essentially it was observed that models with LN LTM require hyper parameter tuning and can be safely constructed with more ambiguous layers when using LN LSM it is predictable to eliminate the ASR model without the use of a pre-prepared scheme and to introduce better adjustments for boundary introduction and greater strength.

Huang et al. [12], a CTC-based S2I framework was actualized that coordinates the presentation of cutting edge, conventional Cascade SLU frameworks. Two methods to improve the S2I framework were examined. The recovery of the proposed execution is 80% impeded because of the utilization of restricted purpose named speech. End-to-end communication in language understanding frameworks require matched speech and semantic comment information, this framework improves the blunder pace of goal grouping by over 60%, and combined speech to-aim information more than 80% can accomplish improvement.

Mukherjee et al. [13] made text-to-speech synthesizer that changes over content into conveyed word, by exploring and taking care of it utilizing natural language processing (NLP) and a while later re-establishing digital signal processing (DSP) advancement to change over this readied content into the depiction of the component. This synthesizer changes over inputted text into incorporated discourse and peruses it out to the client that can be saved as an mp3 file.

Chang et al. [14] have been proposed to end the multi-channel multi-lingual speech cognition model known as MIMO-Speech. Apparently, the model takes the multi-speaker speech recorded by a speaker bundle as data and collects the text for each speaker. MIMO-speech, which initially plans to expand seq2seq to manage multi-channel input and multi-channel output, is planning a nervous system (seq2seq), called multi-channel multi-speaker. MIMO-speech is a prepared nerve that must be modified through the ASR standard.

Jayakumari et al. [15] Hidden Markov Model (HMM) indicator is utilized to change text over to speech effectively. The proposed strategy utilizes the Hidden Markov model to change text over to speech in an effective way. HMM is utilized as an indicator to address the articulation speed to improve quality. The precision of the framework is improved well when contrasted with the traditional strategies. By utilizing the proposed strategy, the accuracy of the structure is improved by a factor of 6% contrasted and the conventional framework.

3 Types and Applications

3.1 First Order HMM

A Hidden Markov model (HMM) is a measurable build that can be utilized to tackle arrangement issues that have an innate state arrangement portrayal. The model can be pictured as an interlocking arrangement of states.

These states are associated by a bunch of progress probabilities, which demonstrate the likelihood of voyaging between two given states. A cycle starts in some state, at that point at discrete time stretches, the measure moves to another state as directed by the change probabilities [16].

3.2 Higher Order HMM

The fundamental contrast between the conventional HMM and a higher-request HMM is that in the hidden layer, the state progress likelihood is represented by the nth request higher-request Markov model $a_{it-n+1}, \cdots, i_{t+1} = A[q_{t+1} = R_{it+1} | q_t = R_{it}, \cdots, q_{t-n+1} = R_{it-n+1}]$ [11].

3.3 Hidden Semi Markov Model

It is a factual model with a similar structure as a hidden Markov model aside from that the inconspicuous cycle is semi Markov instead of Markov. This implies that the likelihood of there being an adjustment in the hidden state relies upon the measure of time that has slipped by since the passage into the present status. This is rather than hidden Markov models where there is a steady likelihood of changing state given endurance in the state up to that time [9].

3.4 Factorial HMM

Hidden Markov models are speculation of combination models. Whenever step, the likelihood thickness over the observables characterized by an HMM is a combination of the densities characterized by each state in the fundamental Markov model. Another speculation of blend models, the cooperative vector quantize (CVQ), gives a characteristic formalism to conveyed state portrayals in HMMs. Factorial concealed Markov models join the state change structure of HMMs with the circulated representations of CVQs [17].

3.5 Second Order HMM

The second order HMM is an expansion to the first order HMM which contains extra data about past movement that prompts better exactness results. Second order HMM depends on one perception in a current state and the progress likelihood work dependent on two past states [18].

3.6 *Layered HMM*

LHMM is a precise model discovered from the Hidden Markov model (HMM). A layered Hidden Markov model (LHMM) comprises of M levels of HMMs, where the HMMs on level $j + 1$ relate to perception images or likelihood generators at level j. Each level j of the LHMM comprises of G_j HMMs running in parallel [18].

3.7 *Autoregressive HMM*

It is a combination of autoregressive time arrangement and hidden Markov chains. Perceptions are created by two or three autoregressive time courses of action, while the switches between each autoregressive time plan are obliged by a hidden Markov chain [19].

3.8 *Hierarchal HMM*

In the hierarchical hidden Markov model (HHMM), each state is viewed as an independent probabilistic model. All the more definitely each condition of the HHMM is itself a HHMM. This infers that the conditions of the HHMM transmit successions of perception images instead of single perception images similar to the case for the standard HMM states [19, 20].

4 Distinguishedness of the Related Work

4.1 *Comparative Study of the HMM Models*

The most significant pros and cons of the HMM models has been discussed in Table 1.

4.2 *Featured Based Significancy of the Developed HMM Models*

5 Conclusion

From the last few decades, the speech recognition has become prominent field of the research. Due to the daily advancement of the technological landscape, it is

Table 2 Developed model of first order HMM

Name of HMM model	Developed model	Features	Limitations	References
First order HMM	Isolated word speech recognition	recognize speech even with discretionarily picked boundaries	Various keys are utilized independently and later consolidated which influence the preparing time	[19, 21]
	Speech emotion recognition	Improved performance of emotion recognition system for different SNR levels	It is not efficient for reducing car noise	[28]
	Acoustic recognition model for English speech	Improves the acknowledgment rates and shows better flexibility	Still needs an enhancement	[2]

Table 3 Developed model of higher order HMM

Name of HMM model	Developed model	Features	Limitations	References
Higher order HMM	High-order HMM-estimation model	Using this procedure one can handle the three well issues for any HOHMM by using a first-request HMM plan	Obtained issues resolving with the help of FOHMM	[19, 22]
	Automatic classification of heartbeat using ECG signals	Learned-HOHMM helps to decode and classify new heartbeats of unspecified types	For heterogeneity, it requires efficient training model	[19, 29]
	Piecewise linear processes and application to speech recognition	A parameter estimation method based on the expectation maximization algorithm was derived for piecewise straight cycles are proposed to all the more likely estimated the conduct of a genuine cycle	The error rates were approximately Same as before using piecewise linear processes	[20]

Table 4 Developed model of hidden semi Markov model [27]

Name of HMM model	Developed model	Features	Limitations	References
Hidden semi Markov model	Optimal detection and error exponents	The implicit ideal probability ratio test demonstrated that it contained a straightforward repetition of the measure equal to the amount of diffusion of the length of the two states	The Neyman-Pearson error type is directly equivalent to the top Lyapunov example for recurrence	[17, 19]
	Feature selection and parameter estimation algorithm	New boundaries, feature saliencies, are acquainted with the model and used to choose features that recognize states	HSMM is depend on the MAP for vigorous change in the training data and modeling assumption	[15]
	Audio Segmentation and clustering	Real-time segmentation for music notes and acoustic sequences	Algorithm doesn't have convergence analysis The EM Framework does not support automated structure discovery, audio monitoring systems, and music sequencing	[18]

drawing the attention of the researcher to design more robust framework which has the ability to detect the speech efficiently. This paper primarily presented the hypothesis of the hidden Markov models and its sort and application. The hidden Markov models play a quite significant role in speech recognition due to its learning model [3]. In addition, it has classified into different types due to their dynamic upgradation as per the requirements. The primary objective of this article is to identify the most efficient HMM models which can be used to implement the robust speech recognition model such as only the few suitable models are obtained like factorial HMM(FHMM), Layered HMM(LHMM), Hierarchal HMM(HHMM) etc. for speech recognition. There are several models have already been developed to deal with the speech recognition by ignoring the relative consequences such as detection rate percentage, noises, security concern, etc. Conclusively, a fusion based HMM model can be significantly assist to develop a more robust speech recognition model with greater precision rate and has the potential to resolve the major consequences get identified during the study.

Table 5 Developed model of factorial HMM

Name of HMM model	Developed model	Features	Limitations	References
Factorial HMM	Load identification	Viterbi algorithm along with FHMM is more efficient than equivalent HMM. Solutions are available for both online and offline Load identification	It is insensitive for stability and accuracy of powered data	[30]
	Audio source separation	Solving problems of BSS, VAD, DE reverberation And DOA estimation simultaneously	FHMM does not hold Weighted sum model due to reverberation and noise component	[23]
	Infinite factorial unbounded state HMM	Supports any number of states in the Markov chain and developed two Markov Chain Monte Carlo algorithm	At this point when the power grid of any city is used to a significant extent, it does not have separate data for each household	[24]
	Energy disaggregation model based on human behavior	Time-subordinate state change framework and the unsure controller exhibit a solid capacity to distinguish Without separating both cyclic and non-cyclic machines into assemblies	Does not support self-adopting feature Lack of flexibility to look after new customer or Migrant Human Behaviors	[31]

Table 6 Developed model of second order HMM

Name of HMM model	Developed model	Features	Limitations	References
Second order HMM	Web service fault	Effective diagnose technology for fault detection and diagnosis	Their determination correctness's actually stay to rise, and their analysis types should be expanded	[25, 30]
	Web service selection	Solve the measurement problem of the web service	A solitary worth has its impediment, which Dismisses the impact of different boundaries	[27]

Table 7 Developed model of layered HMM

Name of HMM model	Developed model	Features	Limitations	References
Layered HMM	Motion intention recognition	LHMM is better than HMM	LHMM has second most probable larger state	[21]
	Dual-layer PM variable flux hybrid memory machine	This lowers the required polarizing current level and inverter rating	Bigger than normal inverter power rating and gigantic pounding power issue during the polarization cycle, which is troublesome for cost-tricky present-day applications	[28]
	LNLSTM for hybrid-HMM and end-to-end ASR	Global joined Norm variant gives 10% improvements in word error rates	Increase in variance epsilon effect the performance	[11]

Table 8 Developed model of auto–regressive HMM

Name of HMM model	Developed model	Features	Limitations	References
Autoregressive HMM	User independent estimations model for gait events with minimal sensor data	It reduces the amount of sensor data	Limits the appropriateness of these outcomes to this present reality control framework	[19]
	Voice Pathology analysis	This method helps for automatic topology generation for voice pathology detection	Optimal criteria rather than the MDL is not investigated	[20]
	MVA processing for Arabic speech recognition	MVA method Can make noise more robust and achieve higher accuracy rates	Vocabularies are limited RASTA-PLP, MFCC, and PNCC MVA is better with technology than baseline	[2]

Table 9 Developed model of non-stationary HMM

Name of HMM model	Developed model	Features	Limitations	References
Non-stationary HMM	Speech enhancement method for stationary and non-stationary noise environments	Upgrade conspire performs far superior to the reference techniques in non-stationary conditions, while giving best in class execution for stationary conditions	Performance is better only in non-stationary environment	[2, 19]
	Speech enhancement in non-stationary noise environments	SARHMM method solves the inherent problem of the ARHMM The precise displaying of the addition differences of the Discourse and commotion can assume a significant job in speech upgrade in non-stationary noise environment	SARHMM strategy don't display expressly the arbitrary varieties of discourse and Noise gains	[3]

Table 10 Developed model of hierarchal HMM

Name of HMM model	Developed model	Features	Limitations	References
Hierarchal HMM	3D Gesture Recognition	Performance that can tackle both temporal hierarchy and an entropy model	Supervised poselet path encoding has to be improved	[31]
	3D skeleton based online gesture segmentation and recognition	It helps to improve accuracy and speed	For heterogeneous inputs it requires efficient training model	[25]
	Hierarchical reinforcement learning framework toward multi-agent navigation	Decoupling the objective driven and impact evasion errands, Prompting a quicker and steadier model to be prepared	Navigation scene is simple Heterogeneous shape should be included	[27]

References

1. Mukherjee P, Santra S, Bhowmick S, Paul A, Chatterjee P, Deyasi A (2018) Development of GUI for text-to-speech recognition using natural language processing, IEEE. 978–1–5386–5550–4
2. Ting W (2019) An acoustic recognition model for english speech based on improved HMM algorithm. In: 11th international conference on measuring technology and mechatronics automation, IEEE. 978–1–7281–2165–9
3. Chen S, Gao F, Liu T (2017) Load identification based on factorial hidden Markov model and online performance analysis. In: 13th IEEE conference on automation science and engineering, IEEE. 978–1–5090–6781–7
4. Jayaraman S, Choudhury T, Kumar P (2017) Analysis of classification models based on cuisine prediction using machine learning. Int Conf Smart Technol Smart Nation (SmartTechCon) 2017:1485–1490
5. Choudhury T, Kumar V, Nigam D (2015) An innovative and automatic lung and oral cancer classification using soft computing techniques. Int J Comput Sci Mob Comput 4(12):313–323
6. Choudhury T, Kumar V, Nigam D (2015) Intelligent classification and clustering of lung and oral cancer through decision tree and genetic algorithm. Int J Adv Res Comput Sci Softw Eng 5(12):501–510
7. Chandel K, Kunwar V, Sabitha S, Choudhury T, Mukherjee S (2016) A comparative study on thyroid disease detection using K-nearest neighbor and Naive Bayes classification techniques. CSI Trans ICT 4(2–4):313–319
8. Dhamija J, Choudhury T, Kumar P, Tetarbe A (2018) Classification of swine flu patients using gene expression:"For a safer future.". In: 2018 International conference on communication, computing and internet of things (IC3IoT), pp 504–507
9. Lee LM, Jean FR (2016) High-order hidden Markov model for piecewise linear processes and applications to speech recognition. J Acoust Soc Am 140(2):204–210
10. Nilu S, Raees K, Shree PR (2012) Applications of speaker recognition. Procedia Enghttps://doi.org/10.1016/j.proeng.2012.06.363
11. Hadar U, Messer H (2009) High-order hidden markov models—estimation and implementation, IEEE. 978–1–4244–2710–9
12. Zeineldeen M, Zeyer A, Schluter R, Ney H (2020) Layer-normalized LSTM for hybrid-HMM and end-to-end ASR. In: International conference on acoustics, speech and signal processing, IEEE. 978–1–5090–6631–5
13. Chang X, Zhang W, Qian Y, Roux JL, Watanabe S (2019) Mimo-speech: end-to-end multi-channel multi-speaker speech recognition. In: Automatic speech recognition and understanding, IEEE. 978–1–7281–0306–8
14. Huang Y, Kuo H-K, Thomas S, Konsy Z, Audhkhasi K, Kingsbury B, Hooryy R, Picheny M (2020) Leveraging unpaired text data for training end-to-end speech-to-intent system. In: International conference on acoustics, speech and signal processing, IEEE. 978–1–5090–6631–5
15. Bietti A, Bach F, Cont A (2015) An online em algorithm in hidden (Semi) Markov models for audio segmentation and clustering. In: International conference on acoustics, speech and signal processing, IEEE. 978–1–4673–6997–8
16. Manogaran G, Vijayakumar V, Varatharajan R, Kumar PM, Sundarasekar R, Hsu CH (2018) Machine learning based big data processing framework for cancer diagnosis using hidden Markov model and GM clustering. Wirel Pers Commun 102(3):2099–2116
17. Wang X, Wang J, Shi D, Khodayar ME (2018) A factorial hidden markov model for energy disaggregation based on human behavior analysis, IEEE. 978–1–5386–7703–2
18. Jia Z, Xing X (2015) Second-order HMM-based diagnosis method for web service fault. In: Fifth international conference on instrumentation and measurement, computer, communication and control, IEEE. 978–1–4673–7723–2
19. Mor B, Garhwal S, Kumar A (2020) A systematic review of hidden Markov models and their applications. Archiv Comput Methods Eng. https://doi.org/10.1007/s11831-020-09422-4

20. Bajovic D, He K, Stankovic L, Vukobratovic D, Stankovic V (2018) Optimal detection and error exponents for hidden semi-Markov models, IEEE.https://doi.org/10.1109/JSTSP.2018.2851506
21. Ang F, Tsutsui H, Miyanaga Y (2015) Incorporation of time-varying LP cepstral features in HMM-based isolated word speech recognition, IEEE. 978–1–4673–7488–0
22. Lee L-M, Fu-Rong Jean DDO (2016) High-order hidden Markov model for piecewise linear processes and applications to speech recognition, IEEE, Acoustical Society of America
23. Jayakumari J, Femina JA (2019) An improved text to speech technique for tamil language using hidden Markov model. In: 7th International conference on smart computing and communications, IEEE. 978–1–7281–1557–3
24. Deng F, Changchun Bao W, Kleijn B (2015) Sparse hidden markov models for speech enhancement in non-stationary noise environments (IEEE). Digital Object Identif. https://doi.org/10.1109/TASLP.2015.2458585
25. Chen H, Liu X, Shi J, Zhao G (2020) Temporal hierarchical dictionary guided decoding for online gesture segmentation and recognition. IEEE Trans Image Process Digital Object Identif. https://doi.org/10.1109/TIP.2020.3028962
26. Karita S, Watanabe S, Iwata T, Delcroix M, Ogawa A, Nakatani T (2019) Semi-supervised end-to-end speech recognition using text-to-speech and autoencoders. In: International conference on acoustics, speech and signal processing, IEEE. 978–1–5386–4658–8
27. Ding W, Li S, Qian H, Chen Y (2018) Hierarchical reinforcement learning framework towards multi-agent navigation. In: International conference on robotics and biomimetics, IEEE. 978–1–7281–0377–8
28. Chenchah F, Lachiri Z (2016) Speech emotion recognition in noisy environment. In: 2nd International conference on advanced technologies for signal and image processing, IEEE. 978–1–4673–8526–8
29. Adams S, Beling PA, Cogill R (2016) Feature selection for hidden Markov models and hidden semi-Markov models, IEEE. https://doi.org/10.1109/ACCESS.2016.2552478
30. Lu Y, Jia Z, Li X, Xing X (2018) A second-order hidden markov model based web services selection. In: 7th Data driven control and learning systems conference, IEEE, 978–1–5386–2618–4
31. Chen H, Liu X, Zhao G (2018) Temporal hierarchical dictionary with HMM for fast gesture recognition. In: 24th International conference on pattern recognition, IEEE. 978–1–5386–3788–3

Intelligent Hand Cricket

Anuj Kinge, Nilima Kulkarni, Aditya Devchakke, Aditya Dawda, and Ankit Mukhopadhyay

Abstract Gesture recognition is a technology that uses optical character recognition to identify gestures, hand movements, etc. It has seen applications on various platforms like VR, AR, gaming consoles (Xbox Kinect), etc. The game of hand cricket has been very popular among school-going children. The game will allow the person to play a game of hand cricket against the computer. The system will use gesture recognition in real-time similar to that found in gaming consoles. Libraries like OpenCV help in real-time computer vision, convolutional neural network or CNN help in analysing the image. In our proposed work, we have developed a model that could recognize the gestures of a user correctly. To achieve this, we have implemented a classification model using an image classifier which groups a linear stack of layers into a Keras model. We have created our data set which consists of 1800 images in total. For every gesture, we have taken the 300 images using the webcam and converted those images into the binary image which consists of pixels that have only two colours.

Keywords Hand cricket · Hand gestures · Convolution neural networks · Classification · Computer vision · Human–computer interaction (HCI)

1 Introduction

Hand cricket is an immensely popular game which is played between two people. It is a derivative of the most common rock, paper and scissors. The school-going children love this game thanks to it being so easy to play. Now in this game, two players show scores using their respective fingers. If the two scores are equal, the

A. Kinge (✉) · N. Kulkarni · A. Devchakke · A. Dawda · A. Mukhopadhyay
Department of Computer Science and Engineering, MIT School of Engineering, MIT Arts Design and Technology University, Pune, India

N. Kulkarni
e-mail: nilima.kulkarni@mituniversity.edu.in

J. M. R. S. Tavares et al. (eds.), *Cyber Intelligence and Information Retrieval*, Lecture Notes in Networks and Systems 291,
https://doi.org/10.1007/978-981-16-4284-5_33

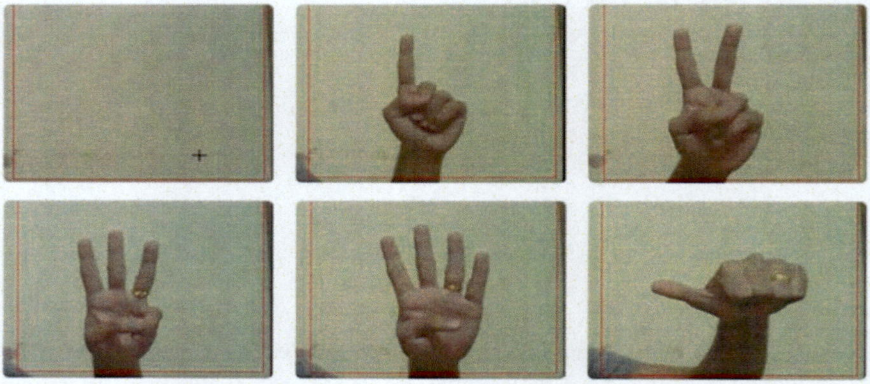

Fig. 1 Hand gestures (0, 1, 2, 3, 4, 6)

batsman is declared out. Else the score is added to the batsman's total score. We need just two players, in this case, the user and the AI system.

One of the aspects of body language is gestures. As per the PRISMA statement, it has been stated that hand gestures are one kind of non-verbal communication [1]. This method is used stand-alone or with another communication (parallel) method such as speech [2]. They are conveyed through the placement of the finger, or shape constructed by the hand and central of the palm. They can be static or dynamic. The main approach to hand gestures can be done with the help of computer vision in our case.

In the approach of camera-based recognition, visibility is the main component. Capturing image from the stream then detecting skin and differentiate it from the background as skin could also have the same colour as the background so all these things make this vision-based complex. But it is still quite popular because chances of errorless and efficiency are high. After taking input from either by these two, we save that inputs to databases by assigning some task. The hand gestures for 0, 1, 2, 3, 4 and 6 are shown in Fig. 1.

Now when we give some inputs to the system by hand gestures as shown in Fig. 1, then the first machine capture command as an image compare this with the database, and if any image found in the database, then the task assigned to that will be performed [3].

To date, there are many research works done on hand gesture recognition. The way of implementing the techniques differ as per the system requirements. However, almost every model has its pros and cons.

The method in [1] implemented the gesture detection using neural network, HMM and fuzzy c-means clustering using Euclidean distance metric. In [4], multiple CNN used where CNN1 gained 96%, while CNN2, CNN3 and CNN4 gained around 96%. It was seen that the model proposed by Oudah et al. [5] gained 100% train accuracy and 99% validation accuracy.

2 Literature Survey

Author name	Method	Observation
Rafiqul Zaman Khan [6]	Gesture detection using neural network, HMM and fuzzy c-means clustering	Hand gesture classification is done through Euclidean distance metric
Zhi-hua Chen [7]	Finger segmentation for real-time gesture detection of hand	The input binary image is in two parts the white pixel of the hand region and black pixel for the background behind the hand. Detection is based on the number of fingers in the white pixel region
Munir Oudah [5]	Deep learning convolutional neural network (ADCNN) is used for better hand classification	The ADCNN result in 100% training data and 99% testing data with 15598 s as the execution time
Raimundo F. Pinto [4]	Morphological operation, counter extraction and polygon approximation, ANN, CNN	The average result of all the CNN is 96%, while CNN1(two-layered) gave 94.7% and CNN2,3,4 gave an accuracy of around 96%
Jay Prakash [3]	Hand gesture detection using models like hidden Markov model, YUV colour space and cam shift algorithm, 3d hand model-based approach, using time flight camera and Naïve Bayes' classifier	Various hand gesture recognition algorithms and models are discussed, and their pros and cons are analysed to find which algorithm and model should be used in a particular scenario
Gatagat [8]	Features of the hand are detected by using the centroid of hand and Euclidean distance	Shape parameter-based hand gesture recognition is done using image pre-processing and segmentation, orientation detection, features extraction, classification and bits generation
Sharma [9]	CNN like Alex Nets, GoogLeNet and ResNet50 used to find real-world object detection from a video feed	The accuracy of all the three CNN, i.e. Alex Nets, GoogLeNet and ResNet50 is checked by using 100 test images from CIF AR-100 and CIF AR-10 test data sets

3 Proposed System

In today's era, CNN is one of the most popular deep learning neural networks which widely used in the field of computer vision. CNN can achieve great results in image classification, object detection and identification. In CNN, instead of typical general

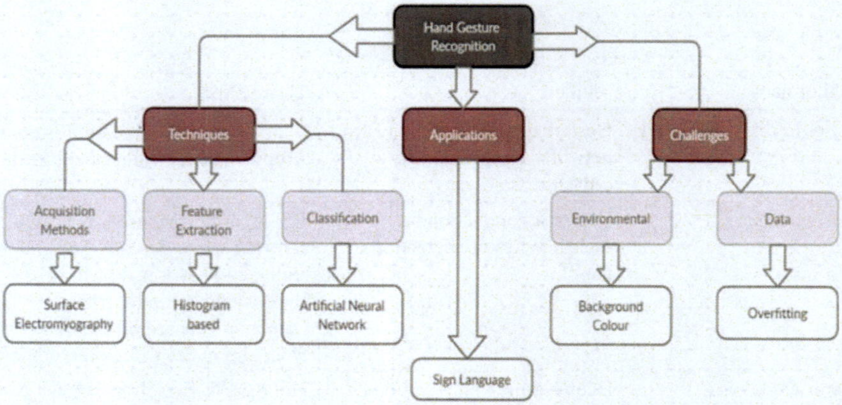

Fig. 2 Hand gesture recognition methods

matrix multiplication, convolution is used in its layers. CNN consists of input layers, output layers and some hidden layers.

CNN mainly consists of three layers for building its architecture, convolutional layer, pooling layer and fully connected layers. Pooling layer is used for spatial dimension reduction. The convolutional layer should have convolutional kernels of fixed height and width, input and output channels where convolutional filters must be equal to the depth of the input feature map. CNN has ReLU, which is a nonlinear function used to introduce nonlinearity. In this way, CNN converts the original image by considering every layer.

In the proposed system, hand gesture recognition is a crucial step. Literature survey reveals that there are many methods for hand gesture recognition as shown in Fig. 2. Each research question is illustrated in this figure, along with the suggested subcategories and the most common technique or problem resulting in them.

In the proposed paper, neural networks are used for hand gesture recognition. As indicated in Fig. 2, hand gesture recognition steps are

Step 1: Acquiring images through the videos.

Step 2: Feature extraction.

Step 3: Classification.

Step 4: Gesture detection.

3.1 Data Set Creation

In the proposed hand cricket method, we have created data set. The data set contains 1800 images. Images are acquired through the camera (5 Mega Pixels), and the images (64px X 64px, 96 DPI, 3 KB size) are fed to the model for detection. The images are taken under normal lighting conditions. Before detection, the insightful information of the image is being served to the model, and this process is called feature extraction. Once the information is gathered, then the classification model comes into the picture. A classification model is used to recognize the gestures of a user correctly. The model groups a linear stack of layers into a Keras model. Figure 3 depicts our proposed system.

Following are the steps which are used to build a classification model.

1. Flattening the dimensions of an image into 1D
2. Normalizing the image pixel values
3. Encoding the categorical column
4. Building a sequential model with appropriate layers
5. Feeding this refined information to the model for training
6. Making predictions

Figure 4 depicts our convolution neural network (CNN). Here, we are taking the input image (user input of the hand gesture) by using an integrated webcam/camera. The special tools can be used for the input such as gloves to detect the movement of

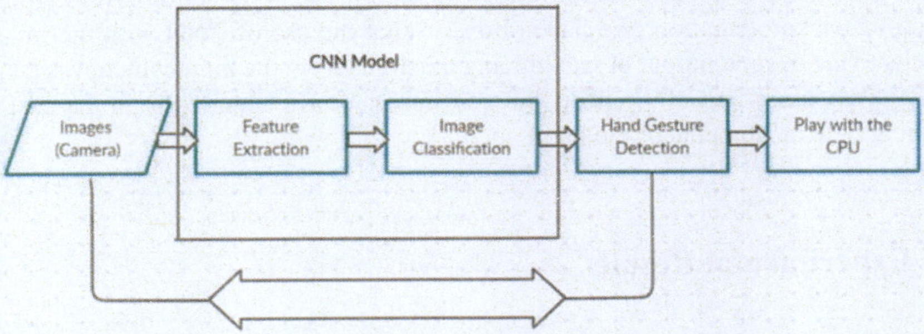

Fig. 3 Proposed system

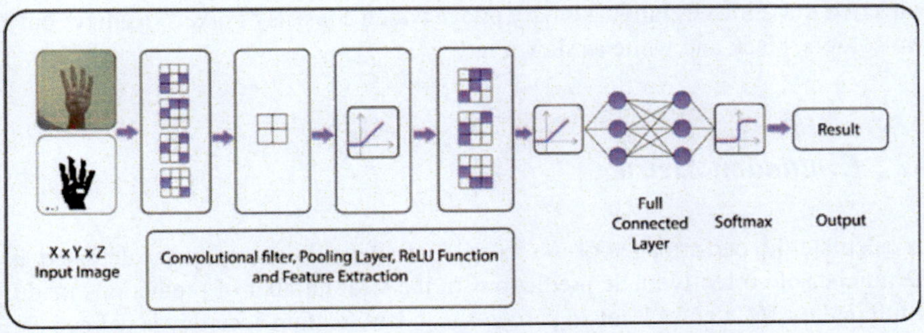

Fig. 4 Convolutional neural network architecture with an example

hand by the user. This is done by the OpenCV library. Convolutional neural network (CNN) extracts the features of the input image and classifies the hand gestures. This can be done using the multi-scale colour feature or recognizing the black and white areas in the input image.

Clustering algorithms are capable of treating the fingers as a cluster and removing the empty spaces present between them. Features needed to be extracted are finger movement or current finger status, alignment of the finger and palm. From the system/computer's view, the performance of hand gesture recognition is affected by the surroundings of the environment (light, background, distance) and the position, direction of the hand.

Three types of layers of CNN are convolution, pooling and the classification layer. It is possible to extract the features by the CNN model. In the convolution layer, every neuron associated with kernel window is convolved with the input image during the CNN training and classification. This kernel is composed of the weights of each neuron.

ReLU is used to replace negative values to zero as the new output images (one for each neuron) contain negative values. Pooling layer decreases the dimensionality of the feature map which ultimately reduces the network training time. During the CNN training, the values of kernel filters applied in the convolution layers are the result of backpropagation. We have observed that the use of CNN with the image results to the desired output of recognizing the digits from the input which was taken by the webcam. The CNN reflects the gestures that have been learnt in the past in training new movements of the hand.

4 Experimental Results

4.1 Data Set

In this proposed work, we have generated our data set which consists of 1800 images in total. For every gesture, we have taken the 300 images using the webcam and converted those images into the binary image which consists of pixels that have only two colours, black and white as shown in Fig. 5.

4.2 Evaluation Metrics

To calculate the performance of the model, classification accuracy is calculated as the number of correctly made predictions to the total number of predictions made, multiplied by 100 to turn it into a percentage. Classification accuracy is defined as:

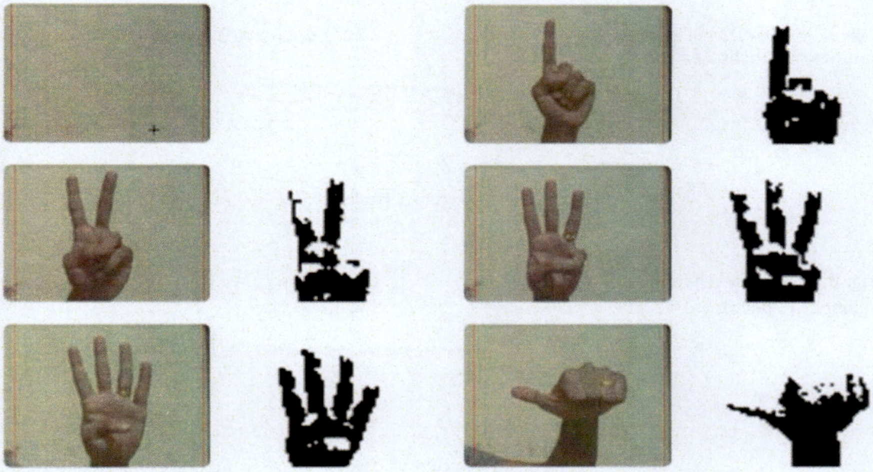

Fig. 5 Hand gesture images and region of interest (ROI)

$$\text{Accuracy} = 1/m \sum_{i=1}^{m} Xi,$$

where $Xi = 1$, if predicted label $\hat{y}(i)$ equals to the true label $y(i)$ for example i, otherwise $Xi = 0$ [10].

4.3 Performance Analysis

In this work, we have fed 80% data set to the model for training while 20% data set for validation. We have attained different accuracy for the different number of epochs.

Steps per epoch = 1000.
Validation Steps = 20% of the data set = 200.
Average Train Accuracy = 96.77%, Average Validation Accuracy = 93.52%
Average Train Accuracy = 98.21%, Average Validation Accuracy = 94.70%

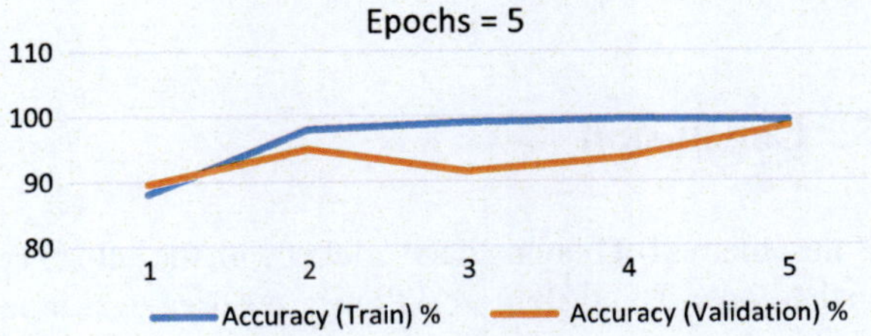

Fig. 6 Accuracy when the number of epochs = 5

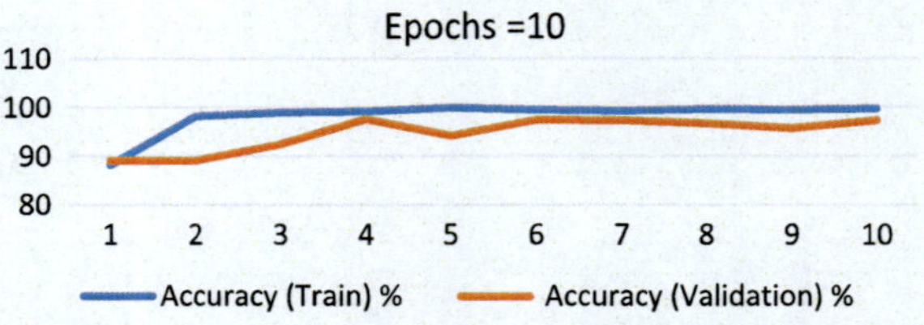

Fig. 7 Accuracy when the number of epochs = 10

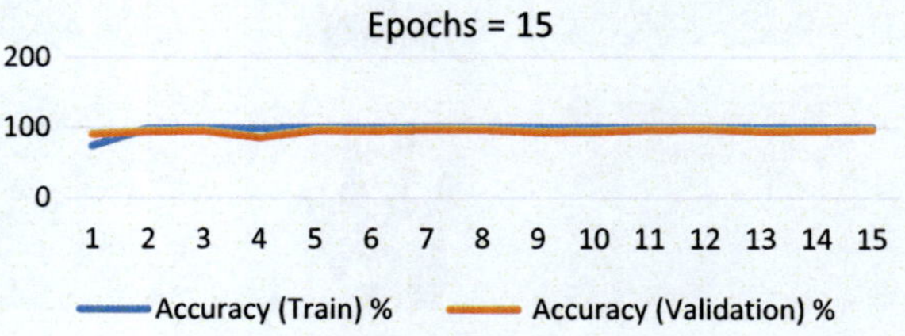

Fig. 8 Accuracy when the number of epochs = 15

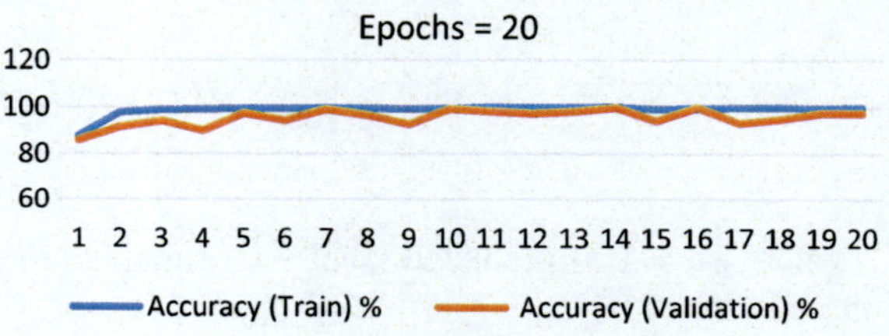

Fig. 9 Accuracy when the number of epochs = 20

Average Train Accuracy = 97.76%, Average Validation Accuracy = 95.38%

Average Train Accuracy = 98.90%, Average Validation Accuracy = 96.60%

As seen in Figs. 6 and 7, our model is under fitting for 5 and 10 epochs, respectively, because the line of validation varies greatly. In Fig. 9, we can see that for 20 epochs, our model seems to be over fitting. In Fig. 8, we can see that model with 15 epochs is performing quite well as compared to the rest of the models. The line of validation of this model perfectly fits our model.

Hence, we have used a model with 15 epochs. We have attained 97.76% of training accuracy and 95.38% of validation accuracy.

We have encountered some challenges during the implementation of the system such as skin tone, lighting condition, etc. However, most of the times the prediction is done correctly.

5 Conclusion

A new method of playing hand cricket with the help of computers has been introduced in this paper. The system workflow is: the user gives input by making the hand gesture, the hand region turns out much darker than the background, then the system scans the gesture using the camera, deduces it and passes it into the programme. The proposed system achieves 97.76% of training accuracy and 95.38% of validation accuracy as seen in the result. With the help of better images for data and a much more robust algorithm, it can be made more efficient. The system proves the effectiveness of CNN image processing. This work proves that machine learning (ML) and deep learning (DL) can not only make human lives easier but also bring some joy and entertainment.

References

1. Moher D, Liberati A, Tetzlaff J, Altman DG (2009) The PRISMA group preferred reporting items for systematic reviews and meta-analyses: the PRISMA statement. PLoS Med 6(7):e1000097. https://doi.org/10.1371/journal.pmed.1000097
2. Nguyen HT, (2008) Adam K (2004) In: Gesture: visible action as utterance. UK: Cambridge University Press. pp 400. $32.32, Hardcover, ISBN 0–521–83525–9, Critical Inquiry in Lang Stud 5(1):72–77. https://doi.org/10.1080/15427580701340790
3. Prakash J, Gautam UK (2019) Hand gesture recognition. Int J Recent Technol Eng (IJRTE) 7(6C) ISSN: 2277–3878
4. Pinto RF, Borges CDB, Almeida AMA, Paula IC (2019) 10 October 2019 by Hindawi Limited in Journal of Electrical and Computer Engineering. J Electri Comput Eng 2019:1–12. https://doi.org/10.1155/2019/4167890
5. Oudah M, Al-Naji AA, Chahl J (2020) Hand gesture recognition based on computer vision: a review of techniques. J Imaging 6:73. https://doi.org/10.3390/jimaging6080073
6. Khan RZ, Ibraheem N (2012) Hand gesture recognition: a literature review. Int J Artif Intell Appl (IJAIA) 3:161–174. https://doi.org/10.5121/ijaia.2012.3412
7. Chen Z-H, Kim J-T, Liang J, Zhang J, Yuan Y-B (2014) Real-time hand gesture recognition using finger segmentation. Sci World J. 267872:9. https://doi.org/10.1155/2014/267872
8. Kaur P, Ganguly P, Verma S, Bansal N (2018) Bridging the communication gap: with real time sign language translation. In: 2018 IEEE/ACIS 17th international conference on computer and information science (ICIS), Singapore, Singapore, pp 485–490. https://doi.org/10.1109/ICIS.2018.8466546
9. Sushma L, Lakshmi KP (2020) An analysis of convolution neural network for image classification using different models. Int J Eng Res Technol (IJERT) 09(10)
10. Rahul B, Suyash G, Nilima K (2020) Diabetic retinopathy stage classification (May 2, 2020). In: 2nd International conference on communication and information processing (ICCIP) 2020, Available at SSRN https://ssrn.com/abstract=3645460 or https://doi.org/10.2139/ssrn.3645460

CNN Based Facial Expression Recognition System Using Deep Learning Approach

Hitesh Kumar Sharma, Tanupriya Choudhury, Adarsh Kandwal, Anurag Mor, Preeti Sharma, Md. Ezaz Ahmed, and Prashant Ahlawat

Abstract In this fast-growing world of 21st century Machine Learning (ML) is one of the most powerful tools for solving a variety of real-life problems. Machine Learning helps both normal as well as differently-abled people. Facial Expression Recognition (FER) system can be widely used in different variety of research areas, such as diagnosis of mental disease and human physiological interaction detection. Machine Learning can automate processes by learning from different datasets. ML is commonly used for tasks like prediction, detection, regression and classification. The core idea of ML is to learn from previous and produce results accordingly. This result may be the output of new input or predictions based on past actions. With this, Deep Learning a subset of ML made possible to learn human facial expression from an image dataset and identify the facial expression performed in real life. The advancement in facial recognition systems helps in the purpose of diagnosis and also market feedback. But, due to various complexities and noise factors in facial expression images, there is still a need for improvement in this field. With different technological advancements in hardware and software, Facial Expression Recognition systems have been developed to help and support real-world applications.

Keywords Facial Expression Recognition (FER) · Convolutional Neural Network (CNN) · Machine Learning · Deep Learning

H. K. Sharma (✉) · T. Choudhury · A. Kandwal (✉) · A. Mor · P. Sharma
School of Computer Science, University of Petroleum and Energy Studies (UPES), Energy Acres, Bidholi, Dehradun, Uttarakhand 232007, India

Md. E. Ahmed
CS Department, Saudi Electronic University, Al Madina, Saudi Arabia
e-mail: m.ezaz@seu.edu.sa

P. Ahlawat
Manipal University, Jaipur, India

J. M. R. S. Tavares et al. (eds.), *Cyber Intelligence and Information Retrieval*, Lecture Notes in Networks and Systems 291,
https://doi.org/10.1007/978-981-16-4284-5_34

1 Introduction

In today's world, most of our time is spent in interaction with digital devices like computers and mobile. Maximum devices to which we interacted most of the time in a day are camera-enabled devices and these are very closely attached to our facial expression. The applications (software) that we use to interact with these systems are static in nature and cannot respond to us based upon our verbal and facial expressions. It will be new advancement in these systems and applications if they can interact with us by understanding our facial expressions and act accordingly. It will improve fast communication and also improve human attachment with robots. There are many ways to express human emotions like hand gestures, facial expressions, etc. However, it has been proved that 55% accuracy is provided by facial expressions. Facial expression is more accurate than any other human expression representation way. In many research Facial Expression is considered a more accurate way of the actual representation of human emotions. Facial Expressions are one of the most important aspects of human communication. The face is responsible for communicating emotions. Facial Expression Recognition system is a challenging field in HCI- Human Computer Interaction and computer vision. This system enables many sectors to make systems for the hospitals to read the mental state of the patient by reading the facial expression, this also helps the event planners to get their visitors feedbacks through the installation of such models, it will also help marketers in getting their customer's feedback. But, recognizing facial expression is complex due to diversity in shape, size, conditions of light and background noises. This model can be useful in mental disease diagnosis, security, getting market feedback and many other areas. This might be very useful for differently-abled people as they can share their emotions with their expressions. With help of thoughts from these people, companies can work on improving their services towards these people. Robots can easily interact with humans in a more friendly manner when they can easily analyze the facial expressions and get their emotional state. In this way, they can be used in the health care system to detect human mental state. An efficient FER can also be used in computer graphics in modelling human faces with accurate measured facial expressions.

2 Literature Review

HCI (Human Computer Interaction) is an advanced domain of computer science. One of the sub-domain of this domain is to identify human emotions through their facial expressions. Human Facial Expression Recognition System (HFERS) plays an important role to fulfil this goal. This expression recognition system will improve efficiency to recognize feeling and emotions of a Human and helps to improve Human Computer Interaction (HCI). This a very attractive and a high potential field of research in Artificial Intelligence, Machine Learning and Computer Vision [1]. A facial expression of a human, generated due to a change in emotional expression in act of some activities performed or by some internal feelings. Identification of human facial expressions is to recognize these expressions using AI-enabled expert system in all circumstances. Researchers in this field identified the potential of research in this field and proposed some modern expert systems used to solve these problems [2]. Facial expression is considered one of the most important features of the human body to identify his/her actual feelings and activities [3 ,4]. Li an Jain, in their work that change in expressions shows human interaction, state of mind and intention for self and society. An expert system used for automatic recognition of facial expression has lots of applications in this era. Animated Movies, Real-time advertisement, Health-care, Gaming, Entertainment, sociable robotics and many other areas are the core potential sectors for these expert systems [5]. Interaction between Human and Robot (HRI) is an emerging technology used for developing an interactional environment for a computational and expert system to understand emotional gestures and interact through some channels for communication between robot and human. It is used for responding to the user based upon his/her emotional gesture [6]. Human creates many gestures on his/her face using eye, nose, mouth hands, etc. [7,8]. There are many methods already discovered to identify these gestures and act accordingly. PCA (Principal Component Analysis) [9] and LDA (Linear Discriminant Analysis) [10,11,12] are some high accuracy methods that have been used in many expert systems for facial gesture recognition systems. A psychologist named Ekman, who worked on facial gesture recognition, said that facial expression on a human face disappears in a fraction of a second [13]. He proposed that facial expression recognition should be fast enough to differentiate expression in both short and long duration [14, 15]. Facial gesture recognition has been successfully identified using 2D images in Convolution Neural Network (CNN) [15]. This method has been successful in various fields such as object recognition, EEG signal classification, character recognition, document recognition, handwritten digit recognition and facial expression recognition.

3 Proposed CNN Model for Facial Expression Recognition

In this section we have provided the basic architecture of CNN, details of the dataset used and the architecture of the CNN model designed.

3.1 Convolutional Neural Network (CNN)

A Convolutional Neural Network (CNN, also called as ConvNet) is a class of deep neural networks [16,17], which is most commonly applied to analyze visual imagery. CNN can be applied in medical image analysis, image and video recognition, image classification, recommender systems, natural language processing and financial time series. A Convolutional Neural Network based Facial Recognition system is a Deep

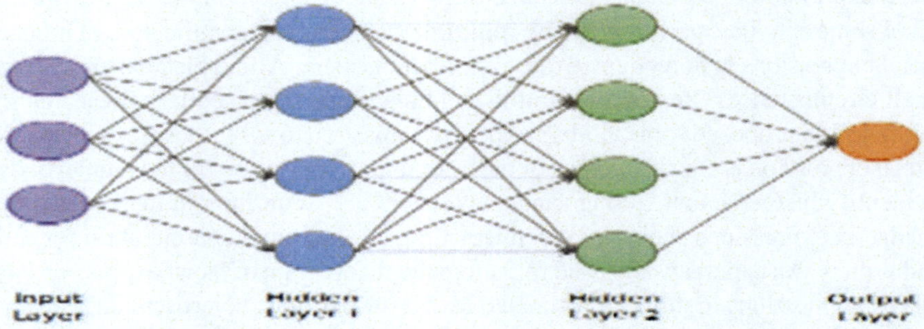

Fig. 1 Example of convolutional neural network

Learning algorithm that can take in an input image, assign learnable weights and biases to various aspects in the image and be able to differentiate one facial expression from the other [7, 18]. The CNN is much faster and data pre-processing is least required as compared to other classification ML algorithms. The main role of the CNN is to reduce the images into a form that is easier to process and without losing features that are important for getting a good prediction.

A simple and very basic example of a multi-layer CNN model is shown in Fig. 1. It consists of an input layer, an output layer and two hidden layers. Hidden layers are used for the extraction and learning features of the dataset. There can be n number of hidden layers in a CNN model. As we go deep in the number of hidden layers more features are extracted by the model. CNN models are known as the best models for image classification. CNN applies various weights and biases on raw pixels of the image and learns different features and after learning these models can classify images [7, 19]. The layers that are designed in a CNN model is categories into three main categories:

- Convolutional Layer
- Max Pool Layer or Pooling Layer
- Fully connected Layer or Dense Layer

These layers of a CNN has shown in Fig. 2. Convolution layers apply a specific filter also known as kernels to an image. In this layer of CNN, a set of different

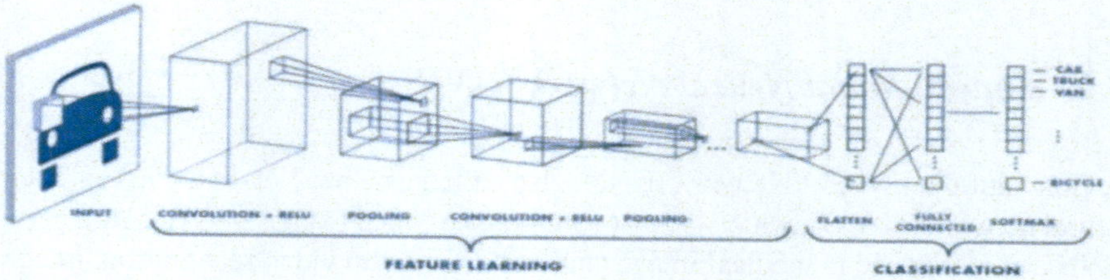

Fig. 2 Hidden layers in deep learning-based CNN model

mathematical operations are performed to produce a feature map of a single value. ReLU is the most popular activation function used in this layer. Pooling layer is used for the down-sample of the extracted image from the Convolution layer. It is used to reduce dimensions in order to minimize the processing time. Max pooling is mainly used in this layer. Dense or Fully connected layer is finally used for classification. The N-Dimension array is flattened to a 1-D array and N Number of outputs are produce based on the number of classifications required. It is a mathematical operation that takes two inputs, which are image matrix and a filter or kernel [20]. In Fig. 3 we have shown the mathematical operations.

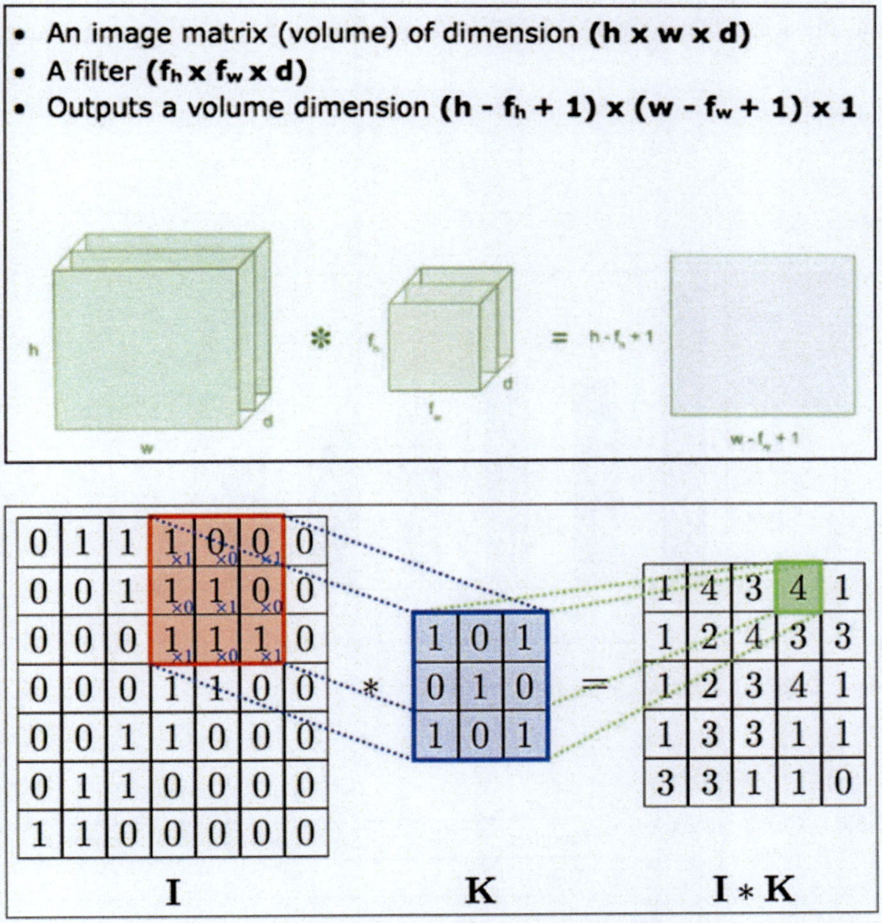

Fig. 3 Mathematical operation in a CNN model

3.2 Facial Expression Recognition Dataset Specifications

For this work, we have used the Facial expression Dataset given by Pierre-Luc Carrier and Aaron Courville in their paper. This dataset is available on Kaggle [21]. It consists of 35,887 images with different facial expressions of many peoples. The dataset consists of 7 labels that are 7 basic emotions that a human can express. In Fig. 4 the distribution of labels in the dataset has shown (Table 1).

Due to limiting computation power, we have taken 7 basic emotions that a human can express. We have taken samples of random people for different expressions. We have trained our model on 28,709 datasets and tested it on 7178. After 15 epochs the training accuracy was found to be 67.5% and for the validation set, it was found to be 48.3%.

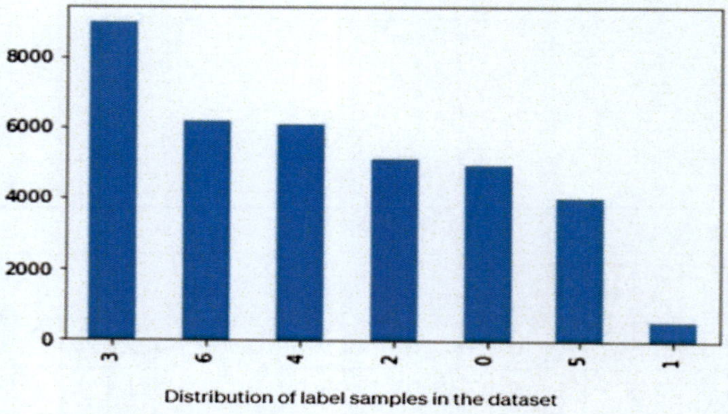

Fig. 4 Distribution of labels in the dataset

Table 1 Classification used for every emotion

Emotion	Label used
Angry	00
Disgust	01
Fear	02
Happy	03
Sad	04
Surprise	05
Neutral	06

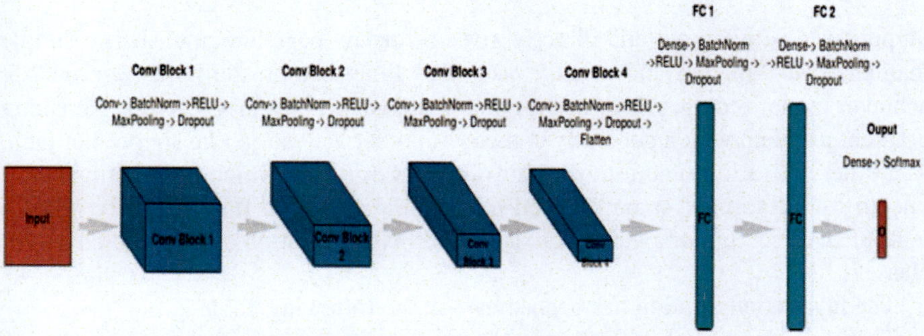

Fig. 5 Proposed CNN model for facial expression recognition

3.3 *Proposed CNN Model Architecture*

The flow diagram and layered configuration architecture is shown in Fig. 5 and Table 2. There is a dropout of 25% after every convolution layer so that there is no overfitting in the training phase.

As shown in Fig. 5, our dataset already contains images in grayscale and are pre-processed before feeding them into our model for the training phase. The dimension of each image is compressed to 32*32 for fast processing. Then the pre-processed image is passed to the neural network for training and feature learning and classification of images in 7 different classes as per the data provided. The model accuracy is checked using training data and was found to be 67.5%. Our convolutional neural network is a sequential model which is basically a linear stack of layers. It contains many layers each with a different function and mechanism. The very first layer which is close to the input layer is the conv2d layer, it helps to bring out unique features of images, by repeatedly using conv2d layer over an input we get a map of activation known as a feature map which indicates the position and strength of a unique feature in an input image. The conv2d in begging have less grasp over a feature than the ones in the last of the model.

In CNN, Filters detect unique features such as edges in an image by detecting various intensities of pixels of the image. The number of filters used in CNN is basically in the power of two which commonly involves 32, 48,128,256,64 and 1024. We also have selected the size of the filter, N*N-dimension of filters are used over an input where N is an odd number. The most common filter size used are 3*3 and 2*2, high dimension filters such as 8*8 are very rarely used.

Another layer used is the Max-Pooling layer which is used to extract the prominent features from the previous features map. It basically involves applying a 2d filter over each channel and selecting the maximum element from the region or the feature map. Max pooling is used to reduce the dimension of an image so that the performance of the model can be increased and processing time can be decreased.

And, the activation function we have used in the Conv2d layer is ReLu (Rectified Linear activation function), basically it is a piecewise linear function that only gives output for positive input and directly gives 0 for any negative input. It is a simple mathematical function and usually takes less time to train the model and also it achieves higher accuracy. This activation function also helps to resolve the vanishing gradient problem which generally comes with other activation like sigmoid or tanh. The other activation function used for output is SoftMax which is sometimes also known as softargmax or normalized exponential function. It is basically used to normalize the output of a network to a probability distribution over predicted output classes.

The layer configuration has been shown in the following Table 2.

There is a dropout of 25% after every max-pooling layer so that there is no overfitting in our model. We have also used successive Conv2d for feature extraction and to achieve better accuracy. Final output layer contains 7 classes as per our image categorization.

Table 2 Layer configuration of proposed CNN

Basic CNN model configuration	
Layers	Layer shape details
optimizer	adam
loss	sparse_categorical_crossentropy
metrics	['accuracy']
Conv2D Layer	48 filters, 3 × 3 Filter size and ReLU Activation
Max-Pooling2d Layer	2 × 2 kernel size
Dropout	25%
Conv2D Layer	128 filters, 5 × 5 Filter Size and ReLU Activation
Max-Pooling2d Layer	2 × 2 kernel
Dropout	25%
Conv2D Layer	64 filters,3 × 3 Filter size and ReLu activation
Max-Pooling2d Layer	2 × 2 kernel size
Dropout	25%
Flatten layer	4608 Neurons
Dense layer	256 Neurons
Batch normalization	Relu activation
Dropout	25%
Dense layer	64 Neurons
Batch normalization	Relu activation
Dropout	25%
Output layer	Softmax 7 classes

4 Implementation of Proposed CNN Model

In this section, we have defined the implementation of our proposed model. The dataset used for learning hand gestures is taken from Kaggle [21]. The image dataset is captured using a Leap motion sensor. These are a set of infrared images. The specifications of the image dataset are given in Table 3.

7178 sample images are used for training our model due to resource limitations. Some randomly selected images are shown below in Fig. 6.

Table 3 Facial expression recognition dataset specification

Total images in dataset	35,887
Total hand gesture types recorded	7 (given in Table 1)
Total classes	7
Raw image dimension	32*32
Total images used for our model	35,887
No. of images used for training	28,709
No. of images used for testing	7178

Fig. 6 Randomly selected image for training

The summary of the model trained on the sampled images is given here. The layers configuration details are explained in Table 4.

The proposed model contains a total of 4, 474, 75 parameters and using these our model has been trained.

Table 4 The layers configuration details

Model Layer	Shape of Output	Parameters Nos
Conv2D (Convolution Layer)_1	(No Param, 32, 32, 48)	480
Batch_Normalization Layer_1	(No Param, 32, 32, 48)	256
Activation Layer_1	(No Param, 32, 32, 48)	0
Max-Pooling layer (Max_Pool2D)_1	(No Param, 24, 24, 48)	0
Drooping Out Layer_1	(No Param, 24, 24, 48)	0
conv2d_1 (Conv2D)_2	(No Param, 24, 24, 128)	204,928
Batch_Normalization Layer_2	(No Param, 24, 24, 128)	64
Activation Layer_2	(No Param, 24, 24, 128)	0
Max-Pooling layer (Max_Pool2D)_2	(No Param, 12, 12, 128)	0
Drooping Out Layer_2	(No Param, 12, 12, 128)	0
Conv2D (Convolution Layer)_2	(No Param, 12, 12, 64)	590,336
Batch_Normalization Layer_3	(No Param, 12, 12, 64)	2032
Activation Layer_3	(No Param, 12, 12, 64)	0
Max-Pooling layer (Max_Pool2D)_3	(No Param, 6, 6, 64)	0
Drooping Out Layer_3	(No Param, 6, 6, 64)	0
Conv2D (Convolution Layer)_2	(No Param, 6, 6, 64)	2,359,808
Batch_Normalization Layer_3	No Param, 6, 6, 64)	2032
Activation Layer_4	(No Param, 6, 6, 64)	0
Max-Pooling layer (Max_Pool2D)_4	(No Param, 3, 3, 64)	0
Drooping Out Layer_4	(No Param, 3, 3, 64)	0
Flatten Layer	(No Param, 4608)	0
Dense Layer_1	(No Param, 256)	1,179,904
Batch_Normalization Layer_5	(No Param, 256)	1024
Activation Layer_5	(No Param, 256)	0
Drooping Out Layer_5	(No Param, 256)	0
Dense Layer_2	(No Param, 64)	131,584
Batch_Normalization Layer_6	(No Param, 64)	2032
Activation Layer_6	No Param, 64)	0
Drooping Out Layer_6	(No Param, 64)	0
Dense Layer_3	(No Param, 7)	3591

5 Experimental Results

Here the results of our proposed model are presented. The following statistical parameters are evaluated for testing the performance of our model.

Precision: Precision provides us the information about the proportion of positive identifications that was actually correct.

$$\mathbf{Precision} = \mathbf{TP}/(\mathbf{TP} + \mathbf{FP})$$

Recall: Recall provides us the information about the proportion of actual positives that were identified correctly.

$$\mathbf{Recall} = \mathbf{TP}/(\mathbf{TP} + \mathbf{FN})$$

F1- Score: The F1 score is interpreted as a weighted average of the precision and recall, where F1 score of value 1 is best and a score of value 0 is worst.

$$\mathbf{F1} = \mathbf{2}^{*}\left[\left(\mathbf{Recall}^{*}\mathbf{Precision}\right)/(\mathbf{Recall} + \mathbf{Precision})\right]$$

Accuracy: Accuracy is a metric for evaluating classification models. Accuracy gives the information about the fraction of predictions that our model got right. In percentage;

$$\mathbf{Accuracy} = \mathbf{100}^{*}[(\mathbf{TP} + \mathbf{TN})/(\mathbf{TP} + \mathbf{TN} + \mathbf{FP} + \mathbf{FN})]$$

Here, TP = True Positive, TN = True Negative, FP = False Positive and FN = False Negative. TP,TN,FP and FN are decided based on the following matrix.

The confusion matrix for our proposed model is given in Table 5. Vertical columns represent predicted results and horizontal rows represent actual images. The heat map of the confusion matrix is given in Fig. 7.

The results of the training are shown in Fig. 8a and b. The accuracy of training and validation becomes constant after 14 epochs and loss of validation is nearly equal to 1.

To test the output from our designed model we gave two videos as input to our model and the model determined the facial expression of the faces on the videos. The results (screenshots of respective videos) are shown in Fig. 9. The output is displayed in a browser with help of FLASK library.

Table 5 The confusion matrix for our proposed model

	Actual positive (1)	Actual negative (0)
Predicted positive (1)	True positive (TP)	True negative (TN)
Predicted negative (0)	False negative (FN)	False positive (FP)

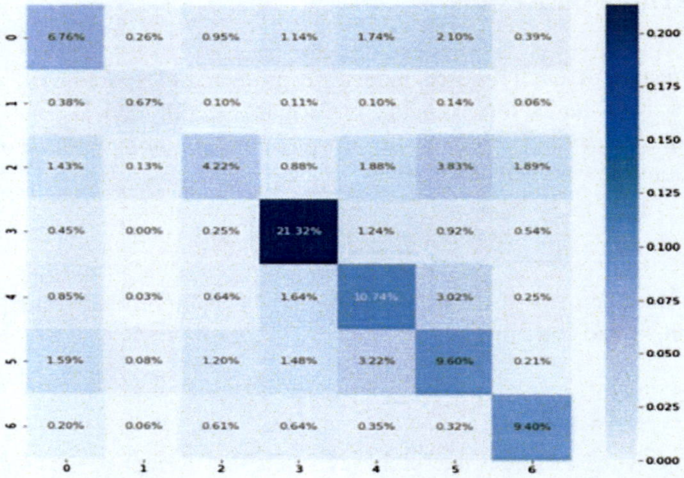

Fig. 7 Heat map of the confusion matrix for our proposed model

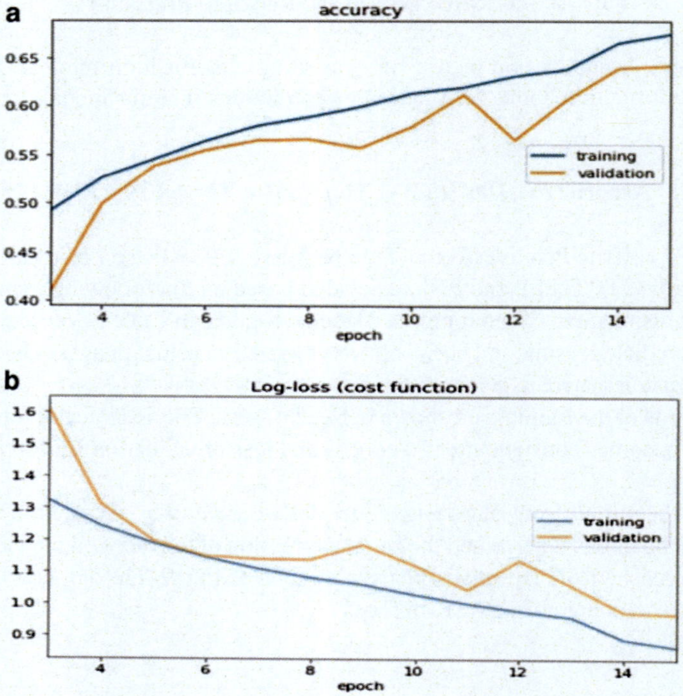

Fig. 8 **a** and **b** Accuracy and cost function graph

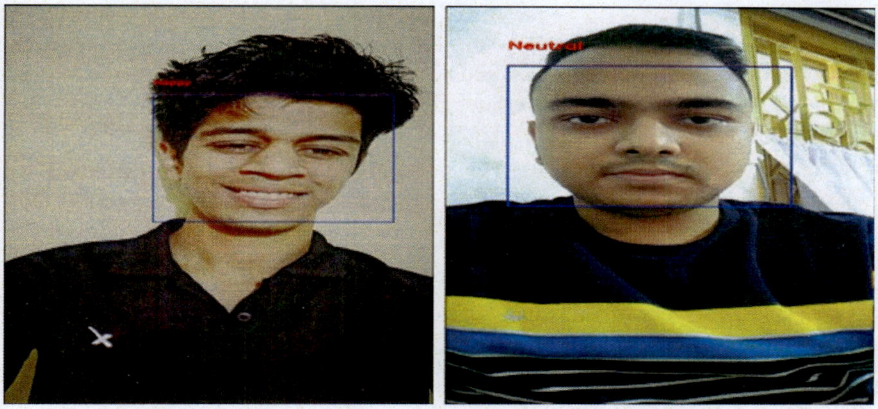

Fig. 9 Test result of our proposed model

6 Application Areas

Facial Expression Recognition system has many application areas. This can be used in health sectors in psychology and neurosciences, security, getting customer feedback, designing of 3-D models in computer graphics (shown in Fig. 10). Applications are not only limited to these fields, but research works are going on in this field to get better advancement.

Some of the applications of Facial Expression Recognition System are:

- Health Sectors: Facial Expression recognition system plays an important role in treating mental disease. With the help of this system, a human's state of mind and emotions can be read by reading his/her facial expression. This has helped a lot in the treatment of mental diseases.
- Computer Graphics: This FER system has helped a lot in designing models with exact facial expressions like humans. With accurate measurement and geometry using the FER system, a human's facial expression can now be seen in virtual models of humans.

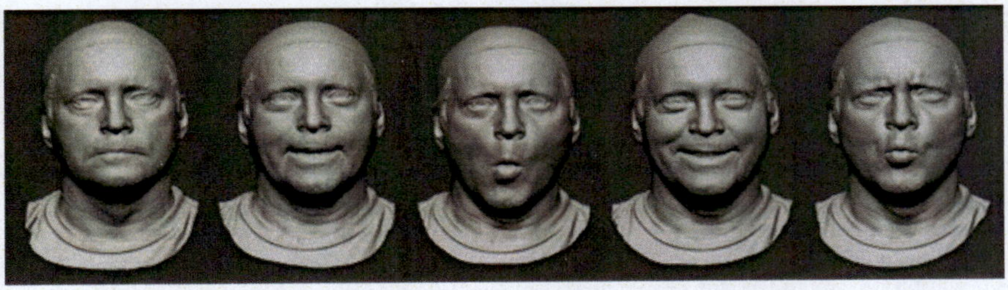

Fig. 10 3D models designed with facial expressions

- Receiving customer feedback: Using the FER system the marketers and event managers can get the feedback of customers or guests by reading facial expressions. This feedback data can be later used to improve the services that are being provided.
- Robotics: FER system can be used by robots in better understanding the emotional state of the person with whom the robot is interacting, this will make the behaviour of the robot more friendly.

There are many more application areas where the Facial Expression Recognition system proved its success and help to eliminates various problems. This will be used in many more areas as and when better models are developed.

7 Conclusion

Facial Expressions are a way of communication for humans. This is also one of the ways of non-verbal communication. This can be used by both normal and differently-abled people for expressing their views and thoughts. This system has the potential in minimizing the communication barrier between computers and humans by converting the facial expression data into computer understandable form. By achieving higher accuracy this can be used in many sectors in solving different kinds of problems. In our designed model the training accuracy achieved is 67.5% and validation accuracy achieved is 48.3%. The confusion matrix shown is representing that the designed model classified all tested images to their respected labels.

Future work in this field will lead to greater accuracy and better features. More different types of datasets will be used for better accuracy and recognition. This will be helpful in getting better results also in noisy images. Due to some limitations, we have used less number of images but in future, we will implement this on high computing larger dataset.

References

1. Mantecón T, del Blanco CR, Jaureguizar F, García N (2016) Hand gesture recognition using infrared imagery provided by leap motion controller. In: International conference on advanced concepts for intelligent vision systems, ACIVS 2016, Lecce, Italy, 24–27 October 2016, pp 47–57. https://doi.org/10.1007/978-3-319-32680-2_5
2. Khanchi I, Ahmed E, Sharma HK (2019) Automated framework for real-time sentiment analysis. In: International conference on next generation computing technologies (NGCT-2019)
3. Patni JC, Sharma HK (2019) Air quality prediction using artificial neural networks. In: International conference on automation computational and technology management (ICACTM)
4. Sharma HK, Singh T, Kshitiz K, Singh H, Kukreja P (2017) Detecting hate speech and insults on social commentary using nlp and machine learning. Int J Eng Technol Sci Res 4(12):279–285
5. Pardeshi V, Sagar S, Murmurwar S, Hage P (2017) Health monitoring systems using IoT and Raspberry Pi—a review. In: 2017 International conference on innovative mechanisms for industry applications (ICIMIA), Bangalore, pp 134–137

6. Mathan Kumar K, Venkatesan RS (2014) A design approach to smart health monitoring using android mobile devices. In: IEEE international conference on advanced communication control and computing technologies (ICACCCT), pp 1740–1744
7. Alani AA et al (2018) Hand gesture recognition using an adapted convolutional neural network with data augmentation. In: 24th IEEE international conference on information management
8. Mukhopadhyay SC (2015) Wearable sensors for human activity monitoring: a review. IEEE Sensors J 15(3):1321–1330
9. Djuknic GM, Richton RE (2001) Geolocation and assisted GPS. Computer 34.2:123–125
10. Sanders G, Thorens L, Reisky M, Rulik O, Deylitz S (2003) GPRS networks. Wiley, Hoboken, NJ
11. Yasen M, Jusoh S (2019) A systematic review on hand gesture recognition techniques, challenges and applications. Peer J Comput Sci
12. Ahlawat P, Biswas SS (2020) Sensors based smart healthcare framework using internet of things (IoT). Int J Sci Technol Res 9(2):1228–1234
13. Taneja S, Ahmed E, Patni JC (2019) I-Doctor: an IoT based self patient's health monitoring system. In: 2019 International conference on innovative sustainable computational technologies, CISCT 2019
14. Zengeler N et al (2019) Hand gesture recognition in automotive human-machine interaction using depth cameras. Sensors 19(1):59. https://doi.org/10.3390/s19010059
15. Szegedy C, Liu W, Jia Y, Sermanet P, Reed S, Anguelov D, Erhan D, Vanhoucke V, Rabinovich A (2015) Going deeper with convolutions. In: 2015 IEEE conference on computer vision and pattern recognition (CVPR)
16. Verma A, Shukla A, Choudhury T, Chauhan A (2019) Recommendation system for prediction of tumour in cells using machine learning approach. In: international conference on recent developments in science, engineering and technology, pp 212–222
17. Singh UV, Gupta E, Choudhury T (2019) Detection of rheumatoid arthritis using machine learning. Int Conf Comput Intell Knowled Econ (ICCIKE) 2019:25–29
18. Nasr-Esfahani1 E et al. (2016) Hand gesture recognition for contactless device control in operating rooms. http://arxiv.org/abs/1611.04138
19. Khanna A, Sah A, Choudhury T (2020) Intelligent mobile edge computing: a deep learning based approach. In: International conference on advances in computing and data sciences, pp 107–116
20. Kumar P, Choudhury T, Rawat S, Jayaraman S (2016) Analysis of various machine learning algorithms for enhanced opinion mining using twitter data streams. Int Conf Micro-Electron Telecommun Eng (ICMETE) 2016:265–270
21. Molchanov P, Gupta S, Kim K, Kautz J (2015) Hand gesture recognition with 3D convolutional neural networks. In: 2015 IEEE conference on computer vision and pattern recognition workshops (CVPRW)

CNN-Based Handwritten Mathematical Symbol Recognition Model

Sakshi, Chetan Sharma, and Vinay Kukreja

Abstract Mathematical expression or formula recognition system is evolving field in pattern recognition. Various methods are implemented by different researchers on wide areas. Handwritten mathematical expression field evolved in last decades rapidly. This paper aims to target and represent CNN as an application of the deep learning method for recognizing the handwritten mathematical symbols. The current trend of the deep learning architectures witnesses CNN is one of the most prominent and widely used techniques that have been successfully implemented in NLP, computer vision, and pattern recognition. The kinds of deep learning models have proven to produce a recognizable state of the art. In this paper, we recognize the handwritten mathematical symbols using CNN. This paper comprises stages involved in recognition process, challenges and proposed methodology used by the author to conduct their experiment and further result and future scope is discussed. Dataset used for this experiment is downloaded from the public available platform (HasyV2 Dataset 2018. A dataset of size 369 classes comprising 168,223 images has been used for experimentation, and the results relieve the accuracy of 76.17%. For further improvement, an extra dense net (fully connected layer) has been used. The DuosdenseNet model has been designed; the accuracy of recognizable results shows a decent work for consideration.

Keywords Convolution neural network · Handwritten mathematical symbol recognition · Math symbols · Handwriting recognition

---

Sakshi · V. Kukreja
Chitkara University Institute of Engineering and Technology, Chitkara University, Punjab, India
e-mail: sakshi@chitkara.edu.in

V. Kukreja
e-mail: vinay.kukreja@chitkara.edu.in

C. Sharma (✉)
Chitkara University, Solan , Himachal Pradesh, India
e-mail: chetan.sharma@chitkarauniversity.edu.in

J. M. R. S. Tavares et al. (eds.), *Cyber Intelligence and Information Retrieval*,
Lecture Notes in Networks and Systems 291,
https://doi.org/10.1007/978-981-16-4284-5_35

1 Introduction

Mathematical formula or expression recognition systems have symbol recognition as the core phase and essential component of the entire recognition process. Mathematical expressions are the essential part of almost every scientific disciplines, and recognition of these is vital problem among the researchers. The search and retrieval operations on the digital mathematical text are more challenging than other handwritten script data. The emerging trends of pen-based and other smart devices have necessitated and captivated the interest in recognizing the mathematical text from the handwritten source. The existing systems are still miles away from an ideal recognition rate, so are the emerging practice in recognition procedures are influenced by machine learning and deep learning architectures [2].

2 Stages in the Recognition

On the broader scale, the math symbol recognition is a phase and step in the recognition of mathematical expressions, which is broadly posed to be performed in two stages: symbol recognition and structural analysis [3]. This symbol recognition is the basis of the structural analysis [4]. It comprises two stages in the recognition process: symbol segmentation and isolated symbol recognition. Various techniques and methods can independently drive the recognition strategy for the completion of these two sub stages. The literature witnesses several stances where each phase is performed using grammar-driven parsing methods, fuzzy logic, and other machine learning and deep learning models.

3 Challenges

The apparent reasons for the main challenges in this domain are two significant characteristics of math symbols:

(a) The 2-dimensional nature of the mathematical texts is one of the factors responsible for ambiguities and complexities incurred in the process of recognition.
(b) The existing similarities in shape and style of many symbols make them look alike and challenging to recognize.
(c) The wide variety of symbols and variations in the spatial relationships demands the recognition system's exclusive training.
(d) Twisting semantics of the mathematical signs and operator (part of symbols) are one of the leading difficulties in the task of recognition [5].

4 Literature Review and Related Work

CNN has been prominently used for the task of recognition since long. Convolutional neural networks (CNNs) can be inferred as the hierarchical representation of input data that facilitates categorization. The authors [6] proposed their work on speech recognition to understand the affect-salient features for speech emotion recognition (SER) with the help of semi-CNN. The author used two stages to train the semi-CNN. Firstly, the author used the unlabeled data to learn sample features by contacting CNN with reconstruction penalization. The extracted features are used further as input to the semi-CNN to make machine-understandable affect silent, discriminative features with the help of novel objective function. The different researchers use CNN to present complex image features, and it has many applications in various fields like iris, face, and much more recognition. CNN shows its excellent performance in ImageNet, pattern recognition, and other computer vision tasks [7] explored the performance of advanced pre-trained CNNs on iris recognition. The author considers two datasets on the iris, which is ND-cross sensor 2013 and CASIA-Iris, and used off the shelf CNN feature to train generic objects which are exceptionally well in representing the iris images. This technique effectively extracts discriminative visual features from iris images and achieves good results in iris recognition.

The author [8] used handwritten quadratic equations to conduct their study and used CNN to classify the specific character. The recognition of the handwritten quadratic equation was accomplished using a group of a handwritten quadratic equation, and a single quadratic equation that is considered to recognize and make a solution for that equation. The author analyzed quadratic equations with the help of compact horizontal projection and combined connected component analysis. CNN is used to classify the particular character, and the correctly detected character is considered the solution for a specific equation considered. The author concludes that their results through an experiment performed are well sufficient for the proposed system.

The author [9] conducted their study on handwritten mathematical digits recognition and used the dataset of 32,400 images to experiment. Dataset mostly consists of Bengali numerical digits. The author's considered dataset includes digit, algebra operator, set-symbols, calculus, and many more. MathNET is the proposed model by the author for recognizing the handwritten digits and other 44 math symbols. The author considered two types of accuracy from the proposed model: train and other are validation accuracy, 96.01, and 96.50%, respectively, which is very goofing precision in recognizing symbols.

In this study, author assessed the faster R-CNN object detection algorithm as a method of symbol detection in the field of handwritten graphics. Using the various patterns of handwritten data evaluation, author raised issues regarding the nature of the handwritten data. The experiment is done using the publically available flowcharts and CROHME 2016 dataset. According to the author's results, faster R-CNN is well performed on both the dataset and enables the possibility to develop the method to detect the symbols in handwritten graphics [10].

The author uses a combinational classifier CNN and LSTM, to conduct their study. According to the author, these both classifiers are refined neural networks and better classifiers to recognize the character in online and offline handwritten data. The author improves the accuracy of CNN and LSTM [11] by introducing the dropout and gradient-based local features technique. The author uses the Mathbrush dataset to conduct their study and conclude that the recognition rate is higher when the combination is used rather than an individual classifier.

The authors [12] conduct their study to recognize the offline handwritten mathematical expressions by mean of CNN for classification. The study starts with collecting handwritten math symbols and expressions, and further, they are preprocessed to conduct segmentation of the collected data. For training purposes, using CNN, handwritten math symbols are used, and each single-segmented character is sent to the CNN classifier to recognize in which category the input symbols lie. The author's proposed model can realize the offline math symbols, and CNN will classify the characters with the softmax function's help. All the recognized symbols and expression are further converted to Latex sequence as an output of the proposed system.

Handwritten mathematical expressions are grouped and identified as in cluster, and clustering is one of the most acceptable methods in the recognition process. The author [13] conducts their study to represent the HMEs. This CNN-based method is used. Different characters and symbols are used in an expression, and every symbols is of variable scale which is located and classified by combining the features from a multi-scale CNN. Weakly supervised training is used in this study which is combined with symbols attention, and this combined technique improves localization and classification predictions. Second, a multilevel spatial distance between two representations for clustering HMEs was also proposed. CROHME 2016 and 2019 dataset is used to conduct the experiments and reveal favorable results 0.99 and 0.96 in purity, respectively.

5 Proposed Methodology

In this study, the first dataset is identified to conduct our experiment. Dataset is collected from [1]. The convolution neural network algorithm is chosen to conduct our experiment, as this algorithm is the most populated recognition method [14]. CNN is always used in the recognition field to achieve the highest accuracy rate of handwritten character recognition [15, 16]. Section 5.1 elaborates on the dataset considered for the experiment, and the Sect. 5 depicts the recognition model and stages involved in the considered model. The CNN model is used to train and test the math character dataset images described in Sect. 5.4. Finally, Sect. 6 presents the conclusion and future scope of this study.

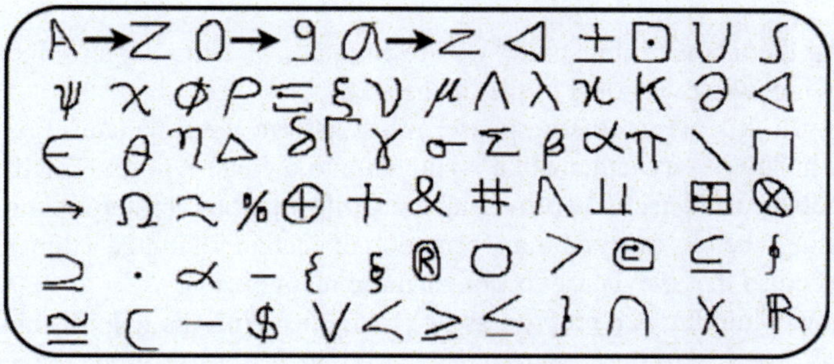

Fig. 1 Dataset character images

5.1 Dataset Depiction

Dataset is downloaded from the [1], which is publically available on kaggle.com. Dataset for handwritten math symbols is named HASYv2, a collection of handwritten alphanumeric chars that includes numbers, letters, mathematical, and scientific symbols used in the field of mathematics in our daily life. The dataset contains 369 symbols classes, and there are 1,68,233 png images in a total of these symbols classes. Each image retrieved from the dataset is 32px × 32 px. A dataset is a derivative of the HWRT dataset developed by Thoma [17]. Each image's dimension is stored in a bounding box of 32 pixels, and each image is then centered to this bounding box. Dataset information is stored in the .csv file, including the symbol, Id, symbol, and image. Symbols classes used in the dataset include Latin uppercase and lowercase letters (A-Z, a-z), numerals (0–9), 32 different arrows, fractals, brackets, etc., other mathematical characters. Some of the character used in the dataset is shown in Fig. 1.

5.2 Preprocessing

Preprocessing is done on the collected dataset. When the data is collected, several preprocessing techniques can be applied to the collected dataset to feed the data into the system for recognition purposes.

Reducing Noise: When an image is clicked or collected, it may contain some noise. To remove noise, various techniques are used by researchers like linear, nonlinear, adaptive, and median filtering [18]. Every technique has its own rules to remove or reduce the noise from the image.

Binarization: Binarization is the process of converting grayscale images into black and white images. Firstly, the threshold value is calculated using the grayscale image and checked whether the pixel has any particular gray value. This is also one of the preprocessing techniques. The adaptive threshold method is used to obtain a binary image with pixel intensity values 0 or 1. Davoula method is used by Chauhan et al.

[19] to find the adaptive threshold. This technique is used to acquire the threshold where in window size is equal to the image size.

Scaling: In this preprocessing phase, we transform the raw dataset of obtained images of handwritten mathematical symbols into a clean typeset. The dataset can be made polished and useful in providing more information for interpretation of data. Under scaling, the resizing and transformation of each input image's dimensions are accomplished so that the dataset becomes more recognizable.

Thinning: Thinning is a preprocessing technique. With the help of thinning, the thickness is diminish of the foreground accordingly. It is also applied to handwritten math characters to get the skeletal image of any HME thinning algorithm. This process is also taking care to enhance recognition speed [20].

5.3 Segmentation

Segmentation is the process used in the field of pattern recognition to smoothing the recognition process. All images collected in this field are in the form of digital form, which further divides into the number of segments (set of pixels). This is used to transform the image into some other meaningful form so that it is easy to analyze the image. Segmentation is used to detect the objects and boundaries in any image [21]. In conclusion, segmentation is the process of labeling each pixel of the image to get their characteristics. Majorly segmentation is divided into three categories structural, stochastic, and hybrid segmentation techniques [22, 23]. Tremendous information is stored in any image, but we cannot extract exact information from a particular image without segmentation. The objective of segmentation in handwritten math symbols or expression recognition is to refine the inputted ME in a way that individual characters and symbols could be passed as inputs to the further stages of the recognition process.

5.4 Classification

Classification is the process of categorized the same class characters in the same group. In this study, the CNN classifier is used on 369 different classes of handwritten mathematical symbols. Duodensnet CNN is proposed in this study which is shown in Fig. 3. Proposed model consists of convolution, max pooling, flatten, softmax, and dense layers which help in the recognition process. As it is known, the input is processed through several layers of convolutional setup. The input is first passed through the convolutional 2D layer, and then the image becomes abstracted to a feature map of size (3), with shape (107,665)) $\times$ feature map height (3) $\times$ feature map width (3), constituting the filter size of $3 \times 3 \times 3$. The constructed convolutional layer within a neural network possesses the convolution kernel which is described by height and width, and it also depends upon the number of input and output channels.

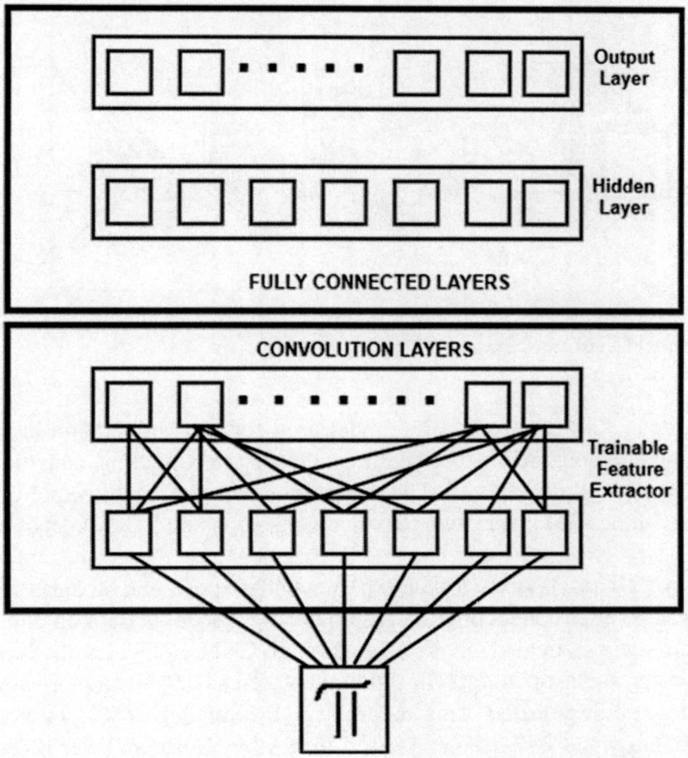

Fig. 2 CNN architecture

In CNN model, the input image is passed to the filter which is of $3 \times 3 \times 3$ of size. The basic architecture is represented through Fig. 2. Filter size of the layer is of 3×3 or 2×2. ReLu is the activation function is applied after the CNN layers, and in this function, the negative values are substituted with 0, and there is no change when positive values are considered. Output size of the image in pooling layer is 50% of the taken input image after applying the sub-sampling. Layers are used to extract the features and after that extracted features are applied to fully connected layers of the model. The input neurons are connected to every output neurons, and in softmax, activation function is passed for recognition.

5.5 *Training and Testing CNN Model*

The CSV file containing the dataset of 369 classes and images (168,233 images) was read through the CSV reader. The total dataset was spitted into two parts, and the ratio of 80:20 was considered for training and testing the model. That constitutes that the 1,07,665 images were used for training the CNN model. The shape of the

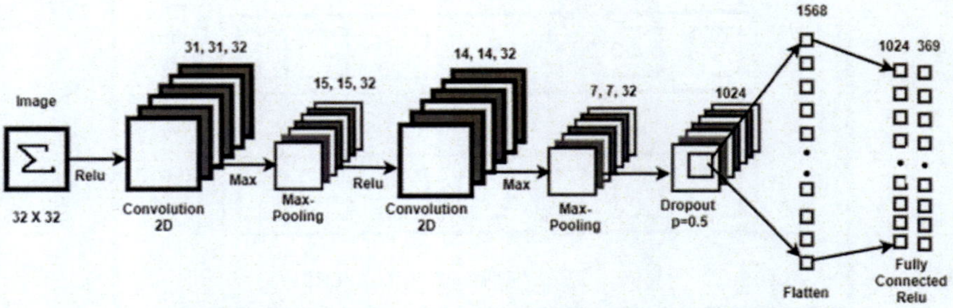

Fig. 3 Proposed DuodenseNet model

input images is 32 * 32 pixels. The model used for experimentation in this study implemented the model using four significant layers: max pooling, convolutional 2d layer, flattening, and dense layer. The recognition systems' architecture constitutes the two convolution 2D layers, two max pooling layers, and 2 Densenet as presented in Fig. 3.

The HasyV2 dataset is used for the experimental purpose, and model is train on the dataset symbols after conducting the preprocessing. Model is train on 369 different handwritten mathematical symbol classes through CNN. CNN classifier is executed with 20 epochs, and dropout rate for training is taken as 0.5. Testing or validation of the dataset symbols is tested using the trained CNN model. 80% data of dataset is used for training, and 20% of data is used for testing. Softmax layer is used in the classifier. The working principle for the deployed softmax layer is as follow

$$S(a = x|b) = \frac{e^{b^T m_c} + n_c}{\sum_j e^{b^T m_i} + n_i}$$

In above equation, S is considered as the class x to be predicted, b is used as the vector of the symbol to be recognized when fully connected layer is passed, weights are represented by m, and n is represented for bias for i various classes. The layers, output shape, and parameters in CNN trainable model are shown in Table 1.

Table 1 CNN trainable model

Layer (type)	Output shape	Param #
Convolution 2D-1	(None, 31, 31, 32)	416
Max pooling Layer-1	(None, 15, 15, 32)	0
Convolution 2D-2	(None, 14, 14, 32)	4128
Max pooling layer -2	(None, 7, 7, 32)	0
Flatten	(None, 1568)	0
Dense-1	(None, 1024)	1,606,656
Dropout	(None, 1024)	0
Dense-2	(None, 369)	378,225

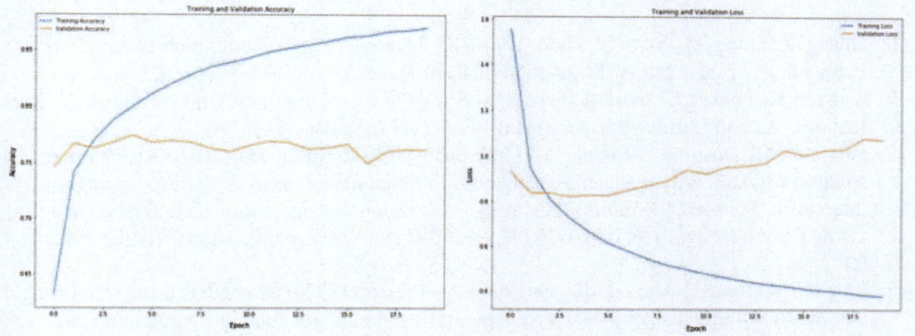

Fig. 4 Training and validation graph

The shape of the convolutional 2D layer and max pooling is (31,31,32) and (15,15,32), which is further diminished by passing through the chosen filters for extraction as shown in Fig. 2. The dropout layer with the value 0.5 is used to prevent the overfitting of the system. The epochs and accuracy values during training and validation are depicted through Fig. 4.

6 Conclusion

The researcher in this work has proposed a system that will recognize the handwritten mathematical symbol. This system created in this work DuodenseNet will be best for isolated characters. Many sub-processes are undertaken, and the developed CNN model produced an accuracy of 76.17%. In the future, merged or connected or joined symbols and many other preprocessing and hybrid segmentation techniques can be employed, potentially improving.

References

1. HasyV2 Dataset (2018)
2. Kukreja V, Dhiman P (2020) A deep neural network based disease detection scheme for Citrus fruits. In: 2020 International conference on smart electronics and communication (ICOSEC), pp 97–101
3. Chan KF, Yeung DY (2000) Mathematical expression recognition: a survey. Int J Doc Anal Recogn (iJDAR) 3(1):3–15. https://doi.org/10.1007/PL00013549
4. Hu L, Zanibbi R (2011) HMM-based recognition of online handwritten mathematical symbols using segmental K-means initialization and a modified pen-up/down feature. In: Proceedings of the international conference on document analysis and recognition, ICDAR, pp 457–462. https://doi.org/10.1109/ICDAR.2011.98
5. Shinde S, Waghulade R (2016) Handwritten mathematical expressions recognition using back propagation artificial neural network. Commun Appl Electron 4(7):1–6. https://doi.org/10.5120/cae2016652125

6. Huang Z, Dong M, Mao Q, Zhan Y (2014) Speech emotion recognition using CNN. In: Proceedings of the 22nd ACM international conference on multimedia, pp 801–804
7. Nguyen K, Fookes C, Ross A, Sridharan S (2017) Iris recognition with off-the-shelf CNN features: A deep learning perspective. IEEE Access 6:18848–18855
8. Hossain MB, Naznin F, Joarder YA, Zahidul Islam M, Uddin MJ (2018) Recognition and solution for handwritten equation using convolutional neural network. In: Proceedings Jt. 7th international conference informatics, electronics vision 2nd international conference imaging, vision pattern recognition, ICIEV-IVPR, pp 250–255. https://doi.org/10.1109/ICIEV.2018.8640991
9. Shuvo SN, Hasan F, Ahmed MU, Hossain SA, Abujar S (2020) MathNET: using CNN bangla handwritten digit, mathematical symbols, and trigonometric function recognition. In: Soft computing techniques and applications, Springer, pp 515–523
10. Julca-Aguilar FD, Hirata NST (2018) Symbol detection in online handwritten graphics using faster R-CNN. In: 2018 13th IAPR international workshop on document analysis systems (DAS), pp 151–156
11. Dai HAIN, Le Duc ANH, NAKAGAWA M (2014) Combination of LSTM and CNN for recognizing mathematical symbols. In: Proceedings of the 17th information-based induction sciences workshop
12. Drsouza L, Mascarenhas M (2018) Offline handwritten mathematical expression recognition using convolutional neural network. In: Proceedings of international conference on information, communication, engineering and technology, ICICET, pp 1–3. https://doi.org/10.1109/ICICET.2018.8533789
13. Nguyen CT, Khuong VTM, Nguyen HT, Nakagawa M (2020) CNN based spatial classification features for clustering offline handwritten mathematical expressions. Pattern Recognit Lett 131:113–120. https://doi.org/10.1016/j.patrec.2019.12.015
14. Kukreja V, Kumar D, Kaur A et al. (2020) GAN-based synthetic data augmentation for increased CNN performance in vehicle number plate recognition. In: 2020 4th international conference on electronics, communication and aerospace technology (ICECA), pp 1190–1195
15. Ciresan DC, Meier U, Gambardella LM, Schmidhuber J (2011) Convolutional neural network committees for handwritten character classification. In: 2011 International conference on document analysis and recognition, pp 1135–1139
16. Ahranjany SS, Razzazi F, Ghassemian MH (2010) A very high accuracy handwritten character recognition system for Farsi/Arabic digits using convolutional neural networks. In: 2010 IEEE fifth international conference on bio-inspired computing: theories and applications (BIC-TA), pp 1585–1592
17. Thoma M (2017) The hasyv2 dataset. arXiv Prepr. arXiv1701.08380
18. Hambal AM, Pei Z, Ishabailu FL (2017) Image noise reduction and filtering techniques. Int J Sci Res 6(3):2033–2038
19. Chauhan S, Sharma E, Doegar A et al. (2016) Binarization techniques for degraded document images—a review. In: 2016 5th international conference on reliability, infocom technologies and optimization (Trends and Future Directions)(ICRITO), pp 163–166
20. Zhang G, Ni G (2008) A new fast and effective recognition method based construction shape for printed digital. IJCSNS 8(5):309
21. Kaur D, Kaur Y (2014) Various image segmentation techniques: a review. Int J Comput Sci Mob Comput 3(5):809–814
22. Khokher MR, Ghafoor A, Siddiqui AM (2013) Image segmentation using multilevel graph cuts and graph development using fuzzy rule-based system. IET Image Process 7(3):201–211
23. Dey V, Zhang Y, Zhong M (2010) A review on image segmentation techniques with remote sensing perspective. vol 38. na Vienna, Austria

A Concise Review of Acute Myeloid Leukemia Recognition Using Machine Learning Techniques

Ashwini P. Patil

Abstract Acute myeloid leukemia is one of the fatal cancers and requires efficient as well as accurate solution for early detection of the disease. Analyses of blood and bone marrow smear images are standard and efficient method in field of leukemia detection. Various machine learning techniques are proposed to develop methods, models, and decision support systems to assist pathologist in decision making. In recent years, there is lot of research contribution toward building deep learning models. This paper reviews various literatures corresponding to recognition of acute myeloid leukemia in microscopic blood and bone marrow smear images using machine learning techniques. This review presents summary of all the methodologies, datasets and results of the reviewed research articles and also highlights some of the limitation, issues, and challenges faced by researchers in this field.

Keywords Acute myeloid leukemia · Classification · Image processing · Image segmentation · Deep learning

1 Introduction

Leukemia is life threatening malignant cancer caused by DNA mutation in the stem cells. Stem cells are produced by bone marrow which gets further matured into platelets, white blood cells, and red blood cells. When a person is suffering from leukemia, stem cells remain in immature state. These immature and non-functional cells are also referred as blast cells which grow in number rapidly replacing all the healthy cells in our bone marrow and blood.

Based on cell lineage and maturity level of cells, leukemia is divided into four main types, i.e., AML—acute myeloid or myelocytic leukemia, ALL—acute lymphoid or lymphocytic leukemia, CML—chronic myeloid or myelocytic leukemia and CLL—chronic lymphocytic or lymphoid leukemia. Acute lymphoid leukemia is a common leukemia diagnosed in children below 14 years old, and the second most common

A. P. Patil (✉)
Department of Computer Applications, CMR Institute of Technology, Bengaluru, India

J. M. R. S. Tavares et al. (eds.), *Cyber Intelligence and Information Retrieval*,
Lecture Notes in Networks and Systems 291,
https://doi.org/10.1007/978-981-16-4284-5_36

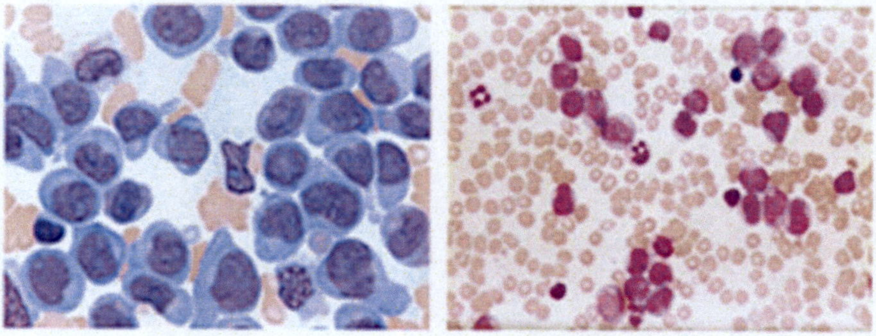

(a) (b)

Fig. 1 Examples of acute myeloid leukemia images: **a** Bone marrow smear image with AML blast **b** Blood smear image with AML blast

cancer is acute myeloid leukemia, and it is mostly diagnosed in adults; acute leukemia is further divided into subtypes according to FAB classification. Lymphoid leukemia is subdivided into L1, L2, and L3, and myeloid leukemia is subdivided into 8 subtypes from M0 to M7. In 1990, 18% of AML cases were reported worldwide. In 2012, AML cases estimated to be 3,51,965, and in 2017, percentage of cases increased to 23.1% [1]. The survival rate of AML patients is 5 years. In USA alone, estimated deaths in 2020 are 11,180, which is 1.8% of all the cancers [2].

Diagnosis of acute myeloid leukemia at early stage is very crucial. In spite of lot of advance technologies, microscopic analyses of blood or bone marrow smear images are still considered as a standard technique to detect leukemia. There is lot of research work conducted to automate the process to analyze microscopic digital bone or blood smear images. Leukemia is field which requires automated and efficient techniques or models for early detection and diagnosis. Figure 1 represents examples of acute myeloid leukemia images.

This paper reviews some of the studies and research work in area of image processing for acute myeloid leukemia. The reviewed articles are selected from year 2015 to 2021. The paper is organized as follows: Sect. 1 introduces the study; Sect. 2 presents the review of methodology and summary of reviewed methodology, dataset, and results; Sect. 3 provides the conclusion and specifies some of the limitations and research issues.

2 Review Methodology

Identification and classification of acute myeloid leukemia involve the similar steps of medical image processing, i.e., acquiring images, preprocessing images segmentation, extraction of feature, and finally classification. Figure 2 shows image processing stages for leukemia detection.

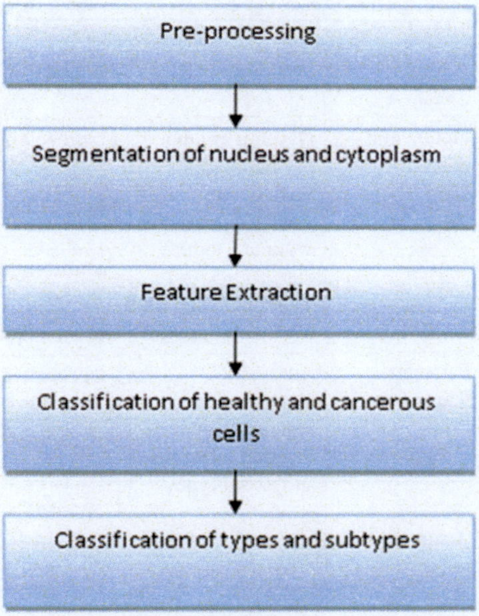

Fig. 2 Image processing stages for acute leukemia detection

2.1 Methodology

First step is to acquire microscopic bone marrow or blood smear images. In the reviewed literature, very few studies have used public dataset. Most commonly used public datasets are ALL-IDB, ASH Image bank, and cancer imaging archive. Preprocessing step focuses on enhancement of bone marrow or blood smear images. This step is needed since the quality of images varies due to camera quality, brightness of the light, and preparation process of smear slides.

In the next step, nucleus and cytoplasm are segmented, and it is considered one of the crucial step because the result of further steps depends upon accuracy of segmentation. K-means clustering is commonly used techniques for segmentation [3–6]. In feature extraction, various texture, color, and shape features are extracted and are fed to the classifier. Classification includes two stages, first is binary classification to identify if the cell is cancerous cell or healthy cell, and second stage is to recognize the types and subtypes of leukemia. Rawat et al. [7] present detail research work on classification of ALL, AML, and their subtypes. [8, 9] have developed deep learning models to automate leukemia recognition with optimal accuracy. Table 1 presents the comparison of all the reviewed articles.

Table 1 Comparison of reviewed literature

Author , Year	Proposed work	Methods		Dataset
Gutam and Sailaja 2015 [3]	Develop system for automatic classification of AML, increase the accuracy, reduce the over-segmentation of nucleus cluster	Preprocessing	Convert to rgb, median filter	A microscopic blood smear image 90 high-quality images of size 184 × 138 from the ASH Image Bank
		Segmentation	K-means clustering	
		Feature extraction	LDP feature extraction	
		Classification	SVM	
Fatemeh Kazemi et al. [4]	Automatic identification system for acute myeloid leukemia and its subtypes	Preprocessing	Convert rgb image to LAB color	A microscopic blood smear image of size 184 × 138 from Shariati Hospital, Iran
		Segmentation	K-means clustering	
		Post-processing	Canny edge detection, edge enhancement, dilation, and hole-filling	
		Feature extraction	texture features, fractal dimension, shape features, and color features	
		Classification	multiclass SVM	
Esti Suryani et al. [10]	Determine AML subtype M0 and M1 based on the morphology WBC image using image processing method	Preprocessing	median filter, rgb to YCbCr	Total 50 Bone Marrow Images 20 of AML M0 30 of AL M1 from Clinical Pathology Laboratory RSUD Dr Moewardi Surakarta
		Segmentation	Thresholding	
		Feature extraction	WBC diameter and nucleus ratio,	
		Classification	KNN	
Jyoti Rawat et. al [7]	C CAD System recongize myeloid and lymphoid cells and their FAB characterization	Preprocessing	rgb to gray scale, histogram equalization, 2D order statistical filter	420 Blood smear images from American Society of Hematology (ASH)
		Segmentation	Global thresholding, Morphological opening	
		Feature extraction	geometric, chromatic, and texture features	

(continued)

Table 1 (continued)

Author , Year	Proposed work	Methods		Dataset
		Classification	SVM	
Preeti Jagadev et. al [11]	Leukemia and its subtype(ALL, AML, CML, CLL) detection using ML techniques	Preprocessing	CMYK to rgb conversion	(Private Dataset) 220 blood smear images from Goa Medical College
		Segmentation	HSV color-based segmentation, Marker controlled watershed algorithm, K-means clustering,	
		Feature extraction	geometric and texture features	
		Classification	SVM	
Andika Setiawan et al. [5]	Classify AML subtypes M4, M5. M7 with SVM	Preprocessing	Median filtering and global contrast stretching	(Private Dataset) Blood smear images from RSUP Sardjito Yogyakarta
		Segmentation	K-means clustering	
		Feature extraction	Perimeter, nucleus ratio, circularity, mean, area, standard deviation	
		Classification	SVM	
Ahmed Negm et al. [6]	Acute leukemia classification system	Preprocessing	Green component selected from rgb	(Private Dataset and Public dataset) ALL-IDB
		Segmentation	K-means clustering	
		Feature extraction	geometry, statistics, textures, and size ratio	
		Classification	neural network and decision tree	
Satvik Dasariraju et al [12]	ML model to diagnose AML and to detect and classify immature leukocytes	Preprocessing	rgb to LAB color space	(Public Dataset) 18, 365 images of leukocyte from the cancer imaging archive
		Segmentation	Otsu thresholding, binary mask and conversion to gray format	

(continued)

Table 1 (continued)

Author, Year	Proposed work	Methods		Dataset
		Feature extraction	Total 16 cytomorphological features were extracted, Which are categorized as color features, nucleus size, and shape, elliptical features	
		Classification	Random forest algorithm	
Maíla Clro et al. [8]	Developed a convolutional neural network model Alert Net-RWD to diagnosis ALL and AML	Classification	CNN model Alert Net-RWD,	16 datasets with 2415 images,
Laura Boldú et al. [9]	Deep learning model to diagnose acute leukemia in blood smear images	Classification	Deep learning mode ALNet	(Private Dataset) Blood smear images Core Laboratory of the Hospital Clinic of Barcelona

2.2 Results

Above comparison, Table 1 gives detailed analysis of proposed work and methods used. Comparison shows that some of the authors have proposed only AML classification, and some have further classified AML into their subtypes. Gutam et al. [3] have achieved 98% accuracy, [6] have achieved 99.517% of accuracy, and [10] have got very low accuracy of 67.58% in AML classification. Kazemi et al. [4] have got AML classification accuracy of 96% and its subtypes (M2-M5) of 87%. Among all the works, Rawat et al. [7] have achieved very high accuracy of 99.5%. They have used public dataset and have also further classified AML into M1, M2, and M3 with 98.5% accuracy, whereas ALL is further classified into L1, L2, and L3 with 97.5% accuracy. [11 ,12, and 8] have achieved accuracy of 99.5, 93.45, and 97.18%, respectively. Boldú et al. [9] used deep learning mode ALNet to classify myeloid leukemia (Sensitivity- 100%, specificity 92.3%, precision 93.7%) and lymphoid leukemia (Sensitivity- 89%, specificity 100%, precision 100%).

3 Conclusion

This paper presents review of various methodologies implemented to detect and classify acute myeloid leukemia. All the reviewed objectives, proposed methodologies datasets, and results are summarized and exhibited in a tabular format. Research in acute myeloid leukemia is lacking as compared to the research contribution toward acute lymphocytic leukemia. Most of the studies make use of private datasets from local hospitals or pathology centers, and hence, the accuracy and efficiency of result cannot be evaluated and compared with other research works. In the recent years, more studies are focused on using deep learning technology to design a model to diagnose leukemia, and hence, a large benchmark public dataset must be introduced. More research work in the development of automated tool to classify acute myeloid leukemia into its subtypes is also required.

References

1. Dong Y, Shi O et al (2020) Leukemia incidence trends at the global regional and national level between 1990 and 2017. Exp Hematol Oncol 9:14. https://doi.org/10.1186/s40164-020-00170-6
2. "Website" [Online] (2021). Available https://seer.cancer.gov/statfacts/html/amyl.html. Accessed 1 Apr 2021
3. Goutam D et al. (2015) Classification of Acute Myelogenous Leukemia in blood microscopic images using supervised classifier. In: 2015 IEEE international conference on engineering and technology (ICETECH), Coimbatore, India, pp 1–5. https://doi.org/10.1109/ICETECH.2015.7275021
4. Kazemi F, Araabi BN, Najafabadi TA (2016) Automatic recognition of acute myelogenous leukemia in blood microscopic images using K-means clustering and support vector machine. J Med Signals Sens. 6(3):183–93. PMID: 27563575; PMCID: PMC4973462
5. Setiawan A, Ratnaningsih T, Palgunadi S, Harjoko A, Wiharto, Suryani E (2018) Classification of cell types in Acute Myeloid Leukemia (AML) of M4, M5 and M7 subtypes with support vector machine classifier. In: International conference on information and communications technology (ICOIACT), Yogyakarta, Indonesia, pp 45–49. https://doi.org/10.1109/ICOIACT.2018.8350822.
6. Negm AS et al. (2018) A decision support system for Acute Leukaemia classification based on digital microscopic images. Alexandria Eng J 57(4):2319–2332. ISSN 1110–0168. https://doi.org/10.1016/j.aej.2017.08.025
7. Rawat J, Bhadauria HS, Singh A, Devgun JS, Virmani J (2017) Computer assisted classification framework for prediction of Acute Lymphoblastic and Acute Myeloblastic Leukemia. Biocybernet Biomed Eng 37(4):637–654. ISSN 0208–5216. https://doi.org/10.1016/j.bbe.2017.07.003

8. Claro M et al (2020) CNN models for acute Leukemia diagnosis. In: International conference on systems, signals and image processing, Brazil, pp 63–68. https://doi.org/10.1109/IWSSIP48289.2020.9145406
9. Laura B, Acevedo A, Merino A, Molina A, Rodellar J (2021) A deep learning model (ALNet) for the diagnosis of acute leukaemia lineage using peripheral blood cell images. Comput. Methods Programs Biomed 202:105999. ISSN 0169–2607
10. Wiharto ES et al. (2017) Cell identification of Acute Myeloid Leukemia AML M0 and M1 using K-nearest neighbour based on morphological images. In: 2017 International conference on data and software engineering (ICoDSE), Indonesia, Palembang, pp 1–6. https://doi.org/10.1109/ICODSE.2017.8285851
11. Jagadev P et al. (2017) Detection of leukemia and its types using image processing and machine learning. In: International conference on trends in electronics and informatics (ICEI), Tirunelveli, India, pp 522–526. https://doi.org/10.1109/ICOEI.2017.8300983
12. Dasariraju S, Huo M, McCalla S (2020) Detection and classification of immature leukocytes for diagnosis of acute myeloid leukemia using random forest algorithm. Bioeng (Basel) 7(4):120. https://doi.org/10.3390/bioengineering7040120.PMID:33019619;PMCID:PMC7711527

A Novel Approach for Web Mining Taxonomy for High-Performance Computing

Debabrata Samanta, Soumi Dutta, Mohammad Gouse Galety, and Sabyasachi Pramanik

Abstract Web mining is a central part of data analysis. The fetching and discovering knowledge from the different web data in data mining mechanism is more important nowadays. Web usage mining customs data mining practice for the investigation of custom decoration from different data storages. In this article paper, introducing a new approach for web mining taxonomy for high-performance computing. The primary motivation of this research is on the data collection in different real-time web servers for implementation and analysis. This article is focussed the WebLog Expert lite 9.3 tools for our study.

Keywords Web mining · Self-organising map · Document object model · WebLog Expert

1 Introduction

The information extraction procedure, that is useful, from web documents is referred to as web mining [1]. The contents of a web page are incorporated as the content data. Mining of the text helps to classify web pages. Web pages—nodes and hyper-links—edges which connect the similar pages are the main contents of a typical graph. Web structure mining has two broad categories. A hyperlink is a connection between two locations of the webpage [2]. After connecting different parts within the web page,

D. Samanta (✉)
Department of Computer Science, Christ University, Bangalore, India

S. Dutta
Institute of Engineering and Management, Kolkata, India

M. G. Galety
College of Engineering and Technology, Catholic University in Erbil, Erbil, Kurdistan Region, Iraq
e-mail: galety.143@ieee.org

S. Pramanik
Haldia Institute of Technology, Haldia, India

J. M. R. S. Tavares et al. (eds.), *Cyber Intelligence and Information Retrieval*, Lecture Notes in Networks and Systems 291,
https://doi.org/10.1007/978-981-16-4284-5_37

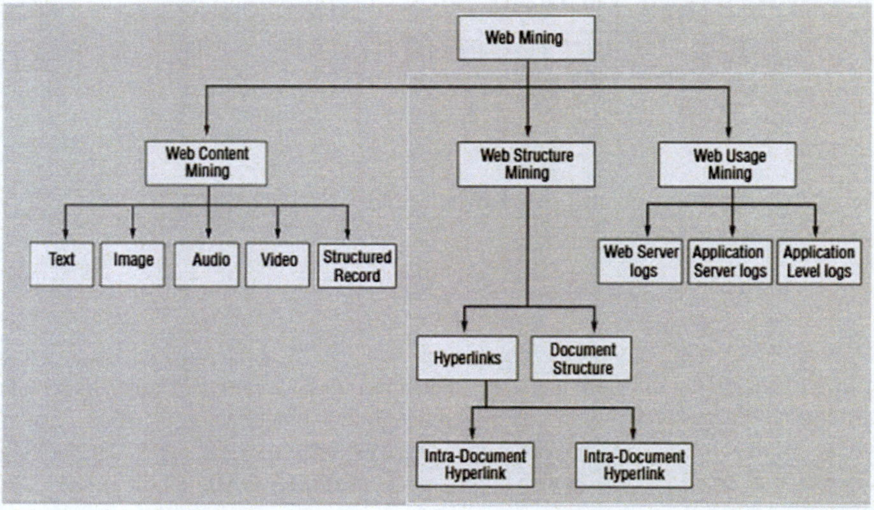

Fig. 1 Basic web mining taxonomy

it is called an intro-document hyperlink, and with touching the separates page, it is called an inter-document hyperlink. A hyperlink has been analysed by two significant bodies; provide an up-to-date survey. It is a tree structure given to the web content of a page based on various tags. Here the focus is on document object model (DOM) from the [3]. Figure 1 represents basic web mining taxonomy.

2 Web Usage Mining

Implementation of the technique for discovering the pattern usage expresses web data in three phases. Association rules, apriori algorithm, sequential rules, self-organising map, genetic algorithm and cluster computing are some of the popular approaches of mining [4, 5]. The self-organizing map (SOM) is a much-accepted network model, which can be categorised for competitive learning networks based on unsupervised learning. This requires minimum human interference and, hence, is used to cluster the data without knowledge of the class. The SOM is used for detecting inherent features of the problem, and it also called SOFM, the self-organizing feature map. Map units, or neurons, form a lattice, and the mapping is onto the plane [6, 7]. The mapping helped to preserve the point distance and mapped to nearby units in the SOM. It helps to analyse high-dimensional and also pre-processed data. Based on the access pattern of the users, the websites are organised so that overload of the search depth for users is reduced. The fittest offspring is chosen for mutation, which is the process for implementing the changes in the genetic material between the parents and their offspring forms [8–10].

Table 1 Sample A translated sequence database

ID For web mining	T-ID	Events
6	Ψ1	bread, milk, oatmeal
1	Ψ2	cereal, cheese, mil,
5	Ψ3	bread, oatmeal, butter
7	Ψ4	bread, milk, cheese, butter, cereal
2	Ψ5	oatmeal
3	Ψ6	milk, bread, butter
4	Ψ7	cereal, milk

3 Background and Related Work

Primary and real data help to mine web content, whereas the second is used for web usage mining (WUM) [11, 12]. The latter uses data proxy, web servers and registered files browser logs or user profiles or any other data that is generated by user interaction. Ecommerce companies act as predators for WUM for tracking the behaviour of the users and using the data for mining. "Collaborative filtering" is also used for identifying similar customers for application of clustering. The recommender for this collaborative system helps to discriminate between customers and their choices. A new algorithm in [13, 14] is used for catering to the rule of association in collaborative recommendation systems. This method is used for reducing the data rating dimension, and then NNs are applied for making collaborative model. In spite of the pattern discoveries, personalisation task is not easy to perform [15–17, 27]. Patterns help in the derivation of average user profiles as the evaluation is done in two steps of clustering and recommender [18]. Weblog analysis (ASCII) contains the Name of the user, IP Address, Access Request, Time Stamp, Error codes, URL that Referred, etc. Transfer log, Agent log, Error log and Referrer log [8] are the server logs [19]. The standard is agent log and transfer. The movement from one page to another is recorded along with the IP address [20, 21]. Table 1 expresses sample A translated sequence database. Figure 2 defines comparison of data from 2007 to 2018.

4 Methods of Parallelism

Several reasons have been pointed out by Zaki for the design of a parallel algorithm to data mining [22, 23]. To begin with: multi-processor CPUs are a better option as it speeds up the system as compared to single processors. Secondly, as a massive amount of data is already stored in the databases, therefore, it is expensive to mine the data in a single system. Hence, tools can be used for parallelisation. Few popular methods are: "task parallelism" (divide and conquer strategy, task queue), "data parallelism" (record based and attribute-based) and "hybrid data/task parallelism" [24, 25].

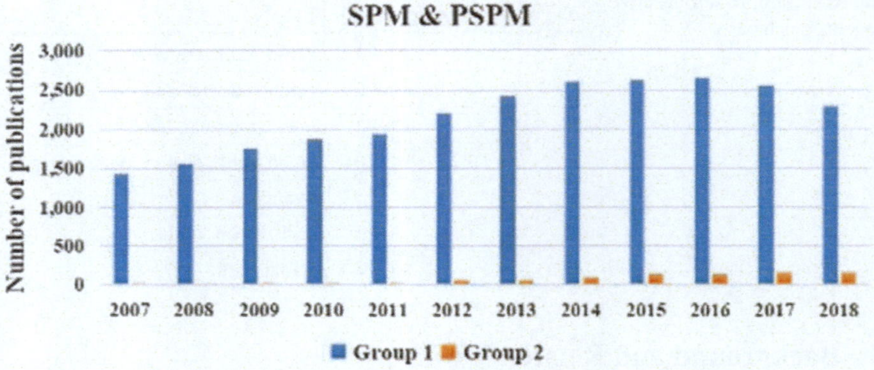

Fig. 2 Comparison of data from 2007 to 2018

(1) Task parallelism1—This divides the search space for separating the processors, leading to multiple works being performed on the data. "Divide and conquer" strategy and "task queue" are the types of approaches. In the first approach, there is division in the search space and the partitions to the specific processor is pre-decided; while in the second approach, portions of the space are assigned whenever it is available.

(2) Data parallelism—In this, the same task runs simultaneously on different dataset. Data-parallel approaches are of two types: record and attribute. In Zaki,G. Md. Gouse, B. Kavitha had pointed out that in record-based data parallelism, the records or attribute lists are horizontally partitioned among the processors.

5 Proposed Work

Genetic algorithm-based approach is the combination of information from content and usage information for providing the relevant search result to the users. Some parameters should be taken into consideration for checking the health of the solution provided [26]. Access frequency measures the frequency of visit, from any id. Several unique visitors—It is decided based on unique visitors. Time duration—This is decided on how much time is spent and how relevant the page is for the user. Several bytes received—The measurement of the data downloaded from the web page reveals its significance. Typical entry and exit points—It is the page where the user starts the search in the web site. The exit point denotes the content that the user is looking for in the page or website. Several advertisements—The importance is represented by the ads presented on it as only frequently visited pages have enough of the ad in it. Several parameters are included in the fitness function. Genetic algorithm is the selection of random population, and then the various applying techniques on many generations until the result are achieved. Table 2 shows sequence Table for our test case. Figure 3 projects final output screen of our test result.

Table 2 Sequence table for our test case

T-ID	Different sequence
Ψ1	< {a,d}, {c}, {d,9}, {9}, fe} >
Ψ2	< {a}, {d}, {c.g}, {e} >
Ψ3	< {a,b}, {c}, Uf.9}, {a,5,9), {e} >
Ψ4	< {b}, (c}, td, f} >
Ψ5	< {a,b}, {c}, {d, f.9}, (g}, fe} >
Ψ6	< {a}, {d},,{e} >
Ψ7	< {a,b}, {c, {a,5,9), {e} >

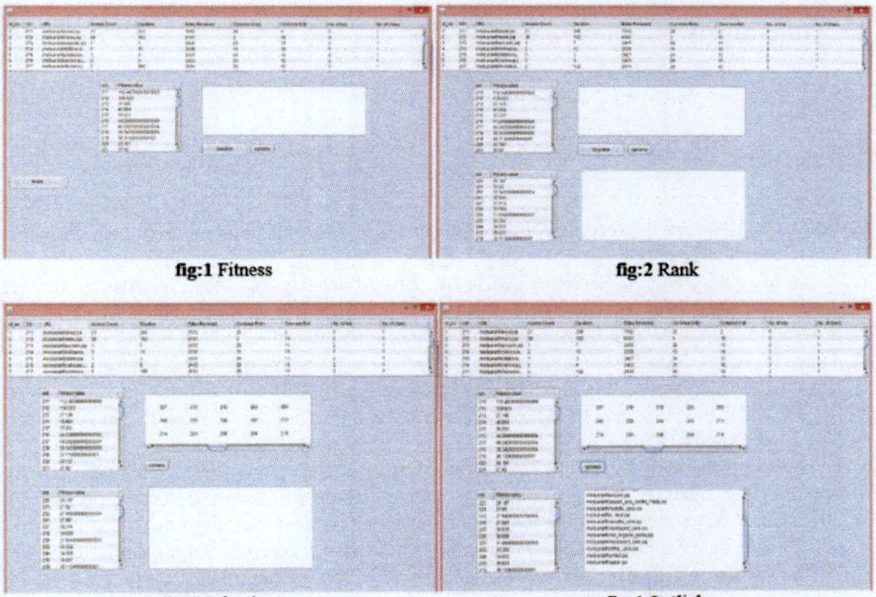

fig:1 Fitness **fig:2** Rank

fig:3 Selection **fig:4** Optlinks

Fig. 3 Final output screen of our test result

6 Advantages of GA Over SOM

This is a natural search optimisation performance, the critical initiative of which is dependent on the process of natural evolution. It is used to solve general optimisations as it is parallel. Fitness function is used optimising a solution and helps in ranking it. Genetic algorithm is used for optimising fitness function and does need to be continuous or differentiable. GA can be applied for discrete services. Secondly, GA is very fruitful for nonlinear function discovery. These are the merits of genetic algorithm, and hence, it is used over SOM.

7 Conclusion

A web page is an advertisement tool, which benefits several commercial sectors, government bodies, institutions …etc. The website designs should comply with W3C standards, and the content should be good enough to impact the visitor's count, which may vary, based on quality and content of the web page provide by any foundation. The server log files should be analysed, and the errors should be detected well in advance. The data of the server of an educational institution were analysed, and it helped for improvising the policies and procedures of the services they provide. Data mining and recognition of the pattern are a prospective field for future research and development to improve the overall health of the commercial, industrial and educational sector.

References

1. Grover L (1996) In: Proceedings of the 28th annual ACM symposium on theory of computing STOC '96, vol 212
2. Zidan M, Abdel-Aty A, El-Sadek A, Zanaty EA, Abdel-Aty M (2017) In: AIP conference proceedings, vol 1905, pp 020005
3. Sagheer A, Zidan M (2013) Autonomous quantum perceptron neural network. arXiv:1312.4149, 1
4. Einstein A, Podolsky B, Rosen N (1935) Phys Rev 47(10):777
5. Lu CY, Zhou XQ, G¨uhne O, Gao WB, Zhang J, Yuan ZS, Goebel A, Yang T, Pan JW (2007) Nat. Phys 3(2):91
6. Kanti PS, Debabrata S (2019) Impact of climate changes on rural livelihoods and re-orienting the situation through human resources towards a sustainable society (March 19, 2019). In: Proceedings of international conference on sustainable computing in science, technology and management (SUSCOM), Amity University Rajasthan, Jaipur, India, February 26–28. Available at SSRN: https://ssrn.com/abstract=3355347 or https://doi.org/10.2139/ssrn.3355347
7. Pradhan KN, Siddappa M, Kavitha S, Debabrata S (2019) Analysis and improvement of wireless network security based on biometrics (March 20, 2019). In: Proceedings of international conference on sustainable computing in science, technology and management (SUSCOM), Amity University Rajasthan, Jaipur-India, February 26–28. Available at SSRN: https://ssrn.com/abstract=3356360 or https://doi.org/10.2139/ssrn.3356360
8. Karthikeyan MP, Samanta D, Banerjee A, Roy A, Inokawa H (2021) Design and development of terahertz medical screening devices. In: Chakraborty M, Jha RK, Balas VE, Sur SN, Kandar D (eds) Trends in wireless communication and information security. Lecture notes in electrical engineering, vol 740. Springer, Singapore. https://doi.org/10.1007/978-981-33-6393-9_40
9. Althar RR, Samanta D (2021) Building intelligent integrated development environment for IoT in the context of statistical modeling for software source code. In: Kumar R, Sharma R, Pattnaik PK (eds) Multimedia technologies in the internet of things environment. Studies in Big Data, vol 79. Springer, Singapore. https://doi.org/10.1007/978-981-15-7965-3_7
10. Podder SK, Samanta D, Gurunath R (2020) Impact of business analytics for smart education system and management functions. In: Sharma H, Govindan K, Poonia R, Kumar S, El-Medany W (eds) Advances in computing and intelligent systems. Algorithms for intelligent systems. Springer, Singapore. https://doi.org/10.1007/978-981-15-0222-4_45

11. Sarkhel D, Deka D, Samanta D, Kumudavalli M, Le DN (2020) GUI-based percentage analysis for curing breast cancer survivors. In: Satapathy S, Bhateja V, Nguyen B, Nguyen N, Le DN (eds) Frontiers in intelligent computing: theory and applications. Advances in intelligent systems and computing, vol 1013. Springer, Singapore. https://doi.org/10.1007/978-981-32-9186-7_33
12. Praveen B, Samanta D, Prasad G, Ranjith Kumar C, Prasad MLM (2020) Protecting medical research data using next gen steganography approach. In: Jain L, Peng SL, Alhadidi B, Pal S (eds) Intelligent computing paradigm and cutting-edge technologies. ICICCT 2019. Learning and analytics in intelligent systems, vol 9. Springer, Cham. https://doi.org/10.1007/978-3-030-38501-9_34
13. Guha A, Samanta D (2020) Real-time application of document classification based on machine learning. In: Jain L, Peng SL, Alhadidi B, Pal S (eds) Intelligent computing paradigm and cutting-edge technologies. ICICCT 2019. Learning and analytics in intelligent systems, vol 9. Springer, Cham. https://doi.org/10.1007/978-3-030-38501-9_37
14. Chakrabarti S, Samanta D (2016) Image steganography using priority-based neural network and pyramid. In: Shetty N, Prasad N, Nalini N (eds) Emerging research in computing, information, communication and applications. Springer, Singapore. https://doi.org/10.1007/978-981-10-0287-8_15
15. Samanta D., Sanyal G. (2012) Novel Shannon's Entropy Based Segmentation Technique for SAR Images. In: Venugopal K.R., Patnaik L.M. (eds) Wireless Networks and Computational Intelligence. ICIP 2012. Communications in Computer and Information Science, vol 292. Springer, Berlin, Heidelberg. https://doi.org/10.1007/978-3-642-31686-9_22
16. Hossain M.A., Samanta D., Sanyal G. (2012) Extraction of Panic Expression from Human Face Based on Histogram Approach. In: Venugopal K.R., Patnaik L.M. (eds) Wireless Networks and Computational Intelligence. ICIP 2012. Communications in Computer and Information Science, vol 292. Springer, Berlin, Heidelberg. https://doi.org/10.1007/978-3-642-31686-9_48
17. Samanta D, Sanyal G (2011) Development of edge detection technique for images using adaptive thresholding. In: Venugopal KR, Patnaik LM (eds) Computer networks and intelligent computing, ICIP 2011. Communications in computer and information science, vol 157. Springer, Berlin, Heidelberg. https://doi.org/10.1007/978-3-642-22786-8_85
18. Horodecki R, Horodecki P, Horodecki M, Horodecki K (2009) Rev Mod Phys 81:865
19. Biswal AK, Singh D, Pattanayak BK, Samanta D, Yang M-H (2021) IoT-based smart alert system for drowsy driver detection. Wireless Commun Mobile Comput 6627217:13. https://doi.org/10.1155/2021/6627217
20. Maheswari M, Geetha S, Selva Kumar S, Karuppiah M, Samanta D, Park Y (2021) PEVRM: probabilistic evolution based version recommendation model for mobile applications. IEEE Access. https://doi.org/10.1109/ACCESS.2021.3053583
21. Gomathy V, Padhy N, Samanta D et al (2020) Malicious node detection using heterogeneous cluster based secure routing protocol (HCBS) in wireless adhoc sensor networks. J Ambient Intell Human Comput 11:4995–5001. https://doi.org/10.1007/s12652-020-01797-3
22. Sivakumar P, Nagaraju R, Samanta D et al (2020) A novel free space communication system using nonlinear In GaAsP microsystem resonators for enabling power-control toward smart cities. Wireless Netw 26:2317–2328. https://doi.org/10.1007/s11276-019-02075-7
23. Khamparia A, Singh PK, Rani P, Samanta D, Khanna A, Bhushan B (2020) An internet of health things-driven deep learning framework for detection and classification of skin cancer using transfer learning. Trans Emerg Tel Tech. e3963. https://doi.org/10.1002/ett.3963
24. Althar RR, Samanta D (2021) The realist approach for evaluation of computational intelligence in software engineering. Innov Syst Softw Eng. https://doi.org/10.1007/s11334-020-00383-2
25. Guha A, Samanta D (2021) Hybrid approach to document anomaly detection: an application to facilitate RPA in title insurance. Int J Autom Comput 18:55–72. https://doi.org/10.1007/s11633-020-1247-y
26. Debabrata S, Sivaram M, Ahmed R , Boopathi CS, Amiri IS, Preecha Y (2020) Distributed feedback laser (DFB) for signal power amplitude level improvement in long spectral band. J Opt Communhttps://doi.org/10.1515/joc-2019-0252

27. Hassanien, AE, Chakrabarti S, Bhattacharya A, Peng SA, Dutta S (Eds) (2020) Semantic Analysis in web-based learning and teaching platform [Special issue]. Int J Web-Based Learn Teach Technol 17(3)

Advance Computing

Performance Evaluation and Comparison of Various Personal Cloud Storage Services for Healthcare Images

Moumita Roy and Monisha Singh

Abstract In recent times, usage of personal cloud storage services for storing e-health records in on a rise. This is due to the constant accessibility, easy sharing, and safe storage of the data at a nominal cost. In this paper, we have analyzed the performance of four personal cloud storage services: Google Drive, Dropbox, Sync.com, and Icedrive using medical image data files of various sizes. The parameters checked were number of packets transmitted during file upload and duration of time to upload, download, and delete the files. The results show us a comparative analysis of the personal cloud storage services based on the parameters and also help us identify certain gaps for the future.

Keywords Personal cloud storage · Healthcare image · Google Drive · Dropbox · Performance analysis

1 Introduction

In recent times, personal cloud storage (PCS) services are used mostly by individual users for storing personal data in one secure place and retrieving and managing them based on requirement. The primary convenience of using PCS services is the constant availability and ready accessibility of the data from personal devices like mobile phones, minimizing data loss, and enabling easy sharing of the data. PCS also allows synchronizing of the files stored in them with the help of cloud servers. The remote storage space provided by the service providers is huge, and the cost involved is negligible or even free sometimes. Due to the simple and cost-effective way in which data can be stored, accessed and shared, the number of users using

M. Roy (✉)
Department of Computer Science, Christ Deemed To Be University, Bengaluru, India

M. Roy · M. Singh
Department of Computer Applications, CMR Institute of Technology, Bengaluru, India
e-mail: monisha.singh@christuniversity.in

J. M. R. S. Tavares et al. (eds.), *Cyber Intelligence and Information Retrieval*, Lecture Notes in Networks and Systems 291,
https://doi.org/10.1007/978-981-16-4284-5_38

PCS services are growing at a rapid rate. Due to the increasing demand, the number of service providers is also on a rise in the market.

One very common usage of PCS is storing of electronic healthcare data, Personal Health Records (PHRs) or images related to health care as they are of utmost importance to the users or patients. Healthcare data is personal and sensitive, and users cannot afford to lose them. PCS is safe and, thus, is ideal for healthcare data. Frequent sharing of such data may also be required with medical practitioners, and so, synchronization is also required. Image data takes lot of space so it also makes sense to use remote storage to store them and keep personal devices uncluttered.

One very common usage of PCS is storing of electronic healthcare data, Personal Health Records (PHRs) or images related to health care as they are of utmost importance to the users or patients. Healthcare data is personal and sensitive, and users cannot afford to lose them. PCS is safe and, thus, is ideal for healthcare data. Frequent sharing of such data may also be required with medical practitioners, and so, synchronization is also required. Image data takes lot of space so it also makes sense to use remote storage to store them and keep personal devices uncluttered.

2 Literature Review

PCS services enable users to store personal and private data, synchronize, and access them at any time from any device in a cost-effective way [1]. These services are gaining popularity in the recent times due to the convenience of use and economic benefits. These platforms work with the help of multiple virtual machines using remote resources for computation and storage. Platforms like Google Drive, Dropbox, OneDrive, Icedrive, Box, Kingsoft, Sync.com, etc., are some of the popular PCS service providers in the present times. Akter et. al. [6] have compared performance of Google Drive, Dropbox, OneDrive, and Box using network analyzing tool Wireshark and got the output in favor of Dropbox.

A. **Personal Cloud Storage Platforms**

Cloud computing services can be distinguished based on the services delivered like infrastructure, computation, or software; or the model of deployment used like public, private, or hybrid. Personal cloud is of hybrid model that is deployed using both public and private cloud partially [2]. Personal cloud storage platforms provide on-demand cloud services to store and manage user's personal data through individual devices like mobile phones. Open REST APIs and client implementations of the APIs are provided by personal clouds. These are used by the developers to develop applications based on the stored data [3]. The data transfer performance depends on several factors like the service provider, geographic location, type traffic, time of the day, etc. The primary reason behind the tremendous popularity of PCS, not forgetting about the limitations, is user experience. The factors that add up to the experience are constant data accessibility and share-ability from personal Internet-connected devices at a very much affordable cost [4]. The limitations primarily include privacy and security of the data stored.

B. Software Platforms

In this paper, we have focused on performance of four different PCS software: Google Drive, Dropbox, Sync.com and Icedrive. Here in this part, let us dive into some details about them.

Google Drive

Google Drive is a data storage system that users can access using their google mail account. It is one of the most popular personal cloud platforms having number of users approximately around 2 billion by the end of 2020. One primary reason for the popularity is the users are given their first 15 GB of storage for free. Beyond that, additional space upto 16 TB can be given to the user based on a subscription. Secondly, it is accessible from any personal device (mobile, laptop, desktop, etc.) having iOS or Android for mobiles or Mac or Windows for computers. Linux-based OS is not supported, and devices using Linux need to depend on external programs [5]. The website allows users to view MS Office files, and to edit them, the files need to be opened as Google Doc. Online and simultaneous file editing and synchronization are possible if the file is shared with multiple collaborators who are all Google Drive users. Sharing is allowed as a viewer or editor, depending on whether to give write permission to the user with whom the file is being shared. Sharing can be done by sharing a link or by email notification [6].

Dropbox

Dropbox is another extremely user-friendly PCS platform that allows its users to store data up to 2 GB for free and add up to 20 GB of free space by getting 500 MB for each friend referral [8]. Unlike Google Drive, Dropbox is accessible from systems having Linux-based OS along with Mac and Windows systems. Along with computers and mobile phones, Dropbox applications also run-on Kindle. It also supports many file formats like MS Office, image, pdf, javascript, audio, video, zip files, etc., but it does not allow online editing like Google Drive. Files need to be downloaded in order to be edited. For a document that is publicly shared in Dropbox, sharing is possible using links with someone who is not a Dropbox user [9]. As shown in Fig. 1, by the end of 2019, Dropbox served about 600 million users in which 14.3 paid a subscription. At a time, Dropbox allows its users to log into three different devices parallelly.

Sync.com

Sync.com is our next PCS that gives an extremely easy platform to store, access, and share data at a competitive cost as other market giants like Google Drive and Dropbox. Each user gets storage space of 5 GB space as free trial which is more than double of the free space provided by Dropbox, and then upto 1 TB with a subscription of $5 per month, though the minimum expense would be of $49 as only annual contracts are allowed [10]. Compatibility covers major OS like Mac

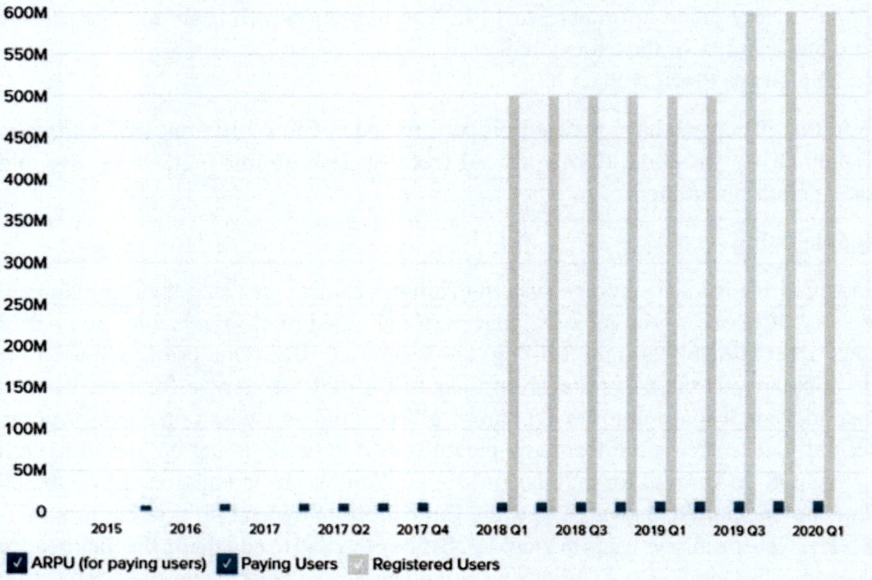

Fig. 1 Dropbox user statistics over last 5 years [7]

and Windows, web-based applications, and Android and Apple mobile handheld devices. Real-time synchronization of files is possible. Data stored is kept secure using end-to-end encryption making it more secure than other platforms, but that makes the synchronization process slower. No third-party app access is possible, and like Google Drive, Sync.com also does not support Linux, but it allows users to connect to upto five different devices at the same time [11].

Icedrive

Icedrive is one of the latest PCS inclusions in the marketspace and is gaining rapid popularity for providing upto 10 GB of free storage to the users and the next 5 TB for $1.64 only [12]. The interface is very simple and user-friendly. Security features are robust with zero-knowledge encryption. All major operating systems like Windows, Mac, and even Linux are compatible to the desktop version, and the mobile app works with Android and iOS. File preview is possible, but collaboration is still a drawback, and upload links are not available.

A comparative summary of benefits and drawbacks for all the four PCS services is shown in Table 1.

Table 1 Benefits and drawbacks of PCS services

Software name	Benefits	Drawbacks
Google drive	• Ease of document access and share • Office compatible	• Cannot upload from device to cloud automatically
Dropbox	• Access latest version of files	• Only user can see the files
Sync.com	• Easy file and folder sharing	• Annual contracts only • Slow synchronization
Icedrive	• Cheapest	• Does not support FTP and SFTP

Table 2 Datasets used for PCS service performance evaluation

Dataset	#1	#2	#3	#4	#5	#6	#7
File Type	JEPG	JEPG	JEPG	JEPG	JEPG	JEPG	JEPG
#Files	1	1	2	17	20	925	1860
Data size	1 MB	5 MB	10 MB	50 MB	100 MB	500 MB	1 GB

3 Experimental Setup

In this paper, we have used dataset containing JPEG files of COVID-19 patient chest x-ray image [15]. Seven folders of different sizes are used. Table 2 has the detail outline of the dataset. A network analyzer tool Wireshark [13, 14] is used to analyze the packets sent in the Internet traffic by the four PCS services while uploading the test data and also to measure the time duration for the upload, download, and deletion of the data. A Windows desktop machine is used to perform the experiment.

4 Results

The results of the experiment conducted on the seven different sized data folders containing JPEG files by uploading, downloading, and deleting them from the four different PCS services are shown in tabular and graphical forms below. Table 3 and Fig. 2 help compare the storage platforms based on number of packets transferred. Similarly, Table 4 and Fig. 3 are for upload time, Table 5 and Fig. 4 are for deletion time, and Table 6 and Fig. 5 are for download time, respectively.

5 Summary

The analysis results show that Google Drive and Icedrive have the highest number of packets, while Sync.com has average and Dropbox has lowest number of packets sent to the Internet traffic while uploading the data files. Time duration analysis shows that

Table 3 Comparison of number of packets transferred during upload in each PCS

File Size (MB)	Google Drive	DropBox	Sync.com	Icedrive
1	3670	73	1017	1116
5	7343	82	4244	4084
10	9110	134	8108	8099
50	45,892	302	43,903	48,581
100	88,905	582	83,517	97,546
500	514,138	3231	517,660	565,396
1000	1,088,433	69,272	1,095,273	1,214,931

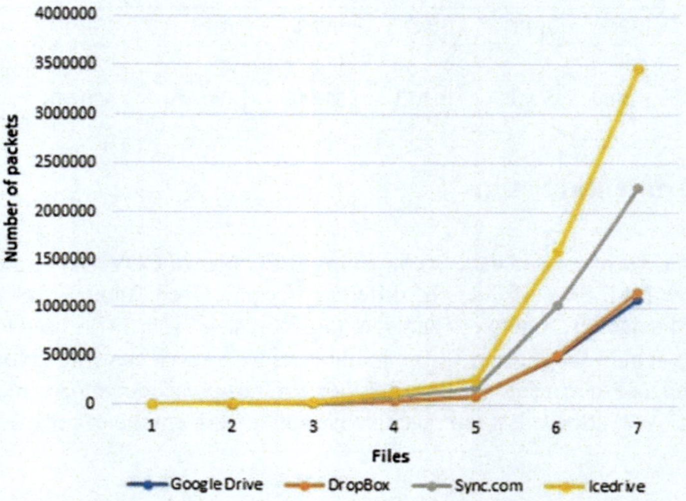

Fig. 2 Analysis of number of data packets used during file upload by each PCS service

Sync.com takes the highest time, and Google Drive takes an average duration, while Dropbox and Icedrive take lowest time in uploading, downloading, and deleting the

Table 4 Comparison of uploading time (in seconds) taken by each PCS

File Size (MB)	Google Drive	DropBox	Sync.com	Icedrive
1	13.84	10.72	27.01	9.1
5	15.37	12.85	36.04	10.07
10	19.91	12.69	60	14.5
50	93.15	23.34	251.4	23.8
100	103.65	29.13	525.84	37.14
500	580.22	113.54	3185.21	196.34
1000	1116.02	4837.84	6432.96	502.59

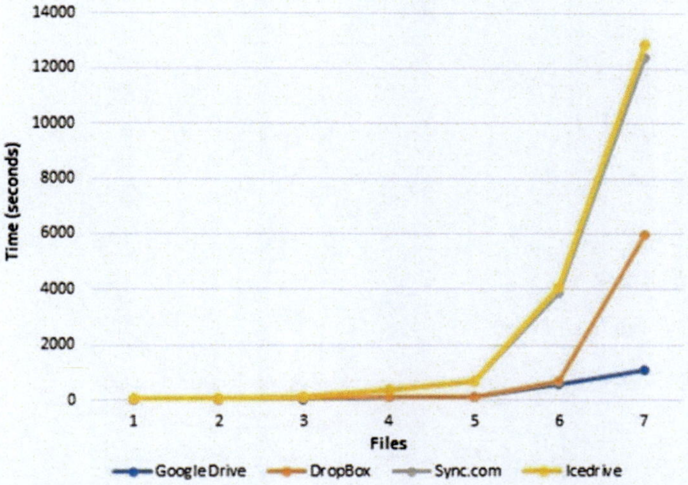

Fig. 3 Analysis of data upload time duration for each PCS service

Table 5 Comparison of deletion time taken by each PCS

File size (MB)	Google Drive	DropBox	Sync.com	Icedrive
1	11.32	10.02	24.74	7.42
5	13.83	13.06	33.01	11.44
10	17.76	13.12	62.93	12.76
50	94.23	24.29	249.76	24.13
100	103.65	29.13	525.84	37.14
500	580.22	113.54	3185.21	196.34
1000	1116.02	4837.84	6432.96	502.59

data files. We can conclude that Dropbox shows a better performance like that in [6], but Icedrive is emerging as a strong contender.

6 Conclusion

In this paper, the analysis is done of the possibilities and the drawbacks of four different PCS services while managing medical image data. The PCS services taken into consideration are Google Drive, Dropbox, Sync.com, and Icedrive. The test data included seven folders of size 1, 5, 10, 50, 100, 500, and 1 GB. The analysis scrutinized the capabilities of PCS services in terms of data packets sent to Internet traffic

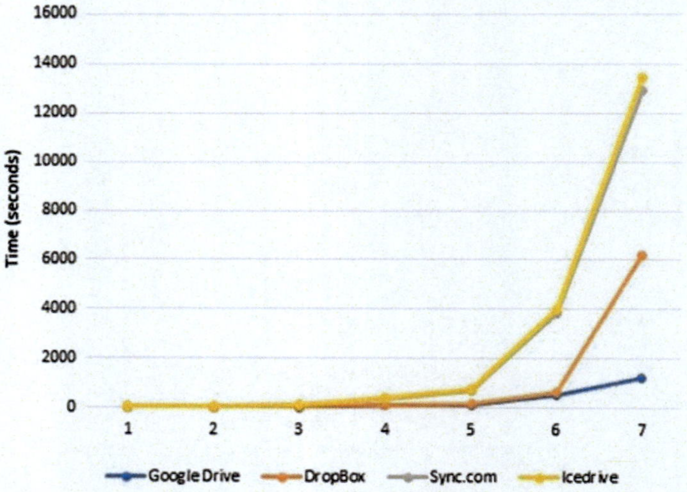

Fig. 4 Analysis of data deletion time duration for each PCS service

Table 6 Comparison of time taken to download files by each PCS

File size (MB)	Google Drive	DropBox	Sync.com	Icedrive
1	15.84	9.31	26.01	9.5
5	17.37	14.85	34.04	13.07
10	20.01	15.69	64	13.5
50	100.15	27.34	259.4	26.8
100	140.65	32.13	542.84	39.14
500	529.22	121.54	3172.21	178.34
1000	1241.02	4952.84	6732.96	528.59

during file upload and the time duration taken by them while uploading, downloading, and deleting the files. The analysis result shows that Google Drive and Icedrive send highest number of packets, while Dropbox sends lowest number of packets to the Internet traffic while uploading the files. During uploading, downloading, and deleting the files, Dropbox and Icedrive take the least time, while Sync.com takes the highest time. The overall performance of Dropbox is superior compared to the other PCS services. This experiment included only image data. Our future aim is to include other data types and also perform qualitative tests to investigate other parameters of the PCS services.

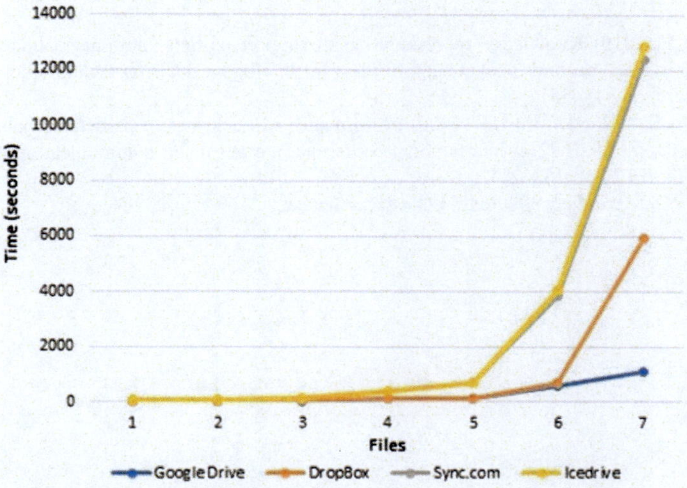

Fig. 5 Analysis of data download time duration for each PCS service

References

1. Gracia-Tinedo R, Artigas MS, Moreno-Martínez A, Cotes C, López PG (2013) Actively measuring personal cloud storage. In: 2013 IEEE sixth international conference on cloud computing, Santa Clara, CA, pp 301–308. https://doi.org/10.1109/CLOUD.2013.25
2. Na S, Park J, Huh E (2010) Personal cloud computing security framework. In:2010 IEEE Asia-Pacific services computing conference, Hangzhou, pp 671-675. https://doi.org/10.1109/APSCC.2010.117
3. Gracia-Tinedo R, Artigas MS, López PG (2013) Cloud-as-a-Gift: effectively exploiting personal cloud free accounts via REST APIs. In: 2013 IEEE sixth international conference on cloud computing, Santa Clara, CA, pp 621–628. https://doi.org/10.1109/CLOUD.2013.47
4. Karel V, DeMoor K, Lieven M (2013) Use-and QoE-related aspects of personal cloud applications: an exploratory survey. 36–37. https://doi.org/10.1109/QoMEX.2013.6603202
5. Quick D, Raymond Choo K-K (2014) Google drive: forensic analysis of data remnants. J Netw Comput Appl 40:179–193. ISSN 1084–8045. https://doi.org/10.1016/j.jnca.2013.09.016
6. Akter M, Gani A, Rahman Md, Hassan M, Almogren A, Ahmad S (2018) Performance analysis of personal cloud storage services for mobile multimedia health record management. IEEE Access 6:1–1. https://doi.org/10.1109/ACCESS.2018.2869848
7. https://businessquant.com/dropboxs-paying-users-growth
8. Das N, Basu S, Bit SD (2020) Efficient dropbox deployment toward improving post-disaster information exchange in a smart city. <i>ACM Trans Spatial Algor Syst</i> 6, 2, 9:18. https://doi.org/10.1145/3373645
9. Idilio D, Mellia M, Munafo MM, Sperotto A, Sadre R, Pras A (2012) Inside dropbox: understanding personal cloud storage services. In: Proceedings of the 2012 internet measurement conference, pp 481–494
10. Bhat WA, Jalal MF, Khan SS, Shah FF, Wani MA (2019) Forensic analysis of Sync. com and flipdrive cloud applications on android platform. Forensic Sci Int 302:109845
11. Levin SP, Levin M (2019) Managing ideas, people, and projects: organizational tools and strategies for researchers. Iscience 20:278–291
12. Ilieva G, Yankova T, Hadjieva V, Doneva R, Totkov G (2020) Cloud service selection as a fuzzy multi-criteria problem. TEM J 9(2):484

13. Sikos LF (2019) Knowledge representation to support partially automated honeypot analysis based on Wireshark packet capture files. In: Intelligent decision technologies, Springer, Singapore, pp 345–351
14. HyunHo K, Lee HJ, Lim HT (2020) Performance of packet analysis between observer and wireshark. In: 2020 22nd International conference on advanced communication technology (ICACT), IEEE, pp 268–271
15. https://github.com/ieee8023/covid-chestxray-dataset

An Approach of Trustworthy Supply Chain Management of Ventilator and PPE in COVID-19 Pandemic Through Blockchain Technology

A. R. Sathya and Barnali Gupta Banik

Abstract The rapid outbreak of the pandemic COVID-19 has uncovered how our healthcare systems are confined to handle situation like emergency resilience, mitigation of risks, and supply chain management. One of the major problems during this pandemic is the growing demand of personal protective equipment (PPE) and significant medical supplies resulting in serious shortages for health care and frontline workers. But there has been several fake information being spread during this pandemic creating panic among the public and no system to authenticate the information. The lack of transparency of information along with the difficulty of accurate tracking of goods movements within the supply chain needs a robust traceability system. So, for efficient planning of operational processes and resource arrangements, we need to adhere to innovative technologies like blockchain. It is known that blockchain plays a significant role in healthcare systems. It improves the regulatory approvals and provides a prompt data management in clinical trials. Similarly, in supply chain, it streamlines the communication between different stakeholders, assures safe, transparent, and secure exchange of data among the collaborators. Thus, a blockchain-based tracking system will help the public to receive reliable and authentic information from government agencies. This paper analyzes the different blockchain-based applications and diverse prospects for handling this pandemic situation in a systematic way. It also proposes a blockchain-based supply chain framework for managing ventilators and PPE kits using Ethereum smart contracts.

Keywords COVID-19 · PPE · Supply chain · Traceability · Blockchain · Ethereum · Smart contracts

A. R. Sathya (✉) · B. G. Banik
Department of Computer Science and Engineering, Koneru Lakshmaiah Education Foundation, Deemed to be University, Hyderabad, Telangana 500075, India

B. G. Banik
e-mail: barnali.guptabanik@ieee.org

J. M. R. S. Tavares et al. (eds.), *Cyber Intelligence and Information Retrieval*, Lecture Notes in Networks and Systems 291,
https://doi.org/10.1007/978-981-16-4284-5_39

1 Introduction

COVID-19 showed us the limitations of current medical and supply chain systems to encounter the increasing demand for PPE kits. The supply chain process is greatly disturbed by the factors like panic purchase or misuse of PPEs leading many to risk lives [1]. The unexpected increase in supply for PPE, like face shields, surgical masks, gloves, gowns, and goggles, has caused massive production and a subsequent surge in costs, making disproportionate distribution of PPEs. As the frontline workers are at higher risk to expose to the virus, they should wear proper PPEs to prevent themselves as well as from further transmission of virus. But the sudden outbreak of pandemic caused a reduced stock, and healthcare workers are forced to repeat the products that are meant for single use. The important challenge during COVID-19 is to confirm sufficient number of PPEs, ventilators, and other medical supplies are available as and when needed. Therefore, in order to combat this situation, a reliable and robust tracking supply chain system for PPE and ventilator is required.

2 Background

2.1 *COVID-19 Crisis and Supply Chain*

This corona virus has affected millions of people's lives across the globe and created a global medical emergency. Not only healthcare systems, COVID-19 affected every other domain like global economy, trade, and finance. Many important events like Dubai expo [2], Olympics [3], had to be canceled seeing the situation. Statistics shows that the range between $2 trillion and $4.1 trillion could be the global cost of corona virus [4]. As a result, scientists around the world have been working hard to find a cure and forecast the prospects for a cure. Back in 2018, the US Department of Health and Human Services (HHS) and the Johns Hopkins Center for Health security predicted the shortage of ventilator. It is expected that up to 742,500 individuals may require mechanical ventilation during a serious respiratory pandemic such as COVID-19. As the situation was out of control, the supply and managing of important medical supplements were handled by HHS and Federal Emergency Management Agency (FEMA) [5]. There is no standard mechanism exists among the hospitals, manufacturers, and other stakeholders to even communicate the supply and demand-related information. In this existing process, few were using emails, few using spreadsheets, and few over phone sharing the demand and supply to state governments which in turn will be provided to FEMA. And then, based on the information provided FEMA conducts a manual process to coordinating the demands and supplies of the state. Because there were many different systems used to collect information and processing that information manually to manage the need and supplies, there has been a delay and deficiency in the supply. Therefore, a tightly integrated mechanism where all the stake holders can share their requirements and supplies is needed.

On the other hand, this pandemic can be considered data driven as every data points like patients tested, patients affected, patients not affected, patients recovered, beds, ventilators, and PPE kits shortfalls have to be reported to track the progress of the situation both by public and officials in real-time [6]. However, if the data is not accurate and verified, the decision made can be wrong. In order to provide an accurate information, the need for a reliable tracking system is inevitable.

Therefore, advanced technologies like artificial intelligence (AI), blockchain, deep learning, and machine learning can help to handle the situation in a structured way. However, blockchain in specific due to its unique characteristics like security, accuracy, transparency, and immutability assures that information over it as authentic and valid.

2.2 *Supply Chain Change Due to COVID-19*

The extent of shortages and disruptions caused by COVID-19 is huge, and certainly, healthcare systems were not prepared for that as supply chains are mostly not on real-time basis. Further medical supply chains have had limited regional and vendor diversification. Moreover, medical supply chains in general are not automated and follow non-digitalized processes due to which no demand analysis could be made in case of unexpected changes in demand or changes. Therefore, the stakeholders suggested few of the changes to improve the stability of healthcare supply chain. The notable changes are shown in Fig. 1 [7].

As medical supply chain does not follow much of digital supply network, it was difficult to cope-up with emergency situations like this pandemic. Hence, embracing new technologies can help the supply network participants to get more reliable information about supply, demand, quality, tracking, etc. E-procurements with several non-medical companies helped to connect with more suppliers; so that in case of shortages, it can be addressed more swiftly. Some procurements even helped in predicting the shortage of medical supplies and manufacture the products accordingly [8]. Several public health organizations across the globe came together to manage shortage of medical supplies. Some initiatives were made to even share

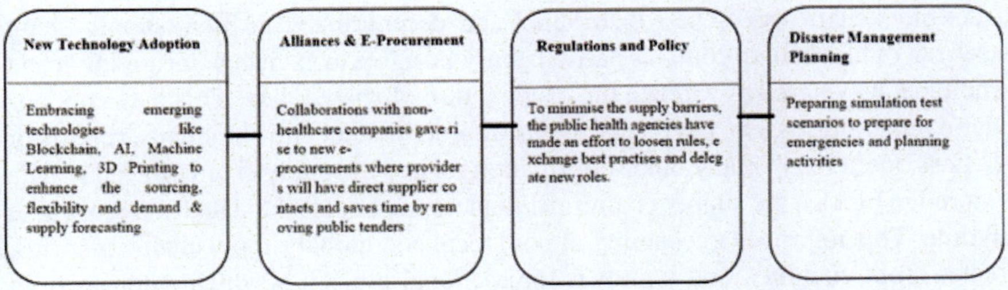

Fig. 1 Supply chain changes due to COVID-19

best practices followed by countries in managing the pandemic. This pandemic also exposed unpreparedness of medical and supply chain areas. Thus, a global disaster management planning committee has been formulated for emergency preparedness and plan for future [9].

2.3 Blockchain Technology versus Medical Supply Chain

The COVID-19 emergency has showed the lack of supply chain readiness across industries. A proper operational healthcare system needs an efficient supply chain system [10]. An efficient supply chain should be capable of delivering important health supplementary like PPEs, ventilators, vaccines, and other medicines in a faster and secure way especially during pandemic outbreaks. Also, lockdowns and shutdowns have made the supply chain process even more challenging. Furthermore, there were many fake PPE despite COVID-19. Many of them had to be thrown away as they failed the quality and safety test [11, 12].

Blockchain can very well fit in the supply chain requirement of safe and transparent way of data exchange as it has built-in features like security (Cryptography), transparency, distributed, and reliable solution at low cost (as it removes intermediaries) [13]. And using Ethereum smart contracts to automate the activities makes it even more easy with not much human intervention. Some of the COVID-19 use cases using blockchain is summarized in below Fig. 2.

To bring any new medicine or medical device into market, it has to be tested completely and show the efficiency and monitor if any side effects appear in clinical trials. It needs a fair and transparent system to carry out the process. Similarly, medical supply chain where the needs and demands of medical supplements can be captured, addressed, and managed in a structured way. Data collection is another important aspect in COVID-19 as it is essential to collect data about infection, symptoms, and taking them for research. When a person is affected, the contacts of them are to be tracked and identified. At the same time, this data has to be accurate and privacy of the data also to be maintained. Therefore, the need for a reliable, transparent, and privacy preserving system is inevitable, and blockchain fills that gap by providing all the needed features.

As blockchain works in a distributed and decentralized fashion, supply chain based out of blockchain contains participants who agree to exchange information and simultaneously agree to validate the exact nature of shared data. This data contains sales data, digital assets, payments, status of procurement and shipments, reports of business operations, or any other information related to supply chain. Since the data is stored in blocks, the supply chain stakeholders can access the data from anywhere anytime. This improves traceability at both local and global supply chain scenarios, creates more visibility, and improves transactional efficiency which would help in preparedness for emergencies. In addition, blockchain also provides multi-tier supply chain linkages starting from suppliers to consumers [14]. The traditional process of supply chain with respect to PPEs and ventilators is showed in Fig. 3.

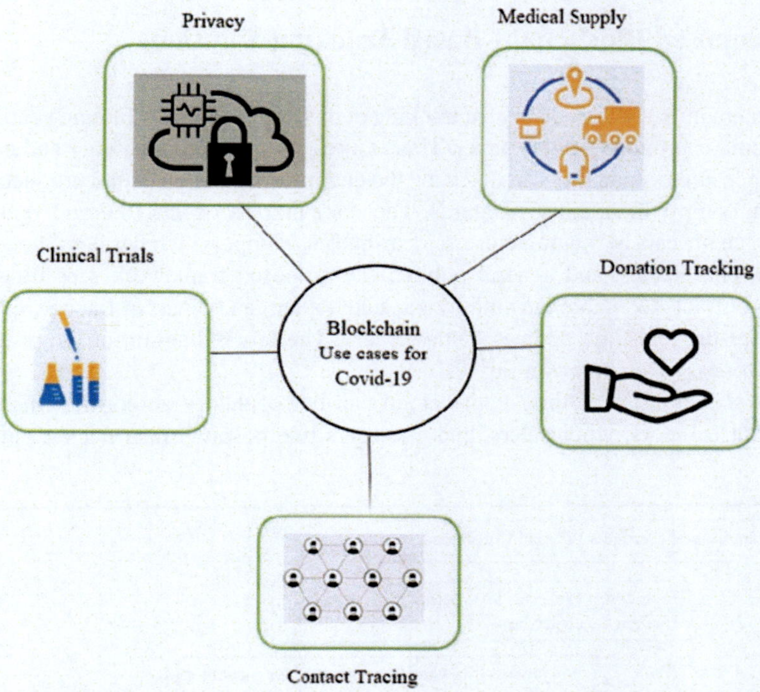

Fig. 2 Blockchain use cases in COVID-19

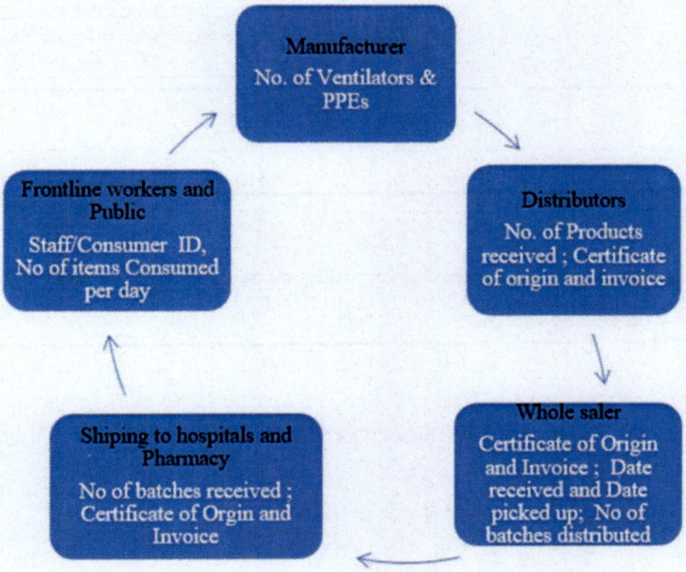

Fig. 3 Supply chain process for healthcare devices

3 Proposed Blockchain-Based Tracking Solutions

A blockchain-based framework for tracking of medical devices, PPEs, and ventilators in specific is proposed in this paper. This assures end-to-end traceability and quality assurance at the same time. In this way, the end user is guaranteed the products sold to them comply the quality standards. The stake holders of this PPE and ventilator supply chain can be manufacturers, distributors, shippers, wholesalers, hospitals, clinics, pharmacies, and general public. The proposed framework uses Ethereum smart contracts to enable the authorized stakeholder can be part of the network and can trace the real-time updates of the orders. The block diagram of the proposed framework is given below in Fig. 4.

The stakeholders of this supply chain can be regulatory authorities, manufacturers, distributors, wholesalers, and end-users like hospital, pharmacies, patients,

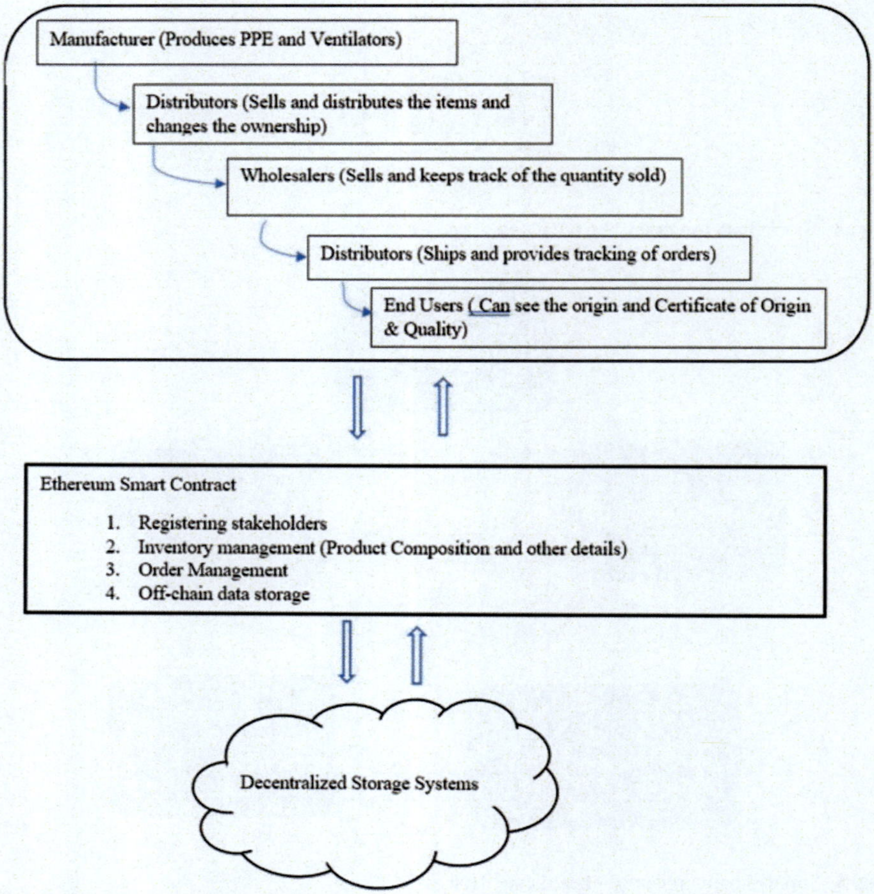

Fig. 4 Blockchain-based framework for PPE and Ventilator supply chain

and public. The role and activities of each stake holder are: (i) the regulatory agency is the administrative authorities who enrolls different stakeholders in the blockchain network. They also are responsible for the quality check of drug, PPE, ventilators, or other medical device and guarantee that they follow regulatory guidelines. (ii) The manufacturer is the one who produces the needed medical supplies using the raw materials. Also does the market study, finds the needed items, gets the raw materials and ensures sufficient quantity produced based on customer needs. It is also to be ensured safety guidelines are adhered and should be able to predict if any shortages might occur in future. (iii) Distributors are the intermediate entity between the manufacturer and the consumers. Once the product is manufactured, the distributor collects the products and distributes either to a wholesaler or to the consumers like hospitals. Sometimes distributors may be repacking the supplies to ensure utmost safety. (iv) the wholesaler is responsible to collect the products in large quantity and satisfy the needs of end customer. (v) Once the order is placed by healthcare service provider, the wholesaler requests the shipment companies to safely transmit the ordered items with a shipping fee agreed by the customer and the wholesaler. (vi) Once the shipment is done, real-time updates of the shipment will be provided by the shipment provider. (vii) The end-users can be hospitals, clinics, and pharmacies who are in need of PPEs and other essential medical devices for their consumers. They are responsible for keeping track of the items and place the demand information based on the availability and need.

Ethereum smart contracts are software code developed to execute functions automatically without any third-party interference. All vital functionalities of the stake holders like (i) registration of stake holders into the network, (ii) the users can make their demand requests, and they can even see the composition of the medical supplies before placing the orders. iii) The manufacturers can forecast the demand and produce the supply and distribute them, and once the order is placed, the complete order details can be accessed and tracked by the consumers. A smart contract can do all of the mentioned without the need for any intermediaries [15]. A decentralized storage system can be used to store data, and all the nodes in the network can get access to it. The decentralized system can be IPFS, MongoDB, Swarm, Filecoin, or any decentralized database. The combined use of blockchain and decentralized storage is considered to be more efficient as the blockchain will store only indexed information.

In this way, transparency for the entire process can be achieved, and every stakeholder will get a real-time update of the products which may not be possible in traditional supply chain processes [16].

3.1 Benefits of Blockchain-Based Solutions

Blockchain is adopted in various sectors of supplychain for process optimization and improving the efficiencies of several other factors like strengthening the relationship of core business stakeholders, attracting more clients, enhanced tracking,

Table 1 Benefits of blockchain-based solutions

Existing methods	Blockchain-based method
Slow processes and involves lot of manual activities	Automatic execution and are quite faster
No tracebility and very difficult in case of theft, loss, or damaging of items	Highly traceable
No transparency	Completely transparent
Loss of revenue and reputation due to lack of traceabilty and transparency	Strengthens customer relationship and provides trust
Higher IT transaction cost	Reduces IT transaction cost

and most importantly transparency which is lacking in the existing manual process. The advantages of a blockchain-based solution over existing methods are briefed in the Table 1.

4 Conclusion

In this paper, the crisis of COVID-19 analyzed and how healthcare sectors and supply chain domains are impacted specifically in tracking and managing the medical devices like PPE and ventilators are briefly explained. It is also discussed how blockchain can be adopted to overcome the challenges, and a blockchain-based supply chain framework is proposed to track the PPE and ventilators. The framework includes verification and validations of data using smart contracts and external storage systems to store the data records. Therefore, it is commercially feasible to adopt a blockchain-based approach for PPE supply chains offers a simplified, stable, trusted, and transparent form of communication between different stakeholders.

References

1. Park C-Y, Kim K, Roth S (2020) Global shortage of personal protective equipment amid covid-19: supply chains, bottlenecks, and policy implications
2. BIE Member States confirm one-year postponement for Expo (2020). Expo 2020 Dubai UAE. https://www.expo2020dubai.com/en/whats-new/expo-stories/bie-announcement. Accessed 25 May 2020
3. Ramsay G (2020) Japanese PM and IOC chief agree to postpone 2020 Olympics until 2021. CNN, 24 March 2020. https://edition.cnn.com/2020/03/24/sport/olympics-postponement-tokyo-2020-sptintl/index.html. Accessed 25 May 2020
4. Alegado S (2020) Global cost of coronavirus may reach $4.1 trillion, adb says. Bloomberg (2020). Accessed 15 August 2020
5. https://www.fema.gov/news-release/2020/03/27/fema-administrator-march-27-2020-letter emergency-managers-requesting-action

6. WHO (2019) Rolling updates on coronavirus disease (COVID-19).World Health Organization. https://www.who.int/emergencies/diseases/novel-coronavirus-2019/events-as-they-happen. Accessed 25 May 2020
7. https://decisionresourcesgroup.com/downloads/4-ways-medtech-supply-chains-changed-due-covid-19/
8. Digital procurement marketplace connects hospitals responding to coronavirus outbreak (2020) Supply Chain Dive. March 26
9. Top 3 supply chain management changes hospitals must make post-Covid-19. Medcity News. June 4 (2020). 13. COVID-19 pandemic shows we must rebuild our national emergency supply chain. The Hill. April 28, 2020
10. Houston L, Probst Y, Humphries A (2015) Measuring data quality through a source data verification audit in a clinical research setting. Stud Health Technol Inform
11. Rana SS (2020) Rise of counterfeit ppe in india amid covid-19. ICLG.com (2020). Accessed 19 Aug 2020
12. Lagasse J (2020) Healthcare industry is grappling with the emergence of counterfeit ppe in the Covid-19 battle. Healthcare Finance (2020). Accessed 19 Aug 2020
13. Zheng Z, Xie S, Dai H, Chen X, Wang H (2017) An overview of blockchain technology: architecture, consensus, and future trends. In: IEEE 6th international congress on big data, Honolulu
14. Kouhizadeh M, Sarkis J (2018) Blockchain practices, potentials, and perspectives in greening supply chains. Sustainability 10(10):3652
15. Parizi RM, Dehghantanha A et al. (2018) Smart contract programming languages on blockchains: an empirical evaluation of usability and security. In: International conference on blockchain, Springer, pp 75-91
16. Ilhaam O, Mazin D, Raja J, Khaled S, Mohammed O, Junaid A (2020) Blockchain-based supply chain traceability for COVID-19 PPE. TechRxiv. Preprint. https://doi.org/10.36227/techrxiv.13227623.v1

The Practical Enactment of Robotics and Artificial Intelligence Technologies in E-Commerce

Sabyasachi Bala, Mohammad Nadeem Khalid, Hardeep Kumar, and Vinod Kumar Shukla

Abstract This paper maps and portrays the current and anticipated relationship between two different consistent pieces of exhibiting science and programming. It takes a gander at the association of cutting-edge displaying and artificial cognizance (computerized reasoning) in the insightful network, proposing, at the same time, an artificial intelligence model that could fit in a couple of parts of cutting edge displaying coherent zone. There are different consistent disseminations concerning artificial mental ability (human-made knowledge) across the requests. Regardless, suggesting automated promoting points of view, this number excess parts nearly nothing. Coherent research on artificial cognizance (computerized reasoning) could benefit from displaying science in different ways. Nowadays, simply a restricted amount of consistent investigation insinuating progressed publicizing and synthetic thinking (human-made brainpower) is related to unequivocal electronic advancing techniques. This paper includes man-made review (man-made knowledge) applications on cutting-edge promoting temperate zone, establishment dissemination, and refers to potential comparative reasons that occur. It gives an artificial intelligence model that could fit in a couple of cutting-edge publicizing occasions. This paper depicts the internet business, advancement circumstance, and possibilities of AI innovation, examines the current circumstance of the use of AI innovation in the field of internet business, fundamentally consider and talk about in detail from the part of the partner of AI, smart coordination, proposal motor, and the ideal valuing application through the exploration of online business insightful activity case—noon.ae, amazon.ae, tests into the significant effect and incredible noteworthiness on the internet business improvement of artificial brainpower.

Keywords Artificial intelligence · Machine learning · E-commerce · E-business

S. Bala
Amity University, Dubai, UAE

S. Bala · M. N. Khalid · H. Kumar · V. K. Shukla (✉)
Department of Engineering and Architecture, Amity University, Dubai, UAE

H. Kumar
e-mail: hkumar@amityuniversity.ae

J. M. R. S. Tavares et al. (eds.), *Cyber Intelligence and Information Retrieval*,
Lecture Notes in Networks and Systems 291,
https://doi.org/10.1007/978-981-16-4284-5_40

1 Introduction

Artificial intelligence [1] is the mechanism of a computer-aided robot or fully controlled digital computer to perform the tasks which are in general, beyond to the human intelligence. It is programmed in the machine or digital computers to behave like human beings and show their performing tasks. It is the power like rational beings and deciding to undertake actions to gain the best possibility of reaching goals [2]. In the modern time, AI has been undergone many industries and gathered boom of many industries such as medical, engineering, and commercial applications such as marketing and sales, and business e-commerce. This comprehensive technology is designed to solve the internet-based commercial activities such as online marketing, sales, e-government activities, purchase, e-bill payments, e-licensing, and online currency investment resulting in brand competitiveness and customer loyalty. The following sections aim to know the current standing and patterns of e-commerce or electronically digital commerce and observe how robotic artificial intelligence influences the activity of e-commerce. Machine learning [3] enables the system to learn by the machine automatically and gather the result by experiencing by specifically writing every instruction for the machine's action and understanding in machine language. It is a computer program that deals with data access and data observation, such as data experience or data commanding to observe and investigate data patterns to develop future development. The main objective is to automatically learn computers by a set of algorithmic programs without human help that acts better than humans.

2 Literature View

Artificial intelligence [4] integrates information technology with analytical solid and computational capability consisting of the Web system, hardware, and software to achieve the desired AI solution. The technology is further modified and restarted as an application of digital information science to promote organization-level and supremacy activities at a competitive platform. Artificial intelligence is the software architecture with highly equipped technology system software to improve environmental technology with working results [5]. Artificial intelligence and machine learning system technology is a subset of an algorithm that always enables deep learning and data processing like a human being and more accurate business knowledge and application. AI incorporates intelligence and has no error or free from error decision making. Here, the working time is guided by state-of-the-art technology and IT systems to minimize the hardness of created problems and maximize perfection with lawfulness. Here, there is no existence of emotional decision-making agendas that influences the problem zone. Here, the decisions are based on statistical analysis.

Machine learning is a kind of man-made intelligence described by Samuel [6] as the limited structure to learn without being randomly adjusted and the same as how a human behaves and expresses. Use of AI can be followed to confirm the market designs or any other occasions from educational assortments to reasonably envision the fundamental pieces of information [7], responses, and reactions so that sponsors can fathom the hidden driver and likelihood of repeating exercises and adjust ideal game plans. Given the business usages of AI, counts are in sample supply [8], going from K-suggests gathering, neural networks, decision trees, reinforcement learning, etc.

Generally, gigantic data is described as 3 V's explicit volume, arrangement, and speed [9] and implies a promoter's ability to aggregate and section an enormous volume of data [10] with unimportant manual work utilizing development. Advancing divisions appropriately use this clever data for re-trying their thing, offering customer-driven necessities. Top-tier IT structure engaged relationships by joining a blend of logical, mechanical assemblies, simple techniques, data interfaces, portrayal moves close, and programming examination [11]. This gigantic data gets changed to sensitive data [12], which is then used for making regard settling on business decisions and imperfections.

Marketing in progressing as indicated by the authoritative point of view can be portrayed as a turn of events, set of establishments, and cycles for making, presenting, passing on, and trading responsibilities that impulse clients, customers, embellishments, and society unhindered [12]. Marketing from the client's point of view will be depicted as developing steps toward achieving the required target of essential worth expansion to related assistants at a foreordained worth/cost. Marketing as per social viewpoint is defined as a social cycle by which people and parties get what they need a great deal through making, offering, and straightforwardly trading things and associations of tremendous worth with others. Key purchaser markets in presence are customers showing, business progressing (industrial marketing), global publicizing, and non-advantage advancing [13].

3 Marketing-Mix

The fundamental business model consists of factors that an organization or a company controls to influence the consumers to buy the products, called Marketing Mix, which consists of product, price, place, and promotions. The dealer uses different publicizing blends (4P's) as estimated by McCarthy [14], which can be masterminded explicitly by product, price, place, and promotion. In any case, as publicizing condition transformed more toward the client as a ruler moves close, extra client-driven [15] models were outlined. One such model was proposed by [16] wherein purchaser abrogating thing, cost supplanting cost, comfort abrogating spot and correspondence overriding movement. Another customer-focused model (solution, information, value, and access) propounded by [17] contrasted dealer-arranged and client-oriented game plan, the cost with client respect, a place with client access, and movement with client data.

Post globalization, as desired business areas get consolidated [18] with their worldwide accomplices, the ascent of cutting-edge exhibiting gadgets completely changed how business measures were streamlined. The regular advances were benchmarked with top-tier development guards, promoting assignments was benchmarked with top-level competitors. As the information open to customers gets quadrupled, improving their base of comparison and giving the dataset for thing assurance [19], customization has taken center stage at no other time and is obligatory united into business and thing portfolios. With the availability of various informational collections about buyer buying behavior and emotional cycle, the STP (segmentation, targeting, and arranging) targets of affiliations end up being more customer-driven and demand-synchronized. Loyalty in this electronic period is not limited to setting your things in customers' minds regardless of the substance of customers. In this thing, organic framework with millions of items and brand decisions, they would easily have the option to get spellbound and diverted by elective brands, lifestyles, and example changes. Standard promoting practices of propelling a host of things through the newspaper, radio, and TV are taking the brands behind the stages. Because of reduced customer thought and offering a way to deal with on the Web and progressed exhibiting headway stages [20] in the like of substance publicizing, Web crawler improvement, Web crawler advancing, online media progression, electronic media indicating. Social media promoting channels are simply expanding customer buying powers as a consistently growing number of clients are voicing their sentiments about association and checks and related thing credits.

An adjustment in standpoint, as an ordinary asset, is driven by organizational thinking and offering the way to deal with contemporary AI-controlled game plans [21]. With an example of progress in financial endeavors of significantly reserved examiners, loyalty credits of customers likewise, capacity skilling of legitimate HR in conclusion in accomplices compensating associations with their merited hold assets and profit. Computerized reasoning-driven stages with expanded progressed limits are filled with business model central purposes of proactive variety and perfect use compensating the related accomplices with improved essential concern, advantages, and future turn of events potential outcomes.

4 Marketing Application Related and AI-Oriented

Man-made knowledge close by subsets of AI moreover, effective learning are unraveling how associations are driven around the world. The speed of imaginativeness have caused unbelievable game(s) plans which are portraying the advancement of the 21st century. Joining AI-engaged headways can redesign affiliations data to take care of limits and make the dynamic cycle more far-reaching and agent [22]. This suggests that the stages can recognize sharp thoughts and subjects across gigantic educational assortments, incredibly brisk and give some wisest reactions to help progressive basic reasoning and consistent, robust process. Moreover, AI plans interpret inclining that a human, which prepares these stages to understand open-source content like online media, ordinary language taking care of, and email responses. Reproduced insight game plans can dive significantly into watchword look, social profiles, and related online data for astounding models and results.

Relation with Consumers: Man-made intelligence can assume a tremendous job in client adjusting and maintenance through giving 24 h administration overseeing 85% of client commitment by 2020 [23]. As AI bots will control immediate to-purchaser commitment roads (visit bots and call-bots), leaving aside adequate time for key dynamic to gifted administrators. Artificial intelligence-empowered client co-operations will improve convenient inquiry the executives, what's more, disbursal with no time limitations (24*7) and holding upcycle. Available to come into work line and home-visit holding up of the organization's administration workforce adds dissatisfaction in clients and a negative effect on the brand picture and estimation of multinationals.

Business Corporate Company: How helpful it is for business associations to be overseen via robotized measures aiding routine dynamic and everyday tasks. Notwithstanding the time and cash investment funds by limiting asset wastage, associations can likewise pick up as far as mastery and specialization in the workplace and upper hand at commercial center. Man-made intelligence stages admit to a whole web of client information [24] data and search chronicles; essential touch focuses, the purpose of deals and purpose of procurement making them capable in the craft of information division's more, investigation than their human partners.

Stakeholders: It is undeniably more effective to enhance social models and government exercises by AI-controlled instruments propelling sensibility, straightforwardness, and duty in social plans and following social new developments. Similarly, by enabling sending of AI engaged enrolling in critical parts of the government affiliations like appraisal deception recognizable proof, network security, advanced bad behavior, and advanced battling [25] (Bhatia, Gautam, and Tripathi, 2018), geographic after of disease scenes and the Census office for inspecting people designs [26] would be of monstrous help.

5 How Does Artificial Intelligence Act on e-commerce?

AI Assistants and Chatbots: A chatbot is an application programming that copies man-made reasoning or copies a speeching (or words) sharing an individual, taking everything into account, language through demonstrating programs, battles, versatile applications, and correspondence. They are crucial considering how an application is led by and gigantic proposed as likely the most frustrating and considering how sides promise clarifications of shaped exertion among machines and individuals. They help brands with seeing purchaser requests: as made as voice. Besides, they're utilized giving these suggestion through NLP. These applications smooth out relationships among affiliations and people, improving client transparency. At that specific time, they could affiliate new occasions to restore the customers' operational sufficiency and responsibility measure by diminishing the critical cost of client keep up. The online business models, are dependent for the help, from AI and its tools. Before, these were, in a general sense, offered standard answers; as of now, they've changed into a sharp enemy of electronic application that acknowledges different issues that the visitors need to battle with. Particular Web-shopping territories at long last have application is to help people settle on buying decisions. Truly, even applications as Facebook Courier have chatbots through which clients can banter with the shipper site also assist with the buying cycle. These bots pass on by utilizing either text or talk or both of enthusiastic circle [27].

Customer Relationship Management and Artificial Intelligence: Man-made care and CRM are a staggering blend. As the website is correct now, the foundation of contemporary publicizing and procedures works out; CRM has brought to make a central purpose a ceaselessly making stream of markers notwithstanding data focuses about likely results, customers, and even affiliations. Alluding to that, an expert discovers the proportion of that data, consistently on the fly, moreover, pick a sharp choice on unequivocally how to utilize that data in their talking best is, no statement with a two-sided undercurrent proposed a disturbing sell. Sales reps are utilized to highlight what's also watching out for them to get information experts ordinarily met with pretty blended plans. In express models, it can end up forestalling delegates who'd ordinarily be excellent in made by theirs from managing their responsibility capably. The fairness of man-made perception in CRM follows simulated intelligence of the examination. It makes cheerful counsel about a buyer or perhaps significant dependent on completing the message about that individual the contraption has collected.

AI targeting sales achievements: The pay drives the world, as we are cautious; online business is fundamentally sensibly about making bargains. PC-based approach will help in the entire customer's experience measure from getting the conceivable buyer to passing on after strategy affiliations. Unquestionably, man-made data may be utilized to make particular paychecks suitable for thing experience leads by offering specific information into bargains skims, which were passed on thing bargains data, standard postal mail business affiliations, etc. Considering these affiliations can improve resource contact to make a great pipeline, look at pack execution, what's more, achieve valuable outcomes. This very PC-based information helps in thing bargains forecasting:

- AI could finish standard positions, making bargains specialists attain zero in on crucial work.
- It gives substantial assistance to sales reps.
- It is a circumstance to oversee more data.

Customer Service: In prospering economy, the customer keeps up is the focal key to exceptional results, and different understanding leeds you to fantastic customer help which is delivered with perfection. AI attracts and relationship to weight on what's squeezing:

- Fulfillment of the clients.
- Observing their necessities.

- Offering little appreciation to the open gateway, it comes up.

Man-made intellectual ability to approach help to give vast customer keep up. This is unavoidably changed over in strategy volume. In a flavor, the customer keeps up environ a congruity is struck between the machine and the male. Best of man-settled on mindfulness decisions are used while simultaneously ace information is utilized to guarantee the custom setting. The conversation bots and PC-based information techniques are capable of level-one help obligation as they respond to the conventional customer worries, regardless of whether they are not dazzling concerning tangled customer unequivocal solicitations. As the affirmation of in-person affiliations will fall in advancing toward times, the centrality of mechanized thinking in Web business will rise enormously.

Automation: Man-made awareness and computer-based intelligence do not recommend that robots are ruling. Still, some vibe has these impressions. AI is simply a commitment to supply the response with AIR for the individual requirements, and which is unreasonably paying towards the constraint. By utilizing AI, agents can interface with customers in a better manner. It can likewise help save them time and wipe out disappointment related to enumerating something in and repeating the moment subtleties.

Use of AI in Marketing: Controlling appearing and inducing spending plans totally as by and large as conceivable requires full-scale space flourishing. This requires extra, what's much more, usually not exactly ideal expenses for a business that is tiny houses. Fortunately, at current events, small business establishments can apply even direct cash-related plans unequivocally by the reasonableness of man-made scholarly fitness approaches foreseen appearing. In general, small firms give up their lifting and publicizing philosophies to PC-based data programs and figures. That supervises showing rehearses across various channels (for instance, Bing, AdWords, Facebook), surveying publicizing execution, and giving arranged proposals as regards most reachable spending distributions concerning PPC, for instance, pay per click. By utilizing man-made data publicizing moves close, affiliations can consider the superb inducing procedure and tremendously limit CPC (click per call).

Practical function of e-commerce application system: Now AI method becomes more technologically enriched, many more commercial activities are applied practically to make the system more human-friendly. As an example (www.noon.ae), it is a regional e-commerce site, especially Gulf country-based, and is widely used in the UAE, the KSA, Oman, Qatar, etc. It also acts as a domestic e-commerce model such as OTO, which is called online to offline. From trusting on the e-commerce platform to user or consumer portal, the platform serves as new retail, delivery, or shipment and deliver errands. In the shipment or logistics phase, the domestic or regional intra-city logistics mode, the delivery system consists of time-to-time delivery with full security of the product [28].

The ultimate business consists of three stages. Those are the user or the consumer selects an item through smartphone or pc or laptop and clears payment (payment may be online or cash on delivery), dispatches, and shipment of the courier. The aim is to develop a selected item, making more value for consumers; core competitiveness is the internet or Wi-Fi smart. At noon, generally, browse items and select system by laptop internet browsing and intelligent algorithm functioning in GCC or any place within GCC. It consists of a recommendation system and a scheduling system. Through these two systems, some movable data such as order log, description of the courier, courier and time track item to know the accurate status of an item with efficient delivery. In the meantime, from data hub as user info, item info will be sent, merchants details with preference scene, and real-time availability to deliver to get positive feedback from the consumer. Hence, the mean waiting time can be reduced, and order at delivery process efficiency can be increased significantly.

Scheduling Process: It is the key part of the delivery that consists of dispatching and way of planning. Previously, the person who is dispatching the order, who makes 1000 orders a day and millions of weeks, makes a genuine e-commerce schedule. Now, the modern time OTO (online to offline) system counts many data and orders that need an intelligent and automated scheduling system and requires tracking the right order for the right person's door. With the provision of GPS navigation system, information will be gathered for the whole system.

Astute recommendation system: Noon.ae can recognize data, and this procedure can detect user sequence analysis. For difficult matters such as user, timings, and place, different orders are made to satisfy different users' needs (Table 1).

Intelligent Recommendation System: These models are on an offline and real-time basis. It indicates merchant mining, label, and attributes. Real-time data shows real-time capacity, real-time inventory, human behavior, and operational data. Offline data gives users long-term behavior, long-term portrait, real-time data, and analyses massive data of buyers, their access to noon.ae, user models and order summery and way to make future development [29].

Table 1 Various levels and responsibility

Application-level	Merchant recommendation, food recommendation, food suggestion
User modeling layer	The online platform, fact analysis, real-time platform
Strategy layer	Delivery strategy, improved pin ratio
Algorithm layer	Ranking, recommendation
Merchants model layer	The offline message, real-time info
Data processing layer	Hadoop, spark, mpi

6 Growth of Artificial Intelligence (AI) in E-commerce Globally

Today, especially in twenty-first-century, artificial intelligence technology has entered in daily life commercial process. It is also referred as one of the smart technology.

Consumers refer to smart product recommendation: More intelligent suggestions prompt more deals, particularly for more giant retailers with millions, maybe countless things. Indeed, man-made intelligence empowered customized item suggestions for online customers is expanding change rates by 915%. Amazon has verification, expressing that item upselling and strategically pitching is liable for a significant 35% of its general revenues. Amazon utilizes a mix of cooperative separating and next-in-grouping models to make expectations on items that clients may require straightaway. Alibaba is likewise using man-made intelligence calculations to drive keen item and search proposals by examining client perusing and communications with the site.

Utilizing man-made intelligence to Give Visual Pursuit, a More Normal Client Experience: One of the more stimulating pieces of man-made insight is the enablement of visual request using pictures. It gives the customer an incredibly improved—and more typical—way to deal with the search for things. If a customer sees a dress she likes with visual pursuit, she does not have to endeavor to depict it or even know the brand deliberately. She ought to snap a photograph, and shrewd picture glancing through development will pull up the same or near things. As examined, Amazon has named their visual interest thing StyleSnap.com. Starting late, an article in usmagazine.com demonstrated a use case for this part that solidly imitated average customer lead. It used an innocuous Instagram photo of Jennifer Aniston to find close unmistakable matches to the top she was wearing in the picture. It should be notified that the photo was not a thing exceptional, nor did it show a great deal of what she was wearing. Anyway, StyleSnap.com passed on a reasonable selection of choices. Certifiable customers would not hold on for a perfect photo; they will act rashly reliant on what they find at that point, and recreated knowledge can manage this lead.

Simulated intelligence Partner for Client care to Lift Deals: A portion of the time, customers essentially need a little help to make a purchase. Man-made cognizance engineers have a mind-blowing answer for this: automated salesmen. They come as chatbots, humble aides, and even voice-request accomplices. Distant aides show remarkable assurance in helping customers find what they need. The North Face uses IBM's mimicked knowledge plan called Watson to enable online clients to locate their optimal coat. Customers answer tends to like "where and when will you use your coat?" through voice input. The reproduced insight programming investigates numerous things to find great matches subject to ceaseless customer information and its investigation, like atmosphere conditions in the area. The more factors the accomplice needs to work with, the more changed an answer it can give.

Voice-engaged organizations are also in this class by a similar token. Amazon's high-level right hand Alexa uses NLP to sort out how people talk (and Alexa is upheld to help customers find and buy things by bantering with customers). Now, the Google Duplex contraption is making capacities like making fundamental food thing records spoken by a customer and, regardless, setting orders for them.

Dynamic Esteeming to Give Ultra-Changed Cutoff points Ceaselessly: Dynamic assessing is a procedure whereby retailers change the expense of the thing reliant on market revenue ceaselessly. It uses the customer data, genuine assessing data, and arrangements trade data to predict when to restrict, what to restrict, and logically register the base proportion of markdown expected to ensure trade. The result is the ability to pass on at the last possible second, altered markdown to a Web business client, with the most important possible advantage to the brand. Amazon is the market boss in this space, at present, seeing immense accomplishment in applying dynamic assessing. They change costs as expected, which is on numerous occasions more than Walmart and Best Buy, and has incited an advantage increment of 25%.

Man-made intelligence for Advancing Stock and Distribution centre Administration and Coordination: The profitable stock organization is connected to keeping up the right stock level that can fulfill market revenue without adding to sit stock. Progressed collaborations are connected to diminishing resource use in movement and transport time. China's JD.com has a basic association with Siasun Robot and Automation Co Ltd., conveying man-made knowledge and robotization to seven collaboration environments responsible for working 209 stockrooms. Reports ensure that online solicitations were duplicated from 2014 to 2015 (a total of 1.26 billion), and 85% of those solicitations were passed on within two days. Alibaba claims that strong collaborations have achieved a 10% decline in-vehicle use and a 30% reduction in development distances. They use a collaboration part, Cainiao, who uses reproduced knowledge to map the best movement courses.

Computerized reasoning can handle a Variety of Complex Online business Issues: There is no limitation to such an issue that man-created insight can settle, at whatever point applied precisely. Let us assume Flipkart, a primary Indian Web business association with 39.5% of the Indian online business market. In India, unlike in western countries, postal area plans are not standardized. People follow neighborhood shows and consistently use close-by names rather than genuine ones. This can provoke major determined issues, critical costs because of misdelivered demands, and defenseless shoppers. Flipkart used man-made reasoning and ML to deal with this intriguing issue. Data analysts at Flipkart figured an area course of action structure, using significant learning and computer-based intelligence, which has a precision speed of 98%. It is expecting a critical occupation in ensuring right and helpful transports.

Web business Specialists can Change the Game With reproduced knowledge and ML: These couple of models are ways that Web business originators save or make a large number of dollars for associations. However, there is a considerable amount more than artificial intelligence planners can do in the online business space. Moreover, it is not just the strong hitters like Amazon or Alibaba.com that are getting in on the game. Undoubtedly, even likely, the most standard online business stages, which are used by SMEs all over the world, are using PC-based insight and Machine Learning to serve their customers, more promptly. Along these lines, the sky's the limit. Web business advancement is not booked to upset any time soon. As a general rule, the lifestyle of transport may have been exacerbated by the COVID-19 situation. Soon, it will be something past generally electronic business associations that ought to use man-made consciousness and ML answers for updated errands, extended arrangements, and better customer experience. [Basic reference: Amazone, Tractica, Servion, Investopedia, Emerj, Research by my business.inc, Flipkart, SearchEngineLand] [30].

7 Conclusion

Currently, e-commerce media and technological e-commerce companies such as www. noon.ae, www.amazon.ae are progressively doing business with artificial intelligence aid and have constituted their e-commerce platforms to increase their profits with other rivals. In UAE, many e-commerce rivals exist, such as Zomato, Talabat, Careem, Uber Eats, Amazone.ae, el grocer, and they have also introduced their commercial robot with AI. In logistics, they have installed a recommendation engine that consists of a visual artificial intelligence station to track the images and word and speech tracing system thus, we speak to get the exact item. Some superior technologies have been introduced to achieve continuous development on research, such as deep learning, word recognition via voice search, biometrics and picture and video recognition system, and automatic processing method such as the NLP system, which helps an intelligent developer. AI methods are booming at a fast pace everywhere and all possible commercial applications and satisfy all other e-commerce demands such as electricity and telephone bill payment with the fastest possible customer satisfaction. It is also an integrated tool for the transformation of e-commerce. Nowadays, from grocery shopping to electric bill payment to medicine delivery everywhere, AI creates better integrated CRM and other promotional activities.

References

1. Athota L, Shukla VK, Pandey N, Rana A (2020) Chatbot for healthcare system using artificial intelligence. In: 2020 8th international conference on reliability, infocom technologies and optimization (trends and future directions) (ICRITO). Noida, India, 2020, pp 619–622. https://doi.org/10.1109/ICRITO48877.2020.9197833
2. Girdher S (2019) Role of artificial intelligence in transforming e-commerce sector by shaffy girther. Res View J 04(06). ISSN 2455–3085
3. Artificial intelligence enabled marketing solutions. A Review by Sushant Kumar Vishnoi, Teena Bagga, Aarushi Sharma And Saadat, Nasir Wani

4. Freeman C (1989) Technolgy policy and economic performance, Pinter Publishers, London & New York
5. Vishnoi, Sushant Kumar, et al. (2018) Artificial Intelligence enabled marketing solutions: A Review. Indian J Econo Bus 167–177
6. Samuel A (1959) Some studies in machine learning using the game of checkers. IBM J Res Develop 3:210–229
7. Bluma AL (1995) Selection of relevant features and examples in machine learning. Elsevier, pp 245–271
8. Ray S (2017) Analytics Vidhya—learn everything about analytics. [Online] https://www.analyticsvidhya.com. Accessed on March 2021
9. Laney D (2001) 3D data management: controlling data colume, velocity and variety. Meta Delta
10. Amir MH (2014) Beyond the hype: big data concepts, methods, and analytics. Int J Inf Manage 35:137–144
11. Roski J, Bo-Linn GW, Andrews TA (2014) Creating value in health care through big data: opportunities and policy implications. Health Aff 33:1115–1122
12. Farell D (2002) How IT enables productivity growth. McKinsey Global Institute, San Francisco, pp 313–324
13. American Marketing Association. (2013) Retrieved from https://www.ama.org/AboutAMA/Pages/Press-Release.aspx. Accessed on March 2021
14. Kotler P, Keller KL, Koshy A, Jha M (2009) Marketing management. In: Kotler P, Keller KL, Koshy A, Jha M (eds) Marketing management: a south indian perspective, 13th Edn. Pearson Publications, p 11
15. Londhe DB (2014) Marketing mix for next generation marketing. In: Symbiosis Institute of Management Studies Annual Research Conference (SIMSARC13*)*, pp 335–340
16. Lauterborn R (1990) New marketing litany. In: Advertising age, pp 26–26
17. Dev CS, Schultz DE (2005) A customer focussed approach can bring the current marketing mix into the 21st century. Mark Manage 14:18–24
18. Coviello N, Milley R, Marcolin B (2001) Understanding IT-enabled interactivity in contemporary marketing. J Interact Mark 15:18–33
19. Goldsmith R (1999) The personalised marketplace: beyond the 4Ps. Mark Intell Plan 17:178–185
20. Mahadevan B (2000) Business models for internet-based E-Commerce: an anatomy. Calif Manage Rev 42:55–69
21. Thielscher M (2011) General game playing in AI research and education. In: Proceedings of the 34th annual german conference on advances in artificial intelligence. Springer-Verlag, Berlin, Heidelberg, pp 26–37
22. Brooks RA (1991) Intelligence without representation. Artif Intell 47:139–159
23. Fowler W (2018) Inside big data. Retrieved from https://insidebigdata.com/2018/08/31/ai-taking-lead-generation-nearest-future. Accessed on March 2021
24. Henderson J, Venkatraman H (1999) Strategic alignment: leveraging information technology for transforming organizations. IBM Syst J 38:472–484
25. Bhatia R, Gautam S, Tripathi A (2018) Digital security. Privacy and Emerging trends in cybercrime, Noida
26. Bryant RE, Katz RH, Lazowska ED (2008) Big-data computing: creating a revolutionary breakthroughs in commerce, science, and society
27. ISSN:2455–3085 (ONLINE), Research Review International Journal of Multidisciplinary, Role of Artificial Intelligence in Transforming E-commerce Sector, Shaffy Girder, *Assistant Professor, Panjab University Constituent College(sikhwala) (India)*
28. Application of AI in e-commerce, doi: https://doi.org/10.1088/1742-6596/1302/3/032030, 1 Shandong agriculture and engineering university, jinan, shandong, China , 2 Shandong normal university, jinan, shangdong, China , 3 Sinopec lubricating oil co. LTD, jinan, shangdong, China , E-mail: song@sdaeu.edu.cn

29. Song X, Yang S, Huang Z, Huang T (2019) The application of artificial intelligence in electronic commerce. J Phys: Conf Ser 1302:032030. https://doi.org/10.1088/1742-6596/1302/3/032030
30. IOP Conference Series: Journal of Physics; Ritu Punhani et al (2021) J Phys: Conf Ser 1714:012026
31. Wanganoo L, Shukla VK (2020) Real-time data monitoring in cold supply chain through NB- IoT. In: 11th international conference on computing, communication and networking technologies (ICCCNT), pp 1–6. https://doi.org/10.1109/ICCCNT49239.2020.9225360

Feasibility Study of Software Engineering Aspects of Bigdata Analytics Applications for Academicians

Nivedita Kasturi, R. B. Geeta, Goldina Ghosh, and S. G. Totad

Abstract Aim of the work is to find research scope specifically for academicians with little or no big data application development knowledge. Observation made are based on research conducted by academicians that are only in terms of theoretical information, and there is no real-time implementation so that work can be validated. Researcher from industry and academicians in collaboration with industry are having real-time implementation for the methods proposed. The observation shows that real-time implementations are done by researcher with big data application development knowledge so we tried to address the issues related to conduction of research in this area by academicians, and also, we tried to provide set of objectives which may help novice researcher. Researcher will face challenges while developing big data applications using software engineering methodologies, as big data application is completely different from traditional application. To accomplish the work, survey has been done in the field of software engineering aspects of big data analytics applications and verifying how feasible it is to conduct research in this area without much experience. The results show that academicians can conduct research in this field with collaboration with software industry so that their work is validated with real-time implementation. The researcher with no or little knowledge in big data application development can follow guidelines provided in research outcome section in this paper so that they can proceed with their research easily. The whole work tried to address the difficulty faced by researcher in conduction of research in this field.

Keywords Big data applications · Data analytics · Software engineering · Phases of software development · Requirement engineering · Design · Architecture · Quality assurance

N. Kasturi · R. B. Geeta · S. G. Totad
KLE Technological University, Hubli, Karnataka, India
e-mail: totad@kletech.ac.in

G. Ghosh (✉)
Institute of Engineering & Management, Kolkata, West Bengal, India
e-mail: goldina.ghosh@iemcal.com

J. M. R. S. Tavares et al. (eds.), *Cyber Intelligence and Information Retrieval*,
Lecture Notes in Networks and Systems 291,
https://doi.org/10.1007/978-981-16-4284-5_41

1 Introduction

Software engineering is a part of all the applications whether it is a traditional windows application, glossy Web application, current big data applications, IoT application, and even artificial intelligent application.

Data analytics and big data application deal with three characteristics that is volume, variety, and velocity. If any of the big data applications need to be built, then it has to address all their characteristics in the framework activities of software engineering. A framework activity refers to planning, communication, analysis and design, construction and testing, and maintenance and support. Design frameworks [30] mentioned for these applications need to be tested with industry experts and specifies the debugging process [22] and also difficulties involved in it.

Gorton et al. [24] provide the overview of how these applications have evolved without any standard process being developed. Big data analytics had created a tremendous improvement in the business analysis and decision-making software, based on the past and current data that exists for a particular business. For this it is required to address how to build a big data application with the quality factor in mind [41].

Software engineering process models are followed to build the application with better quality. But big data application mainly concentrates on the infrastructures to store and manage data while less attention is given to application development [6]. Hence, there is a need to address all the framework activities in the software engineering process for big data application development and also address the challenges faced while maintaining these kinds of applications [5].

Out of all the framework activities in the software engineering process [37], study is done on the requirement analysis and design as well as on maintenance and support of data analytics applications. Effective work needs to be done at early stages of software development as cost of the defect missed at requirement analysis is much higher than the cost of defect missed at implementation [20, 45]. Huge data is being generated from many areas and applications are being developed for these areas. Now the question that needs to be addressed is how can maintenance and support be provided for such applications [42]. Most of the current domains are using big data analytics applications [27], as this kind of applications are most suitable for current environment and makes user life easy. It is required to find how much it is feasible for a new researcher to conduct research on this field and come up with good strategies for data analytics applications concerning software engineering phases [4].

The trend of big data and data analytics has caused tremendous change in all domains [36] like health care, banking and finance, education, marketing, and many more [2]. It is the job of the researcher to find excellent techniques for application development in big data analytics. Our research interest was to find good techniques for developing big data applications and when we proceeded our work with the literature survey to know existing methodologies, we found that it is extremely difficult to understand the domain of application development of big data analytics, if a researcher has not developed an application in big data analytics or a similar domain.

It is required that at least the co-researcher should have known developing big data analytics applications [3, 38, 43].

Following section will give an insight into research conducted in this area. Academicians have proposed many frameworks for big data analytics applications but they lack actual implementation in real time situations so it is not sure that these frameworks will work or no. Similarly work done by industry and in collaboration with industry are having actual implementation and these researchers have mentioned challenges that still exists in this area as well as how important it is to have experience in building big data application if anyone wants to conduct research and propose some standard software engineering process.

The rest of the work is organized in the following sections: Related work is summarized in Sect. 2, Research Method along with Search Strategy, Selection process discussed in Sects. 3 and 4 has Results and Discussions and Sect. 5 consists of Outcomes of the research conducted and Sect. 6 presents Conclusion and Future Work.

2 Related Work

This section discusses the research carried in the direction of Software engineering for big data applications by different researchers. Important points investigated in the related works are how authors have conducted research, how they proceeded and what is the level of difficulty they faced during the development of big data applications. Paper [23] discusses different challenges and difficulty they faced when finding new techniques for different phases of software development life cycle for big data analytics applications.

A thesis work of Vijay Dipti defines how software engineering practices can be related to big data application development and the author conducted the survey of this domain and selected 170 papers which were most relevant and mentioned the gaps exiting between software engineering practices and big data applications development [19]. Big data software is different from other software, and its characteristics make it different. It is possible to engineer big data applications. Carlos et al. [35] discuss all the phases of software concerning big data application and they also mentioned it is possible to engineer these kinds of the system but there is no engineered community that has mastered it. Paper also discusses why there is a need for good collaboration between big data developers and academic researchers. A magazine article by Ion Gorton briefly discusses big data system challenges, and data-intensive systems need different features to sustain for the long term. The author also discusses how important is a software architecture for the building big data software [24].

Yuri et al. [18] propose a framework and components for big data architecture, and it specifies all the requirements for big data systems and how science and industry can be benefited by the new tools developed for big data technology. Authors also mentioned what are the challenges they faced while designing the ecosystems for big data applications. Bagriyanik et al. [7] provide a systematic literature review for how much work has been done in the field of big data related to software engineering and they concluded with an opinion that there is a need for much more work in this field. Each phase of software engineering is mapped with big data system developments, and research challenges in each phase are being categorized in [30] with analysis of gap existing. Epic is a project which has helped in understanding big data system and also helped in addressing two issues related to datasets, how to collect data and store effectively as data grows [5].

Data science projects are the new domain of application in the application domain and its process models are still in infancy stage of evolution so there is a need for good and standard practices for the field of big data software engineering [17]. Large datasets are needed to get accurate results and it also costs in terms of storage, time, and cost. Banerjee and Cukic [8] discuss the storage requirement for large report storage systems and also provide a cost estimation of using the cloud services for large datasets. A new branch of big data analytics to help software development is software analytics. It is to help and improve each phase of the software development life cycle [1]. Workloads of big data in 2015 increased by 40 pbytes of new disk capacity daily. The rate of growth of data repositories is exponential. Big data systems with different technologies making the project life cycle of these applications shorter and it is already seen that computational resources will be considered under the commodities and organizations are already in this path. Data-intensive applications moved from SQL database technology to NoSQL [47].

Architecture is an important part of the whole application and using architecture as a solution, many issues can be addressed related to the distributed system environment. To achieve the required quality, three things need to be considered, those are distributed environment, data storage, and deployment environment with loose coupling which improves the elasticity of system [25]. Many researchers tried to address the issues of 5Vs by developing a case study [13] and a firm with academic researcher have used the existing software engineering practices like RAD with modification specific to architecture have developed the nine projects and validated their approach for developing the big data systems [12]. These are some examples of case study but still, there is a need for addressing challenges mentioned in [36] which required still robust architecture and process models for big data frameworks [27]. Big data systems by its nature are heterogeneous and when different sets of data exist in different environments and these need to be considered together to get appropriate results and when different environments exist together then is a need for pipelining the data coming from these systems [31, 46].

Big data software frameworks have been developed for solving the issues related to big data systems. Some of the software frameworks are Hadoop, Spark, and Storm for the development of big data systems [27]. There are many software frameworks that exist for helping the developer in developing big data systems, and there are systems which are used for performance benchmarking systems for big data applications but there are pitfalls for these systems and all these pitfalls needs to be considered while developing the big data applications [15]. Component-based software development has got more importance in all kinds of application development and similar techniques can be applied to big data application [11]. Industrial big data application needs more structured process model for developing these applications with short time to market. Most important part of application is requirement analysis and validation are being addressed in [48] with development of product.

References [23, 26] provides a systematic mapping of software engineering phases to big data application development but work does not address how feasible it is for a academician to conduct research in this field without any industry collaboration and it does not provide what are the important research questions that can addressed in collaboration with industry person who has sound knowledge of developing big data applications. Following section provides the insight into complete process of survey conduction and article selection procedure.

3 Research Method

The work mainly focuses on extracting the current work in the domain of software engineering for big data analytics applications and how feasible it is for a novice researcher to conduct research in this field if they do not have knowledge of big data application development. Our work also tried to extract the important research question and research objectives which will act research guide for new researcher.

The research method followed to conduct the survey is based on PRISMA model which is a standard procedure. All the articles are selected iteratively using the keyword search from Google Scholar and articles selected are from IEEE, science Direct, ACM and Taylor and Frances. For search strategy year is the first filter as our work is intended to cover 2015 to 2019 year then we applied search string filters as shown in the Sects. 3.1 and 3.2.

3.1 Search Strategy

This section discusses search strategy used for conduction of survey. The search strings (S) are as follows:

S1: Big data and software engineering, **S2**: big data application development, **S3**: requirement engineering for big data application, **S4**: relation of big data and software engineering process, **S5**: an architecture for big data, **S6**: process for big data analytics software, **S7**: SDLC , **S8**: Software implementation of Big data, **S9**: testing methods, **S10**: Quality assurance, **S11**: application development, **S12**: Data-intensive applications, **S13**: Big data architecture, **S14**: design issues of big data application.

3.2 Selection Process

By using the above search strings, research articles were collected but all were not appropriate for the researches, so following elimination criteria are chosen:

1. Research articles where big data concepts are used to improve existing software engineering processes like defect predictions in software code or requirements analysis, design, etc.
2. Research articles which are not in English.
3. Research papers with only theoretical concepts.

All the remaining research articles got after the elimination process has been considered for our study.

In Fig. 1 let total number of articles be *N* and it includes research publication, book chapter and thesis related to the chosen domain that is software engineering for bigdata applications. Duplicate articles may exist after searching using search strings from **S1** to **S14** from Sect. 3.2 where **Nd** is the number of non-duplicate articles out of **N**. In next step, screening takes place, first with respect to abstract, introduction and conclusions and second with respect to full text based on the article's relevancy to domain. **Nrelevant** is the number of articles fetched in first screening

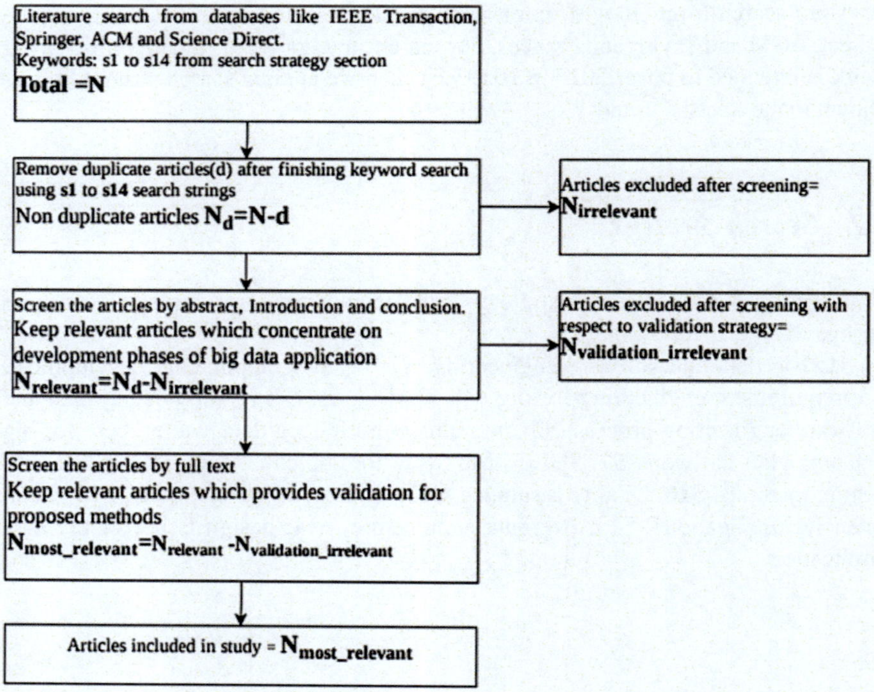

Fig. 1 Flow diagram of research method

and **Nmost_relevant** in second screening. **Nirrelevant** is number of articles removed after first screening and similarly **Nvalidation_irrelevant** is articles removed after second screening. Last step of whole process fetches **Nmost_relevant** articles which have validation for provided methodologies.

Figure 1 describes the whole process. The PRISMA flow diagram [32] is considered but modified according to the strategy chosen for the research work in this paper.

Considering a general definition of filters, Eq. 1 has been derived:

$$Nmost\_relevant = N - (d + Nirrelevant + Nvalidation\_irrelevant) \quad (1)$$

where **N** is the total number of articles which are selected, using search strings **S1** to **S14**. It is a selection of subset of articles from whole. By using subset filter equation from standard mathematics, filters are applied in many domains like model theory, topology, probability theory, etc. In general filters can be defined as non-empty and partially ordered set which includes the elements that are fulfilling certain given criterion. Criteria can be like comparing with some elements or part of another set. In Eq. 2, criterion selected set of search strings. By considering the general filter equation, the following equation has been modified to suite the requirement of work with respect to search strings.

$$x \epsilon N | n = \{S1, S2, S3, S4, S5, S6, S7, S8, S9, S10, S11, S12, S13, S14\} \quad (2)$$

where x is an article chosen from a pool of articles, such that the selected article satisfies the search string criteria and article belongs to N. N is the set of articles which are selected from search space such that all articles satisfy the criteria of search strings from **S1** to **S14**.

Next result section shows how many research articles are published in past few years and how important it is to have industry collaboration or big data application development knowledge to validate proposed methodologies. All proposed methodologies need to be tested for real-time scenarios else these works will be of no use if they are just in theory.

4 Result and Discussion

When big data analytics took its stand in information technology, researchers wanted to extract 100% profit out of it, and in this direction, they started to build the tools which will help in doing so. But later the software engineers felt the potential research that can be taken place in developing big data systems using standard practices of software engineering knowledge areas and even some felt why not improve practices of software engineering using big data analytics. We can use the existing error logs for software built and maintained for predicting the same kind of errors appearing again and again [21, 29, 39, 40].

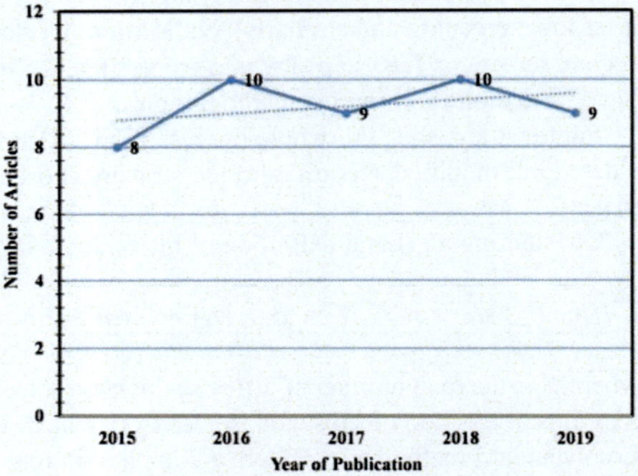

Fig. 2 Articles published from 2015 to 2019

All the results of result section are built using the number of a research article that is published by many researchers in the field of Software engineering for Big data analytics software [3, 9, 10].

The research articles related to software engineering for big data applications are being considered for feasibility study, other research articles which belongs to the domain of big data for software engineering will not belong to the domain. Juneja and Kaur [26] considered the articles in the same direction but also included some papers which considered agile methods of software engineering. In the analysis, we have considered research article of all methodologies in software engineering for big data applications.

We have specifically considered big data analytics application in Fig. 2 result. The result shows how many numbers of articles are published in year 2015 to 2019. It is observed that number of researches conducted in this field increased from 2015 to 2016 and then there is slight decrease and then again it increased. In 2019, specifically papers were collected related to work done on software engineering phases related to big data applications and review papers were excluded. The main aim is to concentrate on software engineering for big data applications, so while exploring the domain, we got the results which are very important in analysis.

Figure 3 gives the analysis of the research interest of academicians, industry, and both. In the year 2015, nearly 13 articles from academicians are published similarly for the next year people with academic background have shown more interest compared to the industry. Academician in collaboration with industry the article published was very less. When the domain was surveyed, it was detected that the collaborative research would fetch, process models, which will surely increase the accuracy of big data systems, and if an academician is researching this filed and if he proposes some standard practices then it can be tested by implementing a real-time industry project. It will be a verification for proposed method.

Survey is done for only three phases of software development life cycle (SDLC). They are requirement's engineering, design and architecture, and quality assur-

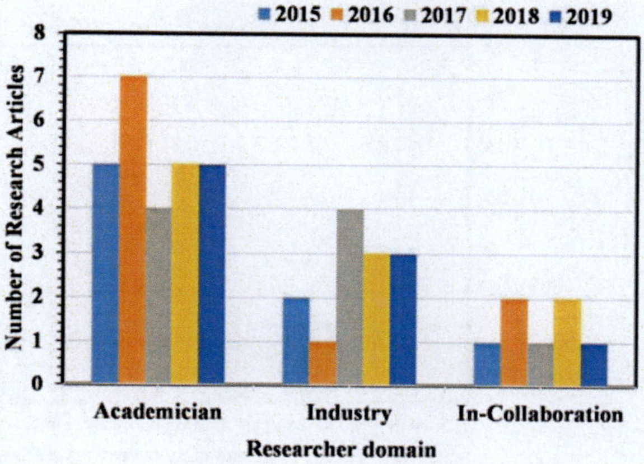

Fig. 3 Research articles published by academicians and industry persons

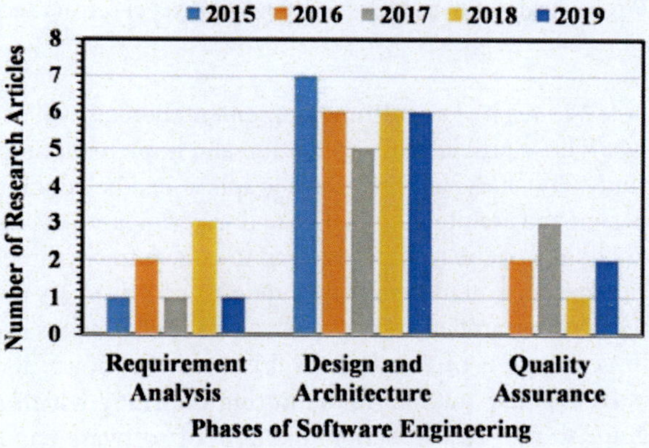

Fig. 4 Research articles of different phases of software engineering with respect to years

ance because implementation phase had been considered by most of the industry researcher to develop big data application but other phases like requirement engineering, design and quality analysis is least considered. Survey conducted mainly helps understanding the need for more research in these phases of software engineering for big data applications [44].

Figure 4 shows the number of research articles published in the year 2015 to 2019 in each of the phases. It is observed that there is much more work done on design and architecture compared to requirement engineering and Quality Assurance [28, 33].

There is a need for more research on remaining fields as requirement engineering is the base for whole software development and it is known that cost of defect introduced during RE is much more that Implementation. Quality assurance is the last phase that ensures the quality of the product.

Figure 5 shows total number of research article in requirement engineering, design and quality assurance. When domain was investigated for articles related to all phases of big data applications, there were many research articles but only some articles like

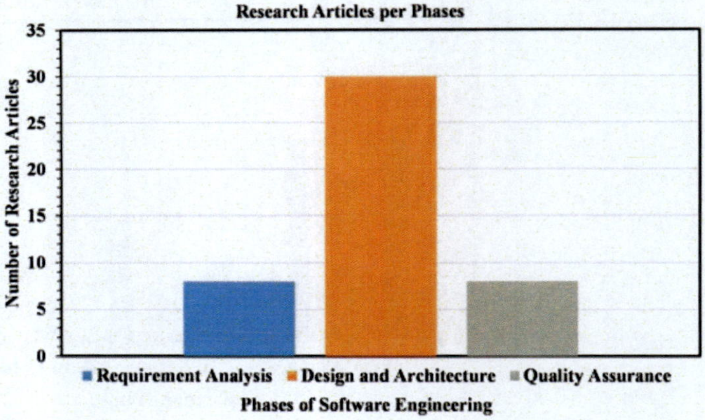

Fig. 5 Total research articles of different phases of software engineering

[14, 34, 42] had a real-time implementation as well as validation of proposed methods. Those articles with validation and implementation are considered in analysis of study. By analyzing the above graph, it can be concluded that more work related to design and architecture had been done compared to requirement analysis and quality assurance so the researcher need to concentrate on these fields.

Next section discusses the outcome of the survey on feasibility study which helps the new researchers to know what are the challenges exists and how they can proceed in conduction of research in this field. Section 5 also discusses general research outcomes that authors found during the study which may act as a research plan or guide to new researchers in the field of software engineering for big data analytics applications.

5 Research Outcome

5.1 Research Questions

After doing a through survey some questions may come and they are as follows:

1. What are the challenges in the software engineering aspect of big data analytics application?
2. How to map software engineering development phases to big data analytics application?
3. What are the research opportunities in big data analytics application and software engineering?
4. How to conduct requirements engineering for big data analytics applications?

5. How to design for big data analytics applications and which architecture is suitable?
6. How to conduct quality assurance for big data analytics applications?

As a result of the above question, very few papers were available as very little work has been conducted in this area and most of the papers were for the fifth question. Many design and architecture techniques can be found as suggested by many researchers based on the big data characteristics like volume, variety, velocity, etc.

5.2 Proposed Research Objectives

The research objectives are framed after collecting the data from existing literature and considering the gaps mentioned in the research articles. These research objectives act as guidelines to researcher who wants to conduct research in the field of software engineering for big data application development:

1. Extract the domain knowledge of how software engineering practices have been used in developing big data analytics systems
    - Identify applications which need software engineering practices or principles
    - Extract the knowledge of how these applications being developed (Existing process)
    - Identify what needs to be improved for these applications.
    - Identify how these applications are tested for their respective best features.
2. Identify the factors/attributes which are suitable for big data analytics applications
    - From the current work on application development of big data analytics application, choose the factors which they considered for development purpose
    - Select factors which are most suitable for the chosen application
    - Factors here refer to reusability, app development time, reliability, security, and memory used and cost-effectiveness of the system.
3. Identify the software engineering knowledge area which can be considered for big data analytics applications (knowledge area: refers to requirement engineering, architecture, design, etc.)
    - Extract the knowledge regarding existing requirement engineering process models for existing applications. (data-intensive applications)
    - Identify whether existing models fit in for big data analytics applications.
    - Find a gap between existing requirement models with the requirements of big data analytics applications.
    - Study regarding existing data architecture models and identify major design works existing in data architecture.

- Based on the factors decided on second objective, decide how to choose an appropriate design model for specific applications from exiting.
- After identifying the most successful architecture design from software engineering practices and apply it to the big data analytics application chosen in objective 1.

4. Verification and validation of chosen models with the exiting other models.

   - Requirements will be tested based on the functional and non-functional requirements of an application.
   - Verify the stated functional requirements are met by testing (manual verification)
   - Check how to verify the non-functional requirements (only for chosen factors)

All the above objectives will help a new researcher to conduct research in the software engineering for big data applications.

6 Conclusion and Future Work

The main aim was to find the research scope in the field of big data by using software engineering concepts specifically for academicians. In order to perform such an activity, certain articles published from 2015 to 2019 was taken under consideration. The survey was conducted on the basis of PRISM model. The outcome of the model was performed by applying some searching strategies which comprised of "search strings" and also the unwanted strategies were eliminated.

The major finding of this research work was research articles published by academicians and industry people for each year starting from 2015 to 2019. It was observed that among all these years in 2016 research articles published by academicians were maximum where as in 2017 research article published by industry people was maximum. The important observation was that models proposed by academicians were not implemented in real-time scenarios so it is difficult to validate them for their correctness but people from industry has validated their models with real time implementations. Even the research articles of different phase of software engineering were evaluated. From this analysis it was found that "Design and Architecture" had a great prospect for all the years in comparison to "Requirement Analysis and Quality Assurance."

Hence, it can be concluded that if any researcher wants to conduct research in this field with sound knowledge in software engineering and big data application development then he must concentrate on the two fields and they are requirement analysis and quality assurance of big data applications. The researcher from academic background should try to implement their proposed models in real time scenarios for validation purpose.

References

1. Abdellatif TM, Capretz LF, Ho D (2015) Software analytics to software practice—a systematic literature review. In: IEEE/ACM 1st international workshop on Big Data software engineering. https://doi.org/10.1109/BIGDSE.2015.14
2. Al-Jaroodi J, Hollein B, Mohamed N (2017) Applying software engineering processes for big data analytics applications development. In: IEEE 7th annual computing and communication workshop and conference. https://doi.org/10.1109/CCWC.2017.7868456
3. Amershi S, Begel A, Bird C, Deline R et al (2019) Software engineering for machine learning: a case study. In: International conference on software engineering: software engineering in practice. https://doi.org/10.1109/ICSE-SEIP.2019.00042
4. Anders A, Bjorn B, Luka C-F, Jan B (2018) Software engineering challenges of deep learning. In: Euromicro conference on software engineering and advanced applications. https://doi.org/10.1109/SEAA.2018.00018
5. Anderson KM (2015) Embrace the challenges software engineering in a big data world. In: IEEE/ACM 1st international workshop on Big Data software engineering. https://doi.org/10.1109/BIGDSE.2015.12
6. Ankit D, Sunil S (2017) A review paper on software engineering areas implementing data mining tools and techniques. Int J Comput Intelligence Res 13(4):559–574
7. Bagriyanik S, Karahoca.A, (2016) Big data in software engineering: a systematic literature reviews. Global J Information Technol 6(1):107–116
8. Banerjee S, Cukic B (2015) On the cost of mining very large open source repositories. In :IEEE/ACM 1st international workshop on Big Data software engineering. https://doi.org/10.1109/BIGDSE.2015.16
9. Basanta-Val P, Audsley NC, Wellings AJ et al (2016) Architecting time-critical Big-Data systems. IEEE Trans Big Data. https://doi.org/10.1109/TBDATA.2016.2622719
10. Cerqueus T, de Almeida EC, Scherzinger S (2015) Safely managing data variety in Big Data software development. In: IEEE/ACM 1st international workshop on Big Data software engineering. https://doi.org/10.1109/BIGDSE.2015.9
11. Chatzipetrou P, Alégroth E, Papatheocharous E et al (2018) Component selection in software engineering—which attributes are the most important in the decision process. In: 44th Euromicro conference on software engineering and advanced applications. https://doi.org/10.1109/SEAA.2018.00039
12. Chen H-M, Kazman R, Haziyev S (2016) Strategic prototyping for developing big data systems. In: IEEE Software. https://doi.org/10.1109/MS.2016.65
13. Chen H-M, Kazman R, Haziyev S et al (2015) Big Data system development: an embedded case study with a global outsourcing firm. In: IEEE/ACM 1st international workshop on Big Data software engineering. https://doi.org/10.1109/BIGDSE.2015.15
14. Chen S, Bronevetsky G, Peng L et al (2016) Software engineering for Big Data systems. IEEE Softw 33(2):32–35. https://doi.org/10.1109/MS.2016.47
15. Chen H, Kazman R, Garbajosa J, Gonzalez E (2016) Toward Big Data value engineering for innovations. In: IEEE/ACM 2nd international workshop on big data software engineering. https://doi.org/10.1109/BIGDSE.2016.016
16. Darlan A, Madhavj Nazim H (2017) Towards a Big Data requirements engineering artefact model in the context of Big Data software development projects. In: IEEE international conference on Big Data. https://doi.org/10.1109/BigData.2017.8258521
17. DeLine R (2015) Research opportunities for the Big Data era of software engineering. In: IEEE/ACM 1st international workshop on Big Data software engineering. https://doi.org/10.1109/BIGDSE.2015.13
18. Demchenko Y, Ngo C, Membrey P (2013) Architecture framework and components for the Big Data eEcosystem. System and Network Engineering Group
19. Dipti Kumar V, Alencar P (2016) Software engineering for big data projects: domains, methodologies and gaps. In: IEEE international conference on Big Data. https://doi.org/10.1109/BigData.2016.7840938

20. Eichelberger H, Qin C, Schmid K (2017) Experiences with the model-based generation of Big Data pipelines. Big Data Mgt Syst Bus Industrial Appl 49–56
21. Fahmideh M, Beydoun G (2019) Big data analytics architecture design—an application in manufacturing systems. J Comput Industrial Eng 128:948–963. https://doi.org/10.1016/j.cie.2018.08.004
22. Gulzar MA (2018) Interactive and automated debugging for Big Data analytics. In: IEEE/ACM 40th International Conference on Software Engineering: Companion (ICSE Companion), pp 509–511
23. Hummel O, Eichelberger H, Giloj A et al (2018) A collection of software engineering challenges for Big Data system development. In: 44th Euromicro conference on software engineering and advanced applications. https://doi.org/10.1109/SEAA.2018.00066
24. Ian Gorton AB, Bener AM (2016) Software engineering for Big Data systems. IEEE Softw 33(2):32–35. https://doi.org/10.1109/MS.2016.47
25. Ian G, John K (2015) Distribution, data, deployment: software architecture convergence in Big Data systems. IEEE Softw 32(3):78–85. https://doi.org/10.1109/MS.2014.51
26. Juneja P, Kaur P (2019) Software engineering for Big Data application development: systematic literature survey using snowballing. In: International conference on computing, power and communication technologies, pp 492–496
27. Karakaya Z (2017)Software engineering issues in big data application development. In: International conference on computer science and engineering. https://doi.org/10.1109/UBMK.2017.8093547
28. Klein J, Buglak R, Blockow D et al (2016) A reference architecture for big data systems in the national security domain. In: BIG Data Software Engineering. IEEE/ACM 2nd international workshop on Big Data Software Engineeringe. https://doi.org/10.1109/BIGDSE.2016.017
29. Kleppmann M (2017) Designing data-intensive applications: the big ideas behind reliable, scalable, and maintainable systems. O'Reilly Media, Inc
30. Madhavji NH, Miranskyy A, Kontogiannis K (2015) Big Picture of Big Data software engineering with example research challenges. In: IEEE/ACM 1st international workshop on Big Data software engineering. https://doi.org/10.1109/BIGDSE.2015.10
31. Martínez FS, Andreas J, Liliana G, Maria VA (2018) A quality model for actionable analytics in rapid software development. In: Euromicro conference on software engineering and advanced applications. https://doi.org/10.1109/SEAA.2018.00067
32. Moher D, Liberati A, Tetzlaff J et al (2009) Preferred reporting items for systematic reviews and meta-analyses: the PRISMA statement. PLoS Med. https://doi.org/10.1371/journal.pmed1000097
33. Naseer A, Alkazemi BY, Waraich EU (2016) A big data approach for proactive healthcare monitoring of chronic patients. In: International Conference on Ubiquitous and Future Networks (ICUFN), pp 943–945
34. O'Donovan P, Leahy K, Bruton K et al (2015) Big data in manufacturing: a systematic mapping study. J Big Data 2. https://doi.org/10.1186/s40537-015-0028-x
35. Otero CE, Peter A (2014) Research directions for engineering Big Data analytics software. IEEE Intelligent Syst. https://doi.org/10.1109/MIS.2014.76
36. Philip Chen CL, Zhang Chun-Yang (2014) Data-intensive applications, challenges, techniques and technologies: a survey on Big Data. J Information Sci. https://doi.org/10.1016/j.ins.2014.01.01
37. Pressmen RS (2010) Software engineering: a practionerś approach. McGraw-Hill, New York
38. Rosenthal S, Mcmillan S, Matthew EG (2015) Developer toolchains for large-scale analytics: two case studies. In: International conference on Big Data. https://doi.org/10.1109/BigData.2015.7363887
39. Sandhya Rani K, Eesha P, Mina M (2020) A conceptual framework for Big Data requirements engineering. J Comput Sci Information Technol https://doi.org/10.5121/csit.2020.100608
40. Scavuzzo M, Di Nitto E, Ardagna D (2018) Experiences and challenges in building a data intensive system for data migration. In: IEEE/ACM 40th international conference on software engineering. https://doi.org/10.1145/3180155.3182534

41. Schram A, Anderson K (2015) Design challenges/solutions for environments supporting the analysis of social media data in crisis informatics research. In: International Conference on System Sciences (HICSS). https://doi.org/10.1109/HICSS.2015.29
42. Shapira G, Chen Y (2015) Common pitfalls of benchmarking Big Data systems. IEEE Trans Services Comput. https://doi.org/10.1109/TSC.2015.2494595
43. Stol KJ, Ralph P, Fitzgerald B (2016) Grounded theory in software engineering research: a critical review and guidelines. In: IEEE/ACM 38th international conference on software engineering. https://doi.org/10.1145/2884781.2884833
44. Vergilio T, Ramachandran M, Mullier D (2020) Requirements engineering for large-scale Big Data applications. In: Ramachandran M, Mahmood Z (eds) Software Engineering in the era of cloud computing. Springer International Publishing, pp 51–84
45. Villanustre F (2015) Industrial Big Data analytics lessons from the trenches. In: IEEE/ACM 1st international workshop on Big Data software engineering. https://doi.org/10.1109/BIGDSE.2015.8
46. Wu D, Liming Z, Xu X et al (2016) A pipeline framework for heterogeneous execution environment of Big Data processing. In: IEEE Software. https://doi.org/10.1109/MS.2016.35
47. Yashaswini J, Geeta RB, Totad Shashikumar G, Prasad Reddy PVGD (2017) Mobile agent based frequent pattern mining for distributed databases. In: International conference on intelligent computing and communication. https://doi.org/10.1007/978-981-10-7245-1_9
48. Zhao H, Hou J (2018) Design concerns for industrial Big Data system in the smart factory domain: from product lifecycle view. In: International Conference on Engineering of Complex Computer Systems (ICECCS). https://doi.org/10.1109/ICECCS2018.2018.00036

Security Concerns in IoT Systems and Its Blockchain Solutions

Rajat Verma, Namrata Dhanda, and Vishal Nagar

Abstract IoT is an ecosystem that deals with electronic objects and performs various operations such as interaction among themselves, controlling operations, monitoring operations that may or may not involve human beings. The evolution of IoT made a significant contribution to the world, and as a result, IoT is omnipresent. Because of this omnipresent attribute, IoT will be a family of 75 billion (approx.) connected devices by the year 2025, and as this time will pass, this number of connected devices will grow exponentially. This will cause enormous data generation, which will be sent and received through the Internet. This will surely cause diverse security, privacy, and interoperability issues that will influence the lives of the people in one or the other way. Dealing with these issues is extremely important since if carelessness will be there, then this data can put a negative impact on the country's economy. The conventional approaches that have been a primary solution in the IoT spectrum are outdated and require modern and smart solutions. This is so because the orthodox solutions of IoT are somehow responsible for the birth of diverse concerns of IoT. This paper focuses on these issues and challenges of IoT and highlights the importance of solving these issues by a decentralized peer-to-peer (P2P) network technology blockchain.

Keywords Blockchain · Decentralization · Internet of Things (IoT) · Network security · Privacy preservation

1 Introduction

IoT is capable enough of communicating and connecting with numerous devices at a particular time. As stated in research, it was estimated that by the year 2020, 50 billion

---

R. Verma (✉) · N. Dhanda
Department of Computer Science and Engineering, Amity University, Uttar Pradesh, Lucknow Campus, India

V. Nagar
Department of Computer Science and Engineering, Pranveer Singh Institute of Technology, Kanpur, India

J. M. R. S. Tavares et al. (eds.), *Cyber Intelligence and Information Retrieval*, Lecture Notes in Networks and Systems 291,
https://doi.org/10.1007/978-981-16-4284-5_42

devices (approximate) will be a part of the connected world of IoT [1]. According to various research communities, IoT has diversified definitions. The term 'things' in IoT corresponds to any sensor in networks, any device, and a human being. Just as the Internet has diversified the world and its working, similarly IoT has the ability to do the same or even more [2]. IoT is an amalgamation of many technologies thus making it a multidimensional field that can contribute to research in many ways. Some technologies that constitute the IoT are information technology, actuator technology, etc. [3]. The extensive increment in the connected devices which are dependent on the centralized authenticator may give birth to many security, privacy, interoperability, and legal issues for the user. Due to this impact on the security aspect of IoT, the focus on the security domain of IoT should remain intact. The current scenario of IoT dependability and security is still in its introduction phase [4]. In contrast, there is state-of-the-art technology blockchain which has gained its significance from the last 12 years. It is an immutable, transparent, peer-to-peer network (P2P network) that aggregates the diverse consensus mechanisms, encryption techniques, and time series data that provides adequate verification as well as traceability of data [5]. There is one limitation of using blockchain. It consumes enormous resources and is computationally very expensive and also includes high consumption of bandwidth. It also includes a high delay [6]. To perfectly aggregate blockchain technology into the security aspect of IoT, there is a requirement of a lightweight security mechanism [7]. As blockchain technology consumes huge resources, it might be unsuitable to make it work on every system it requires. Therefore, there is a need to design an optimized framework of blockchain. Although here, we will consider how blockchain can abolish the privacy and security concerns of IoT [8]. The three important attributes of blockchain are immutability, transparency, and decentralization. This paper will focus on these three parameters.

2 Internet of Things

IoT is an immense web of connected things and people. IoT is a notion of connecting a device to the Internet or already connected devices [9].

2.1 Characteristics

The characteristics of IoT are as follows:

1. Intelligence/SMART [10].
2. Connectivity [1, 11].
3. Sensing [12].
4. Dynamicity.
5. Heterogeneity [13].

2.2 *Security and Privacy Issues in the Internet of Things (Research Gaps)*

With these extensively connected devices by the year 2025, i.e., about 75 billion [11], a huge amount of data will rise that will be exchanged through the Internet. Although, the traditional IoT follows a centralized architecture. This centralized IoT is one of the primitive security issues of IoT. Centralized IoT is illustrated in Fig. 1.

Centralized architecture is depicted as a single point of failure. It means the entire system will be shut when the central administrator is on the failure side [14]. The other privacy issue of IoT depicts the case of fabricated data, transparency aspect as well as privacy expectation of users [15–18]. The concept of fabricated or manipulated data is an old scenario of IoT.

2.2.1 Scenario: Illustration [Fabricated Data]

A person named 'X,' went to a car showroom to purchase a car. X deposited the amount of money to the company and got the keys of the car. The showroom handed a certificate of ownership to Mr. X. Now, typically the owner of the car is Mr. X. The information about his ownership was stored in a conventional ledger or a centralized database or in particular a notebook.

The attacker has some low resources and by that, he was able to attack the database (centralized) which was stored at one specific location, or in the other scenario, he just tore the pages of the book where the record of Mr. X was there. Now, Mr. X cannot claim that he was the real holder of the car as this information is missing from the company's database.

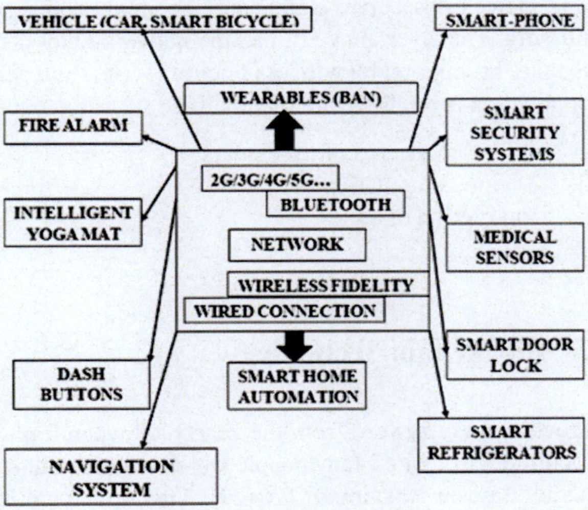

Fig. 1 Centralized IoT architecture

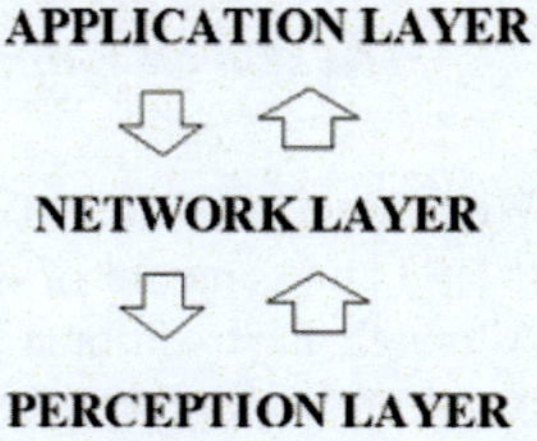

Fig. 2 Three-layer architecture: IoT

Apart from these general issues, the issues of IoT also occur at its different layers in its architecture that is mentioned in the next section.

2.2.2 Three-Layer Architecture of Internet of Things (IoT)

Few concerns are present in the layers of IoT architecture. Three-layer IoT architecture is highlighted in Fig. 2.

Apart from the three-layered scenario, the architecture of IoT can also be visualized as a four-layer IoT model or a five-layer IoT model. In the four-layer IoT model, a support layer is there, which is added between perception layer and the network layer. In a five-layer IoT architecture, the network layer is replaced with the transport layer in combination with the processing layer, and a layer that is present above the application layer is added as the business layer.

In the perception layer, many terminals are there that opens diverse services and allow numerous amounts of susceptibility attacks like distributed denial of service (DDoS) attacks to happen [19]. Some other issues that occur in the perception layer are the terminals which are linked through a network layer device 'gateway' to diverse standards of IEEE. For example, IEEE 802.15.4, IEEE 802.15.1, IEEE 802.11 [20–22]. Here, the security parameters are weak, and a strong login password is also missing. The layer above it, i.e., the network layer takes care of the security issues that are brought up by wireless fidelity (Wi-Fi) and cellular communications. In the application layer, there are mainly three security concerns [23–25]. They are:

1. Privacy concerns of user's data.
2. Secure computation.
3. Encryption of data.

3 Blockchain Technology

It became a buzzword from the year 2008 when it played a major role in the cryptocurrency Bitcoin. Many people were behind it, but it was represented by a pseudo name 'Satoshi Nakamoto' [26–29]. Traditionally, this technology blockchain was

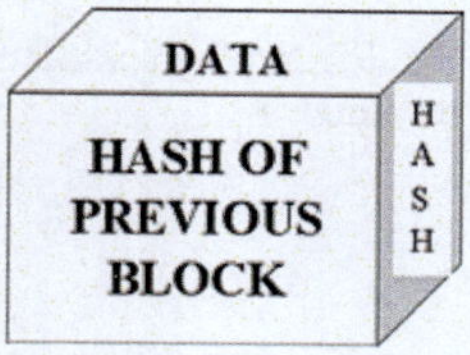

Fig. 3 Layout of a block in blockchain

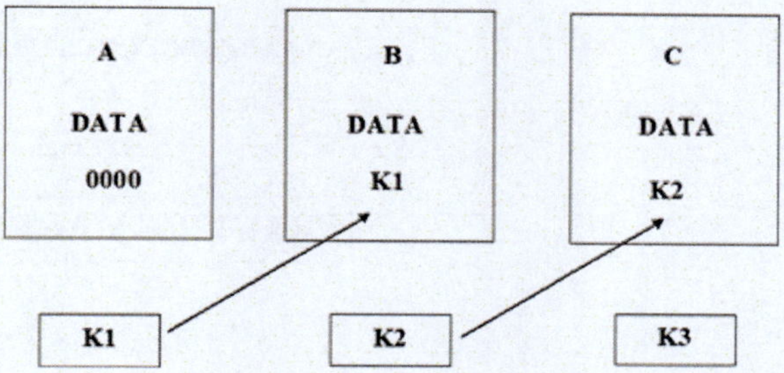

Fig. 4 A simple blockchain

securing the transactions, but in modern use, any type of data can be kept in the chain of blocks and this form of input data in blockchain is under research.

The blockchain is a decentralized P2P network that contains data, facts, and figures especially termed as records, and follows immutability, transparency, and decentralization [30, 31]. A block in a blockchain comprises three entities namely data (facts and figures), the previous block's cryptographic hash value, and the current block's cryptographic hash value. Now, if SHA-256 (hashing technique) is applied, then the cryptographic hash which will be of a fixed length of 64 hexadecimal characters will be generated [32]. The layout of a block is depicted in Fig. 3.

If the blocks are connected in a sequence, it forms a blockchain. If there are other blocks that are not the constituent of the longest chain or the main chain, then they are called orphan blocks. In a blockchain, the initial block is termed as Genesis block [8]. The blockchain layout is depicted in Fig. 4.

In the figure mentioned below, three blocks are present in a blockchain environment, namely A, B, and C. A is known as Genesis block which has a particular hash value depicted as K1. The first block A has a previous hash value of 0000 along with some data. Block B has data in addition to the previous block's hash value as K1 and current hash value as K2. A similar scenario takes place with block C.

3.1 *Role of Cryptographic Hash*

In blockchain concern, one can conclude that the data cannot be modified in an unethical way. That means the data is secured by a cryptographic hash. If one attempts to change the information, the entire cryptographic fixed-length bit string (hash value) will change and every entity in the blockchain will come to know about the tampering

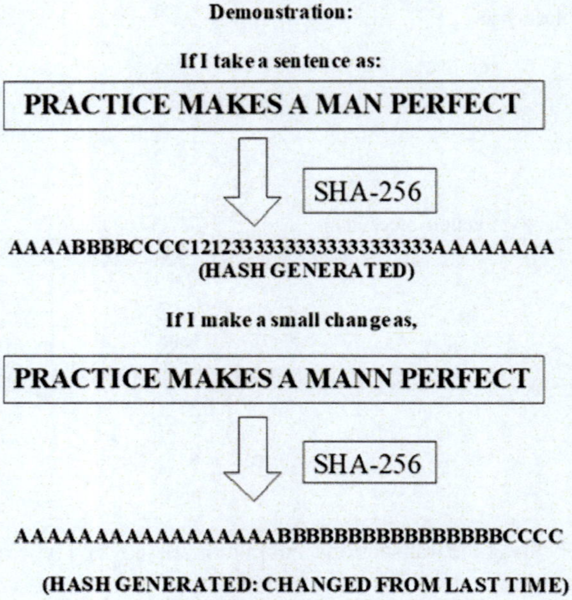

Fig. 5 Cryptographic hash: working

of data that have been tried. The working of the cryptographic hash is shown below in Fig. 5.

3.2 *Characteristics of Blockchain*

Few characteristics of blockchain technology are mentioned below:

1. Tamperproof/Immutable [33].
2. Decentralization.
3. Intensifies Security [34].
4. Distributed Ledger.
5. Consensus Algorithms [35–37].
6. Quick-Settlement.

3.3 *Concepts Involved in Blockchain*

Few concepts of blockchain technology are mentioned below:

(1) **Smart-Contract** [38, 39].
(2) **Token**: This corresponds to the unit of ownership, which can be exchanged among parties. It is represented digitally [40].

(3) **Miners**: They are the validators of the transactions and with the help of them, the problem of double-spending can be avoided [41].
(4) **Valid-Transactions** [42].
(5) **Permissioned Blockchain** [43].
(6) **Permission-less Blockchain** [44].

3.4 Probable Blockchain Solutions for the Internet of Things [Material and Methods]

The initial problem in IoT is of centralization that can be handled with a concept of blockchain technology, i.e., decentralization. If this concept of decentralization happens in IoT, it will be recognized as a decentralized IoT network which is represented in Fig. 6.

This decentralized IoT network will abolish the central point of failure while applying the attributes of blockchain. Further, it also abolishes the problem of fabricated data. A proposed approach, concerning the problem, is depicted in Sect. 2.2.1. Now, suppose the record of the vehicle (car) purchased by Mr. X. is available in a blockchain ledger. Let's assume the record is present in the 4th node of the chain of blocks. The attacker attacks the 4th node and tries to manipulate the contents present in the node by changing its hash value. Now, since it is the blockchain mechanism, the next node, i.e., the 5th node has an entity that contains the previous node's hash value as depicted in the block's structure. Now, since the cryptographic hash value of the 4th node and the 5th node's previous hash value do not match, so the intruder has to change the contents of the 5th node also. If the intruder manipulates the contents of the 5th node, then the previous hash value of the 6th node will not be equal to the current hash value of the 5th node and this cycle goes on until the end of the chain. Now, if the attacker has the assets to alter the entire contents of the chain, then the interloper has to change the contents of all the chains which are available in a

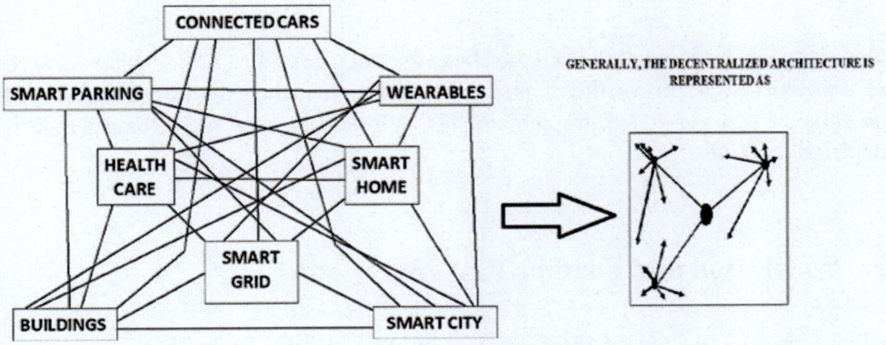

Fig. 6 Decentralized IoT network

Table 1 Issues raised by IoT devices

S. No.	Issue	Recommendation/assumptions
Issue 1	Centralization	This issue can be solved by applying blockchain as it follows decentralization
Issue 2	Fabricated data	This issue can be solved by applying blockchain as it follows decentralization and immutability
Issue 3	Perception layer: multiple terminals	This issue can be solved by applying blockchain as it follows strict rules and protected by cryptographic functions that use secure hashing algorithms (SHA) and Elliptic Curve Digital Signature Algorithm (ECDSA)
Issue 4	Application layer: privacy concern of user's data, secure computation, encryption of data	
Issue 5		
Issue 6		
Issue 7	Data attacks	This issue can be solved by applying blockchain as in-between writing is prohibited in blockchain

decentralized P2P network. This is somehow impossible to do. Hence, the blockchain follows the process of non-fabricated data.

This approach of immutable data follows the approach of privacy preservation along with the CIA triad. Privacy preservation in the sense that if the interloper manipulates the details, then the entire hash value will be altered, and by that the principle of anonymity is followed. Since the information is privacy protected, thus follows confidentiality. Information is correct and available to everyone follows the concept of integrity and availability [45–48].

The summarized table for the issues in IoT is illustrated below for easy understanding.

4 Results and Discussion

One thing can be concluded from the above textual matter, i.e., if blockchain is used as a primary solution for the diverse problems of IoT, then the effectiveness will be enhanced. A variety of problems of IoT can be solved by blockchain which is depicted in Table 1.

5 Conclusion and Further Research Scopes

IoT is a tremendous aspect of the artificial intelligence industry as it has revolutionized every attribute of society from the past decade by providing many advancements and automating manual operations. However, these billions of devices that are linked in a balanced manner will surely surpass a total of approximately 75 billion shortly and will generate data that will be exchanged through the Internet by following the recommendations of the World Wide Web Consortium (W3C). This enormous data will become a key factor to give rise to various security, privacy, and interoperability issues of IoT that have been discussed in this paper. The different concerns of IoT can be addressed by an immutable peer-to-peer technology blockchain that has been depicted here in a detailed manner. The different characteristics of IoT and blockchain are also highlighted in this paper. The probable solutions and approaches of blockchain that can solve the diverse issues of IoT are also shown in this paper.

Additionally, will find some more open issues of IoT whose perspective can be enhanced by blockchain. It is the next big thing as depicted by the popular press; also, it furnishes substantial potential in the field of research and development (R&D). Additionally, focusing on the programming and hardware aspect, it has to be depicted that how both of these technologies can be integrated using one set of programs.

References

1. Khan MA, Salah K (2018) IoT security: review, blockchain solutions, and open challenges. Futur Gener Comput Syst 82:395–411
2. Atzori L, Iera A, Morabito G (2010) The internet of things: a survey. Comput Netw 54(15):2787–2805
3. Giusto D, Iera A, Morabito G, Atzori L (eds) (2010) The internet of things: 20th Tyrrhenian workshop on digital communications. Springer Science & Business Media
4. Qiu H, Wang X, Xie F (2017) A survey on smart wearables in the application of fitness. In: 2017 IEEE 15th international conference on dependable, autonomic and secure computing, 15th international conference on pervasive intelligence and computing, 3rd international conference on big data intelligence and computing and cyber science and technology congress (DASC/PiCom/DataCom/CyberSciTech). IEEE, pp 303–307
5. Nayak P (2017) Internet of things services, applications, issues, and challenges. In: Handbook of research on advanced wireless sensor network applications, protocols, and architectures, IGI Global, pp 353–368
6. Verma R, Dhanda N, Nagar V (2020) Addressing the issues and challenges of internet of things using blockchain technology. Int J Adv Sci Technol 29(05):10074–10082
7. Bhushan B, Sahoo C, Sinha P, Khamparia A (2021) Unification of blockchain and internet of things (BIoT): requirements, working model, challenges and future directions. Wireless Netw 27(1):55–90
8. Singh S, Singh N (2016) Blockchain: future of financial and cyber security. In: 2016 2nd international conference on contemporary computing and informatics (IC3I). IEEE, pp 463–467
9. Borgia E (2014) The internet of things vision: key features, applications and open issues. Comput Commun 54:1–31
10. Collina M, Corazza GE, Vanelli-Coralli A (2012) Introducing the QEST broker: scaling the IoT by bridging MQTT and REST. In: 2012 IEEE 23rd international symposium on personal, indoor and mobile radio communications-(PIMRC). IEEE, pp 36–41
11. Bera A (80) Insightful Internet of things statistics (Infographic). White Paper. Retrieved from https://safeatlast.co/blog/iot-statistics/#gref
12. Mainetti L, Patrono L, Vilei A (2011) Evolution of wireless sensor networks towards the internet of things: a survey. In: SoftCOM 2011, 19th international conference on software, telecommunications and computer networks. IEEE, pp 1–6

13. Razzaque MA, Milojevic-Jevric M, Palade A, Clarke S (2015) Middleware for internet of things: a survey. IEEE Internet Things J 3(1):70–95
14. Heer T, Garcia-Morchon O, Hummen R, Keoh SL, Kumar SS, Wehrle K (2011) Security challenges in the IP-based internet of things. Wireless Pers Commun 61(3):527–542
15. Chin WL, Li W, Chen HH (2017) Energy big data security threats in IoT-based smart grid communications. IEEE Commun Mag 55(10):70–75
16. Derhamy H, Eliasson J, Delsing J (2017) IoT interoperability—on-demand and low latency transparent multiprotocol translator. IEEE Internet Things J 4(5):1754–1763
17. Davies N, Taft N, Satyanarayanan M, Clinch S, Amos B (2016) Privacy mediators: helping IoT cross the chasm. In: Proceedings of the 17th international workshop on mobile computing systems and applications, pp 39–44
18. Hodkowski WA (1997) Future of internet security: how new technologies will shape the internet and affect the law, vol 13. The. Santa Clara Computer & High Tech., LJ, p 217
19. Bhattacharjya A, Zhong X, Wang J, Li X (2019) Security challenges and concerns of internet of things (IoT). Cyber-physical systems: architecture security and application. Springer, Cham, pp 153–185
20. Olyaei BB, Pirskanen J, Raeesi O, Hazmi A, Valkama M (2013) Performance comparison between slotted IEEE 802.15. 4 and IEEE 802.1 lah in IoT based applications. In: 2013 IEEE 9th international conference on wireless and mobile computing, networking and communications (WiMob). IEEE, pp 332–337
21. Rahman T, Chakraborty SK (2018) Provisioning technical interoperability within ZigBee and BLE in IoT environment. In: 2018 2nd international conference on electronics, materials engineering and nano-technology (IEMENTech). IEEE, pp 1–4
22. Tian L, Famaey J, Latré S (2016) Evaluation of the IEEE 802.11 ah restricted access window mechanism for dense IoT networks. In: 2016 IEEE 17th international symposium on a world of wireless, mobile and multimedia networks (WoWMoM). IEEE, pp 1–9
23. Jing Q, Vasilakos AV, Wan J, Lu J, Qiu D (2014) Security of the internet of things: perspectives and challenges. Wireless Netw 20(8):2481–2501
24. Ali M, Khan SU, Vasilakos AV (2015) Security in cloud computing: opportunities and challenges. Inf Sci 305:357–383
25. Upmanyu M, Namboodiri AM, Srinathan K, Jawahar CV (2010) Blind authentication: a secure crypto-biometric verification protocol. IEEE Trans Inf Forensics Secur 5(2):255–268
26. Swan M (2015) Blockchain: blueprint for a new economy. O'Reilly Media, Inc.
27. Nakamoto S (2019) Bitcoin: A peer-to-peer electronic cash system. Manubot
28. Nakamoto S, Bitcoin A (2008) A peer-to-peer electronic cash system. Bitcoin. URL https://bitcoin.org/bitcoin.pdf. p 4
29. Gupta SS (2017) Blockchain. IBM Onlone http://www.IBM.COM
30. Mamoshina P, Ojomoko L, Yanovich Y, Ostrovski A, Botezatu A, Prikhodko P, Izumchenko E, Aliper A, Romantsov K, Zhebrak A, Ogu IO (2018) Converging blockchain and next-generation artificial intelligence technologies to decentralize and accelerate biomedical research and healthcare. Oncotarget 9(5):5665
31. Angrish A, Craver B, Hasan M, Starly B (2018) A case study for blockchain in manufacturing:"FabRec": a prototype for peer-to-peer network of manufacturing nodes. Procedia Manuf 26:1180–1192
32. Appel AW (2015) Verification of a cryptographic primitive: SHA-256. ACM Trans Program Lang Syst (TOPLAS) 37(2):1–31
33. Emmadi N, Narumanchi H (2017) Reinforcing immutability of permissioned blockchains with keyless signatures' infrastructure. In: Proceedings of the 18th international conference on distributed computing and networking, pp 1–6
34. Huang X, Xu C, Wang P, Liu H (2018) LNSC: a security model for electric vehicle and charging pile management based on blockchain ecosystem. IEEE Access 6:13565–13574
35. Mingxiao D, Xiaofeng M, Zhe Z, Xiangwei W, Qijun C (2017) A review on consensus algorithm of blockchain. In: 2017 IEEE international conference on systems, man, and cybernetics (SMC).IEEE, pp 2567–2572

36. Nguyen GT, Kim K (2018) A survey about consensus algorithms used in blockchain. J Inf Proces Syst 14(1):101–128
37. Zheng Z, Xie S, Dai H, Chen X, Wang H (2017) An overview of blockchain technology: architecture, consensus, and future trends. In: 2017 IEEE international congress on big data (BigData congress). IEEE, pp 557–564
38. Kosba A, Miller A, Shi E, Wen Z, Papamanthou C (2016) Hawk: the blockchain model of cryptography and privacy-preserving smart contracts. In: 2016 IEEE symposium on security and privacy (SP). IEEE, pp 839–858
39. Mohanta BK, Panda SS, Jena D (2018) An overview of smart contract and use cases in blockchain technology. In: 2018 9th international conference on computing, communication and networking technologies (ICCCNT). IEEE, pp 1–4
40. Westerkamp M, Victor F, Küpper A (2018) Blockchain-based supply chain traceability: token recipes model manufacturing processes. In: 2018 IEEE international conference on internet of things (iThings) and IEEE green computing and communications (GreenCom) and IEEE cyber, physical and social computing (CPSCom) and IEEE smart data (SmartData). IEEE, pp 1595–1602
41. Xiong Z, Feng S, Niyato D, Wang P, Han Z (2018) Optimal pricing-based edge computing resource management in mobile blockchain. In: 2018 IEEE international conference on communications (ICC). IEEE, pp 1–6
42. Aydinli K (2019) Design and implementation of a scalable IoT-based blockchain (Doctoral dissertation, PhD thesis, Master Thesis, CSG@ IfI, University of Zürich, Switzerland, Supervisors)
43. Zhu Y, Qin Y, Zhou Z, Song X, Liu G, Chu WCC (2018) Digital asset management with distributed permission over blockchain and attribute-based access control. In: 2018 IEEE international conference on services computing (SCC). IEEE, pp 193–200
44. Bahri L, Girdzijauskas S (2019) Trust mends blockchains: living up to expectations. In: 2019 IEEE 39th international conference on distributed computing systems (ICDCS). IEEE, pp 1358–1368
45. Gaetani E, Aniello L, Baldoni R, Lombardi F, Margheri A, Sassone V (2017) Blockchain-based database to ensure data integrity in cloud computing environments
46. Dorri A, Kanhere SS, Jurdak R, Gauravaram P (2017) Blockchain for IoT security and privacy: the case study of a smart home. In 2017 IEEE international conference on pervasive computing and communications workshops (PerCom workshops). IEEE, pp 618–623
47. Talal M, Zaidan AA, Zaidan BB, Albahri AS, Alamoodi AH, Albahri OS, Alsalem MA, Lim CK, Tan KL, Shir WL, Mohammed KI (2019) Smart home-based IoT for real-time and secure remote health monitoring of triage and priority system using body sensors: Multi-driven systematic review. J Med Syst 43(3):42
48. Ghuli P, Kumar UP, Shettar R (2017) A review on blockchain application for decentralized decision of ownership of IoT devices. Adv Comput Sci Technol 10(8):2449–2456

Real-Time Tracking System for Object Tracking Using Internet of Things (IoT)

Hitesh Kumar Sharma, Tanupriya Choudhury, Adarsh Kandwal, Anurag Mor, Preeti Sharma, Md. Ezaz Ahmed, and Prashant Ahlawat

Abstract In the twenty-first century, IoT and machine learning made possible and affordable to track the belonging objects on our mobile screen. Due to daily busy schedules, we generally miss our important belongings at many such places where frequent visit is not possible. Tracking of our goods and belongings is an open challenge in real-time situations. Technologies are matured enough to make this possible in real-time. In this research work, we have integrated IoT, AIML and GPS together in a Web-based and mobile-based app to solve this real-time issue. A low energy consumption device has been developed in this work which makes it more durable and environment friendly. The user can see his/her object on Google Map in real-time which minimizes the probability of miss or stolen his/her item or object. Advanced technologies with open-source tools have been integrated in this system to make it a high-performing system. An Android-based mobile application will also developed to track and monitor the bag/vehicle on mobile screen.

Keywords Internet of Things (IoT) · Object tracking · GPS · GPRS · Cloud computing · Sensors

1 Introduction

In a day, many people travel from one place to another place. While traveling, many people may carry valuable items. There might be a chance of losing/stolen of the items [1]. No airline service provider or any other travel service provider provides the

H. K. Sharma (✉) · T. Choudhury · A. Kandwal · A. Mor · P. Sharma
School of Computer Science, University of Petroleum andEnergy Studies (UPES), Energy Acres, Bidholi, Dehradun, Uttarakhand 248007, India

Md. E. Ahmed
Department of Computer Science, Saudi Electronic University, Al Madina, Kingdom of Saudi Arabia
e-mail: m.ezaz@seu.edu.sa

P. Ahlawat
Manipal University Jaipur, Jaipur, India

J. M. R. S. Tavares et al. (eds.), *Cyber Intelligence and Information Retrieval*, Lecture Notes in Networks and Systems 291,
https://doi.org/10.1007/978-981-16-4284-5_43

real-time tracking feature/service to track the passenger's bags if they got lost/stolen. Similarly, in vehicles, no vehicle company is providing any tracking system to track the vehicle when it is stolen [2, 3]. We know that generally the cost of vehicle will be high, so losing the vehicle cause people more stress. Traditionally, we were lacking in such kind of device or system which can be used for real-time tracking of objects. It will be helpful to track stolen or missed objects. In today's scenario, if a bag or a vehicle has been stolen by someone, then there is no other way to find the missed or stolen object. Only civil police are the option to find the object. The probability of finding missed item with the help of police is very low. Police is also not capable to handle these kinds of situations due to unavailability of such devices in the current world. But using the proposed device, police can track down the position of their missed bag or stolen vehicle and the same can be easily tracked using GPS [4, 5].

To overcome these types of problems, we build a real-time tracking system using GPS. In this system, we have used GPS-enabling API to integrate GPRS. GSM module will be used to find location and tracking the object where the device will be fitted. Using cloud APIs services, we will send the captured data from GPS module to the Google map and visualize the real-time path of the object on Web/mobile app [6, 7]. During travel, this device can be placed or attached with items we carry with us. This device which will have GSM module will be attached in the object like bags or suitcase which make it trackable. The object will get an ID so that it can be monitored on Google Map [8–10]. Using the shortest path algorithm, we can easily find the minimum path of the object from the person. In our research work, we have also deployed a direction feature. A navigation option is provided in our proposed system, enabling that the shortest path will be displayed on Web/app dashboard to reach the object in case of miss/stolen in real-time.

In similar way, the same device can be fixed in two-wheeler vehicles. User can track anytime the real-time position of his/her vehicle from any distance with the help of Android mobile app. It will also help the user to track the vehicle in case of stolen. Our research work can be used for also various scenarios like tracking lost pets, small children's, etc.

2 Literature Review

As we all know that security has been becoming an important issue everywhere. It is becoming necessary nowadays as the possibilities of intrusion are increasing day by day. As we know that security cannot be guaranteed [11, 12]. Using more number of tools and technologies is not the solution for guaranteed security. The effective use of available tools and technologies in properly integrated manner will be more effective. Here, in our research work, security system has been designed that has a special feature, which makes the person or passenger aware of his being bag being stolen/lost or vehicle being stolen. We have designed an Internet of things-based real-time tracking system using GPS. Arduino Uno device is used, which is considered one of modern programmable devices and provide us an interface for accessing other modules [13–15]. In last so many years, many systems with advanced controller has been developed to prevent unauthorized access. We as humans have a tendency of locking whatever is deemed precious to us, be it our belongings in the locker, our houses, other institutional buildings (schools, offices, etc.). Hence, it is a high need of having some smart and cost-effective system to guard our belongings and make them safe from theft [16]. In the more conventional ways, people try to safeguard their belongings by employing a watch dog or maybe security personals [17, 18]. In this literature survey, we find that billions of dollars are generally spent worldwide to make our belongings more secured from unauthorized access. Manpower is installed to protect them from others to being theft. Implementing real-time monitoring device with each object can reduce this cost and manpower overhead also. Hence, to achieve the safety, we have designed an Internet of things-based real-time tracking system using GPS. The sole purpose of using Arduino Uno in our research work is that it has been the open nature of this very module and can provide an effective solution for IoT[19–21]-based applications. The software used on Arduino Uno is entirely available as on open source and the hardware design information is readily available under creative commons licenses for new designers as ourselves, be it schematics, PCB layouts etc. In practice, it is very easy to adapt the software and the hardware and then contribute what you do back on the Arduino research work as a whole.

3 Objectives

The objectives and their flow has shown in Fig. 1.

Objective A: Collecting the components for real-time tracking system.
Objective B: Building the device.
Objective C: Writing business logic using Arduino software (IDE).
Objective D: Development of Web portal and an Android application.
Objective E: Device testing.

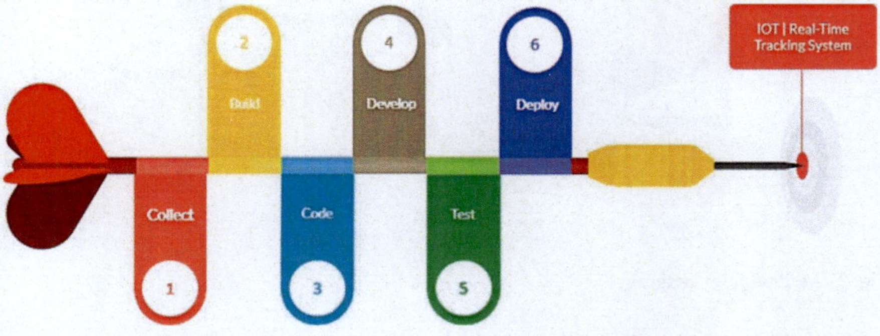

Fig. 1 Flow diagram of objectives

Objective F: Deploying of Web portal on Microsoft Azure/Amazon Web services.

4 Methodology

As shown in Fig. 2, for the creation of real-time tracking system. A system or device needs to fulfill the above-said six objectives. We have used IoT to enable computing device named Arduino Uno and GPS-enabled module for collecting data from the object that needs to be tracked. Collected data need to be stored on cloud storage for anytime availability. To access the data from cloud and projecting on a navigation map, a Web and mobile app-based dashboard is created to track the real-time position of given object.

4.1 Collecting the Components for Real-Time Tracking System

For the development of real-time tracking system, we do require hardware modules and sensors to get the real-time position. We will be requiring Arduino Uno Board (either 3.3 or 5 V), GPS Module Sim 28 M, GSM Module Sim 900A, Battery (12 V | 1A), Power Bank, Antenna, GPS Receiver, Jumper Wires, ThingSpeak API's and Microsoft Azure/AWS subscription. So, the first step is to collect/purchase these components/subscriptions and then go for next phase.

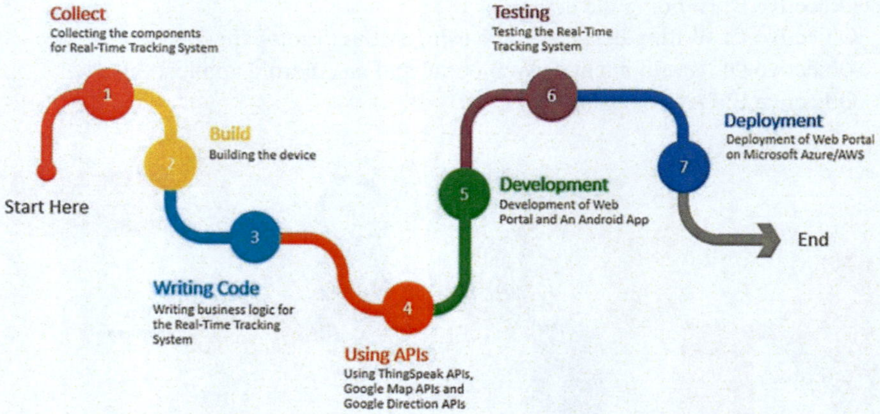

Fig. 2 Methodology diagram

4.2 Building the Device

We will place the device in a bag or vehicle to track its position in real-time and then use it for analytics, mapping or directions. We will use ThingSpeak Cloud API's to track the bag or vehicle. We will be using a GPS module which will provide us the exact position of bag or vehicle. We will use GSM module to send the latitude and longitude coordinates to the ThingSpeak channel via GPRS. And to make connection between all the components, we will use jumper wires.

4.3 Writing Business Logic Using Arduino Software (IDE)

To get the current location of travel bag or vehicle, we need to write the code or the business logic (Arduino programming) for the device using GPS, which will provide us the exact position of bag or vehicle and GSM will send it over GPRS to the ThingSpeak server. We will upload the code in Arduino Uno, in order to interact with GPS and GSM modules.

4.4 Using API

We will be using Google Map API's and Google Map direction services, through which we can display the location on map and shortest path to reach the stolen/lost travel bag or stolen vehicle.

4.5 Development of Web Portal and an Android Based Mobile App

In this research work, we will design and develop a Web-based and mobile-based application which will have two frames. One frame will show that real-time position of stolen object and the second frame will show the shortest path of reaching to that object from current position. Both Web and mobile applications will have same layouts. Mobile app will help to track the object using a hand-held device which will be easy to carry. We will deploy the Android application on the Google Play Market for the ease of user.

4.6 Testing the Device

After the development of real-time tracking system, we will do the testing. We will provide different test cases to the device, which will tell us that, how much our device is efficient, robust and accurate.

4.7 Web Application Deployment on Microsoft Azure/Amazon Web Service

On completing testing, the device the whole system has been deployed on cloud platform for 24*7 availability. Amazon Web Service cloud-enabled platform has been chosen for this purpose. A JavaScript-based Web portal and an Android-based mobile application is also developed to visualization the position on a map for tracked object. The users will be charged as per the usage of cloud services.

5 Implementation Levels

1. For the development of real-time tracking system, we do require hardware modules and sensors to get the real-time position. We are requiring Arduino Uno Board (either 3.3 V or 5 V), GPS Module Sim 28 M, GSM Module Sim 900 A, Battery (12 V | 2A), Power Bank, Antenna, Jumper Wires, ThingSpeak API's and Microsoft Azure/AWS subscription. So, first step is to collect/purchase these components/subscriptions and then go for next phase.
2. We will place the device in a bag or vehicle to track its position in real-time and then use it for analytics, mapping or directions. We will use ThingSpeak Cloud API's to track the bag or vehicle. We will be using a GPS module which will provide us the exact position of bag or vehicle. We will use GSM module to send the latitude and longitude coordinates to the ThingSpeak channel via GPRS. And to make connection between all the components, we will use jumper wires.
3. To get the current location of travel bag or vehicle, we need to write the code or the business logic (Arduino Programming) for the device using GPS, which will provide us the exact position of bag or vehicle and GSM will send it over GPRS to the ThingSpeak server. We will upload the code in Arduino Uno in order to interact with GPS and GSM module. We are using Google Map API's and Google Map Direction Services through which we can display the location on map and shortest path to reach the stolen/lost travel bag or stolen vehicle.
4. We will develop an Android application as well as Web portal, which will have two frames. One frame will be used to visualize the live tracking of stolen object.

And other frame will be used to show the shortest direction to the bag or vehicle using Google Direction Service.

5. After the development of real-time tracking system, we have done testing. We have provided different test case to the device, which tell us that, how much our device is efficient, robust and accurate.
6. After that, we will deploy the Web portal on cloud computing platforms like Amazon Web Service or Microsoft Azure by running the service on a VM.

6 Hardware Used and Circuit Diagram

The complete hardware setup is shown in Fig. 3. Various modules used in this device has also been shown in this diagram.

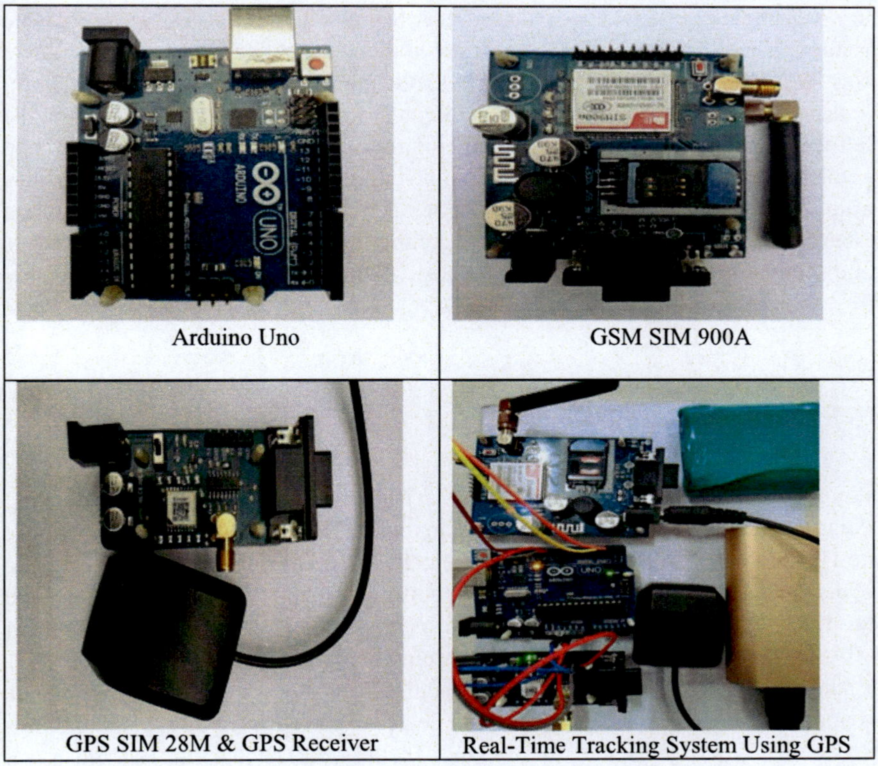

Fig. 3 Hardware components and circuit diagram

7 Conclusion

The work carried out in this research and implementation is that we have used a cost-effective model with minimum components and with latest approach to develop a system which will be used to solve a real-time issue of society. IoT-enabled sensors used in this device are very cost-effective. The machine learning algorithms are also less resource consuming, GPS systems is energy efficient. So this system is cost and energy efficient. Similarly, it was difficult for civilian police also for finding the bag or vehicle. In this research work, we have integrated successfully the technical components of IoT and AIML. Using IoT smart sensors, we have developed a system which will solve this real-time issue with many of the busy people in their daily life. It will be a great contribution to the world and the Internet of things world community so that someone can do more research on this research work and will be able to build even more secure, efficient and robust module or device which will help the consumer in a great way. It will be a nightmare to bag lifter or vehicle lifter as having our research work or device installed in their bag or vehicle. Because having device installed, it will let consumers to track down their lost/stolen bag or stolen vehicle using Web portal as well as an Android application. By doing this research work, we can understand that how we can implement an Internet of things open-source platform in maintaining the latitude and longitude coordinates of the lost/stolen bag or stolen vehicle. Microsoft Azure is a public cloud computing platform. It provides various cloud computing services, which includes computing, data analytics, storage resources and network resources. Cloud computing is a technology which delivers computing resources to the clients on request. As this Web app will be available 24×7.

8 Future Work

Future prospectus of this research work is that it can be extended for real-time tracking system in Airline Travel Sector using some more advanced sensors and some more latest technologies so that it would automatically sense when the airplane accelerates, breaks and use that to stop and start the reporting of its position. This will make the real-time tracking system even much more energy-efficient solution as the GSM SIM 900A device can be turned off and the Arduino Uno put to sleep.

Another feature that can be installed is that in future, we will make our device even more compact and small. So that our device can keep easily in any small things like handbags, pet's belts, etc., for tracking. Then it won't be just limited to the travel sector, which means it could be used everywhere, where the need to locate something. This will make the real-time tracking system even more robust, flexible and precise.

References

1. Al-Mazloum A, Omer E, Abdullah MFA (2013) GPS and SMS-based child tracking system using smart phone. Int J Electr, Robot, Electron Commun Eng 7(2):171–174
2. Lee S, Tewolde G, Kwon J (2014) Design and implementation of vehicle tracking system using GPS/GSM/GPRS technology and Smartphone application. Internet of things (WF-IoT), 2014 IEEE world forum on. IEEE
3. Noppadol C, Russameesawang A, Keeratiwintakorn P (2008) Real-time tracking management system using GPS, GPRS and Google Earth, ECTICON 2008. In: 5th international conference. pp 14–17. ISBN: 978-1-4244-2101-5
4. Khan A, Mishra R (2012) GPS-GSM based tracking system. Int J Eng Trends Technol 3(2):161–164
5. Sharma HK, Shastri A, Biswas R (2012) A framework for automated database tuning using dynamic SGA parameters and basic operating system utilities. Database Syst J 3(4)
6. Kshitiz K, Shailendra (2018) NLP and machine learning techniques for detecting insulting comments on social networking platforms. In: International conference on advances in computing and communication engineering (ICACCE) 2018. IEEE, pp 265–272
7. Sharma HK (2019) Air quality prediction using artificial neural networks. In: International conference on automation computational and technology management (ICACTM)
8. Khanchi I, Ahmed E (2019) Automated framework for real-time sentiment analysis. In: International conference on next generation computing technologies (NGCT-2019)
9. Singh T, Kshitiz K, Singh H, Kukreja P (2017) Detecting hate speech and insults on social commentary using nlp and machine learning. Int J Eng Technol Sci Res 4(12):279–285
10. Ghribi B, Logrippo L (2000) Understanding GPRS: the GSM packet radio service. Comput Netw 34:763–779
11. Djuknic GM, Richton RE (2001) Geolocation and assisted GPS. Computer 34(2):123–125
12. Ahlawat P, Biswas SS (2020) Sensors based smart healthcare framework using internet of things (IoT). Int J Sci Technol Res 9(2):1228–1234
13. Sanders G, Thorens L, Reisky M, Rulik O, Deylitz S (2003) GPRS networks. Wiley, Hoboken, NJ
14. Hoff S, Meyer M, Schieder A (1998) A performance evaluation of Internet access via the general packet radio service of GSM. In Proceedings of. 48th IEEE Vehicular technology conference, Ottawa, vol 3, pp 1760–1764
15. Gharge S, Chhaya M, Chheda G, Deshpande J (2012) Real time bus monitoring system using GPS. An Int J Eng Sci Technol 2(3):441–448
16. Chandurkar S, Mugade S, Sinha S, Borkar P (2013) Implementation of real time bus monitoring and passenger information system. Int J Sci Res Publ 3(5):1–5
17. Krishnan DSR, Gupta SC, Choudhury T (2018) An IoT based patient health monitoring system. Int Conf Adv Comput Commun Eng (ICACCE) 2018:1–7
18. Sharma A, Choudhury T, Kumar P (2018) Health monitoring & management using IoT devices in a cloud based framework. Int Conf Adv Comput Commun Eng (ICACCE) 2018:219–224
19. Obheroi KS, Chaurasia A, Choudhury T, Kumar P (2018) Economical home monitoring system using IOT. In: Proceedings of the second international conference on computational intelligence and informatics, pp 627–637
20. Sharma HK, Taneja S, Ahmed E, Patni JC (2019) I-Doctor: An IoT based self patient's health monitoring system. In: 2019 international conference on innovative sustainable computational technologies, CISCT 2019
21. Sharma HK, Khanchi I, Agarwal N, Seth P, Ahlawat P (2019) Real time activity logger: a user activity detection system. Int J Eng Adv Technol 9(1):1991–1994

Empirical Analysis on Consensus Algorithms of Blockchain

Shivani Wadhwa and Gagandeep

Abstract Nowadays, the rapid growth in the count of cryptocurrencies has been observed. Blockchain is the core technology used in the development of cryptocurrencies. In addition, blockchain also supports various security applications as it has converted the centralized structure of the network into the decentralized structure. Various consensus algorithms exist for deciding how to reach an agreement to append block into the verified and distributed ledger. However, there is a lack of empirical analysis on different consensus algorithms based on various parameters. To support detailed analysis, this study first discusses the most widely used consensus algorithms based on proof-based consensus, voting-based consensus and directed acyclic graph. Further, the analysis on these algorithms has been performed using tolerated power of adversary, verification speed and throughput as performance parameters. This analysis would help the reader to select the best consensus algorithm for a specific application.

Keywords Blockchain · Consensus algorithm · Performance

1 Introduction

Blockchain has brought a revolution in the era of Internet. According to the Gartner prediction, the overall quality of data of the companies using blockchain smart contracts will be increased by approx 50% by 2023 [1]. The blockchain is a distributed storage system that resembles a chain, with each block representing transaction/event information as well as the hash function of the previous block which links the chain cryptographically [2]. It is a distributed digital ledger which records all form of transactions that exist between the two partners in a network in an efficient, persistent and verifiable mode. It provides strong data integrity to the transactions being recorded without the involvement of a third party [3].

S. Wadhwa (✉) · Gagandeep
Department of Computer Science, Punjabi University, Patiala, India

J. M. R. S. Tavares et al. (eds.), *Cyber Intelligence and Information Retrieval*,
Lecture Notes in Networks and Systems 291,
https://doi.org/10.1007/978-981-16-4284-5_44

Researchers of the blockchain mainly focus on three aspects. The first aspect of the blockchain is the cryptocurrency like Bitcoin, Peercoin, Blackcoin, Monero, Litecoin, etc. The second aspect of the blockchain includes the application of the blockchain where it can be applied like a smart meter, smart city, health care, education, supply chain management, etc. The third aspect includes the concepts which are the foundation of blockchain-like, consensus mechanism, prevention from attacks, smart contracts, etc.

1.1 *Working of Blockchain*

Various users exist in the network which gives rise to various numbers of transactions in the decentralized and distributed network [4]. When any user initializes the network, that transaction is sent to the pool where various transactions are kept together. Bunch of transactions are taken into consideration, and then consensus algorithm is applied on these to reach agreement. Different nodes in the network start working on the consensus algorithm to generate a block. Most of the consensus algorithms give reward to the block for producing the successful block. Once the successful block is released by the node, and then it sends that block to other nodes for verification purpose. When all nodes validate the block, then that block is appended in the blockchain. All nodes in the network update the blockchain possessed by them after appending newly generated block into their existing blockchain. Figure 1 shows the sequence of events for appending the block in the blockchain.

The rest of the paper would be organized as follows: Sect. 2 presents the overview of various consensus algorithms which are most widely used for appending block in the blockchain. Section 3 discusses the different parameters which are considered for analysing the performance of consensus algorithms. Section 3 provides the analysis of considered algorithms on performance parameters. Section 4 provides the conclusion of the whole study.

Fig. 1 Working of blockchain

2 Overview of Consensus Algorithms

In an untrustworthy and distributed framework, blockchain innovation guarantees that all the nodes accomplish agreement on the regular record utilizing a specific consensus algorithm. The consensus mechanism is essential in a blockchain technology as it is a key component of the solution for achieving integrity in the distributed environment. Two important properties of good consensus algorithm are persistence and liveness. Persistent property ensures the stable state of the data by providing consistent response. Liveness property infers that all nodes participating in the consensus mechanism agree with the decision being made [5]. These properties confirm that digital ledger is robust, permanent, secure and authentic. Few important consensus algorithms are proof of work (PoW), practical byzantine fault tolerance (PBFT), delegated practical byzantine fault tolerance (DPBFT), proof of stake (PoS), delegated proof of stake (DPoS) and many more. Table 1 illustrates the brief overview of various consensus algorithms. The detailed discussion on these algorithms is given below:

1. **Proof of Work**: In 2008, Satoshi Nakamoto used the PoW in Bitcoin to generate tamperproof and consistent transactions [6]. The main technology incorporated in the bitcoin is the blockchain. PoW involves solving the complex computational puzzle in such a way that the computed hash of the block equals to or less than the constraint value. All miners start working on the block, only one miner will win, and resources consumed by others are simply wasted. Speed and success rate of PoW is highly dependent on the computational capabilities of the miner's hardware [7].

2. **Proof of Stake**: Energy-inefficient PoW consensus algorithm has given rise to proof of stake. PoS came into existence in 2011 for Peercoin cryptocurrency. Possession of stake and age are the main factors on which the creator of the next block depends. Quasi-random process is used to select the creator of the next block which ultimately depends on the stakes possessed by particular node [8]. Nodes having majority of the stakes can make the blockchain centralized.

3. **Delegated Proof of Stake**: The concept of DPoS was introduced in 2014 by Daniel Larimer in Bitshares. In this approach, nodes participate in voting to select the witnesses of the next block [9]. The number of nodes is very limited who have right to witness the next block. After successful creation of the block, the incentive is generally shared with the nodes who voted for the witness.

4. **Practical Byzantine Fault Tolerance**: PBFT is considered as an energy-efficient consensus approach with high throughput. Due to the existence of malicious nodes in the network, PBFT uses the concept of byzantine general problem for the creation of consortium blockchain. In decentralized system, all nodes participate in

Table 1 Brief overview of consensus algorithms

Consensus Algorithm	Type of Consensus	Type of Blockchain	Cryptocurrency	Merits	Suitability
PoW	Proof-based	Public	Bitcoin	Avoids double spending	Depends on computational power
PoS	Proof-based	Public	Peercoin	Consumes less computational power	Depends on stakes
DPoS	Proof-based	Public	Bitshares	High throughput as compared to PoS	Depends on stakes of witnesses
PoL	Proof-based	Public	Not found	Low latency with deterministic confirmation time	Needs SGX-enabled CPU's for computation purposes
PBFT	Voting-based	Private or consortioum	Cosmos (Atom)	Low latency, high throughput and consumes less computation	Suitable for low-scale networks
Raft	Voting-based	Private or consortioum	Not found	Fault tolerant	Suitable to apply in actual systems
Ripple	Proof-based	Public	XRP ripple	Decentralized, avoids double spending	Suitable for achieving decentralized networks
Tangle	Directed acyclic graph	Private	IoTA	Low latency, high throughput, scalable and consumes less computation	Best suitable for IoT devices

voting to find the creator of the next block. To reach the consensus, two-third of the nodes must approve that block otherwise consensus is not reached [10].

5. **Proof of Luck**: Multinovic and Warren in 2017 proposed consensus algorithm for the implementation in trusted execution environment with XGS devices. Generation of the highest luck value is the main criteria to decide that block of which node will be appended into the blockchain. The value of luck number varies between 0 and 1. To reach consensus, all nodes must agree that the main chain will be the blockchain with the summation of largest luck value [11]. Node may not be lucky if there is no synchronization between its clock and network's clock.

6. **Raft:** Raft algorithm was proposed by Stanford's Ongaro and others in 2013 [12]. Quorum uses the raft consensus algorithm to tolerate crashes. Raft works on the simple assumption that atleast $n/2 + 1$ of the total nodes should work normally. Follower, candidate and leader are three types of verifying nodes in this consen-

sus algorithm. Two types of message passing take place between these nodes, i.e. RequestVote and AppendEntries. Leaders are elected on the basis of randomized timers [13].

7. **Ripple**: Schwartz and others proposed the ripple consensus algorithm in 2014 for consortium blockchain. It basically combines the various trusted subnetworks which are part of larger network. In this network, server nodes participate in consensus and client nodes are for transferring the funds. In case of query, each ripple server maintains the Unique Node List (UNL). UNL is used to determine that the transaction should be added into the ledger or not. To reach the consensus even if 20% of the total nodes are faulty, then also agreement can be reached correctly [14].

8. **Tangle**: Tangle is proposed by the Popov in 2016 for IoT devices. The consensus algorithm of tangle is probabilistic and follows the concept of directed acyclic graph.The graph of the tangle is represented by transactions initiated by the users. To add the transaction in the tangle, node must approve two unapproved transactions. The tangle user also approves the transaction by solving the PoW puzzle. Honest participants are rewarded with the incentive.

3 Discussion on Performance Analysis of Considered Consensus Algorithms

Features of good consensus algorithm are efficiency, convenience and its secure nature. Many consensus algorithms exist for providing consensus algorithm with good mix of these features.

1. **Tolerated power of adversary**: This is the main factor to determine security level of the consensus algorithm. This parameter evaluates the definite percentage of the malicious attackers that may enter the network. In the consensus algorithms taken into consideration, the highest rate is 51% in case of PoS and DPoS and least is 20% in case of ripple as shown in Fig. 2. Although there can be variation in this percentage in some cases, most of the time this percentage is recorded. As the tolerated power of adversary increases, the performance of consensus algorithm also increases.

2. **Verification Speed**: This factor is used to determine that how much time is required to validate the transaction into the block. Two major factors affecting this parameter are as follows: the type of blockchain and the speed of the computing resource. Voting-based consensus algorithms generally require less time for verification as compared to proof-based consensus algorithms (refer Fig. 3). Due to this reason, voting-based consensus algorithms are mostly used in pri-

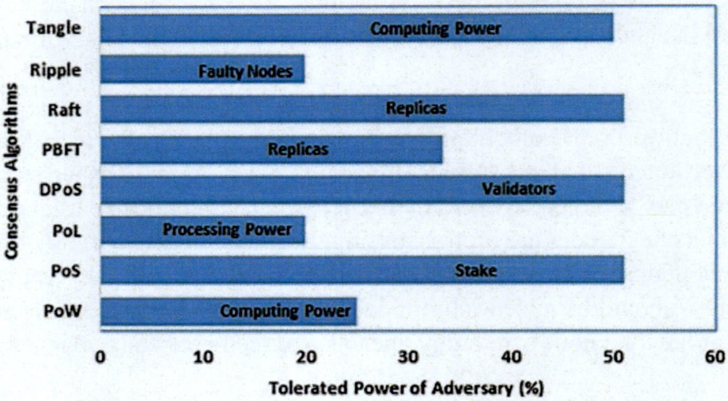

Fig. 2 Tolerated adversary

Fig. 3 Verification speed

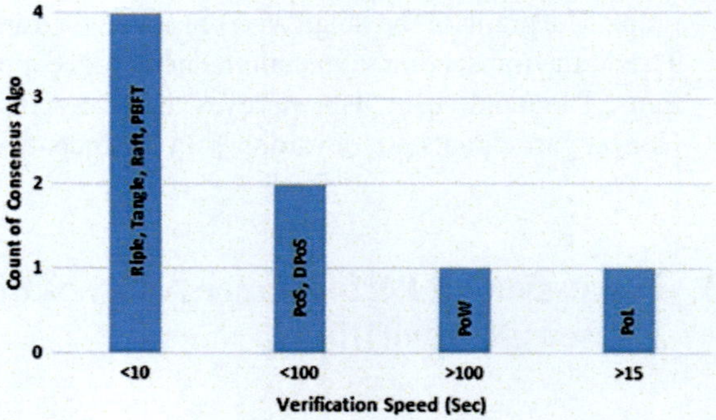

vate blockchain, consortioum blockchain, zkCrowd [15], etc. Tangle also shows good verification speed in smaller networks. Verification rate of traditional PoW is very low which gives rise to other consensus algorithms.

3. **Throughput**: Throughput is considered as the important performance metric of the blockchain. Throughput is evaluated on the basis of count of successful transactions per second that occurred after the deployment of the first transaction. High throughput is achieved in the blockchain with the less block interval time [16]. Blockchain sharding is used nowadays to increase the throughput [17]. High throughput is observed in case of raft and least in case of PoW as illustrated in Fig. 4. Tangle achieves better throughput in networks with less than 250 nodes. Although PoW has least throughput, its variants or hybrid approaches obtain much better throughput than PoW.

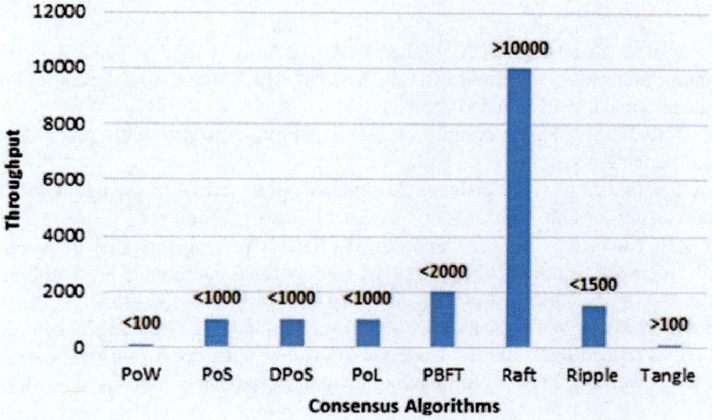

Fig. 4 Throughput

4 Conclusion

Blockchain technology will impact a lot in digitizing the different scenarios. Consensus algorithms are used to manage the transactions in the digital ledger. In this paper, consensus algorithms based on proof-based consensus, voting-based consensus and directed acyclic graph are discussed in detail. Factors affecting tolerated power of adversary, verification speed and throughput of different consensus algorithms are the main performance parameters discussed in this paper. Although a lot of research work has been done to improve these performance parameters, still consensus algorithm with good performance parameters for particular scenarios like IoT, mobile devices does not exist as they have limited resources with them. As a future work, we will tend to improve the performance of these parameters under the limited constraints.

References

1. https://www.ledgerinsights.com/gartner-by-2023-blockchain-can-improve-data-quality-by-50/
2. Xu G, Liu Y, Khan PW (2019) Improvement of the DPoS consensus mechanism in Blockchain based on vague sets. IEEE Trans Industrial Informatics 16(6):4252–4259
3. Zheng Z, Xie S, Dai HN, Chen X, Wang H (2018) Blockchain challenges and opportunities: a survey. Int J Web Grid Serv 14(4):352–375
4. Saleem A, Agarwal AK (2016) Analysis and design of secure web services. In: Proceedings of fifth international conference on soft computing for problem solving. Springer, Singapore, pp 453–462
5. Zhang R, Xue R, Liu L (2019) Security and privacy on blockchain. ACM Comput Surv (CSUR) 52(3):1–34
6. Nakamoto S, Bitcoin A (2008) A peer-to-peer electronic cash system. Bitcoin 4. https://bitcoin.org/bitcoin.pdf

7. Zheng Z, Xie S, Dai H, Chen X, Wang H (2017) An overview of blockchain technology: architecture, consensus, and future trends. In: 2017 IEEE international congress on big data (BigData congress). IEEE, pp 557–564
8. King S, Nadal S (2012) Ppcoin: peer-to-peer crypto-currency with proof-of-stake. self-published paper, August, 19, p 1
9. Schuh F, Larimer D (2017) Bitshares 2.0: general overview. Accessed June 2017 (Online). Available: http://docs.bitshares.org/downloads/bitshares-general.pdf
10. Chondros N, Kokordelis K, Roussopoulos M (2012) On the practicality of practical byzantine fault tolerance. In: ACM/IFIP/USENIX international conference on distributed systems platforms and open distributed processing. Springer, Berlin, pp 436–455
11. Milutinovic M, He W, Wu H, Kanwal M (2016) Proof of luck: an efficient blockchain consensus protocol. In: Proceedings of the 1st workshop on system software for trusted execution, pp 1–6
12. Ongaro D, Ousterhout J (2013) In search of an understandable consensus algorithm (extended version)
13. Huang D, Ma X, Zhang S (2019) Performance analysis of the raft consensus algorithm for private blockchains. IEEE Trans Syst Man Cybernetics Syst 50(1):172–181
14. Schwartz D, Youngs N, Britto A (2014) The ripple protocol consensus algorithm. Ripple Labs Inc White Paper 5(8):151
15. Zhu S, Cai Z, Hu H, Li Y, Li W (2019) zkCrowd: a hybrid blockchain-based crowdsourcing platform. IEEE Trans Industrial Informatics 16(6):4196–4205
16. Xiao Y, Zhang N, Lou W, Hou YT (2020) A survey of distributed consensus protocols for blockchain networks. IEEE Commun Surv Tutorials 22(2):1432–1465
17. Li S, Yu M, Yang CS, Avestimehr AS, Kannan S, Viswanath P (2020) Polyshard: coded sharding achieves linearly scaling efficiency and security simultaneously. IEEE Trans Information Forensics Security 16:249–261

City Traffic Speed Characterization Based on City Road Surface Quality

Ratna Mandal, Soumi Dutta, Rupayan Banerjee, Sujoy Bhattacharya, Ritusree Ghosh, Sougata Samanta, and Tiyasa Saha

Abstract Road surface condition detection not only helps us to improve road surface quality, but also it helps us to implement one of the main aspects of driverless car by maintaining proper speed limit, determining the actual road condition. But preparatory to monitor and analyse the road condition, we need some of the sensors to be available in the vehicle itself or in the road to detect those sensor data for our research work. But to make things simpler, we use our smart phone sensors to encounter those sensors data. As every person carries at least their mobile phone with them while travelling, we have used a sensor logging app to log primarily the GPS and accelerometer data from our smartphone sensors. After getting the data, we have used a plotter to visualize the data, i.e. from which point to which point we are monitoring the data and our GPS footprint trail helped us to do that. After that our main challenge is to get the accelerometer data without any inconsistency, because of the re-orientation issue while capturing the accelerometer data. In order to calculate the road surface index, we need only Z axis value of accelerometer data to be processed, and this road surface index is the metric to measure the road surface quality. Conforming to address whether the road is good, bad or excellent, the road surface index will be used. In this paper, we focus specifically on the sensing component, which uses the accelerometer, GPS sensors in the smart phones to detect potholes and bumps to monitor road surface conditions.

Keywords City traffic speed · Smartphone sensor · Road surface measurement

R. Mandal (✉) · S. Dutta · R. Banerjee · S. Bhattacharya · R. Ghosh · S. Samanta · T. Saha
Institute of Engineering and Management, Calcutta, India
e-mail: ratna.mandal@iemcal.com

S. Dutta
e-mail: soumi.dutta@iemcal.com

J. M. R. S. Tavares et al. (eds.), *Cyber Intelligence and Information Retrieval*, Lecture Notes in Networks and Systems 291,
https://doi.org/10.1007/978-981-16-4284-5_45

1 Introduction

"*You can't understand a city without using its public transportation system.*" —Erol Ozan

Quality of traffic management system highly depends on the estimation of different traffic-related parameters like speed, volume, flow to visualize real-time traffic conditions. These systems provide the basis of services like route recommendations along with driving directions, showing rerouting options, providing transit information, assistance in locating parking lots, etc. Some systems also allow users to report and localize several traffic instances like road accidents, hazards, weather conditions, police warnings, etc., on digital map.

The systems mentioned above rely on GPS trails which is obtained by intelligent sensing of mobility data of commuters. However, such systems suffer from some limitations, like inadequate traffic information in remote areas, high dependence on only real-time mobility data and individual effect on driving pattern which are real barriers of the development of any sophisticated traffic monitoring system. However, studies suggest that vehicle speed is significantly affected by the roughness of the road surface and the intensity of light at night. Vehicles may be forced to travel at a slower speed due to extremely poor road surface conditions. Therefore, existing systems fail to infer the proper reasons for lower-speed levels of traffic.

Measurement of road surface condition is an integral part for the development of smart city transportation service. Rapid urbanization process, inadequate road space and shortfall of budget affect the overall maintenance process of road surface quality across the suburban cities of low economic countries. Therefore, building up a good quality of transportation is challenging as it is highly affected by poor road condition. People have contributed a lot to measure the roughness of road surface by intelligent sensing of accelerometer data of traveller along particular road segments. Notably, some of the previous works like [15, 20, 21] have used contextual information through multi-modal sensing for analysing road conditions and traffic behaviour. Measurement of international roughness index using z-axis accelerometer and GPS have shown the accuracy of IRI measurement can be enhanced by low-cost measurement system. Road surface condition detection not only helps us to improve road surface quality but also it helps us to implement one of the main aspects of driverless car by maintaining proper speed limit, determining the actual road condition.

Regular and seamless monitoring of city road surface condition across the city is required in order to build a good traffic estimation system. An method of image analysis from live video requires expensive installation of roadside cameras or collection of sensor logs from fixed roadside infrastructure requires costly deployment of roadside infrastructure which also needs to rebuild while city road repairing work goes on. Therefore, collection of well spatially and temporarily distributed data in low-cost way across the city is non-trivial. People have used smartphone sensor data, i.e. accelerometer to measure road roughness, but no existing work provides a straight way analysis of variation of speed with road roughness condition.

Therefore, in this paper, our contribution lies broadly in three directions: (1) collection of accelerometer and GPS sensor data from smartphone by a customized android application installed in smartphone of group of volunteers for two-wheeler mode of transport (i.e. we have carried out the data acquisition process by two-wheeler only) for two routes, (2) development of web-based tool to visualize the raw mobility data on digital map and to identify those particular road segments where the vehicles have moved with slower speed pattern and (3) analysis of correlation between raw accelerometer data and speed of vehicles for both jerky and smooth road segments with a measurement of road surface index from z-axis value of accelerometer sensor data.

The rest of this paper is organized as follows: Sect. 2 gives an description of overall in-house data set collected by group of volunteers for a suburban city of India. Section 3 presents the web-based tool for visualization and analysis of raw GPS trails and for validating the collected ground truth data. Section 4 establishes the motivation of exploring the correlation of vehicle speed with raw accelerometer data, and Sect. 5 explains the measurement of road surface index along with experimental results which conforms the aforesaid fact.

2 Data Acquisition

To avoid the problem of heterogeneous and unstructured data, we have carried out the process of data acquisition in a controlled environment where a group of five volunteers have participated to collect the sensory data from smartphone by installing a customized android application in their smartphone while travelling along two or three city road segments by two wheeler across a suburban city of India. They have continued the data collection process for 7 days covering the distance of 15 km, where almost 150 m road segment was broken and full of potholes. The application captured logs from various sensors like GPS (sampled at 1 Hz) and IMU (accelerometer sampled at 197 Hz) embedded in the smartphones.

Besides the spatial diversities like jerky road or smooth road surface, etc., we also captured data across different time zones starting from 9 a.m. till 7 p.m., each day. For this, we planned the data collection in different time intervals like—, 9 a.m. to 1 p.m.—*Morning*, 1 p.m. to 5 p.m.—*Afternoon* and 5 p.m. to 9 p.m.—*Evening*.

After collecting the initial raw data, we found that out of the entire data collected for 20 d, a part of the data (approx. 20%) contains missing or corrupted sensor logs. Thus, to perform further analysis and evaluation of the system, we dropped the corrupted parts of the data and considered only the data from those trips where the sensors have been appropriately logged.

3 Data Visualization

Visualization of raw mobility data on digital map is an integral part of data analysis. The overall process of data collection is carried out by a group of volunteers. While collecting the smartphone data by android application, the volunteers have tagged the locations where the road surface conditions are significantly bad and full of potholes and broken patches. Hence, intuitively, along those road segments, the speed of vehicles should be slower compared to regular road segments. But, individual tagging of data set is highly dependent on human perception and can vary from person to person. These variations can hamper the actual nature of road surface and finally perturbing the correlation between vehicle speed and road roughness. We have collected the ground truth information about the quality of road surface for each of the road segments from offline interaction with local people and have recorded the exact coordinate values along with the length of bad road patches for future analysis.

We have developed a web-based visualization tool as shown in Fig. 1 to visualize the raw mobility trails on digital map. We have utilized the Google API to plot each of the GPS points and recorded at 1 s interval on Google Map. This tool shows all those locations where bus has stopped or moved with a slow speed range. The reason of slow speed nature may be a designated bus stop, a traffic signal, road congestion or damaged road condition. In this paper, we are only focussed to explore the bad road patches and to find out its effect on speed of vehicle. From the observation of slow speed natures and the corresponding ground truth data of bad road segments, we can easily validate the actual slow speed values for damaged road patches which will help for further processing of the data.

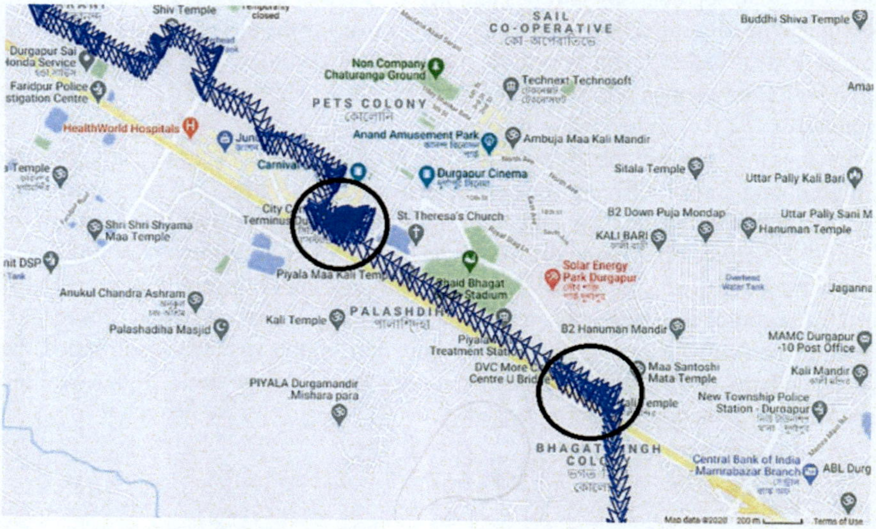

Fig. 1 Visualization of mobility data on google map

4 Methodology

We first start by analysing the collected raw accelerometer data and the instantaneous speed value of vehicle. The sampling frequency of accelerometer sensor is 50 Hz. The vertical acceleration (i.e. along the Z-axis) is required to estimate the road surface condition. In general, the smartphone (i.e. probing device) and its (x, y, z) axes could be in an arbitrary orientation with respect to the vehicle and its (X, Y, Z) axes. Furthermore, this orientation could change over time as the phone is moved around. But, in our study, we have collected the data by group of expert volunteers, and smartphone was fixed on vehicle dashboard so that it cannot be displaced causing change in orientation.

From collected GPS trails of vehicles, we have calculated the instantaneous speed of vehicle. Z-axial acceleration values lie within the range from −40 to 40, and instantaneous speed value ranges from 0 to 30 km/h as shown in Fig. 2a, b. Figure 2c clearly shows the accelerometer signature for a jerky road segments, where the instant speed of a vehicle shows a bouncy nature compared to regular road segments. This fact gives clear indication that instant speed is strongly correlated with the raw z-axis acceleration.

This observation leads us to define the metric of road surface index (R_I) to estimate the condition of the road surface which is a significant factor for slow speed or stay of a bus. Buses typically slow down when the road condition is poor. Moreover, the poorer the condition of the road surface, the longer is the travel time. Therefore, the quality of the road surface is an important reason for the bus stay. R_I gives a quantitative measure for inferring the road surface condition of a given road segment. It may be noted that the impact of poor road surface condition impacts significantly the Z-axis signature of the accelerometer. Experiments show that in a poor road

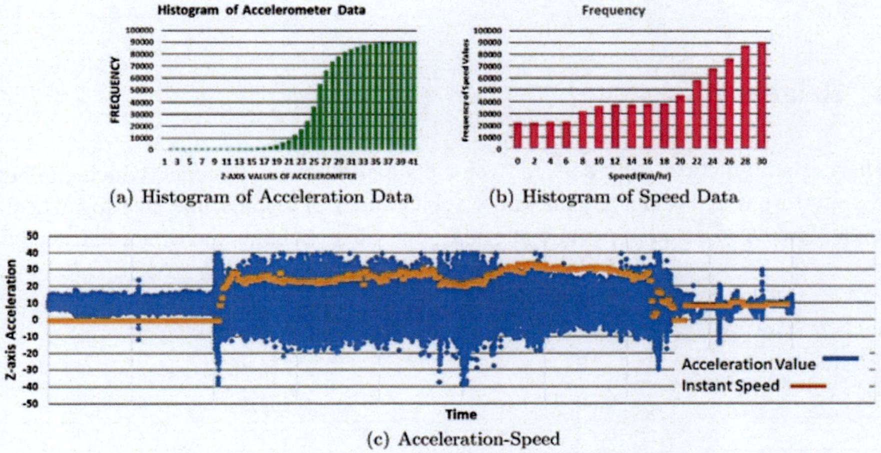

(a) Histogram of Acceleration Data (b) Histogram of Speed Data

(c) Acceleration-Speed

Fig. 2 **a** Frequency of acceleration data for different ranges. **b** histogram of speed data. **c** correlation of instant speed and acceleration data having bumps, potholes or a broken patch, the value of Z-axis signal varies within a certain range as shown in Fig. 2c.

$$R_I(\Delta_i) = \sum_{i=1}^{\Gamma} d_i \tag{1}$$

We have used the Z-axis acceleration, generated for 150 m road segment, assuming that it might impact the stop behaviour. We have calculated the sum of distances (d_i), covered with a particular Z-axis acceleration value (in a discrete scale). Γ is the maximum Z-axis acceleration value as shown in Eq. 1.

5 Experimental Results

We have classified the road segments into different category based on road surface index. Road surface classes are classified into four classes as shown in Table 1. In Table 1, we have shown different classes of road surface classes along with its length of road and total percentage of that particular road surface from overall data set.

Figure 3 shows for R_I 4 (i.e. very poor) category of road segments, the mean speed is 10 km/h and the variation of minimum and maximum speed range is very less, i.e. −10 km/h to +10 km/h. from mean speed. But, for road surface category R_I 1 (i.e. excellent road surface), the mean speed is 40 km/h, which denotes very fast speed class as per our observation. Figure 3 also presents the correlation between mean speed of vehicles and road surface condition. Therefore, this feature can be a prominent feature to identify the road segments where normal flow of traffic can be affected due to bad road condition.

6 Related Works

There is a high correlation between vehicle's inertial measurements (which can be captured by accelerometer) and direct vehicle control signals like steering wheel angle, brake/acceleration pedal operations, etc. This phenomenon of accelerometer

Table 1 Road surface classification based on road surface index

Road surface class	RI	Length of road
RS1	$0 < R_I < 2$	102.3
RS2	$2 < R_I < 4$	57.8
RS3	$4 < R_I < 9$	63.58
RS4	>9	34.67

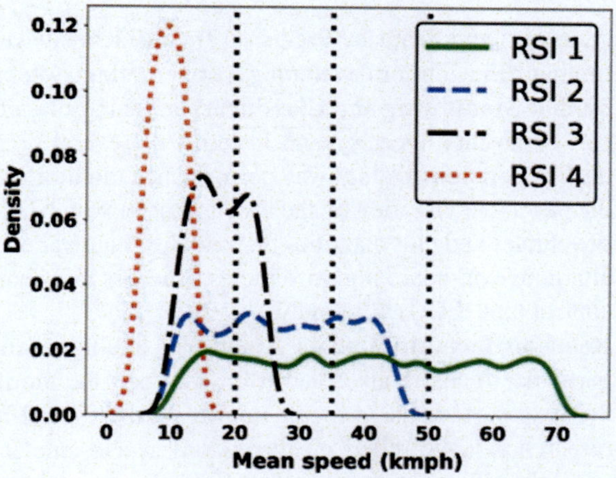

Fig. 3 Distribution of mean speed for different road categories

sensor motivates the researchers to utilize accelerometer sensor in various contexts of intelligent transportation system as follows:

Energy-Aware Navigation Chin Woo Tan [22] has proposed a navigation model without using gyroscope, whereas Rohit Verma has contributed in proposing a navigation system with minimal GPS usage of GPS sensor [25].

Accident Prevention and Detection People have worked a lot in the prevention and detection of accidents by continuously monitoring the acceleration of three axes. D.Selvathi [19], Bruno Fernandes [8], Rajesh Kannan Megalingam [14], Adnan Bin Faiz [6], Nicky Kattukkaran [11], Md. Syedul Amin [4], Chris Thompson [23], Hamid M. Ali [3] and Jules White [26] have identified the abrupt change in acceleration reading for any of the three axes, to be the event of an accident and have proposed several safety measurements.

Driving Behaviour Analysis Not only monitoring the movement of the vehicle, but a good analysis of driving behaviour can also avoid the chances of accidents because aggressive driving is one of the prominent reasons behind most of the accident cases. Vygandas Vaitkus [24] has classified rash driving and regular driving pattern using three-axis accelerometer data as input of the classifier algorithm. Apart from the driver's performance, the efficient driver profiling and driver rating system have different key factors like vehicle health and road anomalies, i.e. presence of bumps, potholes, broken road surface and absence streetlights. Being motivated by this fact, researchers worldwide have emphasized using the smartphone inertial sensor (i.e. accelerometer) to infer the contextual driving behaviour. Many works have focused on continuous monitoring of lateral, longitudinal and vertical acceleration and identification of any abrupt changes in three-axial reading of vehicle-mounted accelerometer sensors or smartphones carried by passengers. These changes in axial reading of accelerometer capture different driving events like lane change, sudden brake, high speed, etc., which clearly indicate aggressive driving or bad road condition or poor health of the vehicle. A handful of work Nadeem Akhtar [1], Thunyasit Pholprasit [16], Chalermpol Saiprasert [18], Johan W. Joubert [10], Munshi Yusuf

Alam [2] and Bratislav Predic [17] considers city road conditions while profiling driving behaviour or developing a driver rating system.

Vehicle Monitoring and Classification Vehicle classification is one major important part of an intelligent system in terms of several useful applications like efficient traffic operation and highway performance monitoring; Wenteng Ma [12] measures the pavement vibration by the accelerometer which helps to detect the axle of wheels of vehicles and thus classifies the vehicles. Zaldivar et al. [27] use smartphones as an alternative on-board unit in vehicles to access information in the vehicle's electronic control unit (ECU) wirelessly.

Road Surface Monitoring Among all intelligent transportation-related works, a particular focus is also placed on the road monitoring system. The condition of the road surface has a major impact on the overall transportation service. Again, researchers have relied on three-axial accelerometer sensor data to capture road anomalies like the presence of potholes, bumps or broken road surface from specific driving signatures. Eriksson et al. [5] developed one of the first road condition monitoring systems that detect and map road anomalies, such as potholes, using an accelerometer and GPS. Mednis et al. [13] and Ghose et al. [9] later both developed road condition monitoring applications for a smartphone that sends sensor data to a remote server and alerts drivers of potholes in the road. Mohan et al. [15] developed a road and traffic monitoring system, named Nericell, which also employs smartphone sensors to detect certain conditions. In addition to the driver behaviour monitoring feature, Fazeen et al. [7] also added a road condition characterization and mapping feature to their system that uses a smartphone's GPS and accelerometer.

7 Conclusion and Future Works

On the move, automatic detection of road surface condition can be utilized in developing a sophisticated transportation service which also aims towards the development of auto driving or driverless car in context of developing countries. Here, we have prepared one customized android application to collect the mobility data and accelerometer data. The application was installed in volunteer's smartphone to collect data for road segments of suburban city of India. However, to make the systems more robust, the following limitations need to be addressed. We have collected data primarily for single mode of transport, i.e. two wheeler but data from other modes of transport should be investigated to have better estimation of road surface condition across the city.

References

1. Akhtar N, Pandey K, Gupta S (2014) Mobile application for safe driving. In: 2014 fourth international conference on communication systems and network technologies. IEEE, pp 212–216
2. Alam MY, Saurav S, Mandal R, Saha S, Nandi S, Chakraborty S (2017) A fair and effective driver rating system for developing regions. In: 2017 9th international conference on communication systems and networks (COMSNETS). IEEE, pp 454–459
3. Ali HM, Alwan ZS Car accident detection and notification system using smartphone, vol 4, pp 620–635
4. Amin MS, Reaz MBI, Bhuiyan MAS, Nasir SS (2014) Kalman filtered gps accelerometer-based accident detection and location system: a low-cost approach. Current Sci 1548–1554
5. Eriksson J, Girod L, Hull B, Newton R, Madden S, Balakrishnan H (2008) The pothole patrol: using a mobile sensor network for road surface monitoring. In: Proceedings of the 6th international conference on mobile systems, applications, and services, pp 29–39
6. Faiz AB, Imteaj A, Chowdhury M (2015) Smart vehicle accident detection and alarming system using a smartphone. In: 2015 international conference on computer and information engineering (ICCIE). IEEE, pp 66–69
7. Fazeen M, Gozick B, Dantu R, Bhukhiya M, González MC (2012) Safe driving using mobile phones. IEEE Trans Intell Transp Syst 13(3):1462–1468
8. Fernandes B, Alam M, Gomes V, Ferreira J, Oliveira A (2016) Automatic accident detection with multi-modal alert system implementation for its. Veh Commun 3:1–11
9. Ghose A, Biswas P, Bhaumik C, Sharma M, Pal A, Jha A (2012) Road condition monitoring and alert application: using in-vehicle smartphone as internet-connected sensor. In: 2012 IEEE international conference on pervasive computing and communications workshops. IEEE, pp 489–491
10. Joubert JW, De Beer D, De Koker N (2016) Combining accelerometer data and contextual variables to evaluate the risk of driver behaviour. Transp Res part F: Traff Psychol Behav 41:80–96
11. Kattukkaran N, George A, Haridas TM (2017) Intelligent accident detection and alert system for emergency medical assistance. In: 2017 international conference on computer communication and informatics (ICCCI). IEEE, pp 1–6
12. Ma W, Xing D, McKee A, Bajwa R, Flores C, Fuller B, Varaiya P (2013) A wireless accelerometer-based automatic vehicle classification prototype system. IEEE Trans Intell Transp Syst 15(1):104–111
13. Mednis A, Strazdins G, Zviedris R, Kanonirs G, Selavo L (2011) Real time pothole detection using android smartphones with accelerometers. In 2011 international conference on distributed computing in sensor systems and workshops (DCOSS). IEEE, pp 1–6
14. Megalingam RK, Nair RN, Prakhya SM (2010) Wireless vehicular accident detection and reporting system. In: 2010 International conference on mechanical and electrical technology. IEEE, pp 636–640
15. Mohan P, Padmanabhan VN, Ramjee R (2008) Nericell: rich monitoring of road and traffic conditions using mobile smartphones. In: Proceedings of the 6th ACM conference on embedded network sensor systems, pp 323–336
16. Pholprasit T, Choochaiwattana W, Saiprasert C (2015) A comparison of driving behaviour prediction algorithm using multi-sensory data on a smartphone. In: 2015 IEEE/ACIS 16th international conference on software engineering, artificial intelligence, networking and parallel/distributed computing (SNPD). IEEE, pp 1–6
17. Predic B, Stojanovic D (2012) Localized processing and analysis of accelerometer data in detecting traffic events and driver behaviour. J UCS 18(9):1152–1176
18. Saiprasert C, Thajchayapong S, Pholprasit, Tanprasert C (2017) Driver behaviour profiling using smartphone sensory data in a V2I environment. In: 2014 international conference on connected vehicles and expo (ICCVE). IEEE, pp 552–557

19. Selvathi D, Pavithra P, Preethi T (2017) Intelligent transportation system for accident prevention and detection. In: 2017 international conference on intelligent computing and control systems (ICICCS). IEEE, pp 442–446
20. Sen R, Raman B, Sharma P (2010) Horn-ok-please. In: Proceedings of the 8th international conference on mobile systems, applications, and services, pp 137–150
21. Sen R, Siriah P, Raman B (2011) Roadsoundsense: acoustic sensing based road congestion monitoring in developing regions. In :2011 8th annual IEEE communications society conference on sensor, mesh and Ad Hoc communications and networks. IEEE, pp 125–133
22. Tan C-W, Park S (2005) Design of accelerometer-based inertial navigation systems. IEEE Trans. Instrum. Measurem. 54(6):2520–2530
23. Thompson C, White J, Dougherty B, Albright A, Schmidt DC (2010) Using smartphones to detect car accidents and provide situational awareness to emergency responders. In: International conference on mobile wireless middleware, operating systems, and applications, Springer, pp 29–42
24. Vaitkus V, Lengvenis , G. Žylius, Driving style classification using long-term accelerometer information. In 2014 19th international conference on methods and models in automation and robotics (MMAR). IEEE, pp 641–644
25. Verma R, Shrivastava A, Mitra B, Saha S, Ganguly N, Nandi S, Chakraborty S (2016) UrbanEye: an outdoor localization system for public transport. In: IEEE INFOCOM 2016-the 35th annual IEEE international conference on computer communications. IEEE, pp 1–9
26. White J, Thompson C, Turner H, Dougherty B, Schmidt DC (2011) Wreckwatch: automatic traffic accident detection and notification with smartphones. Mob Netw Appl 16(3):285
27. Zaldivar J, Calafate CT, Cano JC, Manzoni P (2011) Providing accident detection in vehicular networks through OBD-II devices and Android-based smartphones. In: 2011 IEEE 36th conference on local computer networks. IEEE, pp 813–819

An Overview of Recent Trends in OCR Systems for Manuscripts

Aditi Moudgil, Saravjeet Singh, and Vinay Gautam

Abstract OCR systems capable of recognizing characters have gained maximum concentration of researchers these days, especially when it comes to recognizing ancient documents. Digitization of these documents to make them readable and also searching from paper-based data becomes a great challenge. Literature shows that there are a lot of OCR systems that use different feature extraction and segmentation techniques to calculate the recognition accuracy. This paper focuses on a variety of classifiers and feature extraction techniques used to segment characters individually to make them readable by machines. The purpose of recognition of these documents is to preserve the heritage and valuable information contained in ancient manuscripts written in different languages.

Keywords Ancient scripts · ANN · KNN · OCR · SVM · MLP

1 Introduction

Historical documents contain a lot of invaluable information for society as well as for individuals. Due to the deteriorating factor of these manuscripts, preserving the digital copies of these documents is very essential. So, was the need to develop the OCR system which was capable of recognizing the characters. But developing an OCR which can be made to recognize the manuscripts is a big challenge, where paper-based society is being transformed into a paperless society. Ancient scripts can be in the form of stone inscriptions, old-printed books, yellow pages or typewritten text. Documents written in the fifteenth to eighteenth century are referred. Work-

A. Moudgil · S. Singh (✉) · V. Gautam
Chitkara University Institute of Engineering and Technology, Chitkara University, Punjab, India
e-mail: saravjeet.singh@chitkara.edu.in

A. Moudgil
e-mail: aditi.modgil@chitkara.edu.in

V. Gautam
e-mail: vinay.gautam@chitkara.edu.in

J. M. R. S. Tavares et al. (eds.), *Cyber Intelligence and Information Retrieval*,
Lecture Notes in Networks and Systems 291,
https://doi.org/10.1007/978-981-16-4284-5_46

ing of an OCR is generally based on three models: line segmentation model, word segmentation model and character segmentation model. Ancient script recognition contributes very less for the literature, giving a scope to enhance the contribution for this field. In this paper, different classification methods showing the best accuracy have been considered [1]. A lot of techniques have been used for feature extraction and classification to obtain the best accuracy. The conversion of digitized information into textual form becomes a challenging task but mandatory task so as to make it readable by any machine doing ventures of a million page per second. When it comes to analysing manuscripts which are not even sometimes in readable form and the text in those manuscripts is deteriorated, it becomes even more challenging to get that precious text and to make the masses of text available. For aged documents character recognition is an art of segregating characters from the complete word due to the fact that aged documents sometimes look blur, over inking or sometimes missing characters become some of the reasons which make it difficult to make it readable. Also various classifiers like ANN, CNN, feed forward neural network, SVM, K-NN, MMD, MLP and AdaBoost have already been used but further combination of any of the above mentioned features and classifiers can be used to optimize the results of character recognition. Variability of data sets can also be considered as one of the factor to improve character recognition rate. Figure 1 shows a sample Devanagari ancient document image.

1.1 Various Issues Encountered in OCR

1. Touching characters and overlapping characters create segmentation difficulties.
2. Degraded documents, i.e. degradation due to ink stains, yellow pages, over inking, torn pages, etc., make the designing difficult.
3. Faded ancient documents create difficulties in recognition.
4. Database needs to be manually collected making it a difficult task.
5. Noise in the documents makes pre-processing difficult.

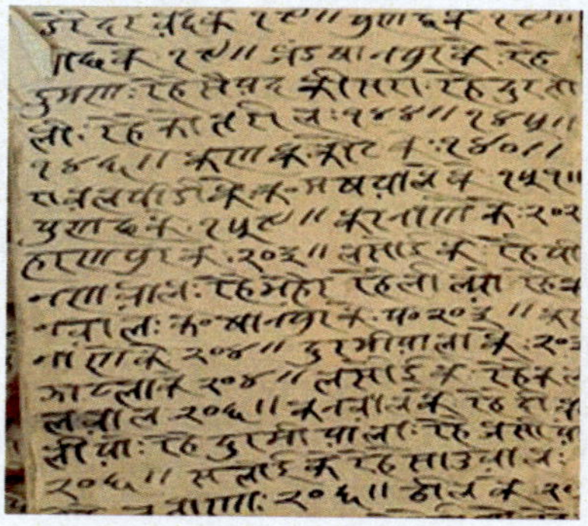

Fig. 1 Devanagari ancient sample document

2 Literature Studied

Jindal et al. in 2008 worked on degraded Gurmukhi documents, and accuracy was calculated using structural feature extraction techniques [2]. Arora et al. in 2007 implemented a hybrid approach which was a combination of structural and intersectional feature extraction techniques [3]. Correspondingly,, shirorekha was removed from Devanagri ancient documents, and a success rate of 89.12% was achieved. Lehal in 2009 implemented a Gurmukhi-based OCR system. In this paper, binary tree and KNN classifiers were worked upon along with structural feature extraction techniques [4].

Additionally, SVM was used in combination of some statistical and structural feature extraction techniques to achieve an accuracy rate of 98.18%. The fourth case considered in this paper was SVM in a combination of gabor filters. Shao et al. in 2014 presented a robust technique for character recognition of Chinese handwritten characters [5]. The author performed the restoration of gradient features to preserve the originality of the features. The data set used was a mixture of character images of two types: omitted stroke and added stroke noise images. Triggs and Dalal in 2005 introduced a novel technique of HOG feature detector technique to calculate the accuracy [6]. Singh et al. in 2012 implemented a gabor filter extraction technique on isolated Gurmukhi handwritten characters [7]. Fivefold cross-validation technique was used in combination with RBF-SVM classifier. Singh and Lehal in 2014 used different feature extraction techniques like directional, DCT, zoning and gradient feature extraction technique to achieve better accuracy [8]. Also classifiers like ANN, KNN and SVM were used. Multiple fold technique was used to achieve the best results. Khanduja and Deepti in 2016 presented a hybrid approach by combining mathematical models with structural features and obtained higher accuracy rates. Devanagri manuscripts consisting of 6152 characters were collected from museums and libraries, and the implementation of multiple classification techniques was implemented on segmented characters [9]. Also the features like open end points, centroid, horizontal peak extent and vertical peak extend were used for feature extraction. Sonika Rani presented a system to extract characters and developed an OCR that was able to recognize 5484 characters from ancient manuscripts along with the extraction of features like DCT zigzag features and a combination of classification techniques have been used like SVM, decision tree, Naïve Bayes, ensemble method, etc. [10]. Sonika et al. in 2017 presented an iterative algorithm for complete segmentation of Devanagri scripts and worked on 100 pages collected from museums and libraries and performed an experiment based on the flow of water droplets to recognize words, lines and characters [11]. Sonika Narang in 2019 presented a paper in which the author designed an ancient manuscript recognition system able to recognize the Devanagri script which in these days were found in fragile condition. Devanagri scripts can be analysed on the basis of three conjuncts, 11 vowels and 33 consonants used to write multiple languages like Hindi, Sanskrit, Marathi and Nepali [9]. In this paper, author proposed a system which was based on the following steps: digitization, pre-processing, segmentation, feature extraction and classification. M.

K Jindal in 2019 worked on the data set consisting of 100 pages written in Devanagri and followed multiple steps to make the ancient data readable with the highest accuracy and reliability [10]. With the pre-segmented set of images, the steps followed for this work include feature extraction followed by the classification method applied, which are responsible for maximizing the accuracy, specificity and sensitivity of the data. In the last step, author tried to improve the accuracy by using AdaBoost and bagging methods to improve the recognition results. Implementing this method separates out the high- and low-frequency components, out of which a few low-frequency components are meaningful which are selected as features. Classification is a very significant method which identifies to which class an image will belong to. In this paper, the author presents a new technique for skewed word detection and correction. If the aspect ratio of the word comes out to be less than 0.7, then word was considered as skewed, and for its correction different algorithms were followed [12]. The word was rotated clockwise and anticlockwise, and finally the height was calculated. The point where minimum height was calculated was noted, and same was followed for word width also. Bansal et al. in 2001 presented a Devanagari OCR which used core strip region coverage, zero crossing, vertical bar features and structural feature extraction techniques along with some classification technique decision tree classifier for obtaining the accuracy [2].

The extensive literature survey can prove to be beneficial for many currently running applications like postal address recognition, car number plate recognition, etc. So, finally, we can conclude that manuscript recognition is still a lacking filed in the area of OCR and can be used for multiple applications.

3 Detailed Working of OCR

The main stages in the development of OCR involve data set collection, preprocessing, segmentation, feature extraction and classification.

Steps involved in OCR are as follows:

- **Image Acquisition**: For image acquisition, a data set of old manuscripts is collected for the language that is to be worked upon followed by the digitization of the images, i.e. converting the images into the electric format. This process can be achieved by scanning the images or capturing images using a camera.
- **Image pre-processing**: It takes in three steps, wherein the first step enhancement of an image is done using any inbuilt tool. The second step involves the cleaning of the image by removing the noise from the image, and in the third step image is corrected for any slant lines or skewness correction is done.
- **Segmentation**: For image segmentation, the image is first segmented into lines, followed by word segmentation, and lastly the character segmentation is done to achieve accuracy for an OCR system. Segmentation takes place in the following steps:

  - **Page Segmentation**: There might be content zones and non-content zones in a report picture. Page division is utilized to section content zones from non-content areas, recognizable proof of the number of segments in content zones.
  - **Line Segmentation**: Line segmentation means splitting the manuscripts into lines in which top-down approach uses the concept of larger manuscripts which are further split into smaller pieces which makes it easy to segment lines, whereas bottom-up approach uses vice versa concept where smaller pieces are segmented for lines of text and are gathered to form a larger manuscript, thus providing better results. Top-down and bottom-up approaches are clubbed together into hybrid methods which gives the best results.
  - **Word Segmentation**: Word segmentation is a process of separation of words from lines that were segmented in previous section. Specifically word is a collection of multiple characters.
  - **Character segmentation**: Character segmentation is the last step performed by OCR in which each character is segmented followed by the recognition of characters.

- **Feature Extraction**: There are a lot of feature extraction techniques that can be used for our OCR system. Feature extraction techniques can be broadly classified as structural features, statistical features and global transformation and moments. Mainly, the feature extraction techniques are classified under these three categories, i.e. zoning, aspect ratio, directional features, projection features, etc.
- **Classification**: The unknown region of the pattern is determined using classification. There are a lot of classification techniques used for classification, e.g. MLP, CNN, SVM, ANN, binary tree, KNN, etc.
- **Post-processing**: Samples are seen for any errors (error detection and correction are done). Correspondingly, accuracy is measured for the collected samples.

4 Feature Extraction Techniques

- **DCT features**: DCT zigzag features are used for compressing the originally sized images. The data of images is compressed into frequency-level components (high frequency and low frequency). It collects all the low-frequency coefficients in the upper left corner and all the high-frequency components in the bottom right corner, out of which the meaningful data is converted into features. Multiple variants of DCT can be studied such as zigzag DCT coefficient.
- **HOG features**: HOG features are generally considered for shape and size features [15] and were initially used for still images.
- **Gradient Features**: The intensity function of each pixel of the image and the how the image gets darker or lighter and in which direction is determined by the gradient function. The direction of the greatest change in the intensity value is measured by using gradient features. Variability of gradient features can be used in the field of handwriting recognition.

- **Zoning Features**: This feature extraction method is the most prominent feature extraction used for handwriting recognition. This technique divides the image into zones. From each zone, some features are extracted. Generally, the foreground pixels are considered as the features of each zone and correspondingly a feature vector are achieved.

5 Classification Techniques

- **Multilayer Perceptron**: As the name indicates, it is a multilayer feed forward kind of network where data is flown between the input and the output layer. Also it allows the concept of backpropagation to solve nonlinearly separable problems. More than one layer is used between the output and the input layer, where the data flows in forward direction through the consecutive layers.
- **ANN**: This network consists of tightly interconnected artificial neurons, capable of performing computations in parallel. Input data is received by input layer further passed on to the hidden layers, and correspondingly the output is received by the output layer. This is also known as a self-trained [11] network. The associated connections between the nodes are assigned some weights, and the modifications are done on the basis of input and output provided to the network. Each neuron receives an input having a specific value on the basis of which output is produced based on the input values.
- **CNN**: It is a backpropagation-based deep learning approach, which deals with high-dimensional complex data. Higher accuracies can be achieved using this algorithm, whereas adding the number of layers makes this algorithm more efficient. CNNs are based on the hierarchical structure of MLPs where alternatively succeeding layers are designed. Higher-level features are learnt from these layers, and the final classification results are produced by the last layer. Sub-sampling and convolutional operations are provided by the succeeding layers.
- **SVM**: SVM is a supervised learning technique which is based on statistics. Data is assigned to multiple classes. SVM can also be used in combination with other techniques [14]. Generally, SVM classifier has been used with the same data set. It can also be used with linear, polynomial and RBF kernels.
- **Random Forests** This technique is based on decision tree which helps to completely understand the inner model of the network, thus representing a white box model. Data segregation takes place based on the question series where leaf node is designated for new data allocation. The complete process starts from the root node of the tree, and later the data is split into features. The nodes are split into sub-nodes, and this process is repeated iteratively. Table 1 shows some of the feature extraction and classification techniques.

Table 1 Trending techniques used in OCR

Ref no	Feature extraction technique	Classification technique	Data set used	Accuracy analysis (%)
[1]	Structural features	SVM and KNN	Gurmukhi degraded manuscripts	83.60
[2]	Strip wise core region covering, vertical stripe feature, horizontal crossings, counting the number of vertex point positions, structural features	Decision tree classifier	Printed Devanagari text	93
[3]	Intersectional features and structural features	Feed forward neural network-based classifier	Devanagri handwritten characters	89.12
[4]	Structural features, gabor filters and statistical features	KNN, SVM and binary tree classifier	Gurmukhi script	98.18
[6]	Gradient features	Modified quadratic discriminant function—Extended version	Handwritten Chinese character	93.50
[8]	Gabor filters	SVM-RBF classifier	Gurmukhi handwritten isolated characters	94.29
[10]	Gabor filters, zoning features, distance directional distribution, DCT features and transition features	KNN, ANN and SVM	Devanagri characters	99.42
[2]	Curve fitting using some mathematical equations and structural features	Not specified	Devanagri characters	93.40
[14]	HOG features	Support Vector machine	Any random database	89
[15]	Zoning features	KNN, SVM, SVM with RBF kernel	Gurmukhi handwritten text	72.5, 73.02, 95.04

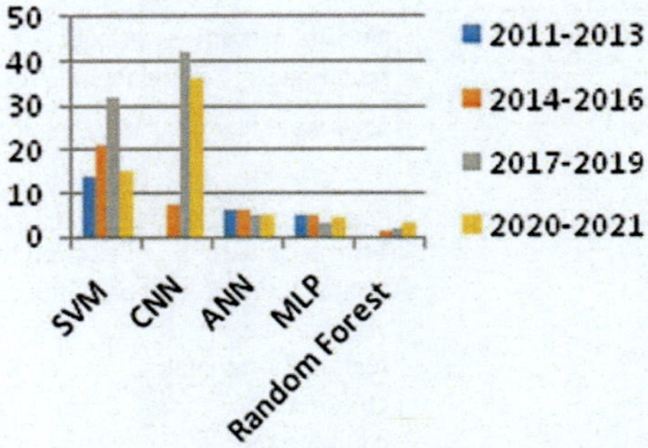

Fig. 2 Year wise classification technique analysis

6 Discussion and Analysis

The graph shown in Fig. 2 represents the multiple classification techniques used in different time slots, i.e. 2011–2013, 2014–2016, 2017–2019 and 2020–2021.

One can easily interpret from this graph that since 2011, the use of SVM technique is increasing, and this technique is used even today. CNN is also an emerging technique these days, and since the past decade the use of CNN is also increasing. The studies show that ANN is being used from 2011 till day. MLP and random forest techniques also played a vital role in classifying the handwritten text. So, the studies reveal that these techniques can even be used today for any kind of classification to be performed on historic manuscripts.

7 Conclusion

OCR is a process that helps in the identification of characters either from handwritten or printed documents. As shown in literature, there are a lot many OCR systems that work on the concept of deep learning and use various kinds of feature extraction, segmentation and classification techniques along with the accuracy measurement techniques mainly in ancient documents. As reported by many authors, the research in this topic is still in its initial stage. The literature presented in this report focuses on various classification and feature extracted techniques along with the language for which the OCR has been designed for. The literature focuses on various application areas of the OCR briefly. There is no standard database available to work upon for Devanagri manuscripts. Already work has been done on basic characters which can further be extended to modifiers and conjuncts. Lacking behind is the need to design novel OCR for manuscript recognition.

References

1. Kumar M, Sharma RK, Jindal MK (2012) Offline handwritten Gurmukhi character recognition: Study of different feature-classifier combinations. In: Proceeding of the workshop on Document Analysis and Recognition, pp 94–99
2. Bansal V, Sinha MK (2001) A complete OCR for printed Hindi text in Devanagari script. In: Proceedings of 6th international conference on document analysis and recognition. IEEE Computer Society, p 0800
3. Arora S, Bhatcharjee D, Nasipuri M, Malik L (2007) A two stage classification approach for handwritten Devnagari characters. In: International conference on computational intelligence and multimedia applications (ICCIMA 2007). IEEE, vol 2, pp 399–403
4. Lehal GS (2009) Optical character recognition of Gurmukhi script using multiple classifiers. In: Proceedings of the international workshop on multilingual OCR, pp 1–9
5. Shao Y, Wang C, Xiao B (2015) A character image restoration method for unconstrained handwritten Chinese character recognition. Int J Document Anal Recog (IJDAR) 18(1):73–86
6. Dalal N, Triggs B (2005) Histograms of oriented gradients for human detection. In: 2005 IEEE computer society conference on computer vision and pattern recognition (CVPR'05). IEEE, vol 1, pp 886–893
7. Hussain J (2018) A hybrid approach handwritten character recognition for Mizo using artificial neural network. In: 2018 international conference on advanced computation and telecommunication (ICACAT). IEEE, pp 1–6
8. Singh J, Lehal GS (2014) Comparative performance analysis of feature (S)-classifier combination for Devanagari optical character recognition system. Int Jo Adv Comput Sci Appl (IJACSA) 15
9. Narang S, Jindal MK, Kumar M (2019) Devanagari ancient documents recognition using statistical feature extraction techniques. Sdhan 44(6):1–8
10. Narang SR, Jindal MK, Kumar M (2019) Devanagari ancient character recognition using DCT features with adaptive boosting and bootstrap aggregating. Soft Comput 23(24):13603–13614
11. Narang SR, Jindal MK, Kumar M (2019) Drop flow method: an iterative algorithm for complete segmentation of Devanagari ancient manuscripts. Multimedia Tools Appl 78(16):23255–23280
12. Mahto MK, Bhatia K, Sharma RK (2015) Combined horizontal and vertical projection feature extraction technique for Gurmukhi handwritten character recognition. In: 2015 international conference on advances in computer engineering and applications. IEEE, pp 59–65
13. Singh P, Budhiraja S (2011) Feature extraction and classification techniques in OCR systems for handwritten Gurmukhi Script-a survey. Int J Eng Res Appl (IJERA) 1(4):1736–1739
14. Narang SR, Jindal MK, Sharma P (2018) Devanagari ancient character recognition using HOG and DCT features. In: 2018 5th international conference on parallel, distributed and grid computing (PDGC). IEEE, pp 215–220
15. Khanduja D, Nain N, Panwar S (2015) A hybrid feature extraction algorithm for devanagari script. ACM Trans Asian Low-Resour Lang Inf Process (TALLIP) 15(1):1–10

Towards Integration of Blockchain and Machine Learning Technologies for Security in Smart Cities

Shivani Wadhwa, Divya Gupta, Aditi Moudgil, and Shalli Rani

Abstract Today, the evolution of the Internet of things (IoT) technology has proved a great support to various applications. The communication among heterogeneous network devices in huge traffic environments always deals with security and privacy concerns. The traditional security approaches are not suitable for these IoT devices. The emerging technologies (such as artificial intelligence, blockchain, machine learning, fog computing) may provide efficient security solutions when embedded with any particular IoT application. In this paper, the authors first discussed various security solutions proposed by different research groups to provide security in IoT networks using blockchain and machine learning technologies. Further, the background of these technologies provides more insight into the working of these technologies. The security in smart cities (an IoT application) with the integration of blockchain and machine learning is presented using a use case scenario of a web trading platform. At last, the open research challenges in the implementation of this integration are explained.

Keywords IoT security · Blockchain · Machine learning · Smart cities

1 Introduction

Over the last several decades, the massive growth in the IoT technology has provided a great support for communication over the Internet [1]. Several devices such as sensors, actuators and data collectors are deployed in a different environment to gather

S. Wadhwa · D. Gupta (✉) · A. Moudgil · S. Rani
Chitkara University Institute of Engineering and Technology, Chitkara University, Punjab, India
e-mail: divya.gupta@chitkara.edu.in

S. Wadhwa
e-mail: shivani.wadhwa@chitkara.edu.in

A. Moudgil
e-mail: aditi.modgil@chitkara.edu.in

S. Rani
e-mail: shalli.rani@chitkara.edu.in

J. M. R. S. Tavares et al. (eds.), *Cyber Intelligence and Information Retrieval*,
Lecture Notes in Networks and Systems 291,
https://doi.org/10.1007/978-981-16-4284-5_47

the information based on some events. With the benefits associated in use of this technology, the count of Internet-connected devices is increasing each passing day. As per the reports generated by CISCO, the count of Internet-connected devices will be almost three times the global population by the year 2023 [2]. This tremendous growth in connected devices is due to the large range of applications dealing in the context of IoT and is not limited to only smart grid, agriculture automation, natural disaster detection, supply chain, smart homes, intelligent transportation systems, smart cities and e-health monitoring systems. The connectivity of these huge number of devices on the Internet while managing the large volume of traffic produced by these IoT devices due to its limited constraints (such as the battery, storage, processing capability) imposes challenges in its implementation. Several studies have focused on the implementation of IoT with limited resources and therefore suggested various solutions in different application domains. Moreover, maintaining end-to-end communication of the user data with all the necessary security and privacy concerns (mainly confidentiality, authentication and integrity) is yet another issue in IoT. Although some research works targeting security in IoT are already present, still, some advancements can be made by integrating IoT security with new leading technologies. To work in this direction, this study considers two emerging technologies such as blockchain and machine learning for providing security in specific IoT scenarios such as smart cities. The security in smart cities will lead to efficient communication in various fields as represented in Fig. 1. This conceptual shift with the integration of blockchain and machine learning in smart cities would pave a new way for secured communication among requesters and providers [3]. The following are the contributions made in this paper:

- The authors first discuss various IoT security solutions proposed by different researchers using blockchain and machine learning technology, respectively.
- An overview of these emerging technologies in relation to IoT and its integartion is explained further.
- The paper presents a use case scenario to represent how the integration of blockchain and machine learning can be used in IoT security.
- The open research challenges in implementation of this integration are also explained to support future research in this direction .

The remaining of this paper is structured as follows: the related work of the IoT security in context of machine learning and blockchain technologies is discussed in Sect. 2. The background of these emerging technologies as well as its combination is discussed in Sect. 3. Section 4 demonstrates a use case scenario for supporting security in the integration of blockchain and machine learning. The various research challenges in the implementation of this integration are presented in Sect. 5. Finally, the conclusion of this study is explained in Sect. 6.

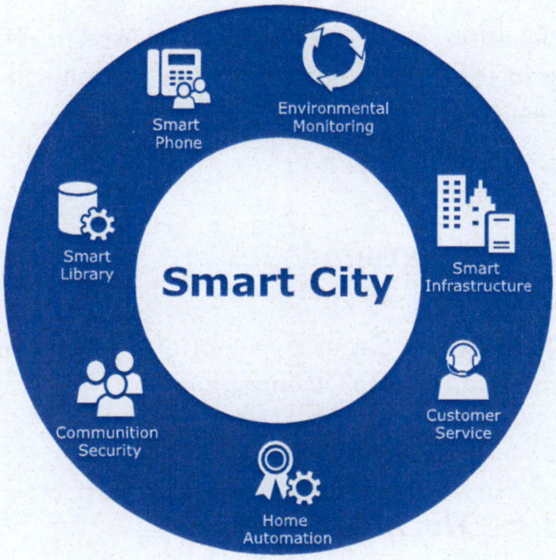

Fig. 1 The different domains of smart city

2 Literature Review

Hyunil Kim et al. proposed a new mechanism for blockchain networks by using distributed machine learning. Error-based aggregation techniques and stochastic gradient descent method are used as core methods to provide security and privacy to permissioned blockchain, i.e. hyper ledger fabric. This method provides prevention from attack, lowers the transaction delay and lowers the computation complexity [4]. Sudeep Tanwar et al. explain various machine learning algorithms for blockchain technology which can provide resilient against attack [5]. Different techniques like support vector machine, convolutional neural networks and deep learning algorithms are discussed to access the attacks on the blockchain networks. Experiment's implementation has been verified on the energy trading system. Somdip Dey presented a paper for attacks on consortium blockchain because there may exist collusion between different institutions. Supervised ML algorithms are used to detect the collusion in the blockchain networks. Different version of proof-of-work is proposed in the framework [6]. A. Besir Kurtulmus proposed DanKu protocol which runs in five stages for successful completion of smart contract in Ethereum [7]. Smart contracts are written by using solidity language. The proposed protocol improved the GPU mining strategy by using machine learning models.

Xuhui Chen et al. introduced a LearningChain for privacy preservation in decentralized blockchain network and secure machine learning methods [8]. L-nearest aggregation algorithm is used to protect from byzantine attacks. Implementation is done on Ethereum platform which proved that proposed technique is very efficient. Anne H. Ngu et al. proposed middleware-based design of IoT [9]. Each and every layer of IoT is explained in detail. Security and adaptability issues that arise in the IoT middleware layer are also discussed. Arsalan Mosenia et al. explained security threats mostly found on edge side in the model. Possible solutions are also implemented in this paper [10]. Eklas Hossain et al. presented the paper which shows how the different ML methods can be used to generate useful information in case of smart grid [11]. Also the authors have given some algorithms to protect data from cyber security menacing.

3 Background

This section discusses the brief overview of considered technologies along with its integration in IoT technology.

3.1 Machine Learning

Machine learning deals in the study where machines learn on the basis of experience and the usage of data. Explicitly programming is not done to train the system rather it uses training data to do so. Various disciplines like philosophy, psychology, statistics, probability and artificial intelligence integrate together to provide acceptance to machine learning algorithms. Main applications of machine learning algorithm include data mining, email filtering, human face recognition, manufacturing process, data mining and computer vision.

Learning approaches mostly include supervised learning, unsupervised learning , semi-supervised learning, deep learning, self-learning, feature learning, reinforcement learning, sparse dictionary learning and robot learning. Few applications include the use of combination of approaches to give better accuracy. Most widely used models of machine learning are decision trees, support vector machines, Bayesian networks, regression analysis, training models, genetic algorithms, etc. [12].

3.2 Blockchain

Blockchain is considered as the ledger which contains all transactions in the form of records. All blocks of the blockchain are cryptographically linked, i.e. hash of the previous block is contained in the next block (refer Fig. 2). Size of the chain grows as more blocks are appended into the chain. To verify and validate the block, consensus algorithms are applied. Main characteristics of blockchain are decentralization, tamper-proof, auditability, persistency, privacy, anonymity and immutability distributed [13].

The fields of the block are block version, Merkle tree root hash, time stamp, nonce, hash of the previous block and parent block hash. Block size and size of transactions are the main factors for deciding the maximum number of transactions that a block can hold.

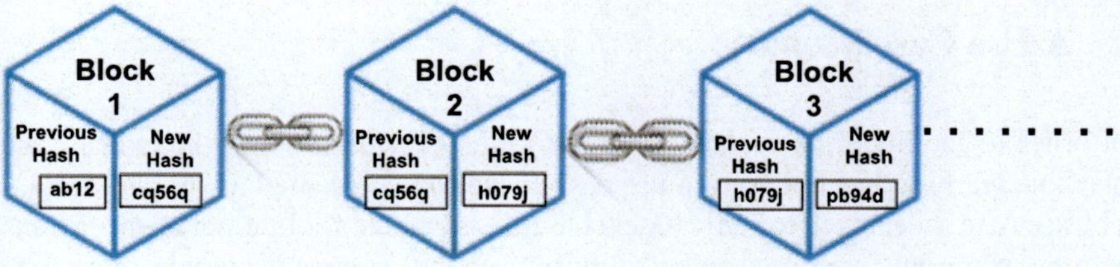

Fig. 2 Appending blocks in blockchain

3.3 Integration of IoT, Machine Learning and Blockchain

Internet of things (IoT) consists of various heterogeneous devices which have their own unique identity and are connected to each other in a collaborative manner [14]. These devices generally have processors with limited capability, small memory and low power. Combining IoT with machine learning can provide better data prediction, data analysis and data visualization. Data generated by IoT devices can be utilized in appropriate manner by machine learning algorithms. Union of machine learning and blockchain can prevent attacks on the blockchain and also helps in collision detection. Security and decentralization can be provided to data generated by IoT devices through blockchain. Integration of IoT, machine learning and blockchain is an emerging area for facilitating security to the smart cities as illustrated in Fig. 3.

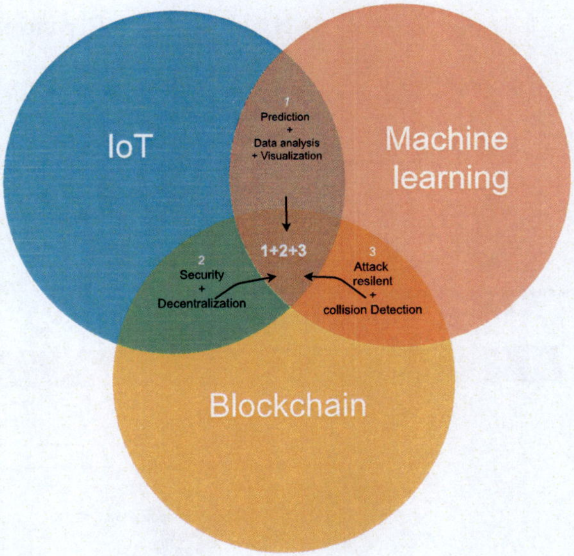

Fig. 3 Combined functionalities of different technologies

4 A Use Case Scenario

In order to facilitate the security in smart cities considering both blockchain and machine learning technologies, a use case scenario is presented as shown in Fig. 4. The use case is designed to study blockchain-based model for Internet services transactions using any digital currency. Any transaction processing within this model makes use of both blockchain and machine learning to complete the payment. The use of consensus algorithm such as byzantine fault-tolerance support high system throughput. The hash functions are used to create a new block. The designed model is just like an intrusion detection system (IDS) which keeps an eye on recurrent neural networks for detecting any false and fraud transactions.

This model starts its processing when demand for Internet services is initiated by any user (such as smart home, smart transportation, e-health care, UAVs) [15]. The user demands services from Internet service provider (ISP). Among all the ISPs present, the user can connect to any one of them based on the charges offered to provide services. Blockchain due to its decentralized approach contains all the transactions performed between users and ISPs. Each ISP publishes his price for providing services to user. Any user wishes to avail services as per the price offered by ISP will make payment transaction using blockchain network. The authentication of user is validated on blockchain network before offering services. On validating the user identity, the ISP provides its services to the user and receives digital currency

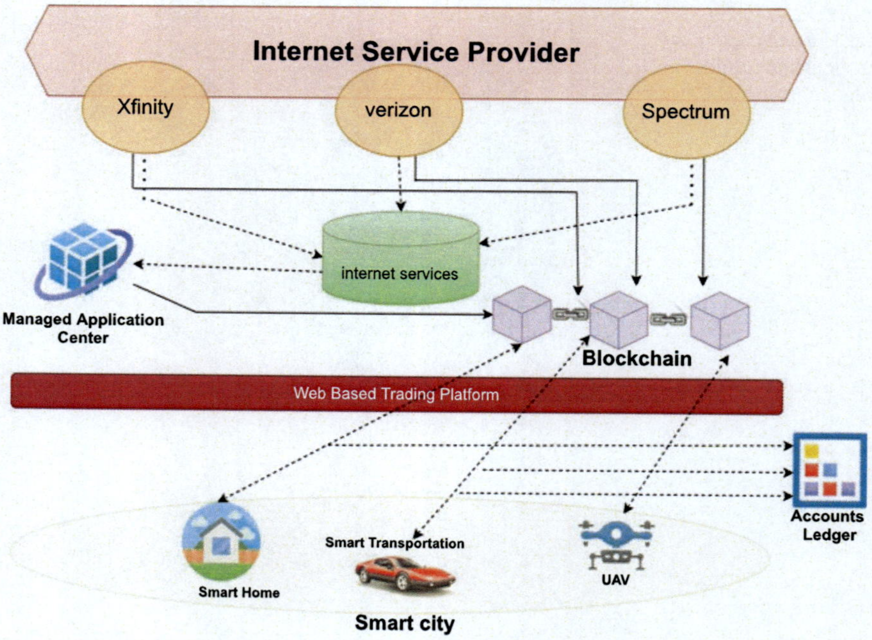

Fig. 4 Use case scenario of web-based trading platform

in its account. The analysis using machine learning algorithms can be performed on this model to predict the frequency of future requests from the users for ISPs. This IDS model will also help in detecting the fraud transactions as well as any attack on this network.

5 Research Challenges

Although various solutions have been offered for providing security to IoT with the use of blockchain and machine learning concepts, still various challenges do exist in the implementation of this integration. The research in this field is in its infancy and demands various future solutions. This section discusses the various open challenges of applying blockchain and ML in the IoT smart cities.

5.1 Storage

The data produced by the smart cities will ultimately increase the size of the blockchain. All these blocks must be stored by all network nodes due to its decentarlized facility. However, storing this huge volume is a challenge for network node due to limited capacity constraint. Moreover, a lot of node's computation power go in waste to store irrelevant data. In addition, the immutable nature of blockchain further offers various storage challenges.

5.2 Privacy and Security

The data stored in the blockchain is provided to all the nodes present in the network which may affect the privacy concern of an individual. This challenge needs a lot of attention; otherwise, privacy of the nodes will be threatened. This challenge can also have impact on prediction capability of the machine learning algorithms. Attacks on the consensus algorithm of the blockchain are very important concern to be considered.

5.3 Implementation

Smart devices generally produce huge amount of data. Implementation of bulk data on blockchain requires good Internet service. Appending a block and including relevant transaction in a block are a major challenge.

5.4 Architecture

Blockchain is mostly used nowadays to provide security to the IoT devices. However, the architecture of blockchain is complex to understand which may result into system performance degradation for some specific applications.

6 Conclusion

Secure communication is always in demand for IoT networks. Emerging technologies such as blockchain and machine learning have proved their effectiveness in various domains. This paper worked towards providing security in IoT smart city applications by utilizing both above-said technologies. Subsequently, the security in smart cities has been explained by designing a use case scenario of a web-based trading platform in integration with blockchain and machine learning. This paper also discussed various open challenges faced in implementing this integration which can further be used as a base for future research in the security domain.

References

1. Saleem A, Agarwal AK (2016) Analysis and design of secure web services. In: Proceedings of fifth international conference on soft computing for problem solving. Springer, Singapore, pp 453–462
2. https://www.cisco.com/c/en/us/solutions/collateral/executive-perspectives/annual-internet-report/white-paper-c11-741490.html
3. Hakak S, Khan WZ, Gilkar GA, Imran M, Guizani N (2020) Securing smart cities through blockchain technology: architecture, requirements, and challenges. IEEE Netw 34(1):8–14
4. Kim H, Kim SH, Hwang JY, Seo C (2019) Efficient privacy-preserving machine learning for blockchain network. IEEE Access 7:136481–136495 (IEEE)
5. Tanwar S, Bhatia Q, Patel P, Kumari A, Singh PK, Hong WC (2019) Machine learning adoption in blockchain-based smart applications: the challenges, and a way forward. IEEE Access 8:474–488
6. Dey S (2018) Securing majority-attack in blockchain using machine learning and algorithmic game theory: A proof of work. In: 2018 10th computer science and electronic engineering (CEEC), IEEE, pp 7–10
7. Besir Kurtulmus A, Daniel K (2018) Trustless machine learning contracts; Evaluating and exchanging machine learning models on the Ethereum Blockchain. arXiv e-prints, pp arXiv-1802
8. Chen X, Ji J, Luo C, Liao W, Li P (2018) When machine learning meets blockchain: a decentralized, privacy-preserving and secure design. In: 2018 IEEE international conference on Big Data (Big Data). IEEE, pp 1178–1187
9. Ngu AH, Gutierrez M, Metsis V, Nepal S, Sheng QZ (2016) IoT middleware: a survey on issues and enabling technologies. IEEE Internet Things J 4(1):1–20
10. Mosenia A, Jha NK (2016) A comprehensive study of security of internet-of-things. IEEE Trans Emerging Topics Comput 5(4):586–602

11. Hossain E, Khan I, Un-Noor F, Sikander SS, Sunny MSH (2019) Application of big data and machine learning in smart grid, and associated security concerns: a review. IEEE Access 7:13960–13988
12. Shanthamallu US, Spanias A, Tepedelenlioglu C, Stanley M (2017) A brief survey of machine learning methods and their sensor and IoT applications. In: 2017 8th International Conference on Information, Intelligence, Systems & Applications (IISA). IEEE, pp. 1–8
13. Zheng Z, Xie S, Dai H, Chen X, Wang H (2017) An overview of blockchain technology: architecture, consensus, and future trends. In: 2017 IEEE international congress on big data (BigData congress). IEEE, pp 557–564
14. Outchakoucht A, Hamza ES, Leroy JP (2017) Dynamic access control policy based on blockchain and machine learning for the internet of things. Int J Adv Comput Sci Appl 8(7):417–424
15. Chhabra R, Verma S, Krishna CR (2017) A survey on driver behavior detection techniques for intelligent transportation systems. In: 2017 7th international conference on cloud computing, data science and engineering-confluence. IEEE, pp 36–41

Computational Intelligence

Intention to Purchase Online Luxury Watches Among Indian Consumers in the New Normal Mode

Soumik Das, Rabin Mazumder, and Shamindra Nath Sanyal

Abstract People purchase luxury products to showcase their wealth, success and exclusivity in the social world. The satisfaction is of great importance to them as it has a direct link with money spent on the luxury items. In this study, we tried to analyse purchasing intentions of luxury watches in the new normal mode. The study is done in few selected cities of India. We circulated online questionnaire to 600 online shoppers by convenience sampling, and 481 data were received. We considered functionalism, experientialism, symbolic interactionism, social influence, dimensions of vanity and hedonism apart from income. The data was analysed for simple frequencies, factor analysis, hierarchical regression and correlation. The results indicated that vanity attainment, functional value, hedonism and household earning have a direct influence on luxury purchase. These factors may be further explored for analysing the other cities not considered in this study.

Keywords Luxury · Purchase intention · Online purchase · Watches

1 Introduction

E-commerce is a vital tool for maintaining revenue, engaging with consumers. When the physical stores around the globe are more or less closed, online shopping plays a pivotal role. Accelerate digital investments and shift marketing budgets to online platforms, focusing rather than branding on customer activation. Consider collaborations with reputable e-retailers as well, in addition to improving your own websites.

S. Das (✉) · R. Mazumder
Institute of Engineering and Management, Kolkata, India
e-mail: soumik.das@iemcal.com

R. Mazumder
e-mail: rabin.mazumder@iemcal.com

S. N. Sanyal
Bennett University, Greater Noida, India
e-mail: shamindra.sanyal@bennett.edu.in

J. M. R. S. Tavares et al. (eds.), *Cyber Intelligence and Information Retrieval*,
Lecture Notes in Networks and Systems 291,
https://doi.org/10.1007/978-981-16-4284-5_48

Not only does digital marketing help improve online sales, but it can also enable customers to visit stores once they reopen [1].

For luxury marketers, e-commerce is never an easy option. It is like tasting the apple in the Garden of Eden, seductive and risky, to sell online luxury products many years ago when advertisers were arguing for luxury e-commerce brands [2–4]. Many luxury marketers thought it was not appropriate for luxury brands to go online because luxury buyers are typically older and less adventurous than regular shoppers, unable to buy online luxury products [4]. In addition, the Internet did not appear to be compatible with luxury brands that were considered exclusive, glamorous, luxurious and timeless as an accessible, affordable, free and fast-medium [3]. Since they are distinctive, high quality, elegant and exclusive, luxury goods are desirable [5]. These can once again be divided into luxury pieces that are soft and heavy [6]. Soft luxuries include objects such as fashion and accessories, while watches and jewellery are considered hard luxuries. By 2025, the global market for this kind of luxury will be about EUR 320–330 billion. The online platform will continue to move northward and estimated to be around 30% of the market by 2025 [5], with double-digit growth already experienced in 2019. With the younger generations becoming the remaining luxury market, this goes hand in hand with Gen Y and Gen Z.

More academicians are enthusiastic about e-commerce and believe it may help luxury enterprises expand as the online world and luxury industry flourish [7–9]. The research topic nowadays is not whether or not luxury businesses should adopt e-commerce, but rather how premium e-commerce companies act. Some studies are looking into how luxury brands keep their sense of luxury by increasing user experience [10], adapting conventional sensory shopping experiences to online shopping [3, 9], and identifying and tracking online luxury customers using modern information technologies (e.g. personalisation and targeting).

2 Theoretical Framework AND Hypothesis

In the luxury industry, buyers are the real owners. We presume that consumers would prefer being back to the normal life post the pandemic circumstances are under control. The luxury industry must prepare for these situations and any other undue circumstances in near future [11]. Today, conventional luxury brands are evolving. Preliminary acquisition of self-online retail site for digital strategy and e-commerce. Why, then, are some of the biggest brands finally investing in brand-own online retailing? [12]. Research found that multi-brand luxury goods stores now account for 78% of global online sales.

The conventional luxury industry, even in India, is now playing catch up to respond to the response of affluent consumers digital behaviours are changing [7]. In addition, digital is increasingly affecting the decision to purchase high-net-worth people. India is home to the world's second-largest digital base, as per the latest McKinsey Survey. It is expected to reach 800 million internet users by 2023, out of about 600 million internet users currently. With the democratisation of the Internet, the internet is growing.

Luxury brand goods are pricey and are mostly part of the niche market. They are highly active in knowledge collection and process of decision making for customers [13]. For several different grounds, they are bought and decorated by citizens. Product characteristics derive from cultural principles. Purchasing luxury watches will reflect wealth, stability and achievement. This study tries to find out the purchase intention on online luxury watches among Indian consumers. In its perceived intentions to buy luxury brand products, we also investigate the impact of wealth. The goal is to investigate the degree to which the object of buying is affected by the practical, experiential, symbolic, social power and character of vanity dimensions and how hedonism has an impact on the online luxury purchase.

The online luxury market in India is ready to bloom, with the amount of HNIs, growing brand recognition and an increasing number of New-Age customers. To wake up and take the digital path, time is ripe for Indian luxury brands. The way forward is digital, yet the road to digitalisation is full of obstacles, but incentives are enticing enough to walk this daunting path. Take the right steps, get your act together and build your presence in one of the fastest rising luxury markets in the world [14].

At least three motivations are outlined for the acquisition of luxury brands [15]: experientialism, symbolic interactionism and functionalism. These were examined in Taiwan [16]. We ask the same questions in the Indian sense that do consumers like to buy luxury goods at all, given the socio-economic situation? For this study, these become our questions. As the key object of research, we use luxury watches, i.e. wristwatches, for men and women.

The Fishbein model, which is highly supported by empirical evidence and utilised in consumer preferences and intentions investigations, was developed by Fishbein and Ajzen in 1975 [17]. As indicated by this model, behavioural intentions are affected by subjective standards, behavioural beliefs and perceived behavioural regulation. External perceived behavioural control refers to the perception by a person of the degree to which external obstacles can be managed. Price is the most significant impediment. The conclusion in the study stated that once they gain perceived strength, the purchase intention branded apparel will be established [18]. This shows that in order to determine the intent of purchasing luxury products, people need ample disposable income.

H1: Income is positively associated with the purpose of buying luxury goods online.

The emphasis on the purpose of consumers to buy was due to the positive effect on the actions of an individual [16–18] and has implication in the sense of brand consumption [15]. The interrelationships between people, product and brand are of great importance in understanding the luxury consumption behaviour [4]. These three criteria differentiate luxury and non-luxury brands from each other. These theories are then used to test ideas advocated [16].

i. Functionalism.
ii. Experientialism.

iii. Symbolic Interactionism.

H2. The understanding of online luxury products is associated favourably with the intention of purchasing.
H2a. The interpretation of functional value is positively associated with online purchasing intention.
H2b. The interpretation of experience worth is associated favourably with the intention of purchasing online.
H2c. The interpretation of symbolic meaning is positively associated with online purchasing intention.
H3. Social effect is associated favourably with the intention of online purchasing.
H4. Vanity is closely correlated with the purpose of shopping online.
H4a. Physical vanity is favourably related to the online intention to buy.
H4b. The vanity of achievement is closely correlated with the purpose of shopping online.
H5. Hedonism is positively related with intention of online purchasing.

3 Methodology

3.1 Designing of Questionnaire

This research adopts the online questionnaire survey method for collection of data. Due to the pandemic situation physically filling up, the questionnaire was an issue. The questionnaire comprises of basic questions on demography and a five-point Likert scale was used to measure the intent to purchase. The questions were modified from the research on antecedents of the intention to buy luxury brands [16]. We focus on luxury brands available in India that are globally renowned, emulating the research of [16]. This is to make sure the items are open and familiar to the participants.

3.2 Methodology of Sampling

In the absence of a sampling frame, the method of convenience sampling was used. From 16th August 2020 to 20th September 2020, total samples collected were 481 which included Indians aged 25 years and above. Samples were collected via online mode from the following cities namely Bangalore, Delhi, Gurgaon and Mumbai.

First, the respondents were told about the objectives of the survey and permission was taken that the data would be solely for research and analysis purposes. A Google Form was distributed to the database obtained through peer referencing. They were requested to fill the form with their own perception and not get influenced by the thought process or suggestions of others.

4 Results and Analysis

In four steps, data review was conducted. The foremost step is the verification of data for appropriateness of the further study. We then analyse the data to identify the respondent's profile, including the mean ratings of the respondent.

Sample measurements were performed using measured scales. In order to test the results with respect to the threshold factor loadings that were accompanied by a hierarchical regression analysis, we also conducted a factor analysis.

4.1 Encounters with the Purchasing of Luxury Goods

The participants were asked about their online purchasing experience and some awareness of some well-known luxury wristwatch brands. The reference pieces are luxury wristwatches for men and women. Table 1 charts down the respondents' experience of purchasing luxury goods online for themselves and their other relationships. 62.6% were respondents who had witnessed the purchase of online luxury watches. Among these, 37.8% bought these luxury brand items for themselves, while 17.2% purchased them for their spouses. Luxury brand purchases by 19.6% of respondents are those who purchased gifts for extended family members. 8.3% of respondents are buying as presents while 1.6% is buying as gifts for others.

The most bought labels are Omega, Longines, Rolex and Tag Heuer. Other purchases are Rado, Seiko, Breitling, Audemars Piguet, Patek Philippe, Jaeger-LeCoultre, Hublot and Cartier. Approximately 27.8% of the respondents are involved in online luxury watch purchases apart from the list given. They have also been asked to state these purchases of brands. These brands are listed in Table 2.

In the Kaiser-Meyer-Olkin (KMO) calculation [19] the factor features of the data were verified by the sphericity and sampling test Bartlett. The results of the two tests are shown in Table 3.

The KMO index value for factor analysis is not less than 0.6, which is to be respected. The Bartlett sphericity test has shown that it is statistically significant at $\alpha = 0.05$ using $\mu 2$-statistics. We certify that the data can be evaluated using a methodology for factor analysis.

Table 1 Online purchasing of luxury wrist watches

Purchasing of luxury wrist watches	Percent
Purchasing of wrist watches	62.6
Purchasing for own use	37.8
Purchasing for spouse	17.2
Purchasing for family members	19.6
Purchasing for friends	8.3
Purchasing for others	1.6

Table 2 Incidence of luxury brand purchase of wrist watches

Wrist watches	Percent
Omega	31.5
Longines	25.3
Seiko	18.3
Rado	11.9
Fossils	18.2
Breitling	10
Audemars Piguet	9.3
Patek Philippe	12.3
Rolex	29.6
Hublot	17.8
Jaeger-LeCoultre	5.9
Cartier	8.8
TAG Heuer	19.7
Others	27.8

Factor analysis

Table 3 Evaluation for running factor analysis suitability

Online luxury purchase dimensions	Kaiser-Meyer-Oikin (KMO) measure of sampling adequacy	Bartlett's test of spherity (χ^2)
Intention to purchase	0.7	567.8
Value function dimension	0.7	515.2
Dimension of experiential value	0.9	1010.6
Dimension of symbolic value	0.8	1108.8
Social influence	0.9	1148.9
Physical vanity	0.8	1087.6
Vanity attainment	0.8	534.4
Value of hedonism	0.7	536.3

All Bartlett's test of sphericity (χ^2) are statistically significant at $\alpha = 0.05$

4.2 *Test of Reliability*

Each component of the intention to buy online luxury includes three to seven constructs. Table 4 shows the Cronbach's Alpha statistics for determining internal consistency. Cronbach's Alpha values are greater than 0.7, indicating that there is strong internal consistency above that value. All dimensions have reasonable inter-item correlations between 0.2 and 0.4, allowing the Alpha of Cronbach to be less than 0.7 [19].

Table 4 Values of Cronbach's alpha according to the various luxury purchase dimensions

Luxury purchase dimensions	Cronbach's alpha
Intention to purchase	0.869
Value function dimension	0.790
Dimension of experiential value	0.878
Dimension of symbolic value	0.779
Social influence	0.867
Vanity physical	0.882
Vanity attainment	0.812
Value of hedonism	0.815

4.3 Factor Analysis Results

There is a greater linear relationship between household income than between personal income and purchasing intent. It ensures that when buying luxury goods, customers often rely on household profits.

Mixed strengths are seen by other correlation coefficients but are statistically relevant at alpha = 0.05. The outcome shows that the study of the factors is generally sound. One of the factors advocated by the author [16] is that consumers have their own understanding of online luxury goods purchases. Second, customers may indicate whether they wish to buy online luxury items. As per the earlier researches which includes the work of Hung, the components that have contributed are shown in Table 5. The load factor values for the different measured components are relatively high, suggesting that Indian customers are highly interested in purchasing online luxury brands in the recent scenario. Both factor loadings are better than 0.6 and their own values are better than unity, a threshold value used by the researcher [16].

Table 5 Component matrix for value function dimension

Components	Factor loading
I purchase luxury brand watches because they are handcrafted	0.701
I purchase luxury brand watches because they have the best design	0.819
I purchase luxury brand watches because they are sophisticated	0.789
I purchase luxury watches as it is superior than the others	0.869

Extraction method: Principal component analysis. Eigenvalue: 2.487

4.4 Value Function Dimension

The value function dimension is shown in Table 5, an immeasurable attribute but indicated by the product's functionality. This emphasises that items which are hand-crafted have more value to the customers as relate to high sophistication, consistency and supremacy. The loading factor is better than 0.7, while the corresponding value is 2.487.

Feature value dimension is the first factor considered to understand the influence of online purchasing intension of luxury products. It indicates that consumers give value to these products as they are special, desirable, rare, beautiful and precious. Factor loadings measured for all these characteristics in luxury goods are approximately 0.8 except for being handmade. Therefore, one reason why luxury goods are consumed by Indian consumers is because of their practical value.

4.5 Dimension of Experiential Value

Customer experience provides them with satisfaction by using luxury brand items. The purchase of this product is due to the product being unique, attractive, splendid, exquisite and rare. Factor loadings, as shown in Table 6, are all better than 0.8 with a related value of 3.67. The components are responsible for 68% of the variation. For Indian online luxury product customers, dimension of experiential value is therefore considered to be extremely important.

Table 6 Component matrix for dimension of experiential value

Components	Factor loading
I purchase luxury brand watches as it is unique	0.858
I purchase luxury brand watches as it is attractive	0.835
I purchase luxury brand watches as it is rare	0.836
I purchase luxury brand watches as it is splendid	0.835
I purchase luxury brand watches as it is exquisite	0.825

Method of extraction: Principal component analysis Eigen value: 3.67

Table 7 Component matrix for social influence

Components	Factor loading
It is important to know what others think of people using such items prior to buying an online luxury watch	0.868
I like to know what luxury branded watches gives others positive impression	0.837
It is vital to know what kind of individuals purchase those products before buying a luxury branded watches	0.825
I prefer to pay attention to other luxury brand watches that others purchase	0.789
Prior purchasing, it is important to know the luxury brands that will be impressive to others	0.709
My friends and I prefer to buy luxury things from the same brand	0.689
I consciously discourage the use of luxury branded watches that are not trendy	0.608

Method of extraction: Principal component analysis Eigen value: 3.67

4.6 Dimension of Symbolic Value

By using luxury brand items, the customer experience provides them with satisfaction. This product was purchased because it is unique, attractive, splendid, exquisite and rare. Table 6 shows that all factor loadings are greater than 0.8, with a related value of 3.67. The components account for 68% of the variation. Dimension of experiential value is thus extremely important for Indian online luxury product customers.

4.7 Dimension of Social Influence

The relation between an individual customer and the environment is social influence. It has turned to be important for them that they use these international luxury brands. The exposure to the world has increased in form of globalisation and foreign education leading to widening knowledge on luxury brands. As shown in Table 7, the loading factor being on a higher note and a statistically significant correlation of 0.41 illustrate their responses to the seven questions that reflect this dimension. The recommendation for online luxury brand marketers is that they can use more online ads to communicate with customers. When online has won the lions' share, this has become increasingly possible internationally.

4.8 Vanity Physical

Vanity physical refers to customers becoming mindful of the way they view themselves. They would be humiliated in this sense if they are not in proper attire for the

Table 8 Component matrix of vanity physical

Components	Factor loading
Looking good is very important for me	0.879
My appearance is an important factor myself	0.885
I'm going to make an effort to look good	0.886
I put great importance on my presence	0.815
I feel uncomfortable when I am not looking best among people around me	0.678

Method of extraction: Principal component analysis

occasions, which also apply to wearing luxury items. It has a correlation of 0.281 which is statistically significant in context to Indian consumers' buying intentions. In Table 8, all of the different characteristics provided to the customers met the analytical factor threshold value of 0.6 with an independent value of 3.72.

4.9 *Vanity Attainment*

Vanity achievement refers to the pride that one retains in career achievement, schooling, life, etc. Its association with the purpose of purchasing is 0.389. For the other features, factor loadings are more than 0.6 in one and better than 0.8 in other cases.

4.10 *Value of Hedonism*

Hedonism points out that the use of a luxury commodity achieves self-pleasure and internal gratification. Luxury goods can thus be viewed as hedonic items to which consumers are emotionally drawn and can find self-pleasure by using them [20]. A measure of 0.671 factor loading indicating that it is of importance to them. The Eigenvalue associated with it was 3.016. The dimension contributed to 63% of the variance in intention to purchase.

4.11 *Regression Analysis*

To evaluate the data with the scale scores on purchase intention, a hierarchical regression was performed. The presence of the moderating factor can be assessed by this regression analysis approach. The moderating variables in our case are the mean sum scores for physical vanity, attainment vanity and hedonism. Regression is carried out in two steps in this study. In the first stage, moderating variables are used as explanatory variables (Model 1) as with purchase intentions being the dependent variable (Fig. 1). The calculation of the complete model is followed by Model 2. Tables 9 and 10 provide the outcome of the regression analyses.

For Model 1 the R-square is relatively small, suggesting that only a trivial proportion of the difference in purchasing intention was explained by Vanity Physical, Vanity Achievement and Hedonism. Model 2 reveals 38.4% of all models, that is, differences in purchase intent. The reduction in R-Square is therefore 23% when the Vanity Physical, Attainment-Vanity and Hedonism are excluded. As shown by a statistically significant shift in the F-value, this is a significant contribution.

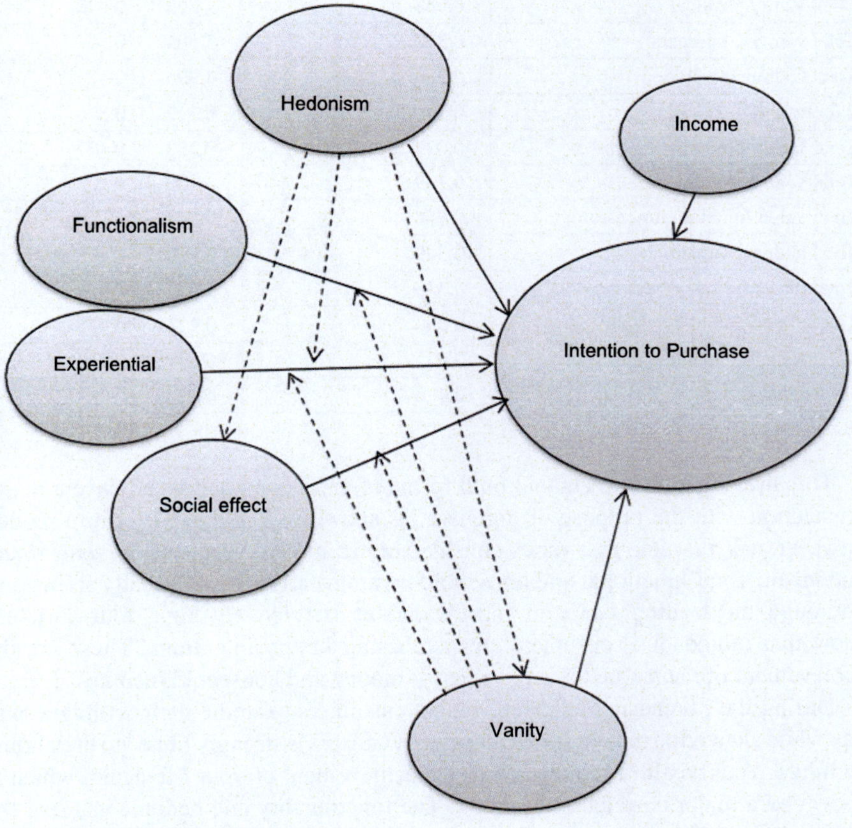

Source: Author (2021), adapted from Hung *et al.* (2011)

---------> Factors that are Moderating

⟶ Factors Influencing Directly

Fig. 1 Proposed model on purchase intention of Luxury Products.

Table 9 Change statistic due to models 1 and 2

Model	R-square	R-square change	F-change	Sig. F-change
Model 1	0.164	0.164	34.415	0.000
Model 2	0.394	0.230	31.112	0.000

Table 10 Results of the hierarchical regression analysis on intention to purchase as the dependent variable

Independent variables	Regression coefficients	*t*-value	Sig. (model)
1 (constant)	3.536	4.199	0
Total vanity physical	0.088	1.803	0.072
Total vanity attainment	0.327	5.594	0
Total hedonism value	0.336	4.996	0
2 (constant)	0.136	0.312	0.876
Total vanity physical	−0.032	−0.753	0.477
Total vanity attainment	0.189	3.803	0
Total value function dimension	0.358	5.347	0
Total hedonic value	0.349	4.996	0
Total dimension of experiential value	0.33	1.351	0.523
Total dimension of symbolic value	0.03	0.223	0.825
Total social influence dimension	0.031	0.649	0.523
Household earning	0.496	4.526	0

This investigation reveals that both forms of vanity and hedonism have a major connection with the purpose of purchase, as shown in Table 9. The entire model, however, has a *t*-value that varies considerably from zero. Vanity achievement and hedonism. Total functional and household earning statistics vary greatly from zero. Although the loading factor of most variables is relatively high, four variables show that indigenous consumers are purchasing luxury intentions. These are the achievement of vanity, hedonism, value for money and household incomes.

During the purchasing research, Indian customers examine their willingness to pay while answering the questions intuitively, and this is strongly linked to their home earnings. Your wealth is connected to the achievement of your life again, which is shown by a major correlation of 0.251. The functionality and hedonic value of the item they bought are also considered. It should be costly to invest your income on these items.

We are changing our model based on the above findings. Vanity attainment, household earning, value function and hedonic value influence Indian customers to buy online luxury wristwatches. Figure 2 depicts the model. Vanity and Hedonism are not treated as a moderating factor in the model, but explicitly affects the purpose of purchasing. The coefficient of regression is also not substantially different from

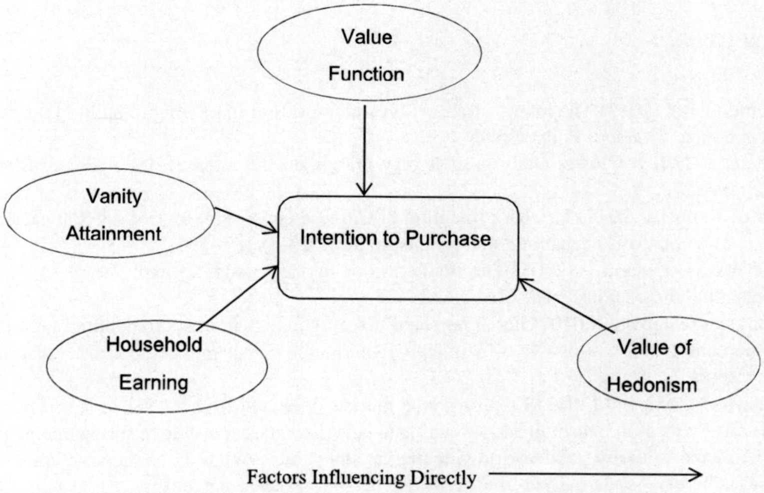

Fig. 2 Modified model of intention to purchase online luxury products

zero, indicating that it has no effect on the purpose of purchasing. From a study [11], earning is also found to be significant.

5 Conclusions

In the components of certain purchasing-intent dimensions, the relatively high loading factor does not mean that it is a clear proposition that subsequent transactions will result. Sales will be one of the potential factors for future transactions. With increase in income, consumers are linked to the aim of purchasing. Other factors that are not associated with sales can lead people to fail to create buying intentions. That is the reason why respondents refused to take part in the survey. It of no use for them to engage in a survey on online luxury items since they have no way of purchasing them. For this reason, vanity attainment can have a statistically relevant *t*-statistic instead of vanity physical for the regression coefficient. The attainment of vanity and hedonism is related to wealth and therefore affordability. The customers of the luxury product will want to experience the consistency, high price and rarity. The online mode of collecting data is also a hindrance as luxury being a high priced item there could have been few more queries answered while physically filling the questionnaire.

References

1. Kapferer J-N (2012) The luxury strategy: break the rules of marketing to build luxury brands. Kogan Page, London; Philadelphia
2. Geerts A (2013) Cluster analysis of luxury brands on the internet. Int J Manag Mark Res 6(2):79–82
3. Jin B, Kang JH (2011) Purchase intention of Chinese consumers toward a US apparel brand: a test of a composite behavi model. J Consum Mark 28(3):187–199
4. Vickers SJ, Renand F (2003) The marketing of luxury goods: an exploratory study—three conceptual dimensions. Mark Rev 3:459–478
5. Bain and Company (2020) Global personal luxury goods market set to contract between 20–35 percent in 2020. https://www.bain.com/about/media-center/press-releases/2020/spring-luxury-report/
6. Bain and Company (2012) Luxury goods market predicted to grow six to seven percent in 2012, defying global turmoil and spreading to new markets, according to spring update of Bain & Company's luxury goods worldwide market study. http://www.bain.com/about/press/pressreleases/luxury-goods-market-predicted-to-grow-six-to-seven-percent-in-2012.aspx#.T_AqbaZQZs8.mailto
7. Going digital: The only option for Indian luxury market: retail https://retail.economictimes.indiatimes.com/e/retail-tech-summit?ag=retailskinspk&utm_source=etretail-com&utm_medium=banner-skinning-banner&utm_campaign=featured-4065-campaig403. February 2020
8. Guercini S, Runfola A (2015) Internationalization through E-Commerce. In: The Case of MultiBrand Luxury Retailers in the Fashion Industry. Wiley
9. Nwankwo S, Hamelin N, Khaled M (2014) Consumer values, motivation and purchase intention for luxuary goods. J Retail Consum Serv 21:735–744
10. Beuckels E, Hudders L (2016) An experimental study to investigate the impact of image interactivity on the perception of luxury an an online shopping context. J Retail Consum Serv 33:135–142
11. McKinsey & Company (2020) A perspective for the luxury-goods industry during—and after—coronavirus
12. Türk B, Scholz M, Berresheim P (2012) Measuring service quality in online luxury goods retailing. J Electron Commer Res 13(1):88
13. Hennigs N, Wiedmann K-P, Klarmann D-OC (2012) Luxury brands in the digital age-exclusivity versus ubiquity. Mark Rev St Gallen 29(1):30–35
14. The Economic Times https://retail.economictimes.indiatimes.com/
15. Berthon PR, Pitt L, Parent M, Berthon J-P (2009) Aesthetics and ephemerality: observing and preserving the luxury bran. Manage Rev 52(1):45–66
16. Hollensen S (2011) Global marketing, 5th edn. FT Prentice-Hall, London
17. Ajzen (1991) The theory of planned behaviour. Organ Behav Human Decis Process 50:179–211
18. Hung K-P, Chen AH, Peng N, Hackley C, Tiwsakul RA, Chou C-L (2011) Antecedents luxury brand purchase intentio. J Prod Brand Manage 20(6):457–467
19. Pallant J (2020) SPSS survival manual, 7th edn. Allen & Unwin, NY
20. Wiedmann K-P, Hennigs N, Siebels A (2009) Value-Based Segmentati

Implementing Virtual Reality in Entertainment Industry

Saniya Zubair Ahmed Ansari, Vinod Kumar Shukla, Komal Saxena, and Bethoven Filomeno

Abstract This paper talks about how information technology can be used to create a system for the entertainment industry, which can help combat the losses incurred during the pandemic. IT is the abbreviated form of information technology. It deals with computer mechanisms such as networking, software, and hardware. Many firms these days opt for IT-based operations as it makes everything more convenient. It is used in a variety of fields like gaming, sports, education, and even entertainment. Among the vast list of subcategories in information technology, one is slowly gaining momentum, the Virtual Reality or VR industry. The paper also proposes a model to capture various images for further uses in Virtual Reality. Virtual Reality is a simulation in which a person can use a 3D surrounding using multiple peripherals. Depending upon its efficiency, it gives the user a real-life experience in the comfort of their seats.

Keywords Virtual Reality · Hologram · Information technology · Smart entertainment · Image recognition · 3D image rendering

S. Z. A. Ansari
Bachelor of Information Technology, Amity University, Dubai, UAE
e-mail: saniyaA1@amitydubai.ae

V. K. Shukla (✉)
Department of Engineering and Architecture, Amity University, Dubai, UAE
e-mail: vshukla@amityuniversity.ae

K. Saxena
Amity Institute of Information Technology, Amity University, Noida, India
e-mail: ksaxena1@amity.edu

B. Filomeno
Department of Humanities Arts and Applied Sciences, Amity University, Dubai, UAE
e-mail: bfilomeno@amityuniversity.ae

J. M. R. S. Tavares et al. (eds.), *Cyber Intelligence and Information Retrieval*, Lecture Notes in Networks and Systems 291,
https://doi.org/10.1007/978-981-16-4284-5_49

1 Introduction

Recently, this technology has been making its way into the mainstream market through the entertainment industry. In the past, Virtual Reality (VR) had only been the subject of an experiment, but nowadays, it is making a mark in the mainstream industry. Almost 50% of the VR industry lies in the entertainment field. Experts say that in coming years, investors will have a vast choice of VR products. An increase in VR entertainment experiences will engage more consumers and bring success that is more commercial. Currently, the one field VR dominates in is gaming, but it progresses to other realms and fields. While VR is blamed for creating a less social environment, it can also unite people virtually. Communication in such worlds is similar to real-life, but it gives flexibility in the choice of environment, mobility, and ways to spend time. A study in social interaction at Neurons Incorporation showed that 59% of people consider VR to communicate with friends and family living far. For example, Facebook is promoting its VR platform Facebook Space. Microsoft also recently acquired the social platform Altspace VR.

1.1 Types of Virtual Reality

Virtual Reality is a completely immersive digital experience, unlike other technological devices. Virtual Reality allows a user to experience something first-hand by giving immersive three-dimensional expertise, which can be very different from the real world. With that in mind, there are three types of Virtual Reality based on the level of immersion. (Fig. 1):

- ***Non-Immersive Virtual Reality***: These types of VRs are not considered as VR by common people but do give a virtual experience. They are usually equipped with input and output devices. However, it still provides a glimpse into a new world. Video games are an example of Non-Immersive Virtual Reality.
- ***Semi-Immersive Virtual Reality***: This type of VR allows a part virtual environment. Though it will give users the experience of a virtual environment through simulation, it also allows them to be connected to their physical surroundings. It provides realism through three-dimensional graphics, which provide a more immersive experience. Most education institutes use this type as it gives a more interactive learning method. Recently, rehabilitation centers

Fig.1 Types of virtual reality

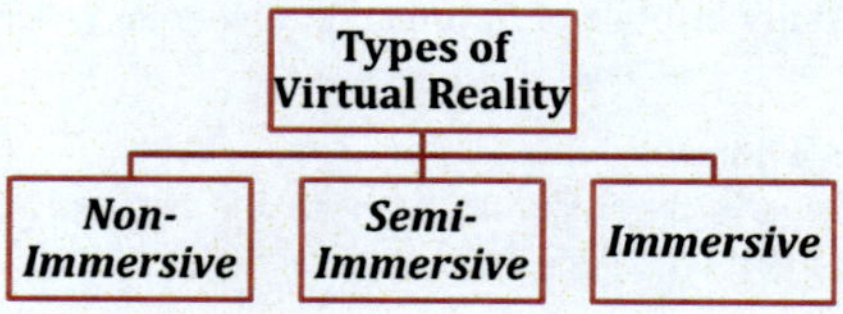

are also implementing semi-immersive technologies to simulate the real-world environment.

- ***Immersive Virtual Reality***: This type gives the most realistic experience in terms of simulation. It uses a head mount display (HMD) to display images. The headsets provide very high image resolution with a wide range of views. The display splits into two different sides to give the maximum real-life experience as possible. This type of Virtual Reality is mainly being used in gaming, but its scope is extending to medical field [1].

1.2 Virtual Reality in the Mainstream Industry

Virtual Reality is a common name in side-entertainment industries like gaming. Consumer-based Virtual Reality Gear was released in the 1990s. In 2010, many more prominent companies like Sony and HTC also began to invest in it, thereby opening up new application development scope [2]. Multidimensional cinema is being used for movies, documentaries, theater, and art exhibitions. Not only that, it is being used in many arcades and amusement parks too. Rollercoasters and similar rides are becoming more thrilling with the addition of Virtual Reality in their mechanism [3]. With the surge of the COVID-19 pandemic, all daily on-site activities have shifted to online platforms. Hence, the entertainment industry is being affected. According to the observer, the entertainment industry is set to lose 160 billion dollars if the pandemic continues over five years. The world's highest populated country China was under lockdown, and it has affected the entertainment industries of all countries around the globe. Hollywood itself has suffered a loss of a hundred million dollars of sale tickets. Not only that, the pandemic hit the USA in March 2020, there is a further possible loss of 20 billion dollars. Theater is estimated to hit harder, with a potential loss of 24 billion dollars. Therefore, Virtual Reality needs to be utilized before more losses occur [4].

2 COVID-19 and Entertainment Industry

According to the World Health Organization, coronavirus is caused by SARS-CoV-2, which spreads from one person to another only when they are in close proximity. The virus spreads from person to person through mouth or nose in small liquid particles as the person coughs, laughs, sings, sneezes, or even breathes heavily. Aerosol transmission occurs in particular situations, most crowded, indoor spaces. Examples of such places are restaurants, pubs, clubs, malls, offices, and places of worship, gyms, and standards [5]. According to BBC health, our body starts releasing chemicals that are called interferons when threatened. It serves as a warning sign. However, the novel coronavirus can turn off the warning. According to Professor Lehner, "it does so well you won't even know you are ill." Blood clotting is much unexpected during COVID, with many doctors not being able to save a patient due to immediate clotting. "Clotting chemicals rise by 200, 300% and so on higher than normal when in Covid," said Professor Beverly Hunt from King's College London. This virus can cause sudden inflammation in patients, making the immune system weaker and overdriven, which harms the patient's harmful consequences. That is why lockdowns were announced globally to stop it from spreading further [6].

2.1 A Solution to Support Both the Sides

Since the entertainment industry is more focused on revenues by designing and selling artists' images, we would like to propose communication between the artist's admirers and artists at the comfort of their seats. This solution makes entertainment more accessible and is very safe as there are no crowds or external people involved. It uses a basic system of image recognition and holographic projection to connect two foreign media.

2.2 The Components of the System

The system comprises cameras, face recognition systems, three-dimensional image rendering software, laptops, and a hologram device. The system also includes the basic technology of face recognition, which uses the following algorithms:

- ***Scale-invariant Feature Transform (SIFT)***:It can detect and describe particular images using image graphing and mapping techniques. It is used to characterize images that are not high resolution or blurred as such occurrences are shared in real life [7].
- ***Speeded-Up Robust Features (SURF)***: It uses wavelet analysis in both horizontal and vertical planes for the size of 6 s. Gaussian weights are also implemented. The wavelet response can be found using an integrated series of images very conveniently [8].
- ***Linear Discriminant Analysis (LDA)***:This dimension reduction technology is essential in machine learning, bioinformatics, information retrieval, and data mining. The main idea is to remove the repetitive and redundant data by changing the data space to smaller dimensions. Multidimensional data has features that correspond to one another. It allows only essential data to be featured [9].

The software will be programmed into a credit card-sized, commonly used computer called the Raspberry Pi. The latest version of Raspberry Pi has both Internet and Bluetooth connection facilities, making it accessible for the project. It was released in February 2016 with a 1.2 GigaHertz 64-bit quad-core processor. Additional features include Power over Ethernet, USB boot, and Network boot. Next to be used is the image recognition software, which uses the following algorithm:

The relative change in position of the two images is recovered. Next, motion and calibration are calculated, along with the three-dimensional structure of the features. This is due to the missing prior calibration. Using this, the distortions and disruptions can be removed, which could lead to a reconstruction precise at the global scale. It allows more flexibility in the process not to require focal lengths and calibrated or even be changed. Since a pixel correlated to a ray in space and projected light from this ray can be predicted using coordinates and estimation in the change of position, a pixel's search can be restricted to a single line [10]. A **laser** is much of a purer light than the normal visible light we are accustomed to. In a normal light beam, the light waves are scattered about in space and are oriented randomly. Light is similar to scattered dust particles, moving about in undetermined directions while the laser is coherent: they all stay in the same line like marching bands in synchronization. When a laser is split to create a hologram, the light beams travel in identical ways. There is a photographic plate that is used to project the rays.

When the rays combine on the plate, the beam has traveled from a different path, and the light rays will be disturbed. This generates the shape of the object. Since the light rays were aligned in a straight line, the distortion shows where the object deflected the light. This gives the outline of the corresponding object. In a different sense, if you combine the two orientations of rays and calculate the change in positions, you can determine the object's exact dimensions. This information is embedded into the photographic plate by laser beams. So a hologram is a permanent record of something that looks from every angle. Every point in a hologram catches the light so that it traveled from every point of the object. That means wherever you look at in a hologram, you see precisely the points where the light hit the object. Since human eyes also function in the same way, viewing light hit at different points of an object creates the illusion of the object being present. So, as you move your head around the object, the holographic generated image seems to change just like a real object. And that is the reason why holograms appear to be three-dimensional [11].

3 Model for Implementing Virtual Reality

The system's mechanism starts in the Concert Venue, where the artist's stage is set up in an isolated environment. Three cameras are set up according to the following formula:

- For three cameras to be set up so that they can get 360 views of the whole stage, we need to consider it as a surrounding circle (Fig. 2).
- Since the total angle of a circle = 360°.
- Three cameras are required to be set up from the circle at an angle of 360/3 = 120.

Draw a perpendicular from each radius of the circle and join the points. The resulting triangle is an equilateral triangle, with each angle equal to 60° (Fig. 3).

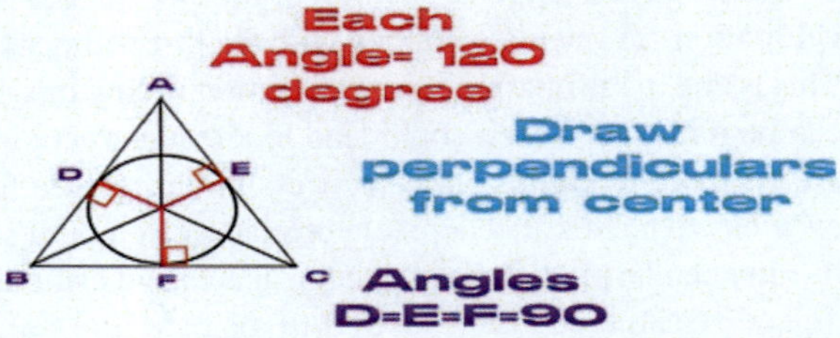

Fig. 2 Triangle diagram

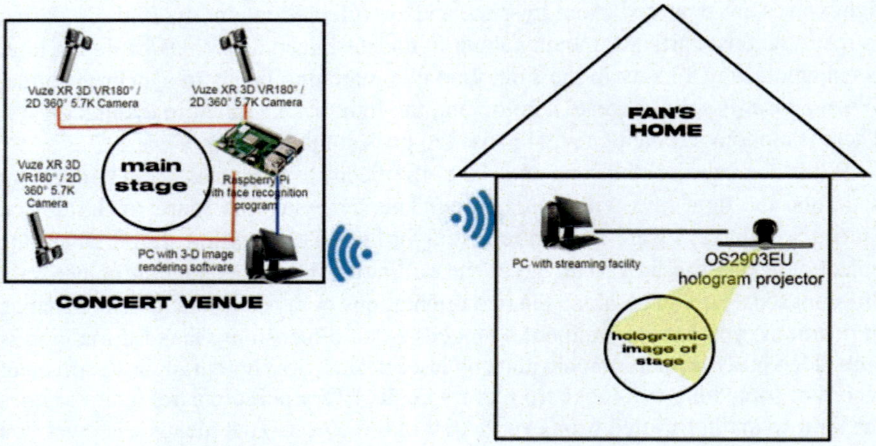

Fig. 3 Proposed model of the system

When an artist performs on the stage, the Vuze XR 3D VR180°/2D 360° 5.7 K Camera captures 5.7 K resolution images of the stage while tracking real-time movements. Each frame of the artist's stage is captured and transferred to the Raspberry Pi, connected to the common network. Here is how Raspberry Pi can be connected to a wireless network. The Pi can also be used as a bridged access point to connect to a wireless network within an existing Ethernet network. This can extend the network to wireless computers and devices. A wireless access point can be created using the premade wireless features of the Raspberry Pi 4 or Raspberry Pi 3 using suitable Universal Serial Bus dongles that support access point modes.

3.1 Prerequisite

1. Ensure you have administrative access to your Raspberry Pi. The IP address of your Raspberry Pi on the network will probably change after installation.
2. Be ready to add a screen and keyboard in case you lose contact with your Raspberry Pi after installation.

3. Connect your Raspberry Pi to the Ethernet network and boot the Raspberry Pi OS.
4. Ensure the Raspberry Pi OS on your Raspberry Pi is up-to-date and reboot if packages were installed in the process.

This situation is particular to wireless LAN interfaces [12–14].

3.2 *Define the Bridge Device IP Configuration*

The bridge device itself needs an IP address so that you can reach your Raspberry Pi on the network. dhcpcd, the DHCP client on the Raspberry Pi, automatically requests an IP address for every active interface. Sudo nano/etc./dhcpcd.

3.3 *Go to the End of the File and Add the Following: interface br0*

Save the file to complete the machine's IP configuration.

3.4 *Ensure Wireless Operation*

The Linux OS helps users comply with these rules by allowing applications to be configured with a two-letter Wi-Fi country code.

3.5 *Configure the Access Point Software*

Create the hostapd configuration file, located at hostapd/hostapd. Conf, to add the various parameters for your new wireless network. Sudo nano/etc./hostapd/hostapd. Add the information below to the configuration file [15].

The Raspberry Pi is connected to the working Internet network and is loaded with the face recognition program. The *scale-invariant feature transform* algorithm is used to detect and make out the different features of the idol's face, to save the details of their features in the database to ensure that only the artist's face is captured. *Speeded-up robust features* use its wavelet analysis to determine whether the object that is the artist moving across the same plane is the same person. It minimizes a large Gaussian plane to a smaller one that ensures the correctly targeted face is captured. The *linear Discriminant analysis* allows the algorithm to sort various images and remove the redundant ones. For example, if the artist moves around the stage faster than the frame speed, then two images will be generated. LDA will remove any redundant images or glitches generated during operation [16].

All the saved images are stored as a backup in the SD card inserted in the Raspberry Pi. The data stored in the Raspberry Pi is transferred to the main control computer with higher performance stability and has the three-dimensional image rendering software installed, where each image is carefully analyzed. It is then layered to create a three-dimensional model of the performing artist. The resulting image is once again combined frame-by-frame to generate the artist's replica video stream performing on the stage.

The corresponding video is then sent to a streaming facility where it is streamed like a normal video stream. The streaming device is connected to a hologram projector. The hologram projector uses the method of projecting laser beams on an object, and the resultant light rays are projected on the floor, which acts as the photographic plate. The image quality depends on the smoothness of the floor, which will be an optional accessory that will be proposed in the future. For now, if the floor is smooth, the image will be clear, but if the floor is irregular, the image will not be of high quality. The result would be an audience watching their favorite artist's concert show in the comfort of their homes, without worrying about the crowd or any threat to their health amid a pandemic.

3.6 Cost of the Mechanism

Since this prototype is designed for the entertainment industry, high-quality products were used to ensure maximum efficiency. The hologram projector costs AED 668, VUZE camera costs 1752 and the Raspberry Pi costs AED 225. The remaining components are standard, and both the client and the organizer may have access to it. The quality of a basic good quality laptop is AED 1250. Installation can cost AED 200 at maximum. The total cost amounts to AED 6349, which is an average cost for an entertainment company. Even low-scale companies can afford the whole system to present their entertainers. All cost is based on international online market till January 2021.

The paper proposes a plan to bring mainstream entertainment to anyone's homes' comfort without sacrificing any of the characteristics. It presents an integrated system using various IT components like VR, holograms, and 3D image rendering to transform reel into real. The system is cost-effective, compact, and accessible to common people, unlike previously proposed solutions. It promises to provide full-fledged entertainment, even during the pandemic.

4 Conclusion

Virtual Reality was always sought to be used only for gaming and leisure time purposes. However, its scope goes beyond gaming and arcade. With technology developing per second, more uses of Virtual Reality are coming into the limelight. VR has the most potential to develop among the other information technology field. It is being used in education, e-sports, medical establishments, and now even the mainstream entertainment industry. It is a versatile technology paired with other flourishing technologies to create many newer technologies. Like this prototype using video cameras, face recognition algorithms, hologram algorithms, and Raspberry Pi, we can create an entirely different system using these technologies. Many countries are under the economy's loss due to the pandemic as their workforce is restricted to work from home. Virtual Reality can help bring relief in their system, as it promotes working from home in an isolated area. Not only that, but the entertainment industry is also on the verge of losing billions of dollars, which could affect world economic status as a whole. Some necessary losses could be covered up by setting up this system of presenting artists to their fans in a paid special concert environment. This technology could potentially save many people from losing their only source of income, entertainment.

References

1. Radar H (2020) The 3 types of virtual reality. Heizen Radar
2. Comparison of VR headsets: project morpheus vs. Oculus Rift vs. HTC Vive. Data Reality. Archived from the original on 20 August 2015. Retrieved 15 August 2015
3. Kevin K (2016) The untold story of magic leap, the world's most secretive startup. WIRED. Retrieved 13 March 2017
4. Katz B (2020) Entertainment industry staring at $160B loss over 5 years due to COVID-19. OBSERVER
5. WHO Team (2020) Coronavirus disease (COVID-19): How is it transmitted?" Who.org
6. Gallagher J (2020) Covid: Why is coronavirus such a threat?. BBC health
7. Weitz E (2016) SIFT—scale-invariant feature transform. Impressum
8. Mordvintsev A, Abid K (2013) Introduction to SURF (Speeded-Up Robust Features) open CV
9. Tyagi N (2019) Introduction to linear discriminant analysis in supervised learning. Analytics steps infomedia LLP 2020
10. Pollefeys M, Van Gool L How computers can automatically build realistic 3D models from images acquired with a hand-held camera. Katholieke Universiteit Leuven
11. Woodford C (2008, 2017, November 2020). Holograms. Retrieved from https://www.explainthatstuff.com/holograms.html. Accessed on Feb 2021
12. Shukla VK, Kohli A, Shaikh FA (2020) IOT based growth monitoring on moringa oleifera through capacitive soil moisture sensor. In: 2020 seventh international conference on information technology trends (ITT). Abu Dhabi, United Arab Emirates, pp 94–98. https://doi.org/10.1109/ITT51279.2020.9320884
13. Madana AL, Shukla VK (2020) Conformity of accident detection using drones and vibration sensor. In: 2020 8th international conference on reliability, infocom technologies and optimization (Trends and Future Directions) (ICRITO). Noida, India, pp. 192–197. https://doi.org/10.1109/ICRITO48877.2020.9197783

14. Nanda I, Sahithi C, Swath M, Maloji S, Shukla VK (2020) IIOT based smart crop protection and irrigation system. In: 2020 seventh international conference on information technology trends (ITT). Abu Dhabi, United Arab Emirates, pp. 118–125. https://doi.org/10.1109/ITT51279.2020.9320783
15. Epoch 1970, Hughes J, Wright M, Lynn H, Scheller A (2020) Setting up a Raspberry Pi as a bridged wireless access point. Raspberry Pi Foundation
16. Gupta S (2018) Understanding image recognition and its uses. E infochips

Women in Information Technology: How Organizationally Committed They Are

Rooprekha Baksi Maiti and Shamindra Nath Sanyal

Abstract The study's aim is to look into the factors that influence organizational commitment among working women in the information technology (IT) industry, as well as how these factors can be used to determine organizational commitment among women in the IT industry. Working women from seven IT companies in Kolkata were interviewed qualitatively for this report, which included 35 participants. For maximum substantiation of the result, data triangulation was established which was followed by testing Cohen's kappa to justify the pattern of synchronism between author and coder. Six constructs were identified in this study that are expected to affect working women's organizational commitment in the IT sector. The constructs were motivation toward work, role of emotional intelligence, importance of incentives, impact on psychological well-being and loneliness, organizational culture, and family duties and responsibilities. More than 85% of the respondents complied to the constructs incorporated in the study which influence organizational commitment of working women in IT industry.

Keywords Organizational commitment · Working women · Psychological well-being and loneliness · Family responsibilities · Cohen's kappa

1 Introduction

Organizational commitment is a psychological and independent attachment of the worker with the organization [1]. Commitment is also involved with "identification and attachment" of the employee with his place of work [2]. Researcher came up with four different variables which reflect organizational commitment that are job

R. B. Maiti (✉)
Amity University Kolkata, Kolkata, India
e-mail: rbaksi@kol.amity.edu

S. N. Sanyal
Bennett University, Greater Noida, India
e-mail: shamindra.sanyal@bennett.edu.in

J. M. R. S. Tavares et al. (eds.), *Cyber Intelligence and Information Retrieval*, Lecture Notes in Networks and Systems 291,
https://doi.org/10.1007/978-981-16-4284-5_50

characteristics, personal characteristics, structural characteristics, and work experiences [3]. The explorers of organizational commitment proclaimed that employees with a higher commitment would pursue to work with the company or the institution even during any organizational commotion [4].

Organizational commitment can be escalated through satisfaction attained in job, rate of turnover, being absent from work and organizational citizenship behavior (OCB) [5]. The researchers also note that commitment, accompanied by a low turnover rate and a large rate of organizational citizenship behavior, would be higher if an employee is happy with his job.

Involvement of women in the job sector is increasing every day, and the role of women has become very promising for the country's growth. In addition to being the backbone of a family's survival, women handle job pressure fairly [6].

In the present study, organizational commitment of the women employees in the IT area is studied in an analytical way. Women play multiple roles that include a wife and a mother and also carry out several household activities. To become successful in the field of work, one needs to display productivity, attendance, and performance in their workplace [6]. Due to the narrow-mindedness and shallow perspective of the society, male employees attain exceeding moral support compared to women at their place of work, and hence, researches prove that low moral support commitment of men is found higher than female [7]. Social support from family and peers enhances organizational commitment of women employees as they can balance work and family life [8, 9].

The key goals of the present study are twofold: firstly, to define the factors that influence organizational commitment among working women in IT sector, as well as how these factors can be used to determine organizational commitment among women in the IT industry.

2 Literature Review

Researchers note that an organization's prospect and success depend not only on employee competencies, but also on how an organization encourages employee commitment [10]. Employee commitment is the crux of human resource management [11] which is accompanied by organizational goals [12].

Job motivation encourages a person to initiate actions to achieve goals or psychological satisfaction [13]. The primary motto and key to success is the commitment of employees and motivated workforce in the dynamic area of work. The fulfillment of needs creates an obligation for employees to reciprocate, which contributes to greater commitment to the organization [14, 15]. A group of employees in an organization observed a strong association between need achievement and work motivation with organizational commitment [16]. According to researchers, motivation at various levels is vital for employee satisfaction and organizational commitment, both of which are critical for a company's performance [17].

Emotional intelligence is referred as a kind of intelligence which encompasses the awareness of an individual regarding his emotions and feelings and moreover his control over emotions [18]. A change from studying cognitive intelligence to emotional intelligence has taken place in the modern age [19] and further research notes that the existence of emotional intelligence, along with cognitive intelligence, is the secret to success for many people [20]. Significant relationship was established between emotional intelligence and organizational commitment [21, 22].

Incentives are referred to as tangible or intangible rewards that motivate workers to perform in a defined way [23], and others focused more on monetary incentives given after an achievement [24]. Researchers found out that incentives and job values are closely linked to organizational commitment [25]. Information exchange, task orientation, organizational integrity, training processes, and incentives are some aspects of organizational dynamics that have a positive impact on organizational commitment [26].

COVID-19 has been studied for its economic impact on the general population, as well as its impact on well-being, fear, and anxiety-prone behavior [27]. A case study of three individuals in the USA showed that they were all struggling to run their families and pay their bills during the pandemic [28]. Researches indicated an association between gender, emotions, and organizational commitment [29, 30]. Psychological well-being is characterized as the subjective experience of a positive feeling of valence, including calmness or happiness, followed by a sense of excitement [31]. It is often correlated with a positive effect, satisfaction, a feeling of happiness, and optimism.

Organizational culture may refer to norms, notion, values, and beliefs that affect organizational behavior [32]. Compared to men, women view the corporate culture from a different context. Research has shown that gender differentiation, lack of role models, and mentoring are a few factors for women's diverse views of organizational culture [33].

When it comes to organizational commitment of working women, family responsibilities play a vital role. Research has shown a difference in accountability between males and females. A research to assess the job performance and work attitudes of women [34] found that women with children displayed lower levels of organizational commitment than women without children. Family obligations can be managed easily when working hours are flexible, which also improves organizational commitment and job satisfaction [35]. In general, women face many challenges in order to balance both professional and family life, which relies on flexible working hours, organizational culture, and managers' behavior [36].

3 Research Methodology

This research was carried out using the qualitative research approach by applying content analysis technique. In the beginning, a pilot study was carried out with 13 working women in the IT industry in Kolkata. Based on the outcomes of the pilot survey, a semi-structured questionnaire was created to assess the factors affecting working women's organizational commitment in the IT field. Since it was unendurable to meet the respondents in person during the pandemic, data was obtained through individual telephonic in-depth interviews. The total interview was recorded with the respondents' permission. Individuals were chosen using a qualitative sampling frame to ensure that they possessed a diverse set of demographic and professional traits. Quota sampling method was applied to select the organizations, followed by the employees. Data were collected from seven different IT companies in Kolkata. Since work cultures vary by location, and each company has its own pay scales and job terms and conditions, different IT firms were chosen. All respondents were given a clear idea of the subject before the interview. Table 1 shows the demographic features of the current sample.

When researchers can devote adequate attention to thorough and rich content analysis and reporting, a sample size of 35 is important [37]. Following the data collection, a coding manual based on the responses given by the participants was developed. The coding was done by the first author, who then delegated another set of coding manual to an independent coder. Following each construct, a table of commonly accepted items by respondents was developed. The pairs of responses were arranged in a two-way contingency table for each item. There were six constructs in the present analysis, consisting of 14 items in total. The authors and the coder chose the same object and rated it the same way. Cohen's kappa was used to assess inter-rater reliability for such variables. The kappa coefficient is used to express how well each construct agrees with each other.

Table 1 Demographic features of the respondents

Demographic features	Specification	Participant number	Frequency (%)
Age	20–30	3	8.50
	30–40	12	34
	40–50	16	45
	50–60	4	11
Education level	Undergraduate	11	31
	Post-graduate	8	23
	Professional	16	46
Work experience (in years)	0–5	5	14
	5–10	8	23
	10–15	16	45
	Above 15	11	31
Monthly income	US$ 400–600	9	26
	US$ 600–800	14	40
	US$ 800–1000	4	11
	US$ > 1000	8	23

4 Data Analysis and Discussion

4.1 Work Motivation

Maslow's Hierarchy of Needs theory is important in the field of organizational behavior when it comes to job motivation. Work motivation enhances the effectiveness of the employees, which improves their level of commitment [38]. Most participants (68%) agreed that the work motivation is instrumental in shaping organizational commitment. They also claimed that if an employee was empowered to work productively, their organizational commitment would be on a higher note. Job satisfaction determines the fundamental behavioral pattern of an employee in the organization. It is also claimed that organizational commitment is simply an extension of job satisfaction, which entails the employee's positive attitude to her organization [39]. Some respondents think that due to the lack of job satisfaction, the propensity for employees to leave job will be higher.

Job security is another significant component of organizational dynamics that often motivates employees. Job security is the expectation of an employee about her continuity in a job situation [40]. Lack of a suitable job position, low promotion opportunities, poor working conditions, all contribute to job insecurity among employees. All the participants agreed that job security is essential for organizational commitment. For women, job security is extremely important because many employees are single mothers and they need to run the family.

To summarize, all the respondents acceded that the components under the construct work motivation, i.e., job satisfaction and job security do play a vital role in organizational commitment of the female employees working the IT.

4.2 Role of Emotional Intelligence

Research suggests that emotional intelligence differs between working and non-working women, and working women are more fulfilled and happier with their lives and are prepared to take on various types of challenges [41]. Majority of the participants (80%) agreed upon the importance of emotional intelligence in their workplace.

4.3 Importance of Incentives

Incentives are designed to enhance employee efficiency regardless of its form, which increases the abilities of individuals and helps to boost their potential [42]. The desire to work hard is fueled by a mixture of monetary and non-monetary rewards. Each participant accepted that incentives increased organizational commitment. Compensation plans should be more accurate for working women because when they work extra hours or on holidays then they have to skillfully balance their family chores.

In summary, all the respondents agreed that the items under the construct financial incentives and non-financial incentives do play a vital role in organizational commitment of the female employees of IT firms in Kolkata.

4.4 Impact of Psychological Well-being and Loneliness of the Female Employees during COVID-19.

Under this construct, the items were anxiety and fear, stress tolerance level, coping strategies, self-care, and positivity. Psychological well-being became a very important aspect during the pandemic. Hence, the views of the working women in IT sectors were taken into consideration. When asked about self-care and positivity, the respondents replied that it is an important dimension of maintaining good mental health, but after managing office and home without domestic help, self-care was rarely possible.

In short, it can be interpreted that participants have different perspectives on psychological well-being and its impact on their organizational commitment.

4.5 Organizational Culture

A correlation between organizational culture, employee behavior and the company's productivity were documented [43]. The aim of this study was to learn about the perspectives of working women in the IT sector on organizational culture and commitment. The elements included in this factor were type of culture, office environment, and co-worker relationships.

All participants accepted that the environment and organizational culture should function smoothly and be acceptable for them. A threatening environment will reduce organizational commitment and job security. However, in terms of working relationships, an interesting conclusion was drawn. Most respondents said that a good and friendly relationship with colleagues increases organizational commitment, but some respondents, on the contrary, disagreed with this idea. They claimed that an employee's commitment is based on her love and association with the organization and does not rely on their relationship with the colleagues.

In short, it can be interpreted that participants have different views on organizational culture and its effect on organizational commitment.

Table 2 Two-way contingency table of observation by the author and the coder

Author	Coder	Column1	Column2
	Yes	No	Total
Yes	21	9	30
No	4	15	19
Total	25	24	49

4.6 *Family Duties and Responsibilities*

Maintaining a fine balance between family responsibilities and work responsibilities has been highlighted worldwide and employees, especially women employees face a lot of difficulties in both these areas due to changes in work dynamics. Studies clarify that a person's job and family are the two most important areas, and satisfaction from each of these areas takes a lot of time and energy simultaneously [44]. A balanced work life and family life reduces conflict and increases organizational commitment [45]. The authors tried to concentrate more intensively on this idea to examine the mindset of the female participants of IT sector. The situation has become more complex while working from home especially during the pandemic situation. The items which were included under this construct were child rearing, looking after the elderly members of the house, household activities, and spending quality time with the family. At times project deadlines do not allow the employees to provide quality time to the family which culminates to guilt feeling. Organizational commitment increases when personal and professional lives are balanced.

To sum up, it was observed that participants have a heterogeneous perspective on family responsibilities and organizational commitment.

Table 2 displays the two-way contingency table that was generated based on the authors' and coder's independent coding of the themes.

The Cohen's kappa value of 0.671 (at 95% confidence level, $p = 0.000$) indicates a "substantial" [46] agreement between the authors' themes and the independent coder. This result shows the research's validity in terms of the respondents' acknowledgment of construct-wise themes.

5 Implications

The study's aim is to look into the factors that influence organizational commitment among working women in the information technology (IT) industry, as well as how these factors can be used to determine organizational commitment among women in the IT industry. The relationships between six different parameters thought to be responsible for enhancing organizational commitment were examined in this research, as well as the various sub-parameters that are also responsible for enhancing commitment. Respondents adopted the proposal of an optimized combination of the six factors, which is equally applicable to consider the relationship between the factors and organizational commitment, while finding ways to change the six influencing factors to enhance the value of enhancing organizational commitment.

The concept of importance of emotional intelligence shapes employee attitude for communicating their idea about and acceptance of organizational commitment, while family responsibilities, especially related to the work from home scenario during the pandemic situation, leads to coping with a more complex situation to perceive organizational commitment. The other factors, namely work motivation, role of incentives, psychological well-being, and loneliness during COVID-19 and organizational culture exerted a substantial influence on the organizational commitment of women employees working in IT firms.

Nevertheless, the current research work has produced a unique systematic approach that involves six themes, two of which are relatively new to the study of female employee of IT sector organizational commitment.

This study holds a theoretical influence in the area of organizational commitment, and moreover, the organizational commitment of female employees of IT sector. As women have to take on both family and work obligations, evaluating their level of commitment, which also affects organizational growth, is very critical. In general, the working women agree that they have to take extra pressure to maintain a balanced work family life.

A significantly high Cohen's kappa value suggests a strong degree of agreement between the themes selected separately by the authors and the independent coder. This connection aided in the improvement of the current study's qualitative investigation.

6 Limitations and Future Research

As every other analysis, shortcomings exist here and are required to be taken into account. First, a robust quantitative research is needed to further endorse the proposal. Second, future studies will benefit from the inclusion of women workers from a wider range of organizations. Organizational culture is such an influential factor that can make a substantial change in the statements of the employees. While great care was taken in selecting the companies so that there is little overlap in their organizational contexts, it is possible that not enough time was spent building trust with managers to allow for full and honest representation. Thirdly, the information was not gathered in person, which may impact the overall efficiency of the procedure. Responses might have been different if more time was given for rapport building with the participants.

The present study was conducted on few IT companies of Kolkata. The mentality of working women living in other metropolitan and emerging cities is believed to reflect various perspectives. A number of factors can affect organizational commitment, but only six have been discussed here. More studies should be done in the future in fields that have a direct effect on organizational commitment.

References

1. Marique G, Stinglhamber F (2011) Identification to proximal targets and affective organizational commitment. J Pers Psychol 10(3):107–117
2. Iqbal S, Ehsan S, Rizwan M, Noreen M (2014) The impact of organizational commitment, job satisfaction, job stress and leadership support on turnover intention in educational institutes. Int J Hum Res Stud 4(2):181
3. Mowday RT, Porter LW, Steers RM (2013) Employee—organization linkages: the psychology of commitment, absenteeism, and turnover. Academic press
4. Meyer JP, Allen NJ (1997) Commitment in the workplace: theory, research, and application. Sage publications
5. Chan WY, Lau S, Nie Y, Lim S, Hogan D (2008) Organizational and personal predictors of teacher commitment: the mediating role of teacher efficacy and identification with school. Am Educ Res J 45(3):597–630
6. Ambad SNA, Rimin F, Harbi A (2017) Determinant of organizational commitment among women employees in Malaysia. Int J Acad Res Bus Soc Sci 7(10):2222–6990
7. Ramadoss K, Rajadhyaksha U (2012) Gender differences in commitment to roles, work-family conflict and social support. J Soc Sci 33(2):227–233
8. Allen TD, Herst DE, Bruck CS, en Sutton M (2000) Consequences associated with work-to-family conflict: a review agenda for future research 278–308
9. Greenhaus C, Shaw 2003, Greenhaus JH, Collins KM, Shaw JD (2003) The relation between work–family balance and quality of life. J Vocat Behav 63(3):510–531
10. Bakan I, Büyükbeşe T, Erşahan B (2011) An investigation of organizational commitment and education level among employees. Int J Emerg Sci 1(3):231–245
11. Gbadamosi G, Ndaba J, Oni F (2007) Predicting charlatan behaviour in a non-Western setting: lack of trust or absence of commitment? J Manage Dev 26(8):753–769
12. Khalili A, Asmawi A (2012) Appraising the impact of gender differences on organizational commitment: empirical evidence from a private SME in Iran. Int J Bus Manage 7(5):100
13. George L, Sabapathy T (2011) Work motivation of teachers: relationship with organizational commitment. Can Soc Sci 7(1):90–99
14. Maiti RB, Sanyal SN (2018) Optimizing the role of organizational commitment: a qualitative study in the school education sector. Int J Organ Anal 26(4):669–690
15. García-Chas R, Neira-Fontela E, Varela-Neira C, Curto-Rodríguez E (2019) The effect of political skill on work role performance and intention to leave: a moderated mediation model. J Leadersh Organ Stud 26(1):98–110
16. Warsi S, Fatima N, Sahibzada SA (2009) Study on relationship between organizational commitment and its determinants among private sector employees of Pakistan. Int Rev Bus Res Papers 5(3):399–410
17. Tella A, Ayeni CO, Popoola SO (2007) Work motivation, job satisfaction, and organisational commitment of library personnel in academic and research libraries in Oyo State, Nigeria. Libr philos Pract 9(2):1–16
18. Kumari P, Priya B (2015) The role of emotional intelligence in organisational commitment: a study of banking sector. GJRA-Global J Res Anal 4(11):59–62
19. Javidparvar L, Hosseini TA, Berjisian R (2013) The relationship between emotional intelligence and leadership performance in primary schools managers of Isfahan. Int J Sci Res Publ 3(8):1–10
20. Aghabozorgi A, Mehni MAK, Alipour O, Azizi B (2014) Impact of emotional intelligence on organizational commitment of nurses in the public hospitals of Sanandaj. Indian J Fundam Appl Life Sci 4(3):120–127
21. Petrides KV, Furnham A (2006) The role of trait emotional intelligence in a gender-specific model of organizational variables 1. J Appl Soc Psychol 36(2):552–569
22. Alikhani M, Aghili R, Arzhang N, Yadolahi A, Shafahi S (2015) Examining the relationship between emotional intelligence, organizational commitment and job satisfaction among managers. GMP Rev 16:272–276

23. Collins A, Bronte-Tinkew J, Burkhauser M (2008) Using incentives to increase participation in out-of-school time programs. Child Trends, Washington, DC, pp 1–7
24. Banjoko SA (2006) Managing corporate reward systems. Pumark Nigeria Limited, Lagos
25. Oliver N (1990) Work rewards, work values, and organizational commitment in an employee-owned firm: evidence from the UK. Hum Relat 43(6):513–526
26. Cheah CS, Chong VSW, Yeo SF, Pee KW (2016) An empirical study on factors affecting organizational commitment among generation X. Procedia Soc Behav Sci 219:167–174
27. Shigemura J, Ursano RJ, Morganstein JC, Kurosawa M, Benedek DM (2020) Public responses to the novel 2019 coronavirus (2019-nCoV) in Japan: mental health consequences and target populations. Psychiatry Clin Neurosci 74(4):281
28. Machokoto W (2020) A Commitment under challenging circumstances: analysing employee commitment during the fight against Covid-19 in the Uk. Int J Adv Res 8:516–522
29. Franzway S (2000) Women working in a greedy institution: Commitment and emotional labour in the union movement. Gend Work Organ 7(4):258–268
30. Singh V, Vinnicombe S (2000) Gendered meanings of commitment from high technology engineering managers in the United Kingdom and Sweden. Gend Work Organ 7(1):1–19
31. Boehm JK, Kubzansky LD (2012) The heart's content: the association between positive psychological well-being and cardiovascular health. Psychol Bull 138(4):655
32. Aktaş E, Çiçek I, Kıyak M (2011) The effect of organizational culture on organizational efficiency: the moderating role of organizational environment and CEO values. Procedia Soc Behav Sci 24:1560–1573
33. Jandeska KE, Kraimer ML (2005) Women's perceptions of organizational culture, work attitudes, and role-modeling behaviors. J Manag Issues 461–478
34. Campbell DJ, Campbell KM, Kennard D (1994) The effects of family responsibilities on the work commitment and job performance of non-professional women. J Occup Organ Psychol 67(4):283–296
35. Scandura TA, Lankau MJ (1997) Relationships of gender, family responsibility and flexible work hours to organizational commitment and job satisfaction. J Organ Behav: Int J Ind, Occup Organ Psychol Behav 18(4):377–391
36. Shiva G (2013) A study on work family balance and challenges faced by working women. IOSR J Bus Manag 14(5):1–4
37. Marshall B, Cardon P, Poddar A, Fontenot R (2013) Does sample size matter in qualitative research? a review of qualitative interviews in IS research. J Comput Inf Syst 54(1):11–22
38. Gagné M, Deci EL (2005) Self-determination theory and work motivation. J Organ Behav 26(4):331–362
39. Ćulibrk J, Delić M, Mitrović S, Ćulibrk D (2018) Job satisfaction, organizational commitment and job involvement: the mediating role of job involvement. Front Psychol 9:132
40. Akpan CP (2013) Job security and job satisfaction as determinants of organizational commitment among university teachers in cross river state, Nigeria. Br J Educ 1(2):82–93
41. Dani S, Singhai M A study of emotional intelligence of working women and housewives. Editorial Board 4(2):14–26
42. Al-Nsour M (2012) Relationship between incentives and organizational performance for employees in the Jordanian Universities. Int J Bus Manage 7(1):78–89
43. Warrick DD (2017) What leaders need to know about organizational culture. Bus Horiz 60(3):395–404
44. Rathi N, Barath M (2013) Work-family conflict and job and family satisfaction: Moderating effect of social support among police personnel. Equality, Divers Incl: Int J 32(4):438–454
45. Casper WJ, Harris C, Taylor-Bianco A, Wayne JH (2011) Work–family conflict, perceived supervisor support and organizational commitment among Brazilian professionals. J Vocat Behav 79(3):640–652
46. Landis JR, Koch GG (1977) The measurement of observer agreement for categorical data. Biometrics 159–174

Comparative Analysis of Brain Tumor Segmentation with Fuzzy C-Means Using Multicore CPU and CUDA on GPU

Sahana, S. Sowmya, and V. Narendra

Abstract Magnetic resonance imaging is widely applied in medical practice. It has become a difficult task to divide the brain's image into distinct groups due to the symbiosis of intensity and noise. In recent years, due to the enhanced soft tissue contrast of non-invasive imaging and magnetic resonance imaging (MRI) images, MRI-based brain tumor segmentation studies are gaining more attention. With nearly two decades of development, innovative approaches to use computer-aided techniques to the field of brain tumors are becoming more mature and approaching common clinical applications. In order to enhance the segmentation performance of MRI brain images, fuzzy C-means (FCM) method based on similarity measurement is implemented in this paper. However, high computational requirements when working with big datasets are the principal problem with these algorithms. GPU today plays a major role in implementing time-consuming algorithms to decrease the complexity of time. With the use of FCM algorithm in GPU reduces the time required for processing the large amount of data, and it is 19 times better than serial execution.

Keywords MRI · FCM · GPU · High computation · Time · Brain tumor segmentation

1 Introduction

Medical imagery is now most important scientific diagnostic and treatment approaches for human brain diseases over the last decades. The diagnostic imaging system such as magnetic resonance imaging is constantly advancing and widely used. MRI is able to give non-invasive, detailed MRI of brain with high contrast and

Sahana (✉) · S. Sowmya · V. Narendra
Department of Computer Science and Engineering, Shri Madhwa Vadiraja Institute of Technology and Management Bantakal, Udupi, Karnataka, India
e-mail: sahanakaranth.cs@sode-edu.in

S. Sowmya
e-mail: sowmya.cs@sode-edu.in

J. M. R. S. Tavares et al. (eds.), *Cyber Intelligence and Information Retrieval*, Lecture Notes in Networks and Systems 291,
https://doi.org/10.1007/978-981-16-4284-5_51

high spatial resolution between different soft tissues. MRI has gained widespread attention for these advantages and is widely used in clinical applications.

Brain tumor is a growth or mass of abnormal cells in the brain. It destroys brain cells and increases softness in the brain. The general tendency is that humans with brain tumors have seen tremendous rise in recent years, due to change in the lifestyle.

When analyzing MRI of brain, a precise division of the anatomical structure and determination of the exact volume of tissue can help identify and diagnose diseases such as schizophrenia, dementia, and multiple sclerosis with morphological changes in brain tissue. MRI of brain is usually characterized by artifacts such as unknown acquisition noise, intensity asymmetry, and partially impact effects. These artifacts make the task of dividing MR brain images difficult. Therefore, there is a fierce competition for computer-aided segmentation which can achieve accurate and optimal results.

Several techniques such as atlas-based techniques, learning-based methods, active form-based methods, statistical parametric mapping (SPM) based on classical voxel, FMRIB automated segmentation tools for segmentation of MR brain images method, and fuzzy clustering-based techniques were suggested. Fuzzy clustering-based algorithms are widely studied and used in the brain images section of MR. On the one hand, tissue distribution in MRI of brain is overlapping and complex, meaning that the information has fuzzy property. It is appropriate to use clustering techniques. Also, fuzzy clustering enables a pixel to have more than one group (or classes) with different members. Such soft separation has an added advantage when processing MRI of brain, because of the partial volume effect the pixel intensity in MRI of brain may be a complex of multiple tissues. Hence, this paper focuses mainly on methods of clustering-based segmentation based on fuzzy C-means (FCM).

The computation time of fuzzy C-means method should be improved. It is possible to enhance the performance of fuzzy C-means method by using parallel implementation of CPU and GPU.

This method is able to identify tumor-induced MR images and the segmentation of that tumorous region. This uses all GPU cores with organized threads, grids, and blocks. The proposed method improves the performance in terms of speed.

2 Background Works

This section focuses on some research work that improved the popular FCM algorithm's segmentation output. This also briefly highlights several methods for enhancing the time complexity of the algorithm for processing the image that was carried out by previous researchers.

Fuzzy-C for the segmentation of synthetic-aperture radar image, means non-local spatial data was developed [1]. Fuzzy means splitting is noise-sensitive. The accuracy of the separation is low due to the nature of spatial distortion in the partitioning of synthetic-aperture radar images. Non-native spatial information for overcoming the

FCM disadvantage is added in the partition. This approach is tested on synthetic-aperture radar images and has a performance gain of 6% relative to current techniques.

A detailed review of cutting-edge brain tumor resection techniques based on MRI is given [2]. Due to the isolation of non-invasive and fine soft tissue from the MR Image, present brain tumor resection procedures execute MRI. It also uses grouping and clustering techniques with specific features and consideration of local spatial details. These methods are intended to give the doctor a tentative test, tumor tracking, and surgical preparation recommendation.

This paper [3] describes the type of performance gain in implementing the basic and essential image processing technique which is achieved by converting the serial programming model, the first-order edge detector operator, to the parallel programming model. The image processing algorithms technique is a kind of algorithm that works substantially in achieving the best pay for CUDA. In general, image processing algorithms mean that one type of computation, as we have seen in edge detection, is often heavily repetitive. Most of those methods are processed into one another individually.

Exact brain tumor segmentation is essential tasks in the production and analysis of medical images since it deals with segmenting tumor region data from brain MRI sequences [4]. A significant technique of segmentation of brain tumors from the MRI image sequences is done by classifying the local window followed by clustering of fuzzy C-means in parallel. GPU built image segmentation using first-order edge detection operators.

The distinct FCM clustering technique to improve contrast enhancement and separate the three main human—white matter, gray matter, and CSF underlying areas—was introduced [5]. The newly introduced technique—"FAAGKFCM" proposed novel objective method, which replaces the usual Euclidean distance metric with the GRBF kernel to use a new WG image.

FRFCM image segmentation technique is introduced to increase the performance and segmentation quality and minimize the impact of image noise [6]. In this, local spatial data of the images was used to enhance the dividing effect. Due to magnetic resonance which is capable of reducing noise while maintaining object appearance, the trade-off between noise reduction and defense against diffusion is easily accomplished. Additionally, this can give good reconstructed results for a variety of images contaminated with noise.

A new method for segmentation of images by mixing the FCM algorithm with a rough-set theory was introduced [7]. In this, the value table of the attributes is built on the basis of the FCM partitioning results with different clustering numbers, and the images are divided into number of smaller regions based on the attribute values' integral relation. Then, by subtracting the value the weight values of each attributes are obtained. Later it is used as the basis for calculating the difference between regions. The similarity of each region is estimated by the similarity relationship defined by the region of difference. Lastly, the relationship described by equality is used to combine regions and complete the image segmentation.

The super-pixel-based fast FCM (SFFCM) algorithm for segmentation of the image color was established [8]. This algorithm is more advanced than other clustering techniques, as it delivers the precise segmentation results and takes short period runtime.

The PFCM method-based similarity measurement with local label data for brain tumor segmentation of the MRI is proposed [9]. This method depends on PFCM, which can detect outliers in the image and defines similarity measurements. Based on similarity of the intraclass and similarity of the interclass, the proposed method can group linear and nonlinear data types. The method proposed may also solve the issue of cluster size sensitivity. In addition, the proposed approach uses local label information to retain more clarity of the image and to reduce the noise effect.

A non-intuitive method of clustering FCM, namely ICFFCM for segmentation of the MRFF images, has been proposed [10]. ICFFCM does not have cluster centers. The clustering is accomplished by adding the proposed similarity of pixel to pixel and the similarity of pixel to cluster. In order to reduce the influence of disturbance on MRI of brain, the resemblance of pixels to pixels considers the spatial values and intensity values.

It has been studied that the fuzzy C-means algorithm and parallel implementation of that algorithm can be used to perform detection and segmentation of tumors in brain MRI study. The fuzzy C-means takes more computation time than other clustering techniques. So it should be improved in terms of its computation time. It is possible to enhance the performance of this algorithm by using parallel implementation of CPU and GPU.

Various approaches have been proposed for retinal detection and segmentation of OD over the last few years. Among all these approaches, it can be classified mainly into two groups, i.e., location and segmentation methods, respectively, based on localizing the OD region from the fundus image and another on estimating the boundary of OD for segmenting it.

3 Methodology

A comparative analysis is carried out using three processors. The performance of the fuzzy C-means algorithm was evaluated by conducting experiments using Python and CUDA on single-core CPU, multicore CPU, and GPU. Experiment was conducted for 250 input images, and the times taken by all the three systems were recorded.

3.1 Single-Core CPU

The whole implementation is carried out in a single thread of CPU. Reading the input image, segmentation (processing), and the producing the output all take place in a single thread. In single-core processing, the process takes place in sequential order

Table 1 Time taken for single-core CPU for different number of images

No. of images	Serial execution
1	8
2	19
4	40
10	90
20	185
50	480
100	970
150	1332
250	2488

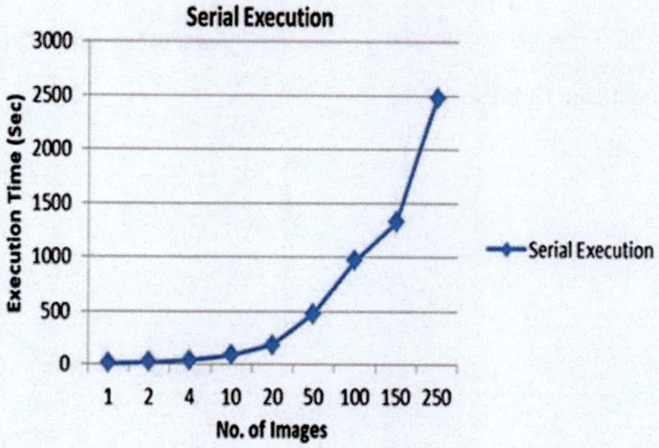

Fig. 1 Number of images verses time taken for serial execution

one after the other. Table 1 shows the serial execution for different number of images and time taken in seconds.

Number of images vs time taken for serial execution is shown in Fig. 1

3.2 *Multicore CPU*

The whole implementation is carried out in a multiple thread of CPU. Taking the input image, segmentation (processing), and the producing the output can take place in a multiple thread simultaneously. In multicore processing, multiple images will be processed simultaneously inside different threads. Table 2 shows the multicore CPU execution for different number of images and time taken in seconds.

Number of images vs time taken for multicore CPU execution is shown in Fig. 2

This performs the same function on both single-core CPU and multicore CPU to check the efficiency of processing speed in it. When single image is passed, the processing speed of single-core processor is faster than multicore processor. In

Table 2 Time taken for multicore CPU for different number of images

No. of images	Multiprocessing
1	11
2	15
4	21
10	40
20	78
50	193
100	389
150	561
250	956

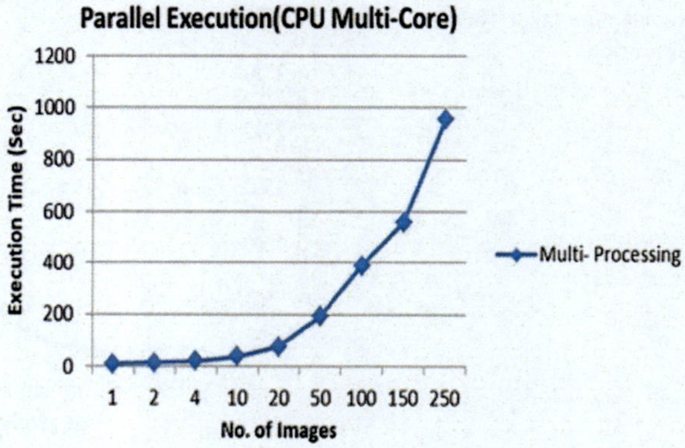

Fig. 2 Number of images verses time taken for multicore CPU execution

multicore processing thread overlapping takes place for single image due to which time taken is more than single-core processor.

The time comparison for different CPU threads for 20 images given in Table 3.

Number of threads vs time taken for 20 images in multicore CPU execution is shown in Fig. 3

Table 3 Time taken with different number of threads

No. of threads	Time (secs)
1	227
2	143
3	112
4	95
6	97
8	95
10	95

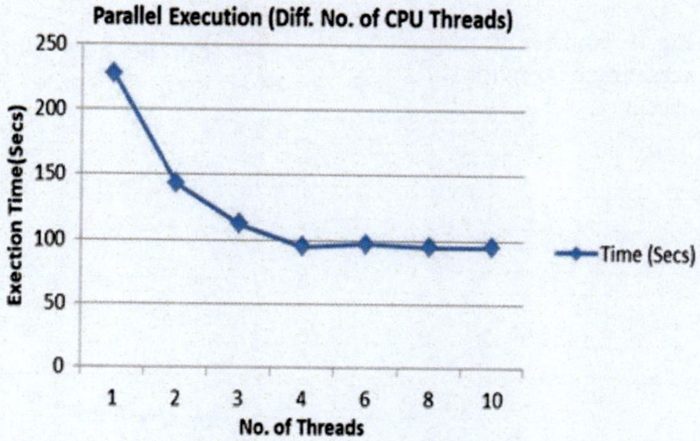

Fig. 3 Number of threads verses time taken for 20 images in multicore CPU

Since the experiment is performed on a dual-core CPU time taken for processing does not change after four threads.

3.3 *GPU Processing*

The entire implementation is divided into two phases. First phase executes on host, i.e., CPU which is responsible for input and output operations and another one executes on device, i.e., GPU which is responsible for processing the brain tumor segmentation. Table 4 shows the GPU processing for different number of images and time taken in seconds.

Number of images vs time taken for GPU execution is shown in Fig. 4

Table 5 shows comparative analysis using three different processors.

Number of images vs time taken for three different executions is shown in Fig. 5

Table 4 Time taken by GPU for different number of images

No. of images	GPU processing
1	1
2	2
4	3
10	5
20	10
50	27
100	54
150	78
250	133

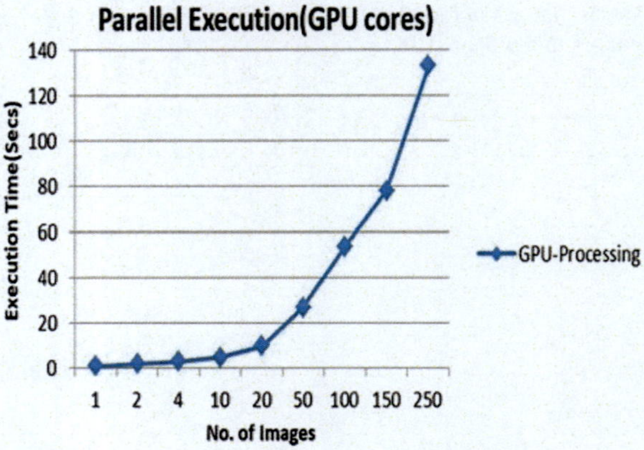

Fig. 4 Number of images verses time taken for GPU execution

Table 5 Time taken by three different processors

No. of images	Serial execution	Multiprocessing	GPU processing
1	8	11	1
2	19	15	2
4	40	21	3
10	90	40	5
20	185	78	10
50	480	193	27
100	970	389	54
150	1332	561	78
250	2488	956	133

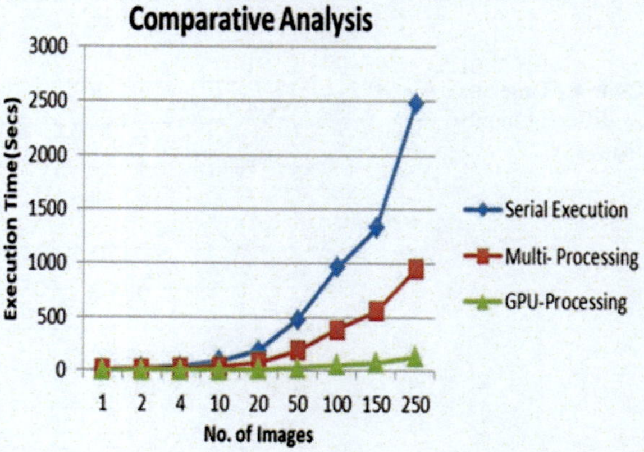

Fig. 5 Number of images verses time taken for three different executions

4 Conclusion

Till now, several techniques exist for brain tumor segmentation. Implementing fuzzy C-means clustering on GPU using CUDA improves the performance in terms of speed. This method is able to identify tumor-induced MR images and the segmentation of the tumorous region. This method also make use of all GPU cores with organized threads, grids, and blocks efficiently. The experimental results show that the performance improvement can be achieved using GPU cores in comparison with single-core CPU and multicore CPU. The speed up achieved by executing fuzzy C-means clustering on GPU using CUDA is 18.7X. For future viewpoint, we can incorporate the machine learning algorithms to detect which type of tumor for the already segmented image. The addition of machine learning for tumor detection with parallel processing will produce accurate result with high efficiency.

References

1. Vimala M, Meena Prakash R, Shantha Selva Kumari R (2015) Segmentation of sar images using fuzzy c means wit non-local spatial information. ARPN J Eng Appl Sci 10. ISSN 1819-6608
2. Liu J, Li M, Wang J, Wu F, Liu T, Pan Y (2015) A survey of MRI-based brain tumor segmentation methods. Tsinghua Sci Technol 19:578–595. ISSNll1007-0214ll04/10ll
3. Saxena S, Sharma N, Sharma S (2016) GPU constructed image segmentation using first order edge detection operators in CUDA environment. J Chem Pharm Res 8(2):379–387. ISSN: 0975-7384
4. Saxena S, Shama S (2018) Brain tumor segmentation by texture feature extraction with the parallel implementation of fuzzy C-Means using CUDA on GPU. In: Fifth international conference on parallel, distributed and grid computing
5. Nongmeikapam K, Kumar WK, Singh AD (2018) Fast and automatically adjustable GRBF kernel based fuzzy C-Means for cluster-wise coloured feature extraction and segmentation of MR images. IET Image Process 12:513–524. https://doi.org/10.1049/iet-ipr.2017.1102
6. Lei T, Jia X, Zhang Y, He L, Meng H, Nandi AK (2018) Significantly fast and robust fuzzy C-Means clustering algorithm based on morphological reconstruction and membership filtering. IEEE Trans Fuzzy Syst 26:3027–3041. https://doi.org/10.1109/TFUZZ.2018.2796074
7. Huang H, Meng F, Zhou S, Jiang F, Manogaran G (2019) Brain image segmentation based on FCM clustering algorithm and rough set. IEEE Access 7. https://doi.org/10.1109/ACCESS.2019.2893063
8. Lei T, Jia X, Zhang Y, Liu S, Meng H, Nandi AK (2019) Superpixel-based fast fuzzy C-Means clustering for color image segmentation. IEEE Trans Fuzzy Syst 27:1753–1766. https://doi.org/10.1109/TFUZZ.2018.2889018
9. Bai X, Zhang Y, Liu H, Chen Z (2019) Similarity measure-based possibilistic FCM with label information for brain MRI segmentation. IEEE Trans Cybern 49:2618–2630. https://doi.org/10.1109/TCYB.2018.2830977
10. Bai X, Zhang Y, Liu H, Wang Y (2019) Intuitionistic center-free FCM clustering for MR brain image segmentation. IEEE J Biomed Health Inf 23:2039–2051

Survey on Podcasting to Improve Teaching Learning Process

A. Priyadharshini, Ashwini Doke, M. Shahina Parveen, and Y. Swathi.Y

Abstract A media file or a series of files that are distributed over Internet and played back on users' computer, mobile phone, or portable media players is termed as a podcast. The survey is about perceptions and impact of podcast on teaching–learning practices. It especially focuses on teaching–learning process. Findings propose that the podcast contents are referred by students especially for assessment preparation and revision. In addition to that podcast helps course instructor to change from traditional teaching methods to recent way of teaching. This paper is result of initial survey of ongoing research in podcasting. We hope to make use of this knowledge to create a design that will bridge a gap and missing links identified.

Keywords Podcasting in engineering education · Podcasting · Educational technology

1 Introduction

1.1 Podcast Definition

It follows push technology, as content is automatically pushed into the subscriber's system there is no need to check for the new content periodically [1]. Nowadays, audio and video applications got momentous acceptance in education industry. Podcast is a new tool that is used to renovate higher education in various fields such as engineering, nursing, geography, psychology, business, and law. It is mainly used as

A. Priyadharshini (✉) · A. Doke · M. S. Parveen · Y. Swathi.Y
Department of Information Science and Engineering, CMR Institute of Technology, Bengaluru, India
e-mail: priyadharshini.a@cmrit.ac.in

M. S. Parveen
e-mail: shahina.m@cmrit.ac.in

Y. Swathi.Y
e-mail: swathi.y@cmrit.ac.in

J. M. R. S. Tavares et al. (eds.), *Cyber Intelligence and Information Retrieval*, Lecture Notes in Networks and Systems 291,
https://doi.org/10.1007/978-981-16-4284-5_52

interview guide [2]. Podcasts can be used as an audiovisual newsletter for a particular course. [3]. Senior students can create podcasts for junior students [4]. 5 minutes of presentation summary can be developed [5]. A short podcast to summarize past and future lecture content as well as reminding students regarding upcoming events and deadlines if any [6]. 3–5 minutes podcast on content has been created by students [4], and for revision, a 5 minutes podcast was developed by students [7]. Watching the videos repeatedly will help. Advantage is flexibility in study [1].

2 Literature Survey

This section describes the survey of literature done in podcasting. The outcome of podcasting is evaluated in three different ways in the paper [2] as follows: (1) Based on the course, questionnaires are formed and targeted toward focus groups and individual interview. (2) Evaluation based on the material access and performance of students. (3) Using made-up use case study.

Advantages of podcast are listed as follows [8]: flexibility, portability, convenience, can listen anytime and anywhere, can do multitasking while listening, can repeatedly listen, playback speed control, automatic RSS subscription, can independently select what to listen, useful for auditory learners and can serve distant students.

Content preparation of podcast needs to be done carefully as there is difference in the learning requirement of learn [9]. Instead of dependent on written notes one can make use of podcast for revision [7]. Need for podcast: (a) watching radio/TV preferably in move, (b) extension of classroom on the Internet, (c) catching up with the classes missed. But this may mislead the students toward skipping classes and watching podcasting which is unjustified.

2.1 Assessing Students

There are two broad categories under students' assessment: (1) No effect or inconclusive results and 2. Positive effects. Based on the assessment survey, the researches [10–12] and [13] have concluded that there is no effect after following podcast, whereas research [14] has concluded that 42 students have experienced higher efficiency in their work.

In another research [15], students of two different classes are taken into consideration, whereas in one class traditional teaching was followed, and in another class, recorded lectures followed by the teacher's feedback on the recorded session were followed. The result shows that students who belong to the second category of class scored high marks compared to the class that followed traditional teaching. In this podcasting, type of teaching workload seems to be higher than in conventional teaching [16].

Table 1 Podcasts listener's location

S. no.	Location	Percentage of students (%)
1	Home	83
2	In transit	14
3	Somewhere else	3

In another work [17], 20 short podcasts have been prepared in technical materials that belong to computer science and engineering, but only 7% fewer students have passed. Students of two different classes [18] one with podcast and one without podcast were taken into account; finally, the podcasting class experienced 12% more top grades and even grade average was higher, and in addition, 2/3 of the students admitted that podcast was helpful to score better grades. In paper [19], two student groups were taken into account, Group1 (count = 32) listened a 25 min lecture and Group2 (count = 34) listened the recording of PowerPoint presentation of the same lecture. Both student groups were assessed by MCQ, and the result shows that the average score of Group1 is 71.24% and Group2 is 62.47%.

2.2 *Impact on Teaching Practice*

Paper [20] suggests to replace the existing in-class teaching methodology to podcasts and to make use of the class time for hands-on in laboratory, which will inculcate experimental learning in students. In addition, researches [21, 17] suggest using compact presentations on theories and methods by leaving more time for classroom discussions. Postgraduate students of University of Sydney, Australia have taken up a questionnaire on podcast [1].

As per the responses, 76% of students have listened to their lecturer's podcasts on their home computers. Student's statements were as follows: (1) Some students have taken notes while listening to the podcasts. (2) Some students listened the podcast while lying on the bed. 2/3 of the students have listened the podcasts for revision. ½ of the students have referred the podcasts for complex content. ½ of the students have listened to podcasts to prepare for assessment. The number of students who have taken the questionnaire is 532. Table 1 depicts the location of listening of podcasts with the percentage of listeners, Table 2 depicts the student's use of podcasting along with percentage, and Table 3 depicts the students' perception about podcasting along with the percentage.

2.3 *Lecturer Perceptions*

According to lecturer, podcasting is used to share information about a particular assessment; time to answer student's queries has been saved; instead of composing

Table 2 Student's use of podcasting

S. no.	Use	Percentage (%)
1	Lecture preparation	13
2	Referring complex content	9
3	Revision	35
4	To replace lectures	3
5	To reinforce information for assessment	55

Table 3 Students' perception about podcasting

S. no.	Perception	Percentage (%)
1	Podcast has enhanced the learning	86
2	Podcast helped to actively engage in learning	79
3	Podcast is used to make learning in student's own time	81
4	Used in more units of study	86

everything in the mail, the lecturer can just write an e-mail and redirect to the podcast; it enables to concentrate on pedagogical design of the content instead of repeating the content [1].

Podcast can be used for various reasons ranging from education, library building, social aspects, entertainment, etc. It is recommended that the advertisers should only use their ads during the starting and ending of the podcast session [10].

We have done initial basic survey on how companies are growing by using SAAS as business model as per Table4.

It has been observed that [22] Enterprise SaaS revenue crossed the $100 billion-mark in 2019 and is growing at an average rate of 30% per annum.

3 Conclusion

Education sector is more profited by the evolvement of podcast in teaching learning process. As it follows learn anywhere and anytime style, students are also getting benefited. In this paper, the survey has been conducted on the existing educational podcast and the findings are as follows: It impacts student's level of understanding, the test grades of students who listened podcast are comparatively higher and the course instructor can use the lecture time to train the students on hands-on which has more impact compare to delivering the content in traditional way. In future, it has been planned to develop interesting short podcasts which require less time to download compared to lengthy podcast. The design and the quality of the podcast have impact over the effectiveness of the podcast which will be time and cost effective for education industry.

Table 4 Revenue generated by SAAS as business model

Company	Launched date	Products	Prizing model	Revenue approx.
G suite Google	First time it launched on 2006 Aug as Google app. But later rebrand as G suite on Sep 27	Gmail, calendar, sites, sheets, forms, slides,	Uses trial ware	\$2.6
Netflix	Aug 1997	Online movies	Initial trial ware, subscriptions	\$20
Drop box	2007	Drop box + with 2 TB, drop box + with 3 TB, Drop box business, drop box enterprise	Trial ware	\$1.90
Sales force	2004	Sales cloud, Service cloud, marketing cloud and commerce cloud	Trial ware	\$17.1
Amazon Prime Video	2005 first time in USA later in 2007 other countries	TV series, blockbuster movies, regional movies, music	Trial ware	\$19.21
Spotify	2006	music streaming	Free as well as premium	€6.76

References

1. Sutton-Brady C, Scott KM, Taylor L, Carabetta G, Clark S (2009) The value of using short-format podcasts to enhance learning and teaching. ALT-J 17(3):219–232. https://doi.org/10.1080/09687760903247609
2. Heilesen SB (2010) What is the academic efficacy of podcasting? Comput Educ 55(3):1063–1068. https://doi.org/10.1016/j.compedu.2010.05.002
3. Bell T, Cockburn A, Wingkvist A, Green R (2007) Podcasts as a supplement in tertiary education: an experiment with two Computer Science courses. In: Parsons D, Ryu H (eds) Mobile learning technologies and applications (MoLTA) 2007. Massey University, Auckland
4. Lee MJW, McLoughlin C, Chan A (2008) Talk the talk: learner-generated podcasts as catalysts for knowledge creation. Br J Edu Technol 39(3):501–521
5. Calder A (2006) If you can't beat 'em, podcast 'em. In: Paper presented at the 9th Pacific Rim–first year in higher education conference, 29 June–1 July, University of Technology, Townsville, Queensland. http://www.fyhe.qut.edu.au/past_papers/2006/Nuts%20&%20Bolts/Calder.pdf
6. Bell T, Cockburn A, Wingkvisit A, Green R (2007) Podcasts as a supplement intertiary education: An experiment with two computer science courses. In: Parsons D, Ryu H (eds) Proceedings of the mobile learning technologies and applications (MoLTA) conference. Massey University, Auckland, New Zealand, pp 70–77

7. Evans C (2008) The effectiveness of m-learning in the form of podcast revision lectures in higher education. Comput Educ 50(2):491–498
8. Jowitt A (2008) Perceptions and usage of library instructional podcasts by staff and students at New Zealand's Universal College of Learning (UCOL). Ref Serv Rev 36(3):312–336
9. Hurst W, Welte M, Jung S (2007) An evaluation of the mobile usage of e-lecture podcasts. In: Proceedings of the 4th international conference on mobile technology, applications, and systems and the 1st international symposium on computer human interaction in mobile technology. ACM, Singapore, pp 16–23
10. Steven McClung PhD, Kristine Johnson MS (2010) Examining the motives of podcast users. J Radio Audio Media 17(1):82–95. https://doi.org/10.1080/19376521003719391
11. Baker R, Harrison J, Thornton B, Yates R (2008) An analysis of the effectiveness of podcasting as a supplemental instructional tool: a pilot study. Coll Teach Methods Styles J 4(3):49–54
12. Abt G, Barry T (2007) The quantitative effect of students using podcasts in a first year undergraduate exercise physiology module. Bioscience Education e-Journal, 10. Apple. (2010). Why podcasting matters. Available from http://www.apple.com/education/podcasting/. Accessed 15 March 2010
13. Hodges CB, Stackpole-Hodges CL, Cox KM (2008) Self-efficacy, self-regulation, and cognitive style as predictors of achievement with podcast instruction. J Edu Comput Res 38(2):139–153
14. Lakhal S, Khechine H, Pascot D (2007) Evaluation of the effectiveness of podcasting in teaching and learning. In: Paper presented at the world conference on e-learning in corporate, government, healthcare, and higher education 2007. Quebec City, Canada
15. Carle AC, Jaffee D, Miller D (2009) Engaging college science students and changing academic achievement with technology: a quasi-experimental preliminary investigation. Comput Educ 52(2):376–380
16. Kurtz BL, Fenwick JB, Ellsworth CC (2007) Using podcasts and tablet PCs in computer science. In: Dicheva D, Pauca VP, Burg J (eds) Proceedings of the 45th annual southeast regional conference. ACM, New York, pp 484–489
17. Smith G, Fidge C (2008) On the efficacy of prerecorded lectures for teaching introductory programming. In: Proceedings of the tenth conference on Australasian computing education, vol 78. Australia Australian Computer Society, Inc., Wollongong, NSW, pp 129–136
18. McCombs S, Liu Y (2007) The efficacy of podcasting technology in instructional delivery. Int J Technol Teach Learn 3(2):123–134
19. McKinney D, Dyck JL, Luber ES (2008) ITunes University and the classroom: can podcasts replace professors? Comput Educ 52(3):617–623
20. Gannod GC, Burge JE, Helmick MT (2008) Using the inverted classroom to teach software engineering. In: Proceedings of the 30th international conference on software engineering. Germany ACM, Leipzig, pp 777–786
21. McKinney AA, Page K (2009) Podcasts and video streaming: useful tools to facilitate learning of path physiology in undergraduate nurse education? Nurse Educ Pract 9(6):372–376
22. https://blog.hubspot.com/service/top-saas-companies

Physiological Signals Based Anxiety Detection Using Ensemble Machine Learning

Vikas Khullar, Raj Gaurang Tiwari, Ambuj Kumar Agarwal, and Soumi Dutta

Abstract Generalized Anxiety disorder (GAD) is a neurological disorder that is mentioned in the Diagnostic and Statistical Manual of Mental Disorders, fifth edition (DSM-V), as well as its prevalence is also on a high note. Throughout the current situation, it's indeed necessary to identify the anxious state to investigate relevant incidents. The aim of this paper is to propose an automatic and intelligent anxiety identification system focused on physiological signals. In methodology, the dataset was collected and cleaned by null removal, duplicate removal, etc. After data pre processing, base machine learning algorithms were compared with ensemble learning models to identify better in terms of metrics.

Keywords Anxiety · Stress · Disorder · Artificial intelligence · Machine learning · Ensemble learning

1 Introduction

GAD (Generalized Anxiety Disorder) is a psychological disorder that impacts people at any age. Constant worry, restlessness, impaired concentration, feeling quickly tired, irritability and some other signs of GAD are also described in the Diagnostic and Statistical Manual of Mental Disorders, fifth edition (DSM-V). In certain cases, GAD coexists with panic disorder, social anxiety disorder, mood disorder and other mental illnesses. Living with GAD may have an effect on everyday tasks such as academic performance, the wellbeing of others, a sense of trust, or other physical complexities,

V. Khullar (✉) · R. G. Tiwari · A. K. Agarwal
Chitkara University Institute of Engineering and Technology, Chitkara University, Punjab, India

R. G. Tiwari
e-mail: rajgaurang@gmail.com

S. Dutta
Institute of Engineering and Management, Kolkata, India
e-mail: soumi.dutta@iemcal.com

J. M. R. S. Tavares et al. (eds.), *Cyber Intelligence and Information Retrieval*, Lecture Notes in Networks and Systems 291,
https://doi.org/10.1007/978-981-16-4284-5_53

among several other factors. The treatment of a diverse variety of mental disorders has been found to be aided by a wide range of technical interventions.

Physiological signals [1] have the strength to provide a precise reading of our body's internal reactions to any raised emotion and some non-invasive biological signals may be used as stress biomarkers. De Santos Sierra et al. have introduced a stress detection method based on physiological evidence. The physiological data used was obtained by non-invasive methods and further trained and tested using a fuzzy logic-based method. The experiment carried out resulted in more than 90% accurate recognition of the stressed condition using the Heart Rate (HR) and GSR data. Author had recommended HR and GSR for real-time stress detection applications [2]. The impact of music on emotions had been studied by Goshvarpour et al. They had studied emotion recognition using Matching Pursuit algorithm and wavelet dictionaries using Electrocardiogram (ECG) and Galvanic Skin Response (GSR) signals with healthy subjects in a musical environment. The calculated emotion was with 100% detection rate in terms of valence and arousal state identification [3].

The combination of sensing technologies and machine learning algorithms have collectively supported stress detection devices. Domínguez-Jiménez et al. had worked for identification of amusement, sadness and neutral as emotions from HR and GSR [4]. Collected signals were analyzed using frequency and time domain to explore features that classified using SVM and Random Forest (RF) classifiers. As result, 100% of accuracy was noted on the test data set. Dillen et al. [5] had analyzed passenger stress during different styles using GSR, HR and eye movement. Further experiments conducted by Ihmig et al. experimented for anxiety level detection using statistical and supervised learning models of machine learning. The training of supervised machine learning models was conducted using collected physiological signals for ECG, GSR and respiration rate. During diverse testing, the trained model resulted in 89.8% of accuracy for binary classification using Bagged trees [6]. Additionally, Rodríguez-Arce et al. had analyzed the anxiety and stress of education institutions to improve the performance of low performing and school dropout students. For data collection, the proposed system was implemented using Arduino along with low-cost sensors related to HR, Skin Temperature (ST) and Oximeter. The collected features were further classified using K-Nearest Neighbour (KNN) and SVM classifiers and approximately 90% accuracy was achieved [7]. Some of the studies were also conducted to identify the reliability and cost-effective non-invasive physiological signals in concern to anxiety detection. Petrescu et al. had proposed the rules for the effective acquisition and processing of physiological signals to measure user anxiety levels [8]. Nath et al. [9] had proposed a Reliable and cost-effective stress detection system using machine learning approaches to control anxiety-related episodes. Author had worked with GSR and Photoplethysmograph (PPG) as biomarkers for the classification of stress levels. Approximately 92% accurate results along with 96% of f1-score had been achieved.

Surveyed literature had mostly resulted in an ensemble or composite methods with an upper edge over base machine learning models. Ensemble algorithms were implemented over the National Epidemiological Survey dataset to identify alcohol-related anxiety or other conditions [10]. Here author had operated with logistic regression, least absolute shrinkage and selection operator (LASSO), RF, Bayesian main-terms logistic regression, generalized additive modelling, adaptive splines, k-nearest neighbours, linear support vector machines and linear discriminant analysis. LASSO algorithm had resulted in best prediction accuracy near 82%. A comparative analysis between different ensemble models was conducted for the WESAD stress dataset prediction related dataset. In this study, Nonlinear Auto-Regressive networks with exogenous inputs (NARX) model performed better than RF and Least-Squares Gradient Boosting (LSBoost) models [11]. A combined clustering and classification techniques had been implemented using "incremental fuzzy cluster ensemble method based on rough set theory (IFCERS)" to handle uncertainty and overlapping datasets efficiently in comparison to other ensemble and base models [12–14]. Zhang et al. compared DT, KNN, LR, SVM models with ensemble model and suggested ensemble model with a higher metric value of area under curve (AUC) [15].

The role of wearable and non-invasive device development is very crucial as there is no need to put it into the human body. Betti et al. [16] had developed and analyzed wearable ECG, EDA and EEG based sensing for stress detection devices over fifteen participants. By implementing Support Vector Machine (SVM) using collected data, the author had achieved 86% of accurate classification. Wireless Internet connected Things (IoT) architecture has been used for health care data with fast access mobility and envisioned to modern smart health care systems by user-friendly clinical applications [17, 18]. Amiri et al. had proposed an Internet of Things (IoT) based framework named WearSense for detecting stereotypic behaviour by extracting features from three-dimension accelerometers of wristband. The conducted experiments had resulted in 94.6% accurate identification related to the stereotypic behaviour [19]. The complex raw data provided by the sensor of IoT devices have been analyzed by using the multilayer structure of artificial intelligence (Li 2018). Khullar et al. had designed and developed an IoT connected wearable bag to identify and analyze the hyperactive behaviour of ASD individuals [20]. Sundaravadivel et al. [21] had implemented an IoT based framework for stress detection using physiological and behavioural signal identification to reduce suicidal thoughts. In the proposed framework, physiological data were collected using commercial wristbands and transmitted and monitored over internet connected devices.

The objective of this paper was to propose physiological signal based anxiety detection using ensemble learning algorithms with better metrics values. Further in this paper, materials and methods are defined to describe dataset collection, pre-processing and utilizing models. In the end, different algorithms are implemented and analyzed to propose a better one.

2 Materials and Methods

In this section descriptions of utilized datasets and implemented pre-processing techniques were discussed. Followed by prediction ensemble model identification and creation was done in comparison to base machine learning models on the basis of parameters as discussed below in detail.

2.1 Dataset Description and Processing

The utilized dataset was collected from open source available [6, 22]. Dataset includes an electrocardiogram, skin conductance and respiration rate as physiological data of 57 individuals during controlled environments with normal and spider-fearful videos. The measurements of physiological signals were done through standard wearable devices at a 100 Hz sampling rate. The measurement of respiration was in percentage with a value range from −50% to 50%, the measurement of Electrocardiogram was in mV with a value range from −1.5 to 1.5 mV and measurement of GSR was in uS with a value range from −12.6 to 41 uS. Here classification labels are marked in binary One if recording done during Spyder video and marked binary Zero when Normal video shown. The collected dataset is attached with labels in a single data file and split all data into training and testing data. Before splitting null values and duplicate values removal was conducted to clean data. Sample data plots for Respiration, Electrocardiogram and Galvanic Skin Response shown in Fig. 1. Nearly equal distribution of stressed and normal dataset classes of 50,000,000 in total is as shown in Fig. 2.

2.2 Predictive Model Selection

In this section, a brief overview of Ensemble predictive models was discussed in comparison to base machine learning techniques. Machine learning techniques are widely used for predictive analytics. Ensemble learning techniques are a combination of several base machine learning models. In reviewed literature, ensemble learning models had resulted in better results in comparison to base machine learning models. In this work initially, we had implemented experimental testing using base machine learning binary classification models such as Logistic Regression (LR), Quadratic Discriminant Analysis (QDA), K-Neighbours Classifier (KNN), Ada Boost Classifier (AB) and Gaussian Naïve Bayes (GNB). As literature suggested the supremacy of ensemble learning models over classic machine learning models is clearly mentioned. So we had implemented ensemble models such as RF, Extra Trees Classifier (ET), Bagging Classifier with Logistic Regression (B-LR), Bagging Classifier with Decision Tree (B-DT) and Voting Classifier with different base classifiers (VC). The efficiency of implemented algorithms was analyzed using various metrics viz., accuracy, precision, recall, F1-Score, zero-one loss, Matthews correlation coefficient and Kappa statistic.

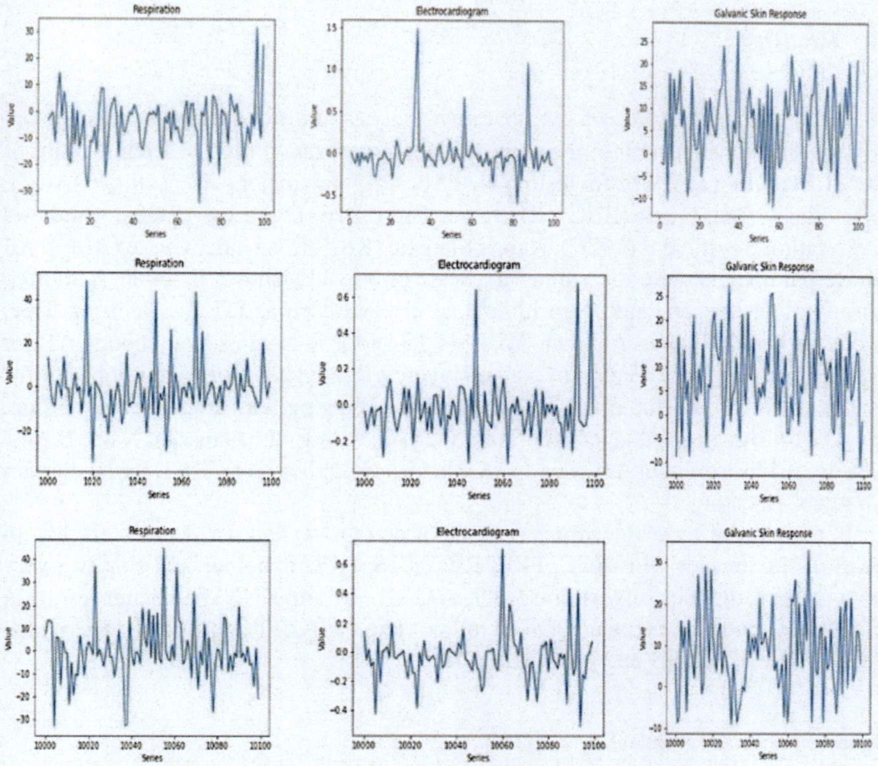

Fig. 1 Respiration, electrocardiogram and galvanic skin response sample data

Fig. 2 Distribution of stressed and normal dataset categories

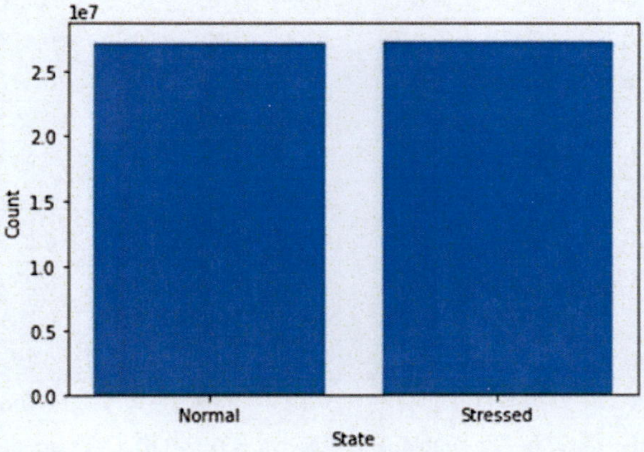

3 Results

In this section K-fold ($K = 5$) comparative analysis of different models was applied to identify a better ensemble or base model. Data were randomly divided into ratios of 70% for training and 30% for testing on the basis of parameters viz. accuracy (ACC), precision (PRE), recall (REC), F1-Score (F1S), Zero-One Loss (ZOL), Matthews Correlation Coefficient (MCC), Kappa Statistic (KS), Area Under Curve (AUC). All above parameters were measured on a scale of 0 to 1 as shown in Table 1. Shorten acronyms of implemented algorithms had utilized such as DT for Decision Tree, QDA for Quadratic Discriminant Analysis, KNN for K-Nearest Neighbours, AB for Ada Boost, GNB for Gaussian Naïve Bayes, LR for Logistic Regression, RF for Random Forest, ET for Extra Trees, B-LR for Bagging with Logistic Regression, BagDT for Bagging with Decision Tree, VOT for Voting with Gaussian Naïve Bayes, Quadratic Discriminant Analysis and K-Nearest Neighbours and HGB for Histogram Gradient Boosting.

In the case of base algorithms (LR, GNB and QDA) had performed very low in most of the metrics with ACC, PRE, REC, F1S, AUC is below 0.61 that is a very poor score. Other base algorithms (such as GNB, DT and KNN) had better results in comparison to LR, GNB and QDA in terms. Here GNB, DT and KNN had provided ACC, PRE, REC, F1S and AUC near 0.75.

Table 1 Metric and model wise results

Tech	Algo	ACC	PRE	REC	F1S	ZOL	MCC	AUC	KP
B	LR	0.55	0.55	0.55	0.55	0.451	0.099	0.559	0.098
E	B-LR	0.55	0.55	0.55	0.55	0.451	0.099	0.559	0.099
B	GNB	0.56	0.58	0.56	0.54	0.438	0.138	0.611	0.125
B	QDA	0.56	0.58	0.56	0.54	0.437	0.139	0.612	0.126
E	AB	0.65	0.65	0.65	0.65	0.354	0.292	0.693	0.292
B	DT	0.76	0.76	0.76	0.76	0.237	0.526	0.764	0.526
E	HGB	0.74	0.75	0.74	0.74	0.258	0.491	0.821	0.485
B	KNN	0.77	0.77	0.77	0.77	0.227	0.545	0.834	0.545
E	VOT	0.77	0.77	0.77	0.77	0.230	0.540	0.843	0.540
E	ET	0.8	0.8	0.8	0.8	0.205	0.592	0.876	0.591
E	RF	0.79	0.79	0.79	0.79	0.209	0.584	0.878	0.582
E	BagDT	0.79	0.79	0.79	0.79	0.208	0.585	0.879	0.584

B—Base, E—Ensemble, DT—Decision Tree, QDA—Quadratic Discriminant Analysis, KNN—K-Neighbours (n_neighbours = 3), AB—Ada Boost, GNB—Gaussian Naïve Bayes, LR—Logistic Regression, RF—Random Forest, ET—Extra Trees, B-LR—Bagging (LogisticRegression, n_estimators = 50), BagDT—Bagging (Decision Tree, n_estimators = 50), VOT—Voting (Gaussian Naïve Bayes, Quadratic Discriminant Analysis, K Neighbours), HGB—Histogram Gradient Boosting

ACC—Accuracy, PRE—Precision, REC—Recall, F1S—F1Score, ZOL—Zero-One Loss, MCC—Matthews Correlation Coefficient, KS—Kappa Statistic, AUC—Area Under Curve

Though some other ensemble techniques (i.e. AB, HGB and VOT) had average results, even B-LR had poor results. However as the literature suggested, outcome results of ensemble techniques are quite improved in comparison to base machine learning algorithms. Here ET, RF, BagDT had better outcomes, more than 0.79 in concern to the metrics viz., ACC, PRE, REC, F1S and AUC. The metric wise comparative graphs are shown in Fig. 3 and pictorial comparison for the AUC for implemented algorithms is also shown in Fig. 4.

Comparative analysis on the basis of MCC and KP were also conducted as shown in Table 1. Similarly, ET, RF and BagDT had results above other implemented algorithms in terms of MCC and KP as 0.58 with the least ZOL near to 0.2.

4 Conclusion

In this paper, ensemble techniques had been proposed for the implementation of binary classification for stress identification using physiological signals of respiration rate, electrocardiogram and galvanic skin response. The data set had included a very high number of rows and 3 columns that is quite a large dataset. Initially, base machine learning algorithms were implemented and then comparatively analyzed with ensemble-based models for classification. Here, extended tree, random forest and bagged decision tree had provided the best outcome with near to 80% positive results. Since the analysis was conducted in this paper, the proposed utilization of extended tree, random forest and bagged decision tree had been recommended for further research. In the future, scalable platforms or big data platforms could be implemented.

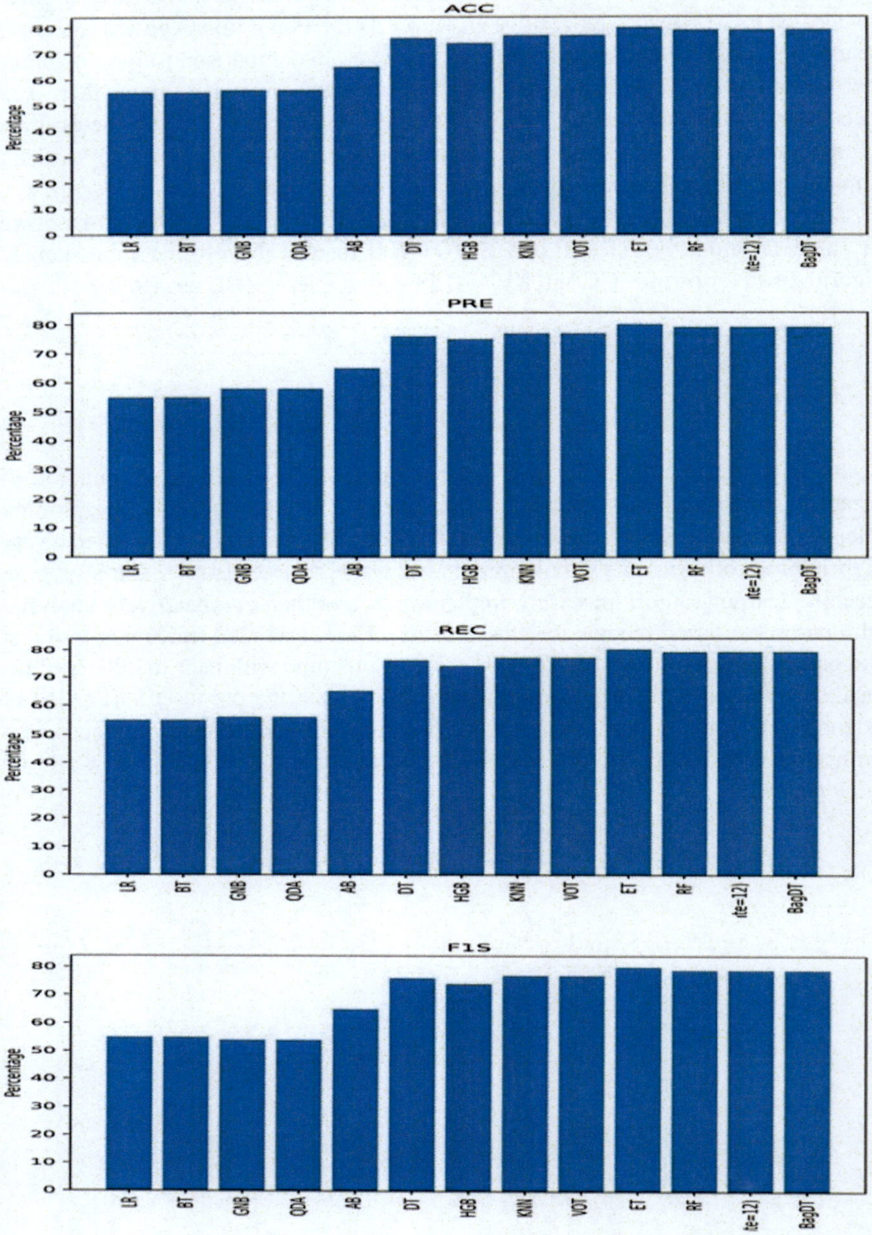

Fig. 3 Metric wise comparative graphs

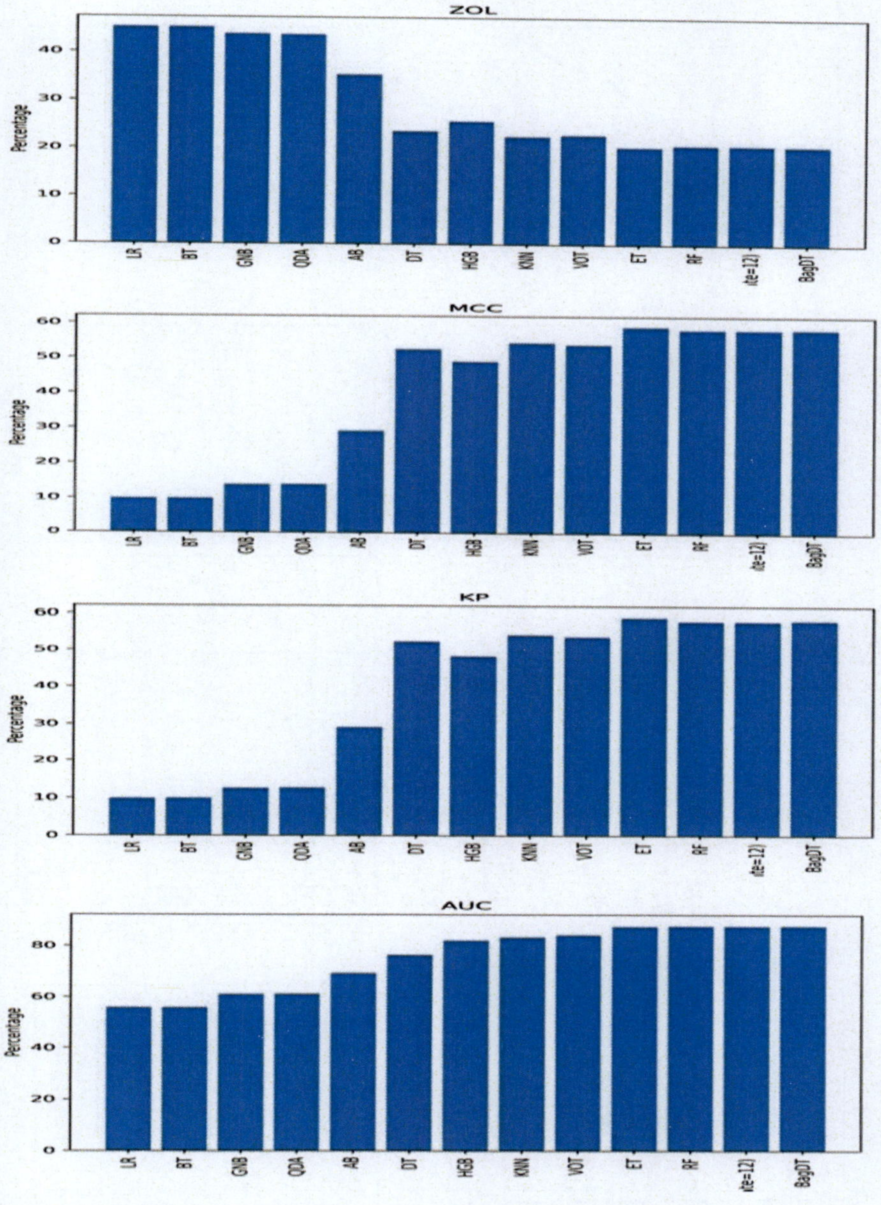

Fig. 3 (continued)

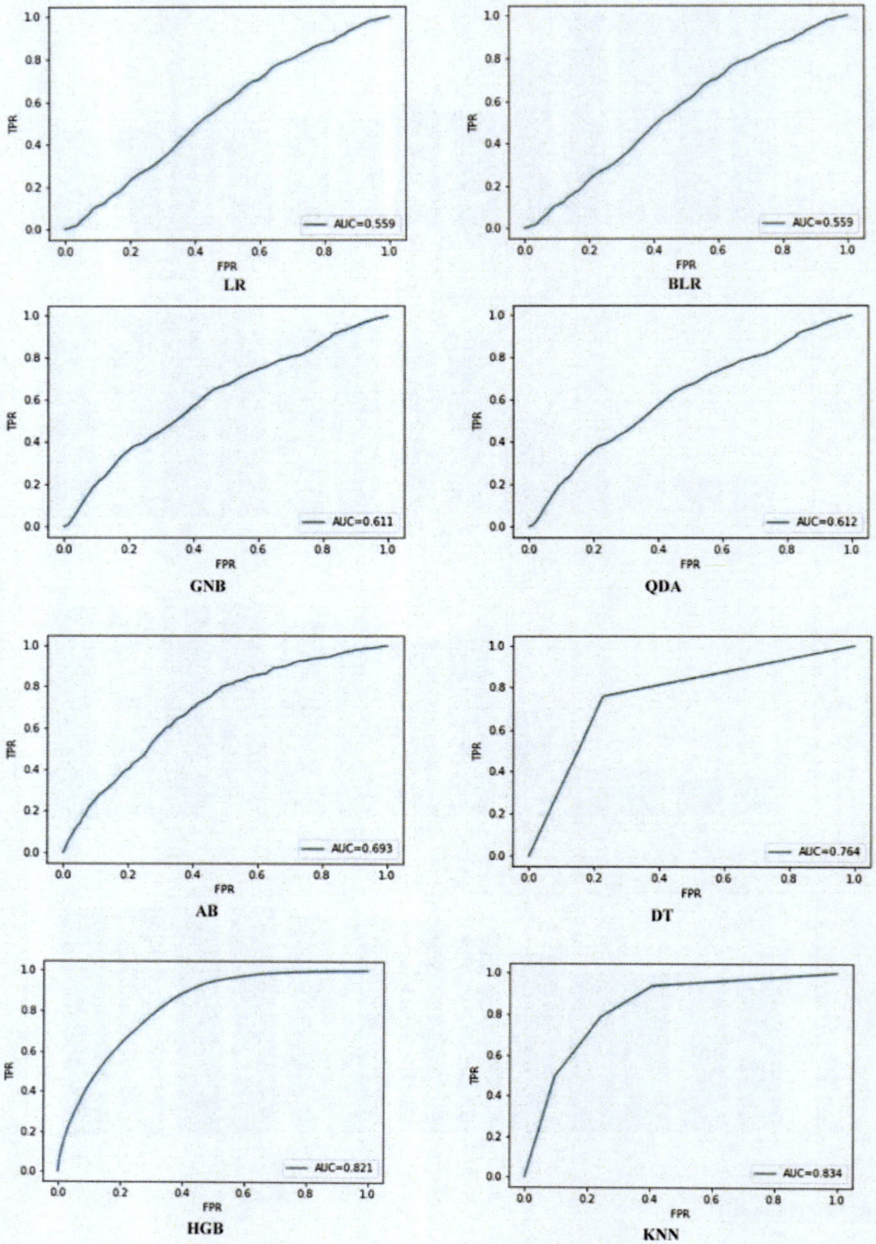

Fig. 4 AUC comparative analysis for implemented models

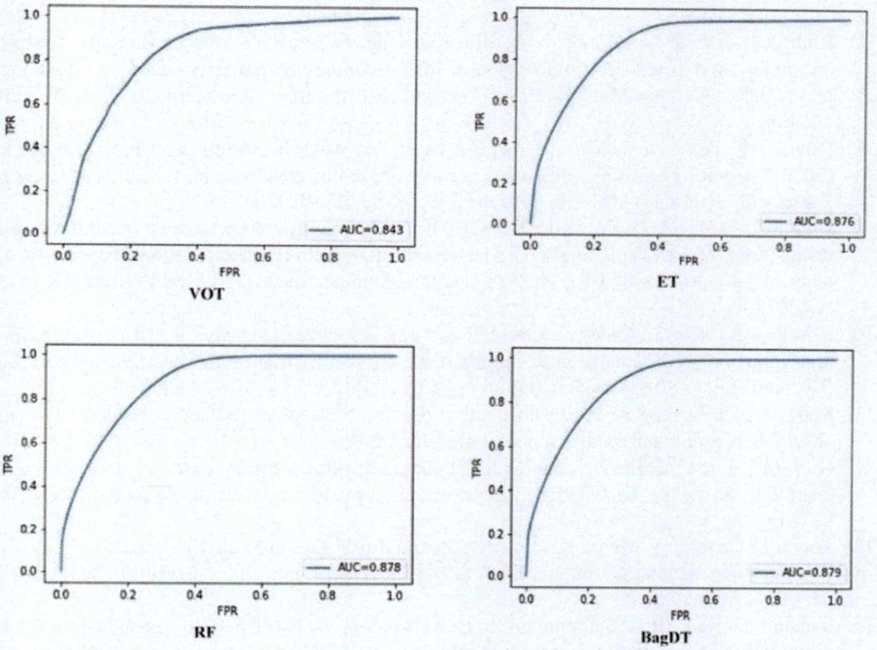

Fig. 4 (continued)

References

1. Bhattacharjee S, Ghatak S, Dutta S, Chatterjee B, Gupta M (2019) A survey on comparison analysis between EEG signal and MRI for brain stroke detection. In: Abraham A, Dutta P, Mandal J, Bhattacharya A, Dutta S (eds) Emerging technologies in data mining and information security. Advances in intelligent systems and computing, vol 814. Springer, Singapore. https://doi.org/10.1007/978-981-13-1501-5_32
2. De Santos Sierra A, Sánchez Ávila C, Guerra Casanova J, Bailador Del Pozo G (2011) A stress-detection system based on physiological signals and fuzzy logic. IEEE Trans Ind Electron 58:4857–4865. https://doi.org/10.1109/TIE.2010.2103538
3. Goshvarpour A, Abbasi A, Goshvarpour A (2017) An accurate emotion recognition system using ECG and GSR signals and matching pursuit method. Biomed J 40:355–368. https://doi.org/10.1016/j.bj.2017.11.001
4. Domínguez-Jiménez JA, Campo-Landines KC, Martínez-Santos JC, Delahoz EJ, Contreras-Ortiz SH (2020) A machine learning model for emotion recognition from physiological signals. Biomed Sig Process Control 55:101646. https://doi.org/10.1016/j.bspc.2019.101646
5. Dillen N, Ilievski M, Law E, Nacke LE, Czarnecki K, Schneider O (2020) Keep calm and ride along: passenger comfort and anxiety as physiological responses to autonomous driving styles. In: Proceedings of the 2020 CHI conference on human factors in computing systems, pp 1–13. https://doi.org/10.1145/3313831.3376247
6. Ihmig FR, Antonio Gogeascoechea H, Neurohr-Parakenings F, Schäfer SK, Lass-Hennemann J, Michael T (2020) On-line anxiety level detection from biosignals: Machine learning based on a randomized controlled trial with spider-fearful individuals. PLoS One 15:1–20. https://doi.org/10.1371/journal.pone.0231517

7. Rodríguez-Arce J, Lara-Flores L, Portillo-Rodríguez O, Martínez-Méndez R (2020) Towards an anxiety and stress recognition system for academic environments based on physiological features. Comput Methods Programs Biomed 190. https://doi.org/10.1016/j.cmpb.2020.105408.
8. Petrescu L, Petrescu C, Mitruţ O, Moise G, Moldoveanu A, Moldoveanu F, Leordeanu M (2020) Integrating biosignals measurement in virtual reality environments for anxiety detection. Sensors (Switzerland) 20:1–32. https://doi.org/10.3390/s20247088
9. Nath RK, Thapliyal H, Caban-Holt A (2020) Validating physiological stress detection model using cortisol as stress bio marker. In: Dig Tech Pap.—2020 IEEE international conference on consumer electronics (ICCE), 2020-Janua, pp 1–5. https://doi.org/10.1109/ICCE46568.2020.9042972.
10. Rosellini AJ, Liu S, Anderson GN, Sbi S, Tung ES, Knyazhanskaya E (2020) Developing algorithms to predict adult onset internalizing disorders: an ensemble learning approach. J Psychiatr Res 121:189–196. https://doi.org/10.1016/j.jpsychires.2019.12.006
11. Martino FDi, Delmastro F High-resolution physiological stress prediction models based on ensemble learning and recurrent neural networks. 2–7
12. Hu J, Li T, Luo C, Fujita H, Yang Y (2017) Incremental fuzzy cluster ensemble learning based on rough set theory. Knowledge-Based Syst 132:144–155. https://doi.org/10.1016/j.knosys.2017.06.020
13. Soumi D, Chandra V, Mehra K, Das AK, Chakraborty T, Ghosh S (2018) Ensemble algorithms for microblog summarization. IEEE Intell Syst 33(3):4–14. https://doi.org/10.1109/MIS.2018.033001411.
14. Samanta D, Rao Althar R, Pramanik S, Dutta S (eds) (2021) Methodologies and applications of computational statistics for machine intelligence. IGI Global. https://doi.org/10.4018/978-1-7998-7701-1
15. Zhang J, Wang Y, Sun Y, Li G (2020) Strength of ensemble learning in multiclass classification of rockburst intensity. Int J Numer Anal Methods Geomech 44:1833–1853. https://doi.org/10.1002/nag.3111
16. Betti S, Lova RM, Rovini E, Acerbi G, Santarelli L, Cabiati M, Ry SD, Cavallo F (2018) Evaluation of an integrated system of wearable physiological sensors for stress monitoring in working environments by using biological markers. IEEE Trans Biomed Eng 65:1748–1758. https://doi.org/10.1109/TBME.2017.2764507
17. Luo E, Bhuiyan MZA, Wang G, Rahman MA, Wu J, Atiquzzaman M (2018) Privacy protector: privacy-protected patient data collection in IoT-based healthcare systems. IEEE Commun Mag 56:163–168. https://doi.org/10.1109/MCOM.2018.1700364
18. Pandia Rajan J, Edward Rajan S (2018) An Internet of Things based physiological signal monitoring and receiving system for virtual enhanced health care network. Technol Heal Care 26:379–385. https://doi.org/10.3233/THC-171173
19. Amiri AM, Peltier N, Goldberg C, Sun Y, Nathan A, Hiremath SV, Mankodiya K (2017) WearSense: detecting autism stereotypic behaviors through smartwatches. J Healthc 5
20. Khullar V, Singh HP, Bala M (2019) IoT based assistive companion for hypersensitive individuals (ACHI) with autism spectrum disorder. Asian J Psychiatr 46:92–102. https://doi.org/10.1016/j.ajp.2019.09.030
21. Sundaravadivel P, Salvatore P, Indic P (2020) M-SID: an IoT-based edge-intelligent framework for suicidal ideation detection. In: IEEE world forum internet things, WF-IoT 2020—symposium proceedings, pp 1–6. https://doi.org/10.1109/WF-IoT48130.2020.9221279.
22. Ihmig FR, Gogeascoechea A, Schäfer S, Lass J, Michael T (2021) Electrocardiogram, skin conductance and respiration from spider-fearful individuals watching spider video clips (v 1.0.0) 1–8

A Survey on Master Data Management Techniques for Business Perspective

Saravjeet Singh and Jaiteg Singh

Abstract Master data (MD) is operational data of the organization which requires minimal updates. Master data is a very crucial element of the organization and is handled using the master data management (MDM). MDM is a process, which is used to enhance the quality of data and process flow in an organization. MDM is used to control the quality of data and creates restrictions on data usage. Master data management helps to maintain the data integrity and provides a single copy of master data. Many organizations are using MDM for business purposes; but still, master data is unable to handle the poor data in the organization. This poor data creates a negative impact on the organization. Due to cost constrain, many small and medium size (SMEs) enterprises are unable to adopt the MDM process. In this paper, we first explain the different aspects of master data, and then we highlight the major studies performed in the field of MDM. Through this study, we provided four major observations in context to the research trends and future research.

Keywords Master data · Organization data · Small and medium size enterprises · Meta data · Data processing

1 Introduction

For an enterprise activities related to quality assessment, updating and enhancement are dependent on master data. Master data (MD) is operational data of the organization which requires minimal updates. Changes in MD are very difficult to implement, and to perform changes in MD, special provision is required. MD is a kind of operational data that has high worth, characterizes center data that helps in a basic dynamic and taking care of business forms over the venture. Master data includes persons details (for example, provider, shopper and controller, accomplices, representative), things (for example items and materials), cost, and other explicit data of an enterprise. Master data management (MDM) is a process used to handle the master data. MDM is

S. Singh (✉) · J. Singh
Chitkara University Institute of Engineering and Technology, Chitkara University, Rajpura, Punjab, India

J. M. R. S. Tavares et al. (eds.), *Cyber Intelligence and Information Retrieval*, Lecture Notes in Networks and Systems 291,
https://doi.org/10.1007/978-981-16-4284-5_54

responsible for the creation, updating, and deletion of the master data. MDM is a very challenging process due to regulatory and constraints associated with MD. Master data management is used for customer relationship management, client integration, employee relationship management, quality management, and other management activities. All the functions associated with MDM are implemented using association rules. These association rules are related to products and procedures. The concept of master data comes into the existence in the late 1990s. In the early 2000s, MDM acts as a research area and different views and implementation MD were presented by the research community [1].

Based on studied literature organizational data is categorized into two basic parts; domain data and meta data. Domain data includes core operational data. It covers complete details of the functional data of organization. Master data provides the information about business prospective. It gives the base to the transactional data. Master is always non-transactional data, and it provides basic attribute of the business. It includes person, place, price, item, and other enterprise specific data. Transactional data is actual business data. It includes all cash flow and business activities. These activities involve sale and purchase history, employee and client transactions, location and product-related transaction. Transactional data is based on master data. Meta data is data about the data. It provides the information about the data used for the operational and domain purpose. It includes all the details related to data generation and documentation. Meta data is further of two types: informational data and operational data. Reference data is fine grain organizational data. It is derived from all other categories of the data. These data sets are used for the reference purpose of back up purpose. Data store in reference space is very less frequently to change.

Master data can be categorized into four basic types. These are party, place, product, and other organization specific data. Party is type of master data, which includes names of the persons or people involved in organization. These covers customer names, supplier name, producer name, employees name, etc. Master data place category covers information related to the location of enterprises, site of the organization, location of the client, address of the organization, client and supplier, etc. Master data under product categories include product details like product name, price, id, value, and category of the product. Apart from this, master data covers enterprise specific data. It includes incense details, variety and guaranty information, contact information, and other legal information's. Master data management is a way to handle, manage, integrate, and unify the referential data so that it can be used for analytical purpose. MDM process involves following five steps:

- **Profile**: MDM process starts with his step. In this step, data from all the sources is assessed and checked for the integrity. Data quality issues are handled in this step.
- **Consolidate**: In this step, data from different sources is first checked for the compatibility.
- **Cleanse and Govern**: In this step, consolidate data is preprocessed and that involves data cleansing and enrichment, addition of missing data, rule definition, privacy policy determination, access permission allocation, and data enrichment.

Table 1 Comparison of master data management commercial available solution

Attribute	SAP netweaver	IBM infosphere	Oracle MDM suite	Informatics MDM
Stable version and year	7.5	11.6	8.2	10.2
Basic components	Business information warehouse, analytical, and visualization tool	Operational and application server, artifacts, inspector, viewer, and pair manager	Apps for logical, operational, and enterprise data, customer, site, and product hub	Database, application, UI, operation reference store, and reference server
Domains	Domain based or domain neutral	Domain based or domain neutral	Domain based	Customer, product, supplier
Method of use	Customer data integration rich product-content management	Application, standard, collaborative	Up sell, on sell, customer management, product management	iPaaS: Data, API and application integration, data catalog

- **Share**: In this step, master data is provided to all silos based on the access right. This step also insures the consistency of the data.
- **Leverage**: Last step is to control the consistency and identity of the data. This step involves updating and maintenance of master data.

MDM provides integrated and single copy of the master data with 360° view. We provided the analysis (Table 1) of these MDM provider based upon their current version, facility and functioning, methods, domain, and invented years. All these major vendors are competitor in market and used by many enterprises for their ERP, CDI, and data management purpose [2–4]. Table 1 provides comparison of frequently used MDM commercial solutions.

2 Current State of Art

Concept of master data was comes into existence in late 90's. As per the studied literature, Siebel used the concept of data organization, and further, this term derived the concept of master data. SAP started working in the domain of master data and after that many big ventures like IBM, Oracle, Infosys, IBM, Google, Informatics, TCS, and so on started working on MDM. Apart from these big venture, research community also participated in this field. According to studied literature, it is observed that big venture like IBM, SAP, etc., provided the solution of MDM for the large enterprises; whereas after decade, MDM solution was also provided for the medium and small size enterprises [5–8]. To highlight the research journey in the field of master data, we provided an analysis of major publication from year 2008 to till data. Table

2 provides the brief information about the prominent research done in the field of master data management.

In current era, data generated by organizations is of big data type so big data have unstructured transactional data and similarly unstructured master data. This unstructured master data or subset of big data requires advance techniques to handle. All the social sites generate big data and manage their data for analysis and business intelligence, and this analysis data includes master data element that requires semantic or no SQL base storage methods. Semantic-based master data management provides a way to handle one of the big data characteristic that is verity [19, 26–30].

MDM is very frequently used for big organizations, but for SMEs, it is not that famous. One of the biggest challenges to implement MDM for SME is its high cost. Few organizations developed their applications to handle the data using MDM approach, but these solutions do not provide 360° view of data. Moreover, some customized solutions provide MDM facilities but fail to cover all aspects of the organization's data. As per the current research trends of network and storage handling, MDM using the cloud (also its related technologies) is also a key field for research.

3 Observations

In this study, we provided different aspect of master data and research done in the field of master data. After performing this study, we come up with following observations. In this paper, we considered publications from year 1998 to 2000 and observed that research in the field of MDM is a continuous process as shown in Fig. 1. Research in the field of MDM started with lower pace and but after few years, this topic attracted many researchers. After more than 20 year, still research is continued in this filed. As per our observation, it is because of vast application of master data in business activities. Almost all applications are using master data and MDM for the business purpose. As the recent study, MDM is used for business intelligence and machine learning techniques are used to provide the solution of the MDM.

Second observation from this study is based on the type of publication. In most of the research areas, research publications are primarily done in journals and conferences but in MDM field industry publications are much higher than other source of publications. Figure 1 shows the publication made in different sources. Third observation is made on the bases of area in which maximum research is performed. For this analysis, we used web crewel to fetch the primary keywords in the publication. We picked the research articles from the Google scholar and performed the analysis using word cloud as shown in Fig. 2. According the mentioned word cloud, it is observed that prime areas for MDM research are business, enterprise, quality, management, reference, unstructured, and many more. So according to this, we can say that maximum research is done in the field of business, and MDM is primarily used for business only.

Next observation from the study is that, MDM is now also evolved in medium and small size (SME) enterprises. Earlier due to high cost, MDM was only used for

Table 2 Prominent research outcome from literature study

Year	Category	Conclusion
2008	Architecture	Gives concept of master data management and its use for customer data integration. This paper explains implementation of MDM and significant of CDI with MDM [9]
2009	Semantic framework	Ontologies are used for the master data management. Link between various data element was established using the RDF files [10]
2010	Medium size enterprise MDM	Provided a framework consist of four different strategies. These stages are based on problem and data [11]
2012	Add-on to IBM Infosphere for unstructured data	Provided architecture to handle the unstructured data. It shows the correlation between unstructured data and text aware master data [12]
2013	Implemented on a project	Provided step by step approach to be followed to implement the MDM process for an organization. Explained the dependencies involved in MDM and how these dependencies can resolved [13]
2014	Decision model for MDM architecture	Provided decision model to identify the appropriate MD architecture the MDM process [14]
2014	Integration between MD and meta data	Provide way to combine the data warehouse meta data and master data [15]
2015	Maturity matrix to access the MD	This study provided a maturity matrix model to assess the master data. This provided key factor used for the assessment of master data [16]
2015	Provided framework to identify data protection issues	This paper highlights the requirement of data protection rules and implemented a solution to identify the data protection issues in government data [17]
2015	Application developed in ArgoUML IDE	Provided method to validate the master data using incremental approach [18]
2015	Semantic framework for food join	Provided semantic framework to handle the master data of food joint of any small to medium size enterprise [19]
2016	Using MDM for business intelligence	Provided an integrating solution for the business intelligence using the master data. Also highlighted the master data activities responsible for the business intelligence [20]

(continued)

Table 2 (continued)

Year	Category	Conclusion
2016	Model for SME using MD API	Provide the reasons why small and medium size enterprise has low cost ERP system and how master data can help to improve this [21]
2017	Challenges related to MD adoption	Highlight the 15 challenges faced by an organization while selecting and adopting MD [22]
2018	Model for finance sector	Provided the maturity model to enhance the performance finance sector using the master data [23]
2019	Framework to identify the challenges	This study provided the challenges faced by government organization while opting master data. Highlighted the factor which differentiate IT and government section for the adoption of MDM [24]
2020	Framework to handle big data	Provided method to evaluate the big data generated using the data network of any organization [25]

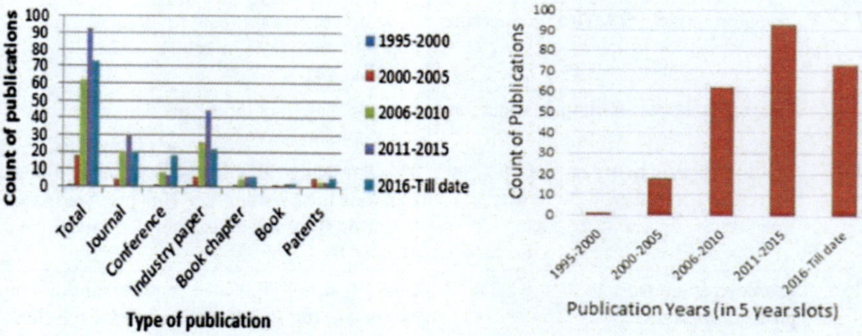

Fig. 1 Publication trends and types from year 1998 to 2020

the large enterprises; but as per recent publication trends, many MDM frameworks were implemented or designed for SMEs. Major concern while design framework for SMEs is efficient and low cost solution. Many semantic and RDF-based solutions provided for the SME's MDM [7, 10, 19, 24].

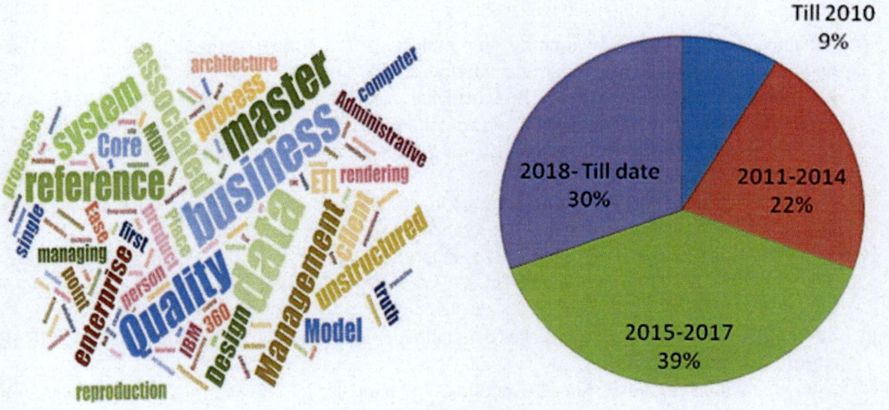

Fig. 2 Frequency of usage of MDM and related elements in research and research trends on SME's

4 Conclusion

Master data is backbone of business, and almost all commercial activities are dependent on master data. To efficiently handle the master data, many solutions and frameworks were designed for MDM by the research and IT industries. These MDM solutions earlier were designed for the large organization, and based on the relational databases but with the fast emergence of unstructured data, these MDM solutions were converted to the hybrid model. These hybrid models were able to handle both structure and unstructured data. Earlier due to high cost SMEs were unable to afford the MDM solution; but based on the recent study, much research is performed for the MDM solution for SMEs. These MDM solutions for SMEs are of low cost and of good efficiency. Semantic and ontological models are also being used by the researcher to make the cost efficient and efficient MDM solution.

References

1. Kokemuller J, Weisbecker A (2009) Master data management: products and research. In: ICIQ, pp 8–18
2. Qodarsih N, Yudhoatmojo SB, Hidayanto AN (2018) Master data management maturity assessment: a case study in the supreme court of the Republic of Indonesia. In: 2018 6th international conference on cyber and IT service management (CITSM), pp 1–7
3. Aditya Rahman A et al (2019) Master data management maturity assessment: a case study of a Pasar Rebo Public Hospital. In: 2019 international conference on advanced computer science and information systems (ICACSIS), pp 497–504
4. Siner V (2020) Master data management in the EU. Comparative analysis of access to base registries
5. Akhmetshin E, Ilyina I, Kulibanova V, Teor T (2019) Special aspects of master data-based integrated management of region reputation in modern IT environment. IOP Conf Ser Mater Sci Eng 497:012022

6. Athanasiadou I (2019) Evaluating the maturity of companies in supplier master data management: the design of a maturity model (2019)
7. Fernando LKW, Haddela PS (2017) Hybrid framework for master data management. In: 2017 seventeenth international conference on advances in ICT for emerging regions (ICTer), pp 1–7
8. Khillari S (2020) Impact of coronavirus on master data management market. Growth, trends and forecast report, 2026
9. Cleven A, Wortmann F (2010) Uncovering four strategies to approach master data management. In: 2010 43rd Hawaii international conference on system sciences, pp 1–10
10. Vilminko-Heikkinen R, Pekkola S (2013) Establishing an organization's master data management function: a stepwise approach. In: 2013 46th Hawaii international conference on system sciences, pp 4719–4728
11. Suram SD, Muppala R (2008) Master data management-CDI. In: TENCON 2008–2008 IEEE region 10 conference, pp 1–6
12. Xiaoyuan, Sun X, Cao F, Ma L, Kanellos N, Zhang K, Pan Y, Wang YY (2009) SMDM: enhancing enterprise-wide master data management using semantic web technologies. In: Proceedings of the VLDB endowment, pp 1594–1597
13. Murthy K et al (2012) Exploiting evidence from unstructured data to enhance master data management. Proc VLDB Endowment 5:1862–1873
14. Spruit M, Pietzka K (2015) MD3M: the master data management maturity model. Comput Hum Behav 51:1068–1076
15. Piedrabuena F, Gonzalez L, Ruggia R (2015) Enforcing data protection regulations within e-government master data management systems. In: ICEIS (3), pp 316–321
16. Baghi E, Schlosser S, Ebner V, Otto B, Oesterle H (2014) Toward a decision model for master data application architecture. In: 2014 47th Hawaii international conference on system sciences, pp 3827–3836
17. Subotic D, Jovanovic V, Poscic P (2014) Data warehouse and master data management evolution—a meta-data-vault approach. Issues Inform Syst 15
18. Myriam LM, Lamolle CLD (2015) Incremental checking of Master Data Management model based on contextual graphs. Enterprise Inform Syst 9(7):681–708
19. Singh J, Singh S (2015) SSMDM: an approach of big data for semantically Master Data Management. In: 2nd international conference on computing for sustainable global development (INDIACom), New Delhi, pp 586–590
20. Haneem F, Kama N, Taskin N, Pauleen D, Bakar NAA (2019) Determinants of master data management adoption by local government organizations: an empirical study. Int J Inf Manag 45:25–43
21. Vilminko-Heikkinen R, Pekkola S (2019) Changes in roles, responsibilities and ownership in organizing master data management. Int J Inf Manag 47:76–87
22. Kekwaletswe RM, Lesole T (2016) A framework for improving business intelligence through master data management. J South Afr Bus Res 2016:1–12
23. Zuniga DV, Cruz RK, Ibanez CR, Dominguez F, Moguerza JM (2018) Master data management maturity model for the microfinance sector in Peru. In: Proceedings of the 2nd international conference on information system and data mining, pp 49–53
24. Park J-K, Gim M-S (2016) Master data management API for assembled ERP tools. J Korea Soc Comput Inform 21:93–101
25. Zhao C, Ren L, Zhang Z, Meng Z (2020) Master data management for manufacturing big data: a method of evaluation for data network. World Wide Web 23:1407–1421
26. Jeong JH, Woo JH, Park JG (2020) Machine learning methodology for management of shipbuilding master data. Int J Naval Archit Ocean Eng 12:428–439
27. Korja K (2019) Developing a product master data management process. Master's thesis
28. Mrigen S, Vikram NB Relevance of master data management in pharmaceutical industries
29. Peeters R, Widlak A (2018) The digital cage: administrative exclusion through information architecture—the case of the Dutch civil registry's master data management system. Gov Inf Q 35:175–183

30. Prokhorov I, Kolesnik N (2018) Development of a master data consolidation system model (on the example of the banking sector). Procedia Comput Sci 145:412–417

Reshaping Education Through Augmented Reality and Virtual Reality

Mahsa Monfared, Vinod Kumar Shukla, Soumi Dutta, and Anjuli Chaubey

Abstract Augmented reality (AR) is the real world to which a virtual world including images sounds, etc. Augmented reality can be defined as a medium connecting digital information with the real world spatially and temporally and is instantly interactive. Virtual reality (VR) is a new technology that allows users to be present and interact with a computer simulation environment. Most virtual reality environments are devices for human-to-computer interaction with computers. Advances in technology in today's world have made a lot of information about our environment turn into a virtual environment. A type of information that has a special place in education today is augmented reality. This form of the virtual world is a living, direct or indirect physical view that adds elements around people's real world. This paper presents the study conducted by data collection and library studies and looks at how augmented reality and virtual reality are used as digital technology in school classrooms around the world. It seems to be one of the most appropriate ways to benefit from education in the age of Covid-19 and the closure of schools in many parts of the world and forcing students to enter the virtual world, which may continue and use these methods.

Keywords Augmented reality · Virtual reality · Education · Technology

M. Monfared
Bachelor of Information Technology, Amity University, Dubai, UAE
e-mail: mahsaM@amitydubai.ae

V. K. Shukla (✉)
Department of Engineering and Architecture, Amity University, Dubai, UAE
e-mail: vshukla@amityuniversity.ae

S. Dutta
Institute of Engineering and Management (IEM), Kolkata West Bengal,, India
e-mail: soumi.dutta@iemcal.com

A. Chaubey
Dr. Harisingh Gour University, Sagar, India

J. M. R. S. Tavares et al. (eds.), *Cyber Intelligence and Information Retrieval*,
Lecture Notes in Networks and Systems 291,
https://doi.org/10.1007/978-981-16-4284-5_55

1 Introduction

We become cyborgs in cyberspace; That is, the space in which the electric computer has transformed humans into virtual humans living in a digital realm. Cyberspace is a metaphor for various technological developments that include the Internet, virtual reality, computer games, and digital databases [1].

Information and communication technology can play a potential role in students' knowledge and skills by applying personalization to learning, a multi-sensory approach, and enhancing interaction with content. Real world virtual content offers a new way to learn and improve common learning quality, making the learning process an interactive, multi-sensory, and enjoyable experience. The opportunities provided by augmented reality technologies offer practical ways to meet the needs of teachers [2]. Augmented reality technology is one of the new learning technologies in the science of educational technology. This technology supports advances in educational sciences as well as advances in educational technology [3].

The twenty-first century is moving in a direction where most jobs will require computer knowledge and skills. Entering this field requires a new type of education that is not in line with the current traditional education, especially what is currently going on in the countries' schools. For some time now, schools in some countries have established e-schools, which took the original model of these schools from the United Kingdom. With this approach, some countries have recently implemented a plan in their schools. Universities have also shifted to another type of education, called virtual education. In innovative schools, the computer influences how teaching and evaluation are done and changes the curriculum. But at the same time, schools' social functions remain in place because they help students in social relationships. In these schools, students learn to process large amounts of information and use this information to learn more. Students can even connect with the world's scientific resources and teachers and students of other schools [4].

It may be possible to take longer steps with the existing conditions using augmented reality and virtual reality technology in education; sometimes, the journey from home to school takes a long time for the student, tedious and boring in the long run. On the other hand, it involves transportation costs and a lot of high costs. Let's put all these issues aside. And help in advancing educational goals. Imagine the use of technology, augmented reality, and virtual reality with courses such as mathematics, geometry, experimental sciences, history, geography, etc. Creating an authentic atmosphere can make the comprehension of the text much more tangible and understandable for the student and the art atmosphere and increase the quality of education. The following section of the paper reflects augmented reality and virtual reality and its applications in the present era, the impact of introducing these two technologies in the application, and the effectiveness of learning programs in education.

2 Augmented Reality (AR)

Augmented reality is a live, direct, or indirect physical view, usually interacting with the user that adds elements to the real world. Augmented reality combines real world entities with computer-generated graphic objects, and its main feature is its promptness that detects, tracks, and interacts with things in a single time. The augmented reality system can be divided into three general parts. The ***first*** part, which is responsible for receiving real environmental information and creates a location system by input sensors such as sound, video, graphics or data. The ***second*** part's task is to receive information from the first part and analyze it and prepare the information needed to add to the real image. The ***third*** section, which finally displays the information. Because smartphones have all three of these components, they are an excellent tool for implementing AR applications [5].

Information is also displayed through augmented reality glasses (augmented reality headsets). These glasses are made by different companies, each of which has a unique name. Google Glass is one of the most attractive wearable products related to the initial production and construction of an augmented reality display and is supported by Google. HoloLens is Microsoft's augmented reality headset. Users who want to talk to others but suffer from long distances can now use this technology to meet and chat face-to-face with the person they want [6].

3 Virtual Reality

Virtual reality is the presentation of the real world in a 3D computer simulator [7]. In other words, virtual reality is a computer simulation of a three-dimensional environment that reconstructs reality using special electronic devices such as a helmet equipped with a sensor and a screen [8]. This concept refers to the complete virtual representation of the real world and the objects within it. In virtual reality, advanced technologies are used to create simulation, interactive, and multidimensional environments [9]. Technically, virtual reality is a computer environment where people operate in a simulated environment and create artificial locations through an interface that stimulates one or more perceptual senses simultaneously. The digital production space is such that user movements are tracked and displayed around it to coordinate user actions [10, 11]. Digitally constructed realities can be as passive as animation or as interactive as virtual reality or video games. This definition confirms three essential features of virtual reality, including immersion, interaction, and imagination [12].

Virtual reality achieves its highest functional efficiency when it helps solve real world problems, offers a creative solution, or enhances human life quality in some way. Virtual reality is the media in the present; Just like the media of music, dance, painting; Which can be used for many purposes. The primary goal of any media is to communicate. Virtual reality can range from purely abstract to fully functional ideas. Like other media, if virtual reality does not help solve problems or provide useful ways to convey messages, ideas or emotions, it is a new emerging technology. It seems that in the near future, this technology become something beyond the media. Recent technological advances show that people's experiences of physical and virtual environments are changing. Virtual reality is likely to play a key role in various industries in the near future [13, 14].

Education Human needs many factors and elements such as objects, other human beings, his natural and social environment, and he/she can meet his social needs about these elements. Thus, in these processes, social phenomena and institutions are created that the social life of human beings is constantly related to these institutions. Education is one of these phenomena and institutions directly related to human nature, personality development, and social life and interacts with social mobility and social change [15].

4 Augmented Reality and Virtual Reality: The Present Era

Augmented reality is a technique that combines a live and instant view with computer-generated virtual images. In this way, a reality is created more than what is seen in appearance. The concept of augmented reality is more about "**reality**" than "**virtual reality.**" Because in augmented reality, there is a reality now, and then facts that are not seen in appearance are added to it. Augmented reality is the surface between the virtual environment (completely virtual) and the perception beyond (completely real).

This technology has been instrumental and widely used in various sciences such as medicine and the military. For example, different doctors see themselves in an operating room environment and perform multiple surgeries virtually, while performing the necessary training by putting special glasses on the eye. Or pilots with this technology can experience the training required in different environments. This experience would probably come at a high cost if it were to be achieved in real life. Psychiatrists use this technology to avoid a unique experience or are afraid of being in a particular environment and treat their patients by placing them in those situations. Augmented reality applications include receiving additional information about a particular restaurant or museum as you approach it. An older example of augmented reality is monitors mounted on people's heads, which provide information about objects in front of a person [16].

Augmented reality includes the following three main features

1. Combining reality and virtual world
2. Immediate reciprocity
3. Create a three-dimensional space

There is no consensus on the above definitions and features, and different people have offered different purposes. In particular, the description of augmented reality is different in other geographical areas and according to the facilities available in those areas. For example, it is possible to use augmented reality facilities on a mobile phone in a geographical location. Still, it is not possible to create a three-dimensional space and interact with other people. Augmented reality is widely used in the non-commercial sciences: for example, in aerospace, this type of technology is used to obtain additional information about the space around a specific spatial area, surgeons use this type of technology for complex and complex surgeries, and soldiers in their wars use this technology to identify the location of enemies.

Despite all the advantages stated for augmented reality, there are many shortcomings and challenges in the implementation and widespread use of this technology. The most critical challenges of advancing this technology are as follows:

- The growth of technology to the extent that it can provide the groundwork for its implementation. Use of appropriate media can bring the ideals of this technology to the forefront [16].

5 Role of Augmented Reality and Virtual Reality in Education

Human beings need to learn the concepts and skills known to the skills of the twenty-first century, such as problem-solving, creativity, creative thinking, systems thinking, etc., to be successful in today's complex world. Therefore, today's intelligent technology, using all the applied sciences, is trying to affect human performance today, which can be improved through education and learning and solve human problems and issues with educational roots in a desirable way. Augmented reality technology is an emerging technology that has just entered the field of education and hopes to create a significant change in the process of teaching, learning, and learning outcomes [6]. One of the advantages of this is that it provides the ability to interact with many students in a three-dimensional environment. This technology (AR), portable and compatible with a variety of hosting scenarios to be used to increase content and education in the traditional classroom, educational supplement in special education classes, content to spread outside the school worldwide, and with other technologies combine to enrich your applications.

Augmented reality technology can bring many benefits, compare different objects in the real world, and enable students to discover information and knowledge of their surroundings. In general, augmented reality is a kind of content-based learning and experience and exploratory learning that may connect new layers of information in the real world. AR allows students to explore the globe interactively and encourages students to collaborate. Structuralists believe that student participation and identity as learners are formed by participation in participatory groups and communities. Structuralism also changed the teacher's role where the responsibility for organizing, integrating, and analyzing content information is in the hands of the learner.

Research on learning technologies has focused on emerging technologies in terms of impact on technology [17]. Augmented reality is a technology that has attracted the attention of those in charge of education around the world. The use of augmented reality in learning has created a new method that enables users to use it without any computer knowledge. Augmented reality brings flexibility to understanding. Augmented reality strengthens communication between learners in different fields and facilitates learning by bringing other disciplines and education [18]. Augmented reality provides a rich visual and verbal learning environment for groups to build knowledge through social interactions. Augmented reality allows students to learn different facts and learn entirely different from traditional classes.

Undoubtedly, students will have better educational, economic, and social opportunities to access computers and information technologies [19, 20]. Augmented educational reality helps students become more aware of the living environment, and the impact of educational simulations on the spread of this be aware and learn. Also, get acquainted with similar IT tools and different methods of collecting, organizing, and presenting information and use these tools independently and in groups. The use of this technology increases students' group interaction and increases their confidence in searching for and accessing information [21].

Augmented reality is gradually gaining popularity in educational communities and has been considered one of the emerging educational technologies. The results of some research in augmented reality indicate the effect of this emerging technology in creating more interaction and activating, satisfying and recognizing the media by students, while learning. This technology has features that can be used in learning as well as designing educational environments.

In connection with augmented reality in education, various researches have been done in real classrooms. With augmented reality technologies, it is possible to combine natural objects with virtual objects and place appropriate information in the natural environment. The use of new augmented reality programs makes it possible to align and coordinate education and entertainment, thus providing new opportunities to support teaching and learning environments in informal settings. Augmented reality allows for the creation of educational experiences based on perception and focuses on the viewer's intellectual and emotional development. The use of augmented reality technology in education to provide lessons to students is one of the best teaching and learning methods. For example, students can view geography, science, and science virtually in a natural environment and easily understand everything. This technology helps students explore and search the real world with virtual elements such as text, video, animation, and image as complementary elements. One of the first applications of augmented reality is to interpret and understand the existing real space with a layer of location-based information. The use of augmented reality in education has made significant progress in recent decades.

On the other hand, with the increase in mobile devices in this field, this technology's introduction is proliferating. Mobile devices allow the use of augmented reality technology in any place, which means that students can learn outside the classroom as much as they learn in the school. With the help of this technology, learners will have no problem visualizing 3D models. With the help of augmented reality, the teacher can increase students' learning power by increasing their attention, concentration, and motivation and helping them better understand the curriculum. Therefore, augmented reality can play an effective role in motivating learning and arousing learners' curiosity.

Virtual reality is one of the newest methods of gaming education. In the virtual environment, all characteristics of activity such as duration, intensity influential of feedback can be changed based on the purpose and ability of individuals. Also, people can view the results of their movement and correct it if necessary. Kell and Summers suggested several years ago regarding the organization of motion that the sound perception system may be superior to the visual system. For example, drummers need to learn the timing and sequence of two-handed movements about audio information to function more efficiently. The audio data role is not limited to the beginning of the movement but can also be used in decision-making processes such as impact selection in racket sports [22].

Virtual reality can be used in a wide range of training topics, from air transport to cross-street. Crossing the street is another skill that can help to have independence in life and be practiced and strengthened as a safe rehabilitation unit. Virtual reality can play an essential role in school homework as an additional factor in a group classroom. Students have learned to understand the concept of space by creating or revealing a sense of place in virtual reality spaces. A number of teachers have linked wearable systems to teaching methods, including the use of Google glasses to visualize and comprehend texts is the best known. Virtual reality education's usual experience often relies on real world implementation of the virtual world to provide a smoother transition from the experimental model similar to reality. In this situation, we can imagine that the real world is under the influence of new technologies, and students with these experiences are presented in the classroom. Most students prefer to learn lessons through video elements and practice a virtual reality. The inclusion of virtual reality wearables in the school enables students to become more familiar with the technologies they use daily and prepare them for the world of digital networking. We can conclude that the closer the virtual reality environment is to reality, and the more students are present and the more motivated. Excited students are to learn with the rich essence of virtual reality [10].

Virtual reality technology has been used as one of the newest technologies today in various fields, universities around the world, including entertainment and education [23]. They have conducted research and studies related to virtual reality-based applications in multiple fields such as machinery, architecture, medicine, physics, design, education, etc [24]. Therefore, it is essential to study the strengths and weaknesses of virtual reality technology to use it properly in various fields, especially in education. In virtual reality technology, the three factors of interaction, imagination, and immersion effectively stimulate the environment for human perception (Fig. 1).

Among these factors, interaction refers to the two-way relationship between humans and the virtual world. Interaction is also focused on optimizing user control. Virtual reality system based on interaction, responds to user input information quickly and based on interaction responds to user input information quickly and accurately, allows users to two-way communication with the computer [25].

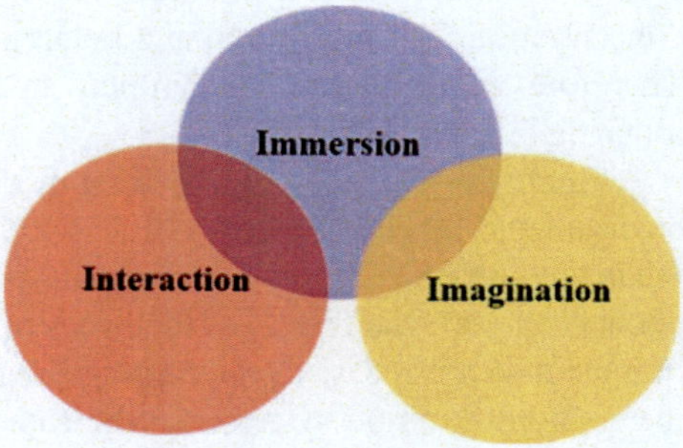

Fig. 1 Factors in in stimulating the environment

Based on this interaction, students' cognitive learning can be improved. The imaging agent also shows how much the user feels like they are in a virtual environment; Even though one knows that one is physically in another setting.

Immersion is one of the important factors in the use of virtual reality technology, the level of user experience as (being in that place) or (losing themselves) (drowning) in the virtual environment and disconnecting from the elements of the physical world it shows [26–28].

6 Virtual Reality and Augmented Reality: Future Ahead

From military training to space exploration and vehicle design, various technologies are ready to evolve in the digital world. Virtual reality and augmented reality still do not live up to their promise [29].

Car construction and design Virtual reality will help car designers and engineers visualize the car model effectively without physically building the model. Technology creates virtual prototypes of car models with complex simulations. Therefore, the performance of the car in actual conditions before construction is examined. For example, Ford uses the Oculus Rift headset in its workflow. Virtual reality technology has been used to create cyberspace by building a car and understanding the vehicle's interior and exterior design before construction. In this way, new ideas and design problems are identified before production.

A new form of journalism: The only way to be able to understand the real news and circumstances of an event is to be present in the situation. Of course, we cannot travel in space and time, but we can fully understand the situation with virtual reality.

Preparing prisoners to enter the world outside prison: A virtual reality project was launched in Colorado for juvenile inmates. This project is done to rehabilitate and rehabilitate prisoners to prepare for the real world. Using Samsung Gear VR, they keep up with the changes in society and have different experiences. So they become acquainted with the changes in the outside world during their imprisonment.

Reduce patients' pain: A new study shows that VR can significantly reduce patients' chronic pain. In this study, pain intensity was recorded in 120 hospitalized patients. Patients were randomly selected to use VR headsets. Virtual reality has been shown to effectively reduce patients' pain.

A new kind of 3D cinema: One of the well-known applications of VR is virtual reality games. VR headsets for long-term gaming still cause dizziness. Gaming companies are trying to combine the gaming environment with 3D cinema and instantaneous movements. The user can interact with the game's virtual characters in the digital world [30].

Rehabilitation of people with disabilities and exposure therapy: In addition to reducing pain, virtual reality technology is used in exposure therapy and treating some diseases such as phobias and post-traumatic stress disorder (PTSD). Research shows that virtual reality exposure therapy helps the user reduce their anxiety about fears and difficult situations. VR is beneficial in rehabilitating people with paralysis. This technology has a significant effect on the user in visualizing and coordinating hands and eyes, restoring brain function, and regaining control of paralyzed organs.

The evolution of medical and military training with VR: Virtual reality will revolutionize the field of healthcare. For example, surgeons practice the surgical procedure before surgery. As a result, virtual reality in medical education makes it easy to visualize and practice. In military training, VR also places soldiers in a simulated combat environment.

Space exploration: NASA researchers use the latest virtual reality technology to remotely control robots on Mars. With the Oculus Rift headset, they allow users to control the robotic arm using hand movements.

7 Conclusion

Today, technology is used by many educators as a teaching aid to make it easier for learners to understand. According to the study and research available, augmented reality is gradually gaining popularity in educational communities and has been considered an emerging educational technology. The results of some research in augmented reality indicate the effectiveness of this emerging technology in creating more interaction and activation, satisfaction, and knowledge of the media by students, while learning. This technology has facilities that can be used in learning as well as designing educational environments, and due to the many benefits of using this technology, its limitations can be ignored, and it is better than this technology to be used alongside other new methods. This technology supports group-based and research-based learning. Augmented reality technology adds excitement and fun to learning activities. Virtual reality technology can be more effective in some components of audience perception. This can express the special place of this technology in future of education. This can also indicate the special place of this technology in future of education. Based on the existing capabilities of virtual reality, it seems that this technology is used in educational topics related to understanding the complexities of the environment, especially in the face of inaccessible and detailed environments that the audience is supposed to feel with their own eyes. Benefited in the environment in which it is present. It also seems that in educational topics that understand the meaning of the environment, virtual reality can be used as an educational supplement.

References

1. Dinloo D, Khademi M (2013) Internet and virtual reality, western tourism. Farvardin 116:115–122
2. Sarami N, Rasti J, Maher M (2015) The application of augmented reality technology as a multi-sensory method in improving dyslexia. In: Second national conference on computer games; opportunities and challenges, Isfahan, University of Isfahan
3. Kabiri AA, Nowruzi N, Hosseini S, Ghasemi A (2017) The need to use information and communication technologies in the philosophy education program for children (P4C). J Edu Technol 4(7 and 6):57–65
4. Niroomand G, Bakhtavari N (2011) The position of new communication technology (smart schools) in education. In: Media studies, vol. 15. Scientific and Research/ISC, pp 95–110. Winter
5. Nodolghi A, Reza M, Khaliq G (2016) Augmented reality in improving educational skills. In: Fifth national conference and fourth international conference on skills and employment
6. Mohebbi F (1398) Augmented reality technology and its application in biology education. Res Biol Edu Fall 3:73–84
7. Çakiroğlu Ü, Gökoğlu S (2019) Development of fire safety behavioral skills via virtual reality. Comput Edu 133:56–68. https://doi.org/10.1016/j.compedu.2019.01.014
8. Zhang M, Zhang Z, Chang Y, Aziz ES, Esche S, Chassapis C (2018) Recent developments in game-based virtual reality educational laboratories using the Microsoft kinect. Int J Emerg Technol Learn 13(1):138–159. https://doi.org/10.3991/ijet.v13i01.7773
9. Sveistrup H (2004) Motor rehabilitation using virtual reality. J Neuroeng Rehabil 1(10):1–8. https://doi.org/10.1186/1743-0003-1-10
10. Sadegh TTDM, Sina KT, Omid H (1398) The study of virtual reality technology capability based on the evaluation of students' visual perception components. In: Higher education, vol. 41. Scientific and Research/ISC, pp 161–199. Spring
11. Innocenti A (2017) Virtual reality experiments in economics. J Behav Exp Econ 69:71–77. https://doi.org/10.1016/j.socec.2017.06.001
12. Bakr AF, El Sayad ZT, Thomas SMS (2018) Virtual reality as a tool for children's participation in kindergarten design process. Alex Eng J 57(4):3851–3861. https://doi.org/10.1016/j.aej.2018.10.003
13. Locke J (1995) Applying virtual reality. IEEE Potentials 14(4):16–18. https://doi.org/10.1109/45.468221
14. Sherman WR, Craig AB (2018) Experience conception and design: applying VR to a Problem. In: Understanding virtual reality, 2nd ed. Elsevier, pp 724–779. https://doi.org/10.1016/B978-0-12-800965-9.00009-X
15. Kaveh M (2012) Pathology of social diseases, vol I, 1st ed. Sociologists Publishing, Tehran
16. (2011) BekBashravieh, Izadpour p 57–59
17. Shukla VK, Verma A (2019) Enhancing LMS experience through AIML base and retrieval base Chatbot using R language. In: 2019 international conference on automation, computational and technology management (ICACTM), London, UK, pp 561–567. https://doi.org/10.1109/ICACTM.2019.8776684.

18. Rajabian Deh Zireh M, Maghami HR, Ismaili Gujar S, Sharifati S The impact of educational augmented reality on lifelong learning and learning performance in students. Learn Edu Technol 9:63–90. Winter
19. Gupta N, Sharma P, Deep V, Shukla VK (2020) Automated attendance system using OpenCV. In: 2020 8th international conference on reliability, Infocom technologies and optimization (Trends and Future Directions) (ICRITO), Noida, India, pp 1226–1230. https://doi.org/10.1109/ICRITO48877.2020.9197936
20. Agarwal P, Shukla VK, Gupta R, Jhamb S (2019) Attendance monitoring system through RFID, face detection and ethernet network: a conceptual framework for sustainable campus. In: 2019 4th international conference on information systems and computer networks (ISCON), Mathura, India, pp 321–325. https://doi.org/10.1109/ISCON47742.2019.9036209
21. Abbasi H (2017) A new perspective on educational technology. J Maturity 1:1–18
22. Lotfi M, Mohammadzadeh H. Sohrabi M (2017) The effect of virtual and real reality practice with and without audio information limitation on learning forehand skills of table tennis. Motor Behav 28:89–108. Summer
23. Dinis FM, Guimaraes AS, Carvalho BR, Martins JPP (2017) Development of virtual reality game-based interfaces for civil engineering education. In: 2017 IEEE global engineering education conference (EDUCON). IEEE, pp 1195–1202. http://refhub.elsevier.com/S22884300(18)30066-6/h0045
24. Satava R (2013) Keynote speaker: virtual reality: current uses in medical simulation and future opportunities and medical technologies that VR can exploit in education and training. In: Virtual reality IEEE, vol 19(4), pp xii-xii. https://ieeexplore.ieee.org/abstract/document/6549339
25. Liang H, Xiaoming B (2013). Application research of virtual reality technology in electronic technique teaching. In: Paper presented at the intelligence computation and evolutionary computation, vol 180, Berlin, Heidelberg, pp 153–159
26. Chen R, Wang X (2008) An empirical study on tangible augmented reality learning space for design skill transfer. Tsinghua Sci Technol 13(1):13–18. http://refhub.elsevier.com/S2288-4300(18)30066-6/h0030
27. D'Souza N, Yoon S, Islam Z (2011) Understanding design skills of the generation Y: an exploration through the VR-KiDS project. Des Stud 32(2):180–209. http://refhub.elsevier.com/S2288-4300(18)30066-6/h0040
28. Burdea GC, Coiffet P (2003) Virtual reality technology. Wiley
29. (2016) Vision of future virtual reality and augmented reality applications. J Power Wearable Technol. Exoskeleton Robots Reference Website
30. Paul PS, Goon S, Bhattacharya A (2012) History and comparative study of modern game engines

Multiregional Segmentation of High-Grade Glioma Using Modified Deep UNET Model with Edge-Detected Multimodal MRI Images

Sonal Gore, Ashwin Mohan, Prajakta Joshi, Prajakta Bhosale, Ashley George, and Jayant Jagtap

Abstract With newly emerging technologies in the field of computer science, there is rising awareness about its applications in the medical sciences. One of such important applications is early and accurate diagnosis of cancer tumor. Brain tumor is a deadly disease and needs to be diagnosed accurately on time. Among many types of brain tumors, high-grade glioma (HGG) is the most belligerent type, and the aggressive nature of tumors affects the survival outcomes of patients. Image processing and deep learning techniques have helped a lot in this endeavor. The proposed work deals with a hybrid methodology that combines simple edge detection technique with deep convolutional neural network to achieve state-of-the-art results. The Brain Tumor Segmentation (BRATS) 2018 dataset is used, which is provided under BRATS contest organized by MICCAI international conference (Medical Image Computing and Computer-Assisted Intervention). The Sobel operator is used for edge detection, and such edge-detected images are further trained using the modified deep UNET (md-UNET)-based model to segment glioma tumor regions into three classes namely—enhancing tumor, non-enhancing necrotic tumor, and edema. The data of 210 patients is used for training the proposed model using train-validate-test split. The model has achieved an accuracy of 98.94%, dice score of 99.03%, and dice loss of 0.0096 for separate test cohort of 6720 multi-modal MRI images of 40 glioma patients.

Keywords MRI · High-grade glioma · Segmentation · Sobel operator · UNET · Enhancing tumor · Non-enhancing tumor · Edema · Necrosis

S. Gore (✉)
Symbiosis International (Deemed University), Lavale, Pune, Maharashtra, India

S. Gore · A. Mohan · P. Joshi · P. Bhosale · A. George
Pimpri Chinchwad College of Engineering, Nigdi, Pune, Maharashtra, India

J. Jagtap
Symbiosis Institute of Technology, Symbiosis International (Deemed University), Lavale, Pune, Maharashtra, India
e-mail: jayant.jagtap@sitpune.edu.in

J. M. R. S. Tavares et al. (eds.), *Cyber Intelligence and Information Retrieval*, Lecture Notes in Networks and Systems 291,
https://doi.org/10.1007/978-981-16-4284-5_56

1 Introduction

A brain tumor is the harmful condition inside the brain to accumulate the unusual development of cells or strong neoplasm. The cancerous tumors with variety of types impact the quality of life up to greater extent [1]. A study conducted by the International Agency for Research on Cancer (IARC) shows that approximately more than 1,26,000 people are diagnosed with brain tumors per year around the world. Nearly 97000 deaths occur due to brain tumors every year [2]. Among various brain tumors, glioma is most aggressive tumor. According to report [3] , the survival rate of glioma patients is very low. In the past few decades, researchers from a wide variety of disciplines like computer science, mathematics, and medical are working to introduce novel methods of detection, classification, and survival predictions of glioma to advance the life quality of cancer patients.

Tumor detection and segmentation is generally carried out using widely used technique termed as magnetic resonance imaging (MRI). MRI with its different modalities help for tumor classification or related predictions non-invasively. Most commonly used MRI modalities are T1-weighted (T1w), T1-weighted with contrast enhanced (T1ce), T2-weighted (T2w), and fluid attenuation inversion recovery (FLAIR). Various 2D or 3D segmentation techniques are proposed to classify the brain cancerous tissues in MR images. Majority of recent techniques have used deep neural network-based models [4, 5], and among these, UNET model is the most popular one [6, 7]. Few earlier techniques have used image processing algorithms related to region or contour detection [8, 9]. The proposed method uses edge detection technique which segments the tumorous tissues from non-tumorous tissues, and it is further followed by multi-region segmentation of glioma tumor into three different regions using modified UNET (md-UNET)-based deep model. These three regions of tumor are namely—edema, enhancing, and non-enhancing tumor which carries different intensity values exhibiting the different features. Such region-wise segmentation definitely further helps to understand the severity and consequences of glioma. This will further facilitate an immediate and most suitable treatment planning for glioma patients. The multi-parametric MRI 2D data is provided as an input to our model to improve the generalizability of model.

2 Related Work

Timely and accurate segmentation of brain tumor is necessary. Many image processing techniques or artificial intelligence (machine learning- and deep learning-based) methods have developed semi-automated or fully automated systems for the tumor segmentation. These automated models help to avoid inter-expert variability in the manual segmentation of tumor. Still there is a scope to design more generalized or more interpretable segmentation methods to facilitate further classification of cancer based on region-specific features.

A Sobel operator-based improved edge detection technique is used for brain tumor segmentation, which has used auto-thresholding method and closed contour algorithm [10]. Sivakumar et al. in [11] presented a computer-aided method with an accuracy of 99.31% that used combination of imroved version of canny edge detection method and enhanced watershed segmentation algorithm. But it has several complex computations involved which makes it tedious. Another notable work [12] combined watershed technique with fuzzy logic to tumor segmentation and classification into benign or malignant type using area of the tumor and color variant contrast. Another study [13] demonstrated an accuracy of 94.2% by using watershed algorithm for tumor segmentation and convolutional neural network (CNN) for classification of tumor and non-tumorous region of multi-modal MRI images of BRATS 2017 dataset. U-Net based architecture, originally proposed by Ronneberger et al., provided more promising results for brain tumor segmentation, and it is an award-winning method at ISBI challenges in the year 2015 [14]. In this regard, Kermi et al. [15] experimented an UNET based model with changes like residual blocks with large padding, no use of pooling layer. It used data augmentation on BRATS 2018 challenge data and yielded mean dice score for complete tumor area, enhancing region, and tumor necrotic core was 0.783, 0.868, and 0.805, respectively. A fully convolutional U-Net-based architecture was designed for multi-regin tumor segmentation and achieved the dice value of 0.83 for complete tumor on BRATS 2015 data [16]. UNET architecture with skip connections provided good results on natural images [17]. This method provides good results on natural images. UNET architecture with five convolutional residual blocks achieved dice coefficient of value 0.9694 using BRATS 2019 challenge data [18].

3 Data Preparation

The imaging data was requested through image processing portal of Center of Biomedical Image Computing and Analytics (CBICA). It is the data released with challenge for multimodal segmentation of brain tumor by Medical Image Computing and Computer-Assisted Intervention (MICCAI) in the year 2018 [19]. The dataset is primarily in 3D format and consists of presurgical MRI scans of 210 high-grade glioma (HGG) patients and 75 low-grade glioma (LGG). The data of each patient contains scans of clinically acquired routine MRI modalities, which consists of T1-weighted (T1w), T1-weighted with contrast enhancement (T1ce), T2-weighted (T2), fluid-attenuated inversion recovery (FLAIR), and it also contains the ground truth of segmented tumor images, which are manually segmented and annotated with four labels by expert neuroradiologist. These annotations include four regions labeled as enhancing tumor (label 4), peritumoral edema (label 2), and the core (label 1) as described in [20]. The core region encompasses necrotic and non-enhancing tumor areas. All these files are provided in NIfTY (Neuroimaging Informatics Technology Initiative) file format. The preprocessed data is distributed as BRATS 2018 challenge data, in which the preprocessing tasks like co-registration, skull stripping, resolution interpolation, etc., are carried out. Since, the presence of multiple tumor regions is more visible with HGGs than LGGs. Therefore, only high-grade glioma is included with our experimentation.

3.1 Data Preprocessing

The BRATS 2018 challenge data is made available in 3D NIfTI format. Every 3D MRI volume is of size 240 × 240 × 155. Each 3D image is sliced into 155 2D slices, each of size 240 × 240, using SimpleITK library available in Python. Only 40 slices covering the maximum tumor area are selected. These 40 slices are further cropped to a size of 192 × 192 to remove unwanted background. Total 8400 2D images of each MRI modality has yielded 33,600 images, which are finally included in our experimentation. The sample preprocessed image of every MRI technique along with its ground truth segmented image is shown in Fig. 1.

4 Method

The proposed work has performed the manual feature extraction by extracting basic feature like edge of tumorous tissue to segment it from non-tumorous tissue. Further, these edge-detected tumorous images are fed as an input to UNET based convolutional neural network model. The architecture of proposed system is shown in Fig. 2.

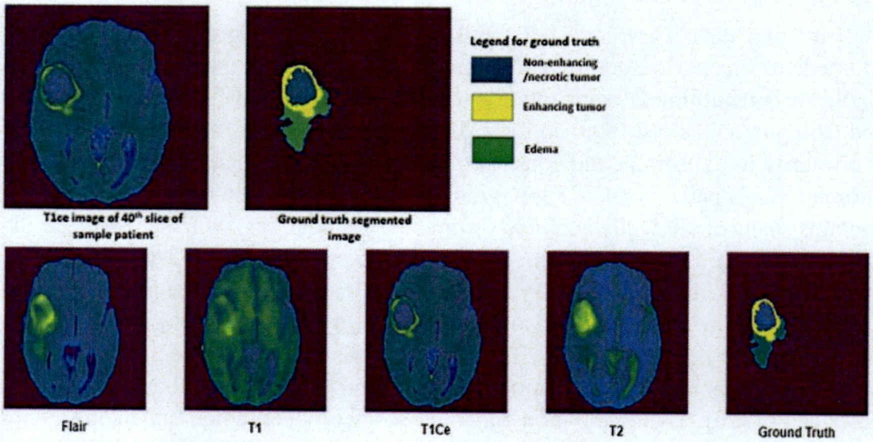

Fig. 1 Sample images of four MRI techniques and ground truth segmentation

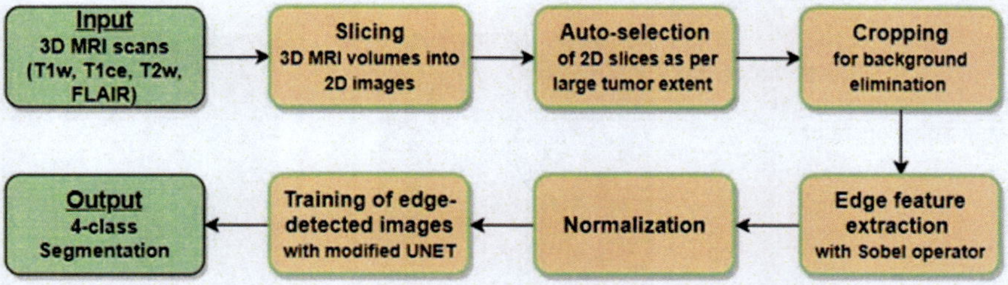

Fig. 2 Flow diagram of proposed system

4.1 Feature Extraction with Edge Detection Method Using Sobel Operator

The edge is a preliminary feature of an image, which help to segment one region from another by recognizing the boundaries with the help of intensity discontinuities in pixel values. A Sobel operator or Sobel filter is an image processing technique used to emphasize edges in an image. It is a discrete differentiation operator, which computes an approximate gradient of image intensity function. It works by convolving the image in both horizontal direction from left to right and vertical direction from top to bottom with a small separable integer-valued filter; hence it is relatively inexpensive in terms of computations. As compared to other edge detection filters, Sobel performs better in terms of good smoothing effect to random noise, which may be inserted due to different artifact effects while acquiring MR images from MRI machine. Secondly, it provides better appearance of tumor region edges due to differential of two rows and two columns. The two filters of size 3×3 used by Sobel operator are shown in Eq. 1, and these are convolved across each image.

$$Gx = \begin{bmatrix} 1 & 0 & -1 \\ 2 & 0 & -2 \\ 1 & 0 & -1 \end{bmatrix}, Gy = \begin{bmatrix} 1 & 2 & 1 \\ 0 & 0 & 0 \\ -1 & -2 & -1 \end{bmatrix} \tag{1}$$

Every image is processed with Sobel operator from Sci-kit learn library, so that the edges are clearly visible to segment tumor region from non-tumorous region. The sample image after applying sobel operator is shown in Fig. 3.

Further, these 2D slices are normalized using an Eq. 2.

$$x' = \frac{(x - \mu)}{\max(x)}, \tag{2}$$

where x' is the normalized value, x is the actual value, μ is the mean of x, $\max(x)$ means the maximum value of x. Then, such preprocessed dataset is trained on U-Net with skip connections.

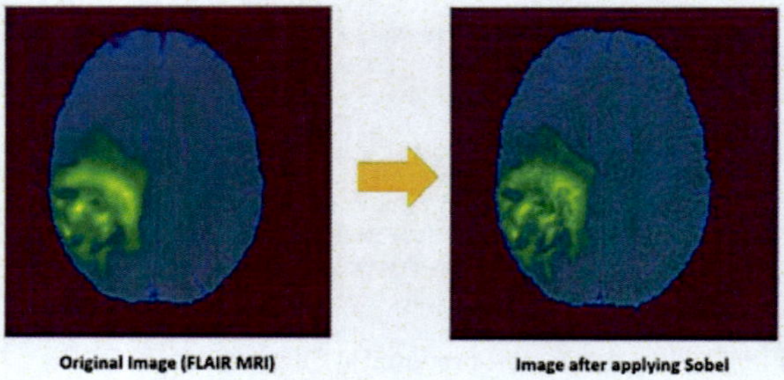

Fig. 3 Edge feature extraction to segment tumor tissue from normal tissue using Sobel operator

4.2 *Auto-feature Extraction of Edge-Detected Images Using Modified Deep U-Shaped Net (md-UNET)based CNN Model*

U-Net is a prevalent architecture majorly used for image segmentation tasks, which is evolved from convolutional neural networks. Long et al. proposed a skip connection-based architecture that has the function of copying the output of past layers to later layers. This helps to utilize the features extracted from earlier layers to the fullest extent by combining (with summation or concatenation) those with input to be fed to far ahead layers of any deep convolutional neural network model [21]. It basically helps to propagate the gradients instead of vanishing those with deeper convolutional layers. In our work, such multiple residual connections are used with original UNET model to combine (with summation) the low-level representation extracted at encoding layer with high-level representation at decoding layer. This makes decoding layers more interpretable to reconstruct the most accurate segmented image at the output of final layer.

U-Net has an encoding and decoding part. Encoding part performs down sampling, and decoding part performs up sampling. Figure 4 shows the schematic representation of modified deep U-shaped network (md-UNET) with multiple skipped connections.

The encoding part consists of five blocks of consecutive convolutional layers. Each block has two 3 × 3 convolutional layers with same padding and activation function as rectified linear unit (ReLU), each followed by a batch-normalization layer and a maximum pooling layer. A dropout layer with dropout rate of 20% is placed after two blocks to avoid overfitting. The sixth block acts as a bridge between the encoding and the decoding part, which consists of a 3 × 3 convolutional layer with same padding and ReLU activation function. The decoding part too consists of five blocks. Each block has a convolutional transpose layer of 3 × 3 voxel with same padding and ReLU activation function merged with skip connection from previous encoder part, followed by a convolution layer 3 × 3 voxel with same padding and ReLU activation. A dropout layer is present after two blocks in the decoding part. Later, the output layer with softmax activation function is used for four-class segmentation. The number of activation filters used are 32, 64, 128, 256, and 512 in five consecutive blocks of encoder as well as decoder path. However, total 1024 activation kernels are used

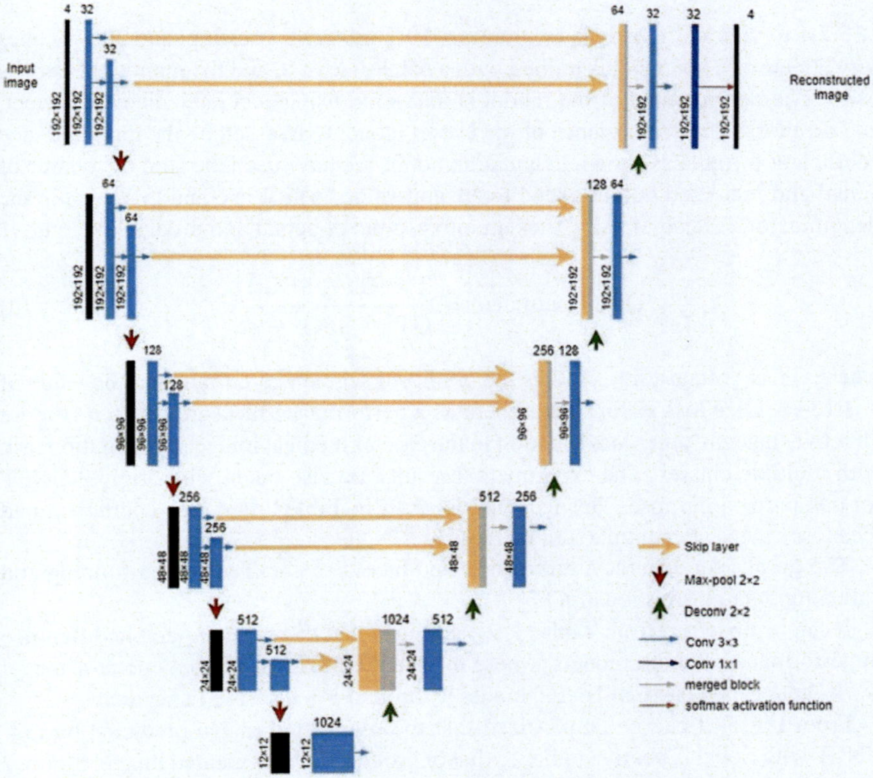

Fig. 4 Modified deep U-shaped network (md-UNET) architecture with multiple skipped connections. Each blue strip represents convolutional layer with multi-feature activation map. The number of feature kernels are written on top of box, and size of input is specified on the lower-left side of each strip. Skip layers are cropped from first five blue-colored boxes of encoder and added with its respective decoding layer

at center of our proposed deep U-shaped CNN model. In total, there are ten blocks each with two residual connections, which comprise of total 21 convolution layers, 10 layers of batch normalization, 5 layers of maximum pooling, and there is finally one convolutional layer at the bridge point of U-shaped network. Adam optimizer is used for optimization at a learning rate of 1e−5, and dice loss is used as a loss function.

5 Experimentation and Results

The proposed experimentation is performed on Google Colaboratory GPU with 25 GB RAM with Keras and Tensorflow library in the backend. The data is split into ratio of 3:1:1 to yield 60% images for training, 20% images for testing, and 20% images for validation. The model is trained with a batch size of 8, and the number of epochs is 30. The performance of the model is measured using accuracy, dice coefficient, and dice loss. The performance of the U-Net model is assessed by the modified dice coefficient formula as shown in equation no. 3. We have used the sum of squares of actual and predicted output added to epsilon value (which is equal to 1e−6) in the denominator, instead of using the sum of absolute of actual and predicted output.

$$\text{Dice Coefficient} = \frac{2 \times |\text{GT} \cap \text{SEG}|)}{\text{GT}^2 + \text{SEG}^2 + \epsilon} \quad (3)$$

where GT is ground truth, SEG is the predicted segmented output, and the value of ϵ is 1e−6. Dice loss is considered here as a performance measure, since it is more effective than any other loss function in the case of medical image segmentation task with multiple classes. The experimental results are also taken with original UNET for comparison purpose. The accuracy is given in Table 1, and dice coefficient and dice loss values are summarized in Table 2.

The graphs for accuracy, dice score, and dice loss obtained during training and validation phases are shown in Fig. 5.

It can be observed from Tables 1, 2, and Fig. 5 that there is comparable difference in performance of both models, where modified UNET with Sobel operator model works better than original UNET model by around 3% increase in accuracy.

From Fig. 6, it can be emphasized that the segmented image predicted by md-UNET with Sobel is nearly similar to that of ground truth segmented image. Figure 7 shows sample image of segmentation output by original UNET and proposed md-UNET with Sobel.

It can be seen that the projected model gives better segmentation output than original UNET.

Table 1 Accuracy of modified U-Net with Sobel operator and of original UNET

Model	Accuracy (%) training	Validation	Testing
Modified UNET with Sobel	99.09	99.08	98.94
Original UNET	95.71	95.67	95.79

Table 2 Dice score, dice loss of modified U-Net with Sobel operator and of original UNET

Model	Dice score training	Validation	Testing	Dice loss training	Validation	Testing
Modified UNET with Sobel	0.9918	0.9917	0.9903	0.0081	0.0082	0.0096
Original UNET	0.9571	0.9567	0.9579	0.0428	0.0432	0.0420

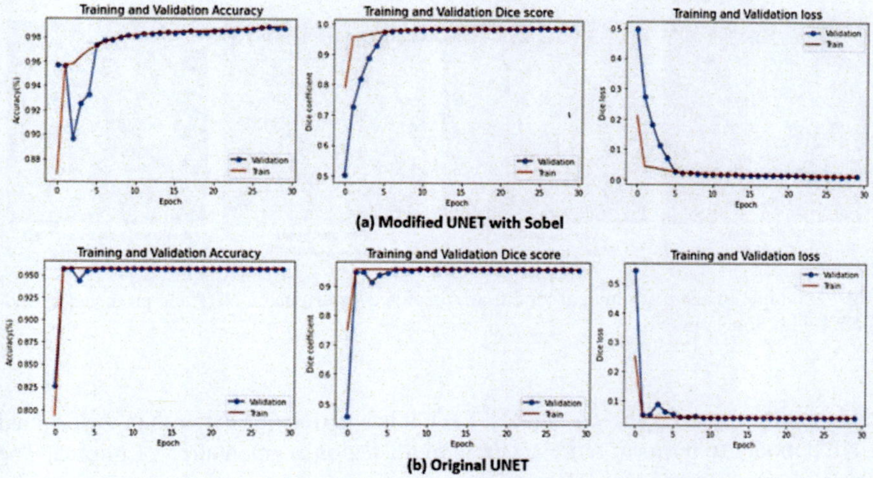

Fig. 5 Graphs for accuracy, dice score, dice loss obtained during training and validation

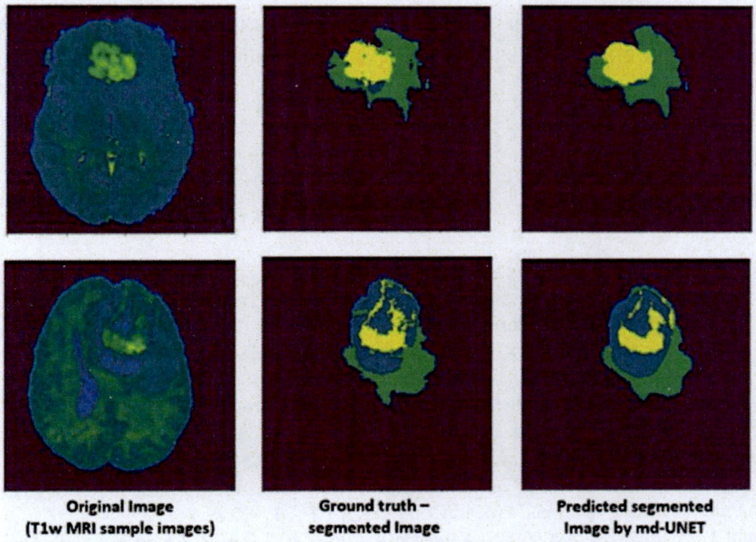

Fig. 6 Sample images: original, ground truth, predicted by md-UNET with Sobel operator

6 Conclusion

The work has proposed here an approach which combines two technologies for segmenting high-grade glioma into multiple regions (enhancing, necrotic, and edema). The experimental results with testing accuracy of 98.94%, dice score of 0.99, dice loss of 0.009 demonstrate that the projected model has performance that is compa-

Fig. 7 Sample images: original, ground truth, predicted by original UNET and predicted by md-UNET with Sobel operator

rable to other state-of-the-art models. Even it has performed better than the original UNET model to perform more accurate multi-region segmentation of glioma. The model combines preliminary feature extraction by using simple, effective Sobel operator with auto-feature extraction of UNET based modified deep convolution neural network model. This model is simple, easy to implement, and computationally efficient. In future, this model can be further enhanced to classify the tumor as per World Health Organization (WHO) specifications and to predict the survival rate of glioma patients.

References

1. Liu R, Page M, Solheim K, Fox S, Chang SM (2009) Quality of life in adults with brain tumors: current knowledge and future directions. Neuro Oncol 11(3):330–339
2. Ferlay J, Shin HR, Bray F, Forman D, Mathers C, Parkin DM (2010) Cancer incidence and mortality worldwide. GLOBOCAN 2008 v2.0 in international agency for research on cancer. Lyon, France
3. Alvarez de Eulate-Beramendi S, Alvarez-Vega MA, Balbin M, Sanchez-Pitiot A, Vallina-Alvarez A, Martino-Gonzalez J (2016) Prognostic factors and survival study in high-grade glioma in the elderly. Br J Neurosurg 30(3):330–336 (2016)
4. Isin A, Direkoglu C, Sah M (2016) Review of MRI-based brain tumor image segmentation using deep learning methods. Proced Comput Sci 102:317–324
5. Bernal J, Kushibar K, Asfaw DS, Valverde S, Oliver A, Marti R, Llado X (2019) Deep convolutional neural networks for brain image analysis on magnetic resonance imaging: a review. Artif Intell Med 95:64–81
6. Zeineldin RA, Karar ME, Coburger J et al (2020) DeepSeg: deep neural network framework for automatic brain tumor segmentation using magnetic resonance FLAIR images. Int J Comput Assist Radiol Surg 15:909–920
7. Lotan E, Jain R, Razavian N, Fatterpekar GM, Lui YW (2019) State of the art: machine Learning applications in Glioma imaging. AJR Am J Roentgenol 212(1):26–37
8. Maiti I, Chakraborty M (2012) A new method for brain tumor segmentation based on watershed and edge detection algorithms in HSV colour model. In: National Conference on Computing and Communication Systems, Durgapur, India, pp 1–5 (2012)
9. Manisha, Radhakrishnan B, Suresh LP (2017) Tumor region extraction using edge detection method in brain MRI images. In: 2017 international conference on circuit, power and computing technologies (ICCPCT), Kollam, India, pp 1–5 (2017)

10. Aslam A, Khan E, Beg M (2015) Improved edge detection algorithm for brain tumor segmentation. Proced Comput Sci 58:430–437
11. Sivakumar V, Janakiraman N (2020) A novel method for segmenting brain tumor using modified watershed algorithm in MRI image with FPGA. Biosyst 198
12. Abid H, Abdulmunim ME (2016) Segmentation brain tumor and diagnosing using watershed algorithm. Am J Eng Res 5(11):31–35
13. Jemimma TA, Vetharaj YJ (2018) Watershed algorithm based DAPP features for brain tumor segmentation and classification. In: 2018 International conference on smart systems and inventive technology (ICSSIT), Tirunelveli, India, pp 155–158
14. Ronneberger O, Fischer P, Brox T (2015) U-Net: convolutional networks for biomedical image segmentation. In: Navab N et al (eds) Medical image computing and computer-Assisted intervention 2015; lecture notes in computer science, vol 9351. Springer, Cham
15. Kermi A, Mahmoudi I, Khadir MT (2018) Deep convolutional neural networks using U-net for automatic brain tumor segmentation in multimodal MRI volumes. In: Crimi A et al (eds) Brainlesion: glioma, multiple sclerosis, stroke and traumatic brain Injuries. Brain Les 2018; Lecture notes in computer science, vol 11384. Springer, Cham
16. Shreyash V, Pankajakshan V (2017) A deep learning architecture for brain tumor segmentation in MRI images. In: IEEE 19th international workshop on multimedia signal processing. Luton, UK, pp 1–6 (2017)
17. Shelhamer E, Long J, Darrell T (2017) Fully convolutional networks for semantic segmentation. IEEE Ann Hist Comput 39(4):640–651
18. Munir K, Frezza F, Rizzi A (2021) Brain tumor segmentation using 2D-UNET convolutional neural network. In: Kose U, Alzubi J (eds) Deep learning for cancer diagnosis. Studies in computational intelligence, vol 908. Springer, Singapore
19. Multimodal brain tumor segmentation challenge (2018). https://www.med.upenn.edu/sbia/brats2018/data.html
20. Menze BH, Jakab A, Bauer S, Kalpathy-Cramer J, Farahani K, Kirby J et al (2015) The multimodal brain tumor image segmentation benchmark (BRATS). IEEE Trans Med Imaging 34(10):1993–2024
21. Drozdzal M, Vorontsov E, Chartrand G, Kadoury S, Pal C (2016) The importance of skip connections in biomedical image segmentation. In: Carneiro G et al (eds) Deep learning and data labeling for medical applications. Lecture notes in computer science, vol 10008, Springer, Cham

Author Index

J. M. R. S. Tavares et al. (eds.), *Cyber Intelligence and Information Retrieval*, Lecture Notes in Networks and Systems 291,
https://doi.org/10.1007/978-981-16-4284-5